W9-CBF-572

Preliminary Edition

Calculus

A Graphing Approach

Volume I

Preliminary Edition

Calculus

A Graphing Approach

Volume I

Ross L. Finney
George B. Thomas, Jr.
Franklin D. Demana
Bert K. Waits

Addison-Wesley Publishing Company

Reading, Massachusetts • Menlo Park, California • New York
Don Mills, Ontario • Wokingham, England • Amsterdam • Bonn
Sydney • Singapore • Tokyo • Madrid • San Juan • Milan • Paris

Please Note: This Preliminary Edition Part 1 of *Calculus: A Graphing Approach* has been prepared as a service to the many schools who have asked to class-test the project. This paperback version includes material most often taught in the first semester or quarter of the calculus sequence. Before the hardback edition is published we will subject this material to many more production procedures and error checks in order to ensure the highest degree of quality and accuracy. Your support and indulgence throughout this process is greatly appreciated.

ISBN 0-201-56903-5

Copyright © 1993 by Addison-Wesley Publishing Company, Inc.

All rights reserved. No part of this publication may be reproduced, stored in a retrieval system, or transmitted, in any form or by any means, electronic, mechanical, photocopying, recording, or otherwise, without the prior written permission of the publisher.
Printed in the United States America.

1 2 3 4 5 6 7 8 9 10-MU-959493

Note to the Instructor

This *Preliminary Edition Part I* of *Calculus: A Graphing Approach* by Ross Finney, George Thomas, Franklin Demana, and Bert Waits provides you with the opportunity to assist the author team in crystallizing their vision of this ground-breaking project. We ask that you consider this project *"work-in-progress"*. It is our hope that through thorough review and class testing you will become partners with the author team and Addison-Wesley in crafting an exciting new generation of calculus materials. We welcome your suggestions, criticism, and praise, for it is only through your scrutiny that this project can reach its fullest potential.

Directly following the Preface you will find tear-out evaluation sheets for your students to complete. We encourage you to forward the evaluations along with your suggestions for improvement to us at the address listed below as you class-test this project.

Thank you for your interest in *Calculus: A Graphing Approach*. We look forward to hearing from you.

Jerome A. Grant
Mathematics Editor
Addison-Wesley Publishing Company
1 Jacob Way
Reading, MA 01867

About The Authors

Ross L. Finney

Ross Finney received his undergraduate degree and Ph.D. from the University of Michigan at Ann Arbor. He taught at MIT from 1980–1990. Currently, he teaches at the Naval Postgraduate School in Monterey, CA, and is involved in a number of mathematical organizations. Ross Finney has worked as a consultant for the Education Development Center in Newton, MA. He also held the position of Director for UMAP (Undergraduate Mathematics and Its Applications Project) from 1977–1984, and was the founding editor of the UMAP Journal. In 1984 he traveled with an MAA (Mathematical Association of America) delegation to China on a teacher's education project through People to People International.

Dr. Finney has co-authored a number of Addison-Wesley textbooks, including *Calculus, Calculus and Analytic Geometry, Elementary Differential Equations with Linear Algebra*, and *The Calculus Toolkit.*

George B. Thomas, Jr.

George Thomas is the author of the first mathematics textbook published by Addison-Wesley — *Calculus and Analytic Geometry* — now a best-selling Calculus text. George Thomas received his Ph.D. from Cornell University and his M.S. and B.S. from the State College of Washington. He taught at Cornell University for 3 years, at MIT for 38 years, and was a visiting Associate Professor at Stanford University for a year while on sabbatical from MIT. In the summer of 1965 he was a lecturer at Birla Institute of Technology and Science in India.

Dr. Thomas has written and co-authored a number of texts for Addison-Wesley including *Calculus, Calculus and Analytic Geometry, Probability: A First Course, Probability with Statistical Applications, Elementary Calculus from an Advanced Viewpoint*, and *Elements of Calculus and Analytic Geometry.*

Franklin D. Demana

Frank Demana received his Masters in Mathematics and Ph.D. in Algebra from Michigan State University. Currently, he is the director of Freshman Programs at Ohio State. As an active supporter of the use of technology to teach and learn mathematics, he is co-director of the very successful C^2PC (Calculator and Computer Precalculus) technology-enhanced curriculum revision project now operating in over 1000 high schools and colleges. He has been the director and co-director of over $10 million of NSF and foundational grant activities in the past 10 years. Along with frequent presentations at professional meetings, he has published a variety of articles in the areas of computer-and calculator-enhanced mathematics instruction.

Dr. Demana has co-authored *Precalculus: A Graphing Approach, Transitions to College Math, Essential Algebra: A Calculator Approach, College Algebra and Trigonometry: A Graphing Approach*, and *College Algebra: A Graphing Approach* for Addison-Wesley.

Bert K. Waits

Bert Waits received his Ph.D. from the Ohio State University in 1969, where he is currently a Professor of Mathematics. Waits is co-director of the C^2PC project along with Frank Demana, and has been co-director or principal investigator on several NSF projects. Waits has published over 50 articles in nationally recognized professional journals. He frequently gives invited lectures, workshops and mini-courses at national meetings of the MAA and NCTM. In 1991, Waits was an invited speaker at the ICTMA conference in the Netherlands, and the Mathematikunterricht und Informatik conference in Wolfbenbüttel, Germany. He has also served as a Visiting Fellow in Mathematics at the Chisolm Institute of Technology (Monash University) in Melbourne, Australia.

Dr. Waits has co-authored *Precalculus: A Graphing Approach, College Algebra and Trigonometry: A Graphing Approach*, and *College Algebra: A Graphing Approach* for Addison-Wesley.

Preface

Calculus: A Graphing Approach grew out of our strong conviction that incorporating technology into instruction makes the successful study of calculus realistic for all students, better prepares them for further study in mathematics and science, and turns calculus into a *Pump* and not a *Filter*. It builds on our ten-year experience in the Calculator and Computer Precalculus (C^2PC) project, a project that pioneered the use of a technology-based approach to mathematics and that integrates the use of numerical, graphical, and algebraic techniques. The approach of this textbook is consistent with the call by national mathematics organizations for reform in calculus that takes advantage of today's technological tools. Moreover, teachers using our technology-based precalculus curriculum reform materials expressed strong interest for similar reform materials in calculus. Their call has motivated and guided our efforts in this new project.

Calculus is the gate through which students wanting advanced training in most scientific, mathematical, and technical fields must pass. It is our strong belief that technology is the key that opens the gate for most students.

The Calculator and Computer Calculus (C^3) Project

The content is easily recognized. We start with this familiar body of calculus material and then make the assumption that computer graphing (function and parametric) and numerical methods for finding derivatives and integrals are available to *all* students on a regular basis for both in-class activites and homework. The tools used to teach and learn the familiar content are new and different.

With technology, we are able to approach problems numerically and graphically. We provide visual support for algebraic and analytic methods. We use visual methods to solve problems and then confirm using algebraic and analytic methods. We also use visual and numerical methods when analytic methods are impossible or very difficult. Technology is used to model and simulate problem situations.

Technology is integrated in *Calculus: A Graphing Approach* but the textbook does not assume the use of any particluar machine or software. Information about certain graphing calculators can be found in the lab manual that accompanies this textbook. We have described the technological approaches in this textbook in a way that is appropriate for all kinds of numerical, graphical, and symbolic capabilites that are available today in software packages and graphing calculators. For example, we zoom-in on graphs, and find derivatives and integrals numerically.

Integration of Technology

The use of a grapher—whether a hand-held graphing calculator or computer graphing software—is not optional. Depending on the particular technology used, it may be necessary to enter certain "toolbox" programs on your graphing calculator. These programs are given in the accompanying lab manual. Technology permits the focus of calculus to be on problem solving and exploration, while building a deeper understanding about algebraic and analytic techniques.

Multiple Representations

We use multiple representations to explore problem situations. First, we find an algebraic representation of the problem. Then we find a complete graph of the algebraic representation and determine which values of the domain of the algebraic representation make sense in the problem situation. Technology allows us to move easily among these representations and exploit connections.

The concept of function is central to the study of calculus. A function can be thought of as a process that produces outputs for associated inputs. The notion can be represented *numerically* as a table of input-output pairs, *graphically* as a plot of output versus input, and *symbolically* as an algebraic representation. With technology we are no longer restricted to a purely algebraic approach to calculus. Now we can also approach concepts of calculus numerically and graphically. Establishing connections among the numerical, graphical, and symbolic approaches establishes a richer understanding about calculus.

Visualization and Numerical Methods

Graphing helps students gain an understanding about functions that can often be confirmed analytically. Graphing can also be used as a powerful problem-solving technique as well as a way to estimate results and support the reasonableness of results obtained analytically.

Numerical methods such as numerical differentiation, numerical integration, and numerical and graphical solution methods for differential equations allow us to remove the need for contrived problems and opens the door for realistic and interesting applications.

Features

New pedagogical features have been incorporated into this textbook.

Explore with a Grapher. This recurring box makes the student an active partner in the development of the mathematics. The Explore activities help the instructor in the modern role of facilitator of learning. The boxes guide the experimental classroom approach that is a consequence of technology.

Sidelight Boxes. Boxes placed in the margin provide commentary on the mathematical development, historical facts, and reminders.

Artwork. We have made a visual distinction between graphs generated with a grapher and hand-sketched art. Art produced with a grapher is outlined with a box and hand-sketched art is not boxed.

Graphing Calculator Laboratory Manual for Calculus. This manual includes instructions for the latest calculators from Casio, Texas Instruments, Sharp, and Hewlett-Packard. The manual provides keystroke-level descriptions for using the technology in calculus, concrete examples, and machine-specific programs that are used in the textbook.

Content

The materials included in Volume I of the Preliminary Edition represent the first six chapters of *Calculus: A Graphing Approach*, or what roughly corresponds to first semester calculus.

Acknowledgments

We would like to thank the many wonderful CalcNet teachers that attended the two-week summer 1991 Calculus institute directed by Gregory Foley of Sam Houston State University in Huntsville, Texas. Several of the CalcNet teachers have used draft copies of these materials and provided important feedback. Special thanks to Ray Barton of Olympus High School in Salt Lake City, Utah for providing many variable suggestions for the textbook and for his work on the laboratory manual. We would also like to thank Morton Brown of the University of Michigan—Ann Arbor for his insightful comments throughout this process. Special thanks also go to David and Mary Winter of Michigan State University in East Lansing, Michigan for their work providing answers for this preliminary edition; to Sherrie Lowery for typing the manuscript; and to Amy Edwards and Greg Ferrar who prepared the artwork for the manuscript.

We would also like to thank Andrew Fisher, Jenny Bagdigian, Laurie Petrycki, Elka Block, Lynn Faber, and Jerome Grant of Addison-Wesley for their hard work in preparing this preliminary edition.

R.L.F. Monterey, CA
G.B.T. Jr. State College, PA
F.D. Columbus, OH
B.K.W. Columbus, OH

Note To The Student

You are about to embark on an exciting journey . The text you'll be using in this course presents a unique combination of traditional calculus material and the latest technology available in mathematics teaching and learning. The goal of this text is to give you a deeper understanding of the mathematical underpinnings of the calculus by allowing you to graphically explore the material. You'll experience math firsthand! And that's a powerful way to learn. After all — seeing is believing!

As this is a preliminary edition, and still "work-in-progress", we welcome any comments you may have as you work through the text. Your feedback will help shape the first edition.

Student Evaluations: On the following pages, you'll find student evaluation questionnaires which we would like you to fill out after you work through *each chapter*. Please feel free to attach additional sheets if necessary. As the questionnaires are printed front-to-back, please submit them to your instructor after you complete *every other* chapter (after completing chapters 2, 4, and 6). Your instructor will forward them on to us. If you have anything else that you'd like us to be aware of, please feel free to send comments directly to:

Andrew Fisher
Marketing Manager
Higher Education Mathematics
Addison-Wesley Publishing Company
1 Jacob Way
Reading, MA 01867 USA

Finney/Thomas/Demana/Waits: *Calculus: A Graphing Approach*

CHAPTER ONE STUDENT EVALUATION

This questionnaire should be completed after you finish chapter ONE. As you work through the chapter, please feel free to make notes to help refresh your memory after you complete the chapter. If you find you need additional space, please attach as many sheets as you feel are necessary. After you have completed the questionnaire for *chapter TWO*, rip this page out of your book, staple any additional comments to it, and hand it to your instructor.

In order to ensure your confidentiality, but to still allow us to track your comments, please fill in the last four digits of your phone number, and your month and day of birth. Also, please fill in the name of your school.

Identity Code: ___ ___ ___ ___ **Birth Date:** ___ ___ - ___ ___ **School Name:** ____________________
(last 4 phone # digits) (month) (day)

1. What is the one aspect you liked best about this chapter? What is the one aspect you liked least? Please answer as descriptively as possible; cite examples when appropriate.

2. Do you find the examples and exercises at section and chapter ends have helped motivate you to learn the mathematics in this chapter? If yes, to what do you attribute this (i.e., interesting applications, variety of exercises)?

3. Do you find the graphing utility material (Explore Boxes, examples illustrating the use of a graphing utility, and exercises requiring the use of a graphing utility) helpful in understanding the mathematical concepts of this chapter? Please explain your response.

4. So that we may gather some feedback on the content and organization of the chapter, please comment on the following aspects of the chapter as far as understandability and usefulness: explanations, illustrations, highlighted material (definitions, theorems, important equations).

5. Would you recommend any other changes to make this chapter better?

May we quote you in advertising? Yes No

OPTIONAL: Name______________________________ Phone (______)________________

May we contact you if we have further questions? Yes No

Finney/Thomas/Demana/Waits: *Calculus: A Graphing Approach*

CHAPTER TWO STUDENT EVALUATION

This questionnaire should be completed after you finish chapter TWO. As you work through the chapter, please feel free to make notes to help refresh your memory after you complete the chapter. If you find you need additional space, please attach as many sheets as you feel are necessary. After you have completed this questionnaire, rip this page out of your book, staple any additional comments to it, and hand it to your instructor.

In order to ensure your confidentiality, but to still allow us to track your comments, please fill in the last four digits of your phone number, and your month and day of birth. Also, please fill in the name of your school.

Identity Code: ___ ___ ___ ___ **Birth Date:** ___ ___ - ___ ___ **School Name:** ____________________
(last 4 phone # digits) (month) (day)

1. What is the one aspect you liked best about this chapter? What is the one aspect you liked least? Please answer as descriptively as possible; cite examples when appropriate.

2. Do you find the examples and exercises at section and chapter ends have helped motivate you to learn the mathematics in this chapter? If yes, to what do you attribute this (i.e., interesting applications, variety of exercises)?

3. Do you find the graphing utility material (Explore Boxes, examples illustrating the use of a graphing utility, and exercises requiring the use of a graphing utility) helpful in understanding the mathematical concepts of this chapter? Please explain your response.

4. So that we may gather some feedback on the content and organization of the chapter, please comment on the following aspects of the chapter as far as understandability and usefulness: explanations, illustrations, highlighted material (definitions, theorems, important equations).

5. Would you recommend any other changes to make this chapter better?

May we quote you in advertising? Yes No

OPTIONAL: Name________________________________ Phone (______)________________

May we contact you if we have further questions? Yes No

Finney/Thomas/Demana/Waits: *Calculus: A Graphing Approach*

CHAPTER THREE STUDENT EVALUATION

This questionnaire should be completed after you finish chapter THREE. As you work through the chapter, please feel free to make notes to help refresh your memory after you complete the chapter. If you find you need additional space, please attach as many sheets as you feel are necessary. After you have completed the questionnaire for *chapter FOUR*, rip this page out of your book, staple any additional comments to it, and hand it to your instructor.

In order to ensure your confidentiality, but to still allow us to track your comments, please fill in the last four digits of your phone number, and your month and day of birth. Also, please fill in the name of your school.

Identity Code: ___ ___ ___ ___ **Birth Date:** ___ ___ - ___ ___ **School Name:** ____________________
(last 4 phone # digits) (month) (day)

1. What is the one aspect you liked best about this chapter? What is the one aspect you liked least? Please answer as descriptively as possible; cite examples when appropriate.

2. Do you find the examples and exercises at section and chapter ends have helped motivate you to learn the mathematics in this chapter? If yes, to what do you attribute this (i.e., interesting applications, variety of exercises)?

3. Do you find the graphing utility material (Explore Boxes, examples illustrating the use of a graphing utility, and exercises requiring the use of a graphing utility) helpful in understanding the mathematical concepts of this chapter? Please explain your response.

4. So that we may gather some feedback on the content and organization of the chapter, please comment on the following aspects of the chapter as far as understandability and usefulness: explanations, illustrations, highlighted material (definitions, theorems, important equations).

5. Would you recommend any other changes to make this chapter better?

May we quote you in advertising? Yes No

OPTIONAL: Name________________________________ Phone (_____)________________

May we contact you if we have further questions? Yes No

Finney/Thomas/Demana/Waits: *Calculus: A Graphing Approach*

CHAPTER FOUR STUDENT EVALUATION

This questionnaire should be completed after you finish chapter FOUR. As you work through the chapter, please feel free to make notes to help refresh your memory after you complete the chapter. If you find you need additional space, please attach as many sheets as you feel are necessary. After you have completed this questionnaire, rip this page out of your book, staple any additional comments to it, and hand it to your instructor.

In order to ensure your confidentiality, but to still allow us to track your comments, please fill in the last four digits of your phone number, and your month and day of birth. Also, please fill in the name of your school.

Identity Code: ___ ___ ___ ___ **Birth Date:** ___ ___ - ___ ___ **School Name:** ____________________

(last 4 phone # digits) (month) (day)

1. What is the one aspect you liked best about this chapter? What is the one aspect you liked least? Please answer as descriptively as possible; cite examples when appropriate.

2. Do you find the examples and exercises at section and chapter ends have helped motivate you to learn the mathematics in this chapter? If yes, to what do you attribute this (i.e., interesting applications, variety of exercises)?

3. Do you find the graphing utility material (Explore Boxes, examples illustrating the use of a graphing utility, and exercises requiring the use of a graphing utility) helpful in understanding the mathematical concepts of this chapter? Please explain your response.

4. So that we may gather some feedback on the content and organization of the chapter, please comment on the following aspects of the chapter as far as understandability and usefulness: explanations, illustrations, highlighted material (definitions, theorems, important equations).

5. Would you recommend any other changes to make this chapter better?

May we quote you in advertising? Yes No

OPTIONAL: Name________________________________ Phone (______)__________________

May we contact you if we have further questions? Yes No

Finney/Thomas/Demana/Waits: *Calculus: A Graphing Approach*

CHAPTER FIVE STUDENT EVALUATION

This questionnaire should be completed after you finish chapter FIVE. As you work through the chapter, please feel free to make notes to help refresh your memory after you complete the chapter. If you find you need additional space, please attach as many sheets as you feel are necessary. After you have completed the questionnaire for *chapter SIX*, rip this page out of your book, staple any additional comments to it, and hand it to your instructor.

In order to ensure your confidentiality, but to still allow us to track your comments, please fill in the last four digits of your phone number, and your month and day of birth. Also, please fill in the name of your school.

Identity Code: ___ ___ ___ ___ **Birth Date:** ___ ___ - ___ ___ **School Name:** ____________________
(last 4 phone # digits) (month) (day)

1. What is the one aspect you liked best about this chapter? What is the one aspect you liked least? Please answer as descriptively as possible; cite examples when appropriate.

2. Do you find the examples and exercises at section and chapter ends have helped motivate you to learn the mathematics in this chapter? If yes, to what do you attribute this (i.e., interesting applications, variety of exercises)?

3. Do you find the graphing utility material (Explore Boxes, examples illustrating the use of a graphing utility, and exercises requiring the use of a graphing utility) helpful in understanding the mathematical concepts of this chapter? Please explain your response.

4. So that we may gather some feedback on the content and organization of the chapter, please comment on the following aspects of the chapter as far as understandability and usefulness: explanations, illustrations, highlighted material (definitions, theorems, important equations).

5. Would you recommend any other changes to make this chapter better?

May we quote you in advertising? Yes No

OPTIONAL: Name______________________________ Phone (_____)________________

May we contact you if we have further questions? Yes No

Finney/Thomas/Demana/Waits: *Calculus: A Graphing Approach*

CHAPTER SIX STUDENT EVALUATION

This questionnaire should be completed after you finish chapter SIX. As you work through the chapter, please feel free to make notes to help refresh your memory after you complete the chapter. If you find you need additional space, please attach as many sheets as you feel are necessary. After you have completed this questionnaire, rip this page out of your book, staple any additional comments to it, and hand it to your instructor.

In order to ensure your confidentiality, but to still allow us to track your comments, please fill in the last four digits of your phone number, and your month and day of birth. Also, please fill in the name of your school.

Identity Code: ___ ___ ___ ___ **Birth Date:** ___ ___ - ___ ___ **School Name:** ____________________
(last 4 phone # digits) (month) (day)

1. What is the one aspect you liked best about this chapter? What is the one aspect you liked least? Please answer as descriptively as possible; cite examples when appropriate.

2. Do you find the examples and exercises at section and chapter ends have helped motivate you to learn the mathematics in this chapter? If yes, to what do you attribute this (i.e., interesting applications, variety of exercises)?

3. Do you find the graphing utility material (Explore Boxes, examples illustrating the use of a graphing utility, and exercises requiring the use of a graphing utility) helpful in understanding the mathematical concepts of this chapter? Please explain your response.

4. So that we may gather some feedback on the content and organization of the chapter, please comment on the following aspects of the chapter as far as understandability and usefulness: explanations, illustrations, highlighted material (definitions, theorems, important equations).

5. Would you recommend any other changes to make this chapter better?

May we quote you in advertising? Yes No

OPTIONAL: Name______________________________ Phone (_____)________________

May we contact you if we have further questions? Yes No

Contents

1 Prerequisites for Calculus 1–1

Overview 1–1
Coordinates and Graphs in the Plane 1–1
Slope, and Equations for Lines 1–14
Functions and Their Graphs 1–30
Shifts, Reflections, Stretches, and Shrinks. Geometric Transformations 1–50
Solving Equations and Inequalities Graphically 1–63
Relations, Inverses, Circles 1–78
A Review of Trigonometric Functions 1–95
Review Questions 1–115
Practice Exercises 1–116

2 Limits and Continuity 2–1

Overview 2–1
Limits 2–1
Continuous Functions 2–13
The Sandwich Theorem and (sin θ)/θ 2–26
Limits Involving Infinity 2–31
Controlling Function Outputs: Target Values 2–46
Defining Limits Formally with Epsilons and Deltas 2–53
Review Questions 2–64
Practice Exercises 2–65

3 Derivatives 3–1

Overview 3–1
Slopes, Tangent Lines, and Derivatives 3–1
Differentiation Rules 3–20
Velocity, Speed, and Other Rates of Change 3–34
Derivatives of Trigonometric Functions 3–51
The Chain Rule 3–58
Implicit Differentiation and Fractional Powers 3–69
Linear Approximations and Differentials 3–79
Review Questions 3–92
Practice Exercises 3–93

4 Applications of Derivatives 4–1

Overview 4–1
Maxima, Minima, and the Mean Value Theorem 4–1
Predicting Hidden Behavior 4–14
Newton's Method and Polynomial Functions 4–26
Rational Functions and Economic Applications 4–42
Radical and Transcendental Functions 4–54
Related Rates of Change 4–65
Antiderivatives, Initial Value Problems, and Mathematical Modeling 4–74
Review Questions 4–88
Practice Exercises 4–89

5 Integration 5–1

Overview 5–1
Calculus and Area 5–1
Formulas for Finite Sums and Area 5–9
Definite Integrals 5–16
Antiderivatives and Definite Integrals 5–29
The Fundamental Theorems of Integral Calculus 5–39
Indefinite Integrals 5–53
Integration by Substitution—Running the Chain Rule Backward 5–64
A Brief Introduction to Logarithms and Exponentials 5–74
Numerical Integration: The Trapezoidal Rule and Simpson's Method 5–94
Review Questions 5–109
Practice Exercises 5–110

6 Applications of Definite Integrals 6–1

Overview 6–1
Areas Between Curves 6–1
Volumes of Solids of Revolution—Disks and Washers 6–11
Cylindrical Shells—an Alternative to Washers 6–25
Lengths of Curves in the Plane 6–34
Areas of Surfaces of Revolution 6–43
Work 6–49
Fluid Pressures and Fluid Forces 6–58
Centers of Mass 6–64
The Basic Idea. Other Modeling Applications 6–76
Review Questions 6–89
Practice Exercises 6–90

Appendixes A–1

Answers A–15

Index I–1

A Brief Table of Integrals T–1

CHAPTER

1

Prerequisites for Calculus

OVERVIEW This chapter reviews the most important things you need to know to start learning calculus and provides an introduction to the use of a graphing utility to investigate mathematical concepts and ideas. The emphasis is on functions and graphs, the main building blocks of calculus.

In calculus, functions are the major tools for describing the real world in mathematical terms, from temperature variations to planetary motions, from business cycles to brain waves, and from population growth to heartbeat patterns. Many functions have particular importance because of the kind of behavior they describe. For example, trigonometric functions describe cyclic, repetitive activity; exponential and logarithmic functions describe growth and decay; and polynomial functions can approximate these and most other functions. We shall examine some of these functions in this chapter and meet others in later chapters.

We shall learn that calculus can help us confirm the graph of a function obtained with a graphing utility, and that it can tell us many things about a graph and its equation that plotting cannot reveal. We assume that you have a graphing utility available at all times.

1.1 Coordinates and Graphs in the Plane

To assign coordinates to a point in a plane, we start with two number lines that cross at their zero points at right angles. Each line represents the real numbers, which are the numbers that can be represented by decimals. Figure 1.1 shows the usual way of drawing the lines, with one line horizontal and the other vertical. The horizontal line is called the **x-axis** and the vertical line the **y-axis**. The point at which the lines cross is the **origin**.

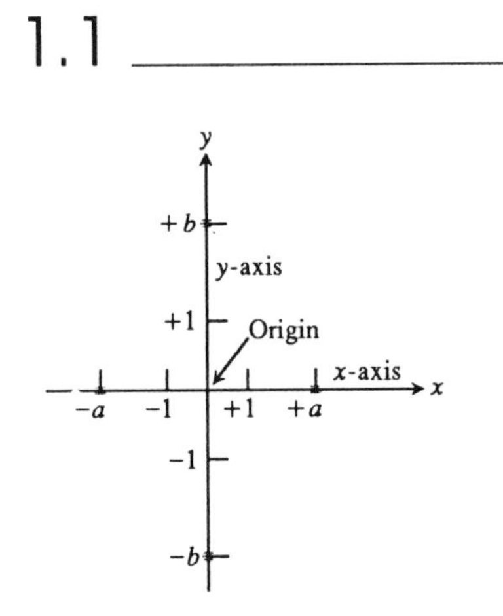

1.1 In Cartesian coordinates, the scaling on each axis is symmetric about the origin.

On the x-axis, the positive number a lies a units to the right of the origin, and the negative number $-a$ lies a units to the left of the origin. On the y-axis, the positive number b lies b units above the origin while the negative number $-b$ lies b units below the origin.

The coordinates defined here are often called *Cartesian* coordinates, after their chief inventor, René Descartes (1596–1650).

With the axes in place, we assign a pair (a, b) of real numbers to each point P in the plane. The number a is the number at the foot of the perpendicular from P to the x-axis. The number b is the number at the foot of the perpendicular from P to the y-axis. Figure 1.2 shows the construction. The notation (a, b) is read "a b."

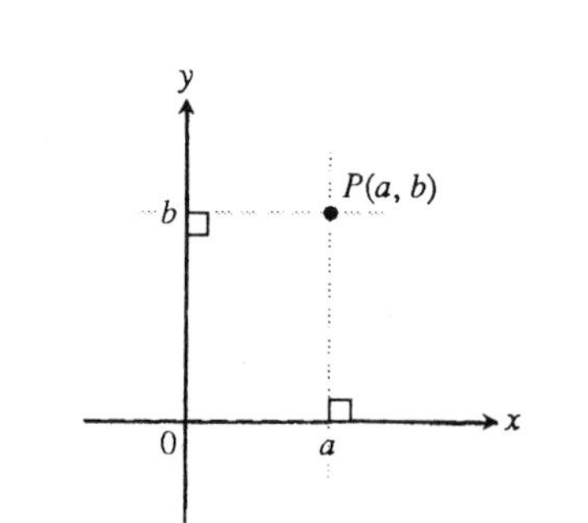

1.2 The pair (a, b) corresponds to the point where the perpendicular to the x-axis at a crosses the perpendicular to the y-axis at b.

The number a from the x-axis is the **x-coordinate** of P. The number b from the y-axis is the **y-coordinate** of P. The pair (a, b) is the **coordinate pair** of the point P. It is an **ordered pair**, with the x-coordinate first and y-coordinate second. To show that P has the coordinate pair (a, b), we sometimes write the P and (a, b) together: $P(a, b)$.

The construction that assigns an ordered pair of real numbers to each point in the plane can be reversed to assign a point in the plane to each ordered pair of real numbers. The point assigned to the pair (a, b) is the point where the perpendicular to the x-axis at a crosses the perpendicular to the y-axis at b. Thus, the assignment of coordinates is a one-to-one correspondence between the points of the plane and the set of all ordered pairs of real numbers. Every point has a pair and every pair has a point, so to speak.

The points on the coordinate axes now have two kinds of numerical labels: single numbers from the axes and paired numbers from the plane. How do the numbers match up? See Fig. 1.3. As you can see, every point on the x-axis has y-coordinate zero and every point on the y-axis has x-coordinate zero. The origin is the point $(0, 0)$.

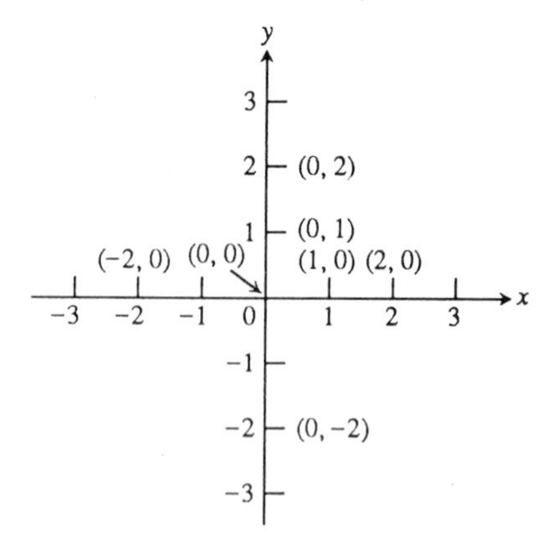

1.3 Points on the axes can be labeled in two ways.

Directions and Quadrants

Motion from left to right along the x-axis is said to be motion in the **positive x-direction**. Motion from right to left is in the **negative x-direction**. Along the y-axis, the positive direction is up, and the negative direction is down.

The origin divides the x-axis into the **positive x-axis** to the right of the origin and the **negative x-axis** to the left of the origin. Similarly, the origin divides the y-axis into the **positive y-axis** and the **negative y-axis**. The axes divide the plane into four regions called **quadrants**, numbered I, II, III, and IV (Fig. 1.4).

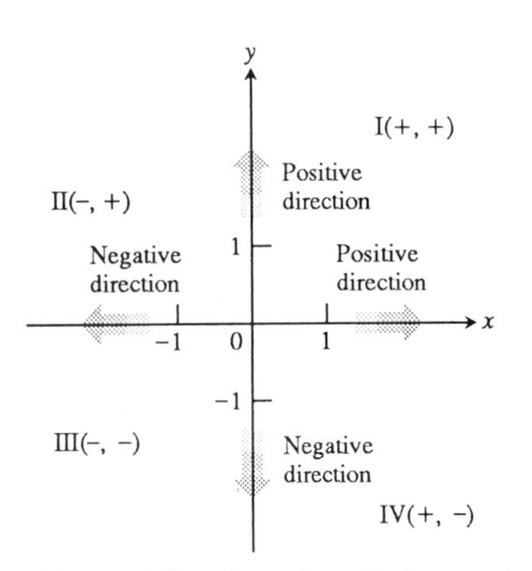

1.4 Directions along the axes: The values of x and y increase in the positive direction and decrease in the negative direction. Roman numerals label the quadrants.

A Word about Scales

When we plot data in the coordinate plane or graph formulas whose variables have different units of measure, we do not need to use the same scale on the two axes. There is no reason to place the two 1's on the axes the same number of millimeters or whatever from the origin.

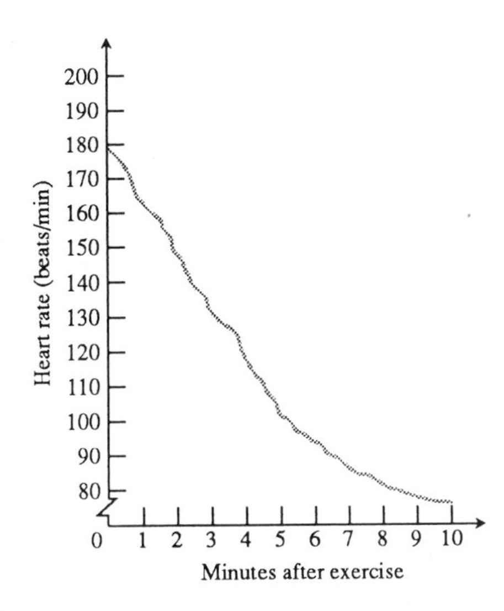

1.5 How the heartbeat returns to a normal rate after running. (Adapted from James F. Fixx's *The Complete Book of Running* (1977). New York: Random House.)

EXAMPLE 1 The graph in Fig. 1.5 shows how long it takes the heart to return to normal after running. The length that shows one minute on the horizontal axis is the same length that shows ten beats on the vertical axis. ■

Sometimes we will assume that the scales on the axes we draw are the same. One unit of distance up and down in the plane will then look the same as one unit of distance right and left. As on a surveyor's map or a scale drawing, line segments that are supposed to have the same length will look as if they do. The measure of angles will also appear to be correct. For example, the angle the line $y = x$ makes with the x-axis will look like a 45° angle.

Graphing Utilities

Devices that automatically produce graphs are called **graphing utilities** or **graphers**. The most practical graphing utilities today are graphing calculators. Computers with graphing software are also graphing utilities. In this book, we assume that you have a graphing utility available at all times.

The display screen of a graphing utility *represents* a rectangular portion of the coordinate plane determined by $x\text{Min} \leq x \leq x\text{Max}$ and $y\text{Min} \leq y \leq y\text{Max}$ called the **viewing rectangle** or **viewing window** **[xMin, xMax] by [yMin, yMax]**. The display screen of a graphing utility consists of a rectangular array of lights called **pixels.** Many popular graphing calculators, including the one we use in this textbook, name each pixel with an ordered pair of real numbers called screen coordinates.

Screen Coordinates

The screen coordinates of the pixels associated with the viewing rectangle [xMin, xMax] by [yMin, yMax] are

$$(x_i,\ y_j) = (x\text{Min} + i\Delta x, y\text{Min} + j\Delta y),$$

where i and j are integers with

$$i = 0, 1, \cdots, N, \quad j = 0, 1, \cdots, M,$$

$$N + 1 = \text{ number of columns of pixels,}$$

$$M + 1 = \text{ number of rows of pixels,}$$

$$\Delta x = \frac{x\text{Max} - x\text{Min}}{N}, \text{ and}$$

$$\Delta y = \frac{y\text{Max} - y\text{Min}}{M}.$$

One popular graphing calculator has $N = 126$ and $M = 62$. With the graphing calculator we use, every point (a, b) of the coordinate plane in the viewing rectangle [xMin, xMax] by [yMin, yMax] corresponds to a unique pixel with screen coordinates as described above. See Fig. 1.6. However, this assignment cannot be reversed because an infinite number of points of the coordinate plane correspond to the same pixel as there are only a *finite* number of pixels on the display screen.

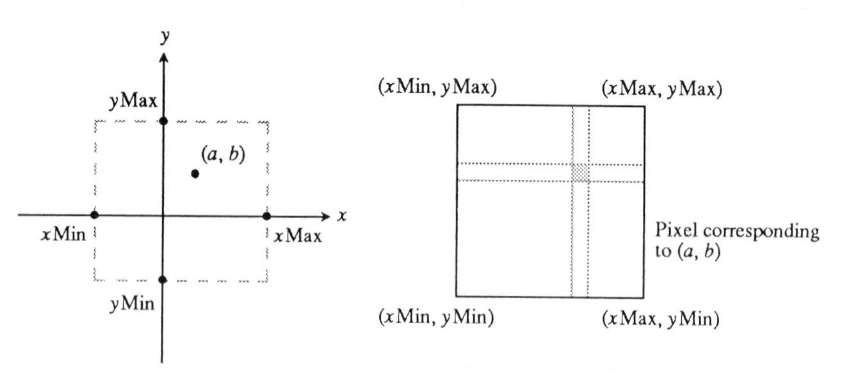

1.6 The viewing rectangle [xMin, xMax] by [yMin, yMax].

The key used to input a viewing rectangle on a graphing utility is usually called the RANGE key.* Many graphing utilities allow the user to choose the *scale marks* on the axes by assigning values to xScl and yScl. The six

* Consult the accompanying *Calculus Graphing Calculator and Computer Graphing Laboratory Manual* or the Owner's Manual for your graphing utility for more information about screen coordinates and viewing rectangle input.

values xMin, xMax, xScl, yMin, yMax, yScl are called the **range** of the viewing rectangle. The following Explore Activity illustrates the graphing utility features described above.

Explore Box

1. Set $x\text{Min} = -10$, $x\text{Max} = 10$, $x\text{Scl} = 1$, $y\text{Min} = -10$, $y\text{Max} = 10$, and $y\text{Scl} = 1$ on the range screen of your graphing utility to enter the $[-10, 10]$ by $[-10, 10]$ viewing rectangle, called the **standard viewing rectangle.**
2. Use the appropriate keys to move the cursor left, right, up, and down and observe the *screen coordinates* of the cursor location.
3. What are the screen coordinates of the four corners of the display screen?
4. How does the viewing rectangle change if xScl or yScl is changed?
5. Determine Δx and Δy for your grapher. If the number of columns of pixels is 127 and the number of rows of pixels is 63, then Δx and Δy are approximately 0.1587 and 0.3226, respectively.
6. Let (x_i, y_j) be a screen coordinate. What is the screen coordinate of the point $(x_i + a, y_j + b)$ of the coordinate plane where $|a| < \frac{\Delta x}{2}$ and $|b| < \frac{\Delta y}{2}$?

Graphs of Equations

The points (x, y) whose coordinates satisfy an equation like $y = x^2$ make up the *graph* of the equation in the xy-plane. Graphs give us a practical way to picture equations as lines or curves.

EXAMPLE 2 Graph the equation $y = x^2$ without a graphing utility.

Solution

STEP 1: Make a table of xy-pairs that satisfy the equation $y = x^2$.

x	$y = x^2$
-2	4
-1	1
0	0
1	1
2	4

STEP 2: Plot the points (x, y) whose coordinates appear in the table.

STEP 3: Sketch a smooth curve through the plotted points. Label the curve with its equation. ≡

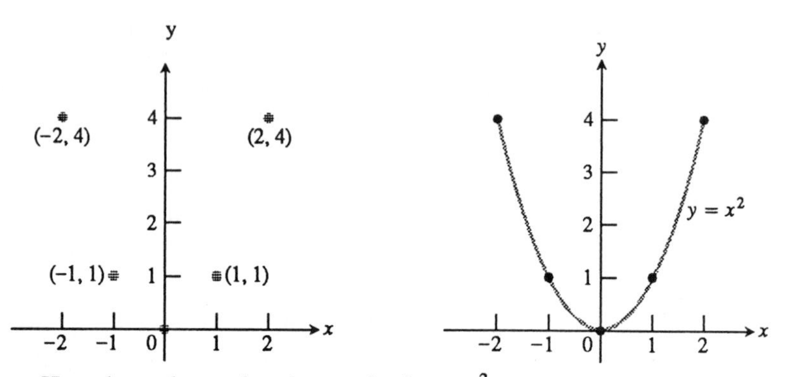

How do we know that the graph of $y = x^2$ doesn't look like one of *these* curves:

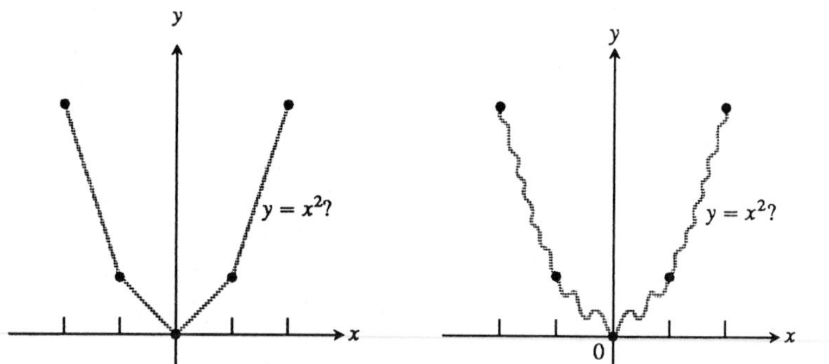

To find out, we might plot more points. However, this is tedious and very time consuming. Graphing utilities are able to quickly plot one point for each column of pixels provided the point lies within the chosen viewing rectangle and the graphs are very reliable. This is one of the many benefits of the use of graphing utilities. Some graphing utilities—including the one we use in this textbook to draw graphs—have an option that allows us to draw a smooth curve through the plotted points (**connected mode**) or simply to show the plotted points (**dot mode**). Figure 1.7 gives *connected* and *dot* mode **computer visualizations** of the graph of $y = x^2$ that strongly support the graph sketched in Step 3 of Example 2 is correct.

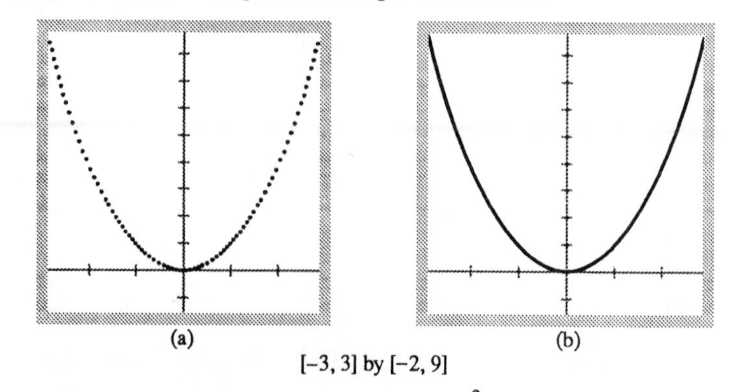

1.7 *Computer visualizations* of the graphs of $y = x^2$ in (a) dot mode and (b) connected mode.

The viewing rectangle used to obtain computer visualizations of graphs will be specified below the figures as in Figure 1.7. Notice that yMin and yMax were chosen in Figure 1.7 so that all points computed by the graphing utility would appear in the viewing rectangle.

In this textbook, you will be asked to produce graphs with and without graphing utilities. We make the following agreement.

Agreement

We use the phrase "sketch a graph of" to mean produce a graph without using a graphing utility. All other phrases such as "draw a graph of" or "graph" will mean you may use any method to obtain the graph.

Intercepts and More about Graphing

Important points to find on an equation's graph are the points where the graph touches or crosses the axes. These points (when they exist) are the graph's **intercepts**.

We find the **x-intercepts** by setting y equal to 0 in the equation and solving for x. We find the **y-intercepts** by setting x equal to 0 and solving for y.

EXAMPLE 3 Find the intercepts of the graph of the equation $y = x^2 - 1$.

Solution

The x-intercepts:

$y = x^2 - 1$ (Write down the given equation.)

$0 = x^2 - 1$ (Let $y = 0$.)

$x^2 = 1$

(Solve for x.)

$x = 1, -1$

The y-intercepts:

$y = x^2 - 1$ (Write down the given equation again.)

$y = (0)^2 - 1$ (Let $x = 0$.)

$y = -1.$ (Solve for y—not much to do in this case.)

There are two x-intercepts ($x = 1$ and $x = -1$) and one y-intercept ($y = -1$).

EXAMPLE 4 Sketch a graph of $y = x^2 - 1$. Support your answer with a graphing utility.

Solution

STEP 1: Find the intercepts (if any). From Example 3, the intercepts are $x = 1, x = -1$, and $y = -1$.

STEP 2: Make a short table of xy-pairs that satisfy the equation $y = x^2 - 1$. Include the intercepts.

x	$y = x^2 - 1$
-3	8
-2	3
-1	0
0	-1
1	0
2	3
3	8

STEP 3: Plot the points (x, y) from the table. Connect them with a smooth curve.

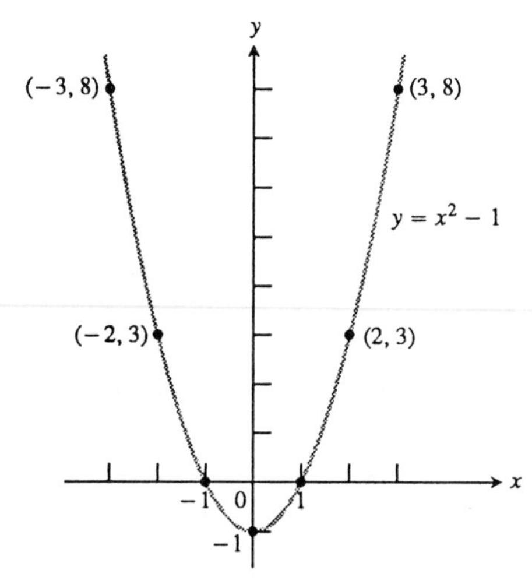

The graph of $y = x^2 - 1$ given in Figure 1.8 supports the sketch in Step 3 above as well as the intercept information in Step 1.

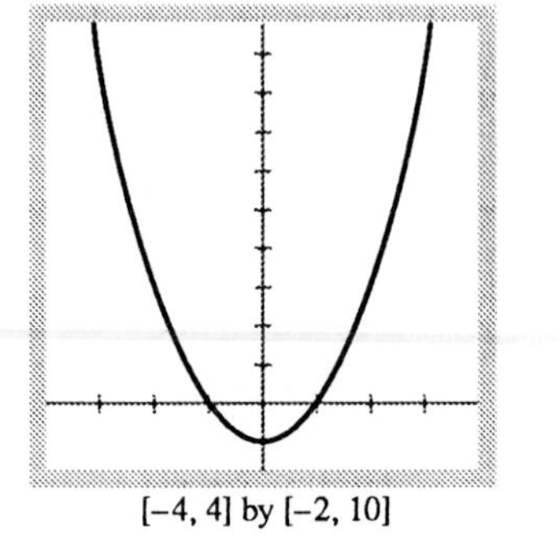

1.8 A computer visualization of the graph of $y = x^2 - 1$.

When we are asked to produce visualizations of graphs, either by hand or with graphing utilities, it is understood that we want the visualizations to *suggest* all the points of the graph and all the important features of the graph. Usually visualizations produced by hand or with graphers cannot show all the points of a graph. However, the visualization in Figure 1.8 suggests that the values of $y = x^2 - 1$ increase without bound for $x > 0$ and increasing, or for $x < 0$ and decreasing. We will say more about this in Chapter 2. Similarly, Figure 1.8 suggests that the graph of $y = x^2 - 1$ has a lowest point (vertex) at $(0, -1)$, and is symmetric with respect to the y-axis. We will say more about symmetry of graphs in Section 1.3, more about graphs of equations of the form $y = ax^2 + bx + c$ (**parabolas**) in Section 1.4, and more about high and low points of graphs in Chapter 3. From this point on we will not distinguish between "graphs" and "visualizations of graphs." We make the following agreement and definition.

Agreement

When we are asked to sketch or draw a graph it is understood that we are asked to produce a visualization of the graph either by hand or with a grapher.

Definition

A graph (or visualization of a graph) is said to be **complete** if it suggests all the points of the graph and all the important features of the graph such as the intercepts.

With the above agreement and definition the graph of $y = x^2 - 1$ given in Figure 1.8 is *complete*. How do we know for sure that the graph is complete? The answer lies in calculus, as we shall see in Chapter 4. There we shall learn to use a marvelous mathematical tool called a *derivative* to find a curve's exact shape between plotted points. Meanwhile, we shall settle for plotting enough points with graphers that produce highly reliable graphs and use the information about quadratic and cubic polynomials given in Figures 1.9 and 1.10 that will be established in Chapter 4.

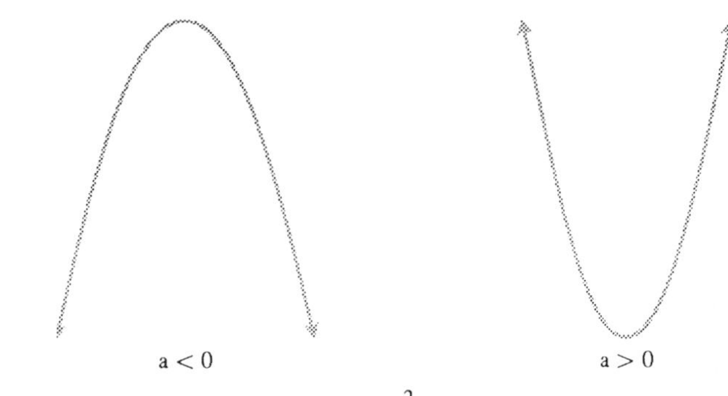

1.9 Possible complete graphs of $y = ax^2 + bx + c, a \neq 0$.

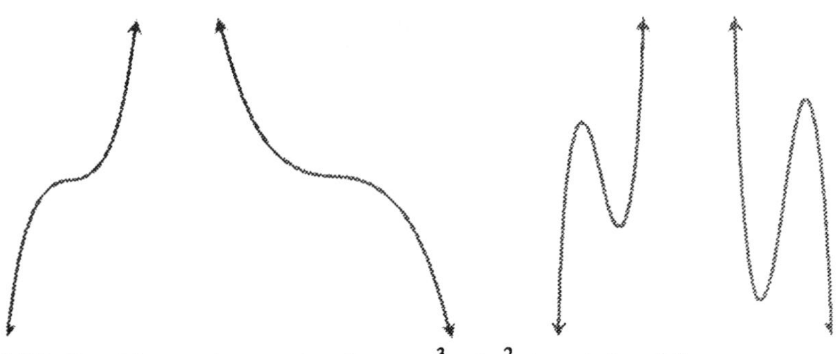

1.10 Possible complete graphs of $y = ax^3 + bx^2 + cx + d, a \neq 0$.

EXAMPLE 5 Which of the viewing rectangles gives a complete graph of $y = -3x^2 + 12x + 5$:

a) $[0, 5]$ by $[0, 5]$

b) $[-10, 10]$ by $[-10, 10]$

c) $[-5, 10]$ by $[-10, 20]$

Solution Figure 1.11 shows the graph of $y = -3x^2 + 12x + 5$ in each of these viewing rectangles. A complete graph must look like the one on the left in Figure 1.9. Thus, only the viewing rectangle in (c) shows a complete graph.

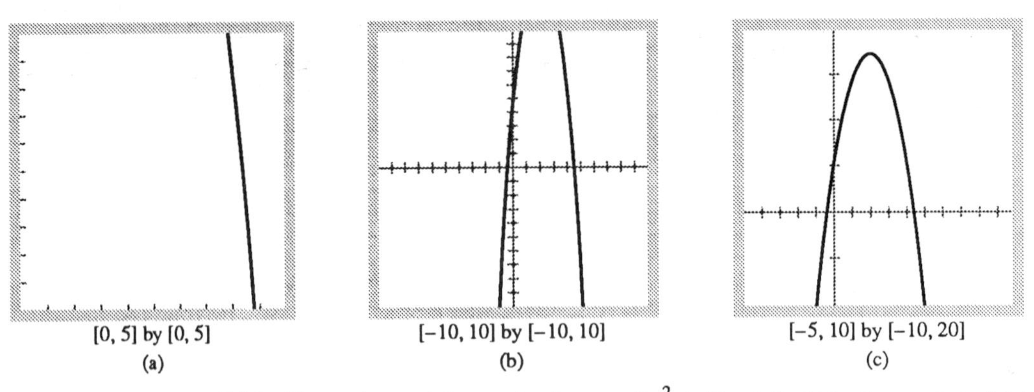

1.11 Three views of the graph of $y = -3x^2 + 12x + 5$. Only the graph in (c) is complete.

■

Example 5 illustrates the importance of choosing an appropriate viewing rectangle in order to display a complete graph of an equation. Also, Fig. 1.11 illustrates that $x\text{Scl}$ and $y\text{Scl}$ will not always be equal to 1 in the graphs shown throughout this book. In Fig. 1.11(a), both $x\text{Scl} = 0.5$ and $y\text{Scl} = 0.5$, whereas in Fig. 1.11(c), $x\text{Scl} = 1$ and $y\text{Scl} = 5$.

EXAMPLE 6 Find the intercepts and draw a complete graph of $y = x^3 - 4x$. Use the graph to estimate the coordinates of any local high or low points.

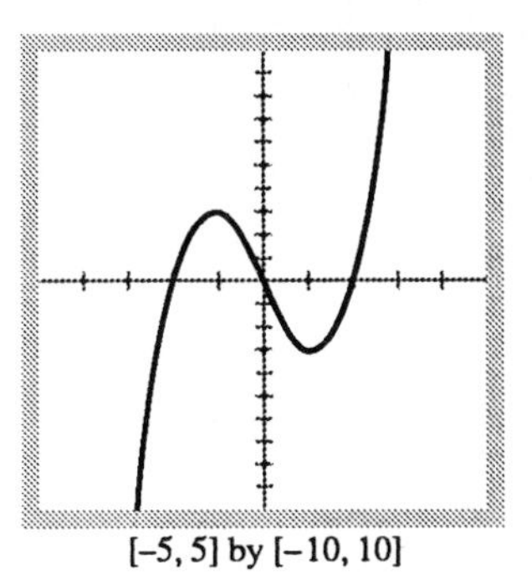
[-5, 5] by [-10, 10]

1.12 A complete graph of $y = x^3 - 4x$.

Solution The graph of $y = x^3 - 4x$ in Figure 1.12 is complete according to Figure 1.10. It appears from Figure 1.12 that $y = 0$, $x = -2$, $x = 0$, and $x = 2$ are the intercepts. We can confirm this information algebraically.
The y-intercepts:

$$y = x^3 - 4x$$
$$y = 0^3 - 4(0) \qquad \text{(Let } x = 0\text{)}$$
$$y = 0$$

The x-intercepts:

$$y = x^3 - 4x$$
$$0 = x^3 - 4x \qquad \text{(Let } y = 0\text{)}$$
$$0 = x(x^2 - 4)$$
$$0 = x(x - 2)(x + 2)$$
$$x = 0, 2, -2$$

We can estimate the coordinates of the local high point near $x = -1$ by reading directly from the graph in Figure 1.12 or by placing the cursor at the point and reading the coordinates to be about $(-1.2, 3.1)$. Similarly, the coordinates of the local low point near $x = 1$ are about $(1.2, -3.1)$. ☰

Finding the x-intercepts in Examples 3 and 6 was easy because the corresponding equations were easy to solve algebraically. In general, finding the x-intercepts will be difficult or impossible algebraically and will require the graphing techniques that will be introduced in Section 1.5.

EXAMPLE 7 Draw a complete graph of $y = x^3 - 2x^2 + x - 30$.

Solution A complete graph must look like one of the four possibilities in Figure 1.10. Figure 1.13 gives three views of the graph of the equation. Only (c) looks like one of the four in Figure 1.10.

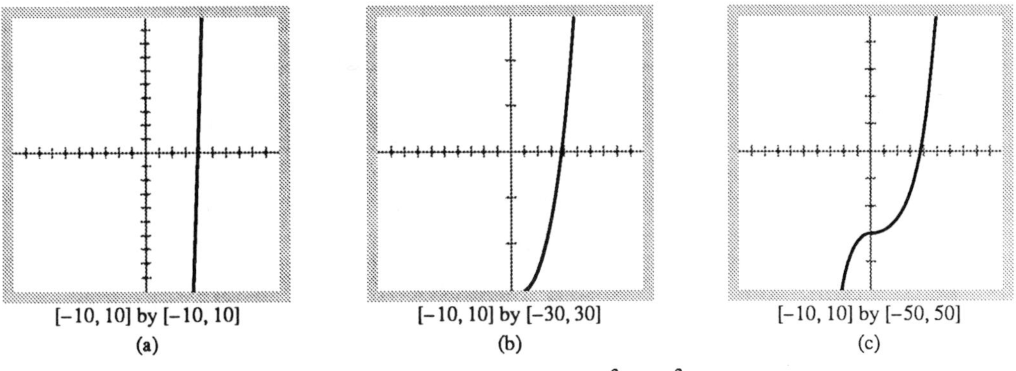
[-10, 10] by [-10, 10] (a) [-10, 10] by [-30, 30] (b) [-10, 10] by [-50, 50] (c)

1.13 Three views of the graph of $y = x^3 - 2x^2 + x - 30$.

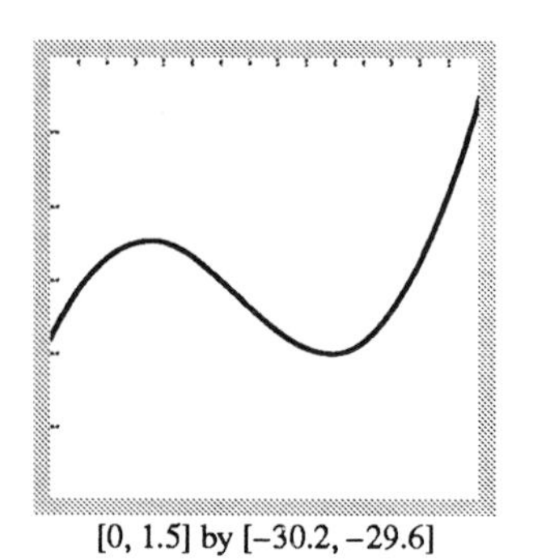

1.14 A closer look at the flat portion of the graph of $y = x^3 - 2x^2 + x - 30$ in Figure 1.13(c).

The relatively flat portion of the graph in Figure 1.13(c) should alert us to the possibility that some of the behavior of the graph may be hidden from view. Figure 1.14 gives a closer look at the flat portion and indicates that a complete graph of this equation has a local high point and a local low point like the third one in Figure 1.10. We can estimate from Figure 1.14 that the coordinates of the local high and low points are about $(0.3, -29.9)$ and $(1, -30)$, respectively.

Notice that the y-intercept is $y = -30$. Any of the three graphs in Figure 1.13 can be used to estimate that the x-intercept is about 4. In Section 1.5, we introduce and use a process called zoom-in to find the x-intercept with more accuracy.

It is not possible to use a single grapher visualization to display the x-intercept, y-intercept, and the local high and low points of the graph of this equation. Thus, Figures 1.13 (c) and 1.14 together constitute a complete graph of this equation illustrating that **sometimes more than one graph is necessary to obtain a complete graph.** ≡

In Chapter 4 we will see how to use calculus to predict the hidden behavior of Figure 1.13(c). Most of the graphs in the book produced by graphers are very reliable. This technology together with the theory of calculus provides us with a powerful technique for the study of graphs and functions in mathematics.

Exercises 1.1

In Exercises 1 and 2, determine which points lie in the indicated viewing rectangle.

1. viewing rectangle: $[-5, 5]$ by $[-10, 10]$
points: $(-6, 2)$, $(0, 6)$, $(5, -5)$, $(-4, 12)$

2. viewing rectangle: $[0, 10]$ by $[0, 100]$
points: $(1.5, 17.5)$, $(0.5, -1)$, $(90, 8)$, $(10, 100)$

In Exercises 3 and 4, determine a viewing rectangle that contains the indicated points.

3. points: $(-17, 3)$, $(21, -12)$, $(18, 76)$

4. points: $(-42, 31)$, $(8, -6)$, $(53, -89)$

In Exercises 5–8, set the range screen on your grapher to enter the indicated viewing rectangle. Choose appropriate values for xScl and yScl, and explain why your choices are reasonable.

5. $[-10, 50]$ by $[-50, 50]$ **6.** $[-1, 1.5]$ by $[1, 2]$

7. $[-1, 0]$ by $[-100, 100]$ **8.** $[-50, 150]$ by $[-2, 2]$

In Exercises 9–14, which of the viewing rectangles give the best complete graph of the indicated equation.

9. $y = 2x^2 - 40x + 150$

a) $[-5, 5]$ by $[-5, 5]$ **b)** $[-10, 10]$ by $[-10, 10]$

c) $[10, 30]$ by $[-50, 100]$

d) $[-100, 100]$ by $[-10, 10]$

e) $[-10, 30]$ by $[-100, 100]$

10. $y = -3x^2 + 9x - 20$

a) $[-10, 10]$ by $[-10, 10]$

b) $[-10, 10]$ by $[-100, 100]$

c) $[2, 10]$ by $[-100, 10]$

d) $[-10, 10]$ by $[-3000, 3000]$

e) $[-300, 300]$ by $[-100, 100]$

11. $y = 20 + 9x - x^3$

a) $[-10, 10]$ by $[-10, 10]$

b) $[-1, 10]$ by $[-30, 40]$

c) $[-5, 0]$ by $[0, 100]$ **d)** $[-5, 5]$ by $[-500, 1000]$

e) $[-10, 10]$ by $[-50, 50]$

12. $y = x^3 - x + 15$

a) $[-5, 5]$ by $[-10, 10]$

b) $[-10, 10]$ by $[-100, 100]$

c) $[-10, 10]$ by $[-30, 30]$

d) $[-5, 5]$ by $[-5, 15]$

e) $[-100, 100]$ by $[-100, 100]$

13. $y = 3x - 800$
   a) $[0, 500]$ by $[0, 500]$ b) $[260, 270]$ by $[-10, 10]$
   c) $[-10, 10]$ by $[-810, -790]$
   d) $[-10, 10]$ by $[-10, 10]$
   e) $[-100, 500]$ by $[-1000, 500]$

14. $y = -2x^2 + 500$
   a) $[-10, 10]$ by $[10, 10]$
   b) $[240, 260]$ by $[-10, 10]$
   c) $[-10, 10]$ by $[490, 510]$
   d) $[-100, 400]$ by $[-200, 800]$
   e) $[0, 400]$ by $[0, 400]$

Find the intercepts and sketch the graph of the equations in Exercises 15–20. Remember, "sketch" means to graph without a graphing utility.

15. $y = x + 1$
16. $y = -x + 1$
17. $y = -x^2$
18. $y = 4 - x^2$
19. $x = -y^2$
20. $x = 1 - y^2$

Draw a complete graph of the equations in Exercises 21–26. Use the graph to estimate the intercepts and then confirm the values algebraically.

21. $y = 3x - 5$
22. $y = 4 - 5x$
23. $y = 10 + x - 2x^2$
24. $y = 2x^2 - 2x - 12$
25. $y = 2x^2 - 8x + 3$
26. $y = -3x^2 - 6x - 1$

Draw a complete graph of the equations in Exercises 27–34. Use the graph to estimate the intercepts and the coordinates of any local high or low points.

27. $y = x^2 + 4x + 5$
28. $y = -3x^2 + 12x - 8$
29. $y = 12x - 3x^3$
30. $y = 2x^3 - 2x$
31. $y = -x^3 + 9x - 1$
32. $y = x^3 - 4x + 3$
33. $y = x^3 + 2x^2 + x + 5$
34. $y = 2x^3 - 5.5x^2 + 5x - 5$

In Exercises 35–38, determine a and b so that the screen coordinates of the viewing rectangle $[-10, a]$ by $[-10, b]$ have the indicated property. The answer depends on the grapher used.

35. $\Delta x = \Delta y = 1$
36. $\Delta x = \Delta y = 0.1$
37. $\Delta x = 0.5, \Delta y = 2$
38. $\Delta x = 2, \Delta y = 10$

Refer to the *Explore* activity in this section. In Exercises 39 and 40, let (a,b) be a point in the coordinate plane and (x_i, y_j) the corresponding screen coordinate.

39. Explain why the error in using x_i as an approximation to a is at most $\Delta x/2$.

40. Explain why the error in using y_j as an approximation to b is at most $\Delta y/2$.

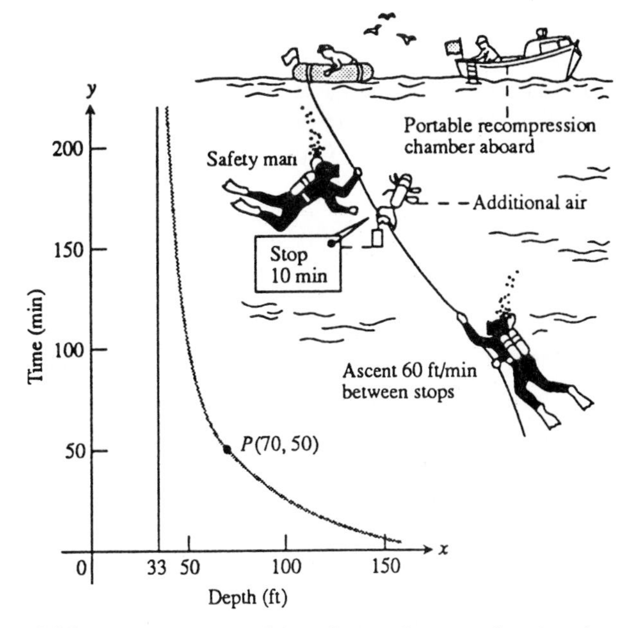

1.15 The coordinates of the points on the curve show how long scuba divers breathing compressed air can safely stay below the surface without a decompression stop. (Data from *U.S. Navy Diving Manual,* NAVSKIPS 250-538.)

Decompression Stops

Scuba divers often have to make decompression stops on their way to the surface after deep or long dives. A stop, which may vary from a few minutes to more than an hour, provides time for the safe release of nitrogen and other gases absorbed by the tissues and blood while the body was under pressure. Dives to 33 ft or less do not require decompression stops. Dives to greater depths can be made without return stops if the diver does not stay down too long. The graph in Fig. 1.15 shows the longest times that a diver breathing compressed air may spend at various depths and still surface directly (at 60 ft/min). The times shown are total lengths of the dives, not just the times spent at maximum depth.

The coordinates of the point P in Fig. 1.15 are $x = 70$ and $y = 50$, so a diver going to 70 ft (but no deeper) can return safely without stopping if the total dive time is 50 min or less. Decompression stops are needed for dives plotted above or to the right of the curve, because their times exceed the corresponding limits set for their depths. Decompression stops are not needed for dives plotted on the curve or for dives plotted below or to the left of it, because the lengths of these dives do not exceed the limits for their depths. Exercises 41–42 are about these time limits.

41. You have been working at 100 ft below the surface for 1 hr. Do you need a decompression stop on the way up?

42. Which of the following dives need decompression stops and which do not?

a) $(40, 100)$ **b)** $(100, 40)$
c) $(70, 100)$ **d)** $(50, 50)$

1.2 Slope, and Equations for Lines

One of the many reasons calculus has proved so useful over the years is that it is the right mathematics for relating a quantity's rate of change to its graph. Explaining this relationship is one of the goals of this book. Our basic plan is first to define what we mean by the slope of a line. Then we define the slope of a curve at each point on the curve. Later we shall relate the slope of a curve to a rate of change. Just how this is done will become clear as the book goes on. Our first step is to find a practical way to calculate the slopes of lines.

Increments

When a particle moves from one point to another in the plane, the net changes in its coordinates are calculated by subtracting the coordinates of the point where it starts from the coordinates of the point where it stops.

EXAMPLE 1 From $A(4, -3)$ to $B(2, 5)$, the net changes in coordinates (Fig. 1.16) are

$$\Delta x = 2 - 4 = -2, \quad \Delta y = 5 - (-3) = 8.$$

The symbols Δx and Δy in Example 1 are read "delta x" and "delta y." They denote net changes or **increments** in the variables x and y. The letter Δ is a capital Greek "dee," for "difference." Neither Δx nor Δy denotes multiplication; Δx is not
"delta times x" nor is Δy "delta times y."

Definition

When a particle moves from (x_1, y_1) to (x_2, y_2), the **increments** are

$$\Delta x = x_2 - x_1 \quad \text{and} \quad \Delta y = y_2 - y_1. \qquad (1)$$

EXAMPLE 2 From $C(5, 6)$ to $D(5, 1)$, the increments are (Fig. 1.16)

$$\Delta x = 5 - 5 = 0, \quad \Delta y = 1 - 6 = -5.$$

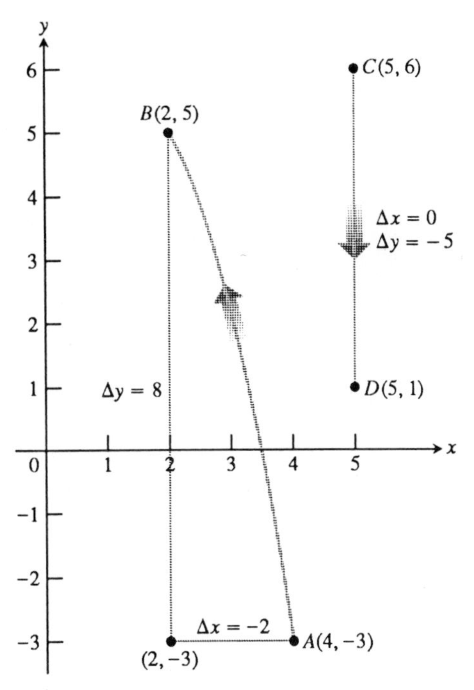

1.16 From A to $B, \Delta x = -2$ and $\Delta y = 8$. From C to $D, \Delta x = 0$ and $\Delta y = -5$.

Slopes of Nonvertical Lines

All lines except vertical lines have slopes. We calculate slopes from changes in coordinates. Once we see how this is done, we shall also see why vertical lines are an exception.

To begin, let L be a nonvertical line in the plane. Let $P_1(x_1,\ y_1)$ and $P_2(x_2,\ y_2)$ be two points on L (Fig. 1.17). We call $\Delta y = y_2 - y_1$ the **rise** from P_1 to P_2 and $\Delta x = x_2 - x_1$ the **run** from P_1 to P_2. Since L is not vertical, $\Delta x \neq 0$ and we may define the **slope** of L to be $\Delta y/\Delta x$, the amount of rise per unit of run. It is conventional to denote the slope by the letter m.

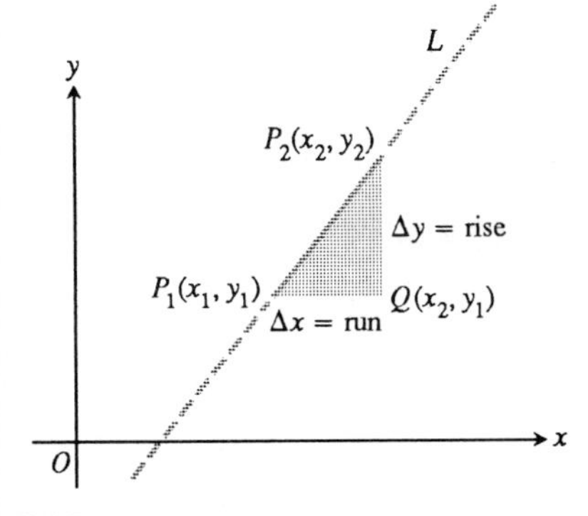

1.17 The slope of the line is

$$m = \frac{\Delta y}{\Delta x} = \frac{\text{rise}}{\text{run}}.$$

Definition

The **slope** of a nonvertical line is $m = \dfrac{\text{rise}}{\text{run}} = \dfrac{\Delta y}{\Delta x} = \dfrac{y_2 - y_1}{x_2 - x_1}$. (2)

Suppose that instead of choosing the points P_1 and P_2 to calculate the slope in Eq. (2), we choose a different pair of points P_1' and P_2' on L and calculate

$$m' = \frac{y_2' - y_1'}{x_2' - x_1'} = \frac{\Delta y'}{\Delta x'}. \tag{3}$$

Will we get the same value for the slope? In other words, will $m' = m$? The answer is yes. The numbers m and m' are equal because they are the ratios

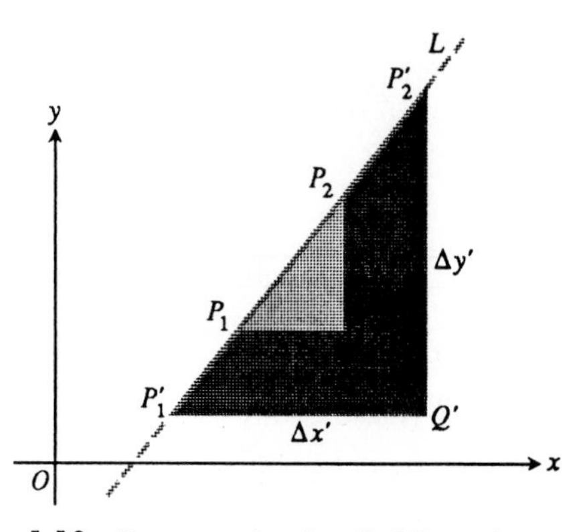

1.18 Because triangles P_1QP_2 and $P'_1Q'P'_2$ are similar, $\Delta y/\Delta x = \Delta y'/\Delta x'$.

of corresponding sides of similar triangles (Fig. 1.18):

$$m' = \frac{\Delta y'}{\Delta x'} = \frac{\Delta y}{\Delta x} = m. \tag{4}$$

The slope of a line depends only on how steeply the line rises or falls and not on the points we use to calculate it.

A line that goes uphill as x increases (Fig. 1.19) has a positive slope. A line that goes downhill as x increases, like the line in Fig. 1.20, has a negative slope. A horizontal line has slope zero. The points on it all have the same y-coordinate, so $\Delta y = 0$.

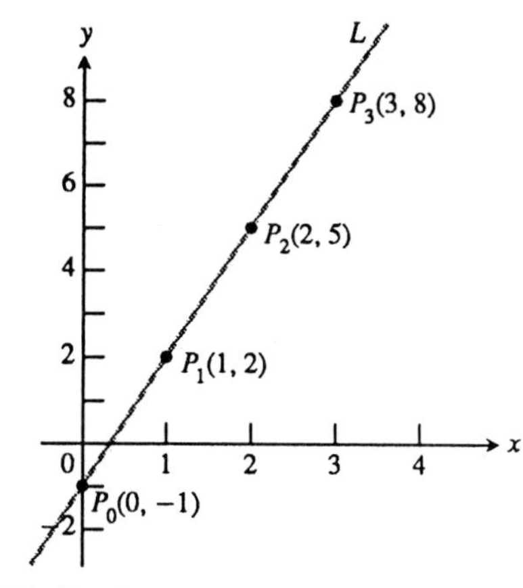

1.19 The slope of this line is

$$m = \frac{\Delta y}{\Delta x} = \frac{(5-2)}{(2-1)} = 3.$$

This means that $\Delta y = 3\Delta x$ for every change of position on the line. (Compare the coordinates of the marked points.)

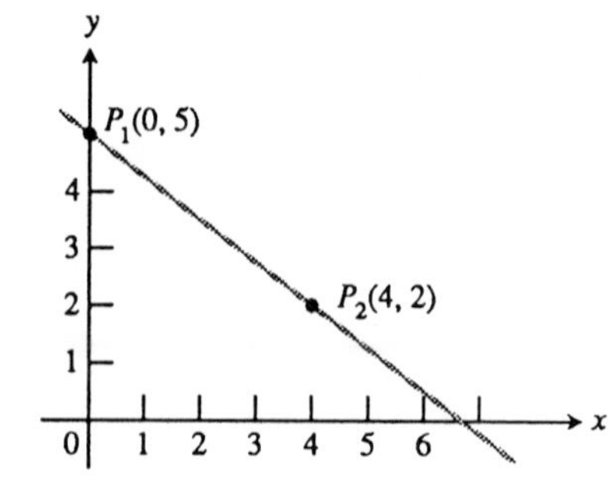

1.20 The slope of this line is

$$m = \frac{\Delta y}{\Delta x} = \frac{(2-5)}{(4-0)} = \frac{-3}{4}.$$

This means that y decreases 3 units every time x increases 4 units.

RAILROADS AND HIGHWAYS

Civil engineers calculate the slope of a roadbed, by calculating the ratio of the distance it rises or falls to the distance it runs horizontally. They call this ratio the **grade** of the roadbed, usually written as a percentage. Along the coast, railroad grades are usually less than 2%. In the mountains, they may go as high as 4%. Highway grades are usually less than 5%.

In analytic geometry we calculate slopes the same way, but we usually do not express them as percentages.

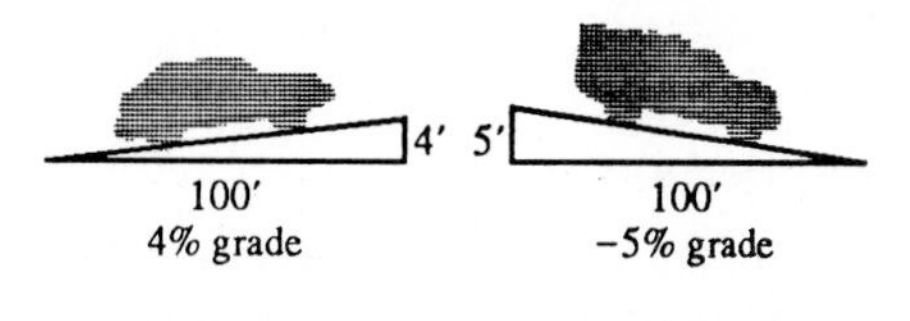

The formula $m = \Delta y/\Delta x$ does not apply to vertical lines because Δx is zero along a vertical line. We express this by saying that vertical lines have no slope or that the slope of a vertical line is undefined.

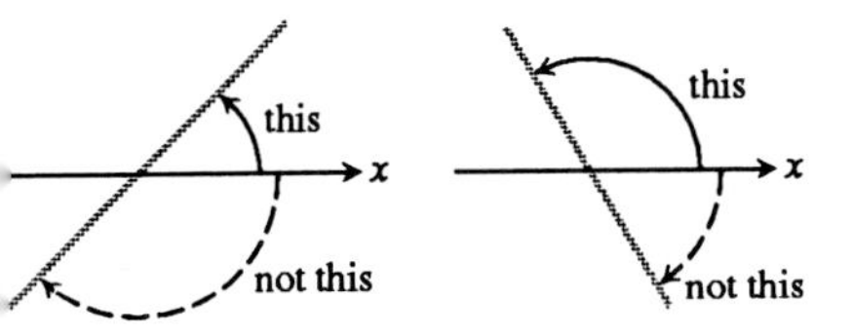

1.21 Angles of inclination are measured counterclockwise from the x-axis.

Angles of Inclination

The **angle of inclination** of a line that crosses the x-axis is the smallest angle we get when we measure counterclockwise from the x-axis around the point of intersection (Fig. 1.21). The angle of inclination of a horizontal line is taken to be $0°$. Thus, angles of inclination may have any measure from $0°$ up to but not including $180°$.

The slope of a line is the tangent of the line's angle of inclination. Figure 1.22 shows why this is true. If m denotes the slope and ϕ the angle, then

$$m = \tan \phi. \tag{5}$$

THE WORLD'S STEEPEST STREETS

The world's steepest street is Baldwin Street in Dunedin, near the southeast corner of New Zealand's South Island; it has a slope of 0.790 and an angle of inclination of 38.3° (as steep as a flight of stairs). The runners-up are both in San Francisco, Filbert Street and 22nd Street, with slopes of 0.613 and angles of 31.5°.

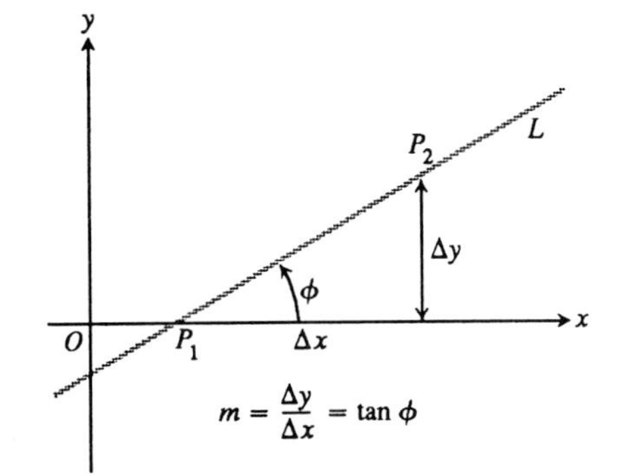

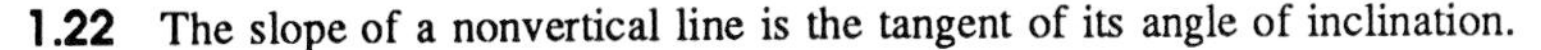

1.22 The slope of a nonvertical line is the tangent of its angle of inclination.

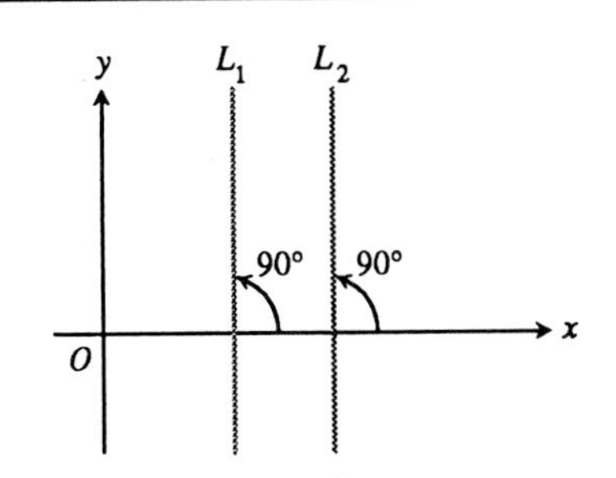

1.23 When $\phi = 90°$, $\tan \phi$ is undefined. Vertical lines have equal angles of inclination but no slope.

EXAMPLE 3 The slopes of lines become increasingly large as their angles of inclination approach 90° from either side, as the following computations show:

ϕ approaching 90° from below	ϕ approaching 90° from above
$\tan 89.9° = 573$	$\tan 90.1° = -573$
$\tan 89.99 = 5730$	$\tan 90.01 = -5730$
$\tan 89.999 = 57296$	$\tan 90.001 = -57296$
$\tan 89.9999 = 572958$	$\tan 90.0001 = -572958$

We say that the slope of a line "becomes infinite" as its angle of inclination approaches 90°. However, vertical lines themselves have no slope (Fig. 1.23).

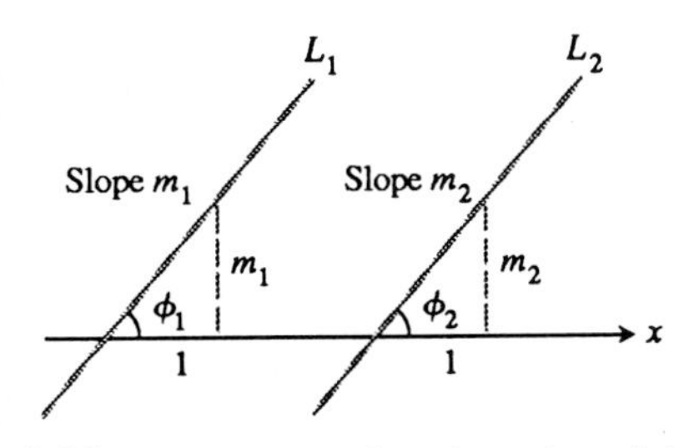

1.24 If $m_1 = m_2$, then $\phi_1 = \phi_2$ and the lines are parallel.

Lines That Are Parallel or Perpendicular

Parallel lines have equal angles of inclination. Hence, if they are not vertical, parallel lines have the same slope. Conversely, lines with equal slopes have equal angles of inclination and are therefore parallel (Fig. 1.24).

If neither of two perpendicular lines L_1 and L_2 is vertical, their slopes m_1 and m_2 are related by the equation $m_1 m_2 = -1$. Figure 1.25 shows why:

$$m_1 = \tan \phi_1 = \frac{a}{h}, \qquad \text{while} \qquad m_2 = -\frac{h}{a}. \tag{6}$$

Hence,

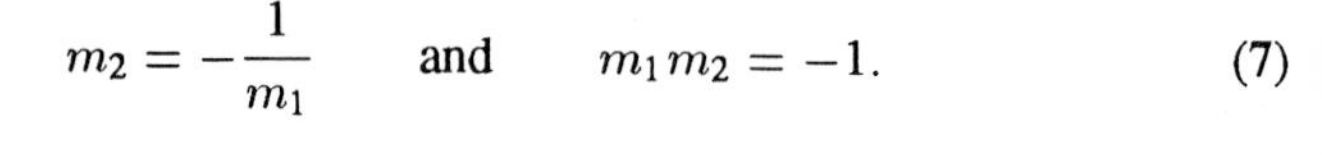

$$m_2 = -\frac{1}{m_1} \qquad \text{and} \qquad m_1 m_2 = -1. \tag{7}$$

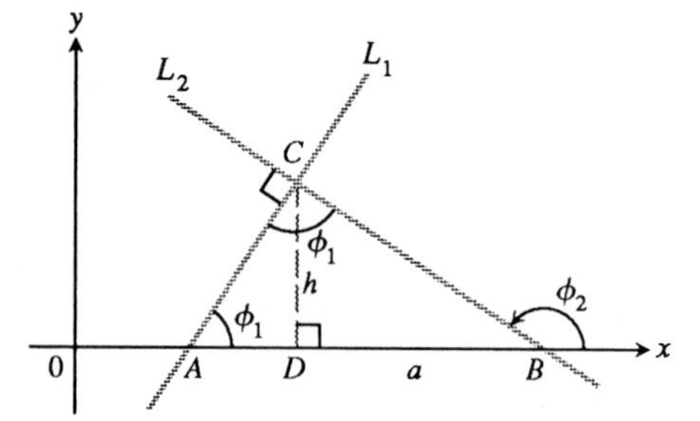

1.25 ΔADC is similar to ΔCDB. Hence ϕ_1 is also the upper angle in ΔCDB. From the sides of ΔCDB, we read $\tan \phi_1 = a/h$.

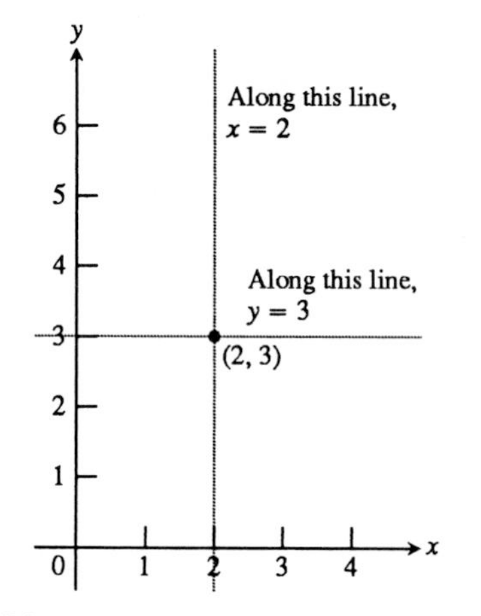

1.26 The standard equations for the horizontal and vertical lines through the point (2, 3) are $x = 2$ and $y = 3$.

EXAMPLE 4 The slope of a line perpendicular to a line of slope $3/4$ is $-4/3$.

Equations for Lines

Definition

> An **equation for a line** is an equation that is satisfied by the coordinates of every point on the line but is not satisfied by the coordinates of points that lie elsewhere.

Horizontal and Vertical Lines

The standard equations for the horizontal and vertical lines through a point (a, b) are simply $y = b$ and $x = a$ (Fig. 1.26). A point (x, y) lies on the horizontal line through (a, b) if and only if $y = b$. It lies on the vertical line through (a, b) if and only if $x = a$.

Point–Slope Equations

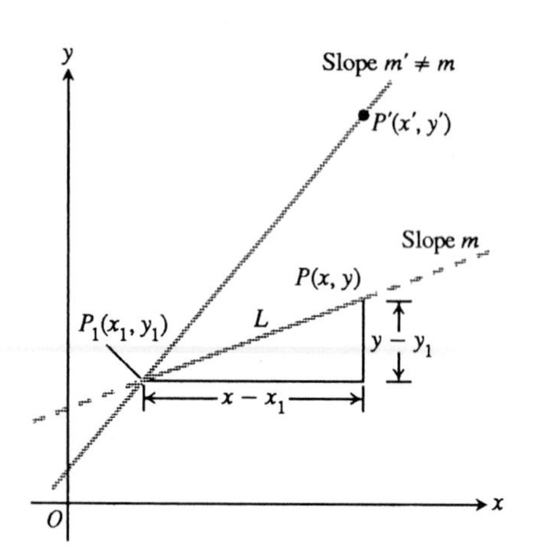

1.27 If L is the line through $P_1(x_1, y_1)$ whose slope is m, then other points $P(x, y)$ lie on this line if and only if slope $PP_1 = m$. This fact gives us the point-slope equation for L.

To write an equation for a line L that is not vertical, it is enough to know its slope m and the coordinates of a point $P_1(x_1, y_1)$ on it. If $P(x, y)$ is any other point on L (Fig. 1.27), then $x \neq x_1$ and we can write the slope of L as

$$\frac{y - y_1}{x - x_1}. \tag{8}$$

We can then set this expression equal to m to get

$$\frac{y - y_1}{x - x_1} = m. \tag{9}$$

Multiplying both sides of Eq. (9) by $x - x_1$ gives us the more useful equation

$$y - y_1 = m(x - x_1). \tag{10}$$

Equation (10) is an equation for L, as we can check right away. Every point (x, y) on L satisfies the equation—even the point (x_1, y_1). What about the points not on L? If $P'(x', y')$ is a point not on L (Fig. 1.27), then the slope m' of $P'P_1$ is different from m, and the coordinates x' and y' of P' do not satisfy Eqs. (9) and (10).

Definition

The equation

$$y - y_1 = m(x - x_1) \tag{11}$$

is the **point-slope equation** of the line that passes through the point (x_1, y_1) with the slope m.

In Example 5 we use a graphing utility to draw the graph of the line. When graphing lines it is important for distance in the x- and y-direction to look the same so that the slope looks correct visually. This means that the screen coordinate parameters Δx and Δy discussed in Section 1.1 must be equal. A viewing rectangle in which $\Delta x = \Delta y$ is called a **square viewing rectangle**. Some graphing utilities have a feature that permits automatic squaring of the viewing rectangle.

EXAMPLE 5 Write an equation for the line that passes through the point (2, 3) with slope −3/2, and sketch a complete graph of the line. Support with a graphing utility.

Solution

$$y - y_1 = m(x - x_1) \qquad \left(\begin{array}{l}\text{Start with the general}\\ \text{point-slope equation, Eq. (11)}\end{array}\right)$$

$$y - 3 = -\frac{3}{2}(x - 2) \qquad (\text{Take } m = -3/2 \text{ and } (x_1, y_1) = (2, 3))$$

$$y = -\frac{3}{2}x + 6$$

This is an equation for the line. Notice that $y = 6$ is the y-intercept and $x = 4$ is the x-intercept. Figure 1.28 gives both a sketched complete graph and a grapher visualization complete graph in a square viewing rectangle of the line. ≡

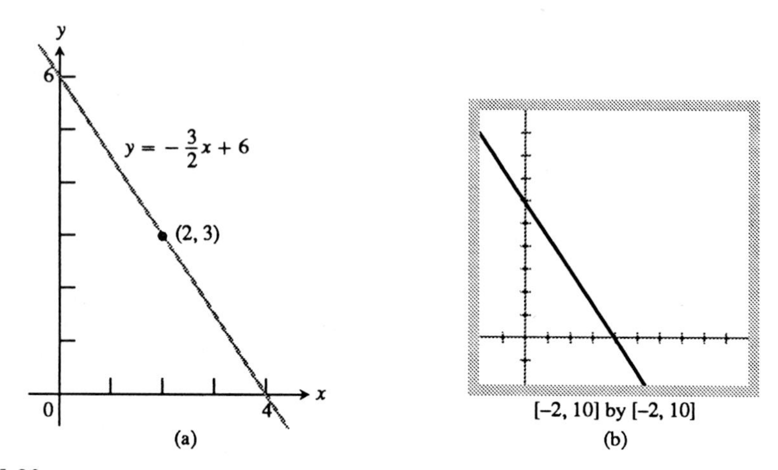

1.28 Complete graphs of the line $y = -3/2x + 6$ of Example 5: (a) sketched and (b) grapher generated.

EXAMPLE 6 Write an equation for the line through $(-2, -1)$ and $(3, 4)$, and draw a complete graph.

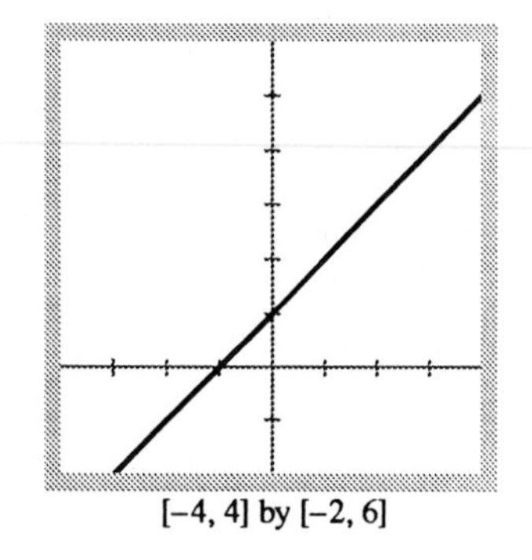

1.29 A complete graph of the line $y = x + 1$ of Example 6.

Solution We first calculate the slope and then use Eq. (11).

$$m = \frac{-1-4}{-2-3} = \frac{-5}{-5} = 1.$$

The (x_1, y_1) in Eq. (11) can be either $(-2, -1)$ or $(3, 4)$:

With $(x_1,\ y_1) = (-2, -1)$	With $(x_1,\ y_1) = (3, 4)$
$y - (-1) = 1 \cdot (x - (-2))$	$y - 4 = 1 \cdot (x - 3)$
$y + 1 = x + 2$	$y - 4 = x - 3$
$y = x + 1$	$y = x + 1$

Same result

Either way, $y = x + 1$ is an equation for the line (Fig. 1.29). 

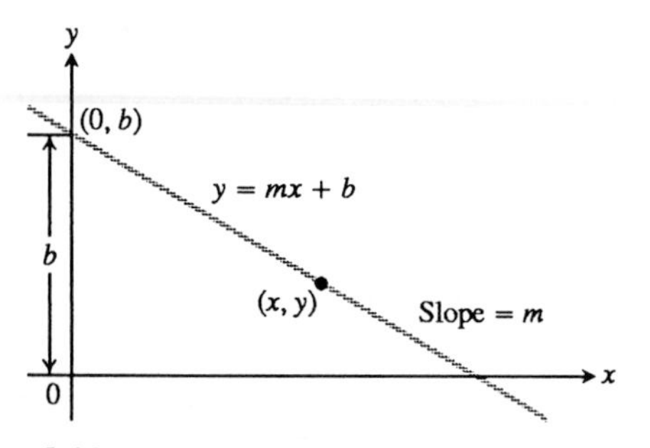

1.30 The line with slope m and y-intercept b has equation $y = mx + b$.

Slope–Intercept Equations

Figure 1.30 shows a line with slope m and y-intercept b. If we take $(x_1,\ y_1) = (0, b)$ in the point-slope equation for the line, we find that

$$y - b = m(x - 0). \tag{12}$$

When rearranged, this becomes

$$y = mx + b. \tag{13}$$

Definition

The equation

$$y = mx + b \tag{14}$$

is the **slope–intercept equation** of the line with slope m and y-intercept b.

EXAMPLE 7 The slope-intercept equation of the line with slope 2 and y-intercept 5 is

$$y = 2x + 5.$$

Explore Activity

1. Draw complete graphs of the lines $y = mx$ in the same square viewing rectangle for $m = 1, 2, 4, 5, -1, -3$.
2. What points do the lines in 1 have in common?
3. Compute the angle of inclination of each of the lines in 1.
4. Draw complete graphs of the lines $y = mx + 3$ in the same square viewing rectangle for $m = 1, 2, 4, 5, -1, -3$.
5. What points do the lines in 4 have in common?
6. Compute the angle of inclination of each of the lines in 4.
7. What can you conclude about the graphs of the lines $y = mx - 5$ for several distinct values of m?

You can conclude from the above Explore Activity that graphs of lines with positive slope rise as x increases and graphs of lines with negative slope fall as x increases.

Explore Activity

1. Draw complete graphs of the lines $y = 2x + b$ in the same square viewing rectangle for $b = 2, 3, -2, -5$.
2. What do the lines in 1 have in common? How are they different?
3. Draw complete graphs of the lines $y = -3x + b$ in the same square viewing rectangle for $b = 2, 3, -2, -5$.
4. What do the lines in 3 have in common? How are they different?
5. What can you conclude about the graphs of the lines $y = 4x + b$ for several distinct values of b?

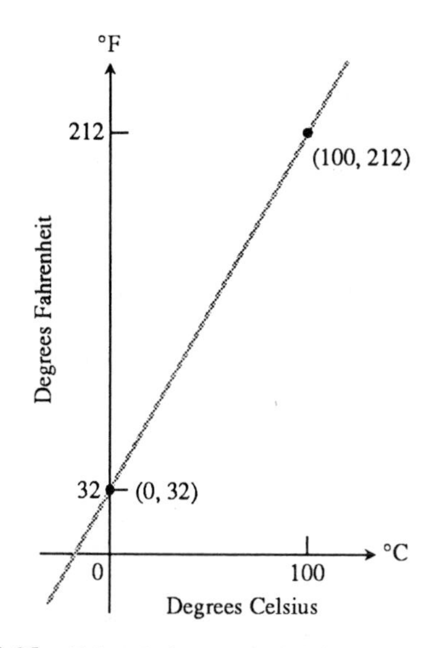

1.31 Fahrenheit versus Celsius temperature.

EXAMPLE 8 Celsius vs. Fahrenheit. The standard equation for converting Celsius temperature to Fahrenheit temperature is a slope-intercept equation. If we plot Fahrenheit temperature against Celsius temperature in a coordinate plane, the points we plot always lie along a straight line (Fig. 1.31). The line passes through the point (0, 32) because $F = 32$ when $C = 0$. It also passes through the point (100, 212) because $F = 212$ when $C = 100$.

This is enough information to make a formula for F in terms of C. The line's slope is

$$m = \frac{212 - 32}{100 - 0} = \frac{180}{100} = \frac{9}{5}.$$

The F-intercept of the line is

$$b = 32.$$

The resulting slope-intercept equation for the line,

$$F = \frac{9}{5}C + 32, \tag{15}$$

is the formula we seek.

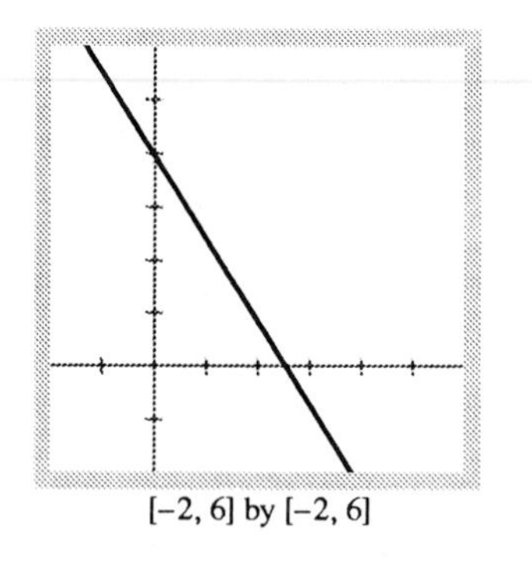

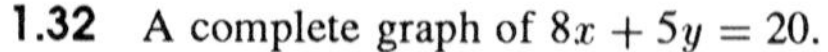

1.32 A complete graph of $8x + 5y = 20$.

EXAMPLE 9 Find the slope and y-intercept and draw a complete graph of the line $8x + 5y = 20$.

Solution Solve the equation for y to put the equation in slope-intercept form. Then read the slope and y-intercept from the equation:

$$8x + 5y = 20$$
$$5y = -8x + 20$$
$$y = -\frac{8}{5}x + 4.$$

The slope is $m = -8/5$. The y-intercept is $b = 4$.

We enter the equation $y = -8/5x + 4$ in our graphing utility to obtain the complete graph in Figure 1.32.

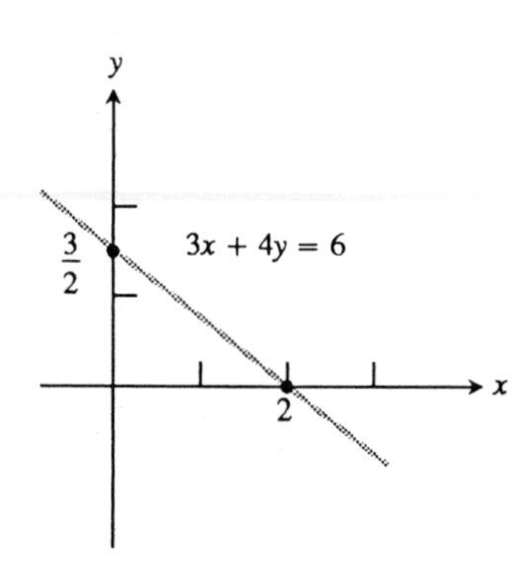

1.33 To sketch a graph of $3x + 4y = 6$, mark the intercepts and draw a line through the marked points.

EXAMPLE 10 Sketch a complete graph of $3x + 4y = 6$.

Solution

STEP 1: Find the x-intercept by setting $y = 0$ to obtain $3x = 6$, or $x = 2$.

STEP 2: Find the y-intercept by setting $x = 0$ to obtain $4y = 6$, or $y = 3/2$.

STEP 3: Plot the intercepts and sketch the line (Figure 1.33).

Notice in Example 9 we needed to solve for y in terms of x in order to use a grapher. In Example 10, to sketch a graph we found the x- and y-intercepts. Generally speaking, it is easier to sketch graphs of lines by hand than to use graphers.

Distance between Points

The coordinates of two points can tell us how far apart the two points are.

If the points lie on a horizontal line, the distance between them is the usual number-line distance between their x-coordinates (Fig. 1.34a).

If the points lie on a vertical line, the distance between them is the number-line distance between their y-coordinates (Fig. 1.34b).

If the line joining the points is not parallel to either coordinate axis, we calculate the distance between the points with the Pythagorean theorem (Fig. 1.34c). The resulting formula works for the other cases as well, so there is only one formula to remember:

Distance Formula for Points in the Plane

The distance between $P(x_1, y_1)$ and $Q(x_2, y_2)$ is

$$d = \sqrt{(x_2 - x_1)^2 + (y_2 - y_1)^2}. \tag{16}$$

EXAMPLE 11 The distance between $P(-1, 2)$ and $Q(3, 4)$ is

$$\sqrt{(3 - (-1))^2 + (4 - 2)^2} = \sqrt{(4)^2 + (2)^2} = \sqrt{20} = \sqrt{4 \cdot 5} = 2\sqrt{5}.$$

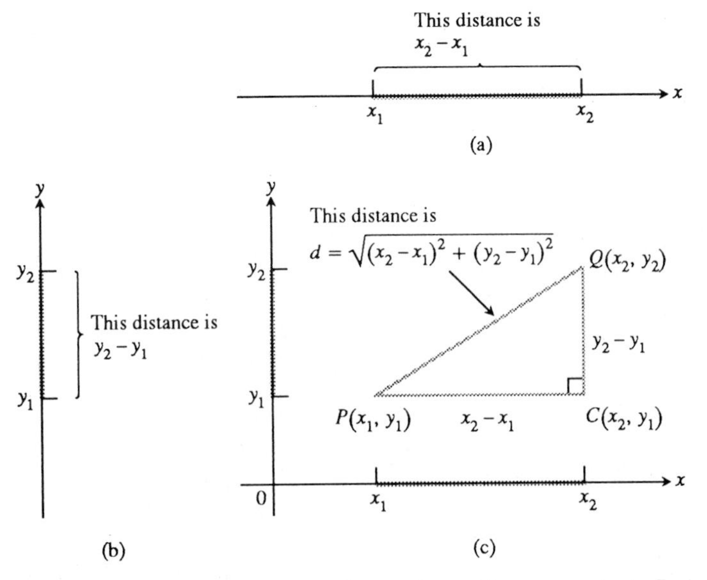

1.34 To calculate the distance between $P(x_1, y_1)$ and $Q(x_2, y_2)$, apply the Pythagorean theorem to triangle PCQ.

Absolute Value

The **absolute value** or **magnitude** of a number x, denoted by $|x|$ (read "the absolute value of x"), is defined by the formula

$$|x| = \begin{cases} x & \text{if } x \geq 0 \\ -x & \text{if } x < 0. \end{cases} \tag{17}$$

The vertical lines in the symbol $|x|$ are called *absolute value bars.*

EXAMPLE 12

$$|3| = 3$$
$$|0| = 0$$
$$|-5| = 5$$

Another way to find the absolute value of a number is to square the number and then take the positive square root:

$$|x| = \sqrt{x^2}. \tag{18}$$

EXAMPLE 13

$$|2| = \sqrt{2^2} = \sqrt{4} = 2$$
$$|-2| = \sqrt{(-2)^2} = \sqrt{4} = 2$$

When we do arithmetic with absolute values, we can always use the following rules:

Arithmetic with Absolute Values

1. $|-a| = |a|$ A number and its negative have the same absolute value.
2. $|ab| = |a|\,|b|$ The absolute value of a product is the product of the absolute values.
3. $\left|\frac{a}{b}\right| = \frac{|a|}{|b|}$ The absolute value of a quotient is the quotient of the absolute values.

EXAMPLE 14

$$|-\sin x| = |\sin x|$$
$$|-2(x+5)| = |-2|\,|x+5| = 2|x+5|$$
$$\left|\frac{3}{x}\right| = \frac{|3|}{|x|} = \frac{3}{|x|}$$

The absolute value of a sum of two numbers is never larger than the sum of their absolute values. When we put this in symbols, we get the important triangle inequality.

The Triangle Inequality

$$|a+b| \le |a| + |b| \text{ for all numbers } a \text{ and } b. \qquad \textbf{(19)}$$

EXAMPLE 15 The number $|a+b|$ is less than $|a|+|b|$ if a and b have different signs. In all other cases, $|a+b|$ equals $|a|+|b|$.

$$|-3+5| = |2| = 2 < |-3| + |5| = 8$$

$$|3+5| = |8| = 8 = |3| + |5|$$

$$|-3+0| = |-3| = 3 = |-3| + |0|$$

$$|-3-5| = |-8| = 8 = |-3| + |-5|.$$

Notice that absolute value bars in expressions such as $|-3+5|$ also work like parentheses: We do the arithmetic inside *before* we take the absolute value.

Absolute Values and Distance

The numbers $|a-b|$ and $|b-a|$ are always equal because

$$|a-b| = |(-1)(b-a)| = |-1|\,|b-a| = |b-a|. \qquad (20)$$

Both $|a-b|$ and $|b-a|$ give the distance between the points a and b on the number line (Fig. 1.35).

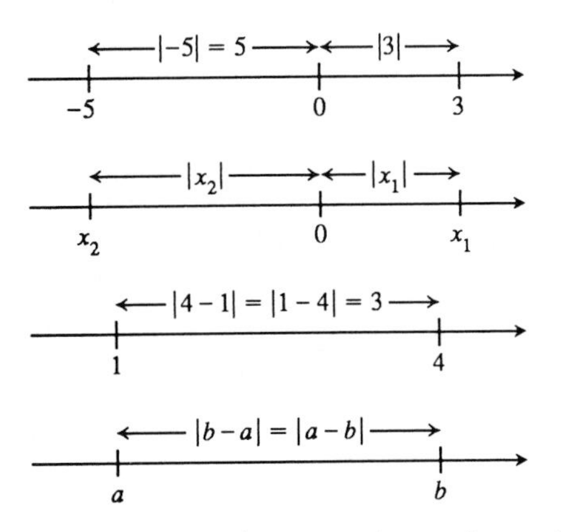

1.35 Absolute values give distances between points on the number line.

Number-line Distance

$|a - b| = |b - a|$ for all numbers a and b. This number is the distance between a and b on the number line.

In Figure 1.34(a) and (b), the distance between the points x_1 and x_2 can be expressed as $|x_1 - x_2|$ or $|x_2 - x_1|$, and the distance between y_1 and y_2 as $|y_1 - y_2|$ or $|y_2 - y_1|$.

The Distance from a Point to a Line

To calculate the distance from a point $P(x_1, y_1)$ to a line L, we find the point $Q(x_2, y_2)$ at the foot of the perpendicular from P to L and calculate the distance from P to Q. The next example shows how this is done.

EXAMPLE 16 Find the distance from the point $P(2, 1)$ to the line $L : y = x + 2$.

Solution We solve the problem in three steps (Fig. 1.36): (1) Find an equation for the line L' through P perpendicular to L; (2) find the point Q where L' meets L; and (3) calculate the distance between P and Q.

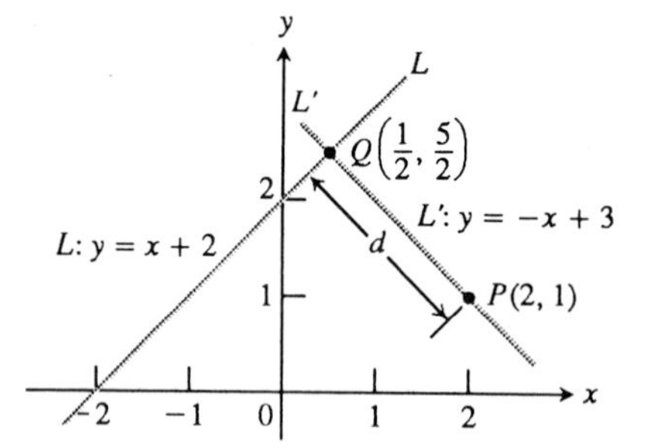

1.36 The distance d from P to L is measured along the line L' perpendicular to L. We can calculate it from the coordinates of P and Q.

STEP 1: We find an equation for the line L' through $P(2, 1)$ perpendicular to L. The slope of $L : y = x + 2$ is $m = 1$. The slope of L' is therefore $m' = -1/1 = -1$. We set $(x_1, y_1) = (2, 1)$ and $m = -1$ in Eq. (11) to find L':

$$y - 1 = -1(x - 2)$$

$$y = -x + 2 + 1$$

$$y = -x + 3.$$

STEP 2: Find the point Q by solving the equations for L and L' simultaneously. To find the x-coordinate of Q, we equate the two expressions for y:

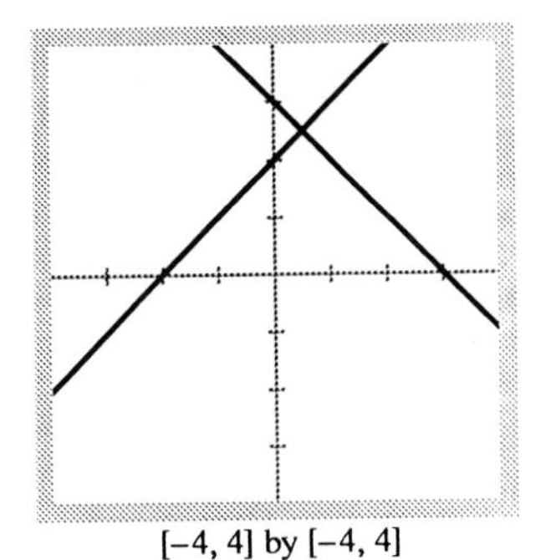

1.37 Point of intersection of the two lines $y = x + 2$ and $y = -x + 3$ of Example 16.

$$x + 2 = -x + 3$$
$$2x = 1$$
$$x = \frac{1}{2}.$$

We can now obtain the y-coordinate by substituting $x = 1/2$ in the equation for either line. We choose $y = x + 2$ arbitrarily and find

$$y = \frac{1}{2} + 2 = \frac{5}{2}.$$

The coordinates of Q are $(1/2, 5/2)$, which is supported by the graph of the two lines in the square viewing rectangle shown in Figure 1.37.

STEP 3: We calculate the distance between $P(2, 1)$ and $Q(1/2, 5/2)$:

$$d = \sqrt{\left(2 - \frac{1}{2}\right)^2 + \left(1 - \frac{5}{2}\right)^2} = \sqrt{\left(\frac{3}{2}\right)^2 + \left(-\frac{3}{2}\right)^2} = \sqrt{\frac{18}{4}} = \frac{3}{2}\sqrt{2}.$$

The distance from P to L is $(3/2)\sqrt{2}$. ∎

The General Linear Equation

The equation

$$Ax + By = C \quad (A \text{ and } B \text{ not both zero}) \tag{21}$$

is called the **general linear equation** because its graph is always a line and because every line has an equation in this form. We shall not take the time to prove this, but notice that all the equations in this section can be arranged in this form. A few of them are in this form already.

Slope

1. The slope of the line through $P_1(x_1, y_1)$ and $P_2(x_2, y_2)$, $x_1 \neq x_2$, is
$$m = \frac{\text{rise}}{\text{run}} = \frac{y_2 - y_1}{x_2 - x_1} = \frac{\Delta y}{\Delta x}.$$
2. $m = \tan\phi$ (ϕ is the angle of inclination).
3. Vertical lines have no slope.
4. Horizontal lines have slope 0.
5. For lines that are neither horizontal nor vertical it is handy to remember:
   - **a)** they are parallel $\Leftrightarrow m_2 = m_1$;
   - **b)** they are perpendicular $\Leftrightarrow m_2 = -1/m_1$.

   (The symbol $\Leftrightarrow$ is read "if and only if.")

Equations for Lines

$x = a$	Vertical line through (a, b)
$y = b$	Horizontal line through (a, b)
$y = mx + b$	Slope–intercept equation
$y - y_1 = m(x - x_1)$	Point–slope equation
$Ax + By = C$	General linear equation (A and B not both zero)

Exercises 1.2

In Exercises 1–4, a particle moves from A to B. Find the net changes Δx and Δy in the particle's coordinates.

1. $A(1, 2), B(-1, -1)$ **2.** $A(-3, 2), B(-1, -2)$

3. $A(-3, 1), B(-8, 1)$ **4.** $A(0, 4), B(0, -2)$

Plot the points A and B in Exercises 5–8. Then find the slope (if any) of the line they determine. Also find the slope (if any) of the lines perpendicular to line AB.

5. $A(1, -2), B(2, 1)$ **6.** $A(-2, -1), B(1, -2)$

7. $A(2, 3), B(-1, 3)$ **8.** $A(1, 2), B(1, -3)$

In Exercises 9–14, find the distances between the given points.

9. $(1, 0)$ and $(0, 1)$ **10.** $(2, 4)$ and $(-1, 0)$

11. $(2\sqrt{3}, 4)$ and $(-\sqrt{3}, 1)$ **12.** $(2, 1)$ and $(1, -1/3)$

13. (a, b) and $(0, 0)$ **14.** $(0, y)$ and $(x, 0)$

Find the absolute values in Exercises 15–20.

15. $|-3|$ **16.** $|2 - 7|$

17. $|-2 + 7|$ **18.** $|1.1 - 5.2|$

19. $|(-2)3|$ **20.** $\left|\frac{2}{-7}\right|$

In Exercises 21–24, find an equation for (a) the vertical line and (b) the horizontal line through the given point.

21. $(2, 3)$ **22.** $(-1, 4/3)$

23. $(0, -\sqrt{2})$ **24.** $(-\pi, 0)$

In Exercises 25–30, write an equation for the line that passes through the point P and has slope m.

25. $P(1, 1), m = 1$ **26.** $P(1, -1), m = -1$

27. $P(-1, 1), m = 1$ **28.** $P(-1, 1), m = -1$

29. $P(0, b), m = 2$ **30.** $P(a, 0), m = -2$

In Exercises 31–36, find an equation for the line through the two points.

31. $(0, 0), (2, 3)$ **32.** $(1, 1), (2, 1)$

33. $(1, 1), (1, 2)$ **34.** $(-2, 1), (2, -2)$

35. $(-2, 0), (-2, -2)$ **36.** $(1, 3), (3, 1)$

In Exercises 37–42, write an equation for the line with the given slope m and y-intercept b. Draw a complete graph.

37. $m = 3, b = -2$ **38.** $m = -1, b = 2$

39. $m = 1, b = \sqrt{2}$ **40.** $m = -1/2, b = -3$

41. $m = -5, b = 2.5$ **42.** $m = 1/3, b = -1$

In Exercises 43–48, find the x- and y-intercepts of the line. Then use the intercepts to sketch a complete graph of the line. Support your answer with a graphing utility.

43. $3x + 4y = 12$ **44.** $x + y = 2$

45. $4x - 3y = 12$ **46.** $2x - y = 4$

47. $y = 2x + 4$ **48.** $x + 2y = -4$

In Exercises 49–54, find an equation for the line through P perpendicular to L. Draw a complete graph of each pair of lines. Then find the distance from P to L.

49. $P(0, 0), L : y = -x + 2$

50. $P(0, 0), L : x + \sqrt{3}y = 3$

51. $P(1, 2), L : x + 2y = 3$

52. $P(-2, 2), L : 2x + y = 4$

53. $P(3, 6), L : x + y = 3$

54. $P(-2, 4), L : x = 5$

In Exercises 55–58, find an equation for the line through P parallel to L. Draw a complete graph of each pair of lines.

55. $P(2, 1), L : y = x + 2$ **56.** $P(0, 0), L : y = 3x - 5$

57. $P(1, 0), L : 2x + y = -2$

58. $P(1, 1), L : x + y = 1$

Increments and Motion

59. A particle starts at $A(-2, 3)$ and its coordinates change by increments $\Delta x = 5, \Delta y = -6$. Find its new position.

60. A particle starts at $A(6,0)$ and its coordinates change by increments $\Delta x = -6, \Delta y = 0$. Find its new position.

61. The coordinates of a particle change by $\Delta x = 5$ and $\Delta y = 6$ as it moves from $A(x, y)$ to $B(3, -3)$. Find x and y.

62. A particle started at $A(1,0)$, circled the origin once counterclockwise, and returned to $A(1,0)$. What were the net changes in its coordinates?

63. What acute angle does the line through the points (0,0) and (1,1) make with the positive x-axis?

64. A rectangle with sides parallel to the axes has vertices at $(3,-2)$ and $(-4,-7)$.

a) Find the coordinates of the other two vertices.

b) Find the area of the rectangle.

65. The rectangle in Fig. 1.38 has sides parallel to the axes. It is three times as long as it is wide. Its perimeter is 56 units. Find the coordinates of the vertices A, B, and C.

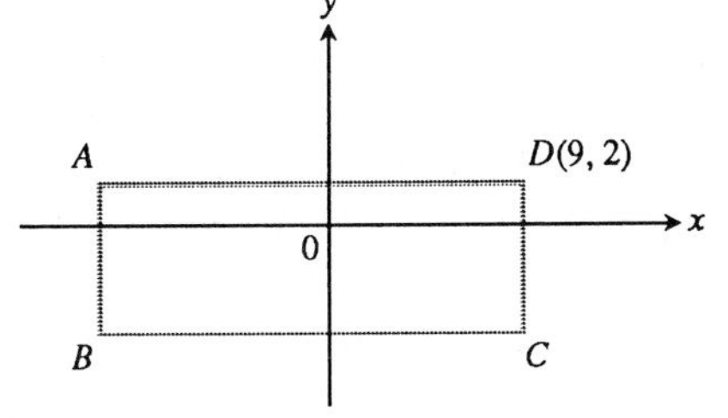

1.38 The rectangle in Exercise 65.

66. A circle in quadrant II is tangent to both axes. It touches the y-axis at (0, 3).

a) At what point does it meet the x-axis?

b) Find the coordinates of the center of the circle.

67. The line through the points (1, 1) and (2, 0) cuts the y-axis at the points (0, b). Find b by using similar triangles.

68. A 90° rotation counterclockwise about the origin takes $(2,0)$ to $(0,2)$ and $(0,3)$ to $(-3,0)$, as shown in Fig. 1.39. Where does the rotation take each of the following points?

a) $(4,1)$
b) $(-2,-3)$
c) $(2,-5)$
d) $(x,0)$
e) $(0,y)$
f) (x,y)
g) What point is taken to $(10,3)$?

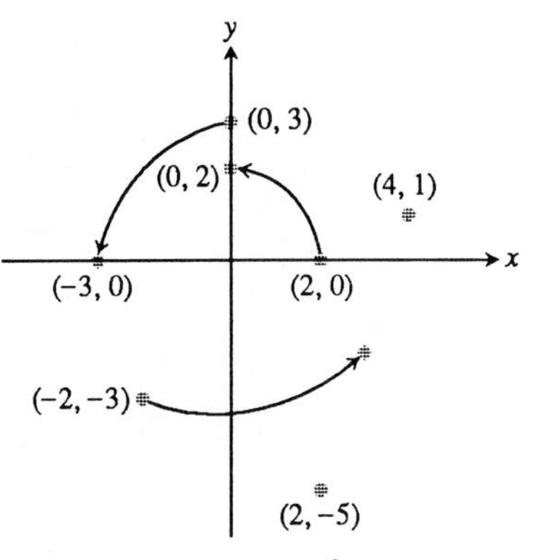

1.39 The points moved by the 90° rotation in Exercise 68.

69. Do not fall into the trap $|-a| = a$. The equation does not hold for all values of a.

a) Find a value of a for which $|-a| \neq a$.

b) For what values of a does the equation $|-a| = a$ hold?

70. For what values of x does $|1-x|$ equal $1-x$? For what values of x does it equal $x-1$?

Write an expression that gives the distance between the points on a number line with coordinates as specified in Exercises 71 and 72.

71. a) x and 3

b) x and -2

72. a) y and -1.3

b) y and 5.5

Applications

73. INSULATION By measuring slopes in Fig. 1.40, find the temperature change in degrees per inch for the following:

a) gypsum wall board

b) fiber glass insulation

c) wood sheathing

74. INSULATION Which of the materials listed in Exercise 73 is the best insulator? The poorest? Explain.

75. PRESSURE UNDER WATER The pressure p experienced by a diver under water is related to the diver's depth d by an equation of the form $p = kd + 1$ (k a constant). When $d = 0$ meters, the pressure is 1 atmosphere. The pressure at 100 meters is about 10.94 atmospheres. Find the pressure at 50 meters.

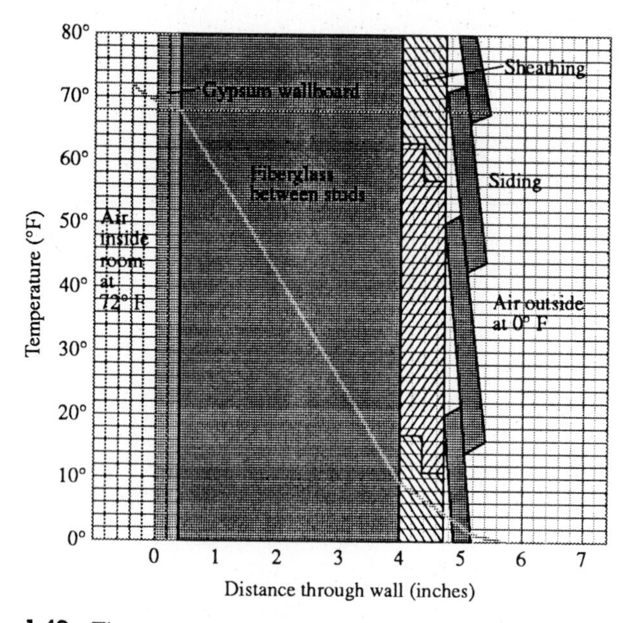

1.40 The temperature changes in the wall in Exercise 73. (Source: *Differentiation*, by W. U. Walton, et al., Project CALC, Education Development Center, Inc., Newton, Mass. (1975), p. 25.)

76. REFLECTED LIGHT A ray of light comes in along the line $x + y = 1$ above the x-axis and reflects off the x-axis. The angle of departure is equal to the angle of arrival (Fig. 1.41). Write an equation of the line along which the departing light travels.

77. FAHRENHEIT VS. CELSIUS Is there a temperature at which a Fahrenheit thermometer gives the same reading as a Celsius thermometer? If so, what is it? (See Example 8.)

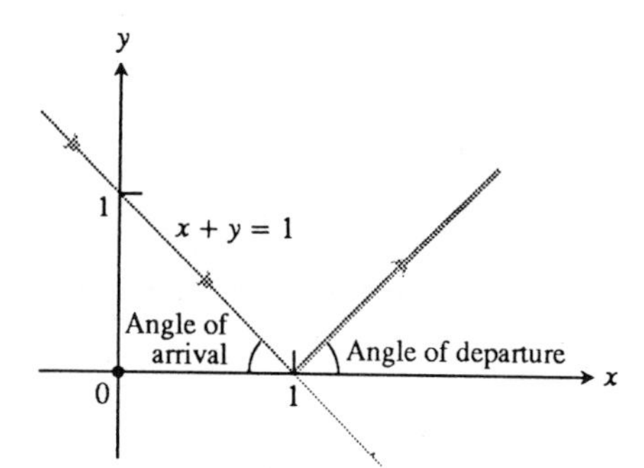

1.41 The path of the light ray in Exercise 76.

78. THE MT. WASHINGTON COG RAILWAY The steepest part of the Mt. Washington Cog Railway in New Hampshire has a phenomenal 37.1% grade. At this point, the passengers in the front of the car are 14 ft above those in the rear. About how far apart are the front and rear rows of seats?

Geometry

79. Three different parallelograms have vertices at $(-1, 1)$, $(2, 0)$, and $(2, 3)$. Sketch them and give the coordinates of the missing vertices.

80. How large a slope can you calculate with your calculator? To find out, continue the list of tangent values in Example 3. The best we could do was $\tan(89.9999999999) = 572957795131$ and $\tan(-89.9999999999) = -572957795131$.

81. For what value of k is the line $2x + ky = 3$ perpendicular to the line $x + y = 1$? For what values of k are the lines parallel?

82. Find the line that passes through the point $(1, 2)$ and the point of intersection of the lines $x + 2y = 3$ and $2x - 3y = -1$.

1.3 Functions and Their Graphs

The values of one variable quantity often depend on the values of another. For example:

The pressure in the boiler of a power plant depends on the steam temperature.

The rate at which the water drains from your bathtub when you pull the plug depends on how deep the water is.

The area of a circle depends on its radius.

In each of these examples, the value of one variable quantity, which we might call y, depends on the value of another variable quantity, which we might call x. Since the value of y in each case is completely determined by the value of x, we say that y is a function of x.

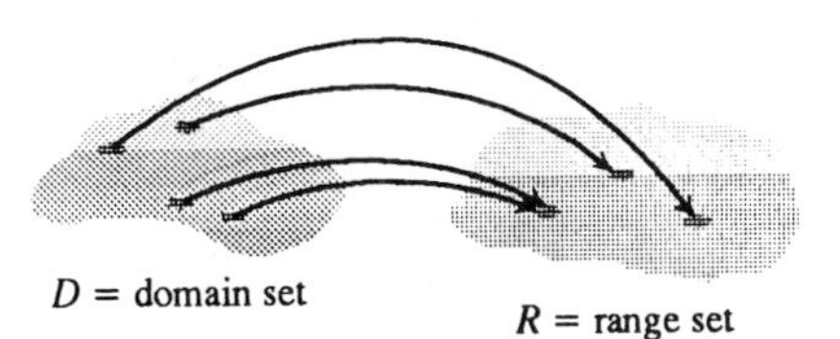

1.42 A function from set D to set R assigns an element of R to each element in D.

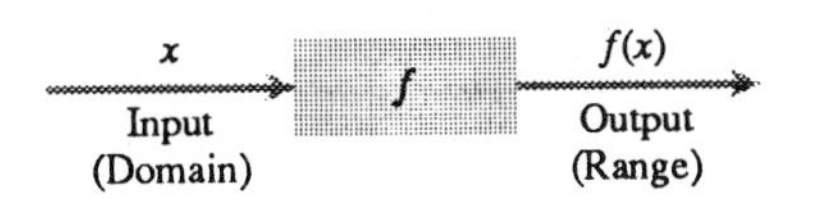

1.43 A flow diagram for a function f.

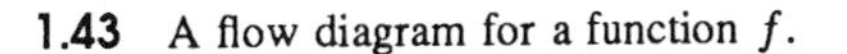

EULER

Leonhard Euler (1707–1783), the dominant mathematical figure of his century, was also an astronomer, physicist, botanist, chemist, and expert in Oriental languages. He was the first scientist to give the function concept the prominence in his work that it has today. The most prolific mathematician who ever lived, his collected books and papers now fill 70 volumes. His introductory algebra text is still read in English translation.

In mathematics, any rule that assigns to each element in one set some element from another set is called a **function**. The sets may be sets of numbers, sets of number pairs, sets of points, or sets of objects of any kind. The sets do not have to be the same. All the function has to do is assign some element from the second set to each element in the first set (Fig. 1.42). Thus a function is like a machine that assigns an output to every allowable input. The inputs make up the function's **domain**; the outputs make up the function's **range** (Fig. 1.43). Do not confuse this use of the word *range* with the use of the [RANGE] key to enter a viewing rectangle on a graphing utility.

Definition

A **function** from a set D to a set R is a rule that assigns a single element of R to each element in D.

The word *single* in the definition of function does not mean that there must be only one element in the function's range, although this can happen for some functions. It means that each input from the domain is assigned exactly one output from the range, no more and no less. In other words, each input appears just once in the list of input-output pairs defined by the function.

Euler invented a symbolic way to say "y is a function of x" by writing

$$y = f(x) \tag{1}$$

which we read "y equals f of x." This notation is shorter than the verbal statements that say the same thing. It also lets us give different functions different names by changing the letters we use. To say that boiler pressure is a function of steam temperature, we can write $p = f(t)$. To say that the area of a circle is a function of its radius, we can write $A = g(r)$. (Here we use a g because we just used f for something else.) We have to know what the variables p, t, A, and r mean, of course, for these equations to make sense.

Real-Valued Functions of a Real Variable

In most of our work, functions have domains and ranges that are sets of real numbers. Such functions are called **real-valued functions of a real variable**, and are usually defined by formulas or equations.

In addition to giving a function a useful name, the notation $y = f(x)$ gives a way to denote specific values of a function. For instance, the value of f at $x = 5$ is $f(5)$ ("f of 5").

EXAMPLE 1 The formula $A = \pi r^2$ gives the area of a circle as a function of its radius, where we would say that $A(r) = \pi r^2$. The area of a circle of radius $r = 2$ is $A(2) = \pi(2)^2 = 4\pi$.

In the context of geometry, the domain of the function $A = \pi r^2$ is the set of all possible radii—in this case the set of all positive real numbers. The range is also the set of positive real numbers. ☰

EXAMPLE 2 The function $y = x^2$. The formula $y = x^2$ defines the number y to be the square of the number x. If $x = 5$, then $y = (5)^2 = 25$.

The domain is the set of allowable x-values—in this case the set of all real numbers. The range, which consists of the resulting y-values, is the set of nonnegative real numbers. This is supported by the graph of $y = x^2$ in Example 2 of Section 1.1. ■

Graphs of Functions

The points in the plane whose coordinates (x, y) are the input-output pairs of a function make up the **graph** of the function. Thus, the graph of a function defined by an equation in x and y is the graph of the equation itself. We obtain visualization of these graphs in the usual way: Either we use a graphing utility to draw the visualization (graph) or we sketch the visualization (graph) by making a table of matching xy-pairs and connecting the points (x, y) with a smooth curve. Again we will use the words *graph* and *visualization of a graph* interchangeably.

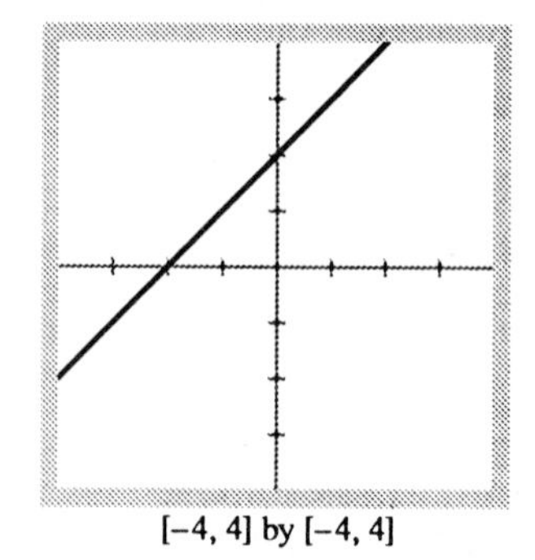

[−4, 4] by [−4, 4]

1.44 A complete graph of the function $y = x + 2$.

EXAMPLE 3 The graph of the function $y = x + 2$ is the line $y = x + 2$. It is the set of points (x, y) in which y equals $x + 2$. Figure 1.44 gives a computer visualization of the complete graph of the function $y = x + 2$. ■

Domains and Ranges Are Often Intervals

The domains and ranges of many functions are intervals of real numbers (Fig. 1.45). The set of all real numbers that lie *strictly between* two fixed numbers a and b is an **open interval**. The interval is "open" at each end because it contains neither of its endpoints. Intervals that contain both endpoints are **closed**. Intervals that contain one endpoint but not both are **half-open**. Half-open intervals could just as well be called half-closed, but no one seems to call them that.

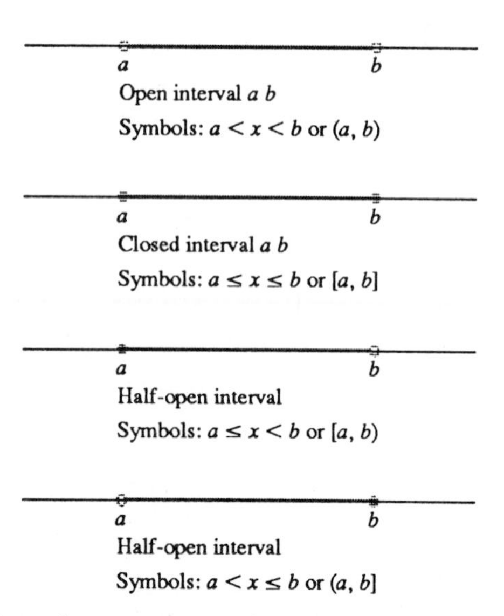

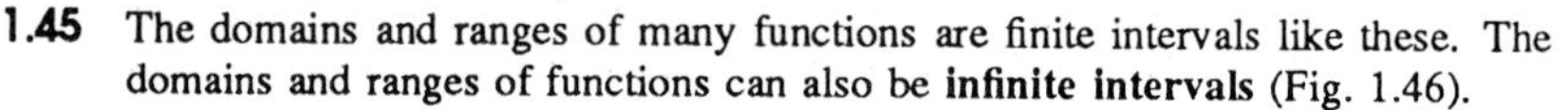

1.45 The domains and ranges of many functions are finite intervals like these. The domains and ranges of functions can also be **infinite intervals** (Fig. 1.46).

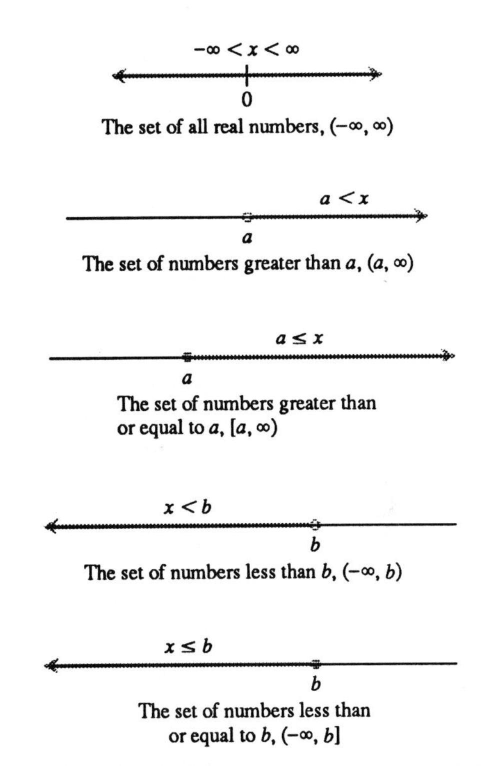

1.46 Rays on the number line and the line itself are called infinite intervals. The symbol ∞ (infinity) in the notation is used merely for convenience; it is not to be taken as a suggestion that there is a number ∞.

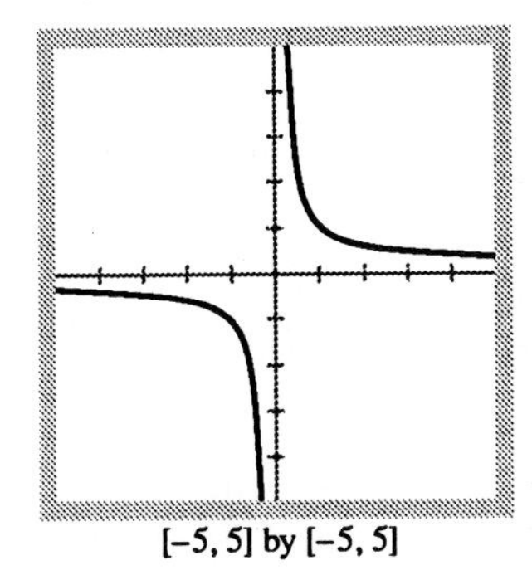

1.47 A complete graph of $y = 1/x$.

EXAMPLE 4 In each of the following functions, the domain is taken to be the largest set of real x-values for which the formula gives real y-values. We shall say more about this convention in a moment.

Function	Domain	Range
$y = \dfrac{1}{x}$	$x \neq 0$	$y \neq 0$
$y = \sqrt{x}$	$0 \leq x$	$0 \leq y$
$y = \sqrt{4 - x}$	$x \leq 4$	$0 \leq y$
$y = \sqrt{1 - x^2}$	$-1 \leq x \leq 1$	$0 \leq y \leq 1$

The formula, or equation $y = 1/x$ gives a real y-value for every x except $x = 0$. Thus, the domain of the function $y = 1/x$ is the set of all real numbers different from 0 as supported by the complete graph in Figure 1.47. This graph also supports that the range of $y = 1/x$ is the set of all real numbers different from 0. To confirm this algebraically requires that we observe that every real number $y \neq 0$ occurs (output) as a reciprocal of a real number (input) x. For example, 3 is the output associated with input 1/3.

The graph of $y = \sqrt{x}$ in Figure 1.48 suggests that both the domain and range of this function are the set of all nonnegative real numbers. The domain can be confirmed by observing that the formula $y = \sqrt{x}$ gives a real y-value only when x is positive or zero. Notice that $y = \sqrt{x}$ is not a real number when

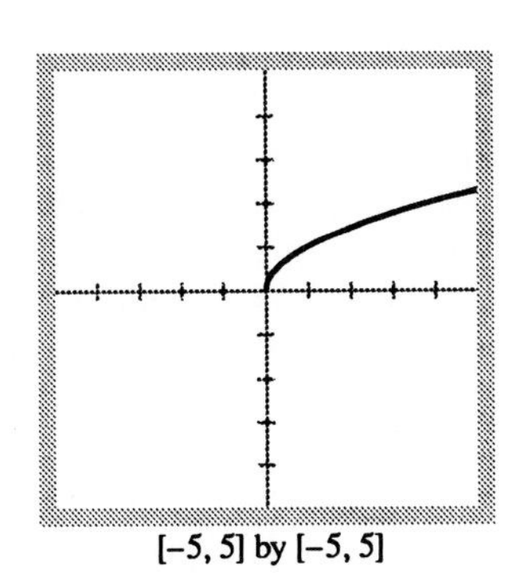

1.48 A complete graph of $y = \sqrt{x}$.

x is negative. (Complex numbers of the form $a + bi$, where $i = \sqrt{-1}$, are excluded from our consideration until Chapter 9.) The range can be confirmed by observing that every nonnegative real number occurs as a y-value.

Notice the graph of $y = \sqrt{4 - x}$ in Figure 1.49 is the graph of $y = \sqrt{x}$ reflected through the y-axis and then shifted right 4 units. We will say more about this later in this section and in Section 1.4. The domain of $y = \sqrt{4 - x}$ appears to be $x \leq 4$ and the range $y \geq 0$. The domain can be confirmed by observing that $4 - x$ cannot be negative, that is, $0 \leq 4 - x$ or $x \leq 4$.

The graph of $y = \sqrt{1 - x^2}$ in Figure 1.50 suggests that the domain is $-1 \leq x \leq 1$ and the range is $0 \leq y \leq 1$. The domain is confirmed by observing that the formula $y = \sqrt{1 - x^2}$ gives a real y-value for every value of x such that $1 - x^2 \geq 0$, that is, for $-1 \leq x \leq 1$. Confirming the range is more difficult. We need to observe that $0 \leq 1 - x^2 \leq 1$ for $-1 \leq x \leq 1$, so that $0 \leq y = \sqrt{1 - x^2} \leq 1$. ≡

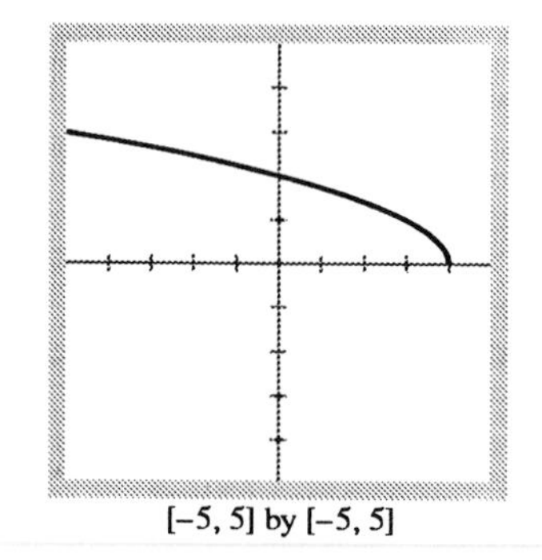
[-5, 5] by [-5, 5]

1.49 A complete graph of $y = \sqrt{4 - x}$.

We can observe from Example 4 that graphs are very helpful in determining domains and ranges of functions, and that confirming this information algebraically can be difficult. We will use graphing utilities together with algebraic procedures when we determine information about functions such as domains and ranges. In Chapters 3 and 4 we will see how to use calculus to confirm that the graphs in Figures 1.47–1.50 are complete.

Independent and Dependent Variables, a Warning about Division by 0, and a Convention about Domains

The variable x in a function $y = f(x)$ is called the **independent variable,** or **argument,** of the function. The variable y, whose value depends on x, is called the **dependent variable.**

We must keep two restrictions in mind when we define functions. First, we *never divide by 0*. When we see $y = 1/x$, we must think $x \neq 0$. Zero is not in the domain of the function. When we see $y = 1/(x - 2)$, we must think "$x \neq 2$."

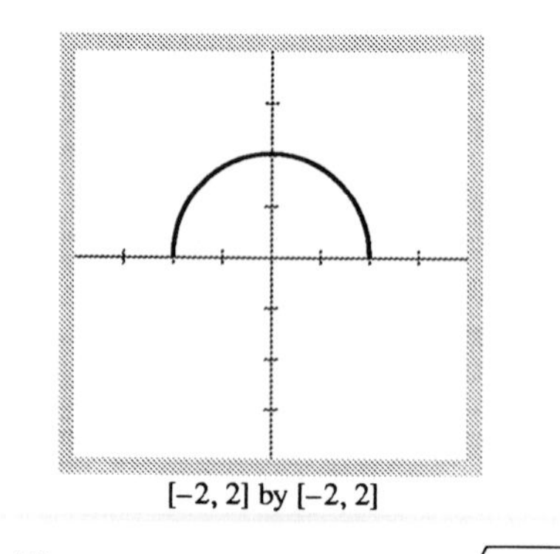
[-2, 2] by [-2, 2]

1.50 A complete graph of $y = \sqrt{1 - x^2}$.

The second restriction is that we shall deal exclusively with real-valued functions (except for a very short while later in the book). We may therefore have to restrict our domains when we have square roots (or fourth roots or other even roots). If $y = \sqrt{1 - x^2}$, we should think "x^2 must not be greater than 1. The domain must not extend beyond the interval $-1 \leq x \leq 1$."

We observe a convention about the domains of functions defined by formulas. If the domain is not stated explicitly, then the domain is automatically the largest set of x-values for which the formula gives real y-values. If we wish to exclude values from this domain, we must say so. The formula $y = x^2$ gives real y-values for every real x. Therefore, writing

$$y = x^2 \quad (x \text{ not restricted})$$

tells everyone that the intended domain is $-\infty < x < \infty$ (Fig. 1.51a). To exclude negative values from the domain, we would limit the equation $y = x^2$ by writing the inequality $x \geq 0$ beside it:

$$y = x^2, \qquad x \geq 0 \text{ (Fig. 1.51b).} \qquad (x \text{ restricted})$$

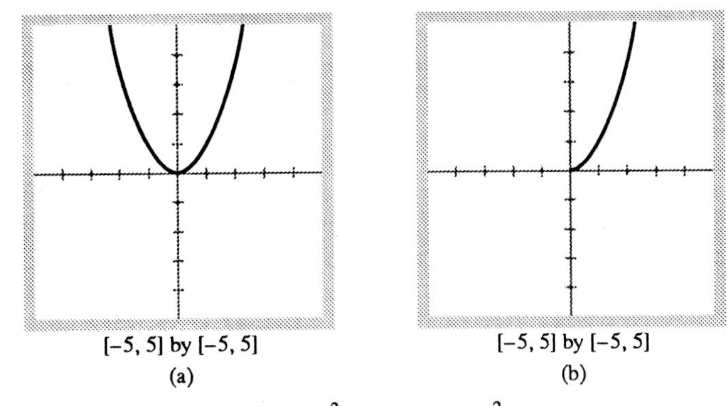

1.51 Complete graphs of (a) $y = x^2$ and (b) $y = x^2, x \geq 0$.

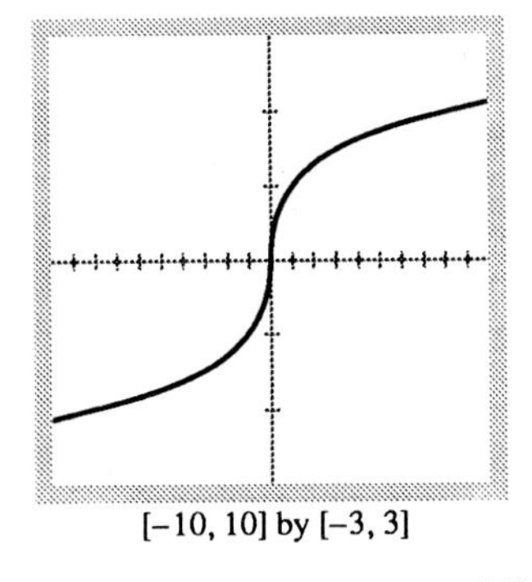

1.52 A complete graph of $y = \sqrt[3]{x}$.

Graphing

Graphing utilities usually produce highly reliable graphs. You probably noticed in Section 1.1 that finding a viewing rectangle(s) that displays a complete graph is sometimes difficult. We can use a graphing utility to explore and find complete graphs when we know or are told what a certain complete graph should look like. In this chapter, we will indicate what complete graphs of some basic functions look like.

EXAMPLE 5 Determine the domain and range and draw a complete graph of each function.
(a) $y = \sqrt[3]{x}$ (b) $y = \sqrt[3]{4 - x^2}$

Solution (a) The formula $y = \sqrt[3]{x}$ gives a real y-value for every real number x, and $y = \sqrt[3]{y^3}$ for every real number y. Thus, the domain and the range of this function is the set of real numbers. The graph of $y = \sqrt[3]{x}$ in Figure 1.52 is complete.

(b) It can be shown that the graph of $y = \sqrt[3]{4 - x^2}$ in Figure 1.53(a) is complete. This graph suggests that the domain is the set of all real numbers, and that the range is the set of all real numbers less than or equal to a where a is about 1.6. To confirm the range, we can start with the graph of $y = 4 - x^2$

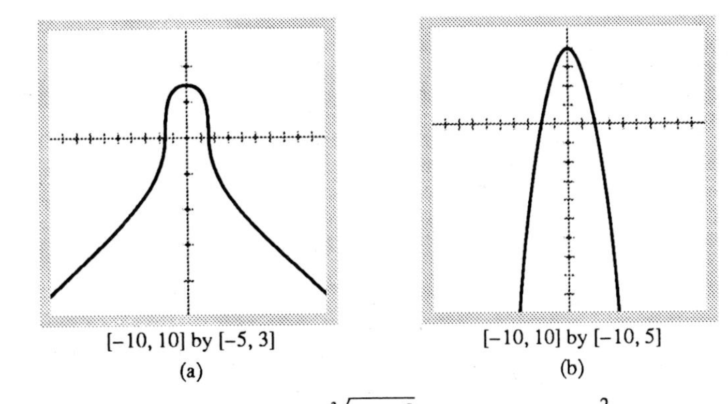

1.53 Complete graphs of (a) $y = \sqrt[3]{4 - x^2}$ and (b) $y = 4 - x^2$.

in Figure 1.53(b) to see that $4 - x^2 \le 4$ for all x and then observe that $\sqrt[3]{4 - x^2} \le \sqrt[3]{4}$. Notice that $\sqrt[3]{4} = 1.5874$ to 4 decimal places. We can see from Figure 1.53(b) that $4 - x^2 < 0$ for $x < -2$ or $x > 2$ so that $\sqrt[3]{4 - x^2} < 0$ for $x < -2$ or $x > 2$. ≡

Figure 1.54 shows complete graphs of some functions frequently used in calculus.

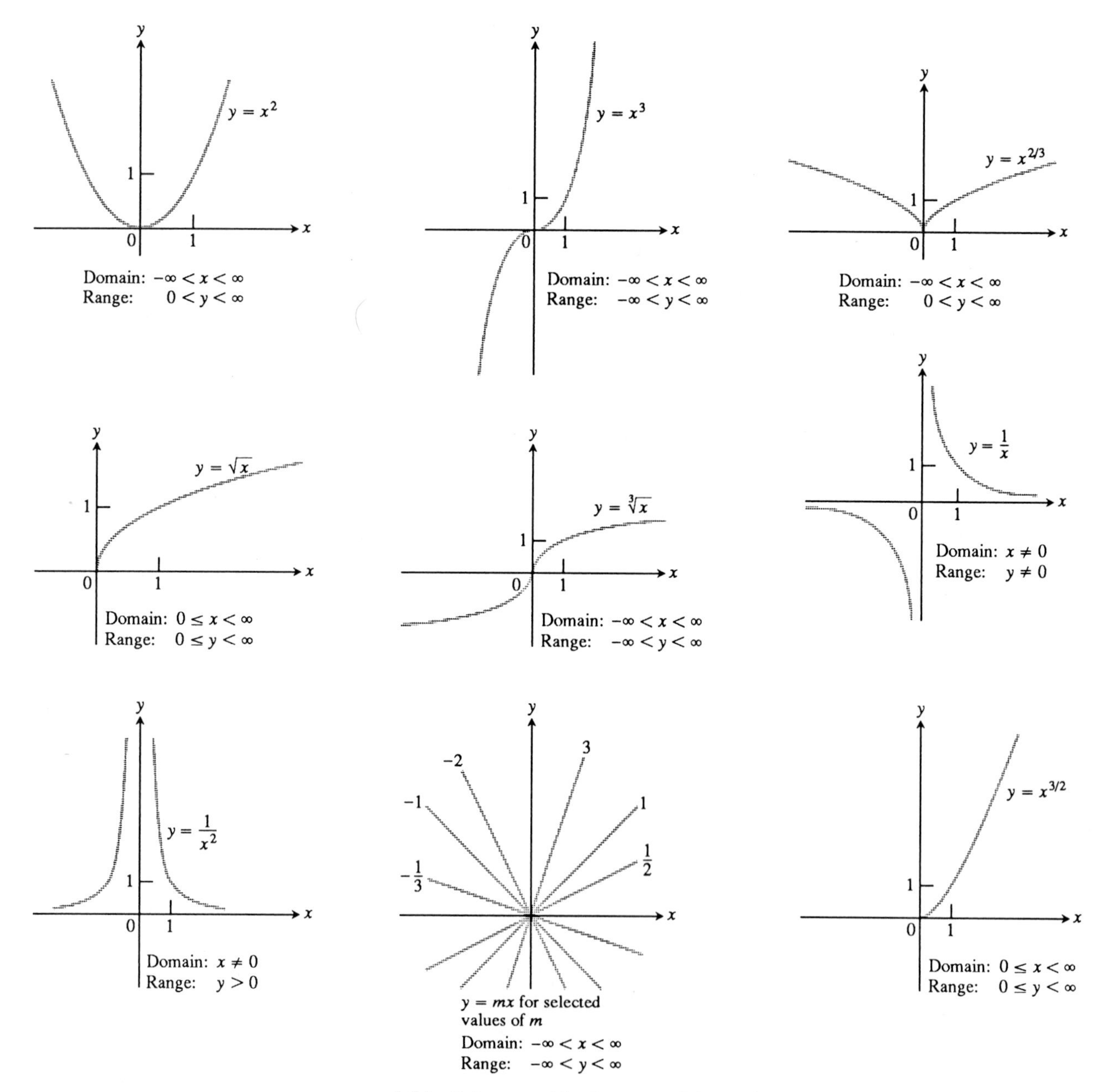

1.54 Reference table of some useful complete graphs.

Symmetry

We can use coordinate formulas to describe important symmetries in the coordinate plane. Figure 1.55 shows how this is done.

The coordinate relations in Figure 1.55 provide the following symmetry tests for graphs.

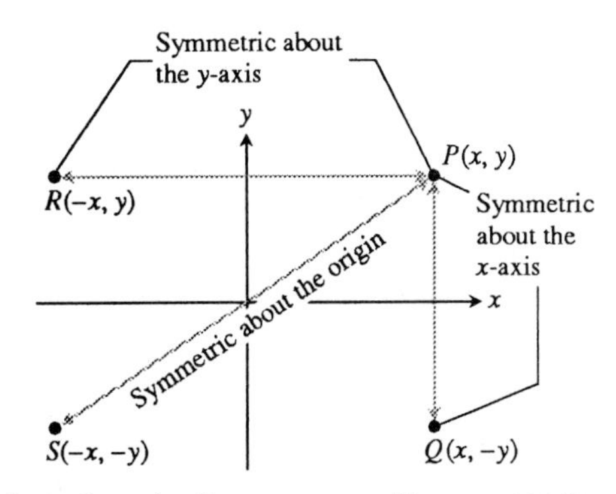

1.55 The coordinate formulas for symmetry with respect to the origin and axes in the coordinate plane.

EXAMPLE 6 Symmetric points:

$P(5, 2)$ and $Q(5, -2)$	symmetric about the x-axis
$P(5, 2)$ and $R(-5, 2)$	symmetric about the y-axis
$P(5, 2)$ and $S(-5, -2)$	symmetric about the origin

Symmetry Tests for Graphs

1. *Symmetry about the x-axis:*
   If the point (x, y) lies on the graph, then the point $(x, -y)$ lies on the graph (Fig. 1.56a).
2. *Symmetry about the y-axis:*
   If the point (x, y) lies on the graph, the point $(-x, y)$ lies on the graph (Fig. 1.56b).
3. *Symmetry about the origin:*
   If the point (x, y) lies on the graph, the point $(-x, -y)$ lies on the graph (Fig. 1.56c).

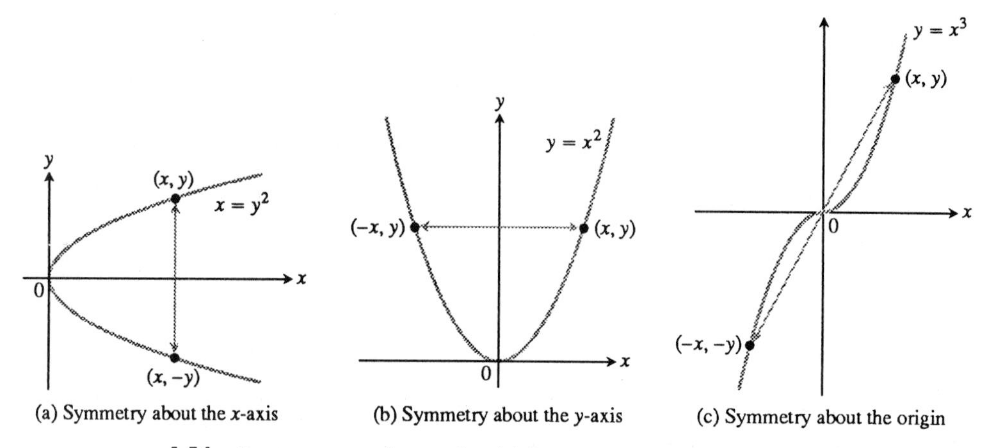

1.56 Symmetry tests for graphs. (a) Symmetry about the x-axis. (b) Symmetry about the y-axis. (c) Symmetry about the origin.

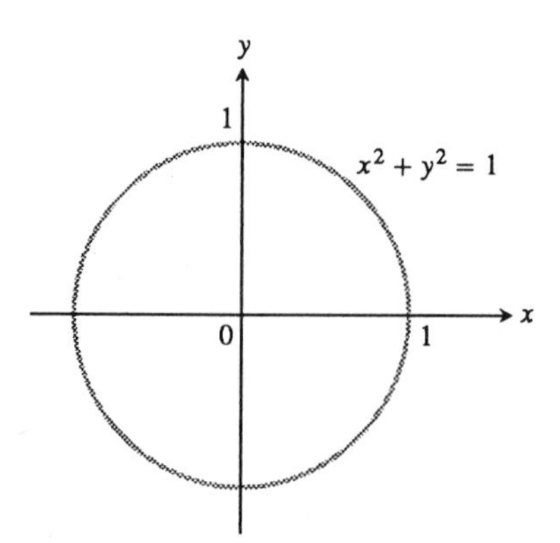

1.57 The graph of the equation $x^2 + y^2 = 1$ is the circle of the radius 1, centered at the origin. It is symmetric about both axes and about the origin. Example 7 shows how to predict these symmetries before you graph the equation.

EXAMPLE 7 The graph of $x^2 + y^2 = 1$ has all three of the symmetries listed above (Fig. 1.57).

1. Symmetry about the x-axis:

(x, y) on the graph $\Rightarrow x^2 + y^2 = 1$ ($\Rightarrow$ means "implies")

$\Rightarrow x^2 + (-y)^2 = 1$ $(y^2 = (-y)^2)$

$\Rightarrow (x, -y)$ on the graph

2. Symmetry about the y-axis:

(x, y) on the graph $\Rightarrow x^2 + y^2 = 1$

$\Rightarrow (-x)^2 + y^2 = 1$ $(x^2 = (-x)^2)$

$\Rightarrow (-x, y)$ on the graph.

3. Symmetry about the origin:

(x, y) on the graph $\Rightarrow x^2 + y^2 = 1$

$\Rightarrow (-x)^2 + (-y)^2 = 1$

$\Rightarrow (-x, -y)$ on the graph

Even Functions and Odd Functions

Definition

A function $y = f(x)$ is an **even** function of x if $f(-x) = f(x)$ for every x in the function's domain. It is an **odd** function of x if $f(-x) = -f(x)$ for every x in the function's domain.

The names even and odd come from powers of x. If y equals an even power of x, as in $y = x^2$ or $y = x^4$, it is an even function of x (because $(-x)^2 = x^2$ and $(-x)^4 = x^4$). If y equals an odd power of x, as in $y = x$ or $y = x^3$, it is an odd function of x (because $(-x)^1 = -x$ and $(-x)^3 = -x^3$).

Saying that a function $y = f(x)$ is even is equivalent to saying that its graph is symmetric about the y-axis. Since $f(-x) = f(x)$, the point (x, y) lies on the curve if and only if the point $(-x, y)$ lies on the curve (Fig. 1.58a).

Saying that a function $y = f(x)$ is odd is equivalent to saying that its graph is symmetric with respect to the origin. Since $f(-x) = -f(x)$, the point (x, y) lies on the curve if and only if the point $(-x, -y)$ lies on the curve (Fig. 1.58b).

The graphs of polynomials in even powers of x are symmetric about the y-axis. The graphs of polynomials in odd powers of x are symmetric about the origin.

EXAMPLE 8 *Even, odd, and neither*

$f(x) = x^2$ — Even function: $(-x)^2 = x^2$ for all x
Symmetry about y-axis

$f(x) = x^2 + 1$ — Even function: $(-x)^2 + 1 = x^2 + 1$ for all x
Symmetry about the y-axis (Fig. 1.59a)

$f(x) = x$ — Odd function: $(-x) = -(x)$ for all x
Symmetry about the origin

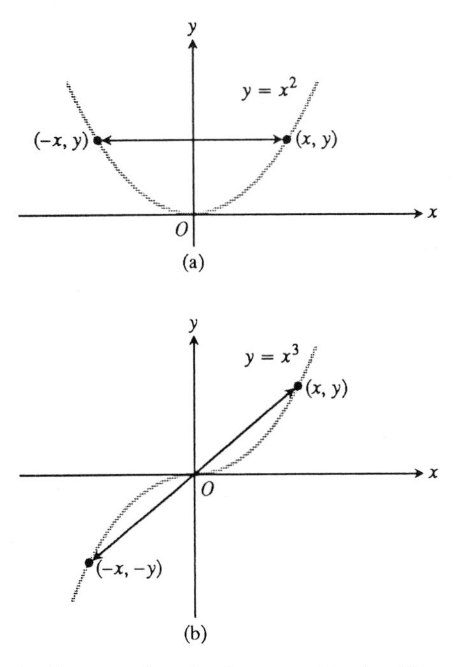

1.58 (a) The graph of an even function is symmetric about the y-axis. (b) The graph of an odd function is symmetric about the origin.

$f(x) = x + 1$	Not odd: $f(-x) = -x + 1$, but $-f(x) = -x - 1$. The two are not equal Not even, either: $(-x) + 1 \neq x + 1$. See Fig. 1.59(b)

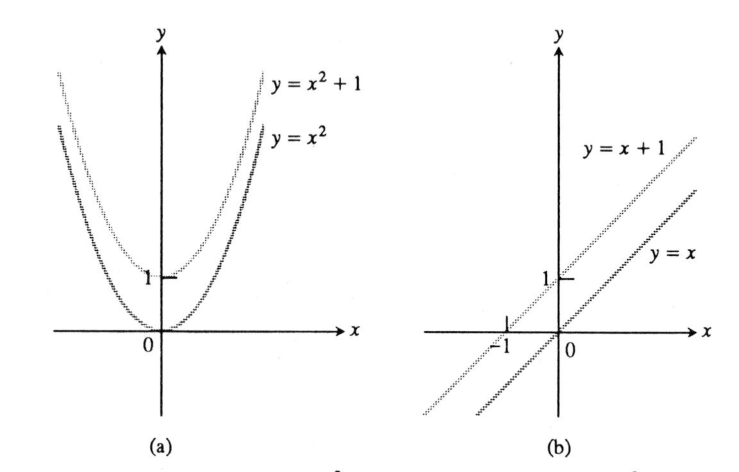

1.59 (a) When we add 1 to $y = x^2$, the resulting function $y = x^2 + 1$ is still even and its graph is still symmetric about the y-axis. (b) When we add 1 to $y = x$, the resulting function $y = x + 1$ is no longer odd. The symmetry about the origin is lost.

Integer-Valued Functions

The greatest integer less than or equal to a number x is called the greatest integer in x. Because each real number x corresponds to only one greatest integer, the greatest integer in x is a function of x. The symbol for it is $\lfloor x \rfloor$, which is read "the greatest integer in x." Figure 1.60 shows the graph of $y = \lfloor x \rfloor$.

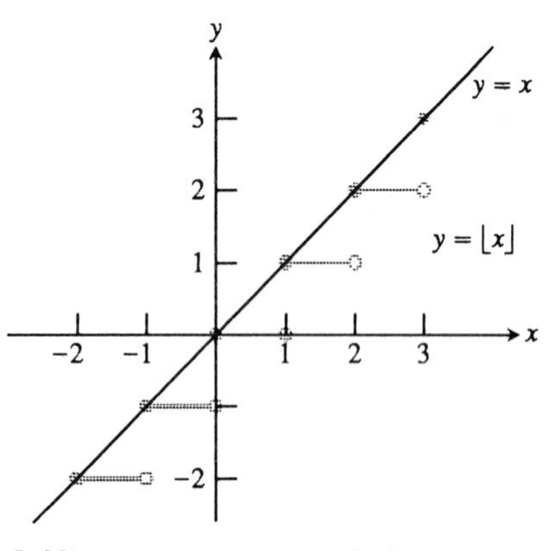

1.60 The graph of $y = \lfloor x \rfloor$ and its relation to the line $y = x$. As the figure shows, $\lfloor x \rfloor$ is less than or equal to x, so it provides an integer floor for x.

EXAMPLE 9 *Values of $y = \lfloor x \rfloor$*

Positive	$\lfloor 1.9 \rfloor = 1$, $\lfloor 2.0 \rfloor = 2$, $\lfloor 2.4 \rfloor = 2$
Zero	$\lfloor 0.5 \rfloor = 0$, $\lfloor 0 \rfloor = 0$
Negative	$\lfloor -1.2 \rfloor = -2$, $\lfloor -0.5 \rfloor = -1$

Notice that if x is negative, $\lfloor x \rfloor$ may have a larger absolute value than x does.

Another common notation for the greatest integer function is $y = [x]$. The notation $y = [x]$ comes from computer science, where it is used to denote the result of rounding x down to the nearest integer. You can think of it as the integer "floor" for x and the notation is chosen to suggest just that:

$$\lfloor x \rfloor \qquad \text{integer floor for } x.$$

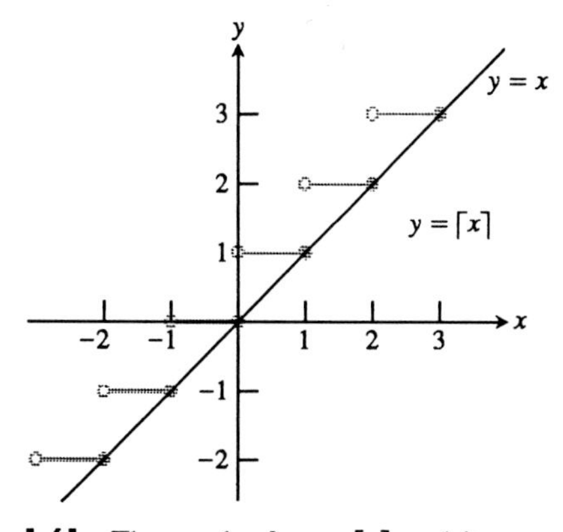

1.61 The graph of $y = \lceil x \rceil$ and its relation to the line $y = x$. As the figure shows, $\lceil x \rceil$ is greater than or equal to x, so it provides an integer ceiling for x.

The companion notation

$$\lceil x \rceil \qquad \text{integer ceiling for } x$$

is used to denote the result of rounding x up to the nearest integer. It denotes the least integer greater than or equal to x (Fig. 1.61).

Notice, for later reference, that the integer floor and ceiling functions exhibit points called *discontinuities*, where the functions jump from one value to the next without taking on the intermediate values. They jump like this at every integer value of x. Some graphing utilities have one or the other of these functions as built-in functions.

Functions Defined in Pieces

While some functions are defined by single formulas, others are defined by applying different formulas to different parts of their domains.

EXAMPLE 10 The values of the function

$$y = f(x) = \begin{cases} -x & \text{if } x < 0 \\ x^2 & \text{if } 0 \le x \le 1 \\ 1 & \text{if } x > 1 \end{cases}$$

are given by the formula $y = -x$ when $x < 0$, by the formula $y = x^2$ when $0 \le x \le 1$, and by the formula $y = 1$ when $x > 1$. The function is *just one function*, however, whose domain is the entire real line (Fig. 1.62). We are able to enter this function directly into the graphing calculator we use in this book to obtain the graph in Figure 1.62. Consult the owner's manual accompanying your graphing utility to see if it can graph functions defined in pieces. If not, then you can use your grapher to obtain the graphs of $y = -x$ for $x < 0$ (Figure 1.63a), $y = x^2$ for $0 \le x \le 1$ (Figure 1.63b), and $y = 1$ for $x > 1$ (Figure 1.63c), and then combine them to sketch the graph shown in Figure 1.64. ■

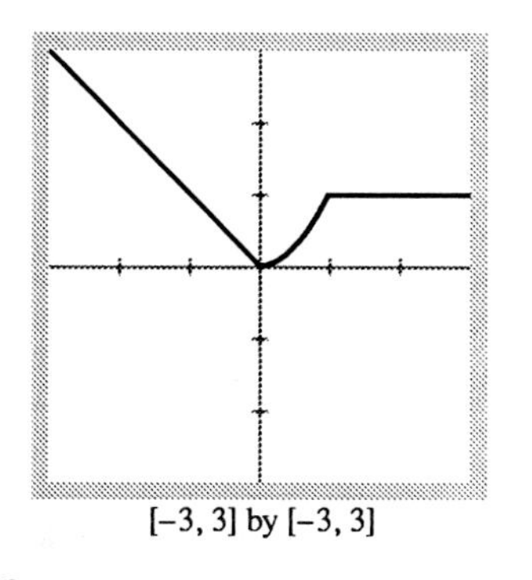

[−3, 3] by [−3, 3]

1.62 A complete graph of the function defined in pieces in Example 10.

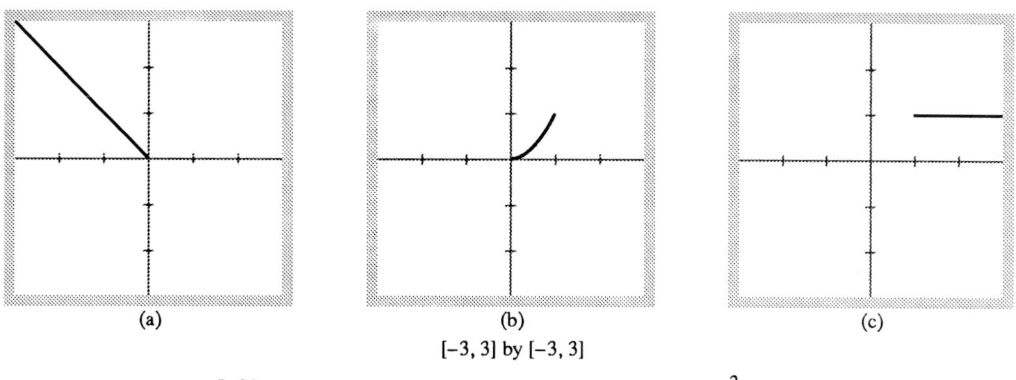

[−3, 3] by [−3, 3]

1.63 The graph of (a) $y = -x$ for $x < 0$, (b) $y = x^2$ for $0 \le x \le 1$, and (c) $y = 1$ for $x > 1$.

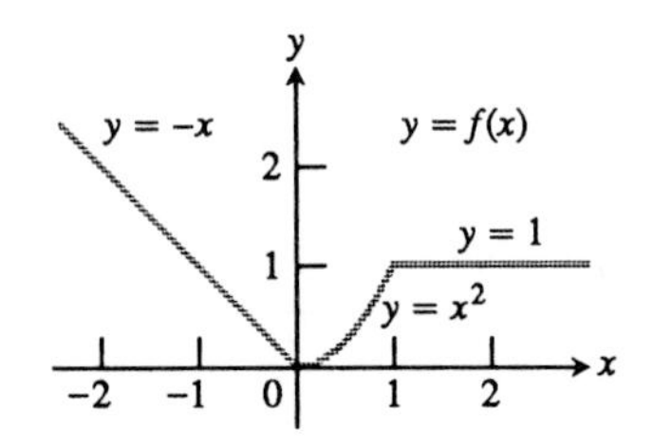

1.64 To sketch the graph of the function $y = f(x)$ shown here, we apply different formulas to different parts of its domain.

EXAMPLE 11 Suppose that the graph of a function $y = f(x)$ consists of the line segments shown in Figure 1.65. Write a formula for f.

Solution We find formulas for the segments from (0, 0) to (1, 1) and from (1, 0) to (2, 1) and piece them together in the manner of Example 10.

Segment from (0, 0) to (1, 1). The line through (0, 0) and (1, 1) has slope $m = (1 - 0)/(1 - 0) = 1$ and y-intercept $b = 0$. Its slope-intercept equation is therefore $y = x$. The segment from (0, 0) to (1, 1) that includes the point (0, 0) but not the point (1, 1) is the graph of the function $y = x$ restricted to the half-open interval $0 \leq x < 1$, namely,

$$y = x, \qquad 0 \leq x < 1.$$

Segment from (1, 0) to (2, 1). The line through (1, 0) and (2, 1) has slope $m = (1 - 0)/(2 - 1) = 1$ and passes through the point (1, 0). The corresponding point-slope equation for the line is therefore

$$y - 0 = 1(x - 1), \qquad \text{or} \qquad y = x - 1.$$

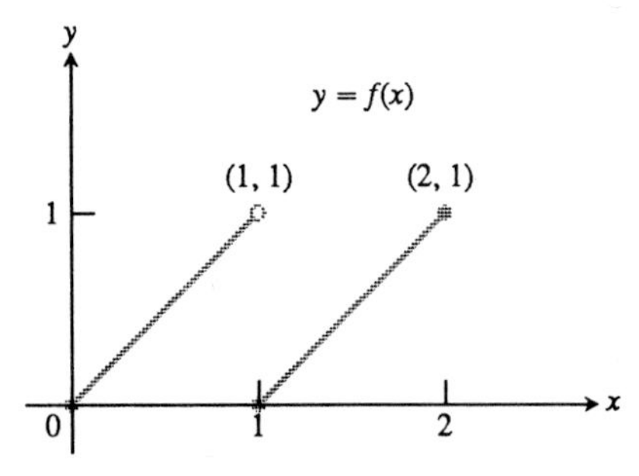

1.65 The graph of the function $y = f(x)$ (Example 11) shown here consists of two line segments. The segment on the left contains the endpoints at the origin (shown by a heavy dot) but does not contain the endpoint (1, 1). The segment on the right contains both endpoints.

The segment from (1, 0) to (2, 1) that includes both endpoints is the graph of $y = x - 1$ restricted to the closed interval $1 \leq x \leq 2$, namely,

$$y = x - 1, \qquad 1 \leq x \leq 2.$$

Formula for the function $y = f(x)$ shown in Fig. 1.65. We obtain a formula for f on the interval $0 \leq x \leq 2$ by combining the formulas we obtained for the two segments of the graph:

$$f(x) = \begin{cases} x & \text{for } 0 \leq x < 1, \\ x - 1 & \text{for } 1 \leq x \leq 2. \end{cases}$$

EXAMPLE 12 The domain of the "step" function $y = g(x)$ graphed in Fig. 1.66 is the closed interval $0 \leq x \leq 3$. Find a formula for $g(x)$.

Solution The graph consists of three horizontal line segments. The left segment is the half-open interval $0 \leq x < 1$ on the x-axis, which we may think of as a portion of the line $y = 0$:

$$y = 0, \qquad 0 \leq x < 1.$$

The second segment is the portion of the line $y = 1$ that lies over the closed interval $1 \leq x \leq 2$:

$$y = 1, \qquad 1 \leq x \leq 2.$$

The third segment is the half-open interval $2 < x \leq 3$ on the line $y = 0$:

$$y = 0, \qquad 2 < x \leq 3.$$

The values of g are therefore given by the three-piece formula

$$g(x) = \begin{cases} 0 & \text{for } 0 \leq x < 1, \\ 1 & \text{for } 1 \leq x \leq 2, \\ 0 & \text{for } 2 < x \leq 3. \end{cases}$$

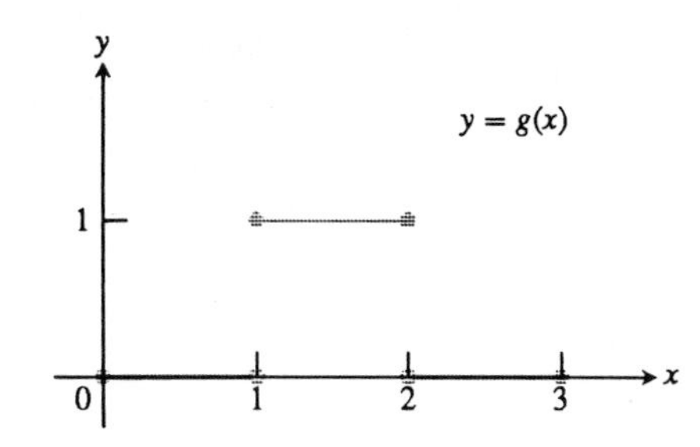

1.66 Functions like the one graphed here are called step functions. Example 12 shows how to write a formula for g.

Absolute Value Function

The function $y = |x|$ is called the *absolute value function.* Its graph lies along the line $y = x$ when $x \geq 0$ and along the line $y = -x$ when $x < 0$ (Fig. 1.67a). The absolute value function is built-in on most graphing utilities as $y = abs(x)$. Figure 1.67(b) supports the sketch of $y = |x|$ in Figure 1.67(a).

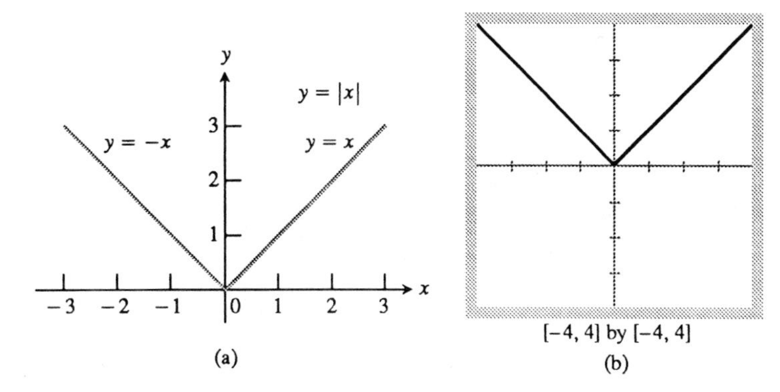

1.67 A complete graph of the absolute value function $|x|$ obtained (a) by sketch and (b) by grapher.

EXAMPLE 13 Draw a complete graph of $f(x) = |x + 1| + |x - 3|$ with a graphing utility and confirm algebraically.

Solution We show that the graph of f in Figure 1.68 is complete. Notice the graph suggests that f can be defined in pieces.

The points where the expressions inside the bars change sign are

$$x = -1 \text{ (for } x + 1)$$

$$x = 3 \text{ (for } x - 3).$$

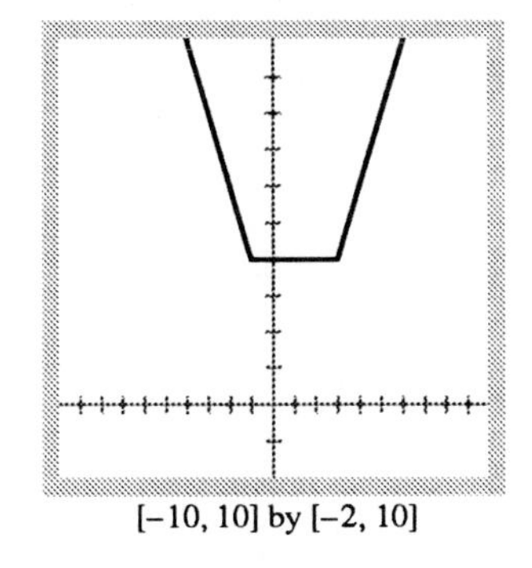

1.68 A complete graph of $f(x) = |x + 1| + |x - 3|$.

These points divide the x-axis into intervals on which we can write absolute-value-free functions for $f(x)$:

For $x < -1$: Here, $x + 1 < 0$ and $x - 3 < 0$, so

$$\begin{aligned} f(x) &= |x+1| + |x-3| \\ &= -(x+1) - (x-3) \\ &= -x - 1 - x + 3 = 2 - 2x. \end{aligned}$$

For $-1 \le x \le 3$: Here, $x + 1 \ge 0$ and $x - 3 \le 0$, so

$$\begin{aligned} f(x) &= |x+1| + |x-3| \\ &= x + 1 - (x-3) \\ &= x + 1 - x + 3 = 4. \end{aligned}$$

For $x > 3$: Here, $x + 1 > 0$ and $x - 3 > 0$, so

$$\begin{aligned} f(x) &= |x+1| + |x-3| \\ &= x + 1 + x - 3 = 2x - 2. \end{aligned}$$

Thus,

$$f(x) = \begin{cases} 2 - 2x, & x < -1 \\ 4, & -1 \le x \le 3 \\ 2x - 2, & x > 3. \end{cases}$$

With the absolute value signs now removed, we can sketch the graph of f in the usual way. ■

To sketch the graph of a formula $y = f(x)$ that contains absolute values, divide the x-axis into intervals on which the absolute values can be removed. Then graph as usual.

Notice that the complete graph in Fig. 1.68 suggests that the equation $|x + 1| + |x - 3| = 0$ has no solution and that the function $f(x) = |x + 1| + |x - 3|$ achieves an absolute minimum value, namely 4, for infinitely many values of x in the interval $-1 \le x \le 3$.

Sums, Differences, Products, and Quotients

The sum $f + g$ of two functions of x is itself a function of x, defined at any point x that lies in both domains. The same holds for the differences $f - g$ and $g - f$, the product $f \cdot g$, and the quotients f/g and g/f, as long as we exclude any points that require division by zero.

EXAMPLE 14 *The sums, differences, products, and quotients of the functions f and g defined by the formulas*

$$f(x) = \sqrt{x} \text{ and } g(x) = \sqrt{1 - x}.$$

Function	Formula	Domain
f	$f(x) = \sqrt{x}$	$0 \le x$
g	$g(x) = \sqrt{1-x}$	$x \le 1$
$f+g$	$(f+g)(x) = f(x) + g(x) = \sqrt{x} + \sqrt{1-x}$	$0 \le x \le 1$
		(The intersection of the domains of f and g)
$f-g$	$(f-g)(x) = f(x) - g(x) = \sqrt{x} - \sqrt{1-x}$	$0 \le x \le 1$
$g-f$	$(g-f)(x) = g(x) - f(x) = \sqrt{1-x} - \sqrt{x}$	$0 \le x \le 1$
$f \cdot g$	$(f \cdot g)(x) = f(x)g(x) = \sqrt{x(1-x)}$	$0 \le x \le 1$
f/g	$\frac{f}{g}(x) = \frac{f(x)}{g(x)} = \sqrt{\frac{x}{1-x}}$	$0 \le x < 1$
		($x = 1$ excluded)
g/f	$\frac{g}{f}(x) = \frac{g(x)}{f(x)} = \sqrt{\frac{1-x}{x}}$	$0 \le x \le 1$
		($x = 0$ excluded)

Figure 1.69(a) shows complete graphs of the functions $f(x) = \sqrt{x}$ and $g(x) = \sqrt{1-x}$. Notice that the domain of f is $0 \le x$ and the domain of g is $x \le 1$. Figure 1.69(b) shows a complete graph of $f + g$ and gives visual support that the domain of $f + g$ is $0 \le x \le 1$, the intersection of the domains of f and g.

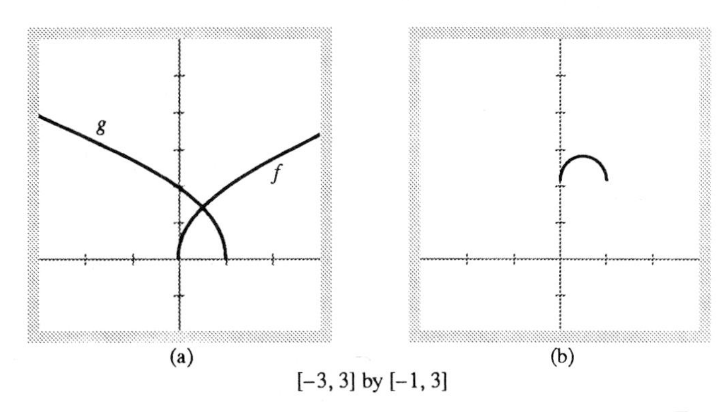

[–3, 3] by [–1, 3]

1.69 Complete graphs of (a) f and g, and (b) $f + g$ where $f(x) = \sqrt{x}$ and $g(x) = \sqrt{1-x}$.

Figure 1.70 gives complete graphs of the functions $f - g$, $g - f$, and fg, and Figure 1.71 gives complete graphs of f/g and g/f. The graphs of f and g can be used to help explain the graphs of $f + g$, $f - g$, $g - f$, $f \cdot g$, f/g, and g/f.

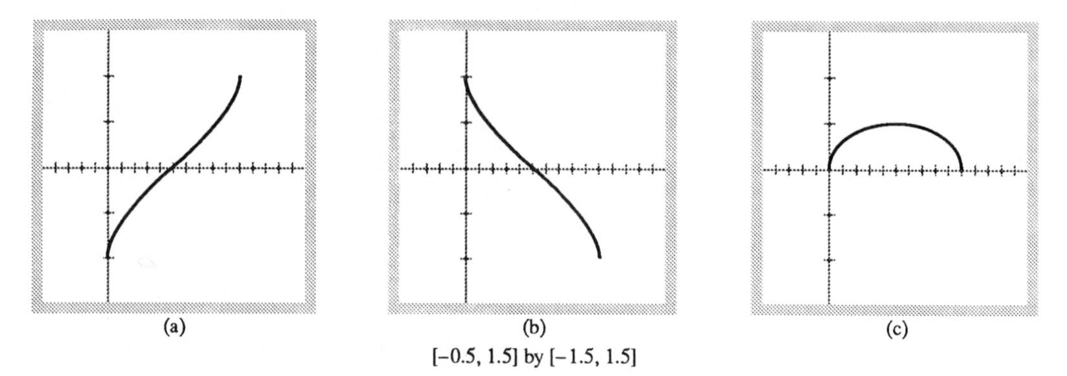

1.70 Complete graphs of (a) $f - g$, (b) $g - f$, and (c) fg where $f(x) = \sqrt{x}$ and $g(x) = \sqrt{1 - x}$.

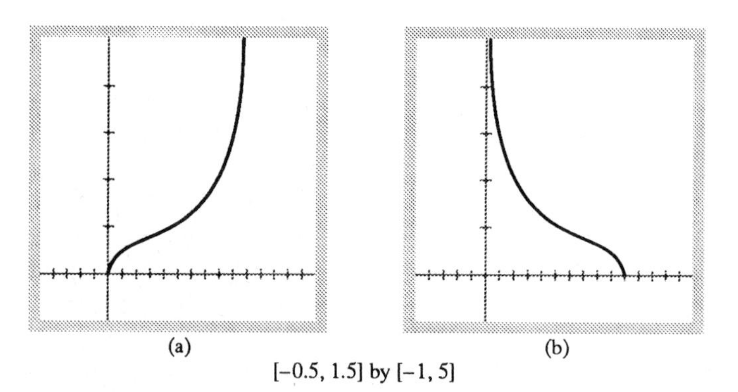

1.71 Complete graphs of (a) f/g and (b) g/f, where $f(x) = \sqrt{x}$ and $g(x) = \sqrt{1 - x}$.

Composition of Functions

Suppose that the outputs of a function g can be used as inputs of a function f. We can then hook g and f together to form a new function whose inputs are the outputs of g and whose outputs are the numbers $f(g(x))$, as in Fig. 1.72. We say that the function $f(g(x))$ (pronounced "f of g of x") is the composite of g and f. It is made by composing g and f in the order of first g, then f. The usual "stand-alone" notation for this composite is $f \circ g$, which is read as "f of g." Thus, the value of $f \circ g$ at x is $(f \circ g)(x) = f(g(x))$.

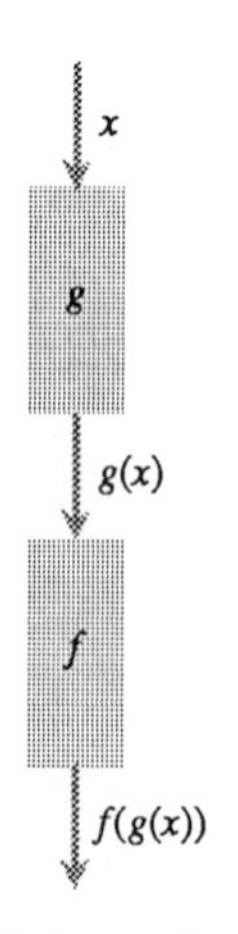

1.72 Two functions can be composed when the range of the first lies in the domain of the second.

EXAMPLE 15 Find a formula for $f(g(x))$ if $g(x) = x^2$ and $f(x) = x - 7$. Then find the value of $f(g(2))$.

Solution To find $f(g(x))$, we replace x in the formula for $f(x)$ by the expression given for $g(x)$:

$$f(x) = x - 7$$

$$f(g(x)) = g(x) - 7 = x^2 - 7.$$

We then find the value of $f(g(2))$ by substituting 2 for x:

$$f(g(2)) = (2)^2 - 7 = 4 - 7 = -3.$$

Changing the order in which functions are composed usually changes the result. In Example 15, we composed $g(x) = x^2$ and $f(x) = x - 7$ in the order of first g, then f, obtaining the function $f \circ g$, whose value at x was $f(g(x)) = x^2 - 7$. In the next example, we see what happens when we reverse the order to obtain the function $g \circ f$.

EXAMPLE 16 Find a formula for $g(f(x))$ if $g(x) = x^2$ and $f(x) = x - 7$. Then find $g(f(2))$.

Solution To find $g(f(x))$, we replace x in the formula $g(x)$ by the expression for $f(x)$:

$$g(x) = x^2$$
$$g(f(x)) = (f(x))^2 = (x - 7)^2.$$

To find $g(f(2))$, we then substitute 2 for x:

$$g(f(2)) = (2 - 7)^2 = (-5)^2 = 25.$$

Explore Box

1. Enter the function $g(x) = x^2$ of Example 15 as $y_1 = x^2$.
2. Some graphing utilities allow the variables used to name functions to also be used as independent variables for other functions.
3. Compare the graph of $y_2 = y_1 - 7$ with the graph of $f(g(x))$ of Example 15.
4. Let $y_3 = x - 7$ and $y_4 = y_3^2$. Compare the graph of y_4 with the graph of $g(f(x))$ of Example 16.

In the notation for composite functions, the parentheses tell which function comes first:

The notation $f(g(x))$ says "first g, then f." To calculate $f(g(2))$, calculate $g(2)$ and then apply f.

The notation $g(f(x))$ says "first f, then g." To calculate $g(f(2))$, calculate $f(2)$ and then apply g.

Exercises 1.3

For each point $P(x, y)$ in Exercises 1–8, use symmetry tests to find the point Q symmetric to P across the x-axis, the point R symmetric to P across the y-axis, and the point S symmetric to P through the origin.

1. $(3, 1)$
2. $(-2, 2)$
3. $(-2, 1)$
4. $(-1, -1)$
5. $(1, -\sqrt{2})$
6. $(-\sqrt{3}, -\sqrt{3})$
7. $(0, \pi)$
8. $(2, 0)$

In Exercises 9–16, find the domain and range of each function and support your answer with a graphing utility.

9. $y = 2 + \sqrt{x - 1}$
10. $y = -3 + \sqrt{x + 4}$
11. $y = -\sqrt{-x}$
12. $y = \sqrt{-x}$

13. $y = 2\sqrt{3 - x}$

14. $y = -3\sqrt{2 - x}$

15. $y = \dfrac{1}{x - 2}$

16. $y = \dfrac{1}{x + 2}$

Determine whether the functions in Exercises 17–26 are even, odd, or neither. Try to answer without writing anything down (except the answer).

17. $y = x^3$

18. $y = x^4$

19. $y = x + 2$

20. $y = x + x^2$

21. $y = x^2 - 3$

22. $y = x + x^3$

23. $y = \dfrac{1}{x^2 - 1}$

24. $y = \dfrac{1}{x - 1}$

25. $y = \dfrac{x}{x^2 - 1}$

26. $y = \dfrac{x^2}{x^2 - 1}$

Use a graph to find the domain and range of the functions in Exercises 27–36. What symmetries described in this section, if any, do the graphs have?

27. $y = x^2 - 9$

28. $y = 4 - x^2$

29. $y = \sqrt[3]{x - 3}$

30. $y = -2\sqrt[3]{x + 2}$

31. $y = 1 + \sqrt[3]{2 - x}$

32. $y = -2 + 5\sqrt[3]{4 - x}$

33. $y = -\dfrac{1}{x}$

34. $y = -\dfrac{1}{x^2}$

35. $y = 1 + \dfrac{1}{x}$

36. $y = 1 + \dfrac{1}{x^2}$

37. Consider the function $y = 1/\sqrt{x}$.
   a) Can x be negative?
   b) Can $x = 0$?
   c) What is the domain of the function?

38. Consider the function $y = \sqrt{(1/x) - 1}$.
   a) Can x be negative?
   b) Can $x = 0$?
   c) Can x be greater than 1?
   d) What is the domain of the function?

Test the equations in Exercises 39–46 to find what symmetries their graphs have, and then draw a complete graph of the equation.

39. $y = -x^2$

40. $x = 4 - y^2$

41. $y = 1/x^2$

42. $y = 1/(x^2 + 1)$

43. $xy = 1$

44. $xy^2 = 1$

45. $x^2y^2 = 1$

46. $x^2 + 4y^2 = 1$

Draw a complete graph of the functions in Exercises 47–58.

47. $y = |x + 3|$

48. $y = |2 - x|$

49. $y = \dfrac{|x|}{x}$

50. $y = \dfrac{|x - 1|}{x - 1}$

51. $y = \dfrac{x - |x|}{2}$

52. $y = \dfrac{x + |x|}{2}$

53. $y = \begin{cases} 3 - x, & x \le 1, \\ 2x, & 1 < x \end{cases}$

54. $y = \begin{cases} 1/x, & x < 0 \\ x, & 0 \le x \end{cases}$

55. $y = \begin{cases} 1, & x < 5, \\ 0, & 5 \le x \end{cases}$

56. $y = \begin{cases} 1, & x < 0, \\ \sqrt{x}, & x \ge 0 \end{cases}$

57. $y = \begin{cases} 4 - x^2 & x < 1, \\ \frac{3}{2}x + \frac{3}{2}, & 1 \le x \le 3, \\ x + 3, & x > 3 \end{cases}$

58. $y = \begin{cases} x^2, & x < 0, \\ x^3, & 0 \le x \le 1, \\ 2x - 1, & x > 1 \end{cases}$

59. For what values of x does (a) $\lfloor x \rfloor = 0$? (b) $\lceil x \rceil = 0$?

60. Does $\lfloor x \rfloor$ ever equal $\lceil x \rceil$? Explain.

61. Graph each function over the given interval.
   a) $y = x - \lfloor x \rfloor, -3 \le x \le 3$
   b) $y = \lfloor x \rfloor - \lceil x \rceil, -3 \le x \le 3$

62. INTEGER PARTS OF DECIMALS When x is positive or zero, $\lfloor x \rfloor$ is the integer part of the decimal representation of x. What is the corresponding description of $\lceil x \rceil$ when x is negative or zero?

63. Find formulas for the functions graphed in the following figures.

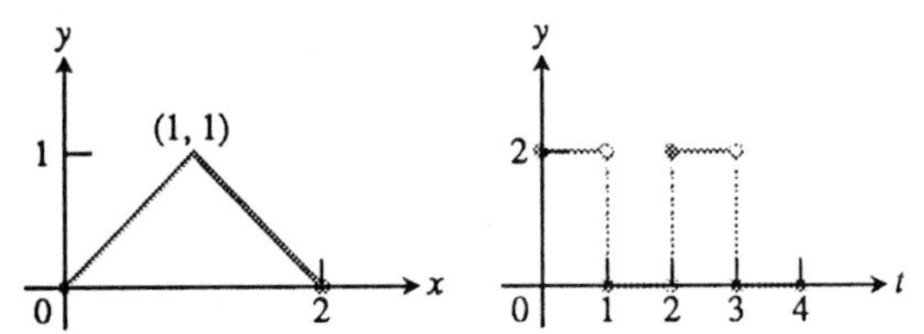

64. Find formulas for the functions graphed in the following figures.

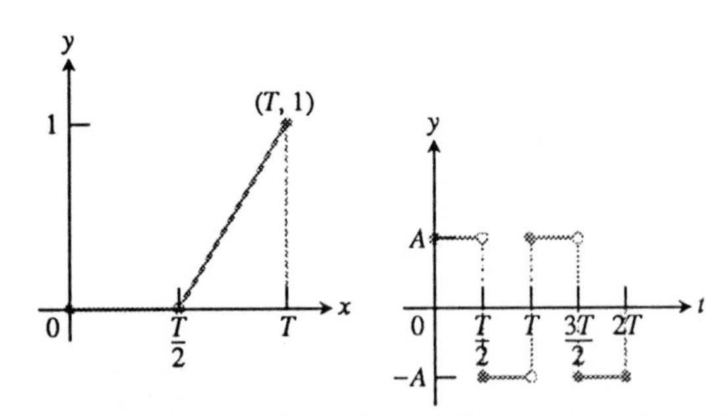

Draw a complete graph of the functions in Exercises 65–68 and confirm algebraically by expressing the function without absolute value symbols.

65. $f(x) = |x + 1| + 2|x - 3|$ (*Hint*: Use three intervals.)

66. $f(x) = |x + 2| + |x - 1|$

67. $f(x) = |x| + |x-1| + |x-3|$ (*Hint*: Use four intervals.)

68. $f(x) = |x+2| + |x| + |x+1|$

In Exercises 69 and 70 find the domains of f and g and then find the corresponding domains and complete graphs of $f+g$, $f-g$, $f \cdot g$, f/g, and g/f.

69. $f(x) = x \quad g(x) = \sqrt{x-1}$

70. $f(x) = \sqrt{x+1}, \quad g(x) = \sqrt{x-1}$

71. If $f(x) = x+5$ and $g(x) = x^2 - 3$, find the following:

a) $f(g(0))$ b) $g(f(0))$

c) $f(g(x))$ d) $g(f(x))$

e) $f(f(-5))$ f) $g(g(2))$

g) $f(f(x))$ h) $g(g(x))$

72. If $f(x) = x+1$ and $g(x) = x-1$, find the following:

a) $f(g(0))$ b) $g(f(0))$

c) $f(g(1))$ d) $g(f(1))$

e) $f(g(x))$ f) $g(f(x))$

73. Copy and complete the following table.

	$g(x)$	$f(x)$	$f \circ g(x)$
a)	$x-7$	$\sqrt{x}$	
b)	$x+2$	$3x$	
c)		$\sqrt{x-5}$	$\sqrt{x^2-5}$
d)	$\dfrac{x}{x-1}$	$\dfrac{x}{x-1}$	
e)		$1+\dfrac{1}{x}$	x
f)	$\dfrac{1}{x}$		x

74. a) If $f(x) = 1/x$, find $f(x+2) - f(2)$.

b) If $F(t) = 4t - 3$, find $F(t+1) - F(1)$.

75. Compare the domains and ranges of the functions $y = \sqrt{x^2}$ and $y = (\sqrt{x})^2$.

76. Find $f(x)$ if $g(x) = \sqrt{x}$ and $(g \circ f)(x) = |x|$.

77. Find $g(x)$ if $f(x) = x^2 + 2x + 1$ and $(g \circ f)(x) = |x+1|$.

78. Find functions $f(x)$ and $g(x)$ whose composites satisfy the two equations

$$(g \circ f)(x) = |\sin x| \text{ and } (f \circ g)(x) = (\sin \sqrt{x})^2.$$

79. THE BEST LOCATION FOR A FACTORY ASSEMBLY TABLE (adapted from *Fantastiks of Mathematiks*, Cliff Sloyer, Janson Publications, Inc., Providence, R.I., 1986). Because of a design change, the parts produced by three machines along a factory aisle (shown here as the x-axis)

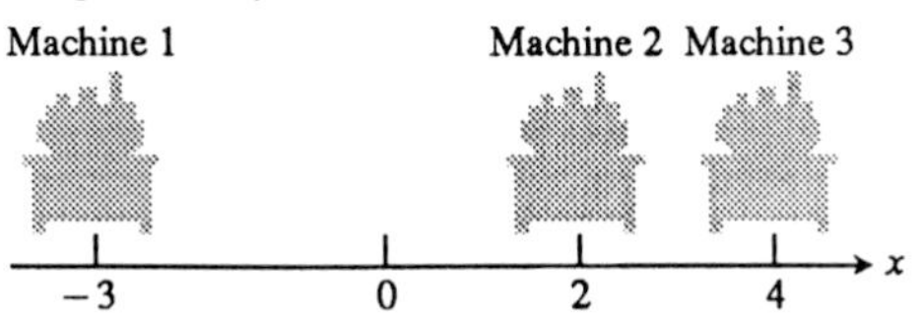

are to go to a nearby table for assembly before they undergo further processing. Each assembly takes one part from each machine, and there is a fixed cost per foot for moving each part. As the plant's production engineer, you have been asked to find a location for the assembly table that will keep the total cost of moving the parts at a minimum.

To solve the problem, you let x represent the table's location and look for the value of x that minimizes the sum

$$d(x) = |x+3| + |x-2| + |x-4|$$

of the distances from the table to the three machines. Since the cost of moving the parts to the assembly table is proportional to the total distance the parts travel, any value of x that minimizes d will minimize the cost.

Complete the job now by graphing $d(x)$ to find its smallest value. Then say where you would put the table.

80. Best location (Continuation of Exercise 79). You solved the table-location problem in Exercise 79 so well that your manager has asked you to solve a similar problem at a neighboring plant. This time there are four machines instead of three

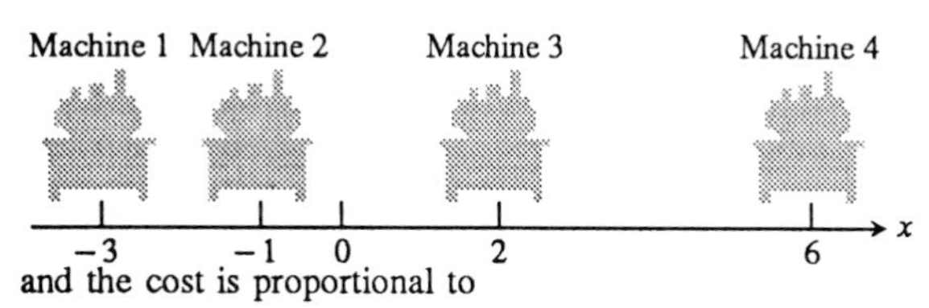

and the cost is proportional to

$$d(x) = |x+3| + |x+1| + |x-2| + |x-6|.$$

Where should the assembly table go now?

81. Best location (Continuation of Exercise 80). As the result of another design change, the assembly in the plant in Exercise 80 is to use twice as many parts from Machine 1 as before and three times as many parts from Machine 3 as before. The total cost of moving parts from the four machines to the assembly table is now proportional to the "weighted" distance

$$d(x) = 2|x+3| + |x+1| + 3|x-2| + |x-6|.$$

What is the minimum value of this new function? Where should the table go?

82. Compare the graphs of $y = \sqrt{x^2}$ and $y = |x|$. Explain.

Determine a viewing rectangle that gives a complete graph of the functions in Exercises 83 and 84.

83. $f(x) = x^{2/3}$ **84.** $g(x) = x^{3/2}$

1.4 Shifts, Reflections, Stretches, and Shrinks. Geometric Transformations.

In this section, we show how to change an equation to move its graph up or down or right or left, vertically stretch or shrink it, or reflect it through the x- or y-axis in the coordinate plane. Each of these changes is called a **geometric transformation** of the previous graph. We practice with parabolas and other basic graphs introduced in the previous sections, but the methods we use apply to other curves as well. We shall have more to say about parabolas when we discuss conic sections in Chapter 10.

How to Shift a Graph

To shift the graph of a function $y = f(x)$ straight up, we add a positive constant to the right-hand side of the formula $y = f(x)$.

EXAMPLE 1 Complete graphs of $y = x^2 + 2$ and $y = x^2 + 4$ are obtained by shifting the graph of $y = x^2$ up 2 units or 4 units, respectively (Fig. 1.73a).

To shift the graph of a function $y = f(x)$ straight down, we subtract a positive constant from the right-hand side of the formula $y = f(x)$.

EXAMPLE 2 Complete graphs of $y = x^2 - 2$ and $y = x^2 - 4$ are obtained by shifting the graph of $y = x^2$ down 2 units or 4 units respectively (Fig. 1.73b).

To shift the graph of $y = f(x)$ to the left, we add a positive constant to x.

EXAMPLE 3 Complete graphs of $y = |x + 2|$ and $y = |x + 4|$ are obtained by shifting the graph of $y = |x|$ left 2 units or 4 units, respectively (Fig. 1.74a).

To shift the graph of $y = f(x)$ to the right, we subtract a positive constant from x.

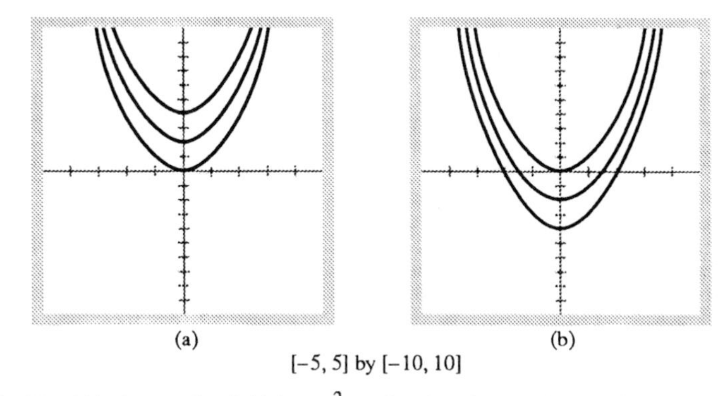

(a) (b)

[−5, 5] by [−10, 10]

1.73 To shift the graph of $f(x) = x^2$ up (or down), we add positive constants to (or subtract them from) the formula for f. Complete graphs of (a) $y = x^2 + c$ for $c = 0, 2, 4$ and (b) $y = x^2 - c$ for $c = 0, 2, 4$.

EXAMPLE 4 Complete graphs of $y = |x-3|$ and $y = |x-5|$ are obtained by shifting the graphs of $y = |x|$ right 3 units or 5 units, respectively (Fig. 1.74b)

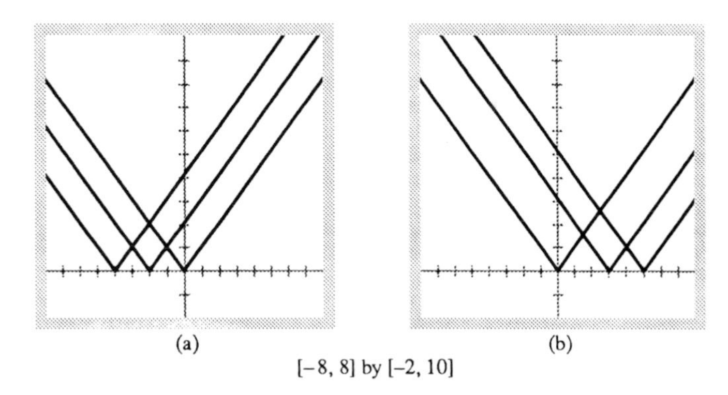

(a) (b)

[−8, 8] by [−2, 10]

1.74 To shift the graphs of $f(x) = |x|$ left (or right), we add positive constants to (or subtract them from) x. Complete graphs of (a) $y = |x + c|$ for $c = 0, 2, 4$ and (b) $y = |x - c|$ for $c = 0, 3, 5$.

EXAMPLE 5 Describe how the graph of $y = 1/(x + 2) - 3$ can be obtained from the graph of $y = 1/x$, and determine its domain and range.

Solution The graph of $y = 1/(x + 2)$ is the graph of $y = 1/x$ shifted left 2 units. The graph of $y = 1/(x + 2) - 3$ is the graph of $y = 1/(x + 2)$ shifted vertically downward 3 units. Thus, the graph of $y = 1/(x + 2) - 3$ can be obtained by shifting the graph of $y = 1/x$ left 2 units and then shifting the resulting graph vertically downward 3 units (Fig. 1.75c). The order can be reversed. The graph of $y = 1/x$ can be shifted vertically downward 3 units to obtain $y = 1/x - 3$ and the resulting graph shifted left 2 units to obtain $y = 1/(x + 2) - 3$. The domain is all $x \neq -2$ and the range is all $y \neq 3$.

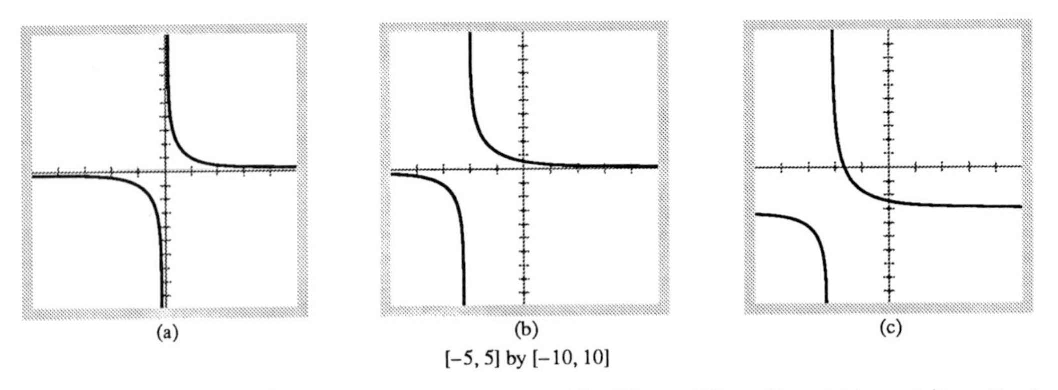

1.75 Complete graphs of (a) $y = 1/x$, (b) $y = 1/(x + 2)$, and (c) $y = 1/(x + 2) - 3$.

> **Shift Formulas ($c > 0$)**
>
> Vertical Shifts
>
> $y = f(x) + c$ or $y - c = f(x)$ — Shifts the graph of f up c units
>
> $y = f(x) - c$ or $y + c = f(x)$ — Shifts the graph of f down c units
>
> Horizontal Shifts
>
> $y = f(x + c)$ — Shifts the graph of f left c units
>
> $y = f(x - c)$ — Shifts the graph of f right c units

The formula for a vertical shift upward is often written as $y - c = f(x)$ instead of $y = f(x) + c$. You can get the new form by subtracting c from both sides of the equation $y = f(x) + c$. But there is also another way to think of it. For $y - c$ to match a given x, the value of y has to be c units larger than before. The graph of y vs. x therefore must be c units higher.

Similarly, the formula for a vertical downward shift of c units is often written as $y + c = f(x)$.

How to Reflect a Graph

To reflect the graph of a function $y = f(x)$ through (or with respect to) the x-axis, we multiply the right-hand side of the formula $y = f(x)$ by -1.

EXAMPLE 6 A complete graph of $y = -\sqrt{x}$ is obtained by reflecting the graph of $y = \sqrt{x}$ through the x-axis (Fig. 1.76b). ☰

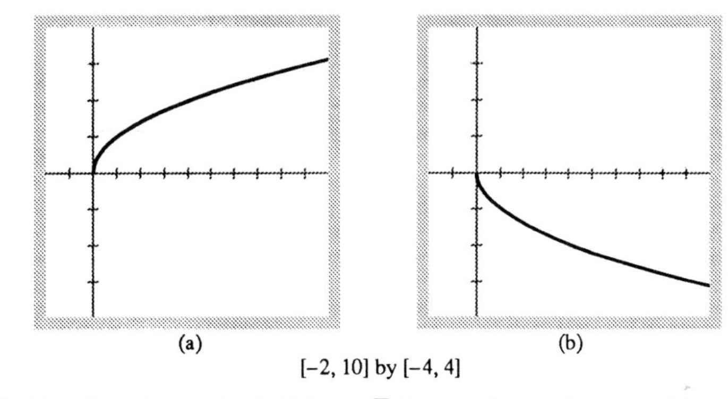

(a) (b)
[−2, 10] by [−4, 4]

1.76 To reflect the graph of $f(x) = \sqrt{x}$ through the x-axis, we multiply the right-hand side of the formula for f by -1. Complete graphs of (a) $y = \sqrt{x}$ and (b) $y = -\sqrt{x}$.

Notice that each graph in Fig. 1.76 is the reflection through the x-axis of the other and that combining the two graphs in Fig. 1.76 produces a new graph that is symmetric with respect to the x-axis.

To reflect the graph of a function $y = f(x)$ through (or with respect to) the y-axis, we multiply x by -1.

EXAMPLE 7 A complete graph of $y = \sqrt[4]{-x}$ is obtained by reflecting the graph of $y = \sqrt[4]{x}$ through the y-axis (Fig. 1.77b). ≡

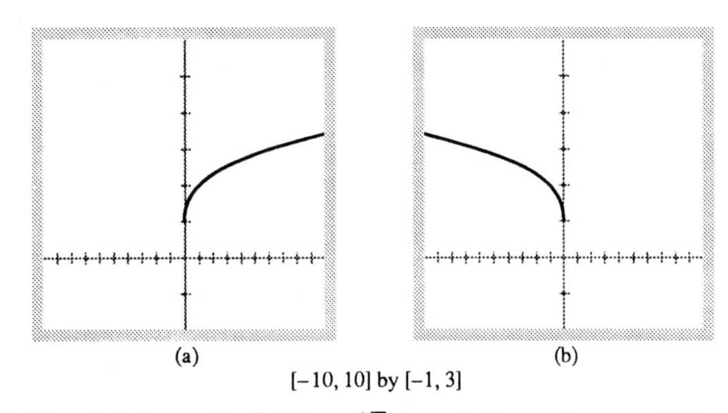

(a) (b)
[−10, 10] by [−1, 3]

1.77 To reflect the graph of $f(x) = \sqrt[4]{x}$ through the y-axis, we multiply x by -1. Complete graphs of (a) $y = \sqrt[4]{x}$ and (b) $y = \sqrt[4]{-x}$.

Notice that each graph in Fig. 1.77 is the reflection through the y-axis of the other and that combining the two graphs in Fig. 1.77 produces a new graph that is symmetric with respect to the y-axis.

Reflection Formulas

With respect to the x-axis

$y = -f(x)$	Reflects the graph of f through the x-axis

With respect to the y-axis

$y = f(-x)$	Reflects the graph of f through the y-axis

How to Vertically Stretch or Shrink a Graph

To vertically stretch the graph of a function $y = f(x)$, we multiply the right-hand side of the formula $y = f(x)$ by a number greater than 1.

EXAMPLE 8 Complete graphs of $y = 2x^2$ and $y = 3x^2$ are obtained by vertically stretching the graph of $y = x^2$ by a factor of 2 or 3, respectively (Fig. 1.78a). ≡

To vertically shrink the graph of a function $y = f(x)$, we multiply the right-hand side of the formula $y = f(x)$ by a number between 0 and 1.

EXAMPLE 9 Complete graphs of $y = 0.5x^2$ and $y = 0.25x^2$ are obtained by vertically shrinking the graph of $y = x^2$ by a factor of 0.5 or 0.25, respectively (Fig. 1.78b). ≡

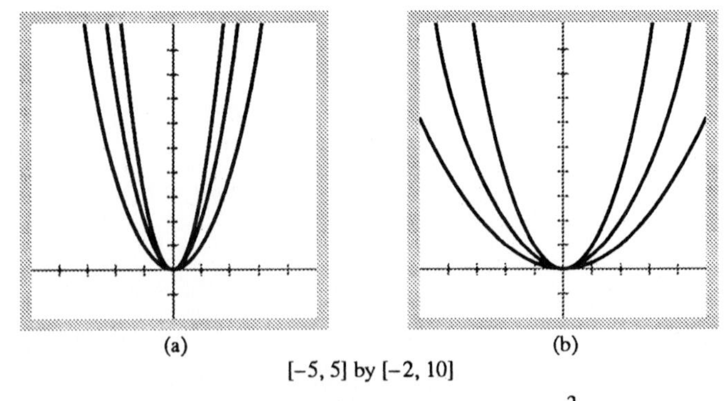

[−5, 5] by [−2, 10]

1.78 To vertically stretch or shrink the graph of $f(x) = x^2$, we multiply the right-hand side of the formula for f by a positive constant. Complete graphs of (a) $y = ax^2$ for $a = 1, 2, 3$, and (b) $y = ax^2$ for $a = 1, 0.5, 0.25$. The parabola $y = ax^2$ widens as a approaches 0 and narrows as a becomes numerically large.

Vertical Stretch or Shrink Formulas ($a > 0$)

$y = af(x)$	$a > 1:$	Vertically stretches the graph of f by a factor of a
	$0 < a < 1:$	Vertically shrinks the graph of f by a factor of a

EXAMPLE 10 Describe how the graph of $y = 2x^3 - 5$ can be obtained from the graph of $y = x^3$.

Solution The graph of $y = 2x^3 - 5$ can be obtained by vertically stretching the graph of $y = x^3$ by a factor of 2 and then shifting the resulting graph vertically downward 5 units (Fig. 1.79c).

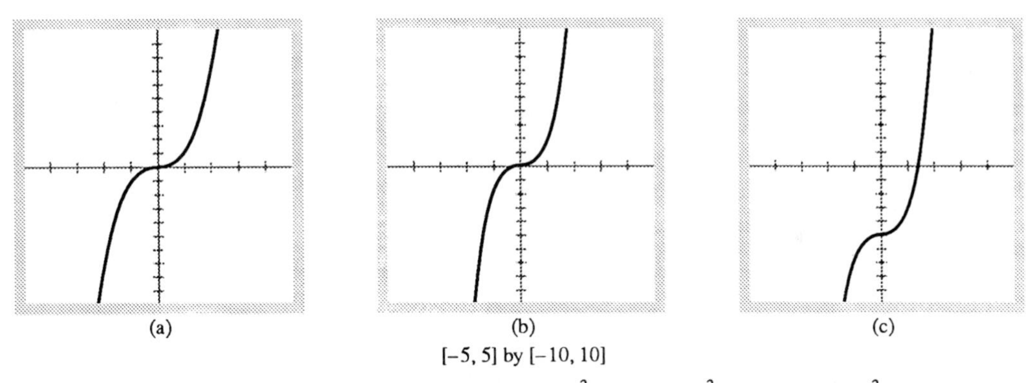

(a) (b) (c)

$[-5, 5]$ by $[-10, 10]$

1.79 Complete graphs of (a) $y = x^3$, (b) $y = 2x^3$, and (c) $y = 2x^3 - 5$.

In Example 10, if we first apply a vertical shift downward of 5 units to $y = x^3$ and then apply a vertical stretch by a factor of 2 to the resulting graph, we obtain the graph of the equation $y = 2(x^3 - 5) = 2x^3 - 10$. We can conclude that, in general, the geometric transformations of vertical stretching and vertical shifting cannot be reversed.

Explore Box

1. Apply a horizontal shift left 2 units to the graph of $y = |x|$ and then a vertical stretch by a factor of 3 to the resulting graph.
2. Write an equation for the final graph in 1.
3. Reverse the order of the geometric transformations applied in 1 and write an equation for the new resulting graph.
4. Compare the equations in parts 2 and 3 and explain any similarities or differences.
5. Repeat parts 1–4 for other pairs of geometric transformations.

EXAMPLE 11 Describe how the graph of $y = -0.5\sqrt[4]{3-x}+1$ can be obtained from the graph of $y = \sqrt[4]{x}$.

Solution First notice that $y = -0.5\sqrt[4]{3-x}+1$ can be written in the form $y = -0.5\sqrt[4]{-(x-3)}+1$. The desired graph can be obtained as follows:

1. Begin with the graph of $y = \sqrt[4]{x}$ (Fig. 1.80a).
2. Shrink vertically by a factor of 0.5 to get $y = 0.5\sqrt[4]{x}$ (Fig. 1.80b).
3. Reflect through the y-axis to get $y = 0.5\sqrt[4]{-x}$ (Fig. 1.80c).
4. Shift right 3 units to get $y = 0.5\sqrt[4]{-(x-3)} = 0.5\sqrt[4]{3-x}$ (Fig. 1.80d).
5. Reflect through the x-axis to get $y = -0.5\sqrt[4]{3-x}$ (Fig. 1.80e).
6. Shift up 1 unit to get $y = -0.5\sqrt[4]{3-x}+1$ (Fig. 1.80f).

The domain is $x \leq 3$ and the range is $y \leq 1$. ■

The Parabola $y = ax^2 + bx + c$

The graph of every equation of the form $y = ax^2 + bx + c$ can be obtained by applying a sequence of geometric transformations (shifts, stretches, shrinks,

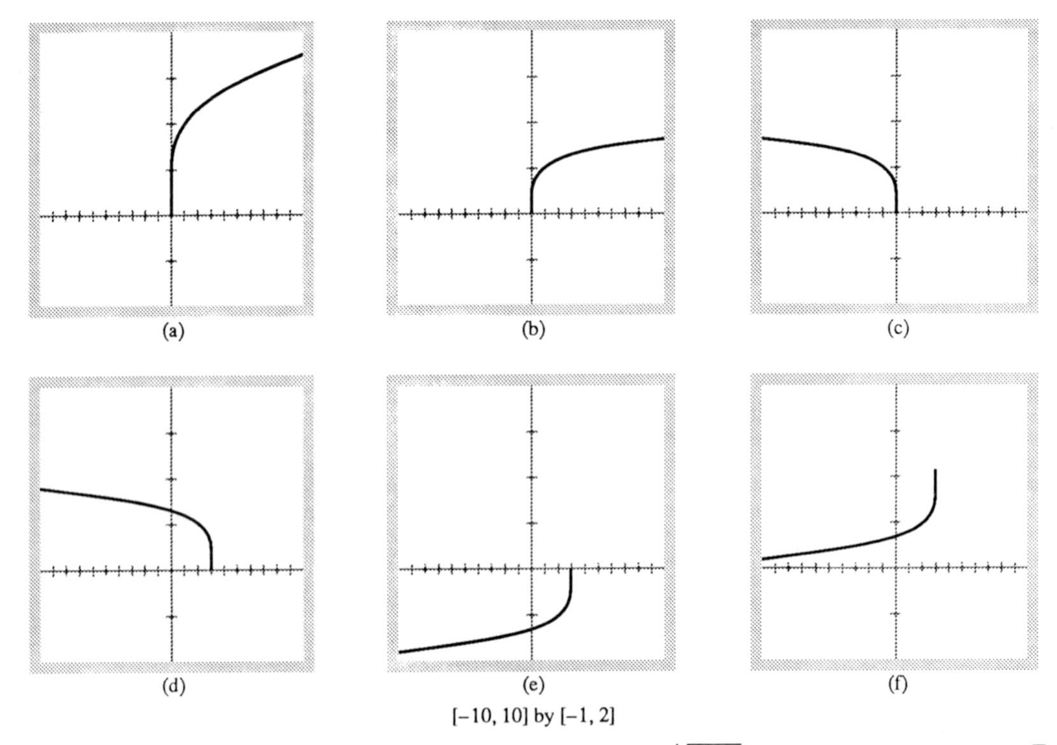

1.80 Transforming $y = \sqrt[4]{x}$ to $y = -0.5\sqrt[4]{3-x}+1$. (a) Begin with $y = \sqrt[4]{x}$. (b) Vertically shrink by 0.5. (c) Reflect through y-axis. (d) Shift right 3 units. (e) Reflect through x-axis. (f) Shift up 1 unit.

reflections) to the graph of $y = x^2$. This follows because $y = ax^2 + bx + c$ can be written in the form $y = a(x + h)^2 + k$ using the process of completing the square, as illustrated in Example 12. The transformed graph will have a vertex and axis of symmetry corresponding to the vertex (0, 0) and axis of symmetry $x = 0$ (y-axis) of the graph of $y = x^2$.

EXAMPLE 12 Describe how the graph of $y = -2x^2 + 12x - 5$ can be obtained from the graph of $y = x^2$. Identify the vertex and axis of symmetry of the graph of $y = -2x^2 + 12x - 5$.

Solution First we complete the square.

$$\begin{aligned} y &= -2x^2 + 12x - 5 \\ &= -2(x^2 - 6x) - 5 \\ &= -2(x^2 - 6x + 9 - 9) - 5 \\ &= -2(x^2 - 6x + 9) + 18 - 5 \\ &= -2(x - 3)^2 + 13 \end{aligned}$$

(Add and subtract the square of 1/2 of the coefficient of x.)

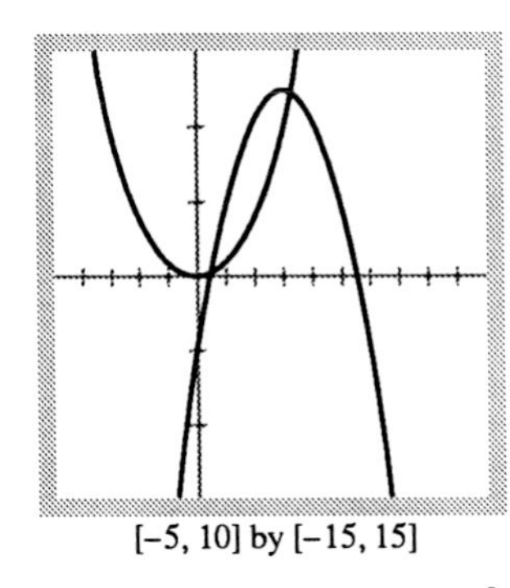
[-5, 10] by [-15, 15]

1.81 Complete graphs of $y = x^2$ and $y = -2x^2 + 12x - 5 = -2(x - 3)^2 + 13$.

The graph of the function $y = -2x^2 + 12x - 5 = -2(x - 3)^2 + 13$ can be obtained from the graph of $y = x^2$ as follows: Shift right 3 units, stretch vertically by a factor of 2, reflect through the x-axis, and finally shift up 13 units (Fig. 1.81).

The vertex is (3, 13), the point corresponding to (0, 0) under the geometric transformations. The axis $x = 0$ (y-axis) of symmetry of the graph of $y = x^2$ corresponds to the line $x = 3$ under the geometric transformations. The line $x = 3$ is the axis of symmetry of the graph of $y = -2x^2 + 12x - 5$. ∎

We can use the process of completing the square to identify the vertex and axis of symmetry of $y = ax^2 + bx + c$.

$$\begin{aligned} y &= ax^2 + bx + c \\ &= a\left(x^2 + \frac{b}{a}x\right) + c \\ &= a\left(x^2 + \frac{b}{a}x + \frac{b^2}{4a^2} - \frac{b^2}{4a^2}\right) + c \\ &= a\left(x^2 + \frac{b}{a}x + \frac{b^2}{4a^2}\right) - \frac{b^2}{4a} + c \\ &= a\left(x + \frac{b}{2a}\right)^2 + c - \frac{b^2}{4a} \end{aligned}$$

(Add and subtract the square of 1/2 of the coefficient of x.)

> **The Graph of $y = ax^2 + bx + c$**
>
> - A parabola that opens upward if $a > 0$ and downward if $a < 0$.
> - The axis of symmetry is $x = -\dfrac{b}{2a}$.
> - The vertex is $\left(-\dfrac{b}{2a}, c - \dfrac{b^2}{4a}\right)$.

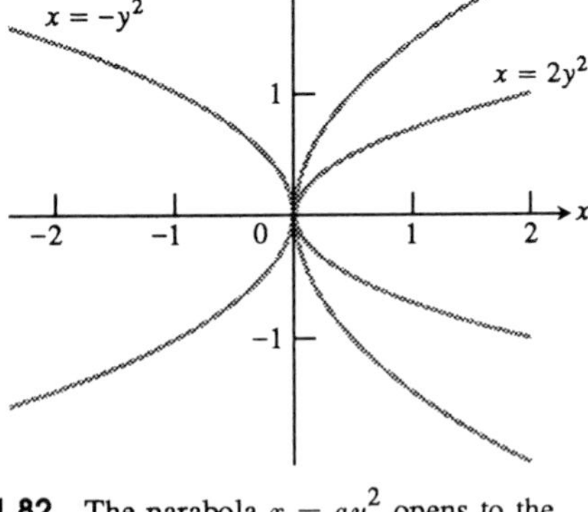

1.82 The parabola $x = ay^2$ opens to the right if $a > 0$ and to the left if $a < 0$.

The above facts confirm the information about possible graphs of $y = ax^2 + bx + c$ given in Figure 1.9 of Section 1.1. In Chapter 4 we will see how to use calculus to identify the vertex of parabolas like these.

Parabolas That Open to the Right or Left

If we interchange x and y in the formula $y = ax^2$, we obtain the equation

$$x = ay^2. \tag{1}$$

With the roles of x and y now reversed, the graph is a parabola whose axis is the x-axis. The vertex still lies at the origin. The parabola opens to the right if $a > 0$ and to the left if $a < 0$ (Fig. 1.82).

EXAMPLE 13 The formula $x = y^2$ gives x as a function of y but does *not* give y as a function of x. If we solve for y, we find that

$$y = \pm\sqrt{x}.$$

For each positive value of x we get *two* values of y instead of the required single value.

When taken separately, however, the formulas

$$y = \sqrt{x} \text{ and } y = -\sqrt{x}$$

do define functions of x. Each formula gives exactly one value of y for each possible value of x. The graph of $y = \sqrt{x}$ is the upper half of the parabola $x = y^2$. The graph of $y = -\sqrt{x}$ is the lower half. The two graphs meet at the origin (Fig. 1.83).

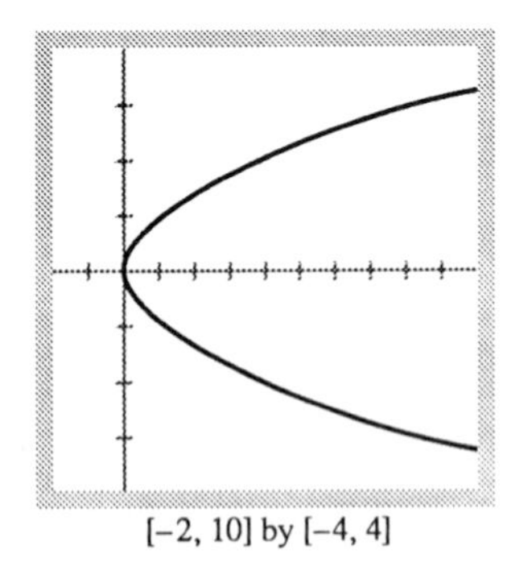

1.83 A computer visualization of the graph of $y^2 = x$ by overlaying the graphs of the two functions $y = \sqrt{x}$ and $y = -\sqrt{x}$.

The techniques used to identify the important features of the graph of the parabola $y = ax^2 + bx + c$ can be extended to the graph of the parabola $x = ay^2 + by + c$.

EXAMPLE 14 Describe how the graph of $x = 3y^2 - 12y + 11$ can be obtained from the graph of $x = y^2$. Identify the vertex and axis of symmetry of $x = 3y^2 - 12y + 11$.

Solution First we complete the square.

$$x = 3y^2 - 12y + 11$$
$$= 3(y^2 - 4y) + 11$$
$$= 3(y^2 - 4y + 4 - 4) + 11 \qquad \text{(Add and subtract the square of 1/2 of the coefficient of } y.)$$
$$= 3(y - 2)^2 - 12 + 11$$
$$= 3(y - 2)^2 - 1$$

The graph of $x = 3y^2 - 12y + 11 = 3(y - 2)^2 - 1$ can be obtained as follows:

1. Begin with the graph of $x = y^2$ (Fig. 1.84a).
2. Stretch horizontally by a factor of 3 to get $x = 3y^2$ (Fig. 1.84b).
3. Shift left 1 unit to get $x = 3y^2 - 1$ (Fig. 1.84c).
4. Shift up 2 units to get $x = 3(y - 2)^2 - 1$ (Fig. 1.84d).

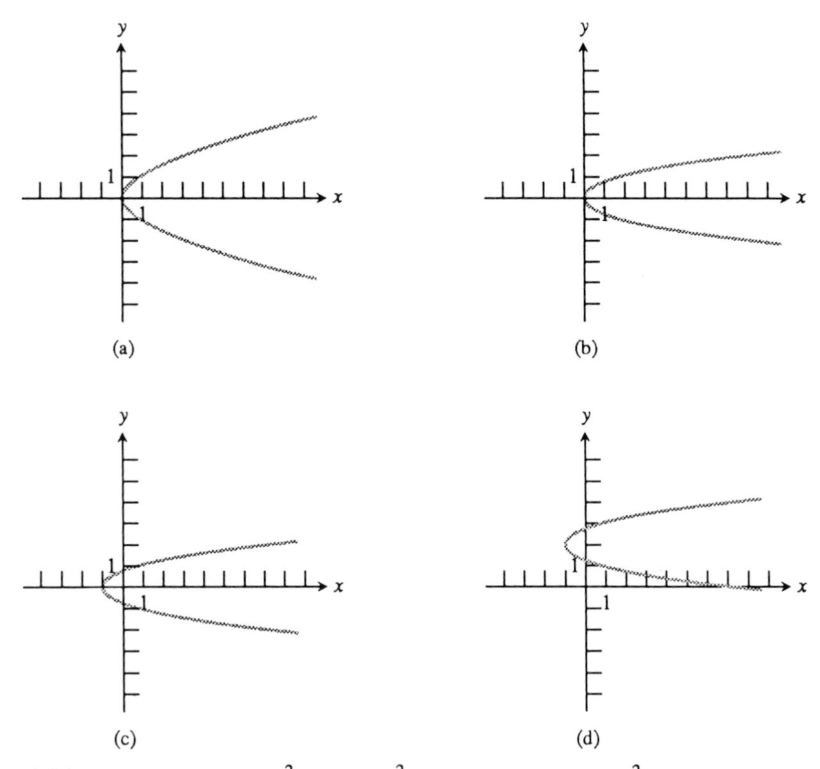

1.84 Transforming $x = y^2$ to $x = 3y^2 - 12y + 11 = 3(y - 2)^2 - 1$. (a) Begin with $x = y^2$. (b) Horizontally stretch by 3. (c) Shift left 1 unit. (d) Shift up 2 units.

To use a grapher to draw the graphs in Figure 1.84, we need to first solve for y in terms of x. For example, in step 4 above we get

$$x = 3(y-2)^2 - 1$$

$$\frac{x+1}{3} = (y-2)^2$$

$$\pm\sqrt{\frac{x+1}{3}} = y - 2$$

$$y = 2 \pm \sqrt{\frac{x+1}{3}}$$

The graph of the function $y = 2 + \sqrt{\frac{x+1}{3}}$ produces the upper half of the parabola, and the graph of the function $y = 2 - \sqrt{\frac{x+1}{3}}$ produces the lower half of the parabola.

If $a > 0$, the graph of $x = ay^2$ is either a horizontal stretch or shrink of the graph of $x = y^2$ in the same way that $y = ax^2$ is either a vertical stretch or shrink of the graph of $y = x^2$. In Section 1.7 we will say more about horizontal stretching.

Exercises 1.4

1. Figure 1.85 shows the graph of $y = x^2$ shifted to two new positions. Write equations for the new graphs.

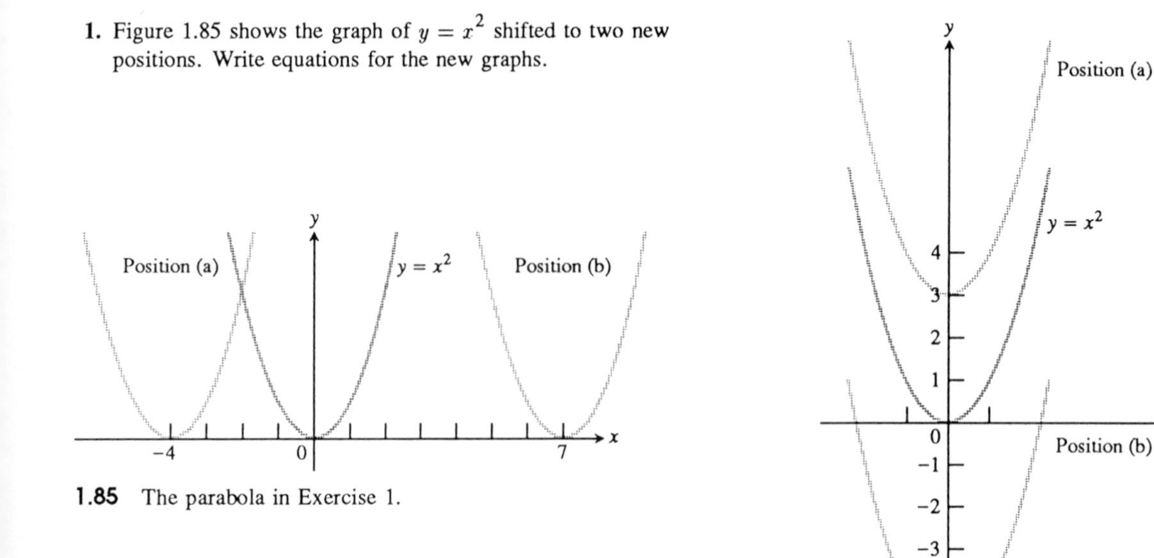

1.85 The parabola in Exercise 1.

1.86 The parabolas in Exercise 2.

2. Figure 1.86 shows the graph of $y = x^2$ shifted to two new positions. Write equations for the new graphs.

3. Match the equations listed below to the graphs in Fig. 1.87.
   a) $y + 4 = (x - 1)^2$ b) $y - 2 = (x - 2)^2$
   c) $y - 2 = (x + 2)^2$ d) $y + 2 = (x + 3)^2$

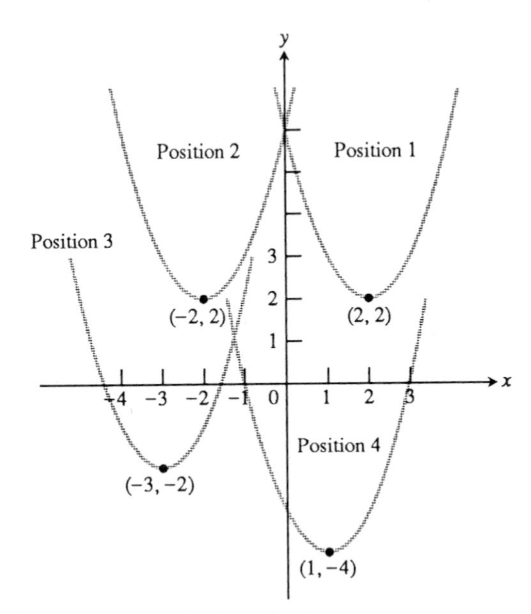

1.87 The parabolas in Exercise 3.

4. Figure 1.88 shows the graph of $y = x^2$ shifted to four new positions. Write an equation for each new graph.

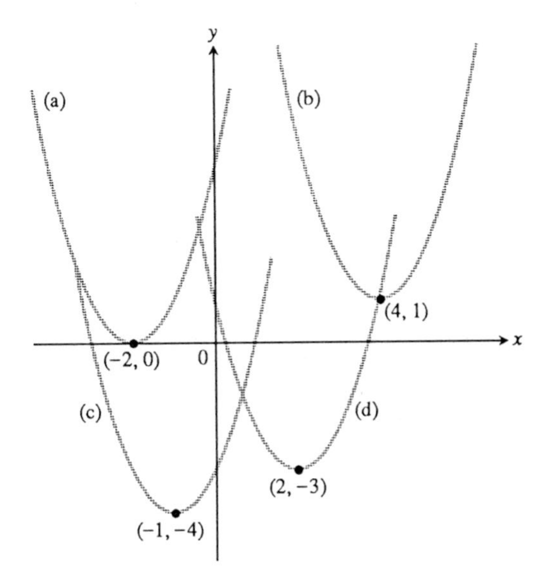

1.88 The parabolas in Exercise 4.

In Exercises 5–14, describe how each graph can be obtained from the graph of $f(x) = |x|, g(x) = 1/x, h(x) = \sqrt{x}$, or $k(x) = x^3$.

5. $y = |x + 4| - 3$
6. $y = |x - 3| + 2$
7. $y = 3\sqrt{-x}$
8. $y = -0.2\sqrt{-x}$
9. $y = \dfrac{2}{x} - 3$
10. $y = -\dfrac{0.5}{x} + 1$
11. $y = -0.5(x - 3)^3 + 1$
12. $y = 3\sqrt{2 - x} - 5$
13. $y = \dfrac{1}{2 - x} + 3$
14. $y = \dfrac{2}{x - 3} - 5$

In Exercises 15–24, sketch a complete graph of each function. Support your answer with a grapher. Determine the domain and range of the function.

15. $y = 4\sqrt[3]{2 - x} - 5$
16. $y = 0.5|x + 3| - 4$
17. $y = 5\sqrt[3]{-x} - 1$
18. $y = -2\sqrt[4]{1 - x} + 3$
19. $y = -\dfrac{1}{(x + 3)^2} + 2$
20. $y = \dfrac{1}{(x - 2)^2}$
21. $y = -2(x - 1)^{2/3} + 1$
22. $y = (x + 2)^{3/2} + 2$
23. $y = 2[1 - x]$
24. $y = [x - 2] + 0.5$

In Exercises 25–28, determine the vertex and axis of symmetry and sketch a complete graph of the parabolas. Support your work with a grapher.

25. $y = -2x^2 + 12x - 11$
26. $y = 3x^2 + 12x + 7$
27. $y = 4x^2 + 20x + 19$
28. $y = -4x^2 + 12x - 3$

Exercises 29–38 specifies the order in which transformations are to be applied to the graph of the given equation. Give an equation for the transformed graph in each case.

29. $y = x^2$, vertical stretch by 3, shift up 4
30. $y = x^2$, shift up 4, vertical stretch by 3
31. $y = \dfrac{1}{x}$, shift down 2, vertical shrink by 0.2
32. $y = \dfrac{1}{x}$, vertical shrink by 0.2, shift down 2
33. $y = |x|$, shift left 2, vertical stretch by 3, shift up 5
34. $y = |x|$, shift right 3, vertical shrink by 0.3, shift down 1
35. $y = x^3$, reflect through x-axis, vertical shrink by 0.8, shift right 1, shift down 2
36. $y = x^3$, vertical stretch by 2, reflect through x-axis, shift left 5, shift down 6
37. $y = \sqrt{x}$, reflect through y-axis, vertical stretch by 5, shift left 6, shift up 5
38. $y = \sqrt[4]{x}$, vertical shrink by 0.7, reflect through y-axis, shift right 8, shift down 7
39. Are the graphs of the equations found in Exercises 29 and 30 the same? Explain any differences.

40. Are the graphs of the equations found in Exercises 31 and 32 the same? Explain any differences.

41. Let $f(x) = \sqrt[3]{x}$.

a) Describe how the graph of $y = \sqrt[3]{-x}$ can be obtained from the graph of f.

b) Describe how the graph of $y = -\sqrt[3]{x}$ can be obtained from the graph of f.

c) Compare the graphs of the functions in (a) and (b). Explain any similarities or differences.

42. Let $f(x) = \sqrt{x}$.

a) Describe how the graph of $y = \sqrt{-x}$ can be obtained from the graph of f.

b) Describe how the graph of $y = -\sqrt{x}$ can be obtained from the graph of f.

c) Compare the graphs of the functions in (a) and (b). Explain any similarities or differences.

In Exercises 43–46, the graph of g is obtained by applying, in order, the given transformations to the graph of $f(x) = |x|$. Find the points on the graph of g that correspond under the sequence of transformations to the points $(-1, 1)$, $(0, 0)$, and $(1, 1)$ on the graph of f.

43. Shift right 3, shift up 2

44. Shift down 4, shift left 1

45. Vertical stretch by 2, reflect through x-axis

46. Vertical shrink by 0.3, shift up 4

A complete graph of f is shown in Figure 1.89. Sketch a complete graph of the functions in Exercises 47–54.

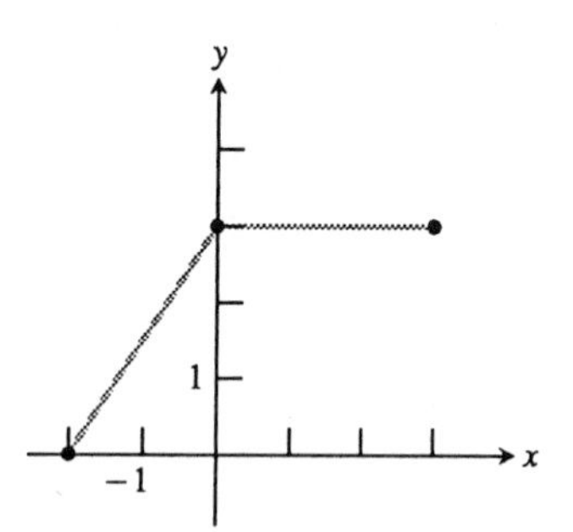

1.89 The graph of f for Exercises 47–54.

47. $y = f(-x)$

48. $y = -f(x)$

49. $y = f(x - 2)$

50. $y = f(x + 3)$

51. $y = 0.5f(x) - 3$

52. $y = -3f(x) + 2$

53. $y = -2f(x + 1) + 3$

54. $y = 0.2f(x - 1) - 1$

55. Sketch a complete graph of $y = \dfrac{2x + 1}{x - 2}$. (*Hint:* Use long division to write $y = 2 + \dfrac{5}{x - 2}$).

56. Sketch a complete graph of $y = \dfrac{x + 1}{x + 3}$. (*Hint:* See Exercise 55).

In Exercises 57–62, use completing the square to rewrite the equation in the form $x = a(x + h)^2 + k$. Then describe how the graph of the equation can be obtained from the graph of $x = y^2$ and sketch a complete graph. Write separate formulas for y in terms of x for the upper and lower branches of the parabolas and support your work with a grapher.

57. $x = y^2 - 6y + 11$

58. $x = y^2 + 4y + 1$

59. $x = 2y^2 + 4y + 1$

60. $x = -3y^2 + 12y - 7$

61. $x = -2y^2 + 12y - 13$

62. $x = 4y^2 + 16y + 9$

63. Let $f(x) = x^2$ and suppose g is obtained by applying the following transformations in some order to f: shift up 3, shift right 2, reflect through the y-axis. How many different graphs can be obtained for g?

64. Let f be any function. Show that:

a) $g(x) = f(x) + f(-x)$ is an even function.

b) $h(x) = f(x) - f(-x)$ is an odd function.

65. The line $y = mx$ is shifted vertically to make it pass through the point (0, b). What is the line's new slope-intercept equation?

66. The line $y = mx$ is shifted horizontally and vertically to make it pass through the point (x_0, y_0). What is the line's new point-slope equation?

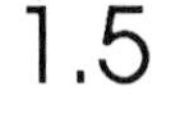

1.5 Solving Equations and Inequalities Graphically

We assume you are familiar with standard algebraic techniques for solving equations and inequalities such as the quadratic formula and factoring. These methods are really not very useful for solving most equations or inequalities (ones like $x^3 - 2x^2 - 7 = 0, \sin x > x$, etc.). In this section, we show how to use a graphing utility to solve equations and inequalities graphically with prescribed accuracy. The computer graphing method discussed in this section is powerful, robust, and general. We introduce algebraic and graphical representations of problem situations, and we discuss the Rational Zeros Theorem, which will be used with graphs to find the rational zeros of polynomial functions.

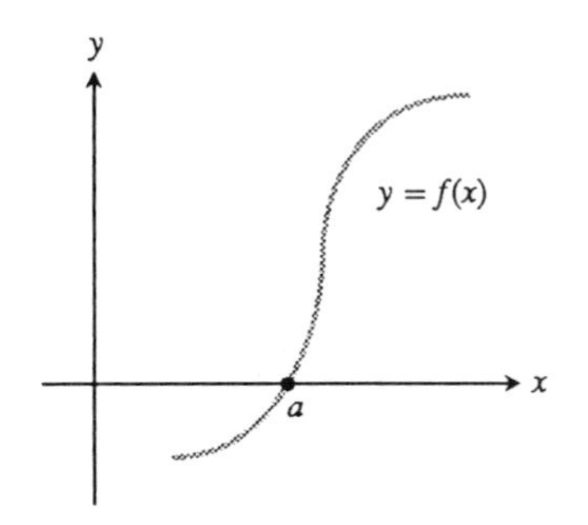

1.90 The real number $x = a$ is a solution to the equation $f(x) = 0$ if and only if the point $(a, 0)$ is on the graph of $y = f(x)$.

Unless we specify otherwise, **solving an equation** will mean finding all its real number solutions. Every equation involving a single variable x can be put in the equivalent form $f(x) = 0$. To use the graph of $y = f(x)$ to help solve the equation $f(x) = 0$, we first observe that $x = a$ is a solution to the equation $f(x) = 0$ if and only if $(a, 0)$ is a point on the graph of $y = f(x)$ (Fig. 1.90). Thus, real solutions to equations $f(x) = 0$ correspond to the x-intercepts of the graph of $y = f(x)$. The key to the success of the graphical method is finding complete graphs.

Solving Equations Graphically Using Zoom-in

A graphing utility can be used to find solutions to a high degree of accuracy using a procedure called zoom-in regardless of the complexity of the problem.* The idea is to "trap" the corresponding x-intercept in a decreasing sequence of viewing rectangles, each new one contained within the previous one, until the last viewing rectangle has enlarged a small enough portion of the graph so that the value of x can be read to the accuracy desired and within the limits of machine precision (Fig. 1.91). Most graphing utilities allow accurate answers to be read to at least 9 or 10 significant digits.

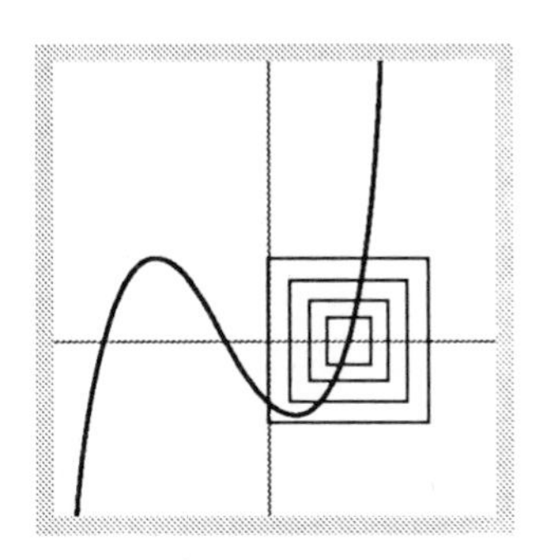

1.91 Zoom-in creates a nested decreasing sequence of viewing rectangles.

EXAMPLE 1 Solve the equation $x^3 - 2x^2 = 1$.

Solution Rewrite the equation in the form $x^3 - 2x^2 - 1 = 0$ and find a complete graph of $f(x) = x^3 - 2x^2 - 1$ (Fig. 1.92). Because the graph in Figure 1.92 is complete, we know that the equation $x^3 - 2x^2 - 1 = 0$ has only one real solution. Next we estimate its value.

* Consult the accompanying *Calculus Graphing Calculator and Computer Graphing Laboratory Manual* or the owner's manual of your graphing utility for more details about zoom-in.

Notice that $f(2) < 0$ and $f(3) > 0$. In Chapter 2 we will see that, as the graph suggests, f has a zero between $x = 2$ and $x = 3$. In fact, we can estimate the zero to be 2.2 because the distance between the scale marks in Fig. 1.91 is 1.

Figure 1.92 suggests a new viewing rectangle to use (Fig. 1.93a). The scale marks are 0.1 unit apart in Fig. 1.93(a), so we can now estimate the solution to be 2.21. In Exercise 51, you will see why it is best to change yMin and yMax as we change xMin and xMax.

Figure 1.93(a) suggests a new viewing rectangle to use (Fig. 1.93b). The scale marks are 0.01 unit apart in Fig. 1.93(b), so we can now estimate the solution to be 2.206. This zoom-in process can be continued until we reach the limit of our grapher's precision. ≡

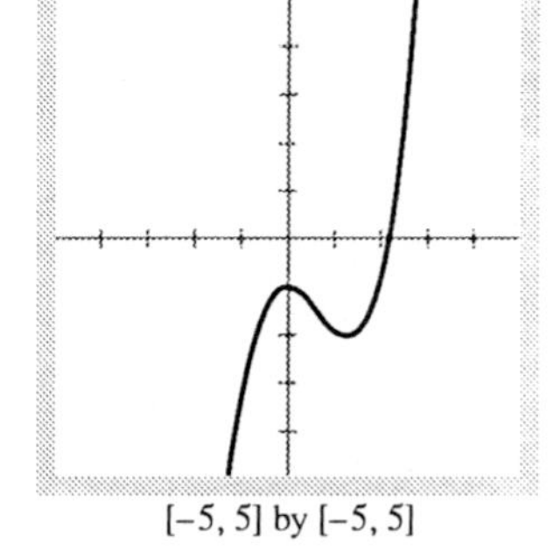

[−5, 5] by [−5, 5]

1.92 A complete graph of $f(x) = x^3 - 2x^2 - 1$ that shows all real solutions to $x^3 - 2x^2 - 1 = 0$.

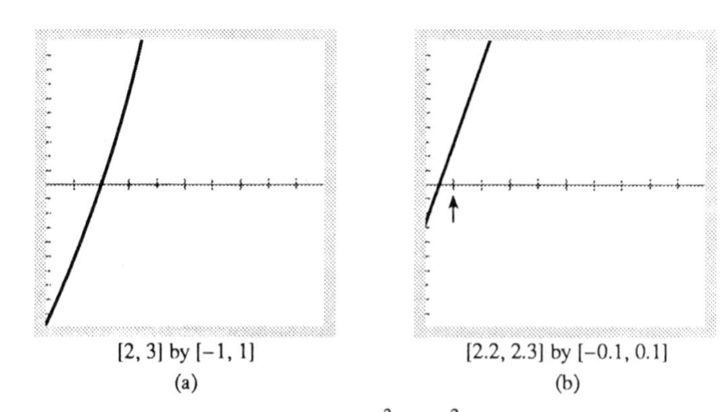

[2, 3] by [−1, 1]
(a)

[2.2, 2.3] by [−0.1, 0.1]
(b)

1.93 Two views of the zero of $f(x) = x^3 - 2x^2 - 1$.

TABLE 1.1 Comparing approximations to the zero of $f(x) = x^3 - 2x^2 - 1$.

x	$f(x) = x^3 - 2x^2 - 1$
2.20	−0.03200
2.205	−0.00328
2.206	0.00248
2.21	0.02566

Analysis of Error

We know from Figure 1.93(b) that the exact solution to $x^3 - 2x^2 - 1 = 0$ is between 2.20 and 2.21. The distance between 2.20 and 2.21 is 0.01 unit, so *any* number in the interval $2.20 \leq x \leq 2.21$ can be reported as the solution with an error of at most 0.01.

Which number should we report as the solution with an error of at most 0.01? Technically speaking, the midpoint 2.205 of the interval $2.20 \leq x \leq 2.21$ will be in error by at most 0.005, half the length of the interval. However, the graph suggests 2.206 is a better answer. Table 1.1 computes the $y = f(x) = x^3 - 2x^2 - 1$ values to five decimals for several approximations to the zero of f and illustrates why our visually suggested observation that 2.206 is a better approximation is correct.

Notice in Table 1.1 that $f(2.206)$ is closer to zero than any other number in the second column. Indeed most graphs produced by graphing utilities are very reliable. Even though some points in intervals may give more accurate estimates of solutions, we make the following definition.

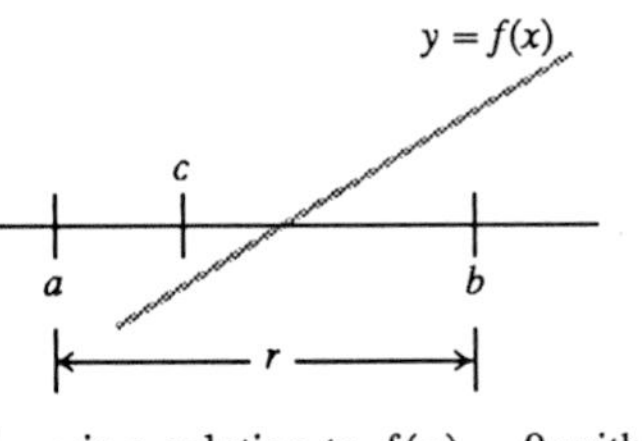

1.94 c is a solution to $f(x) = 0$ with an error of at most r.

Definition

Error of a Solution

Suppose a graph of f crosses the x-axis between two consecutive scale marks a and b and $b - a = r$ (Fig. 1.94). Then, any number c in the interval $[a, b]$ is said to be a **solution with error of at most r** to $f(x) = 0$.

EXAMPLE 2 Solve the equation $x^3 - 2x^2 = 1$ with an error of at most 0.000001.

Solution Continue the process started in Example 1 until we find a viewing rectangle in which we can distinguish consecutive scale marks 0.000001 unit apart (Fig. 1.95).

The graph crosses the x-axis between 2.205569 and 2.205570. Thus the solution is 2.2055694 with an error of at most 0.000001. ☰

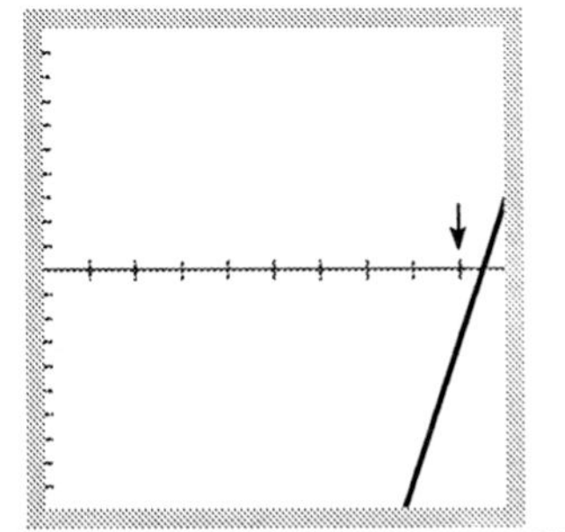

[2.20556, 2.20557] by [−0.00001, 0.00001]

1.95 A graph of $f(x) = x^3 - 2x^2 - 1$ that displays its real zero with an error of at most 0.000001.

Generally it is not necessary to find solutions as accurately as in Example 2. We make the following agreement.

Accuracy Agreement

In this textbook, unless otherwise stated, to *solve an equation* means *to approximate all real solutions with an error of at most 0.01* or *give the exact solution.* Numbers that appear in examples and exercises are assumed to be exact unless otherwise specified.

This agreement works well most of the time. There are, of course, exceptions such as problems where numbers are very large or very small.

Polynomial Equations

An **nth-degree polynomial equation** has the form

$$a_n x^n + a_{n-1} x^{n-1} + \ldots + a_1 x + a_0 = 0,$$

where the a_i's are real numbers with $a_n \neq 0$ and n a positive integer. The equation of Example 1 is a polynomial equation of degree 3, and the equation $x^3 - 2x - 2\cos x = 0$ is not a polynomial equation because of the term $2\cos x$. We assume you know how to algebraically solve polynomial equations of degree 1 and 2 as well as higher degree polynomial equations that can be factored into a product of polynomials each of degree 1 or 2. Remember that $x = a$ is a zero of a polynomial if and only if $(x - a)$ is a factor.

The following theorem allows us to algebraically determine all rational number zeros of a polynomial equation with rational coefficients.

Theorem 1

Rational Zeros Theorem

Suppose all the coefficients in the polynomial equation

$$f(x) = a_n x^n + a_{n-1}x^{n-1} + \cdots + a_1 x + a_0 = 0 \qquad (a_n \neq 0, a_0 \neq 0)$$

are integers.

If $x = c/d$ is a rational zero of f, where c and d have no common factors, then

- c is a factor of the constant term a_0, and
- d is a factor of the leading coefficient a_n.

Example 3 nicely illustrates how computer graphing enhances the application of algebraic procedures.

EXAMPLE 3 Solve $2x^3 - x^2 - 14x - 12 = 0$.

Solution First we use the Rational Zeros Theorem to find any rational zeros of $f(x) = 2x^3 - x^2 - 14x - 12$. Suppose c/d is a rational zero in lowest terms. Then, the list of possibilities for c, d, and c/d are:

$$c: \quad \pm 1, \pm 2, \pm 3, \pm 4, \pm 6, \pm 12$$ (c must be a factor of the constant term -12)

$$d: \quad \pm 1, \pm 2$$ (d must be a factor of the leading coefficient 2)

$$c/d: \quad \pm 1, \pm 2, \pm 3, \pm 4, \pm 6, \pm 12, \pm\frac{1}{2}, \pm\frac{3}{2}.$$

We would compute $f(c/d)$ for each of the 16 possibilities. However, we can use a complete graph of f to reduce this time-consuming and tedious task (Figure 1.96a). We can check that $f(-1) \neq 0$, so any real zero must be between -2 and -1 or between 3 and 4. The only possible rational zero is $-3/2$, and direct computation shows $f(-3/2) = 0$. We can use long division or synthetic division to factor f:

$$\begin{aligned} f(x) &= 2x^3 - x^2 - 14x - 12 \\ &= \left(x + \frac{3}{2}\right)(2x^2 - 4x - 8) \qquad \text{($x = -3/2$ is a zero, so $(x + 3/2)$ is a factor.)} \\ &= \left(x + \frac{3}{2}\right)(2)(x^2 - 2x - 4) \\ &= (2x + 3)(x^2 - 2x - 4) \end{aligned}$$

Now we will use the quadratic formula to find the zeros of the factor $x^2 - 2x - 4$:

$$x = \frac{2 \pm \sqrt{4 + 16}}{2} = \frac{2 \pm 2\sqrt{5}}{2} = 1 \pm \sqrt{5}.$$

Figure 1.96(a) makes the zero at $-3/2$ appear to be a zero of multiplicity 2. The reason is that the real root $1 - \sqrt{5} = -1.236$ is close to $-3/2$. If we zoom-in around $x = -3/2$, we will see that there are two real zeros between -2 and -1 (Fig. 1.96b). Thus, there are three real solutions, one rational and two irrational. ≡

EXPRESSING NUMBERS AS DECIMALS

In this textbook, we write $1 - \sqrt{5} = -1.236$ and $1 + \sqrt{5} = 3.236$ with the understanding that the decimal approximation is accurate to the number of decimal places displayed.

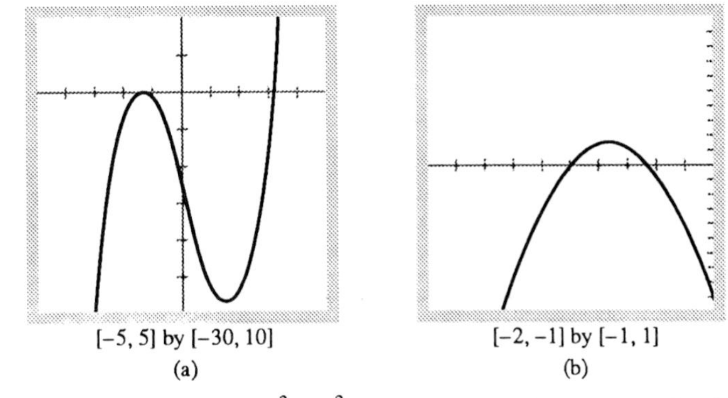

1.96 Two views of $f(x) = 2x^3 - x^2 - 14x - 12$ that show the equation $2x^3 - x^2 - 14x - 12 = 0$ has three real zeros.

Solving Inequalities Graphically Using Zoom-In

Solutions to inequalities usually involve unions of intervals of real numbers. Every inequality involving a single variable x can be put into one of the four equivalent forms: $f(x) < 0, f(x) \leq 0, f(x) > 0, f(x) \geq 0$. To use the graph of $y = f(x)$ to solve any one of the four inequalities we first use the graph to solve the equation $f(x) = 0$. The solutions to the equation give the endpoints of the intervals involved in the solution to the inequality. When we say **solve an inequality** we mean determine the endpoints (solutions to $f(x) = 0$) of the intervals with an error of at most 0.01.

EXAMPLE 4 Solve the inequality $x^3 - 2x^2 < 1$.

Solution Rewrite the inequality in the form $x^3 - 2x^2 - 1 < 0$ and let $f(x) = x^3 - 2x^2 - 1$. This is the function of Example 1, whose complete graph is shown in Figure 1.92. Observe that $x = a$ is a solution to the inequality $f(x) < 0$

if and only if the corresponding point $(a, f(a))$ on the graph of f is *below* the x-axis. Thus, according to our agreement about accuracy, the solution to the inequality $x^3 - 2x^2 - 1 < 0$ could be reported as $-\infty < x < 2.206$, where 2.206 is a solution to the equation $x^3 - 2x^2 - 1 = 0$ with an error of at most 0.01.

In Table 1.1, we showed that $f(2.206) > 0$. Thus, the interval $-\infty < x < 2.206$ includes values of x for which $f(x) > 0$. If we needed to be extra careful, we could replace 2.206 by a slightly smaller number at which f is negative. For example, $f(2.205) < 0$. So we could report the answer as $-\infty < x < 2.205$. Notice 2.205 is a solution to the equation $x^3 - 2x^2 - 1 = 0$ with an error of at most 0.01 and $f(x) < 0$ for all x in $-\infty < x < 2.205$.

It is our intention to report endpoints of intervals involved as solutions to inequalities as agreed. There will be times when we are off a little bit, especially when $\geq$ or $\leq$ are involved and we use zoom-in to find the endpoints. When we need to be more careful we will do so.

EXAMPLE 5 Solve the inequality $2x^3 - x^2 - 14x - 12 \geq 0$.

Solution In Example 3, we showed that $f(x) = 2x^3 - x^2 - 14x - 12$ had three real zeros: $-3/2, 1 - \sqrt{5}, 1 + \sqrt{5}$. We can see from Figure 1.96 that the graph of f is above the x-axis between -1.5 and $1 - \sqrt{5}$ and to the right of $1 + \sqrt{5}$. However, the fact that the endpoints are solutions to the inequality cannot be determined from a graph. This information must be confirmed algebraically. Because of Example 3, we know that the zeros of f are solutions to this inequality. Thus, the solution to the inequality is $-1.5 \leq x \leq 1 - \sqrt{5}$ and $x \geq 1 + \sqrt{5}$. According to our agreement about accuracy we report the answer as $-1.5 \leq x \leq -1.236$ and $x \geq 3.236$, or $[-1.5, -1.236] \cup [3.236, \infty]$, where the endpoints of the intervals are either exact or have an error of at most 0.01.

Equations and Inequalities with Absolute Values

To solve graphically an equation or inequality that contains absolute values requires *no* new techniques. To solve such equations and inequalities algebraically, we write equivalent equations or inequalities without absolute values and then solve as usual.

EXAMPLE 6 Solve the equation $|2x - 3| = 7$ algebraically, and support your work graphically.

Solution The equation says that $2x - 3 = \pm 7$, so there are two possibilities:

$2x - 3 = 7$	$2x - 3 = -7$	(Equivalent equations without absolute values)
$2x = 10$	$2x = -4$	(Solve as usual.)
$x = 5$	$x = -2$	

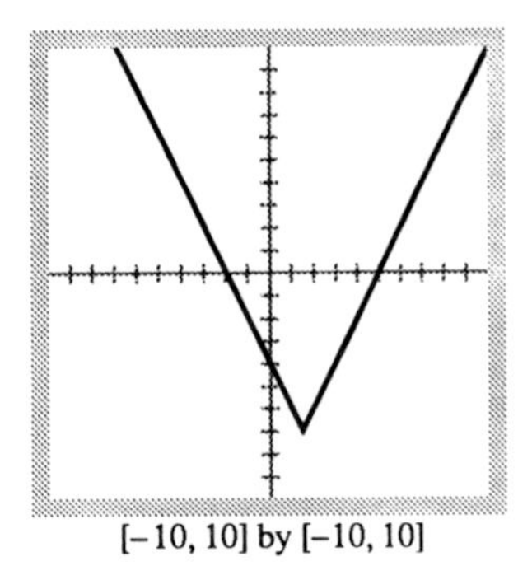

[−10, 10] by [−10, 10]

1.97 A complete graph of $y = |2x - 3| - 7$.

The equation $|2x - 3| = 7$ has two solutions: $x = 5$ and $x = -2$.

The x-intercepts of the graph of the function $y = |2x - 3| - 7$ in Figure 1.97 appear to be $x = 5$ and $x = -2$ supporting the reasonableness of the solutions found algebraically. ≡

Absolute Values and Intervals

The connection between absolute values and distance established in Section 1.2 gives us a new way to write formulas for intervals.

The inequality $|a| < 5$ says that the distance from a to the origin is less than 5. This is the same as saying that a lies between −5 and 5 on the number line. In symbols,

$$|a| < 5 \Leftrightarrow -5 < a < 5. \qquad (1)$$

The set of numbers a with $|a| < 5$ is the open interval from −5 to 5 (Fig. 1.98). The general rule follows.

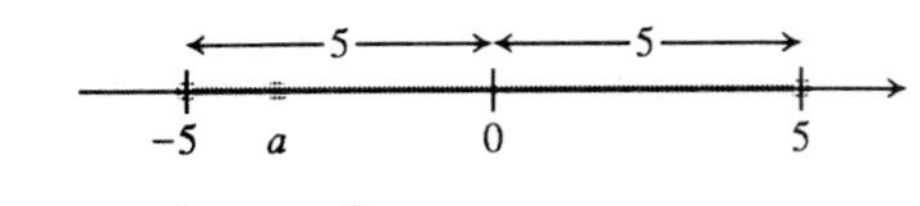

1.98 $|a| < 5$ means $-5 < a < 5$.

> ### Relation between Intervals and Absolute Values
>
> If D is any positive number, then
>
> $$|a| < D \Leftrightarrow -D < a < D, \qquad (2)$$
>
> $$|a| \le D \Leftrightarrow -D \le a \le D. \qquad (3)$$

EXAMPLE 7 Solve the inequality $|x - 5| < 9$ algebraically and support your work graphically.

Solution Change

$$|x - 5| < 9$$

to $-9 < x - 5 < 9$ (Eq. (2) with $a = x - 5$ and $D = 9$)

to $-9 + 5 < x < 9 + 5$ (Adding a positive number to both sides of an inequality gives an equivalent inequality. Adding 5 here isolated the x.)

or $-4 < x < 14$.

The steps we just took are reversible, so the values of x that satisfy the inequality $|x-5| < 9$ are the numbers in the interval $-4 < x < 14$ (Fig. 1.99).

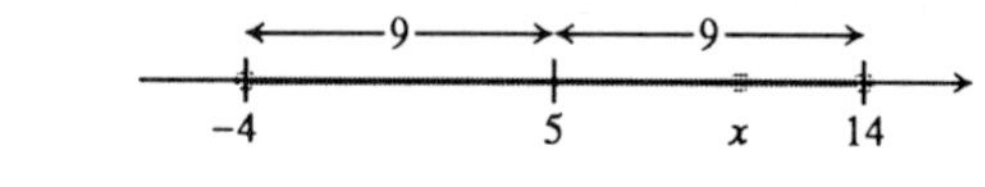

1.99 $|x-5| < 9$ means $-4 < x < 14$.

The graph of $y = |x-5| - 9$ in Figure 1.100 supports the solution found algebraically because the graph appears to be below the x-axis for $-4 < x < 14$.

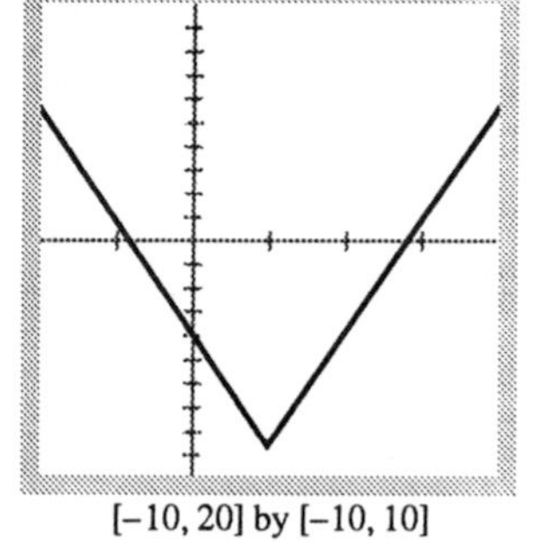

[−10, 20] by [−10, 10]

1.100 A complete graph of $y = |x-5| - 9$.

EXAMPLE 8 Describe the interval $-3 < x < 5$ with an absolute value inequality of the form $|x - x_0| < D$.

Solution We average the endpoint values to find the interval's midpoint:

$$\text{midpoint } x_0 = \frac{-3+5}{2} = \frac{2}{2} = 1. \qquad \text{(midpoint figure from Example 8, } F/T\text{, pg. 59)}$$

The midpoint lies 4 units away from each endpoint. The interval therefore consists of the points that lie within 4 units of the midpoint, or the points x with

$$|x-1| < 4.$$

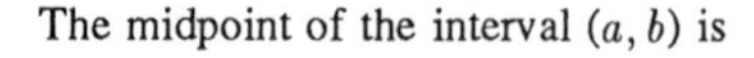

The midpoint of the interval (a, b) is

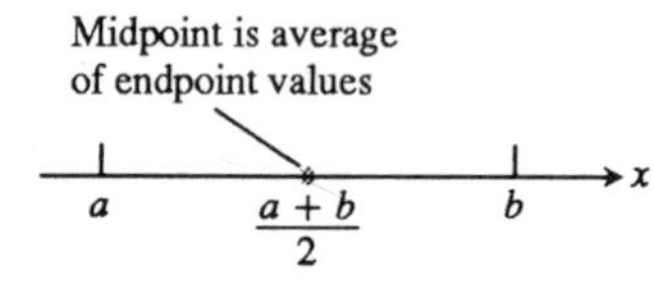

EXAMPLE 9 What values of x satisfy the inequality $\left|\frac{2x}{3}\right| \le 1$?

Solution Change

$$\left|\frac{2x}{3}\right| \le 1$$

to $-1 \le \frac{2}{3} \le 1$ (Eq. (3) with $a = 2x/3$ and $D = 1$)

to $-3 \le 2x \le 3$ (Multiplying both sides of an inequality by a positive number gives an equivalent inequality.)

to $-\frac{3}{2} \le x \le \frac{3}{2}$ (Dividing both sides of an inequality by a positive number gives an equivalent inequality.)

The original inequality holds for x in the closed interval from $-3/2$ to $3/2$ (Fig. 1.101).

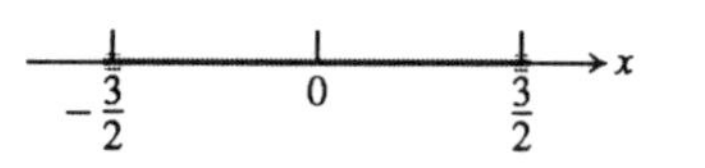

1.101 The inequality $|2x/3| \le 1$ holds on the interval $-3/2 \le x \le 3/2$.

EXAMPLE 10 What values of x satisfy the inequality

$$\left|5 - \frac{2}{x}\right| < 1?$$

Solution Change

$$\left|5 - \frac{2}{x}\right| < 1$$

to $-1 < 5 - \dfrac{2}{x} < 1$ (Eq. (2) with $a = (5 - 2/x)$ and $D = 1$)

to $-6 < -\dfrac{2}{x} < -4$ (Subtracting a positive number, in this case 5, from both sides of an inequality gives an equivalent inequality.)

to $4 < \dfrac{2}{x} < 6$ (Multiplying both sides of an inequality by -1 reverses the inequality.)

to $2 < \dfrac{1}{x} < 3$ (Divide by 2.)

to $\dfrac{1}{3} < x < \dfrac{1}{2}$. (Take reciprocals. When the numbers involved have the same sign, taking reciprocals reverses an inequality.)

The original inequality holds if, and only if, x lies between $1/3$ and $1/2$.

EXAMPLE 11 Support the solution to $|5 - 2/x| < 1$ found algebraically in Example 10 by drawing a complete graph of $y = |5 - 2/x| - 1$.

Solution The techniques of Section 1.4 are useful to help determine the desired complete graph. First, note the graph of $y = 5 - 2/x$ can be obtained from the graph of $y = 1/x$ as follows: Stretch vertically by a factor of 2, reflect through the y-axis, and shift up 5 units (Fig. 1.102a). Next, the graph of $y = |5 - 2/x|$ can be obtained from the graph of $y = 5 - 2/x$ by replacing the portion that is below the x-axis by its reflection through the x-axis (Fig. 1.102b). Finally, the graph of $y = |5 - 2/x| - 1$ is the graph of $y = |5 - 2/x|$ shifted down 1 unit (Fig. 1.102c). We could zoom in to support that the x-intercepts of the graph of $y = |5 - 2/x| - 1$ are $x = 1/3$ and $x = 1/2$. Thus, we have supported the fact that the solution to $|5 - 2/x| < 1$ is $1/3 < x < 1/2$.

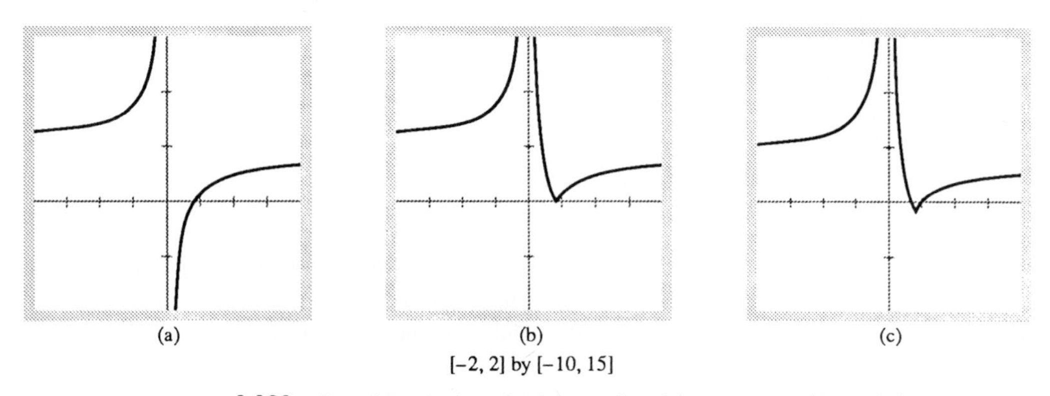

1.102 Complete graphs of (a) $y = 5 - 2/x$, (b) $y = |5 - 2/x|$, and (c) $y = |5 - 2/x| - 1$.

The inequality $|a| > 5$ says that the distance from a to the origin is greater than 5. This is the same as saying that a lies to the left of -5 or to the right of 5 on the number line. In symbols,

$$|a| > 5 \quad \Leftrightarrow \quad a < -5 \text{ or } a > 5. \tag{4}$$

The set of numbers a with $|a| > 5$ is the union of open intervals: $(-\infty, -5) \cup (5, \infty)$ (Fig. 1.103).

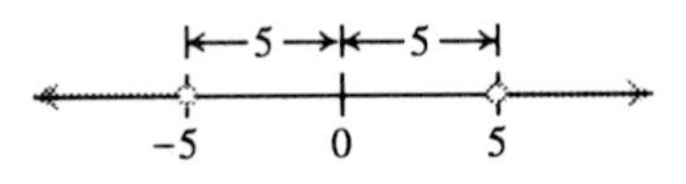

1.103 $|a| > 5$ means $(-\infty, -5) \cup (5, \infty)$.

The general rule follows.

> **Absolute Value Inequalities with $>$ or $\geq$**
>
> If D is any positive number, then
>
> $$|a| > D \Leftrightarrow a < -D \text{ or } a > D, \tag{5}$$
>
> $$|a| \geq D \Leftrightarrow a \leq -D \text{ or } a \geq D. \tag{6}$$

EXAMPLE 12 Solve the inequality $|3/(x-2)| < 1$ algebraically and support your work graphically.

Solution The following inequalities are equivalent provided $x \neq 2$.

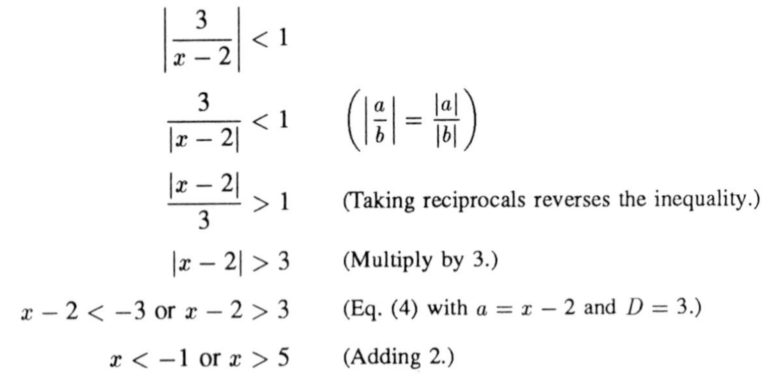

$$\left|\frac{3}{x-2}\right| < 1$$

$$\frac{3}{|x-2|} < 1 \qquad \left(\left|\frac{a}{b}\right| = \frac{|a|}{|b|}\right)$$

$$\frac{|x-2|}{3} > 1 \qquad \text{(Taking reciprocals reverses the inequality.)}$$

$$|x-2| > 3 \qquad \text{(Multiply by 3.)}$$

$$x-2 < -3 \text{ or } x-2 > 3 \qquad \text{(Eq. (4) with } a = x-2 \text{ and } D = 3.)$$

$$x < -1 \text{ or } x > 5 \qquad \text{(Adding 2.)}$$

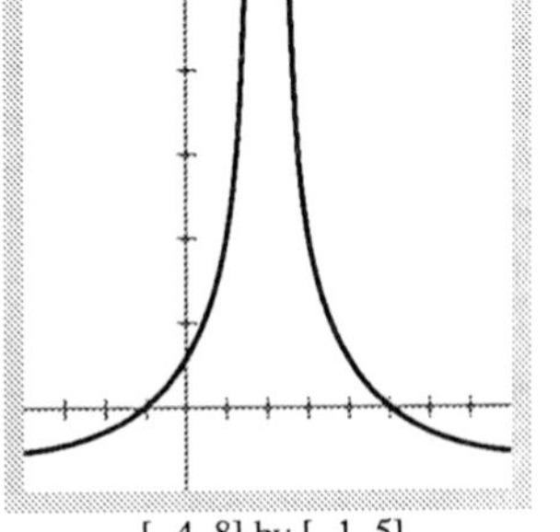

[−4, 8] by [−1, 5]

1.104 A complete graph of $y = |3/(x-2)| - 1$.

The original inequality holds if, and only if, $x < -1$ or $x > 5$.

The graph of $y = |3/(x-2)| - 1$ in Fig. 1.104 is complete and supports the solution found algebraically. ■

Algebraic and Graphical Representations

Consider the following problem situation. A rectangular garden is constructed with one side against an existing wall using 50 feet of fencing. Let x be the

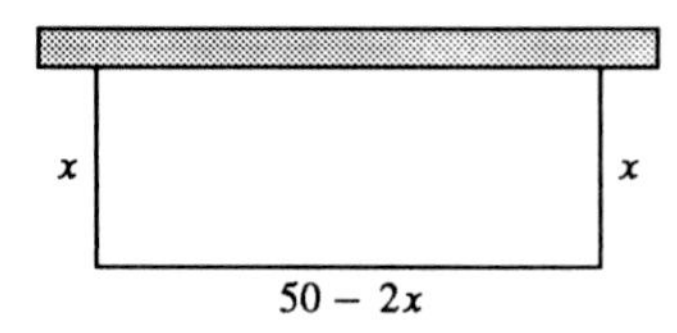

1.105 A garden enclosed using an existing wall and 50 feet of fence.

length of the side of the fence perpendicular to the existing wall (Fig. 1.105). The length of the side parallel to the wall is $50 - 2x$.

The area, A, of the garden can be expressed as a function of x,

$$A(x) = x(50 - 2x).$$

We call $A(x)$ an *algebraic representation.* It is also an *algebraic representation of the area of the garden.* Any complete graph of the algebraic representation $y = A(x)$ is called a *graphical representation of* $y = A(x)$ (Fig. 1.106a).

The axis of symmetry of the parabola $y = A(x)$ must be $x = 12.5$, halfway between the zeros: $x = 0$ and $x = 25$. The coordinates of the vertex are $(12.5, A(12.5)) = (12.5, 312.5)$. Thus, the domain of the function A is the set of all real numbers and the range is $(-\infty, 312.5]$. Notice that the x-coordinate 12.5 of the vertex corresponds to the side length x that produces a garden of maximum possible area 312.5 square feet.

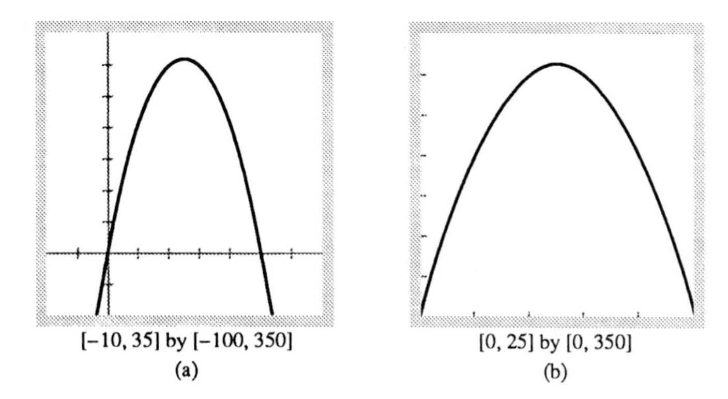

1.106 (a) A complete graph of the algebraic representation $y = x(50 - x)$ of the area of the garden in Fig. 1.105. (b) A graph of the problem situation $y = x(50 - x), 0 \leq x \leq 25$.

Only the values $0 \leq x \leq 25$ in the domain of $y = A(x)$ make sense in the problem situation. If we restrict the domain of an algebraic representation to the values that make sense in the problem situation, then its corresponding graph is called a *graph of the problem situation.* In this case, the graph of $y = x(50 - 2x)$ with x restricted to $0 \leq x \leq 25$ is a graph of the area problem (Fig. 1.106b).

Algebraic and graphical representations of problem situations are not unique. Notice that both $y = x(50 - 2x)$, with the domain the set of all real numbers, and $y = x(50 - 2x), 0 \leq x \leq 25$ are algebraic representations of the problem situation, and their corresponding graphs are graphical representations of the problem situation. However, the domain of the graph of the problem situation is unique.

EXAMPLE 13 Pure acid is added to 40 ounces of a 30% acid solution. Let x be the number of ounces of pure acid added.

a) Determine an algebraic representation $C(x)$ of the concentration of the new mixture.

b) Draw a complete graph of the algebraic representation $y = C(x)$. What are the domain and range of this function?

c) What values of x make sense in the problem situation? Draw a graph of the problem situation.

d) How much pure acid should be added to produce a mixture that is 75% acid?

Solution

a)
$$(0.3)40 = 12 \text{ ounces of pure acid in original mixture}$$
$$x + 12 = \text{ounces of pure acid in final mixture}$$
$$x + 40 = \text{ounces of final mixture}$$
$$C(x) = \frac{x+12}{x+40} = \text{concentration of acid in final mixture}$$

b) The domain of $y = C(x)$ is all real numbers different from -40. We use long division to rewrite $C(x)$.

$$y = \frac{x+12}{x+40}$$
$$y = 1 - \frac{28}{x+40}$$

Now we can see how the graph of $y = C(x)$ is obtained from the graph of $y = 1/x$. So the graph of $y = C(x)$ in Fig. 1.107(a) is complete and the range of $y = C(x)$ is the set of all real numbers different from 1.

c) The values of x that make sense in the problem situation are $0 \le x < \infty$. A graph of the problem situation is given in Fig. 1.107(b).

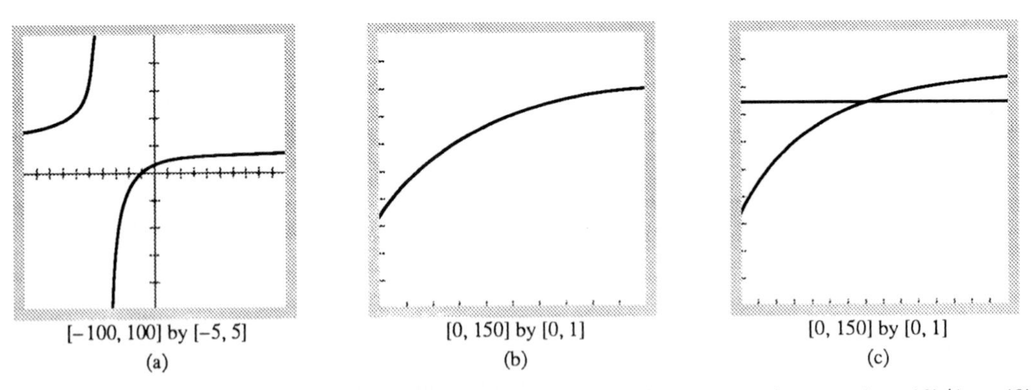

1.107 (a) A complete graph of the algebraic representation $y = (x+12)/(x+40)$ of the mixture problem. (b) A graph of the problem situation of Example 13: $y = (x+12)/(x+40)$, $x \ge 0$. (c) The new solution is 75% acid at the x-coordinate of the point of intersection of $y = (x+12)/(x+40)$ and $y = 0.75$.

d) The x-coordinate of the point of intersection of

$$y = C(x) = \frac{x+12}{x+40}$$

$$y = 0.75$$

gives the amount of pure acid that must be added to produce a 75% mixture (Fig. 1.107c). We could use zoom-in to show that $x = 72$ with an error of at most 0.01. We can also find the solution algebraically.

$$C(x) = 0.75$$

$$\frac{x+12}{x+40} = 0.75$$

$$x + 12 = 0.75x + 30$$

$$0.25x = 18$$

$$x = 72$$

Thus, the solution we found graphically using zoom-in was exact. So 72 ounces of pure acid must be added to 40 ounces of a 30% acid solution to produce a mixture that is 75% acid.

A Word about Solution Methods

Most equations and inequalities cannot be solved using traditional algebraic techniques. These techniques together with the Rational Zeros Theorem even fail with most polynomial equations, because the real roots are usually irrational. However, the graphical solution method is very powerful because it works regardless of the equation. We will see the superiority of the graphical method over and over again throughout this textbook.

Exercises 1.5

1. If $2 < x < 6$, which of the following statements about x are true and which are false?

a) $0 < x < 4$ b) $0 < x - 2 < 4$

c) $1 < \frac{x}{2} < 3$ d) $\frac{1}{6} < \frac{1}{x} < \frac{1}{2}$

e) $1 < \frac{6}{x} < 3$ f) $|x - 4| < 2$

g) $-6 < -x < 2$ h) $-6 < -x < -2$

2. If $-1 < y - 5 < 1$, which of the following statements about y are true and which are false?

a) $4 < y < 6$ b) $|y - 5| < 1$

c) $y > 4$ d) $y < 6$

e) $0 < y - 4 < 2$ f) $2 < \frac{y}{2} < 3$

g) $\frac{1}{6} < \frac{1}{y} < \frac{1}{4}$ h) $-6 < y < -4$

Solve the equations in Exercises 3–8.

3. $|x| = 2$ **4.** $|x - 3| = 7$

5. $|2x + 5| = 4$ **6.** $|1 - x| = 1$

7. $|8 - 3x| = 9$ **8.** $\left|\frac{x}{2} - 1\right| = 1$

Solve the equations in Exercises 9–12 algebraically and support your answers graphically.

9. $6x^2 + 5x - 6 = 0$ **10.** $x^2 - 4x + 1 = 0$

11. $4x^3 - 16x^2 + 15x + 2 = 0$

12. $x^3 - x^2 + x = 0$

Solve the equations in Exercises 13–16 graphically and confirm your answers algebraically.

13. $2x^2 + 7x - 4 = 0$ **14.** $9x^2 - 6x = 4$

15. $18x^3 - 3x^2 - 14x = 4$

16. $x^3 - 2x^2 + 2x = 0$

In Exercises 17–18, find a sequence of four viewing rectangles containing each solution. Choose each sequence to permit the solutions to be read with errors of at most 0.1, 0.01, 0.001, and 0.0001.

17. $x^3 - 2x^2 - 5x + 5 = 0$

18. $x^3 - 4x + 1 = 0$

Solve the equations in Exercises 19–24.

19. $9x^2 - 6x + 5 = 0$

20. $x^2 + x - 12 = 0$

21. $2x^3 - 8x^2 + 3x + 9 = 0$

22. $10 + x + 2x^2 = x^3$

23. $x^3 - 21x^2 + 111x = 71$

24. $x^3 + 19x^2 + 90x + 52 = 0$

The inequalities in Exercises 25–30 define intervals. Describe each interval with inequalities that do not involve absolute values.

25. $|y - 1| \le 2$

26. $|y + 2| < 1$

27. $|3y - 7| < 2$

28. $\left|\frac{y}{3}\right| \le 10$

29. $|1 - y| < \frac{1}{10}$

30. $\left|\frac{7 - 3y}{2}\right| < 1$

Describe the intervals in Exercises 31–34 with absolute value inequalities of the form $|x - x_0| < D$. It may help to draw a picture of the interval first.

31. $3 < x < 9$

32. $-3 < x < 9$

33. $-5 < x < 3$

34. $-7 < x < -1$

Solve the inequalities in Exercises 35–38 algebraically and support your answer graphically.

35. $|x - 5| < 2$

36. $\left|\frac{3x}{2}\right| < 5$

37. $\left|\frac{4}{x - 1}\right| \le 2$

38. $|3x + 2| > 3$

Solve the inequalities in Exercises 39–42 graphically and confirm your answers algebraically.

39. $|x + 3| \le 5$

40. $\left|\frac{2x}{5}\right| \le 1$

41. $\left|\frac{2}{x + 3}\right| < 1$

42. $|3x + 2| \ge 1$

Solve the inequalities in Exercises 43–48.

43. $x^2 + 3x - 10 \le 0$

44. $4x^2 - 8x + 5 > 0$

45. $|2 - 3x| < 4$

46. $\left|5 - \frac{x}{2}\right| \le 1$

47. $x^3 - 6x^2 + 5x + 6 \le 0$

48. $x^3 - 2x^2 - 5x + 20 > 0$

Use zoom-in to find the coordinates of the vertex in Exercises 49–50, each with an error of at most 0.01 and then confirm your answers algebraically.

49. $y = 30x - x^2$

50. $y = -x^2 + 22x - 21$

51. Consider the function $f(x) = x^3 - 2x^2 - 1$ of Example 2.

a) Draw the graph of f in the following viewing rectangles: $[-5, 5]$ by $[-5, 5]$, $[2, 3]$ by $[-5, 5]$, $[2.2, 2.3]$ by $[-5, 5]$, and $[2.2, 2.3]$ by $[-0.1, 0.1]$.

b) Explain why it is necessary to change yMin and yMax when we zoom in on an x-intercept.

52. Solve the equation $x^3 - 2x = 2\cos x$.

53. Explain why $ax^3 + bx^2 + cx + d = 0$ (a, b, c, d real, $a \ne 0$) will always have at least one real zero. Explain how to find all real zeros of any cubic polynomial equation.

54. Explain how the Rational Zeros Theorem can be applied to a polynomial equation with noninteger rational coefficients such as $2x^3 + \frac{1}{2}x^2 - \frac{2}{3}x + 1 = 0$.

55. One hundred feet of fencing is used to enclose a rectangular garden. Let x be the length of one side of the garden.

a) Determine an algebraic representation $A(x)$ for the area of the garden.

b) Draw a complete graph of the algebraic representation $y = A(x)$.

c) What are the domain and range of the algebraic representation $y = A(x)$?

d) What values of x make sense in the problem situation? Draw a graph of the problem situation.

e) Use a graph to determine the dimensions of the garden if the area is 500 ft^2, and confirm your answer algebraically.

f) Determine the possible values of x if the area of the garden is to be less than 500 ft^2.

56. One hundred feet of fencing is used to enclose three sides of a rectangular pasture and the side of a barn closes off the fourth side. Let x be the length of one side of the fence perpendicular to the barn.

a) Determine an algebraic representation $A(x)$ for the area of the pasture.

b) Draw a complete graph of the algebraic representation $y = A(x)$.

c) What are the domain and range of the algebraic representation $y = A(x)$?

d) What values of x make sense in the problem situation? Draw a graph of the problem situation.

e) Use a graph to determine the dimensions of the pasture if the area is 500 ft^2, and confirm your answer algebraically.

f) Determine the possible values of x if the area of the pasture is to be less than 500 ft^2.

57. An 8.5- by 11-in. piece of paper contains a picture with uniform border. The distance from the edge of the paper to the picture is x inches on all sides (Fig. 1.108).

a) Write the area A of the picture as a function of x.

b) Draw a complete graph of the function $y = A(x)$.

c) What are the domain and range of the function $y = A(x)$?

d) What values of x make sense in the problem situation? Which portion of the graph in (b) is a graph of the problem situation?

e) Determine the width of the border if the area of the picture is 60 in^2.

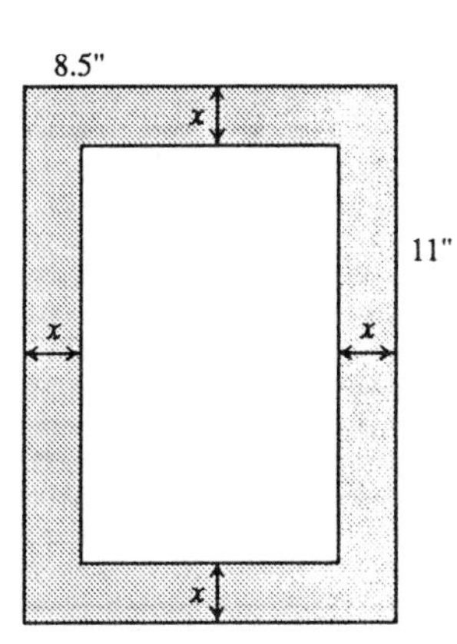

1.108 The figure for Exercise 57.

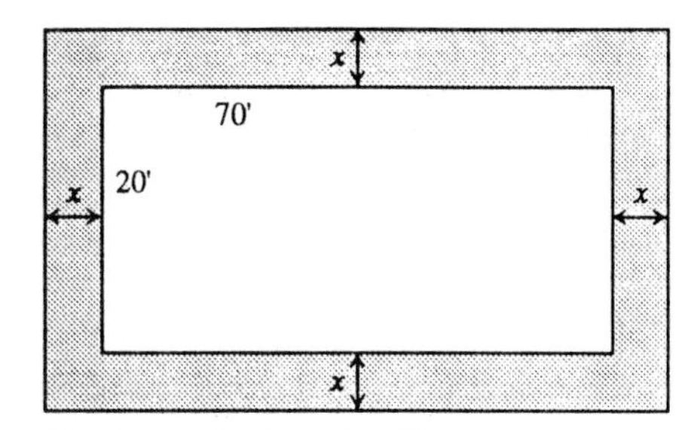

1.109 The figure for Exercise 58.

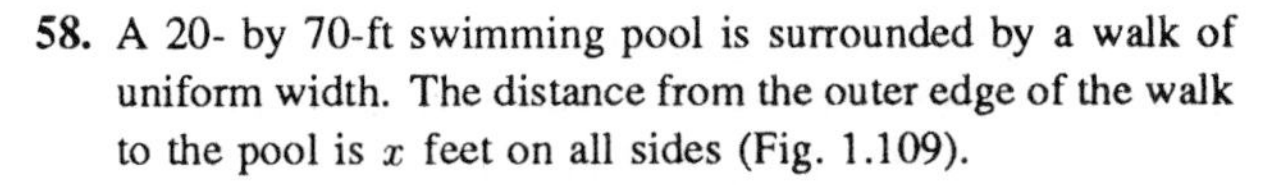

58. A 20- by 70-ft swimming pool is surrounded by a walk of uniform width. The distance from the outer edge of the walk to the pool is x feet on all sides (Fig. 1.109).

a) Write the area A of the sidewalk as a function of x.

b) Draw a complete graph of the function $y = A(x)$.

c) What are the domain and range of the function $y = A(x)$?

d) What values of x make sense in the problem situation? Which portion of the graph in (b) is a graph of the problem situation?

e) Determine the width of the sidewalk if its area is 500 ft^2.

59. Equal squares of side length x are removed from each corner of a 20- inch by 25-inch piece of cardboard. The sides are turned up to form a box with no top (Fig. 1.110).

a) Write the volume V of the box as a function of x.

b) Draw a complete graph of the function $y = V(x)$.

c) What are the domain and range of the function $y = V(x)$?

d) What values of x make sense in the problem situation? Draw a graph of the problem situation.

e) Use zoom-in to determine x so that the resulting box has maximum possible volume. What is the maximum possible volume?

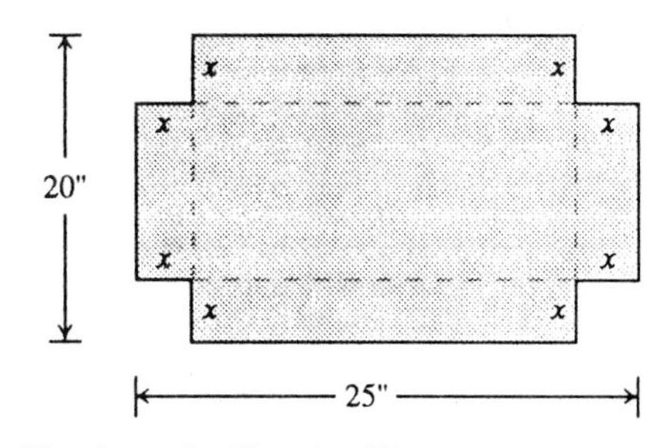

1.110 The figure for Exercise 59.

60. Repeat Exercise 57 for a 25- by 30-in. piece of cardboard.

61. A money box contains 50 coins in dimes and quarters. Let x be the number of dimes in the box.

a) Determine an algebraic representation $V(x)$ for the value of the coins in the box.

b) Draw a complete graph of the algebraic representation $y = V(x)$.

c) What are the domain and range of the function $y = V(x)$?

d) What values of x make sense in the problem situation? Which portion of the graph in (b) is a graph of the problem situation?

e) Determine the number of each coin in the box if the value of the coins is \$9.20.

f) Repeat part (e) if the value of the coins is \$6.25.

62. Sherrie invests \$10,000, part at 6.5% simple interest and the remainder at 8% simple interest. Let x be the amount she invests at 6.5% interest.

a) Determine an algebraic representation $I(x)$ for the total interest received in one year.

b) Draw a complete graph of the algebraic representation $y = I(x)$.

c) What are the domain and range of the function $y = I(x)$?

d) What values of x make sense in the problem situation? Which portion of the graph in (b) is a graph of the problem situation?

e) Determine the amount invested at each rate if Sherrie receives \$766.25 interest in one year.

1.6 Relations, Inverses, Circles

In this section we introduce relations and see that functions are simply relations with specific restrictions. Inverses of relations are introduced and graphed. We will see how to predict when the inverse of a function is itself a function. Exponential and logarithm functions are introduced from a naive point of view with formal development reserved for Chapters 5 and 7.

Recall that a function is a rule that assigns to each element in one set (**domain**) a single element from another set (**range**). If we drop the restriction that only a single element from the range is assigned to each element of the domain, then we have a relation. So a function is a relation.

Definition

> A **relation** from a set D to a set R is a set of ordered pairs of elements where the set of first entries of the ordered pairs is D and the set of second entries of the ordered pairs is R.

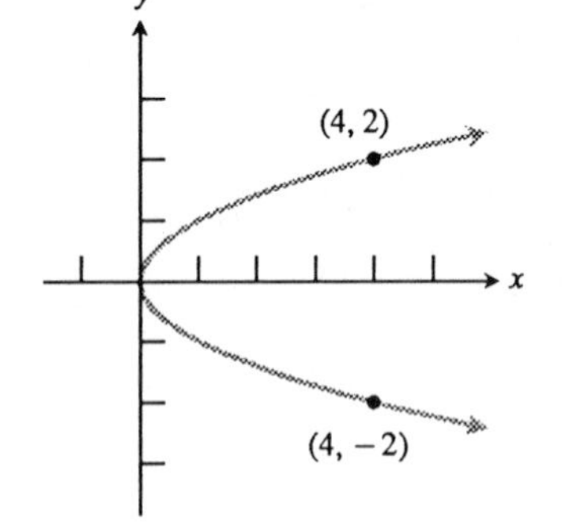

1.111 A complete graph of the relation $x = y^2$.

In most of our work, relations have domains and ranges that are sets of real numbers and are usually defined by formulas or equations. For example, the equation $x = y^2$ defines the relation with domain $x \geq 0$ and range $-\infty < y < \infty$. The domain of a relation defined by an equation in x and y is taken to be the largest set of real x-values for which the corresponding y-values are real. Notice the 4 in the domain of $x = y^2$ is paired with 2 and -2 in the range (Fig. 1.111). The set of points in the coordinate plane that corresponds to the elements of a relation make up the graph of the relation. As we saw in Section 1.4, the graph of $x = y^2$ is a parabola that opens to the right with vertex (0, 0) and axis of symmetry the x-axis.

Vertical Line Test for a Function

The vertical lines $x = a$ with $a > 0$ intersect the graph of $x = y^2$ in Fig. 1.111 in two points, one above the x-axis and the other below. Thus, the domain element a is paired with two elements in the range, $\sqrt{a}$ and $-\sqrt{a}$ in this case. Functions have the property that domain elements (input) are paired with a *single* element in the range (output). So vertical lines intersect graphs of a function in at most one point.

> **Vertical Line Test for a Function**
>
> If every vertical line intersects the graph of a relation in at most one point, then the relation is a function of x.

EXAMPLE 1 Which of the relations graphed in Figure 1.112 are functions?

Solution Parts (b), (c), and (d) are graphs of functions and parts (a) and (e) are not.

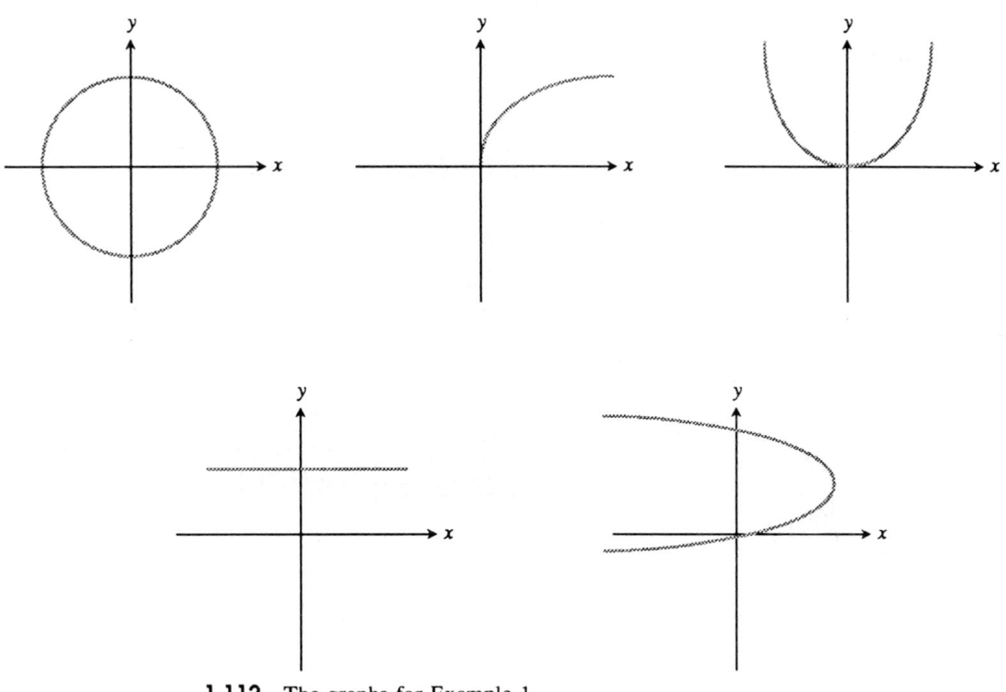

1.112 The graphs for Example 1.

Inverse Relation

Every relation R has associated with it another natural relation called its **inverse.** Basically, the inverse relation is defined by reversing the assignments that produce the relation.

Definition

Inverse Relation

Let R be a relation. The inverse relation R^{-1} of R consists of all those ordered pairs (b, a) for which (a, b) belongs to R. So, domain of R^{-1} = range of R and range of R^{-1} = domain of R.

In Exercise 55 you will be asked to show that the points (a, b) and (b, a) are symmetric with respect to the line $y = x$. This means that the graph of R^{-1} can be obtained by reflecting the graph of R through the line $y = x$. There is also a two-step process for obtaining the graph of R^{-1} from the graph of R: First, rotate the graph of R 90° counterclockwise about the origin and then reflect the resulting graph through the y-axis. In Exercise 56, you will be asked to verify this process for finding the graph of R^{-1}.

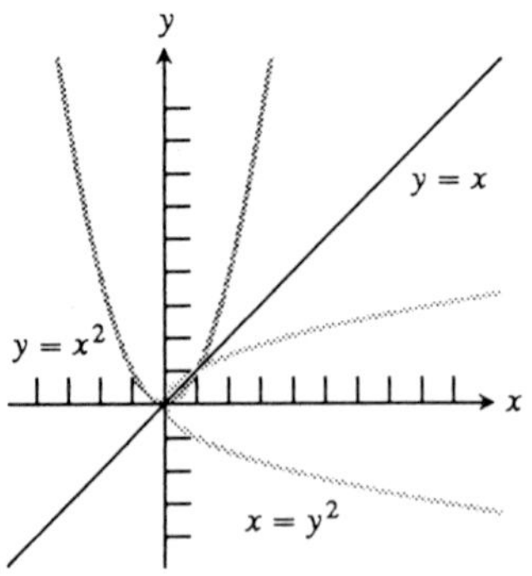

1.113 Complete graphs of the inverse relations $y = x^2$ and $x = y^2$ and the line $y = x$.

EXAMPLE 2 Show that $y = x^2$ and $x = y^2$ are inverse relations.

Solution If we interchange x and y in $y = x^2$ we get $x = y^2$. So $y = x^2$ and $x = y^2$ are inverse relations. Notice that their graphs are symmetric with respect to the line $y = x$ (Fig. 1.113). ≡

When Are Inverse Relations Functions?

We are especially interested in functions whose inverse relations are also functions. Notice that the inverse relation $x = y^2$ of the function $y = x^2$ is not a function because its graph in Fig. 1.111 fails the vertical line test. Suppose the graph of the inverse relation R^{-1} of a relation R satisfies the vertical line test for functions. Then, the symmetry of the graphs of R and R^{-1} with respect to the line $y = x$ gives us a way to test the graph of R to decide if R^{-1} is a function.

Theorem 2

Horizontal Line Test

The inverse relation R^{-1} of the relation R is a function if, and only if, every horizontal line intersects the graph of R in at most one point.

Notice that every horizontal line $y = a$ with $a > 0$ intersects the graph of $y = x^2$ in the two points: $(\sqrt{a}, a)$ and $(-\sqrt{a}, a)$. So the vertical line $x = a$ intersects the graph of $x = y^2$ in the corresponding points under symmetry with respect to the line $y = x$: $(a, \sqrt{a})$ and $(a, -\sqrt{a})$. There is another way to describe functions whose inverses are functions. A function f is said to be **one-to-one** if, and only if, for every pair x_1 and x_2 of distinct values in the domain of f, it is also true that $f(x_1)$ and $f(x_2)$ are distinct.

Theorem 3

Function Test for the Inverse of a Function

The inverse f^{-1} of a function f is a function if, and only if, f is a one-to-one function.

Combining Theorems 2 and 3, we obtain a graphical test of one-to-oneness of a function.

Theorem 4

Graphical Test for One-to-Oneness

The function f is one-to-one if, and only if, every horizontal line intersects the graph of f in at most one point.

EXAMPLE 3 Which of the following functions have inverses that are also functions?

(a) $y = x^3$ (b) $y = \dfrac{1}{x^2}$ (c) $y = -2x + 4$

Solution Complete graphs of the three functions are shown in Fig. 1.114. The graphs of the functions in (a) and (c) satisfy the horizontal line test, but the graph of the function in (b) does not. Thus, by Theorem 2, the inverses of the functions in (a) and (c) are functions, and the inverse of the function in (b) is not. Complete graphs of the three inverses in Fig. 1.115, obtained by reflecting the graphs of the functions through the line $y = x$, support these conclusions. ≡

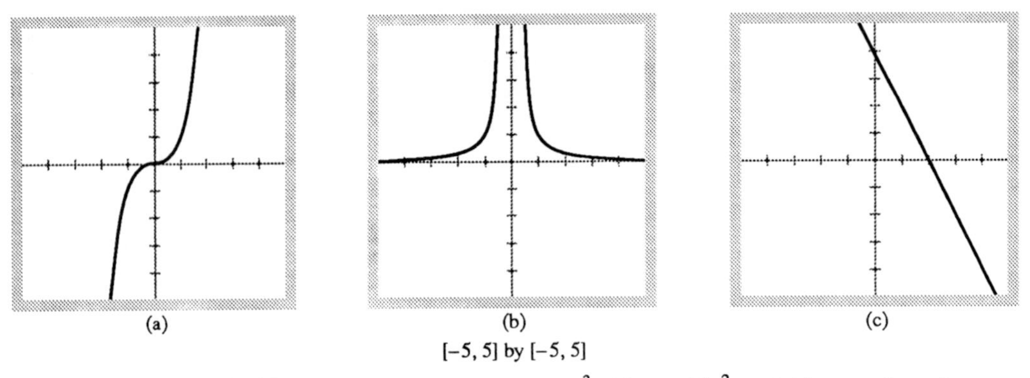

1.114 Complete graphs of (a) $y = x^3$, (b) $y = 1/x^2$, and (c) $y = -2x + 4$.

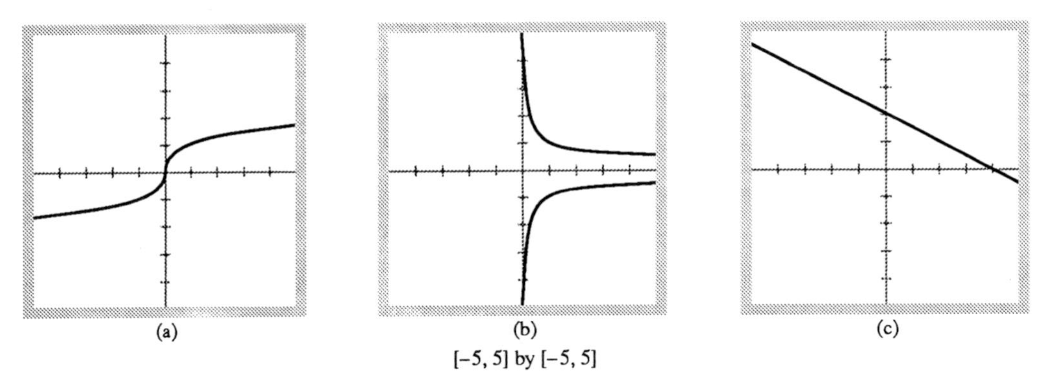

1.115 Complete graphs of the inverses of (a) $y = x^3$, (b) $y = 1/x^2$, and (c) $y = -2x + 4$. Only (a) and (c) are graphs of functions.

Parametric Graphing

Some relations consist of ordered pairs (x, y) where x and y are functions of a third variable, say t. The variable t is called a **parameter** and the equations $x = x(t)$ and $y = y(t)$ are called **parametric equations** for the relation. In this textbook we assume your grapher includes a parametric graphing utility. To use this feature you will need to choose the values of t for which the points $(x(t), y(t))$ are plotted. This is usually done by selecting a value for tMin, tMax, and tStep. Then the points $(x(t), y(t))$ are plotted for

$$t\text{Min},\ t\text{Min} + t\text{Step},\ t\text{Min} + 2t\text{Step},\ \cdots,\ t\text{Min} + kt\text{Step},\ \cdots$$

for all integers k with $t\text{Min} \le t\text{Min} + kt\text{Step} \le t\text{Max}$.

We use a parametric graphing utility to graph a function $y = f(x)$ by setting $x = t$ and $y = f(t)$.

EXAMPLE 4 Use a parametric grapher to support the graph of the inverse of the function $y = x^3$ obtained in Example 3 (Fig. 1.115a).

Solution The graph of the parametric equations

$$x_1(t) = t, y_1(t) = t^3$$

with tMin $= -5$, tMax $= 5$, and tStep $= 0.1$ in the viewing rectangle $[-5, 5]$ by $[-5, 5]$ produces the graph of $y = x^3$ shown in Fig. 1.114(a).

The graph of the inverse of $y = x^3$ is obtained by interchanging x and y in the ordered pairs (x, y). So the graph of the parametric equations

$$x_2(t) = t^3, y_2(t) = t$$

with the same **parametric range settings**

$$-5 \leq t \leq 5, -5 \leq x \leq 5, -5 \leq y \leq 5$$

produces the graph of the inverse of $y = x^3$ shown in Fig. 1.115a. ■

Graphs of inverses of relations defined parametrically can be obtained using a parametric grapher or using one of the two geometric procedures described earlier.

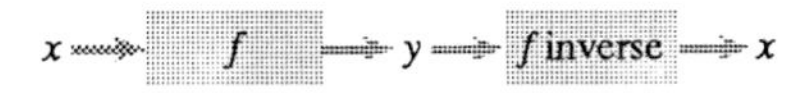

1.116 The inverse of a one-to-one function f sends every output of f back to the input from which it came.

Notice that the functions in Example 3 (a) and (c) are one-to-one functions, and the function in (b) is not (see Fig. 1.114). Since each output of a one-to-one function comes from just one input, a one-to-one function can be reversed to send the outputs back to the inputs from which they came (Fig. 1.116).

As you can see in Fig. 1.117, the result of composing f and f^{-1} in either order is the **identity function**, the function that assigns each number to itself. This gives us a way to test whether two functions f and g are inverses of one another. Compute $f \circ g$ and $g \circ f$. If both composites are identity functions, then f and g are inverses of one another; otherwise they are not. If f squares every number in its domain, g had better take square roots or it isn't the inverse of f.

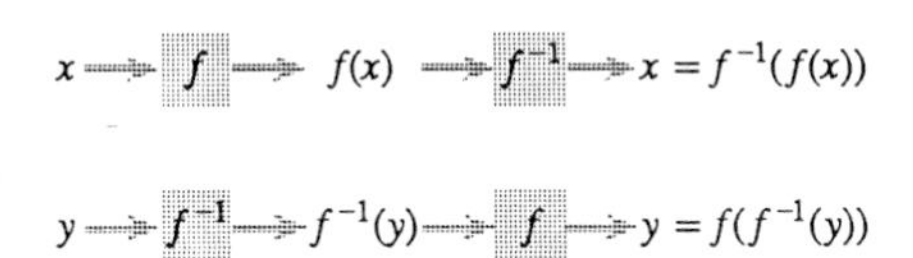

1.117 If $y = f(x)$ is a one-to-one function, then $f^{-1}(f(x)) = x$ and $f(f^{-1}(y)) = y$. Each of the composites $f^{-1} \circ f$ and $f \circ f^{-1}$ is the identity function on its domain.

If the inverse of a function $y = f(x)$ is also a function, we can sometimes find an explicit rule for f^{-1} in terms of x, as Example 4 illustrates.

EXAMPLE 5 Find the inverse of $y = -2x + 4$, expressed as a function of x.

Solution

STEP 1: Switch x and y: $x = -2y + 4$

This is an equation whose graph is the inverse of the function $y = -2x + 4$.

STEP 2: Solve for y in terms of x:

$$x = -2y + 4$$

$$2y = -x + 4$$

$$y = -\frac{1}{2}x + 2$$

The inverse of the function $f(x) = -2x+4$ is the function $f^{-1}(x) = -\frac{1}{2}x+2$. To check, we verify that both composites give the identity function:

$$f^{-1}(f(x)) = -\frac{1}{2}(-2x+4)+2 = x-2+2 = x,$$

$$f(f^{-1}(x)) = -2\left(-\frac{1}{2}x+2\right)+4 = x-4+4 = x.$$

The graph of $f^{-1}(x) = -\frac{1}{2}x+2$ is a straight line with slope $-1/2$ and y-intercept 2 which agrees with the graph of the inverse of $y = -2x+4$ shown in Fig. 1.115(c).

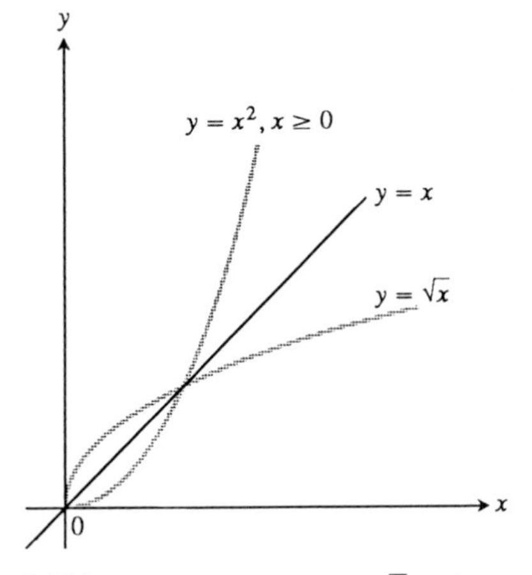

1.118 The functions $y = \sqrt{x}$ and $y = x^2, x \geq 0$, are inverses of one another. Their graphs are symmetric with respect to the line $y = x$.

We can sometimes restrict the domain of a function whose inverse is not a function and create a new function whose inverse is a function.

EXAMPLE 6 Show that the inverse of the function $y = x^2, x \geq 0$ is a function, and express the inverse as a function of x.

Solution The function $y = x^2, x \geq 0$ is one-to-one (Fig. 1.118). Thus, its inverse is a function. Next we find a rule for the inverse.

STEP 1: Switch x and y: $x = y^2, y \geq 0$

STEP 2: Solve for y in terms of x:

$$\sqrt{x} = \sqrt{y^2} = y \quad (\sqrt{y^2} = y \text{ because } y \geq 0)$$

The inverse of the function $y = x^2, x \geq 0$ is the function $y = \sqrt{x}$ (Fig. 1.118).

Notice that unlike the restricted function $y = x^2, x \geq 0$, the unrestricted function $y = x^2$ is not one-to-one and therefore its inverse is not a function.

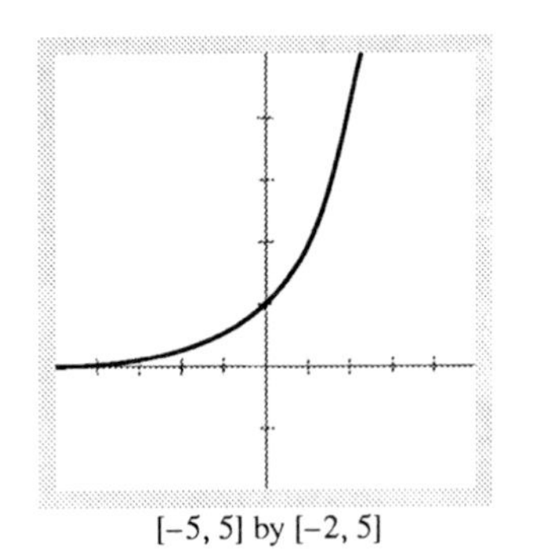

[−5, 5] by [−2, 5]

1.119 A computer visualization of the complete graph of $y = 2^x$.

Exponential and Logarithm Functions

Graphing utilities produce accurate graphs of exponential functions such as $y = 2^x$ (Fig. 1.119). Notice that the domain appears to be $(-\infty, \infty)$ and the range $(0, \infty)$. We have not yet tried to define what we mean by 2^x (or $a^x, a > 0$) for x other than a rational number. What should $2^{\sqrt{3}}$ mean? One calculator gives

$$2^{\sqrt{3}} = 3.32199708548.$$

In Chapters 5 and 7, we will give a precise definition of an exponential function. Until then we will be content with using calculators to find the values of such functions.

Definition

Exponential Function

Let a be a positive real number other than 1. The function $f(x) = a^x$ whose domain is $(-\infty, \infty)$ and whose range is $(0, \infty)$ is the exponential function with base a.

Explore Box

1. Graph the function $y = 2^x, y = 3^x$, and $y = 5^x$ in the same viewing rectangle.
2. For what values of x is it true that $2^x > 3^x > 5^x$?
3. For what values of x is it true that $2^x < 3^x < 5^x$?
4. For what values of x is it true that $2^x = 3^x = 5^x$?
5. Repeat 1–4 for the functions $y = 2^{-x}, y = 3^{-x}$, and $y = 5^{-x}$.
6. Compare the graphs of $y = a^{-x}$ and $y = (1/a)^x$ for $a = 2, 3$, and 5. What would you conjecture for arbitrary $a > 0$?
7. Compare the graphs of $y = a^x$ and $y = a^{-x}$ for $a = 2, 3$, and 5. What would you conjecture for arbitrary $a > 0$?

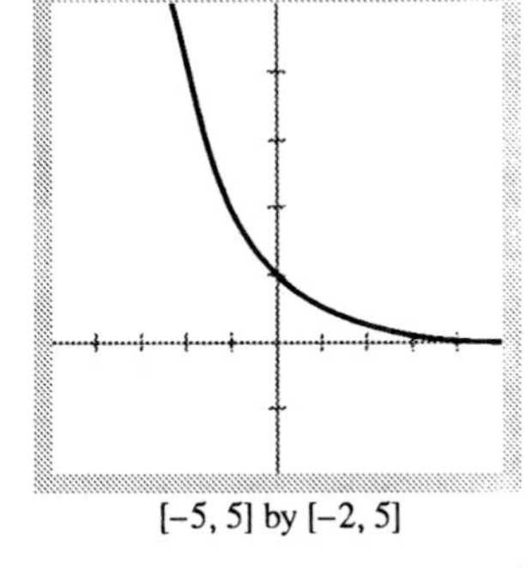
[−5, 5] by [−2, 5]

1.120 A complete graph of the exponential function $y = 0.5^x = 2^{-x}$.

We can draw the following conclusions about exponential functions from the above Explore Activity.

Exponential Functions

- If $a > 1$, the graph of $y = a^x$ has a shape like the graph of $y = 2^x$.
- If $0 < a < 1$, the graph of $y = a^x$ has a shape like the graph of $y = (1/2)^x = 0.5^x$ (Fig. 1.120).
- If $a \neq b$, then $a^x = b^x$ if, and only if, $x = 0$.
- The graph of $y = a^{-x}$ can be obtained by reflecting the graph of $y = a^x$ through the y-axis.

Notice that the two principal models $y = 2^x$ and $y = 2^{-x}$ of exponential functions are one-to-one functions because their graphs satisfy the horizontal line test (see Figs. 1.119 and 1.120). So the inverse relation of an exponential function is also a function. If we interchange x and y in $y = a^x$ we get $x = a^y$. Notice that we have no technique to solve for y in terms of x. For that reason we make the following definition.

Definition

Logarithm Function

Let a be a positive real number other than 1. The function $f(x) = \log_a x$ with domain $(0, \infty)$ and range $(-\infty, \infty)$ is the inverse of the exponential function $y = a^x$, and is called the **logarithm function with base a.**

Figure 1.121 shows complete graphs of $y = \log_2 x$ and $y = \log_{0.5} x$, which are the inverse functions of $y = 2^x$ and $y = 0.5^x$, respectively. In Example 7, we will show how to graph $y = \log_a x$ for any a.

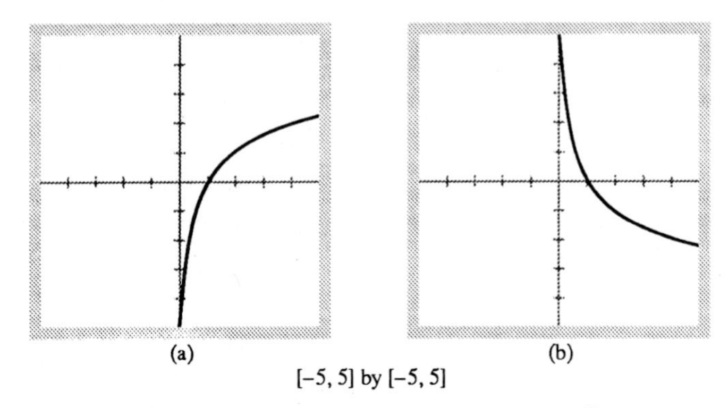

1.121 Complete graphs of (a) $y = \log_2 x$, the inverse of $y = 2^x$ and (b) $y = \log_{0.5} x$, the inverse of $y = 0.5^x$.

There are several immediate consequences of the definition of the logarithm function as the inverse relation of the exponential function: $f(x) = a^x, f^{-1}(x) = \log_a x$.

$$a^0 = 1 \quad \Leftrightarrow \quad \log_a 1 = 0 \tag{1}$$

$$a^1 = a \quad \Leftrightarrow \quad \log_a a = 1 \tag{2}$$

$$x = (f \circ f^{-1})(x) = a^{\log_a x} \tag{3}$$

$$x = (f^{-1} \circ f)(x) = \log_a a^x \tag{4}$$

Equations (1)–(4) have established the following theorem.

Theorem 5

Properties of Logarithms

Let $a > 0$ and $a \neq 1$. Then

a) $\log_a 1 = 0$

b) $\log_a a = 1$

c) $a^{\log_a x} = x$ for every positive real number x

d) $\log_a a^x = x$ for every real number x.

The properties of logarithms given in Theorem 6 are very useful for computation and help us establish the change of base formula in Theorem 7 that will be used to obtain graphs of logarithm functions with a graphing utility.

Theorem 6

Properties of Logarithms

Let a, r, and s be positive real numbers with a $\neq 1$. Then

a) $\log_a rs = \log_a r + \log_a s$

b) $\log_a \dfrac{r}{s} = \log_a r - \log_a s$

c) $\log_a r^c = c \log_a r$ for every real number c.

Proof We prove part (a) and leave (b) and (c) for the exercises. Let $\log_a r = u$ and $\log_a s = v$. Then,

$$r = a^u,\ s = a^v \qquad \text{(Equivalent exponential form)}$$

$$rs = a^u \cdot a^v$$

$$rs = a^{u+v}$$

$$\log_a rs = u + v \qquad \text{(Equivalent logarithmic form)}$$

$$\log_a rs = \log_a r + \log_a s$$

∎

Theorem 7

Change-of-Base Formula

Let a and b be positive real numbers with $a \neq 1$ and $b \neq 1$. Then $\log_b x = \dfrac{\log_a x}{\log_a b}$.

Proof

$$\begin{aligned} y &= \log_b x \\ x &= b^y && \text{(Exponential form)} \\ \log_a x &= \log_a b^y \\ \log_a x &= y \log_a b && \text{(Theorem 6 (c))} \\ y &= \frac{\log_a x}{\log_a b}. \end{aligned}$$

Special Bases for Logarithms

Be careful if you are using a computer-based graphing utility. Some computers only have $y = \ln x$ built in and access it as $y = \log x$.

Traditionally calculators, including graphing calculators, have the logarithm functions with base 10 and base e built-in. We will say more about the important number e in Chapter 5. For now, we can use a calculator to approximate its value. To 10 significant digits

$$e = 2.718281828. \tag{5}$$

We omit the base when using base 10 and replace $\log_e$ by ln:

$$\log_{10} x = \log x \tag{6}$$

$$\log_e x = \ln x \tag{7}$$

The function $y = \ln x$ is usually called the **natural logarithm function**. In Chapter 5 we will see how to define the natural logarithm function using an integral.

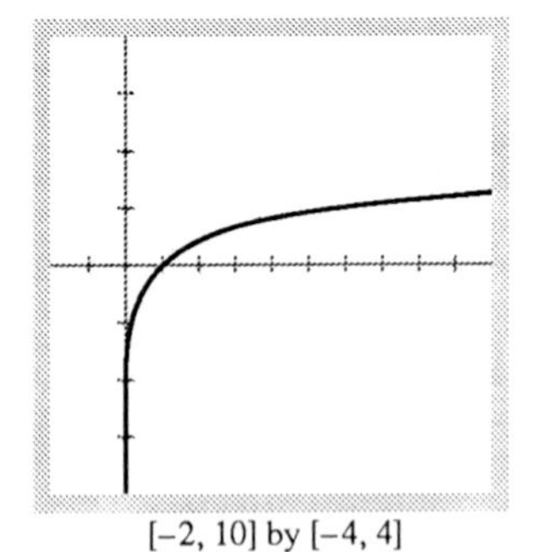

[−2, 10] by [−4, 4]

1.122 A complete graph of $f(x) = \log_5 x$ using either of the representations $f(x) = \log x / \log 5$ or $f(x) = \ln x / \ln 5$.

EXAMPLE 7 Use a graphing utility to draw a complete graph of $f(x) = \log_5 x$. Find the domain and range.

Solution We use the change-of-base formula in Theorem 7 to rewrite $f(x)$ using either log or ln in order to enter f.

$$f(x) = \log_5 x = \frac{\log x}{\log 5} = \frac{\ln x}{\ln 5}$$

Figure 1.122 gives a complete graph of f. Notice that the domain of f is $(0, \infty)$ and the range of f is $(-\infty, \infty)$. All functions $y = \log_a x$ have the same domain and range.

EXAMPLE 8 A complete graph of $f(x) = -1.5(2^{x+1}) - 3$ can be obtained from the graph of $y = 2^x$ as follows: Stretch vertically by a factor of 1.5,

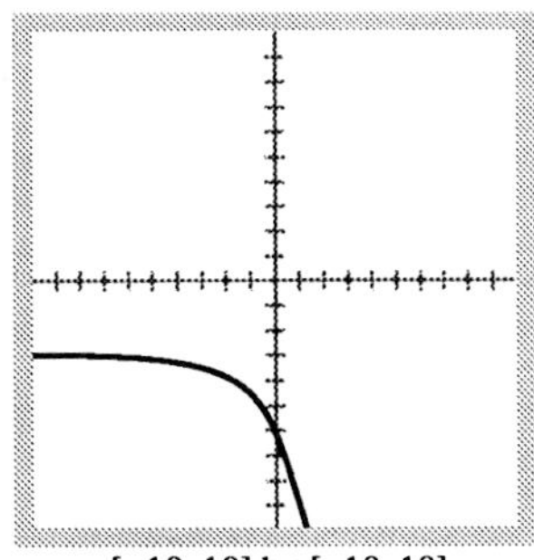

1.123 A complete graph of $f(x) = -1.5(2^{x+1}) - 3$.

reflect through the x-axis, shift left 1 unit, and finally shift down 3 units (Fig. 1.123). The domain of f is $(-\infty, \infty)$ and the range is $(-\infty, -3)$. ☰

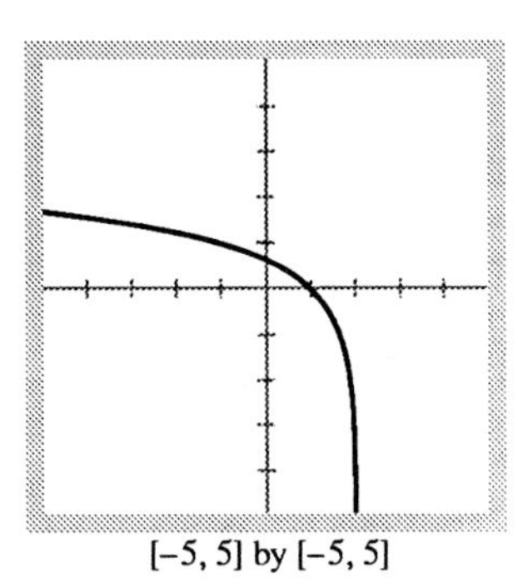

1.124 A complete graph of $f(x) = \log_3(2 - x)$.

EXAMPLE 9 A complete graph of $f(x) = \log_3(2 - x) = \log_3(-(x - 2))$ can be obtained from the graph of $y = \log_3 x$ as follows: Reflect through the y-axis and shift right 2 units (Fig. 1.124). The domain of f is $(-\infty, 2)$ and the range is $(-\infty, \infty)$. ☰

Equations for Circles in the Plane

Definition

A **circle** is the set of points in a plane whose distance from a given fixed point in the plane is a constant. The fixed point is the **center** of the circle. The constant distance is the **radius** of the circle.

To write an equation for the circle of radius a centered at the point $C(h, k)$, we let $P(x, y)$ denote a typical point on the circle (Fig. 1.125). The statement that CP equals a then becomes

$$\sqrt{(x-h)^2 + (y-k)^2} = a \qquad \text{(The distance from } (x, y) \text{ to } (h, k) \text{ equals } a.)$$

or

$$(x-h)^2 + (y-k)^2 = a^2 \qquad \text{(Both sides squared.)} \qquad (8)$$

If $CP = a$, then Eq. (8) holds. If Eq. (8) holds, then $CP = a$. Equation (8) is therefore an equation for the circle.

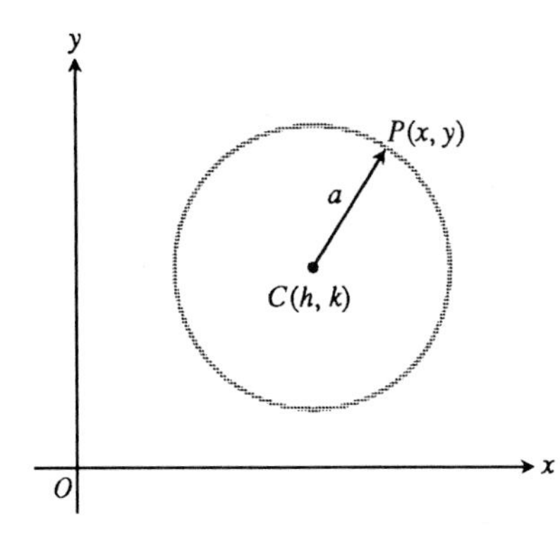

1.125 The standard equation for the circle shown here is $(x - h)^2 + (y - k)^2 = a^2$.

The Standard Equation for the Circle of Radius a Centered at the Point (h, k)

$$(x-h)^2 + (y-k)^2 = a^2 \qquad (9)$$

EXAMPLE 10 The standard equation for the circle of radius 2 centered at the point (3, 4) is

$$(x-3)^2 + (y-4)^2 = (2)^2$$

or

$$(x-3)^2 + (y-4)^2 = 4.$$

There is no need to square out the x and y terms in this equation. In fact, it is better not to do so. The present form reveals the circle's center and radius.

EXAMPLE 11 Find the center and radius of the circle

$$(x-1)^2+(y+5)^2=3.$$

Solution Comparing

$$(x-h)^2+(y-k)^2=a^2 \quad \text{with} \quad (x-1)^2+(y+5)^2=3$$

shows that

$$-h=-1 \quad \text{or} \quad h=1$$

$$-k=5 \quad \text{or} \quad k=-5$$

$$a^2=3 \quad \text{or} \quad a=\sqrt{3}.$$

The center is the point $(h,k)=(1,-5)$. The radius is $a=\sqrt{3}$.

For circles centered at the origin, h and k are 0 and Eq. (9) simplifies to $x^2+y^2=a^2$.

The Standard Equation for the Circle of Radius a Centered at the Origin

$$x^2+y^2=a^2 \qquad (10)$$

The circle of radius 1 unit centered at the origin has a special name, *the unit circle.*

Notice that the circle $(x-h)^2+(y-k)^2=a^2$ is the same as the circle $x^2+y^2=a^2$ with its center shifted from the origin to the point (h,k). The shift formulas we have been using for graphs and functions apply to equations of any kind. Shifts to the right and up are accomplished by subtracting positive values of h and k. Shifts to the left and down are accomplished by subtracting negative values of h and k.

EXAMPLE 12 If the circle $x^2+y^2=25$ is shifted two units to the left and three units up, its new equation is

$$(x-(-2))^2+(y-3)^2=25$$

or

$$(x+2)^2+(y-3)^2=25.$$

As Eq. (9) says it should be, this is the equation of the circle of radius 5 centered at $(h, k) = (-2, 3)$.

The points that lie inside the circle $(x-h)^2+(y-k)^2 = a^2$ are the points less than a units from (h, k). They satisfy the inequality

$$(x-h)^2 + (y-k)^2 < a^2. \tag{11}$$

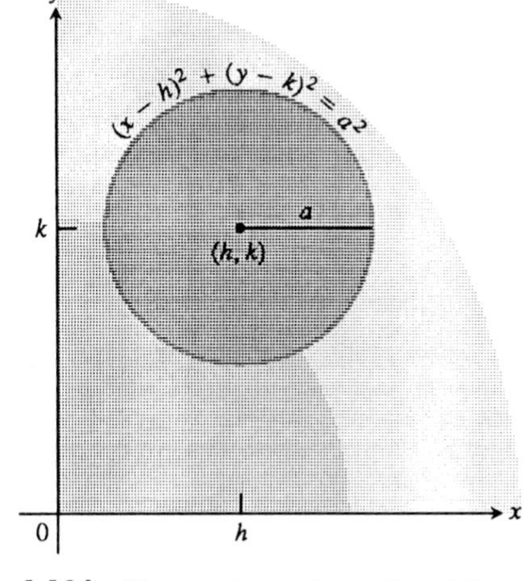

1.126 The interior and exterior of the circle $(x-h)^2 + (y-k)^2 = a^2$.

They make up the region we call the **interior** of the circle (Fig. 1.126).

The circle's **exterior** consists of the points that lie more than a units from (h, k). These points satisfy the inequality

$$(x-h)^2 + (y-k)^2 > a^2. \tag{12}$$

EXAMPLE 13

Inequality	Region
$x^2 + y^2 < 1$	Interior of the unit circle
$x^2 + y^2 \leq 1$	Unit circle plus its interior
$x^2 + y^2 > 1$	Exterior of the unit circle
$x^2 + y^2 \geq 1$	Unit circle plus its exterior

The circle $x^2 + y^2 = 1$ defines a relation that is not a function. If we interchange x and y we get the same equation, so this relation is its own inverse. In Section 1.4 we found that the relation $x = y^2$ could be graphed with a graphing utility. The same is true for circles. If we solve $x^2 + y^2 = 1$ for y we find that

$$y^2 = 1 - x^2 \quad \text{or} \quad y = \pm\sqrt{1-x^2}.$$

Each of the formulas

$$y = \sqrt{1-x^2} \quad \text{or} \quad y = -\sqrt{1-x^2}$$

defines functions of x. Overlaying the graphs of the two functions produces a complete graph of the unit circle (Fig. 1.127c). Notice that the domain and range of the relation $x^2 + y^2 = 1$ are $[-1, 1]$. The domains of $y = \sqrt{1-x^2}$ and $y = -\sqrt{1-x^2}$ are also $[-1, 1]$. However, the range of $y = \sqrt{1-x^2}$ is $[0, 1]$ and the range of $y = -\sqrt{1-x^2}$ is $[-1, 0]$.

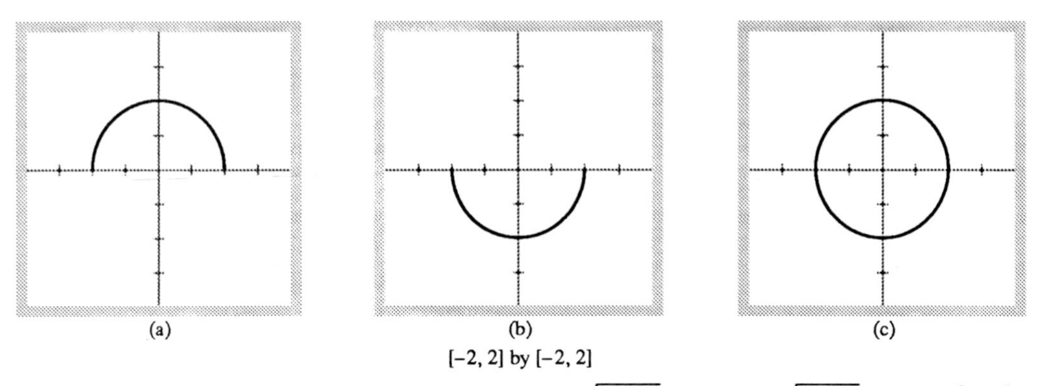

1.127 Complete graphs of (a) $y = \sqrt{1 - x^2}$, (b) $y = -\sqrt{1 - x^2}$, and (c) $x^2 + y^2 = 1$.

EXAMPLE 14 Determine the domain and range and draw a complete graph of the circle $(x - 3)^2 + (y + 4)^2 = 16$.

Solution This circle is centered at $(3, -4)$ and has radius 4. So, the domain is $[-1, 7]$ and the range is $[-8, 0]$. To use a grapher to draw a complete graph, we solve for y.

$$(x - 3)^2 + (y + 4)^2 = 16$$
$$(y + 4)^2 = 16 - (x - 3)^2$$
$$y + 4 = \pm\sqrt{16 - (x - 3)^2}$$
$$y = -4 \pm \sqrt{16 - (x - 3)^2}$$

Figure 1.128 shows a complete graph. ■

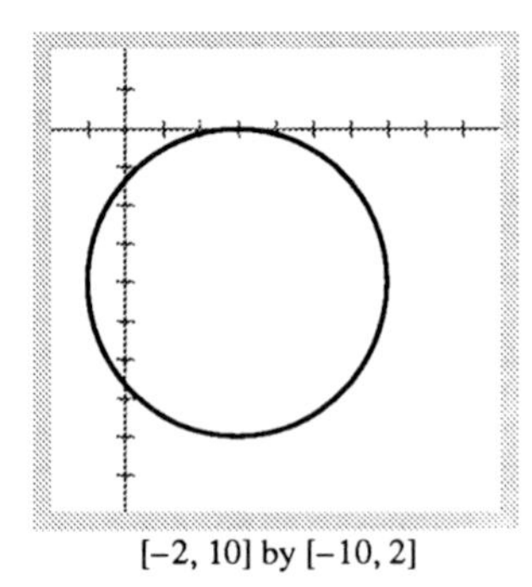

1.128 A computer visualization of the complete graph of circle $(x - 3)^2 + (y + 4)^2 = 16$.

If an equation in x and y can be solved for y to produce one or more formulas, then a grapher can be used to get a complete graph. Of course, we need to know what a complete graph should look like.

USING GRAPHERS TO DRAW CIRCLES

You need to be careful when using graphers to draw circles. The graph may not look like a circle. This will happen if the distance 1 appears to be different on the two axes. To look like a circle, the distance 1 must appear to be the same on each axis.

Some graphing utilities have a feature that allows you to automatically square up a viewing rectangle so that circles will look like circles.

EXAMPLE 15 Draw a complete graph of $\dfrac{x^2}{4} + \dfrac{y^2}{9} = 1$.

Solution In Chapter 10 you will see that a complete graph of this relation is called an *ellipse*. Solve for y:

$$\frac{x^2}{4} + \frac{y^2}{9} = 1$$
$$\frac{y^2}{9} = 1 - \frac{x^2}{4}$$
$$y^2 = \frac{9}{4}(4 - x^2)$$
$$y = \pm\frac{3}{2}\sqrt{4 - x^2}$$

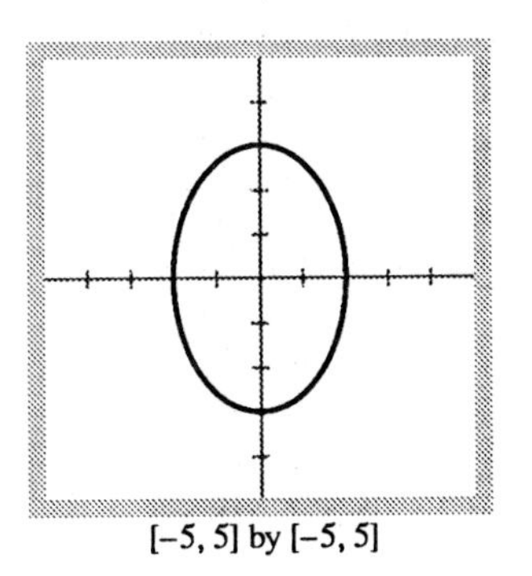
[−5, 5] by [−5, 5]

1.129 A complete graph of the ellipse $\frac{x^2}{4} + \frac{y^2}{9} = 1$.

A complete graph of this relation is shown in Fig. 1.129. Notice that the domain is $[-2, 2]$ and the range is $[-3, 3]$. ≡

The following Explore Activity illustrates some of the conic sections that will be studied in Chapter 10.

> **Explore Box**
>
> The $[-10, 10]$ by $[-10, 10]$ viewing rectangle contains a complete graph of each of the relations. Solve for y and use a grapher to draw a complete graph.
>
> 1. $\frac{x^2}{25} + \frac{y^2}{4} = 1$ (ellipse)
> 2. $\frac{x^2}{16} - \frac{y^2}{9} = 1$ (hyperbola)
> 3. $\frac{y^2}{25} - \frac{x^2}{4} = 1$ (hyperbola)

Exercises 1.6

Which of the functions graphed in Exercises 1–4 are one-to-one?

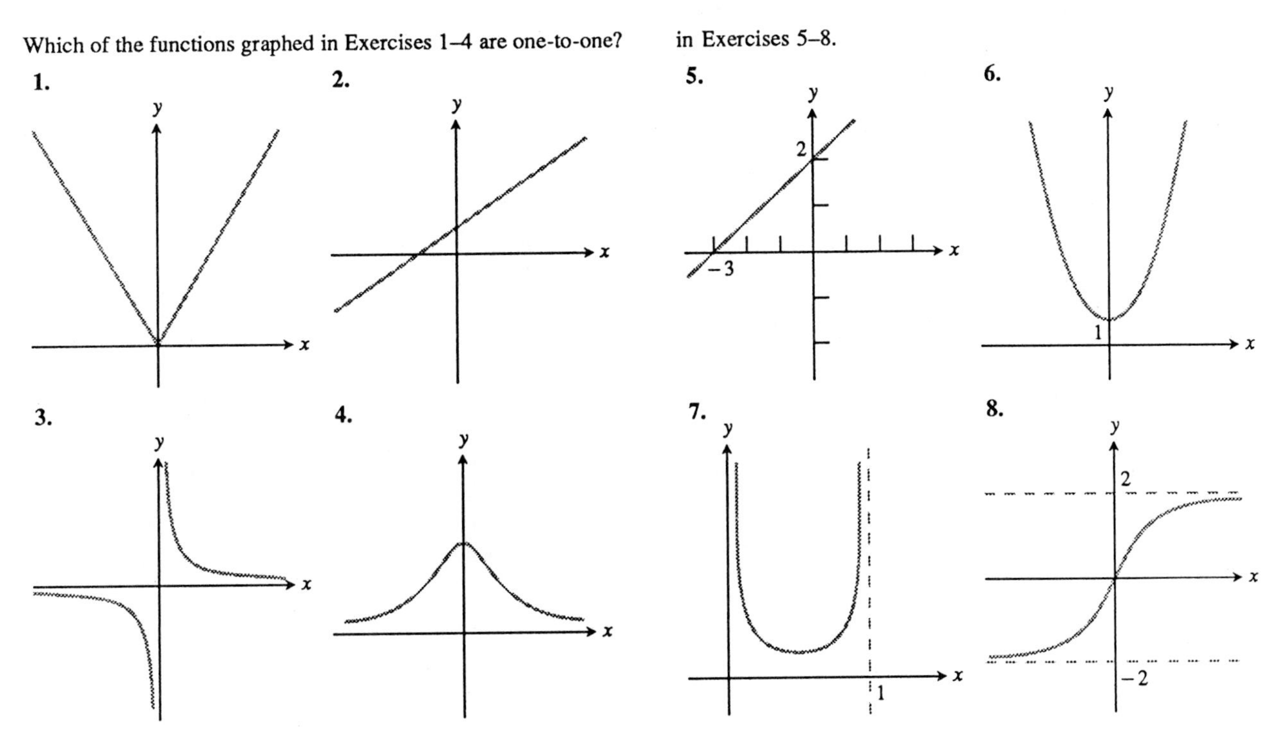

Sketch the graph of the inverse relations of the relations graphed in Exercises 5–8.

Which of the functions in Exercises 9–14 have an inverse relation that is also a function? Draw a graph of the inverse relation and support your answer with a parametric grapher.

9. $y = \dfrac{3}{x-2} - 1$ **10.** $y = x^2 + 5x$

11. $y = x^3 - 4x + 6$ **12.** $y = x^3 + x$

13. $y = \ln x^2$ **14.** $y = 2^{3-x}$

Draw a complete graph of the functions in Exercises 15–18.

15. $y = 2\log_3(x-4) - 1$ **16.** $y = -3\log_5(2-x) + 1$

17. $y = -3\log_{0.5}(x+2) + 2$

18. $y = 2\log_{0.2}(3-x) + 1$

Determine the domain and range and describe how the graphs of the equations in Exercises 19–24 can be obtained from the graph of an appropriate exponential function $y = a^x$, logarithmic function $y = \log_a x$, or circle $x^2 + y^2 = a^2$.

19. $y = -3\log(x+2) + 1$ **20.** $y = 2\ln(3-x) - 4$

21. $y = 2(3^{1-x}) + 1.5$ **22.** $y = -3(5^{x-2}) + 3$

23. $(x+3)^2 + (y-5)^2 = 9$

24. $(x-6)^2 + (y+1)^2 = 25$

In Exercises 25–28, find an equation for the circle with the given center $C(h, k)$ and radius a. Then sketch a complete graph of the circle.

25. $C(0, 2), a = 2$ **26.** $C(-2, 0), a = 3$

27. $C(3, -4), a = 5$ **28.** $C(1, 1), a = \sqrt{2}$

Write equations for the circles in Exercises 29–32.

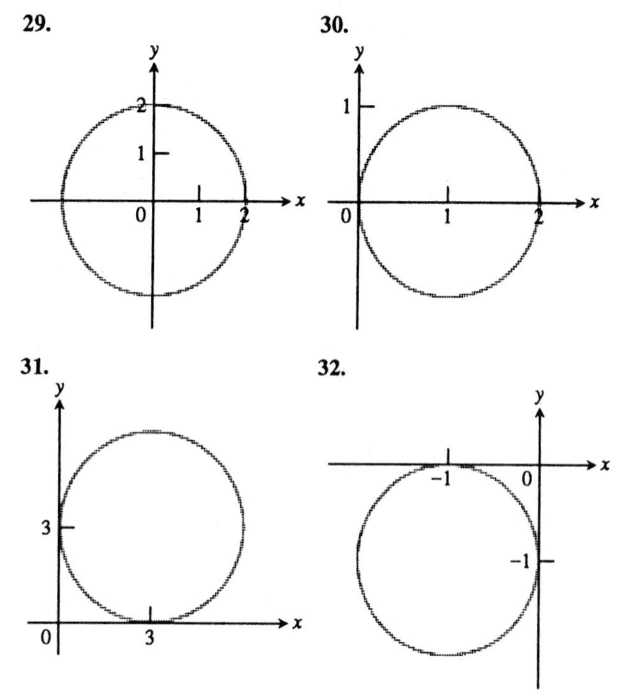

Describe the regions defined by the inequalities and pairs of inequalities in Exercises 33 and 34.

33. a) $x^2 + y^2 > 1$
b) $x^2 + y^2 < 4$
c) the inequalities in (a) and (b) together

34. a) $x^2 + y^2 \geq 1$
b) $x^2 + y^2 \leq 4$
c) the inequalities in (a) and (b) together

35. Write an inequality that describes the points that lie inside the circle with center $C(-2, -1)$ and radius $a = \sqrt{6}$.

36. Write an inequality that describes the points that lie outside the circle with center $C(-4, 2)$ and radius $a = 4$.

In Exercises 37–48, find $f^{-1}(x)$ and show that $f \circ f^{-1}(x) = f^{-1} \circ f(x) = x$. Draw a complete graph of f and f^{-1} in the same viewing rectangle.

37. $f(x) = 2x + 3$ **38.** $f(x) = 5 - 4x$

39. $f(x) = x^3 - 1$ **40.** $f(x) = 2 - x^3$

41. $f(x) = x^2 + 1, x \geq 0$ **42.** $f(x) = x^2, x \leq 0$

43. $f(x) = -(x-2)^2, x \leq 2$

44. $f(x) = (x+1)^2, x \geq -1$

45. $f(x) = \dfrac{1}{x^2}, x \geq 0$ **46.** $f(x) = \dfrac{1}{x^3}, x \neq 0$

47. $f(x) = \dfrac{2x+1}{x+3}, x \neq -3$

48. $f(x) = \dfrac{x+3}{x-2}, x \neq 2$

Solve the equations in Exercises 49–52.

49. $e^x + e^{-x} = 3$ **50.** $2^x + 2^{-x} = 5$

51. $\log_2 x + \log_2(4-x) = 0$

52. $\log x + \log(3-x) = 0$

Use a grapher to obtain complete graphs of the equations in Exercises 53 and 54.

53. $16x^2 - 9y^2 = 144$ **54.** $4x^2 + 9y^2 = 36$

55. Show that the points (a, b) and (b, a) are symmetric with respect to the line $y = x$.

56. Show that the graph of the inverse R^{-1} of a relation R can be obtained by the two-step process: Rotate the graph of R 90° counterclockwise about the origin and then reflect the resulting graph through the y-axis. (*Hint:* The path of a point under this process is $(a, b) \to (-b, a) \to (b, a)$.)

57. Prove parts (b) and (c) of Theorem 6.

58. Use Theorem 7 to show that $\log_b a = \dfrac{1}{\log_a b}$.

59. Explain why each inequality in Example 13 defines a relation that is not a function.

60. Suppose $a \neq 0, b \neq 1$, and $b > 0$. Determine the domain and range of each function.

a) $y = a(b^{c-x}) + d$

b) $y = a \log_b(x - c) + d$

1.7 A Review of Trigonometric Functions

In surveying, navigation, and astronomy, we measure angles in degrees, but in calculus it is usually best to use radians. We shall see why when we study the derivatives of trigonometric functions in Section 3.4. In the present section, we use radians and degrees together so that you can practice relating the two. We show how to horizontally stretch or shrink the graph of a function and use a parametric grapher to graph circles. We also review the trigonometry you will need for calculus and its applications.

Radian Measure

The **radian measure** of the angle ACB at the center of the unit circle (Fig. 1.130) equals the length of the arc that the angle cuts from the unit circle.

If angle ACB cuts an arc $A'B'$ from a second circle centered at C, then circular sector $A'CB'$ will be similar to circular sector ACB. In particular,

$$\frac{\text{Length of arc } A'B'}{\text{Radius of second circle}} = \frac{\text{Length of arc } AB}{\text{Radius of first circle}}. \tag{1}$$

In the notation of Fig. 1.130, Eq. (1) says that

$$\frac{s}{r} = \frac{\theta}{1} = \theta \quad \text{or} \quad \theta = \frac{s}{r}. \tag{2}$$

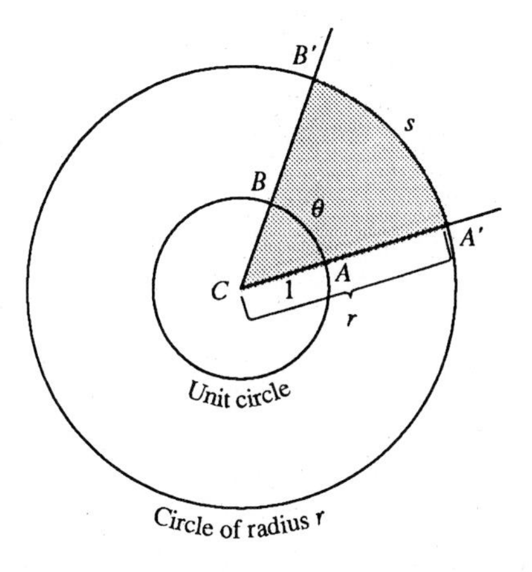

1.130 The radian measure of angle ACB is the length θ of arc AB on the unit circle centered at C. The value of θ can be found from any other circle, however, as the ratio of s to r.

When you know r and s, you can calculate the angle's radian measure θ from this equation. Notice that the units of length for r and s cancel out and that radian measure, like degree measure, is a dimensionless number.

We find the relation between degree measure and radian measure by observing that a semicircle of radius r, which we know has length $s = \pi r$, subtends a central angle of 180°. Therefore,

$$180° = \pi \text{ radians}. \tag{3}$$

We can restate this relation in several useful ways.

Degrees to radians:

- 1 degree makes $\frac{\pi}{180}$ radians (about 0.02 rad)
- To change degrees to radians, multiply degrees by $\frac{\pi}{180}$.

Radians to degees:

- 1 radian makes $\frac{180}{\pi}$ degrees (about 57°)
- To change radians to degrees, multiply radians by $\frac{180}{\pi}$.

EXAMPLE 1 Conversions

Change 45° to radians: $45 \cdot \frac{\pi}{180} = \frac{\pi}{4}\text{rad}$

Change 90° to radians: $90 \cdot \frac{\pi}{180} = \frac{\pi}{2}\text{rad}$

Change $\frac{\pi}{6}$ radians to degrees: $\frac{\pi}{6} \cdot \frac{180}{\pi} = 30°$

Change $\frac{\pi}{3}$ radians to degrees: $\frac{\pi}{3} \cdot \frac{180}{\pi} = 60°$

See Fig. 1.131.

It is important to remember the common conversions in Example 1. For all other conversions we use calculators.

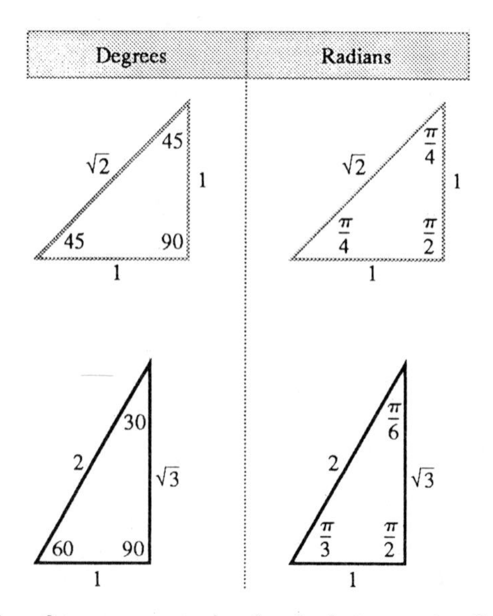

1.131 The angles of two common triangles, in degrees and radians.

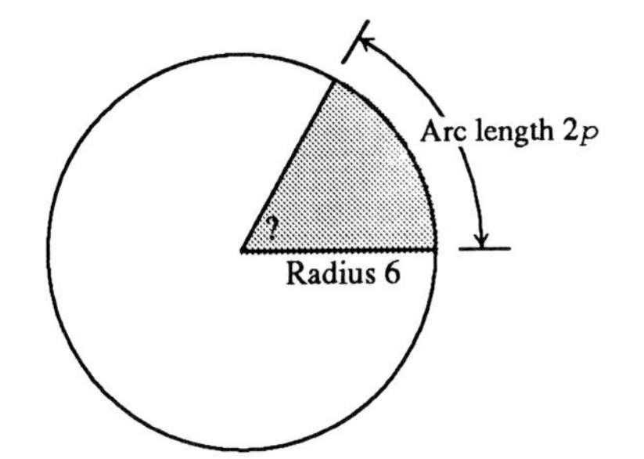

1.132 What is the radian measure of this angle? See Example 3.

EXAMPLE 2 Conversions

Change 26° to radians: $26 \cdot \frac{\pi}{180} = 0.454$ rad

Change 2 radians to degrees: $2 \cdot \frac{180}{\pi} = 114.592°$

EXAMPLE 3 An acute angle whose vertex lies at the center of a circle of radius 6 subtends an arc of length 2π (Fig. 1.132). Find the angle's radian measure.

Solution

$$\theta = \frac{s}{r} = \frac{2\pi}{6} = \frac{\pi}{3} \qquad \text{(Eq. (2) with } s = 2\pi, r = 6.)$$

The angle's measure is $\pi/3$ radians.

The equation $\theta = s/r$ is sometimes written

$$s = r\theta. \tag{4}$$

This equation gives a handy way to find s when you know r and θ.

EXAMPLE 4 An angle of $3\pi/4$ radians lies at the center of a circle of radius 8. How large an arc does it subtend?

Solution

$$s = r\theta = 8 \cdot \frac{3\pi}{4} = 6\pi. \qquad \text{Eq. (4) with } r = 8 \text{ and } \theta = 3\pi/4.$$

The arc is 6π units long.

ROUNDING

When using calculators to do computation, it is common practice to carry full decimal approximations to the end of the computation and then round. In Example 5 we reported the value of θ as 1.745, but we used the full calculator approximation θ when we calculated the value of s.

EXAMPLE 5 How long is the arc subtended by a central angle of 100° in a circle of radius 4?

Solution The equation $s = r\theta$ holds only when the angle is measured in radians, so we must find the angle's radian measure before finding s:

$$\theta = 100\frac{\pi}{180} = 1.745 \text{ rad} \qquad \text{(Convert to radians.)}$$

$$s = r\theta = 4(1.745) = 6.981 \qquad \text{(Then find } s = r\theta.)$$

The arc is 6.981 units long.

When angles are used to describe counterclockwise rotations, our measurements can go arbitrarily far beyond 2π radians or 360°. Similarly, angles that describe clockwise rotations can have negative measures of all sizes (Fig. 1.133).

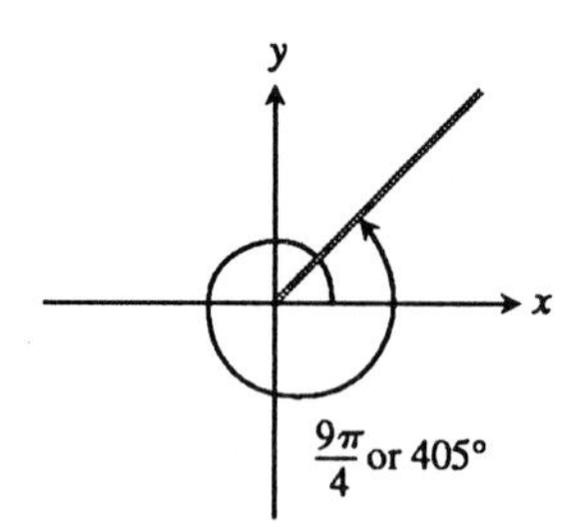

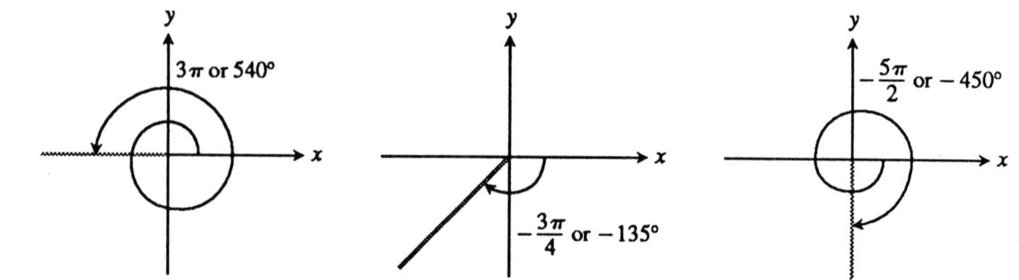

1.133 Angles can have any measure.

The Six Basic Trigonometric Functions

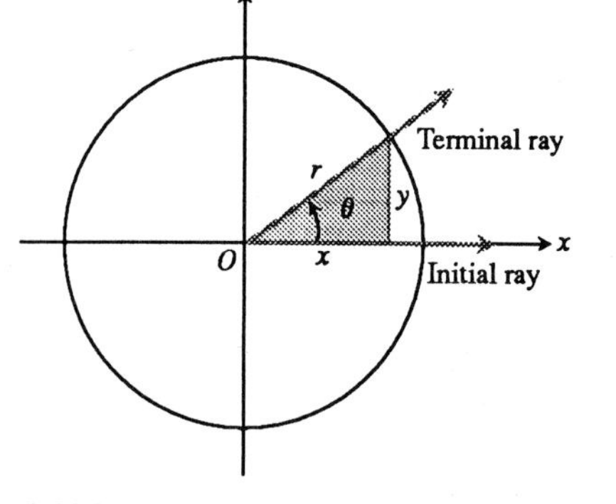

1.134 An angle θ in standard position.

When an angle of measure θ is placed in standard position at the center of a circle of radius r (Fig. 1.34), the six basic trigonometric functions of θ are defined in the following way:

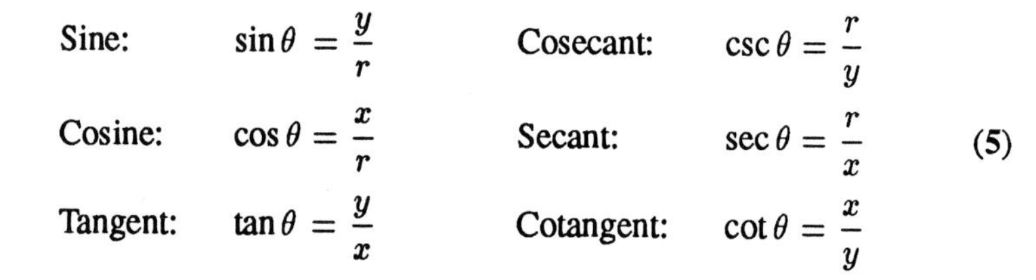

$$\begin{array}{llll} \text{Sine:} & \sin\theta = \dfrac{y}{r} & \text{Cosecant:} & \csc\theta = \dfrac{r}{y} \\ \text{Cosine:} & \cos\theta = \dfrac{x}{r} & \text{Secant:} & \sec\theta = \dfrac{r}{x} \\ \text{Tangent:} & \tan\theta = \dfrac{y}{x} & \text{Cotangent:} & \cot\theta = \dfrac{x}{y} \end{array} \tag{5}$$

As you can see, $\tan\theta$ and $\sec\theta$ are not defined if $x = 0$. In terms of radian measure, this means they are not defined when θ is $\pm\pi/2, \pm 3\pi/2, \ldots$. Similarly, $\cot\theta$ and $\csc\theta$ are not defined for values of θ for which $y = 0$, namely $\theta = 0, \pm\pi, \pm 2\pi, \ldots$. Notice also that

$$\begin{array}{ll} \tan\theta = \dfrac{\sin\theta}{\cos\theta}, & \csc\theta = \dfrac{1}{\sin\theta}, \\ \sec\theta = \dfrac{1}{\cos\theta}, & \cot\theta = \dfrac{1}{\tan\theta} \end{array} \tag{6}$$

whenever the quotients on the right-hand sides are defined.

Because $x^2 + y^2 = r^2$ (Pythagorean theorem),

$$\cos^2\theta + \sin^2\theta = \frac{x^2}{r^2} + \frac{y^2}{r^2} = \frac{x^2 + y^2}{r^2} = 1. \tag{7}$$

The equation $\cos^2\theta + \sin^2\theta = 1$, true for all values of θ, is probably the most frequently used identity in trigonometry.

The coordinates of the point $P(x, y)$ in Fig. 1.134 can be expressed in terms of r and θ as

$$\begin{array}{ll} x = r\cos\theta & (\text{because } x/r = \cos\theta) \\ y = r\sin\theta & (\text{because } y/r = \sin\theta). \end{array} \tag{8}$$

We shall use these equations when we study circular motion and when we work with polar coordinates. Notice that if $\theta = 0$ in Fig. 1.134 then $x = r$ and $y = 0$, so

$$\cos 0 = 1 \qquad \text{and} \qquad \sin 0 = 0.$$

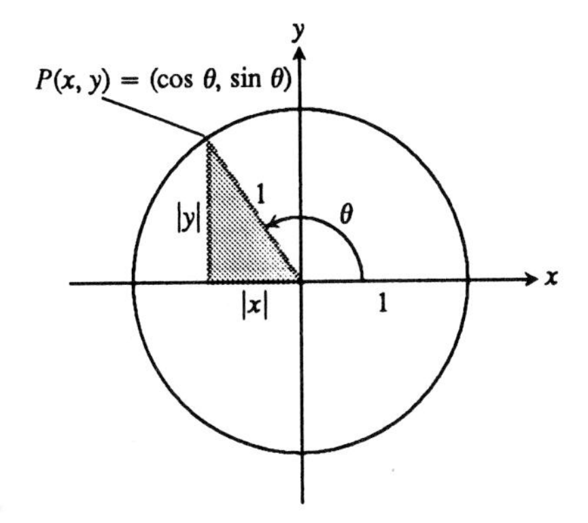

1.135 The acute reference triangle for an angle θ.

If $\theta = \pi/2$, we have $x = 0$ and $y = r$. Hence,

$$\cos\frac{\pi}{2} = 0 \qquad \text{and} \qquad \sin\frac{\pi}{2} = 1.$$

Calculating Sines and Cosines

If the circle in Fig. 1.134 has radius $r = 1$ unit, Eqs. (8) simplify to

$$x = \cos\theta, \qquad y = \sin\theta.$$

We can therefore calculate the values of the cosine and sine from the acute reference triangle made by dropping a perpendicular from the point $P(x, y)$ to the x-axis (Fig. 1.35). The numerical values of x and y are read from the triangle's sides. The signs of x and y are determined by the quadrant in which the triangle lies.

EXAMPLE 6 Find the sine and cosine of $-\pi/4$ radians.

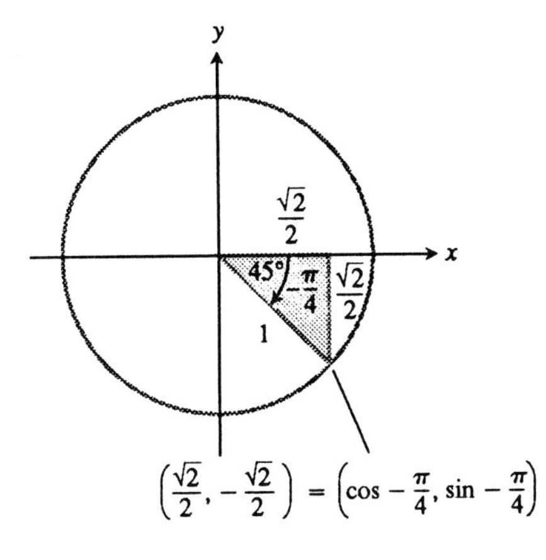

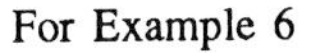
For Example 6

Solution

STEP 1: Draw the angle in standard position and write in the lengths of the sides of the reference triangle.

STEP 2: Find the coordinates of the point P where the angle's terminal ray cuts the circle:

$$\cos -\frac{\pi}{4} = x\text{-coordinate of } P = \frac{\sqrt{2}}{2}.$$

$$\sin -\frac{\pi}{4} = y\text{-coordinate of } P = -\frac{\sqrt{2}}{2}.$$

Table 1.2 gives the values of the sine, cosine, and tangent for selected values. Computations similar to the ones in Example 6 are summarized in Table 1.2.

TABLE 1.2 **Exact values of sin θ, cos θ, and tan θ for selected values of θ**

Degrees	−180	−135	−90	−45	0	45	90	135	180
θ (radians)	$-\pi$	$-3\pi/4$	$-\pi/2$	$-\pi/4$	0	$\pi/4$	$\pi/2$	$3\pi/4$	π
sin θ	0	$-\sqrt{2}/2$	−1	$-\sqrt{2}/2$	0	$\sqrt{2}/2$	1	$\sqrt{2}/2$	0
cos θ	−1	$-\sqrt{2}/2$	0	$\sqrt{2}/2$	1	$\sqrt{2}/2$	0	$-\sqrt{2}/2$	−1
tan θ	0	1		−1	0	1		−1	0

The parametric graphing feature of graphers provide a neat way to compute sines and cosines of angles.

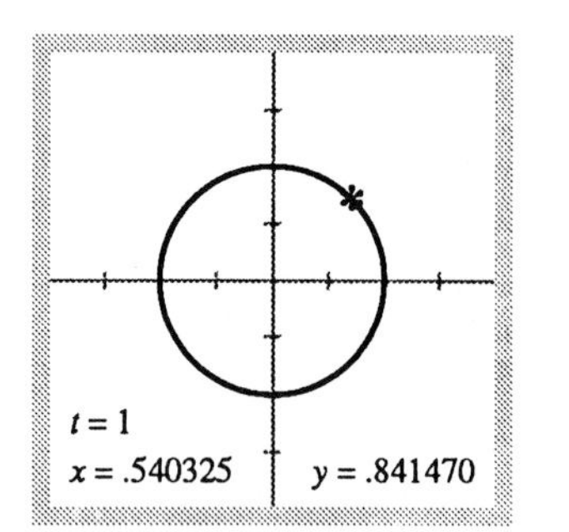

1.136 A complete graph of the unit circle $x^2 + y^2 = 1$ obtained with a parametric grapher. The values of $x = \cos 1$ and $y = \sin 1$ are shown using [TRACE].

EXAMPLE 7 Use a parametric grapher to calculate several values of $\sin\theta$ and $\cos\theta$ for $0 \le \theta \le 2\pi$.

Solution Enter $x(t) = \cos t, y(t) = \sin t$, tMin $= 0$, tMax $= 2\pi$, and tStep $= 0.1$. Choose a square viewing rectangle that includes the unit circle $x^2 + y^2 = 1$. This can be accomplished by choosing a viewing rectangle so that the screen coordinate parameters discussed in Section 1.1 have $\delta x = \delta y = 0.1$. Use the [TRACE] feature to read $x(1) = \cos 1 = 0.5403023$ and $y(1) = \sin 1 = 0.8414710$ (Fig. 1.136). Table 1.3 shows a few more values of $\cos t$ and $\sin t$.

TABLE 1.3 **Some values of cos *x* and sin *x* read using the [TRACE] feature of a parametric grapher.**

t (radians)	$\cos t$	$\sin t$
0.5	0.8775826	0.4794255
1	0.5403023	0.8414710
2	−0.4161468	0.9092974
3.5	−0.9364567	−0.3507832
6.2	0.9965421	−0.0830894

■

Explore Box: Calculating Sines and Cosines

1. Set your parametric grapher in degree mode. Enter $x(t) = \cos t, y(t) = \sin(t)$, tMin $= 0$, tMax $= 360$, and tStep $= 15$.
2. For what values of t are $\cos t$ and $\sin t$ being computed in part 1?
3. Change tMin to -180 and tMax to 180.
4. What is being graphed in part 3?
5. Check the exact values in Table 1.2.
6. Are there other pairs of values for tMin and tMax that give the same graph in part 3?
7. Set your grapher in radian mode. Repeat parts 3–6 using appropriate equivalents of -180 and 180.
8. How can you obtain values of $\sin\theta$ and $\cos\theta$ for some values of θ in $0 \le \theta \le 4\pi$? Explain.

A parametric grapher provides a rapid way to compute sines and cosines. These values can also be computed directly using any scientific calculator.

Most of the time we will use calculators to compute values of trigonometric functions.

CAUTION

When using calculators to compute trigonometric values, be sure your calculator is set in the appropriate mode (degree or radian).

Periodicity

When an angle of measure θ and an angle of measure $\theta + 2\pi$ are in standard position, their terminal rays coincide. The two angles therefore have the same trigonometric-function values:

$$
\begin{aligned}
\cos(\theta + 2\pi) &= \cos\theta \\
\sin(\theta + 2\pi) &= \sin\theta \\
\tan(\theta + 2\pi) &= \tan\theta \\
\cot(\theta + 2\pi) &= \cot\theta \\
\sec(\theta + 2\pi) &= \sec\theta \\
\csc(\theta + 2\pi) &= \csc\theta
\end{aligned}
\tag{9}
$$

Similarly, $\cos(\theta - 2\pi) = \cos\theta$, $\sin(\theta - 2\pi) = \sin\theta$, and so on.

From another point of view, Eqs. (9) tell us that if we start at any particular value $\theta = \theta_0$ and let θ increase or decrease steadily, we see the values of the trigonometric functions start to repeat after any interval of length 2π. We describe this behavior by saying that the six basic trigonometric functions are **periodic** and that they repeat after a fixed **period** of θ-values (in this case 2π).

Definition

> A function $f(x)$ is **periodic** with **period** $p > 0$ if $f(x + p) = f(x)$ for every value of x.

EXAMPLE 8 Equations (9) tell us that the six basic trigonometric functions are periodic with a period of 2π. Other periods include $4\pi, 6\pi$, and so on (positive integer multiples of 2π).

The importance of periodic functions stems from the fact that much of the behavior we study in science is periodic. Brain waves and heartbeats are periodic, as are household voltage and electric current. The electromagnetic field that heats food in a microwave oven is periodic, as are cash flows in seasonal businesses and the behavior of rotational machinery. The seasons are periodic—so is the weather. The phases of the moon are periodic, as are the motions of the planets. There is strong evidence that the ice ages are periodic, with a period of 90,000–100,000 years.

If so many things are periodic, why limit our discussion to trigonometric functions? The answer lies in a surprising and beautiful theorem from advanced calculus that says that every periodic function we want to use in mathematical modeling can be written as an algebraic combination of sines and cosines. Thus, once we learn the calculus of sines and cosines, we will know everything we need to know to model the mathematical behavior of periodic phenomena.

Graphs of Trigonometric Functions

When we graph trigonometric functions in the coordinate plane, we usually denote the independent variable (radians) by x instead of θ. Figure 1.137 shows sketches of the complete graphs of $y = \sin x$, $y = \cos x$, and $y = \tan x$. In the exercises, we will ask you to support these graphs and the graphs in Fig. 1.138 with your grapher.

Notice that the graph of

$$\tan x = \frac{\sin x}{\cos x}$$

"blows up" whenever x nears an odd-integer multiple of $\pi/2$. These are the points for which $\sin x = 1$ and $\cos x = 0$. Notice, too, how the periodicity of the sine, cosine, and tangent appears in the graphs. Choose any starting point, and each graph repeats after an interval of length 2π. We will consider a graph of a periodic function to be complete if it shows at least one period and the period is specified.

Figure 1.138 shows complete graphs of the secant, cosecant, and cotangent functions. Notice again how the periodicity appears in the graphs.

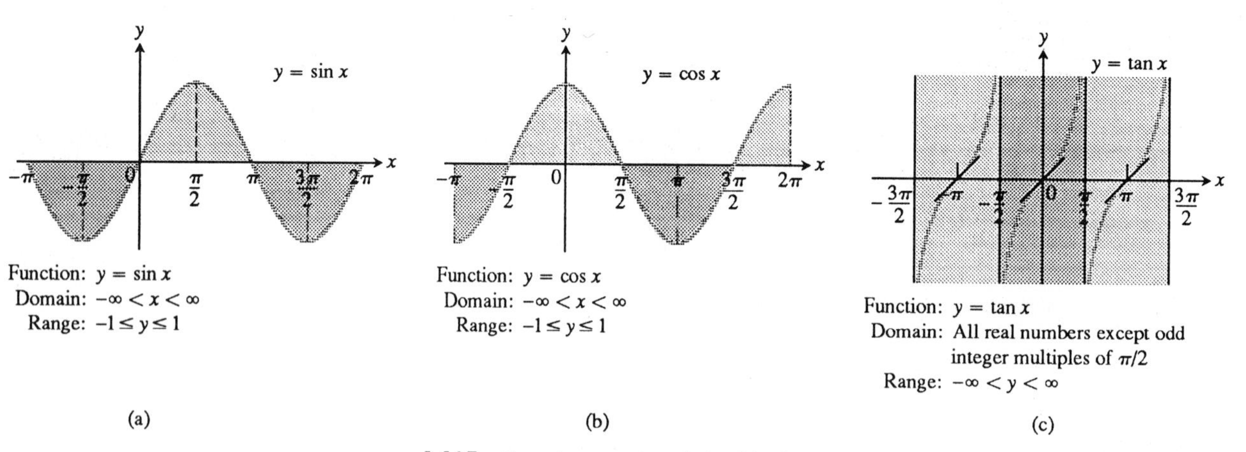

1.137 Complete graphs of the (a) sine, (b) cosine, and (c) tangent as functions of radian measure.

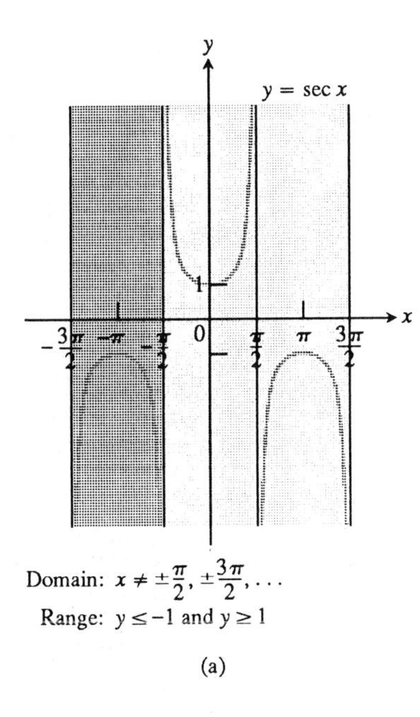

Domain: $x \neq \pm\frac{\pi}{2}, \pm\frac{3\pi}{2}, \ldots$
Range: $y \leq -1$ and $y \geq 1$

(a)

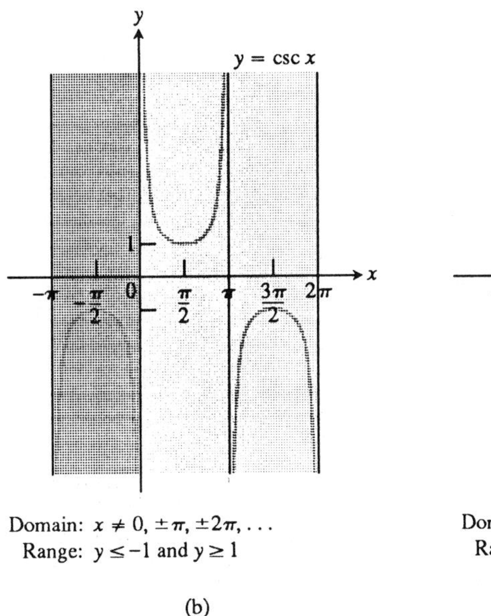

Domain: $x \neq 0, \pm\pi, \pm 2\pi, \ldots$
Range: $y \leq -1$ and $y \geq 1$

(b)

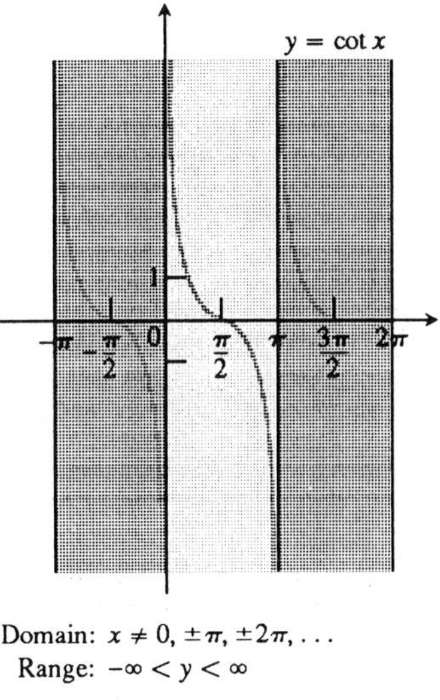

Domain: $x \neq 0, \pm\pi, \pm 2\pi, \ldots$
Range: $-\infty < y < \infty$

(c)

1.138 Complete graphs of the (a) secant, (b) cosecant, and (c) cotangent as functions of radian measure.

The following Explore Activity shows how parametric graphing can be used to investigate the connections between the unit circle definition of the trigonometric values and their graphs.

Explore Box

1. Set your grapher in radian mode and enter the parametric equations

$$x_1(t) = \cos t \qquad x_2(t) = t$$

$$y_1(t) = \sin t \qquad y_2(t) = \sin t$$

with range settings $t\text{Min} = 0, t\text{Max} = 2\pi, x\text{Min} = -3, x\text{Max} = 2\pi, y\text{Min} = -2$, and $y\text{Max} = 2$.

2. Set the [MODE] of your grapher so that the two sets of parametric equations are graphed simultaneously.
3. Describe the graphs of the two sets of parametric equations in part 1. You may need to choose a square viewing rectangle containing the one described in part 1.
4. Use [TRACE] and compare the y-values of the two curves. Explain.
5. Set $x\text{Max} = 4\pi$ and repeat part 4.
6. Change $y_2(t)$ to $\cos t, \tan t, \sec t, \csc t$, and $\tan t$, repeating parts 3–5 each time.

The activity described in the above Explore is sometimes referred to as *unwrapping the trigonometric functions.*

Odd vs. Even

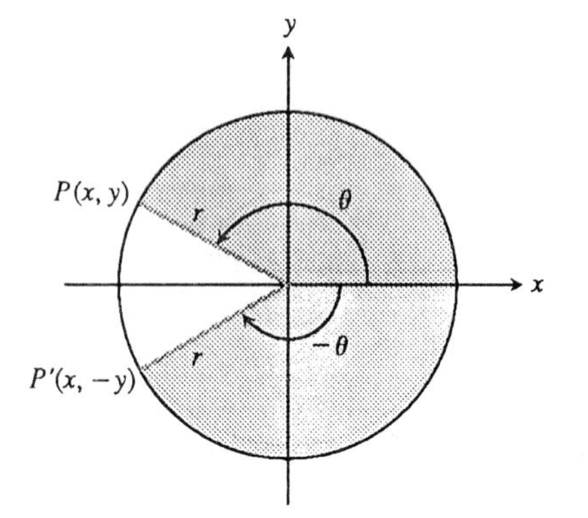

1.139 Angles of opposite sign.

Figure 1.139 shows two angles that have the same magnitude but opposite signs. By symmetry, the points where the terminal rays cross the circle have the same x-coordinate and their y-coordinates differ only in sign. Hence,

$$\cos(-\theta) = \frac{x}{r} = \cos\theta, \qquad \text{(The cosine is an even function.)} \qquad \textbf{(10a)}$$

$$\sin(-\theta) = \frac{-y}{r} = -\sin\theta. \qquad \text{(The sine is an odd function.)} \qquad \textbf{(10b)}$$

Equation (10a) means that reflecting the graph of $f(x) = \cos x$ through the y-axis leaves it unchanged ($f(-x) = f(x)$). Equation (10b) means that reflecting the graph of $g(x) = \sin x$ through the y-axis is the same as reflecting it through the x-axis ($g(-x) = -g(x)$).

EXAMPLE 9

$$\cos\left(-\frac{\pi}{3}\right) = \cos\frac{\pi}{3} = \frac{1}{2}$$

$$\sin\left(-\frac{\pi}{3}\right) = -\sin\frac{\pi}{3} = -\frac{\sqrt{3}}{2}$$

As for the other basic trigonometric functions, the secant is even and the cosecant, tangent, and cotangent are odd. For the secant and tangent,

$$\sec(-\theta) = \frac{1}{\cos(-\theta)} = \frac{1}{\cos\theta} = \sec\theta, \qquad \textbf{(11)}$$

$$\tan(-\theta) = \frac{\sin(-\theta)}{\cos(-\theta)} = \frac{-\sin\theta}{\cos\theta} = -\tan\theta. \qquad \textbf{(12)}$$

Similar calculations show that the cotangent and cosecant are odd, as we ask you to verify in Exercises 73 and 74.

Inverse Trigonometric Functions

Notice that none of the six basic trigonometric functions graphed in Figs. 1.137 and 1.138 is one-to-one. So, the inverses of these functions are not functions. However, in each case the domain can be restricted to produce a new function whose inverse is a function. We define the inverses for $\sin x, \cos x$, and $\tan x$ and give the other three definitions in Chapter 7.

EXAMPLE 10 Each of the following restricted functions is one-to-one:

a) $y = \sin x, -\dfrac{\pi}{2} \leq x \leq \dfrac{\pi}{2}$ (Fig. 1.140a).

b) $y = \cos x, 0 \leq x \leq \pi$ (Fig. 1.140b).

c) $y = \tan x, -\dfrac{\pi}{2} < x < \dfrac{\pi}{2}$ (Fig. 1.140c).

■

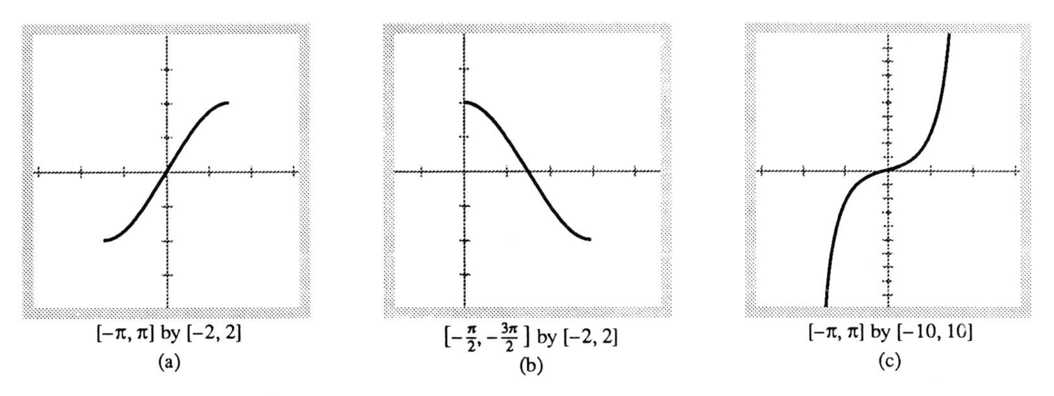

1.140 Complete graphs of the one-to-one functions: (a) $y = \sin x, -\pi/2 \leq x \leq \pi/2$, (b) $y = \cos x, 0 \leq x \leq \pi$, and (c) $y = \tan x, -\pi/2 < x < \pi/2$.

The inverses of the functions graphed in Fig. 1.140 are called the inverse sine, cosine, and tangent functions.

Definitions

The **inverse sine function**, denoted by $y = \sin^{-1} x$, or $y = \arcsin x$, is the function with domain $[-1, 1]$ and range $[-(\pi/2), \pi/2]$ satisfying $x = \sin y$ (Fig. 1.141a).

The **inverse cosine function**, denoted by $y = \cos^{-1} x$, or $y = \arccos x$, is the function with domain $[-1, 1]$ and range $[0, \pi]$ satisfying $x = \cos y$ (Fig. 1.141b).

The **inverse tangent function**, denoted by $y = \tan^{-1} x$, or $y = \arctan x$, is the function with domain $(-\infty, \infty)$ and range $(-\pi/2, \pi/2)$ satisfying $x = \tan y$ (Fig. 1.141c).

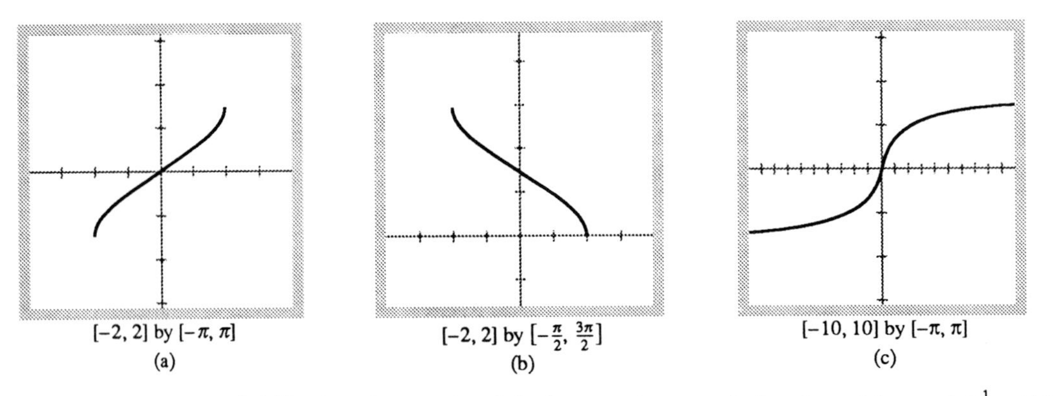

1.141 Complete graphs of the inverse trigonometric functions: (a) $y = \sin^{-1} x$, (b) $y = \cos^{-1} x$, and (c) $y = \tan^{-1} x$.

Scientific calculators have the above inverse functions built in. These functions are useful for finding the possibilities for angles when one of their trigonometric functions is known.

EXAMPLE 11 Solve each of the following equations for x:

a) $\sin x = 0.7$
b) $\tan x = -2.$

Solution

a) Notice that $\theta = \sin^{-1}(0.7)$ gives one angle that is a solution to this equation. We can find that

$$\theta = \sin^{-1}(0.7) = 0.7754$$

with a calculator. The keystrokes needed to access this value are usually [2nd] [SIN], [INV] [SIN], or [SHIFT] [SIN]. Since θ is in the first quadrant, $\pi - \theta$ is in quadrant II and $\sin(\pi - \theta) = \sin\theta$. Thus, $\pi - \theta$ is another solution to this equation (Fig. 1.142).

Any other solution to this equation is coterminal with either θ or $\pi - \theta$. Thus, every solution has one of the two forms:

$$\theta + 2k\pi = 0.7754 + 2k\pi, \text{ or}$$

$$(\pi - \theta) + 2k\pi = 2.3662 + 2k\pi,$$

where k is any integer.

b) First, we find that

$$\tan^{-1}(-2) = -1.1071$$

is in the quadrant IV and is one solution to the equation. Since θ is in quadrant IV, $\theta + \pi$ is in quadrant II and $\tan(\theta + \pi) = \tan\theta$ (Fig. 1.143).

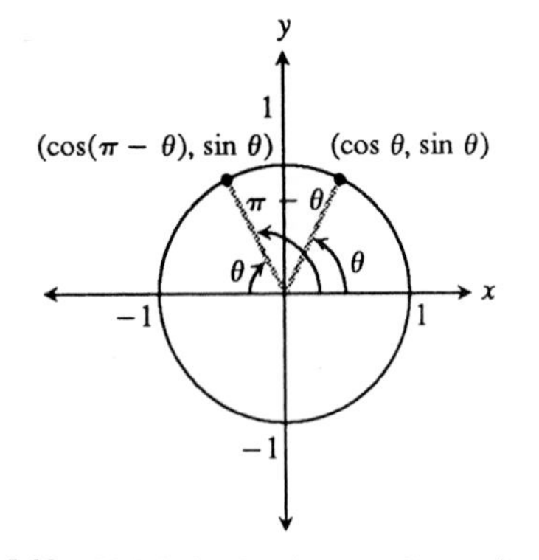

1.142 If θ is in the first quadrant, then $\pi - \theta$ is in the second quadrant and $\sin(\pi - \theta) = \sin\theta$.

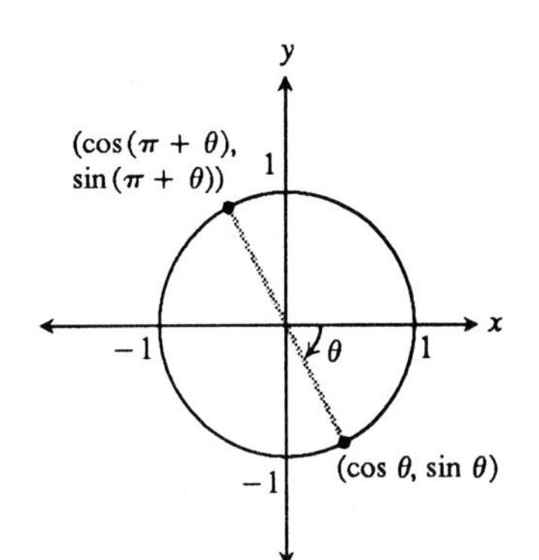

1.143 If θ is in the fourth quadrant, then $\theta + \pi$ is in the second quadrant and $\tan\theta = \tan(\theta + \pi)$.

How to Horizontally Stretch or Shrink a Graph

To horizontally stretch the graph of a function $y = f(x)$, we multiply x by a number between 0 and 1.

EXAMPLE 12 A complete graph of $y = \sin(1/2)x$ is obtained by horizontally stretching the graph of $y = \sin x$ by a factor of $1/(1/2) = 2$.

Solution Figure 1.144 suggests that the period of $y = \sin(1/2)x$ is double the period of $y = \sin x$.

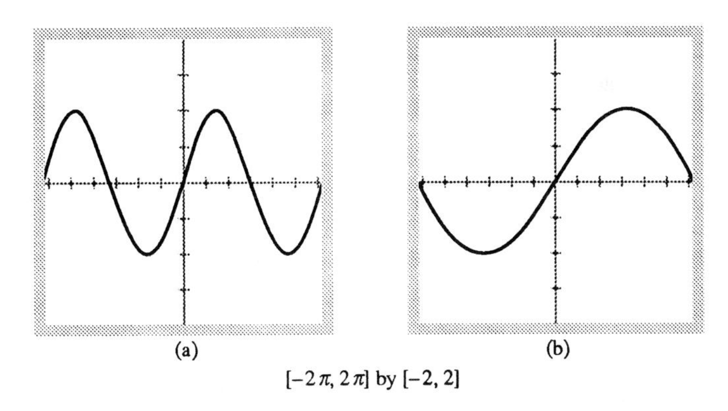

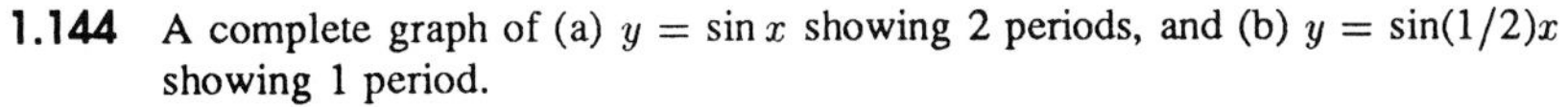

1.144 A complete graph of (a) $y = \sin x$ showing 2 periods, and (b) $y = \sin(1/2)x$ showing 1 period.

To horizontally shrink the graph of a function $y = f(x)$, we multiply x by a number greater than 1.

EXAMPLE 13 A complete graph of $y = \sin 2x$ is obtained by horizontally shrinking the graph of $y = \sin x$ by a factor of 1/2.

Solution Figure 1.145 suggests that the period of $y = \sin 2x$ is 1/2 the period of $y = \sin x$.

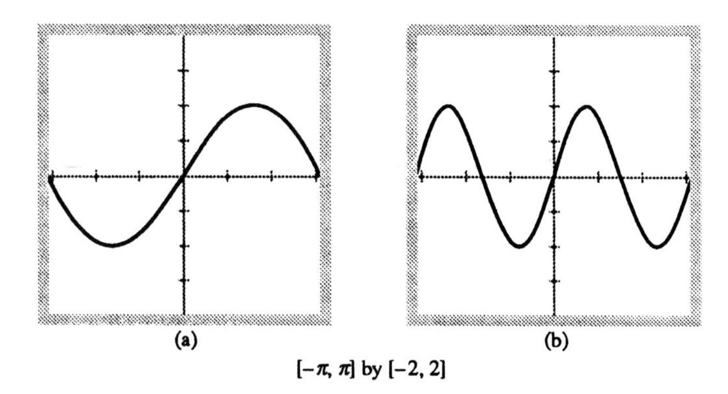

(a) (b)

$[-\pi, \pi]$ by $[-2, 2]$

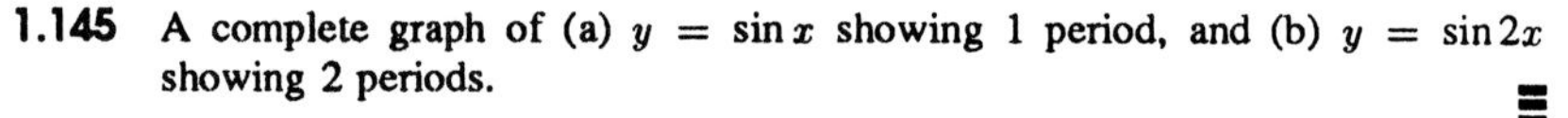

1.145 A complete graph of (a) $y = \sin x$ showing 1 period, and (b) $y = \sin 2x$ showing 2 periods.

Horizontal Stretch or Shrink Formula ($a > 0$)

$y = f(ax)$ $\quad a > 1$: Horizontally shrinks the graph of f by a factor of $1/a$.

$0 < a < 1$: Horizontally stretches the graph of f by a factor of $1/a$.

If f is a periodic function with period p, then the period of $y = f(ax)$ is p/a.

With the addition of the geometric transformation of horizontal stretching and shrinking to the transformations studied in Section 1.4, we can now describe how the graph of $y = af(bx + c) + d$ can be obtained from the graph of $y = f(x)$.

Summary

Graphing $y = af(bx + c) + d = af\left[b\left(x + c/b\right)\right] + d$

The graph of $y = af(bx + c) + d$ can be obtained from the graph of $y = f(x)$ as follows:

1. Vertical stretch or shrink by the factor $|a|$, followed by a reflection through the x-axis if $a < 0$.
2. Horizontal stretch or shrink by the factor $1/|b|$, followed by a reflection through the y-axis if $b < 0$.
3. Horizontal shift $\left|\frac{c}{b}\right|$ units left if $\frac{c}{b} > 0$ or $\left|\frac{c}{b}\right|$ units right if $\frac{c}{b} < 0$.
4. Vertical shift $|d|$ units up if $d > 0$, or $|d|$ units down if $d < 0$.

The following shift formulas show another connection when $f(x)$ is either $\sin x$ or $\cos x$.

Shift Formulas

If you look again at Fig. 1.137, you will see that the cosine curve is the same as the sine curve shifted $\pi/2$ units to the left. Also, the sine curve is the same as the cosine curve shifted $\pi/2$ units to the right. In symbols,

$$\sin\left(x + \frac{\pi}{2}\right) = \cos x, \qquad \cos\left(x - \frac{\pi}{2}\right) = \sin x. \tag{13}$$

Figure 1.146(a) shows the cosine shifted to the left $\pi/2$ units to become the reflection of the sine curve through the x-axis. Next to it, Fig. 1.146(b) shows the sine curve shifted $\pi/2$ units to the right to become the reflection of the cosine curve through the x-axis. In symbols,

$$\cos\left(x + \frac{\pi}{2}\right) = -\sin x, \qquad \sin\left(x - \frac{\pi}{2}\right) = -\cos x. \tag{14}$$

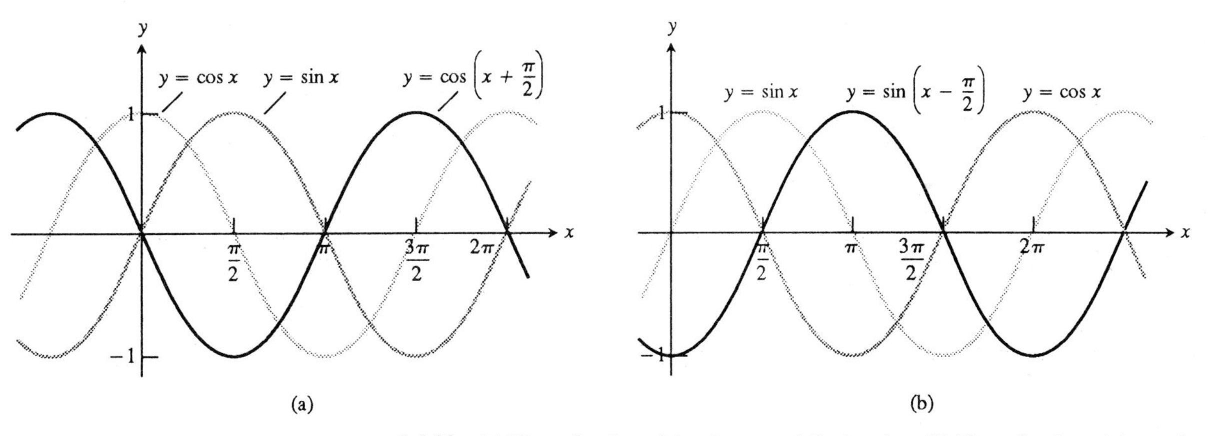

1.146 (a) The reflection of the sine as a shifted cosine, (b) The reflection of the cosine as a shifted sine.

EXAMPLE 14 The builders of the Trans-Alaska Pipeline used insulated pads to keep the heat from the hot oil in the pipeline from melting the permanently frozen soil beneath. To design the pads, it was necessary to take into account the variation in air temperature throughout the year. The variation was represented in the calculations by a *general sine function* of the form

$$f(x) = A \sin\left[\frac{2\pi}{B}(x - C)\right] + D,$$

where $|A|$ is the *amplitude*, $|B|$ is the *period*, C is the *horizontal shift*, and D is the *vertical shift* (Fig. 1.147).

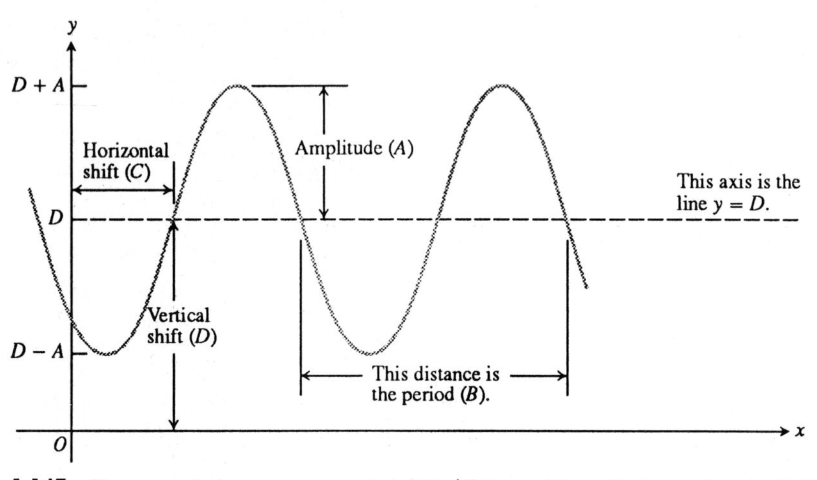

1.147 The general sine curve $y = A\sin[(2\pi/B)(x - C)] + D$, shown for A, B, C, and D positive.

Figure 1.148 shows how we can use such a function to represent temperature data. The data points in the figure are plots of the mean air temperature for Fairbanks, Alaska, based on records of the National Weather Service from 1941 to 1970. The sine function used to fit the data is

$$f(x) = 37\sin\left[\frac{2\pi}{365}(x - 101)\right] + 25,$$

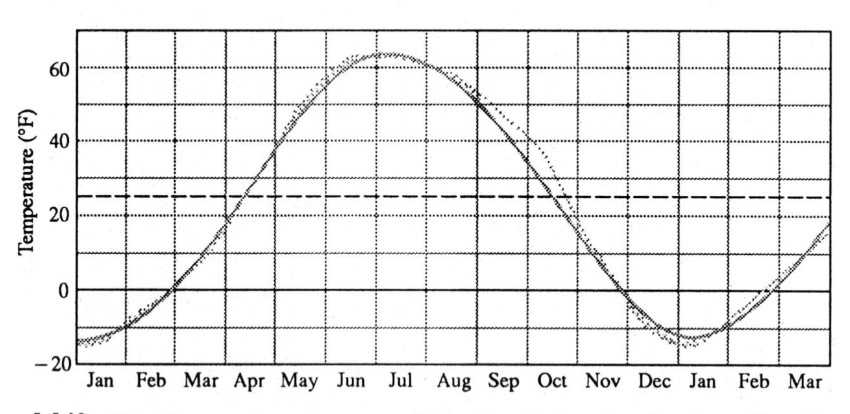

1.148 Normal mean air temperature at Fairbanks, Alaska, plotted as data points. The approximating sine function is

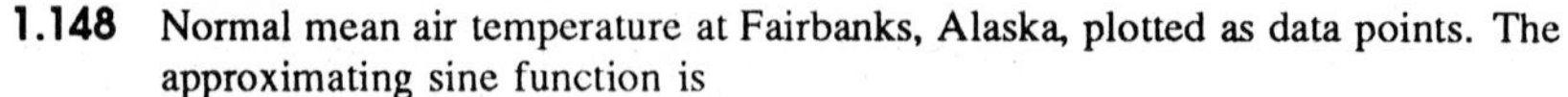

$$f(x) = 37\sin\left[\frac{2\pi}{365}(x - 101)\right] + 25.$$

(*Source:* "Is the Curve of Temperature Variation a Sine Curve?" by B. M. Lando and C. A. Lando, *The Mathematics Teacher*, 7:6, Fig. 2, p. 535 (September 1977).)

where f is temperature in degrees Fahrenheit, and x is the number of the day counting from the beginning of the year. The fit is remarkably good. ≡

Angle Sum and Difference Formulas

As you may recall from an earlier course,

$$\cos(A+B) = \cos A \cos B - \sin A \sin B, \quad (15)$$

$$\sin(A+B) = \sin A \cos B + \cos A \sin B. \quad (16)$$

These formulas hold for all angles A and B.

If we replace B by $-B$ in Eqs. (15) and (16), we get

$$\begin{aligned}\cos(A-B) &= \cos A \cos(-B) - \sin A \sin(-B)\\ &= \cos A \cos B - \sin A(-\sin B)\\ &= \cos A \cos B + \sin A \sin B, \quad (17)\\ \sin(A-B) &= \sin A \cos(-B) + \cos A \sin(-B)\\ &= \sin A \cos B + \cos A(-\sin B)\\ &= \sin A \cos B - \cos A \sin B \quad (18)\end{aligned}$$

Sinusoids

Definition

A **sinusoid** is a function that can be written in the form

$$f(x) = a\sin(bx + c) + d,$$

where a, b, c, and d are real numbers.

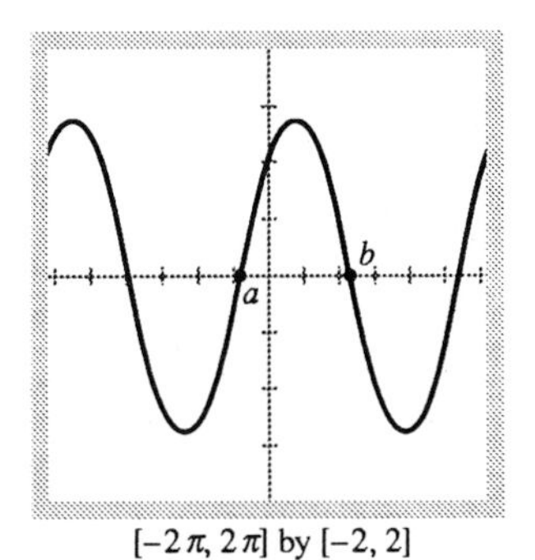

$[-2\pi, 2\pi]$ by $[-2, 2]$

1.149 The graph of $f(x) = \sin x + \cos x$ appears to be a sine curve.

The graph of $f(x) = \sin x + \cos x$ in Figure 1.149 appears to be a sine curve. The following Explore Activity will give additional support for this conjecture.

Explore with a Graphing Utility

Exploring the graph of $f(x) = \sin x + \cos x$

1. Use zoom-in to determine the accuracy of the amplitude of f in Fig. 1.149 as accurately as your grapher will allow. Make a conjecture about the exact value A of the amplitude.
2. Determine the x-intercepts a and b in Fig. 1.149 as accurately as your grapher will allow.
3. Make a conjecture about the exact value of the horizontal shift factor C of the graph of f.
4. Make a conjecture about the exact value of the period B of f.
5. Compare the graphs of $y_1 = f(x)$ and $y_2 = A \sin 2\pi/(B(x + C))$. What might you conjecture?
6. Draw the graph of $y_1 - y_2 + 1$. What might you conjecture?

The above Explore Activity is a special case of the following theorem.

Theorem 8

Sinusoids

For all real numbers a and b, there are real numbers A and α such that

$$a \sin x + b \cos x = A \sin(x + \alpha).$$

EXAMPLE 15 Show that $f(x) = \sin x + \cos x$ is a sinusoid.

Solution

$$\begin{aligned} \sin x + \cos x &= A \sin(x + \alpha) \\ &= A(\sin x \cos \alpha + \cos x \sin \alpha) \qquad \textbf{(16)} \\ &= (A \cos \alpha) \sin x + (A \sin \alpha) \cos x \end{aligned}$$

We have it if we can choose A and α so that

$$A \cos \alpha = 1 \text{ and } A \sin \alpha = 1.$$

Notice

$$\begin{aligned} (A \cos \alpha)^2 + (A \sin \alpha)^2 &= 2 \\ A^2 &= 2. \end{aligned}$$

So, we choose $A = \sqrt{2}$ and $\alpha = \pi/4$. We can use Eq. (16) directly to show that

$$\sin x + \cos x = \sqrt{2} \sin(x + \pi/4).$$

Support can be given for the following identities by comparing graphs of the left- and right-hand sides of the identities.

Double-angle (Half-angle) Formulas

As you will see later on, it is sometimes possible to simplify a calculation by changing trigonometric functions of θ into trigonometric functions of 2θ. There are four basic formulas for doing this, called **double-angle formulas.** The first two come from setting A and B equal to θ in Eqs. (15) and (16):

$$\cos 2\theta = \cos^2\theta - \sin^2\theta \qquad \text{(Eq. (15) with } A = B = \theta) \qquad (19)$$

$$\sin 2\theta = 2\sin\theta\cos\theta \qquad \text{(Eq. (16) with } A = B = \theta) \qquad (20)$$

The other two double-angle formulas come from the equations

$$\cos^2\theta + \sin^2\theta = 1, \qquad \cos^2\theta - \sin^2\theta = \cos 2\theta.$$

We add to get

$$2\cos^2\theta = 1 + \cos 2\theta,$$

subtract to get

$$2\sin^2\theta = 1 - \cos 2\theta,$$

and divide by 2 to get

$$\cos^2\theta = \frac{1 + \cos 2\theta}{2} \qquad (21)$$

$$\sin^2\theta = \frac{1 - \cos 2\theta}{2}. \qquad (22)$$

When θ is replaced by $\theta/2$ in Eqs. (21) and (22), the resulting formulas are called **half-angle formulas**. Some books refer to Eqs. (21) and (22) by this name as well.

Where to Look for Other Formulas

Additional information is available in

1. Appendix 1 of the present book.
2. *CRC Standard Mathematical Tables* (any recent edition). Boca Raton, Florida: CRC Press, Inc.

Exercises 1.7

Exercises 1–4 give angles in degrees. Change them to radians.

1. 510° **2.** 120° **3.** -42° **4.** -150°

Exercises 5–8 give angles in radians. Change them to degrees.

5. 6.2 **6.** $-\frac{\pi}{6}$ **7.** -2 **8.** $\frac{3\pi}{4}$

Exercises 9–14 give angles in radians. Find the sine, cosine, tangent, cotangent, secant, and cosecant of each angle (when defined).

9. a) $\frac{\pi}{3}$ b) $-\frac{\pi}{3}$

10. a) 2.5 b) -2.5

11. a) 6.5 b) -6.5

12. a) 3.7 b) -3.7

13. a) $\frac{\pi}{2}$ b) $\frac{3\pi}{2}$

14. a) 0 b) π

Give the measure of the angles in Exercises 15–18 in radians and degrees. Give exact values whenever possible.

15. $\sin^{-1}(0.5)$

16. $\sin^{-1}\left(-\frac{\sqrt{2}}{2}\right)$

17. $\tan^{-1}(-5)$

18. $\cos^{-1}(0.7)$

Explain how to use a parametric grapher to compute $\sin t$ and $\cos t$ for the values of t specified in Exercises 19 and 20.

19. $0, 0.5, 1, 1.5, \cdots, 6$

20. $0^\circ, 5^\circ, 10^\circ, \cdots, 360^\circ$

21. Choose appropriate viewing rectangles to support the graphs of $y = \sin x, y = \cos x$, and $y = \tan x$ given in Figure 1.137.

22. Choose appropriate viewing rectangles to support the graphs of $y = \sec x, y = \csc x$, and $y = \cot x$ given in Figure 1.138.

23. Choose appropriate viewing rectangles to display two complete periods of $y = \sec x, y = \csc x$, and $y = \cot x$ in degree mode.

24. Choose appropriate viewing rectangles to display two complete periods of $y = \sin x, y = \cos x$, and $y = \tan x$ in degree mode.

In Exercises 25–28, draw the graphs of the functions together over the given intervals of x-values.

25. $y = \sin x$ and $y = \csc x, -\pi \leq x \leq \pi$

26. $y = \cos x$ and $y = \sec x, -\frac{3\pi}{2} \leq x \leq \frac{3\pi}{2}$

27. $y = \cos 4x$ and $y = \cos x, 0 \leq x \leq 2\pi$

28. $y = \sin \frac{x}{4}$ and $y = \sin x, 0 \leq x \leq 8\pi$

Determine the amplitude, period, horizontal shift, and vertical shift and draw a complete graph of the functions in Exercises 29–32.

29. $y = 2\cos \frac{x}{3}$

30. $y = 2\cos 3x$

31. $y = \cot\left(2x + \frac{\pi}{2}\right)$

32. $y = 3\cos\left(x + \frac{\pi}{4}\right) - 2$

In Exercises 33–36, determine the period, domain, and range and draw a complete graph of the function. Describe how the function's graph can be obtained from the graph of one of the six basic trigonometric functions.

33. $y = 3\csc(3x + \pi) - 2$

34. $y = 2\sin(4x + \pi) + 3$

35. $y = -3\tan(3x + \pi) + 2$

36. $y = 2\sin\left(2x + \frac{\pi}{3}\right)$

Solve the equations in Exercises 37–40.

37. $\cos x = -0.7$

38. $\sin x = 0.2$

39. $\tan x = 4$

40. $\sin x = -0.2$

Use a parametric grapher to draw a complete graph of the circles in Exercises 41–44. Specify the viewing rectangle and the range of values for the parameter used.

41. $x^2 + y^2 = 5$

42. $x^2 + y^2 = 4$

43. $(x - 2)^2 + (y + 3)^2 = 9$

44. $(x + 1)^2 + (y + 3)^2 = 16$

45. Prove Theorem 8.

46. Use graphs to support the identities in Equations (19)–(22).

Show that the functions in Exercises 47–50 are sinusoids, $a\sin(bx + c)$. Use a graph to conjecture the values of a, b, and c, and then confirm algebraically.

47. $y = 2\sin x + 3\cos x$

48. $y = \sin x + \sqrt{3}\cos x$

49. $y = \sin 2x + \cos 2x$

50. $y = 2\sin 3x + 2\cos 3x$

51. Which equations have the same graph? Confirm algebraically.

a) $y = \sin$ x

b) $y = \cos x$

c) $y = \sin(-x)$

d) $y = \cos(-x)$

e) $y = -\sin x$

f) $y = -\cos x$

g) $y = \sin\left(x + \frac{\pi}{2}\right)$

h) $y = \sin\left(x - \frac{\pi}{2}\right)$

i) $y = \cos\left(x + \frac{\pi}{2}\right)$

j) $y = \cos\left(x - \frac{\pi}{2}\right)$

k) $y = \cos(x + \pi)$

l) $y = \cos(x - \pi)$

m) $y = \sin(x + \pi)$

n) $y = \sin(x - \pi)$

52. Two more useful identities. Use Eqs. (5) to verify the following identities. Support graphically.

a) $\sec^2\theta = 1 + \tan^2\theta$

b) $\csc^2\theta = 1 + \cot^2\theta$

53. What symmetries do the graphs of the cosine, sine, and tangent have?

54. What symmetries do the graphs of secant, cosecant, and cotangent have?

55. Consider the function $y = \sqrt{(1 + \cos 2x/2)}$.

a) Can x take on any real value?

b) How large can $\cos 2x$ become? How small?

c) How large can $(1 + \cos 2x)/2$ become? How small?

d) What are the domain and range of $y = \sqrt{(1 + \cos 2x/2)}$?

56. Consider the function $y = \tan(x/2)$.

a) What values of $x/2$ must be excluded from the domain of $\tan(x/2)$?

b) What values of x must be excluded from the domain of $\tan(x/2)$?

c) What values does $y = \tan(x/2)$ assume on the interval $-\pi < x < \pi$?

d) What are the domain and range of $y = \tan(x/2)$?

57. Temperature in Fairbanks, Alaska. Find the (a) amplitude, (b) period, (c) horizontal shift, and (d) vertical shift of the general sine function

$$f(x) = 37 \sin \left[\frac{2\pi}{365}(x - 101)\right] + 25.$$

58. Temperature in Fairbanks, Alaska. Use the equation in Exercise 57 to approximate the answers to the following questions about the temperature in Fairbanks, Alaska, shown in Fig. 1.148. Assume that the year has 365 days.

a) What are the highest and lowest mean daily temperatures shown?

b) What is the average of the highest and lowest mean daily temperature shown? Why is this average the vertical shift of the function?

59. What happens if you take $A = B$ in Eq. (17)? Does the result agree with something you already know?

60. What happens if you take $B = \pi/2$ in Eqs. (17) and (18)? Do these results agree with something you already know?

61. What happens if you take $B = \pi/2$ in Eqs. (15) and (16)? Do these results agree with something you already know?

62. What happens if you take $B = \pi$ in Eqs. (15) and (16)? In Eqs. (17) and (18)?

63. Evaluate $\cos 15^\circ$ as $\cos(45^\circ - 30^\circ)$.

64. Evaluate $\sin 75^\circ$ as $\sin(45^\circ + 30^\circ)$.

65. Evaluate $\sin \frac{7\pi}{12}$ (radians) as $\sin\left(\frac{\pi}{4} + \frac{\pi}{3}\right)$.

66. Evaluate $\cos \frac{10\pi}{24}$ (radians) as $\cos\left(\frac{\pi}{4} + \frac{\pi}{6}\right)$.

Use double-angle formulas to find the exact function values in Exercises 67–70 (angles in radians).

67. $\cos^2 \frac{\pi}{8}$

68. $\cos^2 \frac{\pi}{12}$

69. $\sin^2 \frac{\pi}{12}$

70. $\sin^2 \frac{\pi}{8}$

71. The tangent sum formula. The standard formula for the tangent of the sum of two angles is

$$\tan(A + B) = \frac{\tan A + \tan B}{1 - \tan A \tan B}.$$

Derive the formula by writing $\tan(A + B)$ as

$$\frac{\sin(A + B)}{\cos(A + B)}$$

and applying Eqs. (15) and (16).

72. Derive a formula for $\tan(A - B)$ by replacing B by $-B$ in the formula for $\tan(A + B)$ in Exercise 71.

Even vs. Odd

73. a) Show that $\cot x$ is an odd function of x.

b) Show that the quotient of an even function and an odd function is always odd (on their common domain).

c) Describe how the graph of $y = \cot(-x)$ can be obtained from the graph of $y = \cot x$.

74. a) Show that $\csc x$ is an odd function of x.

b) Show that the reciprocal of an odd function (when defined) is odd.

c) Describe how the graph of $y = \csc(-x)$ can be obtained from the graph of $y = \csc x$.

75. a) Show that the product of $y = \sin x \cos x$ is an odd function of x.

b) Show that the product of an even function and an odd function is always odd (on their common domain).

76. a) Show that the function $y = \sin^2 x$ is an even function of x (even though the sine itself is odd).

b) Show that the square of an odd function is always even.

c) Show that the product of any two odd functions is even (on their common domain).

Chapter 1 Review Questions

1. How do you find the distance between two points in the xy-plane? Between a point and a line in the plane? Give examples.

2. Explain the difference between "sketch a graph" and "draw a graph." What are the basic steps in sketching a graph of an equation in x and y? Illustrate them. What does it mean for a graph to be complete? Can a grapher be used to graph any equation in x and y? Give examples.

3. Distinguish between "screen coordinates" and "Cartesian Coordinates." What is a computer visualization of a graph? What is a square viewing rectangle?

4. What coordinate tests determine whether a graph in the xy-plane is symmetric with respect to the coordinate axes or the origin? Give examples.

5. How can you write the equation for a line if you know the coordinates of two points on the line? The line's slope and the coordinates of one point on the line? The line's slope and y-intercept? Give examples.

6. What are the standard equations for lines perpendicular to the coordinate axes?

7. How are the slopes of mutually perpendicular lines related? Give examples.

8. When a line is not vertical, what is the relation between its slope and its angle of inclination?

9. What is a function? Give examples. How do you use a graphing utility to graph a real-valued function of a real variable? What do you have to be careful about?

10. Name some typical functions and draw their graphs.

11. What is an even function? An odd function? What symmetries do the graphs of such functions have? Give examples. Give an example of a function that is neither odd nor even.

12. When is it possible to compose one function with another? Give examples of composites and their values at various points. Does the order in which functions are composed ever matter?

13. How can you write an equation for a circle in the xy-plane if you know its radius a and the coordinates (h, k) of its center? Give examples.

14. What inequality is satisfied by the coordinates of the points that lie inside the circle of radius a centered at (h, k)? What inequality is satisfied by the coordinates of the points that lie outside the circle?

15. The graph of a function $y = f(x)$ in the xy-plane is shifted 5 units to the left, vertical stretched by a factor of 2, reflected through the x-axis, and then shifted 3 units straight up. Write an equation for the new graph.

16. What is a parabola? What are typical equations for parabolas?

17. What does it mean to solve an equation or inequality with an error of at most 0.01? Give examples.

18. Explain the meaning of: algebraic representation, algebraic representation of a problem situation, graphical representation of a problem situation, graph of a problem situation. Give examples.

19. How do you convert between degree measure and radian measure? Give examples.

20. Graph the six basic trigonometric functions as functions of radian measure. What symmetries do the graphs have?

21. What does it mean for a function $y = f(x)$ to be periodic? Give examples of functions with various periods. Name some real-world phenomena that we model with periodic functions.

22. List the angle sum and difference formulas for the sine and cosine functions.

23. List the four basic double-angle formulas for sines and cosines.

24. Define the function $y = |x|$. Give examples of numbers and their absolute values. How are $|-a|$, $|ab|$, $|a/b|$, and $|a+b|$ related to $|a|$ and $|b|$?

25. How are absolute values used to describe intervals of real numbers?

26. What is the inverse of a relation? How are the graphs, domains, and ranges of relations and their inverses related? Give an example.

27. When is the inverse of a relation a function? How do you tell when functions are inverses of one another? Give examples.

Chapter 1 Practice Exercises

In Exercises 1–4, find the points that are symmetric to the given point (a) across the x-axis, (b) across the y-axis, and (c) across the origin.

1. $(1, 4)$ 2. $(2, -3)$ 3. $(-4, 2)$ 4. $(-2, -2)$

Test the equations in Exercises 5–8 to find out whether their graphs are symmetric with respect to the axes or the origin.

5. a) $y = x$ b) $y = x^2$

6. a) $y = x^3$ b) $y = x^4$

7. a) $x^2 - y^2 = 4$ b) $x - y = 4$

8. a) $y = x^{1/3}$ b) $y = x^{2/3}$

Find equations for the vertical and horizontal lines through the points in Exercises 9–12.

9. $(1, 3)$ 10. $(2, 0)$ 11. $(0, -3)$ 12. (x_0, y_0)

In Exercises 13–20, write an equation for the line that passes through point P with slope m. Then use the equation to find the line's intercepts and graph the line.

13. $P(2, 3), m = 2$ 14. $P(2, 3), m = 0$

15. $P(1, 0), m = -1$ 16. $P(0, 1), m = -1$

17. $P(1, -6), m = 3$ 18. $P(-2, 0), m = 1$

19. $P(-1, 2), m = -\frac{1}{2}$ 20. $P(3, 1), m = \frac{1}{3}$

In Exercises 21–24, find an equation for the line through the two points.

21. $(-2, -2), (1, 3)$ 22. $(-3, 6), (1, -2)$

23. $(2, -1), (4, 4)$ 24. $(3, 3)(-2, 5)$

In Exercises 25–28, find an equation for the line with the given slope m and y-intercept b.

25. $m = \frac{1}{2}, b = 2$ 26. $m = -3, b = 3$

27. $m = -2, b = -1$ 28. $m = 2, b = 0$

In Exercises 29–32: (a) Find an equation for the line through P parallel to L. (b) Then find an equation for the line through P perpendicular to L and the distance from P to L.

29. $P(6, 0), L: 2x - y = -2$

30. $P(3, 1), L: y = x + 2$

31. $P(4, -12), L: 4x + 3y = 12$

32. $P(0, 1), L: y = -\sqrt{3}x - 3$

Sketch a complete graph of the equations in Exercises 33–40. Give the domain and range. Support your answer with a grapher.

33. $y = 2x - 3$ 34. $y = |x| - 2$

35. $y = 2|x - 1| - 1$ 36. $y = \sec x$

37. $y = \cos x$ 38. $y = [x]$

39. $x = -y^2$ 40. $y = -2 + \sqrt{1 - x}$

Find the domain and range and draw a complete graph of the functions in Exercises 41–46.

41. $f(x) = x^3 + 8x^2 + x - 37$

42. $f(x) = -1 + \sqrt[3]{1 - x}$

43. $f(x) = \log_7(x - 1) + 1$

44. $f(x) = 3^{2-x} + 1$

45. $f(x) = |x - 2| + |x + 3|$

46. $f(x) = \dfrac{|x - 2|}{x - 2}$

In Exercises 47–50, describe how the graph of f can be obtained from the graph of g.

47. $f(x) = -2(x - 1)^3 + 5, g(x) = x^3$

48. $f(x) = 2\ln(-x + 1) + 3, g(x) = \ln x$

49. $f(x) = 3\sin(3x + \pi), g(x) = \sin x$

50. $f(x) = -2\sqrt[4]{3 - x} + 5, g(x) = \sqrt[4]{x}$

Exercises 51–54 specify the order in which transformations are to be applied to the graph of the given equation. Give an equation for the transformed graph in each case.

51. $y = x^2$, vertical stretch by 2, reflect through x-axis, shift right 2, shift up 3.

52. $x = y^2$, horizontal shrink by 0.5, reflect through y-axis, shift left 3, shift down 2

53. $y = \dfrac{1}{x}$, vertical stretch by 3, shift left 2, shift up 5

54. $x^2 + y^2 = 1$, shift left 3, shift up 5

A complete graph of f is shown in Figure 1.150. Sketch a complete graph of the functions in Exercises 55–58.

55. $y = f(-x)$

56. $y = -f(x)$

57. $y = -2f(x + 1) + 1$

58. $y = 3f(x - 2) - 2$

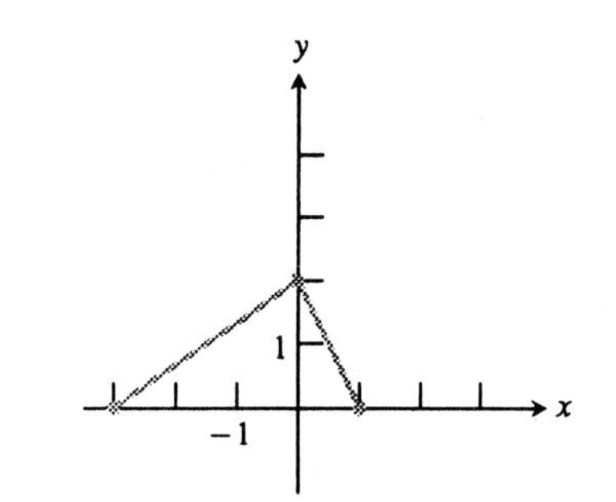

1.150 The graph of f for Exercises 55–58.

Determine the vertex and axis of symmetry and sketch a complete graph of the parabolas in Exercises 59–60. Support your work with a grapher.

59. $y = -x^2 + 4x - 1$ 60. $x = 2y^2 + 8y + 3$

In Exercises 61–66, say whether the functions are even, odd, or neither.

61. a) $y = \cos x$ b) $y = -\cos x$ c) $y = 1 - \cos x$

62. a) $y = \sin x$ b) $y = -\sin x$ c) $y = 1 - \sin x$

63. a) $y = x^2 + 1$ b) $y = x$ c) $y = x(x^2 + 1)$

64. a) $y = x^3$ b) $y = -x$ c) $y = -x^4$

65. a) $y = \sec x$ b) $y = \tan x$ c) $y = \sec x \tan x$

66. a) $y = \csc x$ b) $y = \cot x$ c) $y = \csc x \cot x$

67. Graph the function $y = x - \lfloor x \rfloor$. Is the function periodic? If so, what is its smallest period?

68. Graph the function $y = \lceil x \rceil - \lfloor x \rfloor$. Is the function periodic? If so, what is its smallest period?

Graph the function in Exercises 69–72.

69. $y = \begin{cases} \sqrt{-x}, & -4 \le x \le 0 \\ \sqrt{x}, & 0 \le x \le 4 \end{cases}$

70. $y = \begin{cases} -x - 2, & -2 \le x \le -1 \\ x, & -1 \le x \le 1 \\ -x + 2, & 1 \le x \le 2 \end{cases}$

71. $y = \begin{cases} \sin x, & 0 \le x \le 2\pi \\ 0, & 2\pi < x \end{cases}$

72. $y = \begin{cases} \cos x, & 0 \le x \le 2\pi \\ 0, & 2\pi < x \end{cases}$

Write "two-part" formulas for the functions graphed in Exercises 73 and 74.

73. **74.**

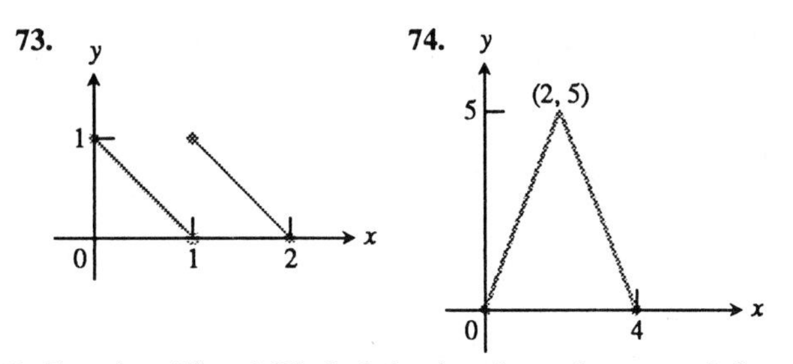

In Exercises 75 and 76, find the domains and ranges of f, g, $f+g$, $f \cdot g$, f/g, g/f. Also, find the domains and ranges of the composites $f \circ g$ and $g \circ f$.

75. $f(x) = \dfrac{1}{x}, g(x) = \dfrac{1}{\sqrt{x}}$

76. $f(x) = \sqrt{x}, g(x) = \sqrt{1-x}$

In Exercises 77–80, write an equation for the circle with the given center (h, k) and radius a.

77. $(h, k) = (1, 1), a = 1$ **78.** $(h, k) = (2, 0), a = 5$

79. $(h, k) = (2, -3), a = \dfrac{1}{2}$ **80.** $(h, k) = (-3, 0), a = 3$

Identify the centers and radii of the circles in Exercises 81–84.

81. $(x-3)^2 + (y+5)^2 = 16$

82. $x^2 + (y-5)^2 = 2$

83. $(x+1)^2 + (y-7)^2 = 121$

84. $(x+4)^2 + (y+1)^2 = 81$

Write inequalities to describe the regions in Exercises 85 and 86.

85. a) The interior of the circle of radius 1 centered at the origin

b) The region consisting of the circle plus its interior

86. a) The exterior of the circle of radius 2 centered at the point (1, 1)

b) The region consisting of the circle plus its exterior.

Solve the equations in Exercises 87–90 algebraically. Support your answer with a grapher.

87. $|x-1| = \dfrac{1}{2}$ **88.** $|2-3x| = 1$

89. $\left|\dfrac{2x}{5} + 1\right| = 7$ **90.** $\left|\dfrac{5-x}{2}\right| = 7$

Describe the intervals in Exercises 91–94 with inequalities that do not involve absolute values.

91. $|x+2| \leq \dfrac{1}{2}$ **92.** $|2x-7| \leq 3$

93. $\left|y - \dfrac{2}{5}\right| < \dfrac{3}{5}$ **94.** $\left|8 - \dfrac{y}{2}\right| < 1$

Solve the equations in Exercises 95–98.

95. $x^3 - 7x^2 + 12x - 2 = 0$

96. $4x^3 - 10x^2 + 9 = 0$

97. $2 + \log_3(x-2) + \log_3(3-x) = 0$

98. $\sin x = -0.7$

Solve the inequalities in Exercises 99–102 algebraically and support your answer graphically.

99. $|1-2x| < 3$ **100.** $\left|\dfrac{2x-1}{5}\right| \leq 1$

101. $\left|\dfrac{3}{x-2}\right| < 1$ **102.** $|2-3x| > 1$

Solve the inequalities in Exercises 103 and 104.

103. $x^3 - 7x^2 + 12x < 2$ (See Exercise 95.) **104.** $4x^3 - 10x^2 + 9 \geq 0$ (See Exercise 96.)

105. Change from degrees to radians:
a) 30° b) 22° c) −130° d) −150°

106. Change from radians to degrees:
a) $\dfrac{3\pi}{2}$ b) −0.9 c) 2.75 d) $-\dfrac{5\pi}{4}$

Find the sine, cosine, tangent, cotangent, secant, and cosecant of the angles in Exercises 107 and 108. The angles are given in radian measure.

107. a) 1.1 b) −1.1 c) $\dfrac{2\pi}{3}$ d) $-\dfrac{2\pi}{3}$

108. a) $\dfrac{\pi}{4}$ b) $-\dfrac{\pi}{4}$ c) 2.7 d) −2.7

109. Graph the following functions side by side over the interval $0 \leq x \leq 2\pi$.
a) $y = \cos 2x$ b) $y = 1 + \cos 2x$ c) $y = \cos^2 x$

110. Graph the following functions side by side over the interval $0 \leq x \leq 2\pi$.
a) $y = \cos 2x$ b) $y = -\cos 2x$
c) $y = 1 - \cos 2x$ d) $y = \sin^2 x$

111. Find $\cos^2 \dfrac{\pi}{6}$

a) by finding $\cos \dfrac{\pi}{6}$ and squaring.

b) by using a double-angle formula.

112. Find $\sin^2 \dfrac{\pi}{4}$

a) by finding $\sin \dfrac{\pi}{4}$ and squaring.

b) by using a double-angle formula.

In Exercises 113 and 114, find $f^{-1}(x)$ and show that $f \circ f^{-1}(x) = f^{-1} \circ f(x) = x$. Draw a complete graph of f and f^{-1} in the same viewing rectangle.

113. $f(x) = 2 - 3x$ **114.** $f(x) = (x+2)^2, x \geq -2$

Draw a graph of the inverse relations of the functions in Exercises 115 and 116. Is the inverse relation a function?

115. $y = x^3 - x$ **116.** $y = \dfrac{x+2}{x-1}$

Give the measure of the angles in Exercises 117 and 118 in radians and degrees.

117. $\sin^{-1}(0.7)$ **118.** $\tan^{-1}(-2.3)$

Draw a complete graph of the function in Exercises 119–122. Explain why your graph is complete.

119. $y = |\cos x|$ **120.** $y = \dfrac{\cos x + |\cos x|}{2}$

121. $y = \dfrac{|\cos x| - \cos x}{2}$ **122.** $y = \dfrac{\cos x - |\cos x|}{2}$

123. A 100-in. piece of wire is cut into two pieces. Each piece of wire is used to make a square wire frame. Let x be the length of one piece of the wire.

a) Determine an algebraic representation $A(x)$ for the total area of the two squares.

b) Draw a complete graph of the algebraic representation $y = A(x)$.

c) What are the domain and range of the algebraic representation $A(x)$?

d) What values of x make sense in the problem situation? Which portion of the graph in (b) is a graph of the problem situation?

e) Use a graph to determine the length of the two pieces of wire if the total area is 400 in^2. Confirm your answer algebraically.

f) Use a graph to determine the length of the two pieces of wire if the total area is a maximum.

124. Equal squares of side length x are removed from each corner of a 20- by 30-in. piece of cardboard, and the sides are turned up to form a box with no top.

a) Write the volume V of the box as a function of x.

b) Draw a complete graph of the algebraic representation $y = V(x)$.

c) What are the domain and range of the algebraic representation $y = V(x)$?

d) What values of x make sense in the problem situation? Draw a graph of the problem situation?

e) Use zoom-in to determine x so that the box has maximum volume. What is the maximum volume?

f) Use a graph to determine the dimensions of a box with volume 750 in^3.

125. Draw a complete graph of the equation $|x| + |y| = 1$.

CHAPTER

2

Limits and Continuity

OVERVIEW This chapter shows how limits of function values are defined and calculated.

Calculus is built on the concept of limit. We will investigate limits graphically and numerically. The rules for calculating limits are straightforward, and most of the limits we need can be found using one or more of direct substitution, graphing, calculator approximation, or algebra. Proving that the calculation rules always work or that approximations are accurate, however, is a more subtle affair that requires a formal definition of limit. We present this definition in Section 2.6 and show there how it is used to justify the rules.

One of the most important uses of limits in calculus is to test functions for continuity. Continuous functions are widely used in science because they serve to model an enormous range of natural behavior. In Section 2.2, we will see what makes continuous functions special. We shall work mainly with continuous functions in this book.

2.1 Limits

If you invest \$100 at a fixed annual rate of 6% compounded k times a year, the interest is added to your account k times a year. If $k = 4$, the interest is compounded quarterly; if $k = 12$, the interest is compounded monthly, and so forth. The amount of interest added at the end of the first period ($1/k$th of a year) is $(0.06/k)(100)$, and the balance in your account is $S_1 = 100 + (0.06/k)(100) = 100\left(1 + 0.06/k\right)$. The amount of interest added at the end of the second period is $(0.06/k)S_1$, and the balance in your account after two periods is $S_2 = S_1 + (0.06/k)S_1 = 100\left(1 + 0.06/k\right)^2$. It can be shown using mathematical induction that the amount in your account at the end of the year is

$$S = 100\left(1 + \frac{0.06}{k}\right)^k$$

if the account earns interest at the fixed rate of 6% compounded k times a year.

For example, if $k = 2$ (compounding semi-annually), then $S = \$106.09$, and if $k = 12$ (compounding monthly), then $S = \$106.17$. It appears that the amount in your account increases as the number of compounding periods increases, giving the impression that more frequent compounding will continue to increase the amount in the account. In fact, some banks advertise paying interest compounded daily or even compounded "continuously." However, there is a limit to how much you will earn, as graphing the compounded interest model demonstrates.

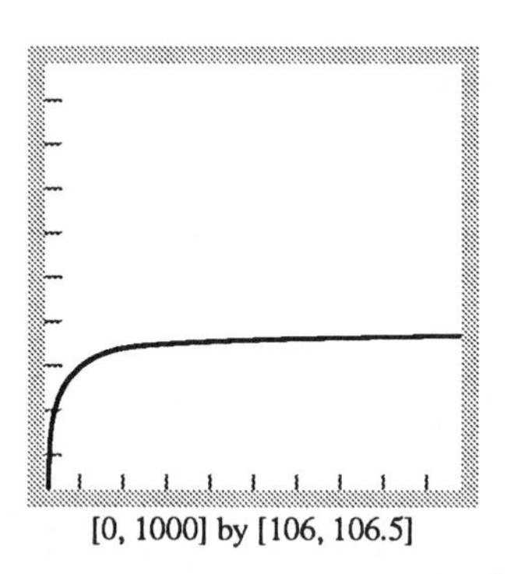

[0, 1000] by [106, 106.5]

2.1 Notice $S = 100\left(1 + 0.06/k\right)^k$ increases as k increases, but less rapidly as k increases.

The graph of the amount S in the account one year later as a function of the number of compounding periods k in Fig. 2.1 suggests that initially ($k \leq 5$) the values of S increase rapidly because the graph is very steep vertically. Then ($k > 5$) the values of S increase less rapidly because the graph is nearly horizontal. The graph of S in Fig. 2.2 appears to be horizontal suggesting that there is a *limit* (about 106.18) to the amount in the account no matter how large k is chosen.

The ideas of *rate of increase* and *limit* of this model will be further explored using the concepts of calculus in Chapter 7—in fact, we will prove the remarkable fact that the amount of the \$100 investment will never exceed $100e^{0.06}$ dollars no matter how large k becomes.

We live in a world of limits. There is a limit to how far a cannon can fire a shell. There is a limit to how fast we can brake for a red light. There is a limit to how much weight we can lift. There is a limit to how long a battery will last. And this is where the calculus comes in. Most of the limits that interest us can be viewed as numerical limits to values of functions. And, as you will see in this and subsequent chapters, graphing is the efficient way to find the limiting values of functions.

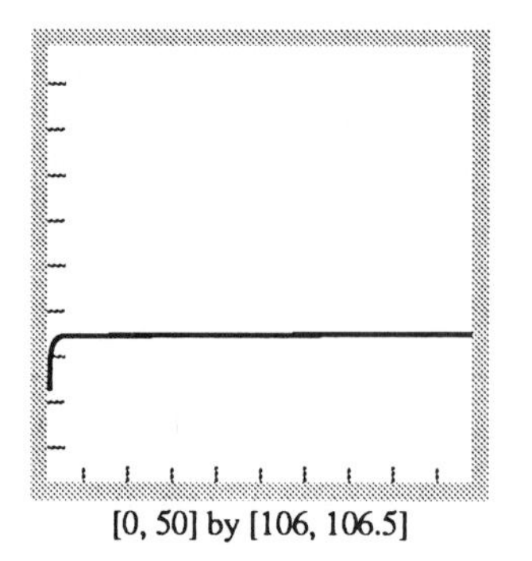

[0, 50] by [106, 106.5]

2.2 As the values of k get very large, the values of $S = 100\left(1 + 0.06/k\right)^k$ approach a limit.

Examples of Limits

To start off on a slightly different track, one of the important things to know about a function f is how its outputs will change when the inputs change. If the inputs get closer and closer to some specific value c, for example, will the outputs get closer to some specific value L? If they do, we want to know that, because it means we can control the outputs by controlling the inputs.

To talk sensibly about this, we need the language of limits. We develop this in two stages. First, we define limit informally, look at examples, and learn the calculation rules that have been discovered over the years to be the most useful ones to know. Then, in Section 2.6, when we have worked with enough examples for the formal definition to make sense, we define limit more precisely and examine the mathematics behind our calculations.

Definition

Informal Definition of Limit

If the values of a function f of x approach the value L as x approaches c, we say f has **limit** L as x approaches c and we write

$$\lim_{x \to c} f(x) = L,$$

(Read "The limit of f of x as x approaches c equals L".)

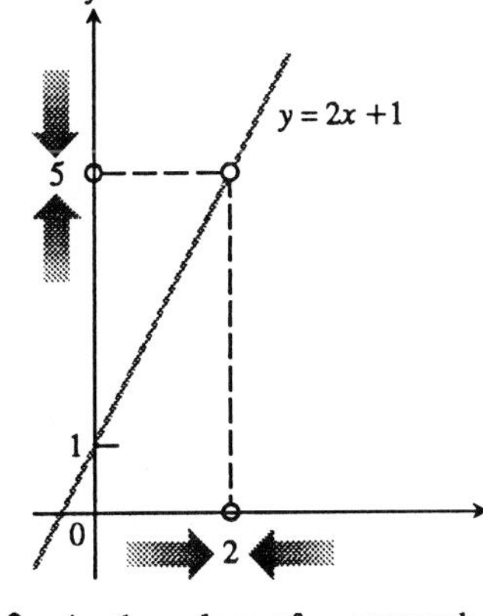

2.3 As the values of x approach 2, the values of $y = 2x + 1$ approach 5.

EXAMPLE 1 As Table 2.1 and Fig. 2.3 suggest,

$$\lim_{x \to 2} (2x + 1) = 5$$

TABLE 2.1 Values of $f(x) = 2x + 1$ as $x \to 2$

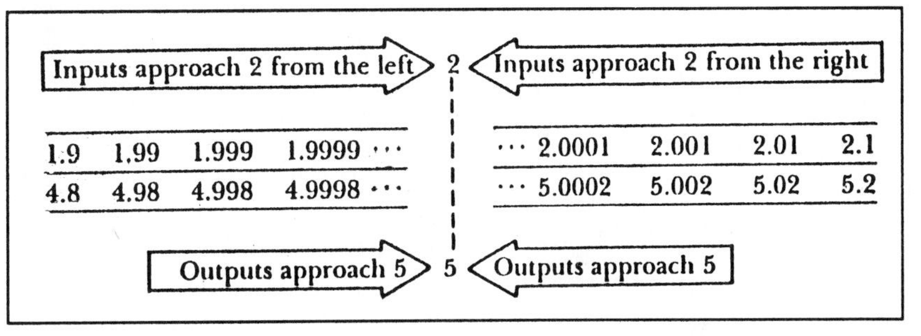

Inputs approach 2 from the left				2	Inputs approach 2 from the right			
1.9	1.99	1.999	1.9999 ···		··· 2.0001	2.001	2.01	2.1
4.8	4.98	4.998	4.9998 ···		··· 5.0002	5.002	5.02	5.2
Outputs approach 5				5	Outputs approach 5			

It turns out that many limits are easy to determine. For example, the limit in Example 1 can be found by *substitution*, that is, $\lim_{x \to 2}(2x + 1) = f(2)$ where $f(x) = 2x + 1$. Unfortunately, some limits we need to compute cannot be found by substitution as Example 2 illustrates.

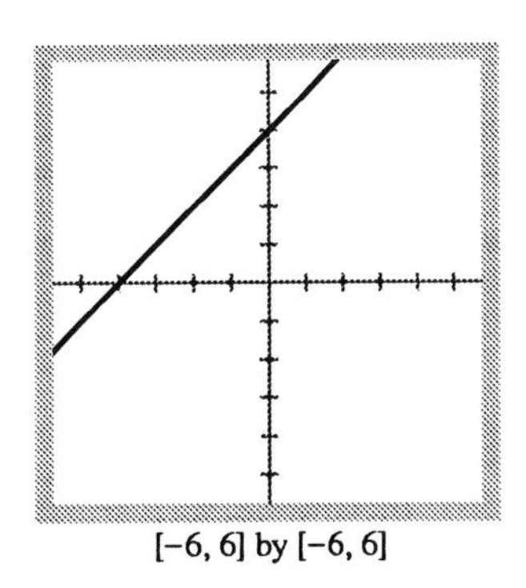

[−6, 6] by [−6, 6]

2.4 Notice that $f(x) = [(x+2)^2 - 4]/x \to 4$ as $x \to 0$ even though $f(0)$ is not defined.

EXAMPLE 2 Determine $\displaystyle\lim_{x \to 0} \frac{(x+2)^2 - 4}{x}$.

Solution To analyze this limit we only need concern ourselves with the values of the fraction for values of x different from 0. If we tried to substitute 0 for x we might conclude the limit does not exist because $0/0$ is undefined. However, this limit *does exist* as suggested by the graph of $f(x) = [(x+2)^2 - 4]/x$ in Fig. 2.4. Indeed, this graph suggests that $\lim_{x \to 0} [(x + 2)^2 - 4]/x = 4$.

This limit can be confirmed algebraically as follows. Because $x \neq 0$,

$$\frac{(x+2)^2-4}{x} = \frac{x^2+4x+4-4}{x} = \frac{x^2+4x}{x} = \frac{x(x+4)}{x} = x+4$$

Thus, $\lim_{x\to 0}[(x+2)^2-4]/x = \lim_{x\to 0}(x+4) = 4$.

Example 2 illustrates an important point about limits: The limit of a function $f(x)$ as x approaches c *never* depends on what happens when $x = c$. The limit, if it exists at all, is entirely determined by the values that f has when $x \neq c$. In Example 2, the quotient $[(x+2)^2-4]/4$ is not even defined at $x = 0$. Yet it has a limit as x approaches 0 and this limit is 4.

EXAMPLE 3 If c is any number, $\lim_{x\to c} 2x^2 = 2c^2$.

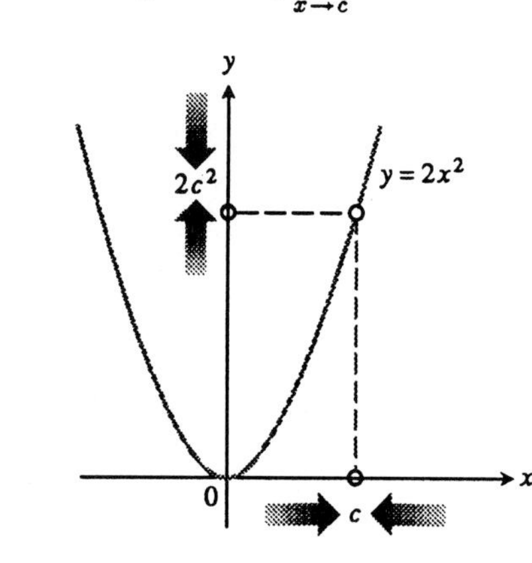

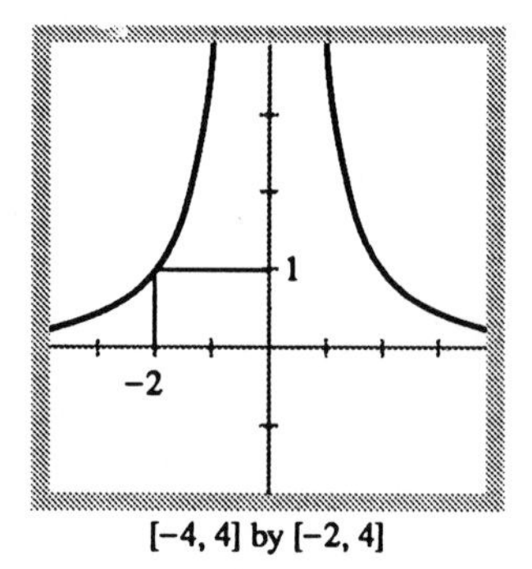

[-4, 4] by [-2, 4]

2.5 As the values of x approach -2, the values of $f(x) = 4/x^2$ approach 1.

EXAMPLE 4 $\lim_{x\to -2} \frac{4}{x^2} = \frac{4}{(-2)^2} = 1.$

The graph of $y = 4/x^2$ in Fig. 2.5 suggests $\lim_{x\to -2} 4/x^2 = 1$.

We could continue to make a huge list of functions and their limits, but we would be missing the boat if we did, because there is a more constructive (and much easier) way to proceed. For instance, once we know the limits for two particular functions as x approaches some value c, we automatically know the limit of their sum—it is the sum of their limits. Similarly, the limit of the difference of two functions is the difference of their limits, and so on, as described in the following theorem.

Properties of Limits

Theorem 1

Properties of Limits

a) If f is the *identity function* $f(x) = x$, then for any value of c

$$\lim_{x \to c} f(x) = \lim_{x \to c} (x) = c.$$

b) If f is the *constant function* $f(x) = k$ (the function whose outputs have the constant value k), then for any value of c

$$\lim_{x \to c} f(x) = \lim_{x \to c} (k) = k.$$

c) If $\lim_{x \to c} f_1(x) = L_1$ and $\lim_{x \to c} f_2(x) = L_2$, then

1. *Sum Rule:* $\lim [f_1(x) + f_2(x)] = L_1 + L_2$

2. *Difference Rule:* $\lim [f_1(x) - f_2(x)] = L_1 - L_2$

3. *Product Rule:* $\lim f_1(x) \cdot f_2(x) = L_1 \cdot L_2$

4. *Constant Multiple Rule:* $\lim k \cdot f_2(x) = k \cdot L_2$ (any number k)

5. *Quotient Rule:* $\lim \dfrac{f_1(x)}{f_2(x)} = \dfrac{L_1}{L_2}$ if $L_2 \neq 0$.

The limits are all taken as $x \to c$, and L_1 and L_2 are real numbers.

In words, the formulas in Theorem 1(c) say:

1. The limit of the sum of two functions is the sum of their limits.
2. The limit of the difference of two functions is the difference of their limits.
3. The limit of a product of two functions is the product of their limits.
4. The limit of a constant times a function is the constant times the limit of the function.
5. The limit of a quotient of two functions is the quotient of their limits, provided the denominator does not tend to zero.

We have included a formal proof of Theorem 1 in Appendix 2. Informally, we can paraphrase the theorem in terms that make it highly reasonable: When x is close to c, $f_1(x)$ is close to L_1 and $f_2(x)$ is close to L_2. Then we naturally think that $f_1(x) + f_2(x)$ is close to $L_1 + L_2$; $f_1(x) - f_2(x)$ is close to $L_1 - L_2$; $f_1(x)f_2(x)$ is close to $L_1 L_2$; $kf_2(x)$ is close to kL_2; and $f_1(x)/f_2(x)$ is close to L_1/L_2 if L_2 is not zero.

What keeps this discussion from being a proof is that the word *close* is vague. Phrases like *arbitrarily close to* and *sufficiently close to* might seem at first to improve the argument, but what are really needed are the formal definitions and arguments developed for the purpose by the great European mathematicians of the nineteenth century. You will see what we mean if you read Section 2.6.

In the meantime, here are some examples of what Theorem 1 can do for us.

EXAMPLE 5 We know from Theorem 1(a,b) that $\lim_{x\to c} x = c$ and $\lim_{x\to c} k = k$. The various parts of Theorem 1(c) now let us combine these results to calculate other limits:

a) $\lim_{x\to c} x^2 = \lim_{x\to c} x \cdot x = c \cdot c = c^2$ — Product

b) $\lim_{x\to c} (x^2+5) = \lim_{x\to c} x^2 + \lim_{x\to c} 5$ — Sum
$= c^2 + 5$ — from (a)

c) $\lim_{x\to c} 4x^2 = 4 \lim_{x\to c} x^2$ — Constant Multiple
$= 4c^2$ — from (a)

d) $\lim_{x\to c} (4x^2-3) = \lim_{x\to c} 4x^2 - \lim_{x\to c} 3$ — Difference
$= 4c^2 - 3$ — from (c)

e) $\lim_{x\to c} x^3 = \lim_{x\to c} x^2 \cdot x = c^2 \cdot c = c^3$ — Product and (a)

f) $\lim_{x\to c} (x^3+4x^2-3) = \lim_{x\to c} x^3 + \lim_{x\to c} (4x^2-3)$ — Sum
$= c^3 + 4c^2 - 3$ — (e) and (d)

g) $\lim_{x\to c} \dfrac{x^3+4x^2-3}{x^2+5} = \dfrac{\lim_{x\to c}(x^3+4x^2-3)}{\lim_{x\to c}(x^2+5)}$ — Quotient
$= \dfrac{c^3+4c^2-3}{c^2+5}.$ — (f) and (b) ■

Example 5 shows the remarkable strength of Theorem 1. From the two simple observations that $\lim_{x\to c} x = c$ and $\lim_{x\to c} k = k$ we can immediately work our way to limits of all polynomials and most **rational functions** (ratios of polynomials). As in Part (f) of Example 5, the limit of any polynomial $f(x)$ as x approaches c is $f(c)$, the number we get when we substitute $x = c$. As in Part (g) of Example 5, the limit of the ratio $f(x)/g(x)$ of two polynomials is $f(c)/g(c)$, provided $g(c)$ is different from 0.

Theorem 2

Limits of Polynomials Can Be Found by Substitution

If $f(x) = a_n x^n + a_{n-1}x^{n-1} + \cdots + a_0$ is any polynomial function, then

$$\lim_{x\to c} f(x) = f(c) = a_n c^n + a_{n-1}c^{n-1} + \cdots + a_0. \qquad (1)$$

Functions with the property $\lim_{x\to c} f(x) = f(c)$ are extremely useful in mathematics, as we shall see in Section 2.2.

Theorem 3

Limits of (Most But Not All) Rational Functions Can Be Found by Substitution

If $f(x)$ and $g(x)$ are polynomials, then

$$\lim_{x \to c} \frac{f(x)}{g(x)} = \frac{f(c)}{g(c)} \tag{2}$$

(provided $g(c) \neq 0$).

EXAMPLE 6

a) $\lim_{x\to 3} x^2(2-x) = \lim_{x\to 3}(2x^2 - x^3) = 2(3)^2 - (3)^3 = 18 - 27 = -9.$

b) Same limit, found another way:
$\lim_{x\to 3} x^2(2-x) = \lim_{x\to 3} x^2 \cdot \lim_{x\to 3}(2-x) = (3)^2 \cdot (2-3) = 9 \cdot (-1) = -9.$

EXAMPLE 7 $\displaystyle \lim_{x\to 2} \frac{x^2+2x+4}{x+2} = \frac{(2)^2+2(2)+4}{2+2} = \frac{12}{4} = 3.$

In Example 7, we can use Eq. (2) to find the limit of $f(x)/g(x)$ because the value of the denominator, $g(x) = x+2$, is different from zero when $x = 2$. However, many interesting limits cannot be found by substitution. In the next example, the denominator is zero when $x = 5$, so we cannot apply Eq. (2) directly. We can rewrite the fraction $f(x)/g(x)$ to find the limit in this case.

EXAMPLE 8

$\displaystyle \lim_{x\to 5} \frac{x^2-25}{3(x-5)}$ — Substitution will not give the limit because $x - 5 = 0$ when $x = 5$.

$\displaystyle = \lim_{x\to 5} \frac{(x+5)(x-5)}{3(x-5)}$ — We factor the numerator to see if $(x-5)$ is a factor. It is. We cancel the $(x-5)$'s, leaving . . .

$\displaystyle = \lim_{x\to 5} \frac{x+5}{3}$ — . . . an equivalent form whose limit we can now find by substitution.

$\displaystyle = \frac{5+5}{3} = \frac{10}{3}.$

In Example 8 we found the limit of a function as x approached a value at which the function was undefined. However, functions need not have a limit at such points as the next example illustrates. More examples of this type of behavior are given in Section 2.4.

EXAMPLE 9 Use a graph to show that $\displaystyle \lim_{x\to 2} \frac{x^3-1}{x-2}$ does not exist.

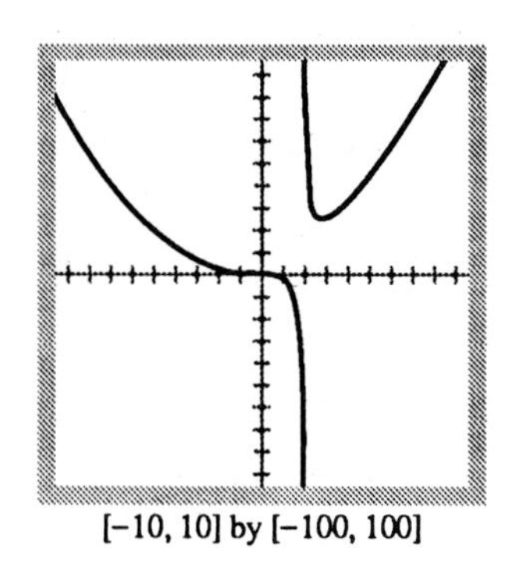

2.6 Notice that as $x \to 2$ the values of $f(x) = (x^3 - 1)/(x - 2)$ do not approach one number.

Solution Notice that the denominator is 0 when replaced by 2. Example 8 shows why this is not enough evidence to conclude that the limit does not exist. The graph in Fig. 2.6 of $f(x) = (x^3 - 1)/(x - 2)$ strongly suggests that as $x \to 2$ (x approaches 2) from either side, the function values do not get close to any one number. This, in turn, suggests that $\lim_{x\to 2}(x^3 - 1)/(x - 2)$ does not exist. ∎

It is also possible for a function to be defined at a value of x and still not have a limit as x approaches that value. In the next example, the function is defined at every value of x, but the function's limit as x approaches 2 is not the same as the function's value at $x = 2$.

EXAMPLE 10 If

$$f(x) = \begin{cases} x, & x \neq 2 \\ 3, & x = 2, \end{cases}$$

then

$$\lim_{x\to 2} f(x) = \lim_{x\to 2}(x) = 2 \quad \text{while} \quad f(2) = 3.$$

As always, the limit is determined by the function's approach behavior, not by what happens at $x = 2$ (Fig. 2.7). ∎

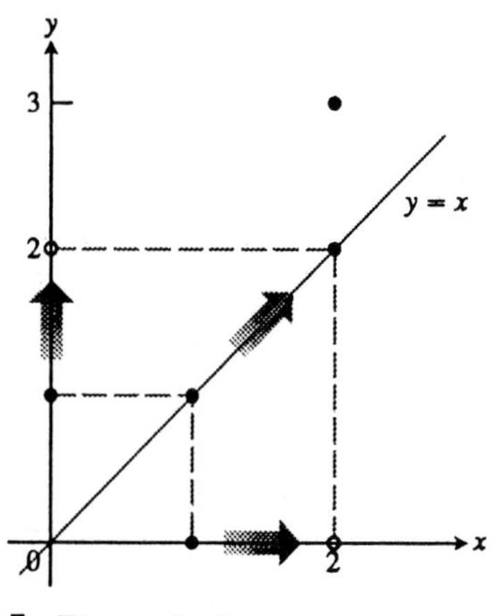

2.7 The graph of

$$f(x) = \begin{cases} x, & x \neq 2 \\ 3, & x = 2. \end{cases}$$

Notice that $f(x) \to 2$ as $x \to 2$ even though $f(2)$ itself is 3.

Right-hand Limits and Left-hand Limits

Sometimes the values of a function $f(x)$ tend to different limits as x approaches a number c from different sides. When this happens, we call the limit of f as x approaches c from the right the **right-hand limit** of f at c, and the limit as x approaches c from the left the **left-hand limit** of f at c.

The notation for the right-hand limit is

$$\lim_{x\to c+} f(x).$$

("the limit of f as x approaches c from the right")
The (+) is there to say that x approaches c through values above c on the number line.

The notation for the left-hand limit is

$$\lim_{x\to c^-} f(x).$$

("the limit of f as x approaches c from the left")
The (−) is there to say that x approaches c through values below c on the number line.

EXAMPLE 11 The greatest-integer function $f(x) = \lfloor x \rfloor$ has different right-hand and left-hand limits at each integer, as we can see in Fig. 2.8,

$$\lim_{x\to 3+} \lfloor x \rfloor = 3 \quad \text{and} \quad \lim_{x\to 3-} \lfloor x \rfloor = 2.$$

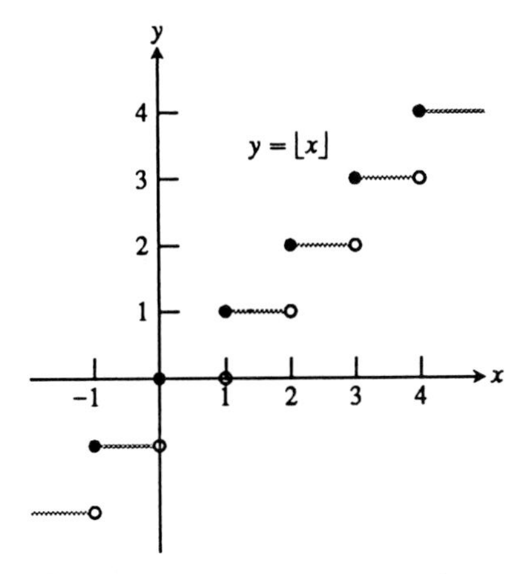

2.8 At each integer, the greatest integer function $y = \lfloor x \rfloor$ has different right-hand and left-hand limits.

The limit of $\lfloor x \rfloor$ as x approaches an integer n from above is n, while the limit as x approaches n from below is $n - 1$. ■

One-sided vs. Two-sided Limits

We sometimes call $\lim_{x \to c} f(x)$ the **two-sided** limit of f at c to distinguish it from the **one-sided** right-hand and left-hand limits of f at c. If the two one-sided limits of $f(x)$ exist at c and are equal, their common value is the two-sided limit of f at c. Conversely, if the two-sided limit of f at c exists, the two one-sided limits exist and have the same value as the two-sided limit.

Relationship between One-sided and Two-sided Limits

A function $f(x)$ has a limit as x approaches c if and only if the right-hand and left-hand limits at c exist and are equal. In symbols,

$$\lim_{x \to c} f(x) = L \quad \Leftrightarrow \quad \lim_{x \to c^+} f(x) = L \quad \text{and} \quad \lim_{x \to c^-} f(x) = L. \tag{3}$$

The implications in Eq. (3) will be proved formally in Section 2.6, but you can see what is going on if you look at the next examples.

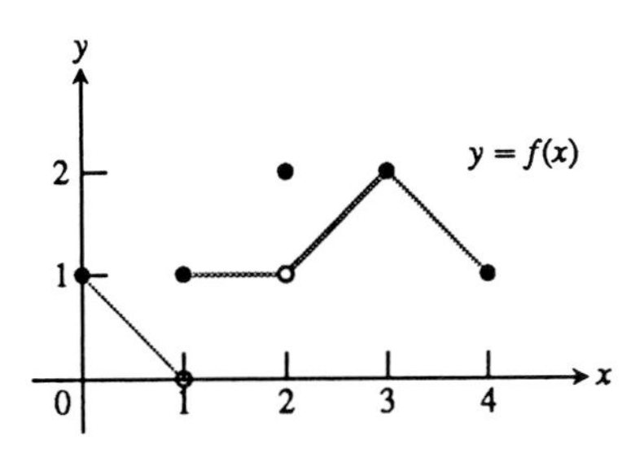

2.9 Example 12 discusses the limit properties of the function $y = f(x)$ graphed here.

EXAMPLE 12 All the following statements about the function $y = f(x)$ graphed in Fig. 2.9 are true.

At $x = 0$: $\lim_{x \to 0^+} f(x) = 1$.

At $x = 1$: $\lim_{x \to 1^-} f(x) = 0$ even though $f(1) = 1$,

$\lim_{x \to 1^+} f(x) = 1$,

$f(x)$ has no limit as $x \to 1$. (The right- and left-hand limits at 1 are not equal.)

At $x = 2$: $\lim_{x \to 2^-} f(x) = 1$,

$\lim_{x \to 2^+} f(x) = 1$,

$\lim_{x \to 2} f(x) = 1$ even though $f(2) = 2$.

At $x = 3$: $\lim_{x \to 3^-} f(x) = \lim_{x \to 3^+} f(x) = \lim_{x \to 3} f(x) = f(3) = 2$.

At $x = 4$: $\lim_{x \to 4^-} f(x) = 1$.

At every other point c between 0 and 4, $f(x)$ has a limit as $x \to c$. ■

EXAMPLE 13 The greatest-integer function $f(x) = \lfloor x \rfloor$ has no limit as x approaches 3. As in Example 11, $\lim_{x \to 3^+} \lfloor x \rfloor = 3$ while $\lim_{x \to 3^-} \lfloor x \rfloor = 2$.

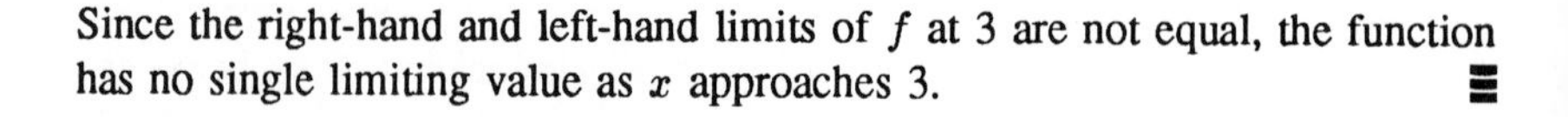

Since the right-hand and left-hand limits of f at 3 are not equal, the function has no single limiting value as x approaches 3. ∎

EXAMPLE 14 Show algebraically that $\lim_{x \to 0} |x| = 0$ and support graphically.

Solution We prove that $\lim_{x \to 0} |x| = 0$ by showing that the right-hand and left-hand limits are both 0:

$$\lim_{x \to 0+} |x| = \lim_{x \to 0+} x = 0. \qquad |x| = x \quad \text{if} \quad x > 0$$

$$\begin{aligned} \lim_{x \to 0-} |x| &= \lim_{x \to 0-} (-x) && |x| = -x \quad \text{if} \quad x < 0 \\ &= -\lim_{x \to 0-} x && \text{A limit property} \\ &= -0 \\ &= 0. \end{aligned}$$

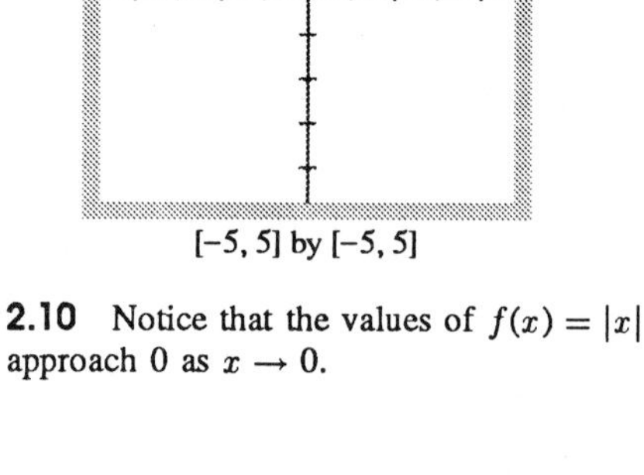

[−5, 5] by [−5, 5]

2.10 Notice that the values of $f(x) = |x|$ approach 0 as $x \to 0$.

The graph of $f(x) = |x|$ in Fig. 2.10 supports $\lim_{x \to 0} |x| = 0$. ∎

The graphs of functions we have seen so far in this section have given correct information about the behavior of functions that could be used to identify or support limits, or to conclude that limits did not exist. Care must be used in drawing conclusions from graphs produced by computers because sometimes these graphs can be misleading as the next example illustrates.

EXAMPLE 15 Investigate the behavior of $f(x) = \dfrac{1 - \cos x^6}{x^{12}}$ as $x \to 0$.

Solution The graph of f in Fig. 2.11 suggests that there is a gap near $x = 0$. If we zoom-in around $x = 0$ we will find that the graph appears to be erratic near $x = 0$. This is *not* true. In Chapter 7 we will see that $\lim_{x \to 0} f(x) = 0.5$. ∎

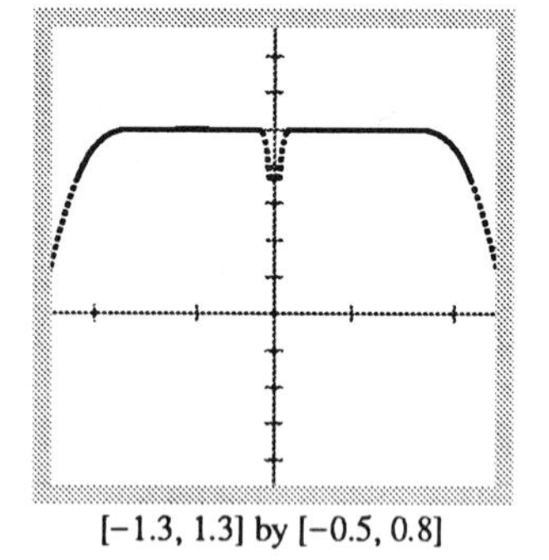

[−1.3, 1.3] by [−0.5, 0.8]

2.11 This graph of $f(x) = (1 - \cos x^6)/x^{12}$ *incorrectly* suggests that there is a gap near $x = 0$.

The grapher used to produce Fig. 2.11 does not have enough precision to produce a correct graph near $x = 0$. Graphers are very useful in mathematics, but cannot be used blindly. Modern technology used in conjunction with the concepts of calculus gives today's students powerful tools to study mathematics.

Exercises 2.1

Find the limits in Exercises 1–14 by substitution and support with a grapher.

1. $\lim_{x \to 2} 2x$
2. $\lim_{x \to 0} 2x$
3. $\lim_{x \to 1} (3x - 1)$
4. $\lim_{x \to 1/3} (3x - 1)$
5. $\lim_{x \to -1} 3x(2x - 1)$
6. $\lim_{x \to -1} 3x^2(2x - 1)$

7. $\lim_{x \to -2} (x+3)^{171}$

8. $\lim_{x \to -4} (x+3)^{1984}$

9. $\lim_{x \to 1} (x^3 + 3x^2 - 2x - 17)$

10. $\lim_{x \to -2} (x^3 - 2x^2 + 4x + 8)$

11. $\lim_{x \to -1} \dfrac{x+3}{x^2+3x+1}$

12. $\lim_{y \to 2} \dfrac{y^2+5y+6}{y+2}$

13. $\lim_{y \to -3} \dfrac{y^2+4y+3}{y^2-3}$

14. $\lim_{x \to -1} \dfrac{x^3-5x+7}{-x^3+x^2-x+1}$

Explain why substitutions will not work to find the limits in Exercises 15–18. Find the limits. Note: Some limits may not exist.

15. $\lim_{x \to -2} \sqrt{x-2}$

16. $\lim_{x \to 0} \dfrac{|x|}{x}$

17. $\lim_{x \to 0} \dfrac{1}{x^2}$

18. $\lim_{x \to 0} \dfrac{(3+x)^2-9}{x}$

Use a grapher to determine the limits in Exercises 19-26 and then confirm algebraically.

19. $\lim_{x \to 1} \dfrac{x-1}{x^2-1}$

20. $\lim_{x \to -5} \dfrac{x^2+3x-10}{x+5}$

21. $\lim_{t \to 1} \dfrac{t^2-3t+2}{t^2-1}$

22. $\lim_{t \to 2} \dfrac{t^2-3t+2}{t^2-4}$

23. $\lim_{x \to 2} \dfrac{2x-4}{x^3-2x^2}$

24. $\lim_{x \to 0} \dfrac{5x^3+8x^2}{3x^4-16x^2}$

25. $\lim_{x \to 0} \dfrac{\frac{1}{2+x}-\frac{1}{2}}{x}$

26. $\lim_{x \to 0} \dfrac{(2+x)^3-2}{x}$

27. a) Graph $f(x) = 100\left(1+0.06/x\right)^x$ and $g(x) = 100e^{0.06}$ in the $[0, 1000]$ by $[106, 106.5]$ viewing window.
 b) Zoom-in around the point on both graphs where x is about 990 until you see they are not the same graphs.
 c) What is the difference between $f(990)$ and $g(990)$? How much money would you have to invest to have a "real" difference (see the beginning of Section 2.1)?

28. a) Graph $f(x) = 100\left(1+0.08/x\right)^x$ and $g(x) = 100e^{0.08}$ in the $[0, 1000]$ by $[108, 108.5]$ viewing window.
 b) Zoom-in around the point on both graphs where x is about 990 until you see they are not the same graphs.
 c) What is the difference between $f(990)$ and $g(990)$? How much money would you have to invest to have a "real" difference (see the beginning of Section 2.1)?

29. Graph $f(x) = [(x+2)^2 - 4]/x$. Magnify the graph around the point $(0, 4)$ as much as possible. What do you observe?

30. Graph $f(x) = [(x+1)^3 - 1]/x$. Magnify the graph around point $(0, 3)$ as much as possible. What do you observe?

31. Graph $f(x) = (x^3 - 1)/(x-2)$. Use tall skinny viewing windows to magnify the graph near $x = 2$. What do you notice about the function as x gets closer to 2 from values greater than 2? From values of x less than 2?

32. Graph $f(x) = \dfrac{1-x^3}{x-2}$. Use tall, skinny viewing windows to magnify the graph near $x = 2$. What do you notice about the function as x gets close to 2 from values greater than 2? From values of x less than 2?

33. Let $f(x) = \begin{cases} 3-x, & x < 2 \\ \dfrac{x}{2}+1, & x > 2 \end{cases}$
 a) Determine a complete graph of f.
 b) Find $\lim_{x \to 2+} f(x)$ and $\lim_{x \to 2-} f(x)$.
 c) Does $\lim_{x \to 2} f(x)$ exist? If so, what is it? If not, why not?

34. Let $f(x) = \begin{cases} 3-x, & x < 2 \\ 2, & x = 2 \\ \dfrac{x}{2}, & x > 2 \end{cases}$
 a) Determine a complete graph of f.
 b) Find $\lim_{x \to 2+} f(x)$ and $\lim_{x \to 2-} f(x)$.
 c) Does $\lim_{x \to 2} f(x)$ exist? If so, what is it? If not, why not?

35. Let $f(x) = \begin{cases} \dfrac{1}{x-1}, & x < 1 \\ x^3-2x+5, & x \geq 1 \end{cases}$
 a) Determine a complete graph of f.
 b) Find $\lim_{x \to 1+} f(x)$ and $\lim_{x \to 1-} f(x)$.
 c) Does $\lim_{x \to 1} f(x)$ exist? If so, what is it? If not, why not?

36. Let $f(x) = \begin{cases} \dfrac{1}{2-x}, & x < 2 \\ 5-x^2, & x \geq 2 \end{cases}$
 a) Determine a complete graph of f.
 b) Find $\lim_{x \to 2+} f(x)$ and $\lim_{x \to 2-} f(x)$.
 c) Does $\lim_{x \to 2} f(x)$ exist? If so, what is it? If not, why not?

37. Let $f(x) = \begin{cases} a-x^2, & x < 2 \\ x^2+5x-3, & x \geq 2 \end{cases}$
 For what values of a does $\lim_{x \to 2} f(x)$ exist?

38. Let $f(x) = \begin{cases} x^3-4x, & x < -1 \\ 2x+a, & x \geq -1 \end{cases}$
 For what value of a does $\lim_{x \to -1} f(x)$ exist?

39. Which of the following statements are true of the function $y = f(x)$ graphed here?

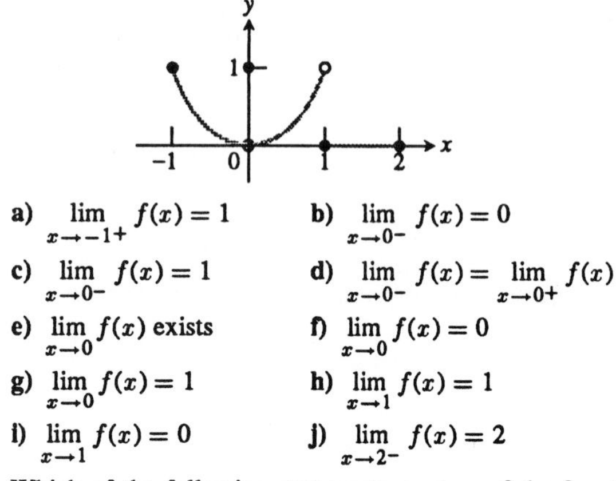

a) $\lim_{x\to -1+} f(x) = 1$

b) $\lim_{x\to 0-} f(x) = 0$

c) $\lim_{x\to 0-} f(x) = 1$

d) $\lim_{x\to 0-} f(x) = \lim_{x\to 0+} f(x)$

e) $\lim_{x\to 0} f(x)$ exists

f) $\lim_{x\to 0} f(x) = 0$

g) $\lim_{x\to 0} f(x) = 1$

h) $\lim_{x\to 1} f(x) = 1$

i) $\lim_{x\to 1} f(x) = 0$

j) $\lim_{x\to 2-} f(x) = 2$

40. Which of the following statements are true of the function graphed here?

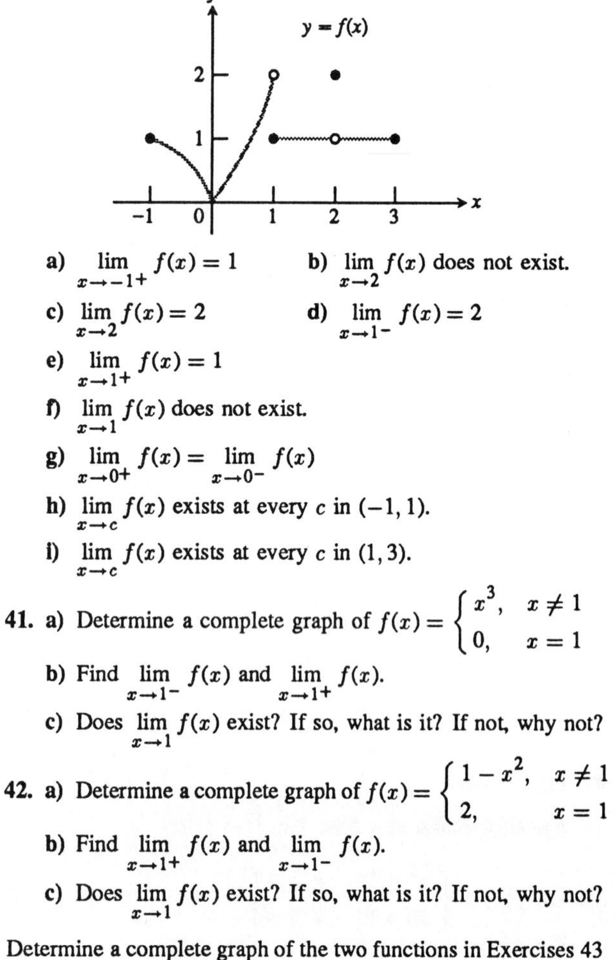

a) $\lim_{x\to -1+} f(x) = 1$

b) $\lim_{x\to 2} f(x)$ does not exist.

c) $\lim_{x\to 2} f(x) = 2$

d) $\lim_{x\to 1-} f(x) = 2$

e) $\lim_{x\to 1+} f(x) = 1$

f) $\lim_{x\to 1} f(x)$ does not exist.

g) $\lim_{x\to 0+} f(x) = \lim_{x\to 0-} f(x)$

h) $\lim_{x\to c} f(x)$ exists at every c in $(-1, 1)$.

i) $\lim_{x\to c} f(x)$ exists at every c in $(1, 3)$.

41. a) Determine a complete graph of $f(x) = \begin{cases} x^3, & x \neq 1 \\ 0, & x = 1 \end{cases}$

b) Find $\lim_{x\to 1-} f(x)$ and $\lim_{x\to 1+} f(x)$.

c) Does $\lim_{x\to 1} f(x)$ exist? If so, what is it? If not, why not?

42. a) Determine a complete graph of $f(x) = \begin{cases} 1 - x^2, & x \neq 1 \\ 2, & x = 1 \end{cases}$

b) Find $\lim_{x\to 1+} f(x)$ and $\lim_{x\to 1-} f(x)$.

c) Does $\lim_{x\to 1} f(x)$ exist? If so, what is it? If not, why not?

Determine a complete graph of the two functions in Exercises 43 and 44. Then answer these questions.

a) At what points c in the domain of f does $\lim_{x\to c} f(x)$ exist?

b) At what points does only the left-hand limit exist?

c) At what points does only the right-hand limit exist?

43. $f(x) = \begin{cases} \sqrt{1 - x^2} & \text{if } 0 \le x < 1, \\ 1 & \text{if } 1 \le x < 2, \\ 2 & \text{if } x = 2. \end{cases}$

44. $f(x) = \begin{cases} x & \text{if } -1 \le x < 0, \text{ or } 0 < x \le 1, \\ 1 & \text{if } x = 0, \\ 0 & \text{if } x < -1, \text{ or } x > 1. \end{cases}$

45. Let $f(x) = \begin{cases} 0, & x \le 0 \\ \sin \dfrac{1}{x}, & x > 0. \end{cases}$

a) Determine a complete graph of f.

b) Does $\lim_{x\to 0+} f(x)$ exist? If so, what is it?

c) Does $\lim_{x\to 0-} f(x)$ exist? If so, what is it?

d) Does $\lim_{x\to 0} f(x)$ exist? If so, what is it? If not, why not?

46. Let $f(x) = \begin{cases} 0, & x = 0 \\ x \sin \dfrac{1}{x}, & x \neq 0 \end{cases}$

a) Determine a complete graph of f.

b) Does $\lim_{x\to 0+} f(x)$ exist? If so, what is it?

c) Does $\lim_{x\to 0-} f(x)$ exist? If so, what is it?

d) Does $\lim_{x\to 0} f(x)$ exist? If so, what is it? If not, why not?

Find the limits of the greatest-integer function in Exercises 47–50.

47. $\lim_{x\to 0+} \lfloor x \rfloor$

48. $\lim_{x\to 0-} \lfloor x \rfloor$

49. $\lim_{x\to 0.5} \lfloor x \rfloor$

50. $\lim_{x\to 2-} \lfloor x \rfloor$

Find the limits in Exercises 51–52.

51. $\lim_{x\to 0+} \dfrac{x}{|x|}$

52. $\lim_{x\to 0-} \dfrac{x}{|x|}$

53. Suppose $\lim_{x\to c} f(x) = 5$ and $\lim_{x\to c} g(x) = 2$. Find

a) $\lim_{x\to c} f(x)g(x)$

b) $\lim_{x\to c} 2f(x)g(x)$

54. Suppose $\lim_{x\to 4} f(x) = 0$ and $\lim_{x\to 4} g(x) = 3$. Find

a) $\lim_{x\to 4} (g(x) + 3)$

b) $\lim_{x\to 4} x f(x)$

c) $\lim_{x\to 4} g^2(x)$

d) $\lim_{x\to 4} \dfrac{g(x)}{f(x) - 1}$

55. Suppose $\lim_{x\to b} f(x) = 7$ and $\lim_{x\to b} g(x) = -3$. Find

a) $\lim_{x\to b} (f(x) + g(x))$

b) $\lim_{x\to b} f(x) \cdot g(x)$

c) $\lim_{x\to b} 4g(x)$

d) $\lim_{x\to b} f(x)/g(x)$

56. Suppose $\lim_{x\to-2} p(x) = 4$, $\lim_{x\to-2} r(x) = 0$, and $\lim_{x\to-2} s(x) = -3$. Find

a) $\lim_{x\to-2} (p(x) + r(x) + s(x))$

b) $\lim_{x\to-2} p(x)\cdot r(x)\cdot s(x)$

Draw a graph and determine the limits in Exercises 57–66.

57. $\lim_{x\to0} x \sin x$

58. $\lim_{x\to0} \frac{1}{x}\sin x$

59. $\lim_{x\to0} x^2 \sin x$

60. $\lim_{x\to0} \frac{1}{x}\sin\frac{1}{x}$

61. $\lim_{x\to0} x^2 \sin\frac{1}{x}$

62. $\lim_{x\to0} (1+x)^{2/x}$

63. $\lim_{x\to0} (1+x)^{3/x}$

64. $\lim_{x\to0} (1+x)^{4/x}$

65. $\lim_{x\to1} \frac{\ln(x^2)}{\ln x}$

66. $\lim_{x\to0} \frac{\sin x}{x}$

2.2 Continuous Functions

Most graphers are designed to plot points and then connect the points with an unbroken curve (connected mode). In other words, outputs are assumed to vary continuously with the inputs and do not jump from one value to another without taking on all the values in between. For example, graphers will sometimes connect the two branches of the graph of $y = 1/x$ suggesting that the function is defined and continuous at $x = 0$. This is why some graphers allow the user to turn off the connecting feature (dot mode) in order to view the plotted points and decide if the function graphed is really continuous.

Continuous functions are the functions we normally use in the equations that describe numerical relations in the world around us. They are the functions we use to find a cannon's maximum range or a planet's closest approach to the sun. They are also the functions we use to describe how a body moves through space or how the speed of a chemical reaction changes with time. In fact, so many observable physical processes proceed continuously that throughout the eighteenth and nineteenth centuries it rarely occurred to anyone to look for any other kind of behavior. It came as quite a surprise when the physicists of the 1920s discovered that the vibrating atoms in a hydrogen molecule can oscillate only at discrete energy levels, that light comes in particles, and that, when heated, atoms emit light in discrete frequencies and not in continuous spectra.

As a result of these and other discoveries, and because of the heavy use of discrete functions in computer science and statistics, the issue of continuity has become one of practical as well as theoretical importance. As scientists, we need to know when continuity is called for, what it is, and how to test for it.

Continuous functions were involved in many of the examples and exercises of Section 2.1. It turns out that a function is continuous at a point c of its domain if the limit as $x \to c$ can be determined by substitution.

The Definition of Continuity

A function $y = f(x)$ that can be graphed over each interval of its domain with one continuous motion of the pen is an example of a **continuous function.** The height of the graph over the interval varies continuously with x. At each

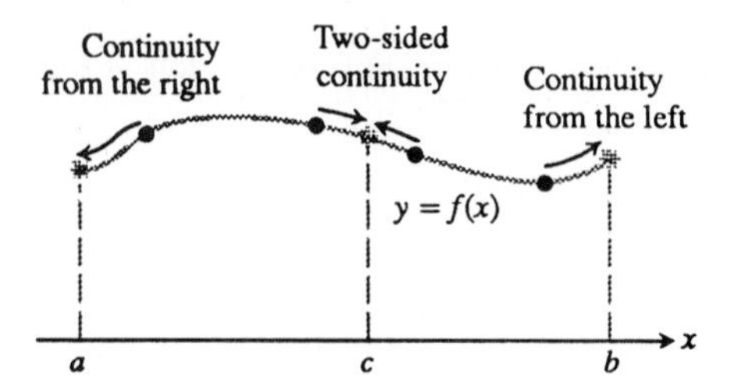

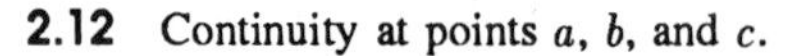
2.12 Continuity at points a, b, and c.

interior point of the function's domain, like the point c in Fig. 2.12, the function value $f(c)$ is the limit of the function values on either side; that is,

$$f(c) = \lim_{x \to c} f(x).$$

The function value at each endpoint is also the limit of the nearby function values. At the left endpoint a in Fig. 2.12,

$$f(a) = \lim_{x \to a+} f(x).$$

At the right endpoint b,

$$f(b) = \lim_{x \to b-} f(x).$$

To be specific, let us look at the function in Fig. 2.13, whose limits we investigated in Example 11 in Section 2.1.

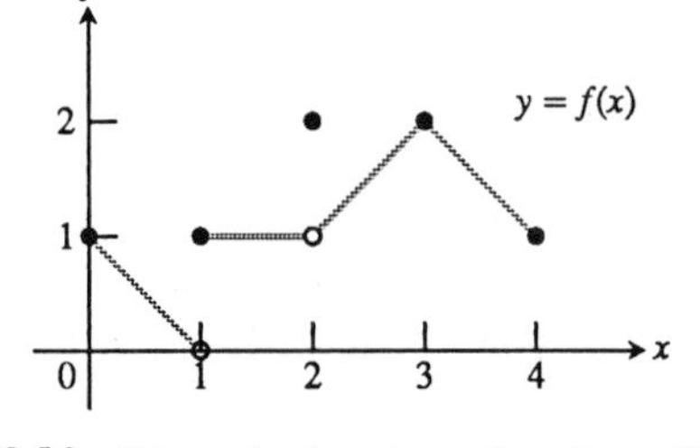

2.13 Discontinuity at $x = 1$ and $x = 2$.

EXAMPLE 1 The function in Fig. 2.13 is continuous at every point in its domain except $x = 1$ and $x = 2$. At these points there are breaks in the graph. Note the relation between the limit of f and the value of f at each point of the function's domain.

Points of discontinuity:

At $x = 1:$ $\lim_{x \to 1} f(x)$ does not exist.

At $x = 2:$ $\lim_{x \to 2} f(x) = 1$, but $1 \neq f(2)$.

Points at which f is continuous:

At $x = 0:$ $\lim_{x \to 0+} f(x) = f(0)$.

At $x = 4:$ $\lim_{x \to 4-} f(x) = f(4)$.

At every point $0 < c < 4$ except $x = 1, 2:$ $\lim_{x \to c} f(x) = f(c)$. ■

We now come to the formal definition of continuity at a point in a function's domain. In the definition we distinguish between continuity at an endpoint (which involves a one-sided limit) and continuity at an interior point (which involves a two-sided limit).

Definitions

Continuity at an Interior Point A function $y = f(x)$ is continuous at an interior point c of its domain if

$$\lim_{x \to c} f(x) = f(c). \qquad (1)$$

Continuity at an Endpoint A function $y = f(x)$ is continuous at a left endpoint a of its domain if

$$\lim_{x \to a^+} f(x) = f(a). \qquad (2)$$

A function $y = f(x)$ is continuous at a right endpoint b of its domain if

$$\lim_{x \to b^-} f(x) = f(b). \qquad (3)$$

Continuous Function A function is continuous if it is continuous at each point of its domain.

Discontinuity at a Point If a function f is not continuous at a point c, we say that f is discontinuous at c and call c a point of discontinuity of f.

How to Test for Continuity at a Point

To test for continuity at a point, we apply the following test.

The Continuity Test

The function $y = f(x)$ is continuous at $x = c$ if and only if *all three* of the following statements are true:

1. $f(c)$ exists (c lies in the domain of f).
2. $\lim_{x \to c} f(x)$ exists (f has a limit as $x \to c$).
3. $\lim_{x \to c} f(x) = f(c)$ (the limit equals the function value).

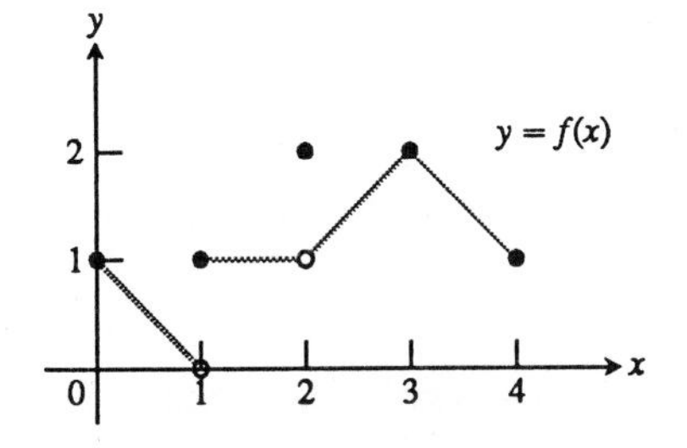

2.14 The function is continuous at $x = 0$, 3, and 4, and discontinuous at $x = 1$ and 2.

(The limit in the continuity test is to be two-sided if c is an interior point of the domain of f; it is to be the appropriate one-sided limit if c is an endpoint of the domain.)

EXAMPLE 2 When applied to the function $y = f(x)$ of Example 1 at the points $x = 0, 1, 2, 3,$ and 4, the continuity test gives the following results. (The graph of f is reproduced here as Fig. 2.14.)

a) f is continuous at $x = 0$ because

i) $f(0)$ exists (it equals 1),

ii) $\lim_{x \to 0+} f(x) = 1$ (f has a limit as $x \to 0^+$),

iii) $\lim_{x \to 0+} f(x) = f(0)$ (the limit equals the function value).

b) f is discontinuous at $x = 1$ because $\lim_{x \to 1} f(x)$ does not exist. The function fails part (2) of the test. (The right-hand and left-hand limits exist at $x = 1$, but they are not equal.)

c) f is discontinuous at $x = 2$ because $\lim_{x \to 2} f(x) \neq f(2)$. The function fails part (3) of the test. This type of discontinuity is called a *removable* discontinuity, because it is possible to redefine f at $x = 2$ that is, ($f(2) = 1$) to make f continuous at $x = 1$. Notice this is not possible at $x = 1$.

d) f is continuous at $x = 3$ because

i) $f(3)$ exists (it equals 2),

ii) $\lim_{x \to 3} f(x) = 2$ (f has a limit as $x \to 3$),

iii) $\lim_{x \to 3} f(x) = f(3)$ (the limit equals the function value).

e) f is continuous at $x = 4$ because

i) $f(4)$ exists (it equals 1),

ii) $\lim_{x \to 4^-} f(x) = 1$ (f has a limit as $x \to 4^-$),

iii) $\lim_{x \to 4^-} f(x) = f(4)$ (the limit equals the function value).

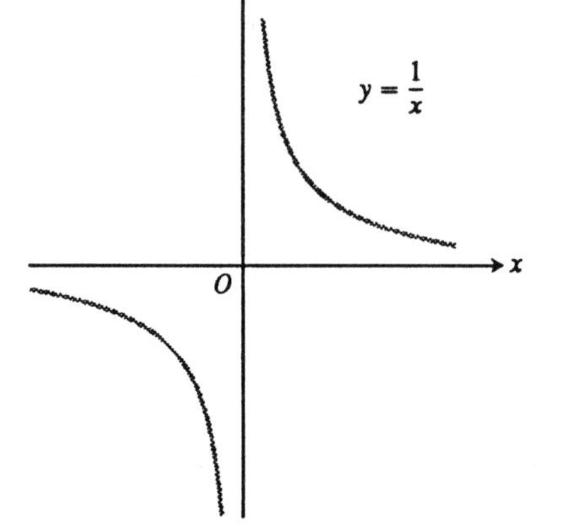

2.15 The function $y = 1/x$ is continuous at every point except $x = 0$.

EXAMPLE 3 The function $y = 1/x$ is continuous because it is continuous at every point of its domain. However, we believe it is much clearer to say that $y = 1/x$ is continuous at every value of x except $x = 0$. The function is not defined at $x = 0$ and therefore fails part (1) of the continuity test at $x = 0$. See Fig. 2.15.

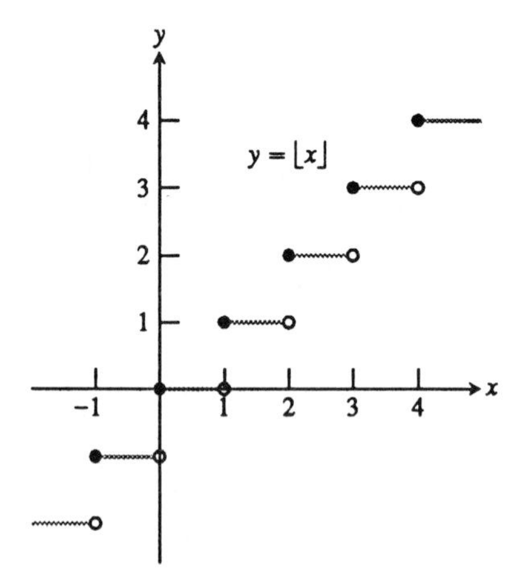

2.16 The greatest integer function $y = \lfloor x \rfloor$ is discontinuous at every integer (but continuous at every other point).

EXAMPLE 4 *The greatest integer function $y = \lfloor x \rfloor$ is discontinuous at every integer.* At every integer, the function fails to have a limit and so fails part (2) of the test (Fig. 2.16).

EXAMPLE 5 *Polynomials are continuous at every point.* In Section 2.1 we saw that $\lim_{x \to c} f(x) = f(c)$ for any polynomial $f(x)$.

EXAMPLE 6 *Rational functions are continuous wherever they are defined.* In Section 2.1 we saw that $\lim_{x \to c} f(x)/g(x) = f(c)/g(c)$ at every point where $g(c) \neq 0$.

EXAMPLE 7 *The transcendental functions like $y = a^x, y = \log_a x, y = \cos x$, and the radical functions $y = \sqrt[n]{x}$, where n is a positive integer greater than 1 are continuous.* These facts are suggested by exploring and magnifying their graphs, and will be proved in later chapters. ▪

EXAMPLE 8 The function $y = |x|$ is continuous at every value of x. It agrees with the continuous (polynomial) function $y = x$ if $x \geq 0$. It agrees with the continuous function $y = -x$ if $x < 0$. And, finally, $\lim_{x \to 0} |x| = 0$, as we saw in Example 14 in Section 2.1. ▪

Algebraic Combinations of Continuous Functions Are Continuous

As you may have guessed, algebraic combinations of continuous functions are continuous at every point at which they are defined. The relevant theorem is this:

Theorem 4

Algebraic Properties of Continuous Functions

If the functions f and g are continuous at $x = c$, then the following combinations are continuous at $x = c$:

1. Sums: $f + g$
2. Differences: $f - g$
3. Products: $f \cdot g$
4. Constant multiples: $k \cdot g$ (any number k)
5. Quotients: f/g (provided $g(c) \neq 0$)

Proof Theorem 4 is a special case of the Limit Property Theorem, Theorem 1 in Section 2.1. If the latter were restated for the functions f and g, it would say that if $\lim_{x \to c} f(x) = f(c)$ and $\lim_{x \to c} g(x) = g(c)$, then

1. $\lim_{x \to c} [f(x) + g(x)] = f(c) + g(c)$,
2. $\lim_{x \to c} [f(x) - g(x)] = f(c) - g(c)$,
3. $\lim_{x \to c} f(x)g(x) = f(c)g(c)$,

4. $\lim_{x \to c} kg(x) = kg(c)$ (any number k),

5. $\lim_{x \to c} \dfrac{f(x)}{g(x)} = \dfrac{f(c)}{g(c)}$ (provided $g(c) \neq 0$).

In other words, the limits of the functions in (1)–(5) as $x \to c$ exist and equal the function values at $x = c$. Therefore, each function fulfills the three requirements of the continuity test at any interior point $x = c$ of its domain. Similar arguments with right-hand and left-hand limits establish the theorem for continuity at endpoints.

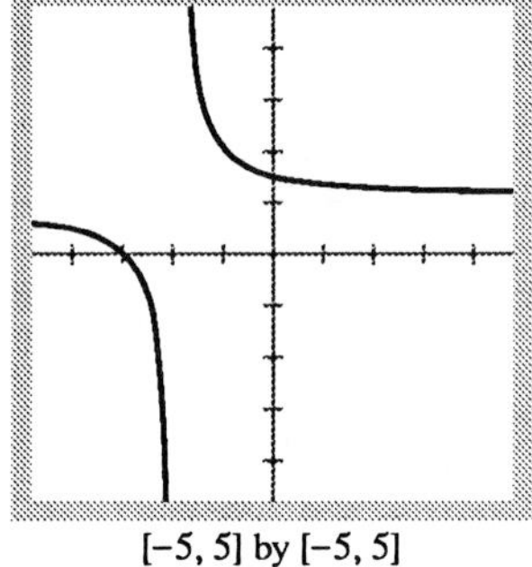

[−5, 5] by [−5, 5]

2.17 The graph of $f(x) = (x^2 + x - 6)/(x^2 - 4)$ suggests that the discontinuity at $x = 2$ is removable.

EXAMPLE 9 The functions

$$f(x) = x^{14} + 20x^4, \qquad g(x) = 5x(2 - x) + 1/(x^2 + 1)$$

are continuous at every value of x. The function

$$h(x) = \frac{x + 3}{x^2 - 3x - 10} = \frac{x + 3}{(x - 5)(x + 2)}$$

is continuous at every value of x except $x = 5$ and $x = -2$.

EXAMPLE 10 *A continuous extension.* The graph of $f(x) = (x^2 + x - 6)/(x^2 - 4)$ in Fig. 2.17 suggests that the discontinuity of f at $x = 2$ is removable. Is it possible to define $f(2)$ in a way that extends f to be continuous at $x = 2$? If so, what value should $f(2)$ have?

Solution Notice f has a discontinuity at $x = 2$ because the denominator is 0 when $x = 2$. However, the graph in Fig. 2.17 appears to be continuous at $x = 2$, that is, it appears that $\lim_{x \to 2} f(x)$ exists, suggesting that the discontinuity at $x = 2$ is removable. Compare the discontinuity at $x = 2$ with the discontinuity at $x = -2$ which is not removable. If the curve is magnified around $x = 2$, the graph will still appear to be continuous at $x = 2$. This is the nature of a removable discontinuity. It *cannot be seen!* Sometimes the hole will be visible on a grapher because of the discrete way graphs are plotted.

To make f continuous at $x = 2$, we need to define $f(2) = \lim_{x \to 2} f(x)$. Fig. 2.17 suggests that $f(2) = 1.25$, and we can confirm this algebraically.

$$\frac{x^2 + x - 6}{x^2 - 4} = \frac{(x - 2)(x + 3)}{(x - 2)(x + 2)}$$

Thus, $\displaystyle \lim_{x \to 2} \frac{x^2 + x - 6}{x^2 - 4} = \lim_{x \to 2} \frac{(x - 2)(x + 3)}{(x - 2)(x + 2)} = \lim_{x \to 2} \frac{x + 3}{x + 2} = \frac{5}{4}.$

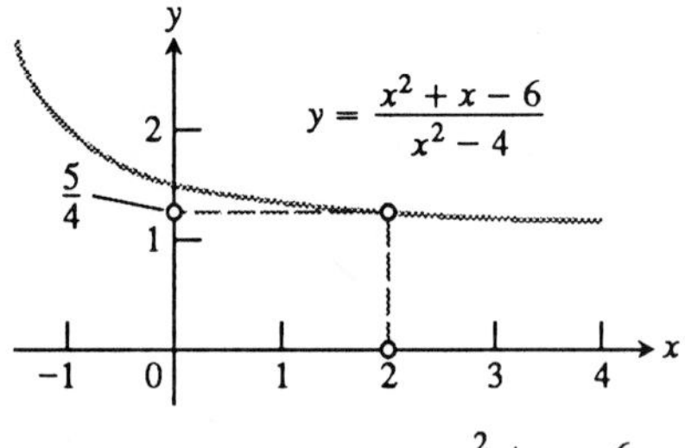

2.18 A graph of $f(x) = \dfrac{x^2 + x - 6}{x^2 - 4}$ illustrating the discontinuity at $x = 2$.

The sketch of f in Fig. 2.18 shows an open circle at $x = 2$ to illustrate that f is discontinuous at $x = 2$. You will usually not see this hole when you use a graphing utility.

The extended function

$$f(x) = \begin{cases} \dfrac{x^2 + x - 6}{x^2 - 4} & \text{if } x \neq 2, \\ \dfrac{5}{4} & \text{if } x = 2, \end{cases} \tag{4}$$

is continuous at $x = 2$ because $\lim_{x \to 2} f(x)$ exists and equals $f(2)$.

The function in Eq. (4) is called the **continuous extension** of the original function to the point $x = 2$. ∎

Explore with a Grapher

Find viewing rectangles in which the hole in the graph of the function f of Example 10 shows up.

Composites of Continuous Functions Are Continuous

All composites of continuous functions are continuous. This means that composites like

$$y = \sin(x^2) \quad \text{and} \quad y = |\cos x|$$

are continuous at every point at which they are defined. The idea is that if $f(x)$ is continuous at $x = c$ and $g(x)$ is continuous at $x = f(c)$, then $g \circ f$ is continuous at $x = c$ (Fig. 2.19).

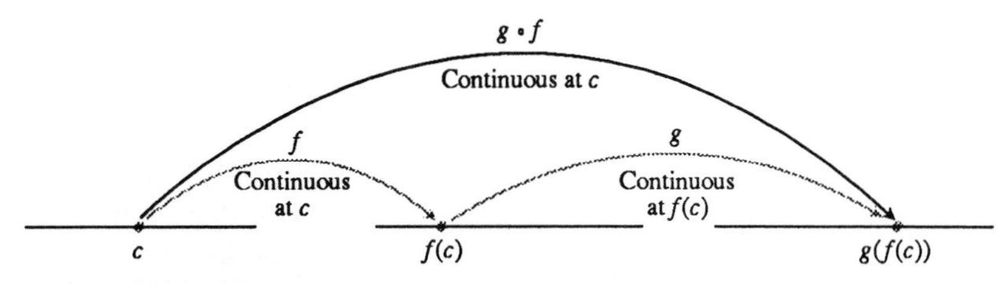

2.19 Composites of continuous functions are continuous.

Theorem 5

If f is continuous at c and g is continuous at $f(c)$, then the composite $g \circ f$ is continuous at c.

For an outline of the proof of Theorem 5 see Problem 6 in Appendix 2.

EXAMPLE 11 Show that the function

$$y = \left| \frac{x \sin x}{x^2 + 2} \right|$$

is continuous at every value of x.

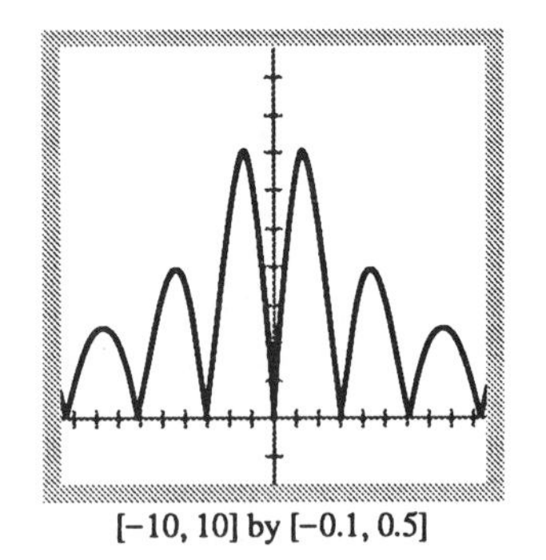

2.20 The graph suggests that $y = \left|(x \sin x)/(x^2 + 2)\right|$ is continuous.

Solution The graph of $y = \left|(x \sin x)/(x^2 + 2)\right|$ in Fig. 2.20 suggests that the function is continuous. We use Theorem 5 to prove this conjecture.

The function y is the composite of the continuous functions

$$f(x) = \frac{x \sin x}{x^2 + 2} \quad \text{and} \quad g(x) = |x|.$$

The function f is continuous by Theorem 4, the function g by Example 8, and their composite $g \circ f$ by Theorem 5.

If a composite function $g \circ f$ is continuous at a point $x = c$, its limit as $x \to c$ is $g(f(c))$. (See Fig. 2.19.)

EXAMPLE 12

a) $\lim_{x \to 1} \sin \sqrt{x - 1} = \sin \sqrt{1 - 1} = \sin 0 = 0$

b) $\lim_{x \to 0} |1 + \cos x| = |1 + \cos 0| = |1 + 1| = 2$

Continuous Functions Have Important Properties

We study continuous functions because they are useful in mathematics and its applications. It turns out that every continuous function is some other function's derivative, as we shall see in Chapter 5. The ability to recover a function from information about its derivative is one of the great powers given to us by calculus. Thus, given a formula $v(t)$ for the velocity of a moving body as a continuous function of time, we shall be able, with the calculus of Chapters 3, 4, and 5, to produce a formula $s(t)$ that tells how far the body has traveled from its starting point at any instant.

In addition, a function that is continuous at every point of a closed interval $[a, b]$ has an absolute maximum value and an absolute minimum value on this interval. We always look for these values when we graph a function, and we shall see the role they play in problem solving (Chapter 4) and in the development of the integral calculus (Chapters 5 and 6).

Finally, a function f that is continuous at every point of a closed interval $[a, b]$ assumes every value between $f(a)$ and $f(b)$. We shall see some consequences of this in a moment. The powerful technique of using zoom-in to solve equations and inequalities introduced in Chapter 1 depends on this important property of continuous functions.

The proofs of these properties require a detailed knowledge of the real number system and we shall not give them here.

Theorem 6

The Max-Min Theorem for Continuous Functions

If f is continuous at every point of a closed interval $[a, b]$, then f takes on both an absolute maximum value M and an absolute minimum value m somewhere in that interval. That is, for some numbers x_1 and x_2 in $[a, b]$ we have $f(x_1) = m$, $f(x_2) = M$, and $m \leq f(x) \leq M$ at every other point x of the interval.

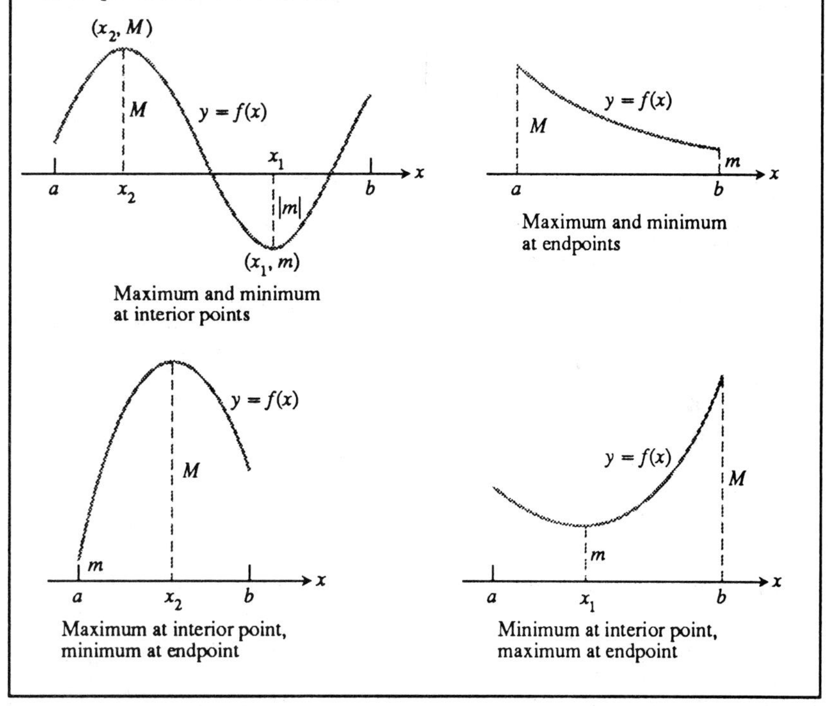

Maximum and minimum at interior points

Maximum and minimum at endpoints

Maximum at interior point, minimum at endpoint

Minimum at interior point, maximum at endpoint

EXAMPLE 13 On the interval $-\pi/2 \leq x \leq \pi/2$, the cosine takes on a maximum value of 1 (once) and a minimum value of 0 (twice). The sine takes on a maximum value of 1 and a minimum value of -1.

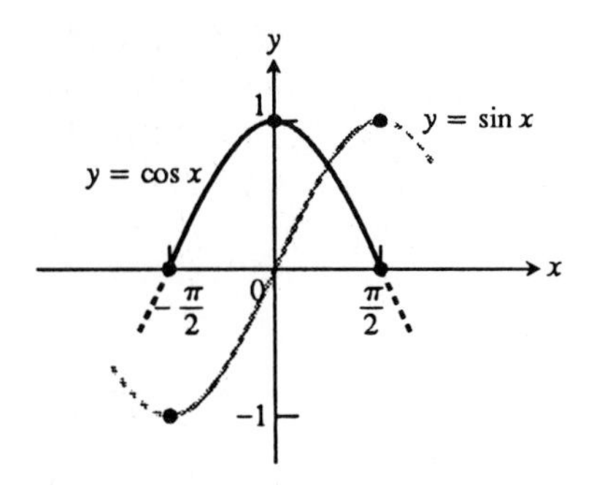

EXAMPLE 14 On an open interval, a continuous function need not have either a maximum or a minimum value. The function $f(x) = x$ has neither a largest nor a smallest value on the interval $0 < x < 1$.

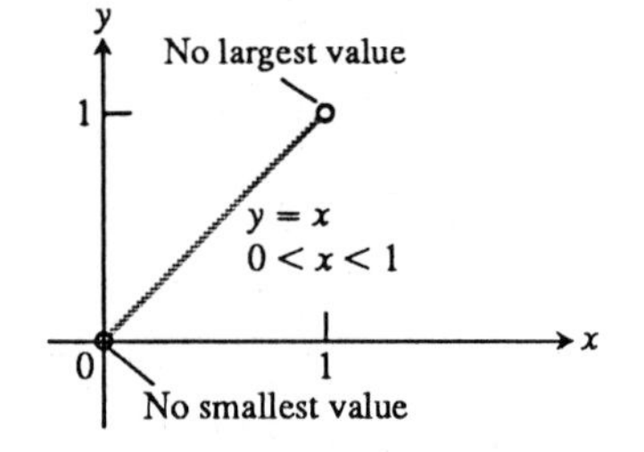

EXAMPLE 15 Even a single point of discontinuity can keep a function from having either a maximum or a minimum value on a closed interval. The function

$$y = \begin{cases} x + 1, & -1 \leq x < 0 \\ 0 & x = 0 \\ x - 1, & 0 < x \leq 1 \end{cases}$$

is continuous at every point of the interval $-1 \leq x \leq 1$ except $x = 0$, yet its graph over the interval has neither a highest nor a lowest point.

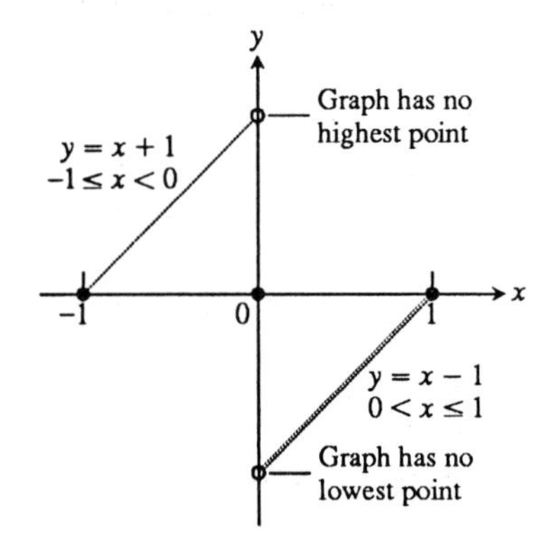

Theorem 7

The Intermediate Value Theorem for Continuous Functions

A function $y = f(x)$ that is continuous on a closed interval $[a, b]$ takes on every value between $f(a)$ and $f(b)$. In other words, if $f(a) \leq y_0 \leq f(b)$ then $y_0 = f(c)$ for some c in $[a, b]$.

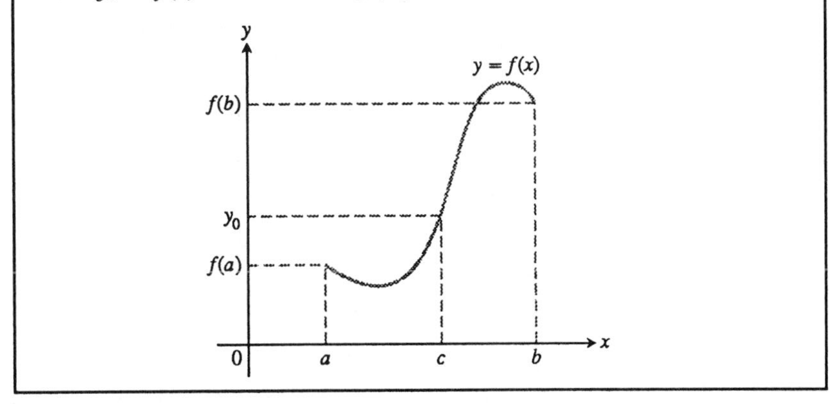

A Consequence for Graphing: Connectivity Suppose we want to graph a function $y = f(x)$ that is continuous throughout some interval I on the x-axis. Theorem 7 tells us that the graph of f over I will never move from one y-value to another without taking on the y-values in between. The graph of f over I will be **connected**: it will consist of a single, unbroken curve, like the graph of $y = \sin x$. The graph of f will not have jumps like the graph of the greatest-integer function (Fig. 2.21), or separate branches like the graph of $y = 1/x$ (Fig. 2.22), or "holes" like the graph of $y = \dfrac{x^2 - 1}{x - 1}$ (Fig. 2.23).

A Consequence for Root Finding Suppose that $f(x)$ is continuous at every point of a closed interval $[a, b]$ and that $f(a)$ and $f(b)$ differ in sign. Then zero lies between $f(a)$ and $f(b)$, so there is at least one number c between a and b where $f(c) = 0$. In other words, if f is continuous and $f(a)$ and $f(b)$

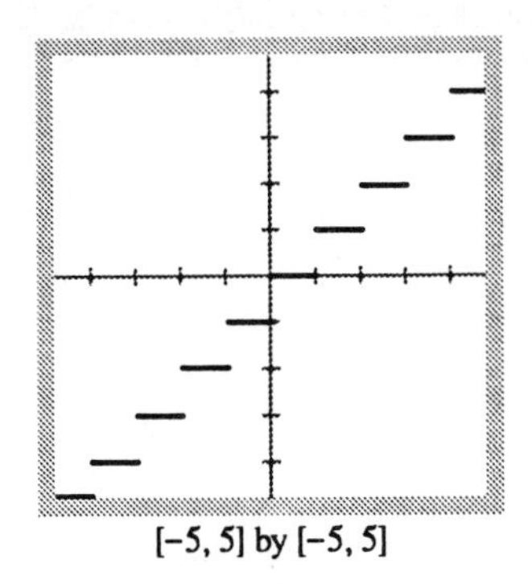

2.21 The graph of $f(x) = \lfloor x \rfloor$ has "jump" discontinuities.

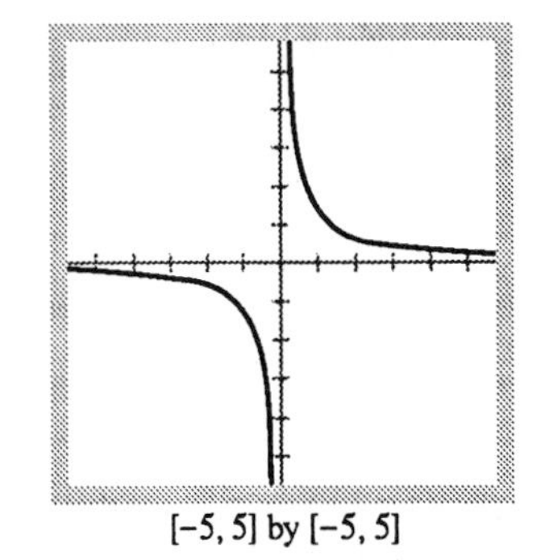

2.22 The graph of $y = 1/x$ has separate branches.

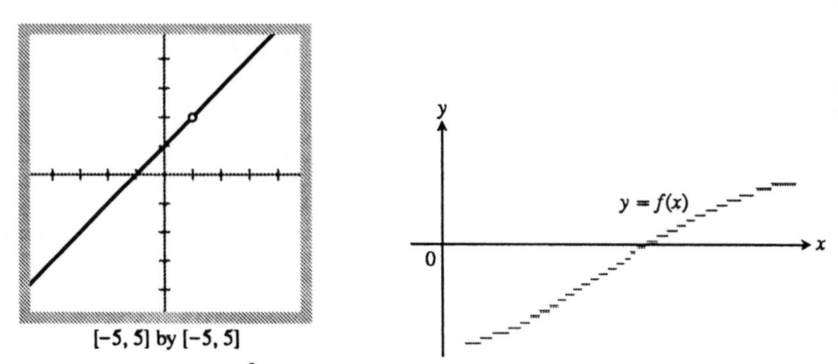

2.23 The graph of $y = (x^2 - 1)/(x - 1)$ has a hole at $x = 1$. The hole was added by hand.

2.24 The graph of a continuous function never has steps like these.

differ in sign, then the equation $f(x) = 0$ has at least one solution in the open interval (a, b).

This is the property of continuous functions that enables us to locate roots using zoom-in. Between any point where the graph lies above the x-axis and any point where it lies below, there must be at least one point where it actually intersects the axis. The graph of a continuous function of x never steps across the x-axis the way the step function in Fig. 2.24 does. When we see the graph of a continuous function cross the axis on a graphing calculator or computer screen, there really is a root there. This is how we found the value of a root in Chapter 1 using zoom-in.

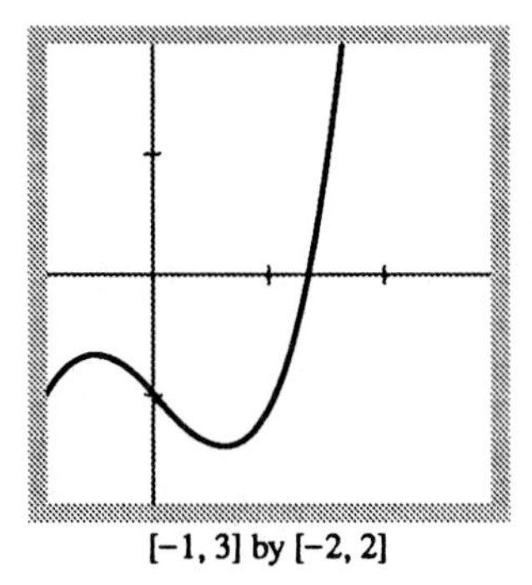

2.25 The graph of $f = x^3 - x - 1$ crosses the x-axis between $x = 1$ and $x = 2$.

EXAMPLE 16 Is any real number exactly 1 less than its cube?

Solution Any such number must satisfy the equation $x = x^3 - 1$ or $x^3 - x - 1 = 0$. Hence, we are looking fo. a zero value of the function $f(x) = x^3 - x - 1$. Figure 2.25 suggests the solution is about 1.3. In Exercise 4.3, we will ask you to find the solution with greater accuracy using zoom-in. ∎

Concluding Remarks

For any function $y = f(x)$ it is important to distinguish between continuity at $x = c$ and having a limit as $x \to c$. The limit, $\lim_{x \to c} f(x)$, is where the function is headed as $x \to c$. Continuity is the property of arriving at the point where $f(x)$ has been heading when x actually gets to c. (Someone is home when you get there, so to speak.) If the limit is what you expect as $x \to c$, and the number $f(c)$ is what you get when $x = c$, then the function is continuous at c if you get what you expect.

Finally, remember the test for continuity at a point:

1. Does $f(c)$ exist?
2. Does $\lim\limits_{x \to c} f(c)$ exist?
3. Does $\lim\limits_{x \to c} f(x) = f(c)$?

For f to be continuous at $x = c$, all three answers must be *yes*.

Exercises 2.2

Exercises 1–6 are about the function graphed in Fig. 2.26, whose formula is

$$f(x) = \begin{cases} x^2 - 1, & -1 \leq x < 0, \\ 2x, & 0 \leq x < 1, \\ 1, & x = 1, \\ -2x + 4, & 1 < x < 2, \\ 0, & 2 < x \leq 3. \end{cases}$$

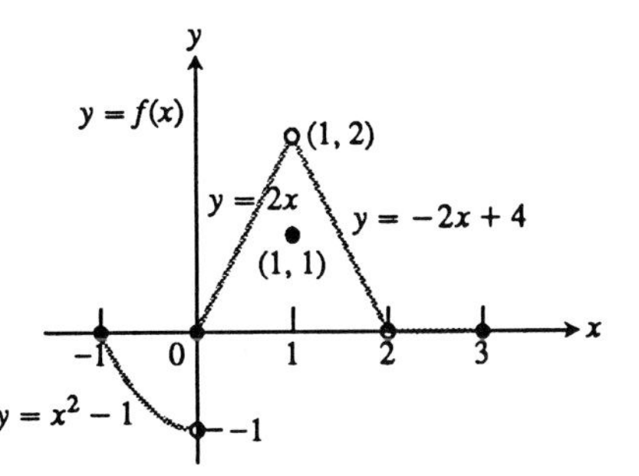

2.26 The function $y = f(x)$ for Exercises 1–6.

1. a) Does $f(-1)$ exist?
   b) Does $\lim_{x \to -1^+} f(x)$ exist?
   c) Does $\lim_{x \to -1^+} f(x) = f(-1)$?
   d) Is f continuous at $x = -1$?
2. a) Does $f(1)$ exist?
   b) Does $\lim_{x \to 1} f(x)$ exist?
   c) Does $\lim_{x \to 1} f(x) = f(1)$?
   d) Is f continuous at $x = 1$?
3. a) Is f defined at $x = 2$? (Look at the definition of f.)
   b) Is f continuous at $x = 2$?
4. At what values of x is f continuous?
5. a) What is the value of $\lim_{x \to 2} f(x)$?
   b) What value should be assigned to $f(2)$ to make f continuous at $x = 2$?
6. To what new value should $f(1)$ be changed to make f continuous at $x = 1$?

At which points are the functions in the following exercises in Section 2.1 continuous?

7. Exercise 33
8. Exercise 34
9. Exercise 35
10. Exercise 36
11. Exercise 39
12. Exercise 40
13. Exercise 41
14. Exercise 42

15. Let $f(x) = \begin{cases} 0, & x < 0 \\ 1, & 0 \leq x \leq 1. \\ 0, & 1 < x. \end{cases}$
    a) Determine a complete graph of f.
    b) At what points is the function continuous?
16. Let $f(x) = \begin{cases} 1, & x < 0 \\ \sqrt{1 - x^2}, & 0 \leq x \leq 1 \\ x - 1, & x > 1 \end{cases}$
    a) Determine a complete graph of f.
    b) Is f continuous? Explain.

Find the points, if any, at which the functions in Exercises 17–30 are *not* continuous.

17. $y = \dfrac{1}{x - 2}$
18. $y = \dfrac{1}{(x + 2)^2}$
19. $y = \dfrac{x + 1}{x^2 - 4x + 3}$
20. $y = \dfrac{x + 3}{x^2 - 3x - 10}$
21. $y = \dfrac{x^3 - 1}{x^2 - 1}$
22. $y = \dfrac{1}{x^2 + 1}$
23. $y = |x - 1|$
24. $y = |2x + 3|$
25. $y = \dfrac{\cos x}{x}$
26. $y = \dfrac{|x|}{x}$
27. $y = \sqrt{2x + 3}$
28. $y = \sqrt[4]{3x - 1}$
29. $y = \sqrt[3]{2x - 1}$
30. $y = \sqrt[5]{2 - x}$
31. The function $f(x)$ is defined by $f(x) = (x^2 - 1)/(x - 1)$ when $x \neq 1$ and by $f(1) = 2$. Is f continuous at $x = 1$? Explain.
32. Define $g(3)$ in a way that extends $g(x) = (x^2 - 9)/(x - 3)$ to be continuous at $x = 3$.
33. Define $h(2)$ in a way that extends $h(x) = (x^2 + 3x - 10)/(x - 2)$ to be continuous at $x = 2$.
34. Define $f(1)$ in a way that extends $f(x) = (x^3 - 1)/(x^2 - 1)$ to be continuous at $x = 1$.
35. Define $g(4)$ in a way that extends $g(x) = (x^2 - 16)/(x^2 - 3x - 4)$ to be continuous at $x = 4$.
36. How should $f(2)$ be redefined in Fig. 2.13 to make the function continuous at $x = 2$?
37. What value should be assigned to a to make the function
    $$f(x) = \begin{cases} x^2 - 1, & x < 3, \\ 2ax, & x \geq 3, \end{cases}$$
    continuous at $x = 3$? Determine a complete graph of f for this value of a.
38. What value should be assigned to b to make the function
    $$g(x) = \begin{cases} x^3, & x < 1/2, \\ bx^2, & x \geq 1/2, \end{cases}$$

continuous at $x = 1/2$? Determine a complete graph of g for this value of a.

Find the limits in Exercises 39–42.

39. $\lim_{x \to 0} \dfrac{1 + \cos x}{2}$

40. $\lim_{x \to 0} \cos\left(1 - \dfrac{\sin x}{x}\right)$

41. $\lim_{x \to 0} \tan x$

42. $\lim_{x \to 0} \sin\left(\dfrac{\pi}{2}\cos(\tan x)\right)$

43. Let $f(x) = x^3 - x - 1$ (see Example 16).

a) Determine the solution to $f(x) = 0$ with an error of at most 10^{-8}.

b) It can be shown that the exact value of the solution in part (a) is

$$\left(\frac{\sqrt{69}}{18} + \frac{1}{2}\right)^{1/3} + \left(\frac{1}{2} - \frac{\sqrt{69}}{18}\right)^{1/3}.$$

Evaluate this exact answer and compare with the value determined in part (a).

44. Let $f(x) = x^3 - 2x + 2$.

a) Determine the solution to $f(x) = 0$ with an error of at most 10^{-4}.

b) It can be shown that the exact value of the solution in part (a) is

$$\left(\frac{\sqrt{57}}{9} - 1\right)^{1/3} - \left(\frac{\sqrt{57}}{9} + 1\right)^{1/3}.$$

Evaluate this exact answer and compare with the value determined in part (a).

45. At what values of x does the function in Fig. 2.13 take on its maximum value? Does the function take on a minimum value? Explain.

46. At what values (if any) does the function in Fig. 2.25 take on a maximum value? A minimum value?

47. Does the function $y = x^2$ have a maximum value on the open interval $-1 < x < 1$? A minimum value? Explain.

48. On the closed interval $0 \le x \le 1$ the greatest-integer function $y = \lfloor x \rfloor$ takes on a minimum value $m = 0$ and a maximum value $M = 1$. It does so even though it is discontinuous at $x = 1$. Does this violate the Max-Min Theorem? Why?

49. A continuous function $y = f(x)$ is known to be negative at $x = 0$ and positive at $x = 1$. Why does the equation $f(x) = 0$ have at least one solution between $x = 0$ and $x = 1$? Illustrate with a sketch.

50. Assuming $y = \cos x$ to be continuous, show that the equation $\cos x = x$ has at least one solution. (*Hint:* Show that the equation $\cos x - x = 0$ has at least one solution.)

2.3 The Sandwich Theorem and $(\sin\theta)/\theta$

One of the most useful facts in calculus is that $\lim_{\theta\to 0}(\sin\theta)/\theta = 1$ when θ is measured in radians. (Unless explicitly stated otherwise, trigonometric arguments will be in radians.) This beautiful and simple result turns out to be the key to measuring the rates at which all trigonometric functions of θ change their values as θ changes, as we shall see in Chapter 3.

This limit is not the kind we can evaluate by substituting $\theta = 0$, that is, f is discontinuous at $\theta = 0$. The graph of $f(\theta) = (\sin\theta)/\theta$ in Fig. 2.27 suggests this discontinuity is *removable* and that $\lim_{\theta\to 0}(\sin\theta)/\theta = 1$. Notice the graph also suggests that $0 < (\sin\theta)/\theta < 1$ for x near 0. Thus the extended function

$$f(\theta) = \begin{Bmatrix} \dfrac{\sin\theta}{\theta}, & \theta \ne 0 \\ 1, & \theta = 0 \end{Bmatrix}.$$

is continuous at every real number θ.

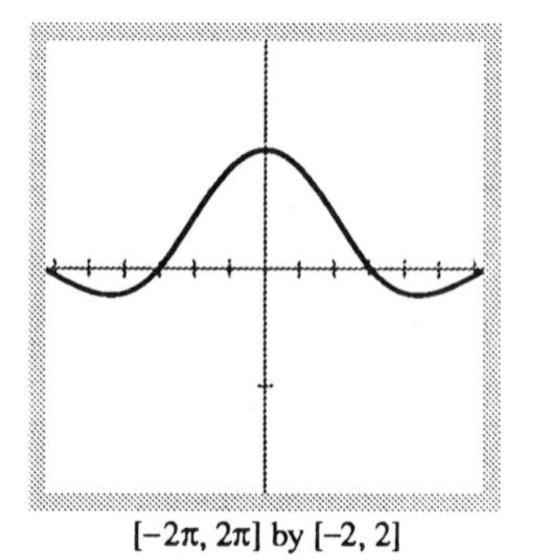

[−2π, 2π] by [−2, 2]

2.27 The graph suggests that the discontinuity of $f(\theta) = (\sin\theta)/\theta$ at $\theta = 0$ is removable.

To prove that $\lim_{\theta\to 0}(\sin\theta)/\theta = 1$, we introduce and apply a powerful theorem called the Sandwich Theorem. What we do is sandwich the fraction $(\sin\theta)/\theta$ between the number 1 and a fraction that is known to approach 1 as θ approaches 0. This tells us that $(\sin\theta)/\theta$ approaches 1 as well.

Theorem 8

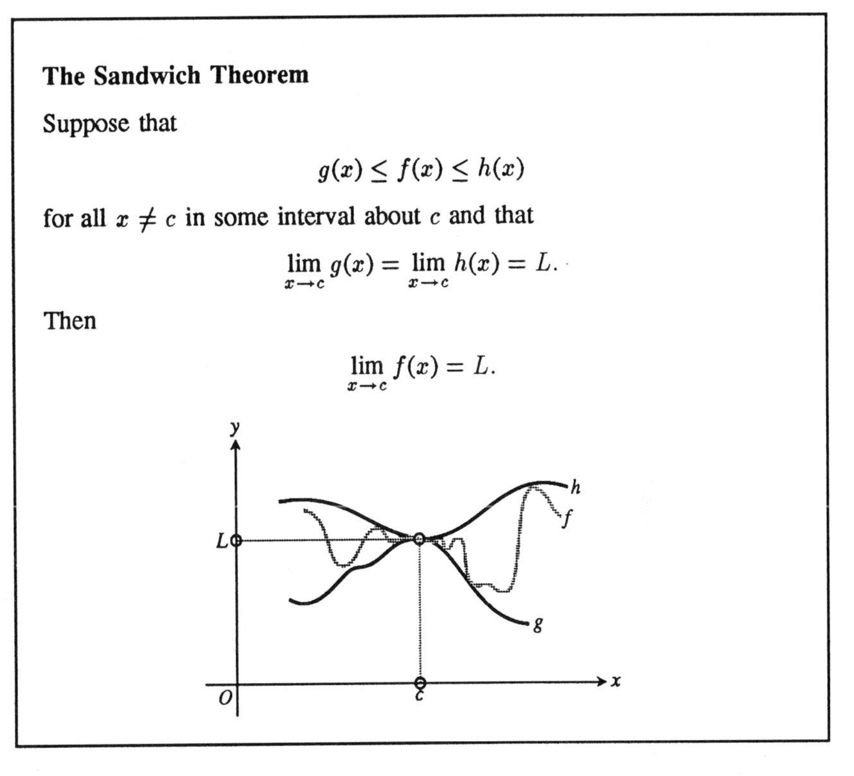

The Sandwich Theorem

Suppose that

$$g(x) \le f(x) \le h(x)$$

for all $x \neq c$ in some interval about c and that

$$\lim_{x\to c} g(x) = \lim_{x\to c} h(x) = L.$$

Then

$$\lim_{x\to c} f(x) = L.$$

The idea is that if the values of f are sandwiched between the values of two functions that approach L, then the values of f approach L, too. We have included a proof of the theorem in Appendix 2.

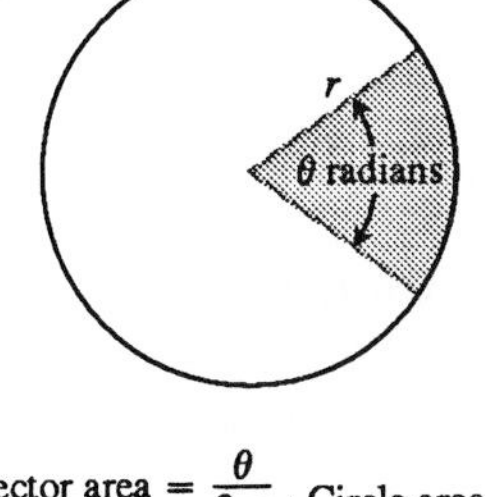

$$\begin{aligned} \text{Sector area} &= \frac{\theta}{2\pi} \cdot \text{Circle area} \\ &= \frac{\theta}{2\pi} \cdot \pi r^2 \\ &= \frac{1}{2} r^2 \theta \quad \text{(Usual form)} \\ &= \frac{\theta}{2} \quad (\text{If } r = 1) \end{aligned}$$

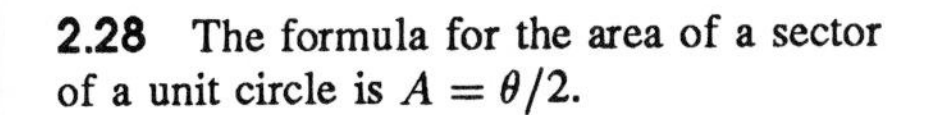

2.28 The formula for the area of a sector of a unit circle is $A = \theta/2$.

EXAMPLE 1 Show that $\lim_{x\to 0} \sin x = 0$, $\lim_{x\to 0} \cos x = 1$, and $\lim_{x\to 0} \tan x = 0$.

Solution To calculate these limits we use the fact that $\sin x$ and $\cos x$ are continuous. In particular, they are continuous at $x = 0$ so

$$\lim_{x\to 0} \sin x = \sin 0 = 0,$$

$$\lim_{x\to 0} \cos x = \cos 0 = 1.$$

Next, $\tan x$ is continuous at $x = 0$ because $\tan x = \sin x / \cos x$ and $\cos 0 = 1 \neq 0$. Thus, $\lim_{x\to 0} \tan x = \tan 0 = 0$. ■

During the proof of $\lim_{\theta\to 0}(\sin\theta)/\theta = 1$ we draw on a formula from geometry that says that the area cut from a unit circle by a central angle of θ radians is $\theta/2$. Figure 2.28 shows where this formula comes from.

EXAMPLE 2 If θ is measured in radians, then

$$\lim_{\theta\to 0} \frac{\sin\theta}{\theta} = 1. \tag{1}$$

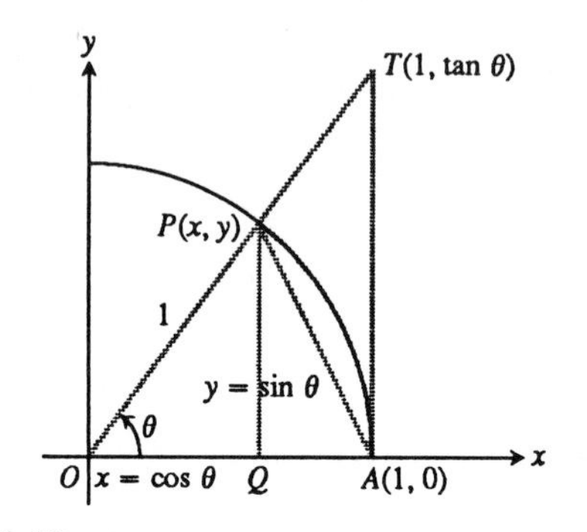

2.29 Area $\Delta OAP <$ area sector $OAP <$ area ΔOAT.

Equation (4) is where the radian measurement comes in: The area of sector OAP is $\theta/2$ only if θ is measured in radians.

Our plan is to show that the right-hand and the left-hand limits are both 1. We will then know that the two-sided limit is 1 as well.

To show that the right-hand limit is 1, we begin with the values of θ that are positive and less than $\pi/2$ (Fig. 2.29). We compare the areas of ΔOAP, sector OAP, and ΔOAT and note that

$$\text{Area } \Delta\, OAP < \text{Area Sector } OAP < \text{Area } \Delta\, OAT. \tag{2}$$

We can express these areas in terms of θ as follows:

$$\text{Area } \Delta\, OAP = \frac{1}{2}\text{base} \times \text{height} = \frac{1}{2}(1)(\sin\theta) = \frac{1}{2}\sin\theta, \tag{3}$$

$$\text{Area sector } OAP = \frac{1}{2}r^2\theta = \frac{1}{2}(1)^2\theta = \frac{\theta}{2}, \tag{4}$$

$$\text{Area } \Delta\, OAT = \frac{1}{2}\text{base} \times \text{height} = \frac{1}{2}(1)(\tan\theta) = \frac{1}{2}\tan\theta, \tag{5}$$

so that

$$\frac{1}{2}\sin\theta < \frac{1}{2}\theta < \frac{1}{2}\tan\theta. \tag{6}$$

The inequality in (6) will go the same way if we divide all three terms by the positive number $(1/2)\sin\theta$:

$$1 < \frac{\theta}{\sin\theta} < \frac{1}{\cos\theta}. \tag{7}$$

We next take reciprocals in (7), which reverses the inequalities:

$$\cos\theta < \frac{\sin\theta}{\theta} < 1. \tag{8}$$

Because $\cos\theta$ approaches 1 as θ approaches 0, the Sandwich Theorem tells us that

$$\lim_{\theta\to0+}\frac{\sin\theta}{\theta} = 1. \tag{9}$$

The limit in Eq. (9) is a right-hand limit because we have been dealing with values of θ between 0 and $\pi/2$, but we obtain the same limit for $(\sin\theta)/\theta$ as θ approaches 0 from the left. For if $\theta = -\alpha$ and α is positive, then

$$\frac{\sin\theta}{\theta} = \frac{\sin(-\alpha)}{-\alpha} = \frac{-\sin(\alpha)}{-\alpha} = \frac{\sin\alpha}{\alpha}. \tag{10}$$

Therefore,

$$\lim_{\theta\to0-}\frac{\sin\theta}{\theta} = \lim_{\alpha\to0+}\frac{\sin\alpha}{\alpha} = 1. \tag{11}$$

Together, Eqs. (9) and (11) imply that $\lim_{\theta\to0}(\sin\theta)/\theta = 1$. ■

Knowing the limit of $(\sin\theta)/\theta$ helps us in calculating a number of related limits.

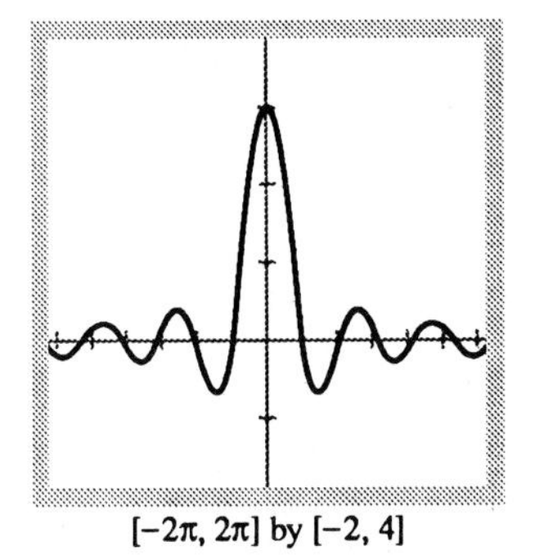
$[-2\pi, 2\pi]$ by $[-2, 4]$

2.30 The graph of $f(x) = (\sin 3x)/x$ suggests $f(x) \to 3$ as $x \to 0$.

EXAMPLE 3 If $a \neq 0$,

$$\lim_{x\to 0} \frac{\sin ax}{ax}$$

$$= \lim_{\theta\to 0} \frac{\sin\theta}{\theta} \qquad \text{Substitute } \theta = ax$$

$$= 1 \qquad \text{Eq. (1)}$$

EXAMPLE 4 Find $\lim_{x\to 0}(\sin 3x)/x$ graphically and then confirm algebraically.

Solution The graph of $f(x) = (\sin 3x)/x$ in Fig. 2.30 suggests that $\lim_{x\to 0}(\sin 3x)/x = 3$. Algebraically,

$$\lim_{x\to 0} \frac{\sin 3x}{x}$$

$$= \lim_{x\to 0} 3\frac{\sin 3x}{3x}$$

$$= \lim_{\theta\to 0} 3 \cdot \frac{\sin\theta}{\theta} \qquad \text{Substitute } \theta = 3x$$

$$= 3 \cdot \lim_{\theta\to 0} \frac{\sin\theta}{\theta}$$

$$= 3 \cdot 1$$

$$= 3$$

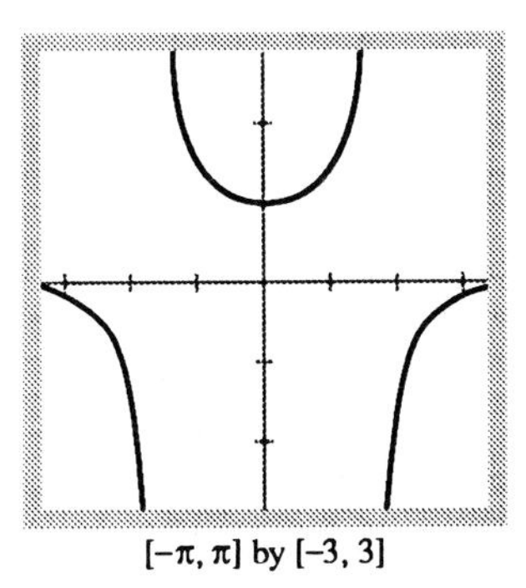
$[-\pi, \pi]$ by $[-3, 3]$

2.31 The graph of $f(x) = (\tan x)/x$ suggests $f(x) \to 1$ as $x \to 0$.

EXAMPLE 5 Find $\lim_{x\to 0} \frac{\tan x}{x}$ graphically and then confirm algebraically.

Solution The graph of $f(x) = (\tan x)/x$ in Fig. 2.31 suggests that $\lim_{x\to 0}(\tan x)/x = 1$.

$$\lim_{x\to 0} \frac{\tan x}{x}$$

$$= \lim_{x\to 0} \frac{\sin x}{x} \frac{1}{\cos x} \qquad \tan x = \frac{\sin x}{\cos x}$$

$$= \lim_{x\to 0} \frac{\sin x}{x} \lim_{x\to 0} \frac{1}{\cos x} \qquad \text{Property of limits}$$

$$= 1 \cdot 1$$

$$= 1$$

Notice in Figs. 2.30 and 2.31 that the graphs suggest that as $x \to 0$ the values of $(\sin 3x)/x$ approach 3 through values less than 3, and values of $(\tan x)/x$ approach 1 through values greater than 1. This important information is not reflected in the algebraic solutions of Examples 4 and 5.

EXAMPLE 6 Find $\lim_{x\to 0} \frac{6\sin 7x}{\sin 9x}$ graphically and then confirm algebraically.

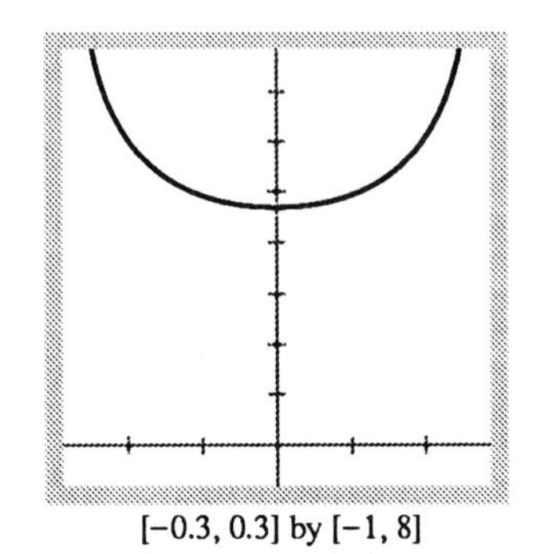

2.32 The graph of $f(x) = (6\sin 7x)/(\sin 9x)$ suggests $f(x) \to 4.7$ as $x \to 0$.

Solution It may take some experimenting to find a graph of $f(x) = (6\sin 7x)/(\sin 9x)$ that gives evidence about the limit of f as $x \to 0$. The graph of f in Fig. 2.32 suggests that $f(x) \to 4.7$ as $x \to 0$. We could zoom-in several times and conjecture that $f(x) \to 4\frac{2}{3}$ as $x \to 0$. Algebraically,

$$\lim_{x\to 0} \frac{6\sin 7x}{\sin 9x}$$

$$= 6 \lim_{x\to 0} \frac{7x\dfrac{\sin 7x}{7x}}{9x\dfrac{\sin 9x}{9x}}$$

$$= 6 \cdot \lim_{x\to 0} \frac{7}{9} \lim_{x\to 0} \frac{\dfrac{\sin 7x}{7x}}{\dfrac{\sin 9x}{9x}}$$

$$= 6 \cdot \frac{7}{9} \cdot 1$$

$$= 4\frac{2}{3}$$

Exercises 2.3

Find the limits in Exercises 1–16 graphically and confirm algebraically.

1. $\lim_{x\to 0} \dfrac{1}{\cos x}$
2. $\lim_{x\to 0} (2\sin x + 3\cos x)$
3. $\lim_{x\to 0} \dfrac{1+\sin x}{1+\cos x}$
4. $\lim_{x\to 0} \dfrac{x^2+1}{1-\sin x}$
5. $\lim_{x\to 0+} \dfrac{x}{\sin x}$
6. $\lim_{x\to 0-} \dfrac{x}{\tan x}$
7. $\lim_{x\to 0} \dfrac{\sin 2x}{x}$
8. $\lim_{x\to 0} \dfrac{x}{\sin 3x}$
9. $\lim_{x\to 0} \dfrac{\tan 2x}{2x}$
10. $\lim_{x\to 0} \dfrac{\tan 2x}{x}$
11. $\lim_{x\to 0} \dfrac{\sin x}{2x^2 - x}$
12. $\lim_{x\to 0} \dfrac{x+\sin x}{x}$
13. $\lim_{x\to 0} \dfrac{\sin^2 x}{x}$
14. $\lim_{x\to 0} \dfrac{2x}{\tan^2 x}$
15. $\lim_{x\to 0} \dfrac{3\sin 4x}{\sin 3x}$
16. $\lim_{x\to 0} \dfrac{\tan 5x}{\tan 2x}$

17. Let $f(x) = \dfrac{\tan 3x}{\sin 5x}$.
    a) Estimate $\lim_{x\to 0} f(x)$ graphically. Hint: Zoom-in around $x = 0$.
    b) Compare the function values of f near 0 to the limit found in part (a).
    c) Find the exact value of $\lim_{x\to 0} f(x)$ algebraically.

18. Let $f(x) = \dfrac{\cot 3x}{\csc 2x}$.
    a) Estimate $\lim_{x\to 0} f(x)$ graphically. Hint: Zoom-in around $x = 0$.
    b) Compare the function values of f near 0 to the limit found in part (a).
    c) Find the exact value of $\lim_{x\to 0} f(x)$ algebraically.

19. *Sandwich Theorem.* The inequality
$$1 - \frac{x^2}{6} < \frac{\sin x}{x} < 1$$
holds when x is measured in radians and $-1 < x < 1$. Use this inequality to calculate $\lim_{x\to 0}(\sin x)/x$. In Chapter 9, we shall see where this inequality comes from.

20. *Sandwich Theorem.* As we saw in Example 2 and again in Exercise 19, we can sometimes use the Sandwich Theorem to calculate the limit of a fraction whose numerator and denominator both approach zero. Another example is the fraction
$$f(x) = \frac{x\sin x}{2 - 2\cos x},$$

which satisfies the inequality

$$1 - \frac{x^2}{6} < \frac{x \sin x}{2 - 2\cos x} < 1$$

when x is an angle in radians close to zero (Fig. 2.33). Use this inequality to find $\lim_{x\to 0} f(x)$. Inequalities like this come from infinite series (Chapter 9).

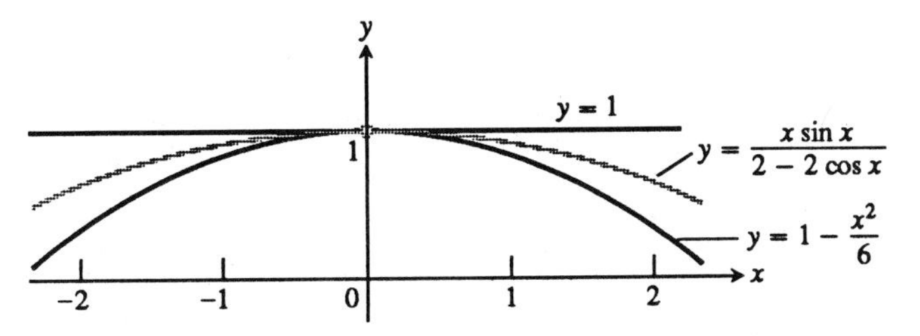

2.33 For all $x \neq 0$, the graph of $f(x) = (x \sin x)/(2 - 2\cos x)$ lies between $y = 1$ and $y = 1 - x^2/6$ (Exercise 20).

21. a) Estimate the value of

$$\lim_{x\to 0} \frac{1 - \cos x}{x^2}.$$

b) *Sandwich Theorem.* Use the inequality

$$\frac{1}{2} - \frac{x^2}{24} < \frac{1 - \cos x}{x^2} < \frac{1}{2}$$

(Fig. 2.34) to find the exact value of the limit in (a). Inequalities like this come from infinite series (Chapter 9).

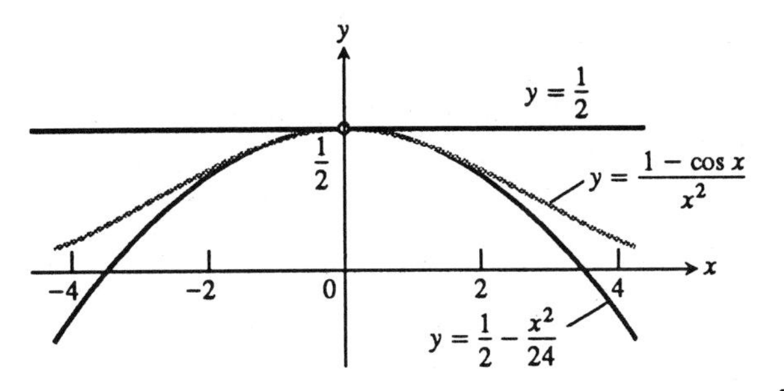

2.34 For all $x = 0$, the graph of $f(x) = (1 - \cos x)/x^2$ lies between the line $y = 1/2$ and the parabola $y = (1/2) - x^2/24$ (Exercise 21).

22. *Sandwich Theorem.* The inequality

$$1 < \frac{\tan x}{x} < 1 + x^2$$

holds when x is measured in radians and $-1 < x < 1$. Use the inequality to calculate $\lim_{x\to 0}(\tan x)/x$.

23. Show that $\lim_{x\to 0}(\cos x)/x$ does not exist.

24. The area formula $A = (1/2)r^2\theta$ derived in Fig. 2.28 for radian measure has to be changed if the angle is measured in degrees. What should the new formula be?

2.4 Limits Involving Infinity

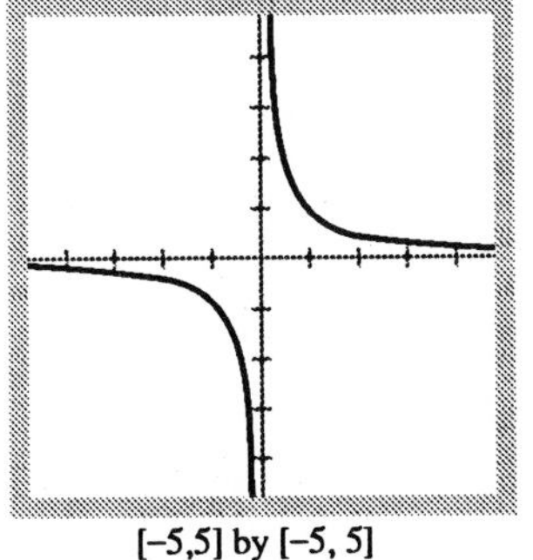

[−5,5] by [−5, 5]

2.35 The graph of $f(x) = 1/x$.

In this section, we describe what it means for the values of a function to approach infinity and what it means for a function $f(x)$ to have a limit as x approaches infinity. Although there is no real number *infinity*, the word *infinity* is useful for describing how some functions behave when their domains or ranges exceed all bounds. Graphers will be very helpful in this section.

Limits as $x \to \infty$ or $x \to -\infty$

The function

$$f(x) = \frac{1}{x} \tag{1}$$

is defined for all real numbers except $x = 0$.

As Fig. 2.35 suggests,

a) $1/x$ is small and positive when x is large and positive.

b) $1/x$ is large and positive when x is small and positive.

c) $1/x$ is large and negative when x is small and negative.

d) $1/x$ is small and negative when x is large and negative.

We summarize these facts by writing:

a) As $x \to \infty$, $1/x \to 0$.
b) As $x \to 0$ from the right ($x \to 0^+$), $1/x \to \infty$.
c) As $x \to 0$ from the left ($x \to 0^-$), $1/x \to -\infty$.
d) As $x \to -\infty$, $1/x \to 0$.

The symbol ∞, **infinity**, does not represent any real number. We cannot use ∞ in arithmetic in the usual way, but it is convenient to be able to say things like "the limit of $1/x$ as x approaches infinity is 0."

EXAMPLE 1 $\lim_{x \to \infty} \frac{1}{x} = 0$

As $x \to \infty$, the graph of $1/x$ gets arbitrarily close to the x-axis, in the following sense: No matter how small a positive number you name, the value of $1/x$ eventually gets smaller.

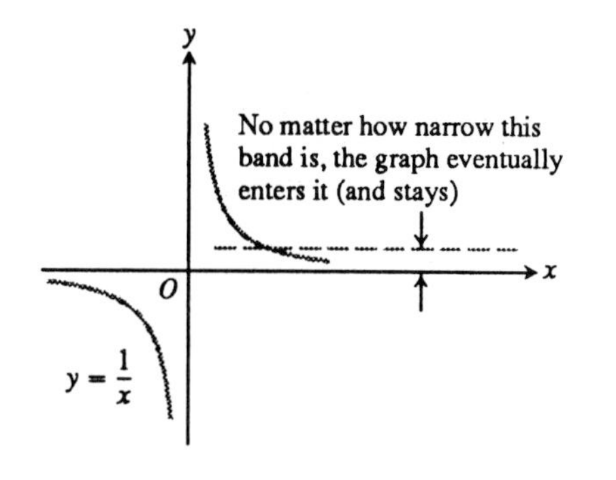

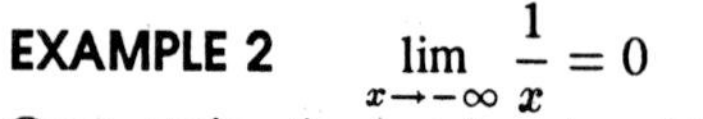

EXAMPLE 2 $\lim_{x \to -\infty} \frac{1}{x} = 0$

Once again, the graph gets arbitrarily close to the x-axis, this time rising toward it from below.

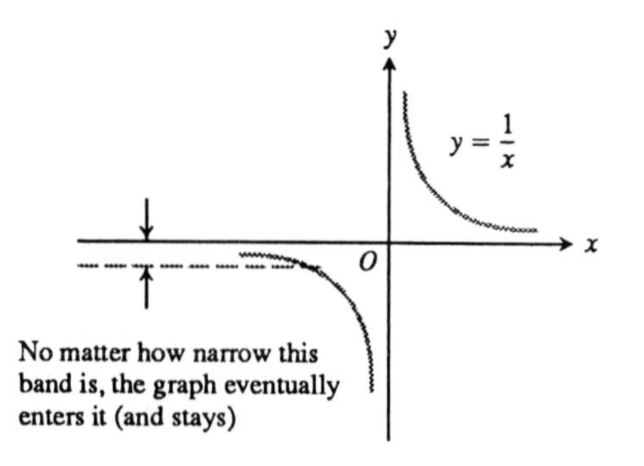

Calculation Rules for Functions with Finite Limits as $x \to \pm\infty$

Our strategy is again the one that worked so well in Section 2.1. We find the limits of two "basic" functions as $x \to \infty$ and $x \to -\infty$ and then, to find everything else, we use a theorem about limits of algebraic combinations. In Section 2.1, the basic functions were the constant function $y = k$ and the identity function $y = x$. Here, the basic functions are $y = k$ and the reciprocal $y = 1/x$. Our presentation continues to be informal.

EXAMPLE 3 If $f(x) = 3$ is the constant function whose outputs have the constant value 3, then

$$\lim_{x\to\infty} f(x) = \lim_{x\to\infty} (3) = 3$$
$$\lim_{x\to-\infty} f(x) = \lim_{x\to-\infty} (3) = 3.$$

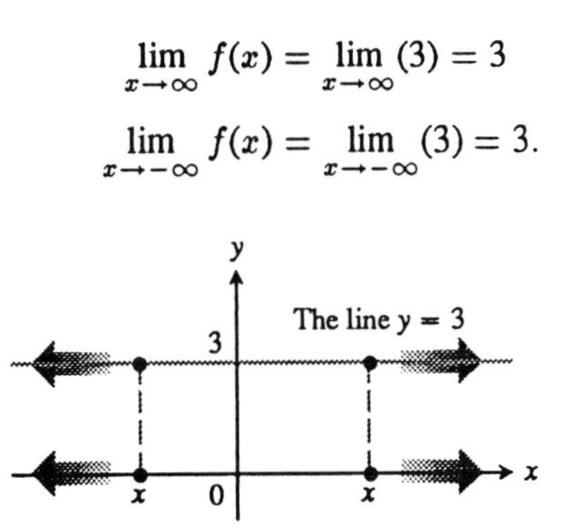

EXAMPLE 4 What we saw in Example 3 holds for any number k: If f is the constant function with $f(x) = k$, then

$$\lim_{x\to\infty} f(x) = \lim_{x\to\infty} (k) = k$$
$$\lim_{x\to-\infty} f(x) = \lim_{x\to-\infty} (k) = k.$$

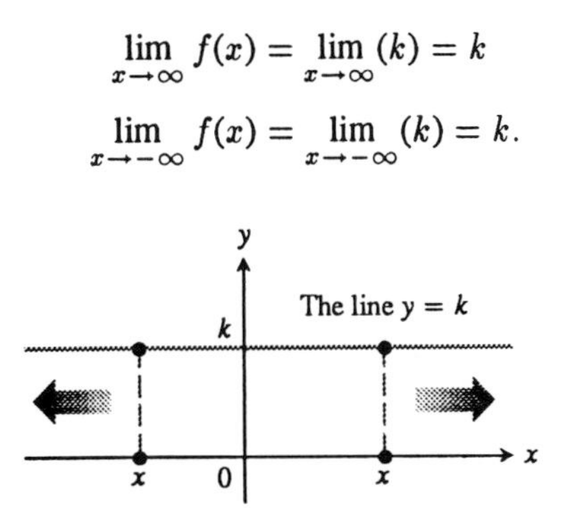

Believe it or not, we now have enough specific information to calculate the limits of a wide variety of rational functions as $x \to \pm\infty$. To do so, we simply use the limit properties listed in the following theorem.

Theorem 9

Properties of Finite Limits as $x \to \pm\infty$

If

$$\lim_{x\to\infty} f_1(x) = L_1 \quad \text{and} \quad \lim_{x\to\infty} f_2(x) = L_2$$

and L_1 and L_2 are (finite) real numbers, then

1. *Sum Rule:* $\lim_{x\to\infty} [f_1(x) + f_2(x)] = L_1 + L_2$
2. *Difference Rule:* $\lim_{x\to\infty} [f_1(x) - f_2(x)] = L_1 - L_2$
3. *Product Rule:* $\lim_{x\to\infty} f_1(x) \cdot f_2(x) = L_1 \cdot L_2$
4. *Constant Multiple Rule:* $\lim_{x\to\infty} k \cdot f_1(x) = k \cdot L_1$ (any number k)
5. *Quotient Rule:* $\lim_{x\to\infty} \dfrac{f_1(x)}{f_2(x)} = \dfrac{L_1}{L_2}$ if $L_2 \neq 0$.

These properties hold for $x \to -\infty$ as well as $x \to \infty$.

These properties are just like the properties we stated in Section 2.1 for limits as $x \to c$, and we use them the same way.

EXAMPLE 5

$$\lim_{x\to\infty}\left(5 + \frac{1}{x}\right) = \lim_{x\to\infty} 5 + \lim_{x\to\infty} \frac{1}{x} \qquad \text{Sum Rule}$$

$$= 5 + 0 = 5. \qquad \text{Known values}$$

EXAMPLE 6

$$\lim_{x\to-\infty} \frac{4}{x^2} = \lim_{x\to-\infty} 4 \cdot \lim_{x\to-\infty} \frac{1}{x} \cdot \lim_{x\to-\infty} \frac{1}{x} \qquad \text{Product Rule}$$

$$= 4 \cdot 0 \cdot 0 = 0. \qquad \text{Known values}$$

EXAMPLE 7

$$\lim_{x\to\pm\infty} \frac{\sin x}{x} = 0$$

Notice first that $\sin x$ lies between -1 and 1. Therefore, for all positive values of x,

$$-\frac{1}{x} \le \frac{\sin x}{x} \le \frac{1}{x},$$

and for all negative values of x,

$$-\frac{1}{x} \ge \frac{\sin x}{x} \ge \frac{1}{x}.$$

The Sandwich Theorem, stated in Section 2.3 for limits as $x \to c$, also holds for limits as $x \to \pm\infty$.

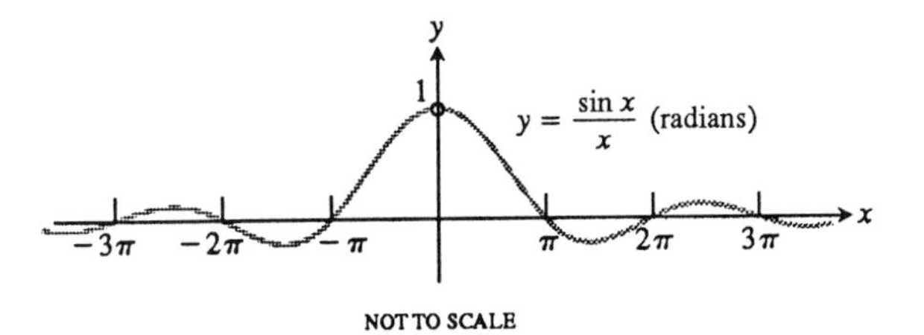

2.36 The graph of $y = (\sin x)/x$ oscillates about the x-axis. The amplitude of the oscillation decreases toward zero as $x \to \pm\infty$.

Because $-1/x$ and $1/x$ both approach 0 as $x \to \infty$, a modified version of the Sandwich Theorem tells us that $(\sin x)/x$ approaches 0 as well (Fig. 2.36). ■

Horizontal Asymptotes and End Behavior Models

The x-axis (the line $y = 0$) appears indistinguishable from the graph of $f(x) = 1/x$ for $|x|$ large, even though the graph of f never touches the x-axis. We say that the x-axis is a horizontal asymptote of f. Notice that

$$\lim_{x\to\infty} \frac{1}{x} = 0 \text{ and } \lim_{x\to-\infty} \frac{1}{x} = 0.$$

Definition

A line $y = b$ is a **horizontal asymptote** of the graph of a function $y = f(x)$ if either

$$\lim_{x\to\infty} f(x) = b \text{ or } \lim_{x\to-\infty} f(x) = b.$$

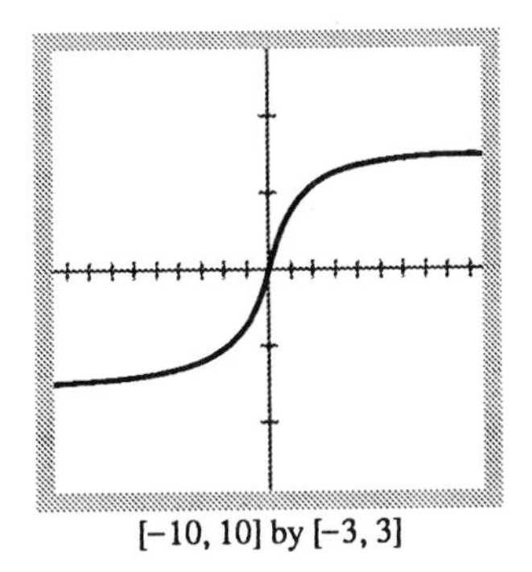

2.37 The function $y = \tan^{-1} x$ has two horizontal asymptotes.

Figure 2.36 illustrates that the x-axis is a horizontal asymptote for the function $f(x) = (\sin x)/x$. Functions can have two horizontal asymptotes as the graph of $y = \tan^{-1} x$ in Fig. 2.37 illustrates. We also call the function $y = 0$ an *end behavior model* for $f(x) = (\sin x)/x$ in the sense that $y = 0$ is a simpler function that behaves virtually the same way as f for $|x|$ large.

Definition

Assume $\lim_{x\to\infty} f(x) \neq 0$ and $\lim_{x\to-\infty} f(x) \neq 0$. The function g is called an **end behavior model** of f if $\lim_{x\to\infty} \frac{f}{g} = 1$ and $\lim_{x\to-\infty} \frac{f}{g} = 1$. If $\lim_{x\to\infty} f(x) = \lim_{x\to-\infty} f(x) = 0$, then $g(x) = 0$ is an **end behavior model** of f.

If only one of $\lim_{x\to\infty} f(x)$ and $\lim_{x\to-\infty} f(x)$ is 0, the above definition can be extended to define an end behavior model of f. We will see many examples of end behavior models in this section.

Limits of Rational Functions as $x \to \pm\infty$

Usually, we first investigate the limit of a rational function as $x \to \pm\infty$ graphically, then (when the limit exists) we confirm algebraically by dividing the numerator and denominator by the highest power of x in the denominator.

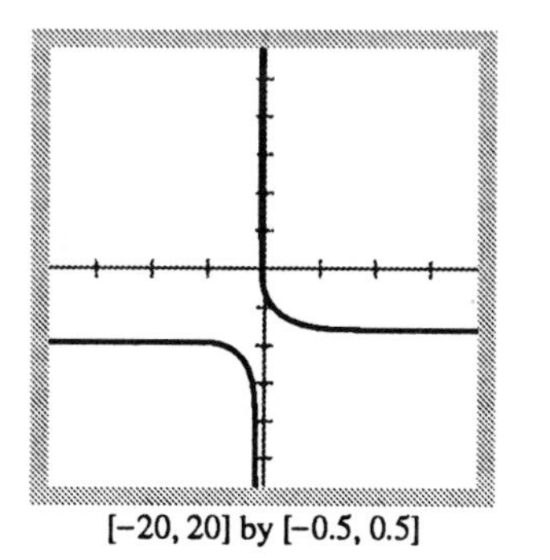

[−20, 20] by [−0.5, 0.5]

2.38 Notice that the values of $f(x) = -x/(7x+4)$ approach a limit as $x \to \infty$ and as $x \to -\infty$.

EXAMPLE 8 *A rational function in which the numerator and denominator have the same degree.* The graph of $f(x) = -x/(7x+4)$ in Fig. 2.38 suggests that $\lim_{x\to\infty} f(x)$ and $\lim_{x\to-\infty} f(x)$ are both about -0.14. Algebraically,

$$\lim_{x\to\infty} \frac{-x}{7x+4} = \lim_{x\to\infty} \frac{-1}{7+(4/x)} \qquad \text{Divide numerator and denominator by highest power of } x \text{ in denominator, in this case } x.$$

$$= \frac{-1}{7+0} \qquad \text{Known values}$$

$$= -\frac{1}{7}$$

$$= -0.142\ \ldots$$

Notice also that $\lim_{x\to-\infty} f(x) = -1/7$. The line $y = -1/7$ is a horizontal asymptote and an end behavior model for f (see Exercise 55).

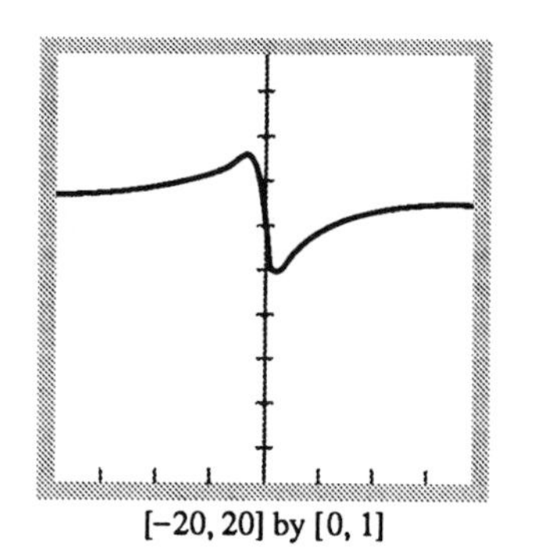

[−20, 20] by [0, 1]

2.39 Notice that the values of $f(x) = (2x^2 - x + 3)/(3x^2 + 5)$ approach a limit as $x \to \infty$ and as $x \to -\infty$.

EXAMPLE 9 *Another rational function in which the numerator and denominator have the same degree.* The graph of $f(x) = (2x^2 - x + 3)/(3x^2 + 5)$ in Fig. 2.39 suggests that $\lim_{x\to\infty} f(x)$ and $\lim_{x\to-\infty} f(x)$ are both about 0.65. Algebraically,

$$\lim_{x\to-\infty} \frac{2x^2 - x + 3}{3x^2+5} = \lim_{x\to-\infty} \frac{2-(1/x)+(3/x^2)}{3+(5/x^2)} \qquad \text{Divide numerator and denominator by } x^2.$$

$$= \frac{2-0+0}{3+0} \qquad \text{Known values}$$

$$= \frac{2}{3}$$

$$= 0.66\ \ldots$$

Notice also that $\lim_{x\to\infty} f(x) = 2/3$. The line $y = 2/3$ is a horizontal asymptote and an end behavior model for f.

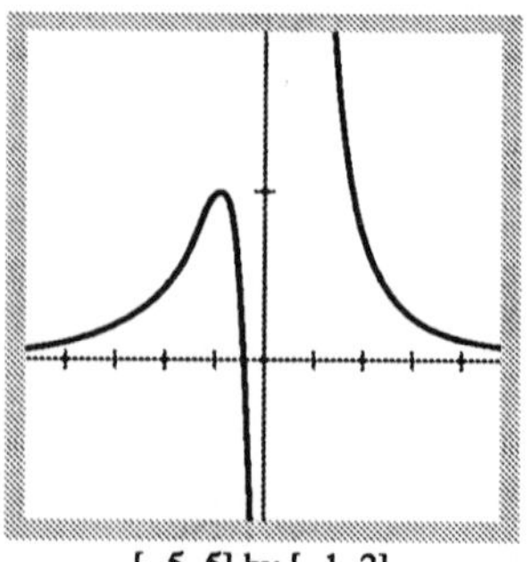

[−5, 5] by [−1, 2]

2.40 Notice that the values of $f(x) = (5x+2)/(2x^3 - 1)$ approach 0 as $x \to \infty$ and as $x \to -\infty$.

EXAMPLE 10 *A rational function in which the degree of the numerator is less than the degree of the denominator.* The graph of $f(x) = (5x+2)/(2x^3-1)$ in Fig. 2.40 suggests that $\lim_{x\to\infty} f(x) = \lim_{x\to-\infty} f(x) = 0$. Algebraically,

$$\lim_{x\to\pm\infty} \frac{5x+2}{2x^3-1} = \lim_{x\to\pm\infty} \frac{(5/x^2)+(2/x^3)}{2-(1/x^3)} \qquad \text{Divide numerator and denominator by } x^3.$$

$$= \frac{0+0}{2-0} \qquad \text{Known values}$$

$$= 0$$

The line $y = 0$ (x-axis) is a horizontal asymptote and an end behavior model of f. ∎

EXAMPLE 11 Sums and Products

a)

$$\lim_{x\to\infty}\left(\frac{-x}{7x+4}+\frac{5x+2}{2x^3-1}\right) = \lim_{x\to\infty}\frac{-x}{7x+4}+\lim_{x\to\infty}\frac{5x+2}{2x^3-1} \qquad \text{Sum Rule}$$

$$= -\frac{1}{7}+0 = -\frac{1}{7}. \qquad \text{Known values}$$

b)

$$\lim_{x\to\infty}\left(\frac{-x}{7x+4}\cdot\frac{5x+2}{2x^3-1}\right) = \lim_{x\to\infty}\frac{-x}{7x+4}\cdot\lim_{x\to\infty}\frac{5x+2}{2x^3-1} \qquad \text{Product Rule}$$

$$= -\frac{1}{7}\cdot 0 = 0. \qquad \text{Known values}$$

We could have multiplied the original fractions together first, to express the product as a single rational function, but it would have taken longer to get the answer that way. ∎

REMINDER

The *degree* of $a_nx^n + a_{n-1}x^{n-1} + \cdots + a_1x + a_0$ is n, the largest exponent. The *leading coefficient* is a_n.

A simple rule for finding limits of rational functions as $x \to \pm\infty$ is this: If the numerator and denominator have the same degree, the limit is the ratio of the leading coefficients. If the degree of the numerator is less than the degree of the denominator, the limit is zero. If the degree of the numerator is greater than the degree of the denominator, the limit is not a real number, as we shall see in a moment.

Lim $f(x) = \infty$ or lim $f(x) = -\infty$

As suggested by the behavior of $1/x$ as $x \to 0$, or x^2 as $x \to \infty$, we sometimes want to say such things as

$$\lim_{x\to c} f(x) = \infty, \qquad (1)$$

$$\lim_{x\to c^+} f(x) = \infty, \qquad (2)$$

$$\lim_{x\to c^-} f(x) = \infty, \qquad (3)$$

$$\lim_{x\to\infty} f(x) = \infty, \qquad (4)$$

$$\lim_{x\to-\infty} f(x) = \infty. \qquad (5)$$

Note, none of the limits exist in the sense of Section 2.1. However, in every instance, we mean that the value of $f(x)$ eventually exceeds any positive real number B. That is, for any real number B no matter how large, the values of f eventually satisfy the condition

$$f(x) > B. \tag{6}$$

Similarly, we write

$$\lim_{x\to c} f(x) = -\infty, \quad \lim_{x\to c^+} f(x) = -\infty, \quad \lim_{x\to c^-} f(x) = -\infty \tag{7}$$

to say that no matter how large the negative number $-B$ may be numerically, the values of f eventually satisfy the condition

$$f(x) < -B. \tag{8}$$

Polynomial End Behavior

Polynomial functions $f(x) = a_n x^n + a_{n-1}x^{n-1} + \cdots + a_1 x + a_0$ of degree 1 or higher have the property that

$$\lim_{x\to\infty} f(x) = \infty \text{ or } -\infty,$$

and

$$\lim_{x\to-\infty} f(x) = \infty \text{ or } -\infty.$$

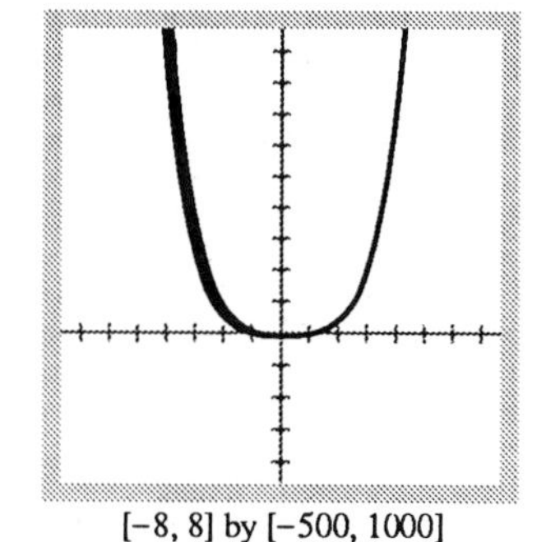

[−8, 8] by [−500, 1000]

2.41 $y = 3x^4$ is an end behavior model of $f(x) = 3x^4 - 2x^3 + 3x^2 - 5x + 6$.

It turns out that for $|x|$ large, the values of $f(x)$ and $a_n x^n$ are approximately the same. More precisely, $y = a_n x^n$ is an *end behavior model* of f, that is,

$$\lim_{x\to\pm\infty} \frac{f(x)}{a_n x^n} = 1 \qquad (a_n \neq 0).$$

EXAMPLE 12 Show that $y = 3x^4$ is an end behavior model of $f(x) = 3x^4 - 2x^3 + 3x^2 - 5x + 6$.

Solution The graphs of $y = 3x^4$ and f in Fig. 2.41 are nearly identical. This is visual evidence that the end behavior of f and $y = 3x^4$ are the same. Algebraically,

$$\begin{aligned}\lim_{x\to\pm\infty} \frac{f(x)}{3x^4} &= \lim_{x\to\pm\infty} \frac{3x^4 - 2x^3 + 3x^2 - 5x + 6}{3x^4} \\ &= \lim_{x\to\pm\infty} \left(1 - \frac{2}{3x} + \frac{1}{x^2} - \frac{5}{3x^3} + \frac{2}{x^4}\right) \\ &= 1.\end{aligned}$$

■

The algebraic techniques of Example 12 can be used to establish the following Theorem.

Theorem 10

End Behavior Models of Polynomial Functions

If $f(x) = a_n x^n + a_{n-1}x^{n-1} + \ldots + a_1 x + a_0$ $(a_n \neq 0)$, then $y = a_n x^n$ is an end behavior model of f.

Thus, there are really only four types of polynomial end behavior models:

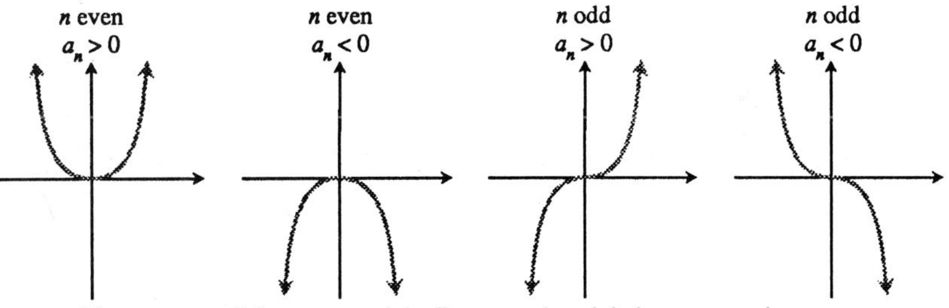

The steepness of the curves and the flatness at the origin increase as n increases

Vertical Asymptotes

Some functions have asymptotes that are vertical lines.

Definition

A line $x = a$ is a **vertical asymptote** of the graph of a function $y = f(x)$ if either $\lim_{x\to a^-} f(x) = \pm\infty$ or $\lim_{x\to a^+} f(x) = \pm\infty$.

EXAMPLE 13 Show that the y-axis is a vertical asymptote of $y = \dfrac{1}{x}$.

Solution

$$\lim_{x\to 0+} \frac{1}{x} = \infty$$

$$\lim_{x\to 0-} \frac{1}{x} = -\infty$$

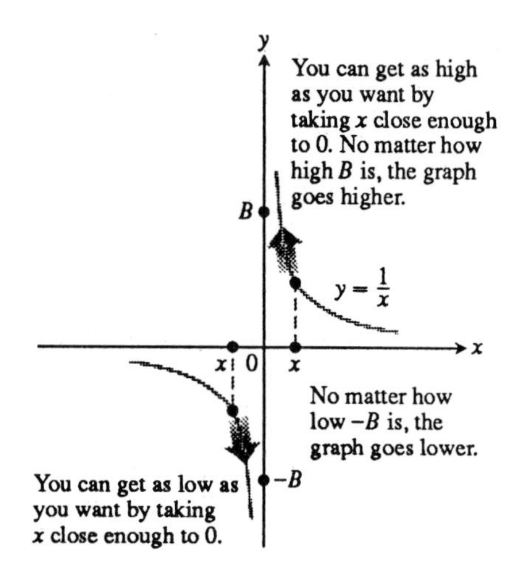

EXAMPLE 14 Show that the y-axis is a vertical asymptote of $y = \dfrac{1}{x^2}$.

Solution

$$\lim_{x\to 0+} \frac{1}{x^2} = \infty$$

$$\lim_{x\to 0-} \frac{1}{x^2} = \infty$$

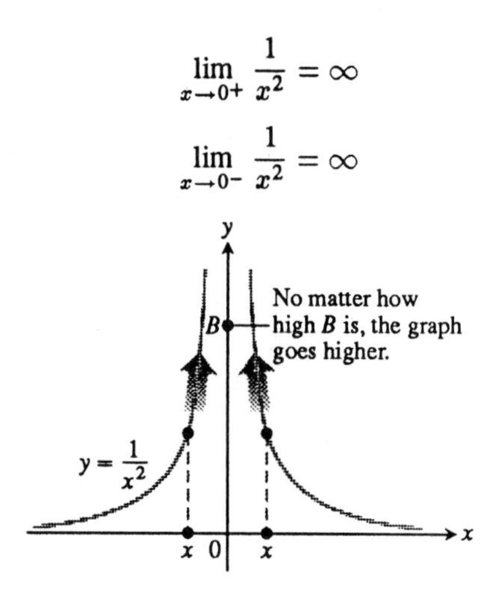

Use your grapher to support the information in the graphs above.

EXAMPLE 15 Determine the vertical asymptotes of $f(x) = \dfrac{1}{x-1}$.

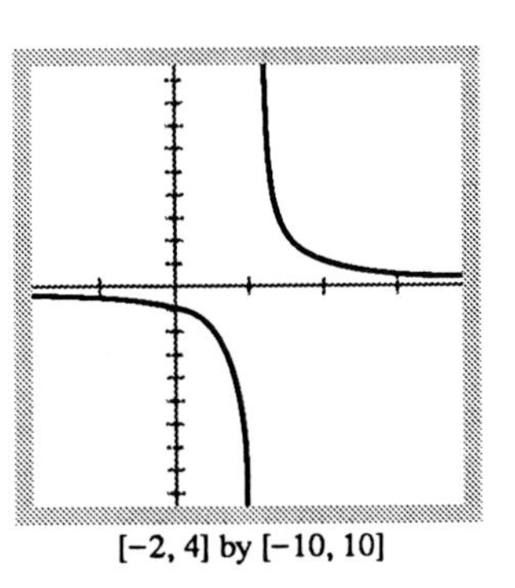

[−2, 4] by [−10, 10]

2.42 It appears that the line $x = 1$ is a vertical asymptote of $f(x) = 1/(x - 1)$.

Solution The graph of f in Fig. 2.42 suggests that $x = 1$ is a vertical asymptote and that $\lim_{x \to 1^+} 1/(x-1) = \infty$ and $\lim_{x \to 1^-} 1/(x-1) = -\infty$. This can be confirmed by investigating the values of f for values of x near 1. ▪

Some functions have several vertical asymptotes as the next example illustrates.

EXAMPLE 16 Determine all vertical asymptotes graphically and then confirm algebraically.

$$f(x) = \frac{x^2 - 4}{x^3 - 2x^2 - 9x + 18}$$

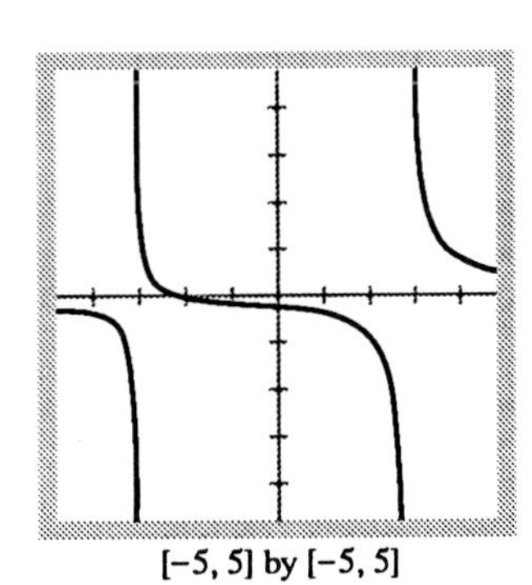

[−5, 5] by [−5, 5]

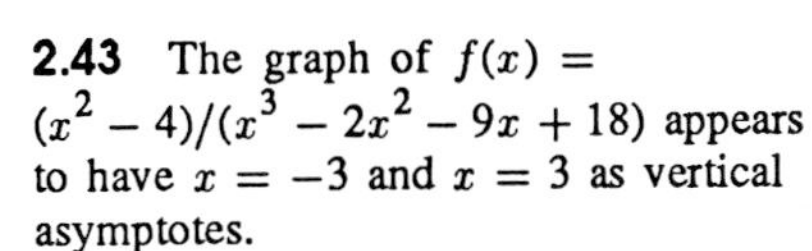

2.43 The graph of $f(x) = (x^2 - 4)/(x^3 - 2x^2 - 9x + 18)$ appears to have $x = -3$ and $x = 3$ as vertical asymptotes.

Solution The graph of f in Fig. 2.43 suggests that the graph of f has two vertical asymptotes, and they appear to be $x = -3$ and $x = 3$. Next, we factor the numerator and denominator of f.

$$\frac{x^2 - 4}{x^3 - 2x^2 - 9x + 18} = \frac{(x-2)(x+2)}{(x-2)(x^2-9)} = \frac{(x-2)(x+2)}{(x-2)(x+3)(x-3)}$$

The discontinuity of f at $x = 2$ is removable, so f does *not* have a vertical asymptote at $x = 2$. We can rewrite f as

$$\frac{x+2}{(x+3)(x-3)}.$$

Now, we can see that the vertical asymptotes are at $x = -3$ and $x = 3$. Notice

$$\frac{x+2}{x+3} \to \frac{5}{6} \text{ and } \frac{1}{x-3} \to \infty \text{ as } x \to 3^+$$

so that

$$f(x) = \frac{x+2}{(x+3)(x-3)} = \frac{x+2}{x+3} \cdot \frac{1}{x-3} \to \infty \text{ as } x \to 3^+.$$

In a similar manner, we can algebraically confirm that $\lim_{x \to 3^-} f(x) = -\infty$, $\lim_{x \to -3^+} f(x) = \infty$, and $\lim_{x \to -3^-} f(x) = -\infty$. ▪

Vertical Asymptotes

If the numerator and denominator of a rational function have no common factors, then factors of the form $x - a$ of the denominator correspond to vertical asymptotes of the form $x = a$.

Examples 15 and 16 illustrate important properties of limits that we list in the following theorem.

Theorem 11

Properties of Infinite Limits

If c and L are real numbers and $\lim_{x\to c} f(x) = L \neq 0$ and $\lim_{x\to c} g(x) = \pm\infty$, then

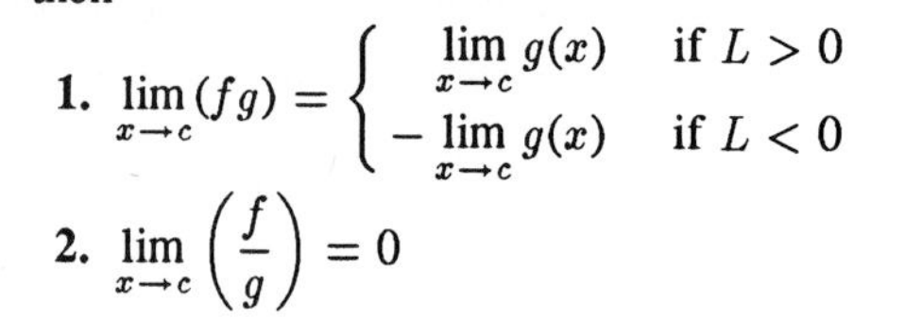

1. $\lim_{x\to c} (fg) = \begin{cases} \lim_{x\to c} g(x) & \text{if } L > 0 \\ -\lim_{x\to c} g(x) & \text{if } L < 0 \end{cases}$

2. $\lim_{x\to c} \left(\frac{f}{g}\right) = 0$

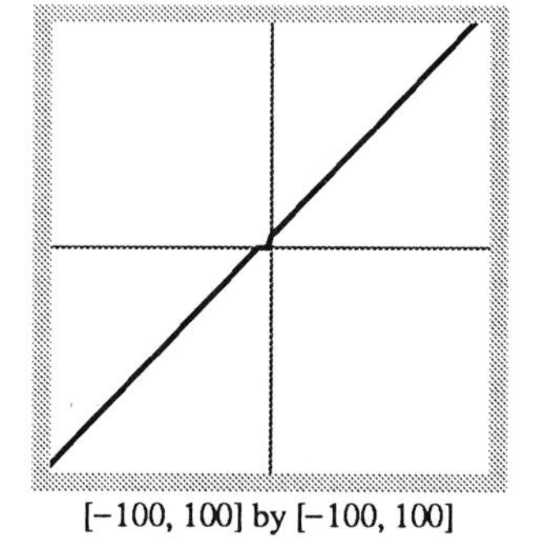

[−100, 100] by [−100, 100]

2.44 It appears that $y = x$ is an end behavior model of $f(x) = (x^2 + 4x + 5)/x$.

Rational Function End Behavior

Let $f(x) = g(x)/h(x)$ where g and h are polynomials. If $\deg g(x) \leq h(x)$ we have seen that f has a limit as $x \to \pm\infty$. Moreover, if $\lim_{x\to\pm\infty} f(x) = L$, then $y = L$ is an end behavior model and a horizontal asymptote of f. The situation for $\deg g > \deg h$ is more complex and more interesting. For example, we know that x^2 is an end behavior model of $x^2 + 4x + 5$. It seems reasonable that x^2/x is an end behavior model of $f(x) = \dfrac{x^2 + 4x + 5}{x}$. In fact,

$$\lim_{x\to\pm\infty} \frac{\frac{x^2+4x+5}{x}}{\frac{x^2}{x}} = \lim_{x\to\pm\infty} \frac{x^2 + 4x + 5}{x^2} = 1$$

confirming that $x = x^2/x$ is an end behavior model of $f(x) = (x^2+4x+5)/x$. A graph of f in a large viewing window also illustrates that x is an end behavior model of f (Fig. 2.44). It can be shown that $y = x + a$ is also an end behavior model of f. So, end behavior model is not unique. However, there is a special end behavior model that can be found using division.

$$\frac{x^2 + 4x + 5}{x} = x + 4 + \frac{5}{x}$$

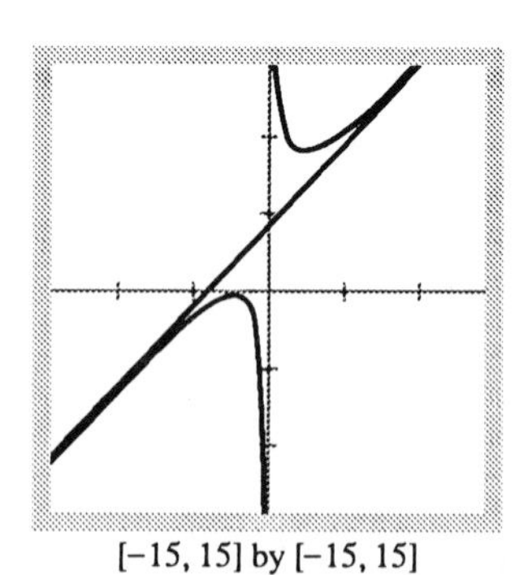

[−15, 15] by [−15, 15]

2.45 The graph of $f(x) = (x^2+4x+5)/x$ approaches the line $y = x + 4$ as $x \to \pm\infty$.

Among the end behavior models $x + a$ of f, it can be shown that $y = x + 4$ best approximates f (Fig. 2.45). The line $y = x + 4$ is sometimes called a slant asymptote of f. We will call it the end behavior asymptote to distinguish it among all the possible end behavior models of f.

Definition

> **End Behavior Asymptote**
>
> Let $f(x) = g(x)/h(x)$ be a rational function with $\deg g > \deg h$. Let $Q(x)$ and $R(x)$ be the quotient and remainder when $g(x)$ is divided by $h(x)$, that is,
>
> $$f(x) = Q(x) + \frac{R(x)}{h(x)}.$$
>
> $Q(x)$ is called the **end behavior asymptote** of f.

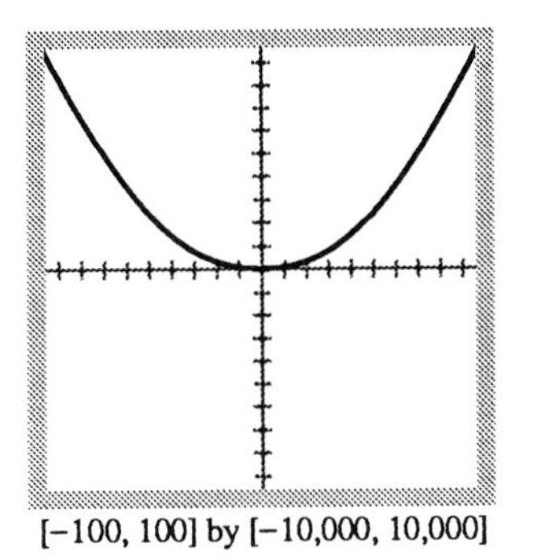

[−100, 100] by [−10,000, 10,000]

2.46 The graph of $f(x) = (x^3 - 4x^2 + 2x + 8)/(x - 2)$ looks like the graph of $y = x^2$ in large viewing windows.

$Q(x)$ is an end behavior model of f and it can be shown that among the end behavior models of f, $Q(x)$ gives the best approximation of f.

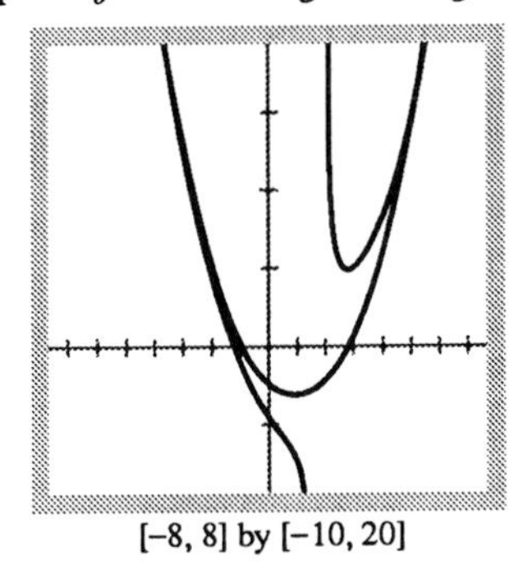

[−8, 8] by [−10, 20]

2.47 The graph of $f(x) = (x^3 - 4x^2 + 2x + 8)/(x - 2)$ approaches the graph of the *end behavior asymptote* $y = x^2 - 2x - 2$ as $x \to \pm\infty$.

EXAMPLE 17 Show graphically that $y = x^2$ is an end behavior model for f. Then, determine the end behavior asymptote algebraically and support graphically.

$$f(x) = \frac{x^3 - 4x^2 + 2x + 8}{x - 2}$$

Solution Notice that x^3 is an end behavior model of the numerator of f, x is an end behavior model of the denominator of f, and $x^2 = x^3/x$ is an end behavior model of f. The graph of f in Fig. 2.46 supports $y = x^2$ as an end behavior model of f. The quotient is $x^2 - 2x - 2$ and the remainder is 4 when $x^3 - 4x^2 + 2x + 8$ is divided by $x - 2$. Thus, $y = x^2 - 2x - 2$ is the end behavior asymptote of f and $f(x) = x^2 - 2x - 2 + 4/(x - 2)$. Figure 2.47 shows that except near the vertical asymptote $x = 2$ of f, $y = x^2 - 2x - 2$ is a good approximation of f. ∎

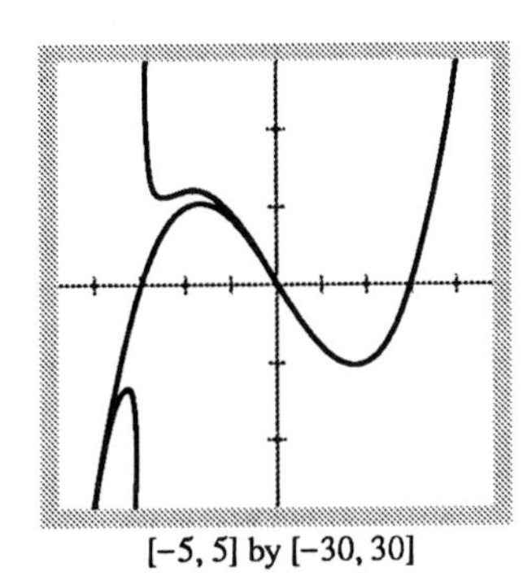

[−5, 5] by [−30, 30]

2.48 The graph of $f(x) = (x^4 + 3x^3 - 9x^2 - 27x + 2)/(x + 3)$ approaches the graph of the *end behavior asymptote* $y = x^3 - 9x$ as $x \to \pm\infty$.

EXAMPLE 18 Determine the end behavior asymptote algebraically and support graphically.

$$f(x) = \frac{x^4 + 3x^3 - 9x^2 - 27x + 2}{x + 3}$$

Solution The quotient is $x^3 - 9x$ and the remainder is 2 when $x^4 + 3x^3 - 9x^2 - 27x + 2$ is divided by $x + 3$. Thus, $x^3 - 9x$ is the end behavior asymptote of f and provides a good approximation to f for values of x away from $x = -3$ (Fig. 2.48). ∎

> **Summary for Rational Functions**
>
> a) $\lim_{x\to\pm\infty} \dfrac{f(x)}{g(x)}$ is 0 if $\deg(f) < \deg(g)$
>
> b) $\lim_{x\to\pm\infty} \dfrac{f(x)}{g(x)}$ is finite if $\deg(f) = \deg(g)$
>
> c) $\lim_{x\to\pm\infty} \dfrac{f(x)}{g(x)}$ is infinite if $\deg(f) > \deg(g)$
>
> However, an end behavior model and the end behavior asymptote give much deeper insight into the behavior of f for values of x large in absolute value.

Changing Variables with Substitutions

Sometimes a change of variable can turn an unfamiliar expression into one whose limit we know how to find. Here are two examples.

EXAMPLE 19

$$\lim_{x\to\infty} \sin\frac{1}{x} = \lim_{\theta\to 0+} \sin\theta \qquad \text{Substitute } \theta = 1/x. \text{ Then } \theta \to 0^+ \text{ as } x \to \infty.$$

$$= 0. \qquad \text{Known value}$$

■

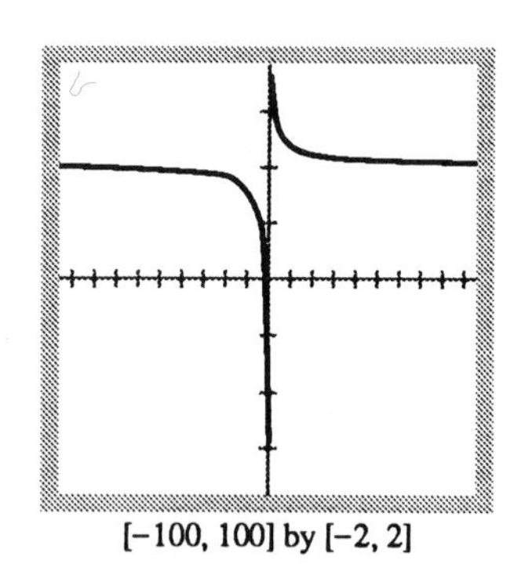

[−100, 100] by [−2, 2]

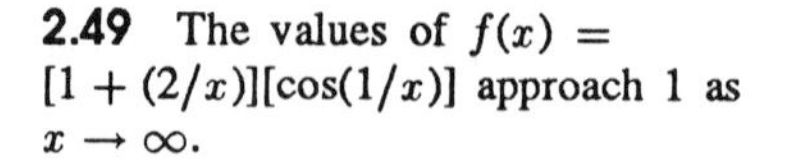

2.49 The values of $f(x) = [1 + (2/x)][\cos(1/x)]$ approach 1 as $x \to \infty$.

EXAMPLE 20 Determine the limit graphically and then confirm algebraically using substitution.

$$\lim_{x\to-\infty} \left(1 + \frac{2}{x}\right)\left(\cos\frac{1}{x}\right)$$

Solution The graph of $f(x) = [1 + (2/x)][\cos(1/x)]$ in Fig. 2.49 suggests that the limit is 1. Algebraically,

$$\lim_{x\to-\infty} \left(1 + \frac{2}{x}\right)\left(\cos\frac{1}{x}\right) = \lim_{\theta\to 0-} (1 + 2\theta)(\cos\theta) \qquad \text{Substitute } \theta = \frac{1}{x}. \text{ Then } \theta \to 0^- \text{ as } x \to -\infty.$$

$$= \lim_{\theta\to 0-} (1 + 2\theta) \lim_{\theta\to 0-} \cos\theta \qquad \text{Product Rule}$$

$$= 1 \cdot 1 = 1. \qquad \text{Known values}$$

■

Exercises 2.4

Use graphs to find the limits of the functions defined by the expressions in Exercises 1–12 (a) as $x \to \infty$ and (b) as $x \to -\infty$.

1. $\dfrac{2x+3}{5x+7}$
2. $\dfrac{2x^3+7}{x^3-x^2+x+7}$
3. $\dfrac{x+1}{x^2+3}$
4. $\dfrac{3x+7}{x^2-2}$
5. $\dfrac{3x^2-6x}{4x-8}$
6. $\dfrac{x^4}{x^3+1}$
7. $\dfrac{1}{x^3-4x+1}$
8. $\dfrac{10x^5+x^4+31}{x^6}$
9. $\dfrac{-2x^3-2x+3}{3x^3+3x^2-5x}$
10. $\dfrac{-x^4}{x^4-7x^3+7x^2+9}$
11. $\left(\dfrac{-x}{x+1}\right)\left(\dfrac{x^2}{5+x^2}\right)$
12. $\left(\dfrac{2}{x}+1\right)\left(\dfrac{5x^2-1}{x^2}\right)$

Find the limits in Exercises 13–20.

13. $\lim_{x\to 2+} \dfrac{1}{x-2}$
14. $\lim_{x\to 2-} \dfrac{1}{x-2}$
15. $\lim_{x\to 2+} \dfrac{x}{x-2}$
16. $\lim_{x\to 2-} \dfrac{x}{x-2}$
17. $\lim_{x\to -3+} \dfrac{1}{x+3}$
18. $\lim_{x\to -3-} \dfrac{1}{x+3}$
19. $\lim_{x\to -3+} \dfrac{x}{x+3}$
20. $\lim_{x\to -3-} \dfrac{x}{x+3}$

Find the end behavior asymptote algebraically and support graphically.

21. $f(x)=\dfrac{x-2}{2x^2+3x-5}$
22. $T(y)=\dfrac{2y+3}{4-y^2}$
23. $g(x)=\dfrac{3x^2-x+5}{x^2-4}$
24. $f(x)=\dfrac{2-3x^2}{5+2x-6x^2}$
25. $f(x)=\dfrac{x^2-2x+3}{x+2}$
26. $f(x)=\dfrac{x^2-3x-7}{x+3}$
27. $g(x)=\dfrac{x^3-2x+1}{x-2}$
28. $g(x)=\dfrac{x^4-2x^2-x+3}{x^2-4}$
29. $f(x)=\dfrac{x}{x^2+3}-\dfrac{x^2+2}{1+x-x^2}$
30. $g(x)=\dfrac{x^2-4}{x+1}+\dfrac{x}{x^2-3x+2}$

In Exercises 31–34, use graphs to find the values of the limits.

31. $\lim_{x\to 2+} \dfrac{1}{x^2-4}$
    $\lim_{x\to 2-} \dfrac{1}{x^2-4}$
    $\lim_{x\to -2+} \dfrac{1}{x^2-4}$
    $\lim_{x\to -2-} \dfrac{1}{x^2-4}$
32. $\lim_{x\to 1+} \dfrac{x}{x^2-1}$
    $\lim_{x\to 1-} \dfrac{x}{x^2-1}$
    $\lim_{x\to -1+} \dfrac{x}{x^2-1}$
    $\lim_{x\to -1-} \dfrac{x}{x^2-1}$
33. $\lim_{x\to -2+} \dfrac{x^2-1}{2x+4}$
    $\lim_{x\to -2-} \dfrac{x^2-1}{2x+4}$
34. $\lim_{x\to 0+} \left(x^2+\dfrac{4}{x}\right)$
    $\lim_{x\to 0-} \left(x^2+\dfrac{4}{x}\right)$

Find the limits in Exercises 35–42.

35. $\lim_{x\to 0+} \dfrac{\lfloor x \rfloor}{x}$
36. $\lim_{x\to 0-} \dfrac{\lfloor x \rfloor}{x}$
37. $\lim_{x\to \infty} \dfrac{|x|}{|x|+1}$
38. $\lim_{x\to -\infty} \dfrac{x}{|x|}$
39. $\lim_{x\to 0+} \dfrac{1}{\sin x}$
40. $\lim_{x\to 0-} \dfrac{1}{\sin x}$
41. $\lim_{x\to (\pi/2)+} \dfrac{1}{\cos x}$
42. $\lim_{x\to (\pi/2)-} \dfrac{1}{\cos x}$
43. Let $f(x)=\begin{cases} \dfrac{1}{x}, & x<0 \\ -1, & x\ge 0. \end{cases}$
    Find $\lim f(x)$ as $x \to -\infty$, 0^-, 0^+, and ∞.
44. Let $f(x)=\begin{cases} \dfrac{x-2}{x-1}, & x\le 0 \\ \dfrac{1}{x^2}, & x>0. \end{cases}$
    Find $\lim f(x)$ as $x \to -\infty$ 0^-, 0^+, and ∞.

Find the limits in Exercises 45-50.

45. $\lim_{x\to \infty} \left(2+\dfrac{\sin x}{x}\right)$
46. $\lim_{x\to -\infty} \dfrac{\sin x}{x}$
47. $\lim_{x\to \infty} \left(1+\cos\dfrac{1}{x}\right)$
48. $\lim_{x\to \infty} x\sin\dfrac{1}{x}$
49. $\lim_{x\to \infty} \dfrac{\sin 2x}{x}$
50. $\lim_{x\to \infty} \dfrac{\cos(1/x)}{1+(1/x)}$

As we mentioned, the Sandwich Theorem in Section 2.3 also holds for limits as $x \to \pm\infty$. Use the theorem to find the limits in Exercises 51 and 52.

51. Find $\lim_{x\to\infty} f(x)$ and $\lim_{x\to -\infty} f(x)$ if

$$\frac{2x^2}{x^2+1} < f(x) < \frac{2x^2+5}{x^2}.$$

52. *The greatest integer function.* Find $\lim_{x\to\infty} \lfloor x \rfloor / x$ and $\lim_{x\to-\infty} \lfloor x \rfloor / x$ given that

$$\frac{x-1}{x} < \frac{\lfloor x \rfloor}{x} \leq 1 \quad (x \neq 0).$$

53. Draw complete graphs of the following functions in the same viewing rectangle: $2x$, $2x^3$, $2x^5$, $2x^7$. Compare their limits as $x \to \pm\infty$, and the steepness of the graphs. How can you distinguish their behavior? Explain.

54. Draw complete graphs of the following functions in the same viewing rectangle: $-3x^2$, $-3x^4$, $-3x^6$, $-3x^8$. Compare their limits as $x \to \pm\infty$, and the steepness of the graphs. How can you distinguish their behavior? Explain.

55. Show that $y = -1/7$ is an end behavior model for the function

$$f(x) = -\frac{x}{7x+4}$$

of Example 8. *Hint:* Show that $\lim_{x\to\pm\infty} f(x)/(-1/7) = 1$.

56. Show that $y = 2/3$ is an end behavior model for the function

$$f(x) = -\frac{2x^2 - x + 3}{3x^2 + 5}$$

of Example 9. *Hint:* Show that $\lim_{x\to\pm\infty} f(x)/(2/3) = 1$.

57. Let $\lim_{x\to c} f(x) = 0$ and $\lim_{x\to c} g(x) = \infty$. Give examples to show that $\lim_{x\to c}(fg)$ can be 0, finite and nonzero, or infinite.

58. Let L be a real number, $\lim_{x\to c} f(x) = L$, and $\lim_{x\to c} g(x) = \pm\infty$. Can $\lim_{x\to c}(f \pm g)$ be determined? Explain.

59. Prove Theorem 10.

60. *Use graphs to find the limits.*

a) $\lim_{x\to\infty} \dfrac{\ln(x+1)}{\ln x}$.

b) Does the 1 in $\ln(x+1)$ really matter? Suppose you have 999 there instead. What do you get for the value of

$$\lim_{x\to\infty} \frac{\ln(x+999)}{\ln x}?$$

c) $\lim_{x\to\infty} \dfrac{\ln x^2}{\ln x}$.

d) $\lim_{x\to\infty} \dfrac{\ln x}{\log x}$.

The behavior you see here will all be explained in Chapter 7.

2.5 Controlling Function Outputs: Target Values

We sometimes want the outputs of a function $y = f(x)$ to lie near a particular target value y_0. This need can come about in different ways. A gas-station attendant, asked for \$5.00 worth of gas, will try to pump the gas to the nearest cent. A mechanic grinding a 3.385-in. cylinder bore will not let the bore exceed this value by more than .002 in. A pharmacist making ointments will measure the ingredients to the nearest milligram. The formal definition of limit given in the next section requires that we control function outputs.

So, the question becomes: How accurate do our machines and instruments have to be to keep the outputs within useful bounds? When we express this question with mathematical symbols, we ask: How closely must we control x to keep $y = f(x)$ within an acceptable interval about some particular target value y_0? We give examples to show how to answer this question.

EXAMPLE 1 Controlling a Linear Function Notice if $x = 4$ and $y = 2x - 1$, then $y = 7$. How close to $x_0 = 4$ must we hold x to be sure that $y = 2x - 1$ lies within 2 units of $y_0 = 7$? Estimate graphically and confirm algebraically.

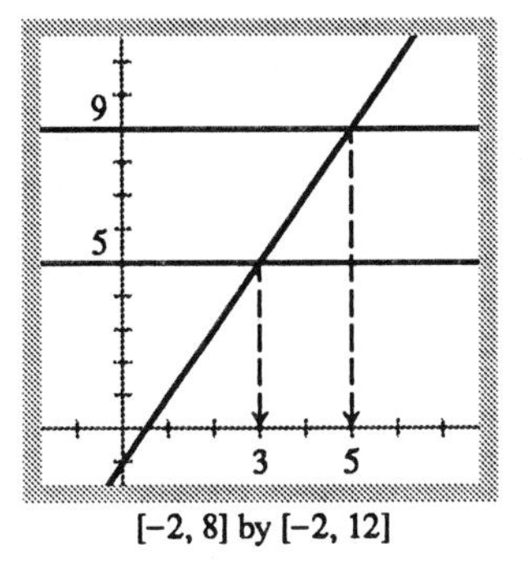

2.50 Keeping x between 3 and 5 will keep $y = 2x - 1$ between $y = 5$ and $y = 9$.

Solution We are asked: For what values of x is $|y - 7| < 2$? Equivalently,

$$|y - 7| < 2$$
$$-2 < y - 7 < 2$$
$$5 < y < 9.$$

We can use the graphs of $y = 2x - 1$, $y = 5$, and $y = 9$ in Fig. 2.50 to estimate that y will be between 5 and 9 if x is between 3 and 5. Thus x must be within 1 unit of $x_0 = 4$. Algebraically, we first express $|y - 7|$ in terms of x:

$$|y - 7| = |(2x - 1) - 7| = |2x - 8|.$$

The question then becomes: What values of x satisfy the inequality $|2x - 8| < 2$? To find out, we change

$$|2x - 8| < 2$$
$$\text{to } -2 < 2x - 8 < 2$$
$$\text{to } 6 < 2x < 10$$
$$\text{to } 3 < x < 5.$$

■

Notice that $x_0 = 4$ is the midpoint of the interval $3 < x < 5$ found in Example 1. Thus, the interval is symmetric about $x = 4$ and can be expressed in the form $|x - 4| < 1$. We could choose smaller intervals about $x_0 = 4$, say $|x - 4| < 0.1$, in which the values of y would still lie within 2 units of $y_0 = 7$. However, we try to choose the interval about x_0 in which the values of y are controlled to be as large as possible. The answers you provide for the exercises of this section should use this convention.

The procedure used in Example 1 will not always lead to an interval symmetric about x_0. However, we can restrict the interval to one that is symmetric about x_0 in which the function outputs are within the prescribed range of the target value y_0. This will be useful in Section 2.6.

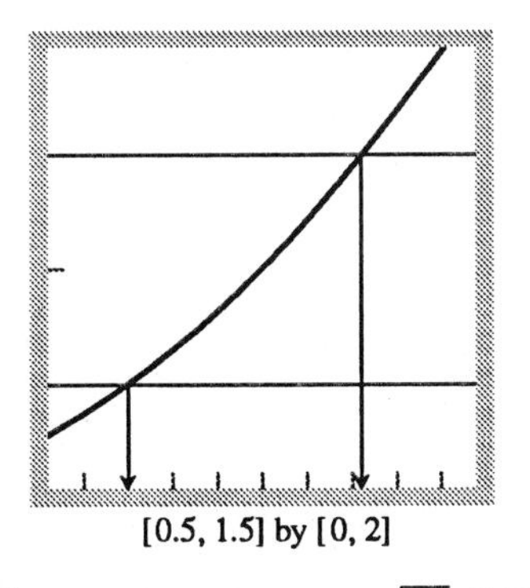

2.51 Keeping x between $\sqrt{0.5}$ (0.71) and $\sqrt{1.5}$ (1.22) will keep $y = x^2$ between $y = 0.5$ and $y = 1.5$.

EXAMPLE 2 Controlling a Quadratic Function Observe if $x = 1$ and $y = x^2$, then $y = 1$. How close to $x_0 = 1$ must we hold x to be sure that $y = x^2$ lies within 0.5 units of $y_0 = 1$? Estimate graphically and confirm algebraically.

Solution Notice that $|y - 1| < 0.5$ is equivalent to $0.5 < y < 1.5$. We need to estimate the x-coordinates of the points of intersection in the first quadrant of the horizontal lines $y = 0.5$ and $y = 1.5$ with the graph of $y = x^2$ (Fig. 2.51). The x-coordinate of the point of intersection of $y = 0.5$ with $y = x^2$ is $\sqrt{0.5}$, the positive solution to $x^2 = 0.5$. Similarly, $\sqrt{1.5}$ is the x-coordinate of the point of intersection of $y = 1.5$ with $y = x^2$. Thus, $0.5 < y < 1.5$ if $\sqrt{0.5} < x < \sqrt{1.5}$, or, accurate to hundredths, $0.71 < x < 1.22$. To be safe, we have rounded up the left endpoint approximation and rounded down the right endpoint. Algebraically,

$$|y - 1| < 0.5$$

$$|x^2 - 1| < 0.5$$

$$-0.5 < x^2 - 1 < 0.5$$

$$0.5 < x^2 < 1.5$$

$$\sqrt{0.5} < x < \sqrt{1.5}$$ For a, b, and c nonnegative, $a < b < c \Rightarrow \sqrt{a} < \sqrt{b} < \sqrt{c}$.

Notice the interval $0.71 < x < 1.22$ of Example 2 is *not* symmetric about $x_0 = 1$.

EXAMPLE 3 Find a symmetric interval about $x_0 = 1$ for which $y = x^2$ lies within 0.5 units of $y_0 = 1$.

Solution For the interval $0.71 < x < 1.22$ of Example 2, the distance 0.22 from $x_0 = 1$ to the right end point 1.22 is smaller than the distance 0.29 from $x_0 = 1$ to the left end point 0.71. A symmetric interval about $x_0 = 1$ is given by

$$1 - 0.22 < x < 1 + 0.22,$$

or equivalently, $$|x - 1| < 0.22$$

[0, 3] by [0, 3]

2.52 Keeping x between 1.75 and 2.25 will keep $y = \sqrt{3x - 2}$ between $y = 1.8$ and $y = 2.2$.

EXAMPLE 4 Controlling a Square Root Function Notice if $x = 2$ and $y = \sqrt{3x - 2}$, then $y = 2$. Find a symmetric interval about $x_0 = 2$ for which $y = \sqrt{3x - 2}$ lies within 0.2 units of $y_0 = 2$. Estimate graphically and confirm algebraically.

Solution We need to determine the values of x for which $\left|\sqrt{3x - 2} - 2\right| < 0.2$, or $1.8 < \sqrt{3x - 2} < 2.2$. Figure 2.52 shows the graphs of $y = 1.8$, $y = 2.2$, and $y = \sqrt{3x - 2}$. Accurate to hundredths, $\sqrt{3x - 2} = 1.8$ if $x = 1.75$ and $\sqrt{3x - 2} = 2.2$ if $x = 2.28$. The end points have been appropriately rounded. Thus, $1.8 < y < 2.2$ if $1.75 < x < 2.28$. If we want a symmetric interval about $x = 2$, then we can restrict $1.75 < x < 2.28$ to $1.75 < x < 2.25$ or $|x - 2| < 0.25$. Notice 0.25 is the distance from $x = 2$ to the nearest endpoint, 1.75, of the interval $1.75 < x < 2.28$. Algebraically,

$$|y - 2| < 0.2$$

$$\left|\sqrt{3x - 2} - 2\right| < 0.2$$

$$1.8 < \sqrt{3x - 2} < 2.2$$

$$(1.8)^2 < 3x - 2 < (2.2)^2$$

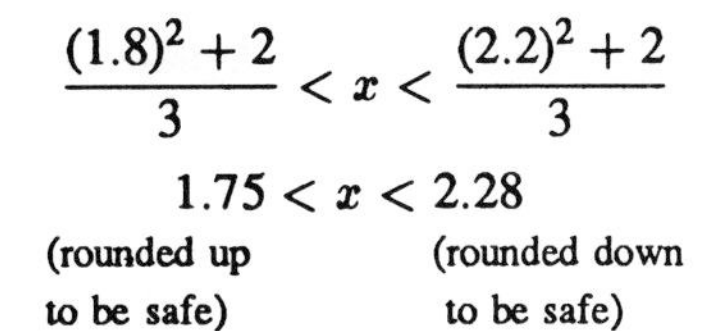

$$\frac{(1.8)^2+2}{3} < x < \frac{(2.2)^2+2}{3}$$

$$1.75 < x < 2.28$$

(rounded up to be safe) (rounded down to be safe)

Alternatively, we can start with $|x-2| < 0.25$ and show that $|y-2| < 0.2$.

$$|x-2| < 0.25$$

$$-0.25 < x-2 < 0.25$$

$$1.75 < x < 2.25$$

$$3(1.75)-2 < 3x-2 < 3(2.25)-2$$

$$3.25 < 3x-2 < 4.75$$

$$\sqrt{3.25} < \sqrt{3x-2} < \sqrt{4.75}$$

$$1.8 < \sqrt{3x-2} < 2.2$$

(rounded down to be safe) (rounded up to be safe)

$$-0.2 < \sqrt{3x-2}-2 < 0.2$$

$$\left|\sqrt{3x-2}-2\right| < 0.2$$

[0, 1] by [0, 1]

2.53 Keeping x between 0.31 and 0.77 will keep $y = \sin x$ between $y = 0.3$ and $y = 0.7$.

EXAMPLE 5 Controlling a Trigonometric Function Observe if $x = \pi/6$ and $y = \sin x$, then $y = 0.5$. How close to $x_0 = \pi/6$ must we hold x to be sure that $y = \sin x$ lies within 0.2 units of $y_0 = 0.5$?

Solution We need to determine the values of x near $\pi/6$ for which $|\sin x - 0.5| < 0.2$ or $0.3 < \sin x < 0.7$. Figure 2.53 shows the graphs of $y = 0.3$, $y = 0.7$, and $y = \sin x$. If x is between 0 and 1 and $\sin x = 0.3$. Then $x = \sin^{-1} 0.3$ or, rounded appropriately to hundredths, $x = 0.31$. Similarly, if $x = \sin^{-1} 0.7$, then, rounded appropriately to hundredths, $x = 0.77$. Thus, $0.3 < y < 0.7$ if $0.31 < x < 0.77$.

In Example 5, it can be shown that $|x-\pi/6| < 0.2$ is a symmetric interval about $x_0 = \pi/6$ for which $y = \sin x$ lies within 0.2 units of $y_0 = 0.5$. To do this we need only estimate and compare $\pi/6 - 0.31$ and $0.77 - \pi/6$.

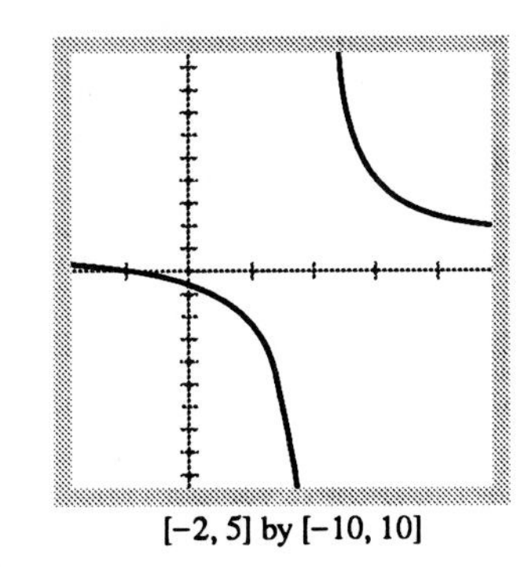

[−2, 5] by [−10, 10]

2.54 A complete graph of $f(x) = (x+1)/(x-2)$.

EXAMPLE 6 Controlling a Rational Function Notice if $x = 3$ and $f(x) = (x+1)/(x-2)$, then $f(3) = 4$. Find an interval about $x_0 = 3$ for which the values of f lie within 0.1 units of 4. Estimate graphically and confirm algebraically.

Solution We need to determine the values of x for which $3.9 < f(x) < 4.1$. Figure 2.54 gives a complete graph of f, and Fig. 2.55 gives the graph of $f, y = 3.9$, and $y = 4.1$ near the point (3,4). To algebraically determine the values of x for which $f(x) = 3.9$ and $f(x) = 4.1$, we need to solve the

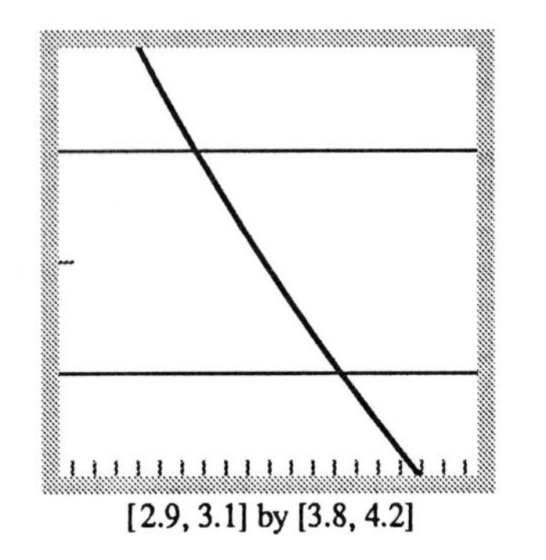

[2.9, 3.1] by [3.8, 4.2]

2.55 Keeping x between 2.97 and 3.03 will keep $f(x) = (x+1)/(x-2)$ between 3.9 and 4.1.

equations

$$\frac{x+1}{x-2} = 3.9 \text{ and } \frac{x+1}{x-2} = 4.1.$$

The solution to the first equation is 3.03 and the solution to the second 2.97. Thus, $f(x) = (x+1)/(x-2)$ is within 0.1 units of 4 if $2.97 < x < 3.03$, or $|x-3| < 0.03$. We can verify directly that $|f(x)-4| < 0.1$ if $|x-3| < 0.03$, provided we first observe that $f(x) = \dfrac{x+1}{x-2} = 1 + \dfrac{3}{x-2}$.

$$|x-3| < 0.03$$

$$2.97 < x < 3.03$$

$$0.97 < x-2 < 1.03$$

$$\frac{3}{0.97} > \frac{3}{x-2} > \frac{3}{1.03}$$

$$\frac{3}{1.03} < \frac{3}{x-2} < \frac{3}{0.97}$$

$$1 + \frac{3}{1.03} < 1 + \frac{3}{x-2} < 1 + \frac{3}{0.97}$$

$$1 + \frac{3}{1.03} < \frac{x+1}{x-2} < 1 + \frac{3}{0.97}$$

$$\frac{3}{1.03} - 3 < \frac{x+1}{x-2} - 4 < \frac{3}{0.97} - 3$$

$$-0.1 < f(x) - 4 < 0.1$$

(rounded down to be safe) (rounded up to be safe)

$$|f(x) - 4| < 0.1$$

■

In the next example, the endpoints of the interval found on the x-axis are negative. To round a negative number up (the left endpoint) we must decrease its absolute value, and to round a negative number down (the right endpoint) we must increase its absolute value. In absolute value, this is the opposite of what we have done in the previous examples with intervals on the positive x-axis.

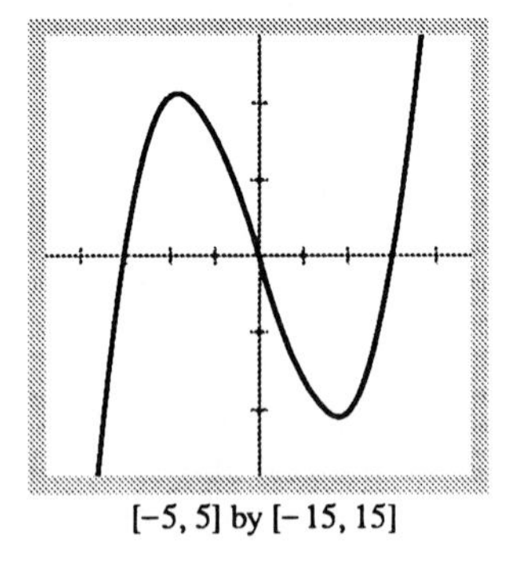

[−5, 5] by [−15, 15]

2.56 A complete graph of $f(x) = x^3 - 9x$.

EXAMPLE 7 Find an interval about $x_0 = -1$ for which $f(x) = x^3 - 9x$ lies within 0.5 units of $f(-1) = 8$.

Solution We need to determine the values of x near $x_0 = -1$ for which $7.5 < f(x) < 8.5$. Figure 2.56 gives a complete graph of f and Fig. 2.57 gives the graph of f, $y = 7.5$ and $y = 8.5$ near the point $(-1, 8)$. We can zoom-in to determine that the x-coordinate of the point of intersection of $y = 7.5$ and $y = f(x)$ is -0.92. Similarly, the x-coordinate of the point of intersection of $y = 8.5$ and $y = f(x)$ is -1.08. Thus, $|f(x) - 8| < 0.5$ if

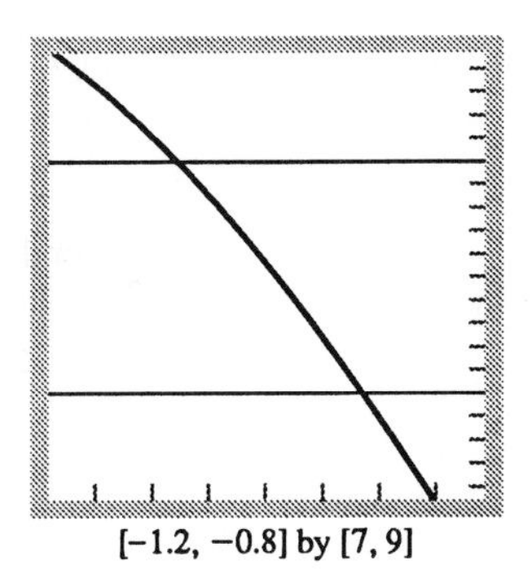

2.57 Keeping x between -1.08 and -0.92 will keep $f(x) = x^3 - 9x$ between 7.5 and 8.5.

$-1.08 < x < -0.92$ or $|x + 1| < 0.08$. The endpoints of the interval have been appropriately rounded.

EXAMPLE 8 Controlling the Area of a Circle In what interval about $r_0 = 10$ must we hold r to be sure that $A = \pi r^2$ lies within π square units of $A_0 = 100\pi$?

Solution We want to know the values of r for which $|A - A_0| < \pi$. To find them, change the inequality from

$$|A - A_0| < \pi$$

to $$|\pi r^2 - 100\pi| < \pi$$

to $$-\pi < \pi r^2 - 100\pi < \pi$$

to $$-1 < r^2 - 100 < 1$$

to $$99 < r^2 < 101$$

to $$\sqrt{99} < r < \sqrt{101}.$$

For a, b, and c nonnegative, $a < b < c \Rightarrow \sqrt{a} < \sqrt{b} < \sqrt{c}$.

The interval of possible radii is the open interval from $r = \sqrt{99}$ to $r = \sqrt{101}$ (Fig. 2.58).

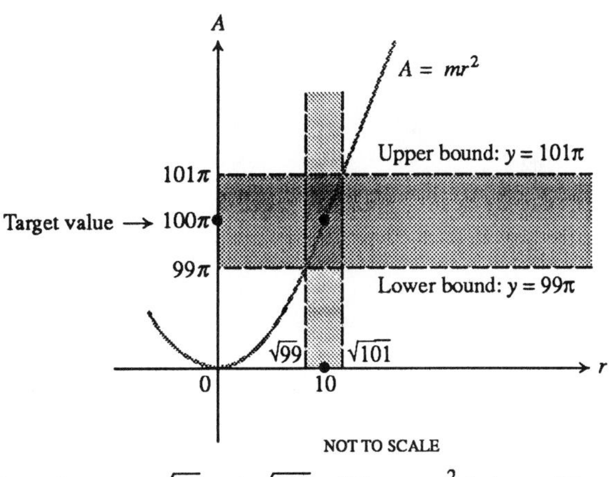

2.58 Keeping r between $\sqrt{99}$ and $\sqrt{101}$ will keep πr^2 between 99π and 101π.

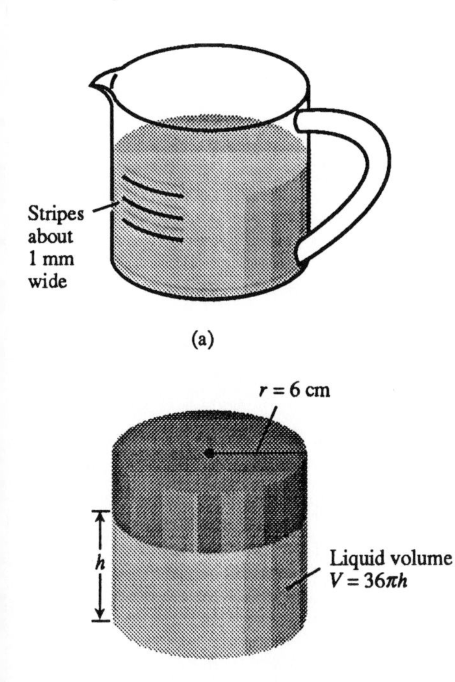

2.59 A typical 1-liter measuring cup (a) modeled in (b) as a right circular cylinder of radius $r = 6$ cm. To get a liter of water to the nearest 1%, how accurately must we measure h? See Example 9.

EXAMPLE 9 Why the Stripes on a 1-liter Kitchen Measuring Cup Are About a Millimeter Wide The interior of a typical 1-L measuring cup is a right circular cylinder of radius 6 cm (Fig. 2.59). The volume of water we put in the cup is therefore a function of the level h to which the cup is filled, the formula being

$$V = \pi(6)^2 h = 36\pi h.$$

How closely do we have to measure h to measure out one liter of water (1000 cm^3) with an error of no more than 1% (10 cm^3)?

Solution In terms of V and h, we want to know in what interval of values to hold h to make V satisfy the inequality

$$|V - 1000| = |36\pi h - 1000| \le 10.$$

To find out, we change

$$|36\pi h - 1000| \le 10$$

to

$$-10 \le 36\pi h - 1000 \le 10$$

to

$$990 \le 36\pi h \le 1010$$

to

$$\frac{990}{36\pi} \le h \le \frac{1010}{36\pi}$$

to

$$8.8 \le h \le 8.9.$$

The interval in which we should hold h is about one tenth of a centimeter wide (one millimeter). With stripes one millimeter wide, we can therefore expect to measure a liter of water with an accuracy of one percent, which is more than enough accuracy for cooking. ≡

Exercises 2.5

1. If $2 < x < 6$, which of the following statements about x are true and which are false?

a) $0 < x < 4$ b) $0 < x - 2 < 4$

c) $1 < \frac{x}{2} < 3$ d) $\frac{1}{6} < \frac{1}{x} < \frac{1}{2}$

e) $1 < \frac{6}{x} < 3$ f) $|x - 4| < 2$

g) $-6 < -x < 2$ h) $-6 < -x < -2$

2. If $-1 < y - 5 < 1$, which of the following statements about y are true and which are false?

a) $4 < y < 6$ b) $|y - 5| < 1$

c) $y > 4$ d) $y < 6$

e) $0 < y - 4 < 2$ f) $2 < \frac{y}{2} < 3$

g) $\frac{1}{6} < \frac{1}{y} < \frac{1}{4}$ h) $-6 < y < -4$

In Exercises 3–10, match each absolute value inequality with the interval it determines.

3. $|x + 3| < 1$ a) $-2 < x < 1$

4. $|x - 5| < 2$ b) $-1 < x < 3$

5. $\left|\frac{x}{2}\right| < 1$ c) $3 < x < 7$

6. $|1 - x| < 2$ d) $-\frac{5}{2} < x < -\frac{3}{2}$

7. $|2x - 5| \le 1$ e) $-2 < x < 2$

8. $|2x + 4| < 1$ f) $-4 < x < 4$

9. $\left|\frac{x - 1}{2}\right| < 1$ g) $-4 < x < -2$

h) $2 \le x \le 3$

10. $\left|\frac{2x + 1}{3}\right| < 1$ i) $-2 \le x \le 2$

The inequalities in Exercises 11–18 define intervals. Describe each interval with inequalities that do not involve absolute values.

11. $|y - 1| \le 2$ **12.** $|y + 2| < 1$

13. $|3y - 7| < 2$ **14.** $|2y + 5| < 1$

15. $\left|\frac{y}{2} - 1\right| \le 1$ **16.** $\left|2 - \frac{y}{2}\right| < \frac{1}{2}$

17. $|1 - y| < \frac{1}{10}$ **18.** $\left|\frac{7 - 3y}{2}\right| < 1$

Describe the intervals in Exercises 19–22 with absolute value inequalities of the form $|x - x_0| < D$. It may help to draw a picture of the interval first.

19. $3 < x < 9$ **20.** $-3 < x < 9$

21. $-5 < x < 3$ **22.** $-7 < x < -1$

23. How close to $x_0 = -1$ must we hold x to be sure that $y = x^2$ lies within 0.5 units of $y_0 = 1$? Estimate graphically and confirm algebraically. (See Example 2.)

24. Find an interval in $[\pi/2, \pi]$ for which $y = \sin x$ lies within 0.2 units of $y_0 = 0.5$. (See Example 5.)

25. Find an interval in $[0, \pi/2]$ for which $y = \cos x$ lies within 0.2 units of $y_0 = 0.4$.

26. Find an interval in $[0, \pi/2]$ for which $y = \tan x$ lies within 0.2 units of $y_0 = 2$.

27. Find an interval in $[-\pi/2, 0]$ for which $y = \cos x$ lies within 0.2 units of $y_0 = 0.4$.

28. Find an interval in $[-\pi/2, 0]$ for which $y = \tan x$ lies within 0.2 units of $y_0 = -2$.

Each of Exercises 29–36 gives a function $y = f(x)$, a number E, and a target value y_0. In what interval must we hold x in each case to be sure that $y = f(x)$ lies within E units of y_0? Estimate graphically and confirm algebraically.

29. $y = x^2, \ E = 0.1, \ y_0 = 100, \ x_0 > 0$

30. $y = x^2, \ E = 0.1, \ y_0 = 100, \ x_0 < 0$

31. $y = \sqrt{x-7}, \ E = 0.1, \ y_0 = 4$

32. $y = \sqrt{19-x}, \ E = 1, \ y_0 = 3$

33. $y = 120/x, \ E = 1, \ y_0 = 5$

34. $y = 1/4x, \ E = 1/2, \ y_0 = 1$

35. $y = \dfrac{3-2x}{x-1}, \ E = 0.1, \ y_0 = -3$

36. $y = \dfrac{3x+8}{x+2}, \ E = 0.1, \ y_0 = 1$

Each of Exercises 37–42 gives a function $y = f(x)$, a number E, and a target value y_0. In what interval must we hold x in each case to be sure that $y = f(x)$ lies within E units of y_0?

37. $y = x^2 - 5, \ E = 0.5, \ y_0 = 11, \ x_0 > 0$

38. $y = x^2 - 5, \ E = 0.5, \ y_0 = 11, \ x_0 < 0$

39. $y = x^3 - 9x, \ E = 0.2, \ y_0 = 5, \ x_0$ near -3

40. $y = x^4 - 10x^2, \ E = 0.2, \ y_0 = -5, \ x_0$ near 1

41. $y = e^x, \ E = 0.1, \ y_0 = 0.5$

42. $y = \ln x, \ E = 0.1, \ y_0 = 2$

Each of Exercises 43–46 gives a function $y = f(x)$, a number E, a point x_0, and a target value y_0. In what interval about x_0 must we hold x in each case to be sure that $y = f(x)$ lies within E units of y_0? Describe the interval with an absolute value inequality of the form $|x - x_0| < D$.

43. $y = x + 1, \ E = 0.5, \ x_0 = 3, \ y_0 = 4$

44. $y = 2x - 1, \ E = 1, \ x_0 = -2, \ y_0 = -5$

45. $y = 2x^2 + 1, \ E = 0.2, \ x_0 = 1, \ y_0 = 3$

46. $y = \sqrt{2x-3}, \ E = 0.2, \ x_0 = 3.5, \ y_0 = 2$

47. *Grinding engine cylinders.* Before contracting to grind engine cylinders to a cross-section area of 9 in^2, you want to know how much deviation from the ideal cylinder diameter of $x_0 = 3.385$ in. you can allow and still have the area come within 0.01 in^2 of the required 9 in^2. To find out, you let $A = \pi(x/2)^2$ and look for the interval in which you must hold x to make $|A - 9| \le 0.01$. What interval do you find?

48. *Manufacturing electrical resistors.* Ohm's law for electrical circuits, like the one shown in Fig. 2.60, states that $V = RI$. In this equation, V is a constant voltage, I is the current in amperes, and R is the resistance in ohms. Your firm has been asked to supply the resistors for a circuit in which V will be 120 volts and I is to be 5 ± 0.1 amperes. In what interval does R have to lie for I to be within 0.1 amperes of the target value $I_0 = 5$?

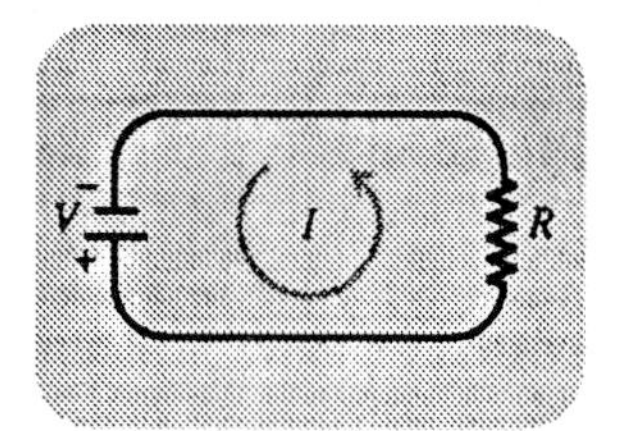

2.60 The circuit in Exercise 48.

2.6 Defining Limits Formally with Epsilons and Deltas

We have spent four sections calculating limits. Our basic tools were graphical estimation, the Sandwich Theorem and the theorems that gave the calculation rules for sums, differences, products, and quotients. With these theorems in hand, we started with sensible assumptions about the limits of constant functions and the identity function and worked our way to limits of rational functions. In every case, the calculations were straightforward and the results made sense.

The only problem is that we do not yet know why the theorems we used to get our results are true. The entire calculus depends on these theorems and we haven't a clue about why they hold. If we were to try to prove them now, however, we would quickly realize that we never said what a limit really is (except by example) and that we do not have a definition good enough to establish even the simple facts that

$$\lim_{x \to x_0} (k) = k \qquad \text{and} \qquad \lim_{x \to x_0} (x) = x_0. \tag{1}$$

To establish these facts and to understand why the limit theorems hold, we must take one final step and define *limit* formally. This section is where we take that step.

As it turns out, the mathematics we need to make the notion of limit precise enough to be useful is the same mathematics we used in Section 2.5 to study target values of functions. There, we had a function $y = f(x)$, a target value y_0, and an upper bound E on the amount of error we could allow the output value y to have. We wanted to know how close we had to keep x to a particular value x_0 so that y would lie within E units of y_0. In symbols, we were asking for a value of D that would make the inequality $|x - x_0| < D$ imply the inequality $|y - y_0| < E$. The number D described the amount by which x could differ from x_0 and still give y-values that approximated y_0 with an error less than E.

In the limit discussions that follow, we shall use the traditional Greek letters δ (delta) and ϵ (epsilon) in place of the English letters D and E. These are the letters that Cauchy and Weierstrass used in their pioneering work on continuity in the nineteenth century. In their arguments, δ meant "différence" (French for *difference*) and ϵ meant "erreur" (French for *error*).

As you read along, please keep in mind that the purpose of this section is not to calculate limits of particular functions. We already know how to do that. The purpose here is to develop a technical definition of limit that is good enough to establish the limit theorems on which our calculations depend.

The Definition of Limit

Suppose we are watching the values of a function $f(x)$ as x approaches x_0 (without x taking on the value x_0 itself). What do we have to know about the values of f to say that they have a particular number L as their limit? What observable pattern in their behavior would guarantee their eventual approach to L?

Certainly we want to be able to say that $f(x)$ stays within one tenth of a unit of L as soon as x stays within a certain radius r_1 of x_0, as shown here:

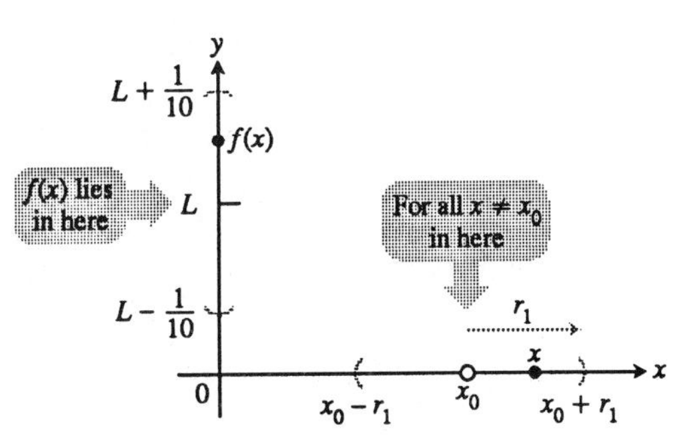

But that in itself is not enough, because as x continues on its course toward x_0, what is to prevent $f(x)$ from jittering about within the interval from $L - 1/10$ to $L + 1/10$ without tending toward L?

We need to say also that as x continues toward x_0, the number $f(x)$ has to get still closer to L. We might say this by requiring $f(x)$ to lie within 1/100 of a unit of L for all values of x within some smaller radius r_2 of x_0:

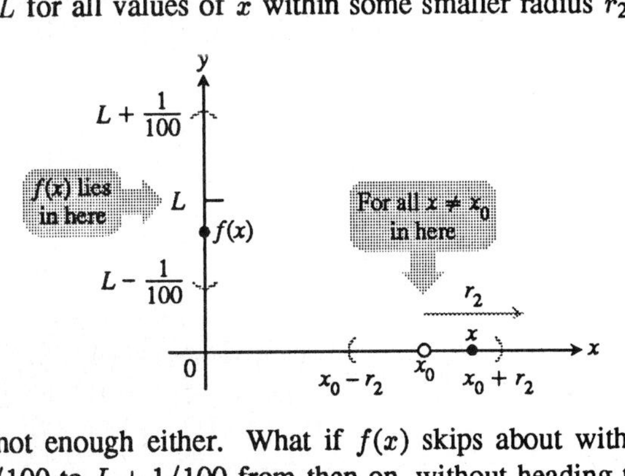

But this is not enough either. What if $f(x)$ skips about within the interval from $L - 1/100$ to $L + 1/100$ from then on, without heading toward L?

We had better require that $f(x)$ lie within 1/1000 of a unit of L after a while. That is, for all values of x within some still smaller radius r_3 of x_0, all the values of $y = f(x)$ should lie in the interval

$$L - \frac{1}{1000} < y < L + \frac{1}{1000},$$

as shown here:

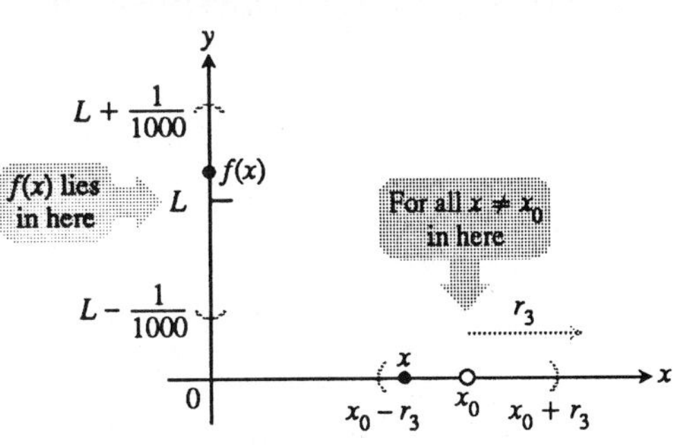

This still does not guarantee that $f(x)$ will now move toward L as x approaches x_0. Even if $f(x)$ has not skipped about before, it might start now. We need more.

We need to require that for *every* interval about L, no matter how small, we can find an interval of numbers about x_0 whose f-values all lie within that interval about L. In other words, given any positive radius ϵ about L, there should exist some positive radius δ about x_0 such that for all x within δ units of x_0 (except $x = x_0$ itself) the values $y = f(x)$ lie within ϵ units of L:

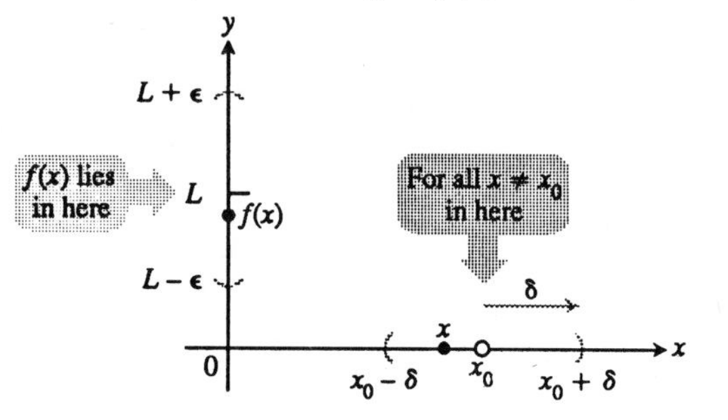

If f satisfies these requirements we will say that

$$\lim_{x \to x_0} f(x) = L.$$

Here, at last, is a mathematical way to say "the closer x gets to x_0, the closer $y = f(x)$ must get to L."

Definition

> The **limit** of $f(x)$ as x approaches x_0 is the number L if the following criterion holds:
>
> Given any radius $\epsilon > 0$ about L there exists a radius $\delta > 0$ about x_0 such that for all x
>
> $$0 < |x - x_0| < \delta \quad \text{implies} \quad |f(x) - L| < \epsilon. \qquad (2)$$

If it turns out that $f(x_0) = L$, then the inequality $0 < |x - x_0| < \delta$ can be replaced by $|x - x_0| < \delta$ because the inequality $|f(x) - L| < \epsilon$ is automatically satisfied for $x = x_0$.

To return to the notions of error and difference, we might think of machining something like a generator shaft to a close tolerance. We try for diameter L, but since nothing is perfect, we must be satisfied to get the diameter $f(x)$ somewhere between $L - \epsilon$ and $L + \epsilon$. The δ is the measure of how accurate our control setting for x must be to guarantee this degree of accuracy in the diameter of the shaft.

Examples—Testing the Definition

Whenever someone proposes a new definition, it is a good idea to see if it gives results that are consistent with past experience. For instance, our experience tells us that as x approaches 1, the number $5x - 3$ approaches $5 - 3 = 2$. If our new definition were to lead to some other result, we would want to throw the definition out and look for a new one. The following three examples are included in part to show that the definition in Eq. (2) gives the kinds of results we want.

EXAMPLE 1 Testing the Definition Show that

$$\lim_{x\to 1}(5x-3)=2.$$

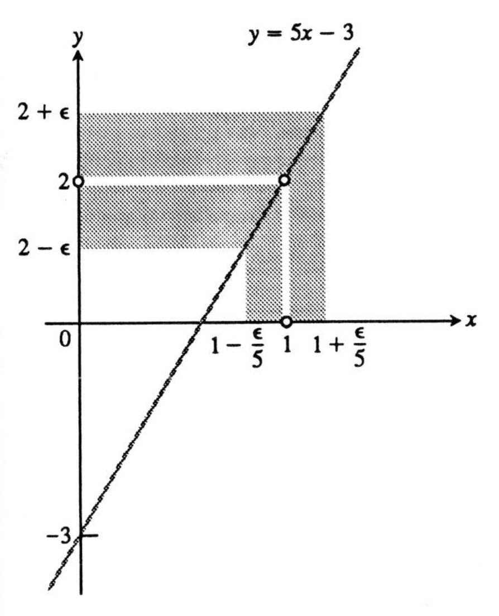

2.61 For the function $f(x) = 5x - 3$, we find that $|x-1| < \epsilon/5$ will guarantee $|f(x)-2| < \epsilon$.

Solution In the definition of limit, we set $x_0 = 1$, $f(x) = 5x-3$, and $L = 2$. To show that $\lim_{x\to 1}(5x-3) = 2$, we need to show that for any number $\epsilon > 0$ there exists a number $\delta > 0$ such that for all x,

$$0 < |x-1| < \delta \quad \Rightarrow \quad |(5x-3)-2| < \epsilon. \qquad (3)$$

(The symbol $\Rightarrow$ is read "implies.") To find a suitable value for δ, we change the ϵ-inequality from

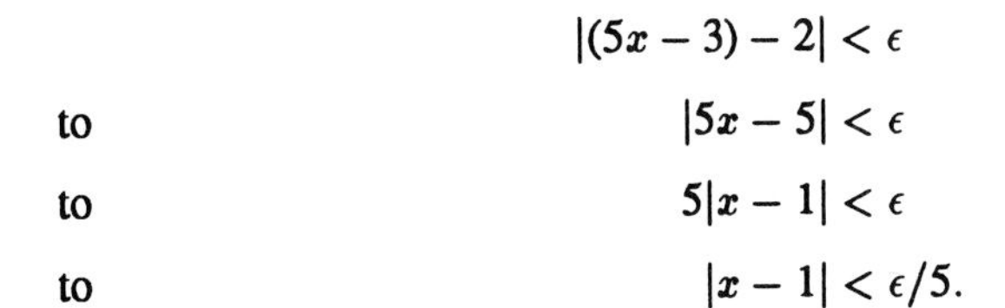

$$|(5x-3)-2| < \epsilon$$

to

$$|5x-5| < \epsilon$$

to

$$5|x-1| < \epsilon$$

to

$$|x-1| < \epsilon/5.$$

Because the above inequalities are equivalent, the last line here tells us that the original ϵ-inequality, and hence the implication in (3), will hold if we choose $\delta = \epsilon/5$ (Fig. 2.61).

The value $\delta = \epsilon/5$ is not the only value that will make the implication in (3) hold. Any smaller positive δ will do as well. The definition does not ask for a "best" δ, just one that will work.

EXAMPLE 2 Establishing a Basic Fact Show that for any number x_0

$$\lim_{x\to x_0}(x) = x_0.$$

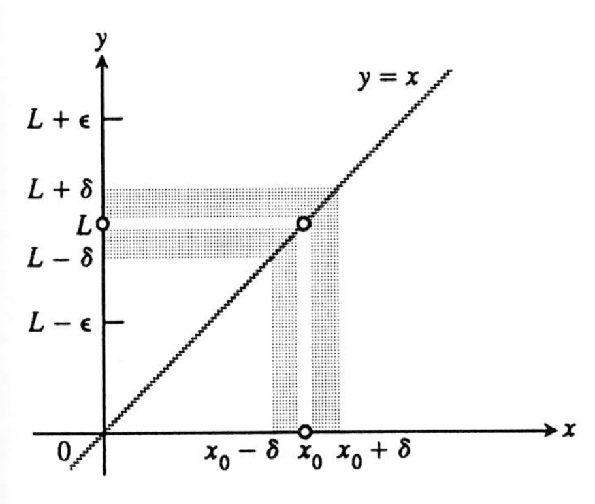

2.62 For the function $f(x) = x$, we find that $|x - x_0| < \delta$ will guarantee $|f(x) - x_0| < \epsilon$ whenever $\delta \le \epsilon$.

Solution In the definition of limit, we set $f(x) = x$ and $L = x_0$. To show that $\lim_{x\to x_0}(x) = x_0$ we must show that for any $\epsilon > 0$ there exists a $\delta > 0$ such that for all x,

$$0 < |x - x_0| < \delta \quad \Rightarrow \quad |x - x_0| < \epsilon.$$

The implication will hold if δ is ϵ itself or any smaller positive number (Fig. 2.62).

EXAMPLE 3 Establishing Another Basic Fact Let $f(x) = k$ be the function whose outputs have the constant value k. Show that for any number x_0

$$\lim_{x\to x_0} f(x) = k.$$

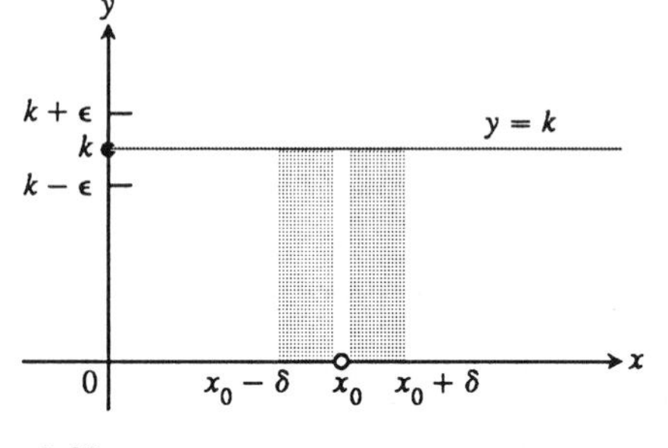

2.63 For the function $f(x) = k$, we find $|f(x) - k| < \epsilon$ for any positive δ.

Solution In the definition of limit, we set $f(x) = k$ and $L = k$. To show that $\lim_{x \to x_0} f(x) = k$, we must show that for any $\epsilon > 0$ there exists a $\delta > 0$ such that for all x,

$$0 < |x - x_0| < \delta \quad \Rightarrow \quad |k - k| < \epsilon.$$

This implication will hold for any positive δ, because $|k - k| = 0$ is less than every positive ϵ for all x (Fig. 2.63).

Finding Deltas for Given Epsilons

Here is a numerical example of finding δ for a given ϵ. This is very similar to what we did in Section 2.5.

EXAMPLE 4 If $f(x) = 1/x$, then $\lim_{x \to 0.5} f(x) = 2$. Notice $f(0.5) = 2$. Let $\epsilon = 0.01$. Find a value of $\delta > 0$ for which $|x - 0.5| < \delta$ implies $|f(x) - 2| < \epsilon$.

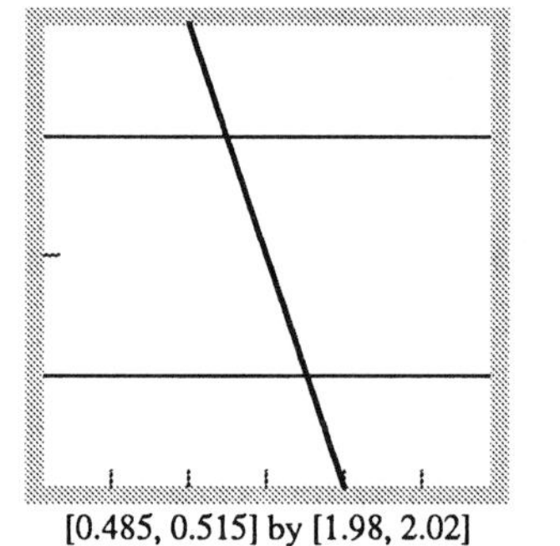

[0.485, 0.515] by [1.98, 2.02]

2.64 Keeping x between 0.498 and 0.502 will keep $f(x) = 1/x$ between $y = 1.99$ and $y = 2.01$.

Solution Notice that $|f(x) - 2| < 0.01$ is equivalent to $1.99 < f(x) < 2.01$. We can use the graph of $f, y = 1.99$, and $y = 2.01$ in Fig. 2.64 to estimate that $|f(x) - 2| < 0.01$ if $0.498 < x < 0.502$, or $|x - 0.5| < 0.002$. Thus, we can choose $\delta = 0.002$. Alternatively, we can use equivalent inequalities to determine δ algebraically.

$$\left| \frac{1}{x} - 2 \right| < 0.01$$

$$1.99 < \frac{1}{x} < 2.01$$

$$\frac{1}{2.01} < x < \frac{1}{1.99}$$

We can use a calculator to show that the distance between $x_0 = 0.5$ and the left endpoint $1/2.01$ is less than the distance between $x_0 = 0.5$ and the right endpoint, $1/1.99$. Accurate to 8 decimal places, the smaller distance is 0.00248756. This estimate has been rounded down so we can choose $\delta = 0.5 - 1/2.01 = 0.00248756$.

In Example 5, we extend Example 4 and find a $\delta > 0$ for arbitrary ϵ.

EXAMPLE 5 Let $f(x) = 1/x$ and $\epsilon > 0$. Find a value of $\delta > 0$ for which $|x - 0.5| < \delta$ implies $|f(x) - 2| < \epsilon$.

Solution First we show that, without any loss of generality, we can assume $\epsilon < 2$. Suppose $\epsilon \geq 2, 0 < \epsilon_1 < 2$, and we can find a $\delta > 0$ such that

$$|x - 0.5| < \delta \quad \Rightarrow \quad |f(x) - 2| < \epsilon_1.$$

Then,

$$|x - 0.5| < \delta \quad \Rightarrow \quad |f(x) - 2| < \epsilon_1 < 2 \leq \epsilon$$

so that

$$|x-0.5|<\delta \quad \Rightarrow \quad |f(x)-2|<\epsilon.$$

Thus, we can assume $0<\epsilon<2$.

The following inequalities are equivalent.

$$\left|\frac{1}{x}-2\right|<\epsilon$$

$$-\epsilon<\frac{1}{x}-2<\epsilon$$

$$2-\epsilon<\frac{1}{x}<2+\epsilon$$

$$\frac{1}{2-\epsilon}>x>\frac{1}{2+\epsilon} \qquad \epsilon<2 \Rightarrow 2-\epsilon>0$$

$$\frac{1}{2+\epsilon}<x<\frac{1}{2-\epsilon}$$

Because $\epsilon>0$, we have $2-\epsilon<2<2+\epsilon$. Thus, $\dfrac{1}{2+\epsilon}<\dfrac{1}{2}<\dfrac{1}{2-\epsilon}$. The distance from $x_0=0.5$ to the left endpoint $1/(2+\epsilon)$ is

$$\frac{1}{2}-\frac{1}{2+\epsilon}=\frac{\epsilon}{2(2+\epsilon)},$$

and the distance from $x_0=0.5$ to the right endpoint $1/(2-\epsilon)$ is

$$\frac{1}{2-\epsilon}-\frac{1}{2}=\frac{\epsilon}{2(2-\epsilon)}.$$

Graphs can be used, if necessary, to show that $\dfrac{\epsilon}{2(2+\epsilon)}$ is smaller than $\dfrac{\epsilon}{2(2-\epsilon)}$ (see Exercise 36). Thus, we can choose $\delta=\dfrac{\epsilon}{2(2+\epsilon)}$. ∎

Locally Straight Functions

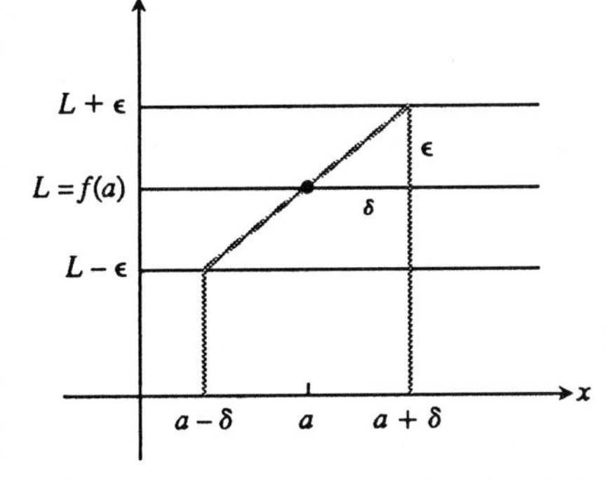

2.65 A zoom-in view of a function f that is locally straight and $\lim_{x\to a} f(x)=L$.

In Examples 1 and 5 we found δ's that corresponded to arbitrary ϵ's. Generally, finding a δ corresponding to an arbitrary ϵ can be very difficult because δ can depend on ϵ in a complicated way. In Chapter 3 we introduce differentiable functions. Functions that are differentiable are locally straight, that is, as we zoom-in, the graph of the function will appear to be a straight line. We can estimate the relationship between δ and ϵ at $x=a$ where $\lim_{x\to a} f(x)=L$ and f is locally straight at $x=a$.

Figure 2.65 shows the graph of such a function near $x=a$ between the horizontal lines $y=L+\epsilon$ and $y=L-\epsilon$. It turns out that if f is differentiable at $x=a$, then f is continuous at $x=a$ so that $L=f(a)$. We choose δ as indicated in Fig. 2.65. This is the largest δ can be so that $0<|x-a|<\delta$ implies $|f(x)-L|<\epsilon$. Let m be the slope of the straight line graph of f in Fig. 2.65. Because m may be negative we have $|m|=\epsilon/\delta$ and $\delta=\epsilon/|m|$. An appropriate estimate for m gives an appropriate estimate for δ as illustrated in the next example.

EXAMPLE 6 If $f(x) = x/(x^2 - 1)$, then $\lim_{x\to 2} f(x) = 2/3$. Let $\epsilon > 0$ and estimate $\delta > 0$ so that $|x - 2| < \delta$ implies $|f(x) - 2/3| < \epsilon$.

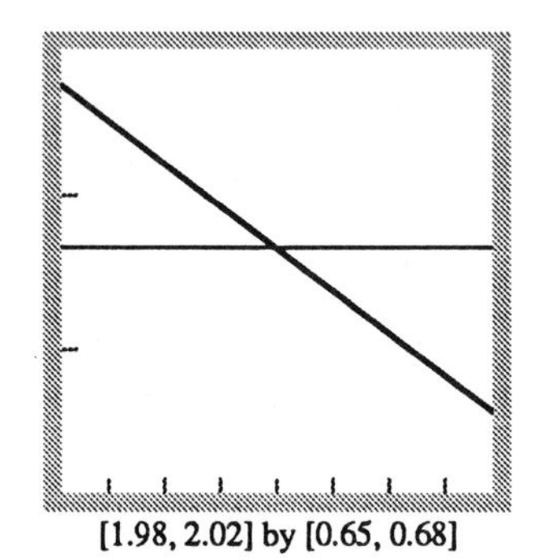

2.66 A zoom-in view of $f(x) = x/(x^2 - 1)$ and $y = 2/3$.

Solution Figure 2.66 gives a zoom-in view of f near the point (2,2/3) that appears to be a straight line, and the graph of $y = 2/3$. We can find the coordinates of two points on the graph of f in Fig. 2.66 to estimate the slope of the straight line graph to be $m = -0.5559$. Thus, $\epsilon/\delta = |m|$ or $\delta = \epsilon/|m|$ (see Fig. 2.65). To be safe we choose a smaller value for δ by replacing $|m|$ by 0.6 and see that $\delta = \epsilon/0.6$, or $\delta = (5/3)\epsilon$.

How Limit Theorems Are Proved

Although we shall not ask you to prove limit theorems yourself, we want to show how a typical proof goes, if only to support our claim that having a precise definition of limit now makes it possible for us to prove the limit theorems on which calculus depends. Our example will be the proof of the sum rule for limits (the first part of Theorem 1(c) from Section 2.1). You can find a proof of the rest of Theorem 1 in Appendix 2.

Theorem 1

Part 1(c):

If $\lim_{x\to x_0} f_1(x) = L_1$ and $\lim_{x\to x_0} f_2(x) = L_2$, then $\lim_{x\to x_0} (f_1(x) + f_2(x)) = L_1 + L_2$.

Proof To show that $\lim_{x\to x_0}(f_1(x) + f_2(x)) = L_1 + L_2$, we must show that for any $\epsilon > 0$ there exists a $\delta > 0$ such that for all x

$$0 < |x - x_0| < \delta \quad\Rightarrow\quad |f_1(x) + f_2(x) - (L_1 + L_2)| < \epsilon. \tag{4}$$

Suppose, then, that ϵ is a positive number. The number $\epsilon/2$ is positive, too, and because $\lim_{x\to x_0} f_1(x) = L_1$ we know that there is a $\delta_1 > 0$ such that for all x

$$0 < |x - x_0| < \delta_1 \quad\Rightarrow\quad |f_1(x) - L_1| < \frac{\epsilon}{2}. \tag{5}$$

Because $\lim_{x\to x_0} f_2(x) = L_2$, there is also a $\delta_2 > 0$ such that for all x

$$0 < |x - x_0| < \delta_2 \quad\Rightarrow\quad |f_2(x) - L_2| < \frac{\epsilon}{2}. \tag{6}$$

Now, either δ_1 equals δ_2 or it doesn't. If δ_1 equals δ_2, the implications in (5) and (6) both hold true for their common value δ. Taken together, (5) and

(6) then say that, for all x, $0 < |x - x_0| < \delta$ implies

$$|f_1(x) + f_2(x) - (L_1 + L_2)|$$
$$= |(f_1(x) - L_1) + (f_2(x) - L_2)|$$
$$\le |(f_1(x) - L_1)| + |(f_2(x) - L_2| \quad \text{Triangle inequality}$$
$$< \frac{\epsilon}{2} + \frac{\epsilon}{2} \quad \text{The implications in (5) and (6) both hold for } \delta \text{ because } \delta = \delta_1 = \delta_2.$$
$$< \epsilon.$$

If $\delta_1 \neq \delta_2$, let δ be the smaller of δ_1 and δ_2. The implications in (5) and (6) then both hold for all x such that $0 < |x - x_0| < \delta$. As before,

$$|f_1(x) + f_2(x) - (L_1 + L_2)| < \epsilon.$$

Either way, we know that given any $\epsilon > 0$ there exists a $\delta > 0$ such that for all x

$$0 < |x - x_0| < \delta \quad \Rightarrow \quad |f_1(x) + f_2(x) - (L_1 + L_2)| < \epsilon.$$

According to the $\epsilon - \delta$ definition of limit, then,

$$\lim_{x \to x_0} (f(x_1) + f(x_2)) = L_1 + L_2.$$

■

The Relation Between One-sided and Two-sided Limits

The formal definitions of right-hand and left-hand limits go like this:

Definition

Right-hand Limit: $\lim_{x \to x_0^+} f(x) = L$

The limit of $f(x)$ as x approaches x_0 from the right is the number L if the following criterion holds:

Given any radius $\epsilon > 0$ about L there exists a radius $\delta > 0$ to the right of x_0 such that for all x

$$x_0 < x < x_0 + \delta \text{ implies } |f(x) - L| < \epsilon. \tag{7}$$

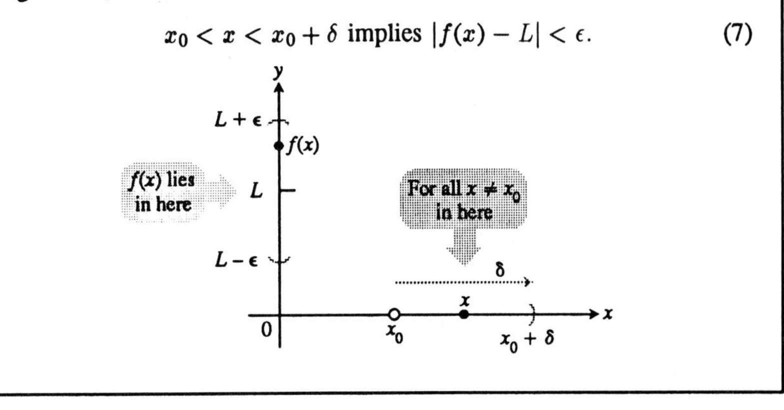

Definition

Left-hand Limit: $\lim_{x \to x_0^-} f(x) = L$

The limit of $f(x)$ as x approaches x_0 from the left is the number L if the following criterion holds:

Given any radius $\epsilon > 0$ about L there exists a radius $\delta > 0$ to the left of x_0 such that for all x

$$x_0 - \delta < x < x_0 \text{ implies } |f(x) - L| < \epsilon. \tag{8}$$

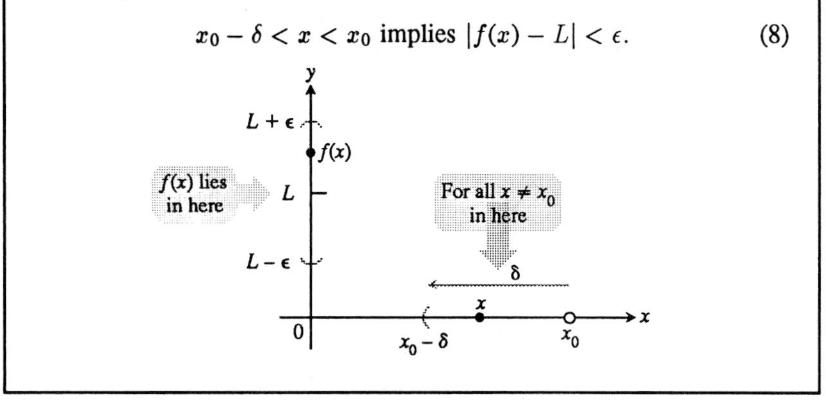

By comparing Eqs. (7) and (8) with Eq. (2), we can see the relation between the one-sided limits just defined and the two-sided limit defined earlier. If we subtract x_0 from the δ-inequalities in Eqs. (7) and (8), they become

$$0 < x - x_0 < \delta \quad \Rightarrow \quad |f(x) - L| < \epsilon \tag{9}$$

and

$$-\delta < x - x_0 < 0 \quad \Rightarrow \quad |f(x) - L| < \epsilon. \tag{10}$$

Together, Eqs. (9) and (10) say the same thing as

$$0 < |x - x_0| < \delta \quad \Rightarrow \quad |f(x) - L| < \epsilon, \tag{11}$$

which is Eq. (2) in the definition of limit. In other words, $f(x)$ has limit L at x_0 if and only if the right-hand and left-hand limits of f at x_0 exist and equal L.

Exercises 2.6

In Exercises 1–4, sketch the interval (a, b) on the x-axis with the point x_0 inside. Then find the largest value of $\delta > 0$ such that $|x - x_0| < \delta$ implies $a < x < b$.

1. $a = 1,\ b = 7,\ x_0 = 5$

2. $a = 1,\ b = 7,\ x_0 = 2$

3. $a = -7/2,\ b = -1/2,\ x_0 = -3$

4. $a = -7/2,\ b = -1/2,\ x_0 = -3/2$

Use the graphs in Exercises 5–10 to find a $\delta > 0$ such that for all x,

$$0 < |x - x_0| < \delta \quad \Rightarrow \quad |f(x) - L| < \epsilon.$$

5.

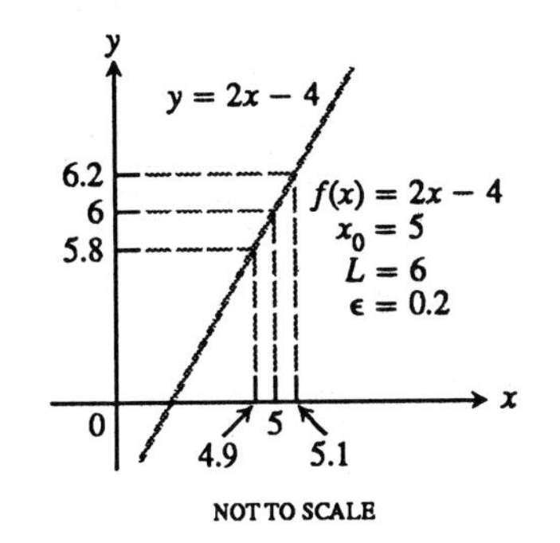

6.

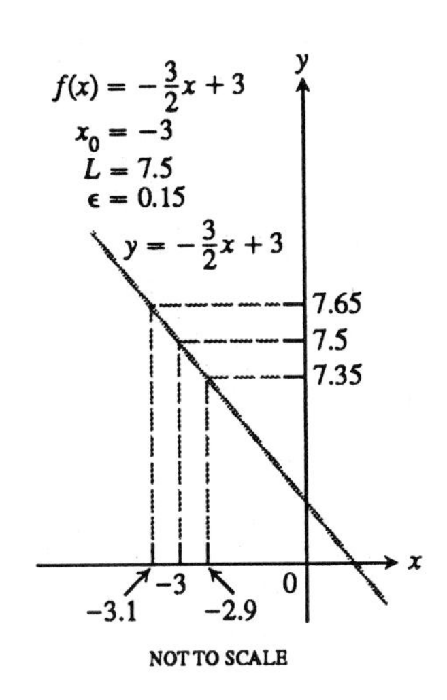

7.

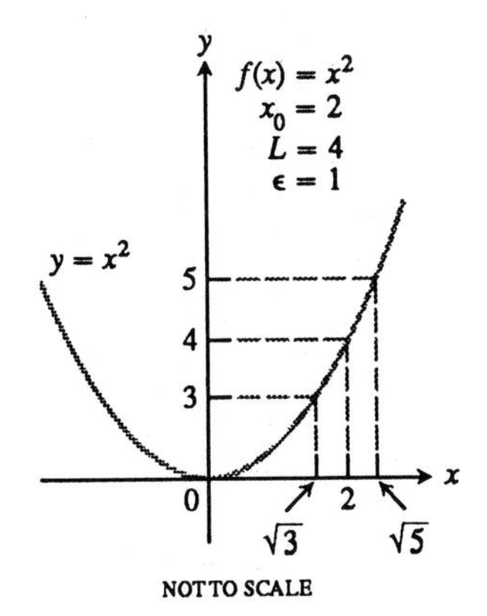

8.

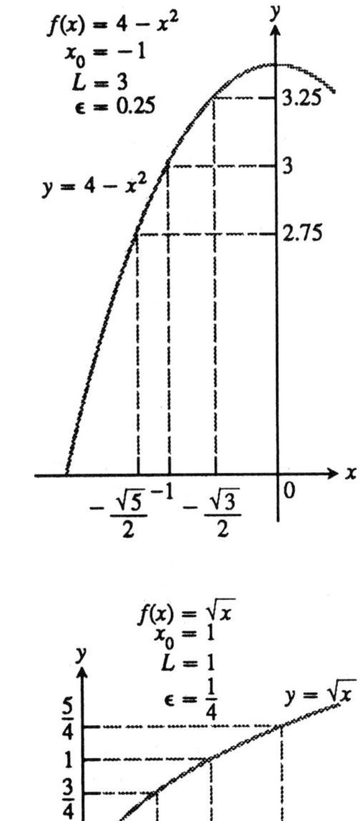

9.

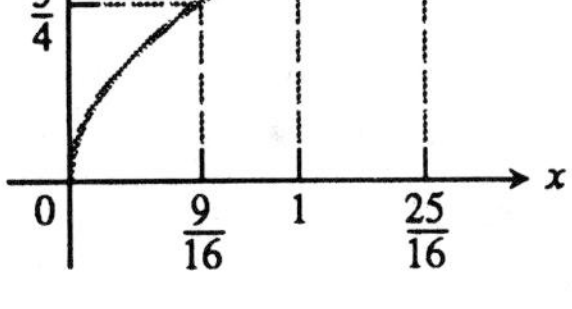

10.

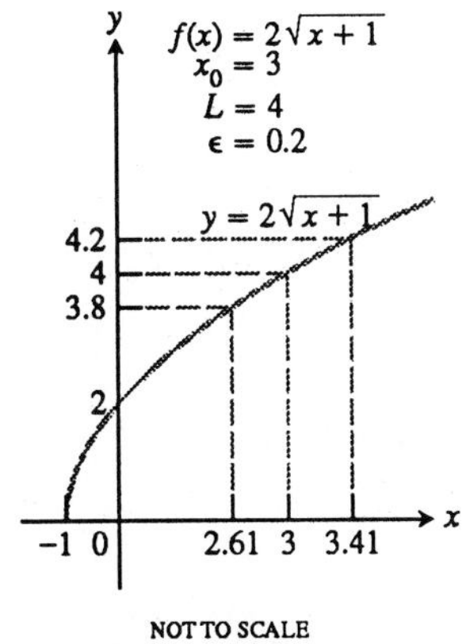

Each of Exercises 11–18 gives a function $f(x)$, a point x_0, and a positive number ϵ. Find $L = \lim_{x \to x_0} f(x)$. Then find a number $\delta > 0$ such that for all x,

$$0 < |x - x_0| < \delta \quad \Rightarrow \quad |f(x) - L| < \epsilon.$$

11. $f(x) = 2x + 3,\ x_0 = 1,\ \epsilon = 0.01$

12. $f(x) = 3 - 2x,\ x_0 = 3,\ \epsilon = 0.02$

13. $f(x) = \dfrac{x^2 - 4}{x - 2}, x_0 = 2, \epsilon = 0.05$

14. $f(x) = \dfrac{x^2 + 6x + 5}{x + 5}, x_0 = -5, \epsilon = 0.05$

15. $f(x) = \sqrt{x - 7}, x_0 = 11, \epsilon = 0.01$

16. $f(x) = \sqrt{1 - 5x}, x_0 = -3, \epsilon = 0.5$

17. $f(x) = 4/x, x_0 = 2, \epsilon = 0.4$

18. $f(x) = 4/x, x_0 = 1/2, \epsilon = 0.04$

In Exercises 19 and 20, find the largest $\delta > 0$ such that for all x,

$$0 < |x - 4| < \delta \quad \Rightarrow \quad |f(x) - 5| < \epsilon.$$

19. $f(x) = 9 - x$; $\epsilon = 0.01, 0.001, 0.0001$, arbitrary $\epsilon > 0$

20. $f(x) = 3x - 7$; $\epsilon = 0.003, 0.0003$, arbitrary $\epsilon > 0$

The functions in Exercises 21–28 are locally straight. Let $L = \lim_{x \to x_0} f(x)$, $\epsilon > 0$, and estimate $\delta > 0$ so that $|x - x_0| < \delta$ implies $|f(x) - L| < \epsilon$. You may assume ϵ is small, say $\epsilon < 0.01$.

21. $f(x) = \sin x, x_0 = 1$

22. $f(x) = \tan x, x_0 = 1$

23. $f(x) = \cos x, x_0 = 1$

24. $f(x) = \sec x, x_0 = 4$

25. $f(x) = x^3 - 4x, x_0 = 0.5$

26. $f(x) = 9x - x^3, x_0 = 2.5$

27. $f(x) = \dfrac{x}{x^2 - 4}, x_0 = -1$

28. $f(x) = \dfrac{2x}{5 - x^2}, x_0 = -1$

29. Given $\epsilon > 0$, find an interval $I = (5, 5 + \delta)$, $\delta > 0$, such that if x lies in I then $\sqrt{x - 5} < \epsilon$. What limit is being verified?

30. Given $\epsilon > 0$, find an interval $I = (4 - \delta, 4)$, $\delta > 0$, such that if x lies in I then $\sqrt{4 - x} < \epsilon$. What limit is being verified?

31. Graph the function

$$f(x) = \begin{cases} 4 - 2x, & x < 1, \\ 6x - 4, & x \geq 1. \end{cases}$$

Then, given $\epsilon > 0$, find the largest δ for which $f(x)$ lies between $y = 2 - \epsilon$ and $y = 2 + \epsilon$ for x in the interval $I = (1 - \delta, 1 + \delta)$.

32. Let $f(x) = |x - 5|/(x - 5)$. Find the set of x-values for which

$$1 - \epsilon < f(x) < 1 + \epsilon, \text{ for } \epsilon = 4, 2, 1, \text{ and } 1/2.$$

33. Define what it means to say that $\lim_{x \to 2} f(x) = 5$.

34. Define what it means to say that $\lim_{x \to 0} g(x) = k$.

35. Let $0 < \epsilon < 2$. Use graphs to support that $\dfrac{1}{2 + \epsilon} < \dfrac{1}{2} < \dfrac{1}{2 - \epsilon}$. (See Example 5.)

36. Let $0 < \epsilon < 2$. Use graphs to support that $\dfrac{\epsilon}{2(2 + \epsilon)} < \dfrac{\epsilon}{2(2 - \epsilon)}$, then verify algebraically. (See Example 5.)

37. Suppose $0 < \epsilon < 4$. Find the largest $\delta > 0$ with the property that $|x - 2| < \delta$ implies $|x^2 - 4| < \epsilon$. What limit is being verified? What happens to δ as ϵ decreases toward 0? Graph δ as a function of ϵ.

38. Suppose $0 < \epsilon < 1$. Find the largest $\delta > 0$ with the property that $|x - 3| < \delta$ implies $\left|\dfrac{2}{x - 1} - 1\right| < \epsilon$. What limit is being verified? What happens to δ as ϵ decreases toward 0? Graph δ as a function of ϵ.

Review Questions

1. You have been asked to calculate the limit of a function $f(x)$ as x approaches a finite number c. What theorems are available for calculating the limit? Give examples to show how the theorems are used.

2. What is the relation between one-sided and two-sided limits? How is this relation sometimes used to calculate a limit or to prove that a limit does not exist? Give examples.

3. You used a graphing calculator to estimate $\lim_{\theta \to 0} (\sin \theta)/\theta$ and found the graph of $f(\theta) = (\sin \theta)/\theta$ to be the horizontal line $y = 0.0174532925 \approx \pi/180$ in the figure below. What did you do wrong?

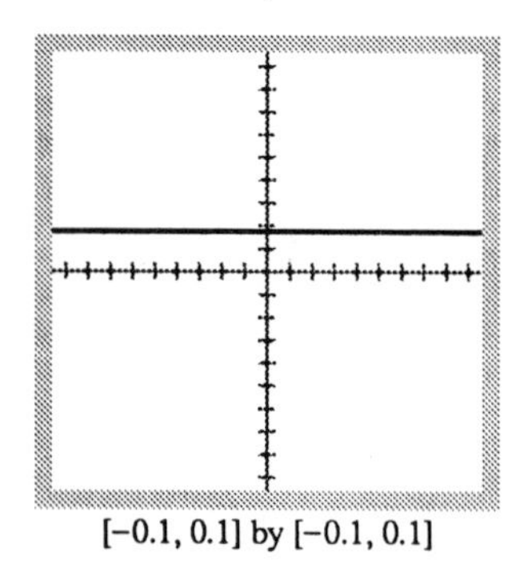

[−0.1, 0.1] by [−0.1, 0.1]

For Exercise 3

4. What is the procedure for finding the limit of a rational function of x as x approaches $\pm\infty$? When is the limit

zero? Finite and different from zero? Infinite?

5. What test can you apply to find out whether a function $y = f(x)$ is continuous at point $x = c$? Give examples of functions that are continuous at $x = 0$. Give examples of functions that fail to be continuous at $x = 0$ for various reasons. (They don't have to be examples from the book. You may make up your own.)

6. What can be said about the continuity of polynomial functions and rational functions?

7. What can be said about the continuity of composites of continuous functions?

8. What are the important theorems about continuous functions? Can functions that are not continuous be expected to have the properties guaranteed by these theorems? Give examples.

9. What are the formal definitions of (two-sided) limit, right-hand limit, and left-hand limit?

10. Discuss how a grapher can be used to *help* determine the limit of a function. Explain why this is not conclusive.

11. Define horizontal and vertical asymptote.

12. Give an example of a function with a removable discontinuity.

13. Define end behavior and end behavior model for a function.

14. Define end behavior asymptote for a rational function.

15. How are absolute values used to describe intervals of real numbers?

16. Show by example how absolute values are used to control function values.

Practice Exercises

Find the limits in Exercises 1–20. Some limits may not exist.

1. $\lim_{x\to -2} x^2(x+1)$

2. $\lim_{x\to 3} (x+2)(x-5)$

3. $\lim_{x\to 3} \frac{x-3}{x^2}$

4. $\lim_{x\to -1} \frac{x^2+1}{3x^2-2x+5}$

5. $\lim_{x\to -2} \left(\frac{x}{x+1}\right)\left(\frac{3x+5}{x^2+x}\right)$

6. $\lim_{x\to 1} \left(\frac{1}{x+1}\right)\left(\frac{x+6}{x}\right)\left(\frac{3-x}{7}\right)$

7. $\lim_{x\to 4} \sqrt{1-2x}$

8. $\lim_{x\to 5} \sqrt[4]{9-x^2}$

9. $\lim_{x\to 1} \frac{x^2-1}{x-1}$

10. $\lim_{x\to -5} \frac{x^2+3x-10}{x+5}$

11. $\lim_{x\to 2} \frac{x-2}{x^2+x-6}$

12. $\lim_{x\to 1} \frac{x^2-2x+1}{x^3-2x^2+x}$

13. $\lim_{x\to 0} \frac{(1+x)(2+x)-2}{x}$

14. $\lim_{x\to 0} \frac{\frac{1}{2+x}-\frac{1}{2}}{x}$

15. $\lim_{x\to \infty} \frac{2x+3}{5x+7}$

16. $\lim_{x\to -\infty} \frac{2x^2+3}{5x^2+7}$

17. $\lim_{x\to -\infty} \frac{x^2-4x+8}{3x^3}$

18. $\lim_{x\to \infty} \frac{1}{x^2-7x+1}$

19. $\lim_{x\to -\infty} \frac{x^2-7x}{x+1}$

20. $\lim_{x\to \infty} \frac{x^4+x^3}{12x^3+128}$

Find the limits in Exercises 21–24.

21. $\lim_{x\to 3+} \frac{1}{x-3}$

22. $\lim_{x\to 3-} \frac{1}{x-3}$

23. $\lim_{x\to 0+} \frac{1}{x^2}$

24. $\lim_{x\to 0-} \frac{1}{|x|}$

Find the limits in Exercises 25–28.

25. $\lim_{x\to 0} \frac{\sin 2x}{4x}$

26. $\lim_{x\to 0} \frac{x+\sin x}{x}$

27. $\lim_{x\to 0} \frac{\sin^3 2x}{x^3}$

28. $\lim_{x\to 0} \frac{2\csc 5x}{\csc 3x}$

29. Let $f(x) = \frac{\sec 2x \csc 9x}{\cot 7x}$

a) Estimate $\lim_{x\to 0} f(x)$ graphically.

b) Compare the function values of f near 0 to the limit found in part (a).

c) Describe (informally, no proof required) the behavior of f near 0. Hint: Restrict x to $[-0.2, 0.2]$.

d) Find the exact value of $\lim_{x\to 0} f(x)$ algebraically.

30. Let $f(x) = \frac{\csc 5x \sec 8x}{\csc 8x \sec 3x}$

a) Estimate $\lim_{x\to 0} f(x)$ graphically.

b) Compare the function values of f near 0 to the limit found in part (a).

c) Describe (informally, no proof required) the behavior of f near 0. Hint: Restrict x to $[-0.2, 0.2]$.

d) Find the exact value of $\lim_{x\to 0} f(x)$ algebraically.

In Exercises 31 and 32 find the limits.

31. a) $\lim_{x\to -2+} \frac{x+3}{x+2}$ b) $\lim_{x\to -2-} \frac{x+3}{x+2}$

32. a) $\lim_{x\to 2+} \frac{x-1}{x^2(x-2)}$ b) $\lim_{x\to 2-} \frac{x-1}{x^2(x-2)}$

c) $\lim_{x\to 0+} \frac{x-1}{x^2(x-2)}$ d) $\lim_{x\to 0-} \frac{x-1}{x^2(x-2)}$

33. Let

$$f(x) = \begin{cases} 1, & x \le -1 \\ -x, & -1 < x < 0 \\ 1, & x = 0 \\ -x, & 0 < x < 1 \\ 1, & x \ge 1 \end{cases}$$

a) Determine a complete graph of f.

b) Find the right-hand and left-hand limits of f at -1, 0, and 1.

c) Does f have a limit as x approaches -1? 0? 1? If so, what is it? If not, why not?

d) At which of the points $x = -1, 0, 1$, if any, is f continuous?

34. Repeat the questions in Exercise 33 for the function

$$f(x) = \begin{cases} 0, & x \le -1 \\ |2x|, & -1 < x < 1 \\ 0, & x = 1 \\ 1, & x > 1. \end{cases}$$

35. Let $f(x) = \begin{cases} |x^3 - 4x|, & x < 1 \\ x^2 - 2x - 2, & x \ge 1. \end{cases}$

a) Determine a complete graph of f.

b) Find the right-hand and left-hand limits of f at 1.

c) Does f have a limit as x approaches 1? If so, what is it? If not, why not?

d) At what points is f continuous? Why?

e) At what points is f not continuous? Why?

36. Let $f(x) = \begin{cases} 1 - \sqrt{3 - 2x}, & x < 3/2 \\ 1 + \sqrt{2x - 3}, & x \ge 3/2 \end{cases}$

a) Determine a complete graph of f.

b) Find the right-hand and left-hand limits of f at 3/2.

c) Does f have a limit as x approaches 3/2? If so, what is it? If not, why not?

d) At what points is f continuous? Why?

e) At what points is f not continuous? Why?

37. Let $f(x) = \begin{cases} -x, & x < 1 \\ x - 1, & x > 1. \end{cases}$

a) Graph f.

b) Find the right-hand and left-hand limits of f at $x = 1$.

c) What value, if any, should be assigned to $f(1)$ to make f continuous at $x = 1$?

38. Repeat Exercise 37 for the function

$$f(x) = \begin{cases} 3x^2, & x < 1 \\ 4 - x^2, & x > 1. \end{cases}$$

Find the points, if any, at which the functions in Exercises 39 and 40 are *not* continuous.

39. $f(x) = \dfrac{x + 1}{4 - x^2}$

40. $f(x) = \sqrt[3]{3x + 2}$

Find the end behavior asymptotes in Exercises 41–44 algebraically and support graphically.

41. $f(x) = \dfrac{2x + 1}{x^2 - 2x + 1}$

42. $g(x) = \dfrac{2x^2 + 5x - 1}{x^2 + 2x}$

43. $h(x) = \dfrac{x^3 - 4x^2 + 3x + 3}{x - 3}$

44. $T(x) = \dfrac{x^4 - 3x^2 + x - 1}{x^3 - x + 1}$

45. Suppose that $f(x)$ and $g(x)$ are defined for all x and that $\lim_{x \to c} f(x) = -7$ and $\lim_{x \to c} g(x) = 0$. Find the limit as $x \to c$ of the following functions.

a) $3f(x)$ b) $(f(x))^2$ c) $f(x) \cdot g(x)$

d) $\dfrac{f(x)}{g(x) - 7}$ e) $\cos(g(x))$ f) $|f(x)|$

46. Suppose that $f(x)$ and $g(x)$ are defined for all x and that $\lim_{x \to 0} f(x) = \frac{1}{2}$ and $\lim_{x \to 0} g(x) = \sqrt{2}$. Find the limits as $x \to 0$ of the following functions.

a) $-g(x)$ b) $g(x) \cdot f(x)$ c) $f(x) + g(x)$

d) $1/f(x)$ e) $x + f(x)$ f) $\dfrac{f(x) \cdot \sin x}{x}$

47. Use the inequality

$$0 \le \left|\sqrt{x} \sin \frac{1}{x}\right| \le \sqrt{x}$$

to find $\lim_{x \to 0} \sqrt{x} \sin(1/x)$.

48. Use the inequality

$$0 \le \left|x^2 \sin \frac{1}{x}\right| \le x^2$$

to find $\lim_{x \to 0} x^2 \sin(1/x)$.

Use the Sandwich Theorem to find the limits in Exercises 49 and 50.

49. $\lim_{x \to \infty} \left(\dfrac{\sin x}{\sqrt{x}}\right)$

50. $\lim_{x \to \infty} \dfrac{\cos x}{\sqrt{x}}$

Given that $\lim_{x \to \infty} (\sin x)/x = \lim_{x \to \infty} (\cos x)/x = 0$, find the limits in Exercises 51 and 52.

51. $\lim_{x \to \infty} \dfrac{x + \sin x}{x}$

52. $\lim_{x \to \infty} \dfrac{x + \sin x}{x + \cos x}$

53. Let $f(x) = \begin{cases} \dfrac{x^2 + 2x - 15}{x - 3}, & x \ne 3 \\ k, & x = 3. \end{cases}$

What value, if any, should be assigned to k to make f continuous at $x = 3$?

54. Let $f(x) = \begin{cases} \dfrac{\sin x}{2x}, & x \ne 0 \\ k, & x = 0. \end{cases}$

What value, if any, should be assigned to k to make f continuous at $x = 0$?

55. The function $y = 1/x$ does not take on either a maximum or a minimum on the interval $0 < x < 1$ even though the function is continuous on this interval. Does this contradict the Max-Min Theorem for continuous functions? Why?

56. What are the maximum and minimum values of the function $y = |x|$ on the interval $-1 \le x < 1$? Note that the interval is not closed. Is this consistent with the Max-Min Theorem for continuous functions? Why?

57. True, or false? If $y = f(x)$ is continuous, with $f(1) = 0$ and $f(2) = 3$, then f takes on the value 2.5 at some point between $x = 1$ and $x = 2$. Explain.

58. Show that there is at least one value of x for which $x + \cos x = 0$.

The Definition of Limit

59. Define what it means to say that

$$\lim_{x \to 1} f(x) = 3.$$

60. Define what it means to say that

$$\lim_{x \to 0} \frac{\sin x}{x} = 1.$$

Wrong descriptions of limit. Show by example that the statements in Exercises 61 and 62 are wrong.

61. The number L is the limit of $f(x)$ as x approaches x_0 if $f(x)$ gets closer to L as x approaches x_0.

62. The number L is the limit of $f(x)$ as x approaches x_0 if, given any $\epsilon > 0$, there is a value of x for which $|f(x) - L| < \epsilon$.

Each of Exercises 63–68 gives a function $y = f(x)$, a number E, and a target value y_0. In each case, find an interval of x-values for which $y = f(x)$ lies within E units of y_0. (Recall our agreement to try to choose the interval as large as possible.) Then find an interval which can be described with an absolute value inequality of the form $|x - x_0| < D$.

	$f(x)$	E	y_0
63.	$\sqrt{x+2}$	1	4
64.	$\sqrt{\dfrac{x+1}{2}}$	1/2	1
65.	$\dfrac{x-1}{x-3}$	0.1	2
66.	$\dfrac{x-1}{x-3}$	0.1	-2
67.	$x^3 - 4x$	0.1	4
68.	$x^3 - 4x$	0.1	$1 (-1 < x_0 < 0)$

69. The function $f(x) = 2x - 3$ is continuous at $x = 2$. Given a positive number ϵ, how small must δ be for $|x - 2| < \delta$ to imply $|f(x) - 1| < \epsilon$?

70. The function $f(x) = |x|$ is continuous at $x = 0$. Given a positive number ϵ, how small must δ be for $|x - 0| < \delta$ to imply $|f(x) - 0| < \epsilon$?

Each of Exercises 71–74 gives a function $f(x)$, a point x_0, and a positive number ϵ. Find $L = \lim_{x \to x_0} f(x)$. Then find a number $\delta > 0$ such that for all x

$$0 < |x - x_0| < \delta \Rightarrow |f(x) - L| < \epsilon.$$

71. $f(x) = 5x - 10$, $x_0 = 3$, $\epsilon = 0.05$

72. $f(x) = 5x - 10$, $x_0 = 2$, $\epsilon = 0.05$

73. $f(x) = \sqrt{x - 5}$, $x_0 = 9$, $\epsilon = 1$

74. $f(x) = \sqrt{2x - 3}$, $x_0 = 2$, $\epsilon = 1/2$

The functions in Exercises in 75 and 76 are locally straight at x_0. Let $L = \lim_{x \to 0} f(x)$, $\epsilon > 0$, and estimate $\delta > 0$ so that $|x - x_0| < \delta$ implies $|f(x) - L| < \epsilon$. You may assume ϵ is small, say $\epsilon < 0.01$.

75. $f(x) = \dfrac{x^2 - x}{x + 2}$, $x_0 = 5$ 76. $f(x) = \dfrac{x - 1}{x^2 + 3x}$, $x_0 = -5$

77. *Controlling the flow from a draining tank.* Torricelli's law says that if you drain a tank like the one in Fig. 2.67, the rate y at which the water runs out is a constant times the square root of the water's depth. As the tank drains, x decreases, and so does y, but y decreases less rapidly than x. The value of the constant depends on the size of the exit valve.

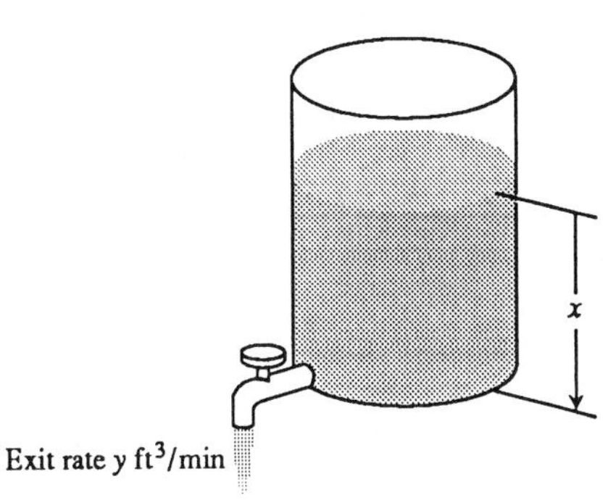

2.67 The tank in Exercise 77

Suppose that for the tank in question, $y = \sqrt{x}/2$. You are trying to maintain a constant exit rate of $y_0 = 1$ ft^3/min by refilling the tank with a hose from time to time. How deep must you keep the water to hold the rate to within 0.2 ft^3/min of $y_0 = 1$? Within 0.1 ft^3/min of $y_0 = 1$? In other words, in what interval must you keep x to hold y within 0.2 (or 0.1) units of $y_0 = 1$?

Remark: What if we want to know how long it will take the tank to drain if we do not refill it? We cannot answer such a question with the usual equation time = amount/rate, because the rate changes as the tank drains. We could always open the valve, sit down with a watch, and wait; but with a large tank or a reservoir that might take hours or even days. With calculus, we will be able to find the answer in just a minute or two, as you will see if you do Exercise 45 in Section 4.7.

78. *Dimension changes in equipment.* As you probably know, most metals expand when heated and contract when cooled, and people sometimes have to take this into account in their work. Boston and Maine Railroad crews try to lay track at temperatures as close to 65°F as they can, so the track won't expand too much in the summer or shrink too much in the winter. Surveyors have to correct their measurements for temperature when they use steel measuring tapes.

The dimensions of a piece of laboratory equipment are often so critical that the machine shop in which it is made has to be held at the same temperature as the laboratory where a part is to be installed. And, once the piece is installed, the laboratory must continue to be held at that temperature.

A typical aluminum bar that is 10 cm wide at 70°F will be

$$y = 10 + (t - 70) \times 10^{-4}$$

centimeters wide at a nearby temperature t. As t rises above 70, the bar's width increases; as t falls below 70, the bar's width decreases.

Suppose you had a bar like this made for a gravity-wave detector you were building. You need the width of the bar to stay within 0.0005 cm of the ideal 10 cm. How close to 70°F must you maintain the temperature of your laboratory to achieve this? In other words, how close to $t_0 = 70$ must you keep t to be sure that y lies within 0.0005 of $y_0 = 10$?

CHAPTER

3

Derivatives

OVERVIEW Derivatives are the functions we use to measure the rates at which things change. We define derivatives as limiting values of average changes, just as we define slopes of curves as limiting values of slopes of secants. Now that we can calculate limits, we can calculate derivatives.

The notion of derivative is one of the most important ideas in calculus. Any subject area that uses calculus, and most subjects do, has applications of derivatives. This chapter shows how to accurately approximate the derivative of a given function using the numerical derivative and how to determine exact derivatives quickly, without having to do tedious limit calculations. An important result is the Chain Rule, which shows how to calculate derivatives of composite functions. Graphs of numerical derivatives will be used to support the determination of exact derivatives. We will see that the ability to use a computer to graph numerical derivatives is as useful as determining exact derivatives.

Applications of derivatives will be discussed mainly in the next chapter. However, to give some idea of the usefulness and importance of derivatives, some applications are introduced in this chapter.

3.1 Slopes, Tangent Lines, and Derivatives

In this section, we get our first view of the role calculus plays in describing how rapidly things change. Our point of departure is the coordinate plane of Descartes and Fermat. The plane is the natural place to draw curves and calculate the slopes of lines and it is from the slopes of lines that we find the slopes of curves. Once we can do that, we can do two really important things: we can find tangent lines for curves and we can find formulas for rates of change.

These formulas for the rates at which functions change define new functions called *derivatives*. Derivatives are first calculated with limits. In this section we develop the ideas of slope and derivative, and see how some typical calculations go. We also introduce the numerical derivative and explore graphs of derivatives.

Average Rates of Change

We encounter average rates of change in such forms as average speeds (distance traveled divided by elapsed time, say, in miles per hour), growth rates of populations (in percent per year), and average monthly rainfall (in inches per month). The **average rate of change** in a quantity over a period of time is the amount of change divided by the time it takes.

Experimental biologists often want to know the rates at which populations grow under controlled laboratory conditions. Figure 3.1 shows data from a fruit fly-growing experiment, the setting for our first example.

EXAMPLE 1 The average growth rate of a laboratory population

The graph in Fig. 3.1 shows how the number of fruit flies *(Drosophila)* grew in a controlled 50-day experiment. The graph was made by counting flies at regular intervals, plotting a point for each count, and drawing a smooth curve through the plotted points.

There were 150 flies on day 23 and 340 flies on day 45. This gave an increase of $340 - 150 = 190$ flies in $45 - 23 = 22$ days. The average rate of change in the population from day 23 to day 45 was therefore

$$\textit{Average rate of change:}\quad \frac{\Delta p}{\Delta t} = \frac{340 - 150}{45 - 23} = \frac{190}{22} \approx 9 \text{ flies/day.} \tag{1}$$

The average rate of change in Eq. (1) is also the slope of the secant line through the two points

$$P(23, 150) \quad \text{and} \quad Q(45, 340)$$

on the population curve. (A line through two points on a curve is called a **secant** to the curve.) We can calculate the slope of the secant PQ from the coordinates of P and Q:

$$\textit{Secant slope:}\quad \frac{\Delta p}{\Delta t} = \frac{340 - 150}{45 - 23} = \frac{190}{22} \approx 9 \text{ flies/day.} \tag{2}$$

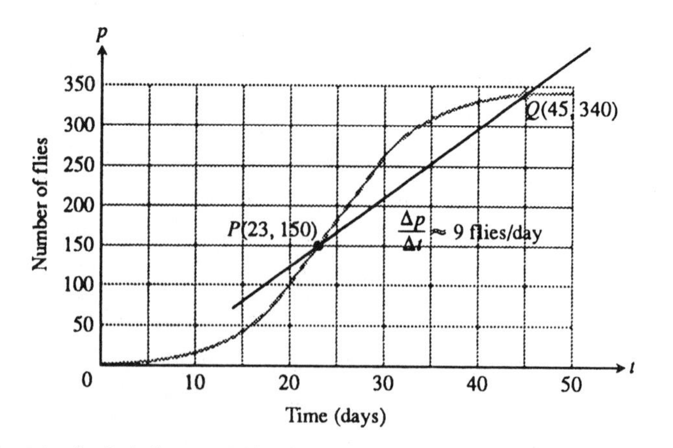

3.1 Growth of a fruit fly population in a controlled experiment. (*Source:* A. J. Lotka, *Elements of Mathematical Biology*. Dover, New York (1956), p. 69.)

By comparing Eqs. (1) and (2) we can see that the average rate of change in (1) is the same number as the slope in (2), units and all. *We can always think of an average rate of change as the slope of a secant line.*

HOW DO YOU FIND A TANGENT TO A CURVE?

It is hard to overestimate how important the answer to this question was to the scientists of the early seventeenth century. In optics, the angle at which a ray of light strikes the surface of a lens is defined in terms of the tangent to the surface. In physics, the direction of a body's motion at any point of its path is along the tangent to the path. In geometry, the angle between two intersecting curves is the angle between their tangents at the point of intersection.

In addition to knowing the average rate at which the population grew from day 23 to day 45, we may also want to know how fast the population was growing on day 23 itself. To find out we can watch the slope of the secant PQ change as we back Q along the curve toward P. The results for four positions of Q are shown in Fig. 3.2.

In terms of geometry, what we see as Q approaches P along the curve is this: The secant PQ approaches the tangent line AB that we drew by eye at P. This means that within the limitations of our drawing, the slopes of the secants approach the slope of the tangent, which we calculate from the coordinates of A and B to be

$$\frac{350-0}{35-15} = 17 \text{ flies/day}.$$

In terms of population change, what we see as Q approaches P is this: The average growth rates for increasingly smaller time intervals approach the slope

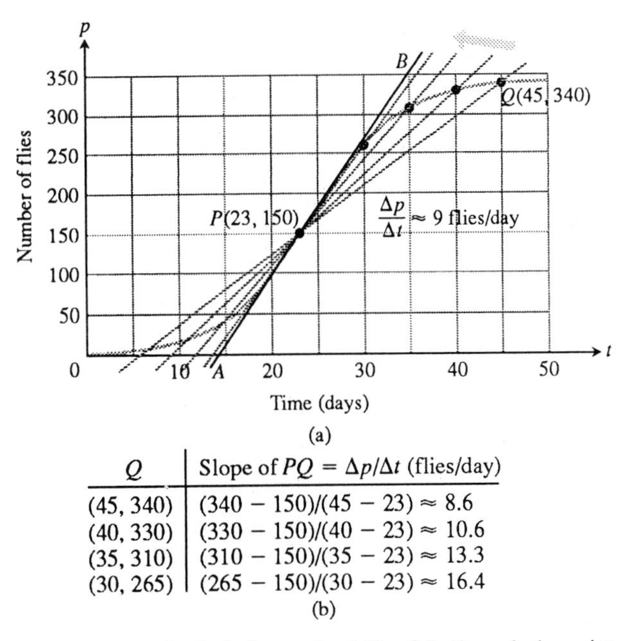

Q	Slope of $PQ = \Delta p/\Delta t$ (flies/day)
(45, 340)	$(340-150)/(45-23) \approx 8.6$
(40, 330)	$(330-150)/(40-23) \approx 10.6$
(35, 310)	$(310-150)/(35-23) \approx 13.3$
(30, 265)	$(265-150)/(30-23) \approx 16.4$

(b)

3.2 (a) Four secants to the fruit fly graph of Fig. 3.1, through the point $P(23, 150)$. (b) The slopes of the four secants.

of the tangent to the curve at P (17 flies per day). The slope of the tangent line is therefore the number we take as the rate at which the fly population was changing on day $t = 23$.

Defining Slopes and Tangent Lines

The moral of the fruit fly story would seem to be that we should define the rate at which the value of the function $y = f(x)$ is changing with respect to x at any particular value $x = x_1$ to be the slope of the tangent to the curve $y = f(x)$ at $x = x_1$. But how are we to define the tangent line at an arbitrary point P on the curve and deduce its slope from the formula $y = f(x)$? The problem here is that we only know one point. Our usual definition of slope requires two points.

The solution that Fermat finally found in 1629 proved to be one of that century's major contributions to calculus. We still use his method of defining tangents to produce formulas for slopes of curves and rates of change. It goes like this:

1. We start with what we *can* calculate, namely the slope of a secant through P and a point Q nearby on the curve.
2. We find the limiting value of the secant slope (if it exists) as Q approaches P along the curve.
3. We take this number to be the slope of the curve at P and define the tangent to the curve at P to be the line through P with this slope.

Explore with a Grapher

1. Graph $y = x^2$ and $y = 4x - 4$ in the same *square* viewing window. Does the line $y = 4x - 4$ appear to be tangent to the graph of $y = x^2$ at the point $(2, 4)$?
2. Zoom-in at the point $(2, 4)$ several times. What happens? Why?

EXAMPLE 2 Find the slope of the parabola $y = x^2$ at the point $P(2, 4)$. Write an equation for the tangent to the parabola at this point.

Solution We begin with a secant line that passes through $P(2, 4)$ and a neighboring point $Q(2 + h, (2 + h)^2)$ on the curve (Fig. 3.3). We then write an expression for the slope of the secant line and find the limiting value of this slope as Q approaches P along the curve.

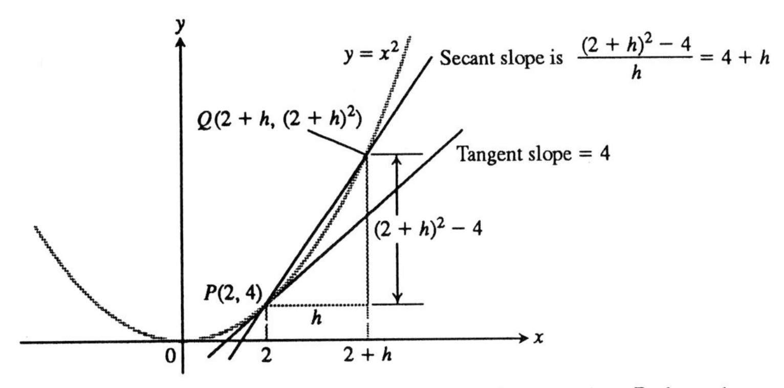

3.3 The slope of the secant PQ approaches 4 as Q approaches P along the curve.

The slope of the secant PQ is

$$\begin{aligned} \text{Secant slope} &= \frac{\Delta y}{\Delta x} = \frac{(2+h)^2 - 2^2}{h} \\ &= \frac{h^2 + 4h + 4 - 4}{h} \\ &= \frac{h^2 + 4h}{h} = h + 4. \end{aligned}$$

The limit of the secant slope as Q approaches P along the curve is

$$\lim_{Q \to P} (\text{Secant slope}) = \lim_{h \to 0} (h + 4) = 4.$$

The tangent to the parabola at P is the line through P with slope 4:

Point : $(2, 4)$

Slope : $m = 4$

Equation : $y - 4 = 4(x - 2)$

$y = 4x - 8 + 4$

$y = 4x - 4.$

The mathematics we just used to find the slope of the parabola $y = x^2$ at the point $P(2, 4)$ will find the slope of the parabola at any other point, too. Here's how it works.

EXAMPLE 3 Find the slope of the parabola $y = x^2$ at any point on the curve.

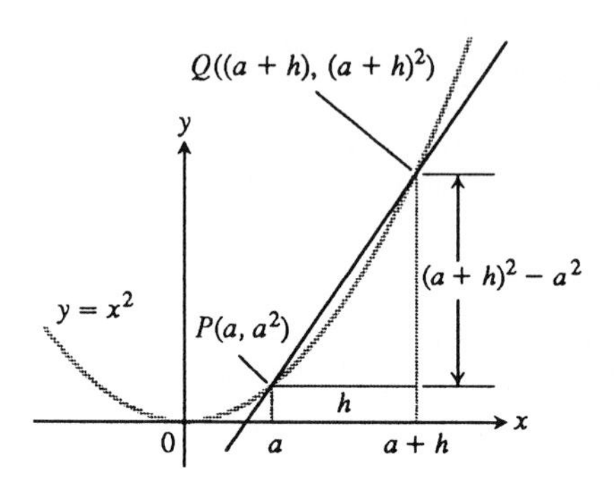

3.4 The slope of the secant PQ is $[(a_1 + h)^2 - x_1^2]/h = 2x_1 + h$.

Solution Let $P(a, a^2)$ be the point. In the notation of Fig. 3.4, the slope of the secant line through P and any nearby point $Q(a+h, (a+h)^2)$ is

$$\begin{aligned}\text{Secant slope} &= \frac{\Delta y}{\Delta x} = \frac{(a+h)^2 - a^2}{h} \\ &= \frac{h^2 + 2ah + a^2 - a^2}{h} \\ &= \frac{h^2 + 2ah}{h} = h + 2a.\end{aligned}$$

The limit of the secant slope as Q approaches P along the curve is

$$\lim_{Q \to P} (\text{Secant slope}) = \lim_{h \to 0} (h + 2a) = 2a.$$

Since a can be any value of x, at any point (x, y) on the parabola, the slope is

$$m = 2x.$$

When $x = 2$, for example, the slope is 4, as in Example 2. ▪

The next example shows how to use the slope formula $m = 2x$ from Example 3 to find equations for tangent lines.

EXAMPLE 4 Find equations for the tangents to the curve $y = x^2$ at the points $(-1/2, 1/4)$ and $(1, 1)$.

Solution We use the slope formula $m = 2x$ from Example 3 to find the point-slope equation for each line.

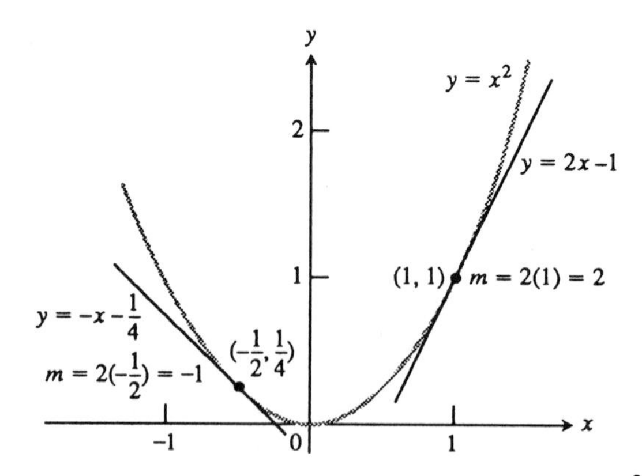

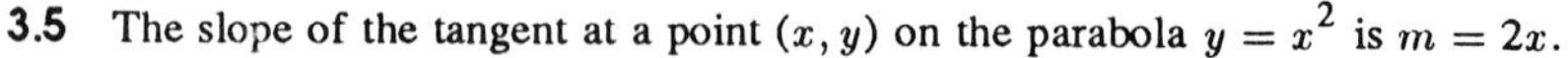
3.5 The slope of the tangent at a point (x, y) on the parabola $y = x^2$ is $m = 2x$.

Tangent at $(-1/2, 1/4)$	Point:	$(-1/2, 1/4)$
	Slope:	$m = 2x = 2(-1/2) = -1$
	Equation:	$y - 1/4 = -1(x - (-1/2))$
		$y - 1/4 = -x - 1/2$
		$y = -x - 1/4$
Tangent at $(1, 1)$	Point:	$(1, 1)$
	Slope:	$m = 2x = 2(1) = 2$
	Equation:	$y - 1 = 2(x - 1)$
		$y - 1 = 2x - 2$
		$y = 2x - 1$

See Fig. 3.5.

The Derivative of a Function

The function $m = 2x$ that gives the slope of the parabola $y = x^2$ at x is the derivative of the function $y = x^2$.

To find the derivative of an arbitrary function $y = f(x)$ (when the function has one—we'll come back to that), we simply repeat for f the steps we took in Examples 2 and 3 for x^2. We start with an arbitrary point $P(x, f(x))$ on the graph of f, as in Fig. 3.6. The slope of the secant line through P and a nearby point $Q(x + h, f(x + h))$ is then

$$\text{Secant slope} = \frac{\Delta y}{\Delta x} = \frac{f(x+h) - f(x)}{h}. \tag{3}$$

The slope of the graph at P is the limit of the secant slope as Q approaches P along the graph. We find this limit from the formula for f by calculating the limit

$$\lim_{h \to 0} \frac{f(x+h) - f(x)}{h}. \tag{4}$$

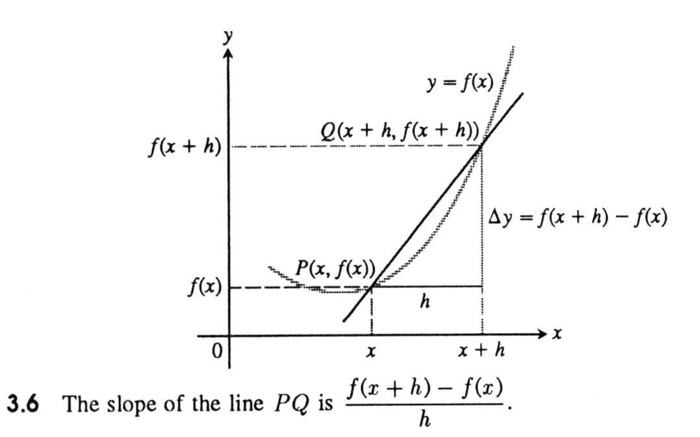

3.6 The slope of the line PQ is $\dfrac{f(x+h) - f(x)}{h}$.

The limit in Eq. (4) is itself a function of x. We denote it by f' ("f prime") and call it the derivative of f. Its domain is a subset of the domain of f. For most of the functions in this book, f' will be defined at all or all but a few of the points where f is defined.

Definition

The Derivative of a Function

The **derivative** of a function f is the function f' whose value at x is defined by the equation

$$f'(x) = \lim_{h \to 0} \frac{f(x+h) - f(x)}{h}. \tag{5}$$

The fraction $(f(x+h) - f(x))/h$ is the **difference quotient** for f at x.

Differentiable at a Point A function that has a derivative at a point x is said to be **differentiable at x.**

Differentiable Function A function that is differentiable at every point of its domain is called **differentiable.**

The Slope and Tangent When the number $f'(x)$ exists it is called the **slope** of the curve $y = f(x)$ at x. The line through the point $(x, f(x))$ with slope $f'(x)$ is the **tangent** to the curve at x.

The derivative of a function f, that is,

$$\lim_{h \to 0} \frac{f(x+h) - f(x)}{h},$$

involves the difference quotient function

$$D(h) = \frac{f(x+h) - f(x)}{h},$$

which is *discontinuous* at $h = 0$. Thus the limit at 0 *cannot* be evaluated by substituting 0 for h in the difference quotient. Later in this section you will apply the difference quotient limit definition to determine derivatives of functions.

The most common notations for the derivative of a function $y = f(x)$, besides $f'(x)$, are

y' ("y prime")	(Nice and brief)
$\dfrac{dy}{dx}$ ("$dydx$")	(Names the variables and has a "d" for derivative)
$\dfrac{df}{dx}$ ("$dfdx$")	(Emphasizes the function's name)
$D_x(f)$ ("Dx of f")	(Emphasizes the idea that taking the derivative is an operation performed on f)
$\dfrac{d}{dx}(f)$ ("ddx of f")	(Ditto).

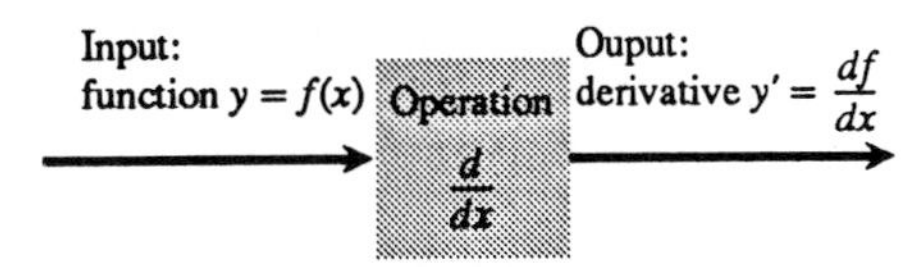

3.7 Flow diagram for the operation of taking a derivative with respect to x.

We also read dy/dx as "the derivative of y with respect to x" and df/dx as "the derivative of f with respect to x." See Fig. 3.7. Example 3 showed that $f'(x) = 2x$ for $f(x) = x^2$ using a slope of secant line limit argument. You can verify that the difference quotient limit definition of the derivative will give the same result.

The Slopes of Lines

We now have two definitions for the slope of a line $y = mx + b$: the number m and, at each point, the derivative of the function $f(x) = mx + b$. Whenever we bring in a new definition, it is a good idea to be sure that the new and old definitions agree on objects to which they both apply. We do this in the next example.

EXAMPLE 5 The derivative of the function $f(x) = mx + b$ is the slope of the line $y = mx + b$.

Solution The idea is to show that the derivative of $f(x) = mx + b$ has the constant value m. To see that this is so, we calculate the limit in Eq. (5) with $f(x) = mx + b$. The calculation takes four steps:

STEP 1: Write out $f(x)$ and $f(x + h)$:

$$f(x) = mx + b$$

$$f(x + h) = m(x + h) + b = mx + mh + b.$$

STEP 2: Subtract $f(x)$ from $f(x + h)$:

$$f(x + h) - f(x) = mh.$$

STEP 3: Divide by h:

$$\frac{f(x + h) - f(x)}{h} = \frac{mh}{h} = m.$$

STEP 4: Take the limit as $h \to 0$:

$$\lim_{h \to 0} \frac{f(x + h) - f(x)}{h} = \lim_{h \to 0} (m) = m.$$

The derivative of f does indeed have the constant value m. ■

Numerical Derivatives

In this textbook, we assume you have an electronic tool that can accurately approximate the derivative of any differentiable function at a given value in its domain. We will denote a numerical approximation of $f'(a)$ by

$$\text{NDer}\,(f(x), a).$$

It is understood that the particular approximation depends on the computation device and method used. We write $f'(a) = \text{NDer}\,(f(x), a)$ with the understanding that the equal sign denotes *approximately* equal. In some cases, $\text{NDer}\,(f(x), a)$ will exist when $f'(a)$ does not exist so care must be used in interpreting numerical results.

Our graphing calculator uses the symmetric difference quotient method (6) to determine $\text{NDer}\,(f(x), a)$. It can be proven that the symmetric difference quotient limit

$$\lim_{h \to 0} \frac{f(a+h) - f(a-h)}{2h} \tag{6}$$

is equal to $f'(a)$, whenever $f'(a)$ exists.

The name *symmetric* is given to this method because it represents the slope of a secant line that extends h units on either side of $(a, f(a))$. (See Fig. 3.8). You will see in Exercise 25 that the symmetric difference quotient method is a better way to approximate the derivative than the standard difference quotient limit definition. Most calculators with numerical derivative functionality allow the user to select a tolerance h. In fact, the tolerance is used as the value of h in the symmetric difference quotient approximation of $f'(a)$. That is,

$$f'(a) \approx \frac{f(a+h) - f(a-h)}{2h}. \tag{7}$$

In this text we will use the symmetric difference quotient method with $h = 0.01$ to compute $\text{NDer}\,(f(x), a)$. Thus

$$\text{NDer}\,(f(x), a) = \frac{f(a+0.01) - f(a-0.01)}{0.02} \tag{8}$$

For example, if $f(x) = x^2$ we compute $\text{NDer}\,(x^2, 2)$ by evaluating the right-hand side of (7).

$$\begin{aligned} \text{NDer}\,(x^2, 2) &= \frac{(2.01)^2 - (1.99)^2}{.02} \\ &= 4 \text{ (exactly)} \end{aligned}$$

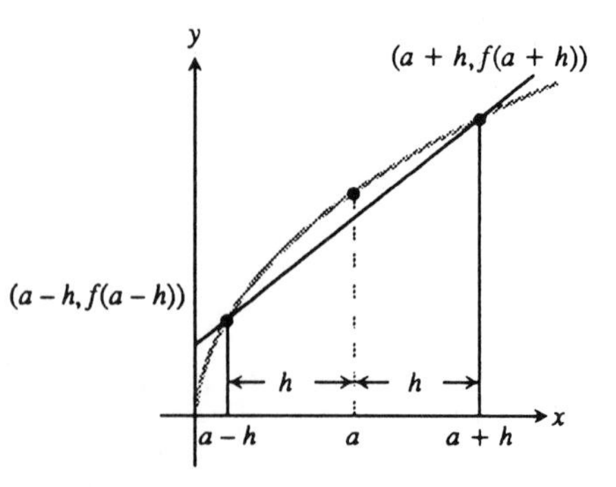

3.8 A symmetric secant line.

We state the following interesting theorem about NDer computations for quadratic functions, which you will prove in Exercise 32.

Theorem 1

> If $f(x) = ax^2 + bx + c$, $a \neq 0$, then NDer $(f(x), x)$ is *exactly* $f'(x)$.

It will be shown later in this chapter that $D_x(x^3) = 3x^2$. So if $f(x) = x^3$, $f'(2) = 12$. Compare this result with the next example.

EXAMPLE 6 If $f(x) = x^3$, compute NDer $(x^3, 2)$.

Using (7),

$$\text{NDer}\,(x^3, 2) = \frac{(2.01)^3 - (1.99)^3}{0.02}$$
$$= 12.0001.$$

The last example suggests that NDer is very accurate when $h = 0.01$. This is usually the case. From now on in this text we assume you will use the NDer feature of your graphing calculator or computer when asked to compute NDer $(f(x), a)$. The NDer feature can be used to compute numerical derivatives of very complicated functions as the next example illustrates.

EXAMPLE 7 Let

$$f(x) = \frac{2^{x+1}x^5 - \sin\left(\frac{x+1}{x-2}\right)}{\sqrt[3]{x^3 - 2x + 5}}.$$

Compute NDer $(f(x), -1)$.

Solution Our calculator gives

$$\text{NDer}\,(f(x), -1) = 2.584469517.$$

It can be shown that $f'(-1) = 2.58416625008$ is accurate to eleven decimal places. So NDer $(f(x), -1)$ in Example 7 is a very accurate approximation of $f'(-1)$. Using the rules introduced later in this chapter, the exact derivative of the function in Example 7 could be determined. However, it is a very tedious process, and in many cases, accurate approximations of derivatives are all that is needed.

Graphs of Derivatives

In this text we assume you have an electronic tool that will quickly produce graphs of the numerical derivative, that is, graphs of $y =$ NDer $(f(x), x)$. We will see that the ability to obtain accurate *graphs* of $y = \text{NDer}(f(x), x)$ is often more useful than determining exact derivatives.

ACCURACY AGREEMENT

Henceforth, all computations using NDer will be displayed accurate to two decimal places.

Explore with a Grapher

1. Graph $y = \text{NDer}(f(x), x)$ for each of the following cubic functions:

$$f_1(x) = 3x^3 - 2x^2 + 5x - 6, \quad f_2(x) = -x^3 + x^2 + 2x + 5,$$
$$f_3(x) = x^3 - 4x - 5, \quad f_4(x) = -2x^3 - 3x^2 + 5x - 10$$

2. Graph $y = \text{NDer}(f(x), x)$ for each of the following quartic functions:

$$f_1(x) = 4x^4 - 3x^3 + 2x^2 - x + 10, \quad f_2(x) = -x^4 + 2x^3 - 3x + 5$$
$$f_3(x) = 2x^4 + x^2 - 2x + 5, \quad f_4(x) = -2x^4 + x^3 - 2x^2 - 6x + 7$$

3. Based on Activity 1, the graph of $y = \text{NDer}(\text{cubic polynomial}, x)$ is a polynomial of what degree?
4. Based on Activity 2, the graph of $y = \text{NDer}(\text{quartic polynomial}, x)$ is a polynomial of what degree?
5. Make a conjecture: based on Activities 3 and 4, the graph of the derivative of an nth degree polynomial is a polynomial of what degree?

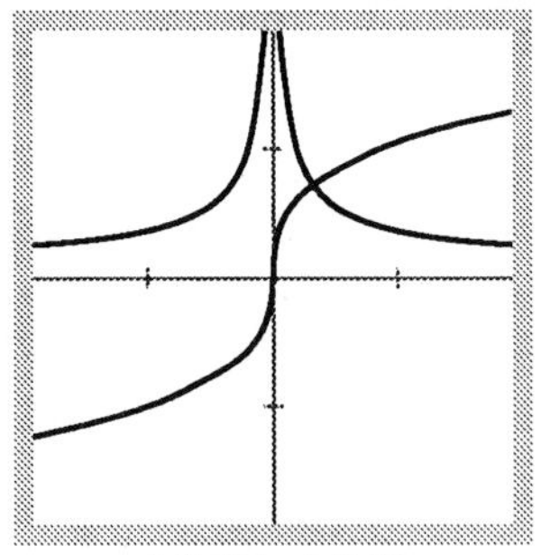

[−2, 2] by [−2, 2]

3.9 The graph of $y = \sqrt[3]{x}$ and $y = \text{NDer}(\sqrt[3]{x}, x)$. $D_x\sqrt[3]{x}$ does not exist at $x = 0$.

The student should be aware that sometimes NDer computations produce wrong results. For example, the derivative of the function $y = \sqrt[3]{x}$ at $x = 0$ does not exist. However, $\text{NDer}(\sqrt[3]{x}, 0) = 21.54$ on one graphing calculator. Figure 3.9 shows how the graph of $y = \text{NDer}(\sqrt[3]{x}, x)$ appears to have an infinite discontinuity at $x = 0$ (i.e. $\lim_{x\to 0}[\text{NDer}(\sqrt[3]{x}, x)] = \infty$). In this case the *graph* suggests the correct fact that $D_x\sqrt[3]{x}$ does not exist at 0. Can you explain why $D_x\sqrt[3]{x}$ does not exist at 0? *Hint*: What is the slope of the tangent line at $x = 0$?

The Fruit Fly Problem Revisited The graph of $P(t) = 350/(1 + 8^{(25-t)/10})$ in Fig. 3.10 is a very close approximation to the graph of the fruit fly population in Fig. 3.1. It is easy to compute that $\text{NDer}(P(t), 23) = 17.43$. This compares very favorably with the slope estimate of 17 given in the beginning of the section, the rate at which the fly population was changing on day 23. The curve in Fig. 3.10 is called a *logistics growth curve*.

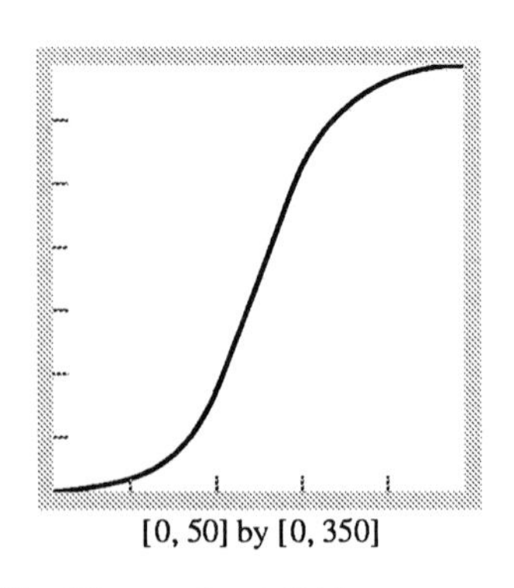

[0, 50] by [0, 350]

3.10 The graph of $P(t) = 350/(1 + 8^{(25-t)/10})$

Typical Exact Derivative Calculations

EXAMPLE 8 Find $f'(x)$ if $f(x) = 1/x$. Graph $y = f'(x)$ and compare it to the graph of $y = \text{NDer}\left(\frac{1}{x}, x\right)$.

Solution We take $f(x) = 1/x$, $f(x+h) = 1/(x+h)$, and form the difference quotient

$$\frac{f(x+h)-f(x)}{h} = \frac{\frac{1}{x+h}-\frac{1}{x}}{h} = \frac{1}{h}\cdot\frac{x-(x+h)}{x(x+h)} = \frac{1}{h}\cdot\frac{-h}{x(x+h)} = \frac{-1}{x(x+h)}. \tag{9}$$

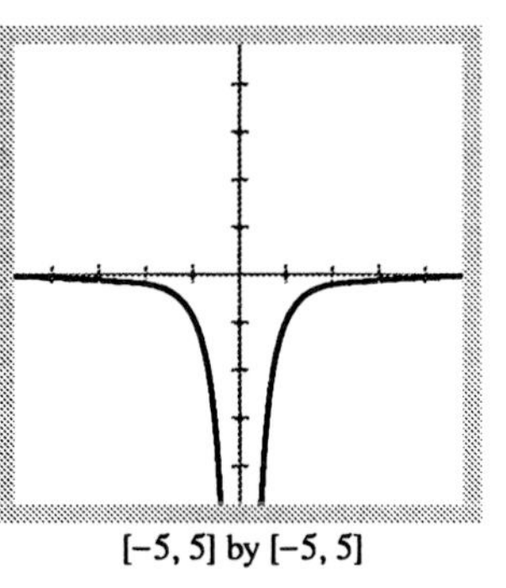

[−5, 5] by [−5, 5]

3.11 The graph of $y = \text{NDer}(1/x, x)$ appears to be the same as the graph of $f'(x) = -(1/x^2)$.

We then take the limit as $h \to 0$:

$$f'(x) = \lim_{h\to 0}\frac{f(x+h)-f(x)}{h} = \lim_{h\to 0}\frac{-1}{x(x+h)} = \frac{-1}{x(x+0)} = -\frac{1}{x^2}.$$

The graph of $f'(x) = -(1/x^2)$ and $y = \text{NDer}(1/x, x)$ seem to be identical (see Fig. 3.11).

Example 8 shows how the graph of $y = \text{NDer}(1/x, x)$ provides strong support that the manipulations used to determine $f'(x) = -(1/x^2)$ are correct.

The algebraic steps we use to calculate $f'(x)$ directly from the limit definition are always the same:

1. Write out $f(x)$ and $f(x+h)$.
2. Subtract $f(x)$ from $f(x+h)$.
3. Divide by h.
4. Take the limit as $h \to 0$.

EXAMPLE 9 Show that the derivative of $y = \sqrt{x}$ is $dy/dx = 1/(2\sqrt{x})$.

Solution We use Eq. (5) with $f(x+h) = \sqrt{x+h}$ and $f(x) = \sqrt{x}$ to form the quotient

$$\frac{f(x+h)-f(x)}{h} = \frac{\sqrt{x+h}-\sqrt{x}}{h}. \tag{10}$$

Rationalizing the numerator in Eq. (10), we find

$$\frac{\sqrt{x+h}-\sqrt{x}}{h} = \frac{\sqrt{x+h}-\sqrt{x}}{h}\cdot\frac{\sqrt{x+h}+\sqrt{x}}{\sqrt{x+h}+\sqrt{x}} = \frac{(x+h)-x}{h(\sqrt{x+h}+\sqrt{x})} = \frac{1}{\sqrt{x+h}+\sqrt{x}}. \tag{11}$$

Now as h approaches 0, the denominator in the final form approaches $\sqrt{x} + \sqrt{x} = 2\sqrt{x}$. The quotient $1/(2\sqrt{x})$ is defined for $x > 0$. Therefore,

$$\frac{dy}{dx} = \lim_{h\to 0}\frac{\sqrt{x+h}-\sqrt{x}}{h} = \lim_{h\to 0}\frac{1}{\sqrt{x+h}+\sqrt{x}} = \frac{1}{2\sqrt{x}} \quad (x \neq 0). \tag{12}$$

> **Explore with a Grapher**
>
> Graph $y = \text{NDer}(\sqrt{x}, x)$ and compare it with the graph of $dy/dx = 1/(2\sqrt{x})$. What do you see? What does it mean?

EXAMPLE 10 Find an equation for the tangent to the curve $y = \sqrt{x}$ at $x = 1.5$.

Solution The slope at $x = 1.5$ is the value of the function's derivative there.

$$\text{NDer}(\sqrt{x}, 1.5) = 0.41$$

Thus the tangent is the line through the point $(1.5, \sqrt{1.5})$ with slope 0.41.

Point : $(1.5, \sqrt{1.5}) = (1.5, 1.22)$

Slope : 0.41

Equation : $y - 1.22 = 0.41(x - 1.5)$

$$y = 0.41(x - 1.5) + 1.22$$

The tangent is the line $y = 0.41(x - 1.5) + 1.22$. The graphs in Fig. 3.12 support this result. ∎

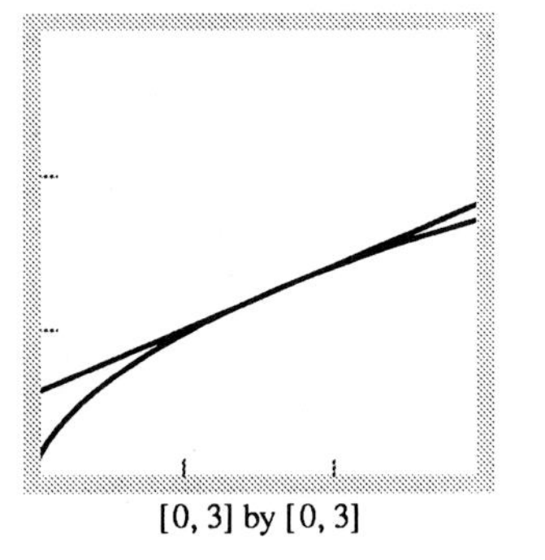

[0, 3] by [0, 3]

3.12 The curve $y = \sqrt{x}$ and its tangent line $y = 0.41(x - 1.5) + 1.22$ at $(1.5, \sqrt{1.5})$.

Analytically, in Example 10,

$$\frac{dy}{dx} = \frac{1}{2\sqrt{x}}.$$

So the tangent line is $y = (1/2\sqrt{1.5})(x - 1.5) + \sqrt{1.5}$. Compare its graph with the tangent line graph in Figure 3.12.

The question of when a function is a derivative is one of the central questions in all calculus, and Newton and Leibniz's answer to this question revolutionized the world of mathematics. We shall see what their answer was when we get to Chapter 5.

Differentiable Functions Are Continuous

A function is continuous at every point at which it has a derivative.

Theorem 2

> If f has a derivative at $x = c$, then f is continuous at $x = c$.

Proof Our task is to show that $\lim_{x \to c} f(x) = f(c)$ or, equivalently, that

$$\lim_{x \to c} [f(x) - f(c)] = 0. \tag{13}$$

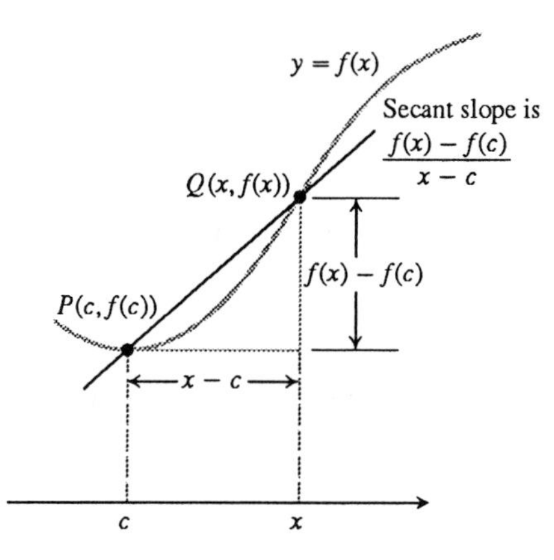

3.13 Figure for the proof that a function is continuous at every point at which it has a derivative.

To this end, we let $P(c, f(c))$ be a point on the graph of f and let $Q(x, f(x))$ be a point nearby (Fig. 3.13). The slope of the secant PQ is

$$\text{Secant slope} = \frac{f(x) - f(c)}{x - c}.$$

By definition, the derivative of f at c is the limiting value of this slope as Q approaches P along the curve, which means in this case, the limit as $x \longrightarrow c$:

$$f'(c) = \lim_{x \to c} \frac{f(x) - f(c)}{x - c}. \tag{14}$$

Why should the mere existence of this limit imply that $[f(x) - f(c)] \longrightarrow 0$ as $x \longrightarrow c$? Because, with the denominator $x - c$ going to zero, the quotient can have a finite limit only if the numerator goes to zero at the same time. Indeed, this is exactly what we find if we apply the Limit Product Rule from Chapter 2:

$$\lim_{x \to c} [f(x) - f(c)] = \lim_{x - c} \left[(x - c) \frac{f(x) - f(c)}{x - c}\right] \qquad \text{Algebra}$$

$$= \lim_{x - c} (x - c) \cdot \lim_{x \to c} \frac{f(x) - f(c)}{x - c} \qquad \text{Limit Product Rule}$$

$$= 0 \cdot f'(c) = 0. \qquad \text{Known values}$$

Differentiable on a Closed Interval—One-sided Derivatives

A function $y = f(x)$ is *differentiable on a closed interval* $[a, b]$ if it has a derivative at every interior point and if the limits

$$\lim_{h \to 0+} \frac{f(a + h) - f(a)}{h} \qquad \text{Right-hand derivative at } a$$

$$\lim_{h \to 0-} \frac{f(b + h) - f(b)}{h} \qquad \text{Left-hand derivative at b}$$

exist at the endpoints. In the right-hand derivative, h is positive and $a + h$ approaches a from the right. In the left-hand derivative, h is negative and $b + h$ approaches b from the left (Fig. 3.14).

Right-hand and left-hand derivatives may be defined at any point of a function's domain.

The usual relation between one-sided and two-sided limits holds for derivatives. A function has a (two-sided) derivative at a point if and only if the function's right-hand and left-hand derivatives are defined and are equal at that point.

EXAMPLE 11 The function $f(x) = |x|$ has no derivative with respect to x at $x = 0$ even though it has a derivative with respect to x at every other point. The reason is that the right-hand and left-hand derivatives of $f(x) = |x|$ are not equal at $x = 0$:

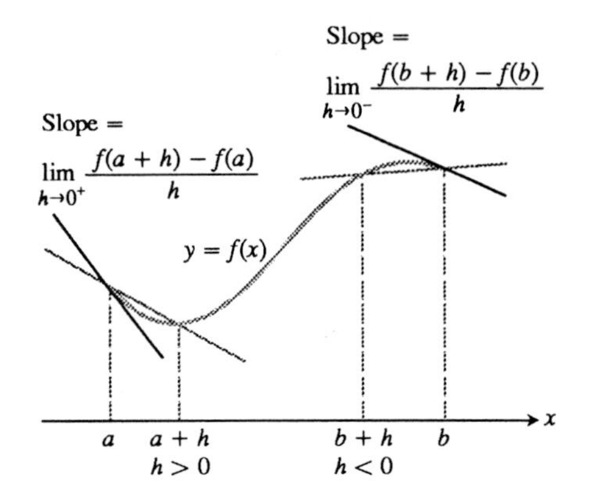

3.14 Derivatives at endpoints are one-sided limits.

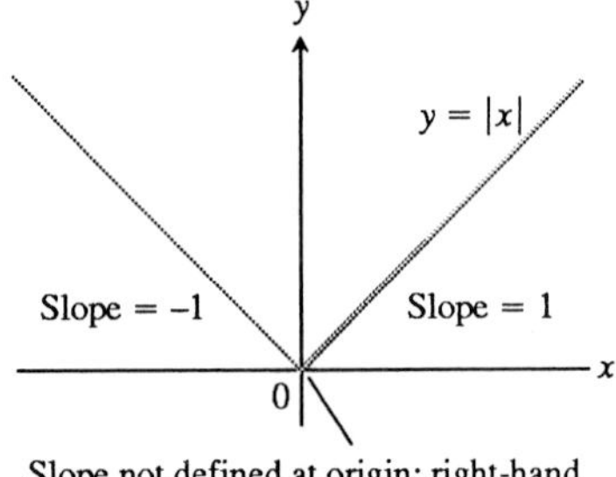

3.15 Slope not defined at origin: right-hand derivative $\neq$ left-hand derivative. The absolute value function has a derivative at every point except the origin.

$$
\begin{aligned}
\lim_{h\to 0+} \frac{|0+h|-|0|}{h} &= \lim_{h\to 0+} \frac{|h|}{h} \\
&= \lim_{h\to 0+} \frac{h}{h} \qquad \text{Because } h > 0 \\
&= \lim_{h\to 0+} 1 = 1,
\end{aligned}
$$

$$
\begin{aligned}
\lim_{h\to 0-} \frac{|0+h|-|0|}{h} &= \lim_{h\to 0-} \frac{|h|}{h} \\
&= \lim_{h\to 0-} \frac{-h}{h} \qquad |h| = -h \text{ because } h < 0 \\
&= \lim_{h\to 0-} (-1) = -1.
\end{aligned}
$$

It is also easy to use a slope argument to show that $f'(0)$ does not exist. To the right of the origin, $|x| = x$, and the slope of the curve is 1 at every point. To the left of the origin, $|x| = -x$, and the slope is -1 at every point (Fig. 3.15).

EXAMPLE 12 Show that the following function has no derivative at $x = 0$.

$$
y = \begin{cases} x^2, & x \le 0 \\ 2x, & x > 0 \end{cases}
$$

Solution There is no derivative at the origin because the right-hand and left-hand derivatives are different there. The slope of the parabola on the left is $2(0) = 0$ (from Example 3). The slope of the line on the right is 2.

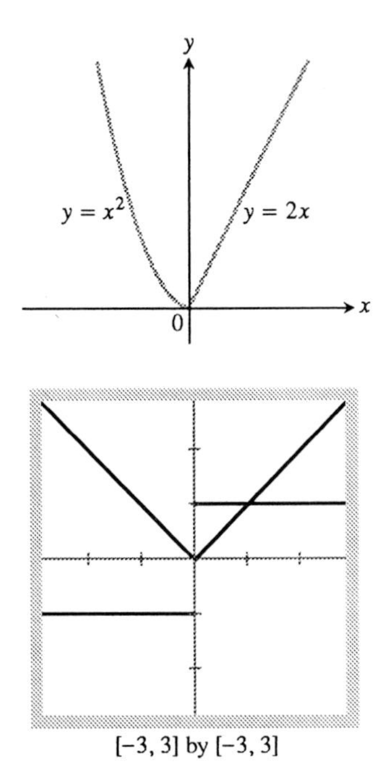

3.16 The graphs of $f(x) = |x|$ and $y = \text{NDer}(|x|, x)$ supports the claim that $D_x f$ does not exist at 0.

The previous two examples provide counterexamples to the converse of Theorem 2. The converse of Theorem 2 is this statement:

If f is continuous at $x = c$, then $f'(c)$ exists.

This is a false statement! For example if $f(x) = |x|$ then $f'(0)$ does not exist; yet f is continuous at $x = 0$. See Fig. 3.16.

Continuous curves that fail to have a tangent anywhere play an important role in chaos theory, in part because there is no way to measure the length of such a curve. We'll see what length has to do with derivatives when we get to Section 6.4.

HOW ROUGH CAN THE GRAPH OF A CONTINUOUS FUNCTION BE?

The absolute value function fails to be differentiable at a single point. Using a similar idea, we can use a saw-tooth graph (Fig. 3.17) to define a continuous function that fails to have a derivative at infinitely many points.

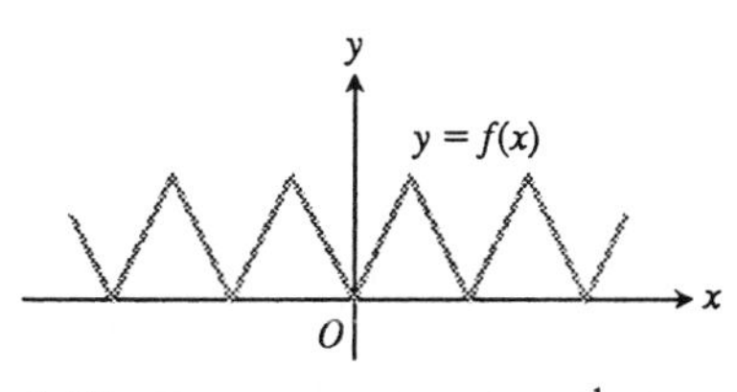

3.17 The graph of $y = |\sin^{-1}(\sin \pi x)|$

But can a *continuous* function fail to have a derivative at *every* point? The answer, surprisingly enough, is yes, as Karl Weierstrass (1815–1897) found in 1872. One of his formulas (there are many like it) was

$$f(x) = \sum_{n=0}^{\infty} \left(\frac{2}{3}\right)^n \cos(9^n \pi x),$$

a formula that expresses f as an infinite sum of cosines with increasingly shorter periods. By adding wiggles to wiggles infinitely many times, so to speak, the formula produces a continuous graph that is too bumpy to have a tangent anywhere. For example, the graph of the first four terms of Weierstrass's function

$$y = \cos(\pi x) + \frac{2}{3}\cos(9\pi x)$$

$$+ \left(\frac{2}{3}\right)^2 \cos(9^2 \pi x) + \left(\frac{2}{3}\right)^3 \cos(9^3 \pi x)$$

is shown in Fig. 3.18. It already has many wiggles and is very bumpy! In the exercises you will graph the first eight terms of the Weierstrass function. Can you imagine what the graph of the first 100 terms of the Weierstrass function would look like?

When Does a Function *Not* Have a Derivative at a Point?

As we know, a function has a derivative at a point x_0 if the slopes of the secant lines through $P(x_0, f(x_0))$ and a nearby point Q on the graph approach a limit as Q approaches P. Whenever the secants fail to take up a limiting position as Q approaches P, the derivative does not exist. Typically, a function whose graph is otherwise smooth will fail to have a derivative at a point where the graph has

1. A *corner* where the one-sided derivatives differ. For example, let $f(x) = \begin{cases} x, & x < 0 \\ -x^2, & x \geq 0. \end{cases}$ In Exercise 33 you will graph both $y = f(x)$ and $y = \text{NDer}(f(x), x)$.

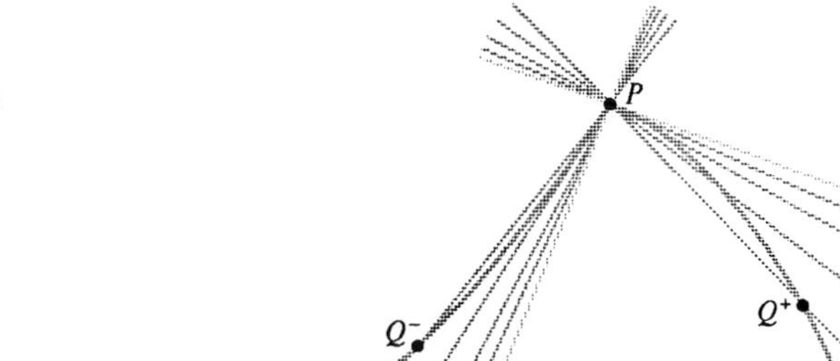

2. A *jump* that makes one of the one-sided derivatives infinite (here the left derivative). For example, let $f(x) = \begin{cases} -\sqrt{|x|}, & x < 0 \\ 1 - x^2, & x \geq 0. \end{cases}$ In Exercise 34 you will graph both $y = f(x)$ and $y = \text{NDer}(f(x), x)$.

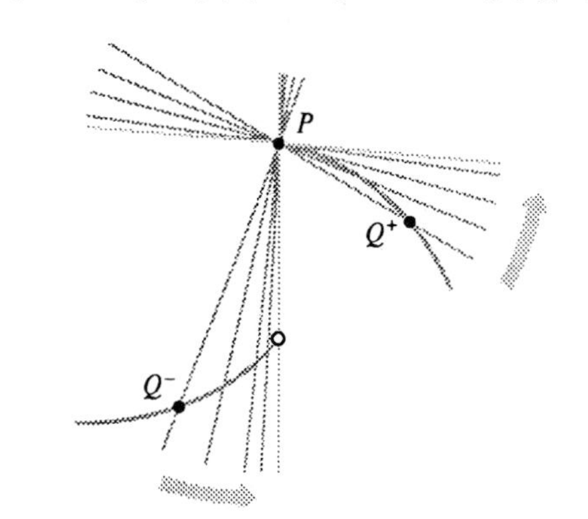

3. A *vertical* tangent. For example let $f(x) = -\sqrt[3]{|x|}$. In Exercise 35 you will graph both $y = f(x)$ and $y = \text{NDer}(f(x), x)$.

A function will also fail to have a derivative at any point of discontinuity.

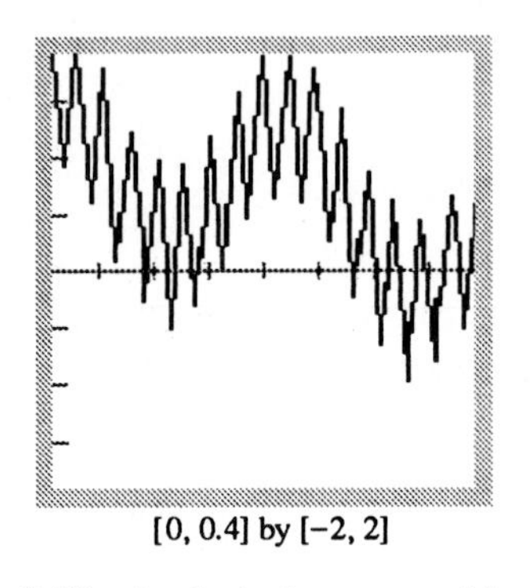

3.18 A wiggly, bumpy graph!

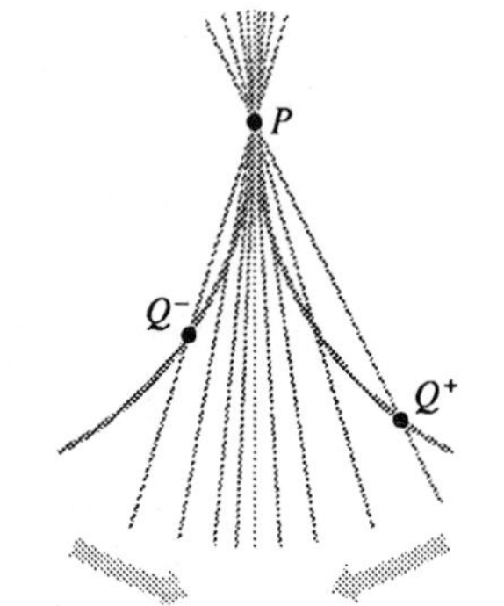

This is from the *contrapositive* of Theorem 2: if f is not continuous at $x = c$, then $f'(c)$ does not exist. The contrapositive (if *not* Q, then *not* P) is always true for an "if P, then Q" theorem. Thus, the greatest-integer function $y = \lfloor x \rfloor$ is not differentiable at any integer value of x. On the other hand, a function like $y = |x|$ can be continuous at a point and still not have a derivative there, as we saw in Example 11. The function $f(x) = 1/x$ is differentiable. However, $f'(0)$ does not exist because 0 is not in the domain (f is discontinuous at $x = 0$).

What Functions Are Differentiable?

Most of the functions we have worked with so far are differentiable. Polynomials are differentiable, as are rational functions and trigonometric functions. Composites of differentiable functions are differentiable, and so are sums, differences, products, powers, and the quotients of differentiable functions, where defined. We shall explain all this as the chapter continues.

Exercises 3.1

In Exercises 1–10, use Eq. (5) to find the derivative dy/dx of the given function $y = f(x)$ and then draw its complete graph. Use the graph of $y = \text{NDer}(f(x), x)$ to support the derivative you determine. Then find the slope of the curve $y = f(x)$ at $x = 3$ and write an equation for the tangent there.

1. $y = 2x^2 - 5$

2. $y = x^2 - x$

3. $y = \dfrac{2}{x}$

4. $y = \dfrac{1}{x+1}$

5. $y = x + \dfrac{9}{x}$

6. $y = x - \dfrac{1}{x}$

7. $y = \sqrt{2x}$

8. $y = \sqrt{2x+3}$

9. $y = \dfrac{1}{\sqrt{x}}$

10. $y = \dfrac{1}{\sqrt{2x+3}}$

In Exercises 11–16, find an equation for the tangent to the curve at the given point $(x, f(x))$. Use $\text{NDer}(f(x), x)$ to compute the slope of the tangent line. Then graph the curve and tangent in the same viewing window.

11. $y = x^2 + 1$, $x = 2$.

12. $y = 2x^3 - 5x - 2$, $x = 1.5$.

13. $y = \sqrt{4 - x^2}$, $x = -1$.

14. $y = (x - 1)^3 + 1$, $x = 2.5$.

15. $y = \dfrac{x^2 - 4}{x^2 + 1}$, $x = 2$.

16. $y = x \sin x$, $x = 2$.

Which of the graphs in Exercises 17–20 suggest a function $y = f(x)$ that is

a) Continuous at every point of its domain?

b) Differentiable at every point of its domain?

c) Both (a) and (b)?

d) Neither (a) nor (b)?

Explain in each case.

17. 18.

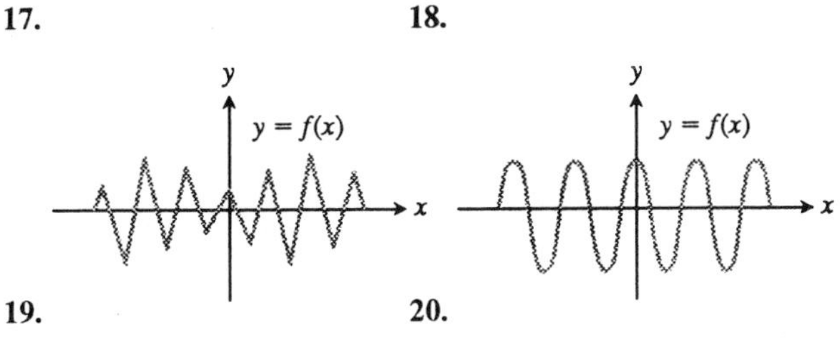

19. 20.

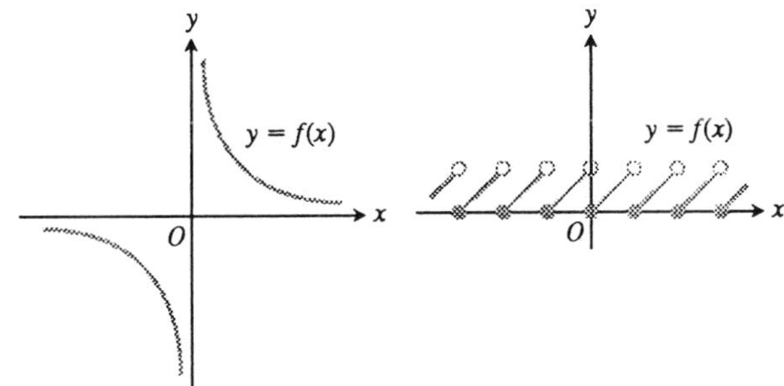

Use right-hand and left-hand derivatives to show that the functions graphed in Exercises 21–22 are not differentiable at the indicated point P.

21. 22.

23. & 24. Write each function in Exercises 21–22 as a piecewise function and graph the function and its numerical derivative in the viewing rectangle $[-3,3]$ by $[-3,3]$.

25. Let

$$D(h)=\frac{f(a+h)-f(a)}{h}$$

and

$$S(h)=\frac{f(a+h)-f(a-h)}{2h}.$$

Complete the following tables for the specified value of a for each of the following differentiable functions: $f_1(x)=3x^2-2x$, $f_2(x)=2x^3+4x$, $f_3(x)=\sin x$, and $f_4(x)=1/x^2$.

a) $a=2$

h	$D(h)$	$S(h)$
-0.1		
0.1		
-0.01		
0.01		
-0.001		
0.001		
-0.0001		
0.0001		

b) $a=0$

h	$D(h)$	$S(h)$
-0.1		
0.1		
-0.01		
0.01		
-0.001		
0.001		
-0.0001		
0.0001		

c) Make a conjecture about the exact value of the derivatives of each function. Which method $D(h)$ (difference quotient) or $S(h)$ (symmetric difference quotient) is closest to your conjecture for the same value of h?

d) Explain the pitfall of using the $S(h)$ method to approximate the derivative of $f_4(x)=1/x^2$ at $x=0$.

26. Let $f(x)=\sqrt[3]{\dfrac{x^2\sin(\tan x)}{x^2\cdot 2^x+x^5-4x^2}}$. Compute $f'(-1)$, $f'(0)$, $f'(1.5)$, and $f'(3.5)$.

27. Let $g(x)=\dfrac{x^8-2x+5}{(x^2+3)^4(2+\sin x)}$. Compute $g'(-2)$, $g'(0)$, $g'(1.5)$, and $g'(5)$.

28. (Discussion) Explain why you must interpret the results of Exercises 26 and 27 carefully. Hint: Draw the graph of $y=f(x)$ given in Exercise 26 in the $[-5,5]$ by $[-0.5,0.5]$ viewing window. What is the domain of f?

29. Graph the first eight terms of the Weierstrass function in the standard viewing window. Zoom-in several times. How wiggly and bumpy is this graph? Specify a viewing window in which the displayed portion of the graph is smooth.

30. a) Graph $y=\sqrt[3]{x-2}$ and $f(x)=\text{NDer}(y,x)$ in the $[-1,4]$ by $[-2,2]$ viewing window.

b) Where does $D_x y$ fail to exist? Why? How does the graph of $f(x)=\text{NDer}(y,x)$ help answer this question?

c) For what values of x is $D_x y>0$? Is the curve $y=\sqrt[3]{x-2}$ increasing or decreasing for all x? For what value of x is the tangent line slope to the curve $y = \sqrt[3]{x-2}$ positive?

d) Explain the connection, if any, among the answers to the questions in part c.

31. Let $f(x) = \begin{cases} -x^2, & x < 0 \\ 4 - x^2, & x \geq 0 \end{cases}$

a) Draw the graph of $y = \text{NDer}\,(f(x), x)$ in $[-5, 5]$ by $[-3, 3]$.

b) Describe the graph in part a. Write a linear function algebraic representation for $y = \text{NDer}\,(f(x), x)$.

c) What is $\text{NDer}\,(f(x), 0)$? What is $f'(0)$? (Be careful!)

32. Let $f(x) = ax^2 + bx + c$, $a \neq 0$. Prove that $[f(x+h) - f(x-h)/2h] = 2ax + b$ provided that $h \neq 0$. (This means that the symmetric difference quotient method produces the *exact* derivative value for quadratic functions.)

33. Let $f(x) = \begin{cases} x, & x < 0 \\ -x^2, & x \geq 0. \end{cases}$
Graph $y = f(x)$ and $y = \text{NDer}\,(f(x), x)$ in the same viewing window. What is $f'(0)$?

34. Let $f(x) = \begin{cases} -\sqrt{|x|}, & x < 0 \\ 1 - x^2, & x \geq 0. \end{cases}$
Graph $y = f(x)$ and $y = \text{NDer}\,(f(x), x)$ in the same viewing window. What is $f'(0)$?

35. Let $f(x) = f(x) = -\sqrt[3]{|x|}$. Graph $y = f(x)$ and $y = \text{NDer}\,(f(x), x)$ in the same viewing window. What is $f'(0)$?

36. a) Draw the graph of $y = a^x$, and $y = \text{NDer}\,(a^x, x)$ in the $[-1, 2]$ by $[-2, 8]$ viewing window for $a = 0.5$, $a = 0.75$, $a = 1$, $a = 1.5$, $a = 2$ and $a = 3$.

b) Determine a value for a so that the two graphs in part a are identical.

c) Find a nontrivial function $y = f(x)$ with the property that $f(x) = f'(x)$ for each value of x in the domain of f.

37. a) Draw the graph of $y = \sin x$ and $y = \text{NDer}\,(\sin x, x)$ in the $[-10, 10]$ by $[-2, 2]$ viewing window.

b) Make a conjecture about $D_x(\sin x)$.

c) Test your conjecture by graphing your conjecture together with $y = \text{NDer}\,(\sin x, x)$.

38. Let $f(x) = \sqrt[5]{x}$. What are $f/(0)$ and NDer $(f, 0)$? Explain.

39. Let $f(x? = |x|$. What are $f/(0)$ and NDer $(f, 0)$? Explain.

3.2 Differentiation Rules

The process of calculating a derivative is called differentiation. The goal of this section is to show how to differentiate functions rapidly—without having to apply the definition each time. It will then be an easy matter to calculate the velocities, accelerations, and other important rates of change we will encounter in Section 3.3.

Integer Powers, Multiples, Sums, and Differences

The first rule of differentiation is that the derivative of every constant function is zero. In short,

RULE 1

Derivative of a Constant If c is a constant, then

$$\frac{d}{dx}(c) = 0.$$

The reason for this rule is the following calculation. If $f(x) = c$ is a function with a constant value c, then

$$\lim_{h\to 0} \frac{f(x+h) - f(x)}{h} = \lim_{h\to 0} \frac{c-c}{h} = \lim_{h\to 0} 0 = 0.$$

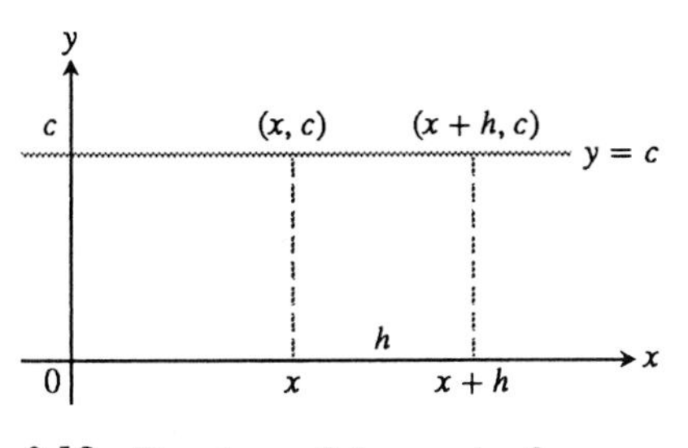

3.19 The slope of the graph of $y =$ constant is zero at every point.

This is another way to say that the values of constant functions never change and that the slope of a horizontal line is zero at every point. See Fig. 3.19.

> **Explore with a Grapher**
>
> 1. Graph $y = \text{NDer}\,(x^3, x)$ in the $[-5, 5]$ by $[-5, 5]$ viewing. Overlay graphs of the form $y = nx^2$ for $n = 1, 2, 3, 4, 5$. Which graph matches the graph of the derivative of $f(x) = x^3$?
> 2. Try the same idea for the derivative of $g(x) = x^4$. Overlay graphs of $y = nx^3$, for various values of n, on the graph of $y = \text{NDer}\,(x^4, x)$. Which graph matches the graph of the derivative of $g(x) = x^4$?
> 3. Can you generalize your results of Explore 1 and 2?

The Exploration suggests the next rule about derivatives of positive integer powers of x.

RULE 2

> **Power Rule for Positive Integer Powers of x** If n is a positive integer, then
>
> $$\frac{d}{dx}(x^n) = nx^{n-1}. \tag{1}$$

To apply the power rule, we subtract 1 from the original exponent (n) and multiply the result by n.

EXAMPLE 1

$$\frac{d}{dx}(x) = \frac{d}{dx}(x^1) = 1 \cdot x^0 = 1$$

$$\frac{d}{dx}(x^2) = 2x^1 = 2x$$

$$\frac{d}{dx}(x^3) = 3x^2$$

$$\frac{d}{dx}(x^4) = 4x^3$$

$$\frac{d}{dx}(x^5) = 5x^4$$

Proof of Rule 2 We set $f(x) = x^n$ and find the limit as h approaches zero of

$$\frac{f(x+h) - f(x)}{h} = \frac{(x+h)^n - x^n}{h}. \tag{2}$$

Since n is a positive integer, we can apply the algebra formula

$$a^n - b^n = (a-b)(a^{n-1} + a^{n-2}b + \cdots + ab^{n-2} + b^{n-1}),$$

with $(a-b) = h$, $a = x+h$, and $b = x$, to replace the expression $(x+h)^n - x^n$ in Eq. (2) by a form that is divisible by h. The resulting division tells us that

$$\begin{aligned}\frac{f(x+h) - f(x)}{h} &= \frac{(x+h)^n - x^n}{h} \\ &= \frac{(h)[(x+h)^{n-1} + (x+h)^{n-2}x + \cdots + (x+h)x^{n-2} + x^{n-1}]}{h} \\ &= \underbrace{[(x+h)^{n-1} + (x+h)^{n-2}x + \cdots + (x+h)x^{n-2} + x^{n-1}]}_{n \text{ terms, each with limit } x^{n-1} \text{ as } h \to 0}.\end{aligned} \tag{3}$$

Hence

$$\frac{d}{dx}(x^n) = \lim_{h \to 0} \frac{f(x+h) - f(x)}{h} = nx^{n-1}.$$

The next rule says that when a differentiable function is multiplied by a constant its derivative is multiplied by the same constant.

RULE 3

The (Constant) Multiple Rule If u is a differentiable function of x and c is a constant, then

$$\frac{d}{dx}(cu) = c\frac{du}{dx}. \tag{4}$$

In particular, if n is a positive integer then

$$\frac{d}{dx}(cx^n) = cnx^{n-1}. \tag{5}$$

EXAMPLE 2 The derivative

$$\frac{d}{dx}(7x^5) = 7 \cdot 5x^4 = 35x^4$$

says that if we stretch the graph of $y = x^5$ by multiplying each y-coordinate by 7, then we multiply the slope at each point by 7. See Fig. 3.20.

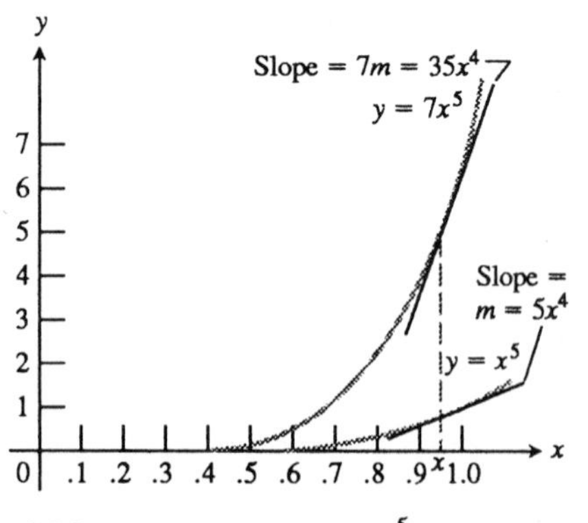

3.20 The graphs of $y = x^5$ and the stretched curve $y = 7x^5$. Multiplying the y-coordinates by 7 multiplies the slopes by 7.

EXAMPLE 3 *(A useful special case)* The derivative of the negative of any function is the negative of the function's derivative:

$$\begin{aligned}\frac{d}{dx}(-u) &= \frac{d}{dx}(-1\cdot u)\\ &= -1\cdot\frac{d}{dx}(u) \qquad \text{Eq. (4) with } c=-1\\ &= -\frac{du}{dx}.\end{aligned}$$

Proof of Rule 3

$$\begin{aligned}\frac{d}{dx}cu &= \lim_{h\to 0}\frac{cu(x+h)-cu(x)}{h} &&\text{Derivative definition applied to } f(x)=cu(x)\\ &= c\lim_{h\to 0}\frac{u(x+h)-u(x)}{h} &&\text{Limit property}\\ &= c\frac{du}{dx}. &&u \text{ is differentiable.}\end{aligned}$$

The next rule says that the derivative of the sum or difference of two differentiable functions is the sum or difference of their derivatives.

RULE 4

The Sum and Difference Rule If u and v are differentiable functions of x then their sum and difference are differentiable at every point where u and v are both differentiable. At such points,

1. $\frac{d}{dx}(u+v)=\frac{du}{dx}+\frac{dv}{dx}$, 2. $\frac{d}{dx}(u-v)=\frac{du}{dx}-\frac{dv}{dx}$

Similar equations hold for more than two functions, as long as the number of functions involved is finite.

EXAMPLE 4

1.
$$\begin{aligned}y &= x^4+12x\\ \frac{dy}{dx} &= \frac{d}{dx}(x^4)+\frac{d}{dx}(12x)\\ &= 4x^3+12\end{aligned}$$

2.
$$\begin{aligned}y &= \frac{7x^2}{3}-5\\ \frac{dy}{dx} &= \frac{d}{dx}\left(\frac{7x^2}{3}\right)-\frac{d}{dx}(5)\\ &= \frac{7}{3}\cdot 2x-0=\frac{14}{3}x\end{aligned}$$

3.
$$\begin{aligned} y &= x^3 + 3x^2 - 5x + 1 \\ \frac{dy}{dx} &= \frac{d}{dx}(x^3) + \frac{d}{dx}(3x^2) - \frac{d}{dx}(5x) + \frac{d}{dx}(1) \\ &= 3x^2 + 3 \cdot 2x - 5 + 0 \\ &= 3x^2 + 6x - 5 \end{aligned}$$

Notice that we can differentiate any polynomial term by term, the way we differentiated the polynomials in Example 4.

Proof of Rule 4 To prove Part 1, we apply the derivative definition with $f(x) = u(x) + v(x)$:

$$\begin{aligned} \frac{d}{dx}[u(x) + v(x)] &= \lim_{h \to 0} \frac{[u(x+h) + v(x+h)] - [u(x) + v(x)]}{h} \\ &= \lim_{h \to 0} \left[\frac{u(x+h) - u(x)}{h} + \frac{v(x+h) - v(x)}{h}\right] \\ &= \lim_{h \to 0} \frac{u(x+h) - u(x)}{h} + \lim_{h \to 0} \frac{v(x+h) - v(x)}{h} \\ &= \frac{du}{dx} + \frac{dv}{dx}. \end{aligned}$$

The proof of Part 2 is similar.

EXAMPLE 5 Finding horizontal tangents Does the curve $y = x^4 - 2x^2 + 2$ have any horizontal tangents? If so, where?

Solution The horizontal tangents, if any, occur where the slope dy/dx is zero. To find these points, we

1. Calculate dy/dx : $\quad \frac{dy}{dx} = \frac{d}{dx}(x^4 - 2x^2 + 2) = 4x^3 - 4x,$

2. Solve the equation $\frac{dy}{dx} = 0$ for x:
$$\begin{aligned} 4x^3 - 4x &= 0 \\ 4x(x^2 - 1) &= 0 \\ x &= 0, 1, -1. \end{aligned}$$

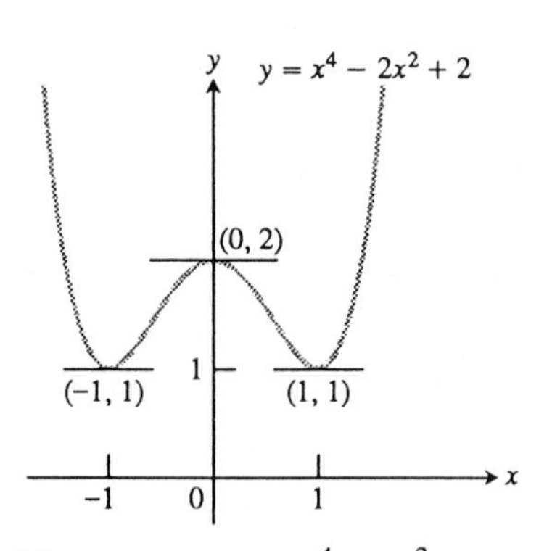

3.21 The curve $y = x^4 - 2x^2 + 2$ and its horizontal tangents. The tangents were located by setting dy/dx equal to zero and solving for x, as in Example 5.

The curve in question has horizontal tangents at $x = 0, 1$, and -1. The corresponding points on the curve (found from the equation $y = x^4 - 2x^2 + 2$) are (0, 2), (1, 1), and (−1, 1). See Fig. 3.21.

The successful solution of Example 5 depends on both our ability to correctly apply the differentiation rules *and* to solve the equation $dy/dx = 0$ by *factoring*. Usually we are not so fortunate as the next example illustrates.

EXAMPLE 6 Where does the curve $f(x) = x^4 - x^3 + x^2 - 2x + 6$ have horizontal tangent lines, if any?

It is easy to determine that $f'(x) = 4x^3 - 3x^2 + 2x - 2$. However solving the equation $f'(x) = 0$ is another matter. It can be shown that there are no rational zeros using the rational zeros theorem. Thus the real solutions are *irrational.* Try finding them analytically. Using zoom-in you can confirm that $f'(x) = 0$ when $x = 0.852$ with an error of less than 0.01 (Fig. 3.22). It can be shown that there are no other real solutions of $f'(x) = 0$. Thus there is only one place of horizontal tangency and it occurs at the point $(0.852, f(0.852))$. Figure 3.23 supports this result.

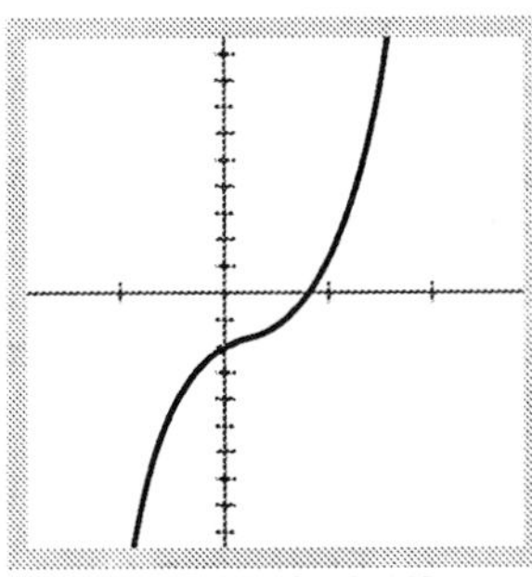

[−2, 3] by [−10, 10]

3.22 A complete graph of $y = f'(x) = 4x^3 - 3x^2 + 2x - 2$.

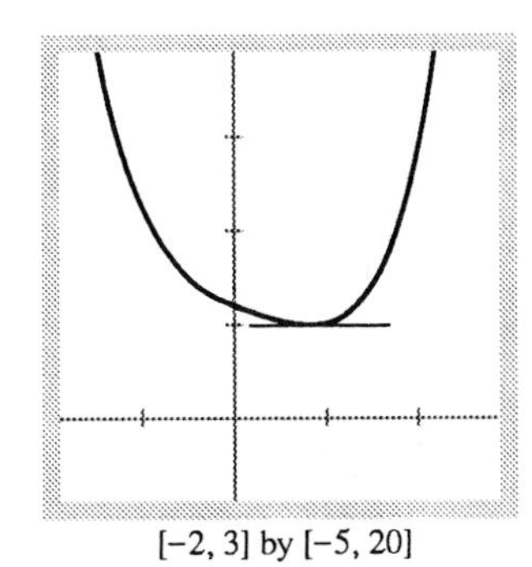

[−2, 3] by [−5, 20]

3.23 A complete graph of $f(x) = x^4 - x^3 + x^2 - 2x + 6$ showing its only horizontal tangent at $x = 0.852$.

How to Solve $f'(x) = 0$

There are at least five general ways to solve $f'(x) = 0$ given a differentiable function $y = f(x)$.

1. Determine the exact derivative $f'(x)$ and find the exact solutions of $f'(x) = 0$ (usually by factoring).
2. Graph $y = \text{NDer}(f(x), x)$. Determine a complete graph and use zoom-in to solve $f'(x) = 0$ to the desired accuracy.
3. Graph the exact derivative $y = f'(x)$. Determine a complete graph and use zoom-in to solve $f'(x) = 0$ to the desired accuracy.
4. (If availiable on your electronic tool.) Use the SOLVER feature of your computer or graphing calculator tool to numerically solve $\text{NDer}(f(x), x) = 0$. Note: Effective SOLVERS first draw the graph to prompt the user to determine good initial approximations.
5. Use a Computer Algebra System like *Derive*™ or *Mathematica*™ to apply a combination of methods 1 and 4 to solve the equation $f'(x) = 0$.

Method 1 is rarely useful because most derived equations $f'(x) = 0$ do not have exact solutions or have exact solutions that are tedious to find (e.g., using the Cardan formulas to find exact solutions of cubic equations). In this text we assume you will use Method 2 anytime unless stated otherwise. Method 3 is the same as Method 2 if $f'(x)$ is easy to determine. Method 4 is acceptable if the SOLVER feature is available to you. It simply automates the graphing zoom-in procedure. Method 5 is optional.

Second and Higher Order Derivatives

The derivative

$$y' = \frac{dy}{dx}$$

is the *first derivative* of y with respect to x. The first derivative may also be a differentiable function of x. If so, its derivative,

$$y'' = \frac{dy'}{dx} = \frac{d}{dx}\left(\frac{dy}{dx}\right) = \frac{d^2y}{dx^2},$$

is called the *second derivative* of y with respect to x. If y'' ("y double prime") is differentiable, its derivative,

$$y''' = \frac{dy''}{dx} \qquad \text{("}y\text{ triple prime"),}$$

is the *third derivative* of y with respect to x. The names continue as you imagine they would, with

$$y^{(n)} = \frac{d}{dx}y^{(n-1)} \qquad (\text{“}y\text{ super n”})$$

denoting the n*th derivative* of y with respect to x.

EXAMPLE 7 The first four derivatives of $y = x^3 - 3x^2 + 2$ are

First derivative : $y' = 3x^2 - 6x$

Second derivative : $y'' = 6x - 6$

Third derivative : $y''' = 6$

Fourth derivative : $y^{(4)} = 0.$

The function has derivatives of all orders, but the fifth and subsequent order derivatives are all zero.

In this text, we assume your graphing calculator or computer can easily compute values of the *second* derivative of a differentiable function. We use the notation

$$\text{NDer}\,(f(x), a)$$

to denote the numerical second derivative. Actually $\text{NDer}\,(f(x), a) = \text{NDer}$ $(\text{NDer}\,(f(x), (x), a)$. See Exercises 27–32. We use the same tolerance $h = 0.01$ for NDer. The accuracy agreement about reporting answers accurate to two decimal places also applies. We assume your graphing calculator or computer can easily graph $y = \text{NDer}\,(f(x), x)$.

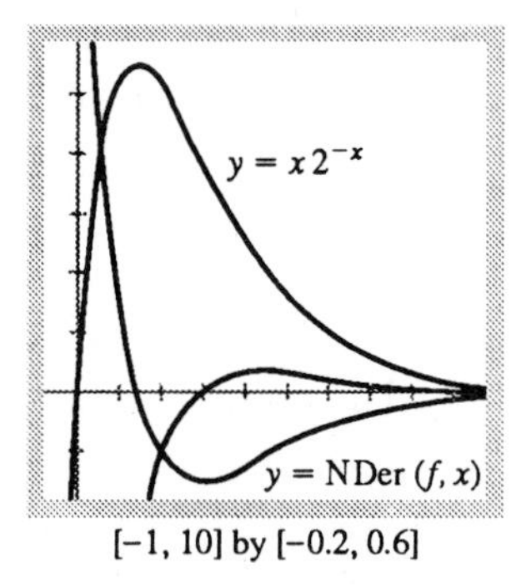

[−1, 10] by [−0.2, 0.6]

3.24 The graphs of $f(x) = x2^{-x}, y = \text{NDer}\,(f, x)$ and $y = \text{NDer}\,(f, x)$.

EXAMPLE 8 Let $f(x) = x2^{-x}$. Graph $f, y = \text{NDer}\,(f(x), x)$ and $y = \text{NDer}(f(x), x)$ in the $[-1, 10]$ by $[0.2, -0.6]$ viewing rectangle. The solution is displayed in Fig. 3.24.

Notice how the x-intercept of the graph of $y = \text{NDer}\,(x2^{-x}, x)$ can be used to locate the horizontal tangent line to the graph of $y = \text{NDer}\,(x2^{-x}, x)$. Can you explain why? We will see in the next chapter how the graphs of $y = f'(x)$ and $y = f''(x)$ can be used to predict the behavior of the graph of $y = f(x)$.

Products

While the derivative of the sum of two functions is the sum of their derivatives and the derivative of the difference of two functions is the difference of their derivatives, the derivative of the product of two functions is *not* the product of their derivatives. The derivative of a product is a sum of *two* products, as we now explain.

RULE 5

The Product Rule The product of two differentiable functions u and v is differentiable and

$$\frac{d}{dx}(uv) = u\frac{dv}{dx} + v\frac{du}{dx}. \tag{6}$$

As with the Sum and Difference Rule, the Product Rule is understood to hold only at values of x where u and v both have derivatives. At such a value of x, the derivative of the product uv is u times the derivative of v plus v times the derivative of u.

Proof of Rule 5

$$\frac{d}{dx}(uv) = \lim_{h\to 0} \frac{u(x+h)v(x+h) - u(x)v(x)}{h}.$$

To change this fraction into an equivalent one that contains difference quotients for the derivatives of u and v, we subtract and add $u(x+h)v(x)$ in the numerator. Then,

$$\begin{aligned}
\frac{d}{dx}(uv) &= \lim_{h\to 0} \frac{u(x+h)v(x+h) - u(x+h)v(x) + u(x+h)v(x) - u(x)v(x)}{h} \\
&= \lim_{h\to 0}\left[u(x+h)\frac{v(x+h)-v(x)}{h} + v(x)\frac{u(x+h)-u(x)}{h}\right] \\
&= \lim_{h\to 0} u(x+h)\cdot \lim_{h\to 0}\frac{v(x+h)-v(x)}{h} + v(x)\cdot \lim_{h\to 0}\frac{u(x+h)-u(x)}{h}.
\end{aligned}$$

As h approaches zero, $u(x+h)$ approaches $u(x)$ because u, being differentiable at x, is continuous at x. The two fractions approach the values of du/dx at x and dv/dx at x. In short,

$$\frac{d}{dx}(uv) = u\frac{dv}{dx} + v\frac{du}{dx}.$$

EXAMPLE 9 Find the derivative of $y = (x^2+1)(x^3+3)$.

Solution From the Product Rule with

$$u = x^2 + 1, \qquad v = x^3 + 3,$$

we find

$$\begin{aligned}
\frac{d}{dx}[(x^2+1)(x^3+3)] &= (x^2+1)(3x^2) + (x^3+3)(2x) \\
&= 3x^4 + 3x^2 + 2x^4 + 6x \\
&= 5x^4 + 3x^2 + 6x.
\end{aligned}$$

This particular example can be done as well (perhaps better) by multiplying out the original expression for y and differentiating the resulting polynomial. We do that now as a check. From

$$y = (x^2 + 1)(x^3 + 3) = x^5 + x^3 + 3x^2 + 3,$$

we obtain

$$\frac{dy}{dx} = 5x^4 + 3x^2 + 6x,$$

in agreement with our first calculation.

There are times, however, when the Product Rule *must* be used, as the next example shows.

EXAMPLE 10 Let $y = uv$ be the product of the functions u and v, and suppose that

$$u(2) = 3, \qquad u'(2) = -4, \qquad v(2) = 1, \qquad \text{and} \qquad v'(2) = 2.$$

Find $y'(2)$.

Solution From the Product Rule, in the form

$$y' = (uv)' = uv' + uv',$$

we have

$$y'(2) = u(2)v'(2) + v(2)u'(2) = (3)(2) + (1)(-4) = 6 - 4 = 2.$$

Quotients

Just as the derivative of the product of two differentiable functions is not the product of their derivatives, the derivative of the quotient of two functions is not the quotient of their derivatives. What happens instead is this:

RULE 6

The Quotient Rule At a point where $v \neq 0$, the quotient $y = u/v$ of two differentiable functions is differentiable and

$$\frac{d}{dx}\left(\frac{u}{v}\right) = \frac{v\frac{du}{dx} - u\frac{dv}{dx}}{v^2}. \tag{7}$$

As with the earlier combination rules, the Quotient Rule holds only at values of x at which u and v both have derivatives.

Proof of Rule 6

$$\frac{d}{dx}\left(\frac{u}{v}\right) = \lim_{h\to 0} \frac{\frac{u(x+h)}{v(x+h)} - \frac{u(x)}{v(x)}}{h}$$

$$= \lim_{h\to 0} \frac{v(x)u(x+h) - u(x)v(x+h)}{hv(x+h)v(x)}$$

To change the last fraction into an equivalent one that contains the difference quotients for the derivatives of u and v, we subtract and add $v(x)u(x)$ in the numerator. This allows us to continue with

$$\frac{d}{dx}\left(\frac{u}{v}\right) = \lim_{h\to 0} \frac{v(x)u(x+h) - v(x)u(x) + v(x)u(x) - u(x)v(x+h)}{hv(x+h)v(x)}$$

$$= \lim_{h\to 0} \frac{v(x)\frac{u(x+h)-u(x)}{h} - u(x)\frac{v(x+h)-v(x)}{h}}{v(x+h)v(x)}.$$

Taking the limit in the numerator and denominator now gives the Quotient Rule. ∎

EXAMPLE 11 Find the derivative of $y = \dfrac{x^2-1}{x^2+1}$.

Solution We apply the Quotient Rule with $u = x^2 - 1$ and $v = x^2 + 1$:

$$\frac{dy}{dx} = \frac{(x^2+1)\cdot 2x - (x^2-1)\cdot 2x}{(x^2+1)^2}$$

$$= \frac{2x^3 + 2x - 2x^3 + 2x}{(x^2+1)^2}$$

$$= \frac{4x}{(x^2+1)^2}.$$

∎

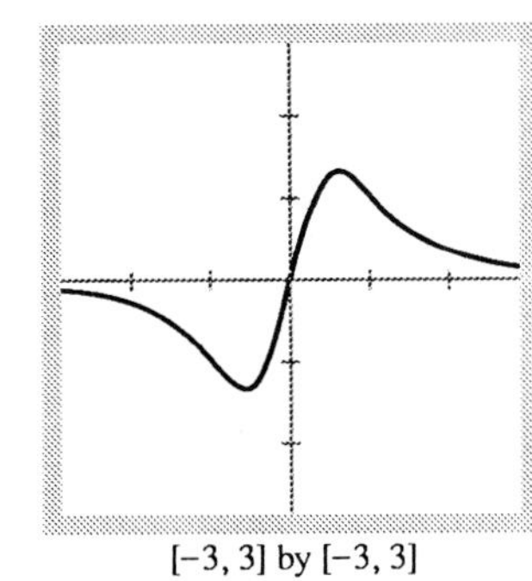

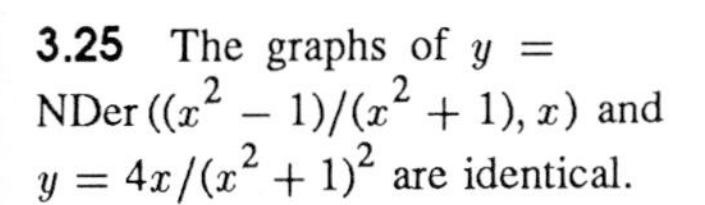

[−3, 3] by [−3, 3]

3.25 The graphs of $y =$ NDer $((x^2-1)/(x^2+1), x)$ and $y = 4x/(x^2+1)^2$ are identical.

Observing that the graphs of $y = \text{NDer}(f(x), x)$ and the exact derivative $y = f'(x)$ are identical provides powerful support that the manipulations used to obtain the exact derivative $f'(x)$ in Example 11 are correct (Fig. 3.25).

Graphs provide strong support for correctness of algebraic manipulations and analytic methods. However, we must be careful when we try to draw conclusions from graphs as the next Explore illustrates.

Explore with a Grapher

Graph $y = \sin x$ and $y = x$ in the $[-0.1, 0.1]$ by $[-0.1, 0.1]$ viewing window. Observe that the graphs appear to be identical. Does this mean $\sin x = x$? Explain.

Negative Integer Powers of *x*

The rule for differentiating negative powers of x is the same as the rule for differentiating positive powers of x.

RULE 7

Power Rule for Negative Integer Powers of x If n is a negative integer and $x \neq 0$, then

$$\frac{d}{dx}(x^n) = nx^{n-1}. \tag{8}$$

Proof of Rule 7 The proof uses the Quotient Rule in a clever way. If n is a negative integer, then $n = -m$ where m is a positive integer. Hence, $x^n = x^{-m} = 1/x^m$ and

$$\frac{d}{dx}(x^n) = \frac{d}{dx}\left(\frac{1}{x^m}\right)$$

$$= \frac{x^m \cdot \frac{d}{dx}(1) - 1 \cdot \frac{d}{dx}(x^m)}{(x^m)^2}$$ Quotient Rule with $u = 1$ and $v = x^m$

$$= \frac{0 - mx^{m-1}}{x^{2m}}$$ Since $m > 0$, $\frac{d}{dx}(x^m) = mx^{m-1}$.

$$= -mx^{-m-1}$$

$$= nx^{n-1}.$$ Changing back to n

EXAMPLE 12

$$\frac{d}{dx}\left(\frac{1}{x}\right) = \frac{d}{dx}(x^{-1}) = (-1)x^{-2} = -\frac{1}{x^2}$$

$$\frac{d}{dx}\left(\frac{4}{x^3}\right) = 4\frac{d}{dx}(x^{-3}) = 4(-3)x^{-4} = -\frac{12}{x^4}$$

EXAMPLE 13 Find an equation for the tangent to the curve

$$y = x + \frac{2}{x}$$

at the point (1, 3).

Solution First, NDer $(x + 2/x, 1) = -1.00020002$. This suggests the slope at $x = 1$ is -1, which we confirm analytically.

The slope of the curve is

$$\frac{dy}{dx} = \frac{d}{dx}(x) + 2\frac{d}{dx}\left(\frac{1}{x}\right)$$

$$= 1 + 2\left(-\frac{1}{x^2}\right)$$ From Example 12

$$= 1 - \frac{2}{x^2}.$$

The slope at $x = 1$ is

$$\left.\frac{dy}{dx}\right|_{x=1} = \left[1 - \frac{2}{x^2}\right]_{x=1} = 1 - 2 = -1.$$

The line through (1, 3) with slope $m = -1$ is

$$y - 3 = (-1)(x - 1) \quad \text{Point-slope equation}$$

$$y = -x + 1 + 3$$

$$y = -x + 4.$$

See Fig. 3.26.

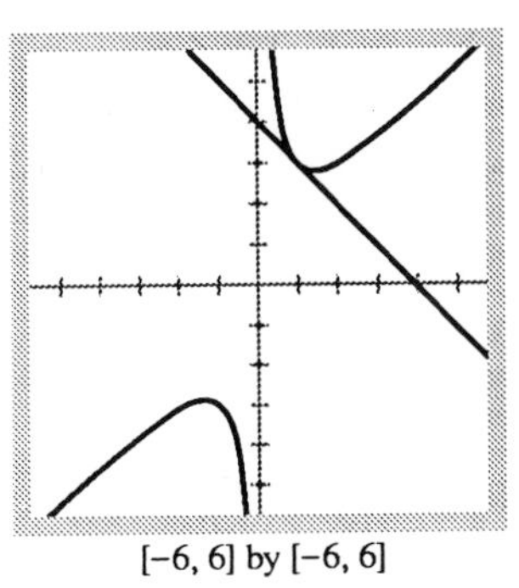

[−6, 6] by [−6, 6]

3.26 The slope of the tangent to the curve $y = x + (2/x)$ at the point $(1, 3)$ is found by evaluating y' at $x = 1$.

Choosing Which Rules to Use

The choice of which rules to use in solving a differentiation problem can make a difference in how much work you have to do. Here is an example.

EXAMPLE 14 Do not use the Quotient Rule to find the derivative of

$$y = \frac{(x-1)(x^2-2x)}{x^4}.$$

Instead, expand the numerator and divide by x^4:

$$y = \frac{(x-1)(x^2-2x)}{x^4} = \frac{x^3 - 3x^2 + 2x}{x^4} = x^{-1} - 3x^{-2} + 2x^{-3}.$$

Then use the Sum and Power Rules:

$$\frac{dy}{dx} = -x^{-2} - 3(-2)x^{-3} + 2(-3)x^{-4} = -\frac{1}{x^2} + \frac{6}{x^3} - \frac{6}{x^4}.$$

Differentiation Rules versus Numerical Derivatives

In many cases, the numerical derivative or the graph of the numerical derivative is all that is needed to solve problems where the derivative is required. Of course, the issue of accuracy is crucial. There is often no need to do the manipulations to find exact derivatives, particularly if the function is complicated. Thus the application of the numerical derivative is often a more appropriate choice, as is illustrated in the next example.

We will see in Chapter 4 that the second derivative of a function is very important in describing the complete behavior of a function.

EXAMPLE 15 Let $f(x) = \dfrac{5x}{x^2+4}$. Solve the inequality $f''(x) > 0$.

Solution The graphing method is easy. $f''(x) > 0$ precisely for the values of x where the graph of $y = f''(x)$ is *above* the x-axis. So we need the graph of $y = \text{NDer}\left(5x/(x^2+4), x\right)$. Figure 3.27 shows two views of the graph of the second derivative. The graph on the left is *not* a complete graph of $y = f''(x)$. For example it is not clear if there are 1, 2, 3, or more zeros.

The figure on the right shows there is a zero of $f''(x)$ between 3 and 4. In the exercises, you will see that if f is an even function, then f' is odd, and if f is odd, then f' is even. In this case, f'' is odd since f is odd. Since $y = f''(x)$ is *odd*, there is another zero between -4 and -3. Zoom-out can be used to show there are only three real zeros. Zoom-in can be used to solve the inequality $f''(x) > 0$. The solution consists of the intervals $(-3.464, 0)$ and $(3.464, \infty)$. That is, for $-3.464 < x < 0$ or $x > 3.464$, the second derivative of $f(x) = 5x/(x^2 + 4)$ is positive.

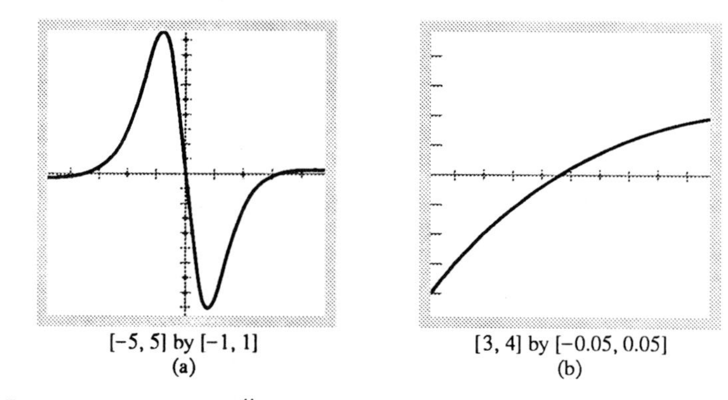

3.27 Two views of $y = f''(x)$.

The exact second derivative of the function in the last example is $f''(x) = [10x(x^2 - 12)]/(x^2 + 4)^3$. Confirm this yourself using the quotient rule. It is easy to see using a sign chart method that the exact solution of $f''(x) > 0$ consists of the intervals $(-\sqrt{12}, 0)$ and $(\sqrt{12}, \infty)$. Why? Compare this to the solution found in the last example. Which method do you prefer? What if the numerator of $f''(x)$ was not factorable? For example, try to determine *analytically* where $f''(x) > 0$ for $f(x) = (2x - 5)/(3x^2 + 4)$. That is, try to find the *exact* solution to $f''(x) > 0$ (see Exercise 53).

Exercises 3.2

In Exercises 1–8, find dy/dx and d^2y/dx^2 when y equals the given expression. Try to answer without writing anything down (except the answer).

1. x

2. $-x$

3. $-x^2 + 3$

4. $\frac{x^3}{3} - x$

5. $2x + 1$

6. $x^2 + x + 1$

7. $\frac{x^3}{3} + \frac{x^2}{2} + x$

8. $1 - x + x^2 - x^3$

In Exercises 9–12, find $y' = dy/dx$ and $y'' = d^2y/dx^2$ when y equals the given expression.

9. $x^4 - 7x^3 + 2x^2 + 15$

10. $5x^3 - 3x^5$

11. $4x^2 - 8x + 1$

12. $\frac{x^4}{4} - \frac{x^3}{3} + \frac{x^2}{2} - x + 3$

Find all the nonzero derivatives of the functions in Exercises 13–16.

13. $y = x^2 - x$

14. $y = \frac{x^3}{3} + \frac{x^2}{2} - 5$

15. $y = \frac{x^4}{2} - \frac{3}{2}x^2 - x$ **16.** $y = \frac{x^5}{120}$

In Exercises 17–20, find dy/dx when y equals the given expression. Find each derivative two ways: (a) by applying the Product Rule and (b) by multiplying the factors to produce a sum of simpler terms to differentiate. In these examples, (b) tends to be faster, but it is not always possible to multiply first this way. For functions like $y = x \sin x$, encountered later, the Product Rule offers the only way for exact solutions.

17. $(x+1)(x^2+1)$ **18.** $(x+1)(3-x^2)$

19. $(x-1)(x^2+x+1)$ **20.** $\left(x+\frac{1}{x}\right)\left(x-\frac{1}{x}\right)$

In Exercises 21–26, find dy/dx when y equals the given expression. Support your answer by graphing $y = dy/dx$ and $y = \text{NDer}(y, x)$ in the same viewing window.

21. $\frac{x-1}{x+7}$ **22.** $\frac{2x+5}{3x-2}$

23. $(1-x)(1+x^2)^{-1}$ **24.** $(2x-7)^{-1}(x+5)$

25. $\frac{x^2}{1-x^3}$ **26.** $\frac{x^2-1}{x^2+x-2}$

In Exercises 27–32, find $y' = dy/dx$ and $y'' = d^2y/dx^2$ when y equals the given expression. Support your answers by graphing $y = dy/dx$ and $y = \text{NDer}(y, x)$ and by graphing $y = d^2y/dx^2$ and $y = \text{N}_2\text{Derv}(y, x)$ in the same viewing window.

27. $\frac{3}{x^2}$ **28.** $-\frac{1}{x}$

29. $\frac{5}{x^4}$ **30.** $-\frac{3}{x^7}$

31. $x + 1 + \frac{1}{x}$ **32.** $\frac{12}{x} - \frac{4}{x^3} + \frac{1}{x^4}$

In Exercises 33–36, find dy/dx when y equals the given expression. Support your answers by graphing $y = dy/dx$ and $y = \text{NDer}(y, x)$ in the same viewing window.

33. $\frac{x^3+7}{x}$ **34.** $\frac{x^2+5x-1}{x^2}$

35. $\frac{(x-1)(x^2+x+1)}{x^3}$ **36.** $\frac{(x^2+x)(x^2-x+1)}{x^4}$

For each of the following functions $y = f(x)$, determine the equation of the tangent line at $(a, f(a))$, and graph both function and the tangent line in the same viewing window.

37. $f(x) = x3^{-0.2x}$, $a = 1$

38. $f(x) = \frac{\sin x}{x}$, $a = \pi$

39. $f(x) = \frac{x+3}{x^3-2x+5}$, $a = 0$

40. $f(x) = \sqrt[3]{\frac{x-1}{x^2+5}}$, $a = 2$

Determine complete graphs of $y = f'(x)$ and $y = f''(x)$ for each of the following functions $y = f(x)$.

41. $y = \frac{2^x}{x^2-1}$ **42.** $y = \frac{3^x}{4-x^2}$

43. $y = \sqrt[3]{\frac{x+3}{x-5}}$ **44.** $y = \sqrt[3]{\frac{x+1}{x^2+2}}$

45. $y = x \sin x$ **46.** $y = x^2 \sin x$

47–52. For each function in Exercises 41–46, solve $f'(x) = 0$ and $f''(x) > 0$.

53. Let f be the function $f(x) = (2x-5)/(3x^2+4)$. Determine $f''(x)$. Try to solve $f''(x) > 0$ *exactly*. Solve $f''(x) > 0$ using any appropriate method.

54. Let K, m, n, a, and b be positive constants. Explain why horizontal tangent lines to the graphs of $y = Ka^x/(m + nb^x)$ and $y = a^x/(m + nb^x)$ have the same x-coordinate. Determine the x-coordinate of all points where the tangent lines are horizontal for $a = 2$, $m = 2$, $n = 1$, and $b = 3$. Draw complete graphs for $K = 1$, $K = 2$, and $K = 3$ in the same viewing window.

55. Suppose u and v are functions of x that are differentiable at $x = 0$ and that

$$u(0) = 5, \quad u'(0) = -3, \quad v(0) = -1, \quad v'(0) = 2.$$

Find the values of the following derivatives at $x = 0$.

a) $\frac{d}{dx}(uv)$

b) $\frac{d}{dx}\left(\frac{u}{v}\right)$

c) $\frac{d}{dx}\left(\frac{v}{u}\right)$

d) $\frac{d}{dx}(7v - 2u)$

56. Which of the following numbers is the slope of the line tangent to the curve $y = x^2 + 5x$ at $x = 3$?

a) 24

b) $-5/2$

c) 11

d) 8

57. Which of the following numbers is the slope of the line $3x - 2y + 12 = 0$?

a) 6

b) 3

c) 3/2

d) 2/3

58. Find the equation of the line perpendicular to the tangent to the curve $y = x^3 - 3x + 1$ at the point $(2, 3)$.

59. Find the tangents to the curve $y = x^3 + x$ at the points where the slope is 4. What is the smallest slope on the curve? At what value of x does the curve have this slope?

60. Find the points on the curve $y = 2x^3 - 3x^2 - 12x + 20$ where the tangent is parallel to the x-axis.

61. Find the x- and y-intercepts of the line that is tangent to the curve $y = x^3$ at the point $(-2, -8)$.

62. Find the tangents to *Newton's Serpentine,*

$$y = \frac{4x}{x^2 + 1}$$

at the origin and the point $(1, 2)$.

63. Find the tangent to the *Witch of Agnesi*

$$y = \frac{8}{4 + x^2}$$

at the point $(2, 1)$.
There is a nice story about the name of this curve in the historical note on Agnesi in Chapter 10.

When we work with functions of a single variable in mathematics, we normally call the independent variable x and the dependent variable y. Applied fields use many different letters, however. Here are some examples. In these cases the exact derivative is very useful.

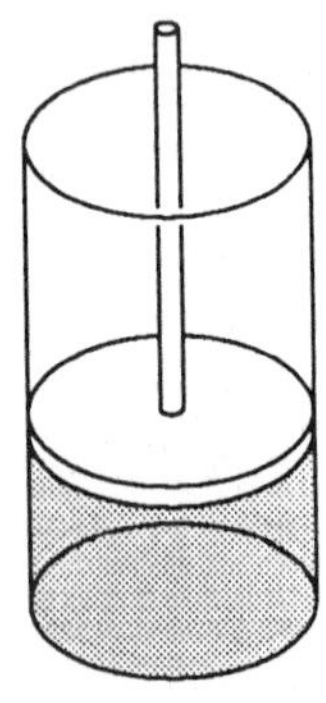

64. *Cylinder pressure.* If the gas in a cylinder is maintained at a constant temperature T, the pressure P is related to the volume V by a formula of the form

$$P = \frac{nRT}{V - nb} - \frac{an^2}{V^2}$$

in which a, b, n, and R are constants. Find dP/dV.

65. *Free fall.* When a rock falls from the rest near the surface of the earth, the distance it covers during the first few seconds is given by the equation

$$s = 4.9t^2.$$

In this equation, s is the distance in meters and t is the elapsed time in seconds. Find ds/dt and d^2s/dt^2.

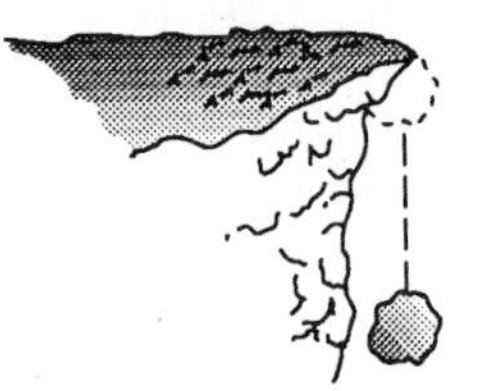

66. *The body's reaction to medicine.* The reaction of the body to a dose of medicine can often be represented by an equation of the form

$$R = M^2\left(\frac{C}{2} - \frac{M}{3}\right),$$

where C is a positive constant and M is the amount of medicine absorbed in the blood. If the reaction is a change in blood pressure, R is measured in millimeters of mercury. If the reaction is a change in temperature, R is measured in degrees, and so on.

Find DR/dM. This derivative, as a function of M, is called the sensitivity of the body to the medicine. In Chapter 4, we shall see how to find the amount of medicine to which the body is most sensitive. (*Source: Some Mathematical Models in Biology,* Revised Edition, R. M. Thrall, J. A. Mortimer, JK. R. Rebman, R. F. Baum, eds., December 1967, PB-202 364, p. 221; distributed by N.T.I.S., U.S. Department of Commerce.)

67. Show that if f is an even function, then f' is an odd function.

68. Show that if f is an odd function, then f' is an even function.

3.3 Velocity, Speed, and Other Rates of Change

In this section, we see how derivatives provide the mathematics we need to understand the way things change in the world around us. With derivatives, we can describe the rates at which water reservoirs empty, populations change, rocks fall, the economy changes, and an athlete's blood sugar varies with exercise. We begin with free fall, the kind of fall that takes place in a vacuum near the surface of the earth.

Free Fall

Near the surface of the earth, all bodies fall with the same constant acceleration. The distance a body falls after it is released from rest is a constant multiple of the square of the time elapsed. At least, that is what happens when the body falls in a vacuum, where there is no air to slow it down. The square-of-time rule also holds for dense, heavy objects like rocks, ball bearings, and steel tools during the first few seconds of their fall through air, before their velocities build up to where air resistance begins to matter. When air resistance is absent, or insignificant, and the only force acting on a falling body is the force of gravity, we call the way the body falls *free fall.*

The equation we write to say that the distance an object falls from rest is proportional to the square of the time elapsed is

$$s = \frac{1}{2}gt^2. \tag{1}$$

In this equation, s is distance, t is time, and g, as we shall see in a moment, is the constant acceleration given to an object by the force of gravity.

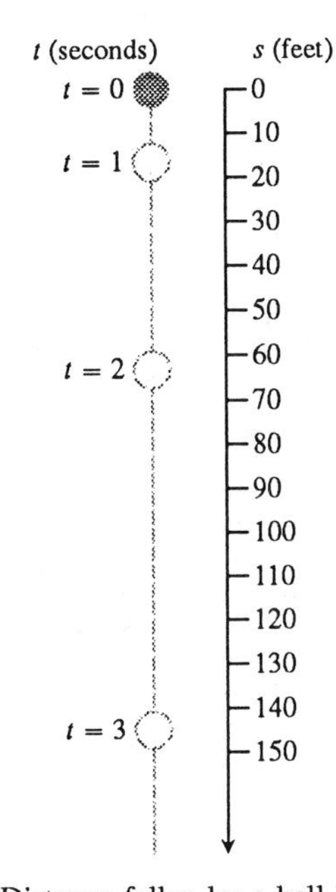

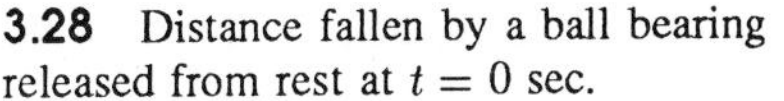
3.28 Distance fallen by a ball bearing released from rest at $t = 0$ sec.

EXAMPLE 1 The value of g in the equation $s = (1/2)gt^2$ depends on the units used to measure t and s. With t in seconds (the usual unit),

$$g = 32 \text{ ft/sec}^2 \qquad s = \frac{1}{2}(32)t^2 = 16t^2 \qquad (s \text{ in feet}),$$

$$g = 9.80 \text{ m/sec}^2 \qquad s = \frac{1}{2}(9.80)t^2 = 4.9t^2 \qquad (s \text{ in meters}),$$

$$g = 980 \text{ cm/sec}^2 \qquad s = \frac{1}{2}(980)t^2 = 490t^2 \qquad (s \text{ in centimeters}).$$

The abbreviation ft/sec^2 is read "feet per second squared" or "feet per second per second." The other units for g are "meters per second squared" and "centimeters per second squared." ■

EXAMPLE 2 Figure 3.28 shows the free fall of a heavy ball bearing released from rest at time $t = 0$. During the first 2 sec, the ball falls

$$s(2) = 16(2)^2 = 16 \cdot 4 = 64 \text{ ft}.$$

■

EXAMPLE 3 How long did it take the ball bearing in Fig. 3.28 to fall the first 100 feet?

Solution From Example 1 the free-fall equation for s in feet and t in seconds is

$$s = 16t^2.$$

To find the time it took the ball bearing to cover the first 100 ft, we substitute $s = 100$ and solve for t.

$$100 = 16t^2$$

$$t^2 = \frac{100}{16}$$

$$t = \frac{5}{2} \quad \text{Time increases from } t = 0 \text{ so we ignore the negative root.}$$

It took the ball 2.5 seconds to fall the first 100 feet.

Parametric Graphing

As indicated in Section 1.6, we assume your grapher includes a *parametric* graphing utility. A curve C defined parametrically by

$$C = \{(x(t), y(t)) : t\min \leq t \leq t\max\}$$

is a relation, that is, simply a subset of the coordinate plane. The variable t is called a parameter and the equations $x = x(t)$ and $y = y(t)$ are called parametric equations. Most curves defined parametrically are not functions (see Exercise 20). The vertical line $x = a$ can be defined parametrically by the parametric equations $x(t) = a$, $y(t) = t$.

The TRACE feature of graphing calculators and some computer software can be used to simulate motion as illustrated in the next example.

EXAMPLE 4 Graph the parametric equations $x(t) = 1$, $y(t) = -16t^2$ in the $[0, 3]$ by $[-200, 0]$ viewing window for $0 \leq t \leq 4$ with tstep= 0.01. Use the TRACE feature of your graphing calculator to approximate the time it takes the ball bearing in Fig. 3.28 to fall 100 feet, 125 feet, 150 feet and 180 feet (Fig. 3.29).

The ball bearing falls to 100 feet in about 2.5 seconds, 125 feet in about 2.79 seconds, 150 feet in about 3.06 seconds, and 180 feet in about 3.35 seconds.

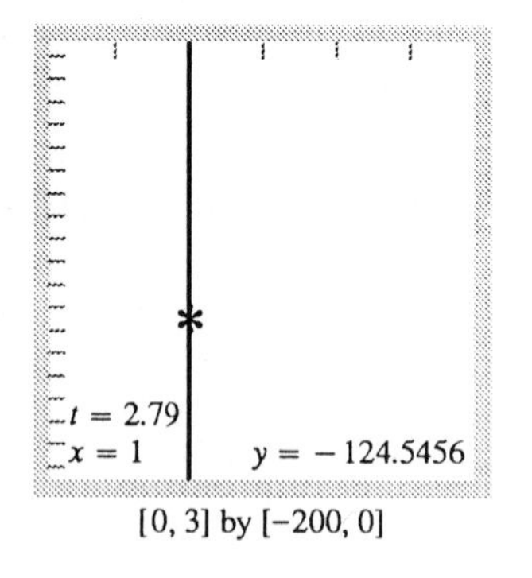

[0, 3] by [−200, 0]

3.29 A simulation of the path of the ball bearing of Example 4.

Velocity

Suppose we have a body moving along a coordinate line and we know that its position at time t is $s = f(t)$.

Position at time t ... and at time $t + \Delta t$

$s = f(t)$ $s + \Delta s = f(t + \Delta t)$ s

As the body moves along, it has a velocity at each particular instant and we want to find out what that velocity is. The information we seek must somehow be contained in the formula $s = f(t)$, but how do we find it?

We reason like this: In the interval from any time t to the slightly later time $t + \Delta t$, the body moves from position $s = f(t)$ to position

$$s + \Delta s = f(t + \Delta t). \tag{2}$$

The body's net change in position, or *displacement*, for this short time interval is

$$\Delta s = f(t + \Delta t) - f(t). \tag{3}$$

The body's average velocity for the time interval is Δs divided by Δt.

Definition

The **average velocity** of a body moving along a line from position $s = f(t)$ to position $s = f(t + \Delta t)$ is

$$v_{av} = \frac{\text{displacement}}{\text{travel time}} = \frac{\Delta s}{\Delta t} = \frac{f(t + \Delta t) - f(t)}{\Delta t}. \tag{4}$$

To find the body's velocity at the exact instant t, we take the limit of the average velocity over the interval from t to $t + \Delta t$ as the interval gets shorter and shorter and Δt shrinks to zero. Here is where the derivative comes in. For, as we now know, this limit is the derivative of f with respect to t.

Definition

Instantaneous velocity is the derivative of position. If the position function of a body moving along a line is $s = f(t)$, the body's velocity at time t is

$$v(t) = \frac{ds}{dt} = \lim_{\Delta t \to 0} \frac{f(t + \Delta t) - f(t)}{\Delta t}. \tag{5}$$

When people use the word **velocity** alone, they usually mean instantaneous velocity.

EXAMPLE 5 Fig. 3.30 is a time-to-distance graph of a 1989 Ford Thunderbird SC. The slope of the secant PQ is the average speed for the ten-second interval from $t = 5$ to $t = 15$ sec, in this case 40 m/sec or 144 km/h. The slope of the tangent at P is the speedometer reading at $t = 5$ sec, about 20 m/sec or 72 km/h. The car's top speed is 235 km/h (146 mph). (*Source: Car and Driver*, March 1989.)

EXAMPLE 6 Velocity during free fall

Free fall equation from Example 1	Corresponding velocity equation
$s = \frac{1}{2}gt^2$	$v = \frac{ds}{dt} = gt$
$s = 16t^2$	$v = 32t$
$s = 4.9t^2$	$v = 9.8t$
$s = 490t^2$	$v = 980t$

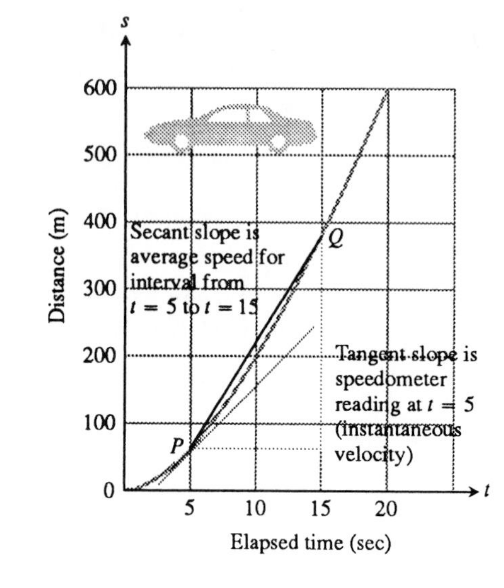

3.30

EXAMPLE 7 From Example 6, the velocity of the falling ball bearing t seconds after release is $v = 32t$ ft/sec (Fig. 3.28).

$$\text{At } t = 2: \qquad v = 32(2) = 64 \text{ ft/sec},$$

$$\text{At } t = 3: \qquad v = 32(3) = 96 \text{ ft/sec}.$$

Any function $y = f(x)$ can be graphed parametrically by graphing the parametric equations $x = t$, $y = f(t)$. Computer graphing can be used to nicely illustrate the connection between the path of a moving object and its displacement (position) as a function of time t as the next example shows.

EXAMPLE 8 A dynamite blast blows a heavy rock straight up with a launch velocity of 160 ft/sec (about 109 mph) (Fig. 3.31a). It reaches a height of $s = 160t - 16t^2$ ft after t sec.

a) Use a graphing utility to simulate the path of the rock and the graph of the rock's height as a function of time t.

b) How high does the rock go?

c) How fast is the rock traveling when it is 256 ft above the ground on the way up? On the way down?

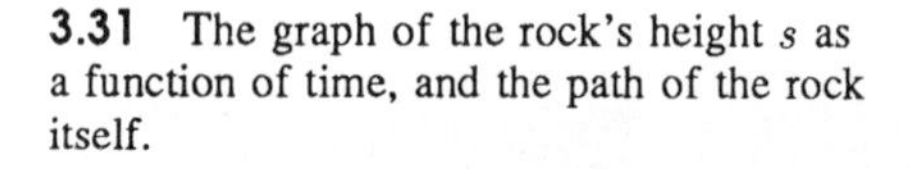
3.31 The graph of the rock's height s as a function of time, and the path of the rock itself.

Solution First assume that the rock goes straight up along the line $x = 3$. Then the path of the rock is given parametrically by

$$x_1(t) = 3$$

$$y_1(t) = 160t - 16t^2.$$

The graph of the *height function* $s(t) = 160t - 16t^2$ is the graph of the parametric equations

$$x_2(t) = t$$
$$y_2(t) = 160t - 16t^2.$$

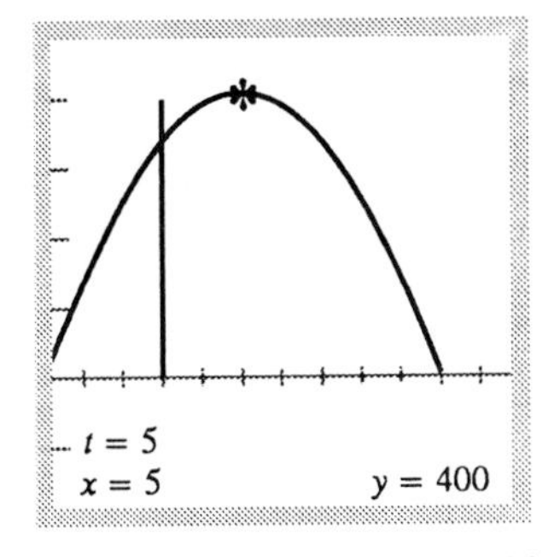

3.32 Two views of the rock problem of Example 8.

Set your grapher to graph *simultaneously* and graph both pairs of parametric equations in $[0, 12]$ by $[-150, 500]$ for $0 \le t \le t\,\mathrm{max}$. What value for t max should we use? Try different values and see what happens! Using "guess and check" and TRACE, you can quickly determine that the rock's maximum height is 400 ft. and it hits the ground after 10 seconds (see Fig. 3.32).

It can also be easily determined in this manner that $y = 256$ when $t = 2$ or 8. The velocity of the rock is given by

$$v(t) = \frac{ds}{dt} = 160 - 32t \quad \text{ft/sec.}$$

So $v(2) = 96$ ft./sec. and $v(8) = -96$ ft/sec. This means the velocity of the rock when it is 256 ft. above the ground is 96 ft/sec. (on the way up), and -96 ft/sec (on the way down). The velocity function can be graphed parametrically by

$$x_3(t) = t$$
$$y_3(t) = 160 - 32t.$$

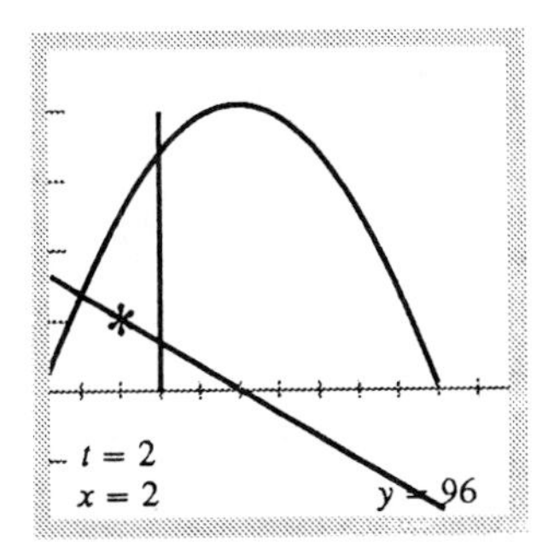

3.33 The path of the rock, the height function, and the velocity function.

A graph of all three curves, (the path of the rock, the height function, and the velocity function) is shown in Fig. 3.33. Notice the velocity is zero at the same time the height is maximum.

Why is the downward velocity negative? It has to do with how we set up the coordinate system. Since s measures height from the ground up, changes in s are positive as the rock rises and negative as the rock falls (Fig. 3.31). ≡

Analytically we can confirm the results obtained by graphing parametric equations to solve Example 8. The maximum height occurs when $v(t) = 0$. So

$$v(t) = \frac{ds}{dt} = 160 - 32t = 0 \Leftrightarrow t = 5.$$

Thus the maximum height is $s(5) = 400$ feet. Furthermore $s(t) = 160t - 16t^2 = 256$ exactly when $t = 2$ and $t = 8$ (by factoring) so $v(2) = 96$ and $v(8) = -96$ are the exact desired velocities.

Speed

If we drive over to a friend's house and back at 30 mph, say, the speedometer will show 30 on the way over but it will not show -30 on the way back. The speedometer always shows speed, and speed is the absolute value of velocity. Speed measures the rate of forward progress regardless of direction.

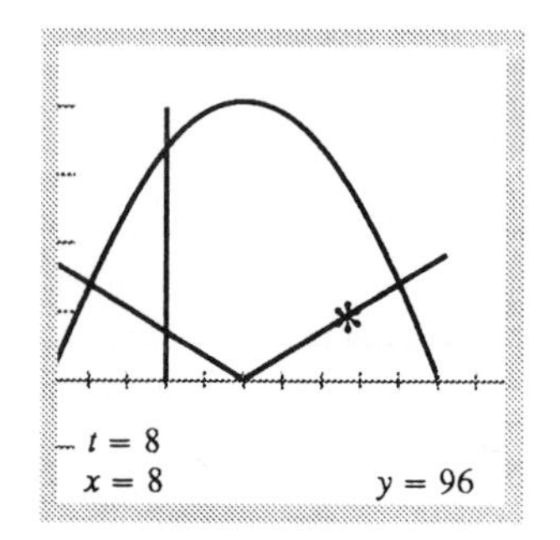

3.34 The graph of the speed of the rock, the height function, and the path of the rock.

Definition

> **Speed** is the absolute value of velocity.

EXAMPLE 9 When the rock in Example 8 passed the 256-ft mark, its forward speed was 96 ft/sec on the way up and $|-96| = 96$ ft/sec again on the way down. ☰

The graph of the speed function $y = |v(t)| = |160 - 32t|$ is V-shaped (Fig. 3.34).

Acceleration

In studies of motion along a coordinate line, we usually assume that the body's position function $s = f(t)$ has a second derivative as well as a first. The first derivative gives the body's velocity as a function of time; the second derivative gives the body's acceleration. Thus the velocity is how fast the position is changing and the acceleration is how fast the velocity is changing. The acceleration tells how quickly the body picks up or loses speed.

Definition

> **Acceleration** is the derivative of velocity. If a body's position at time t is $s = f(t)$, then the body's acceleration at time t is
>
> $$a = \frac{dv}{dt} = \frac{d^2s}{dt^2}. \tag{7}$$

EXAMPLE 10 The acceleration of the rock in Example 8 is

$$a = \frac{dv}{dt} = \frac{d}{dt}(160 - 32t) = 0 - 32 = -32 \text{ ft/sec}^2.$$

The minus sign confirms that the acceleration is downward, in the negative s direction. Whether the rock is going up or down, it is subject to the same constant downward pull of gravity. ☰

Other Rates of Change

Average and instantaneous rates give us the right language for many other applications.

EXAMPLE 11 The number of gallons $g(t)$ of water in a reservoir at time t(min) can be regarded as a differentiable function of t. If the volume of water changes by the amount Δg in the interval from time t to time $t + \Delta t$, then

$$\frac{\Delta g}{\Delta t} \text{ (gal/min)} = \text{average rate of change for the time interval,}$$

$$\frac{dg}{dt} = \lim_{\Delta t \to 0} \frac{\Delta g}{\Delta t} = \text{instantaneous rate of change at time } t.$$

Although it is natural to think of rates of change in terms of motion and time, there is no need to be so restrictive. We can define the average rate of change of any function over any interval of its domain as the change in the function divided by the length of the interval. We can then go on to define the instantaneous rate of change as the limit of average change as the length of the interval goes to zero.

Definition

Rates of Change The **average rate of change** of a function $f(x)$ over the interval from x to $x + h$ is

$$\text{Average rate of change} = \frac{f(x+h) - f(x)}{h}.$$

The **(instantaneous) rate of change** of f at x is the derivative

$$f'(x) = \lim_{h \to 0} \frac{f(x+h) - f(x)}{h},$$

provided the limit exists.

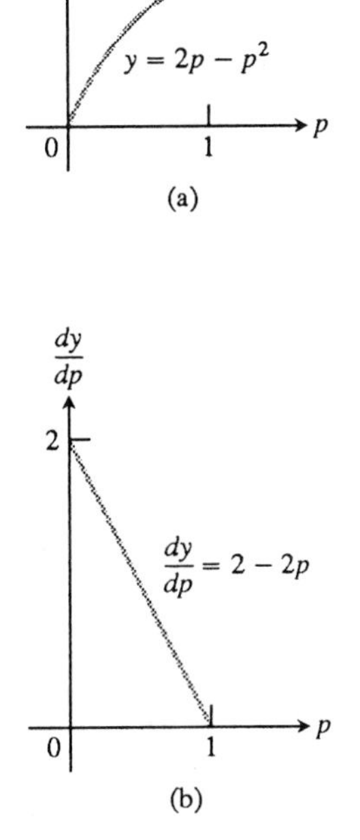

3.35 (a) The frequency $y = 2p - p^2$ of smooth skin in garden peas as a function of the prevalence p of the dominant smooth-skin gene. (b) The graph of dy/dp shows, among other things, how much y will change in response to a small change in p—a lot, if p is near 0, but not much if p is near 1.

Notice the use of the word *instantaneous* even when x does not represent time. It is conventional to do that.

EXAMPLE 12 The Austrian monk Gregor Johann Mendel (1822–1884), working with garden peas and other plants, provided the first scientific explanation of hybridization. His careful records showed that if p (a number between zero and one) is the frequency of the gene for smooth skin in peas (dominant) and $(1 - p)$ is the frequency of the gene for wrinkled skin in peas, then the proportion of smooth-skinned peas in the population at large is

$$y = 2p(1 - p) + p^2 = 2p - p^2.$$

The graph of y versus p in Fig. 3.35(a) suggests that the value of y is more sensitive to a change in p when p is small than it is to a change in p when p is large. Indeed, this is born out by the derivative graph in Fig. 3.35(b), which shows that dy/dp is close to 2 when p is near zero and close to 0 when p is near 1.

We will be able to say more about how sensitive functions are to changes in their variables when we get to Section 3.7. ∎

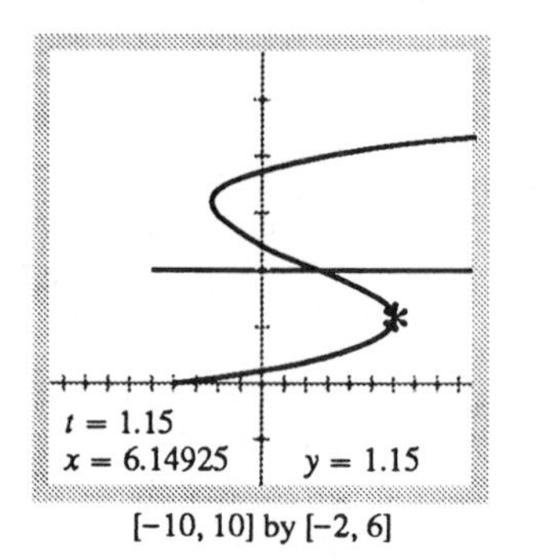

[−10, 10] by [−2, 6]

3.36 TRACE shows that the particle turns around at about $t = 1.15$ seconds.

EXAMPLE 13 The position (x-coordinate) of a particle moving on the line $y = 2$ is given by $s(t) = 2t^3 - 13t^2 + 22t - 5$ where t is time in seconds. a) Describe the motion of the particle for $t \geq 0$. b) When does it speed up? c) When does it slow down? d) When does it change direction? e) When is it at rest? f) Describe the velocity and speed of the particle.

Solution Let $x_1(t) = 2t^3 - 13t^2 + 22t - 5$ and $y_1(t) = 2$. Further let $x_2(t) = x_1(t)$ and $y_2(t) = t$. Figure 3.36 shows both the path of the particle $(x_1(t), y_1(t))$ on the line $y = 2$ and an expanded view of the path $(x_2(t), y_2(t))$ for $0 \leq t \leq 5$ and tstep = 0.05.

Now let $x_3(t) = t$ and $y_3(t) = x_1(t)$. The graph of the path of the particle $(x_1(t), y_1(t))$ and the graph of the particle's *position as a function of time* $(x_3(t), y_3(t))$ is shown in Fig. 3.37.

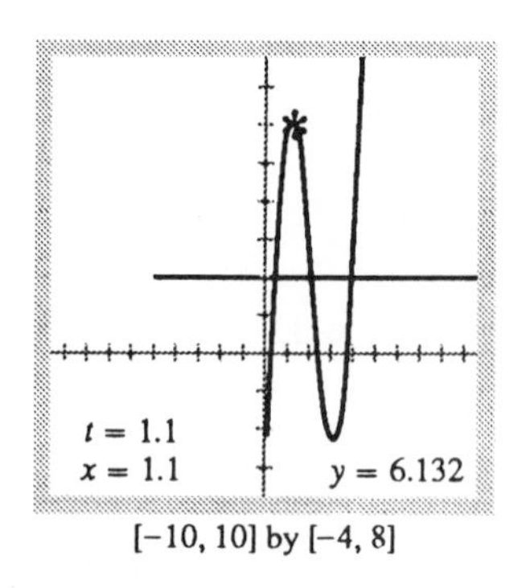

[−10, 10] by [−4, 8]

3.37 The distance of the particle from (0, 2) and its path on the line $y = 2$.

The particle starts out ($t = 0$) at the point $(-5, 2)$ and moves to the right along the line $y = 2$ until it stops ($s'(t) = 0$) at about $t = 1.15$ secs ($x = 6.15$). In DOT mode the distance between consecutive points from $t = 0$ to $t = 1.15$ is decreasing. (Try graphing in DOT mode. Experiment by trying different tstep increments.) This means the particle is slowing down from its initial velocity at $t = 0$ to velocity 0 units/sec at $t = 1.15$. Why? Graph the velocity function $s'(t) = 6t^2 - 26t + 22$ (choose $x_2(t) = t$, $y_2(t) = 6t^2 - 26t + 22$) to see that $s'(t)$ is positive and decreasing for $t = 0$ to $t = 1.15$ (Fig. 3.38). The other zero of $s'(t)$ is at $t = 3.18$.

From $t = 1.15$ to $t = 3.18$ the particle is moving to the *left* so the velocity is *negative.* The velocity decreases from 0 units/sec to -6.17 units/sec (the value of the second coordinate at the vertex of the graph in Fig. 3.38) and then increases back to 0 units/sec. However, the *speed* (choose $x_3(t) = T$, $y_3(t) = \text{abs}\, y_2(t)$) which is the absolute value of the velocity increases from 0 to a local maximum of 6.17 and then decreases to 0 (Fig. 3.39) in the $(1.15, 3.18)$ interval.

The velocity and speed both increase for t in $(3.18, \infty)$ and both decrease for t in $(-\infty, 1.15)$. If we start at any value of t less than 1.15 the particle slows down to 0 units/sec at $t = 1.15$, speeds up to 6.17 units/sec at $t = 2.15$, slows down to 0 units/sec at $t = 3.18$, and speeds up for t greater that 3.18. The particle moves to the right until $t = 1.15$, then moves left from $t = 1.15$ to $t = 3.18$, and finally moves to the right for t greater than 3.18. It is very instructive to simultaneously graph the expanded motion of the particle $x_1(t) = s(t), y_1(t) = t)$ and the velocity function $(x_2(t) = t, y_2(t) = s'(t))$. Try it. ∎

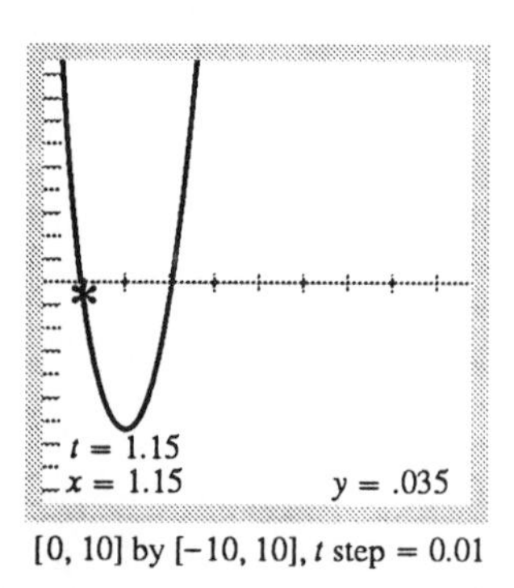

[0, 10] by [−10, 10], t step = 0.01

3.38 The graph of the velocity of the moving particle.

Derivatives in Economics

Economists often call the derivative of a function the **marginal value** of the function.

EXAMPLE 14 **Marginal cost.** Suppose it costs a company $c(x)$ dollars to produce x tons of steel in a week. It costs more to produce $x + h$ tons per week, and the cost difference, divided by h, is the average increase in cost per ton per week:

$$\frac{c(x+h)-c(x)}{h} = \text{average increase in cost.} \tag{8}$$

The limit of the ratio as $h \to 0$ is the marginal cost when x tons of steel are produced:

$$c'(x) = \lim_{h\to 0} \frac{c(x+h)-c(x)}{h} = \text{marginal cost.} \tag{9}$$

How are we to interpret this derivative? First of all, it is the slope of the graph of c at the point marked P in Fig. 3.40. But there is more.

Figure 3.41 shows an enlarged view of the curve and its tangent at P. We can see that if the company, currently producing x tons, increases production by one ton, then the incremental cost $\Delta c = c(x+1) - c(x)$ of producing that one ton is approximately $c'(x)$. That is,

$$\Delta c \approx c'(x) \qquad \text{when} \qquad \Delta x = 1. \tag{10}$$

Herein lies the economic importance of marginal cost. It estimates the cost of producing one unit beyond the present production level; that is, it approximates the cost of producing one more car, one more radio, one more washing machine, whatever.

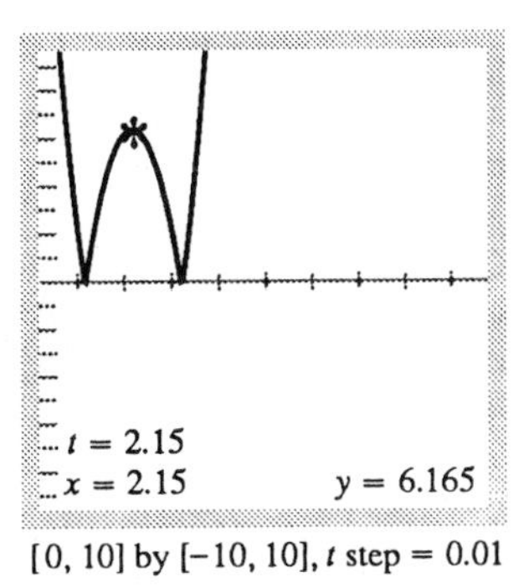

[0, 10] by [−10, 10], t step = 0.01

3.39 The graph of the speed of the particle.

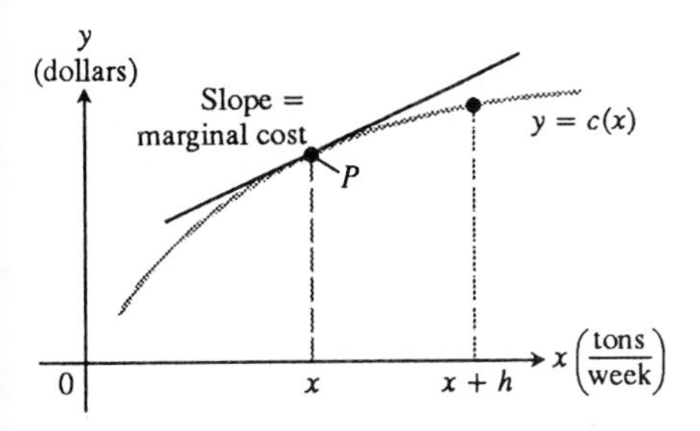

3.40 Weekly steel production: $c(x)$ is the cost of producing x tons per week. The cost of producing an additional h tons is $c(x+h) - c(x)$.

EXAMPLE 15 **Marginal cost (continued).** Suppose it costs

$$c(x) = x^3 - 6x^2 + 15x$$

dollars to produce x stoves and your shop is currently producing 10 stoves a day. About how much extra will it cost to produce one more stove a day?

Solution The cost of producing one more stove a day when 10 are produced is about $c'(10)$. Since

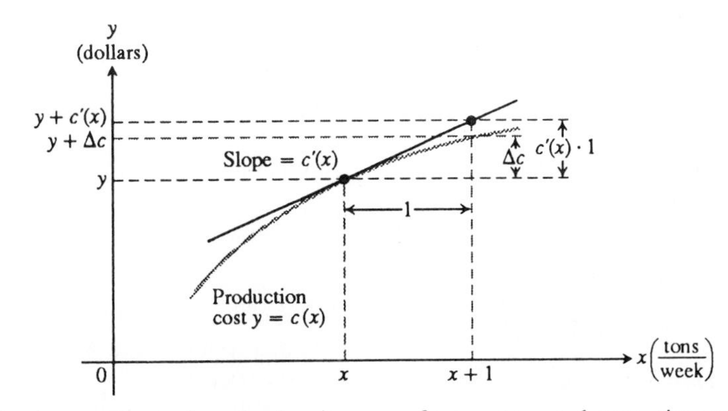

3.41 As weekly steel production increases from x to $x+1$ tons, the cost curve rises by the amount Δc. The tangent line rises by the amount slope · run $= c'(x)\cdot 1 = c'(x)$. Since $\Delta c/\Delta x \approx c'(x)$, we have $\Delta c \approx c'(x)$ when $\Delta x = 1$.

$$\begin{aligned} c'(x) &= \frac{d}{dx}(x^3 - 6x^2 + 15x) \\ &= 3x^2 - 12x + 15, \\ c'(10) &= 3(100) - 12(10) + 15 \\ &= 195. \end{aligned}$$

The additional cost will be about \$195.

Of course, the actual additional cost to increase production from 10 to 11 cabinets is $C(11) - C(10) = 770 - 550 = \$220$. We will see a more realistic application of marginal change in Example 17.

EXAMPLE 16 **Marginal revenue.** If

$$r(x) = x^3 - 3x^2 + 12x$$

gives the dollar revenue from selling x thousand candy bars, the marginal revenue when x thousand are sold is

$$\begin{aligned} r'(x) &= \frac{d}{dx}(x^3 - 3x^2 + 12x) \\ &= 3x^2 - 6x + 12. \end{aligned}$$

As with marginal cost, the marginal-revenue function estimates the increase in revenue that will result from selling one additional unit. If you currently sell 10 thousand candy bars a week, you can expect your revenue to increase by about

$$\begin{aligned} r'(10) &= 3(100) - 6(10) + 12 \\ &= \$252 \end{aligned}$$

if you increase sales to 11 thousand bars a week.

The actual revenue increase when production is increased from 10 thousand to 11 thousand candy bars is easily computed as $r(11) - r(10) = \$280$. The importance of marginal change is better illustrated by the next example.

Estimating f from the Graph of f'

Sometimes only the marginal change function is known; that is, only dy/dx is known for many values of x, not $y = f(x)$. The question is, can f be determined?

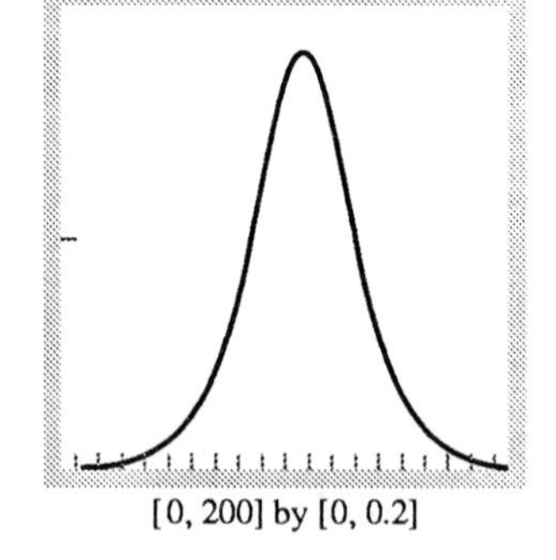

3.42 A marginal profit graph.

EXAMPLE 17 It has been observed that the daily marginal profit of a company producing and selling x widgets per day is given by the graph of $y = M(x)$ shown in Fig. 3.42.

What is the graph of $y = P(x)$, the profit function? We assume $D_x(P(x)) = M(x)$ and $P(0) = 0$. In Exercise 21 you will see why it is important to establish an initial value of P. The marginal profit graph shows that the profit is increasing very slowly for each additional 1 unit of production when x is between about 0 and 40 units and between about 160 and 200 units. Why? Furthermore, the maximum marginal increase in profit is about \$0.17 when the production level is about 106 units. What can we tell about the profit function? Could we determine a graph of this profit function? For now, these are difficult questions. In the next several chapters, we will develop the concepts and tools that will provide easy answers to these questions. See Exercise 22 for the actual profit function that results in the marginal profit graph in Fig. 3.42. In the next example, we investigate a related, but easier question, estimating the graph of f' from a graph of f. ≡

Estimating f' from a Graph of f

When we record data in the laboratory or in the field, we are often recording the values of a function $y = f(x)$. We might be recording the pressure in a gas as a function of volume, or the size of a population as a function of time. To see what the function looks like, we usually plot the data points and fit them with a curve. Even if we have no formula for the function $y = f(x)$ from which to calculate the derivative $y' = f'(x)$, we can still graph f': We estimate the slopes on the graph of f and plot these slopes. The following examples show how this is done and what we can learn from the graph of f'.

EXAMPLE 18 On April 23, 1988, the human-powered airplane *Daedalus* flew a record-breaking 119 km from Crete to the island of Santorini in the Aegean Sea, southeast of mainland Greece. During the 6-hr endurance tests before the flight, researchers monitored the prospective pilots' blood-sugar concentrations. The concentration graph for one of the athlete-pilots is shown in Fig. 3.43(a), where the concentration in milligrams/deciliter is plotted against time in hours.

The graph is made of line segments connecting data points. The constant slope of each segment gives an estimate of the derivative of the concentration between measurements. We calculated the slopes from the coordinate grid and plotted the derivative as a step function in Fig. 3.43(b). To make the plot for the first hour, for instance, we observed that the concentration increased from about 79 mg/dl to 93 mg/dl. The net increase was $\Delta y = 93 - 79 = 14$ mg/dl. Dividing this by $\Delta t = 1$ hr gave the rate of change:

$$\frac{\Delta y}{\Delta t} = \frac{14}{1} = 14 \text{ mg/dl per hour.}$$

≡

When we have so many data that the graph we get by connecting the data points looks like a smooth curve, we may also wish to plot the derivative as a smooth curve. The next example shows how this is done.

EXAMPLE 19 Graph the derivative of the function f in Fig. 3.44(a).

Solution First, we draw a pair of coordinate axes, marking the horizontal axis in x-units and the vertical axis in slope units (Fig. 3.44(b)). Next, we

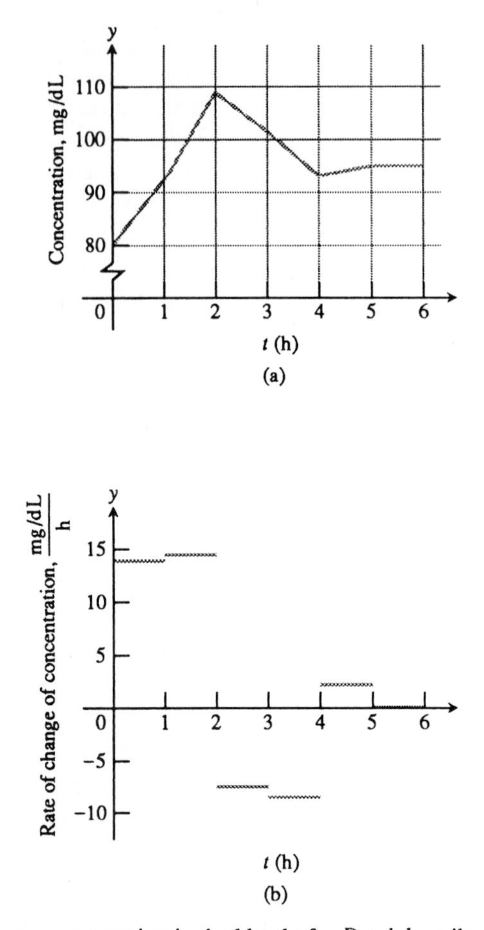

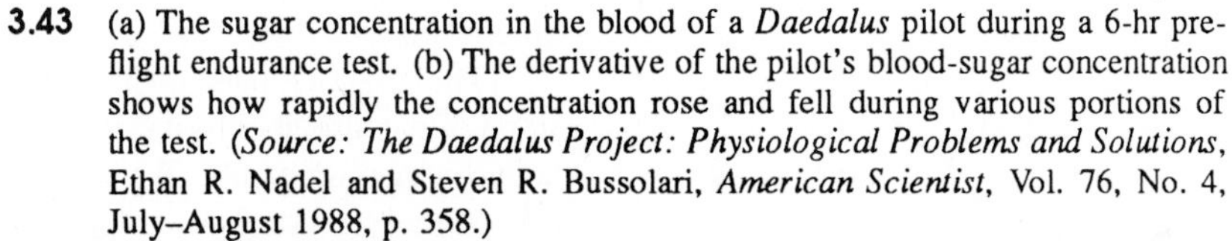

3.43 (a) The sugar concentration in the blood of a *Daedalus* pilot during a 6-hr preflight endurance test. (b) The derivative of the pilot's blood-sugar concentration shows how rapidly the concentration rose and fell during various portions of the test. (*Source: The Daedalus Project: Physiological Problems and Solutions*, Ethan R. Nadel and Steven R. Bussolari, *American Scientist*, Vol. 76, No. 4, July–August 1988, p. 358.)

estimate the slope of the graph of f in y-units per x-unit at frequent intervals, plotting the corresponding points against the new axes. We then connect the plotted points with a smooth curve.

From the graph of $y' = f'(x)$ we can see at a glance

1. Where f's rate of change is positive, negative, or zero.
2. The rough size of the growth rate at any x and its size in relation to the size of $f(x)$.
3. Where the rate of change itself is increasing or decreasing.

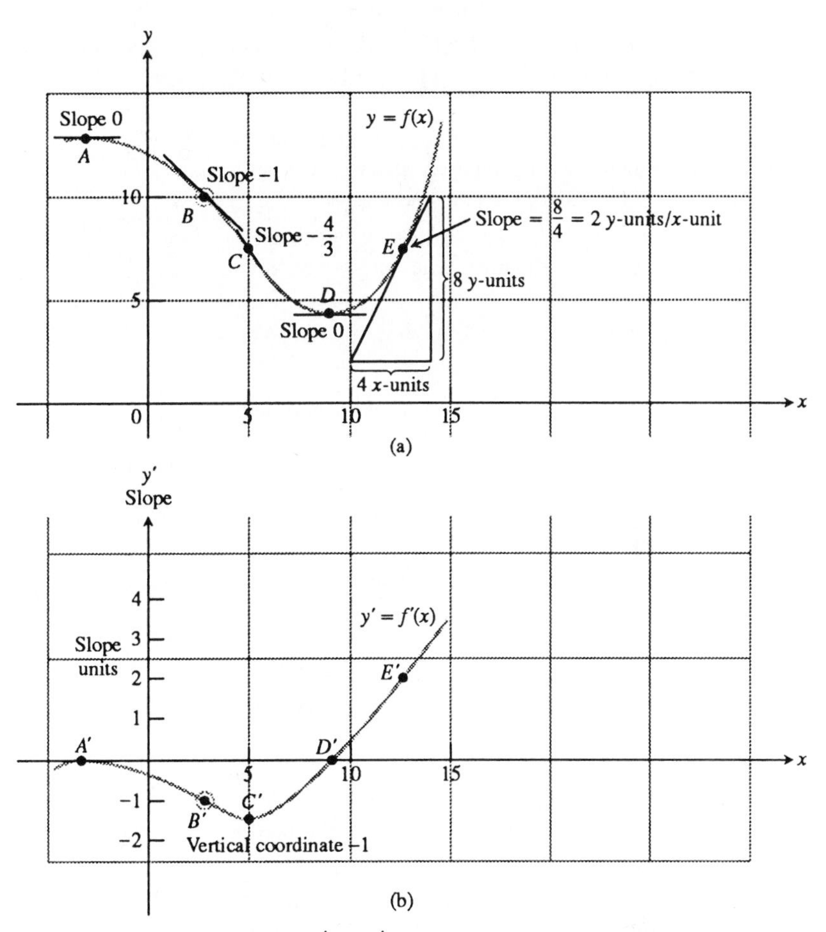

3.44 We made the graph of $y' = f'(x)$ in (b) by plotting slopes from the graph of $y = f(x)$ in (a). The vertical coordinate of B' is the slope at B, and so on. The graph of $y' = f'(x)$ is a visual record of how the slope of f changes with x.

Exercises 3.3

The equations in Exercises 1–6 give the position $s = f(t)$ of a particle moving along the line $y = 3$; s in meters and t in seconds.

a) Use a parametric graphing utility to simulate the motion of the particle for the indicated values of t. Describe the path of the particle.

b) Determine the position of the particle at $t = 0$, $t = 1$, $t = 2$, and $t = 3$ seconds.

c) Determine where and at what time the direction of the particle changes (if any). Also determine the velocity and acceleration at these times.

d) Determine the total distance the particle travels for the specified t interval.

e) Simultaneously graph both the path of the particle and its position as a function of time.

f) Simultaneously graph both the path of the particle, the velocity function $y = v(t) = s'(t)$, and the acceleration function $y = a(t) = s''(t)$. When is the particle at rest?

1. $s = t^2 - 3t + 2, \quad 0 \le t \le 5$
2. $s = 5 + 3t - t^2, \quad 0 \le t \le 5$
3. $s = t^3 - 6t^2 + 7t - 3, \quad 0 \le t \le 5$
4. $s = 4 - 7t + 6t^2 - t^3, \quad 0 \le t \le 5$
5. $s = t \sin t, \quad 0 \le t \le 15$
6. $s = 5 \sin\left(\frac{2t}{\pi}\right), \ 0 < t \le 2\pi$
7. The equations for free fall at the surfaces of Mars and Jupiter (s in meters, t in seconds) are Mars, $s = 1.86t^2$; Jupiter,

$s = 11.44t^2$. How long would it take a rock falling from rest to reach a velocity of 16.6 m/sec (about 100 km/h) on each planet?

8. A rock thrown vertically upward from the surface of the moon at a velocity of 24 m/sec (about 86 km/h) reaches a height of $s = 24t - 0.8t^2$ meters in t seconds.

a) Find the rock's velocity and acceleration. (The acceleration in this case is the acceleration of gravity on the moon.)

b) How long did it take the rock to reach its highest point?

c) How high did the rock go?

d) How long did it take the rock to reach half its maximum height?

e) How long was the rock aloft?

9. Devise a computer simulation of the problem situation in Exercise 8 that supports the answers obtained analytically.

10. On the earth, in the absence of air, the rock in Exercise 8 would reach a height of $s = 24t - 4.9t^2$ meters in t seconds. How high would the rock go?

11. A 45-caliber bullet fired straight up from the surface of the moon would reach a height of $s = 832t - 2.6t^2$ feet after t seconds. On the earth, in the absence of air, its height would be $s = 832t - 16t^2$ feet after t seconds. How long would it take the bullet to get back down in each case?

12. Devise a computer simulation of the problem situation in Exercise 11 that supports the answers obtained analytically.

13. When a bactericide was added to a nutrient broth in which bacteria were growing, the bacterium population continued to grow for a while, but then stopped growing and began to decline. The size of the population at time t (hours) was $b(t) = 10^6 + 10^4t - 10^3t^2$. Find the growth rates at $t = 0$, $t = 5$, and $t = 10$ hours.

14. The number of gallons of water in a tank t minutes after the tank has started to drain is $Q(t) = 200(30 - t)^2$. How fast is the water running out at the end of 10 min? What is the average rate at which the water flows out during the first 10 min?

15. *Marginal cost.* Suppose that the dollar cost of producing x washing machines is $c(x) = 2000 + 100x - 0.1x^2$.

a) Find the average cost of producing 100 washing machines.

b) Find the marginal cost when 100 washing machines are produced.

c) Show that the marginal cost when 100 washing machines are produced is approximately the cost of producing one more washing machine after the first 100 have been made, by calculating the latter cost directly.

16. *Marginal revenue.* Suppose the revenue from selling x custom-made office desks is

$$r(x) = 2000\left(1 - \frac{1}{x+1}\right)$$

dollars.

a) Draw a complete graph of r. What values of x make sense in this problem situation?

b) Find the marginal revenue when x desks are produced.

c) Use the function $r'(x)$ to estimate the increase in revenue that will result from increasing production from 5 desks a week to 6 desks a week.

d) Find the limit of $r'(x)$ as $x \to \infty$. How would you interpret this number?

17. The position of a body at time t sec is $s = t^3 - 6t^2 + 9t$ m. Find the body's acceleration each time the velocity is zero.

18. The velocity of a body at time t sec is $v = 2t^3 - 9t^2 + 12t - 5$ m/sec. Find the body's speed each time the acceleration is zero.

19. Determine complete graphs of the following parametric equations.

a) $x(t) = 3t - 5\sin t$
$y(t) = 3t - 5\cos t$

b) $x(t) = \dfrac{6\cos t}{4 - 3\cos t}$
$y(t) = \dfrac{6\sin t}{4 - 3\cos t}$

c) $x(t) = 2 - 7\cos t$
$y(t) = -2 + 3\cos t$

20. Which of the graphs in Exercise 19 are graphs of functions?

21. Let $f'(x) = 3x^2$.

a) Compute the derivatives of $g(x) = x^3$, $h(x) = x^3 - 2$, and $t(x) = x^3 + 3$.

b) Graph the numerical derivation of g, h and t.

c) Characterize the *family* of functions that have the property that $f'(x) = 3x^2$.

d) Is there a unique f such that $f'(x) = 3x^2$ *and* $f(0) = 0$? What is it?

e) Is there a unique g such that $g'(x) = 3x^2$ *and* $g(0) = 3$? What is it?

22. The monthly profit (in thousands of dollars) of a software company is given by

$$P(x) = \frac{10}{1 + 50 \cdot 2^{5-0.1x}}$$

where x is the number of software packages sold.

a) Draw a complete graph of $y = P(x)$.

b) What values of x make sense in the problem situation?

c) Draw a graph of $y = P'(x)$ (Use $y = \text{NDer}(P(x), x)$.). Compare with the graph of $y = M(x)$ in Example 17.

d) What is the profit when the marginal profit is maximum? What is the marginal profit when 50 units are produced? 100 units, 125 units, 150 units, 175 units, and 300 units?

e) What is $\lim_{x \to \infty} P(x)$? What is the maximum profit possible?

f) Is there a practical explanation to the maximum profit question? Explain your reasoning.

23. When a model rocket is launched, the propellant burns for a few seconds, accelerating the rocket upward. After burnout, the rocket coasts upward for a while and then begins to fall. A small explosive charge pops out a parachute shortly after the rocket starts down. The parachute slows the rocket to keep it from breaking when it lands.

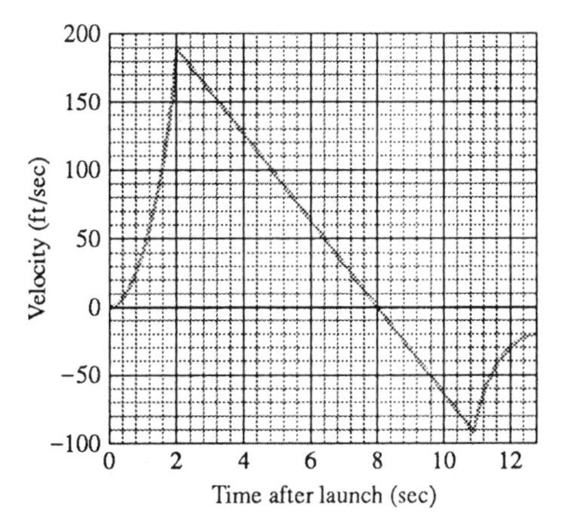

3.45 Velocity of the model rocket in Exercise 23.

Figure 3.45 shows velocity data from the flight of a model rocket. Use the data to answer the following.

a) How fast was the rocket climbing when the engine stopped?

b) For how many seconds did the engine burn?

c) When did the rocket reach its highest point? What was its velocity then?

d) When did the parachute pop out? How fast was the rocket falling then?

e) How long did the rocket fall before the parachute opened?

f) When was the rocket's acceleration greatest? When was the acceleration constant?

24. *Pisa by parachute (continuation of Exercise 23).* Mike McCarthy parachuted from the top of the Tower of Pisa on August 5, 1988. Make a rough sketch to show the shape of the graph of his downward velocity during the jump.

Exercises 25 and 26 are about the graphs in Fig. 3.46. The graphs in part (a) show the numbers of rabbits and foxes in a small arctic population. They are plotted as functions of time for 200 days. The number of rabbits increases at first, as the rabbits reproduce. But the foxes prey on the rabbits and, as the number of foxes increases, the rabbit population levels off and then drops. Figure 3.46(b) shows the graph of the derivative of the rabbit population. We made it by plotting slopes, as in Example 19.

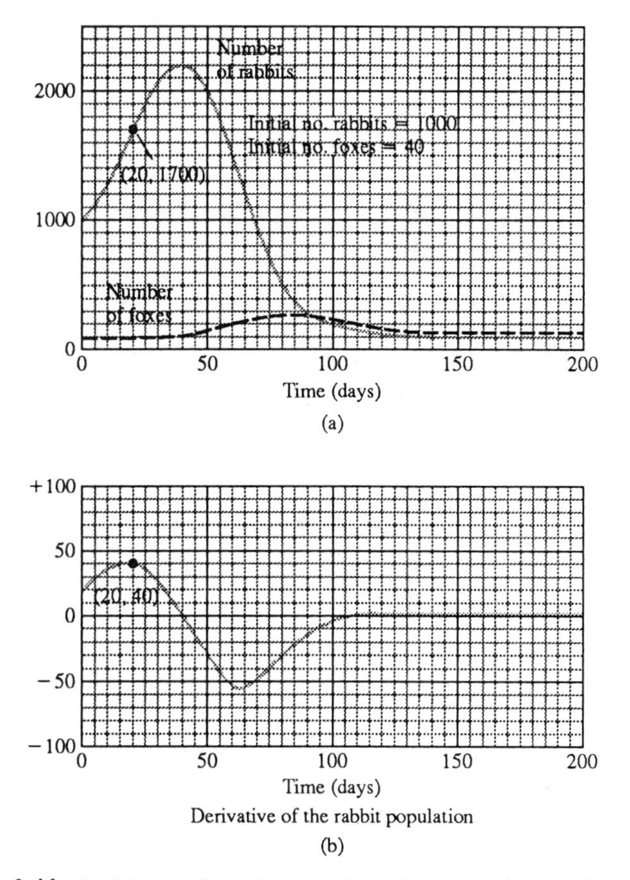

3.46 Rabbits and foxes in an arctic predator–prey food chain. *(Source: Differentiation,* by W. U. Walton et al., Project CALC, Education Development Center, Inc., Newton, Mass. (1975), p. 86.)

25. a) What is the value of the derivative of the rabbit population in Fig. 3.46 when the number of rabbits is largest? Smallest?

b) What is the size of the rabbit population in Fig. 3.46 when its derivative is largest? Smallest?

26. In what units should the slopes of the rabbit and fox population curves be measured?

Match the graphs of the functions in Exercises 27–30 with the graphs of the derivatives from the following list:

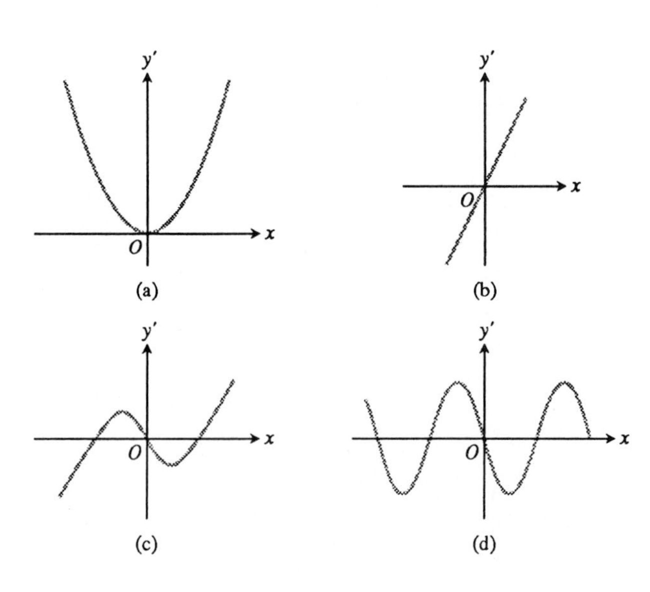

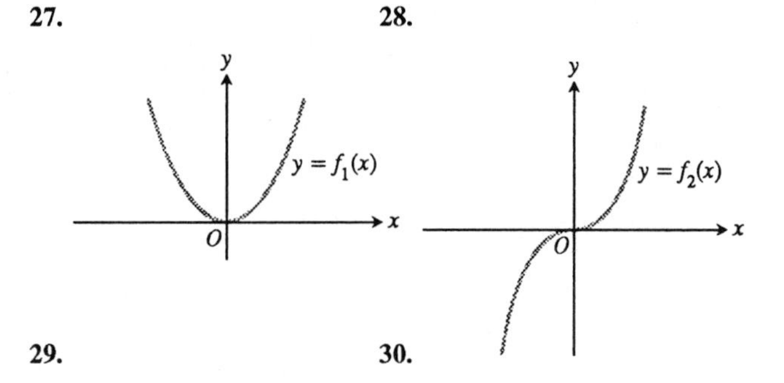

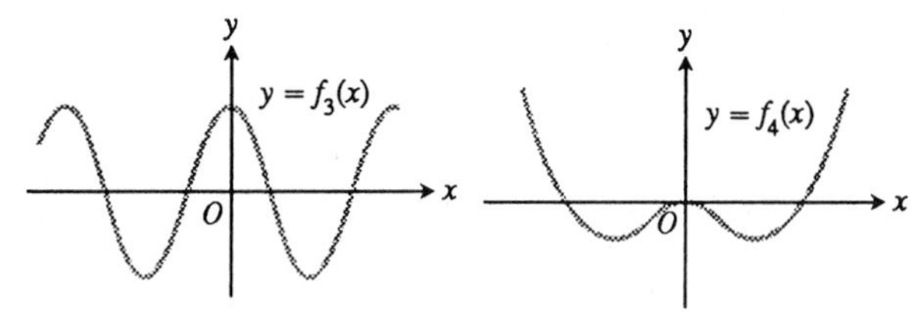

31. The graph of the function $y = f(x)$ in Fig. 3.47 is made of line segments joined end to end.

a) Graph the function's derivative.

b) At what values of x between $x = -3$ and $x = 7$ is the derivative not defined?

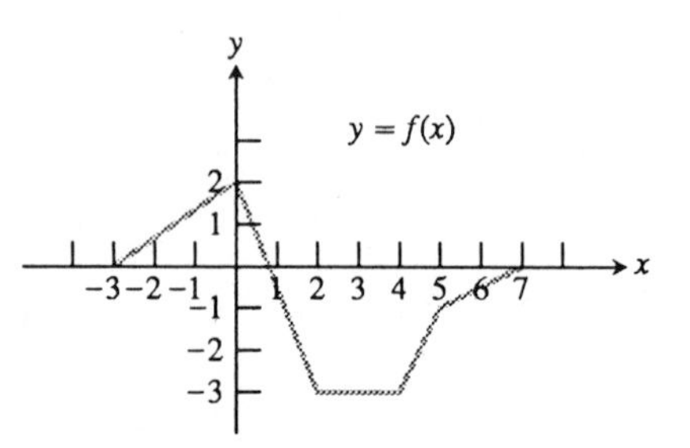

3.47 The graph for Exercise 31.

32. *Growth in the economy.* The graph in Fig. 3.48 shows the values of the U.S. Gross National Product (GNP) $y = f(t)$ for the years 1983–1988

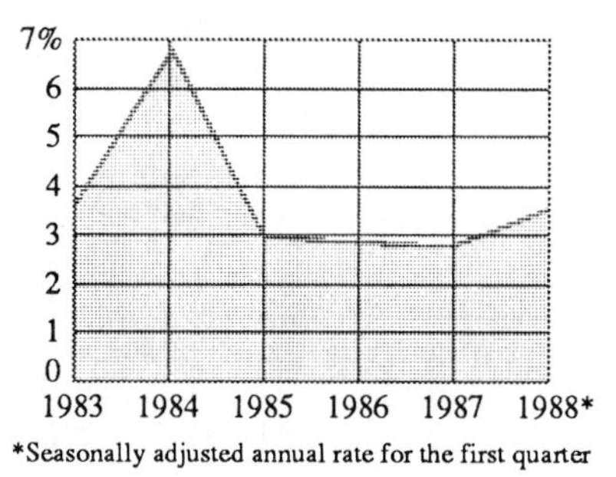

3.48 The graph for Exercise 32.

(*Source: The Wall Street Journal*, July 11, 1988, page 1.) Graph the function's derivative (where defined).

33. *Fruit flies (Example 1, Section 3.1 continued).* Populations starting out in closed environments grow slowly at first, when there are relatively few members, then more rapidly as the number of reproducing individuals increases and resources are still abundant, then slowly again as the population reaches the carrying capacity of the environment.

a) Use the graphical technique of Example 19 to graph the derivative of the fruit fly population introduced in Section 3.1. The graph of the population is reproduced here as Fig. 3.49. What units should be used on the horizontal and vertical axes for the derivative's graph?

b) During what days does the population seem to be increasing fastest? Slowest?

34. At what time is the particle of Example 13 at the point (5, 2)?

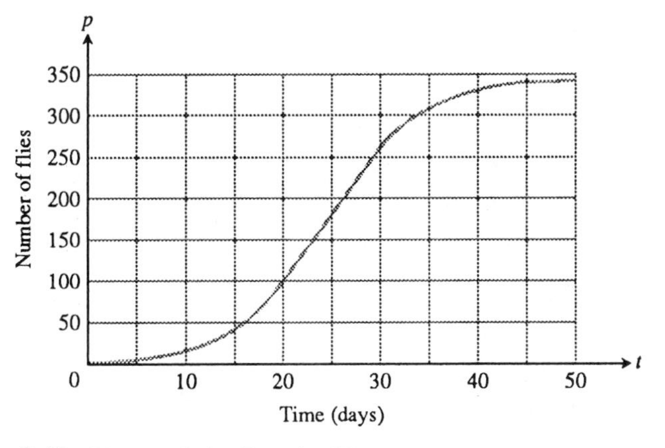

3.49 The graph for Exercise 33.

3.4 Derivatives of Trigonometric Functions

As we mentioned in Section 1.7, trigonometric functions are important because so many of the phenomena we want information about are periodic (heart rhythms, earthquakes, tides, weather). Continuous periodic functions can always be expressed in terms of sines and cosines, so the derivatives of sines and cosines play a key role in describing and predicting important changes. This section shows how to differentiate the six basic trigonometric functions.

Figure 3.50 suggests the graph of $y = \text{NDer}(\sin x, x)$ is the graph of $y = \cos x$. This further suggests the remarkable result that the derivative of the sine function is the cosine function. The proof of this fact involves several steps as we now illustrate.

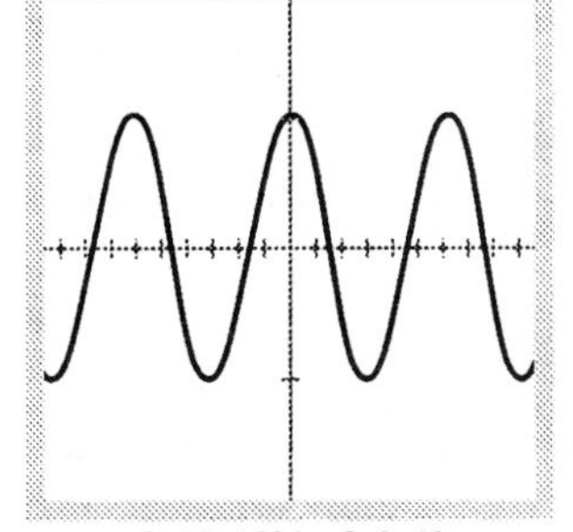

3.50 The graph of $y = \text{NDer}(\sin x, x)$ is identical to the graph of $y = \cos x$.

A New Limit

We begin by showing that

$$\lim_{h \to 0} \frac{\cos h - 1}{h} = 0. \tag{1}$$

We could have done this earlier, but we didn't need it then.

To prove Eq. (1), we divide the identity

$$\frac{1 - \cos 2\theta}{2} = \sin^2 \theta \quad \text{Section 1.7, Eq. 22}$$

by θ so that we get

$$\frac{1 - \cos 2\theta}{2\theta} = \frac{\sin \theta}{\theta} \cdot \sin \theta.$$

Letting $\theta \longrightarrow 0$ then gives

$$\lim_{\theta\to 0}\frac{1-\cos 2\theta}{2\theta} = \lim_{\theta\to 0}\frac{\sin\theta}{\theta}\cdot\lim_{\theta\to 0}\sin\theta \quad \text{Limit Product Rule}$$

$$= 1\cdot 0 = 0. \quad \text{Known values}$$

Replacing 2θ by h gives Eq. (1).

The Derivative of the Sine

The derivative of $y = \sin x$ is the limit

$$\frac{dy}{dx} = \lim_{h\to 0}\frac{\sin(x+h)-\sin x}{h}. \tag{2}$$

To calculate this limit, we combine Eq. (1) with two results from our earlier work:

1. $\sin(x+h) = \sin x\cos h + \cos x\sin h$ — Section 1.7, Eq. 16,
2. $\lim_{h\to 0}\frac{\sin h}{h} = 1$ — Section 2.3.

Then, taking all limits as $h \to 0$, we have

$$\begin{aligned}\frac{dy}{dx} &= \lim\frac{\sin(x+h)-\sin x}{h}\\ &= \lim\frac{\sin x\cos h + \cos x\sin h - \sin x}{h}\\ &= \lim\frac{\sin x(\cos h - 1) + \cos x\sin h}{h}\\ &= \lim\sin x\cdot\lim\frac{\cos h - 1}{h} + \lim\cos x\cdot\lim\frac{\sin h}{h}\\ &= \sin x\cdot 0 + \cos x\cdot 1\\ &= \cos x.\end{aligned} \tag{3}$$

In short,

$$\frac{d}{dx}\sin x = \cos x. \tag{4}$$

The sine and its derivative obey all the usual differentiation rules.

EXAMPLE 1

a) $y = x^2 - \sin x$ $\quad \dfrac{dy}{dx} = 2x - \dfrac{d}{dx}(\sin x)$ $\quad$ Difference Rule

$$= 2x - \cos x$$

b) $y = x^2 \sin x$ $\quad \dfrac{dy}{dx} = x^2 \dfrac{d}{dx}(\sin x) + 2x \sin x$ $\quad$ Product Rule

$$= x^2 \cos x + 2x \sin x$$

c) $y = \dfrac{\sin x}{x}$ $\quad \dfrac{dy}{dx} = \dfrac{x \cdot \frac{d}{dx}(\sin x) - \sin x \cdot 1}{x^2}$ $\quad$ Quotient Rule

$$= \frac{x \cos x - \sin x}{x^2}$$

In case you are still wondering why calculus uses radian measure when the rest of the world seems to use degrees, the answer is provided by the argument that the derivative of the sine is the cosine. The derivative of $\sin x$ is $\cos x$ *only* if x is measured in radians. The argument requires that when h is a small increment in x,

$$\lim_{h \to 0} (\sin h)/h = 1.$$

This is true only for radian measure, as we discussed in Section 2.3.

Now that we know that the sine is differentiable, we also know that it is continuous. The same holds for the other trigonometric functions in this section. Each one is differentiable at every point in its domain and is therefore continuous at every point in its domain.

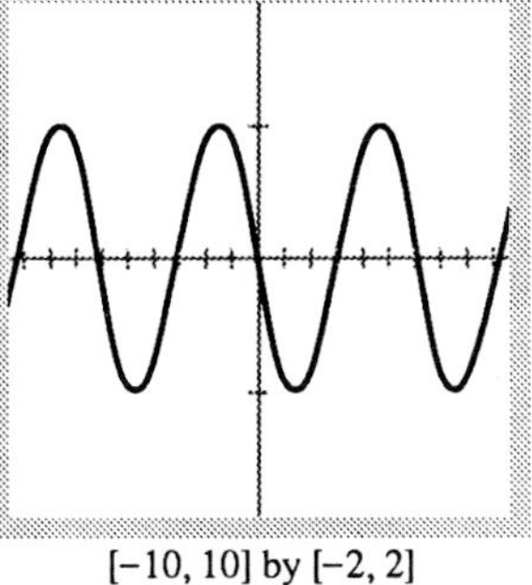

[−10, 10] by [−2, 2]

3.51 The graph of $y = \text{NDer}(\cos x, x)$ and $y = -\sin x$ are identical.

The Derivative of the Cosine

The graph of $y = \text{NDer}(\cos x, x)$ in Fig. 3.51 strongly suggests the fact that

$$\frac{d}{dx} \cos x = -\sin x .$$

You will prove this result in the Exercises.

$$\frac{d}{dx} \cos x = -\sin x. \qquad (5)$$

Notice the minus sign. The derivative of the sine is the cosine, but the derivative of the cosine is *minus* the sine.

EXAMPLE 2

a) $y = 5x + \cos x$

$$\frac{dy}{dx} = \frac{d}{dx}(5x) + \frac{d}{dx}(\cos x) \qquad \text{Sum Rule}$$
$$= 5 - \sin x$$

b) $y = \sin x \cos x$

$$\frac{dy}{dx} = \sin x \frac{d}{dx}(\cos x) + \cos x \frac{d}{dx}(\sin x) \qquad \text{Product Rule}$$
$$= \sin x(-\sin x) + \cos x(\cos x)$$
$$= \cos^2 x - \sin^2 x$$

c) $y = \dfrac{\cos x}{1 - \sin x}$

$$\frac{dy}{dx} = \frac{(1 - \sin x)\frac{d}{dx}(\cos x) - \cos x \frac{d}{dx}(1 - \sin x)}{(1 - \sin x)^2} \qquad \text{Quotient Rule}$$
$$= \frac{(1 - \sin x)(-\sin x) - \cos x(0 - \cos x)}{(1 - \sin x)^2}$$
$$= \frac{1 - \sin x}{(1 - \sin x)^2} \qquad \sin^2 x + \cos^2 x = 1$$
$$= \frac{1}{1 - \sin x}$$

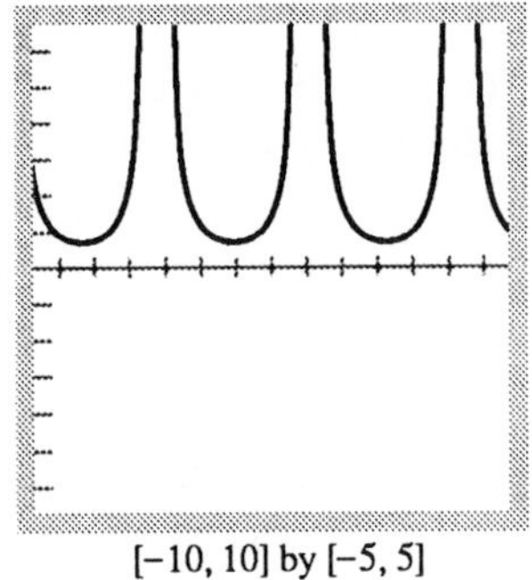

[−10, 10] by [−5, 5]

3.52 The graph of $y =$ NDer $(\cos x/(1 - \sin x), x)$.

Remember after any exact derivative manipulation you can check your answer by graphing the numerical derivative and comparing it with the graph of your answer. For example, the graph of $y =$ NDer $(\cos x/(1 - \sin x), x)$ in Fig. 3.52 appears to be the same as the graph of $dy/dx = 1/(1 - \sin x)$ and thus provides strong support for the answer in Example 2(c).

Simple Harmonic Motion

The motion of a weight bobbing up and down on the end of a spring is a *simple harmonic motion.* The next example describes a case in which there are no opposing forces like air friction to slow the motion down.

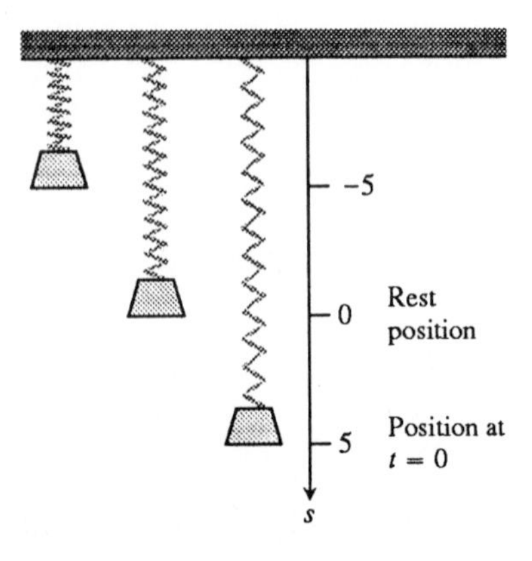

3.53 The weighted spring in Example 3.

EXAMPLE 3 A weight hanging from a spring (Fig. 3.53) is stretched 5 units beyond its rest position and released at time $t = 0$ to bob up and down. Its position at any later time t is

$$s = 5 \cos t.$$

What are its velocity and acceleration at time t?

Solution We have

Position: $\qquad s = 5 \cos t$

Velocity: $\dfrac{ds}{dt} = \dfrac{d}{dt}(5\cos t) = 5\dfrac{d}{dt}(\cos t) = -5\sin t$

Acceleration: $\dfrac{dv}{dt} = \dfrac{d}{dt}(-5\sin t) = -5\dfrac{d}{dt}(\sin t) = -5\cos t.$

Here is what we can learn from these equations:

1. As time passes, the weight moves up and down between $s = 5$ and $s = -5$ on the s-axis. The amplitude of the motion is 5. The period of the motion is 2π.
2. The function $\sin t$ attains its greatest magnitude (1) when $\cos t = 0$, as the graphs of the sine and cosine show. Hence, the weight's speed, $|v| = 5|\sin t|$, is greatest when $\cos t = 0$, that is, when $s = 0$.

   The weight's speed is zero when $\sin t = 0$. This occurs when $\cos t = \pm 1$, at the end points of the interval of motion.
3. The acceleration, $a = -5\cos t$, is zero only at the origin, where $\cos t = 0$. When the weight is anywhere else, the spring is either pulling on it or pushing on it. The acceleration is greatest in magnitude at the points farthest from the origin, where $\cos t = \pm 1$.

Explore with a Grapher

Simultaneously graph the parametric equations

$$x_1(t) = -1$$
$$y_1(t) = 5\cos t$$
$$x_2(t) = t$$
$$y_2(t) = -5\sin t$$

for $0 \le t \le 3\pi$. Use the TRACE key to explore the path of the weight on the spring and the velocity function from Example 3.

The Derivatives of the Other Basic Functions

Because $\sin x$ and $\cos x$ are differentiable functions of x, the related functions

$$\tan x = \frac{\sin x}{\cos x} \qquad \sec x = \frac{1}{\cos x}$$
$$\cot x = \frac{\cos x}{\sin x} \qquad \csc x = \frac{1}{\sin x}$$

are differentiable at every value of x at which they are defined. Their derivatives, calculated from the quotient rule, are given by the following formulas.

$$\frac{d}{dx}\tan x = \sec^2 x \qquad (6) \qquad\qquad \frac{d}{dx}\sec x = \sec x \tan x \qquad (7)$$

$$\frac{d}{dx}\cot x = -\csc^2 x \qquad (8) \qquad\qquad \frac{d}{dx}\csc x = -\csc x \cot x \qquad (9)$$

Notice the minus signs in the equations for the cotangent and cosecant. To show how a typical calculation goes, we derive Eq. (6). The other derivations are left as exercises.

THE WORD *NORMAL*

When analytic geometry was developed in the seventeenth century, European scientists still wrote about their work and ideas in Latin, the one language that all educated Europeans could read and understand. The word *normalis*, which scholars used for perpendicular in Latin, became *normal* when they discussed geometry in English.

EXAMPLE 4 Find dy/dx if $y = \tan x$.

Solution

$$\frac{d}{dx}\tan x = \frac{d}{dx}\left(\frac{\sin x}{\cos x}\right) = \frac{\cos x \frac{d}{dx}(\sin x) - \sin x \frac{d}{dx}(\cos x)}{\cos^2 x}$$

$$= \frac{\cos x \cos x - \sin x(-\sin x)}{\cos^2 x} = \frac{\cos^2 x + \sin^2 x}{\cos^2 x}$$

$$= \frac{1}{\cos^2 x} = \sec^2 x$$

EXAMPLE 5 Find y'' if $y = \sec x$.

Solution

$$y = \sec x$$

$$y' = \sec x \tan x \qquad \text{(Eq. 7)}$$

$$y'' = \frac{d}{dx}(\sec x \tan x)$$

$$= \sec x \frac{d}{dx}(\tan x) + \tan x \frac{d}{dx}(\sec x)$$

$$= \sec x(\sec^2 x) + \tan x(\sec x \tan x)$$

$$= \sec^3 x + \sec x \tan^2 x$$

EXAMPLE 6

a) $\frac{d}{dx}(3x + \cot x) = 3 + \frac{d}{dx}(\cot x) = 3 - \csc^2 x$

b) $\frac{d}{dx}\left(\frac{2}{\sin x}\right) = \frac{d}{dx}(2\csc x) = 2\frac{d}{dx}(\csc x)$

$$= 2(-\csc x \cot x) = -2\csc x \cot x$$

EXAMPLE 7 Find the equation of the line that is normal (perpendicular to the tangent) to the curve $y = x \sin x$ at $x = 4$.

Solution The slope of the tangent line to the curve at $(4, 4\sin 4) = (4, -3.03)$ is

$$\text{NDer}(x \sin x, 4) = -3.37.$$

The normal (perpendicular) has slope $-1/-3.37 = 0.30$ and the equation of the normal line to the curve at $(4, -3.03)$ is

$$y = 0.30(x - 4) - 3.03 \quad \text{or}$$

$$y = 0.30x - 4.23.$$

The graphs in Fig. 3.54 support this result. ∎

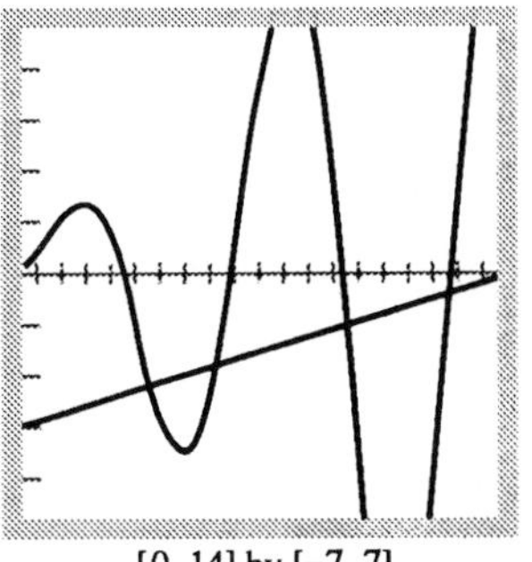

[0, 14] by [-7, 7]

3.54 The graph of $y = x \sin x$ and its normal line $y = 0.3x - 4.23$ at $x = 4$.

Exercises 3.4

In Exercises 1–24, find dy/dx when y equals the given expression. Support your answer by graphing the numerical derivative and comparing with the graph of the exact derivative.

1. $1 + x - \cos x$

2. $2 \sin x - \tan x$

3. $\frac{1}{x} + 5 \sin x$

4. $x^2 - \sec x$

5. $\csc x - 5x + 7$

6. $2x + \cot x$

7. $x \sec x$

8. $x \csc x$

9. $x^2 \cot x$

10. $4 - x^2 \sin x$

11. $3x + x \tan x$

12. $x \sin x + \cos x$

13. $\sin x \sec x$

14. $\sec x \csc x$

15. $\tan x \cot x$

16. $\cos x(1 + \sec x)$

17. $\frac{4}{\cos x}$

18. $5 + \frac{1}{\tan x}$

19. $\frac{\cos x}{x}$

20. $\frac{2}{\csc x} - \frac{1}{\sec x}$

21. $\frac{x}{1 + \cos x}$

22. $\frac{\sin x + \cos x}{\cos x}$

23. $\frac{\cot x}{1 + \cot x}$

24. $\frac{\cos x}{1 + \sin x}$

25. Find y'' if $y = \csc x$.

26. Find $y^{(4)} = d^4y/dx^4$ if
  a) $y = \sin x$,
  b) $y = \cos x$.

In Exercises 27–30, find equations for the lines that are tangent and normal to the curve $y = f(x)$ at the given point $(x, f(x))$. Support your answers with graphs using square windows.*

27. $y = \sin x$, $x = 0$

28. $y = \tan x$, $x = 0$

29. $y = 2\sin^2 x$, $x = 2$

30. $y = \frac{2 + \cot x}{x}$, $x = 1$

31. Prove $D_x \cos x = -\sin x$.

32. Show that the graphs of $y = \sec x$ and $y = \cos x$ have horizontal tangents at $x = 0$.

33. Show that the graphs of $y = \tan x$ and $y = \cot x$ never have horizontal tangents.

Do the graphs of the function in Exercises 34–37 have any horizontal tangents in the interval $0 \le x \le 2\pi$? If so, where? If not, why not?

34. $y = x + \sin x$

35. $y = 2x + \sin x$

36. $y = x + \cos x$

37. $y = x + 2\cos x$

38. *Simple harmonic motion.* The equations in (a) and (b) give the position $s = f(t)$ of a body moving along a coordinate line. Find each body's velocity, speed, and acceleration at time $t = \pi/4$.

  a) $s = 2 - 2\sin t$

  b) $s = \sin t + \cos t$

39. *Running machinery too fast.* Suppose that a mass like a piston is moving straight up and down and that its position at time t seconds is

$$s = A \cos(2\pi bt).$$

The value of A in this equation is the amplitude of the motion, and b is the frequency (number of times the piston moves up and down each second). What effect does doubling the frequency have on the piston's velocity and acceleration? (Once you find out, you will know why machinery breaks when you run it too fast.)

40. Use a parametric graphing utility to simulate the motion of

---

* Consult the *Graphing Calculator and Computer Graphing Laboratory Manual* for details on how to set square windows.

the body and the position of the body and the velocity both as a function of time t.

41. Find equations for the lines that are tangent and normal to the curve $y = \sqrt{2}\cos x$ at the point $(\pi/4, 1)$.

42. Find the points on the curve $y = \tan x$, $-\pi/2 < x < \pi/2$, where the tangent is parallel to the line $y = 2x$.

Find equations for the horizontal tangents to the graph in Exercises 43 and 44. Draw graphs that support your answers.

43. $y = \cot x - \sqrt{2}\csc x,\ 0 < x < \pi$

44. $y = \tan x + 3\cot x - 3,\ 0 < x < \pi/2$

45. Graph $y = \tan x$ and its derivative together over the interval $-\pi/2 < x < \pi/2$.

46. Graph $y = \cot x$ and its derivative together for $0 < x < \pi$.

47. Although $\lim_{h\to 0}(1 - \cos h)/h = 0$, it turns out that

$$\lim_{h\to 0}\frac{1-\cos h}{h^2} \neq 0.$$

Determine the limit.

48. Derive Eq. (7) by writing $\sec x = 1/\cos x$ and differentiating with respect to x.

49. Derive Eq. (8) by writing $\cot x = (\cos x)/(\sin x)$ and differentiating with respect to x.

50. Derive Eq. (9) by writing $\csc x = 1/\sin x$ and differentiating with respect to x.

3.5 The Chain Rule

We now know how to differentiate $\sin x$ and x^2-4, but how do we differentiate a composite like $\sin(x^2 - 4)$? The answer is, with the Chain Rule, which says that the derivative of the composite of two differentiable functions is the product of their derivatives. The Chain Rule is probably the most widely used differentiation rule in mathematics. This section describes the rule and how to use it.

Introductory Examples

A few examples will show what is going on.

EXAMPLE 1 The function $y = 6x - 10 = 2(3x - 5)$ is the composite of the functions $y = 2u$ and $u = 3x - 5$. How are the derivatives of these three functions related?

Solution We have $\dfrac{dy}{dx} = 6$, $\dfrac{dy}{du} = 2$, $\dfrac{du}{dx} = 3$. Since $6 = 2\cdot 3$,

$$\frac{dy}{dx} = \frac{dy}{du}\frac{du}{dx}.$$

■

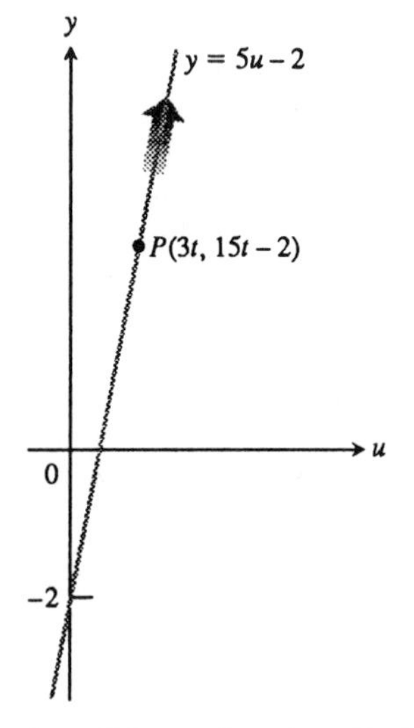

3.55 A particle moving along the line $y = 5u - 2$ in such a way that $u = 3t$.

EXAMPLE 2 A particle moves along the line $y = 5u-2$ in the uy-plane in such a way that its u-coordinate at time t is $u = 3t$. Find dy/dt (Fig. 3.55).

Solution As a function of t,

$$y = 5u - 2 = 5(3t) - 2 = 15t - 2.$$

Therefore,

$$\frac{dy}{dt} = \frac{d}{dt}(15t - 2) = 15.$$

Notice that $dy/du = 5, du/dt = 3$, and

$$\frac{dy}{dt} = 5 \cdot 3 = \frac{dy}{du}\frac{du}{dt}.$$

EXAMPLE 3 The function $y = 9x^2 + 6x + 1 = (3x + 1)^2$ is the composite of $y = u^2$ and $u = 3x + 1$. The derivatives involved are

$$\frac{dy}{dx} = \frac{d}{dx}(9x^2 + 6x + 1) = 18x + 6 = 6(3x + 1) = 6u$$

$$\frac{dy}{du} = \frac{d}{du}(u^2) = 2u$$

$$\frac{du}{dx} = \frac{d}{dx}(3x + 1) = 3.$$

Once again.

$$\frac{dy}{dx} = \frac{dy}{du}\frac{du}{dx}.$$

EXAMPLE 4 In the gear train in Fig. 3.56, the ratios of the radii of gears A, B, and C are 3:1:2. If gear A turns x times, then gear B turns $u = 3x$ times and gear C turns $y = u/2 = (3/2)x$ times. In terms of derivatives,

$$\frac{dy}{du} = \frac{1}{2} \qquad C \text{ turns at half } B\text{'s rate.}$$

$$\frac{du}{dx} = 3 \qquad B \text{ turns at three times } A\text{'s rate.}$$

In this example, too, we can calculate dy/dx by multiplying dy/du by du/dx:

$$\frac{dy}{dx} = \frac{3}{2} = \frac{1}{2} \cdot 3 = \frac{dy}{du}\frac{du}{dx}. \qquad \begin{array}{l} C \text{ turns at three-halves } A\text{'s rate,} \\ \text{three-halves of a turn for } A\text{'s one.} \end{array}$$

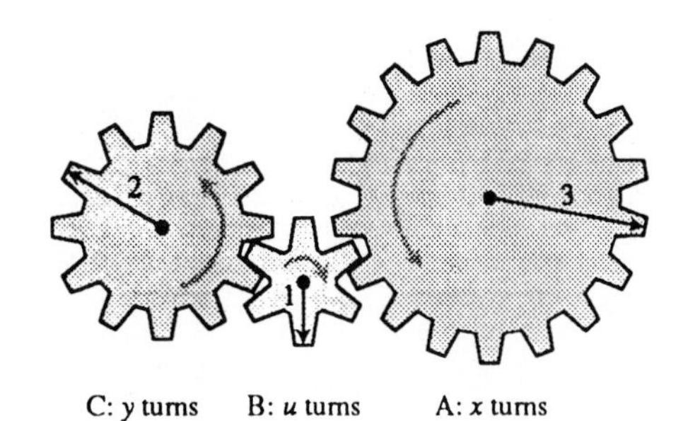

3.56 When wheel A takes x turns, wheel B takes u turns and wheel C takes y turns. By comparing circumferences we see that $dy/du = 1/2$ and $du/dx = 3$. What is dy/dx?

The Chain Rule

The preceding examples all work because the derivative of a composite $f \circ g$ of two differentiable functions is the product of their derivatives evaluated at appropriate points. This is the observation we state formally as the Chain Rule. As in Section 1.3, the notation $f \circ g$ ("f of g") denotes the composite of the functions f and g, with f following g. The value of $f \circ g$ at a point x is $(f \circ g)(x) = f(g(x))$.

The Chain Rule (First Form)

Suppose that $f \circ g$ is the composite of the differentiable functions $y = f(u)$ and $u = g(x)$. Then $f \circ g$ is a differentiable function of x whose derivative at each value of x is

$$(f \circ g)'_{\text{at } x} = f'_{\text{at } u=g(x)} \cdot g'_{\text{at } x}. \tag{1}$$

In short,

$$(f \circ g)'(x) = f'(g(x)) \cdot g'(x). \tag{2}$$

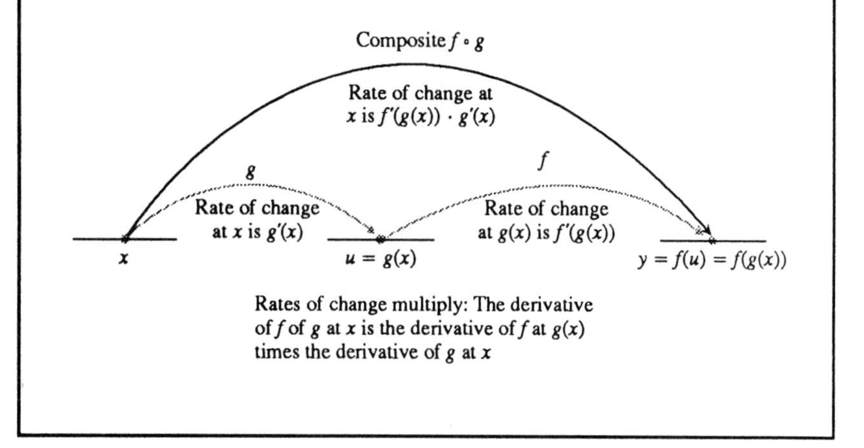

Rates of change multiply: The derivative of f of g at x is the derivative of f at $g(x)$ times the derivative of g at x

Equations (1) and (2) name the function involved as well as the dependent and independent variables. Once we know what the functions are, as we usually do in any particular example, we can get by with writing the Chain Rule a shorter way.

Chain Rule (Shorter Form)

If y is a differentiable function of u and u is a differentiable function of x, then y is a differentiable function of x and

$$\left.\frac{dy}{dx}\right|_{\text{at } x} = \left.\frac{dy}{du}\right|_{\text{at } u(x)} \cdot \left.\frac{du}{dx}\right|_{\text{at } x}. \tag{3}$$

Equation (3) still tells how each derivative is to be evaluated. When we don't need to be told that, we can get along with an even shorter form.

Chain Rule (Shortest Form)

If y is a differentiable function of u and u is a differentiable function of x, then

$$\frac{dy}{dx} = \frac{dy}{du}\frac{du}{dx}. \qquad (4)$$

You might think it would be a relatively easy matter to prove the Chain Rule by starting with the derivative definition the way we started the proofs of the Product and Quotient Rules. Unfortunately, this is the way the *hard* proof starts. The (relatively) easy proof starts with an equation we won't get to until Section 3.7. We have therefore placed the proof of the Chain Rule in Appendix 3, to be looked at later. We'll direct your attention to it when the time comes.

Like different instruments in a doctor's bag, each form of the Chain Rule makes some task a little easier. We shall use them all in the examples that follow. But remember—they all express the same one rule: The derivative of a composite of differentiable functions is the product of their derivatives.

EXAMPLE 5 If $f(u) = \sin u$ and $u = g(x) = x^2 - 4$, find $(f \circ g)'$ at $x = 2$.

Solution Equation (1) for the Chain Rule gives

$$\begin{aligned}(f \circ g)'_{\text{at } x=2} &= f'_{\text{at } u=g(2)} \cdot g'_{\text{at } x=2} \\ &= \frac{d}{dx}(\sin u)_{\text{at } u=0} \cdot \frac{d}{dx}(x^2 - 4)_{\text{at } x=2} \\ &= \cos u\,\big|_{u=0} \cdot 2x\big|_{x=2} \\ &= 1 \cdot 4 = 4.\end{aligned}$$

Explore with a Grapher

Let $y_1 = x^2 - 4$, $y_2 = \sin y_1$. Confirm that $y_2 = f \circ g$ from Example 5. Graph y_2 and $y_3 = \text{NDer}(y_2, x)$ in the $[-3, 3]$ by $[-4, 4]$ viewing window. Use TRACE to explore the values of $(f \circ g)'$ for $-3 \leq x \leq 3$. How many horizontal tangent lines are there for $-3 \leq x \leq 3$?

EXAMPLE 6 Find dy/dx at $x = 0$ if $y = \cos u$ and $u = \pi/2 - 3x$.

Solution With Eq. (3) this time,

$$\left.\frac{dy}{dx}\right|_{x=0} = \left.\frac{dy}{du}\right|_{u=\pi/2} \cdot \left.\frac{du}{dx}\right|_{x=0}$$

$$= -\sin u \,\big|_{u=\pi/2} \cdot (-3) = 3 \sin \frac{\pi}{2} = 3 \cdot 1 = 3.$$

It is more direct in Example 6 to write $y = \cos(\pi/2 - 3x)$. Then

$$\frac{dy}{dx} = -\sin\left(\frac{\pi}{2} - 3x\right) \cdot \frac{d}{dx}\left(\frac{\pi}{2} - 3x\right)$$

$$= 3 \sin\left(\frac{\pi}{2} - 3x\right),$$

and at $x = 0$

$$\frac{dy}{dx} = 3 \sin\left(\frac{\pi}{2} - 3 \cdot 0\right) = 3.$$

EXAMPLE 7 Express dy/dx in terms of x if $y = u^3$ and $u = x^2 - 1$.

Solution

$$\frac{dy}{dx} = \frac{dy}{du}\frac{du}{dx} \qquad \text{Eq. (4)}$$

$$= 3u^2 \cdot 2x$$

$$= 3(x^2 - 1)^2 \cdot 2x \qquad u = x^2 - 1$$

$$= 6x(x^2 - 1)^2$$

Integer Powers of Differentiable Functions

The Chain Rule enables us to differentiate powers like $y = \sin^5 x$ and $y = (2x + 1)^{-3}$ because these powers are composites:

$$y = \sin^5 x \text{ is } u^5 \quad \text{with} \quad u = \sin x,$$

$$y = (2x + 1)^{-3} \text{ is } u^{-3} \quad \text{with} \quad u = 2x + 1.$$

If u is any differentiable function of x and $y = u^n$, then the Chain Rule in the form

$$\frac{dy}{dx} = \frac{dy}{du} \cdot \frac{du}{dx}$$

gives

$$\frac{dy}{dx} = \frac{d}{du}(u^n) \cdot \frac{du}{dx}$$

$$= nu^{n-1}\frac{du}{dx}. \qquad \text{Differentiating } u^n \text{ with respect to } u \text{ itself gives } nu^{n-1} \text{ by Rule 2 in Section 3.2.}$$

Integer Powers of a Differentiable Function

> If u^n is an integer power of a differentiable function $u(x)$, then u^n is differentiable and
>
> $$\frac{d}{dx}u^n = nu^{n-1}\frac{du}{dx}. \tag{5}$$

EXAMPLE 8

a)

$$\frac{d}{dx}\sin^5 x = 5\sin^4 x \frac{d}{dx}(\sin x) \qquad \text{Eq. (5) with } u = \sin x,\, n = 5$$

$$= 5\sin^4 x \cos x$$

b)

$$\frac{d}{dx}(2x+1)^{-3} = -3(2x+1)^{-4}\frac{d}{dx}(2x+1) \qquad \text{(Eq. (5) with } u = 2x+1,\, n = -3)$$

$$= -3(2x+1)^{-4}(2)$$

$$= -6(2x+1)^{-4}$$

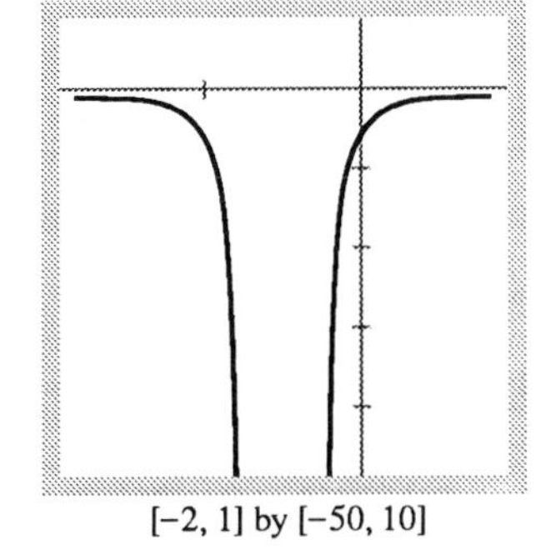

[−2, 1] by [−50, 10]

3.57 The graphs of $y = \text{NDer}((2x+1)^{-3}, x)$ and $y = -6(6x+1)^{-4}$ are the same.

Figure 3.57 supports this computation. ■

The "Inside-Outside" Rule

It sometimes helps to think about the Chain Rule the following way. If $y = f(g(x))$, Eq. (2) tells us that

$$\frac{dy}{dx} = f'(g(x)) \cdot g'(x). \tag{6}$$

In words, Eq. (6) says: To find dy/dx, differentiate the "outside" function f and leave the "inside" $g(x)$ alone; then multiply by the derivative of the inside.

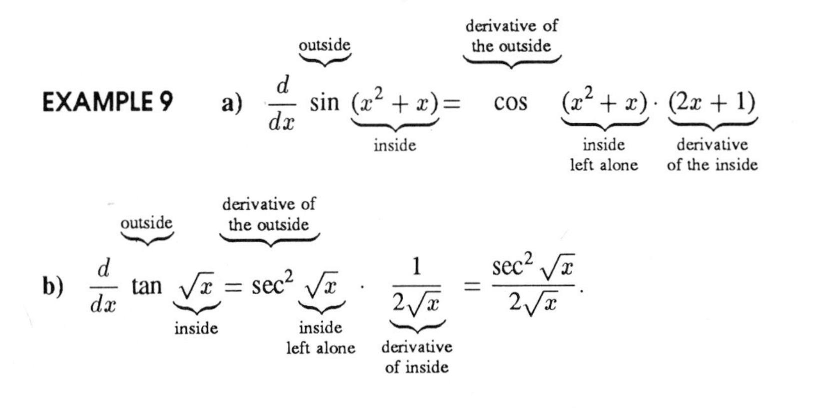

■

Repeated Use

We sometimes have to use the Chain Rule two or more times to get the job done. Here is an example.

EXAMPLE 10

a)

$$\begin{aligned}
\frac{d}{dx}\cos^2 3x &= 2\cos 3x \cdot \frac{d}{dx}(\cos 3x) && \text{Power (Chain) Rule} \\
&= 2\cos 3x(-\sin 3x)\frac{d}{dx}(3x) && \text{Chain Rule again} \\
&= 2\cos 3x(-\sin 3x)(3) \\
&= -6\cos 3x \sin 3x
\end{aligned}$$

b)

$$\begin{aligned}
\frac{d}{dx}\sin(1+\tan 2x) &= \cos(1+\tan 2x)\cdot\frac{d}{dx}(1+\tan 2x) && \text{Chain Rule} \\
&= \cos(1+\tan 2x)\cdot\sec^2 2x\cdot\frac{d}{dx}(2x) && \text{Chain Rule} \\
&= 2\cos(1+\tan 2x)\sec^2 2x
\end{aligned}$$

Derivative Formulas that Include the Chain Rule

Many of the derivative formulas you will encounter in your scientific work already include the Chain Rule.

If f is a differentiable function of u and u is a differentiable function of x, then substituting $y = f(u)$ in the Chain Rule formula

$$\frac{dy}{dx} = \frac{dy}{du}\frac{du}{dx}$$

leads to

$$\frac{d}{dx}f(u) = f'(u)\frac{du}{dx}. \tag{7}$$

When we spell this out for the functions whose derivatives we have studied so far, we get the formulas in Table 3.1

TABLE 3.1 Derivative Formulas That Include the Chain Rule

$$\frac{d}{dx}u^n = nu^{n-1}\frac{du}{dx} \qquad (n \text{ an integer})$$

$\frac{d}{dx}\sin u = \cos u\frac{du}{dx}$	$\frac{d}{dx}\sec u = \sec u \tan u\frac{du}{dx}$
$\frac{d}{dx}\cos u = -\sin u\frac{du}{dx}$	$\frac{d}{dx}\cot u = -\csc^2 u\frac{du}{dx}$
$\frac{d}{dx}\tan u = \sec^2 u\frac{du}{dx}$	$\frac{d}{dx}\csc u = -\csc u \cot u\frac{du}{dx}$

EXAMPLE 11

a)

$$\frac{d}{dx}\sin(-x) = \cos(-x)\frac{d}{dx}(-x) \qquad \left(\frac{d}{dx}\sin u = \cos u\frac{du}{dx} \text{ with } u = -x\right)$$

$$= -\cos(-x)$$

b)

$$\frac{d}{dx}\tan\left(\frac{1}{x}\right) = \sec^2\frac{1}{x}\cdot\frac{d}{dx}\left(\frac{1}{x}\right) \qquad \left(\frac{d}{dx}\tan u = \sec^2 u\frac{du}{dx} \text{ with } u = 1/x\right)$$

$$= \sec^2\frac{1}{x}\cdot\left(-\frac{1}{x^2}\right)$$

$$= -\frac{1}{x^2}\sec^2\frac{1}{x}$$

Figure 3.58 supports this computation.

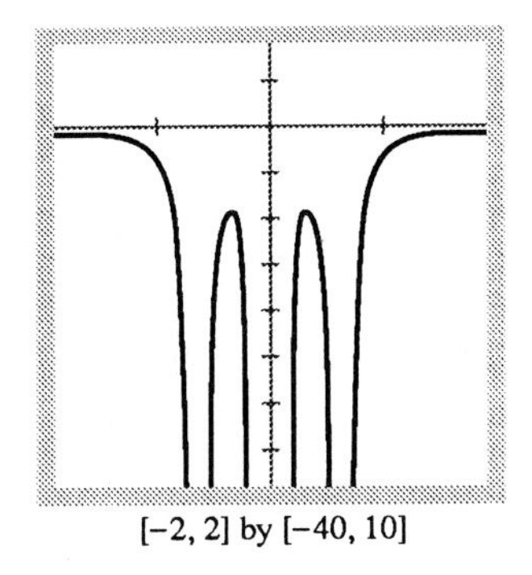

[−2, 2] by [−40, 10]

3.58 The graphs of $y =$ NDer $(\tan(1/x), x)$ and $y = (-1/x^2)\sec^2(1/x)$ are the same.

c)

$$\frac{d}{dx}\cos(\cos x) = -\sin(\cos x)\cdot\frac{d}{dx}(\cos x) \qquad \frac{d}{dx}\cos u = -\sin u\frac{du}{dx} \text{ with } u = \cos x$$

$$= -\sin(\cos x)\cdot(-\sin x)$$

$$= \sin x \sin(\cos x)$$

Melting Ice Cubes

In mathematics, we tend to use letters like f, g, x, y, and u for functions and variables. However, applied fields use letters like V, for volume, and s, for side, that come from the names of the things being modeled. The letters in the Chain Rule then change, too, as in the next example.

EXAMPLE 12 *The melting ice cube.* A melting ice cube lost a fourth of its volume during the first hour. How long will it take this ice cube to melt?

Solution As with all applications to science, we start with a mathematical model. Let us assume that the cube has side length s, so that its volume is $V = s^3$, and that V and s are differentiable functions of time t. Suppose also that the cube's volume decreases at a rate that is proportional to its surface area. This assumption seems reasonable enough when we think that the melting takes place at the surface: Changing the surface area changes the amount of ice exposed to melting. In mathematical terms,

The surface area of a cube of side length s is $6\,s^2$.

$$\frac{dV}{dt} = -k(6s^2).$$

The minus sign is there because the volume is decreasing. We assume that the proportionality factor k is constant. (It probably depends on many things, like the relative humidity of the surrounding air, the air temperature, and the incidence or absence of sunlight, to name only a few.)

Finally, we need at least one more piece of information: How long will it take the ice cube to melt some specific percent? We have nothing to guide us unless we make one or more observations, but now let us assume a particular set of conditions in which the cube lost 1/4 of its volume during the first hour. (You could use letters instead of precise numbers: say $n\%$ in h hr. Then your answer would be in terms of n and h.) Now to work.

Mathematically, we now have the following problem:

$$\textit{Given}: \quad V = s^3 \quad \text{and} \quad \frac{dV}{dt} = -k(6s^2)$$

$$V = V_0 \quad \text{when} \quad t = 0$$

$$V = (3/4)V_0 \quad \text{when} \quad t = 1 \text{ hr}$$

$$\textit{Find:} \quad \text{The value of } t \text{ when } V = 0$$

We apply the Chain Rule to differentiate $V = s^3$ with respect to t:

$$\frac{dV}{dt} = 3s^2\frac{ds}{dt}.$$

We set this equal to the given rate, $\dfrac{dV}{dt} = -k(6s^2)$, to get

$$3s^2\frac{ds}{dt} = -6ks^2,$$

$$\frac{ds}{dt} = -2k.$$

Thus, the side length is *decreasing* at the constant rate of $2k$ units per hour (centimeters, inches, whatever).

Since $ds/dt = -2k$, $s = -2kt + s_0$ where s_0 is the initial length of the side of the cube before the melting began. The key here is to recognize that linear functions ($y = mx + b$) are the only functions with constant derivatives ($y' = m$). This fact will be proved in Chapter 4.

Since $V = s^3$, it follows that

$$V(t) = (-2kt + s_0)^3.$$

Note that $V(0) = s_0^3$, which is the initial volume of the ice cube.

We know that when $t = 1$

$$V(1) = (3/4)\,V(0).$$

Now we want to solve the equation $V(t) = 0$.

First we find k,

$$V(1) = (-2k + s_0)^3 = (3/4)\,V(0) = (3/4)\,s_0^3$$

so

$$k = \frac{s_0 - \sqrt[3]{(3/4)s_0^3}}{2}$$

or

$$k = s_0\left(\frac{1 - \sqrt[3]{3/4}}{2}\right)$$

so

$$V(t) = \left(-2s_0\left(\frac{1 - \sqrt[3]{3/4}}{2}\right)t + s_0\right)^3$$

$$V(t) = s_0^3((\sqrt[3]{3/4} - 1)t + 1)^3.$$

Now $V(t) = 0$ when

$$(\sqrt[3]{3/4} - 1)t + 1 = 0 \qquad (\text{Note}: \quad s_0 \neq 0)$$

or

$$t = \frac{1}{1 - \sqrt[3]{3/4}} = 10.936\ldots.$$

So it will take 9.94 more hours for the ice cube to melt if 1/4 of the cube melts in one hour. ∎

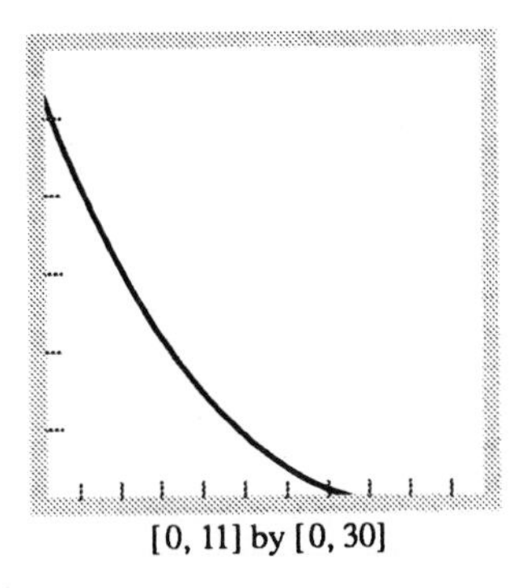

3.59 The volume of a melting ice cube as a function of time.

Notice the solution is independent of s_0. This means the size of the ice cube has no effect on its melting time.

> **Explore with a Grapher**
>
> Figure 3.59 shows a graph of the volume of the melting ice cube as a function of time. Here $s_0 = 3$. When will the volume be 1/2 of the initial volume? 1/4 of the initial volume?

Remark If we were natural scientists who were really interested in testing our model, we could collect some data and compare them with the results of the mathematics. One practical application might lie in analyzing the proposal to tow large icebergs from polar waters to offshore locations near southern California where the melting ice would provide fresh water. As a first approximation, we might assume that the iceberg is a large cube, or a pyramid, or a sphere.

Exercises 3.5

In Exercises 1–18, find dy/dx when y equals the given expression.

1. $\sin(x+1)$
2. $\sin(x/2)$
3. $\cos 5x$
4. $\cos(-2x)$
5. $\sin(2\pi x/5)$
6. $\cos(3\pi x/2)$
7. $\tan(2-x)$
8. $\tan 5(x-1)$
9. $\sec(2x-1)$
10. $\sec(x^2+1)$
11. $\csc(x^2+7x)$
12. $\csc(1-2x)$
13. $\cot(3x+\pi)$
14. $\cot(1/x)$
15. $\sec^2 x - \tan^2 x$
16. $\csc^2 x - \cot^2 x$
17. $\sin^3 x$
18. $\cos^3 x$
19. $(2x+1)^5$
20. $(4-3x)^9$
21. $(x+1)^{-3}$
22. $(2x+1)^{-3}$
23. $\left(1-\frac{x}{7}\right)^{-7}$
24. $\left(\frac{x}{2}-1\right)^{-10}$
25. $\left(1+x-\frac{1}{x}\right)^3$
26. $\left(\frac{x}{5}+\frac{1}{5x}\right)^5$
27. $(x^2+2x+3)^3$
28. $(x^2+2x+3)^{-3}$

In Exercises 29–36, find dy/dx when y equals the given expression. Support your answer with the graph of $y = \text{NDer}(y, x)$.

29. $\cos(\sin x)$
30. $\sin(\sin x)$
31. $1+x+\sec^2 x$
32. $x^2\cos\left(\frac{1}{x}\right)$
33. $(\csc x+\cot x)^{-1}$
34. $-(\sec x+\tan x)^{-1}$
35. $\sin\left(\frac{x-2}{x+3}\right)$
36. $\left(\frac{\sin x}{1+\cos x}\right)^2$

In Exercises 37–42, find dy/dx when y equals the given expression. You will need to use the Chain Rule two or three times in each case.

37. $\sin^2(3x-2)$
38. $\sec^2 5x$
39. $(1+\cos 2x)^2$
40. $(1-\tan(x/2))^{-2}$
41. $\sin(\cos(2x-5))$
42. $(1+\cos^2 7x)^3$

Find y'' in Exercises 43–46.

43. $y=\tan x$
44. $y=9\tan(x/3)$
45. $y=\cot x$
46. $y=\cot(3x-1)$

In Exercises 47–52, find the value of $(f\circ g)'$ at the given value of x.

47. $f(u)=u^5+1,\ u=g(x)=\sqrt{x},\ x=1$

48. $f(u)=1-\frac{1}{u},\ u=g(x)=\frac{1}{1-x},\ x=-1$

49. $f(u)=\cot\frac{\pi u}{10},\ u=g(x)=5\sqrt{x},\ x=1$

50. $f(u)=u+\frac{1}{\cos^2 u},\ u=g(x)=\pi x,\ x=1/4$

51. $f(u)=\frac{2u}{u^2+1},\ u=g(x)=10x^2+x+1,\ x=0$

52. $f(u)=\left(\frac{u-1}{u+1}\right)^2,\ u=g(x)=\frac{1}{x^2}-1,\ x=-1$

What happens if you can write a function as a composite in different ways? Do you get the same derivative each time? The Chain Rule says you should. Try it with the functions in Exercises 53–56.

53. Find dy/dx if $y=\cos(6x+2)$ by writing y as a composite with

a) $y=\cos u$ and $u=6x+2$

b) $y=\cos 2u$ and $u=3x+1$.

54. Find dy/dx if $y=\sin(x^2+1)$ by writing y as a composite with

a) $y=\sin(u+1)$ and $u=x^2$

b) $y=\sin u$ and $u=x^2+1$.

55. Find dy/dx if $y=x$ by writing y as the composite of

a) $y=(u/5)+7$ and $u=5x-35$

b) $y=1+(1/u)$ and $u=1/(x-1)$.

56. Find dy/dx if $y=\sin(\sin(2x))$ by writing y as the composite of

a) $y=\sin u$ and $u=\sin 2x$

b) $y=\sin(\sin u)$ and $u=2x$.

57. Find ds/dt if $s=\cos\theta$ and $d\theta/dt=5$ when $\theta=3\pi/2$.

58. Find dy/dt if $y=x^2+7x-5$ and $dx/dt=1/3$ when $x=1$.

59. What is the largest value the slope of the curve $y=\sin(x/2)$ can ever have?

60. Write an equation for the tangent to the curve $y=\sin mx$ at the origin.

61. Find the lines that are tangent and normal to the curve $y=2\tan(\pi x/4)$ at $x=1$. Support your answer with a graph.

62. *Orthogonal curves.* Two curves are said to cross at right angles if their tangents are perpendicular at the crossing point. The technical word for "crossing at right angles" is *orthogonal.* Show that the curves $y=\sin 2x$ and $y=-\sin(x/2)$ are orthogonal at the origin. Draw both graphs and both tangents in a square viewing window.

63. Suppose that functions f and g and their derivatives have the following values at $x=2$ and $x=3$:

x	$f(x)$	$g(x)$	$f'(x)$	$g'(x)$
2	8	2	$\frac{1}{3}$	-3
3	3	-4	2π	5

Find the values of the following derivatives.

a) $\frac{d}{dx}\{2f(x)\}$ at $x = 2$

b) $\frac{d}{dx}\{f(x) + g(x)\}$ at $x = 3$

c) $\frac{d}{dx}\{f(x) \cdot g(x)\}$ at $x = 3$

d) $\frac{d}{dx}\left\{\frac{f(x)}{g(x)}\right\}$ at $x = 2$

e) $\frac{d}{dx}\{f(g(x))\}$ at $x = 2$

f) $\frac{d}{dx}\{\sqrt{f(x)}\}$ at $x = 2$

g) $\frac{d}{dx}\left\{\frac{1}{g^2(x)}\right\}$ at $x = 3$

h) $\frac{d}{dx}\{\sqrt{f^2(x) + g^2(x)}\}$ at $x = 2$

64. Show that the function $y = \cos(2x + B)$ satisfies the equation

$$y'' = -4y$$

whatever the value of B. Equations like $y'' = -4y$ come up whenever we design loudspeakers and audio amplifiers, and their solutions are always cosines. We shall say more about equations like this in Chapter 16.

65. Suppose that the radius of a soap bubble is increasing at the rate of 1/2 cm/sec. How fast is the volume changing when the radius is 10 cm?

To find out, start with the equation $V = (4/3)\pi r^3$ for the volume of a sphere. Then assume that r is a differentiable function of time t and use the Chain Rule to differentiate both sides of the equation with respect to t. Then substitute $r = 10$ and $dr/dt = 1/2$ to find the value of dV/dt.

66. *Temperatures in Fairbanks, Alaska.* The equation that approximates the average temperature (F°) in Fairbanks, Alaska during a typical 365-day year on day x is

$$y = 37 \sin\left[\frac{2\pi}{365}(x - 101)\right] + 25.$$

a) Draw a complete graph of $y = f(x)$. Assume $x = 0$ is January 1.

b) On what day is the temperature increasing the fastest?

c) About how many degrees per day is the temperature increasing when it is increasing at its fastest?

3.6 Implicit Differentiation and Fractional Powers

Graphing Curves of the Form $F(x, y) = 0$

The equations $x^2 + y^2 = 64$, $y^2 = x$, and $y^5 + \sin xy = 0$ define relations that are *not* functions. The equation $y = x^3 + 2x - 3$ defines a function. Each can be written in the form

$$F(x, y) = 0.$$

For example $x^2 + y^2 = 64$ is

$$F(x, y) = 0 \quad \text{where} \quad F(x, y) = x^2 + y^2 - 64,$$

and $y = x^3 + 2x - 3$ is $F(x, y) = 0$ where $F(x, y) = y - x^3 - 2x + 3$. Some equations of the form $F(x, y) = 0$ can be graphed using a *parametric* graphing utility. Here are three examples in the following explore:

> **Explore with a Grapher**
>
> 1. Any function $y = f(x)$. Graph $x(t) = t$ and $y(t) = f(t)$. Note the function $y = f(x)$ is equivalent to the equation $F(x, y) = 0$ where $F(x, y) = y - f(x)$. Try it for $y = x^3 - 2x$.
> 2. Any ellipse $\dfrac{x^2}{A^2} + \dfrac{y^2}{B^2} = 1$. Graph $x(t) = A\cos t$ and $y(t) = B\sin t$. Try it for $\dfrac{x^2}{4} + \dfrac{y^2}{9} = 1$.
> 3. The relation $y^5 + \sin xy = 0, y \neq 0$. Note that $|y| \leq 1$. Why? (See Exercise 48). Solving for x we get
>
> $$\sin xy = -y^5$$
>
> $$xy = \sin^{-1}(-y^5) + 2k\pi.\ (k \text{ any integer})$$
>
> We graph the part with $k = 0$ parametrically by setting
>
> $$x(t) = \sin^{-1}(-t^5)$$
>
> $$y(t) = t$$
>
> with $-1 \leq t \leq 1$ (Fig. 3.60). For this part $x \leq 0$. Why?

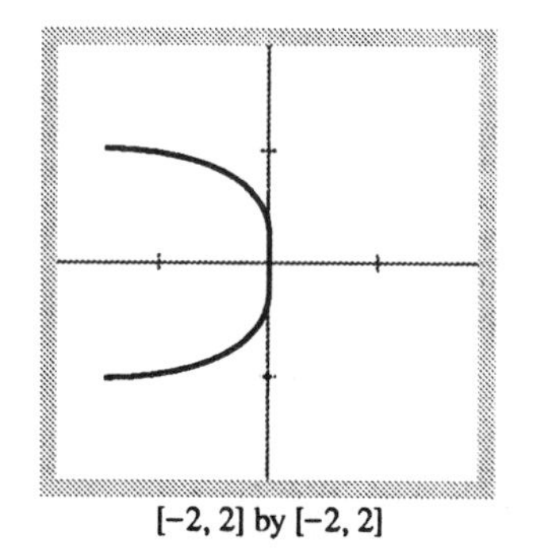

3.60 The graph of part of the relation $y^5 + \sin xy = 0$.

When an equation like $y^5 + \sin xy = 0$ defines y as a differentiable function of x but does not let us solve for y in terms of x, we can still find dy/dx with a technique called implicit differentiation. This section describes the technique and uses it to extend the Power Rule for differentiation to include fractional exponents.

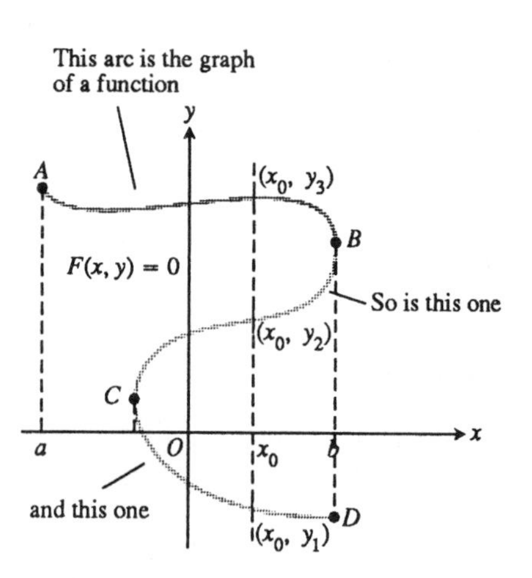

3.61 As a whole, the curve $F(x, y) = 0$ is not the graph of a function of x. Some of the vertical lines that cross it intersect it more than once. However, the curve can be divided into separate arcs that *are* the graphs of functions of x.

Implicit Differentiation

The graph of an equation $F(x, y) = 0$ is not the graph of a function of x if some of the vertical lines that cross it intersect it more than once. For example, the numbers y_1, y_2, and y_3 in Fig. 3.61 all correspond to the same x-value, $x = x_0$. However, various parts of the curve $F(x, y) = 0$ may well be the graphs of functions of x. The arc AB in Fig. 3.61 is the graph of a function of x, and so are arcs BC and CD.

Sometimes we can find explicit formulas for the functions defined by an equation $F(x, y) = 0$ but usually we cannot. The equation $F(x, y) = 0$ has defined the functions *implicitly* but not *explicitly*.

When may we expect the functions defined by an equation $F(x, y) = 0$ to be differentiable? The answer is, when their graphs are smooth enough to have a tangent at every point, as they will, for instance, if the formula for F is an algebraic combination of powers of x and y (a theorem from advanced mathematics).

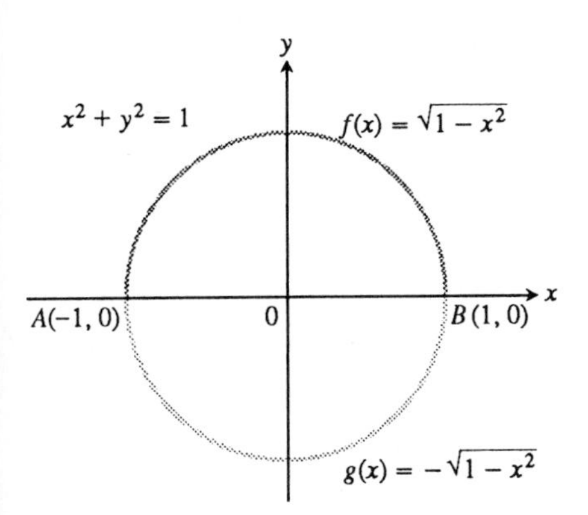

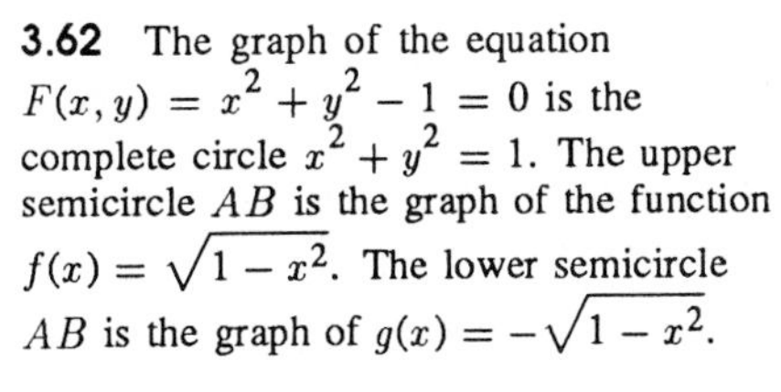
3.62 The graph of the equation $F(x, y) = x^2 + y^2 - 1 = 0$ is the complete circle $x^2 + y^2 = 1$. The upper semicircle AB is the graph of the function $f(x) = \sqrt{1 - x^2}$. The lower semicircle AB is the graph of $g(x) = -\sqrt{1 - x^2}$.

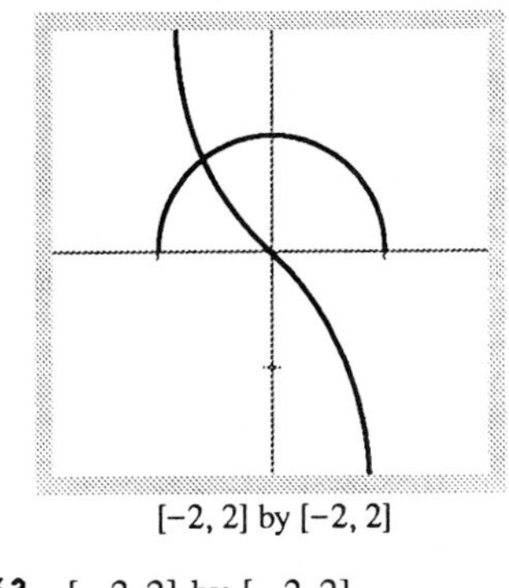

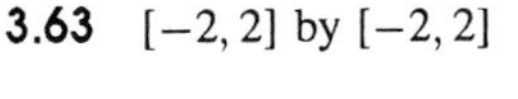
3.63 [−2, 2] by [−2, 2]

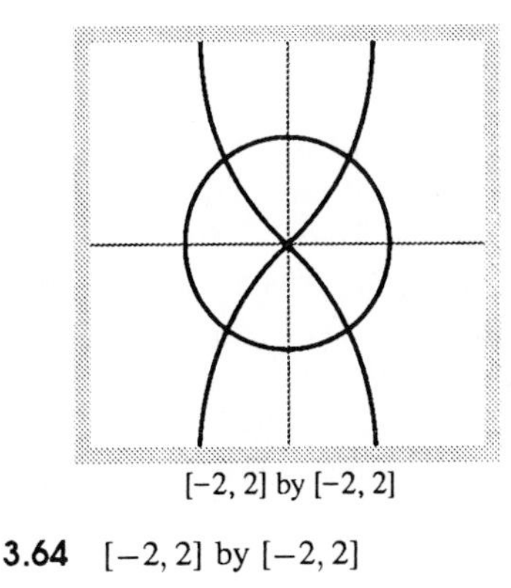

3.64 [−2, 2] by [−2, 2]

EXAMPLE 1 The graph of $F(x, y) = x^2+y^2-1 = 0$ is the circle $x^2+y^2 = 1$. Taken as a whole, the circle is not the graph of any single function of x (Fig. 3.62). Each x in the interval $-1 < x < 1$ gives two values of y, namely $y = \sqrt{1 - x^2}$ and $y = -\sqrt{1 - x^2}$.

The upper and lower semicircles are the graphs of the functions $f(x) = \sqrt{1 - x^2}$ and $g(x) = -\sqrt{1 - x^2}$. These functions are differentiable because they are composites of differentiable functions. The quickest way to find their derivatives, however, is not to differentiate the square-root formulas but to differentiate both sides of the original equation, treating y as a differentiable but otherwise unknown function of x:

$$x^2 + y^2 = 1$$

$$\frac{d}{dx}(x^2) + \frac{d}{dx}(y^2) = \frac{d}{dx}(1)$$

$$2x + 2y\frac{dy}{dx} = 0$$

$$\frac{dy}{dx} = -\frac{x}{y}.$$

This formula for dy/dx is simpler than either of the formulas we would get by differentiating f and g, and holds for all points on the curve above or below the x-axis. It is also easy to evaluate at any such point. At $(\sqrt{2}/2, \sqrt{2}/2)$, for instance,

$$\frac{dy}{dx} = -\frac{\sqrt{2}/2}{\sqrt{2}/2} = -1.$$

≡

To calculate the derivatives of other implicitly defined functions we simply proceed as in Example 1: We treat y as a differentiable (but otherwise unknown) function of x and apply the already familiar rules of differentiation to differentiate both sides of the defining equation, then solve for dy/dx. This procedure is called *implicit differentiation.*

Looking at the circle parametrically provides an alternate approach as illustrated in the next example.

EXAMPLE 2 Graph the circle $x^2 + y^2 = 1$ parametrically. Compute and graph dy/dx.

Solution The graph of $x = \cos t$ and $y = \sin t$ for $0 \leq t \leq 2\pi$ is the circle $x^2 + y^2 = 1$. Now $dx/dt = -\sin t$ and $dy/dt = \cos t$. It will be established later in the text that

$$\frac{dy}{dx} = \frac{dy}{dt} \cdot \frac{dt}{dx} = \frac{dy}{dt} \cdot \frac{1}{\frac{dx}{dt}}.$$

It follows that $dy/dx = -(\cos t/\sin t) = -(x/y)$ as found in Example 1. Figure 3.63 shows that graph of the upper half of the circle with the graph of its derivative $dy/dx = -x/y = -x/\sqrt{1 - x^2}$. Figure 3.64 shows the complete circle and the graph of both derivatives.

Can you explain why the graphs of the derivatives "blow up" at $x = -1$ and $x = 1$? Why are they so different for values close to but smaller than 1?

EXAMPLE 3 Find dy/dx if $2y = x^2 + \sin y$.

Solution

$$2y = x^2 + \sin y$$

$$\frac{d}{dx}(2y) = \frac{d}{dx}(x^2) + \frac{d}{dx}(\sin y) \qquad \text{Differentiate both sides with respect to } x.$$

$$2\frac{dy}{dx} = 2x + \cos y \frac{dy}{dx}$$

$$2\frac{dy}{dx} - \cos y \frac{dy}{dx} = 2x \qquad \text{Collect terms with } dy/dx.$$

$$(2 - \cos y)\frac{dy}{dx} = 2x \qquad \text{Factor out } dy/dx \ldots$$

$$\frac{dy}{dx} = \frac{2x}{2 - \cos y} \qquad \ldots \text{ and divide.}$$

In the exercises, we will ask you to graph the equation $2y = x^2 + \sin y$ and dy/dx.

Implicit Differentiation Takes Four Steps

1. Differentiate both sides of the equation with respect to x.
2. Collect the terms with dy/dx on one side of the equation.
3. Factor out dy/dx.
4. Solve for dy/dx by dividing.

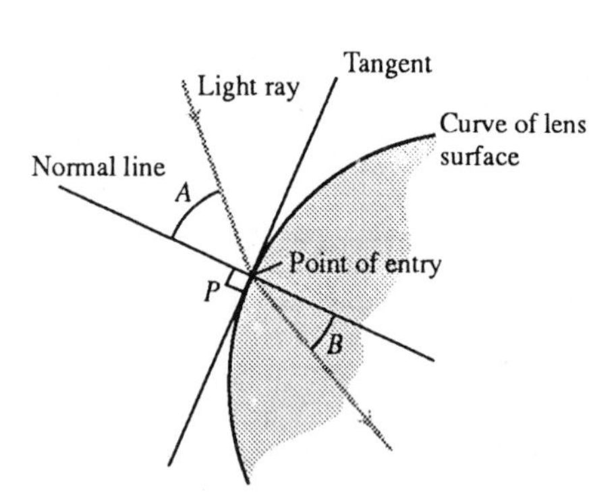

3.65 The profile or cutaway view of a lens, showing the bending (refraction) of a light ray as it passes through the lens surface.

Lenses, Tangents, and Normal Lines

In the law that describes how light changes direction as it enters a lens, the important angles are the angles the light makes with the line perpendicular to the surface of the lens at the point of entry (angles A and B in Fig. 3.65). This line is called the *normal to the surface* at the point of entry. In a profile view of a lens like the one in Fig. 3.65, the normal is the line perpendicular to the tangent to the profile curve at the point of entry.

The profiles of lenses are often described by quadratic curves like the one in Fig. 3.66. When they are, we can use implicit differentiation to find the tangents and normals.

EXAMPLE 4 Find the tangent and normal to the curve $x^2 - xy + y^2 = 7$ at the point $(-1, 2)$. (See Fig. 3.66.)

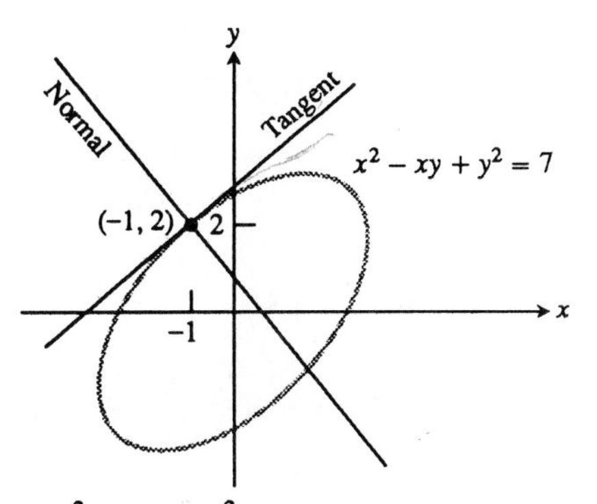

3.66 The graph of $x^2 - xy + y^2 = 7$ is an ellipse. The slope of the curve at the point $(-1, 2)$ is $(dy/dx)_{(-1,2)} = 4/5$.

Solution We first use implicit differentiation to find dy/dx:

$$x^2 - xy + y^2 = 7$$

$$\frac{d}{dx}(x^2) - \frac{d}{dx}(xy) + \frac{d}{dx}(y^2) = \frac{d}{dx}(7)$$ Differentiate both sides with respect to x. . .

$$2x - \left(x\frac{dy}{dx} + y\frac{dx}{dx}\right) + 2y\frac{dy}{dx} = 0$$ . . . treating xy as a product and y^2 as a power.

$$(2y - x)\frac{dy}{dx} = y - 2x$$ Collect terms.

$$\frac{dy}{dx} = \frac{y - 2x}{2y - x}$$ Solve as usual.

We then evaluate the derivative at $x = -1, y = 2$, to obtain

$$\left.\frac{dy}{dx}\right|_{(-1,2)} = \left.\frac{y - 2x}{2y - x}\right|_{(-1,2)}$$

$$= \frac{2 - 2(-1)}{2(2) - (-1)} = \frac{4}{5}.$$

The tangent to the curve at $(-1, 2)$ is

$$y - 2 = \frac{4}{5}(x - (-1))$$

$$y = \frac{4}{5}x + \frac{14}{5}.$$

The normal to the curve at $(-1, 2)$ is

$$y - 2 = -\frac{5}{4}(x + 1)$$

$$y = -\frac{5}{4}x + \frac{3}{4}.$$

Using Implicit Differentiation to Find Derivatives of Higher Order

Implicit differentiation can also produce derivatives of higher order. Here is an example.

EXAMPLE 5 Find d^2y/dx^2 if $2x^3 - 3y^2 = 7$.

Solution To start, we differentiate both sides of the equation with respect to x to find $y' = dy/dx$:

$$2x^3 - 3y^2 = 7$$

$$\frac{d}{dx}(2x^3) - \frac{d}{dx}(3y^2) = \frac{d}{dx}(7)$$

$$6x^2 - 6yy' = 0 \tag{1}$$

$$x^2 - yy' = 0$$

$$y' = \frac{x^2}{y}. \qquad \text{when } y \neq 0.$$

We now apply the Quotient Rule to find y'':

$$y'' = \frac{d}{dx}\left(\frac{x^2}{y}\right) = \frac{2xy - x^2y'}{y^2} = \frac{2x}{y} - \frac{x^2}{y^2}y'. \tag{2}$$

Finally, we substitute $y' = x^2/y$ to express y'' in terms of x and y:

$$y'' = \frac{2x}{y} - \frac{x^2}{y^2}\left(\frac{x^2}{y}\right) = \frac{2x}{y} - \frac{x^4}{y^3}. \tag{3}$$

The second derivative is not defined at $y = 0$ but is given by Eq. (3) when $y \neq 0$. ≡

Fractional Powers of Differentiable Functions

We know that the Power Rule

$$\frac{d}{dx}u^n = nu^{n-1}\frac{du}{dx} \tag{4}$$

holds when n is an integer. Our goal now is to show that it holds when n is a fraction. We will then be able to differentiate functions like

$$y = x^{4/3} \qquad \text{and} \qquad y = (\cos x)^{-1/5}$$

that were beyond our reach before.

HELGA VON KOCH'S SNOWFLAKE CURVE (1904)

Start with an equilateral triangle, calling it Curve 1. On the middle third of each side, build an equilateral triangle pointing outward. Then erase the old middle thirds. Call the expanded curve Curve 2. Now put equilateral triangles, again pointing them outward, on the middle thirds of the sides of Curve 2. Erase the old middle thirds to make Curve 3. Repeat the process, as shown, to define an infinite sequence of plane curves. The limit curve of the sequence is Koch's snowflake curve.

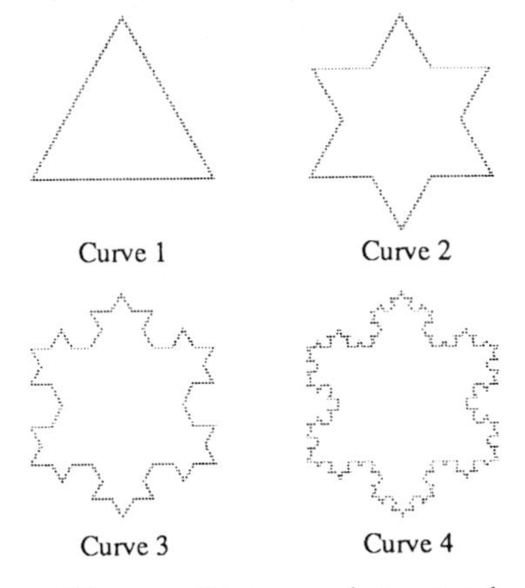

The snowflake curve is too rough to have a tangent at any point. In other words, the equation $F(x, y) = 0$ defining the curve does not define y as a differentiable function of x or x as a differentiable function of y at any point.

The snowflake curve plays an important role in chaos theory. We shall encounter it again when we study length in Section 6.4.

Power Rule for Fractional Exponents

If n is any rational number, then

$$\frac{d}{dx}x^n = nx^{n-1}, \tag{5}$$

provided $x \neq 0$ if $n - 1 < 0$ (i.e., $n < 1$).

If n is a rational number and u is a differentiable function of x, then u^n is a differentiable function of x and

$$\frac{d}{dx}u^n = nu^{n-1}\frac{du}{dx}, \tag{6}$$

provided $u \neq 0$ if $n < 1$.

The restrictions $x \neq 0$ if $n < 1$ and $u \neq 0$ if $n < 1$ are there to protect against inadvertent attempts to divide by zero. There is nothing mysterious about these restrictions. They come up quite naturally in practice, as the next example shows.

EXAMPLE 6

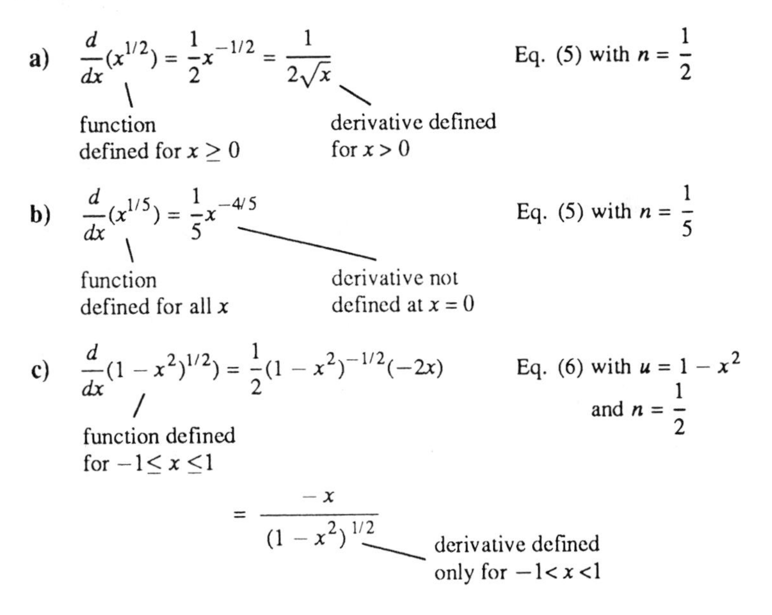

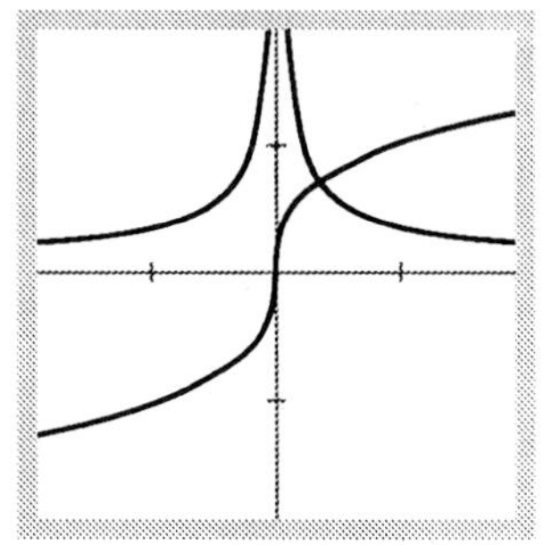

[−2, 2] by [−2, 2]

3.67 The graph of $y = x^{1/3}$ and the graph of $d\left(x^{1/3}\right)/dx = (1/3)\,x^{-2/3}$.

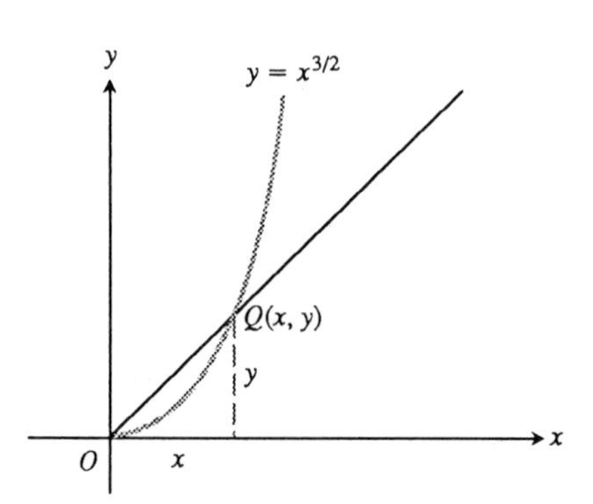

3.68 The graph of $y = x^{3/2}$. The slope of the curve at $x = 0$ is $\lim_{Q \to 0} m_{OQ} = 0$.

> **Explore with a Grapher**
>
> Notice the derivative of $y = x^{1/3}$ is not defined at $x = 0$. (Fig. 3.67) How do you explain this in terms of the slope of the tangent line to the curve at a point on the graph near the origin?

The derivatives of the functions $x^{4/3}$ and $(\cos x)^{-1/5}$ are defined wherever the functions themselves are defined, as we see in the next example.

EXAMPLE 7

a) $\dfrac{d}{dx}x^{4/3} = \dfrac{4}{3}x^{1/3}$

b)
$$\begin{aligned}\frac{d}{dx}(\cos x)^{-1/5} &= -\frac{1}{5}\cos x^{-6/5}\frac{d}{dx}(\cos x)\\ &= -\frac{1}{5}(\cos x)^{-6/5}(-\sin x)\\ &= \frac{1}{5}\sin x(\cos x)^{-6/5}\end{aligned}$$

EXAMPLE 8 Find the derivative of $y = x^{3/2}$ at $x = 0$.

Solution When the graph of a function stops abruptly at a point, as the graph of $y = x^{3/2}$ does at $x = 0$ (Fig. 3.68), we calculate its derivative as a one-sided limit. The Power Rule still applies, giving in this case

$$\left.\frac{dy}{dx}\right|_{x=0} = \left.\frac{3}{2}x^{1/2}\right|_{x=0} = 0.$$

We can see why this equation holds by looking at the geometry of the curve. The slope of a typical secant line through the origin and a point $Q(x, y)$ on the curve is

$$m_{OQ} = \frac{y-0}{x-0} = \frac{x^{3/2}}{x} = x^{1/2}.$$

As Q approaches the origin from the right, m_{OQ} approaches zero, in agreement with the result from the Power Rule.

Proof of the Power Rule for Fractional Exponents We prove Eq. (5) first and then apply the Chain Rule to get Eq. (6).

To prove Eq. (5), let p and q be integers with $q > 0$ and suppose that $y = x^{p/q}$. Then

$$y^q = x^p.$$

This equation is an algebraic combination of powers of x and y, so the advanced theorem we mentioned earlier assures us that y is a differentiable function of x. Since p and q are integers (for which we already have the Power Rule), we can differentiate both sides of the equation with respect to x and obtain.

$$qy^{q-1}\frac{dy}{dx} = px^{p-1}.$$

Hence, if $y \neq 0$,

$$\frac{dy}{dx} = \frac{p}{q}\frac{x^{p-1}}{y^{q-1}} = \frac{p}{q}\frac{x^{p-1}}{(x^{p/q})^{q-1}} = \frac{p}{q}\frac{x^{p-1}}{x^{p-p/q}} = \frac{p}{q}x^{(p/q)-1}.$$

This proves Eq. (5).

To prove Eq. (6) we let $y = u^{p/q}$ and apply the Chain Rule in the form

$$\frac{dy}{dx} = \frac{dy}{du}\frac{du}{dx}.$$

From Eq. (5), $(d/du)u^{p/q} = (p/q)u^{(p/q)-1}$. Hence

$$\frac{dy}{dx} = \frac{p}{q}u^{(p/q)-1}\frac{du}{dx}$$

and we're done.

Exercises 3.6

Find dy/dx in Exercises 1–18.

1. $y = x^{9/4}$
2. $y = x^{-3/5}$
3. $y = \sqrt[3]{x}$
4. $y = \sqrt[4]{x}$
5. $y = (2x+5)^{-1/2}$
6. $y = (1-6x)^{2/3}$
7. $y = x\sqrt{x^2+1}$
8. $y = \dfrac{x}{\sqrt{x^2+1}}$
9. $x^2y + xy^2 = 6$
10. $x^3 + y^3 = 18xy$
11. $2xy + y^2 = x + y$
12. $x^3 - xy + y^3 = 1$
13. $x^2y^2 = x^2 + y^2$
14. $(3x+7)^2 = 2y^3$
15. $y^2 = \dfrac{x-1}{x+1}$
16. $x^2 = \dfrac{x-y}{x+y}$
17. $y = \sqrt{1-\sqrt{x}}$
18. $y = 3(2x^{-1/2}+1)^{-1/3}$

Find dy/dx in Exercises 19–26.

19. $y = \sqrt{1+\cos 2x}$
20. $y = \sqrt{\sec 2x}$
21. $y = 3(\csc x)^{3/2}$
22. $y = [\sin(x+5)]^{5/4}$
23. $x = \tan y$
24. $x = \sin y$
25. $x + \tan(xy) = 0$
26. $x + \sin y = xy$

In Exercises 27–32, use implicit differentiation to find dy/dx and then d^2y/dx^2.

27. $x^2 + y^2 = 1$
28. $x^{2/3} + y^{2/3} = 1$
29. $y^2 = x^2 + 2x$
30. $y^2 + 2y = 2x + 1$
31. $y + 2\sqrt{y} = x$
32. $xy + y^2 = 1$

In Exercises 33–36, find the lines that are (a) tangent and (b) normal to the curve at the given point.

33. $x^2 + xy - y^2 = 1$, $(2, 3)$
34. $x^2 + y^2 = 25$, $(3, -4)$
35. $x^2y^2 = 9$, $(-1, 3)$
36. $y^2 - 2x - 4y - 1 = 0$, $(-2, 1)$
37. Find the two points where the curve $x^2 + xy + y^2 = 7$ crosses the x-axis, and show that the tangents to the curve at these points are parallel. What is the common slope of these tangents?
38. Find points on the curve $x^2 + xy + y^2 = 7$ (a) where the tangent is parallel to the x-axis and (b) where the tangent is parallel to the y-axis. (In the latter case, dy/dx is not defined, but dx/dy is. What value does dx/dy have at these points?)
39. Assume that the equation $2xy + \pi \sin y = 2\pi$ defines y as a differentiable function of x. Find dy/dx when $x = 1$ and $y = \pi/2$.

40. Find an equation for the tangent to the curve $x \sin 2y = y \cos 2x$ at the point $(\pi/4, \pi/2)$.

41. *The Eight Curve.* a) Find the slopes of the figure-eight–shaped curve

$$y^4 = y^2 - x^2$$

at the two points shown on the graph. b) Use a parametric graphing utility to reproduce the curve.

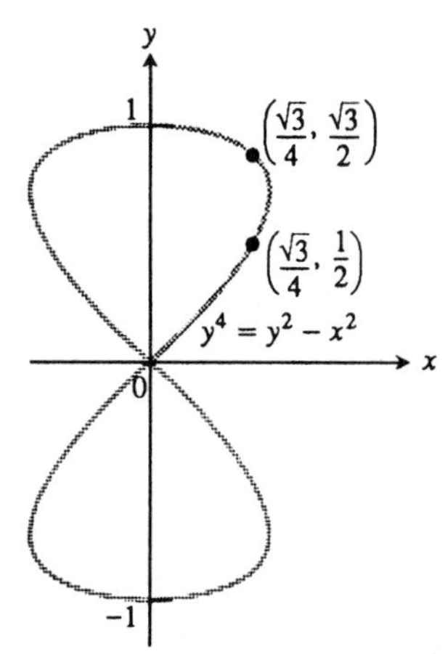

42. *The Cissoid of Diocles (dates from about 200 BC).* a) Find equations for the tangent and normal to the Cissoid of Diocles,

$$y^2(2 - x) = x^3,$$

at the point $(1, 1)$. b) Use a parametric graphing utility to reproduce the curve and the tangent and normal lines at $(1, 1)$.

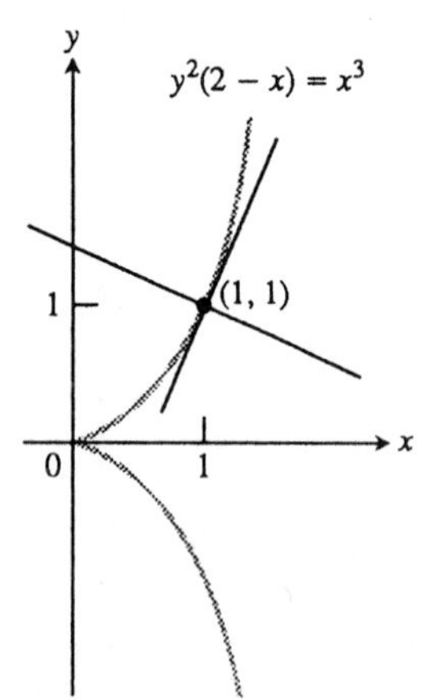

43. Which of the following could be true if $f''(x) = x^{-1/3}$?

a) $f(x) = \frac{3}{2}x^{2/3} - 3$ b) $f(x) = \frac{9}{10}x^{5/3} - 7$

c) $f'''(x) = -\frac{1}{3}x^{-4/3}$ d) $f'(x) = \frac{3}{2}x^{2/3} + 6$

44. *Orthogonal Curves.* Two curves are *orthogonal* at a point of intersection if their tangents there cross at right angles. Show that the curves $2x^3 + 3y^2 = 5$ and $y^2 = x^3$ are orthogonal at $(1, 1)$ and $(1, -1)$. Use a parametric graphing utility to draw the curves and to show the tangent lines.

45. The position of a body moving along a coordinate line at time t is $s = \sqrt{1 + 4t}$, with s in meters and t in seconds. Find the body's velocity and acceleration when $t = 6$ sec.

46. The velocity of a falling body is $v = k\sqrt{s}$ meters per second (k a constant) at the instant the body has fallen s meters from its starting point. Show that the body's acceleration is constant.

47. Use a parametric graphing utility to graph the curve given by the equation in Example 3. Show that it is a function. Graph its derivative.

48. Consider the curve $y^5 + \sin xy = 0$ discussed in Activity 3 of the Explore at the beginning of the section.

a) Show that $-1 \leq y \leq 1$.

b) For the part $x = \dfrac{\sin^{-1}(-y^5)}{y}$ graphed in Fig. 3.60, determine the possible values of x.

c) Use a parametric grapher to graph the part $xy = \sin^{-1}(-y^5) + 2\pi$. What are the possible values of x in this case?

49. Consider the relation $x = \dfrac{\sin^{-1}(-y^5)}{y}$ graphed in Activity 3 of the Explore at the beginning of the section.

a) Find x when $y = -1/2$.

b) Find $\dfrac{dy}{dx}$ using analytic methods and compute its value at $y = -1/2$.

c) Draw the tangent line to the graph in Fig. 3.60 at $y = -1/2$.

50. Refer to Example 4. There are two points at which $\dfrac{dy}{dx}$ does not exist. Find them.

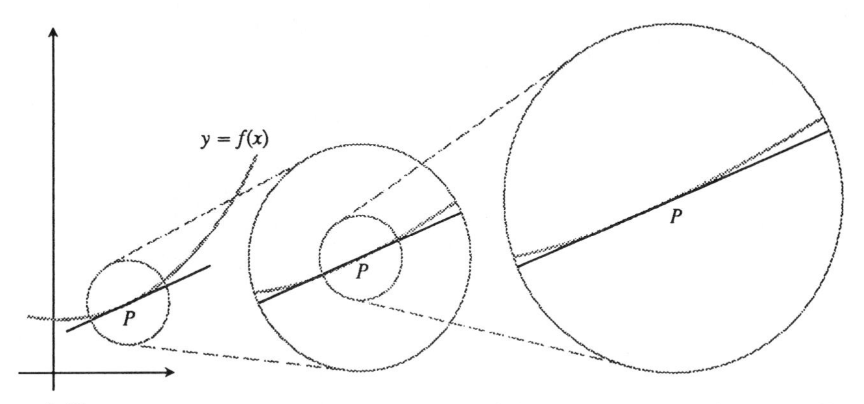

3.69 Successive magnifications show a close fit between a curve and its tangent line.

3.7 Linear Approximations and Differentials

Sometimes we can approximate complicated functions with simpler ones that give the accuracy we want for specific applications without being so hard to work with. It is important to know how to do this, and in this section we study the simplest of the useful approximations. For reasons that will be clear in a moment, the approximation is called a linearization.

We also introduce a new symbol, dx, for an increment in a variable x. This symbol is called the differential of x. In the physical sciences, it is used more frequently than Δx. In mathematics, differentials are used to estimate changes in function values, as we shall see toward the end of this section.

Explore with a Grapher

Graph $f(x) = x^2 - x - 3$ in the standard viewing window. Zoom-in by a factor of 10 at the point $(2, f(2))$. Overlay the line $y - f(2) = f'(2)(x-2)$. Comments? Now zoom-in again by a factor of 10 and then overlay the line $y - f(2) = f'(2)(x - 2)$. Comments?

Linearizations Are Linear Replacement Formulas

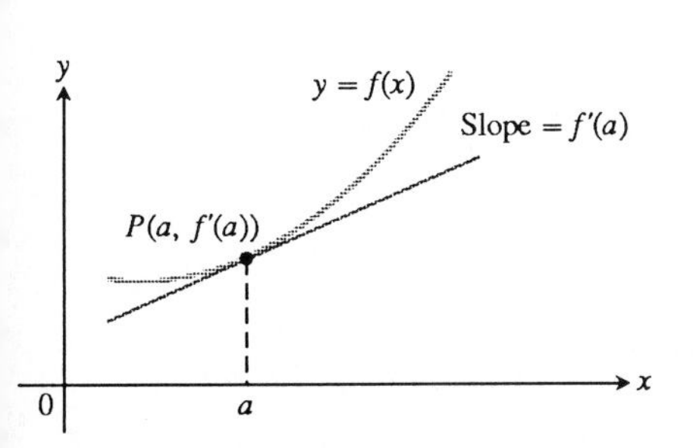

3.70 The equation of the tangent line is $y = f(a) + f'(a)(x - a)$.

As you can see in Fig. 3.69, the tangent to a curve $y = f(x)$ lies close to the curve near the point of tangency. For a brief interval to either side, the y-values along the tangent line give good approximations to the y-values on the curve. Therefore, to simplify the expression for the function near this point, we propose to replace the formula for f over this interval by the formula for its tangent line.

In the notation of Fig. 3.70, the tangent passes through the point $P(a, f(a))$ with slope $f'(a)$, so its point-slope equation is

$$y - f(a) = f'(a)(x - a)$$

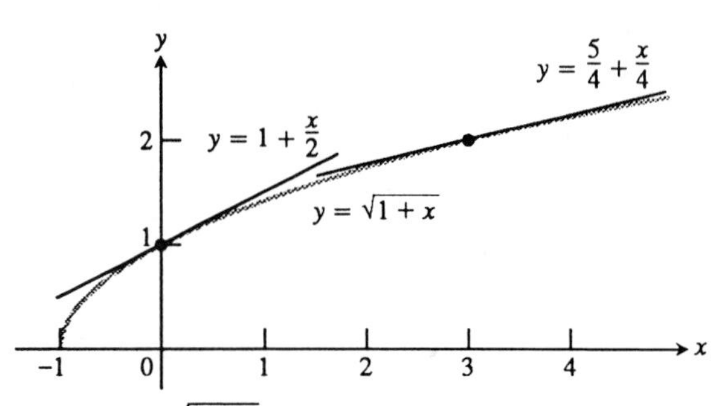

3.71 The graph of $y = \sqrt{1+x}$ and its linearizations at $x = 0$ and $x = 3$.

or

$$y = f(a) + f'(a)(x - a). \tag{1}$$

Thus, the tangent line is the graph of the function

$$L(x) = f(a) + f'(a)(x - a). \tag{2}$$

For as long as the line remains close to the graph of f, $L(x)$ will give a good approximation to $f(x)$.

Definitions

> **Linearization and Standard Linear Approximation** If $y = f(x)$ is differentiable at $x = a$, then
>
> $$L(x) = f(a) + f'(a)(x - a) \tag{3}$$
>
> is the **linearization** of f at a. The approximation
>
> $$f(x) \approx L(x)$$
>
> is the **standard linear approximation** of f at a.

EXAMPLE 1 Find the linearization of $f(x) = \sqrt{1+x}$ at $x = 0$.

Solution We evaluate Eq. (3) for $f(x) = \sqrt{1+x}$ and $a = 0$.

The derivative of f is

$$f'(x) = \frac{1}{2}(1+x)^{-1/2} = \frac{1}{2\sqrt{1+x}}.$$

Its value at $x = 0$ is $1/2$. We substitute this along with $a = 0$ and $f(0) = 1$ into Eq. (3):

$$L(x) = f(a) + f'(a)(x - a) = 1 + \frac{1}{2}(x - 0) = 1 + \frac{x}{2}.$$

The linearization of $\sqrt{1+x}$ at $x = 0$ is $L(x) = 1 + \frac{x}{2}$. See Fig. 3.71. ■

In case you are wondering how close the approximation

$$\sqrt{1+x} \approx 1 + \frac{x}{2}$$

really is, we can try a few values with a calculator:

$$\sqrt{1.2} \approx 1 + \frac{0.2}{2} = 1.10 \qquad \text{Accurate to 2 decimals}$$

$$\sqrt{1.05} \approx 1 + \frac{0.05}{2} = 1.025 \qquad \text{Accurate to 3 decimals}$$

$$\sqrt{1.005} \approx 1 + \frac{0.005}{2} = 1.00250 \qquad \text{Accurate to 5 decimals.}$$

The approximation becomes more accurate as we move toward the center, $x = 0$, and less accurate as we move away. As Fig. 3.71 suggests, the approximation will probably be too crude to be useful if we move out as far, say, as $x = 3$. To approximate $\sqrt{1+x}$ near $x = 3$, we had best find its linearization at $x = 3$.

EXAMPLE 2 Find the linearization of $f(x) = \sqrt{1+x}$ at $x = 3$.

Solution We evaluate Eq. (3) for $f(x) = \sqrt{1+x}, f'(x) = 1/(2\sqrt{1+x})$, and $a = 3$. With

$$f(3) = 2, f'(3) = \frac{1}{2\sqrt{1+3}} = \frac{1}{4},$$

Eq. (3) gives

$$L(x) = 2 + \frac{1}{4}(x-3) = 2 + \frac{x}{4} - \frac{3}{4} = \frac{5}{4} + \frac{x}{4}.$$

Thus, near $x = 3$,

$$\sqrt{1+x} \approx \frac{5}{4} + \frac{x}{4}.$$

At $x = 3.2$, the linearization we just obtained gives

$$\sqrt{1+x} = \sqrt{1+3.2} \approx \frac{5}{4} + \frac{3.2}{4} = 1.250 + 0.800 = 2.050,$$

which differs from $\sqrt{4.2} = 2.04939$ by less than one thousandth. The linearization from Example 1 gives

$$\sqrt{1+x} = \sqrt{1+3.2} \approx 1 + \frac{3.2}{2} = 1 + 1.6 = 2.6,$$

a result that is off by more than 25%. The linearization at $x = 3$ is obviously the one to use for values of $\sqrt{1+x}$ near 3.

Do not be misled by our calculations here into thinking that whatever we do with a linearization is better done with a calculator. In practice, we would never use a linearization to find the value of a particular square root. That is not what linearizations are for. The utility of the linearizations in Examples 1 and 2 lies in their ability to replace the complicated formula $\sqrt{1+x}$ by a simpler formula. If we have to work with $\sqrt{1+x}$ for values of x close to 0, and can tolerate the small amount of error involved, we can safely work with $1+(x/2)$ instead. Of course, we then need to know just how much error there really is. We shall look at that in a moment but the full answer won't come until Chapter 9.

EXAMPLE 3 The most important linearization for replacing roots and powers is

$$(1+x)^k \approx 1+kx \qquad \text{(any number } k\text{).} \tag{4}$$

(See Exercise 22.) The approximation is good for values of x near zero. For instance, when x is numerically small,

$$\begin{aligned}
\sqrt{1+x} &\approx 1+\frac{x}{2} \\
\frac{1}{1-x} &= (1-x)^{-1} \approx 1+(-1)(-x) = 1+x \\
\sqrt[3]{1+5x^4} &= (1+5x^4)^{1/3} \approx 1+\frac{1}{3}(5x^4) = 1+\frac{5}{3}x^4 \\
\frac{1}{\sqrt{1-x^2}} &= (1-x^2)^{-1/2} \approx 1+\left(-\frac{1}{2}\right)(-x^2) = 1+\frac{1}{2}x^2.
\end{aligned} \tag{5}$$

Trigonometric functions have delightfully simple linearizations at the origin.

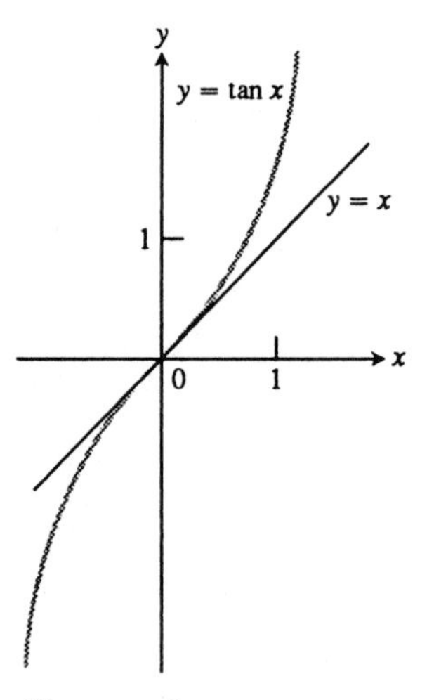

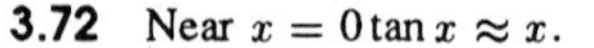
3.72 Near $x=0$ $\tan x \approx x$.

EXAMPLE 4 Find the linearization of $f(x)=\tan x$ at $x=0$.

Solution We use the equation

$$L(x) = f(a) + f'(a)(x-a)$$

with $f(x)=\tan x$ and $a=0$. Since

$$f(0)=\tan(0)=0, \qquad f'(0)=\sec^2(0)=1,$$

we have $L(x)=0+1(x-0)=x$. Near $x=0$ (Fig. 3.72).

$$\tan x \approx x.$$

In Exercises 13 and 14, you will be asked to derive the linearizations of $\sin x$ and $\cos x$ at $x=0$.

Linearizations at $x = 0$

Function $f(x)$	Linearization $L(x)$
$\sin x$	x
$\cos x$	1
$\tan x$	x

Of course, trigonometric functions have linearizations at other points as well, any point where they are differentiable.

EXAMPLE 5 Find the linearization of $f(x) = \cos x$ at $x = \pi/2$.

Solution We use the equation

$$L(x) = f(a) + f'(a)(x - a)$$

with $f(x) = \cos x$ and $a = \pi/2$. Since

$$f(\pi/2) = \cos(\pi/2) = 0, \qquad f'(\pi/2) = -\sin(\pi/2) = -1,$$

the linearization is

$$L(x) = 0 - 1\left(x - \frac{\pi}{2}\right) = -x + \frac{\pi}{2}.$$

See Fig. 3.73.

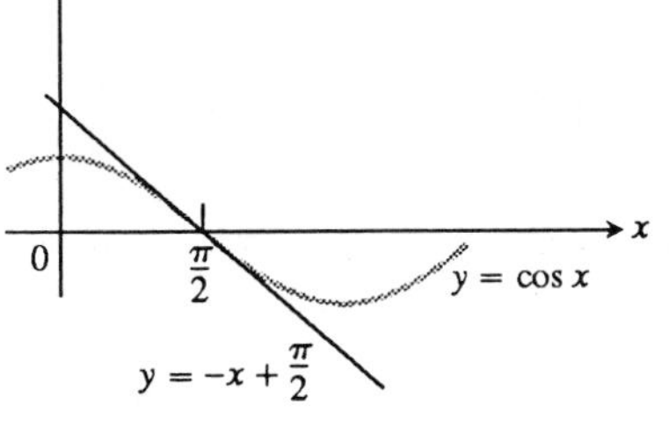

3.73 The graph of $y = \cos x$ and its linearization at $x = \pi/2$. Near $x = \pi/2$, $\cos x \approx -x + (\pi/2)$.

Estimating Change with Differentials

Suppose we know the value of a differentiable function $f(x)$ at a particular point x_0 and want to predict how much this value will change if we move nearby to the point $x_0 + h$. If h is small, f and its linearization L at x_0 will change by nearly the same amount. Since the values of L are always simple to calculate, calculating the change in L gives a practical way to estimate the change in f.

In the notation of Fig. 3.74, the change in f is

$$\Delta f = f(x_0 + h) - f(x_0).$$

The corresponding change in L is

$$\begin{aligned}\Delta L &= L(x_0 + h) - L(x_0) \\ &= f(x_0) + f'(x_0)[(x_0 + h) - x_0] - f(x_0) \qquad (6) \\ &= f'(x_0)h.\end{aligned}$$

The formula for Δf is usually as hard to work with as the formula for f. The formula for ΔL, however, is always simple to work with. As you can see, the change in L is just a constant times h.

The change $\Delta L = f'(x_0)h$ is usually described with the more suggestive notation

$$df = f'(x_0)dx, \qquad (7)$$

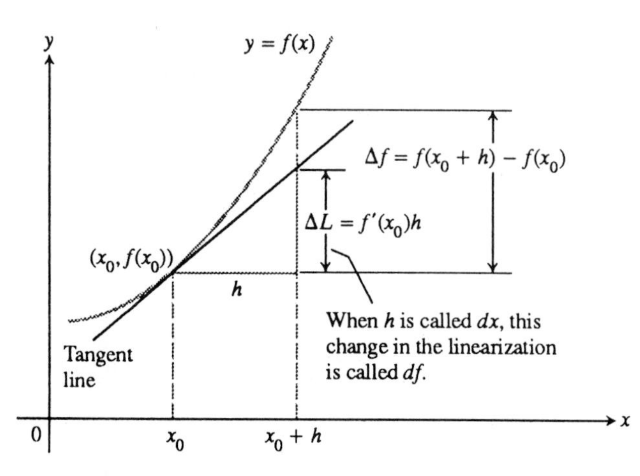

3.74 If h is small, the change in the linearization of f is nearly the same as the change in f.

in which df denotes the change in the linearization of f that results from the change dx in x. We call dx the **differential** of x and df the corresponding **differential** of f.

If $y = f(x)$ and we divide both sides of the equation $dy = f'(x)dx$ by dx, we obtain the familiar equation

$$\frac{df}{dx} = f'(x).$$

This equation now says that we may regard the derivative df/dx as a quotient of differentials. In many calculations, it is convenient to be able to think this way. For example, in writing the Chain Rule as

$$\frac{dy}{dx} = \frac{dy}{du}\frac{du}{dx},$$

we can think of the derivatives on the right as quotients in which the du's cancel to produce the fraction on the left. This gives a quick check on whether we remembered the rule correctly.

EXAMPLE 6 The radius of a circle increases from an initial value of $r_0 = 10$ by an amount $dr = 0.1$ (Fig. 3.75). Estimate the corresponding increase in the circle's area $A = \pi r^2$ by calculating dA. Compare dA with the true change ΔA.

Solution To calculate dA, we apply Eq. (7) to the function $A = \pi r^2$:

$$dA = A'(r_0)dr = 2\pi r_0 dr.$$

We then substitute the values $r_0 = 10$ and $dr = 0.1$:

$$dA = 2\pi(10)(0.1) = 2\pi.$$

The estimated change is 2π square units.

A direct calculation of ΔA gives

$$\Delta A = \pi(10.1)^2 - \pi(10)^2 = (102.01 - 100)\pi = \underbrace{2\pi}_{dA} + \underbrace{0.01\pi}_{\text{error}}.$$

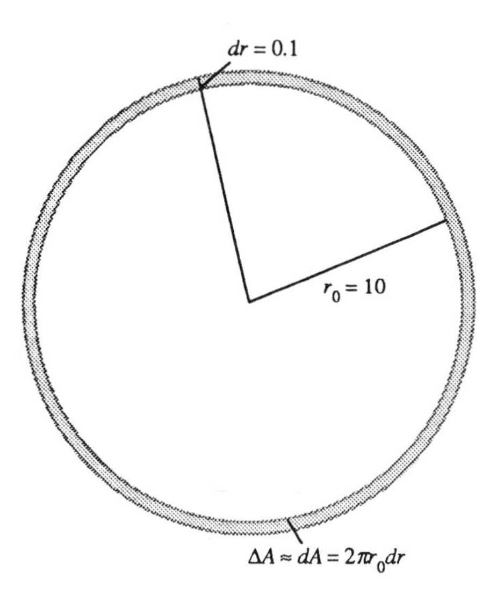

3.75 When dr is small compared to r_0, as it is when $dr = 0.1$ and $r_0 = 10$, the differential $dA = 2\pi r_0 dr$ gives a good estimate of ΔA. See Example 6.

The error in the estimate dA is 0.01π square units. As a percentage of the circle's original area, the error is quite small, as we can see from the following calculation:

$$\frac{\text{Error}}{\text{Original area}} = \frac{0.01\pi}{100\pi} = 0.01\%.$$

Absolute, Relative, and Percentage Change

What is the difference again between Δf and df? The increment Δf is the change in f; the differential df is the change in the linearization of f. Unike Δf, the differential df is always simple to calculate, and it gives a good estimate of Δf when the change in x is small.

As we move from x_0 to a nearby point, we can describe the corresponding change in the value of f in three ways:

	True	**Estimate**
Absolute change	Δf	df
Relative change	$\dfrac{\Delta f}{f(x_0)}$	$\dfrac{df}{f(x_0)}$
Percentage change	$\dfrac{\Delta f}{f(x_0)} \times 100$	$\dfrac{df}{f(x_0)} \times 100$

EXAMPLE 7 Estimate the percentage change that will occur in the area of a circle if its radius increases from $r_0 = 10$ units to 10.1 units.

Solution From the preceding table we have

$$\text{Estimated percentage change} = \frac{dA}{A(r_0)} \times 100.$$

With $dA = 2\pi$ (from Example 6) and $A(r_0) = 100\pi$, the formula gives

$$\frac{dA}{A(r_0)} \times 100 = \frac{2\pi}{100\pi} \times 100 = 2\%.$$

EXAMPLE 8 Suppose the earth were a perfect sphere and we determined its radius to be 3959 ± 0.1 miles. What effect would the tolerance of ± 0.1 have on our estimate of the earth's surface area?

Solution The surface area of a sphere of radius r is $S = 4\pi r^2$. The uncertainty in the calculation of S that arises from measuring r with a tolerance of dr miles is about

$$dS = \left(\frac{dS}{dr}\right) dr = 8\pi r\, dr.$$

With $r = 3959$ and $dr = 0.1$,

$$dS = 8\pi(3959)(0.1) = 9950 \text{ square miles}$$

to the nearest square mile, which is about the area of the state of Maryland. In absolute terms this might seem like a large error. However 9950 mi^2 is a relatively small error when compared to the calculated surface area of the earth:

$$\frac{dS}{\text{Calculated } S} = \frac{9950}{4\pi(3959)^2} \approx \frac{9950}{196{,}961{,}284} \approx .005\%.$$

EXAMPLE 9 About how accurately should we measure the radius r of a sphere to calculate the surface area $S = 4\pi r^2$ within 1% of its true value?

Solution We want any inaccuracy in our measurement to be small enough to make the corresponding increment ΔS in the surface area satisfy the inequality

$$|\Delta S| \le \frac{1}{100} S = \frac{4\pi r^2}{100}. \tag{8}$$

We replace ΔS in this inequality with

$$dS = \left(\frac{dS}{dr}\right) dr = 8\pi r\, dr.$$

This gives

$$|8\pi r\, dr| \le \frac{4\pi r^2}{100} \qquad \text{or} \qquad |dr| \le \frac{1}{8\pi r} \cdot \frac{4\pi r^2}{100} = \frac{1}{2}\frac{r}{100}.$$

We should measure the radius with an error dr that is no more than 0.5% of the true value.

EXAMPLE 10 **Unclogging arteries.** In the late 1830s, the French physiologist Jean Poiseuille ("pwa·*zoy*") discovered the formula we use today to predict how much the radius of a partially clogged artery has to be expanded to restore normal flow. His formula,

$$V = kr^4, \tag{9}$$

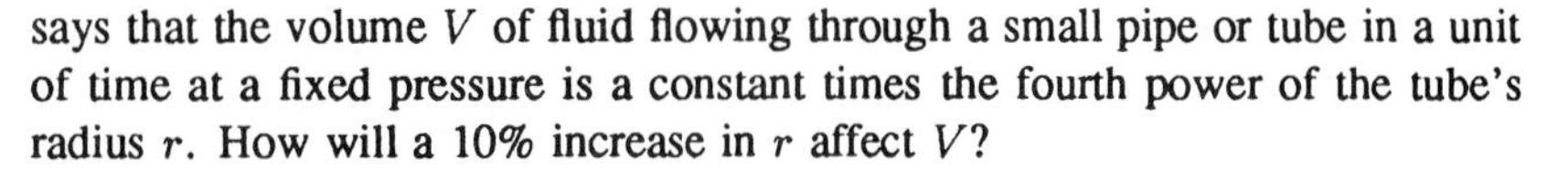

says that the volume V of fluid flowing through a small pipe or tube in a unit of time at a fixed pressure is a constant times the fourth power of the tube's radius r. How will a 10% increase in r affect V?

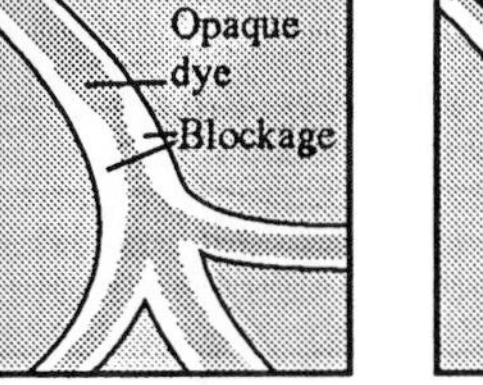

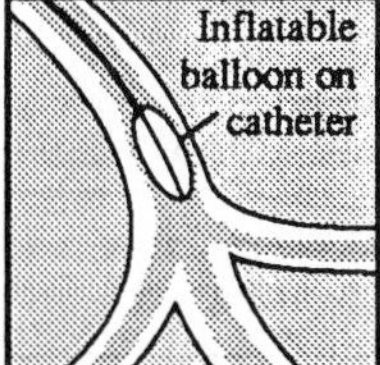

Angiography—An opaque dye is injected into a partially blocked artery to make the inside visible under x-rays. This reveals the location and severity of the blockage.

Angioplasty—A balloon-tipped catheter is inflated inside the artery to widen it at the blockage site.

Solution The differentials of r and V are related by the equation

$$dV = \frac{dV}{dr}dr = 4kr^3dr.$$

Hence,

$$\frac{dV}{V} = \frac{4kr^3dr}{kr^4} = 4\frac{dr}{r}. \quad \text{(Dividing by } V = kr^4.\text{)}$$

The relative change in V is four times the relative change in r, so a 10% increase in r will produce a 40% increase in the flow.

Sensitivity

The equation $df = f'(x)dx$ tells how sensitive the output of f is to a change in input at different values of x. The larger the value of f' at x, the greater is the effect of a given change dx.

EXAMPLE 11 You want to calculate the height of a bridge from the equation $s = 16t^2$ by timing how long it takes a heavy stone you drop to splash into the water below. How sensitive will your calculation be to a 0.1 sec error in measuring the time?

Solution The size of ds in the equation

$$ds = 32t\,dt$$

depends on how big t is. If $t = 2$ sec, the error caused by $dt = 0.1$ is only

$$ds = 32(2)(0.1) = 6.4 \text{ ft.}$$

Three seconds later, at $t = 5$ sec, the error caused by the same dt is

$$ds = 32(5)(0.1) = 16 \text{ ft.}$$

The Error in the Approximation $\Delta f \approx f'(a)\Delta x$

How well does the quantity $f'(a)\Delta x$ estimate the true increment $\Delta f = f(a + \Delta x) - f(a)$? We measure the error by subtracting one from the other:

$$\begin{aligned}\text{Approximation error} &= \Delta f - f'(a)\Delta x \\ &= f(a+\Delta x) - f(a) - f'(a)\Delta x \\ &= \underbrace{\left(\frac{f(a+\Delta x)-f(a)}{\Delta x} - f'(a)\right)}_{\epsilon}\Delta x, \qquad (10) \\ &= \epsilon \cdot \Delta x.\end{aligned}$$

As $\Delta x \to 0$, the difference quotient

$$\frac{f(a+\Delta x)-f(a)}{\Delta x}$$

approaches $f'(a)$ (remember the definition of $f'(a)$), so the quantity in parentheses becomes a very small number (which is why we called it ϵ). In fact,

$$\epsilon \to 0 \qquad \text{as} \qquad \Delta x \to 0.$$

Thus, when Δx is small, the approximation error $\epsilon\Delta x$ is smaller still.

$$\underbrace{\Delta f}_{\substack{\text{true}\\ \text{change}}} = \underbrace{f'(a)\Delta x}_{\substack{\text{estimated}\\ \text{change}}} + \underbrace{\epsilon\Delta x}_{\text{error}} \tag{11}$$

While we do not know exactly how small the error is and will not be able to make much progress on this front until later in Chapter 9, there is something worth noting here, namely the *form* taken by the equation.

If $y = f(x)$ is differentiable at $x = a$, and x changes from a to $a + \Delta x$, the change Δy in f is given by an equation of the form

$$\Delta y = f'(a)\Delta x + \epsilon\Delta x \tag{12}$$

in which $\epsilon \to 0$ as $\Delta x \to 0$.

Surprising as it may seem, just knowing the form of Eq. (12) enables us to bring the proof of the Chain Rule to a successful conclusion. You can find out what we mean by turning to Appendix 3.

Derivatives in Differential Notation

Every formula like

$$\frac{d(u+v)}{dx} = \frac{du}{dx} + \frac{dv}{dx}$$

has a corresponding differential formula like

$$d(u+v) = du + dv$$

TABLE 3.2 Formulas for Differentials

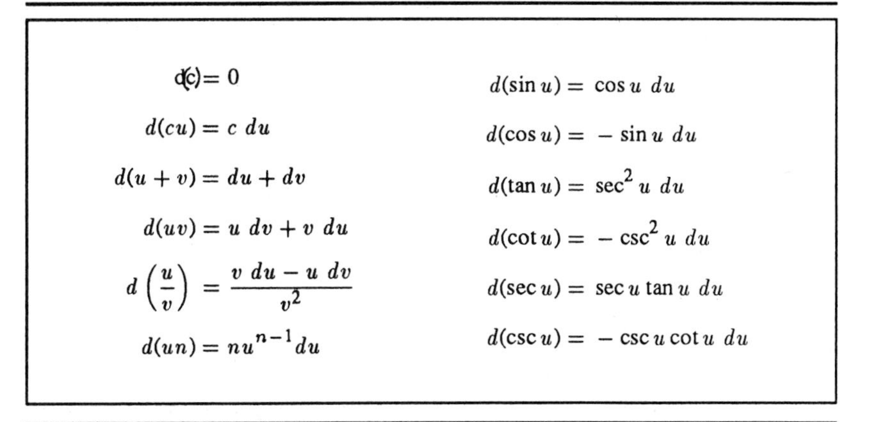

$d(c) = 0$	$d(\sin u) = \cos u\ du$
$d(cu) = c\ du$	$d(\cos u) = -\sin u\ du$
$d(u + v) = du + dv$	$d(\tan u) = \sec^2 u\ du$
$d(uv) = u\ dv + v\ du$	$d(\cot u) = -\csc^2 u\ du$
$d\left(\frac{u}{v}\right) = \frac{v\ du - u\ dv}{v^2}$	$d(\sec u) = \sec u \tan u\ du$
$d(un) = nu^{n-1}du$	$d(\csc u) = -\csc u \cot u\ du$

that comes from multiplying both sides by dx.

To find dy when y is a differentiable function of x, we may either find dy/dx and multiply by dx or use one of the formulas in Table 3.2.

EXAMPLE 12

a) $d(3x^2 - 6) = 6xdx$

b) $d(\cos 3x) = -(\sin 3x)d(3x) = -3 \sin 3xdx$

c)
$$d\frac{x}{x+1} = \frac{(x+1)dx - x\ d(x+1)}{(x+1)^2} = \frac{x\ dx + dx - x\ dx}{(x+1)^2} = \frac{dx}{(x+1)^2}$$

Notice that a differential on one side of an equation always calls for a differential on the other side of the equation. Thus, we never have $dy = 3x^2$ but, instead, $dy = 3x^2dx$.

Exercises 3.7

In Exercises 1–6, find the linearization $L(x)$ of $f(x)$ at $x = a$.

1. $f(x) = x^4$ at $x = 1$

2. $f(x) = x^{-1}$ at $x = 2$

3. $f(x) = x^3 - x$ at $x = 1$

4. $f(x) = x^3 - 2x + 3$ at $x = 2$

5. $f(x) = \sqrt{x}$ at $x = 4$

6. $f(x) = \sqrt{x^2 + 9}$ at $x = -4$

You want linearizations that will replace the functions in Exercises 7–12 over intervals that include the given points x_0. To make your subsequent work as simple as possible, you want to center each linearization not at x_0 but at a nearby integer $x = a$ at which the given function and its derivative are easy to evaluate. What linearization do you use in each case?

7. $f(x) = x^2 + 2x, x_0 = 0.1$

8. $f(x) = x^{-1}, x_0 = 0.6$

9. $f(x) = 2x^2 + 4x - 3, x_0 = -0.9$

10. $f(x) = 1 + x, x_0 = 8.1$

11. $f(x) = \sqrt[3]{x}, x_0 = 8.5$

12. $f(x) = \dfrac{x}{x+1}, x_0 = 1.3$

In Exercises 13–18, find the linearization $L(x)$ of the given function at $x = a$. Then graph f and L together near $x = a$.

13. $f(x) = \sin x$ at $x = 0$

14. $f(x) = \cos x$ at $x = 0$

15. $f(x) = \sin x$ at $x = \pi$

16. $f(x) = \cos x$ at $x = -\pi/2$

17. $f(x) = \tan x$ at $x = \pi/4$

18. $f(x) = \sec x$ at $x = \pi/4$

19. Use the formula $(1 + x)^k \approx 1 + kx$ to find linear approximations of the following functions for values of x near zero. Graph each function and its linearization in $[-2, 2]$ by $[-2, 2]$.

a) $(1 + x)^2$ b) $\dfrac{1}{(1+x)^5}$

c) $\dfrac{2}{1-x}$ d) $(1 - x)^6$

e) $3(1 + x)^{1/3}$ f) $\dfrac{1}{\sqrt{1+x}}$

20. Use the approximation $(1 + x)^k \approx 1 + kx$ to estimate and compare with a calculator value.

a) $(1.002)^{100}$ b) $\sqrt[3]{1.009}$

21. Find the linearization of $f(x) = \sqrt{x+1} + \sin x$ at $x = 0$. How is it related to the individual linearizations for $\sqrt{x+1}$ and $\sin x$?

22. We know from the Power Rule that the equation

$$\frac{d}{dx}(1 + x)^k = k(1 + x)^{k-1}$$

holds for every rational number k. In Chapter 7, we shall show that it holds for every irrational number as well. Assuming this result for now, verify Eq. (4) by showing that the linearization of $f(x) = (1 + x)^k$ at $x = 0$ is $L(x) = 1 + kx$ for any number k.

23. What happens when you take successive square roots of 2 by entering 2 in your calculator and pressing the square-root key repeatedly? Why? Each keypress divides the decimal part of the display approximately in half. The explanation comes from the fact that the linearization of $\sqrt{1+x}$ is $1 + (x/2)$. The x is the decimal part of the display $(1.x)$ and each new square root is about $1 + (x/2)$.
If you have not done so already, enter 2 into your calculator and press $\boxed{\sqrt{\ }}$ repeatedly to see what happens.

24. Repeat Exercise 23 with (a) $\sqrt{0.5}$, and (b) $\sqrt[10]{2}$, and (c) $\sqrt[10]{0.5}$.

In Exercises 25–30, each function $f(x)$ changes value when x changes from x_0 to $x_0 + dx$. Find

a) the change $\Delta f = f(x_0 + dx) - f(x_0)$;

b) the value of the estimate $df = f'(x_0)dx$; and

c) the error $|\Delta f - df|$.

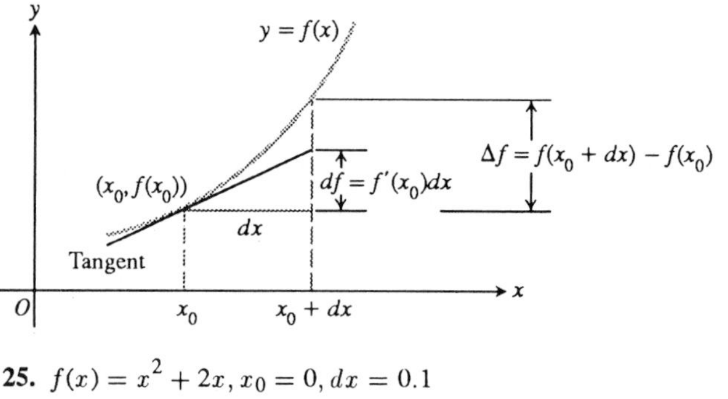

25. $f(x) = x^2 + 2x, x_0 = 0, dx = 0.1$

26. $f(x) = 2x^2 + 4x - 3, x_0 = -1, dx = 0.1$

27. $f(x) = x^3 - x, x_0 = 1, dx = 0.1$

28. $f(x) = x^4, x_0 = 1, dx = 0.1$

29. $f(x) = x^{-1}, x_0 = 0.5, dx = 0.1$

30. $f(x) = x^3 - 2x + 3, x_0 = 2, dx = 0.1$

In Exercises 31–36, write a differential formula that estimates the given change in volume or surface area.

31. The change in the volume $V = (4/3)\pi r^3$ of a sphere when the radius changes from r_0 to $r_0 + dr$.

32. The change in the surface area $S = 4\pi r^2$ of a sphere when the radius changes from r_0 to $r_0 + dr$.

33. The change in the volume $V = x^3$ of a cube when the edge lengths change from x_0 to $x_0 + dx$.

34. The change in the surface area $S = 6x^2$ of a cube when the edge lengths change from x_0 to $x_0 + dx$.

35. The change in the volume $V = \pi r^2 h$ of a right circular cylinder when the radius changes from r_0 to $r_0 + dr$ and the height does not change.

36. The change in the lateral surface area $S = 2\pi rh$ of a right circular cylinder when the height changes from h_0 to $h_0 + dh$ and the radius does not change.

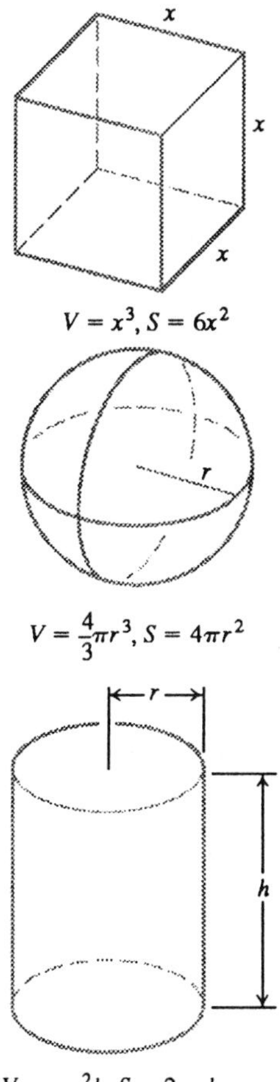

37. The radius of a circle is increased from 2.00 to 2.02 m.

a) Estimate the resulting change in area.

b) Express the estimate in (a) as a percentage of the circle's original area.

38. The diameter of a tree was 10 in. During the following year, the circumference grew 2 in. About how much did the tree's diameter grow? The tree's cross section area?

39. The edge of a cube is measured as 10 cm with an error of 1%. The cube's volume is to be calculated from this measurement. Estimate the percentage error in the volume calculation.

40. About how accurately should you measure the side of a square to be sure of calculating the area within 2% of its true value?

41. The diameter of a sphere is measured as 100 ± 1 cm and the volume is calculated from this measurement. Estimate the percentage error in the volume calculation.

42. Estimate the allowable percentage error in measuring the diameter d of a sphere if the volume is to be calculated correctly to within 3%.

43. The height and radius of a right circular cylinder are equal, so the cylinder's volume is $V = \pi h^3$. The volume is to be calculated from a measurement of h and must be calculated with an error of no more than 1% of the true value. Find approximately the greatest error that can be tolerated in the measurement of h, expressed as a percentage of h.

44. a) About how accurately must the interior diameter of a 10-m-high cylindrical storage tank be measured to calculate the tank's volume to within 1% of its true value?

b) About how accurately must the tank's exterior diameter be measured to calculate the amount of paint it will take to paint the side of the tank within 5% of the true amount?

45. A manufacturer contracts to mint coins for the federal government. How much variation dr in the radius of the coins can be tolerated if the coins are to weigh within 1/1000 of their ideal weight? Assume that the thickness does not vary.

46. *Continuation of Example 10.* By what percentage should r be increased to increase V by 50%?

47. *Continuation of Example 11.* Show that a 5% error in measuring t will cause about a 10% error in calculating $s(t)$ from the equation $s = 16t^2$.

48. *The effect of flight maneuvers on the heart.* The amount of work done by the heart's main pumping chamber, the left ventricle, is given by the equation

$$W = PV + \frac{Vpv^2}{2g},$$

where W is the work per unit time, P is the average blood pressure, V is the volume of blood pumped out during the unit of time, p is the density of the blood, v is the average velocity of the exiting blood, and g is the acceleration of gravity.

When P, V, p, and v remain constant, W becomes a function of g and the equation takes the simplified form

$$W = a + \frac{b}{g} \quad (a, b \text{ constant}) \tag{13}$$

As a member of NASA's medical team, you want to know how sensitive W is to apparent changes in g caused by flight maneuvers, and this depends on the initial value of g. As part of your investigation, you decide to compare the effect on W of a given change dg on the moon, where $g = 5.2$ ft/sec^2, with the effect the same change dg would have on Earth, where $g = 32$ ft/sec^2. You use Eq. (13) to find the ratio of dW_{moon} to dW_{Earth}. What do you find?

Using Linearizations to Solve Equations

49. Let $g(x) = \sqrt{x} + \sqrt{1 + x} - 4$.

a) Find $g(3) < 0$ and $g(4) > 0$ to show (by the Intermediate Value Theorem, Section 2.2) that the equation $g(x) = 0$ has a solution between $x = 3$ and $x = 4$.

b) To estimate the solution of $g(x) = 0$, replace the square roots by their linearizations at $x = 3$ and solve the resulting linear equation.

c) Check your estimate in the original equation.

d) Solve $g(x) = 0$ using graphing zoom-in and compare with (b).

e) Find the exact solutions to $g(x) = 0$.

50. Carry out the following steps to estimate the solution of $2\cos x = \sqrt{1+x}$.

a) Let $f(x) = 2\cos x - \sqrt{1+x}$. Find $f(0) > 0$ and $f(\pi/2) < 0$ to show that $f(x)$ has a zero between 0 and $\pi/2$.

b) Find the linearizations of $\cos x$ at $x = \pi/4$ and $\sqrt{1+x}$ at $x = 0.69$.

c) To estimate the solution of the original equation, replace $\cos x$ and $\sqrt{1+x}$ by their linearizations from (b) and solve the resulting linear equation for x. Check your estimate in the original equation.

d) Solve the original equation using graphing zoom-in.

e) Can you find exact solutions to the original equation?

Derivatives in Differential Form

In Exercises 51–62, find dy.

51. $y = x^3 - 3x$

52. $y = x\sqrt{1-x^2}$

53. $y = 2x/(1+x^2)$

54. $y = (3x^2 - 1)^{3/2}$

55. $y + xy - x = 0$

56. $xy^2 + x^2y - 4 = 0$

57. $y = \sin(5x)$

58. $y = \cos(x^2)$

59. $y = 4\tan(x/2)$

60. $y = \sec(x^2 - 1)$

61. $y = 3\csc(1 - (x/3))$

62. $y = 2\cot\sqrt{x}$

Review Questions

1. What is a derivative? A right-hand derivative? A left-hand derivative? How are they related? Give examples.

2. What geometric significance do derivatives have?

3. How is the differentiability of a function at a point related to its continuity (if any) at a point?

4. Give some examples of differentiable functions.

5. Explain the meaning of NDer (f, x). Is NDer $(f, x) = f'(x)$ always true? Illustrate with examples.

6. When does a function typically *not* have a derivative at a point? Illustrate with a graph.

7. What rules do you know for calculating derivatives? Give examples.

8. Explain how the three formulas

a) $\dfrac{d(x^n)}{dx} = nx^{n-1}$,

b) $\dfrac{d(cu)}{dx} = c\dfrac{du}{dx}$,

c) $\dfrac{d(u+v)}{dx} = \dfrac{du}{dx} + \dfrac{dv}{dx}$

let us differentiate any polynomial.

9. Explain how the graph of $y =$ NDer (f, x) can be used to support the analytic computation of $D_x f(x)$.

10. What formula do we need, in addition to the three listed in Question 7, to differentiate rational functions?

11. What is a second derivative? A third derivative? How many derivatives do the functions you know have? Give examples.

12. When a body moves along a coordinate line and its position $s(t)$ is a differentiable function of t, how do you define the body's velocity, speed, and acceleration? Give an example.

13. Besides velocity, speed, and acceleration, what other rates of change are found with derivatives?

14. What are the derivatives of the six basic trigonometric functions? How does their calculation depend on radian measure?

15. When is the composite of two functions differentiable at a point? What do you need to know to calculate its derivative there? Give examples.

16. What is implicit differentiation and what is it good for? Give examples.

17. What is the linearization $L(x)$ of a function $f(x)$ at a point $x = a$? What is required of f at a for the linearization to exist? How are linearizations used? Give examples.

18. If x moves from x_0 to a nearby value $x_0 + dx$, how do we estimate the corresponding change in the value of a differentiable function $f(x)$? How do we estimate the relative change? The percentage change? Give an example.

19. How are derivatives expressed in differential notation? Give examples.

Practice Exercises

In Exercises 1–34, find dy/dx analytically. Support with a graph of $f(x) = \text{NDer}(y, x)$.

1. $y = x^5 - \frac{1}{8}x^2 + \frac{1}{4}x$
2. $y = 3 - 7x^3 + 3x^7$
3. $y = (x+1)^2(x^2 + 2x)$
4. $y = (2x-5)(4-x)^{-1}$
5. $y = 2\sin x \cos x$
6. $y = \sin x - x\cos x$
7. $y = \frac{x}{x+1}$
8. $y = \frac{2x+1}{2x-1}$
9. $y = (x^3+1)^{-4/3}$
10. $y = (x^2 - 8x)^{-1/2}$
11. $y = \cos(1-2x)$
12. $y = \cot\frac{2}{x}$
13. $y = (x^2 + x + 1)^3$
14. $y = \left(-1 - \frac{x}{2} - \frac{x^2}{4}\right)^2$
15. $y = \sqrt{2u + u^2},\ u = 2x + 3$
16. $y = \frac{-u}{1+u},\ u = \frac{1}{x}$
17. $xy + y^2 = 1$
18. $xy + 2x + 3y = 1$
19. $x^2 + xy + y^2 - 5x = 2$
20. $x^3 + 4xy - 3y^2 = 2$
21. $5x^{4/5} + 10y^{6/5} = 15$
22. $\sqrt{xy} = 1$
23. $y^2 = \frac{x}{x+1}$
24. $y^2 = \sqrt{\frac{1+x}{1-x}}$
25. $y^2 = \frac{(5x^2+2x)^{3/2}}{3}$
26. $y = \frac{3}{(5x^2+2x)^{3/2}}$
27. $y = \sqrt{x} + 1 + \frac{1}{\sqrt{x}}$
28. $y = x\sqrt{2x+1}$
29. $y = \sec(1+3x)$
30. $y = \sec^2(1+3x)$
31. $y = \cot x^2$
32. $y = x^2\cos 5x$
33. $y = \sqrt{\frac{1-x}{1+x^2}}$
34. $y^2 = \frac{x^2-1}{x^2+1}$

35. a) Graph the function

$$f(x) = \begin{cases} x, & 0 \le x \le 1 \\ 2 - x, & 1 < x \le 2. \end{cases}$$

b) Is f continuous at $x = 1$?

c) Is f differentiable at $x = 1$? Explain.

36. a) Find the values of the left-hand and right-hand derivatives of

$$f(x) = \begin{cases} \sin 2x, & x \le 0 \\ mx, & x > 0 \quad (m \text{ constant}) \end{cases}$$

at $x = 0$.

b) For what value of m, if any, is f differentiable at $x = 0$.

37. Find the points on the curve $y = 2x^3 - 3x^2 - 12x + 20$ where the tangent is parallel to the x-axis.

38. The line normal to the curve $y = x^2 + 2x - 3$ at $(1, 0)$ intersects the curve at what other point?

39. The position at time $t \ge 0$ of a particle moving along a coordinate line is

$$s(t) = 10\cos(t + \pi/4).$$

a) Use a parametric graphing utility to simulate the motion of the particle.

b) What is the particle's starting position ($t = 0$)?

c) What are the points farthest to the left and right of the origin reached by the particle?

d) Find the particle's velocity and acceleration at the points in (b).

e) When does the particle first reach the origin? What are its velocity, speed, and acceleration then?

40. On Earth, you can easily shoot a paper clip 64 ft straight up into the air with a rubber band. In t seconds after firing, the paper clip is $s = 64t - 16t^2$ ft above your hand.

a) Use a parametric graphing utility to simulate the position of the paper clip.

b) How long does it take the paper clip to reach its maximum height? With what velocity does it leave your hand?

c) On the moon, the same force will send the paper clip to a height of $s(t) = 64t - 2.6t^2$ ft in t seconds. About how long will it take the paper clip to reach its maximum height and how high will it go?

41. Suppose two balls are falling from rest at a certain height in centimeters above the ground. Use the equation $s = 490t^2$ to answer the following questions:

a) How long did it take the balls to fall the first 160 cm? What was their average velocity for the period?

b) How fast were the balls falling when they reached the 160 cm mark? What was their acceleration then?

42. The following data give the coordinates s of a moving body for various values of t. Plot s versus t on coordinate paper and sketch a smooth curve through the given points. Assuming that this smooth curve represents the motion of the body, estimate the velocity at (a) $t = 1.0$; (b) $t = 2.5$; (c) $t = 2.0$.

s (in ft)	10	38	58	70	74	70	58	38	10
t (in sec)	0	0.5	1.0	1.5	2.0	2.5	3.0	3.5	4.0

43. The graphs in Fig. 3.76 show the distance traveled (miles), velocity (mph), and acceleration (mph/sec) for each second

of a 2-minute automobile trip. Which graph shows
a) distance? **b)** velocity? **c)** acceleration?

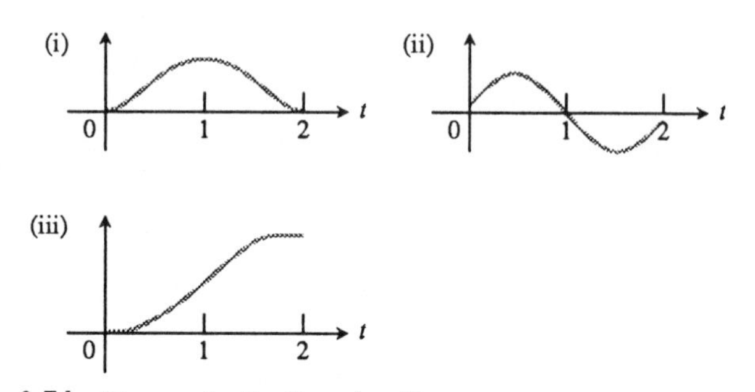

3.76 The graphs for Exercise 43.

44. The graph in Fig. 3.77 shows the position $s(t)$ of a truck traveling on a highway. The truck starts at $t = 0$ and returns 15 hours later at $t = 15$.

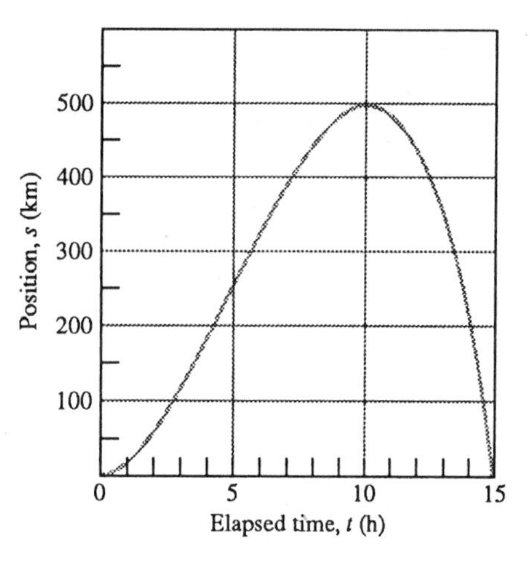

3.77 The time vs. position graph for the truck in Exercise 44.

a) Use the technique described at the end of Section 3.3 to graph the truck's velocity $v = ds/dt$ for $0 \leq t \leq 15$. Then repeat the process, with the velocity curve, to graph the truck's acceleration dv/dt.

b) Suppose $s(t) = 15t^2 - t^3$. Graph ds/dt and d^2s/dt^2 and compare your graphs with those in (a).

45. Use the following information to graph the function $y = f(x)$ for $-1 \leq x \leq 6$.

i) The graph of f is made of line segments joined end to end.

ii) The graph starts at the point $(-1, 2)$.

iii) The derivative of f, where defined, is the step function shown in Fig. 3.78.

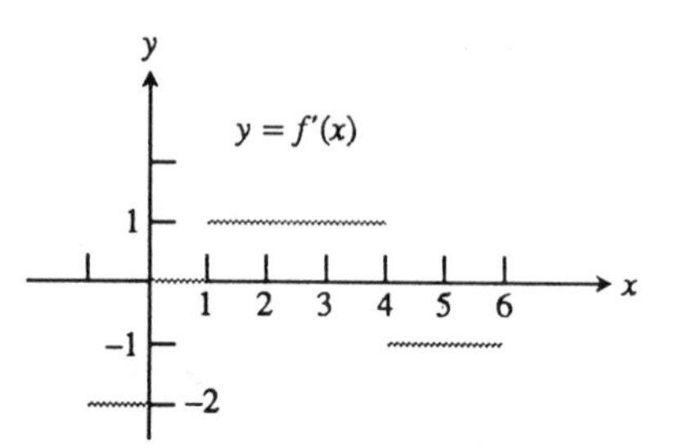

3.78 The graph for Exercise 45.

46. Repeat Exercise 45, supposing that the graph starts at $(-1, 0)$ instead of $(-1, 2)$.

47. If a hemispherical bowl of radius 10 in. is filled with water to a depth of x in., the volume of water is given by $V = \pi[10 - (x/3)]x^2$. Find the rate of increase of the volume per inch increase of depth.

48. A bus will hold 60 people. The number x of people per trip who use the bus is related to the fare charged (p dollars) by the law $p = [3 - (x/40)]^2$. Write an expression for the total revenue $r(x)$ per trip received by the bus company. What number of people per trip will make the marginal revenue dr/dx equal to zero? What is the corresponding fare? (This is the fare that maximizes the revenue, so the bus company should probably rethink its fare policy.)

In Exercises 49 and 50, find an equation for

a) the horizontal tangent and

b) the tangent to the curve at the indicated point P.

49.

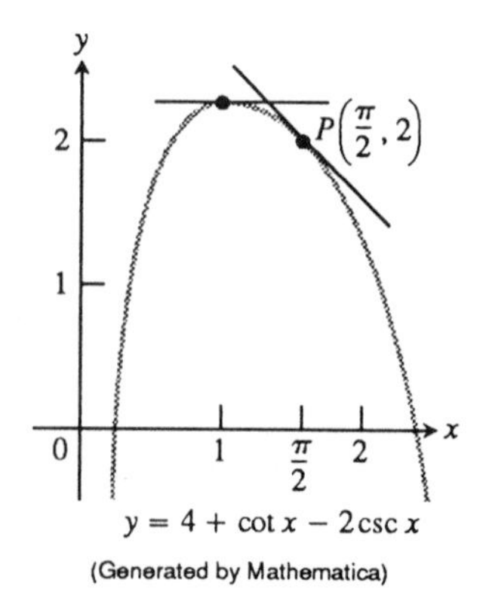

50.

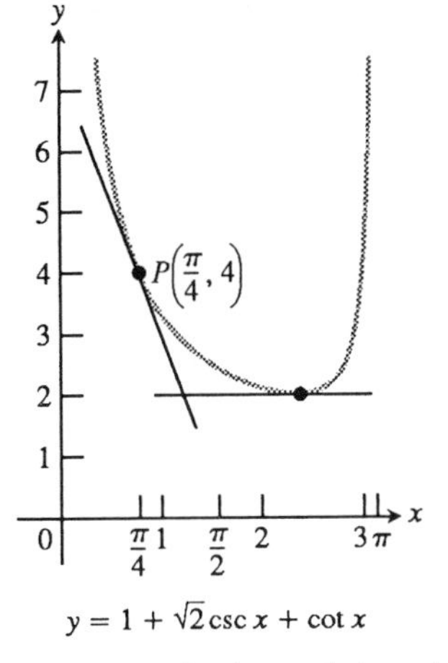

$y = 1 + \sqrt{2}\csc x + \cot x$

51. The accompanying graph of $y = \sin(x - \sin x)$ suggests that the curve might have horizontal tangents at the x-axis. Does it?

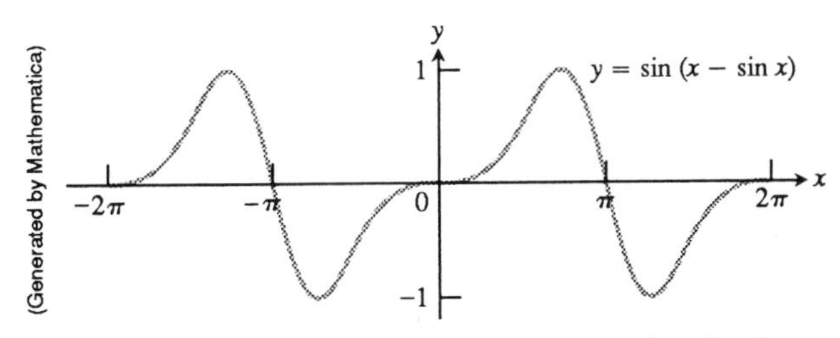

52. The figure shows a boat 1 km offshore, sweeping the shore with a search light. The light turns at the constant rate $d\theta/dt = -3/5$ radians per second. (This rate is called the light's *angular velocity*.)

a) Express x (see the figure) in terms of θ.

b) Differentiate both sides of the equation you obtained in (a) with respect to t. Then substitute $d\theta/dt = -3/5$. This will express dx/dt (the rate at which the light moves along the shore) as a function of θ.

c) How fast (m/sec) is the light moving along the shore when it reaches point A?

d) How many revolutions per minute is 0.6 radian per second?

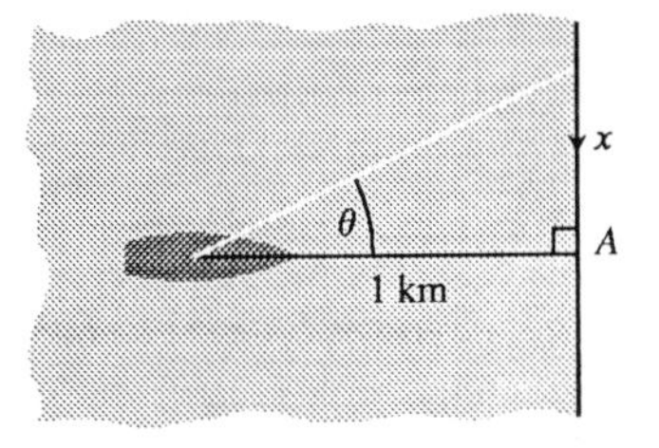

53. Suppose that functions f and g and their derivatives have the following values at $x = 0$ and $x = 1$.

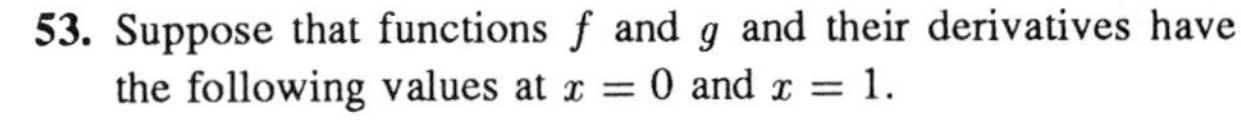

x	$f(x)$	$g(x)$	$f'(x)$	$g'(x)$
0	1	1	5	1/3
1	3	−4	−1/3	−8/3

Find the derivatives of the following combinations at the given value of x.

a) $5f(x) - g(x)$, $x = 1$

b) $f(x)g^3(x)$, $x = 0$

c) $\dfrac{f(x)}{g(x)+1}$, $x = 1$

d) $f(g(x))$, $x = 0$

e) $g(f(x))$, $x = 0$

f) $(x + f(x))^{3/2}$, $x = 1$

g) $f(x + g(x))$, $x = 0$

54. Suppose that $f(x) = x^2$ and $g(x) = |x|$. Then the composites

$$(f \circ g)(x) = |x|^2 = x^2 \quad \text{and} \quad (g \circ f)(x) = |x^2| = x^2$$

are both differentiable at $x = 0$ even though g is not differentiable at $x = 0$. Does this contradict the Chain Rule? Explain.

55. If the identity $\sin(x + a) = \sin x \cos a + \cos x \sin a$ is differentiated with respect to x, is the resulting equation also an identity? Does this principle apply to the equation $x^2 - 2x - 8 = 0$? Explain.

56. Find dy/dt at $t = 0$ if $y = 3\sin 2x$ and $x = t^2 + \pi$.

57. Find ds/du at $u = 2$ if $s = t^2 + 5t$ and $t = (u^2 + 2u)^{1/3}$.

58. Find dw/ds at $s = 0$ if $w = \sin(\sqrt{r} - 2)$ and $r = 8\sin(s + \pi/6)$.

59. Find the points where the tangent to the curve $y = \sqrt{x}$ at $x = 4$ crosses the coordinate axes.

60. What horizontal line crosses the curve $y = \sqrt{x}$ at a 45° angle?

61. Find the lines that are tangent and normal to the curve at the given point.

a) $x^2 + 2y^2 = 9$ at $(1, 2)$

b) $x^3 + y^2 = 2$ at $(1, 1)$

c) $xy + 2x - 5y = 2$ at $(3, 2)$

62. Which of the following statements could be true if $f''(x) = x^{1/3}$?

I. $f(x) = \frac{9}{28}x^{7/3} + 9$

II. $f'(x) = \frac{9}{28}x^{7/3} - 2$

III. $f'(x) = \frac{3}{4}x^{4/3} + 6$

IV. $f(x) = \frac{3}{4}x^{4/3} - 4$

a) I only

b) III only

c) II and IV only

d) I and III only

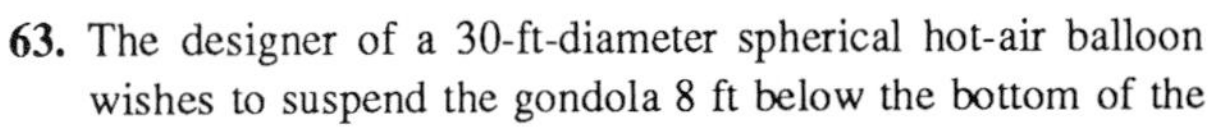

63. The designer of a 30-ft-diameter spherical hot-air balloon wishes to suspend the gondola 8 ft below the bottom of the balloon. Two of the cables are shown running from the top edges of the gondola to their points of tangency, $(-12, -9)$ and $(12, -9)$. How wide must the gondola be?

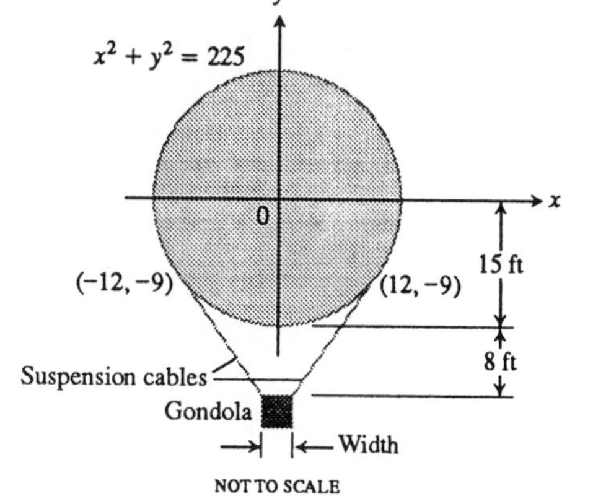

64. What determines the fundamental frequency of a vibrating piano string?
We measure the frequencies at which wires vibrate in cycles (trips back and forth) per second. The unit of measure is a *hertz*: 1 cycle per second. Middle A on a piano has a frequency of 440 hertz. For any given wire, the fundamental frequency y is a function of four variables:

r: the radius of the wire
l: the length
d: the density of the wire
T: the tension (force) holding the wire taut.

With r and l in centimeters, d in grams per cubic centimeter, and T in dynes (it takes about 100,000 dynes to lift an apple), the fundamental frequency of the wire is

$$y = \frac{1}{2rl}\sqrt{\frac{T}{\pi d}}.$$

If we keep all the variables fixed except one, then y can be alternately thought of as four different functions of one variable, $y(r)$, $y(l)$, $y(d)$, and $y(T)$. How would changing each variable then affect the string's fundamental frequency? To find out, calculate $y'(r)$, $y'(l)$, $y'(d)$, and $y'(T)$.

65. Find d^2y/dx^2 by implicit differentiation:

a) $x^3 + y^3 = 1$ b) $y^2 = 1 - \dfrac{2}{x}$

66. a) By differentiating $x^2 - y^2 = 1$ implicitly, show that $dy/dx = x/y$.

b) Then show that $d^2y/dx^2 = -1/y^3$.

67. Find d^2y/dx^2 if

a) $y = \sqrt{2x + 7}$ b) $x^2 + y^2 = 1$

68. If $y^3 + y = 8x - 6$, find d^2y/dx^2 at the point $(1, 1)$.

69. Find the linearizations of

a) $\tan x$ at $x = -\pi/4$ b) $\sec x$ at $x = -\pi/4$.

Graph the curves and linearizations together.

70. A useful linear approximation to

$$\frac{1}{1 + \tan x}$$

at $x = 0$ can be obtained by combining the approximations

$$\frac{1}{1+x} \approx 1 - x \quad \text{and} \quad \tan x \approx x$$

to get

$$\frac{1}{1 + \tan x} \approx 1 - x.$$

Show that this is the standard linear approximation of $1/(1 + \tan x)$.

71. Let $f(x) = \sqrt{1 + x} + \sin x - 0.5$.

a) Find $f(-\pi/4) < 0$ and $f(0) > 0$, to show that the equation $f(x) = 0$ has a solution between $-\pi/4$ and 0.

b) To estimate the solution of $f(x) = 0$, replace $\sqrt{1+x}$ and $\sin x$ by their linearizations at $x = 0$ and solve the resulting linear equation.

c) Check your estimate in the original equation.

d) Solve $f(x) = 0$ using graphing zoom-in. What are the exact solutions?

72. Let

$$f(x) = \frac{2}{1 - x} + \sqrt{1 + x} - 3.1.$$

a) Find $f(0) < 0$ and $f(0.5) > 0$ to show that the equation $f(x) = 0$ has a solution between $x = 0$ and $x = 0.5$.

b) To estimate the solution of the equation $f(x) = 0$, replace $2/(1 - x)$ and $\sqrt{1 + x}$ by their linearizations at $x = 0$ and solve the resulting linear equation.

c) Check your estimate in the original equation.

d) Solve $f(x) = 0$ using graphing zoom-in. What are the exact solutions?

73. Write a formula that estimates the change that occurs in the volume of a right circular cone when the radius changes from r_0 to $r_0 + dr$ and the height does not change.

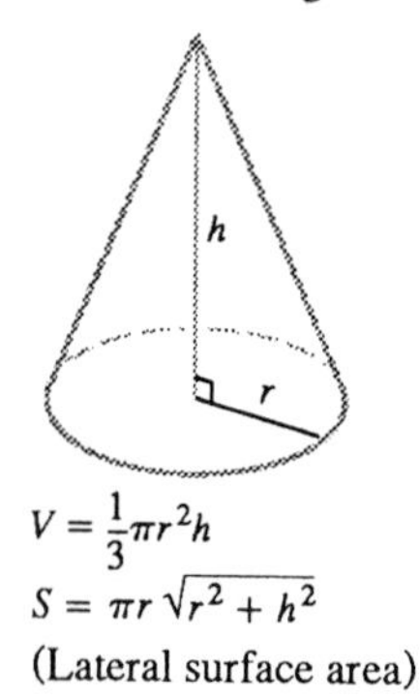

$V = \frac{1}{3}\pi r^2 h$
$S = \pi r\sqrt{r^2 + h^2}$
(Lateral surface area)

74. Write a formula that estimates the change that occurs in the lateral surface area of a cone when the height changes from h_0 to $h_0 + dh$ and the radius does not change.

75. a) How accurately should you measure the edge of a cube to be reasonably sure of calculating the cube's surface area with an error of no more than 2%?

b) Suppose the edge is measured with the accuracy required in (a). About how accurately can the cube's volume be calculated from the edge measurement? To find out, estimate the percentage error in the volume calculation that would result from using the edge measurement.

76. The circumference of the equator of a sphere is measured as 10 cm with a possible error of 0.4 cm. The measurement is then used to calculate the radius. The radius is then used to calculate the surface area and volume of the sphere. Estimate the percentage errors in the calculated values of (a) the radius, (b) the surface area, and (c) the volume.

77. To find the height of a tree, you measure the angle from the ground to the treetop from a point 100 ft away from the base. The best figure you can get with the equipment at hand is $30^\circ \pm 1^\circ$. About how much error could the tolerance of $\pm 1^\circ$ create in the calculated height? Remember to work in radians.

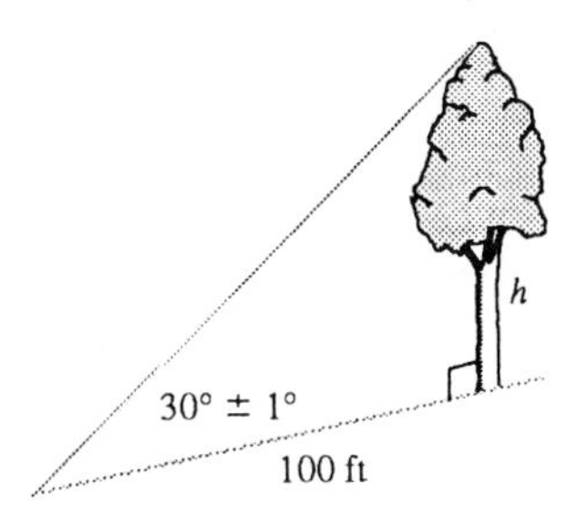

78. To find the height of a lamppost, you stand a 6-ft pole 20 ft from the lamp and measure the length a of its shadow. The figure you get for a is 15 ft, give or take an inch. Calculate the height of the lamppost from the value $a = 15$ and estimate the possible error in the result.

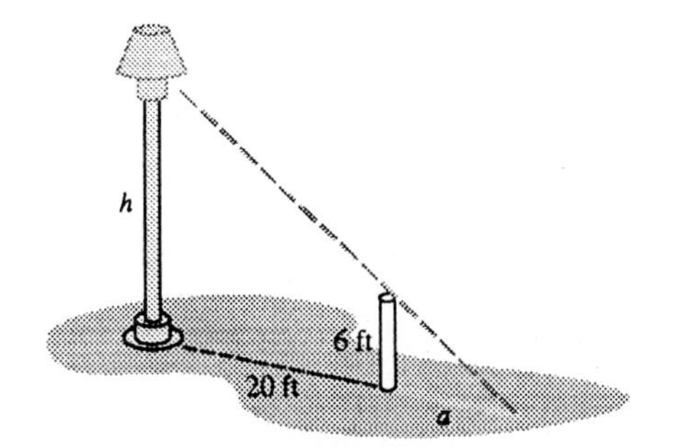

79. *The linearization is the best linear approximation.* Suppose that $y = f(x)$ is differentiable at $x = a$ and that $g(x) = m(x - a) + c$ (m and c constants). If the error $E(x) = f(x) - g(x)$ were small enough near $x = a$, we might think of using g as a linear approximation of f instead of the linearization $L(x) = f(a) + f'(a)(x - a)$. Show that if we impose on g the conditions

1. $E(a) = 0$ — The approximation error is zero at $x = a$.

2. $\lim_{x \to a} \frac{E(x)}{x - a} = 0$ — The error is negligible when compared with $(x - a)$,

then $g(x) = f(a) + f'(a)(x - a)$. Thus, the linearization gives the only linear approximation whose error is both zero at $x = a$ and negligible in comparison with $(x - a)$.

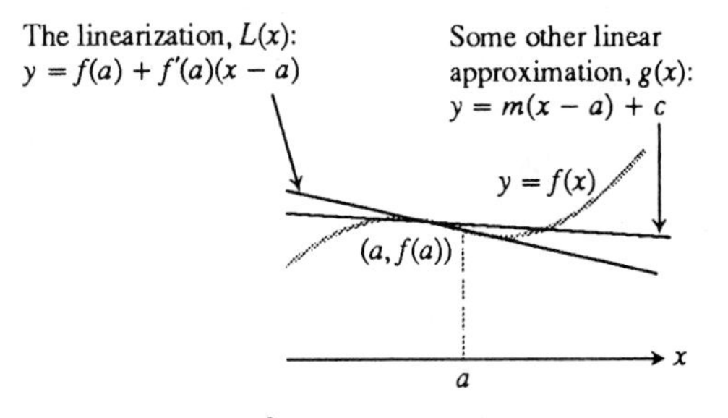

80. The volume $y = x^3$ of a cube with edges of length x increases an amount Δy when x increases by an amount Δx. Show with a sketch how to represent Δy geometrically as the sum of the volumes of

a) three slabs of dimensions x by x by Δx

b) three bars of dimensions x by Δx by Δx

c) one cube of dimensions Δx by Δx by Δx.

The differential formula $dy = 3x^2 dx$ estimates the change in y with the three slabs.

CHAPTER

4

Applications of Derivatives

OVERVIEW In the past derivatives were used to sketch reliable graphs of polynomial, rational, radical, and transcendental functions, and to find the largest and smallest values that a differentiable function assumes on a closed interval. This use of derivatives is less important today because of graphing calculators and computers. However, the techniques introduced in this chapter are still very important from a conceptual point of view. This chapter also shows how derivatives can be used to determine behavior of functions that is sometimes hidden from view in a computer graph, and to confirm information discovered or conjectured graphically. The verification that a graph really looks like what is on the screen must still come from calculus. Even if we have a computer plot hundreds of points, we cannot be sure what the curve does between the plotted points without calculus.

If you know the rate at which a function is changing, this chapter will show that you can often calculate the rates at which functions closely related to it are changing at the same time. If we know the derivative of a function and the value of the function at a particular point, this chapter will show how to find the function. The key to recovering functions from their derivatives is The Mean Value Theorem, a theorem whose corollaries provide the gateway to the so-called integral calculus we shall begin studying in Chapter 5.

Many of the applications in this chapter involve writing down equations that model real world situations we want to find out about and the steps we go through each time to get the answers we want are almost always the same. We close the chapter by showing the role that these steps play in a general process called modeling and how we, as scientists, can use these steps to formulate ideas and test them against reality.

4.1 Maxima, Minima, and the Mean Value Theorem

Differential calculus is the mathematics of working with derivatives. One of the things we can do with derivatives is to algebraically confirm where functions take on their maximum and minimum values. In this section we lay the theoretical ground for finding these extreme values as they are called, and establish the first derivative test for determining when functions are increasing or decreasing on an interval.

We also introduce the Mean Value Theorem, one of the most exciting and influential theorems in calculus.

Maxima and Minima—Relative vs. Absolute

Figure 4.1 shows a point c where a function $y = f(x)$ with domain $[a, b]$ has a maximum value. If we move to either side of c, the function values get smaller and the curve falls away. When we take in more of the curve, however, we find that f assumes an even larger value at d. Thus, $f(c)$ is not the absolute maximum value of f on the interval $[a, b]$ but only a *relative* or *local* maximum value.

Definitions

A function f has a **local maximum** value at an interior point c of its domain if $f(x) \leq f(c)$ for all x in some open interval I about c. The function has an **absolute maximum** value at c if $f(x) \leq f(c)$ for all x in the domain.

Similarly, f has a **local minimum** value at an interior point c of its domain if $f(x) \geq f(c)$ for all x in an open interval I about c. The function has an **absolute minimum** value at c if $f(x) \geq f(c)$ for all x in the domain.

The definitions of local maximum and local minimum are extended to endpoints of the function's domain by requiring the intervals I to be appropriate half-open intervals containing the endpoints.

Notice that an absolute maximum is also a local maximum because it is the largest value in its immediate neighborhood as well as overall. Hence a list of all local maxima will include the absolute maximum if there is one.

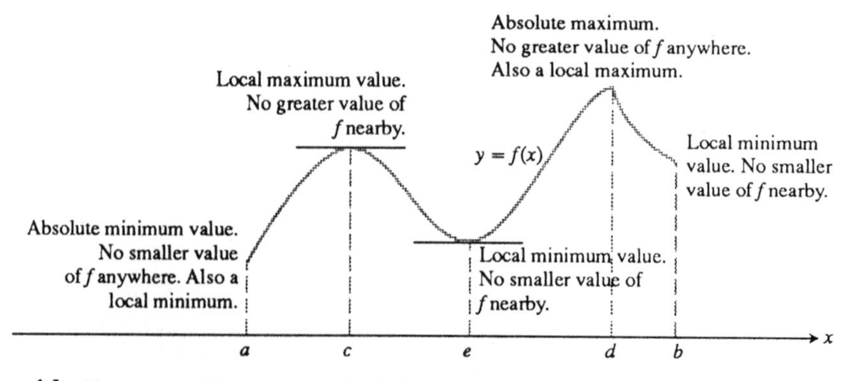

4.1 How to classify maxima and minima.

Similarly, an absolute minimum, when it exists, is also a local minimum. A list of all local minima will include the absolute minimum if there is one.

The First Derivative Theorem

Explore with a Graphing Utility

1. Draw a complete graph of $f(x) = |x^2 - 2x|$.
2. Does f have an extreme value at $x = 0$?
3. Does $f'(0)$ exist? Explain.
4. Repeat parts (2) and (3) at $x = 2$.
5. What does f have at $x = 1$?
6. Does $f'(1)$ exist? Explain.

In Fig. 4.1 two extreme values of f occur at endpoints of the function's domain, one occurs at a point where f' fails to exist, and two occur at interior points where $f' = 0$. This is typical for a function defined on a closed interval. As the following theorem says, a function's first derivative is always zero at an interior point where the function has a local extreme value. Hence the only places where a function f can ever have an extreme value are

1. Interior points where f' is zero,
2. Interior points where f' does not exist,
3. Endpoints of the function's domain.

We shall see the importance of this observation as the chapter continues.

Explore with a Grapher

1. Draw a complete graph of $f(x) = x^3 - 12x$.
2. Compute NDer (f, x) for several values of x near 2, say $x = 1.9, 1.99, 2.1, 2.01$, etc.
3. Repeat part (2) for several values of x near -2.
4. Based on parts (2) and (3) what would you conjecture about $f'(a)$ (if it exists) if f has an extreme value at $x = a$?

Theorem 1

The First Derivative Theorem for Local Extreme Values

If a function f has a local maximum or a local minimum value at an interior point c of an interval where it is defined, and if f' exists at c, then

$$f'(c) = 0.$$

Proof You may not have seen an argument like the one we are about to use, so we shall explain its form first. We want to show that $f'(c) = 0$, and our plan is to do that indirectly by showing first that $f'(c)$ cannot be positive and second that $f'(c)$ cannot be negative either. Why does that show $f'(c) = 0$? Because, in the entire real number system, only one number is neither positive nor negative, and that number is zero.

To be specific, suppose f has a local maximum value at $x = c$, so that $f(x) \leq f(c)$ for all values of x near c (Fig. 4.2). Since c is an interior point of f's domain, the limit

$$\lim_{x \to c} \frac{f(x) - f(c)}{x - c} \tag{1}$$

defining $f'(c)$ is two-sided. This means that the right-hand and left-hand limits both exist at $x = c$, and both equal $f'(c)$.

Let's examine these limits separately. To the right of c, $f(x) \leq f(c)$ so that $f(x) - f(c) \leq 0$. Also $x - c > 0$ so that $(f(x) - f(c))/(x - c) \leq 0$. Thus,

$$\lim_{x \to c+} \frac{f(x) - f(c)}{x - c} \leq 0. \tag{2}$$

Similarly, to the left of c, $f(x) - f(c) \leq 0$ and $x - c < 0$. Thus, $(f(x) - f(c))/(x - c) \geq 0$ and

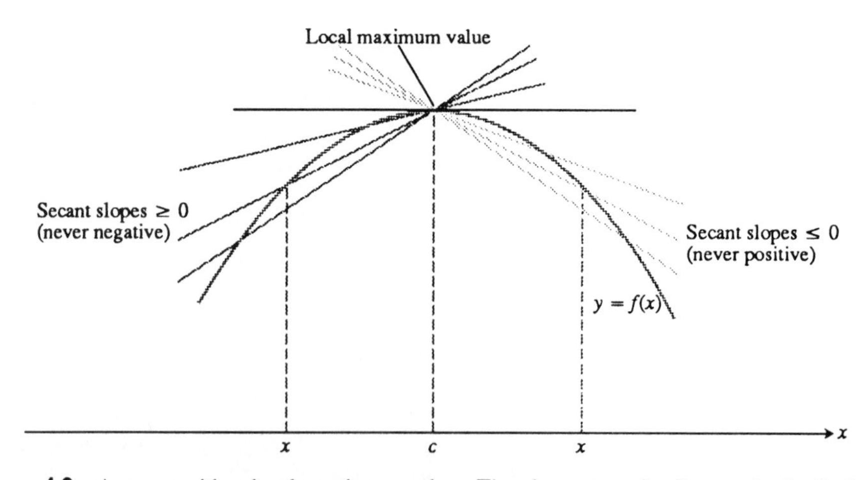

4.2 A curve with a local maximum value. The slope at c, simultaneously the limit of nonpositive numbers and nonnegative numbers, is zero.

$$\lim_{x \to c^-} \frac{f(x) - f(c)}{x - c} \geq 0. \tag{3}$$

The inequality in (2) says that $f'(c)$ cannot be greater than zero, whereas (3) says that $f'(c)$ cannot be less than zero. So $f'(c) = 0$.

This proves the theorem for local maximum values. To prove it for local minimum values, simply replace f by $-f$ and run through the argument again. ∎

Rolle's Theorem

There is strong geometric evidence that between any two points where a smooth curve crosses the x-axis there is a point on the curve where the tangent is horizontal. A 300-year-old theorem of Michel Rolle (1652–1719) assures us that this is indeed the case.

When the French mathematician Michel Rolle published his theorem in 1691, his goal was to show that between every two zeros of a polynomial function there always lies a zero of the polynomial we now know to be the function's derivative. (The version of the theorem we have proved here is not restricted to polynomials.)

Rolle distrusted the new methods of calculus, however, and spent a great deal of time and energy denouncing its use and attacking l'Hôpital's all too popular (he felt) calculus book. It is ironic that Rolle is known today only for his inadvertent contribution to a field he tried to suppress.

Theorem 2

Rolle's Theorem

Suppose that $y = f(x)$ is continuous at every point of the closed interval $[a, b]$ and differentiable at every point of its interior (a, b). If

$$f(a) = f(b) = 0,$$

then there is at least one number c between a and b at which

$$f'(c) = 0.$$

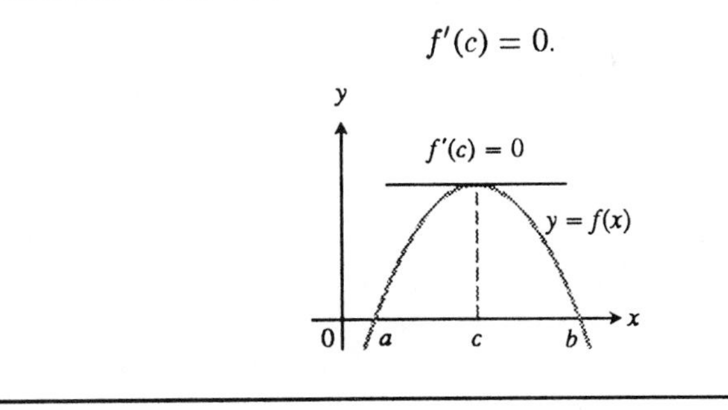

Proof of Rolle's Theorem We know from Section 2.2 that a continuous function defined on a closed interval assumes absolute maximum and minimum values on the interval. The question is, where? Theorem 1 tells us there are only three kinds of places to look:

1. At interior points where f' is zero,
2. At interior points where f' does not exist,
3. At the endpoints of the function's domain, in this case a and b.

By hypothesis, f has a derivative at every interior point. That rules out (2), leaving us with interior points where $f' = 0$ and with the two endpoints a and b.

If either the maximum or the minimum occurs at a point c inside the interval, then $f'(c) = 0$ by Theorem 1, and we have found a point for Rolle's theorem.

If both the maximum and the minimum occur at the endpoints a and b where f is zero, then zero is the maximum value of f as well as the minimum value of f. In other words, for every value of x,

$$0 = \min(f) \leq f(x) \leq \max(f) = 0.$$

So f is zero throughout the interval, and because f has a constant value, its derivative is zero throughout the interval. In other words, $f'(c) = 0$ at every interior point. Either way, we find a point c in (a, b) where $f'(c)$ is zero. This concludes the proof. ∎

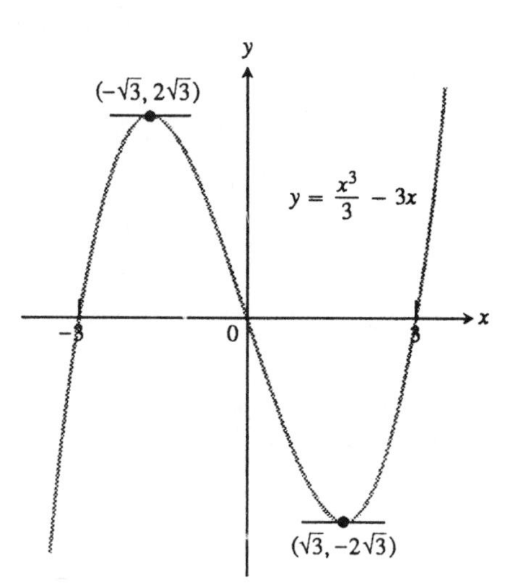

4.3 As predicted by Rolle's theorem, this smooth curve has horizontal tangents between the points where it crosses the x-axis.

EXAMPLE 1 The polynomial function

$$f(x) = \frac{x^3}{3} - 3x$$

graphed in Fig. 4.3 is continuous at every point of the interval $-3 \leq x \leq 3$ and differentiable at every point of the interval $-3 < x < 3$. Since $f(-3) = f(3) = 0$, Rolle's theorem says that f' must be zero at least once in the open interval between $a = -3$ and $b = 3$. In fact, $f'(x) = x^2 - 3$ is zero twice in this interval, once at $x = -\sqrt{3}$ and again at $x = \sqrt{3}$. ∎

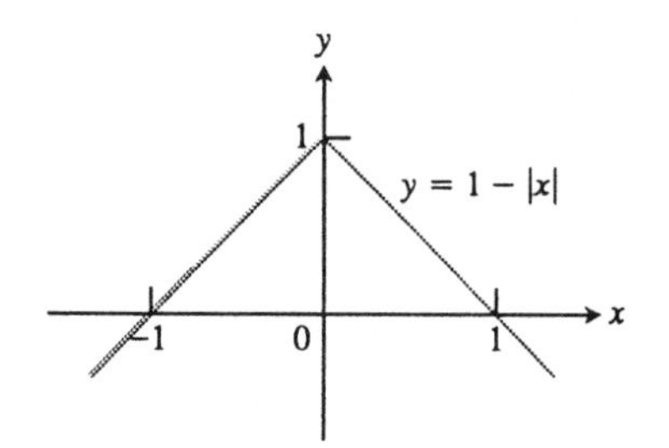

4.4 This curve has no horizontal tangent between the points where it crosses the x-axis.

EXAMPLE 2 As the function $f(x) = 1 - |x|$ shows (Fig. 4.4), the differentiability of f is essential to Rolle's theorem. If we allow even one interior point in (a, b) where f is not differentiable, there may be no horizontal tangent to the curve. ∎

Finding Solutions of Equations

When we solve equations graphically or numerically, it helps to know how many solutions there are in a given interval $[a, b]$. With Rolle's theorem we can sometimes find out.

Suppose, for example, that

1. f is continuous on $[a, b]$ and differentiable on (a, b),
2. $f(a)$ and $f(b)$ have opposite signs,
3. $f' \neq 0$ between a and b.

Then f has exactly one zero between a and b: It cannot have more than one because f' would then have a zero too. Yet it has at least one, by the Intermediate Value Theorem of Section 2.2.

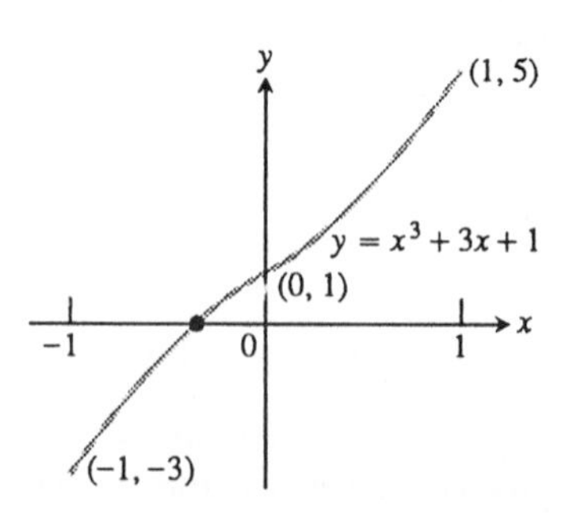

4.5 The only real zero of the polynomial $y = x^3 + 3x + 1$ is the one shown here between -1 and 0.

EXAMPLE 3 The function $f(x) = x^3 + 3x + 1$ is continuous and differentiable on the interval $-1 \leq x \leq 1$, $f(-1) = -3$ and $f(1) = 5$ have opposite signs, and $f'(x) = 3x^2 + 3$ is never zero. Therefore the equation $x^3 + 3x + 1 = 0$ has only one real solution in the interval $-1 \leq x \leq 1$ (Fig. 4.5).

In this case we can also tell that f has only one real zero, period. If f had more than one, f' would have a zero, and it doesn't. ∎

The Mean Value Theorem

The Mean Value Theorem is Rolle's theorem for a chord instead of an interval. You will see what we mean if you look at Fig. 4.6. The figure shows the graph of a differentiable function f defined on an interval $a \le x \le b$. There is a point on the curve where the tangent is parallel to the chord AB.

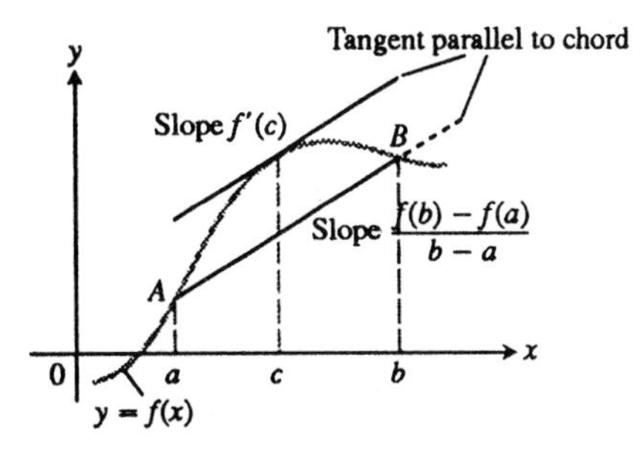

4.6 Geometrically, the Mean Value Theorem says that somewhere between A and B the curve has at least one tangent parallel to chord AB.

In Rolle's theorem, the line AB is the x-axis and $f'(c) = 0$. Here the line AB is a chord joining the endpoints of the curve above a and b, and $f'(c)$ is the slope of the chord.

Theorem 3

The Mean Value Theorem

If $y = f(x)$ is continuous at every point of the closed interval $[a, b]$ and differentiable at every point of its interior (a, b), then there is at least one number c between a and b at which

$$\frac{f(b) - f(a)}{b - a} = f'(c). \tag{4}$$

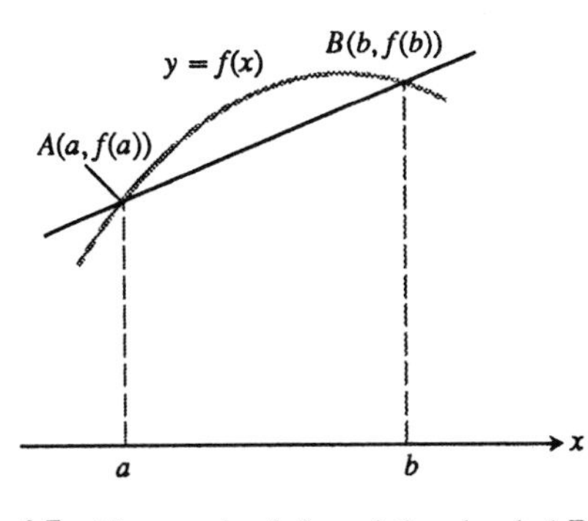

4.7 The graph of f, and the chord AB over the interval $a \le x \le b$.

Proof If we graph f over $[a, b]$ and draw the line through the endpoints $A(a, f(a))$ and $B(b, f(b))$, the figure we get, Fig. 4.7, resembles the one we drew for Rolle's theorem. The difference is that the line AB need not be the x-axis because $f(a)$ and $f(b)$ may not be zero. We cannot apply Rolle's theorem directly to f, but we can apply it to the function that measures the vertical distance between the graph of f and the line AB. This, it turns out, will tell us what we want to know about the derivative of f.

The line AB is the graph of the function

$$g(x) = f(a) + \frac{f(b) - f(a)}{b - a}(x - a) \tag{5}$$

(point–slope equation), and the formula for the vertical distance between the graphs of f and g at x is

$$d(x) = f(x) - g(x) = f(x) - f(a) - \frac{f(b) - f(a)}{b - a}(x - a). \tag{6}$$

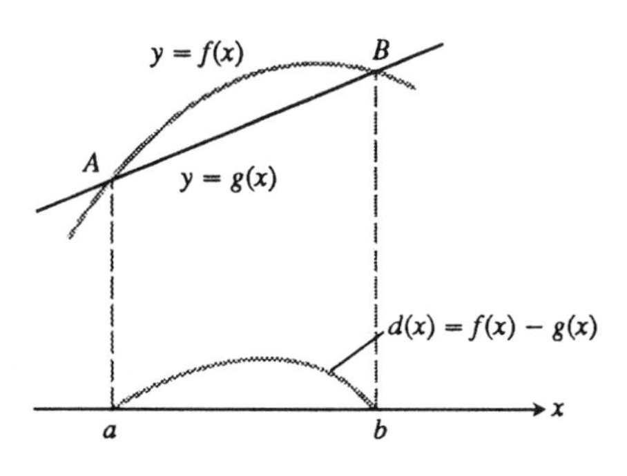

4.8 The chord AB in Fig. 4.17 is the graph of the function $g(x)$. The function $d(x) = f(x) - g(x)$ gives the vertical distance between the graphs of f and g at x.

Figure 4.8 shows the graphs of f, g, and d together.

The function d satisfies the hypotheses of Rolle's theorem on the interval $[a, b]$. It is continuous on $[a, b]$ because f and g are. Both $d(a)$ and $d(b)$ are zero because the graphs of f and g pass through A and B.

Therefore $d' = 0$ at some point c between a and b. To see what this says about f', we differentiate both sides of Eq. (6) with respect to x and set $x = c$. This gives

$$d'(x) = f'(x) - \frac{f(b) - f(a)}{b - a}, \quad \text{Derivative of Eq. (6) ...}$$

$$d'(c) = f'(c) - \frac{f(b) - f(a)}{b - a}, \quad \text{... with } x = c$$

$$0 = f'(c) - \frac{f(b) - f(a)}{b - a}, \quad d'(c) = 0$$

$$f'(c) = \frac{f(b) - f(a)}{b - a}, \quad \text{Rearranged}$$

which is what we set out to prove.

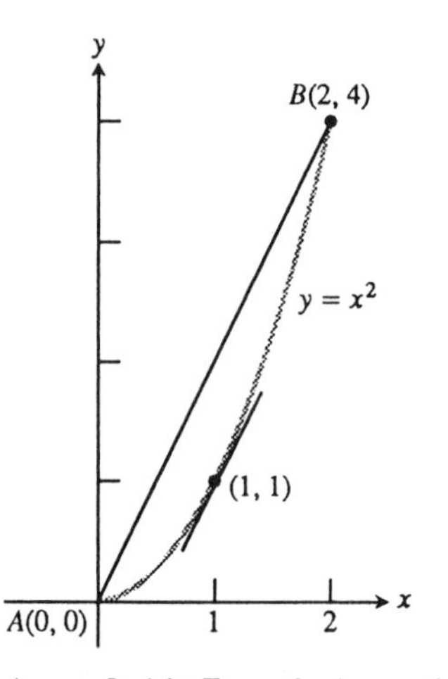

4.9 As we find in Example 4, $x = 1$ is where the tangent is parallel to the chord.

If $f'(x)$ is continuous on $[a, b]$, then the Max–Min Theorem for continuous functions in Section 2.2 tells us that f' has an absolute maximum value max f' and an absolute minimum value min f' on the interval. Since the number $f'(c)$ can neither exceed max f' nor be less than min f', the equation

$$\frac{f(b) - f(a)}{b - a} = f'(c) \tag{7}$$

gives us the inequality

$$\min f' \le \frac{f(b) - f(a)}{b - a} \le \max f'. \tag{8}$$

The importance of the Mean Value Theorem lies in the estimates that sometimes come from Eq. (8) and in the mathematical conclusions that come from Eq. (7), one of which we shall see in a moment.

We usually do not know any more about the number c than the theorem tells us, which is that c exists. In a few cases we can satisfy our curiosity about the identity of c, as in the next example. Keep in mind, however, that our ability to identify c is the exception rather than the rule, and the importance of the Mean Theorem lies elsewhere.

EXAMPLE 4 The function $f(x) = x^2$ (Fig. 4.9) is continuous for $0 \le x \le 2$ and differentiable for $0 < x < 2$. Since $f(0) = 0$ and $f(2) = 4$, the Mean Value Theorem says that at some point c in the interval the derivative $f'(x) = 2x$ must have the value $(4 - 0)/(2 - 0) = 2$. In this (exceptional) case we can identify c by solving the equation $2c = 2$ to get $c = 1$.

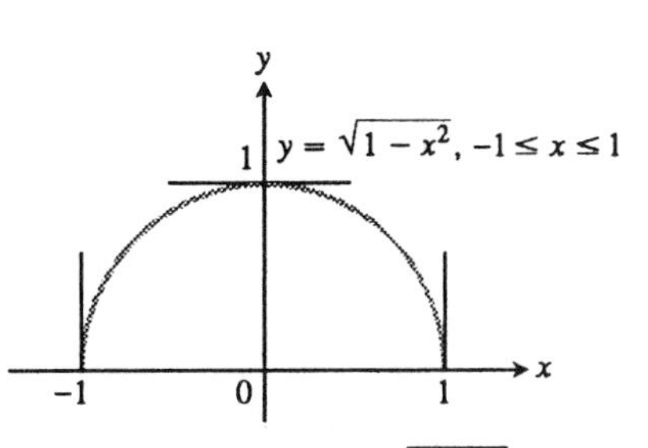

4.10 The function $y = \sqrt{1 - x^2}$ satisfies the hypotheses (and conclusion) of the Mean Value Theorem on the interval $[-1, 1]$ despite the presence of vertical tangents at $x = 1$ and $x = -1$.

EXAMPLE 5 The function $y = \sqrt{1 - x^2}$ (Fig. 4.10) satisfies the hypotheses (and conclusion) of the Mean Value Theorem on the interval $-1 \le x \le 1$. It is continuous on the closed interval, and its derivative

$$y' = \frac{-x}{\sqrt{1 - x^2}}$$

is defined at every interior point. The graph has a horizontal tangent at $x = 0$.

Notice that the function is not differentiable at $x = -1$ and $x = 1$. It does not need to be for the theorem to apply. ▪

Physical Interpretations

If we think of $(f(b) - f(a))/(b - a)$ as the average change in f over $[a, b]$ and $f'(c)$ as an instantaneous change, then the Mean Value Theorem says that the instantaneous change at some interior point must equal the average change over the entire interval.

EXAMPLE 6 If a car takes 8 sec to drive 352 ft, its average velocity for the 8-sec interval is 352/8 = 44 ft/sec, or 30 mph. At some point during the acceleration, the theorem says, the speedometer must read exactly 30.

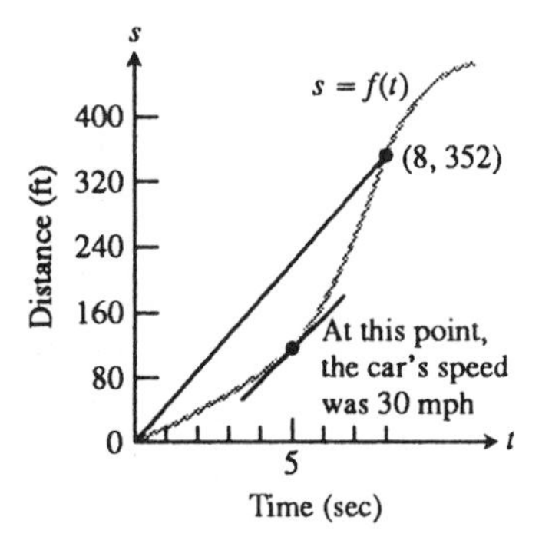

▪

The First Corollary

The Mean Value Theorem is famous for, among other things, three important corollaries. The first, which we shall get to in a moment, says exactly when graphs rise and fall. The second, which we shall come to in Section 4.7, says that only constant functions can have zero derivatives. The third, also in Section 4.7, says that functions with identical derivatives must differ at most by a constant value.

You may have noticed that differentiable functions all seem to increase when their derivatives are positive and to decrease when their derivatives are negative. The corollary we are about to prove says that this is always true. To prove the corollary, we need precise definitions of *increasing* and *decreasing*.

Definitions

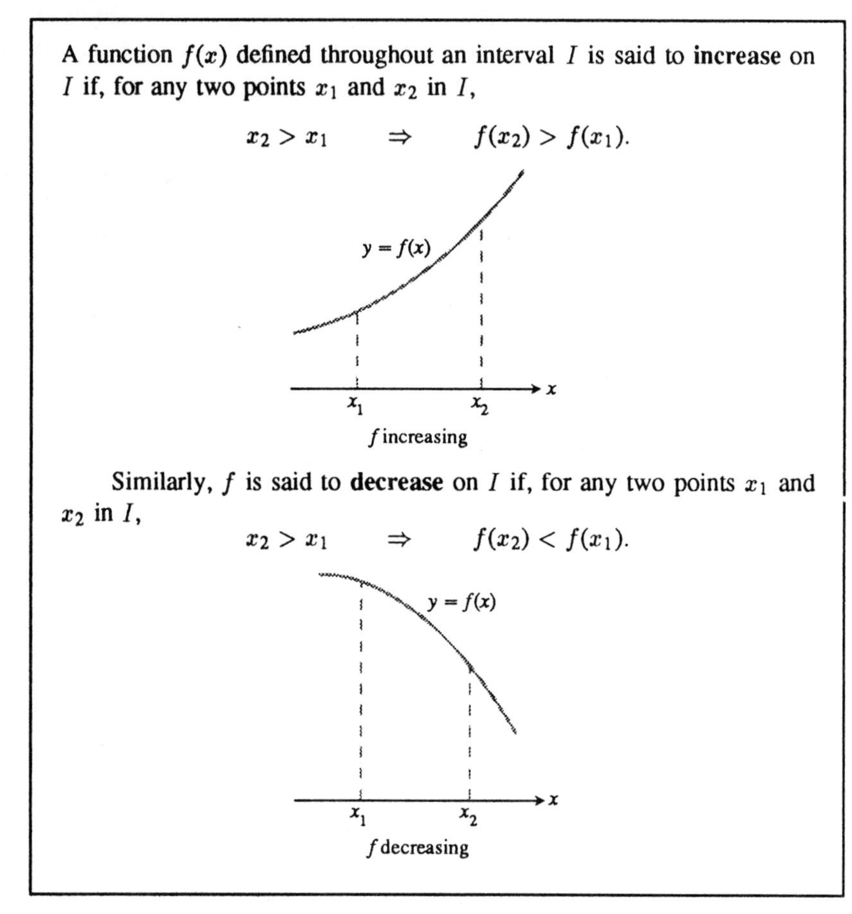

A function $f(x)$ defined throughout an interval I is said to **increase** on I if, for any two points x_1 and x_2 in I,

$$x_2 > x_1 \quad \Rightarrow \quad f(x_2) > f(x_1).$$

f increasing

Similarly, f is said to **decrease** on I if, for any two points x_1 and x_2 in I,

$$x_2 > x_1 \quad \Rightarrow \quad f(x_2) < f(x_1).$$

f decreasing

Explore with a Grapher

1. Determine complete graphs of $f(x) = \dfrac{x^3}{3} - 3x$ and $f'(x) = x^2 - 3$ in the same viewing rectangle.
2. Estimate where f is increasing and where it is decreasing.
3. Estimate where $f'(x)$ is positive and where $f'(x)$ is negative.
4. What might you conjecture about the connection between increasing and decreasing behavior of f and the sign of f'?

Corollary 1

> **The First Derivative Test for Increasing and Decreasing:** f increases when $f' > 0$ and decreases when $f' < 0$.
>
> Suppose that f is continuous at each point of the closed interval $[a, b]$ and differentiable at each point of its interior (a, b). If $f' > 0$ at each point of (a, b), then f increases throughout $[a, b]$. If $f' < 0$ at each point of (a, b), then f decreases throughout $[a, b]$. In either case, f is one-to-one on the interval $[a, b]$.

Proof Let x_1 and x_2 be any two numbers in $[a, b]$ with $x_1 < x_2$. Apply the Mean Value Theorem to f on $[x_1, x_2]$:

$$f(x_2) - f(x_1) = f'(c)(x_2 - x_1) \tag{9}$$

for some c between x_1 and x_2. The sign of the right-hand side of Eq. 9 is the same as the sign of $f'(c)$ because $x_2 - x_1$ is positive. Therefore

$$f(x_2) > f(x_1) \qquad \text{if } f'(x) \text{ is positive on } (a, b)$$

(f is increasing) and

$$f(x_2) < f(x_1) \qquad \text{if } f'(x) \text{ is negative on } (a, b)$$

(f is decreasing). In either case, $x_1 \neq x_2$ implies that $f(x_1) \neq f(x_2)$, so f is one-to-one. ∎

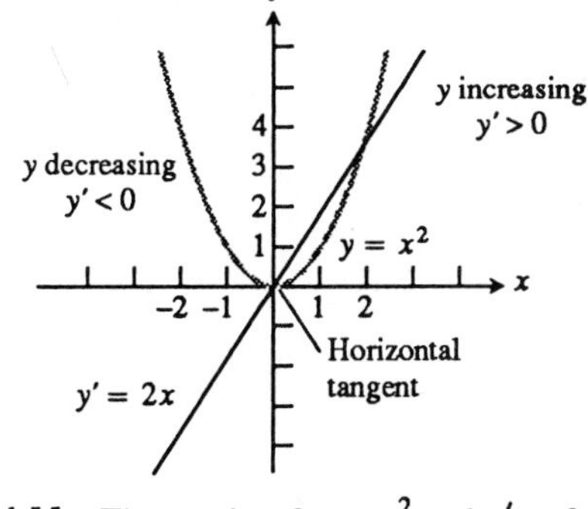

4.11 The graphs of $y = x^2$ and $y' = 2x$.

Notice that we were also able to show that a function has to be one-to-one on any interval where its derivative is positive or its derivative is negative. Knowing this will pay off later on, in Chapter 7.

EXAMPLE 7 The function $y = x^2$ decreases on $(-\infty, 0)$, where the derivative $y' = 2x$ is negative, and increases on $(0, \infty)$, where the derivative is positive. In between, $y' = 0$ and the tangent to the curve is horizontal (Fig. 4.11). ∎

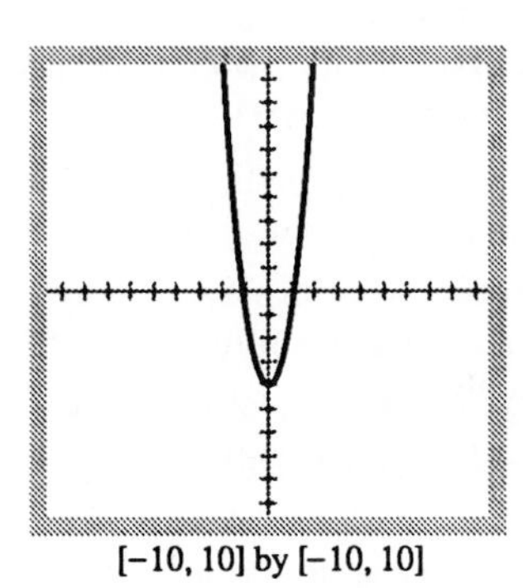

[-10, 10] by [-10, 10]

4.12 A complete graph of $f'(x) = 3x^2 - 4$.

EXAMPLE 8 Let $f(x) = x^3 - 4x$. Determine (a) the intervals on which f is increasing and the intervals on which f is decreasing, (b) the local extrema of f, and (c) the absolute maximum and absolute minimum of f on $[0, 3]$.

Solution

a) The domain of both f and $f'(x) = 3x^2 - 4$ is $(-\infty, \infty)$. Thus, the only possible local extrema of f occur where $f'(x) = 0$. We can see from the graph of f' in Fig. 4.12 that $f'(x) > 0$ for both $x < -\sqrt{4/3} = 1.15$ and $x > \sqrt{4/3} = 1.15$, and $f'(x) < 0$ for $-1.15 < x < 1.15$. Thus f increases in $(-\infty, -1.15)$, decreases in $(-1.15, 1.15)$, and increases in $(1.15, \infty)$.

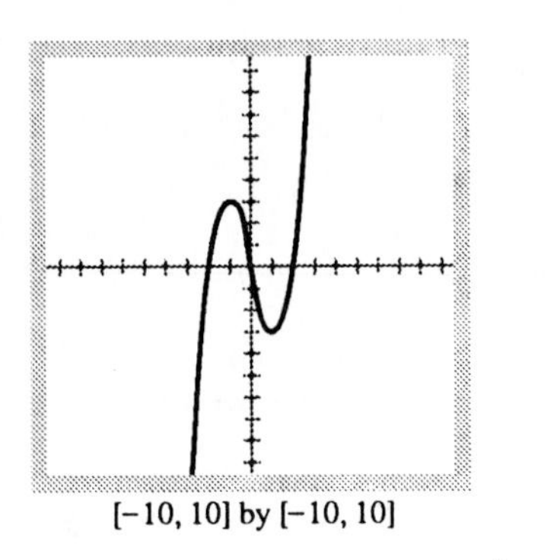

[−10, 10] by [−10, 10]

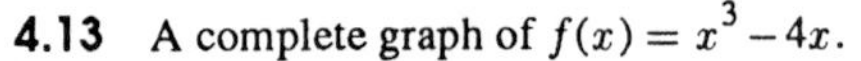

4.13 A complete graph of $f(x) = x^3 - 4x$.

b) The increasing–decreasing behavior of f determined in (a) means that f has a local maximum of 3.08 at $x = -1.15$ and a local minimum of -3.08 at $x = 1.15$ (Fig. 4.13). In the next section we will formalize an algebraic procedure for finding local extrema. We can use zoom-in to support the reasonableness of these values.

c) The value of f at $x = 0$ is 0 and the value of f at $x = 3$ is 15. So on $[0, 3]$, f has an absolute minimum of -3.08 at $x = 1.15$, a local minimum of 0 at $x = 0$, and an absolute maximum of 15 at $x = 3$ (Fig. 4.14).

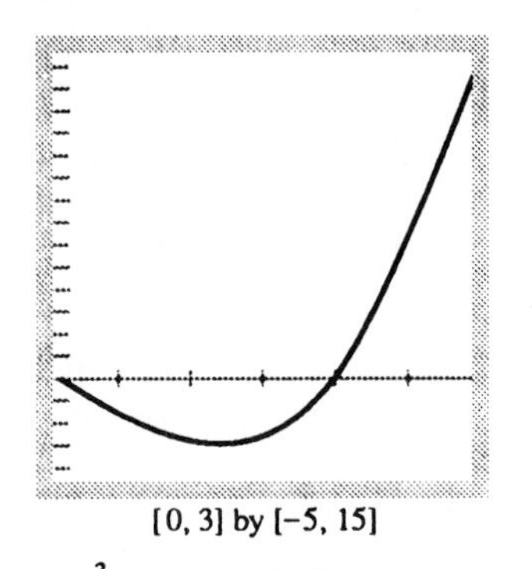

[0, 3] by [−5, 15]

4.14 The graph of $f(x) = x^3 - 4x$ on $[0, 3]$.

Exercises 4.1

Rolle's Theorem. Show that the equations in Exercises 1–4 have exactly one solution in the given interval.

1. $x^4 + 3x + 1 = 0,\ -2 \le x \le -1$
2. $-x^3 - 3x + 1 = 0,\ 0 \le x \le 1$
3. $x - \dfrac{2}{x} = 0,\ 1 \le x \le 3$
4. $2x - \cos x = 0,\ -\pi \le x \le \pi$
5. a) Draw a complete graph of and plot the zeros of each polynomial on a line together with the zeros of its first derivative:
   i) $y = x^2 + 8x + 15$ ii) $y = x^3 - 3x^2 + 4$
   iii) $y = x^3 - 33x^2 + 216x$ iv) $y = x^3 - 4x + 2$
   What pattern do you see?

   b) Use Rolle's theorem to prove that between every two zeros of the polynomial $x^n + a_{n-1}x^{n-1} + \cdots + a_1x + a_0$ there lies a zero of the polynomial
   $$nx^{n-1} + (n-1)a_{n-1}x^{n-2} + \cdots + a_1.$$
6. The function
   $$y = f(x) = \begin{cases} x & \text{if } 0 \le x < 1, \\ 0 & \text{if } x = 1, \end{cases}$$
   is zero at $x = 0$ and at $x = 1$. Its derivative, $y' = 1$, is different from zero at every point between 0 and 1. Why doesn't that contradict Rolle's theorem?
7. a) Draw a complete graph of and plot the zeros of $y = \sin x$ on a line together with the zeros of its first derivative.

   b) Use Rolle's theorem to prove that between every two zeros of $y = \sin x$ there is a zero of $y = \cos x$.
8. a) Draw a complete graph of and plot the zeros of $y = (x^2 - 4x + 2)/(x + 1)$ on a line together with the zeros of its first derivative.

   b) Can you use Rolle's theorem to prove the existence of each zero of the derivative of y?
9. Let $f(x) = |x^3 - 9x|$.

   a) Determine a complete graph of f.

   b) Does $f'(0)$ exist? Explain.

c) Does $f'(-3)$ exist? Explain.

d) Determine all local extrema of f.

e) Are (b)–(d) in conflict with Theorem 1? Explain.

f) Determine the intervals on which f is increasing and the intervals on which f is decreasing?

10. Let $g(x) = (x-2)^{2/3}$.

a) Determine a complete graph of g.

b) Does $g'(2)$ exist? Explain.

c) Determine all local extrema of g.

d) Are (b) and (c) in conflict with Theorem 1? Explain.

e) Determine the intervals on which g is increasing and the intervals on which g is decreasing.

In Exercises 11–18, determine the local extrema of the function, the intervals on which the function is increasing and the intervals on which the function is decreasing, and the absolute maximum and absolute minimum of the function on the specified interval.

11. $f(x) = 5x - x^2$, $[0, 6]$

12. $f(x) = x^2 - x - 12$, $[-4, 4]$

13. $f(x) = \sqrt{x-4}$, $[0, 7]$

14. $f(x) = 4 - \sqrt{x+2}$, $[-2, 5]$

15. $y(x) = x^4 - 10x^2 + 9$, $[-3, 3]$

16. $g(x) = -x^4 + 5x^2 - 4$, $[-3, 3]$

17. $h(x) = x^4 - x^3$, $[-1, 1]$

18. $h(x) = x^3 - x^2$, $[-1, 1]$

The Mean Value Theorem. Find the value or values of c that satisfy the equation

$$\frac{f(b) - f(a)}{b - a} = f'(c)$$

in the conclusion of the Mean Value Theorem for the functions and intervals in Exercises 19–21.

19. $f(x) = x^2 + 2x - 1$, $0 \le x \le 1$

20. $f(x) = x^{2/3}$, $0 \le x \le 1$

21. $f(x) = x + \dfrac{1}{x}$, $\dfrac{1}{2} \le x \le 2$

22. $f(x) = \sqrt{x-1}$, $1 \le x \le 3$

23. SPEEDING A trucker handed in a ticket at a toll booth, showing that in 2 hrs the truck had covered 159 mi on a toll road on which the speed limit was 65 mph. The trucker was cited for speeding. Why?

24. TEMPERATURE CHANGE It took 20 sec for a thermometer to rise from 10°F to 212°F when it was taken from a freezer and placed in boiling water. Show that somewhere along the way the mercury was rising at exactly 10.1°F/sec.

25. TRIREMES Classical accounts tell us that a 170-oar trireme once covered 184 sea miles in 24 hrs. Show that at some point during this feat the trireme's speed exceeded 7.5 knot.

26. Suppose that the derivative of a differentiable function $f(x)$ is never zero on the interval $0 \le x \le 1$. Show that $f(0) \ne f(1)$.

27. Show that for any numbers a and b

$$|\sin b - \sin a| \le |b - a|.$$

28. Suppose that f is differentiable for $a \le x \le b$ and that $f(b) < f(a)$. Show that f' is negative at some point between a and b.

29. Show that $y = 1/x$ decreases on any interval on which it is defined.

30. Show that $y = 1/x^2$ increases on any interval to the left of the origin and decreases on any interval to the right of the origin.

Make the estimates in Exercises 31 and 32 by applying the inequality

$$\min f' \le \frac{f(b) - f(a)}{b - a} \le \max f'.$$

31. Suppose that $f'(x) = 1/(1 + x^4 \cos x)$ for $0 \le x \le 0.1$ and that $f(0) = 1$. Estimate $f(0.1)$.

32. Suppose that $f'(x) = 1/(1 - x^4)$ for $0 \le x \le 0.1$ and that $f(0) = 2$. Estimate $f(0.1)$.

33. Suppose that $f(0) = 3$ and that $f'(x) = 0$ for all x. Use the Mean Value Theorem to show that $f(x)$ must be 3 for all x.

34. Suppose that $f'(x) = 2$ and that $f(0) = 5$. Use the Mean Value Theorem to show that $f(x) = 2x + 5$ at every value of x.

4.2 Predicting Hidden Behavior

Computer-drawn graphs of most functions that appear in this textbook are usually very reliable. In this section we will see how to use calculus to confirm completeness of graphs determined technologically, and to discover behavior hidden from view on a computer graph. The verification that the graph really looks like what's on the screen must still come from calculus. The computer can only suggest what *might* be true. We also introduce the concepts of concavity of graphs, and points where the sense of concavity changes, commonly called points of inflection of graphs. We will see that it is difficult to determine points of inflection graphically.

The First Derivative

When we know that a function has a derivative at every point of an interval, we also know that it is continuous throughout the interval (Section 3.1) and that its graph over the interval is connected (Section 2.2). Thus the graphs of $y = \sin x$ and $y = \cos x$ remain unbroken however far extended, as do the graphs of polynomials. The graphs of $y = \tan x$ and $y = 1/x^2$ break only at points where the functions are undefined. On every interval that avoids these points, the functions are differentiable, so they are continuous and have connected graphs.

We gain additional information about the shape of a function's graph when we know where the function's first derivative is positive, negative, or zero. For, as we saw in Section 4.1, this tells us where the graph is rising or falling or has a horizontal tangent (Fig. 4.15.).

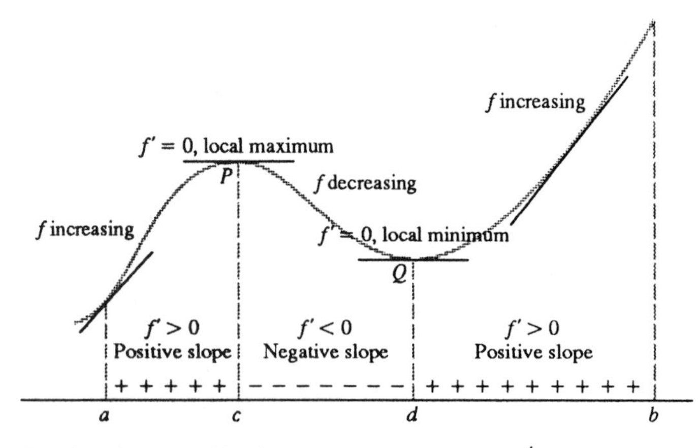

4.15 The function $y = f(x)$ increases on (a, c) where $f' > 0$, decreases on (c, d) where $f' < 0$, and increases again on (d, b). The transitions are marked by horizontal tangents.

There are two things to watch out for here, however. A curve may have a horizontal tangent without having a local maximum or minimum, and a curve may have a local maximum or minimum without having a horizontal tangent.

EXAMPLE 1 The curves $y = x^3$ and $y = -x^3$ have horizontal tangents at the origin without having maxima or minima there.

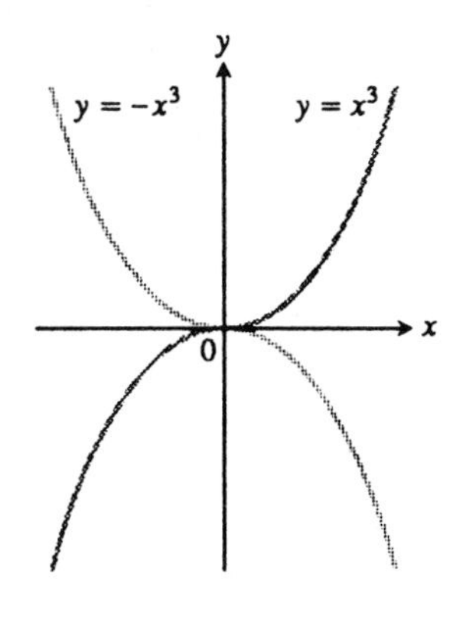

■

EXAMPLE 2 The function $y = |x|$ takes on a minimum value at $x = 0$ without having a horizontal tangent there.

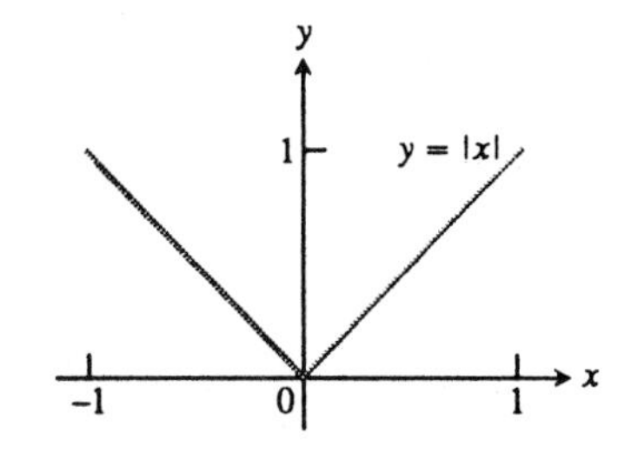

■

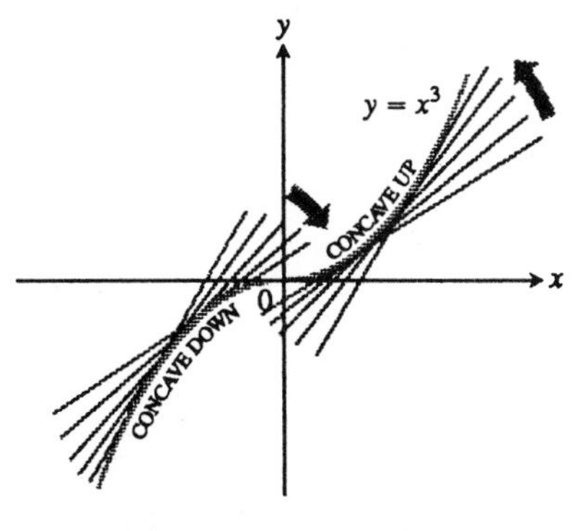

4.16 The graph of $y = x^3$ is concave down on the left, concave up on the right.

Concavity

As you can see in Fig. 4.16, the function $y = x^3$ increases as x increases, but the portions of the curve defined on the intervals $(-\infty, 0)$ and $(0, \infty)$ turn in different ways. If we come in from the left toward the origin along the curve, the curve turns to our right and falls below its tangents. As we leave the origin, the curve turns to our left and rises above its tangents.

To put it another way, the slopes of the tangents to the curve $y = x^3$ decrease as we approach the origin from the left and increase as we move from the origin into the first quadrant.

We say that the curve $y = x^3$ is concave down on the interval $(-\infty, 0)$, where y' decreases, and concave up on the interval $(0, \infty)$, where y' increases.

Explore with a Grapher

1. Write an equation for the tangent line to $f(x) = x^3 - 9x$ at $(a, 0)$.
2. Graph f and its tangent line at $(a, 0)$ for several negative values of a. Near $x = a$ is the graph of f above or below its tangent line?
3. Repeat part (2) for several positive values of a.
4. Overlay the graphs of f and f'. Find the largest interval on which f' is increasing. Find the largest interval on which f' is decreasing.
5. Where is the graph of f concave up? Concave down?

Definition

The graph of a differentiable function $y = f(x)$ is **concave up** on an interval where y' is increasing and **concave down** on an interval where y' is decreasing.

If a function $y = f(x)$ has a second derivative as well as a first (as do most of the functions we deal with in this text), we can apply Corollary 1 of the Mean Value Theorem (Section 4.1) to the function $f' = y'$ to conclude that y' decreases if $y'' < 0$ and increases if $y'' > 0$. We therefore have a test that we can apply to the formula $y = f(x)$ to determine the concavity of its graph. It is called the second derivative test for concavity.

The Second Derivative Test for Concavity

The graph of $y = f(x)$ is
concave down on any interval where $y'' < 0$,
concave up on any interval where $y'' > 0$.

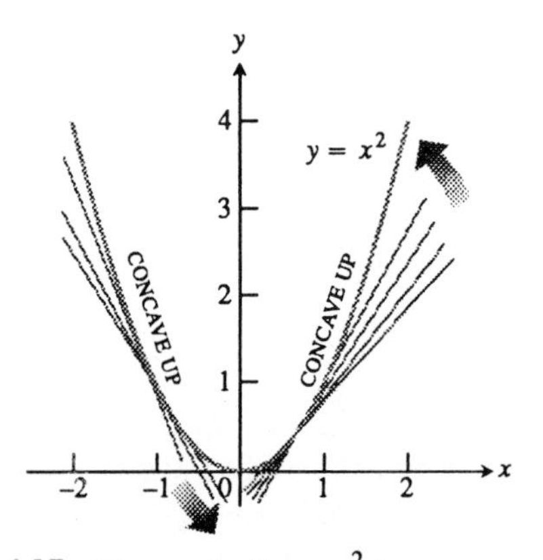

4.17 The graph of $y = x^2$ is concave up. The tangent turns counterclockwise as x increases; y' is increasing.

The idea is that if $y'' < 0$, then y' decreases as x increases and the tangent turns clockwise. Conversely, if $y'' > 0$, then y' increases as x increases and the tangent turns counterclockwise.

EXAMPLE 3 The curve $y = x^2$ (Fig. 4.17) is concave up on the entire x-axis because its second derivative $y'' = 2$ is always positive.

EXAMPLE 4 The curve $y = x^3$ in Fig. 4.16 is concave down to the left of the origin, where its second derivative $y'' = 6x$ is negative, and concave up to the right of the origin, where its second derivative $y'' = 6x$ is positive.

Points of Inflection

When we graph the position of a moving body as a function of time, we want to be able to see where the acceleration changes sign. These points are easy to find because they are the points where the curve changes concavity. In mathematics such points are called points of inflection.

Points of inflection are important in business applications. The growth of an individual company or a new product often follows a *logistics* or *life cycle* curve like the one shown in Fig. 4.18. Sales of a new product will generally grow slowly at first, then the sales will experience a rapid growth. Eventually, the growth of sales slows down again. The function f in Fig. 4.18 is increasing with its rate of increase (f') at first increasing ($f'' > 0$) up to the point of inflection and then decreasing ($f'' < 0$). This is the opposite of what happens in Fig. 4.16.

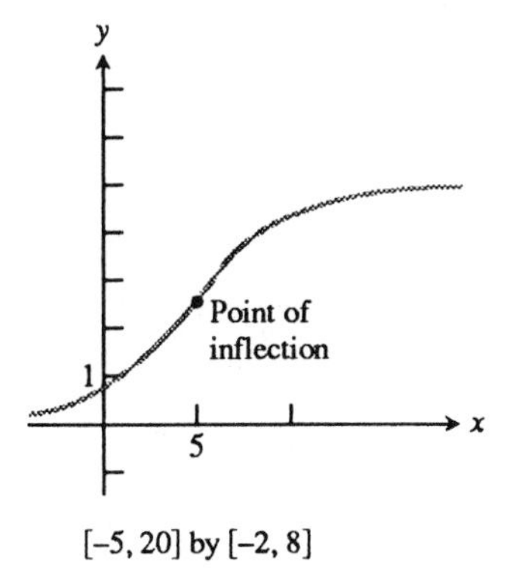

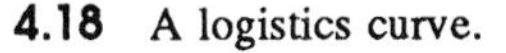

4.18 A logistics curve.

Definition

> A point on the graph of a differentiable function where the concavity changes is called a **point of inflection.**

Thus a point of inflection on a twice-differentiable curve is a point where y'' is positive on one side and negative on the other. At such a point y'' is zero because derivatives have the intermediate value property.

> At a point of inflection on the graph of a twice-differentiable function, $y'' = 0$.

EXAMPLE 5 The graph of the simple harmonic motion $y = \sin x$ shown here changes concavity at $x = 0$ and $x = \pi$, where the acceleration $y'' = -\sin x$ is zero.

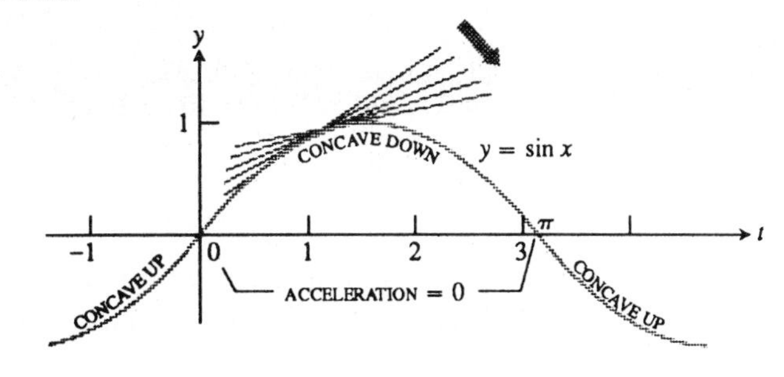

EXAMPLE 6 Inflection points have important applications in some areas of economics. Suppose that the function $y = c(x)$ in Fig. 4.19 is the total cost of producing x

units of something. The point of inflection at P is then the point at which the marginal cost (the approximate cost of producing one more unit) changes from decreasing to increasing.

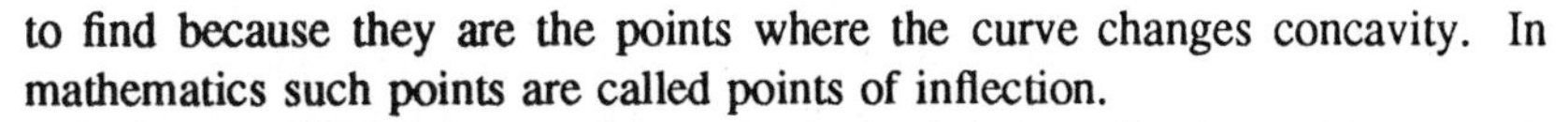

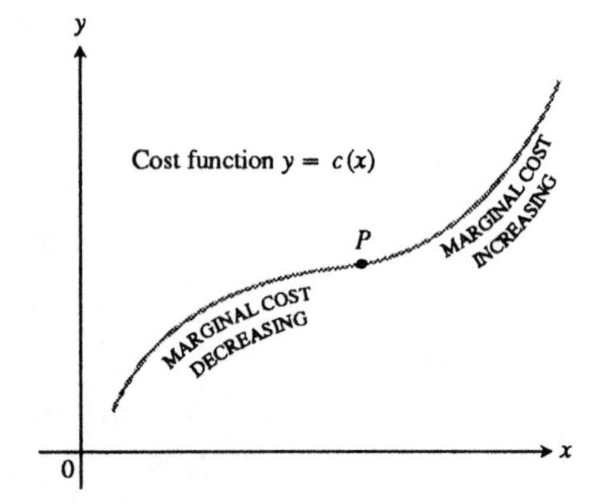

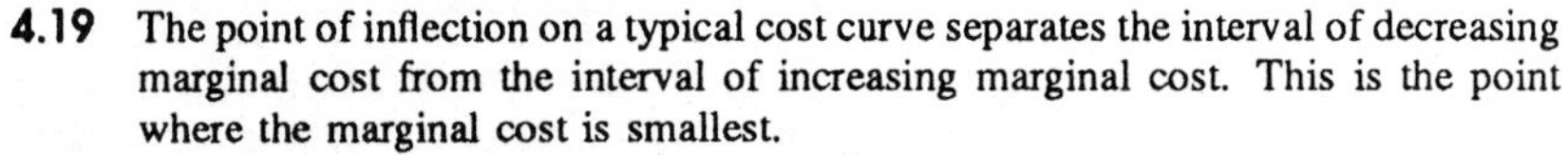

4.19 The point of inflection on a typical cost curve separates the interval of decreasing marginal cost from the interval of increasing marginal cost. This is the point where the marginal cost is smallest.

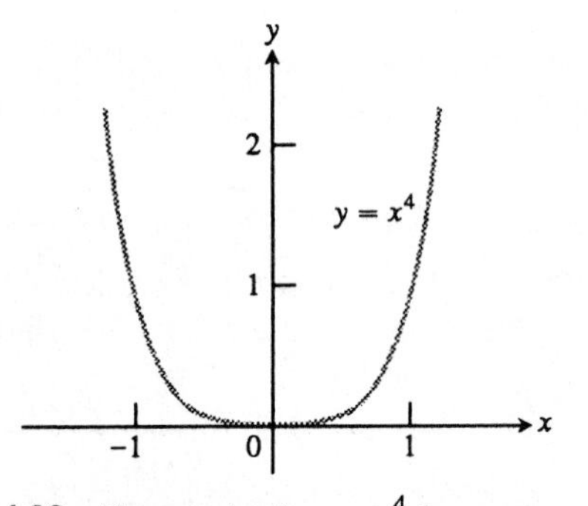

4.20 The graph of $y = x^4$ has no inflection point at the origin, even though $y''(0) = 0$.

It is possible for y'' to be zero at a point that is *not* a point of inflection, and you have to watch out for this. Also, a point of inflection may occur where y'' fails to exist.

EXAMPLE 7 *No inflection where* $y'' = 0$. The curve $y = x^4$ (Fig. 4.20) has no inflection point at $x = 0$. Even though $y'' = 12x^2$ is zero there, it does not change sign.

EXAMPLE 8 *An inflection point where* y'' *does not exist.* The curve $y = x^{1/3}$ (Fig. 4.21) has a point of inflection at $x = 0$, but y'' does not exist there. The formulas for y' and y'' are

$$y' = \frac{1}{3}x^{-2/3}, \qquad y'' = -\frac{2}{9}x^{-5/3}.$$

The curve is concave up for $x < 0$, where $y'' > 0$ and y' is increasing, and concave down for $x > 0$, where $y'' < 0$ and y' is decreasing.

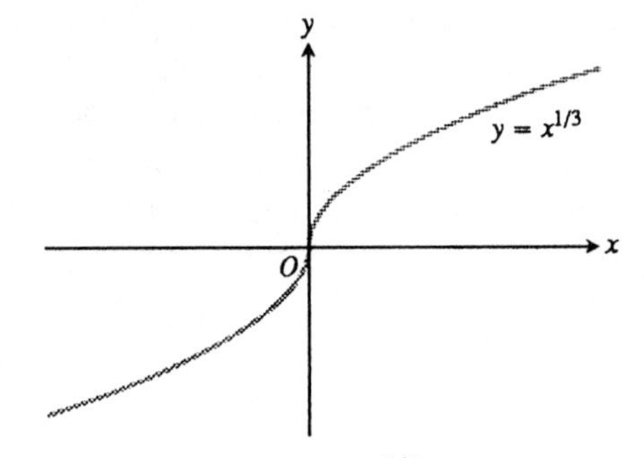

4.21 The graph of $y = x^{1/3}$ shows that a point where y'' fails to exist can be a point of inflection.

Confirming Computer-Generated Graphs

In Chapter 1 we discovered that the graph of $f(x) = x^3 - 2x^2 + x - 30$ shown in Fig. 4.22(a) has hidden behavior. Closer examination of the relatively flat portion of the graph of f reveals two extrema of f (Fig. 4.22(b)). This hidden behavior can be predicted by investigating $f'(x) = 3x^2 - 4x + 1$ (Fig. 4.22(c)). Notice that $f'(x) = (3x - 1)(x - 1)$ has two real zeros and that f' is positive for x in $(-\infty, 1/3) \cup (1, \infty)$ and negative in $(1/3, 1)$. Thus, the graph of f rises as x approaches $1/3$ from the left, falls from $x = 1/3$ to $x = 1$, and rises again to the right of $x = 1$. The sign pattern for f' tells us that the curve has a local maximum at $x = 1/3$, where f' changes from $+$ to $-$, and a local minimum at $x = 1$, where f' changes from $-$ to $+$. Determining the zeros of f' and the sign of f' helps identify local extrema of f.

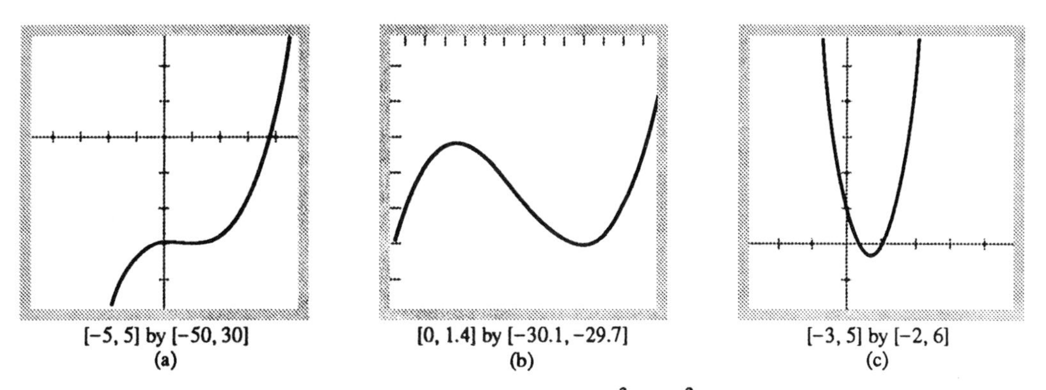

4.22 (a) & (b) Two views of $f(x) = x^3 - 2x^2 + x - 30$. (c) A graph of $f'(x) = 3x^2 - 4x + 1$.

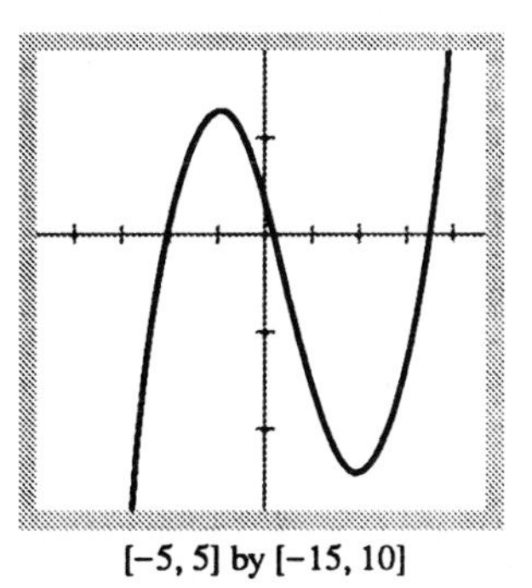

4.23 A complete graph of $f(x) = x^3 - 2x^2 - 7x + 2$.

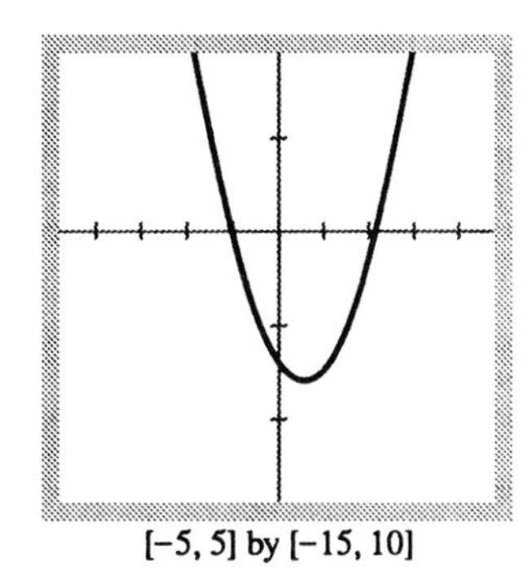

4.24 A complete graph of $f'(x) = 3x^2 - 4x - 7$.

Local Extrema

If $f'(c) = 0$ and if the sign of f' changes at c, then f has a local extrema at $x = c$:

local maximum if f' changes from $+$ to $-$,
local minimum if f' changes from $-$ to $+$,

Calculus can be used to confirm completeness of graphs generated by computer or sketched by hand.

EXAMPLE 9 Draw a complete graph of $f(x) = x^3 - 2x^2 - 7x + 2$. Use analytic calculus to confirm the complete graph and to find the inflection points and local extreme values. Identify the intervals on which the graph is rising, falling, concave up and concave down.

Solution The graph of f in Fig. 4.23 appears to be complete. Notice that

$$f'(x) = 3x^2 - 4x - 7 = (3x - 7)(x + 1)$$

so that f' is zero at $x = -1$ and $x = 7/3$. We can either estimate from the graph or use zoom-in to support the reasonableness of these exact answers. The domain of f and f' are $(-\infty, \infty)$ so that extreme values occur only where $f' = 0$.

Next observe that

$$f'(x) = (3x - 7)(x + 1) > 0 \quad \text{for} \quad (-\infty, -1) \cup (7/3, \infty)$$

$$f'(x) = (3x - 7)(x + 1) < 0 \quad \text{for} \quad (-1, 7/3)$$

which is supported by the graph of f' in Fig. 4.24.

Thus

$$f \text{ rises in } (-\infty, -1),$$

$$f \text{ falls in } (-1, 7/3), \text{ and}$$

$$f \text{ rises in } (7/3, \infty).$$

This sign pattern of f confirms that f has a local maximum of 6 at $x = -1$ and a local minimum of -12.52 at $x = 7/3$. Finally notice that it is difficult to read the coordinate of the point where the graph of f in Fig. 4.23 changes from concave down to concave up, that is, at the associated point of inflection. In Section 3.7 we observed that differentiable functions are locally linear. This means that the tangent line at a point is a good approximation to the function, so that when we use zoom-in the graph of the function appears to be a straight line. Thus, zoom-in will not help us locate the points where graphs change concavity. To find the coordinates of the point of inflection we need the tools of calculus. Notice that

$$f''(x) = 6x - 4$$

is 0 for $x = 2/3$, negative for $x < 2/3$, and positive for $x > 2/3$. Therefore, f is

$$\text{concave down in } (-\infty, 2/3)$$

$$\text{concave up in } (2/3, \infty)$$

and has a point of inflection at $x = 2/3$. The y-coordinate of the point of inflection is -3.26. This analytic analysis confirms that the graph of f in Fig. 4.23 is complete. ≡

Steps in Confirming a Complete Graph of $y = f(x)$

1. Find y' and y''.
2. Find where y' is positive, negative, and zero.
3. Find where y'' is positive, negative, and zero.
4. Compare graphical analysis of steps 1–3 with a graph. Remember that it is difficult to estimate points of inflection from graphs.
5. Correct the graph, if necessary.

The analysis of Example 9 takes on new meaning for the position function of a particle moving along a coordinate line. In Chapter 3 we used parametric graphing to analyze the motion of a particle with distance from the origin satisyfing $y = s(t) = 2t^3 - 13t^2 + 22t - 5$ where t is in seconds.

EXAMPLE 10 Describe the motion of a particle moving on the x-axis with distance from the origin given by $s(t) = 2t^3 - 13t^2 + 22t - 5$.

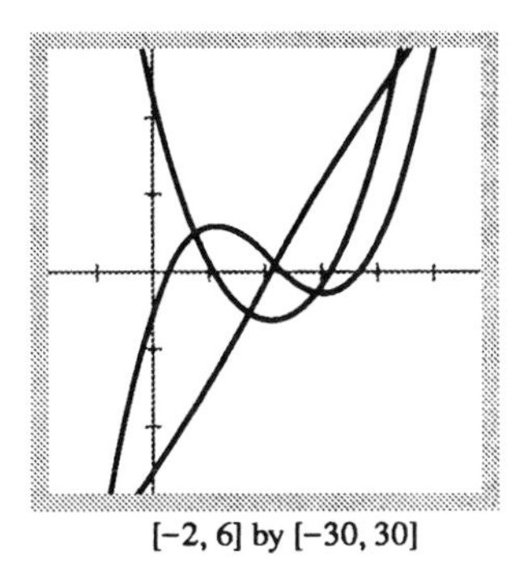

[−2, 6] by [−30, 30]

4.25 Complete graphs of $s(t) = 2t^3 - 13t^2 + 22t - 5$, $v(t) = 6t^2 - 26t + 22$, and $a(t) = 12t - 26$.

Solution The velocity and acceleration of the particle are given by

$$v(t) = s'(t) = 6t^2 - 26t + 22$$

$$a(t) = v'(t) = 12t - 26.$$

Figure 4.25 shows complete graphs of $s(t)$, $v(t)$, and $a(t)$. Notice that $a(t) = 0$ for $t = 13/6$, so $a(t) = v'(t)$ is negative for $t < 13/6$ and $a(t)$ is positive for $t > 13/6$. Thus, $v(t)$ is decreasing for $t < 13/6$ and $v(t)$ is increasing for $t > 13/6$. Also $v(t) = 0$ for $t = 1.15$ and 3.18. Thus, $v(t) = s'(t)$ is negative in $(1.15, 3.18)$ and positive in $(-\infty, 1.15) \cup (3.18, \infty)$. It follows that $s(t)$ is increasing in the intervals $(-\infty, 1.15)$ and $(3.18, \infty)$, and is decreasing in the interval $(1.15, 3.18)$. This means that the particle is moving to the right in $(-\infty, 1.15)$ and in $(3.18, \infty)$; it is moving to the left in $(1.15, 3.18)$.

The particle slows down if $v(t)$ is positive and decreasing, or if $v(t)$ is negative and increasing. Similarly, the particle speeds up if $v(t)$ is negative and decreasing, or if $v(t)$ is positive and increasing.

We can summarize the motion of the particle as follows:

Time	v(t)	Particle Motion
$t < 1.15$	positive and decreasing	slowing down, moving right
$t = 1.15$	0	stops
$1.15 < t < \frac{13}{6}$	negative and decreasing	speeding up, moving left
$t = \frac{13}{6}$	minimum	acceleration 0
$\frac{13}{6} < t < 3.18$	negative and increasing	slowing down, moving left
$t = 3.18$	0	stops
$t > 3.18$	positive and increasing	speeding up, moving right

■

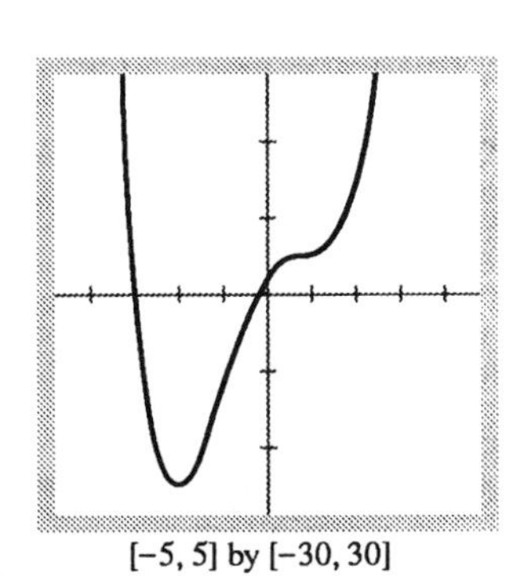

[−5, 5] by [−30, 30]

4.26 A complete graph of $f(x) = x^4 - 6x^2 + 9x + 1$.

EXAMPLE 11 Draw a complete graph of $f(x) = x^4 - 6x^2 + 9x + 1$. Find the inflection points and local extreme values. Identify the intervals on which the graph is rising, falling, concave up and concave down.

Solution We show that the graph of f in Fig. 4.26 is complete. This time we cannot use algebra to easily solve $f'(x) = 4x^3 - 12x + 9 = 0$. However, the graph of f' in Fig. 4.27 supports the completeness of the graph of f in Fig. 4.26.

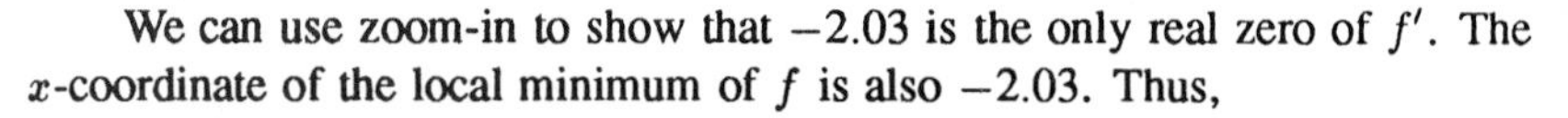

We can use zoom-in to show that -2.03 is the only real zero of f'. The x-coordinate of the local minimum of f is also -2.03. Thus,

$$f' < 0 \text{ and } f \text{ is falling in } (-\infty, -2.03), \text{ and}$$

$$f' > 0 \text{ and } f \text{ is rising in } (-2.03, \infty).$$

The local minimum of f is $f(-2.03) = -25.01$.

It appears from Fig. 4.26 that the concavity of f changes three times. If we try to use zoom-in to find these three points the local straightness of the graph of f hampers accuracy. However, we can solve

$$f''(x) = 12x^2 - 12 = 0$$

to locate the points of inflection

$$f'' > 0 \text{ and } f \text{ concave up in } (-\infty, -1),$$

$$f'' < 0 \text{ and } f \text{ concave down in } (-1, 1), \text{ and}$$

$$f'' > 0 \text{ and } f \text{ concave up in } (1, \infty).$$

Thus, the graph of f has $(-1, -13)$ and $(1, 5)$ as points of inflection.

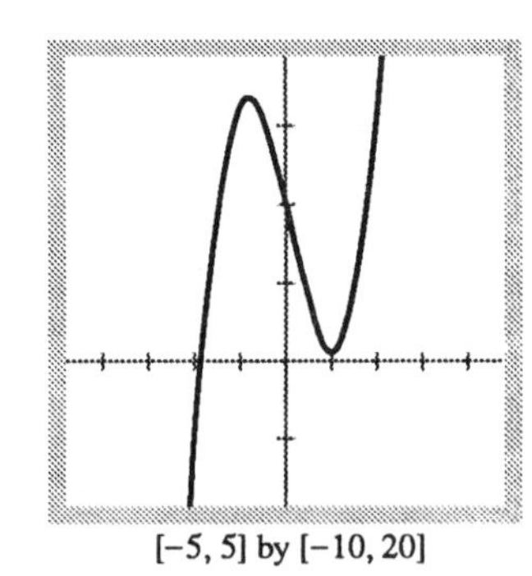

[−5, 5] by [−10, 20]

4.27 A complete graph of $f'(x) = 4x^3 - 12x + 9$.

If we encounter a function whose derivative we do not know or is very complicated, we can use the numerical derivative feature of our graphing utility.

EXAMPLE 12 Draw a complete graph of the logistics curve $f(x) = 1/(0.2 + 2^{-0.5x})$, and find any points of inflection.

Solution This is the curve shown in Fig. 4.28. First notice that $f > 0$ for all x, $\lim_{x\to-\infty} f(x) = 0$, and $\lim_{x\to\infty} f(x) = 1/0.2 = 5$. We show that the graph of f in Fig. 4.28 is complete.

Derivatives of exponential functions will be established later in the text. For now we use the numerical derivative feature to obtain the derivatives of f, all of which exist. It appears that f is always increasing which is supported by the fact that the graph of f' in Fig. 4.29(a) lies above the x-axis. We can also see that the rate of increase of f, namely f', increases up to a local maximum and then decreases, a characteristic feature of logistic curves. Of course, f'' is zero at the local maximum of f', and we can argue that f has a point of inflection at this value of x either from the graph of f' or the graph of f'' in Fig. 4.29(b). Zoom-in on the graph of f'' can be used to show that (4.64, 2.5) is the point of inflection of the graph of f.

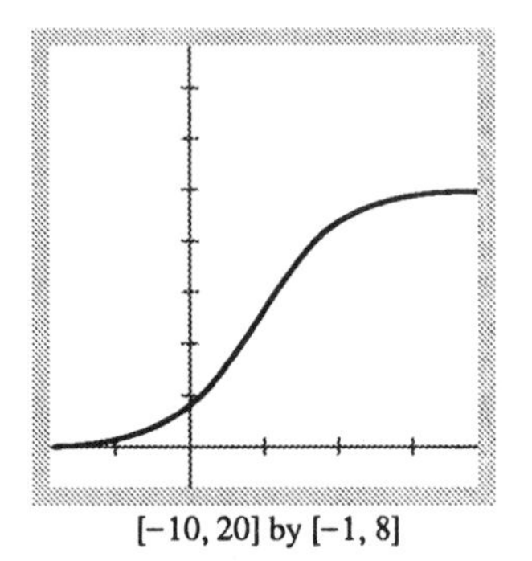

[−10, 20] by [−1, 8]

4.28 A complete graph of the logistics curve $f(x) = 1/(0.2 + 2^{-0.5x})$.

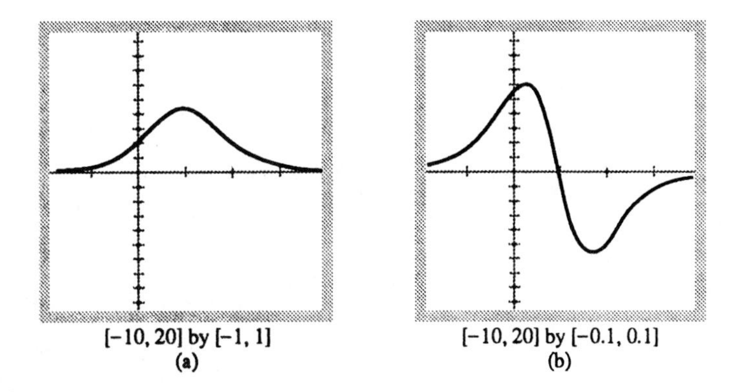

4.29 (a) A graph of $y = \text{NDer}(f, x)$ and (b) $y = \text{N}_2\text{NDer}(f, x)$ where $f(x) = (0.2 + 2^{-0.5x})^{-1}$.

A Useful Shortcut—The Second Derivative Test for Local Maxima and Minima

Instead of looking at how the sign of f' changes at a point where $f' = 0$, we can often use the following test to determine whether there is a local maximum or minimum at the point.

> **The Second Derivative Test for Local Maxima and Minima**
>
> If $f'(c) = 0$ and $f''(c) < 0$, then f has a local maximum at $x = c$.
>
> If $f'(c) = 0$ and $f''(c) > 0$, then f has a local minimum at $x = c$.

Notice that the test requires us to know f'' only at c itself, and not in an interval about c. This makes the test easy to apply. That's the good news. The bad news is that the test fails if $f''(c)$ or if $f''(c)$ fails to exist or is hard to find.

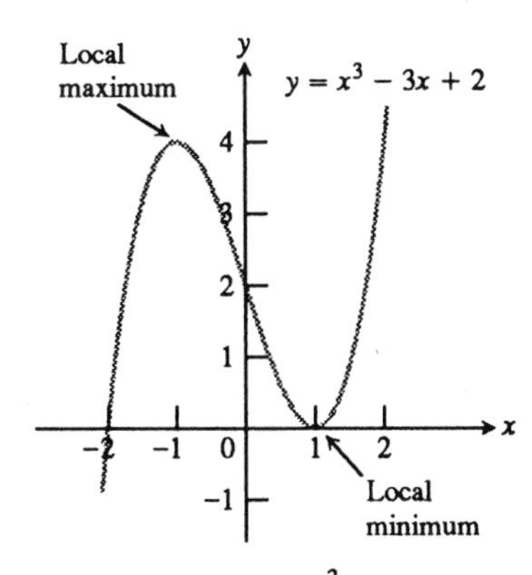

4.30 The graph of $y = x^3 - 3x + 2$.

EXAMPLE 13 Find all maxima and minima of the function

$$y = x^3 - 3x + 2$$

on the interval $-\infty < x < \infty$ analytically.

Solution The domain has no endpoints and the function is differentiable at every point. Therefore extreme values can occur only where the first derivative,

$$y' = 3x^2 - 3 = 3(x - 1)(x + 1),$$

equals zero, which means at $x = 1$ and $x = -1$. The second derivative,

$$y'' = 6x,$$

is positive at $x = 1$ and negative at $x = -1$. Hence $y(1) = 0$ is a local minimum value and $y(-1) = 4$ is a local maximum value (Fig. 4.30.) ≡

EXAMPLE 14 The second derivative test does not identify the local minimum of the function $y = x^4$ at $x = 0$. The second derivative, $y'' = 12x^2$, is zero when $x = 0$, so the test does not apply. ∎

Summary

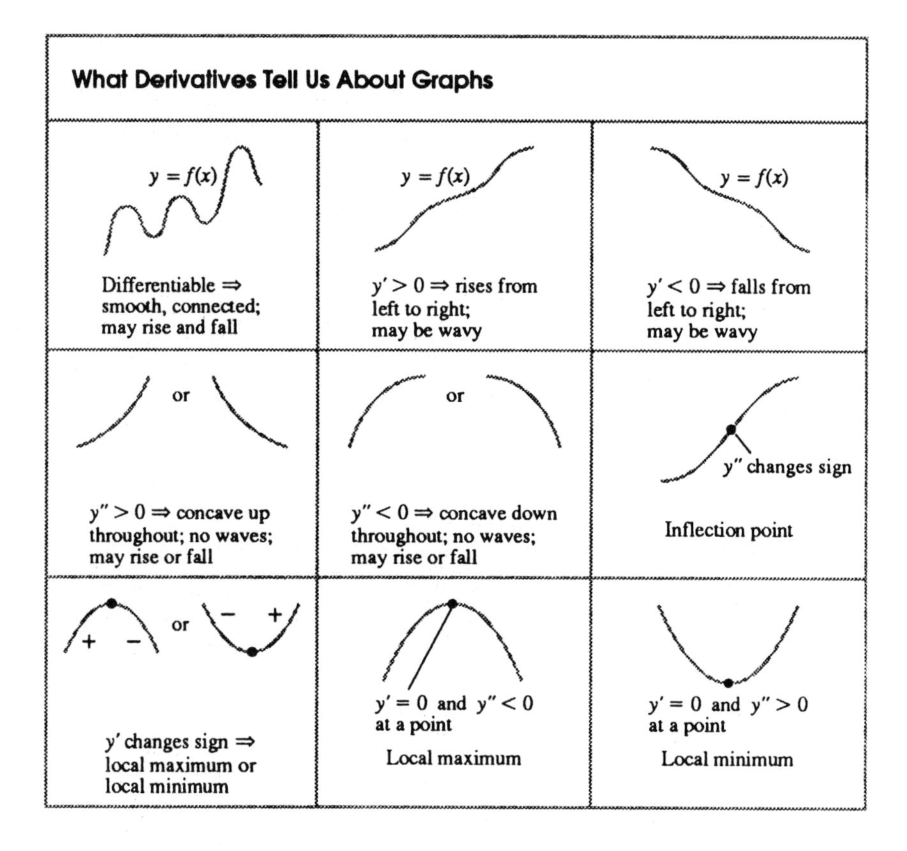

Exercises 4.2

For the functions f graphed in Exercises 1 and 2, identify where the derivative f' is 0, positive, and negative.

1.

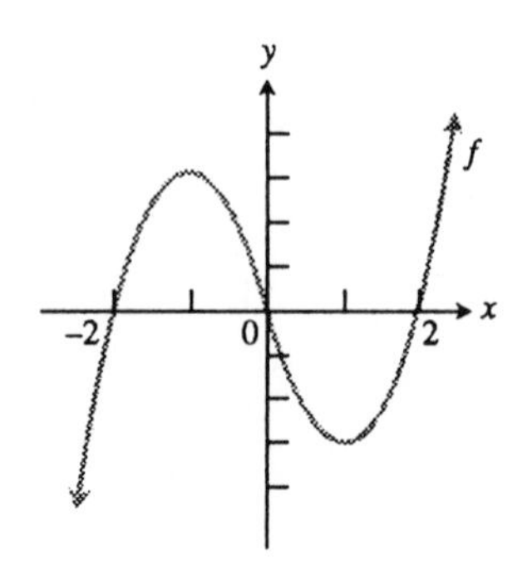

2.

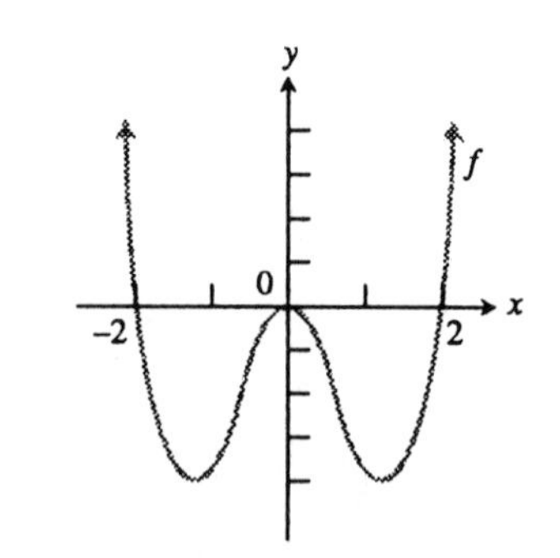

For the functions f whose derivatives are graphed in Exercises 3 and 4, identify the intervals on which the graph of the function

f is rising, falling, and the local extrema of f.

3.

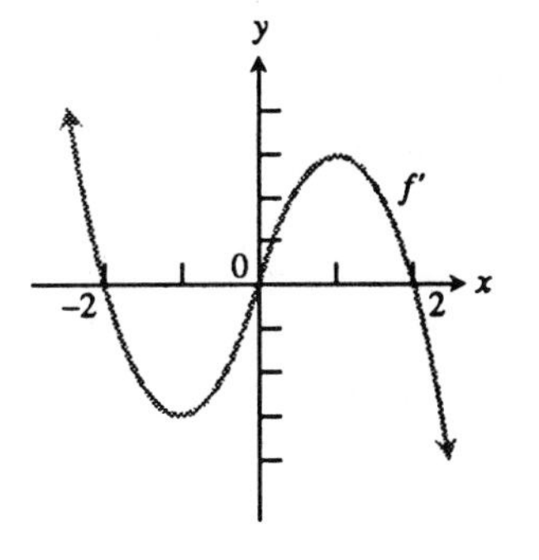

4.

In Exercises 5–10, find the inflection points, local maximum and minimum values, and identify the intervals on which the graphs are rising, falling, concave up, and concave down analytically. Then sketch a complete graph and support your answers with a graphing utility.

5. $y = x^2 - x - 1$ **6.** $y = 4x^2 + 8x + 1$
7. $y = x^3 - 6x^2 + 9x + 1$ **8.** $y = -2x^3 + 6x^2 - 3$
9. $y = 2x^4 - 4x^2 + 1$ **10.** $y = x^4 - 2x^2$

Draw a complete graph of the functions in Exercises 11–16 and confirm that the graphs are complete analytically. Find the inflection points, local maximum and minimum values, and identify the intervals on which the graphs are rising, falling, concave up, and concave down.

11. $y = 2x^3 - 5x^2 + 4x + 10$

12. $y = 4x^3 + 21x^2 + 36x - 20$

13. $y = 3x^4 - x^2 - 10$

14. $y = 20 + 2x^2 - 9x^4$

15. $y = x + \sin x,\ 0 \le x \le 2\pi$

16. $y = x - \sin x,\ 0 \le x \le 2\pi$

Draw a complete graph of the functions in Exercises 17–22. Find the inflection points, local maximum and minimum values, and identify the intervals on which the graphs are rising, falling, concave up, and concave down.

17. $y = x^4 - 8x^2 + 4x + 2$

18. $y = -x^4 + 4x^3 - 4x + 1$

19. $y = -x^4 + 2x^2 - 3x - 2$

20. $y = 2x^4 - x^2 - 3x + 5$

21. $y = \dfrac{5}{1 + 2^{1-0.5x}}$ **22.** $y = \dfrac{-7}{1 + 3^{1-0.5x}}$

Describe the motion of the particle moving on the x-axis with distance from the origin given by the functions in Exercises 23–26.

23. $s(t) = t^2 - 4t + 3$ **24.** $s(t) = 6 - 2t - t^2$
25. $s(t) = t^3 - 3t + 3$ **26.** $s(t) = 3t^2 - 2t^3$

VELOCITY AND ACCELERATION Each of the graphs in Exercises 27 and 28 is the graph of the position function $y = s(t)$ of a body moving back and forth on a coordinate line. At approximately what times is each body's (a) velocity equal to zero? (b) acceleration equal to zero?

27.

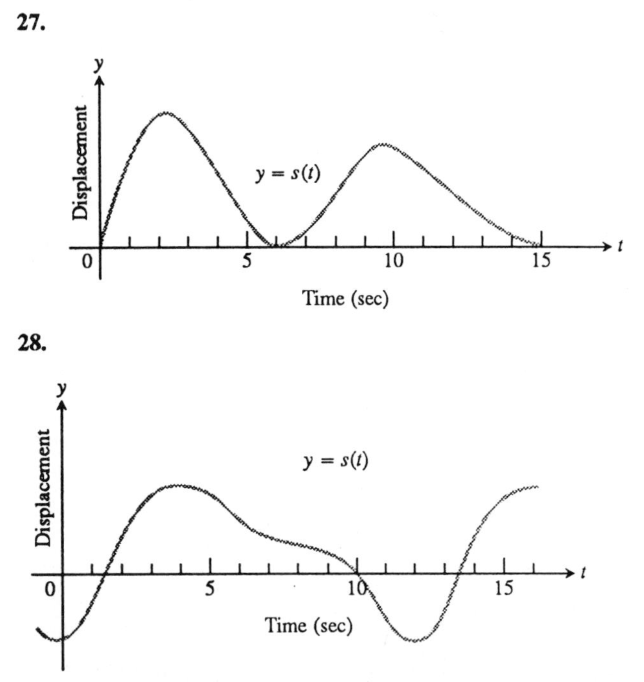

28.

29. Sketch a smooth curve $y = f(x)$ through the origin with the properties that $f'(x) < 0$ for $x < 0$, and $f'(x) > 0$ for $x > 0$.

30. Sketch a smooth curve $y = f(x)$ through the origin with the properties that $f''(x) < 0$ for $x < 0$, and $f''(x) > 0$ for $x > 0$.

31. Sketch a continuous curve $y = f(x)$ having the following characteristics:

$f(-2) = 8$, $f'(2) = f'(-2) = 0$,
$f(0) = 4$, $f'(x) < 0$ for $|x| < 2$,
$f(2) = 0$, $f''(x) < 0$ for $x < 0$,
$f'(x) > 0$ for $|x| > 2$, $f''(x) > 0$ for $x > 0$.

32. Sketch a continuous curve $y = f(x)$ with the following properties. Label coordinates where possible.

x	y	Curve
$x < 2$		Falling, concave up
2	1	Horizontal tangent
$2 < x < 4$		Rising, concave up
4	4	Inflection point
$4 < x < 6$		Rising, concave down
6	7	Horizontal tangent
$x > 6$		Falling, concave down

33. Suppose that the derivative of the function $y = f(x)$ is

$$y' = (x-1)^2(x-2).$$

At what points, if any, does the graph of f have a local minimum, local maximum, or point of inflection?

34. Suppose that the derivative of the function $y = f(x)$ is

$$y' = (x-1)^2(x-2)(x-4).$$

At what points, if any, does the graph of f have a local minimum, local maximum, or point of inflection?

Find the local maximum and minimum values of the functions in Exercises 35 and 36.

35. $y = x + \dfrac{1}{x}$

36. $y = \dfrac{x}{2} + \dfrac{1}{2x-1}$

37. If $f(x)$ is a differentiable function and $f'(c) = 0$ at an interior point c of f's domain, must f have a local maximum or minimum at $x = c$? Explain.

38. If $f(x)$ is a twice-differentiable function and $f'(c) = 0$ at an interior point c of f's domain, must the graph of f have an inflection point at $x = c$? Explain.

39. *Quadratic curves.* True, or false? A quadratic curve $y = ax^2 + bx + c$ never has an inflection point. (*Hint:* What is the corresponding formula for y''?)

40. *Cubic curves.* True, or false? A cubic curve $y = ax^3 + bx^2 + cx + d$, $a \neq 0$, always has one inflection point. (*Hint:* What is the corresponding formula for y''?).

LINEARIZATIONS AT INFLECTION POINTS Linearizations fit particularly well at points of inflection. You will see what we mean if you graph the pair of functions in Exercises 41–44.

41. $f(x) = \sin x$ and its linearization $L(x) = x$ at $x = 0$

42. $f(x) = \sin x$ and its linearization $L(x) = -x + \pi$ at $x = \pi$.

43. *Newton's serpentine.* $f(x) = 4x/(x^2+1)$ and its linearization $L(x) = 4x$ at $x = 0$.

44. *Newton's serpentine.* $f(x) = 4x/(x^2+1)$ and its linearization $L(x) = -(x/2) + 3\sqrt{3}/2$ at the point $(\sqrt{3}, \sqrt{3})$.

4.3 Newton's Method and Polynomial Functions

ALGORITHM, RECURSION, AND ITERATION

It is customary to call a specified sequence of computational steps like the one in Newton's method an *algorithm*. When an algorithm proceeds by repeating a given set of steps over and over, using the answer from the previous step as the input for the next, the algorithm is called *recursive* and each repetition is called an *iteration*. Newton's method is one of the really fast recursive techniques for finding roots.

In this section we use calculus and graphing utilities to provide more detail about polynomial functions including the possible number of local extreme values they can have. Applications whose solutions involve optimizing the values of the corresponding model function will be introduced and then integrated throughout the rest of the chapter. We begin the section by introducing a numerical technique called Newton's method or, as it is more accurately called, the Newton-Raphson method and use it to speed up the process of solving an equation $f(x) = 0$. The method is based on the idea of using a tangent line to replace the graph of $y = f(x)$ near the points where f is zero. Once again, we see that linearization is the key to solving a practical problem.

The Procedure for Newton's Method

1. Guess a first approximation to a root of the equation $f(x) = 0$. A graph of $y = f(x)$ will help.
2. Use the first approximation to get a second, the second to get a third, and so on. To go from the nth approximation x_n to the next approximation x_{n+1}, use the formula

$$x_{n+1} = x_n - \frac{f(x_n)}{f'(x_n)}, \tag{1}$$

where $f'(x_n)$ is the derivative of f at x_n.

We first show how the method works and then go to the theory behind it. Of course, solving $f(x) = 0$ using graphing zoom-in is considerably faster than using Newton's Method with computation done with pencil and paper. The important point about Newton's method is that it can be used as a basis to write a short program to do the arithmetic on your graphing utility. Some graphing utilities have a built-in SOLVER key that uses a numerical technique like Newton's method to solve an equation $f(x) = 0$. Consult the manual accompanying your graphing utility to see if it has such a built-in feature.

In Section 4.1 we determined that the equation $x^3 + 3x + 1 = 0$ had exactly one real root. You can use zoom-in or the SOLVER on your graphing utility to show that this zero is -0.322185354626.

EXAMPLE 1 Use Newton's Method to solve $x^3 + 3x + 1 = 0$.

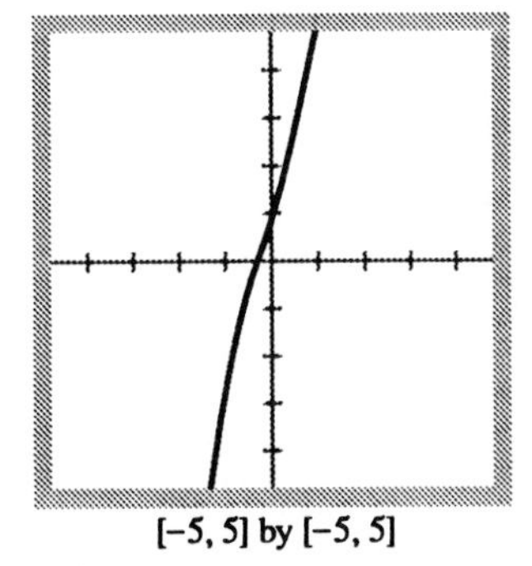

4.31 A complete graph of $f(x) = x^3 + 3x + 1$.

Solution The graph of $f(x) = x^3 + 3x + 1$ in Fig. 4.31 suggests we use $x_0 = -0.3$ as an initial approximation to the solution. With $f'(x) = 3x^2 + 3$, Eq. (1) becomes

$$x_{n+1} = x_n - \frac{x_n^3 + 3x_n + 1}{3x_n^2 + 3}. \tag{2}$$

Starting with $x_0 = -0.3$ we get the following results.

$$x_0 = -0.3$$

$$x_1 = -0.322324159021$$

$$x_2 = -0.322185360251$$

$$x_3 = -0.322185354626$$

The x_n for $n \geq 4$ all appear to be equal to x_3 on the graphing calculator we used to calculate the above values. We conclude that x_3 is the solution. ∎

Explore with a Grapher

Most graphing calculators store the result of a computation in a temporary memory register, usually called ANS.

1. Store −0.3 in ANS and then enter the expression
$$\text{ANS} - \frac{\text{ANS}^3 + 3\,\text{ANS} + 1}{3\,\text{ANS}^2 + 3}$$
which is the right-hand side of Eq. (2) with x_n replaced by ANS.
2. Evaluate the expression in part 1 and compare with the value of x_1 in Example 1.
3. Repeat part 2 two more times and compare with the values of x_2 and x_3. Continue to evaluate. Explain.
4. Repeat parts 1–3 starting with 0 stored in ANS, and compare with the numbers obtained previously.

Newton's method is the method used by most calculators to calculate roots because it converges so fast (more about that later).

What Is the Theory behind the Method?

It is this: We use the tangent to approximate the graph of $y = f(x)$ near the point $P(x_n, y_n)$, where $y_n = f(x_n)$ is small, and we let x_{n+1} be the value of x where that tangent line crosses the x-axis. (We assume that the slope $f'(x_n)$ of the tangent is not zero.) The equation of the tangent is

$$y - y_n = f'(x_n)(x - x_n). \tag{3}$$

We put $y_n = f(x_n)$ and $y = 0$ into Eq. (3) and solve for x to get

$$x = x_n - \frac{f(x_n)}{f'(x_n)}.$$

See Fig. 4.32.

From another point of view, every time we replace the graph of f by one of its tangent lines we are replacing the function by one of its linearizations L. We then solve $L(x) = 0$ to estimate the solution of $f(x) = 0$.

Strengths and Limitations of the Method

Newton's method does not work if $f'(x_n) = 0$. In that case, choose a new starting point. Of course, it may happen that $f(x) = 0$ and $f'(x) = 0$ have a common root. To detect whether this is so, we could first find the solutions of $f'(x) = 0$ and check the value of $f(x)$ at such places. Or we could graph f and f' together on a computer to look for places where the graphs might cross the x-axis together.

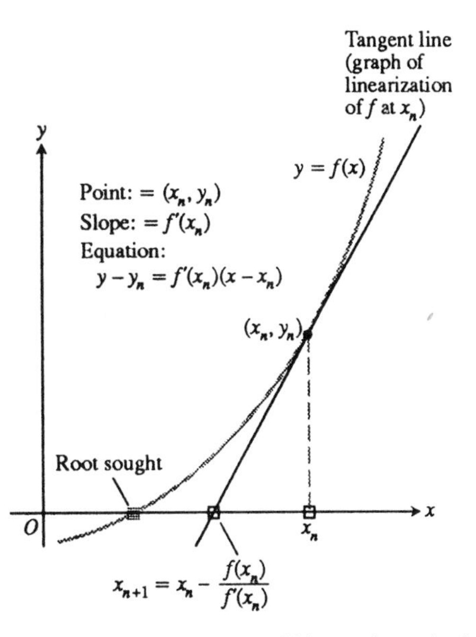

4.32 The geometry of the successive steps of Newton's method. From x_n we go up to the curve and follow the tangent line down to find x_{n+1}.

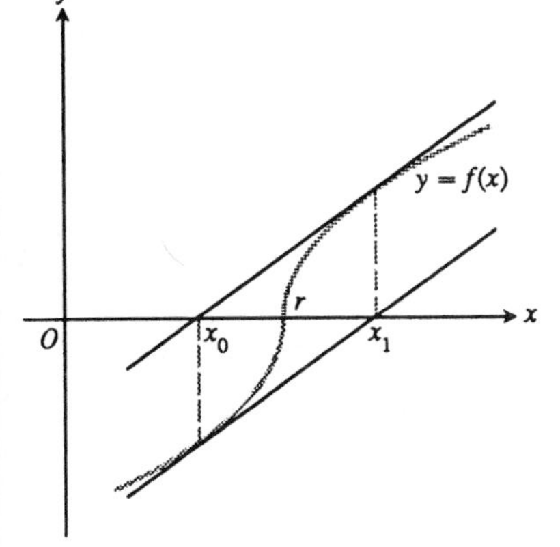

4.33 The graph of a function for which

$$f(x) = \begin{cases} \sqrt{x - r} & \text{for } x \geq r, \\ -\sqrt{r - x} & \text{for } x < r, \end{cases} \tag{4}$$

the graph will be like that shown in Fig. 4.33. If we begin with $x_0 = r - h$, we get $x_1 = r + h$ and Newton's method fails to converge.

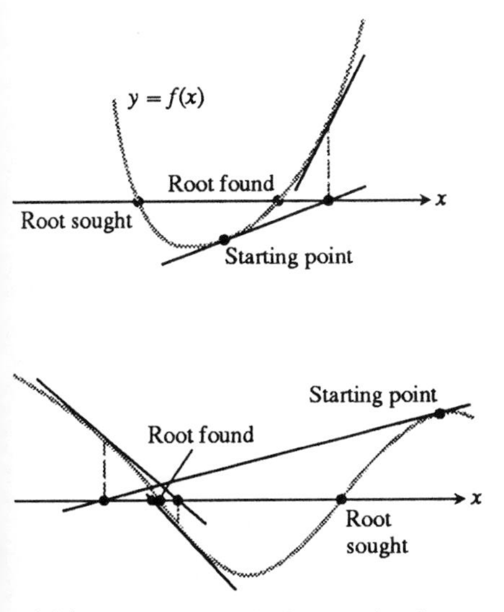

4.34 Newton's method may miss the root you want if you start too far away.

Newton's method does not always converge. For instance, if successive approximations go back and forth between these two values. No amount of iteration will bring us any closer to the root r than our first guess.

If Newton's method does converge, it converges to a root of $f(x)$. However, the method may converge to a root different from the expected one if the starting value is not close enough to the root sought. Figure 4.34 shows two ways this might happen.

When will Newton's method converge? A result from advanced calculus says that if the inequality

$$\left| \frac{f(x)f''(x)}{[f'(x)]^2} \right| < 1 \tag{5}$$

holds for all values of x in an interval about a root r of f, then the method will converge to r for any starting value x_0 in the interval. This is a *sufficient*, but not a necessary, condition. The method can (and does) converge in some cases when there is no interval about r in which the inequality (5) holds. Newton's method will always work if the curve $y = f(x)$ is convex ("bulges") toward the axis in the interval between x_0 and the root sought. See Fig. 4.35.

The speed with which Newton's method converges to a root f is expressed by the advanced-calculus formula

$$|r - x_{n+1}| \leq \frac{1}{2} \frac{\max |f''|}{\min |f'|} (r - x_n)^2, \tag{6}$$

where "max" and "min" refer to the maximum and minimum values f'' and f' have in an interval surrounding r. The formula says that the error at step $n + 1$ is no greater than a constant times the square of the error at step n. That might not seem like much to write home about, but think of what it says. If the constant $(1/2) \max |f''| / \min |f'|$ is less than or equal to 1, and

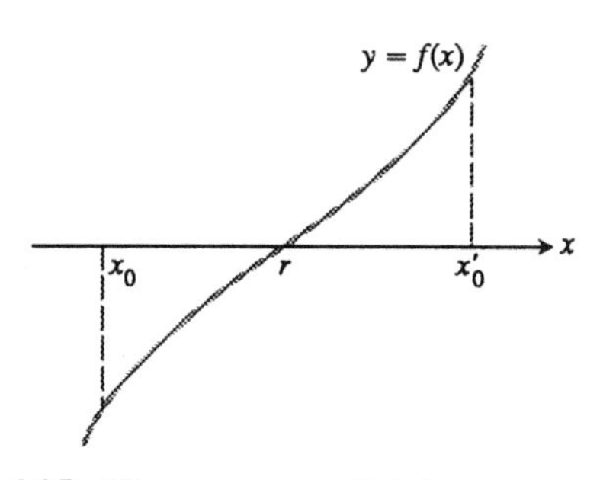

4.35 The curve $y = f(x)$ is convex toward the axis between x_0 and r and between x'_0 and r. Newton's method will converge to r from either starting point.

$|r - x_n| \leq 10^{-3}$, then $|r - x_{n+1}| \leq 10^{-6}$. *In a single step,* the method takes us to within less than one millionth of the root. Little wonder that Newton's method, when it applies, is the method of choice for modern computers.

Polynomial Functions

We state without proof three theorems about the roots of polynomials.

Theorem 4

If the complex number $a + bi$ is a root of a polynomial with real coefficients, then so is its complex conjugate $a - bi$.

Check that both $1+2i$ and $1-2i$ are zeros of the polynomial $x^2 - 2x + 5$.

Theorem 5

Fundamental Theorem of Algebra

Every polynomial function with real coefficients and degree greater than 0 has at least one zero. The zero may be a real number or a nonreal complex number.

Polynomial functions are continuous, so we can use the end behavior and the Intermediate Value Theorem to see that any polynomial of odd degree has at least one real zero. However, some polynomials, for example $x^2 + 1$, have no real zeros. Theorem 5 can be used to establish the following important fact about polynomials.

Theorem 6

A polynomial can be written in the form

$$f(x) = a(x - c_1)(x - c_2) \cdots (x - c_n)$$

where n is the degree of f and the c_i are its zeros.

Notice that the example

$$\begin{aligned} f(x) &= (x+2)(x+2)(x-1)(x-1)(x-1) \\ &= (x+2)^2(x-1)^3 \end{aligned}$$

shows that the c_i in Theorem 6 need not be distinct. In this case we say that -2 is a zero or root of *multiplicity* 2 and 1 is zero of *multiplicity 3*. Counting multiplicities we can conclude from Theorem 6 that a polynomial of degree n has n zeros.

A polynomial function f and its derivative have domain $(-\infty, \infty)$, so extreme values of f can occur only at points where $f' = 0$.

Theorem 7

> The number of extrema of a polynomial function f of degree n can be $n-1$, $n-3$, and so forth.

Proof If f has an extreme value at $x = a$, then $f'(a) = 0$ so a is a real zero of f'. The derivative f' of f is a polynomial of degree $n-1$. By Theorem 6, f' has $n-1$ zeros counting multiplicities and some may not be real. Because of Theorem 4 the possible number of real zeros of f' is $n-1$, $n-3 = (n-1)-2$, and so forth. ∎

A description of possible graphs of polynomial functions of degree 3 and 4 are given in the next two examples. In the exercises you will be asked to give similar descriptions for polynomials of degree 5.

EXAMPLE 2 Describe the possible graphs of $f(x) = ax^3 + bx^2 + cx + d$, where $a \neq 0$.

Solution From Theorem 7 we know that f has either 0 or 2 local extrema in this case.

If $a > 0$,

$$\lim_{x\to-\infty} f(x) = -\infty \quad \text{and} \quad \lim_{x\to\infty} f(x) = \infty,$$

and if $a < 0$,

$$\lim_{x\to-\infty} f(x) = \infty \quad \text{and} \quad \lim_{x\to\infty} f(x) = -\infty.$$

The continuity and end behavior of f together with the Intermediate Value Theorem requires f to take on every real value, that is, the range of f is $(-\infty, \infty)$. Thus, f has at least one real zero and has no absolute maximum or absolute minimum value. Notice that

$$f'(x) = 3ax^2 + 2bx + c$$

$$f''(x) = 6ax + 2b.$$

Since $f''(-b/3a) = 0$ and f'' changes sign at $-b/3a$, f has a point of inflection at $x = -b/3a$. The concavity of f changes from up to down or down to up at $x = -b/3a$.

Notice that the vertex of the parabola $f'(x) = 3ax^2 + 2bx + c$ also occurs at $x = -b/3a$.

Case 1 First assume that $f' \geq 0$ for all x or $f' \leq 0$ for all x. If $f' \geq 0$ for all x, then f is always increasing and the graph of f looks like the graph of $y = x^3$, except that the point of inflection is at $x = -b/3a$. Similarly, if $f' \leq 0$ for all x, then the graph of f looks like the graph of $y = -x^3$. In either case, f has exactly one real zero and no local extreme values (Fig. 4.36).

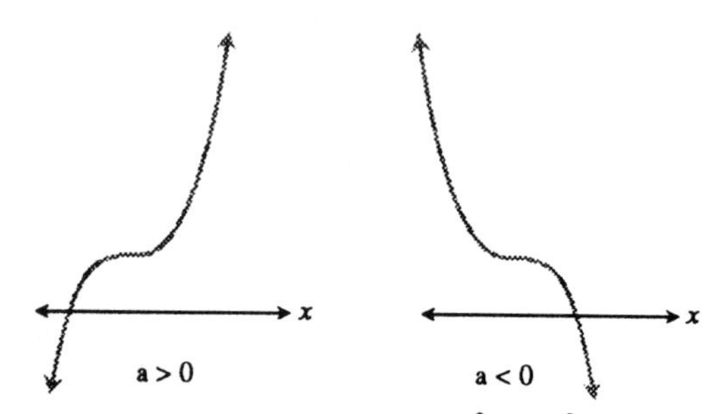

4.36 Two possibilities for the graph of $f = ax^3 + bx^2 + cx + d$ when $f' \geq 0$ or $f' \leq 0$ for all x.

Case 2 Next we assume $f' = 0$ has two distinct real zeros, say x_1 and x_2 with $x_1 < x_2$. If $a > 0$, then $f > 0$ in $(-\infty, x_1) \cup (x_2, \infty)$ and $f' < 0$ in (x_1, x_2); and if $a < 0$, then $f' > 0$ in (x_1, x_2) and $f' < 0$ in $(-\infty, x_1) \cup (x_2, \infty)$. In either case, f has local extreme values at x_1, x_2, and, depending on the location of the x-axis, f has one, two (one of multiplicity 2), or three real zeros (Fig. 4.37).

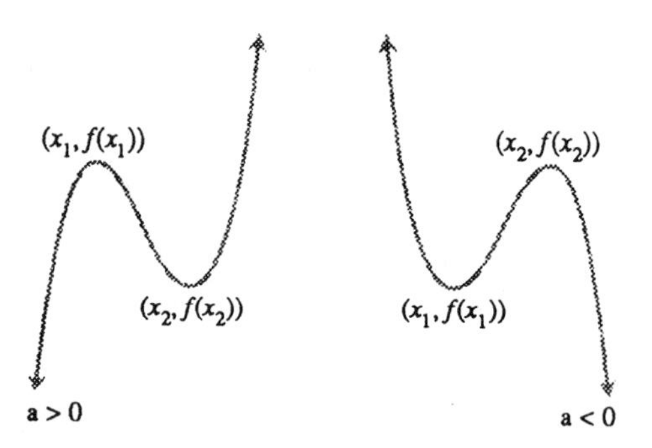

4.37 Two possibilities for the graph of $f = ax^3 + bx^2 + cx + d$ when f' has two distinct real zeros.

■

It is interesting to note that if the horizontal line $y = c$ intersects the graph of f in Fig. 4.37 in three distinct points, then $f(x) - c$ has three distinct real roots. This proves the following theorem.

Theorem 8

> If f is a polynomial of degree 3 with two local extrema, then there are real numbers a and c and three distinct real numbers x_1, x_2, and x_3 such that
>
> $$f(x) = a(x - x_1)(x - x_2)(x - x_3) + c.$$

EXAMPLE 3 Describe the possible graphs of $f(x) = ax^4 + bx^3 + cx^2 + dx + e$, $a \neq 0$.

Solution By Theorem 7, f has either 1 or 3 local extrema, and so has at least one local extrema. The example $y = x^4 + 2$ shows that f need not have any real zeros in this case.

If $a > 0$, then $\lim_{|x|\to\infty} f(x) = \infty$, and if $a < 0$, then $\lim_{|x|\to\infty} f(x) = -\infty$. The continuity and end behavior of f means that either f has an absolute minimum and no absolute maximum ($a > 0$), or f has an absolute maximum and no absolute minimum ($a < 0$). Notice that

$$f'(x) = 4ax^3 + 3bx^2 + 2cx + d$$

$$f''(x) = 12ax^2 + 6bx + 2c.$$

Case 1 First we assume $f'' \geq 0$ for all x or $f'' < 0$ for all x. If $f'' \geq 0$ for all x, then the graph of f is always concave up, and if $f'' \leq 0$ for all x, then the graph of f is always concave down. Now f' is a polynomial of degree 3 and must have a graph like one of the four shown in Figs. 4.36 and 4.37. If f' has a graph like one of the two shown in Fig. 4.37, then its derivative f'' would have two distinct real zeros, one at each of the local extrema of f' violating $f'' \geq 0$ or $f'' < 0$ for all x. Thus, f' must have a graph like one of the two in Fig. 4.36. It follows that f' has one real zero, say at x_1, and either $f' < 0$ in $(-\infty, x_1)$ and $f' > 0$ in (x_1, ∞), or $f' > 0$ in $(-\infty, x_1)$ and $f' < 0$ in (x_1, ∞). Therefore, f has exactly one extremum (at x_1) in this case (Fig. 4.38).

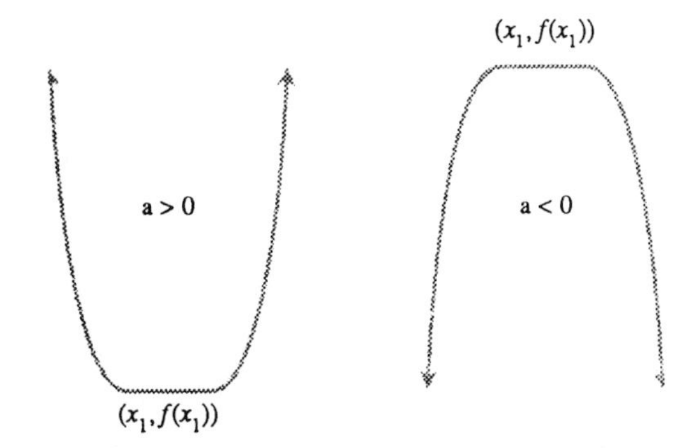

4.38 Two possibilities for the graph of $f = ax^4 + bx^3 + cx^2 + dx + e$ when $f'' \geq 0$ or $f'' \leq 0$ for all x.

Case 2 The remaining possibility is that $f'' = 0$ has two distinct real solutions, say x_1 and x_2 with $x_1 < x_2$. Thus, either $f'' > 0$ in $(-\infty, x_1) \cup (x_2, \infty)$ and $f'' < 0$ in (x_1, x_2), or $f'' < 0$ in $(-\infty, x_1) \cup (x_2, \infty)$ and $f'' > 0$ in (x_1, x_2). It follows that the graph of f is either concave up in $(-\infty, x_1) \cup (x_2, \infty)$ and concave down in (x_1, x_2) or the reverse. Thus, f has two points of inflection. Since f'' changes sign at x_1 and x_2, f' must have local extrema at x_1 and x_2. Because f' is a polynomial of degree 3 with two local extrema the graph of f' must look like one of the two in Fig. 4.37. Now we must take into account the number of real zeros of f'.

Case 2(a) We assume the conditions of Case 2 hold and, in addition, f' has one or two distinct real zeros. The possibilities for f' are shown in Fig. 4.39.

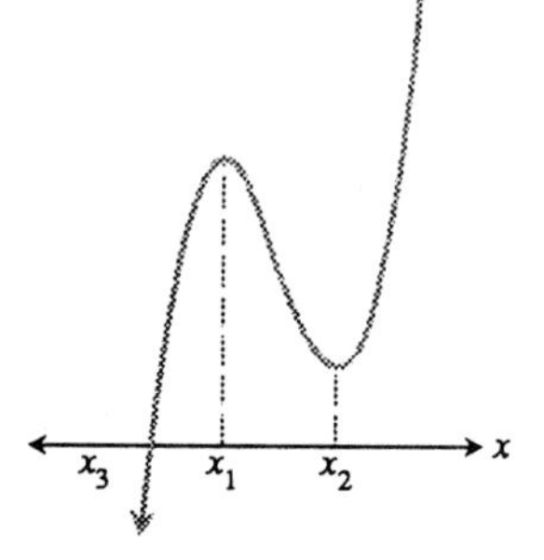

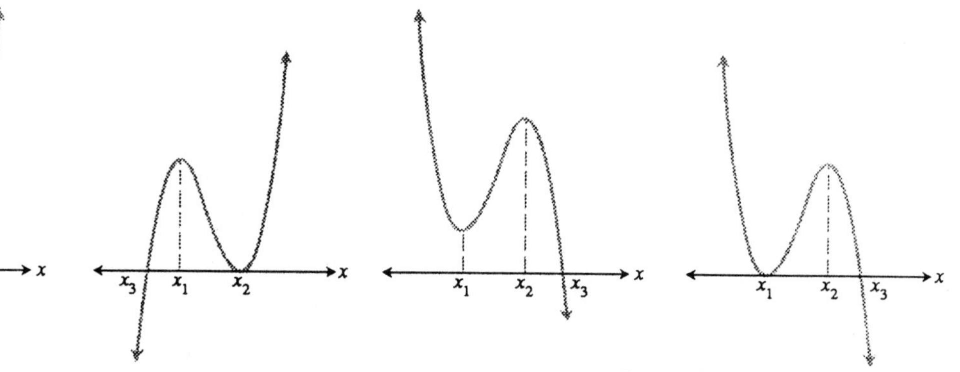

4.39 Four possibilities for the graph of $f' = 4ax^3 + 3bx^2 + 2cx + d$ in the case that f'' has two distinct real zeros at x_1 and f' has one or two real zeros.

Notice that f has a relative extrema at x_3 but not at the other zeros of f' in Fig. 4.38 because the sign of f' does not change sign at these zeros. In this case, f has exactly one extrema and two points of inflection so its graph looks like one of the two in Fig. 4.40.

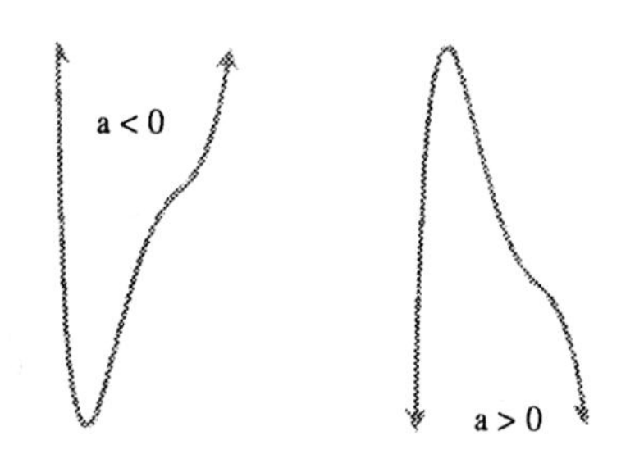

4.40 Two possibilities for the graph of $f = ax^4 + bx^3 + cx^2 + dx + e$ when f'' has two distinct real zeros and f' has one or two real zeros.

Case 2(b) We assume the conditions of Case 2 hold and, in addition, f' has three distinct real zeros. The possibilities for f' are shown in Fig. 4.41.

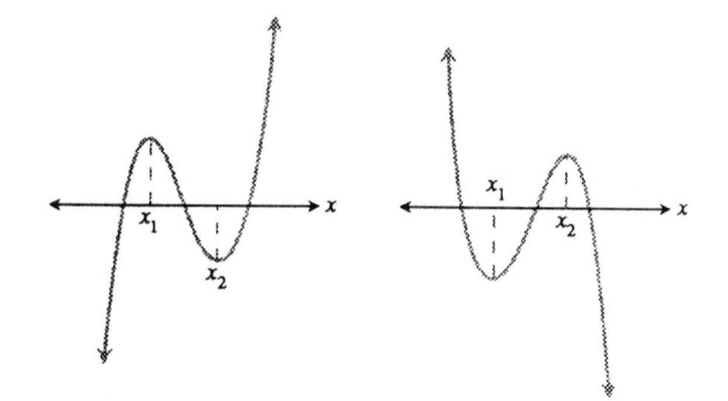

4.41 Two possibilities for the graph of $f' = 4ax^3 + 3bx^2 + 2cx + d$ in the case that f'' has two distinct real zeros at x_1 and x_2 and f' has three distinct real zeros.

Notice that f has relative extrema at each of the three real zeros of f' in Fig. 4.41 because the sign of f' changes at each of the zeros. In this case, f has exactly three extrema and two points of inflection so its graph looks like one of the two in Fig. 4.42. ☰

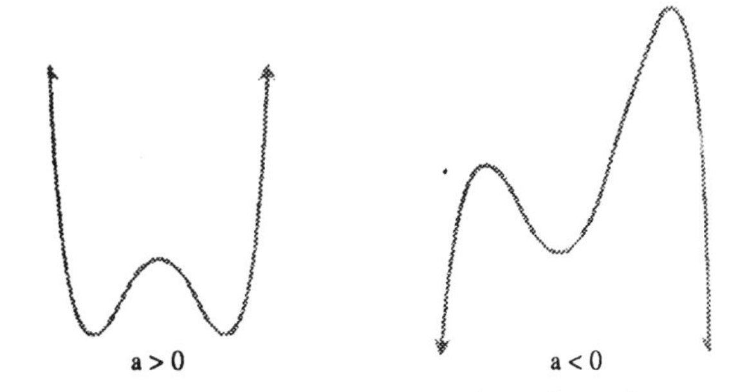

4.42 Two possibilities for the graph of $f = ax^4 + bx^3 + cx^2 + dx + e$ when f'' has two distinct real zeros and f' has three distinct real zeros.

If we are not concerned with details about concavity, we can describe the possible graphs of polynomials with less effort. For example, suppose that g is a polynomial of degree 4 with end behavior $\lim_{|x|\to\infty} g = -\infty$. Then there must be an interval $(-\infty, a)$ in which g is always increasing. Because g cannot be increasing in $(-\infty, \infty)$, the graph of g must rise to a local maximum and then begin to decrease. It may decrease for all x to the right of the local maximum. This means that g has one extrema and its graph looks like the second one in either Fig. 4.38 or 4.40. On the other hand the graph of g may fall until it hits a local minimum and then rise again. Because of the end behavior, the graph of g must eventually fall again. This means g has at least three local extrema, but because of Theorem 7, g has exactly three local extrema and the graph of g falls for all x beyond the third local extrema. In this case, the graph of g looks like the second one in Fig. 4.42.

Optimization

In the mathematical models in which we use differentiable functions to describe the things that interest us, optimization means finding where some function has its greatest or smallest value. What is the size of the most profitable production run? What is the best shape for an oil can? What is the stiffest beam we can cut from a 12-in. log?

Critical Points and Endpoints

Our basic tool is the observation we made in Section 4.1 about local maxima and minima. There we discovered that the extreme values of any function f whatever can occur only at

1. Interior points where $f' = 0$,
2. Interior points where f' does not exist,
3. Endpoints of the function's domain.

None of these points is necessarily the location of an extreme value, but these are the only candidates.

The points where $f' = 0$ or fails to exist are called the **critical points** of f. Thus the only points worth considering in the search for a function's extreme values are critical points and endpoints.

Applied Example from Mathematics

EXAMPLE 4 Products of numbers Find two numbers whose sum is 20 and whose product is as large as possible analytically, and support your answer graphically.

Solution If one number is x, the other is $(10 - x)$. Their product is

$$f(x) = x(20 - x) = 20x - x^2.$$

Figure 4.43 gives a complete graph of f.

We can see from this graph that there is a maximum, and from what we know about parabolas it occurs when $x = 10$. Using calculus we need to evaluate f at the critical points and endpoints. There are no endpoints. The first derivative,

$$f'(x) = 20 - 2x,$$

is defined for all values of x and is zero only at $x = 10$. The value of f at this one critical point is

$$\text{critical point value:} \quad f(10) = 100.$$

We conclude that the maximum value is $f(10) = 100$. The corresponding numbers are $x = 10$ and $20 - 10 = 10$.

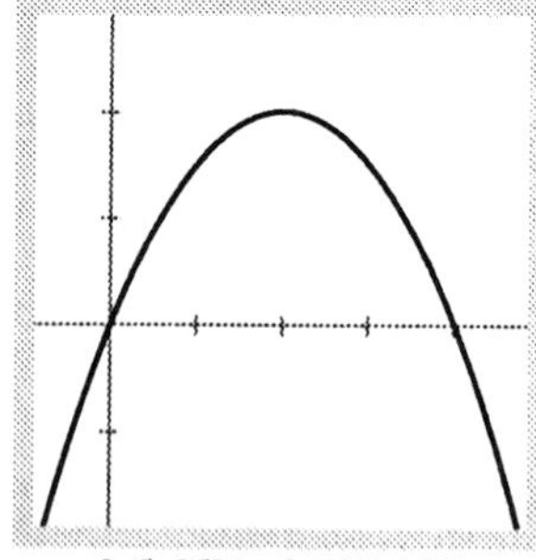

4.43 A complete graph of $f(x) = x(20 - x)$.

Continuous Functions on Closed Intervals

Most applications call for finding the absolute maximum value or absolute minimum value of a continuous function on a closed interval. The Max-Min Theorem in Section 2.2 assures us they exist. The number of points where these values might occur is usually so small that we can simply list the points and compute the corresponding function values to see what the maximum and minimum are and where they are taken on.

Applied Example from Industry

To find the absolute maximum and minimum values of a continuous function on a closed interval, evaluate the function at the critical points and endpoints and take the largest and smallest of these values.

EXAMPLE 5 Metal fabrication. An open-top box is to be made by cutting small congruent squares from the corners of a 20-by-25-in. sheet of tin and bending up the sides (Fig. 4.44). Use analytic calculus to determine how large the squares cut from the corners should be to make the box hold as much as possible, the resulting maximum value, and support your answer graphically.

Solution The height of the box is x and the other two dimensions are $(20 - 2x)$ and $(25 - 2x)$. The volume of the box is

$$V(x) = x(20 - 2x)(25 - 2x) = 4x^3 - 90x^2 + 500x.$$

Figure 4.45(a) gives a complete graph of V.

The dimension $(20 - 2x)$ of the box must be nonnegative. Thus, the only values of x that make sense in the problem are $0 \leq x \leq 10$ (Fig. 4.45(b)). We can use Fig. 4.45(b) to either estimate or use zoom-in to see that the maximum

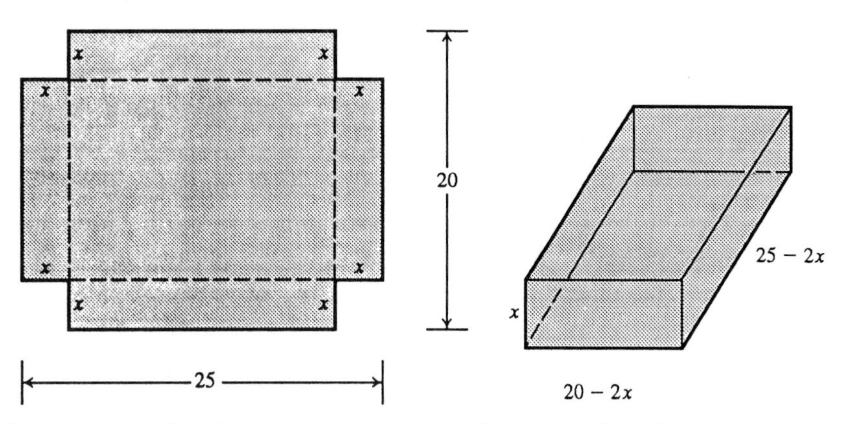

4.44 To make an open box, in (a) squares are cut from the corners of a 20-by-25-in. sheet of tin and in (b) the sides are bent up. What value of x gives the largest volume?

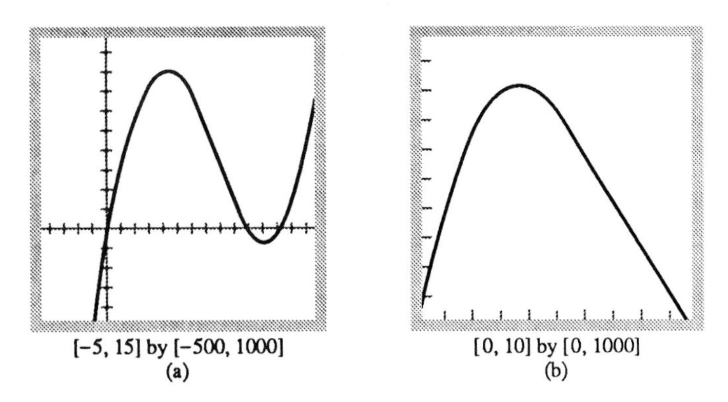

4.45 (a) A complete graph of $V(x) = x(20 - 2x)(25 - 2x)$ and (b) a complete graph of the volume of the box problem in Example 5.

value of V is about 820.5 and occurs when x is about 3.7. Using calculus we need to evaluate V at the critical points and endpoints. The first derivative,

$$V'(x) = 12x^2 - 180x + 500,$$

is defined for all x and is zero at

$$\frac{180 - \sqrt{180^2 - 48(500)}}{24} = 3.68 \quad \text{and} \quad \frac{180 + \sqrt{180^2 - 48(500)}}{24} = 11.32.$$

Only $x = 3.68$ lies in the domain $0 \le x \le 10$ of the problem and makes the critical-point list. The values of V at this one critical point and two endpoints are

$$\text{critical point value: } V(3.68) = 820.53$$

$$\text{endpoint values: } V(0) = 0, \quad V(10) = 0.$$

The maximum volume is 820.53 in^3. The cut-out squares should be 3.68 in. on a side.

Strategy for Solving Max-Min Problems

1. *Draw a picture.* Label the parts that are important in the problem.
2. *Write an algebraic representation.* Write an algebraic representation for the quantity whose maximum or minimum value you want. If you can, express the quantity as a function of a single variable, say $y = f(x)$. This may require some algebra and the use of information from the statement of the problem. Note the domain in which the values of x are to be found.
3. *Test the critical points and endpoints.* The extreme value of f will be found among the values f takes on at the endpoints of the domain and at the points where f' is zero or fails to exist. List the values of f at these points. If f has an absolute maximum or minimum on its domain, it will appear on the list. You may have to examine the sign pattern of f' or the sign of f'' to decide whether a given value represents a maximum, a minimum, or neither.
4. Support your answers graphically.

Exercises 4.3

In Exercises 1–8 use Newton's method to find all real solutions.

1. $x^2 + x - 1 = 0$. Confirm your answers using the quadratic formula.
2. $x^3 + x - 1 = 0$. Support your answers using zoom-in.
3. $x^4 + x - 3 = 0$
4. $2x - x^2 + 1$
5. $x^4 - 2x^3 - x^2 - 2x + 2 = 0$
6. $2x^4 - 4x^2 + 1 = 0$
7. $9 - \frac{3}{2}x^2 + x^3 - \frac{1}{4}x^4 = 0$
8. $x^5 - 5x^3 + 4x + 5 = 0$
9. Use Newton's method to find the positive fourth root of 2

by solving the equation $x^4 - 2 = 0$. Start with $x_0 = 1$ and find x_2.

10. Use Newton's method to find the negative fourth root of 2 by solving the equation $x^4 - 2 = 0$. Start with $x_0 = -1$ and find x_2.

11. Suppose your first guess in using Newton's method is lucky, in the sense that x_0 is a root of $f(x) = 0$. What happens to x_1 and later approximations?

12. You plan to estimate $\pi/2$ to five decimal places by solving the equation $\cos x = 0$ by Newton's method. Does it matter what your starting value is? Explain.

13. OSCILLATION Show that if $h > 0$, applying Newton's method to

$$f(x) = \begin{cases} \sqrt{x}, & x \geq 0 \\ \sqrt{-x}, & x < 0 \end{cases}$$

leads to $x_1 = -h$ if $x_0 = h$ and to $x_1 = h$ if $x_0 = -h$. Draw a picture that shows what is going on.

14. APPROXIMATIONS THAT GET WORSE AND WORSE Apply Newton's method to $f(x) = x^{1/3}$ with $x_0 = 1$, and calculate x_1, x_2, x_3, and x_4. Find a formula for $|x_n|$. What happens to $|x_n|$ as $n \to \infty$? Draw a picture that shows what is going on.

In Exercises 15–20 find the inflection points, local maximum and minimum values, and identify the intervals on which the graphs are rising, falling, concave up, and concave down analytically. Then sketch a complete graph and support your answers with a graphing utility. Indicate the number of real roots.

15. $y = x^3 - 3x^2 + 5x - 4$
16. $y = \frac{1}{3}x^3 - 2x^2 + 4x + 8$
17. $y = x^3 - 2x^2 - 3x + 8$
18. $y = x^3 + 10x^2 - 23x + 12$
19. $y = 12 + x - 4x^3$
20. $y = 2x^3 - x^2 - 14x - 12$

Draw a complete graph of the functions in Exercises 21–24 and confirm that the graphs are complete analytically. Find the inflection points, local maximum and minimum values, and identify the intervals on which the graphs are rising, falling, concave up, and concave down. Indicate the numbers of real zeros.

21. $y = 20 - 3x - \frac{1}{3}x^3$
22. $y = -9x^3 + 4x - 15$
23. $y = x^4 + x^2 + x + 8$
24. $y = x^4 - 8x^3 + 17x^2 - 10x - 1$

Draw a complete graph of the functions in Exercises 25–36. Find the inflection points, local maximum and minimum values, and identify the intervals on which the graphs are rising, falling, concave up, and concave down. Indicate the number of real zeros.

25. $y = 4x^3 - 17x^2 + 8x - 1$
26. $y = -x^3 - 3x^2 + 10x + 3$
27. $y = 10 + \frac{8}{3}x^3 - x^4$
28. $y = \frac{1}{3}x^4 + \frac{4}{3}x^3 + 2x^2 + 2x - 10$
29. $y = x^4 - 2x^2 + x + 20$
30. $y = 20 - 4x + 2x^2 + 3x^3 - \frac{9}{4}x^4$
31. $y = -x^4 - 7x^3 - 9x^2 + 7x + 12$
32. $y = x^4 - 8x^3 + 14x^2 + 8x - 5$
33. $y = -x^5 - x^4 + 11x^3 + 9x^2 - 18x + 5$
34. $y = x^5 - \frac{5}{2}x^4 - 5x^2 + 1$
35. $y = 3x^5 + \frac{15}{4}x^4 + 10x^3 + \frac{15}{2}x^2 + 15x + 1$
36. $y = \frac{1}{5}x^5 - \frac{9}{4}x^4 + \frac{19}{3}x^3 + \frac{9}{2}x^2 - 16x + 1$

37. The sum of two nonnegative numbers is 20. Find the numbers if the sum of their squares is to be as large as possible.

38. Show that among all rectangles with an 8-ft perimeter, the one with the largest area is a square.

39. The figure below shows a rectangle inscribed in an isosceles right triangle whose hypotenuse is 2 units long.

a) Express the y-coordinate of P in terms of x. (You might start by writing an equation for the line AB.)

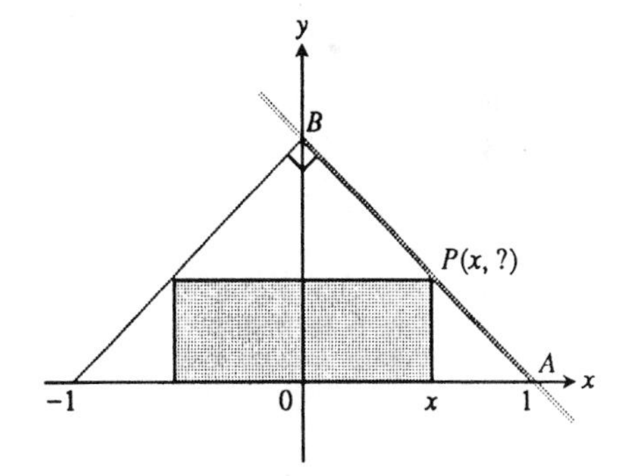

b) Express the area of the rectangle in terms of x.

c) What is the largest area the rectangle can have?

40. A rectangle has its base on the x-axis and its upper two vertices on the parabola $y = 12 - x^2$. What is the largest area the rectangle can have?

41. You are planning to make an open rectangular box from an 8-in. × 15-in. piece of cardboard by cutting squares from the corners and folding up the sides. What are the dimensions of the box of largest volume you can make this way?

42. A rectangular plot of farmland will be bounded on one side by a river and on the other three sides by a single-strand electric fence. With 800 m of wire at your disposal, what is the largest area you can enclose?

43. The height of an object moving vertically is given by

$$s = -16t^2 + 100t + 200,$$

with s in feet and t in seconds. Find

a) the object's velocity when $t = 0$,

b) its maximum height, and

c) its velocity when $s = 0$.

44. The height of an object moving vertically is given by

$$s = -16t^2 + 96t + 112,$$

with s in feet and t in seconds. Find (a) the object's velocity when $t = 0$, (b) its maximum height, and (c) its velocity when $s = 0$.

45. The U.S. Postal Service will accept a box for domestic shipment only if the sum of the length and girth (distance around) does not exceed 108 in. Find the dimensions of the largest acceptable box with a square end.

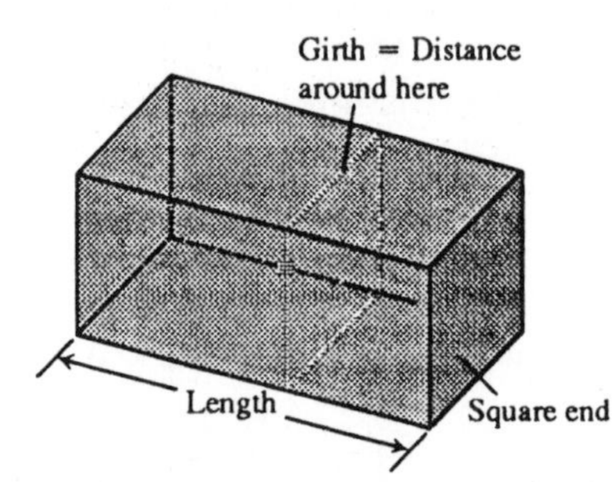

46. THE SONOBUOY PROBLEM From C. O. Wilde's *The Contraction Mapping Principle*, UMAP Unit 326 (Arlington, Mass.: COMAP, Inc.). In submarine location problems it is often necessary to find the submarine's closest point of approach (CPA) to a sonobuoy (sound detector) in the water. Suppose that the submarine travels on a parabolic path $y = x^2$ and that the buoy is located at the point $(2, -1/2)$.

a) Show that the value of x that minimizes the square of the distance, and hence the distance, between the points (x, x^2) and $(2, -1/2)$ in the figure below is a solution of the equation $x = 1/(x^2 + 1)$.

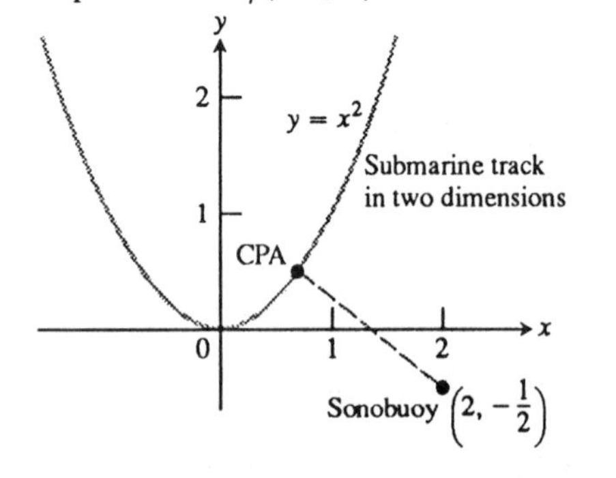

b) Use Newton's method to solve the equation in part (a) to five decimal places.

47. Compare the answers to the following two construction problems.

a) A rectangular sheet of perimeter 36 cm and dimensions x cm by y cm is to be rolled into the cylinder shown in part (a) of the figure below. What values of x and y give the largest volume?

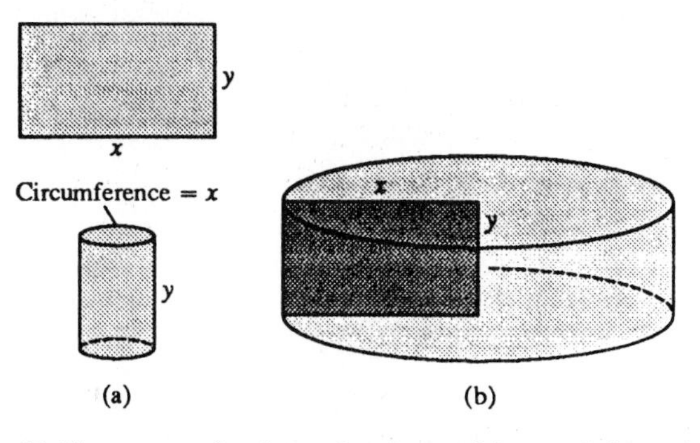

b) The rectangular sheet of perimeter 36 cm and dimensions x by y is to be revolved about one of the sides of length y to sweep out the cylinder in part (b) of the figure above. What values of x and y give the largest volume?

48. A right triangle whose hypotenuse is $\sqrt{3}$ meters long is revolved about one of its legs to generate a right circular cone. Find the radius, height, and volume of the cone of greatest volume that can be made this way.

49. A spherical floating buoy has a radius of 1 meter and a density $\frac{1}{4}$ that of sea water. By Archimedes' law the volume of the displaced water is equal to the volume of the submerged portion of the buoy (see figure below). Find the depth x that the buoy sinks in sea water.

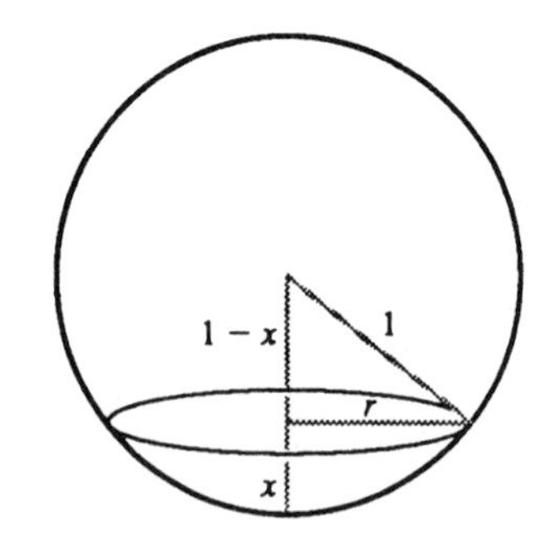

50. Repeat Exercise 49 assuming the buoy has a density $\frac{1}{3}$ that of sea water.

51. What values of a and b make

$$f(x) = x^3 + ax^2 + bx$$

have

a) a local maximum at $x = -1$ and a local minimum at $x = 3$?

b) a local minimum at $x = 4$ and a point of inflection at $x = 1$?

52. Find the volume of the largest right circular cone that can be inscribed in a sphere of radius 3.

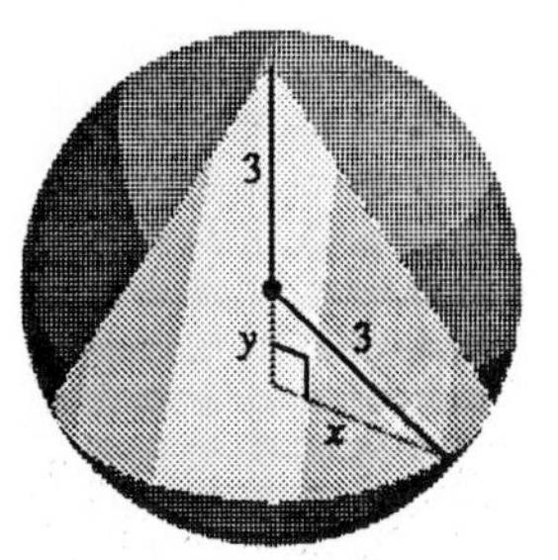

53. THE STRENGTH OF A BEAM The strength of a rectangular beam is proportional to its width times the square of its depth. Find the dimensions of the strongest beam that can be cut from a 12-in.-diameter log.

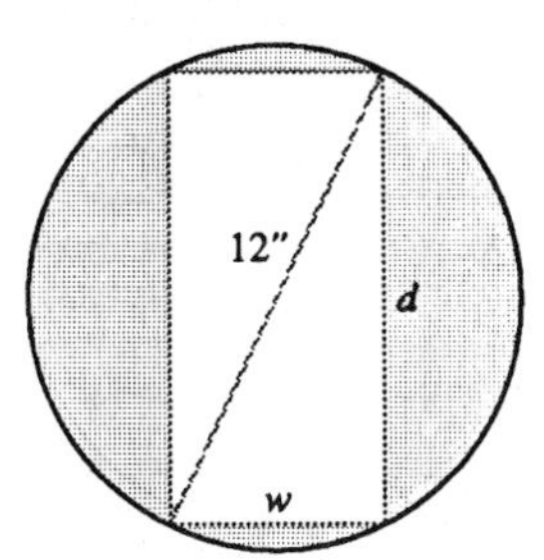

54. THE STIFFNESS OF A BEAM The stiffness of a rectangular beam is proportional to its width times the cube of its depth. Find the dimensions of the stiffest beam that can be cut from a 12-in.-diameter log. Compare your answers with the dimensions of the beam in Exercise 53.

55. TIN PEST Metallic tin, when kept below 13° C for a while, becomes brittle and crumbles to a gray powder. Tin objects eventually crumble to this gray powder spontaneously if kept in a cold climate for years. The Europeans who saw the tin organ pipes in their churches crumble away years ago called the change *tin pest* because it seemed to be contagious. And indeed it was, for the gray powder is a catalyst for its own formation.

A *catalyst* for a chemical reaction is a substance that controls the rate of the reaction without undergoing any permanent change in itself. An *autocatalytic reaction* is one whose product is a catalyst for its own formation. Such a reaction may proceed slowly at first if the amount of catalyst present is small, and slowly again at the end, when most of the original substance is used up. But in between, when both the substance and its product are abundant, the reaction proceeds at a faster pace.

In some cases it is reasonable to assume that the rate $v = dx/dt$ of the reaction is proportional both to the amount of the original substance present and to the amount of product. That is, v may be considered to be a function of x alone, and

$$v = kx(a - x) = kax - kx^2,$$

where

$x =$ the amount of product,
$a =$ the amount of substance at the beginning, and
$k =$ a positive constant.

At what value of x does the rate v have a maximum? What is the maximum value of v?

56. Suppose that at time t the position of a particle moving on the x-axis is $x = (t - 1)(t - 4)^4$.

a) When is the particle at rest?

b) During what time interval does the particle move to the left?

c) What is the fastest the particle goes while moving to the left?

Medicine

57. HOW WE COUGH When we cough, the trachea (TRAY-*kee-uh*, windpipe) contracts to increase the velocity of the air going out. This raises the questions of how much it should contract to maximize the velocity and whether it really contracts that much when we cough.

Under reasonable assumptions about the elasticity of the tracheal wall and about how the air near the wall is slowed by friction, the average flow velocity v can be modeled by the equation

$$v = c(r_0 - r)r^2 \text{ cm/sec}, \quad \frac{r_0}{2} \le r \le r_0,$$

where r_0 is the rest radius of the trachea in centimeters and c is a positive constant whose value depends in part on the length of the trachea.

Show that v has its maximum value when $r = (2/3)r_0$, that is, when the trachea is about 33% contracted. The remarkable fact is that X-ray photographs confirm that the trachea contracts about this much during a cough.

58. SENSITIVITY TO MEDICINE (continuation of Exercise 66, Section 3.2) Find the amount of medicine to which the body is most sensitive by finding the value of M that maximizes the derivative dR/dM.

59. The derivative of a polynomial is $p'(x) = (x+1)(x-1)^2(x-5)^3$. What is the degree of p? Discuss the graph of p. Use technology to determine where the graph of p is concave up.

60. Prove that if a polynomial of degree 4 has three local extrema, then there are real numbers a and c and four distinct real numbers x_1, x_2, x_3, and x_4 such that

$$f(x) = a(x - x_1)(x - x_2)(x - x_3)(x - x_4) + c.$$

61. CURVES THAT ARE ALMOST FLAT NEAR THE ROOT Some curves are so flat that Newton's method stops some distance from the root. Try Newton's method on $f(x) = (x-1)^{40}$ with a starting value of $x_0 = 2$. How close does the computer come to the root $x = 1$?

62. FINDING A ROOT DIFFERENT FROM THE ONE SOUGHT All three roots of $f(x) = x^4 - x^2$ can be found by starting Newton's method near $x = -\sqrt{2}/2$.

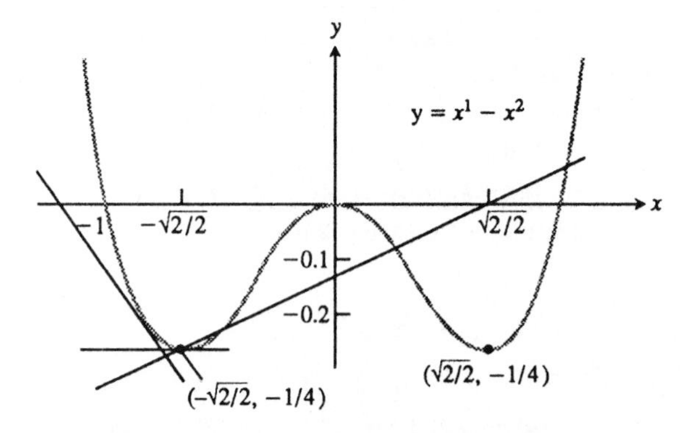

The method will find the root $x = -1$ if x_0 is far enough to the left of $-\sqrt{2}/2$. If x_0 is too close to $-\sqrt{2}/2$, the computer will encounter a zero slope or a value of x_1 too large to handle. There is a zone just to the right of $-\sqrt{2}/2$ where values of x_0 lead to $x = 1$ instead of $x = 0$.

Likewise, selected values of x_0 near $\sqrt{2}/2$ will lead to $x = 1$, $x = 0$, $x = -1$, or no root at all.

Try it.

63. a) Explain why the following four statements ask for the same information:

i) Find the roots of $f(x) = x^3 - 3x - 1$.

ii) Find the x-coordinates of the intersections of the curve $y = x^3$ with the line $y = 3x + 1$.

iii) Find the x-coordinates of the points where the curve $y = x^3 - 3x$ crosses the horizontal line $y = 1$.

iv) Find the values of x where the derivative of $g(x) = (1/4)x^4 - (3/2)x^2 - x + 5$ equals zero.

b) Sketch the graph of $f(x) = x^3 - 3x - 1$ over the interval $-2 \le x \le 2$.

c) Find the roots of $f(x) = x^3 - 3x - 1$ to five decimal places.

64. The curve $y = \tan x$ crosses the line $y = 2x$ somewhere between $x = 0$ and $x = \pi/2$. Where?

65. Estimate π to five decimal places by applying Newton's method to the equation $\tan x = 0$ with $x_0 = 3$. Remember to use radians.

66. LOCATING A PLANET To calculate a planet's space coordinates, we have to solve equations like $x = 1 + 0.5\sin x$. Graphing the function $f(x) = x - 1 - 0.5\sin x$ suggests that the function has a root near $x = 1.5$. Use Newton's method to find the solution accurate to five decimal places.

4.4 Rational Functions and Economic Applications

In this section we use calculus and graphing utilities to provide more detail about rational functions. Applications whose models involve rational functions will be introduced. We also illustrate how calculus makes an important contribution to economic theory.

Let $f(x) = p(x)/h(x)$ be a rational function with $p(x)$ and $h(x)$ polynomials with no common factors. In Section 2.4 we rewrote f in the form

$$f(x) = q(x) + \frac{r(x)}{h(x)} \quad (r(x) = 0 \text{ or } \deg r(x) < \deg h(x))$$

where $q(x)$ and $r(x)$ are the quotient and remainder, respectively, when $p(x)$ is divided by $h(x)$. The zeros of $h(x)$ give rise to vertical asymptotes of f, and $q(x)$ is the end behavior asymptote of f. This means that, except near the vertical asymptotes of f, $q(x)$ is a very good approximation to $f(x)$. In Example 1, $q(x) = 0$.

EXAMPLE 1 Draw a complete graph of

$$f(x) = \frac{3x+2}{2x^2+5x-3} = \frac{3x+2}{(2x-1)(x+3)}.$$

Use analytic calculus to confirm the complete graph and to find the inflection points and local extremum values. Identify the intervals on which the graph is rising, falling, concave up, and concave down.

Solution We show that the graph of f in Fig. 4.46(a) is complete. Notice that f has $x = -3$ and $x = 1/2$ as vertical asymptotes, and

$$\lim_{x\to -3^-} f(x) = -\infty, \quad \lim_{x\to -3^+} f(x) = \infty, \quad \lim_{x\to \frac{1}{2}^-} f(x) = -\infty, \quad \lim_{x\to \frac{1}{2}^+} f(x) = \infty.$$

Also $\lim_{|x|\to\infty} f(x) = 0$. It appears that f is concave down in $(-\infty, -3)$, concave up in $(1/2, \infty)$, and goes from concave up to concave down in $(-3, 1/2)$. So f should have a single point of inflection in $(-3, 1/2)$. Also f has no local extremum values and is decreasing in the three intervals $(-\infty, -3)$, $(-3, 1/2)$, and $(1/2, \infty)$ that make up the domain of f.

We can use the quotient rule to show that

$$f'(x) = -\frac{6x^2 + 8x + 19}{(2x^2 + 5x - 3)^2}.$$

Notice that f' is a rational function with the same vertical asymptotes as f. It can be shown that the numerator of f', $6x^2 + 8x + 19$, is greater that 0 for all x. This means that $f' < 0$ for all x confirming the decreasing behavior of f and that f has no extremum values. We actually used the numerical derivative feature to produce the graph of f' and f'' in Fig. 4.46(b) and (c). Usually the graphs of f' and f'' are enough to draw the necessary conclusions about the graph of f so that we can avoid the tedious computation of exact derivatives. We can use zoom-in or a root SOLVER to show that $f'' = 0$ at $x = -1.05$ (Fig. 4.46(c)). Thus, $f'' < 0$ in $(-\infty, -3)$ and $(-1.05, 1/2)$, and $f'' > 0$ in $(-3, -1.05)$ and $(1/2, \infty)$. It follows that f is concave down in $(-\infty, -3)$ and $(-1.05, 1/2)$ and f is concave up in $(-3, -1.05)$ and $(1/2, \infty)$.

We can show that

$$f''(x) = \frac{2(12x^3 + 24x^2 + 114x + 107)}{(2x^2 + 5x - 3)^3},$$

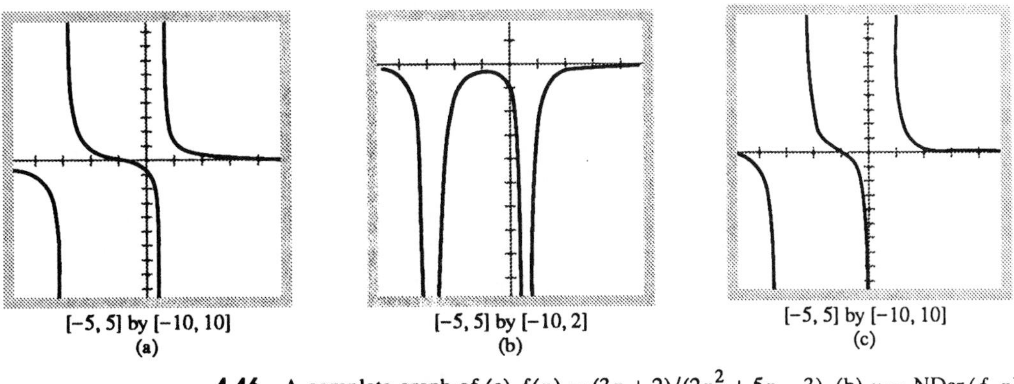

[−5, 5] by [−10, 10]
(a)

[−5, 5] by [−10, 2]
(b)

[−5, 5] by [−10, 10]
(c)

4.46 A complete graph of (a) $f(x) = (3x + 2)/(2x^2 + 5x - 3)$, (b) $y = \text{NDer}(f, x)$, and (c) $y = \text{N}_2\text{Der}(f, x)$.

and that the polynomial of degree 3 in the numerator of f'' has exactly one real root which is consistent with the graph of f'' in Fig. 4.46(c). ∎

Graphs produced by graphing utilities are usually very accurate. This means we can save time normally spent in algebraic computation of derivatives by instead using the graphs of the derivatives to support the information determined by the graph of the function as illustrated in Example 2.

EXAMPLE 2 Draw a complete graph of

$$f(x) = \frac{x^4 + x^3 - 6x^2 + 6}{x^2 + x - 6} = x^2 + \frac{6}{(x-2)(x+3)}.$$

Find the points of inflection, the local extremum values, and identify the intervals on which the graph is rising, falling, concave up, and concave down.

Solution Two views help clarify the behavior of f (Fig. 4.47). Notice that f has vertical asymptotes at $x = -3$ and $x = 2$. Also $y = x^2$ is the end behavior asymptote of f.

Notice that f has five local extrema, three local minimums, and two local maximums. The graph of f is concave up in $(-\infty, -3)$ and $(2, \infty)$. There appear to be two points of inflection in the interval $(-3, 2)$. We use two views of the graph of f' to give more information about the graph of f (Fig. 4.48).

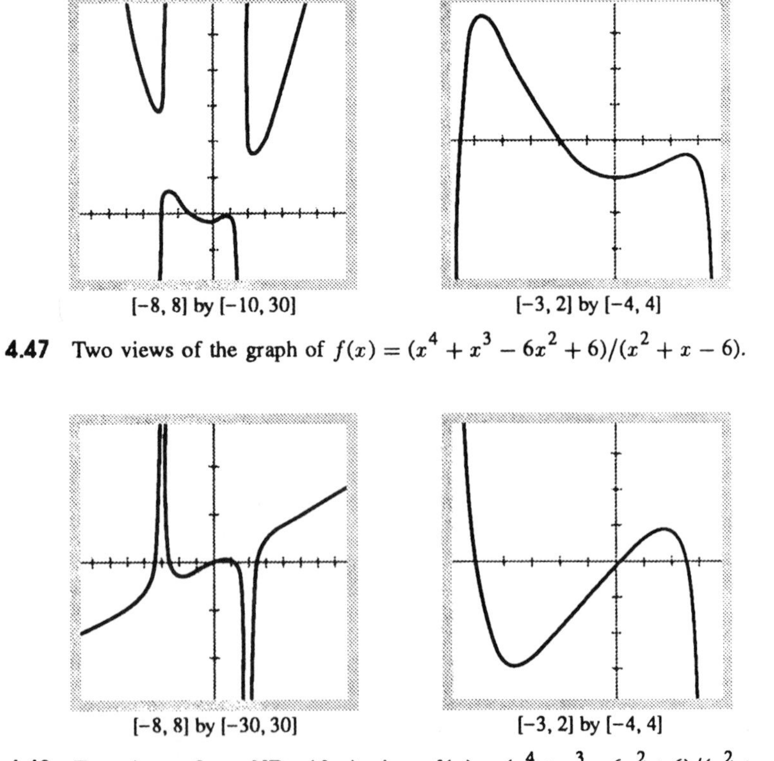

4.47 Two views of the graph of $f(x) = (x^4 + x^3 - 6x^2 + 6)/(x^2 + x - 6)$.

4.48 Two views of $y = \text{NDer}(f, x)$ where $f(x) = (x^4 + x^3 - 6x^2 + 6)/(x^2 + x - 6)$.

We can see from Fig. 4.48 that $f' = 0$ has five solutions and that f' changes sign at each of these zeros. We can use zoom-in or a root SOLVER to determine that f has

a relative minimum of 14.33 at $x = -3.42$,

a relative maximum of 3.59 at $x = -2.51$,

a relative minimum of -1.01 at $x = 0.10$,

a relative maximum of -0.30 at $x = 1.34$, and

a relative minimum of 8.43 at $x = 2.49$.

Thus, f is decreasing in

$$(-\infty, -3.42), \quad (-2.51, 0.10), \quad (1.34, 2), \quad \text{and} \quad (2, 2.49),$$

and f is increasing in

$$(-3.42, -3), \quad (-3, -2.51), \quad (0.10, 1.34), \quad \text{and} \quad (2.49, \infty).$$

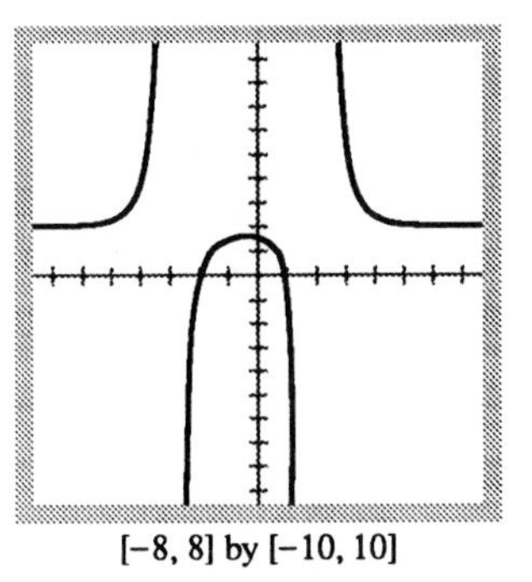

[−8, 8] by [−10, 10]

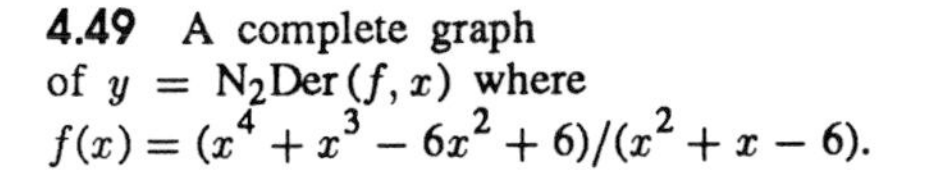

4.49 A complete graph of $y = \text{N}_2\text{Der}(f, x)$ where $f(x) = (x^4 + x^3 - 6x^2 + 6)/(x^2 + x - 6)$.

Next we use a graph of f'' to establish the concavity of f (Fig. 4.49).

We can show that $f'' = 0$ at $x = -1.93$ and $x = 0.93$. Thus, $(-1.93, 2.30)$ and $(0.93, -0.56)$ are pointsof inflection of the graph of f. Further, the graph of f is concave up in

$$(-\infty, -3), \quad (-1.93, 0.93), \quad \text{and} \quad (2, \infty),$$

and concave down in

$$(-3, -1.93), \quad \text{and} \quad (0.93, 2).$$

It can be shown that

$$f'(x) = 2x - \frac{6(2x+1)}{(x^2+x-6)^2}, \quad \text{and}$$

$$f''(x) = 2 + \frac{12(3x^2+3x+7)}{(x^2+x-6)^3}.$$

These expressions can be used to confirm the graphs in Figs. 4.48 and 4.49, and to analytically confirm the information determined above. ■

The following guidelines help summarize what we have learned in Examples 1 and 2.

Sketching the Graph of a Rational Function

1. Determine the vertical asymptotes.
2. Determine the limit from the left and the right at each vertical asymptote.
3. Sketch the end behavior asymptote.
4. Modify the sketch in part (3) to be consistent with part (2).
5. Use the graph of the first derivative (or analyze the exact derivative analytically) to check the rise, fall, and extrema values of the sketch in part (4). Modify, if necessary.
6. Use the graph of the second derivative (or analyze the exact derivative analytically) to check the concavity of the sketch in part (4). Modify, if necessary.

We return to the study of optimization problems.

EXAMPLE 3 Product design You have been asked to design a one-liter oil can shaped like a right circular cylinder. Determine what dimensions will use the least material analytically. Support your answer with a graph.

Solution We picture the can as a right circular cylinder with height h and diameter $2r$ (Fig. 4.50). If r and h are measured in centimeters and the volume is expressed as 1000 cm^3, then r and h are related by the equation

$$\pi r^2 h = 1000. \tag{1}$$

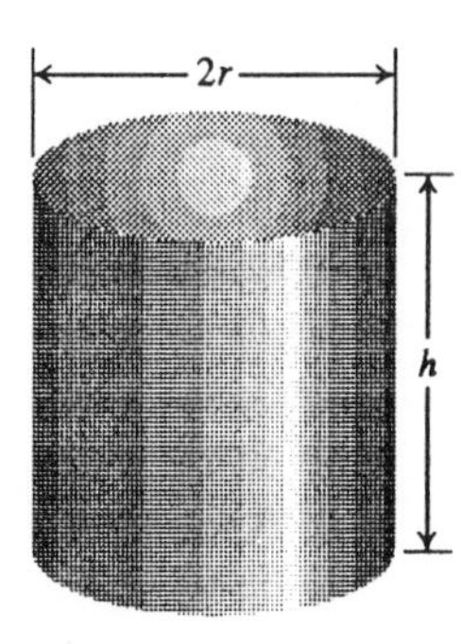

4.50 This one-liter can uses the least material when $h = 2r$ (Example 3).

How shall we interpret the phrase "least material"? One possibility is to ignore the thickness of the material and the waste in manufacturing. Then we ask for dimensions r and h that make the total surface area

$$A = \underbrace{2\pi r^2}_{\text{Cylinder ends}} + \underbrace{2\pi r h}_{\text{Cylinder wall}} \tag{2}$$

as small as possible while satisfying the constraint $\pi r^2 h = 1000$. (Exercise 34 describes one way we might take waste into account.)

What kind of oil can do we expect? Not a tall, thin one like a 6-ft pipe, nor a short, wide one like a covered pizza pan. We expect something in between.

We are not quite ready to follow the procedure of the earlier examples because Eq. (2) gives A as a function of two variables and our procedure calls for A to be a function of a single variable. However, Eq. (1) can be solved to express either r or h in terms of the other.

Solving for h is easier, so we take

$$h = \frac{1000}{\pi r^2}. \tag{3}$$

This changes the formula for A to

$$A = 2\pi r^2 + 2\pi r h = 2\pi r^2 + 2\pi r \frac{1000}{\pi r^2} = 2\pi r^2 + \frac{2000}{r}. \tag{4}$$

Our mathematical goal is to find the minimum value of A on the open interval $r > 0$ using analytic calculus. To do so, we will use the second derivative as well as the first. Here the model is a rational function of r instead of a function of x, but we analyze it the same way.

Since A is differentiable throughout its domain, and its domain has no endpoints, it can have a minimum value only where its first derivative is zero.

$$A = 2\pi r^2 + \frac{2000}{r}$$

$$\frac{dA}{dr} = 4\pi r - \frac{2000}{r^2} \qquad \text{Differentiate.}$$

$$4\pi r - \frac{2000}{r^2} = 0 \qquad \text{Set equal to zero.}$$

$$4\pi r^3 = 2000 \qquad \text{Rearrange.}$$

$$r = \sqrt[3]{\frac{500}{\pi}} = 5.42. \qquad \text{Solve for } r.$$

So something happens at $r = \sqrt[3]{500/\pi}$, but what?

To find out, we calculate d^2A/dr^2:

$$\frac{dA}{dr} = 4\pi r - \frac{2000}{r^2},$$

$$\frac{d^2A}{dr^2} = 4\pi + \frac{4000}{r^3}.$$

This derivative is positive throughout the domain of A. Hence the graph of A is concave up throughout its entire domain, and the value of A at $r = \sqrt[3]{500/\pi}$ is an absolute minimum. Fig. 4.51 shows the graph of A as a function of r.

When $r = \sqrt[3]{500/\pi}$,

$$h = \frac{1000}{\pi r^2} = 2\sqrt[3]{500/\pi} = 2r. \qquad \text{After some arithmetic}$$

Notice how using the exact value of r allows us to determine the relationship $h = 2r$. Thus the most efficient can is one in which the height equals the diameter. So

$$r = 5.42 \text{ cm, and } h = 10.84 \text{ cm.}$$

The value of r can be supported by using graphing zoom-in (see Fig. 4.51).

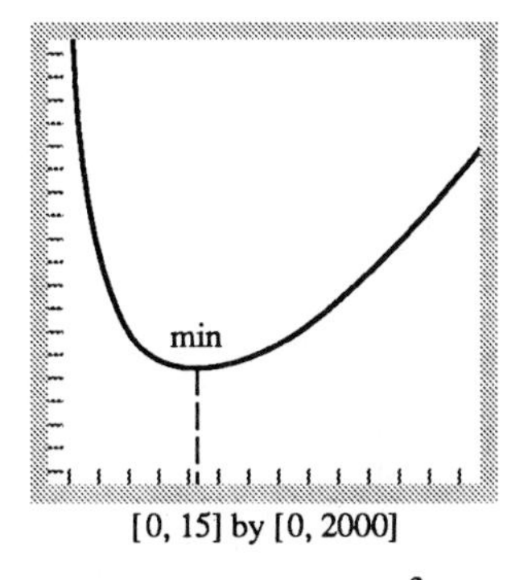

4.51 The graph of $A = 2\pi r^2 + 2000/r$ is concave up for $r > 0$.

EXAMPLE 4 Maximizing profit Suppose a manufacturer can sell x items a week for a revenue of $r(x) = 200x - 0.01x^2$ cents, and it costs $c(x) = 50x + 20{,}000$ cents to make x items. Is there a most profitable number of items to make each week? If so, use analytic calculus to find what it is.

Solution Profit is revenue minus cost, so the weekly profit on x items is

$$p(x) = r(x) - c(x) = 150x - 0.01x^2 - 20,000.$$

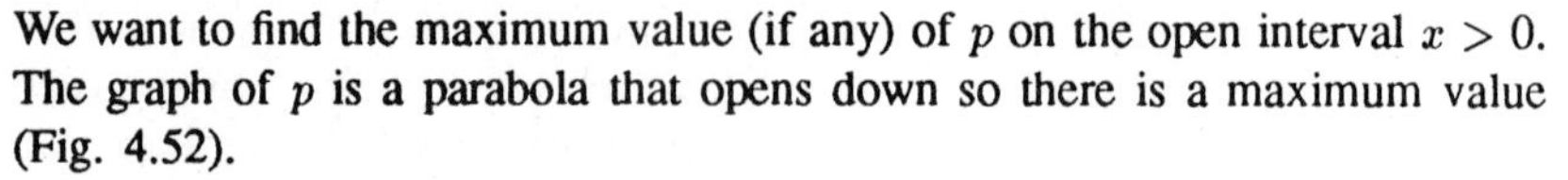

We want to find the maximum value (if any) of p on the open interval $x > 0$. The graph of p is a parabola that opens down so there is a maximum value (Fig. 4.52).

The maximum occurs where

$$\frac{dp}{dx} = 150 - 0.02x = 0,$$

which means at

$$x = \frac{150}{0.02} = 7500.$$

The local maximum is an absolute maximum.

To answer the questions, then, there *is* a production level for maximum profit, and that level is $x = 7500$ units per week. ∎

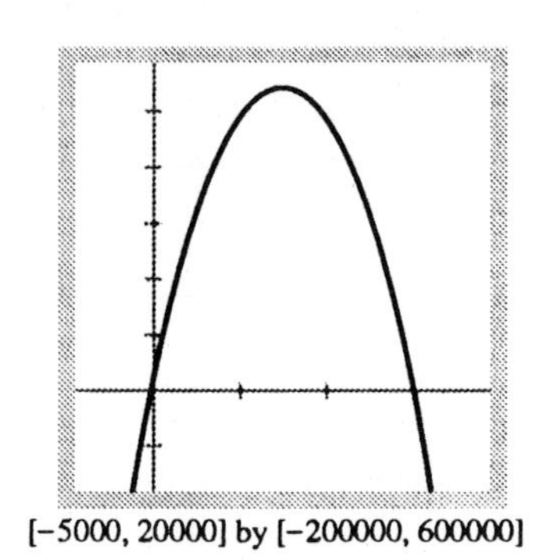

[−5000, 20000] by [−200000, 600000]

4.52 A complete graph of $p(x) = 150x - 0.01x^2 - 20{,}000$.

Cost and Revenue in Economics

Here we want to point out two of the many places where calculus makes an important contribution to economic theory. The first has to do with the relationship between profit, revenue (money received), and cost. Suppose that

$$r(x) = \text{ the revenue from selling } x \text{ items,}$$

$$c(x) = \text{ the cost of producing the } x \text{ items,}$$

$$p(x) = r(x) - c(x) = \text{ the profit from selling } x \text{ items.}$$

The marginal revenue and cost at this production level (x items) are

$$\frac{dr}{dx} = \text{marginal revenue,}$$

$$\frac{dc}{dx} = \text{marginal cost.}$$

The first theorem is about the relationship of p to these derivatives.

Theorem 9

> Maximum profit (if any) occurs at a production level at which marginal revenue equals marginal cost.

Proof of Theorem 9 We assume that $r(x)$ and $c(x)$ are differentiable for all $x > 0$, so if $p(x) = r(x) - c(x)$ has a maximum value, it occurs at a production level at which $p'(x) = 0$. Since $p'(x) = r'(x) - c'(x)$, $p'(x) = 0$ implies

$$r'(x) - c'(x) = 0 \quad \text{or} \quad r'(x) = c'(x).$$

This concludes the proof (Fig. 4.53). ∎

What guidance do we get from this theorem? We know that a production level at which $p'(x) = 0$ need not be a level of maximum profit. It might

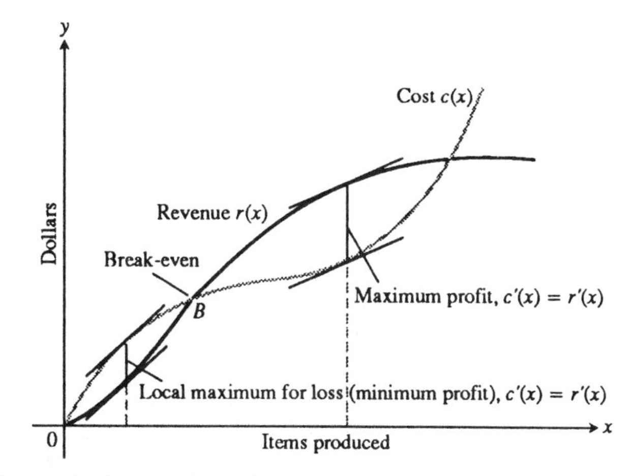

4.53 The graph of a typical cost function starts concave down and later turns concave up. It crosses the revenue curve at the break-even point B. To the left of B, the company operates at a loss. To the right, the company operates at a profit, with the maximum profit occurring where $c'(x) = r'(x)$. Farther to the right, cost exceeds revenue (perhaps because of a combination of market saturation and rising labor and material costs) and production levels become unprofitable again.

be a level of minimum profit, for example. But if we are making financial projections for our company, we should look for production levels at which marginal cost seems to equal marginal revenue. If there is a most profitable production level, it will be one of these.

EXAMPLE 5 Suppose that

$$r(x) = 10x \quad \text{and} \quad c(x) = x^3 - 6x^2 + 15x + 5,$$

where x represents thousands of units. Is there a production level that maximizes profit? If so, what is it?

Solution

$$r(x) = 10x, \quad c(x) = x^3 - 6x^2 + 15x + 5 \qquad \text{Find } r'(x) \text{ and } c'(x).$$

$$r'(x) = 10, \quad c'(x) = 3x^2 - 12x + 15$$

$$3x^2 - 12x + 15 = 10 \qquad \text{Set them equal.}$$

$$3x^2 - 12x + 5 = 0 \qquad \text{Rearrange.}$$

Using the quadratic formula we find that $x = 2 \pm \sqrt{21}/3$ or $x = 0.472$ and $x = 3.528$. The possible production levels for maximum profit are $x = 0.472$ thousand units and $x = 3.527$ thousand units. A quick glance at the graphs in Fig. 4.54 or at the corresponding values of r and c shows $x = 3.528$ to be a point of maximum profit and $x = 0.472$ to be a local maximum for loss. ∎

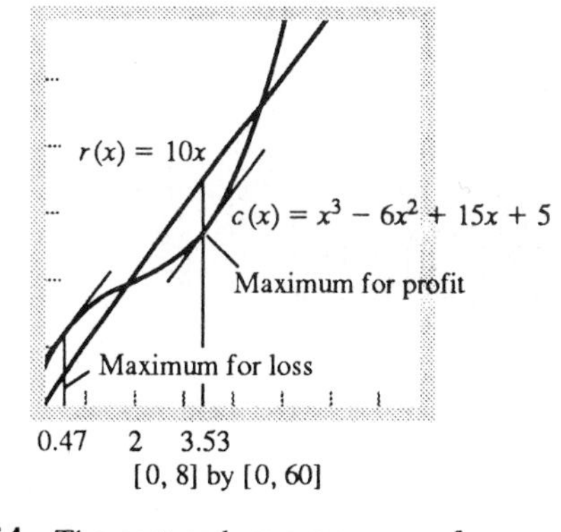

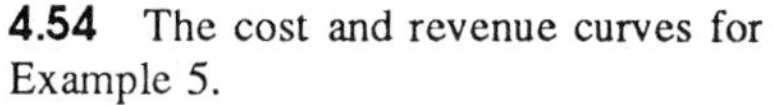

4.54 The cost and revenue curves for Example 5.

Another way to look for optimal production levels is to look for levels that minimize the average cost of the units produced. The next theorem gives a way to find them.

Theorem 10

> The production level (if any) at which average cost is smallest is a level at which the average cost equals the marginal cost.

Proof of Theorem 10 We start with

$$c(x) = \text{ cost of producing } x \text{ items, } x > 0,$$

$$\frac{c(x)}{x} = \text{ average cost of producing } x \text{ items,}$$

assumed differentiable.

If the average cost can be minimized, it will be at a production level at which

$$\frac{d}{dx}\left(\frac{c(x)}{x}\right) = 0$$

$$\frac{xc'(x) - c(x)}{x^2} = 0 \qquad \text{Quotient Rule}$$

$$xc'(x) - c(x) = 0 \qquad \text{Multiply by } x^2$$

$$\underbrace{c'(x)}_{\text{Marginal cost}} = \underbrace{\frac{c(x)}{x}}_{\text{Average cost}}.$$

This completes the proof. ∎

Again we have to be careful about what the theorem does and does not say. It does not say that there is a production level of minimum average cost, but it does say where to look to see if there is one. Look for production levels at which average cost and marginal cost are equal. Then check to see if any of them gives a minimum average cost.

EXAMPLE 6 Suppose $c(x) = x^3 - 6x^2 + 15x + 5$ (x in thousands of units). Is there a production level that minimizes average cost? If so, what is it?

Solution Theorem 10 tells us to look for production levels at which average cost equals marginal cost:

Cost: $c(x) = x^3 - 6x^2 + 15x + 5$

Marginal cost: $c'(x) = 3x^2 - 12x + 15,$

Average cost: $\frac{c(x)}{x} = x^2 - 6x + 15 + \frac{5}{x}$

$$3x^2 - 12x + 15 = x^2 - 6x + 15 + \frac{5}{x} \qquad \text{(MC = AC)}$$

$$2x^2 - 6x - \frac{5}{x} = 0.$$

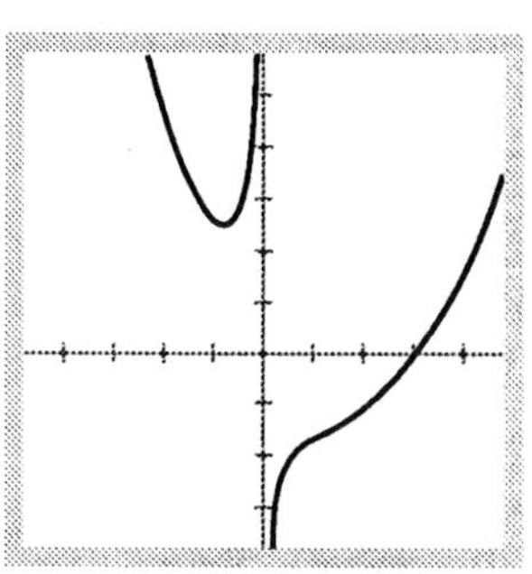

[−5, 5] by [−20, 30]

4.55 A complete graph of $f(x) = 2x^2 - 6x - 5/x$.

Now $f(x) = 2x^2 - 6x - 5/x$ is a rational function with end behavior asymptote $y = 2x^2 - 6x$ and vertical asymptote $x = 0$ (y-axis). Figure 4.55 gives a complete graph of f. Notice that f has 3.238 as its only real zero.

The only production level that might minimize average cost is $c = 3.238$ thousand units.

We check the derivatives:

$$\frac{c(x)}{x} = x^2 - 6x + 15 + \frac{5}{x} \quad \text{(Average cost)}$$

$$\frac{d}{dx}\left(\frac{c(x)}{x}\right) = 2x - 6 - \frac{5}{x^2},$$

$$\frac{d^2}{dx^2}\left(\frac{c(x)}{x}\right) = 2 + \frac{10}{x^3} > 0 \text{ for } x > 0.$$

The second derivative is positive, so the graph of the average cost is concave up. Therefore the value 3.238, where the first derivative is zero, gives an absolute minimum. We can also use a graph of the average cost $y = c(x)/x$ to support that there is an absolute minimum at $x = 3.238$. ■

Modeling Discrete Phenomena with Differentiable Functions

In case you are wondering how we can use differentiable functions $c(x)$ and $r(x)$ to describe the cost and revenue that come from producing a number of items x that can only be an integer, here is the rationale.

When x is large, we can reasonably fit the cost and revenue data with smooth curves $c(x)$ and $r(x)$ that are defined not only at integer values of x but at the values in between. Once we have these differentiable functions, which are supposed to behave like the real cost and revenue when x is an integer, we can apply calculus to come to conclusions about their values. We then translate these mathematical conclusions into inferences about the real world that we hope will have predictive value. When they do, as is the case with the economic theory here, we say that the functions give a good model of reality.

What do we do when our calculus tells us that the best production level is a value of x that isn't an integer, as it did in Example 5 when it said that $x = 2 + \sqrt{21}/3$ thousand units would be the production level for maximum profit? The answer is to use the nearest convenient integer. For $x = 2+\sqrt{21}/3$ thousand, we might use 3528, or perhaps 3520 or 3530 if we ship in boxes of 10.

Exercises 4.4

In Exercises 1–6 find the inflection points, local maximum and minimum values, and identify the intervals on which the graph are rising, falling, concave up, and concave down analytically. Then sketch a complete graphs and support your answers with a graphing utility.

1. $y = \dfrac{x^2 - 1}{x}$

2. $y = \dfrac{x^2 + 4}{2x}$

3. $y = \dfrac{x^4 + 1}{x^2}$

4. $y = \dfrac{x^3 + 1}{x^2}$

5. $y = \dfrac{x}{x^2 - 4}$ **6.** $y = \dfrac{x - 1}{x^3 - 2x^2}$

Draw a complete graph of the functions in Exercises 7–12 and use analytic calculus to confirm that the graphs are complete. Find the inflection point, local maximum and minimum values, and identify the intervals on which the graphs are rising, falling, concave up, and concave down.

7. $y = \dfrac{1}{x^2 - 1}$ **8.** $y = \dfrac{x^2}{x^2 - 1}$

9. $y = -\dfrac{x^2 - 2}{x^2 - 1}$ **10.** $y = \dfrac{x^2 - 4}{x^2 - 2}$

11. $y = \dfrac{x^2 - 4}{x - 1}$ **12.** $y = -\dfrac{x^2 - 4}{x + 1}$

Draw a complete graph of the functions in Exercises 13–26. Find the inflection points, local maximum and minimum values, and identify the intervals on which the graphs are rising, falling, concave up, and concave down.

13. $y = x^2 - x + \dfrac{1}{x + 2}$ **14.** $y = x + \dfrac{4}{x^2}$

15. $y = \dfrac{x + 1}{x^2 + 1}$ **16.** $y = \dfrac{x^2 + 1}{x^3 - 4x}$

17. $y = \dfrac{2x^3 + 3x^2 - 2x + 1}{x^2 + 2x}$

18. $y = \dfrac{3x^3 - 7x^2 - 7x - 1}{x^2 - 3x}$

19. $y = \dfrac{2x^3 - 4x^2 + 3}{x - 2}$

20. $y = \dfrac{-x^3 - 3x^2 + x + 5}{x + 3}$

21. $y = \dfrac{x^4 - 3x^2 + 4x + 1}{x^2 + x - 2}$

22. $y = \dfrac{2x^4 + 3x^3 + x^2 + x - 3}{x^2 + x}$

23. $y = \dfrac{2x^5 - 18x^3 + 2}{x^2 - 9}$

24. $y = \dfrac{x^5 + 3x^4 - 11x^3 - 3x^2 + 9x - 4}{x^2 + 3x - 10}$

25. $y = \dfrac{8}{x^2 + 4}$ (Agnesi's witch)

26. $y = \dfrac{4x}{x^2 + 4}$ (Newton's serpentine)

27. What is the smallest perimeter possible for a rectangle whose area is 16 in^2?

28. You are planning to make an open rectangular box with a square base that will hold a volume of 50 ft^3. What are the dimensions of the box with minimum surface area?

29. A 216-m^2 rectangular pea patch is to be enclosed by a fence and divided into two equal parts by another fence parallel to one of the sides. What dimensions for the outer rectangle will require the smallest total length of fence? How much fence will be needed?

30. THE LIGHTEST STEEL HOLDING TANK Your iron works has contracted to design and build a 500-ft^3, square-based, open-top, rectangular steel holding tank for a paper company. The tank is to be made by welding half-inch-thick stainless steel plates together along their edges. As the production engineer, your job is to find dimensions for the base and height that will make the tank weigh as little as possible. What dimensions do you tell the shop to use?

31. CATCHING RAIN WATER An 1125-ft^3 open-top rectangular tank with a square base x ft on a side and y ft deep is to be built with its top flush with the ground to catch runoff water. The costs associated with the tank involve not only the material from which the tank is made but also an excavation charge proportional to the product xy. If the total cost is

$$c = 5(x^2 + 4xy) + 10xy,$$

what values of x and y will minimize it?

32. You are designing a poster to contain 50 in^2 of printing with margins of 4 in. each at top and bottom and 2 in. at each side. What overall dimensions will minimize the amount of paper used?

33. What are the dimensions of the lightest open-top right circular cylindrical can that will hold a volume of 1000 cm^3? Compare the result here with the result in Example 3.

34. You are designing 1000-cm^3 right circular cylindrical cans whose manufacture will take waste into account. There is no waste in cutting the aluminum for the sides, but the tops and bottoms of radius r will be cut from squares that measure 2 units on a side. The total amount of aluminum used up by each can will therefore be

$$A = 8r^2 + 2\pi rh$$

rather than the $A = 2\pi r^2 + 2\pi rh$ in Example 3. In Example 3 the ratio of h to r for the most economical cans was 2 to 1. What is the ratio now?

35. What value of a makes

$$f(x) = x^2 + \frac{a}{x}$$

have

a) a local minimum at $x = 2$?

b) a point of inflection at $x = 1$?

36. Show that

$$f(x) = x^2 + \frac{a}{x}$$

cannot have a local maximum for any value of a.

37. Show that the function $y = 6x^2 + 8x + 19$, the numerator of f' of Example 1, is greater than 0 for all x.

38. Show that the function $f(x) = x^2 - x + 1$ is never negative.

39. A rectangular sheet of $8\frac{1}{2}$-in. × 11-in. paper is placed on a flat surface, and one of the corners is lifted up and placed on the opposite longer edge. (The other corners are held in their original positions.) With all four corners now held fixed, the paper is smoothed flat (Fig. 4.56). Make the length of the crease as small as possible (call the length L).

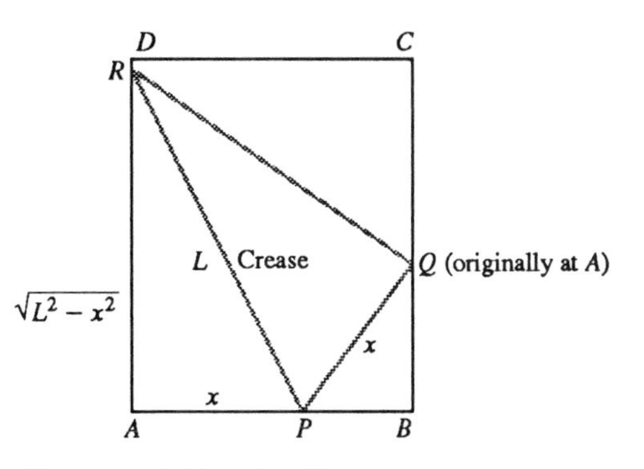

4.56 The paper in Exercise 39.

a) Try it with paper.

b) Show that $L^2 = 2x^3/(2x - 8.5)$.

c) Minimize L^2.

d) Find the minimum value of L.

40. Let $f(x)$ and $g(x)$ be the differentiable functions graphed in Fig. 4.57. Point c is the point where the vertical distance between the curves is the greatest. Show that the tangents to the curves at $x = c$ have to be parallel.

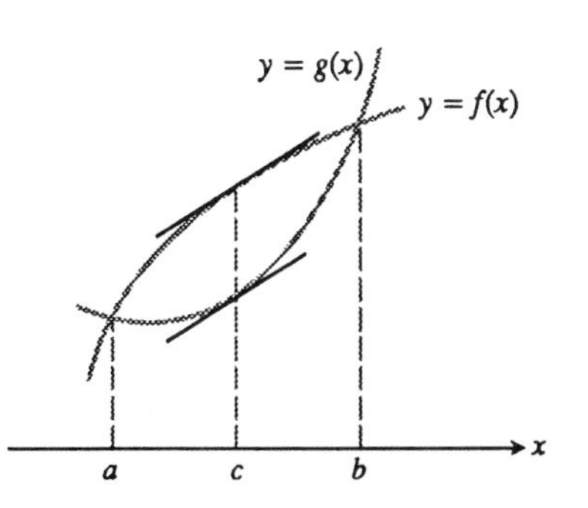

4.57 The graphs for Exercise 40.

Business

41. It costs a manufacturer c dollars each to manufacture and distribute a certain item. If the items sell at x dollars each, the number sold is given by $N = a/(x - c) + b(100 - x)$, where a and b are certain positive constants. What selling price will bring a maximum profit?

42. You operate a tour service that offers the following rates:

a) $200 per person if 50 people (the minimum number to book the tour) go on the tour.

b) For each additional person, up to a maximum of 80 people total, everyone's charge is reduced by $2.

It costs you $6000 (a fixed cost) plus $32 per person to conduct the tour. How many people does it take to maximize your profit?

43. THE BEST QUANTITY TO ORDER One of the formulas for inventory management says that the average weekly cost of ordering, paying for, and holding merchandise is

$$A(q) = \frac{km}{q} + cm + \frac{hq}{2},$$

where q is the quantity you order when things run low (shoes, radios, brooms, or whatever the item might be), k is the cost of placing an order (the same, no matter how often you order), c is the cost of one item (a constant), m is the number of items sold each week (a constant), and h is the weekly holding cost per item (a constant that takes into account things such as space, utilities, insurance, and security). Your job, as the inventory manager for your store, is to find the quantity that will minimize $A(q)$. What is it? (The formula you get for the answer is called the *Wilson lot size formula.*)

44. CONTINUATION OF EXERCISE 43 Shipping costs sometimes depend on order size. When they do, it is more realistic to replace k by $k + bq$, the sum of k and a constant multiple of q. What is the most economical quantity to order now?

Economics

45. Use Theorem 9 to show that if $r(x) = 6x$ and $c(x) = x^3 - 6x^2 + 15x$ are your revenue and cost functions, then your operation will never be profitable and the best you can do is break even (have revenue equal cost).

46. Suppose $c(x) = x^3 - 20x^2 + 20,000x + 1000$ is the cost of manufacturing x items. Use Theorem 10 to find a production level that will minimize the average cost of making x items.

47. Let $f(x)$ be a rational function.

a) Show that the domains of f, f', f'', and so forth are the same.

b) Show that the vertical asymptotes of f, f', f'', and so forth are the same.

48. Let $f(x)$ be a rational function with $g(x)$ its end behavior asymptotes. Show that $g'(x)$ and $g''(x)$ are the end behavior asymptotes of f' and f'', respectively.

4.5 Radical and Transcendental Functions

In this section we use calculus and graphing utilities to expand our understanding about radical and transcendental functions. We use the numerical derivative feature of graphing utilities and delay computations of exact derivatives until later in the book. Applications with these functions as models will be explored.

Radical Functions

The domain and range of $y = \sqrt{x}$ is $[0, \infty)$ if n is even and $(-\infty, \infty)$ if n is odd. The graphs of $y = \sqrt[n]{x}$ with n even all have behavior similar to that illustrated in Example 1.

EXAMPLE 1 Draw a complete graph of $f(x) = \sqrt{x}$ and use analytic calculus to confirm its behavior.

Solution Figure 4.58(a) gives a complete graph of f. We can see that f is increasing in its domain $[0, \infty)$, and has no local extrema. The graph of f has no points of inflection, and is concave down on its domain. The graph of f' and f'' in Figs. 4.58 (b) and (c) confirm this behavior. Notice that

$$f'(x) = \frac{1}{2}x^{-\frac{1}{2}} = \frac{1}{2\sqrt{x}}, \quad \text{and}$$

$$f''(x) = -\frac{1}{4}x^{-\frac{3}{2}} = -\frac{1}{4\sqrt{x^3}}.$$

Neither derivative exists at $x = 0$. Also $f' > 0$ in its domain $(0, \infty)$ and $f'' < 0$ in its domain $(0, \infty)$. ∎

The graphs of $y = \sqrt[n]{x}$ with n odd all have behavior similar to that illustrated in Example 2.

EXAMPLE 2 Draw a complete graph of $f(x) = \sqrt[3]{x}$ and use analytic calculus to confirm its behavior.

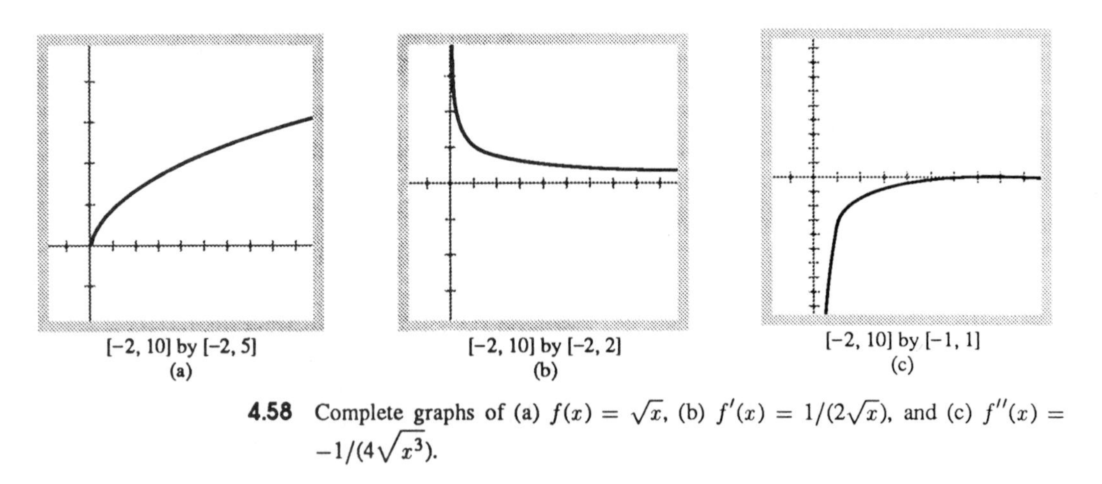

4.58 Complete graphs of (a) $f(x) = \sqrt{x}$, (b) $f'(x) = 1/(2\sqrt{x})$, and (c) $f''(x) = -1/(4\sqrt{x^3})$.

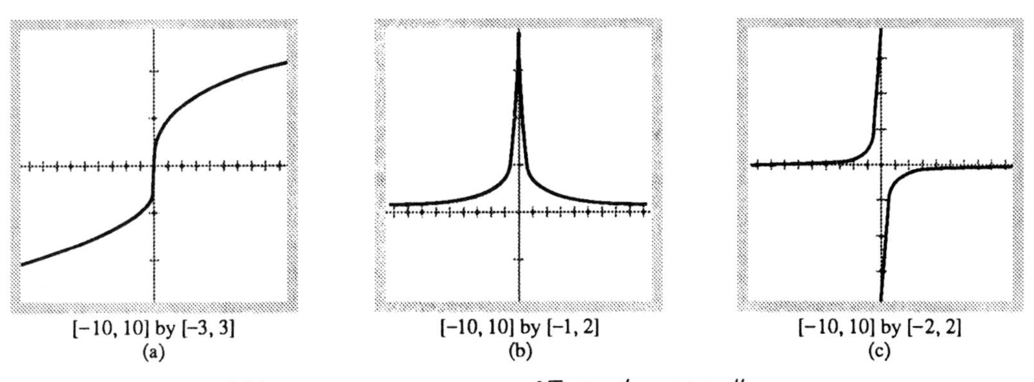

4.59 The graph of (a) $f(x) = \sqrt[3]{x}$, (b) f', and (c) f''.

Solution Fig. 4.59 (a) gives a complete graph of f. Notice that f is increasing in

its domain $(-\infty, \infty)$ and has no local extrema. Differentiating, we have

$$f'(x) = \frac{1}{3}x^{-2/3} = \frac{1}{3\sqrt[3]{x^2}}, \quad \text{and}$$

$$f''(x) = -\frac{2}{9}x^{-5/3} = \frac{-2}{9\sqrt[3]{x^5}}.$$

Neither derivative exists at $x = 0$. The graph of f' in Fig. 4.59(b) shows that $f' > 0$ for $x \neq 0$, confirming that f is always increasing and has no local extrema. The graph of f'' in Fig. 4.59(c) shows that $f'' > 0$ for $x < 0$ and $f'' < 0$ for $x > 0$. Thus, the graph of f is concave up in $(-\infty, 0)$ and concave down in $(0, \infty)$. We do *not* say that f has a point of inflection at $(0, 0)$ because, by definition, f' and f'' must exist at a point of inflection. ■

We must be careful when graphing functions of the form $y = x^{m/n}$ with graphing utilities. For example, entering the exponent in the form 2/3 will sometimes produce an incorrect graph of $y = x^{2/3}$. Try this with your graphing utility and compare with the graph of Example 3. If you get the wrong graph, try entering the function in the form $y = (x^2)^{1/3}$.

EXAMPLE 3 Find the absolute maximum and minimum values of $y = x^{2/3}$ on the interval $-2 \leq x \leq 3$ analytically. Support your answers with a graphing utility.

Solution We evaluate the function at the critical points and endpoints and take the largest and smallest of these values.

The first derivative,

$$y' = \frac{2}{3}x^{-1/3} = \frac{2}{3\sqrt[3]{x}},$$

has no zeros but is undefined at $x = 0$. The values of the function at this one critical point and at the endpoints are

critical point value: $f(0) = 0$

endpoint values: $f(-2) = (-2)^{2/3} = 4^{1/3} = 1.59,$

$f(3) = (3)^{2/3} = 9^{1/3} = 2.08.$

We conclude that the function's maximum value is $9^{1/3}$, taken on at $x = 3$. The minimum value is 0, taken on at $x = 0$. The complete graph of f in Fig. 4.60 supports these conclusions. ■

THE LEAST AND THE GREATEST

Many problems of the seventeenth century that motivated the development of the calculus were maxima and minima problems. Often these problems came from research in physics, such as finding the maximum range of a cannon. Galileo showed that the maximum range of a cannon is obtained with a firing angle of 45 degrees above the horizontal. He also found formulas for predicting maximum heights reached by projectiles fired at various angles to the ground. Pierre de Fermat worked on other problems of maxima and minima, culminating in his principle of least time (see Exercise 67). This was generalized by Sir William Hamilton in his principle of least action—one of the most powerful underlying ideas in physics.

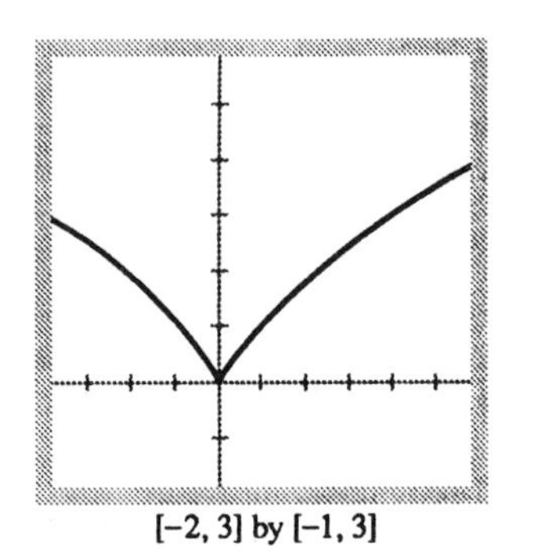

4.60 The graph of $f(x) = x^{2/3}$ showing extreme values on $-2 \leq x \leq 3$.

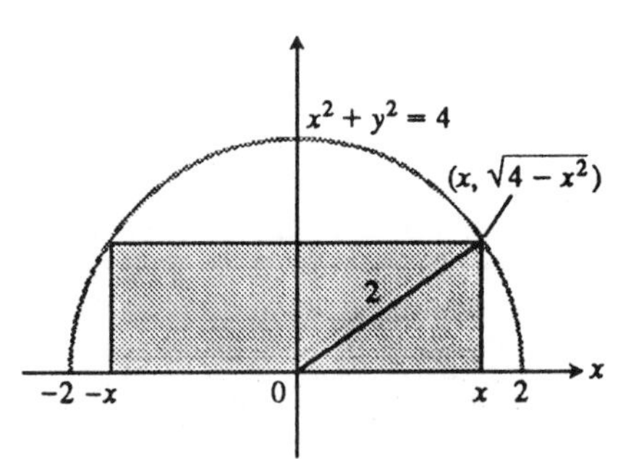

4.61 The rectangle and semicircle in Example 4.

EXAMPLE 4 Geometry A rectangle is to be inscribed in a semicircle of radius 2. Find the largest area the rectangle and its dimensions analytically. Support your answer with a graphing utility.

Solution To describe the dimensions of the rectangle, we place the circle and rectangle in the coordinate plane (Fig. 4.61). The length, height, and area of the rectangle can then be expressed in terms of the position x of the lower right-hand corner:

$$\text{Length: } 2x$$

$$\text{Height: } \sqrt{4 - x^2}$$

$$\text{Area: } 2x \cdot \sqrt{4 - x^2}.$$

Our mathematical goal is now to find the absolute maximum value of the continuous function

$$A(x) = 2x\sqrt{4 - x^2}$$

on the interval $0 \leq x \leq 2$. We do this by examining the values of A at the critical points and endpoints.

The derivative

$$\frac{dA}{dx} = \frac{-2x^2}{\sqrt{4 - x^2}} + 2\sqrt{4 - x^2}$$

is not defined when $x = 2$ and is equal to zero when

$$\frac{-2x^2}{\sqrt{4 - x^2}} + 2\sqrt{4 - x^2} = 0$$

$$-2x^2 + 2(4 - x^2) = 0$$

$$8 - 4x^2 = 0$$

$$x^2 = 2$$

$$x = \pm\sqrt{2}.$$

For $0 \leq x \leq 2$ we therefore have

$$\text{Critical point value: } A(\sqrt{2}) = 2\sqrt{2}\sqrt{4 - 2} = 4,$$

$$\text{Endpoint values: } A(0) = 0, \quad A(2) = 0.$$

The area has a maximum value of 4 when the rectangle is $2x = 2\sqrt{2}$ units long by $\sqrt{4 - x^2} = \sqrt{2}$ units high. (b) The graph in Fig. 4.62(b) supports these conclusions.

Trigonometric Functions

In Chapter 1 and in prior courses you have gained a good understanding of the graphs of $y = af(bx + c) + d$ where f is any one of the six basic trigonometry functions: $\sin x$, $\cos x$, $\tan x$, $\csc x$, $\sec x$, $\cot x$. In the exercises you will use calculus to confirm and deepen your earlier understanding.

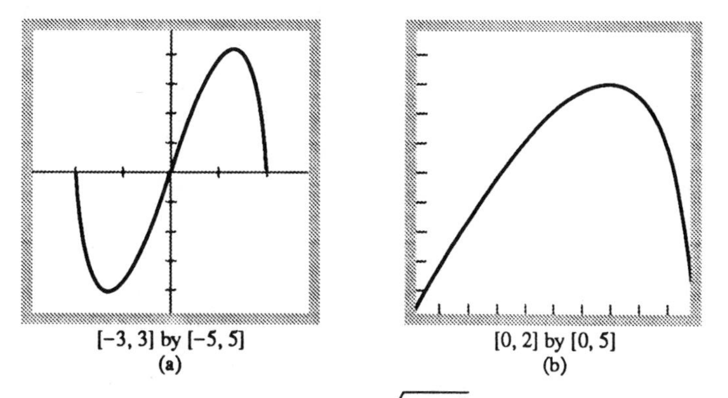

[−3, 3] by [−5, 5]
(a)

[0, 2] by [0, 5]
(b)

4.62 (a) A complete graph of $A(x) = 2x\sqrt{4 - x^2}$. (b) The graph of A in $[0, 2]$.

EXAMPLE 5 Draw a complete graph of $f(x) = \tan x$ and use analytic calculus to confirm its behavior.

Solution We know that f is a periodic function with period π. The complete graph of f in Fig. 4.63(a) shows three periods. The derivatives of f must also be periodic with period π. Notice that

$$f'(x) = \sec^2 x \quad \text{and,}$$

$$f''(x) = 2 \sec x \sec x \tan x = 2 \sec^2 x \tan x.$$

Now, $f'(x) > 0$ for all x so f is increasing in $(-\pi/2, \pi/2)$ and in any interval $(-\pi/2 + k\pi, \pi/2 + k\pi)$ obtained by shifting this interval by an integer multiple k of π (Fig. 4.62(b)). Also f has no local extrema. The graph of f'' in Fig. 4.63(c) confirms that f is concave down in $(-\pi/2, 0)$ and concave up in $(0, \pi/2)$, as well as any intervals obtained by shifting this pair of intervals by an integer multiple of π.

The graph of f has points of inflection at $(k\pi, 0)$ for every integer k. ■

We can conclude from Example 5 that it is enough to describe the behavior

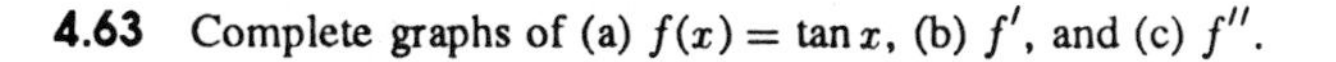

(a)

(b)
[−3π/2, 3π/2] by [−10, 10]

(c)

4.63 Complete graphs of (a) $f(x) = \tan x$, (b) f', and (c) f''.

of a periodic function in one complete period of its graph.

EXAMPLE 6 Draw a complete graph of $f(x) = \tan x - 3\cos x$. Find the inflection points, local extreme, and identify the intervals on which the graph is rising, falling, concave up, and concave down in one complete period.

Solution The period of f is a divisor of 2π because the period of $\tan x$ and $\cos x$ are π and 2π, respectively. The graph of f in Fig. 4.64(a) shows that the period of f is 2π. f has vertical asymptotes at $x = \pi/2$ and $x = 3\pi/2$. There appears to be a local maximum and a local minimum in each of the intervals $\pi/2 < x < 3\pi/2$ and $3\pi/2 < x < 2\pi$. The graph of

$$f'(x) = \sec^2 x + 3\sin x$$

in Fig. 4.64(b) supports this observation and can be used to show that the zeros of f' in $0 \le x \le 2\pi$ are 3.55, 3.98, 5.45, and 5.88. So, f has

a local maximum of 3.19 at $x = 3.55$,

a local minimum of 3.12 at $x = 3.98$,

a local maximum of -3.12 at $x = 5.45$, and

a local minimum of -3.19 at $x = 5.88$.

Thus, f is

increasing in $(0, \pi/2), (\pi/2, 3.55), (3.98, 3\pi/2),$

$(3\pi/2, 5.45)$, and $(5.88, 2\pi)$, and

decreasing in (3.55, 3.98) and (5.45, 5.88).

We can use the graph of

$$f''(x) = 2\sec^2 x \tan x + 3\cos x$$

shown in Fig. 4.64(c) to show that the zeros of f'' in $0 \le x \le 2\pi$ are 3.79 and 5.63. So f is

concave up in $(0, \pi/2)$, $(3.79, 3\pi/2)$, and $(5.63, 2\pi)$, and

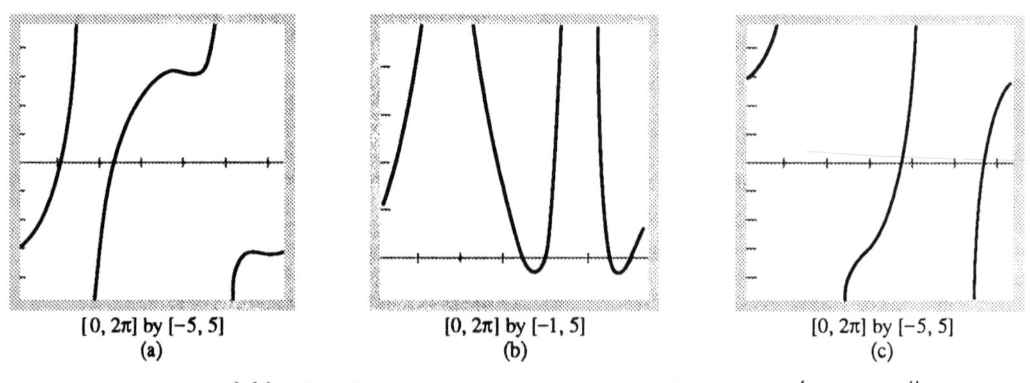

4.64 Complete graphs of (a) $f(x) = \tan x - 3\cos x$, (b) f', and (c) f''.

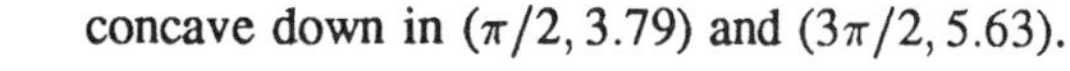

concave down in $(\pi/2, 3.79)$ and $(3\pi/2, 5.63)$.

Finally, f has points of inflection at $x = 3.79$ and 5.63. ∎

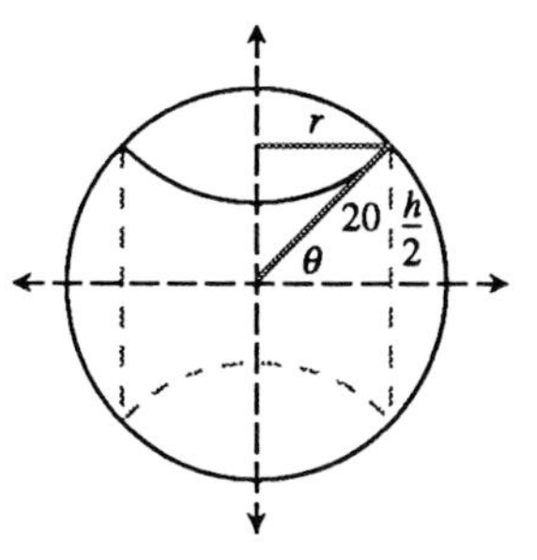

4.65 The sphere of Example 7.

EXAMPLE 7 A hole is drilled through the center of a sphere of radius 20 cm to form two spherical caps and a right circular cylinder as illustrated in Fig. 4.65. Determine the dimensions of the cylinder of largest possible volume that can be cut from the sphere, and the largest possible volume.

Solution Let r, h, and θ be determined as in Fig. 4.65. The volume of the cylinder is $V = \pi r^2 h$. Notice that $\sin\theta = (h/2)/20$ and $\cos\theta = r/20$. Thus, $V = 16{,}000\pi \sin\theta \cos^2\theta$, where $0 \le \theta \le \pi/2$. We need to find the absolute maximum of V on the interval $0 \le \theta \le \pi/2$. We can see from Fig. 4.66 that the absolute maximum occurs at the value of θ in $0 \le \theta \le \pi/2$ for which $V'(\theta) = 0$. This value of θ is 0.62 and the corresponding value of the volume is $19{,}347.19 \text{ cm}^3$. We can use calculus to confirm these values.

$$V'(\theta) = 16{,}000\pi(\cos^3\theta - 2\sin^2\theta\cos\theta) = 0$$

$$2\sin^2\theta\cos\theta = \cos^3\theta$$

$$\tan^2\theta = 1/2 \quad (\text{Since } \cos\theta \pm 0.)$$

$$\tan\theta = 1/\sqrt{2}.$$

The last step follows because $\tan\theta > 0$ in $0 \le \theta \le \pi/2$. Thus, $\theta = \tan^{-1}(1/\sqrt{2}) = 0.6154797087$ and the corresponding value of V is $19{,}347.19322 \text{ cm}^3$. Finally, $r = 16.33$ and $h = 29.09$. ∎

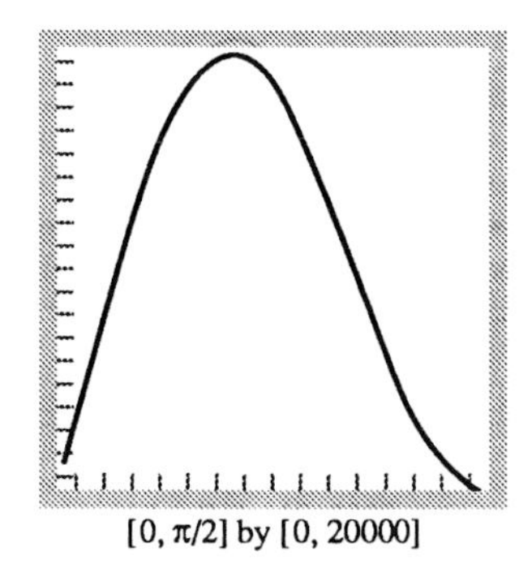

4.66 The graph of $V(\theta) = 16{,}000\pi \sin\theta \cos^2\theta$ in $0 \le \theta \le \pi/2$.

Exponential and Logarithm Functions

In Chapter 1 we defined $y = \log_a x$ as the inverse of the exponential function $y = a^x$, for $a > 0$. The domain of $y = a^k$ is $(-\infty, \infty)$ and the range is $(0, \infty)$. This means that the domain of $y = \log_a x$ is $(0, \infty)$ and its range is $(-\infty, \infty)$. The function $y = a^x$ has behavior similar to $y = 2^x$ if $a > 1$ and similar to $y = 0.5^x$ if $0 < a < 1$.

EXAMPLE 8 Draw a complete graph of (a) $f(x) = 2^x$, and (b) $g(x) = 0.5^x$.

Solution (a) Figure 4.67(a) gives a complete graph of f. The graphs of f' and f'' in Fig. 4.67(b) and (c) allow us to conclude that f is increasing and concave up in its domain $(-\infty, \infty)$, and has no local extrema or points of inflection.

We use the numerical derivative feature to find f' and f'' because we have not yet established exact derivatives of exponential functions. Later we will see that $f'(x) = (\ln 2)2^x$ and $f''(x) = (\ln 2)^2 2^x$. This explains why the graphs of f' and f'' are vertical shrinks of the graph of f.

(b) The graphs of g, g', and g'' in Fig. 4.68 allow us to conclude that g is decreasing and concave down in its domain $(-\infty, \infty)$, and has no local extrema or points of inflection.

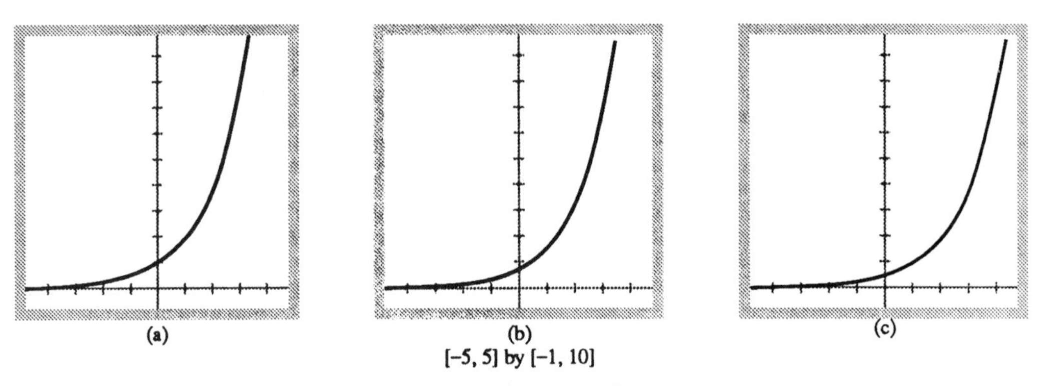

4.67 Complete graphs of (a) $f(x) = 2^x$, (b) NDer (f, x), and (c) N_2Der (f, x).

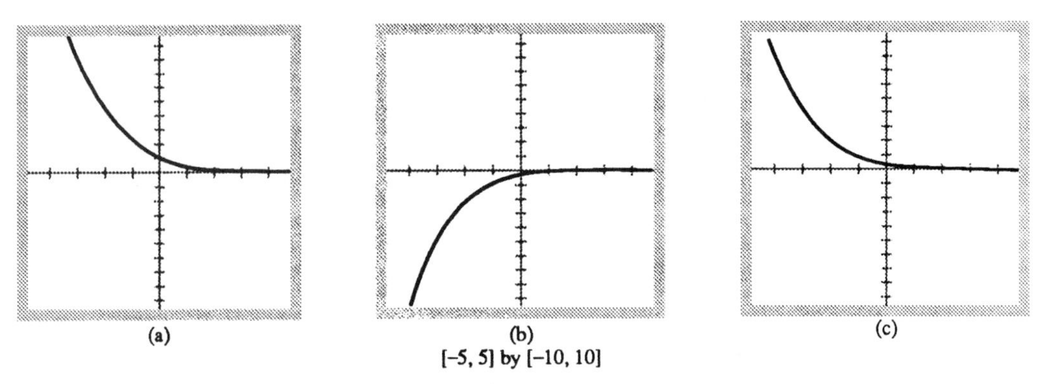

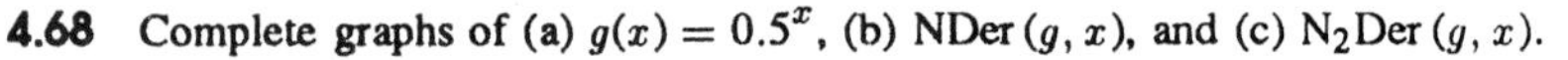
4.68 Complete graphs of (a) $g(x) = 0.5^x$, (b) NDer (g, x), and (c) N_2Der (g, x).

Notice that the graph of g is the reflection in the y-axis of the graph of f. This is because

$$g(x) = 0.5^x = \left(\frac{1}{2}\right)^x = 2^{-x} = f(-x).$$

We will show later that $g'(x) = (\ln 0.5)0.5^x$ and $g''(x) = (\ln 0.5)^2 0.5^x$. This together with $\ln 0.5 < 0$ explains the graph in 4.68(b). ∎

REMINDER

To graph $y = \log_a x$ with a graphing utility we need to use the change of base formula

$$\log_a x = \frac{\log_b x}{\log_b a}$$

established in Chapter 1. Enter $y = \log_a x$ as either

$$y = \log_a x = \frac{\log_{10} x}{\log_{10} a}, \quad \text{or}$$

$$y = \log_a x = \frac{\ln x}{\ln a}.$$

Caution: Some graphing utilities use log, which means logarithm base 10, for ln, which means logarithm base e.

Later we will establish connections between derivatives of functions that are inverses of each other. For now we will use the numerical derivative feature for the graphs of $y = \log_a x$. The function $y = \log_a x$ has behavior similar to $y = \log_2 x$ if $a > 1$ and similar to $y = \log_{0.5} x$ if $0 < a < 1$.

EXAMPLE 9 Draw a complete graph of (a) $f(x) = \log_2 x$, and (b) $g(x) = \log_{0.2} x$.

Solution

a) We write

$$f(x) = \log_2 x = \frac{\ln x}{\ln 2}$$

in order to use our graphing utility. The graphs of f, f' and f'' in Fig. 4.69 allow us to conclude that f is increasing and concave down in its domain $(0, \infty)$, and has no local extrema or points of inflection.

The graphs of $y = 2^x$ and $y = \log_2 x$ are reflections in the line $y = x$ of each other. Compare the behavior of $y = \log_2 x$ with the behavior of $y = 2^x$ given in Example 8.

b) The graphs of g, g', and g'' in Fig. 4.70 allow us to conclude that g is decreasing and concave up in its domain $(0, \infty)$, and has no extrema or points of inflection.

EXAMPLE 10 Draw a complete graph of $f(x) = xe^{-x}$.

Solution The graph of f in Fig. 4.71(a) suggests that $\lim_{x\to\infty} f = 0$ and that we need to take a closer look to the right of $x = 0$. A second view of f in Fig. 4.71(b) shows that f has a local maximum to the right of $x = 0$. This is confirmed by the graph of f' in Fig. 4.72(a) from which we can determine that f' has a zero at $x = 1$.

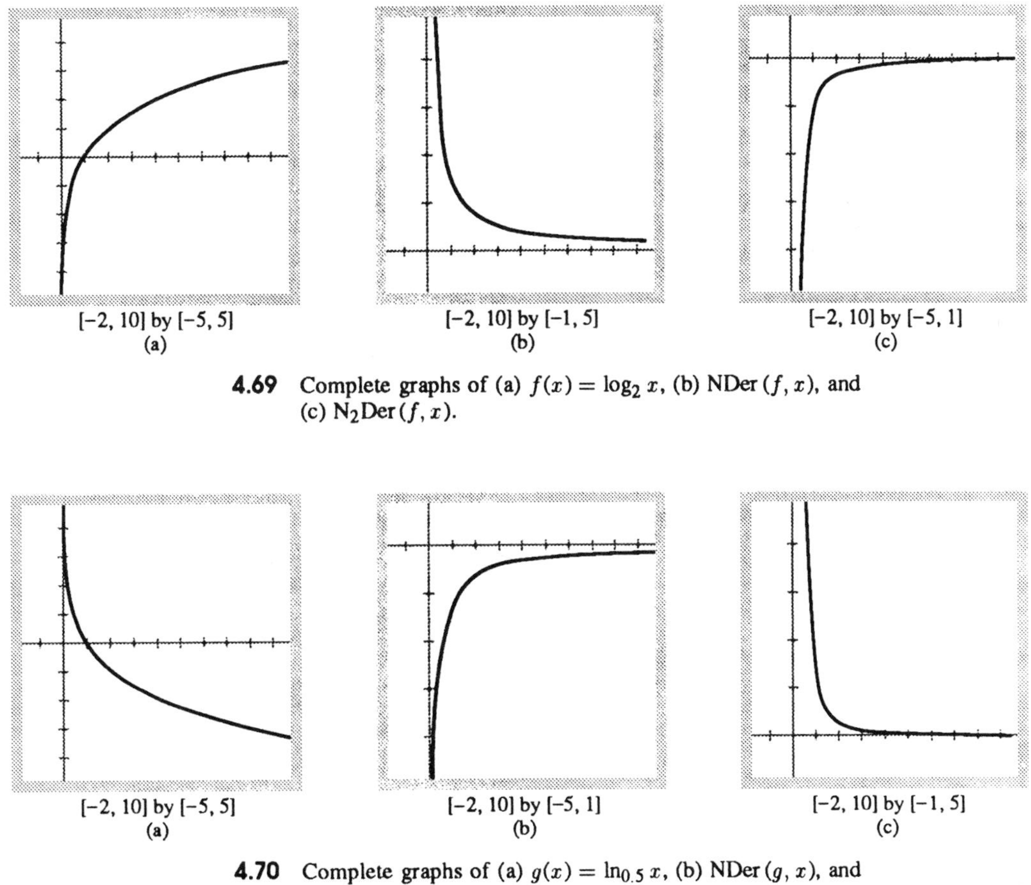

4.69 Complete graphs of (a) $f(x) = \log_2 x$, (b) NDer (f, x), and (c) $\mathrm{N_2Der}(f, x)$.

4.70 Complete graphs of (a) $g(x) = \ln_{0.5} x$, (b) NDer (g, x), and (c) $\mathrm{N_2Der}(g, x)$.

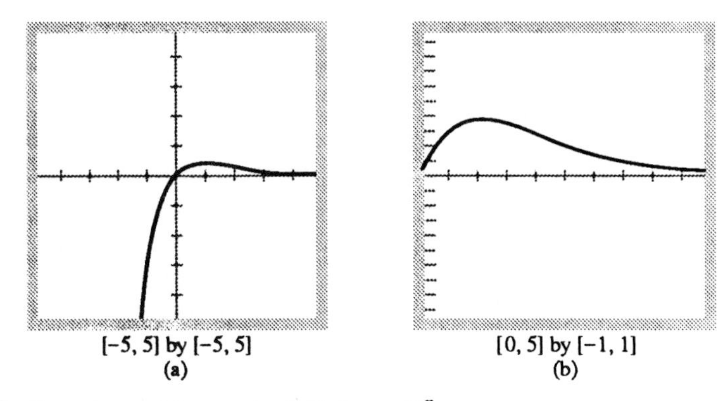

[−5, 5] by [−5, 5]
(a)

[0, 5] by [−1, 1]
(b)

4.71 Two views of the graph of $f(x) = xe^{-x}$.

A second view of the graph of f' in Fig. 4.72(b) helps us determine that f is increasing in $(-\infty, 1)$, decreasing in $(1, \infty)$, and has a local maximum of 0.37 at $x = 1$.

Two views of f'' in Fig. 4.73 help us determine that f has a point of inflection at $x = 2$ and that f is concave down in $(-\infty, 2)$ and concave up in $(2, \infty)$.

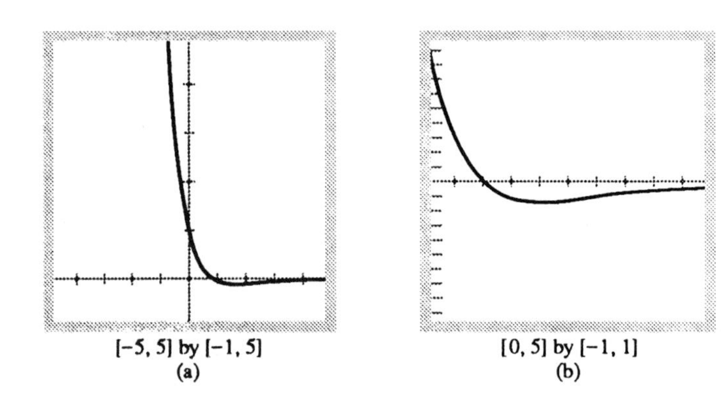

[−5, 5] by [−1, 5]
(a)

[0, 5] by [−1, 1]
(b)

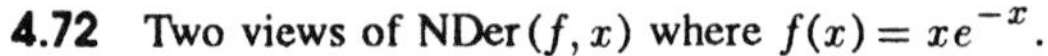

4.72 Two views of NDer(f, x) where $f(x) = xe^{-x}$.

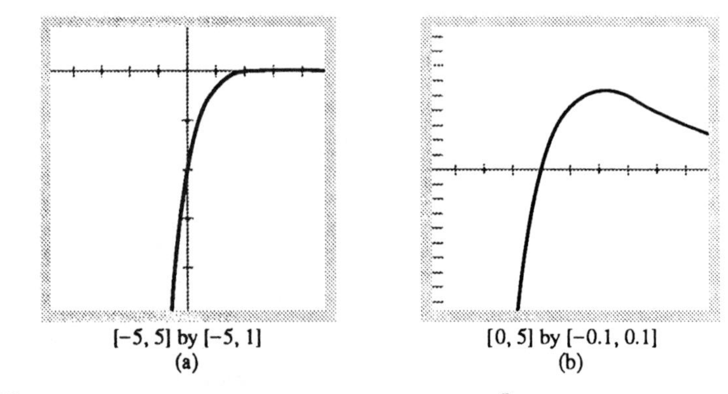

[−5, 5] by [−5, 1]
(a)

[0, 5] by [−0.1, 0.1]
(b)

4.73 Two views of $\text{N}_2\text{Der}(f, x)$ where $f(x) = xe^{-x}$.

EXAMPLE 11 Draw a complete graph of $f(x) = \dfrac{e^x + e^{-x}}{2}$.

Solution The graphs of f, f', and f'' in Fig. 4.74 allow us to conclude that f has a local minimum of 1 at $x = 0$, is decreasing in $(-\infty, 0)$, increasing in $(0, \infty)$, concave up in its domain $(-\infty, \infty)$, and has no points of inflection.

The function of Example 11 is called the *hyperbolic cosine* of x. The hyperbolic functions defined below will be studied in more detail later in the book. In the exercises you will begin to investigate their behavior.

Definition

The hyperbolic functions are defined in a way similar to the trigonometric functions.

$$\text{Hyperbolic sine of } x: \sinh x = \frac{e^x - e^{-x}}{2}$$

$$\text{Hyperbolic cosine of } x: \cosh x = \frac{e^x + e^{-x}}{2}$$

$$\text{Hyperbolic tangent of } x: \tanh x = \frac{\sinh x}{\cosh x} = \frac{e^x - e^{-x}}{e^x + e^{-x}}$$

$$\text{Hyperbolic cotangent of } x: \coth x = \frac{\cosh x}{\sinh x} = \frac{e^x + e^{-x}}{e^x - e^{-x}}$$

$$\text{Hyperbolic secant of } x: \operatorname{sech} x = \frac{1}{\cosh x} = \frac{2}{e^x + e^{-x}}$$

$$\text{Hyperbolic cosecant of } x: \operatorname{csch} x = \frac{1}{\sinh x} = \frac{2}{e^x - e^{-x}}$$

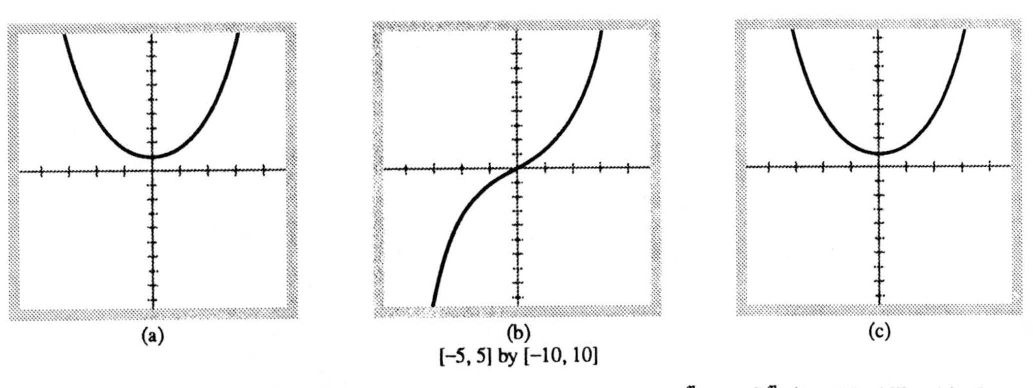

4.74 Complete graphs of (a) $f(x) = (e^x + e^{-x})/2$, (b) NDer(f, x), and (c) $\text{N}_2\text{Der}(f, x)$.

The hyperbolic functions are built-in on some graphing calculations and are important functions. They describe the motions of waves in elastic solids, the shapes of hanging electric power lines, and the temperature distributions in metal cooling fins.

Exercises 4.5

In Exercises 1–8 find the inflection points, local extrema, and identify the intervals on which the graphs are rising, falling, concave up, and concave down analytically—in only one complete period if the function is periodic. Then sketch a complete graph and suppport your answer with a graphing utility.

1. $y\sqrt[3]{x}$

2. $y = \sqrt[5]{1-x}$

3. $y = \sqrt{2x+3}$

4. $y = \sqrt[4]{x-2}+5$

5. $y = 2\sin(3x+5)+3$

6. $y = -3\cos(2x+\pi)+1$

7. $y = \sin 3x + \cos 3x$

8. $y = 5\sin 2x + 3\cos 2x$

Draw a complete graph of the functions in Exercises 9–14 and use analytic calculus to confirm that the graphs are complete. Find the inflection points, local extrema, and identify the intervals on which the graphs are rising, falling, concave up, and concave down—in only one complete period if the function is periodic.

9. $y = \sqrt[3]{3-x}+5$

10. $y = \sqrt[4]{x-4}-2$

11. $y = x^{3/4}$

12. $y = \sqrt[3]{x^2-3x}$

13. $y = 3\csc(2x+\pi)-5$

14. $y = -2\sec(3x+\pi)+7$

Draw a complete graph of the functions in Exercises 15–52. Find the inflection points, local extrema, and identify the intervals on which the graphs are rising, falling, concave up, and concave down—in only one complete period if the function is periodic.

15. $y = x^2 + \sqrt[3]{x}$

16. $y = x^{2/3} - x$

17. $y = (x+2)^{3/5}$

18. $y = (3-x)^{2/3}$

19. $y = \sec 2x$

20. $y = \tan(3x+\pi)$

21. $y = 2e^{3x+1}+5$

22. $y = 2^{x-3}+5$

23. $y = 3\log_2(x+1)$

24. $y = 2\log_3(2-x)$

25. $y = \sqrt{x^2-2x}$

26. $y = \sqrt[3]{x^3 = 4x}$

27. $y = \tanh x$

28. $y = \coth x$

29. $y = \operatorname{sech} x$

30. $y = \operatorname{csch} x$

31. $y = \cosh 2x$

32. $y = \sinh x$

33. $y = \sinh(x-1)$

34. $y = \cosh(2-x)$

35. $y = \sec x \tan x$

36. $y = \csc x \cot x$

37. $y = \sin 2x + \cos 3x$

38. $y = \sin\dfrac{x}{2} + \sin\dfrac{x}{3}$

39. $y = \cot x + 4\sin x$

40. $y = 3\sin x - \cot x$

41. $y = x\ln x$

42. $y = x^2\ln x$

43. $y = x^2\ln|x|$

44. $y = \dfrac{\ln x}{x}$

45. $y = 2^{x^2-1}$

46. $y = 3^{-x^2}$

47. $y = x2^{-x^2}$

48. $y = x3^x$

49. $y = \dfrac{x}{\sin x}$ on $-4\pi \le x \le 4\pi$

50. $y = x^2\sin x$ on $-4\pi \le x \le 4\pi$

51. $y = \dfrac{x-\sin x}{x^2+x}$

52. $y = \dfrac{1-\cos x}{x^2-2x}$

Find the absolute maximum and minimum of the functions in Exercises 53–56 on the specified interval.

53. $y = x^{3/5}$ on $-5 \le x \le 5$

54. $y = x^{4/3}$ on $-3 \le x \le 4$

55. $y = e^x \sin x$ on $-6 \le x \le 6$

56. $y = e^{-x}\cos x$ on $-5 \le x \le 5$

57. The sum of two nonnegative numbers is 20. Find the numbers if one number plus the square root of the other is to be as large as possible.

58. What is the largest possible area for a right triangle whose hypotenuse is 5 cm long?

59. Two sides of a triangle have lengths a and b, and the angle between them is θ. What value of θ will maximize the triangle's area? (*Hint:* $A = (1/2)ab\sin\theta$.)

60. Find the largest possible value of $s = 2x + y$ if x and y are side lengths in a right triangle whose hypotenuse is $\sqrt{5}$ units long.

61. You are planning to close off a corner of the first quadrant with a line segment 20 units long running from $(a, 0)$ to $(0, b)$. Show that the area of the triangle enclosed by the segment is largest when $a = b$.

62. The trough in Fig. 4.75 is to be made to the dimensions shown. Only the angle θ can be varied. What value of θ will give the trough its maximum volume?

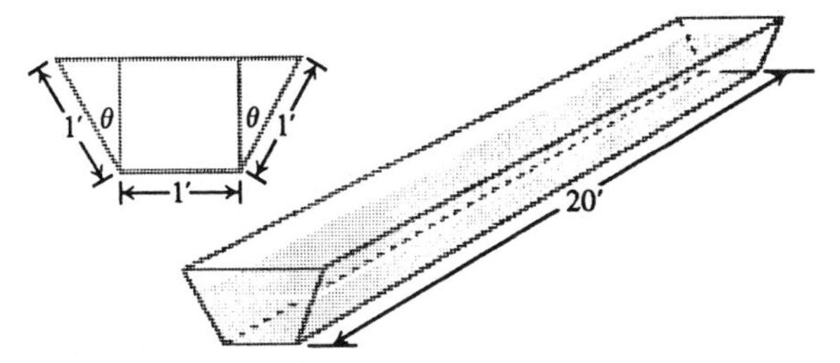

4.75 The trough in Exercise 62.

63. How close does the curve $y = \sqrt{x}$ come to the point $(3/2, 0)$?

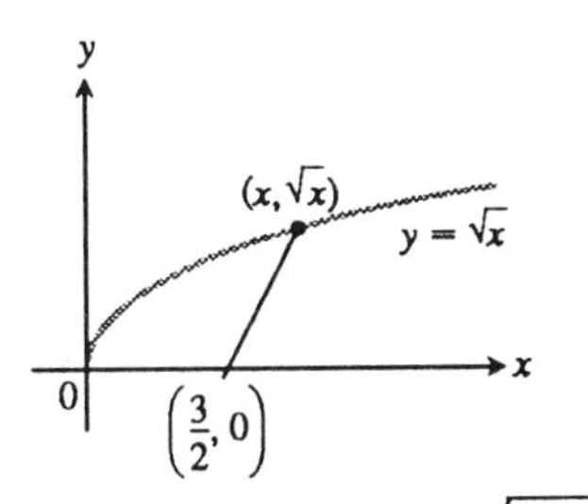

64. How close does the semicircle $y = \sqrt{16 - x^2}$ come to the point $(1, \sqrt{3})$?

65. Is the function $f(x) = 3 + 4\cos x + \cos 2x$ ever negative? How do you know?

66. Is the function $f(x) = (3^x + 3^{-x} - 2)/2$ ever negative? How do you know?

67. Fermat's principle in optics states that light always travels from one point to another along a path that minimizes the travel time. Figure 4.76 shows light from a source A reflected by a plane mirror to a receiver at point B. Show that for the light to obey Fermat's principle, the angle of incidence must equal the angle of reflection.

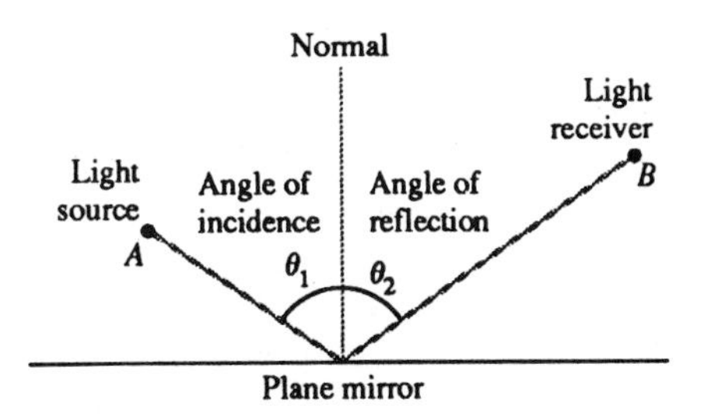

4.76 In studies of light reflection, the angles of incidence and reflection are measured from the line normal to the reflecting surface. Exercise 67 asks you to show that if light obeys Fermat's "least-time" principle, then $\theta_1 = \theta_2$.

68. Jane is 2 miles offshore in a boat and wishes to reach a coastal city 6 mi down a straight shoreline from the point nearest the boat. She can row 2 mph and can walk 5 mph. Where should she land her boat in order to reach the village in the least amount of time?

69. Find the dimensions of a right circular cylinder of maximum possible volume that can be inscribed in a sphere of radius 10 cm.

70. Rework Example 4 by writing the area of the inscribed rectangle as a function of the angle θ that the line segment from the origin to the upper right-hand corner of the rectangle makes with the positive x-axis.

4.6 Related Rates of Change

How fast does the radius change when you blow air into a spherical soap bubble at the rate of 10 cm^3/sec? How fast does the water level drop when a cylindrical tank is drained at the rate of 3 liters/sec?

Questions like these ask us to calculate the rate at which one variable changes from the rate at which another variable is known to change. To calculate that rate, we write an equation that relates the two variables and differentiate it to get an equation that relates the rate we seek to the rate we know.

EXAMPLE 1 The soap bubble How fast does the radius of a spherical soap bubble change when you blow air into it at the rate of 10 cm^3/sec? Support your answer graphically.

Solution We are given the rate at which the volume is changing and are asked for the rate at which the radius is changing.

We think abstractly at first, picturing the bubble as a sphere whose volume V and radius r are differentiable functions of time t. The equation that relates V and r is

$$V = \frac{4}{3}\pi r^3. \tag{1}$$

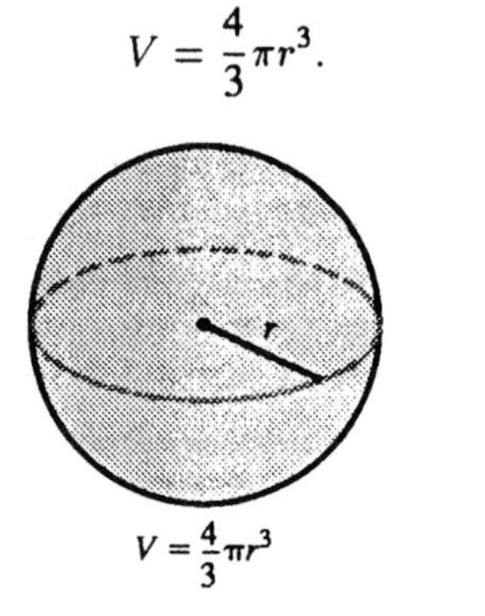

$V = \frac{4}{3}\pi r^3$

To find an equation that relates the rate we seek (dr/dt) to the rate we know (dV/dt), we use the Chain Rule to differentiate both sides of Eq. (1) with respect to t:

$$\frac{dV}{dt} = \frac{d}{dt}\left(\frac{4}{3}\pi r^3\right) = \frac{4}{3}\pi \frac{d}{dt}(r^3) = \frac{4}{3}\pi \cdot 3r^2 \frac{dr}{dt} = 4\pi r^2 \frac{dr}{dt}. \tag{2}$$

We are told that

$$\frac{dV}{dt} = 10. \quad \text{Air is blown in the rate of } 10 \text{ cm}^3/\text{sec.}$$

We substitute this value in Eq. (2) and solve for dr/dt:

$$10 = 4\pi r^2 \frac{dr}{dt} \qquad \frac{dr}{dt} = \frac{10}{4\pi r^2}. \tag{3}$$

We see from Eq. (3) that the rate at which r changes at any particular time depends on how big r is at the time. When r is small, dr/dt will be large; when r is large, dr/dt will be small:

$$\text{At } r = 1 \text{ cm}: \quad \frac{dr}{dt} = \frac{10}{4\pi} \approx 0.8 \text{ cm/sec},$$

$$\text{At } r = 10 \text{ cm}: \quad \frac{dr}{dt} = \frac{10}{400\pi} \approx 0.008 \text{ cm/sec}.$$

Because $dV/dt = 10$ and $V = 0$ when $t = 0$, we can write $V = 10t$. We say more on this in the next section. Substituting this value for V into Eq. (1), we can express r as a function of t.

$$V = \frac{4}{3}\pi r^3$$

$$10t = \frac{4}{3}\pi r^3$$

$$\frac{30t}{4\pi} = r^3$$

$$r = \sqrt[3]{\frac{7.5}{\pi}} t^{1/3}$$

Figure 4.77(a) gives a complete graph of r for $t \geq 0$. We can see that initially the slope of the tangent (dr/dt) is large and decreases over time. We can also use the graph of r' in Fig. 4.77(b) to draw the same conclusion.

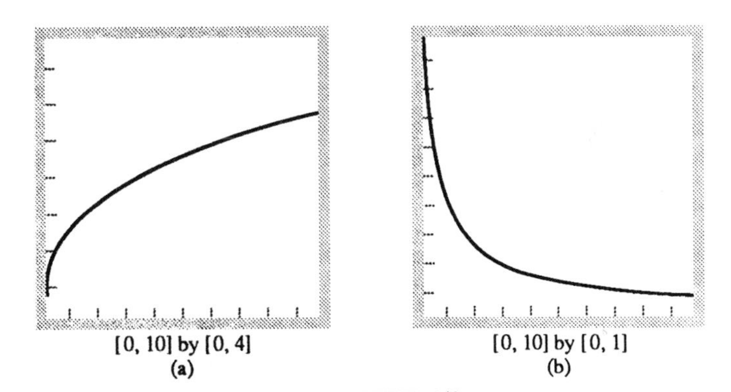

4.77 A complete graph of (a) $r = (\sqrt[3]{7.5/\pi})t^{1/3}$ for $t \geq 0$, and (b) r' for $t \geq 0$.

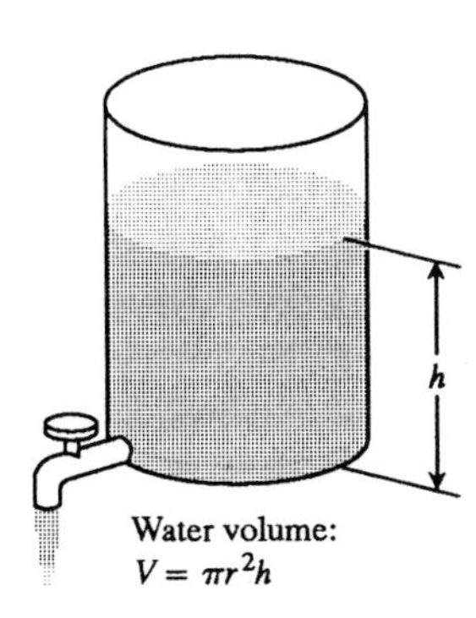

4.78 The cylindrical tank in Example 2.

EXAMPLE 2 The cylindrical tank How fast does the water level drop when a cylindrical tank is drained at the rate of 3 liters/sec?

Solution We draw a picture of a partially filled cylindrical tank, calling its radius r and the height of the water h (Fig. 4.78). We call the volume of water in the tank V.

The radius r is a constant, but V and h change with time. We think of V and h as differentiable functions of time and use t to represent time. The derivatives dV/dt and dh/dt give the rates at which V and h change. We are told that

$$\frac{dV}{dt} = -3, \quad \text{The tank is drained at the rate of 3 liters/sec.}$$

and we are asked for

$$\frac{dh}{dt}. \quad \text{How fast does the water level drop?}$$

To answer the question, we first write an equation that relates V and h:

$$V = \pi r^2 h. \quad \text{The tank is cylindrical.}$$

We then differentiate both sides with respect to t to get an equation that relates dh/dt to dV/dt:

$$\frac{dV}{dt} = \pi r^2 \frac{dh}{dt}.$$

We substitute the known value $dV/dt = -3$ and solve for dh/dt:

$$\frac{dh}{dt} = -\frac{3}{\pi r^2}. \tag{4}$$

The water level is dropping at the constant rate of $3/\pi r^2$. ■

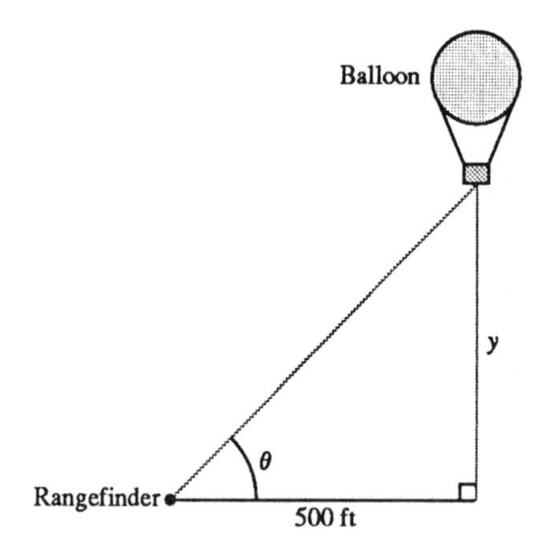

4.79 If $d\theta/dt = 0.14$ when $\theta = \pi/4$, what is the value of dy/dt when $\theta = \pi/4$? See Example 3.

EXAMPLE 3 A rising balloon A hot-air balloon, rising straight up from a level field, is tracked by a range finder 500 ft from the point of lift-off. At the moment the range finder's elevation angle is $\pi/4$, the angle is increasing at the rate of 0.14 radians/min. How fast is the balloon rising?

Solution We answer the question in six steps.

STEP 1: *We draw a picture and name the variables and constants* (Fig. 4.79). The variables in the picture are

$\theta =$ the angle the range finder makes with the ground (radians)

$y =$ the height of the balloon (feet).

We let t represent time and assume θ and y to be differentiable functions of t.

The one constant in the picture is the distance from the range finder to the point of lift-off (500 ft).

STEP 2: *We write down the additional numerical information:*

$$\frac{d\theta}{dt} = 0.14 \text{ rad/min} \qquad \text{when} \quad \theta = \frac{\pi}{4}.$$

STEP 3: *We write down what we are asked to find:* We are asked to find dy/dt when $\theta = \pi/4$.

STEP 4: *We write an equation that relates the variables:* The equation that relates y to θ is

$$\frac{y}{500} = \tan\theta \qquad \text{or} \qquad y = 500\tan\theta.$$

STEP 5: *We differentiate with respect to t to find how dy/dt* (which we want) *is related to $d\theta/dt$* (which we know):

$$\frac{dy}{dt} = 500\sec^2\theta\,\frac{d\theta}{dt}.$$

STEP 6: *We evaluate with $\theta = \pi/4$ and $d\theta/dt = 0.14$ to find dy/dt:*

$$\frac{dy}{dt} = 500(\sqrt{2})^2(0.14) = (1000)(0.14) = 140. \qquad \sec\frac{\pi}{4} = \sqrt{2}$$

At the moment in question, the balloon is rising at the rate of 140 ft/min. ■

> **Strategy for Related Rate Problems**
>
> 1. *Draw a picture and name the variables and constants.* Use t for time. Assume all variables are differentiable functions of t.
> 2. *Write down the numerical information* (in terms of the symbols you have chosen).
> 3. *Write down what you are asked to find* (usually a rate, expressed as a derivative).
> 4. *Write an equation that relates the variables.* You may have to combine two or more equations to get a single equation that relates the variable whose rate you want to the variable whose rate you know.
> 5. *Differentiate* with respect to t to express the rate you want in terms of the rate and variables whose values you know.
> 6. *Evaluate.*

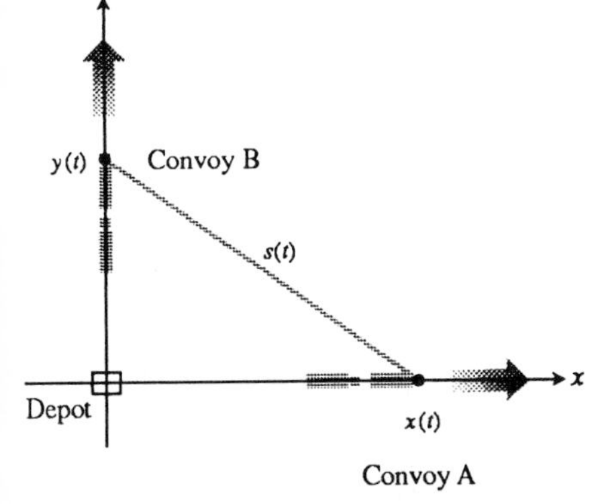

4.80 If you know where the convoys are and how fast they are moving, you can calculate how fast the distance between them is growing (Example 4)

Example 4 serves as a check of the procedures of this section. In part (b) we will see that we can solve this problem from first principles.

EXAMPLE 4 Truck convoys Two truck convoys leave a depot, convoy A traveling east at 40 mph and convoy B traveling north at 30 mph. (a) How fast is the distance between the convoys changing 6 min later, when convoy A is 4 mi from base and convoy B is 3 mi from base? (b) How fast is the distance changing at any time?

Solution (a) We carry out the steps of the basic strategy.

STEP 1: *Picture and variables:* We picture the convoys in the coordinate plane, using the positive x-axis as the eastbound highway and the positive y-axis as the northbound highway (Fig. 4.80). We let t represent time and set

$$x(t) = \text{position of convoy A},$$

$$y(t) = \text{position of convoy B},$$

$$s(t) = \text{distance between convoys}.$$

We assume x, y, and s to be differentiable functions of t.

STEP 2: *Numerical information:* At the time in question,

$$x = 4 \text{ mi}, \qquad y = 3 \text{ mi}, \qquad \frac{dx}{dt} = 40 \text{ mph}, \qquad \frac{dy}{dt} = 30 \text{ mph}.$$

STEP 3: *To find:* $\dfrac{ds}{dt}$.

STEP 4: *How the variables are related:*

$$s^2 = x^2 + y^2 \quad \text{Pythagorean theorem}$$

The equation $s = \sqrt{x^2 + y^2}$ would also work.

STEP 5: *Differentiate with respect to t:*

$$2s\frac{ds}{dt} = 2x\frac{dx}{dt} + 2y\frac{dy}{dt}$$

$$\frac{ds}{dt} = \frac{1}{s}\left(x\frac{dx}{dt} + y\frac{dy}{dt}\right).$$

STEP 6: *Evaluate* with $x = 4, y = 3, (dx/dt) = 40, (dy/dt) = 30$:

$$\frac{ds}{dt} = \frac{1}{\sqrt{4^2+3^2}}(4(40) + 3(30)) = \frac{1}{5}(160 + 90) = \frac{250}{5} = 50.$$

At the moment in question, the distance between the convoys is growing at the rate of 50 mph.

b) Using the formula $d = rt$ we can write $x = 40t$ and $y = 30t$. Thus,

$$s = \sqrt{x^2 + y^2} = \sqrt{(40t)^2 + (30t)^2} = 50t,$$

So that the distance between convoys is changing at the constant rate of 50 mph. If we substitute $s = 50t$, $x = 40t$, $y = 30t$, $dx/dt = 40$, and $dy/dt = 30$ into

$$\frac{ds}{dt} = \frac{1}{s}\left(x\frac{dx}{dt} + y\frac{dy}{dt}\right)$$

we find $ds/dt = 50$ so that the techniques of this section produce the same answer as the formula $d = rt$.

EXAMPLE 5 Relief from a heart attack A heart attack victim has been given a blood vessel dilator to lower the pressure against which the heart has to pump. For a short while after the drug is administered, the radii of the affected blood vessels will increase at about 1% per minute. According to Poiseuille's law, $V = kr^4$ (Section 3.7, Example 10), what percentage rate of increase can we expect in the blood flow over the next few minutes (all other things being equal)?

Solution

STEP 1: *Picture and variables:* We really don't need a picture, and the variables r and V are already named. It remains only to assume that r and V are differentiable functions of time t.

STEP 2: *Numerical information:*

$$\frac{dr/dt}{r} = \frac{1}{100}. \quad (r \text{ increases at } 1\%/\text{min.})$$

STEP 3: *To find:*

$$\frac{dV/dt}{V}.$$ What percentage increase can we expect in the blood flow?

STEP 4: *How the variables are related:* $V = kr^4$.

STEP 5: *Differentiate* (and in this case divide as well) to find how $(dV/dt)/V$ is related to $(dr/dt)/r$:

$$\frac{dV}{dt} = 4kr^3\frac{dr}{dt}$$

$$\frac{dV/dt}{V} = \frac{4kr^3}{kr^4}\frac{dr}{dt} = 4\frac{dr/dt}{r}. \quad \text{Divide by } V = kr^4.$$

STEP 6: *Evaluate* by substituting $(dr/dt)/r = 1/100$ to find $(dV/dt)/V$:

$$\frac{dV/dt}{V} = 4\left(\frac{1}{100}\right) = \frac{4}{100}.$$

The blood flow will increase 4%/min.

Exercises 4.6

1. The radius r and area $A = \pi r^2$ of a circle are differentiable functions of t. Write an equation that relates dA/dt to dr/dt.

2. The radius r and surface area $S = 4\pi r^2$ of a sphere are differentiable functions of t. Write an equation that relates dS/dt to dr/dt.

3. The side length x and volume $V = x^3$ of a cube are differentiable functions of t. Write an equation that relates dV/dt to dx/dt.

4. The radius r and volume $V = (1/3)\pi r^2 h$ of a right circular cone are differentiable functions of t. How is dV/dt related to dr/dt if h is constant?

5. The height h and volume $V = (1/3)\pi r^2 h$ of a right circular cone are differentiable functions of t. How is dV/dt related to dh/dt if r is constant?

6. Let $x(t)$ and $y(t)$ be differentiable functions of t and let $s = \sqrt{x^2 + y^2}$ be the distance between the points $(x, 0)$ and $(0, y)$ in the xy-plane. How is ds/dt related to dx/dt and dy/dt?

7. HEATING A PLATE. When a circular plate of metal is heated in an oven, its radius increases at the rate of 0.01 cm/min. At what rate is the plate's area increasing when the radius is 50 cm?

8. CHANGING VOLTAGE Ohm's law for electrical circuits like the one here states that $V = IR$, where V is the voltage, I is the current in amperes, and R is the resistance in ohms. Suppose that V is increasing at the rate of 1 volt/sec while I is decreasing at the rate of 1/3 amp/sec. Let t denote time in seconds.

   a) What is the value of dV/dt?

   b) What is the value of dI/dt?

   c) What equation relates dR/dt to dV/dt and dI/dt?

   d) Find the rate at which R is changing when $V = 12$ volts and $I = 2$ amp. Is R increasing or decreasing?

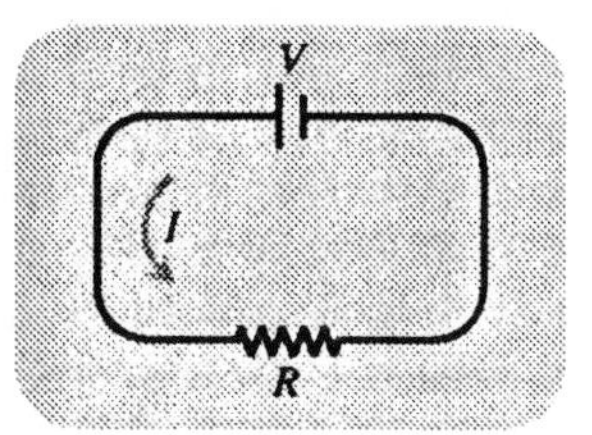

9. CHANGING DIMENSIONS IN A RECTANGLE The length l of a rectangle is decreasing at the rate of 2 cm/sec while the width w is increasing at the rate of 2 cm/sec. When $l = 12$ cm and $w = 5$ cm, find the rates of change of (a) the area, (b) the perimeter, and (c) the lengths of the diagonals of the rectangle. Which of these quantities are decreasing and which are increasing?

10. COMMERCIAL AIR TRAFFIC Two commercial jets at 40,000 ft are flying at 520 mph along straight-line courses that cross at right angles. How fast is the distance between the planes closing when plane A is 5 mi from the intersection point and plane B is 12 mi from the intersection point? How fast is the distance closing at any time?

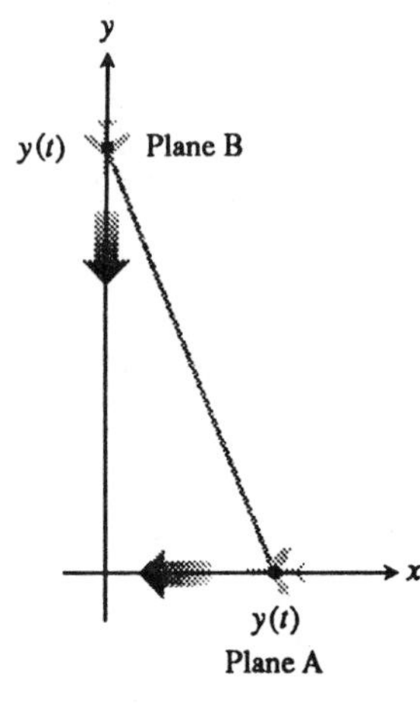

11. A SLIDING LADDER A 13-ft ladder is leaning against a house when its base starts to slide away. By the time the base is 12 ft from the house, the base is moving at the rate of 5 ft/sec. How fast is the top of the ladder sliding down the wall then? How fast is the area of the triangle formed by the ladder, wall, and ground changing?

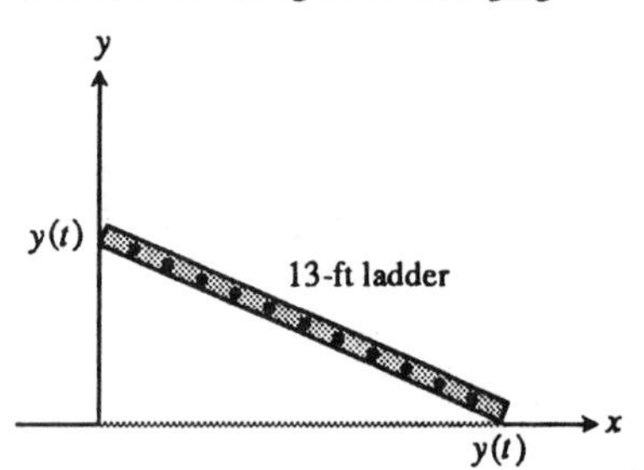

12. A SHRINKING LOLLIPOP A spherical Tootsie Roll Pop you are sucking on is giving up volume at a steady rate of 800 ml/min. How fast will the radius be decreasing when the Tootsie Roll Pop is 20 mm across?

13. BORING A CYLINDER The mechanics at Lincoln Automotive are reboring a 6-in.-deep cylinder to fit a new piston. The machine they are using increases the cylinder's radius one-thousandth of an inch every 3 min. How rapidly is the cylinder volume increasing when the bore (diameter) is 3.80 in.?

14. A GROWING SAND PILE Sand falls from a conveyor belt at the rate of 10 ft^3/min onto a conical pile. The radius of the base of the pile is always equal to half the pile's height. How fast is the height growing when the pile is 5 ft high?

15. A GROWING RAINDROP Suppose that a drop of mist is a perfect sphere and that, through condensation, the drop picks up moisture at a rate proportional to its surface area. Show that under these circumstances the drop's radius increases at a constant rate.

16. THE RADIUS OF AN INFLATING BALLOON A spherical balloon is inflated with helium at the rate of $100\pi ft^3$/min. How fast is the balloon's radius increasing at the instant the radius is 5 ft? How fast is the surface area increasing?

17. HAULING IN A DINGHY A dinghy is pulled toward a dock by a rope from the bow through a ring on the dock 6 ft above the bow. If the rope is hauled in at the rate of 2 ft/sec, how fast is the boat approaching the dock when 10 ft of rope are out?

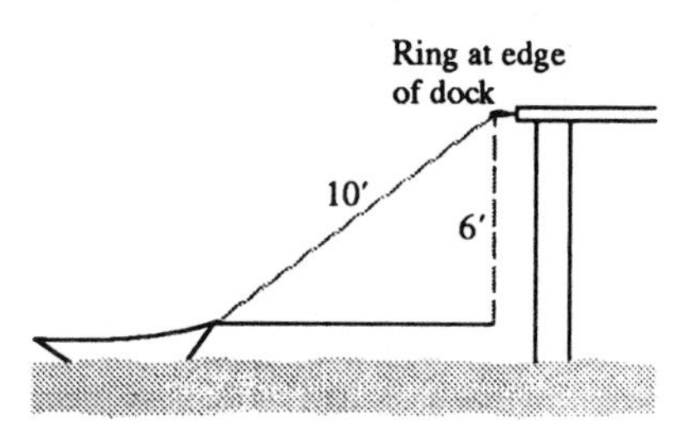

18. A BALLOON AND A BICYCLE A balloon is rising vertically above a level, straight road at a constant rate of 1 ft/sec. Just when the balloon is 65 ft above the ground, a bicycle passes under it, going 17 ft/sec. How fast is the distance between the bicycle and balloon increasing 3 sec later?

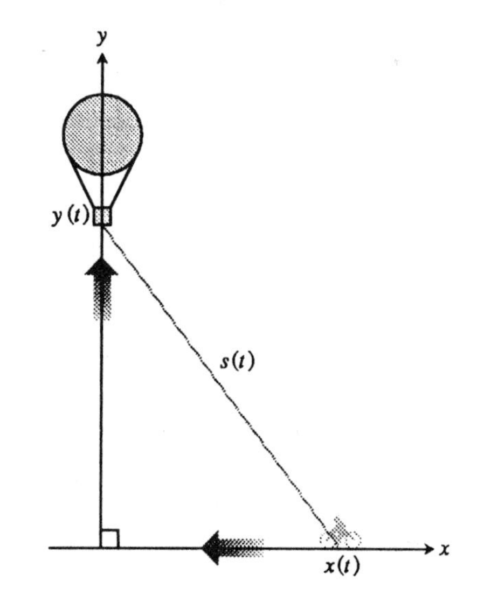

19. MAKING COFFEE Coffee is draining from a full conical filter basket into a cylindrical coffee pot at the rate of 10 in^3/min. How fast is the level in the pot rising when the height of coffee in the filter is 5 in? How fast is the level in the filter cone falling?

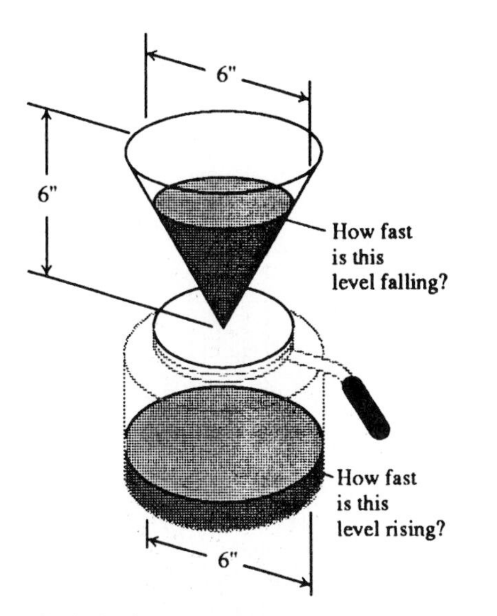

20. BLOOD FLOW Cold water has the effect of contracting the blood vessels in the hands, and the radius of a typical vein might decrease at the rate of 20%/min. According to Poiseuille's law, $V = kr^4$ (see Example 5), at what percentage rate can we expect the volume of blood flowing through that vein to decrease?

21. CARDIAC OUTPUT In the late 1860s, Adolf Fick, a professor of physiology in the Faculty of Medicine in Würtzberg, Germany, developed the method we use today for measuring how much blood your heart pumps in a minute. Your cardiac output as you read this sentence is probably about 7 liters a minute. At rest it is likely to be a bit under 6 L/min. If you are a trained marathon runner running a marathon, your cardiac output can be as high as 30 L/min.

Your cardiac output can be calculated with the formula

$$y = \frac{Q}{D},$$

where Q is the number of milliliters of CO_2 you exhale in a minute and D is the difference between the CO_2 concentration (ml/L) in the blood pumped to the lungs and the CO_2 concentration in the blood returning from the lungs. With $Q = 233$ ml/min and $D = 97 - 56 = 41$ ml/min,

$$y = \frac{233 \text{ ml/min}}{41 \text{ ml/L}} \approx 5.68 \text{ L/min},$$

close to the 6 L/min that most people have at basal (resting) conditions. (Data courtesy of J. Kenneth Herd, M.D., Quillen Dishner College of Medicine, East Tennessee State University.)

Suppose that when $Q = 233$ and $D = 41$, we also know that D is decreasing at the rate of 2 units a minute but that Q remains unchanged. What is happening to the cardiac output?

22. FILLING A CONICAL TANK Water runs into the conical tank shown here at the rate of 9 ft^3/min. How fast is the water level rising when the water is 6 ft deep?

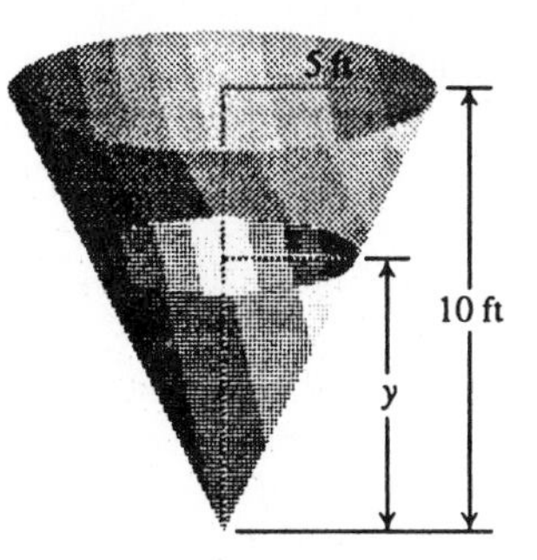

23. MOVING ALONG A PARABOLA A particle moves along the parabola $y = x^2$ in the first quadrant in such a way that its x-coordinate increases at a steady 10 m/sec. How fast is the angle of inclination θ of the line joining the particle to the origin changing when $x = 3$ m? What is the limiting value of $d\theta/dt$ as $x \to \infty$?

24. COST, REVENUE, AND PROFIT A company can manufacture x items at a cost of $c(x)$ dollars, a sales revenue of $r(x)$ dollars, and a profit of $p(x) = r(x) - c(x)$ dollars (everything in thousands). Find the rates of change of cost, revenue, and profit for the following values of x and dx/dt.

a) $r(x) = 9x$, $c(x) = x^3 - 6x^2 + 15x$, and $dx/dt = 0.1$ when $x = 2$

b) $r(x) = 70x$, $c(x) = x^3 - 6x^2 + 45/x$, and $dx/dt = 0.05$ when $x = 1.5$

25. A MOVING SHADOW A man 6 ft tall walks at the rate of 5 ft/sec toward a street light that is 16 ft above the ground. At what rate is the tip of his shadow moving? At what rate is the length of his shadow changing when he is 10 ft from the base of the light?

26. ANOTHER MOVING SHADOW A light shines from the top of a pole 50 ft high. A ball is dropped from the same height from a point 30 ft away from the light. How fast is the shadow of the ball moving along the ground 1/2 sec later? (Assume the ball falls a distance $s = 16t^2$ ft in t sec.)

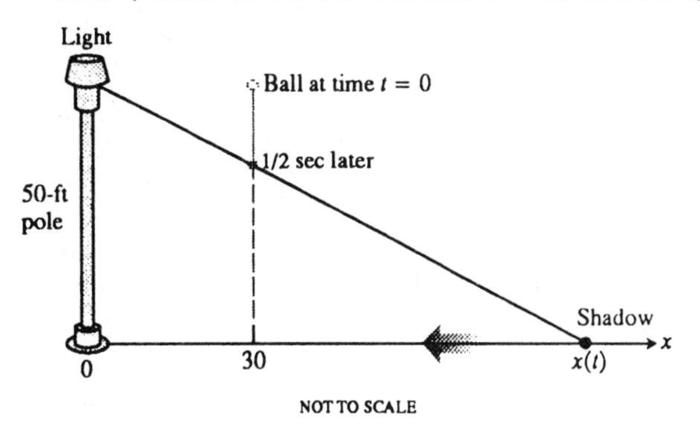

27. FLYING A KITE A girl flies a kite at a height of 300 ft, the wind carrying the kite horizontally away from her at a rate of 25 ft/sec. How fast must she let out the string when the kite is 500 ft away from her?

28. A MELTING ICE LAYER A spherical iron ball 8 in. in diameter is coated with a layer of ice of uniform thickness. If the ice melts at the rate of 10 in^3/min, how fast is the thickness of the ice decreasing when it is 2 in. thick? How fast is the outer surface area of ice decreasing?

29. HIGHWAY PATROL A highway patrol plane flies 3 mi above a level, straight road at a steady ground speed of 120 mph. The pilot sees an oncoming car and determines with radar that the line-of-sight distance from the plane to the car is 5 mi and decreasing at the rate of 160 mph. Find the car's speed along the highway.

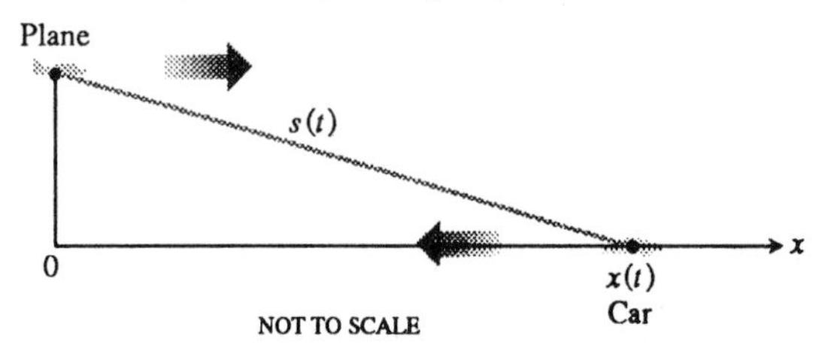

30. THE SUN'S SHADOW On a morning when the sun will pass directly overhead, the shadow of an 80-ft building on level ground is 60 ft long. At the moment in question, the angle θ the sun makes with the ground is increasing at the rate of $0.27°$/min. At what rate is the shadow decreasing? (Remember to use radians. Express your answer in inches per minute, to the nearest tenth.)

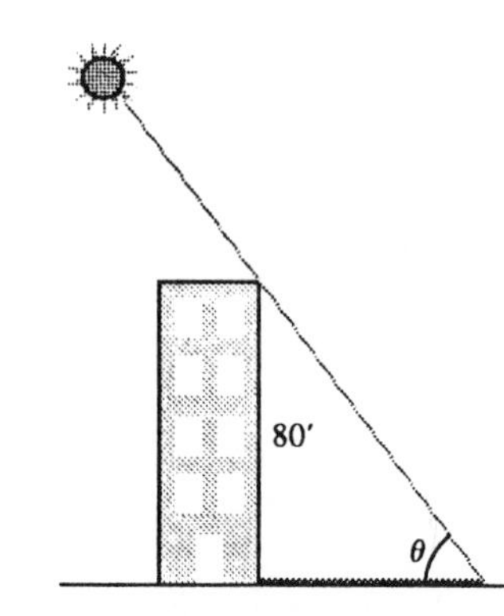

31. SHIPS Two ships are steaming straight away from a point O along routes that make a $120°$ angle. Ship A moves at 14 knots (nautical miles per hour—a nautical mile is 2000 yd). Ship B moves at 21 knots. How fast are the ships moving apart when $OA = 5$ and $OB = 3$? How fast are the ships moving apart at any time?

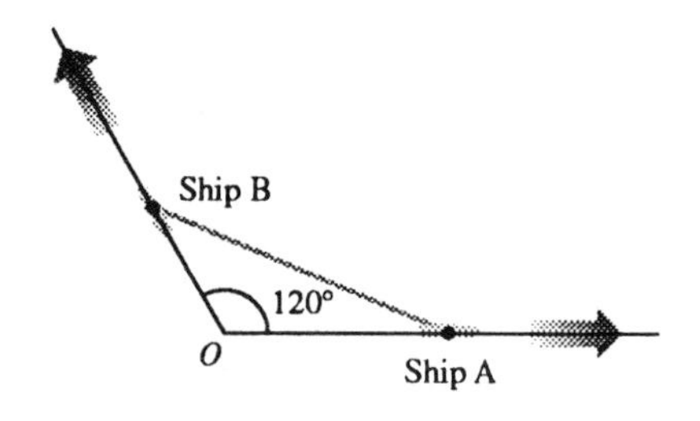

4.7 Antiderivatives, Initial Value Problems, and Mathematical Modeling

One of the early accomplishments of calculus was predicting the future position of a moving body from one of its known locations and a formula for its velocity function. Today we view this as one of a number of occasions on which we recover a function from one of its known values and a formula for its rate of change. It is a routine process today, thanks to calculus, to calculate a factory's future output from its present output and its production rate or to predict a population's future size from its present size and its growth rate.

The process we use in this section to find a formula for a function from one of its known values and its derivative $f(x)$ has two steps. The first is to find a formula that gives all the functions that could possibly have f as a derivative. These functions are the so-called antiderivatives of f, and the formula that gives them all is the general antiderivative of f. The second step is then to use the known function value to select the particular antiderivative we want from the general antiderivative. This process depends on our ability to recognize a function whose derivative is the given function f. In general finding one antiderivative is a very difficult and often impossible task.

There are modern sophisticated Computer Algebra Systems that are designed to produce antiderivatives. However, this software often fails because there are no exact formulas for antiderivatives of many functions.

There are numerical techniques that can be used to approximate and produce graphs of antiderivatives that satisfy the known value. Some graphing calculators have this feature. If yours does, you can use it to check the computations of this section.

Finding a formula that gives all of a function's antiderivatives might seem like an impossible task, but it turns out that if we can find even one of a function's antiderivatives, then we can find them all, thanks to the remaining two corollaries of the Mean Value Theorem of Section 4.1.

The present section begins with these two corollaries. It then shows how to "reverse" known differentiation formulas to find general antiderivatives and uses this technique to solve a number of differential equations. It closes with a brief discussion of mathematical modeling, the process by which we, as scientists, use mathematics to learn about reality.

The Mean Value Theorem's Second and Third Corollaries

The first corollary of the Mean Value Theorem in Section 4.1 gave the first derivative test for rise and fall. The second corollary says that only constant functions can have zero derivatives. The third says that functions with identical derivatives can differ only by a constant value.

Corollary 2

If $F' = 0$, then F is a constant.

If $F'(x) = 0$ for all x in (a, b), then F has a constant value throughout (a, b). In other words, there is a constant C such that $F(x) = C$ for all x in (a, b).

Corollary 2 is the converse of the rule that says the derivative of a constant is zero. While the derivatives of nonconstant functions may be zero from time to time, the only functions whose derivatives are zero throughout an entire interval are the functions that are constant on the interval.

Proof of Corollary 2 We want to show that F has a constant value throughout the interval (a, b). The way we do so is to show that if x_1 and x_2 are any two points in (a, b), then $F(x_2) = F(x_1)$.

Suppose, then, that x_1 and x_2 are two points in (a, b), numbered from left to right, say, so that $x_1 < x_2$. Then F satisfies the hypotheses of the Mean Value Theorem on the closed interval $[x_1, x_2]$. It is differentiable at every point of the interval and hence continuous at every point of the interval. Therefore

$$\frac{F(x_2) - F(x_1)}{x_2 - x_1} = F'(c)$$

at some point c between x_1 and x_2. Since $F' = 0$ throughout (a, b), this translates into

$$\frac{F(x_2) - F(x_1)}{x_2 - x_1} = 0, \qquad F(x_2) - F(x_1) = 0, \qquad \text{and} \qquad F(x_2) = F(x_1).$$

Explore with a Grapher

1. Let $f_1(x) = \sqrt[3]{x+1}$, $f_2(x) = \sqrt[3]{x+1} + 5$, $f_3(x) = \sqrt[3]{x+1} - 6$, $f_4(x) = \sqrt[3]{5x+1}$, and $f_5(x) = 4\sqrt[3]{x+1}$. Compare the graphs of NDer(f_i, x) for $i = 1, 2, 3, 4$, and 5 in $[-10, 10]$ by $[-2, 2]$.
2. Let $g_1(x) = e^x$, $g_2(x) = e^x + 2$, $g_3(x) = e^x - 3$, $g_4(x) = e^{3x}$, and $g_5(x) = 2e^x$. Compare the graphs of NDer(g_i, x) for $i = 1, 2, 3, 4$, and 5 in $[-5, 5]$ by $[-5, 5]$.
3. What might you conjecture based on the observations in parts (1) and (2)?

Corollary 3

Functions with identical derivatives can differ only by a constant.

If $F_1'(x) = F_2'(x)$ at each point x of an open interval (a, b), then there is a constant C such that

$$F_1(x) = F_2(x) + C$$

for all x in (a, b).

Corollary 3 says that the only way two functions can have identical rates of change throughout an interval is for their values on the interval to differ by a constant. For example, we know that the derivative of the function x^2 is $2x$. Therefore every other function whose derivative is $2x$ is given by the formula $F(x) = x^2 + C$ for some value of C. No other functions have $2x$ as their derivative.

Proof of Corollary 3 Since $F_1'(x) = F_2'(x)$ at each point of (a, b), the derivative of the function $F = F_1 - F_2$ at each point is

$$F'(x) = F_1'(x) - F_2'(x) = 0.$$

Therefore F has a constant value C throughout (a, b) (from Corollary 2), so $F_1(x) - F_2(x) = C$. That is, $F_1(x) = F_2(x) + C$ at each point (a, b). ∎

Finding Antiderivatives

As we mentioned in the introduction, a function $F(x)$ is an **antiderivative** of a function $f(x)$ over an interval I if $F'(x) = f(x)$ at every point of I. Once we have found one antiderivative F of f, Corollary 3 of the Mean Value Theorem tells us that all others are given by the formula

$$y = F(x) + C.$$

We call $F(x) + C$ the **general antiderivative** of f over the interval I. Each particular antiderivative of f is given by this formula for some value of C. The constant C is called the **arbitrary constant** in the formula. Thus once we have a particular antiderivative F of f, the general antiderivative of f is F plus an arbitrary constant.

In the examples that follow, the interval I will be the natural domain of f unless we say otherwise.

The use of the letters F and f is conventional in this context, even though F and f are pronounced the same way in normal speech. To distinguish between the two, we recommend saying "cap eff" for F and "little eff" for f.

We can find many of the antiderivatives we need in scientific work by reversing derivative formulas we already know. The next example shows you what we mean.

EXAMPLE 1

Function $f(x)$	General antiderivative $F(x) + C$	Reversed derivative formula
$\cos x$	$\sin x + C$	$\frac{d}{dx} \sin x = \cos x$
$\cos 2x$	$\frac{\sin 2x}{2} + C$	$\frac{d}{dx} \frac{\sin 2x}{2} = \cos 2x$
$3x^2$	$x^3 + C$	$\frac{d}{dx} x^3 = 3x^2$
$\frac{1}{2\sqrt{x}}$	$\sqrt{x} + C$	$\frac{d}{dx} \sqrt{x} = \frac{1}{2\sqrt{x}}$
$\frac{1}{x^2}$	$-\frac{1}{x} + C$	$\frac{d}{dx} \left(-\frac{1}{x}\right) = \frac{1}{x^2}$

∎

There are also some useful general rules.

EXAMPLE 2 General Rules

Function	General antiderivative	Source
1. $k\dfrac{du}{dx}$ (k constant)	$ku + C$	Constant Multiple Rule
2. $\dfrac{du}{dx} + \dfrac{dv}{dx}$	$u + v + C$	Sum Rule
3. $\dfrac{du}{dx} - \dfrac{dv}{dx}$	$u - v + C$	Difference Rule
4. $x^n (n \neq -1)$	$\dfrac{x^{n+1}}{n+1} + C$	Power Rule
5. $\sin kx$	$-\dfrac{\cos kx}{k} + C$	Chain Rule
6. $\cos kx$	$\dfrac{\sin kx}{k} + C$	Chain Rule

Examples 3–5 show how to apply the rules in Example 2. Check the results in one of two ways: take the derivative of the general antiderivative to see if it agrees with the function, or graph the numerical derivative of the general antiderivative for a specific value of C to see if it is the same as the graph of the function.

EXAMPLE 3 We can find antiderivatives term by term.

Function	General antiderivative	Source
$10x$	$5x^2 + C$	(Constant Multiple and Power Rules)
$10x - x^2$	$5x^2 - \dfrac{x^3}{3} + C$	(... along with the Difference Rule ...)
$10x - x^2 + 2$	$5x^2 - \dfrac{x^3}{3} + 2x + C$	(... and Sum Rule ...)

EXAMPLE 4 Fractional powers are handled the same way as integer powers.

Function	General antiderivative	Source
$\sqrt{x} = x^{1/2}$	$\dfrac{x^{3/2}}{3/2} + C = \dfrac{2}{3}x^{3/2} + C$	(Power Rule with $n = 1/2$)
$\dfrac{1}{\sqrt{x}} = x^{-1/2}$	$\dfrac{x^{1/2}}{1/2} + C = 2x^{1/2} + C$	(Power Rule with $n = -1/2$)

EXAMPLE 5 The k in Rules 5 and 6 of Example 2 can be any real number different from zero.

Function	General antiderivative	Source
$6\sin 3x$	$6 \cdot \frac{-\cos 3x}{3} + C = -2\cos 3x + C$	(Rule 5 with $k = 3$)
$5\cos \frac{x}{2}$	$5 \cdot \frac{\sin(x/2)}{1/2} + C = 10\sin \frac{x}{2} + C$	(Rule 6 with $k = \frac{1}{2}$)
$\cos 2\pi x$	$\frac{\sin 2\pi x}{2\pi} + C$	(Rule 6 with $k = 2\pi$)

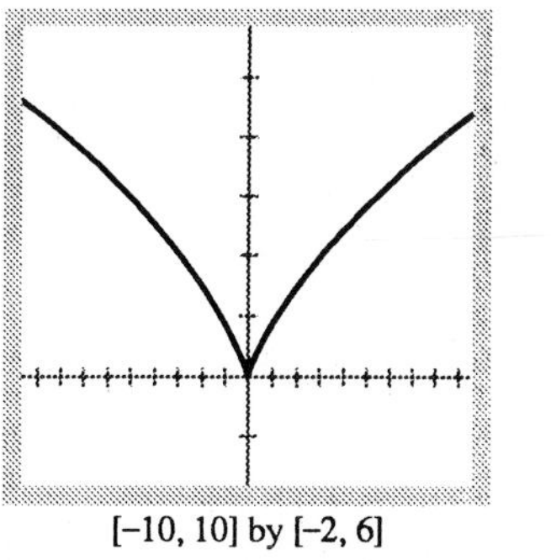
[−10, 10] by [−2, 6]

4.81 Complete graphs of $f(x) = x^{2/3}$ and NDer (F, x) where $F(x) = (3/5)x^{5/3}$.

EXAMPLE 6 Find the general antiderivative of $f(x) = x^{2/3}$ and support the answer graphically.

Solution Rule 4 says that

$$y = \frac{3}{5}x^{5/3} + C$$

is the general antiderivative of f. Let $C = 0$ and $F(x) = \frac{3}{5}x^{5/3}$. Figure 4.81 gives visual support that the graphs of f and the numerical derivative of F are identical.

Differential Equations and Initial Value Problems

The problem of finding a function y of x when we know its derivative

$$\frac{dy}{dx} = f(x)$$

and its value y_0 at a particular point x_0 is called an **initial value problem.** We solve the problem in two steps. First we find the general antiderivative of f,

$$y = F(x) + C. \tag{1}$$

Then we use the fact that $y = y_0$ when $x = x_0$ to find the right value of C. In this case $y_0 = F(x_0) + C$, so $C = y_0 - F(x_0)$, and the solution of the initial value problem is the function

$$y = F(x) + (y_0 - F(x_0)). \tag{2}$$

This function has the right derivative, because

$$\frac{dy}{dx} = \frac{d}{dx}F(x) + \frac{d}{dx}(y_0 - F(x_0)) = f(x) + 0 = f(x). \tag{3}$$

It also has the right value when $x = x_0$, because

$$y\,\Big|_{x=x_0} = F(x_0) + (y_0 - F(x_0)) = y_0. \tag{4}$$

An equation like

$$\frac{dy}{dx} = f(x) \tag{5}$$

that has a derivative in it is called a **differential equation.** Equation (5) gives dy/dx as a function of x. A more complicated differential equation might involve y as well as x:

$$\frac{dy}{dx} = 2xy^2. \tag{6}$$

It might also involve higher-order derivatives:

$$\frac{d^2y}{dx^2} - \frac{dy}{dx} + 5y = 3. \qquad (7)$$

We shall see how to solve equations like (6) and (7) when we get to Chapter 16. For the time being, we shall steer away from such complications.

If your grapher can graph the solution to an initial value problem, then compare the graphs of the solutions to Examples 7–10 to the one produced by your graphing utility.

In the language of differential equations, the general antiderivative $y = F(x) + C$ of the function $f(x)$ is called the **general solution** of the equation $dy/dx = f(x)$. We solve the initial value problem by using the **initial condition** that $y = y_0$ when $x = x_0$ to find the **particular solution** $y = F(x) + (y_0 - F(x_0))$.

EXAMPLE 7 Finding velocity from acceleration. The acceleration of gravity near the surface of the earth is 9.8 m/sec^2. This means that the velocity v of a body falling freely in a vacuum changes at the rate of

$$\frac{dv}{dt} = 9.8 \text{ m/sec}^2.$$

If the body is dropped from rest, what will its velocity be t seconds after it is released?

Solution In mathematical terms, we want to solve the initial value problem that consists of

The differential equation: $\dfrac{dv}{dt} = 9.8$

The initial condition: $v = 0$ when $t = 0$.

To solve it, we first use what we know about antiderivatives to find the general solution of the differential equation $dv/dt = 9.8$:

$$v = 9.8t + C.$$

Then we use the initial condition to find the right value of C for this particular problem:

$$v = 9.8t + C,$$

$$0 = 9.8(0) + C, \qquad (v = 0 \text{ when } t = 0)$$

$$C = 0.$$

The velocity of the falling body t seconds after release is

$$v(t) = 9.8t \text{ m/sec}.$$

EXAMPLE 8 Find the curve whose slope at the point (x, y) is $3x^2$ if the curve is required to pass through the point $(1, -1)$.

Solution In mathematical language, we are asked to solve the initial value problem that consists of

The differential equation: $\dfrac{dy}{dx} = 3x^2$

The initial condition: $y = -1$ when $x = 1$.

The solve it, we first use what we know about antiderivatives to find the general solution of the differential equation:

$$y = x^3 + C.$$

Then we substitute $x = 1$ and $y = -1$ to find C:

$$y = x^3 + C,$$
$$-1 = (1)^3 + C,$$
$$C = -2.$$

The curve we want is $y = x^3 - 2$ (Fig. 4.82).

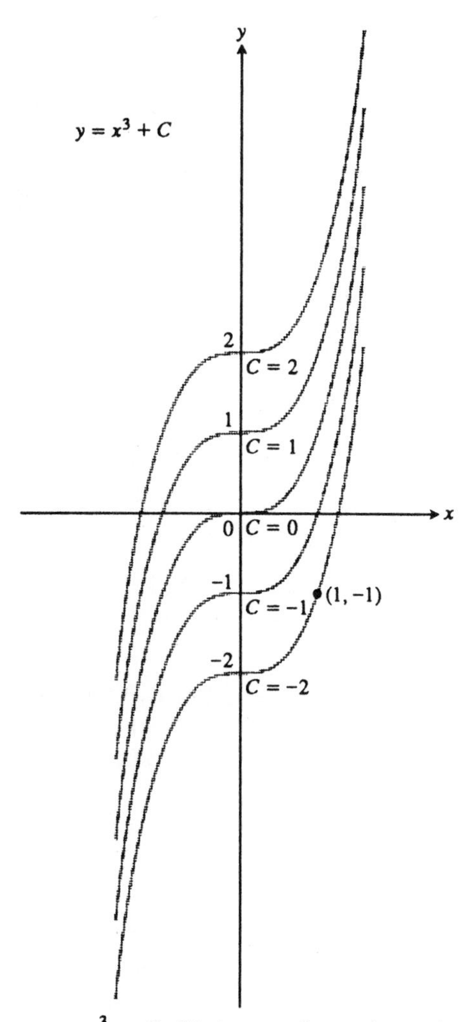

4.82 The curves $y = x^3 + C$ fill the coordinate plane without overlapping. In Example 8 we identify the curve $y = x^3 - 2$ as the one that passes through the point $(1, -1)$.

EXAMPLE 9 Suppose that the marginal cost of manufacturing an item when x thousand items are produced is

$$\frac{dc}{dx} = 10 + 3\sqrt{x}$$

thousand dollars and that the cost of producing 4 thousand items is 60 thousand dollars. What will it cost to produce 9 thousand items?

Solution We have

$$\text{The differential equation:} \quad \frac{dc}{dx} = 10 + 3\sqrt{x}$$

$$\text{The initial condition:} \quad c = 60 \text{ when } x = 4.$$

The general solution of the differential equation is

$$c(x) = 10x + 3\frac{x^{3/2}}{3/2} + C, \quad \text{Power Rule with } n = \frac{1}{2}$$

$$c(x) = 10x + 2x^{3/2} + C.$$

We substitute $x = 4, c(4) = 60$ to find the right value for C:

$$60 = 10(4) + 2(4)^{3/2} + C,$$

$$60 = 40 + 2(8) + C,$$

$$C = 60 - 40 - 16 = 4.$$

The cost of producing x thousand items is therefore

$$c(x) = 10x + 2x^{3/2} + 4.$$

The cost of producing 9 thousand items is

$$c(9) = 90 + 2(9)^{3/2} + 4 = 94 + 2(27) = 148$$

thousand dollars. ≡

Some problems require us to solve two or more differential equations in a row. Here is an example.

EXAMPLE 10 A heavy projectile is fired straight up from a platform 10 ft above the ground, with an initial velocity of 160 ft/sec. Assume that the only force affecting the projectile during its flight is from gravity, which produces a downward acceleration of 32 ft/sec^2. Find an equation for the projectile's height above the ground as a function of time t if $t = 0$ when the projectile is fired.

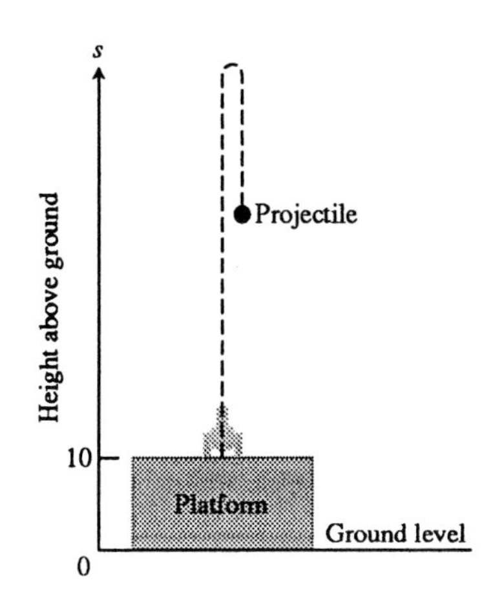

4.83 The sketch for modeling the projectile motion in Example 10.

Solution To model the problem, we draw a figure (Fig. 4.83) and let $s(t)$ denote the projectile's height above the ground at time t. We assume s to be a twice-differentiable function of t, so that

$$v = \frac{ds}{dt} \quad \text{and} \quad a = \frac{dv}{dt} = \frac{d^2s}{dt^2}.$$

Since gravity acts in the negative s direction, the direction of decreasing s in our model, the initial value problem to solve is

$$\text{The differential equation:} \quad \frac{dv}{dt} = -32 \text{ ft/sec}^2$$

$$\text{The initial conditions:} \quad v(0) = 160 \text{ ft/sec}, \qquad s(0) = 10 \text{ ft.}$$

We find v from the equation

$$v = \frac{ds}{dt} = -32t + C_1 \quad \text{General antiderivative of } -32$$

and s from the equation

$$s = -16t^2 + C_1t + C_2. \quad \text{General antiderivative of } -32t + C_1$$

The appropriate values of C_1 and C_2 are determined by the initial conditions:

$$C_1 = v(0) = 160, \qquad C_2 = s(0) = 10.$$

The projectile's height above the ground at time t is

$$s(t) = -16t^2 + 160t + 10.$$

Notice that the formula for s agrees with the one used in prior courses and in Chapter 3.

Mathematical Modeling

The development of a mathematical model usually takes four steps: First we observe something in the real world (a ball bearing falling from rest or the trachea contracting during a cough, for example) and construct a system of mathematical variables and relationships that imitate some of its important features. We build a mathematical metaphor for what we see. Next we apply (usually) existing mathematics to the variables and relationships in the model to draw conclusions about them. After that we translate the mathematical conclusions into "news" about the system under study. Finally we check the news against observation to see if the model has predictive value. We also investigate the possibility that the model applies to other systems. The really good models are the ones that lead to conclusions that are consistent with observation, that have predictive value and broad application, and that are not too hard to use.

The natural cycle of mathematical imitation, deduction, interpretation, and comparison is shown in the following diagrams.

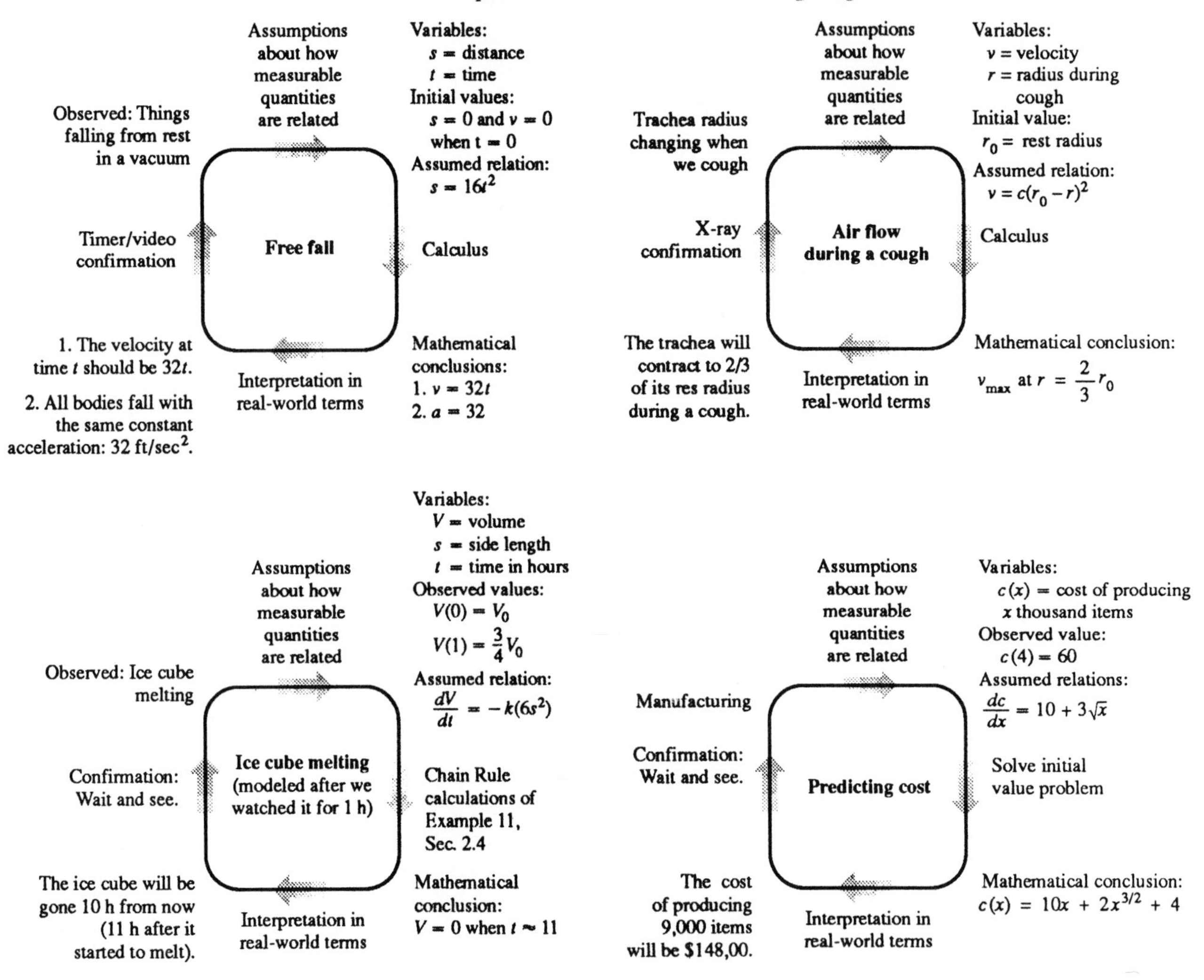

Computer Simulation

When a system we want to study is complicated, we can sometimes experiment first to see how the system behaves under different circumstances. But if this is not possible (the experiments might be expensive, time-consuming, or dangerous), we might run a series of simulated experiments on a computer—experiments that behave like the real thing, without the disadvantages. For example, in Chapter 3 we simulated the motion of objects using the parametric graphing feature of our graphing utility (really a computer). Thus we might model the effects of atomic war, the effect of waiting a year longer to harvest trees, the effect of crossing particular breeds of cattle, or the effect of reducing atmospheric ozone by 1%, all without having to pay the consequences or wait to see how things work out.

We also bring computers in when the model we want to use has too many calculations to be practical any other way. NASA's space-flight models are run on computers—they have to be to generate course corrections on time. If you want to model the behavior of galaxies that contain billions and billions of stars, a computer offers the only possible way. One of the most spectacular computer simulations in recent years, carried out by Alar Toomre at MIT, explained a peculiar galactic shape that was not consistent with our previous ideas about how galaxies are formed. The galaxies had acquired their odd shapes, Toomre deduced, by passing through one another.

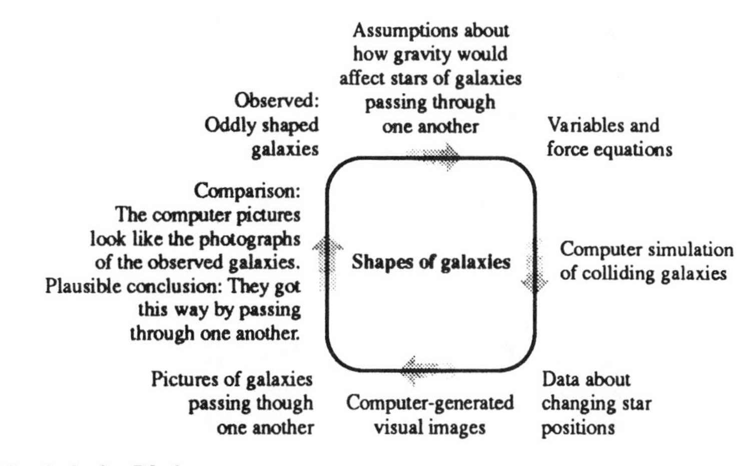

Models in Biology

You may have noticed that we haven't mentioned models in biology yet. The reason is that most mathematical models of life processes use either exponential functions or logarithms, functions whose exact derivatives are not found until Section 5.8. Typical of the models we shall study there is the model for unchecked bacterial growth. The basic assumption is that at any time t the rate dy/dt at which the population is changing is proportional to the number $y(t)$ of bacteria present. If the population's original size is y_0, this leads to the initial value problem

$$\text{Differential equation: } \frac{dy}{dt} = ky$$

$$\text{Initial condition: } y = y_0 \quad \text{when} \quad t = 0.$$

As you will see, the solution turns out to be $y = y_0 e^{kt}$, so the modeling cycle looks like this:

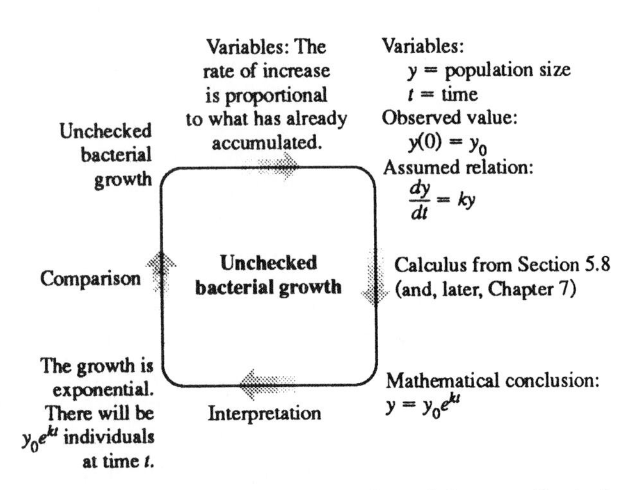

This model is one of the really good models we talked about earlier, because it applies to so many of the phenomena we want to forecast and understand: cell growth, heat transfer, radioactive decay, the flow of electrical current, and the accumulation of capital by compound interest, to mention only a few. We shall see how all of this works by the time we are through with Chapter 7.

Exercises 4.7

Find the general antiderivatives of the functions in Exercises 1–18. Do as many as you can without writing anything down (except the answer). Then support your answers with a graphing utility.

1. a) $2x$ b) 3 c) $2x+3$

2. a) $6x$ b) -2 c) $6x-2$

3. a) $3x^2$ b) x^2 c) x^2+2x+1

4. a) $8x^7$ b) x^7 c) x^7-6x+8

5. a) $-3x^{-4}$ b) x^{-4} c) $x^{-4}+2x+3$

6. a) $\frac{1}{x^2}$ b) $\frac{5}{x^2}$ c) $2-\frac{5}{x^2}$

7. a) $\frac{3}{2}\sqrt{x}$ b) $4\sqrt{x}$ c) $x^2-4\sqrt{x}$

8. a) $\frac{1}{2}x^{-1/3}$ b) $-\frac{1}{2}x^{-3/2}$ c) $-\frac{3}{2}x^{-5/2}$

9. a) $\frac{2}{3}x^{-1/3}$ b) $\frac{1}{3}x^{-2/3}$ c) $-\frac{1}{3}x^{-4/3}$

10. a) $x^{-1/3}$ b) $x^{-2/3}$ c) $x^{-4/3}$

11. a) $-\sin 3x$ b) $3\sin x$ c) $3\sin x - \sin 3x$

12. a) $\pi\cos \pi x$ b) $\frac{\pi}{2}\cos\frac{\pi x}{2}$ c) $\cos\frac{\pi x}{2}$

13. a) $\sec^2 x$ b) $5\sec^2 5x$ c) $\sec^2 5x$

14. a) $\csc^2 x$ b) $7\csc^2 7x$ c) $\csc^2 7x$

15. a) $\sec x\tan x$ b) $2\sec 2x\tan 2x$ c) $4\sec 2x\tan 2x$

16. a) $\csc x\cot x$ b) $8\csc 4x\cot 4x$ c) $\csc 4x\cot 4x$

17. $(\sin x - \cos x)^2$ (*Hint:* $2\sin x\cos x = \sin 2x$)

18. $(1+2\cos x)^2$ (*Hint:* $2\cos^2 x = 1+\cos 2x$)

19. Suppose that $1-\sqrt{x}$ is an antiderivative of $f(x)$ and that $x+2$ is an antiderivative of $g(x)$. Find the *general* antiderivatives of the following functions.

a) $f(x)$ b) $g(x)$ c) $-f(x)$
d) $-g(x)$ e) $f(x)+g(x)$ f) $3f(x)-2g(x)$
g) $x+f(x)$ h) $g(x)-4$

20. Repeat Exercise 19, assuming that e^x is an antiderivative of $f(x)$ and that $x\sin x$ is an antiderivative of $g(x)$.

Solve the initial value problems in Exercises 21–32 for y as a function of x. If possible, support your answers with a graphing utility.

21. $\frac{dy}{dx} = 2x-7,\ y=0$ when $x=2$

22. $\frac{dy}{dx} = 10-x,\ y=-1$ when $x=0$

23. $\frac{dy}{dx} = x^2+1,\ y=1$ when $x=0$

24. $\frac{dy}{dx} = x^2+\sqrt{x},\ y=1$ when $x=1$

25. $\frac{dy}{dx} = -5/x^2, x > 0; y = 3$ when $x = 5$

26. $\frac{dy}{dx} = \frac{1}{x^2} + x, x > 0; y = 1$ when $x = 2$

27. $\frac{dy}{dx} = 3x^2 + 2x + 1, y = 0$ when $x = 1$

28. $\frac{dy}{dx} = 9x^2 - 4x + 5, y = 0$ when $x = -1$

29. $\frac{dy}{dx} = 1 + \cos x, y = 4$ when $x = 0$

30. $\frac{dy}{dx} = \cos x + \sin x, y = 1$ when $x = \pi$

31. $\frac{d^2y}{dx^2} = 2 - 6x, y = 1$ and $\frac{dy}{dx} = 4$ when $x = 0$

32. $\frac{d^3y}{dx^3} = 6; y = 5, \frac{dy}{dx} = 0$, and $\frac{d^2y}{dx^2} = -8$ when $x = 0$

Exercises 33 and 34 give the velocity and initial position of a body moving along a coordinate line. Find the body's position at time t. Simulate the motion with a parametric grapher.

33. $v = 9.8t, s = 10$, when $t = 0$

34. $v = \sin t, s = 0$, when $t = 0$

Exercises 35 and 36 give the acceleration, initial velocity, and initial position of a body moving along a coordinate line. Find the body's position at time t. Simulate the motion with a parametric grapher.

35. $a = 32, v = 20$, and $s = 0$ when $t = 0$

36. $a = \sin t, v = -1$, and $s = 1$ when $t = 0$

37. Find the curve in the xy-plane that passes through the point (9,4) and whose slope at each point is $3\sqrt{x}$.

38. a) Find a function $y = f(x)$ with the following properties:
    i) $\frac{d^2y}{dx^2} = 6x$;
    ii) Its graph in the xy-plane passes through the point (0,1) and has a horizontal tangent there.

    b) How many functions like this are there? How do you know?

39. REVENUE FROM MARGINAL REVENUE Suppose that the marginal revenue when x thousand units are sold is

$$\frac{dr}{dx} = 3x^2 - 6x + 12$$

dollars per unit. Find the revenue function $r(x)$ given that $r(0) = 0$.

40. COST FROM MARGINAL COST Suppose that the marginal cost of manufacturing an item when x thousand items are produced is

$$\frac{dc}{dx} = 3x^2 - 12x + 15$$

dollars per item. Find the cost function $c(x)$ if $c(0) = 400$.

41. On the moon the acceleration of gravity is 1.6 m/sec^2. If a rock is dropped into a crevasse, how fast will it be going just before it hits bottom 30 sec later?

42. A rocket lifts off the surface of Earth with a constant acceleration of 20 m/sec^2. How fast will the rocket be going 1 min later?

43. With approximately what velocity do you enter the water if you dive from a 10-m platform? (Use $g = 9.8$ m/sec^2.)

44. The acceleration of gravity near the surface of Mars is 3.72 m/sec^2. If a rock is blasted straight up from the surface with an initial velocity of 93 m/sec (about 208 mph), how high does it go? (*Hint:* When is the velocity zero?)

45. HOW LONG WILL IT TAKE A TANK TO DRAIN? If we open a valve to drain the water from a cylindrical tank (Fig. 4.84), the water will flow fast when the tank is full but slow down as the tank drains. It turns out that the rate at which the water level drops is proportional to the square root of the water's

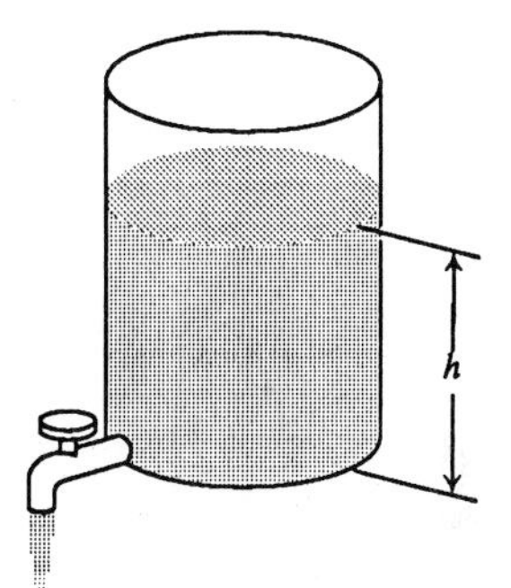

4.84 The tank in Exercises 45 and 46.

depth. In the notation of Fig. 4.84, this means that

$$\frac{dy}{dt} = -k\sqrt{y}. \quad (8)$$

The value of k depends on the acceleration of gravity and the cross-sectional areas of the tank and drain hole. Equation (8) has a minus sign because y decreases with time. To solve Eq. (8), rewrite it as

$$\frac{1}{\sqrt{y}}\frac{dy}{dt} = -k$$

and carry out the following steps.

a) Find the general antiderivative of each side of Eq. (9).

b) Set the antiderivatives in (a) equal and combine their arbitrary constants into a single arbitrary constant. (Nothing is achieved by having two when one will do.) This will give an equation that relates y directly to t.

46. CONTINUATION OF EXERCISE 45 (a) Suppose t is measured in minutes and $k = 1/10$. Find y as a function of t

if $y = 9$ ft when $t = 0$. (b) How long does it take the tank to drain if the water is 9 ft deep to start with?

47. Figure 4.85 shows the graph of a function $y = f(x)$ that solves one of the following initial value problems. Which one? How do you know?

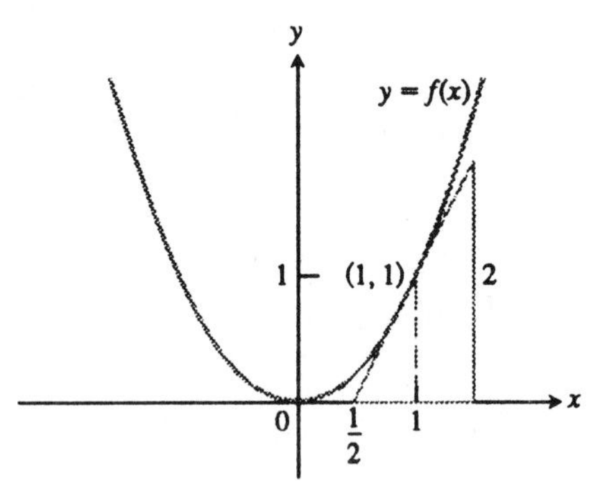

4.85 The function in Exercise 47.

a) $dy/dx = 2x, y(1) = 0$

b) $dy/dx = x^2, y(1) = 1$

c) $dy/dx = 2x + 2, y(1) = 1$

d) $dy/dx = 2x, y(1) = 1$.

Chapter 4 Review Questions

1. Outline a general method for solving related rate problems. Illustrate with an example.
2. State the first derivative test for local extreme values.
3. What does it mean for a function $y = f(x)$ to have an absolute local maximum or minimum value?
4. How do you find the local and absolute maximum and minimum values of a function $y = f(x)$?
5. What are the hypotheses and conclusion of Rolle's theorem? How does the theorem sometimes help you to tell how many solutions an equation has in a given interval?
6. What are the hypotheses and conclusion of the Mean Value Theorem? What physical interpretation does the theorem sometimes have? Give an example.
7. This chapter gives three important corollaries of the Mean Value Theorem. State each one and describe how it is used.
8. How do you test a function to find out where its graph is concave up or concave down? What is an inflection point? What physical significance do inflection points sometimes have?
9. List the steps you would take to confirm a computer-generated graph of a function. How does calculus tell you the shape of the graph between plotted points? Give an example.
10. Give a general description of the class of polynomial functions. Indicate the possible number of real zeros, and the possible number of local extrema.
11. Describe how you would sketch the graph of a rational function.
12. Indicate a reasonable way to describe the behavior of a periodic function.
13. Describe how you would use a graphing utility to draw the graph of $f(x) = \log_a x$.
14. Outline a general method for solving max-min problems. Illustrate with an example.
15. What guidance do you get from calculus about finding production levels that maximize profit? That minimize average manufacturing cost?
16. What is an antiderivative of a function $y = f(x)$? When a function has an antiderivative, how do we find its general antiderivative? Illustrate with an example.
17. What general rules can you call on to help find antiderivatives? Show, by example, how they are used.
18. What is an initial value problem? How do you solve one? Illustrate with an example.
19. How can you predict cost from marginal cost and revenue from marginal revenue?
20. Describe Newton's method for solving equations. Give an example. What is the theory behind the method? What are some of the things to watch out for when you use the method?

Chapter 4 Practice Exercises

1. The radius of a circle is changing at the rate of $-2/\pi$ m/sec. At what rate is the circles's area changing when $r = 10$ m?

2. The coordinate of a particle moving in the metric xy-plane are differentiable functions of time t with $dx/dt = -1$ m/sec and $dy/dt = -5$ m/sec. How fast is the particle approaching the origin as it passes through the point (5, 12)?

3. The volume of a cube is increasing at the rate of 1200 cm^3/min at the instant its edges are 20 cm long. At what rate are the edges changing at that instant?

4. A point moves smoothly along the curve $y = x^{3/2}$ in the first quadrant in such a way that its distance from the origin increases at the constant rate of 11 units per second. Find dx/dt when $x = 3$.

5. Water drains from the conical tank in Fig. 4.86 at the rate of 5 ft^3/mn. (a) What is the relation between the variables h and r in the figure? (b) How fast is the water level dropping when $h = 6$ ft?

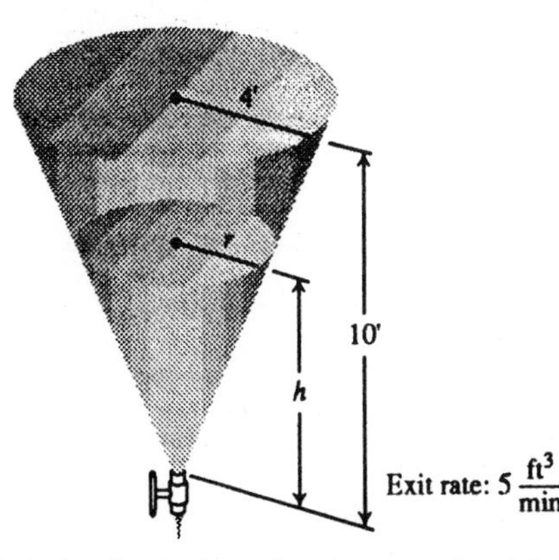

4.86 Exercise 5 asks how fast the water level in this tank is dropping when the water is 6 ft deep.

6. Two cars are approaching an intersection along straight highways that cross at right angles, car A moving at 36 mph and car B at 50 mph. At what rate is the straight-line distance between the cars changing when car A is 5 mi and car B is 12 mi from the intersection? At what rate is the distance changing at any time?

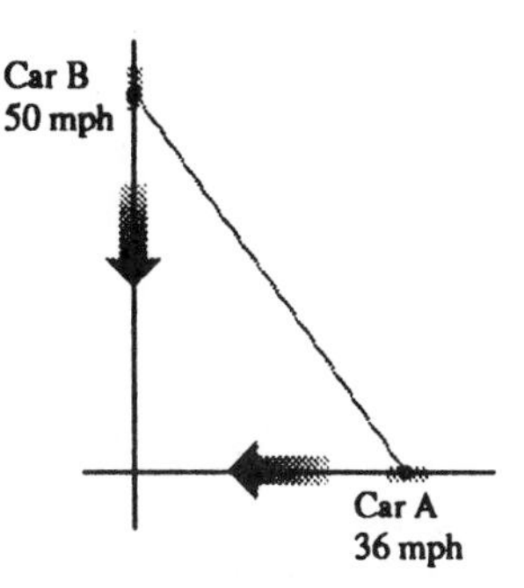

7. You are videotaping a race from a stand 132 ft from the track, following a car that is traveling at 180 mph (264 ft/sec). How fast will your camera angle θ be changing when the car is right in front of you? A half-second later?

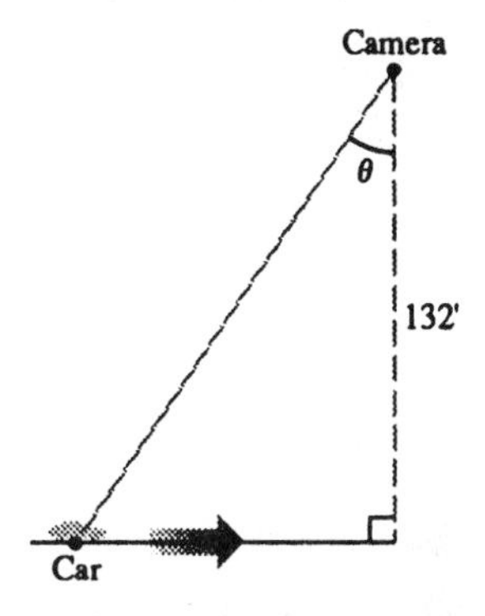

8. As telephone cable is pulled from a large spool to be strung from the telephone poles along a street, it unwinds from the spool in layers of constant radius. If the truck pulling the cable moves at a steady 6 ft/sec (a touch over 4 mph), use the equation $s = r\theta$ to find how fast (radians/sec) the spool is turning when the layer of radius 1.2 ft is being unwound?

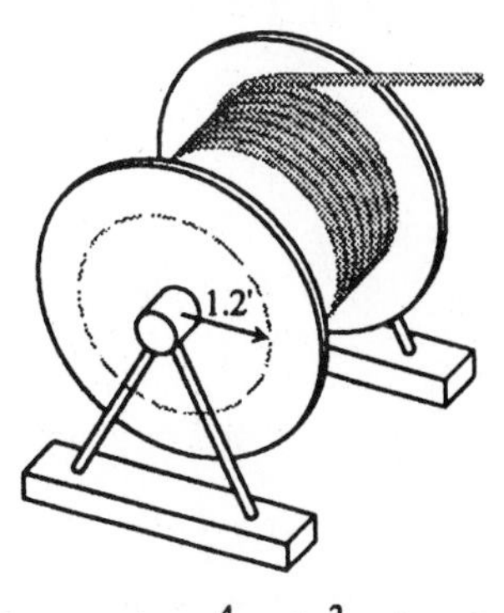

9. Show that the equation $x^4 + 2x^2 - 2 = 0$ has exactly one solution on the interval $0 \le x \le 1$.

10. If $f'(x) \le 2$ for all x, what is the most f can increase on the interval $0 \le x \le 6$?

11. Show that $y = x/(x+1)$ increases on every interval in its domain.

12. Show that $y = x^3 + 2x$ has no maximum or minimum value.

Use calculus in Exercises 13–16 to find the inflection points, local extrema, and identify the intervals on which the graphs are rising, falling, concave up, and concave down. Then sketch a complete graph and support your answer with a graphing utility.

13. $y = -x^3 - 3x^2 - 4x - 2$ **14.** $y = x^3 - 9x^2 - 21x - 11$

15. $y = \sqrt[3]{x-2}$ **16.** $y = \sqrt[4]{1-x}$

Draw a complete graph of the functions in Exercises 17 and 18 and use calculus to confirm that the graphs are complete. Find the inflection points, local extrema, and identify the intervals on which the graphs are rising, falling, concave up, and concave down.

17. $y = 1 + x - x^2 - x^4$ **18.** $y = \frac{2}{3}x^3 + 5x + 20$

Draw a complete graph of the functions in Exercises 19–32.

19. $y = -\frac{8}{3}x^3 + 4x^2 - 2x - 12$

20. $y = -x^4 + 4x^3 - 4x^2 + x + 20$

21. $y = x^4 - \frac{8}{3}x^3 - \frac{x^2}{2} + 1$

22. $y = 4x^5 + 5x^4 + \frac{20}{3}x^3 + 4$

23. $y = -x^5 + \frac{7}{3}x^3 + 5x^2 + 4x + 2$

24. $y = \frac{1}{5}x^5 + \frac{3}{2}x^4 + \frac{5}{3}x^3 - 6x^2 + 3x + 1$

25. $y = \dfrac{5 - 4x + 4x^2 - x^3}{x - 2}$

26. $y = \dfrac{3x^3 - 5x^2 - 11x - 11}{x^2 - 2x - 3}$

27. $y = \log_3 |x|$

28. $y = e^{x-1} - x$

29. $y = x \log(x - 2)$

30. $y = \sin 3x + \cos 4x$

31. $y = \sqrt[4]{x - x^2}$

32. $y = \sinh(x + 2)$

33. Suppose that the first derivative of $y = f(x)$ is

$$y' = 6(x+1)(x-2)^2.$$

At what points, if any, does the graph of f have a local maximum, local minimum, or point of inflection?

34. Suppose that the first derivative of $y = f(x)$ is

$$y' = 6x(x+1)(x-2).$$

At what points, if any, does the graph of f have a local maximum, local minimum, or point of inflection?

35. What value of b will make the graph of $y = x^3 + bx^2 - 5x + 7$ have a point of inflection at $x = 1$?

36. Use the following information to find the values of a, b, and c in the formula

$$f(x) = \frac{x + a}{bx^2 + cx + 2}.$$

a) The values of a, b, and c are either 0 or 1.

b) The graph of f passes through the point $(-1, 0)$.

c) The line $y = 1$ is an asymptote of the graph of f.

37. At which of the five points on the graph of $y = f(x)$ shown here (a) are y' and y'' both negative? (b) is y' negative and y'' positive?

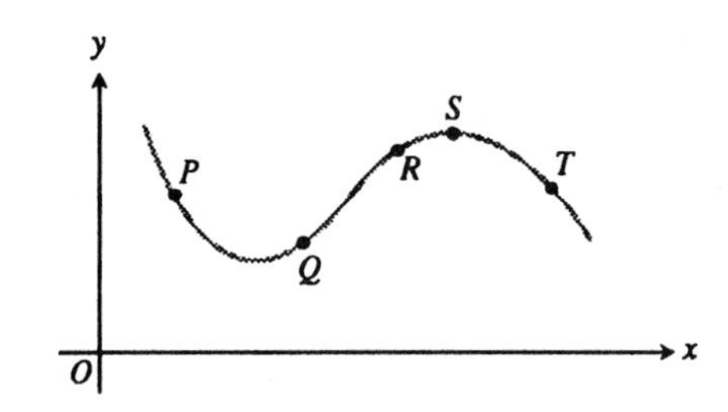

38. Here is the graph of the fruit fly population again. On approximately what day did the population's growth rate change from increasing to decreasing?

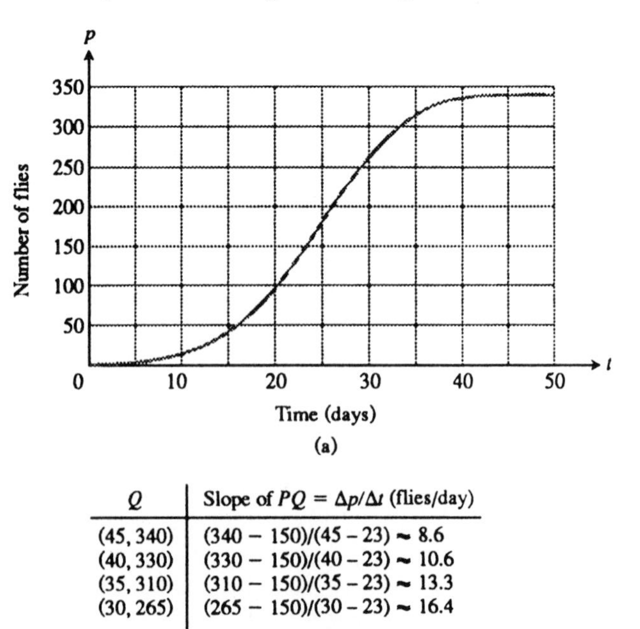

(a)

Q	Slope of $PQ = \Delta p/\Delta t$ (flies/day)
(45, 340)	(340 − 150)/(45 − 23) ~ 8.6
(40, 330)	(330 − 150)/(40 − 23) ~ 10.6
(35, 310)	(310 − 150)/(35 − 23) ~ 13.3
(30, 265)	(265 − 150)/(30 − 23) ~ 16.4

(b)

39. Find the maximum and minimum values of $f(x) = 10 + 20x - 11x^2 - 8x^3 - x^4$ on $-6 \leq x \leq 1$ and say where they are taken on.

41. Find the maximum and minimum values of $f(x) = \sqrt{x} + \cos x$ on $0 \leq x \leq 11$ and say where they are taken on.

Describe the motion of the particle moving on the x-axis with distance from the origin given by the functions in Exercises 41 and 42. Simulate the motion with a parametric grapher.

42. $s(t) = t^3 + t^2 - 6t + 5$

43. $s(t) = 3 + 4t - 3t^2 - t^3$

44. If the perimeter of the circular sector shown here is 100 ft, what values of r and s will give the sector the greatest area?

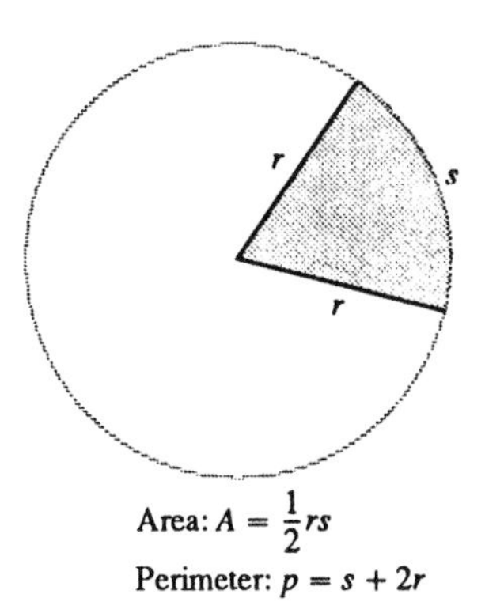

Area: $A = \frac{1}{2}rs$

Perimeter: $p = s + 2r$

45. An isosceles triangle has its vertex at the origin and its base parallel to the x-axis with the vertices above the axis on the curve $y = 27 - x^2$. Find the largest area the triangle can have.

46. Find the dimensions of the largest open storage bin with a square base and vertical sides that can be made from 108 ft^2 of sheet steel. (Neglect the thickness of the steel and assume there is no waste.)

47. A customer has asked you to design an open-top rectangular stainless steel vat. It is to have a square base and a volume of 32 ft^3, to be welded from quarter-inch plate, and to weigh no more than necessary. What dimensions do you recommend?

48. Find the height and radius of the largest right circular cylinder that can be put in a sphere of radius $\sqrt{3}$.

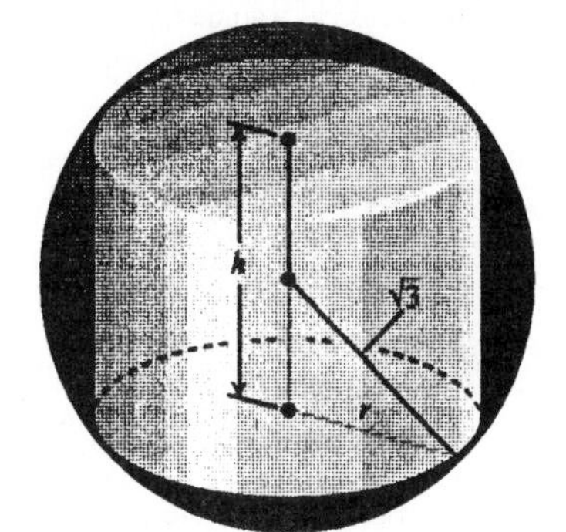

49. Figure 4.87 shows two right circular cones, one upside down inside the other. The two bases are parallel, and the vertex of the smaller cone lies at the center of the larger cone's base. What values of r and h will give the smaller cone the largest possible volume?

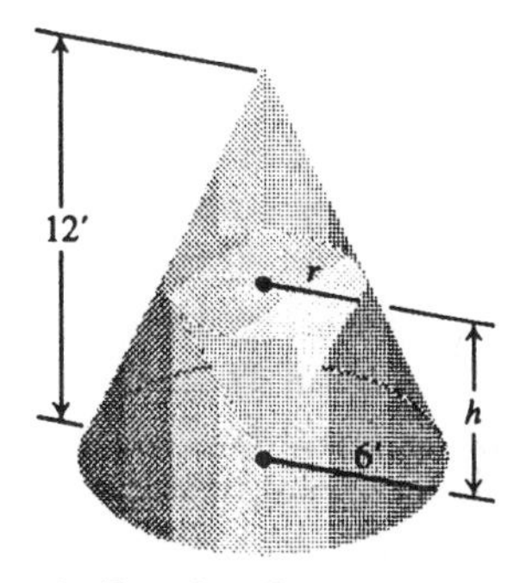

4.87 The cone in Exercise 48.

50. A drilling rig 12 mi off shore is to be connected by a pipe to a refinery on shore, 20 mi down the coast from the rig (Fig. 4.88). If underwater pipe costs \$40,000 per mile and land-based pipe costs \$30,000 per mile, what values of x and y give the least expensive connection?

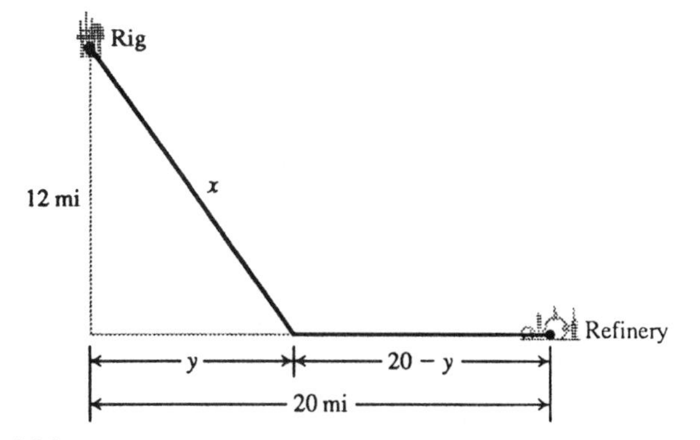

4.88 Diagram for the pipe in Exercise 49.

51. An athletic field is to be built in the shape of a rectangle x units long capped by semicircular regions of radius r at the two ends. The field is to be bounded by a 400-m racetrack. What values of x and r will give the rectangle the largest possible area?

52. Your company can manufacture x hundred grade A tires and y hundred grade B tires a day, where $0 \le x \le 4$ and

$$y = \frac{40 - 10x}{5 - x}.$$

Your profit on Grade A tires is twice your profit on grade B tires. What is the most profitable number of each kind of tire to make?

53. The positions of two particles on the s-axis are $s_1 = \sin t$ and $s_2 = \sin(t + \pi/3)$. What is the farthest apart the particles ever get?

54. The formula $F(x) = 3x + C$ gives a different function for each value of C. All of these functions, however, have the same derivative with respect to x, namely $F'(x) = 3$. Are these the only differentiable functions whose derivative is 3? Could there be any others? Explain.

55. Show that

$$\frac{d}{dx}\left(\frac{x}{x+1}\right) = \frac{d}{dx}\left(-\frac{1}{x+1}\right)$$

even though

$$\frac{x}{x+1} \ne -\frac{1}{x+1}.$$

Doesn't this contradict Corollary 3 of the Mean Value Theorem? Explain.

56. Find the general antiderivatives of the following functions.

a) 0 b) 1 c) x
d) x^2 e) x^{10} f) x^{-2}
g) x^{-5} h) $x^{5/2}$ i) $x^{4/3}$
j) $x^{3/4}$ k) $x^{1/2}$ l) $x^{-1/2}$
m) $x^{-3/7}$ n) $x^{-7/3}$

57. Find the general antiderivatives of the following functions.

a) $\sin x$ b) $\cos x$ c) $\sec x \tan x$
d) $-\csc^2 x$ e) $\sec^2 x$ f) $-\csc x \cot x$

Find the general antiderivatives of the functions in Exercises 57–72. Then support your answer with a graphing utility.

58. $3x^2 + 5x - 7$

59. $\dfrac{1}{x^2} + x + 1$

60. $\sqrt{x} + \dfrac{1}{\sqrt{x}}$

61. $\sqrt[3]{x} + \sqrt[4]{x}$

62. $3\cos 5x$

63. $8\sin(x/2)$

64. $3\sec^2 3x$

65. $4\csc^2 2x$

66. $\dfrac{1}{2} - \cos x$

67. $3x^5 + 16\cos 8x$

68. $\sec\dfrac{x}{3}\tan\dfrac{x}{3} + 5$

69. $1 - \csc x \cot x$

70. $\tan^2 x$ (*Hint:* $\tan^2 x = \sec^2 x - 1$)

71. $\cot^2 x$ (*Hint:* $\cot^2 x = \csc^2 x - 1$)

72. $2\sin^2 x$ (*Hint:* $2\sin^2 x = 1 - \cos 2x$)

73. $\cos^2 x - \sin^2 x$ (*Hint:* $\cos^2 x - \sin^2 x = \cos 2x$)

Solve the initial value problems in Exercises 73–78. If possible, support your answer with a graphing utility.

74. $\dfrac{dy}{dx} = 1 + x + \dfrac{x^2}{2}$, $y = 1$ when $x = 0$

75. $\dfrac{dy}{dx} = 4x^3 - 21x^2 + 14x - 7$, $y = 1$ when $x = 1$

76. $\dfrac{dy}{dx} = \dfrac{x^2+1}{x^2}$, $y = -1$ when $x = 1$

77. $\dfrac{dy}{dx} = \left(x + \dfrac{1}{x}\right)^2$, $y = 1$ when $x = 1$

78. $\dfrac{d^2y}{dx^2} = -\sin x$, $y = 0$ and $\dfrac{dy}{dx} = 1$ when $x = 0$

79. $\dfrac{d^2y}{dx^2} = \cos x$, $y = -1$ and $\dfrac{dy}{dx} = 0$ when $x = 0$

80. Does any function $y = f(x)$ satisfy all of the following conditions? If so, what is it? If not, why not?

a) $d^2y/dx^2 = 0$ for all x
b) $dy/dx = 1$ when $x = 0$
c) $y = 0$ and $x = 0$.

81. Find an equation for the curve in the xy-plane that passes through the point $(1, -1)$ if its slope at x is always $3x^2 + 2$.

82. You sling a shovelful of dirt up from the bottom of a 17-ft hole with an initial velocity of 32 ft/sec. Is that enough speed to get the dirt out of the hole, or had you better duck?

83. The acceleration of a particle moving along a coordinate line is $d^2s/dt^2 = 2 + 6t$ m/sec². At $t = 0$, the velocity is 4 m/sec. Find the velocity as a function of t. Then find how far the particle moves during the first second of its trip, from $t = 0$ to $t = 1$.

84. Show that the equation $x^3 + x - 1 = 0$ has exactly one solution and use Newton's method to find it to three decimal places.

85. *Estimating reciprocals without division.* Newton's method in Section 4.3 can be used to estimate the reciprocal of a positive number a without ever dividing by a, by taking $f(x) = (1/x) - a$. For example, if $a = 3$ the function involved is $f(x) = (1/x) - 3$.

a) Graph $y = (1/x) - 3$. Where does the graph cross the x-axis?

b) Show that the recursion formula in Newton's method in this case is

$$x_{n+1} = x_n(2 - 3x_n),$$

so indeed there is no division.

86. Use Newton's method to find where the curve $y = -x^3 + 3x + 4$ crosses the x-axis.

87. Use Newton's method to solve the equation $\sec x = 4$ on the interval $0 \leq x \leq \pi/2$.

88. Use Newton's method to solve the equation $2\cos x - \sqrt{1+x} = 0$.

89. Find the approximate values of r_1 through r_4 in the factorization

$$8x^4 - 14x^3 - 9x^2 + 11x - 1 = 8(x - r_1)(x - r_2)(x - r_3)(x - r_4).$$

CHAPTER

5

Integration

OVERVIEW This chapter introduces the second main branch of calculus, the branch called integral calculus. Integral calculus is the mathematics we use to find lengths, areas, and volumes of irregular shapes; to calculate the average values of functions; and to predict future population sizes and future costs of living. In this chapter, we set the stage for these and other applications.

The development of integral calculus starts from the calculation of areas. We show how to use antiderivatives in these calculations, and how this technique leads to a natural definition of area as a limit of finite sums. The limits used to define areas are special cases of a kind of limit called a *definite integral.* Presenting the properties of definite integrals and developing numerical methods of computing definite integrals are central goals of this chapter.

The single most important concept in this chapter is the connection between definite integrals and derivatives. The discovery of this connection (called the fundamental theorem of integral calculus) by Leibniz and Newton turned calculus into the most important application of mathematics in the world. The rest of this chapter describes the basic methods for finding antiderivatives and using them to evaluate integrals.

In Section 5.8 we see how to use integrals to define the natural logarithm function. We examine this function and its inverse, the exponential function, in more detail in Chapter 7. The discussion here is to complete our basic picture of how to use definite integrals to define new functions and to illustrate important applications in business, finance, and the life sciences.

In this textbook we will assume your graphing calculator or computer software has the ability to compute definite integrals easily with a high degree of accuracy. Some graphing calculators without a definite integral "key" (e.g. the TI-81) can be easily programmed to accomplish what is needed for this textbook. Furthermore, you will find it desirable that your graphing calculator or computer software has the ability to plot functions *defined* by integrals and to differentiate them numerically.

5.1 Calculus and Area

Integral calculus is the mathematics we use to define and calculate the areas of irregularly shaped regions like the cross sections of machine parts and airplane wings. This section explains what calculus and area have to do with

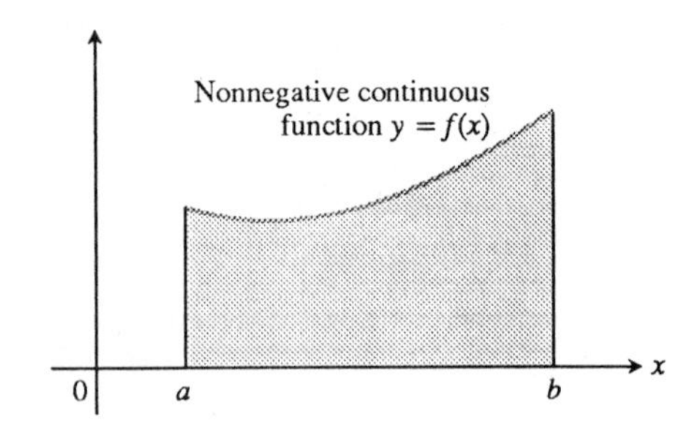

5.1 We can use integral calculus to find the areas of regions like the shaded one here.

one another and shows how our ability to find antiderivatives enables us to calculate the areas of regions like the one in Fig. 5.1.

Regions Bounded by Curves

To find the area of a triangle, we use the formula $A = (1/2)bh$, area equals one-half base times height. To find the areas of more general polygonal regions, we can divide them into triangles, then add the areas of the triangles (Fig. 5.2). But we get stuck if we try to calculate the area of a circle this way. No matter how many triangles we draw inside the circle, their straight edges never quite match the curve of the circle, and some of the circle's interior remains uncovered.

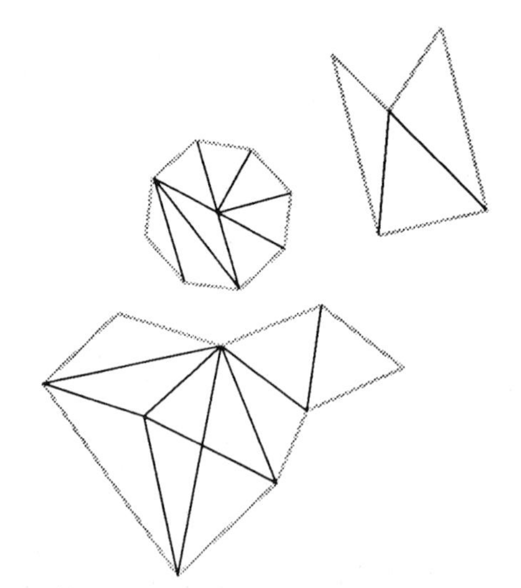

5.2 We find the areas of plane regions with polygonal boundaries by dividing the regions into triangles. The answer is the same for every triangulation.

The Greeks of the fifth century B.C. overcame this problem by filling the circle with an infinite sequence of increasingly fine regular polygons, exhausting the circle's area, so to speak, step by step (Fig. 5.3). They then took the circle's area to be the limit of the areas of these polygons. A decreasing sequence of circumscribed polygons would have worked as well (Fig. 5.4).

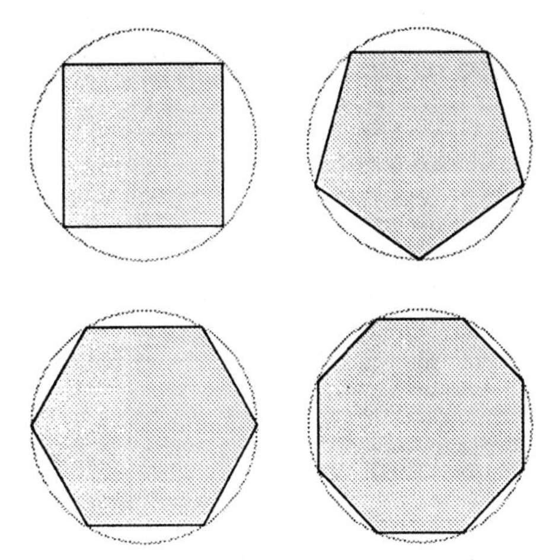

5.3 An increasing sequence of regular polygons inscribed in a circle. Approximations like these were the basis of the method used in classical Greek times to determine the area of a circle.

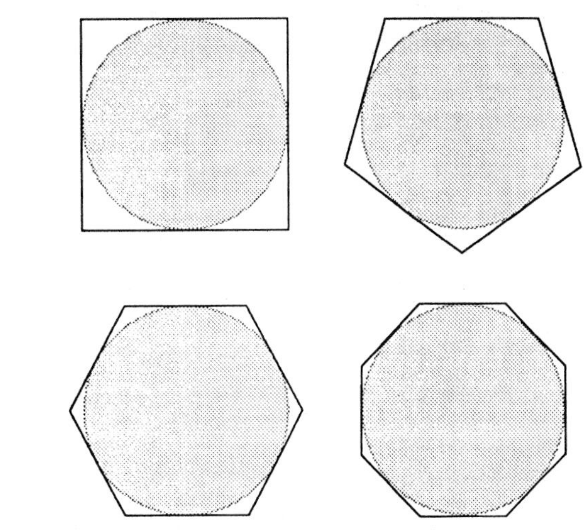

5.4 Circumscribed polygons work too. In this case the polygons' areas approach the circle's area from above instead of below.

The difficulty with applying this approach to more general curves is not the involvement of limits (at least, that's not a difficulty for us). Rather, it is the complication associated with finding workable formulas for the areas of the inscribed polygons, which in an arbitrary curve can assume irregular shapes. We can avoid this difficulty if, instead of working in the abstract plane of Euclidean geometry, we work in the coordinate plane of Descartes and Fermat. For then we can approximate the region under a curve with rectangles whose numerical dimensions, and hence areas, are given by the curve itself.

You will see what we mean if you look at the curve and rectangles in Fig. 5.5. The rectangle areas, added up, approximate the area between the curve $y = f(x)$ and the x-axis over the interval from $x = a$ to $x = b$.

The area of each rectangle, base times height, is the base length times some particular function value, a value we can find from the formula $y = f(x)$.

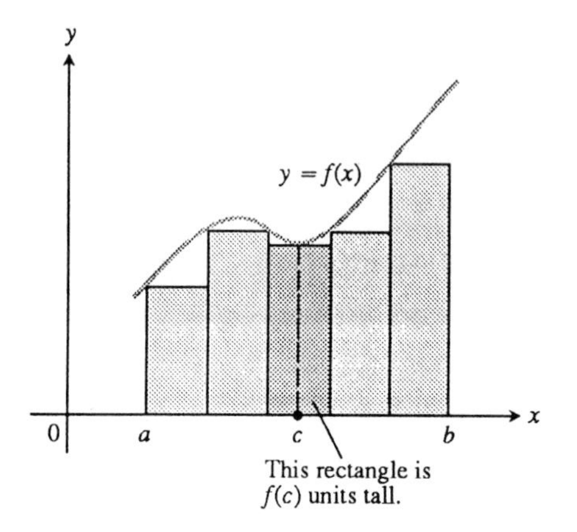

5.5 If we approximate the region under the curve $y = f(x)$ from $x = a$ to $x = b$ with inscribed rectangles that reach from the x-axis up to the curve, then the height of each rectangle is the value of f at some point along the rectangle's base.

Notice how the approximations improve as the rectangles become thinner and more numerous (Fig. 5.6). With each refinement, we get closer to filling up the region whose area we want to find. To finish the job, all we need is

1. A way to write formulas for sums of large numbers of terms, and
2. A way to find the numerical limits of such sums as the number of terms tends to infinity (when the limits exist, that is).

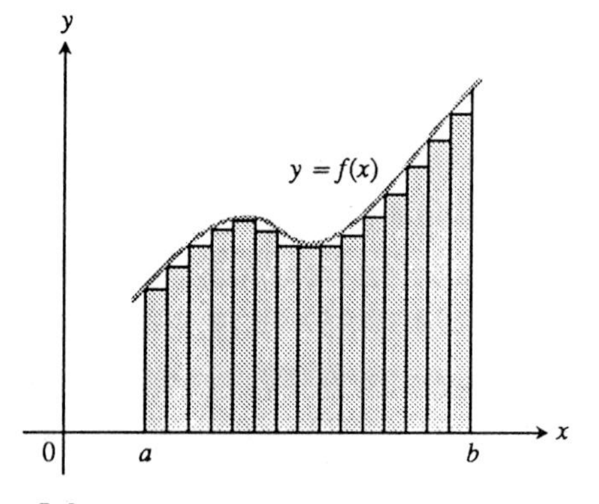

5.6 The more rectangles we use, the better the approximation becomes (provided all the rectangles become narrower as we go along).

As soon as we know how to take these two steps, we shall be able to define and calculate all the areas we want.

What we find, when we take these steps, will also be surprising. We shall be able to do much more than just calculate areas. And, thanks to a great breakthrough in connecting integration and differentiation discovered by Leibniz and Newton, the calculations will be easier. Also, we will illustrate this breakthrough using the power of visualization in a most unusual and instructive manner.

The Area under the Graph of a Nonnegative Continuous Function

Here is a preview of how we shall be able to calculate area once we have made the necessary mathematical arrangements. We all know what we want area to be like, so let us suppose for the moment that the forthcoming mathematical definition in Section 5.3 gives us everything we want. (It will.) Let $y = f(x)$ be a nonnegative continuous function, like the one graphed here, for example:

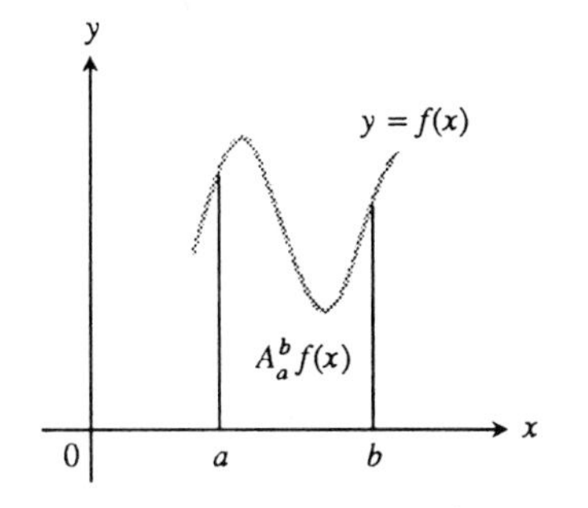

Let $A_a^b f(x)$ denote the area of the region that lies between the curve and the x-axis from a up to b. We'll call this area under the curve from a to b:

$$A_a^b f(x) = \text{the area under the curve from } a \text{ to } b.$$

Let $f(x) = x^2$ and let's try to find $A_0^5 f(x)$, that is, the *area* under the curve $f(x) = x^2$ from $x = 0$ to $x = 5$ (Fig. 5.7a).

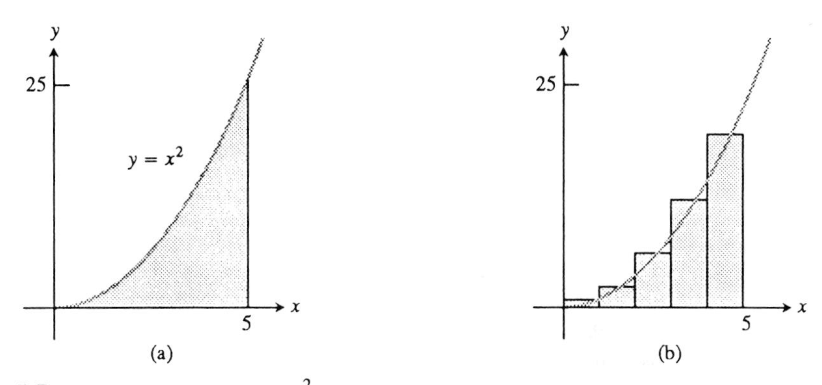

5.7 The area under $f(x) = x^2$ from $x = 0$ to $x = 5$ is (a) shaded and (b) approximated by the area of the five rectangles.

We do not (yet) know a direct way to compute the area. So we will first *approximate* the desired area using a method called the *rectangle approximation method* (RAM). The idea is to approximate the area by using rectangles of known area as shown in the Fig 5.7 (b).

We establish some notation. Let f be a nonnegative continuous function and n the number of rectangles to be used in the approximation of $A_a^b f(x)$. Divide the interval $[a, b]$ into n *subintervals*, each of equal width $\Delta x = (b - a)/n$. The first subinterval is $[x_0, x_1]$ where $x_0 = a$. The next is $[x_1, x_2]$. The kth subinterval is $[x_{k-1}, x_k]$. The last one (the nth one) is $[x_{n-1}, x_n]$ where $x_n = b$ (Fig. 5.8).

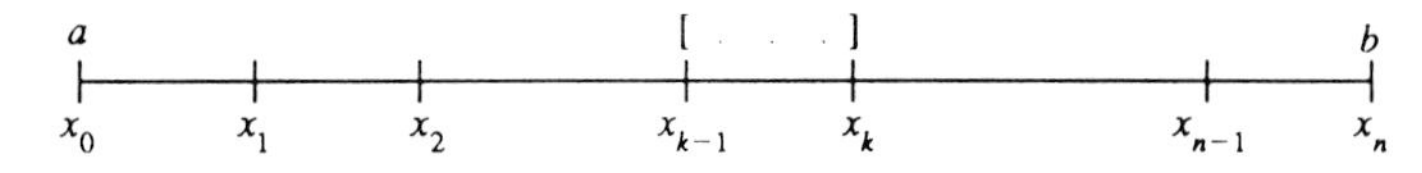

5.8 The kth subinterval of $[a, b]$. Since $\Delta x = (b - a)/n$ is the width of each subinterval, notice that $x_0 = a, x_1 = a + \Delta x, x_2 = a + 2\Delta x, \ldots, x_k = a + k\Delta x, \ldots, x_n = a + n\Delta x = b$.

Next we need a way to determine the *height* of each rectangle. This is easy. For a given subinterval $[x_{k-1}, x_k]$, we have three obvious choices. We can choose the left endpoint (x_{k-1}), or the right endpoint (x_k), or the midpoint $((x_{k-1} + x_k)/2)$, and then compute the corresponding function *value* $(f(x_{k-1}), f(x_k)$ or $f(x_{k-1} + x_k)/2))$ for the height. We use $\text{LR}_n f$, $\text{RR}_n f$ and $\text{MR}_n f$ to denote the RAM determined by the left endpoint, right endpoint and midpoint, respectively, using n approximating rectangles of equal width for the area under the curve $y = f(x)$ from $x = a$ to $x = b$.

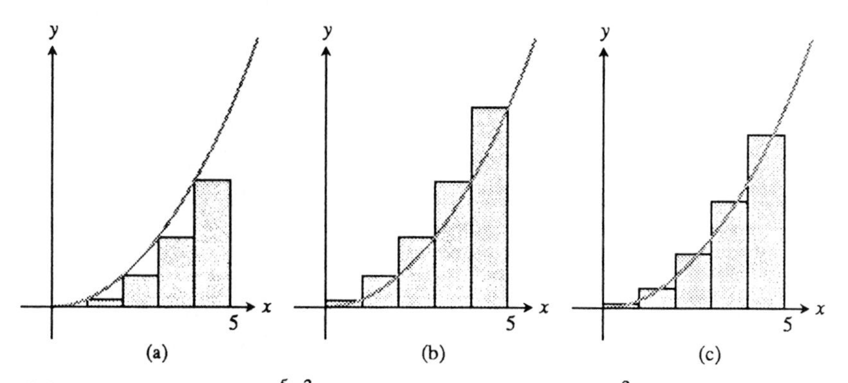

5.9 Approximations to $A_0^5 x^2$, the area under the curve $y = x^2$ from $x = 0$ to $x = 5$, using (a) $\text{LR}_5 x^2$, (b) $\text{RR}_5 x^2$, and (c) $\text{MR}_5 x^2$.

Figure 5.9 shows the three RAM area approximations using 5 subintervals for the area under $f(x) = x^2$ from $x = 0$ to $x = 5$. Next we will actually compute $\text{LR}_n f$, $\text{RR}_n f$, and $\text{MR}_n f$ to numerically approximate the area under f by *adding the areas of each of the 5 rectangles.* From Figure 5.9,

$$\begin{aligned}\text{LR}_5 x^2 &= 1 \cdot 0^2 + 1 \cdot 1^2 + 1 \cdot 2^2 + 1 \cdot 3^2 + 1 \cdot 4^2 \\ &= 0 + 1 + 4 + 9 + 16 = 30\end{aligned}$$

$$\text{RR}_5 x^2 = \text{LR}_5 x^2 + 1 \cdot 5^2 - 1 \cdot 0^2 = 55$$

(Note the easy computation based on the computation for $\text{LR}_5 x^2$)

$$\begin{aligned}\text{MR}_5 x^2 &= 1 \cdot \left(\frac{1}{2}\right)^2 + 1 \cdot \left(\frac{3}{2}\right)^2 + 1 \cdot \left(\frac{5}{2}\right)^2 + 1 \cdot \left(\frac{7}{2}\right)^2 + 1 \cdot \left(\frac{9}{2}\right)^2 \\ &= 41.25.\end{aligned}$$

Explore Box

1. Compute $\text{LR}_n x^2$, $\text{RR}_n x^2$, and $\text{MR}_n x^2$ for $n = 2, 4$, and 6.
2. Compare $\dfrac{\text{LR}_n x^2 + \text{RR}_n x^2}{2}$ with $\text{MR}_n x^2$.
3. Explain why the values in 2 are the same or different.

Notice $\mathrm{LR}_5 x^2 = 30$ is an *under*estimate of the actual area and $\mathrm{RR}_5 x^2$ is an *over*estimate of the actual area. The actual area is exactly $41\frac{2}{3}$ (we will prove this later). So $\mathrm{MR}_5 x^2 = 41.25$ is a good estimate! In the exercise you will be asked to explain why LR_n and RR_n are always underestimates and overestimates, respectively, of $A_0^5 x^2$.

The key to determining a mathematical definition of area is to understand the answer to this question. How can $\mathrm{LR}_n x^2$, $\mathrm{RR}_n x^2$ and $\mathrm{MR}_n x^2$ be made to more closely approximate $A_0^5 x^2$, the actual area under $f(x) = x^2$ from $x = 0$ to $x = 5$? The answer is to make n larger and to use a computer to do the repetitive arithmetic computation! Refer to the *Graphing Calculator Laboratory Manual for Calculus* that accompanies this textbook and enter the simple RAM program for computing $\mathrm{LR}_n f$, $\mathrm{RR}_n f$, and $\mathrm{MR}_n f$. Table 5.1 gives some computer RAM computations for $A_0^5 x^2$ that we will ask you to verify in the exercises.

TABLE 5.1 **RAM Computations for $A_0^5 x^2$.**

n	$\mathrm{LR}_n f$	$\mathrm{RR}_n f$	$\mathrm{MR}_n f$
5	30	55	41.25
10	35.625	48.125	41.5625
25	39.2	44.2	41.65
50	40.425	42.925	41.6625
100	41.04375	42.29375	41.665625
1000	41.6041875	41.7291875	41.66665625

Notice how as n increases the $\mathrm{LR}_n f$ values seem to increase and approach $41\frac{2}{3}$ with each $\mathrm{LR}_n f < 41\frac{2}{3}$, and how the $\mathrm{RR}_n f$ values seem to decrease and approach $41\frac{2}{3}$ with each $\mathrm{RR}_n f > 41\frac{2}{3}$. Indeed, $\mathrm{LR}_n f$ and $\mathrm{RR}_n f$ "squeeze down" on $41\frac{2}{3}$ in this way as n increases. Finally notice that $\mathrm{MR}_n f$ also seems to approach $41\frac{2}{3}$ as n increases but even more rapidly. In fact, for $f(x) = x^2$ on 0 to 5, $\lim_{n\to\infty} \mathrm{LR}_n f = \lim_{n\to\infty} \mathrm{RR}_n f = \lim_{n\to\infty} \mathrm{MR}_n f = 41\frac{2}{3}$. Now it is sensible to define the area under $y = f(x)$ from a to b to be $\lim_{n\to\infty} \mathrm{LR}_n f$, or $\lim_{n\to\infty} \mathrm{RR}_n f$, or $\lim_{n\to\infty} \mathrm{MR}_n f$, if the limits exist. We will see later that if f is continuous, these limits always exist and are equal.

The Area under the Graph of a Nonnegative Continuous Function $y = f(x)$ from $x = a$ to $x = b$, denoted by $A_a^b f(x)$, is

$$A_a^b f(x) = \lim_{n\to\infty} \mathrm{LR}_n f(x) = \lim_{n\to\infty} \mathrm{RR}_n f(x) = \lim_{n\to\infty} \mathrm{MR}_n f(x).$$

Furthermore, if f is increasing on $[a, b]$, then $\text{LR}_n f(x) < A_a^b f(x) < \text{RR}_n f(x)$; and if f is decreasing on $[a, b]$, then $\text{LR}_n f(x) > A_a^b f(x) > \text{RR}_n f(x)$. The next example is typical.

EXAMPLE 1 Estimate the area under $f(x) = x^2 \sin x$ from 0 to 3 using the RAM with $n = 5, 10, 25, 50$, and 100.

Solution First we must verify $x^2 \sin x \geq 0$ for x in $[0, 3]$. A simple algebraic or graphical analysis confirms this.

Next we apply a RAM computer program to obtain Table 5.2. You will be asked to confirm the values in this table in the exercises.

We will see in Chapter 8 (with a lot of effort) that the *exact* area is $-7\cos 3 + 6\sin 3 - 2$ or 5.77666752456 to 12 digits. Notice that $\text{MR}_{100} f(x)$ is "correct" to three decimal places while MR_{1000} is correct to *five* decimal places! The reality is that these computer based numerical methods are extremely powerful and useful.

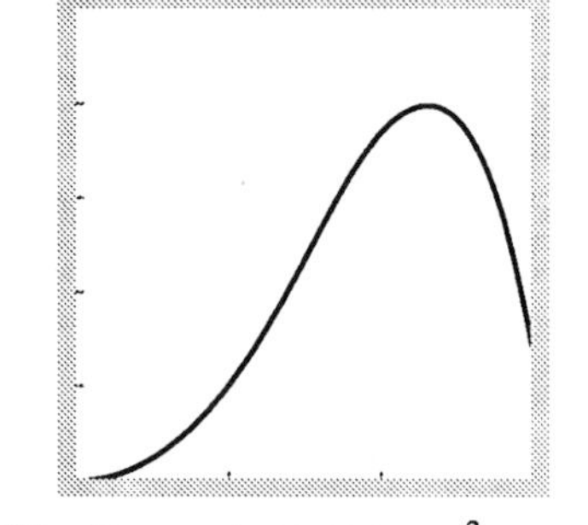

5.10 The graph of $f(x) = x^2 \sin x$ from $0 \leq x \leq 3$.

What Lies Ahead

Let's return to the area under $f(x) = x^2$ from 0 to 5, $A_0^5 x^2$. It should now be clear that $A_0^5 x^2 = \lim_{n\to\infty} (\Delta x \cdot x_0^2 + \Delta x \cdot x_1^2 + \cdots + \Delta x \cdot x_k^2 + \cdots + \Delta x \cdot x_{n-1}^2)$ using the $\text{LR}_n f$ RAM where Δx is the width of one of the n equal subintervals. In the next section we will introduce notation to make writing this sum easy and then actually prove that this limit is *exactly* $41\frac{2}{3}$.

If all we wanted to do was find areas, we would be nearly done now, but the idea of approximating things with small manageable pieces the way we approximated regions with rectangles extends to thousands of other situations. We calculate the volumes of large objects by slicing them like loaves of bread and adding up the volumes of the slices. We find the lengths of curves by approximating small pieces of the curve with line segments and adding the lengths of the line segments. The same idea extends to finding the areas of surfaces and the forces against dams, and to calculating how much work it takes to serve a tennis ball or lift a satellite into orbit. We will see these examples and more in subsequent chapters.

TABLE 5.2 RAM Approximation for $A_0^3(x^2 \sin x)$.

n	$\text{LR}_n f$	$\text{RR}_n f$	$\text{MR}_n f$
5	5.15480..	5.91685..	5.89668..
10	5.52574..	5.90676..	5.80684..
25	5.69078..	5.84319	5.781504
50	5.736146..	5.81235..	5.77787..
100	5.75701..	5.79511..	5.77696..
1000	5.77475..	5.77856..	5.77667..

Exercises 5.1

For each of the following functions $y = f(x)$, consider the area above the x-axis under the graph of $y = f(x)$ from $x = a$ to $x = b$.

a) Make a sketch illustrating the RAM for LR_5, RR_5, and MR_5 showing the five approximating rectangles.

b) Write out by hand LR_5, RR_5, and MR_5, and compute each sum.

1. $y = 6 - x^2$ for $x = 0$ to $x = 2$
2. $y = x^2 + 2$ for $x = -3$ to $x = 2$
3. $f(x) = x + 1$ for $x = 0$ to $x = 5$
4. $f(x) = 5 - x$ for $x = 0$ to $x = 5$
5. $f(x) = 2x^2$ for $x = 0$ to $x = 5$
6. $f(x) = x^2 + 2$ for $x = 1$ to $x = 6$

For each of the following functions, estimate the area above the x-axis under the graph of $y = f(x)$ from $x = a$ to $x = b$ using the RAM for $n = 10$, 100, and 1000. Do all three RAM's: $LR_n f, RR_n f$, and $MR_n f$. First verify each function is non-negative on the specified interval $[a, b]$.

7. $f(x) = x^2 - x + 3$ for $x = 0$ to $x = 3$
8. $f(x) = 2x^2 - 5x + 6$ for $x = -1$ to $x = 4$
9. $f(x) = 2x^3 + 3$ for $x = 0$ to $x = 5$
10. $f(x) = x^3 + x^2 + 2x + 3$ for $x = -1$ to $x = 3$
11. $f(x) = \sin x$ from $x = 0$ to $x = \pi$
12. $f(x) = \cos x$ from $x = 0$ to $x = \frac{\pi}{2}$
13. $f(x) = e^{-x^2}$ for $x = -5$ to $x = 5$
14. $f(x) = 2 + \frac{\sin x}{x}$ for $x = -3$ to $x = 4$
15. For each area problem in Exercises 3, 5, 7, 9, and 11 make a conjecture about the *exact* area.
16. For each area problem in Exercises 4, 6, 8, 10, and 12 make a conjecture about the *exact* area.
17. Confirm the values in Table 5.1.
18. Confirm the values in Table 5.2.
19. In Figure 5.9 we found that $LR_5 x^2$ and $RR_5 x^2$ were underestimates and overestimates, respectively, of $A_0^5 x^2$. Show that $LR_n x^2$ and $RR_n x^2$ are always underestimates and overestimates, respectively of $A_0^5 x^2$.
20. Assume f is an increasing, non-negative function on $[a, b]$. What can you conclude about $LR_n f, RR_n f$, and $A_a^b f$?
21. Let $f(x) = 4 - x^2$. Show that $LR_n x^2 = A_{-2}^2(4 - x^2) = RR_n x^2$ for any positive integer n. Can you generalize this result and explain why it holds?
22. Assume f is a decreasing, non-negative function on $[a, b]$. What can you always conclude about $LR_n f, RR_n f$ and $A_a^b f$?
23. Explain why in Exercise 11 $LR_n \sin x = RR_n \sin x$ for $n = 10, 100$ and 1000.
24. Show that if f is non-negative on $[a, b]$ and the line $y = a + b/2$ is a line of symmetry of the graph of $y = f(x)$ then $LR_n f = RR_n f$ for any positive integer n.

5.2 Formulas for Finite Sums and Area

As we saw in Section 5.1, we need a notation for sums of large numbers of terms. We develop that notation here.

Sigma Notation for Finite Sums

The letter we use to indicate sums is Σ (pronounced "sigma"), the Greek letter for S.

EXAMPLE 1

The sum	In sigma notation	One way to read the notation
$a_1 + a_2$	$\sum_{k=1}^{2} a_k$	The sum of a sub k from k equals 1 to k equals 2.
$a_1 + a_2 + a_3$	$\sum_{k=1}^{3} a_k$	The sum of a sub k from k equals 1 to k equals 3.
$a_1 + a_2 + a_3 + a_4$	$\sum_{k=1}^{4} a_k$	The sum of a sub k from k equals 1 to k equals 4.
$a_1 + a_2 + \cdots + a_n$	$\sum_{k=1}^{n} a_k$	The sum of a sub k from k equals 1 to k equals n.

As you can see, the notation remains compact no matter how many terms are being added. It is just what we need for writing sums that involve millions and millions of terms.

There are many different ways to read the notation $\sum_{k=1}^{n} a_k$, all equally good. Some people say, "Summation from k equals 1 to n of a sub k." Others say, "Summation a k from k equals 1 to n." Still others, "The sum of the a sub k's as k goes from 1 to n," and so on. Take your pick.

Definitions

Sigma Notation for Finite Sums

The symbol

$$\sum_{k=1}^{n} a_k$$

denotes the sum of the n terms

$$a_1 + a_2 + \cdots + a_{n-1} + a_n.$$

The variable k is the **index of summation.** The values of k run through the integers from 1 to n. The a's are the **terms** of the sum; a_1 is the first term, a_2 is the second term, a_k is the **kth term**, and a_n is the nth and last term. The number n is the **upper limit of summation**; the number 1 is the **lower limit of summation.**

Here are some numerical examples.

EXAMPLE 2

The sum in sigma notation	The sum written out—one term for each value of k	The value of the sum
$\sum_{k=1}^{5} k$	$1+2+3+4+5$	15
$\sum_{k=1}^{3} (-1)^k k$	$(-1)^1(1)+(-1)^2(2)+(-1)^3(3)$	$-1+2-3=-2$
$\sum_{k=1}^{2} \frac{k}{k+1}$	$\frac{1}{1+1}+\frac{2}{2+1}$	$\frac{1}{2}+\frac{2}{3}=\frac{7}{6}$

∎

EXAMPLE 3 Find the value of $\sum_{k=1}^{3} \sin\left(\frac{k\pi}{2}\right)$.

Solution

$$\sum_{k=1}^{3} \sin\left(\frac{k\pi}{2}\right) = \sin\left(\frac{1\cdot\pi}{2}\right) + \sin\left(\frac{2\cdot\pi}{2}\right) + \sin\left(\frac{3\cdot\pi}{2}\right) \quad \left(\text{A term for each value of } k\right)$$

$$= 1 + 0 + (-1) = 0 \quad \left(\text{The terms evaluated and added}\right)$$

∎

The lower limit of summation does not have to be 1; it could be some other integer.

EXAMPLE 4

a) $$\sum_{k=0}^{2} \frac{1}{2^k} = \frac{1}{2^0} + \frac{1}{2^1} + \frac{1}{2^2} = \frac{1}{1} + \frac{1}{2} + \frac{1}{4} = \frac{7}{4}$$

b) $$\sum_{k=-3}^{-1} (k+1) = (-3+1) + (-2+1) + (-1+1) = -2 + -1 + 0 = -3$$

∎

Algebra Rules for Finite Sums

When you work with finite sums, you can always use the following rules.

Algebra Rules

1. Sum Rule: $\sum_{k=1}^{n}(a_k + b_k) = \sum_{k=1}^{n} a_k + \sum_{k=1}^{n} b_k$

2. Difference Rule: $\sum_{k=1}^{n}(a_k - b_k) = \sum_{k=1}^{n} a_k - \sum_{k=1}^{n} b_k$

3. Constant Multiple Rule: $\sum_{k=1}^{n} ca_k = c \cdot \sum_{k=1}^{n} a_k$ (Any number c)

4. Constant Value Rule: $\sum_{k=1}^{n} a_k = n \cdot c$ if a_k has the constant value c.

There are no surprises in this list of rules, but the formal proofs require a technique called mathematical induction (Appendix 4).

EXAMPLE 5

a) $\sum_{k=1}^{n}(k - k^2) = \sum_{k=1}^{n} k - \sum_{k=1}^{n} k^2$ (Difference Rule)

b) $\sum_{k=1}^{n} -a_k = \sum_{k=1}^{n} -1 \cdot a_k = -1 \cdot \sum_{k=1}^{n} a_k = -\sum_{k=1}^{n} a_k$ (Constant Multiple Rule)

c) $\sum_{k=1}^{3}(k + 4) = \sum_{k=1}^{3} k + \sum_{k=1}^{3} 4$ (Sum Rule)

$$= (1 + 2 + 3) + (3 \cdot 4) = 6 + 12 = 18$$

Standard Formulas for Sums

Over the years people have discovered a variety of formulas for the values of finite sums. The most famous of these are the formula for the sum of the first n integers (which Gauss discovered at age 5) and the formulas for the sums of the squares and cubes of the first n integers.

The first n integers:	$\sum_{k=1}^{n} k = \frac{n(n+1)}{2}$	(1)
The first n squares:	$\sum_{k=1}^{n} k^2 = \frac{n(n+1)(2n+1)}{6}$	(2)
The first n cubes:	$\sum_{k=1}^{n} k^3 = \left(\frac{n(n+1)}{2}\right)^2$	(3)

Notice the relationship between the first sum and the third.

EXAMPLE 6

a) $\sum_{k=1}^{5} k = 1+2+\cdots+5 = \frac{5(5+1)}{2} = \frac{5\cdot 6}{2} = 15$

b) $\sum_{k=1}^{5} k^2 = 1+4+\cdots+25 = \frac{5(6)(2\cdot 5+1)}{6} = 5\cdot 11 = 55$

c) $\sum_{k=1}^{5} k^3 = 1+8+\cdots+125 = \left(\frac{5(5+1)}{2}\right)^2 = 15^2 = 225$

EXAMPLE 7 Using values from Example 6, we have

a) $\sum_{k=1}^{5} \frac{k^3}{3} = \frac{1}{3}\sum_{k=1}^{5} k^3 = \frac{225}{3} = 75,$

b) $\sum_{k=1}^{5}(k-k^2) = \sum_{k=1}^{5} k - \sum_{k=1}^{5} k^2 = 15 - 55 = -40.$

Computing Area

Now let's return to the computation of $A_0^5(x^2) = \lim_{n\to\infty} \mathrm{RR}_n x^2$. The key here is to recognize that $\Delta x = (5-0)/n = 5/n$ is the width of each of the n rectangles and that the right hand endpoint of the kth interval $[x_{k-1}, x_k]$ is $x_k = 0 + k\Delta x = k5/n$. Thus

$$\begin{aligned}
\mathrm{RR}_n x^2 &= \Delta x \cdot \left(1 \cdot \frac{5}{n}\right)^2 + \Delta x \cdot \left(2 \cdot \frac{5}{n}\right)^2 + \cdots \\
&\quad + \Delta x \cdot \left(k \cdot \frac{5}{n}\right)^2 + \cdots + \Delta x \cdot \left(n \cdot \frac{5}{n}\right)^2 \\
&= \Delta x \sum_{k=1}^{n} \left(k \cdot \frac{5}{n}\right)^2 \\
&= \frac{5}{n} \cdot \frac{5^2}{n^2} \sum_{k=1}^{n} k^2 \qquad \left(\text{Since } \Delta x = \frac{5}{n}\right) \\
&= \frac{5}{n} \cdot \frac{5^2}{n^2} \frac{n(n+1)(2n+1)}{6} \\
&= \frac{125}{6} \frac{n(n+1)(2n+1)}{n^3} \\
&= \frac{125}{6} \frac{2n^3 + 3n^2 + n}{n^3} \\
&= \frac{125}{6} \left(2 + \frac{3}{n} + \frac{1}{n^2}\right)
\end{aligned}$$

As $n \to \infty$, $\mathrm{RR}_n x^2 = 250/6 = 41\frac{2}{3}$ since $\lim_{n\to\infty} (2 + 3/n + 1/n^2) = 2$.

We leave verification of $\lim_{n\to\infty} \mathrm{LR}_n x^2 = \lim_{n\to\infty} \mathrm{MR}_n x^2 = 41\frac{2}{3}$ for the exercises.

Could we determine a formula for $A_0^x(x^2)$ for any value of $x > 0$? Sure. All that is required in the derivation of $A_0^5(x^2)$ is to replace 5 with x. Thus, $\mathrm{RR}_n x^2 = 5/n \cdot 5^2/n^2 \sum_{k=1}^{n} k^2$ becomes

$$\begin{aligned}
\mathrm{RR}_n x^2 &= \frac{x}{n} \cdot \frac{x^2}{n^2} \sum_{k=1}^{n} k^2 \\
&= \frac{x^3}{n^3} \frac{2n^3 + 3n^2 + n}{6} \\
&= \frac{x^3}{6} \left(2 + \frac{3}{n} + \frac{1}{n^2}\right)
\end{aligned}$$

As $n \to \infty$, $\mathrm{RR}_n x^2 = x^3/3$ for any positive value of x. That is, the area under the curve $f(x) = x^2$ from 0 to x is *exactly* $x^3/3$. So

$$A_0^x(x^2) = \frac{x^3}{3}, x \geq 0.$$

As a tantalizing glimpse of what will come in Section 5.4 note that $D_x x^3/3 = x^2$. How remarkable!

Exercises 5.2

Write the sums in Exercises 1–14 without sigma notation. Then evaluate them.

1. $\sum_{k=1}^{4} \frac{1}{k}$
2. $\sum_{k=1}^{4} \frac{12}{k}$
3. $\sum_{k=1}^{3} (k+2)$
4. $\sum_{k=1}^{5} (2k-1)$
5. $\sum_{k=0}^{4} \frac{k}{4}$
6. $\sum_{k=-2}^{2} 3k$
7. $\sum_{k=1}^{2} \frac{6k}{k+1}$
8. $\sum_{k=1}^{3} \frac{k-1}{k}$
9. $\sum_{k=1}^{5} k(k-1)(k-2)$
10. $\sum_{k=0}^{3} (1-k)(2-k)$
11. $\sum_{k=1}^{4} \cos k\pi$
12. $\sum_{k=1}^{3} \sin \frac{\pi}{k}$
13. $\sum_{k=1}^{4} (-1)^k$
14. $\sum_{k=1}^{4} (-1)^{k+1}$

Express the sums in Exercises 15–20 in sigma notation.

15. $1+2+3+4+5+6$
16. $1+4+9+16$
17. $\frac{1}{2}+\frac{1}{4}+\frac{1}{8}+\frac{1}{16}$
18. $1+\frac{1}{2}+\frac{1}{3}+\frac{1}{4}+\frac{1}{5}$
19. $\frac{1}{5}-\frac{2}{5}+\frac{3}{5}-\frac{4}{5}+\frac{5}{5}$
20. $-\frac{1}{5}+\frac{2}{5}-\frac{3}{5}+\frac{4}{5}-\frac{5}{5}$

Use algebra and the formulas in Eqs. (1)–(3) to evaluate the sums in Exercises 21–28.

21. $\sum_{k=1}^{10} k$
22. $\sum_{k=1}^{7} 2k$
23. $\sum_{k=1}^{6} -k^2$
24. $\sum_{k=1}^{6} (k^2+5)$
25. $\sum_{k=1}^{5} k(k-5)$
26. $\sum_{k=1}^{7} (2k-8)$
27. $\sum_{k=1}^{100} k^3 - \sum_{k=1}^{99} k^3$
28. $\left(\sum_{k=1}^{7} k\right)^2 - \sum_{k=1}^{7} k^3$

Find a formula in terms of n for $\mathrm{RR}_n f$ for each of the functions $y = f(x)$ from the specified exercise in Section 5.1. Then determine $\lim_{n\to\infty} \mathrm{RR}_n f$ to find the *exact* area under the curve $y = f(x)$ from $x = a$ to $x = b$.

29. $f(x) = x+1$ for $x = 0$ to $x = 5$ (Exercise 3)
30. $f(x) = 2x^2$ for $x = 0$ to $x = 5$ (Exercise 5)
31. $f(x) = x^2+2$ for $x = 1$ to $x = 4$ (Exercise 6)
32. $f(x) = x^2+x-3$ for $x = 0$ to $x = 3$ (Exercise 7)
33. $f(x) = 2x^3+3$ for $x = 0$ to $x = 5$ (Exercise 9)
34. $f(x) = x^3+x^2+2x+3$ for $x = -1$ to $x = 3$ (Exercise 10)
35. Compute $\mathrm{LR}_n x^2$ for $x = 0$ to $x = 5$. Show that $\lim_{x\to\infty} \mathrm{LR}_n x^2$ is exactly $41\frac{2}{3}$.
36. Show that
    a) $\sum_{k=1}^{n} (2k-1)^2 = \frac{(2n-1)n(2n+1)}{3}$
    b) $\sum_{k=1}^{n} (2k-1)^3 = n^2(2n^2-1)$

    Hint: First show that $\sum_{k=1}^{n} (2k-1)^2 = \sum_{k=1}^{2n} k^2 - \sum_{k=1}^{n} (2k)^2$.
37. Use the result of Exercise 36(a) to show that $\lim_{x\to\infty} \mathrm{MR}_n x^2 = 41\frac{2}{3}$ in the area computation of $A_0^5(x^2)$.
38. Show that $\lim_{x\to\infty} \mathrm{LR}_n x^3$ is exactly 156.25 in the area computation of $A_0^5(x^3)$.
39. Use the result of Exercise 36(b) to show that $\lim_{x\to\infty} \mathrm{MR}_n x^3$ is exactly 156.25 in the area computation of $A_0^5(x^3)$.
40. Find a formula for $A_0^x(x^3)$ for $x \geq 0$.
41. Which of the following express $1+2+4+8+16+32$ in sigma notation?
    a) $\sum_{k=1}^{6} 2^{k-1}$ b) $\sum_{k=0}^{5} 2^k$ c) $\sum_{k=-1}^{4} 2^{k+1}$
42. Which formula is not equivalent to the others?
    a) $\sum_{k=-1}^{1} \frac{(-1)^k}{k+2}$ b) $\sum_{k=0}^{2} \frac{(-1)^k}{k+1}$
    c) $\sum_{k=1}^{3} \frac{(-1)^k}{k}$ d) $\sum_{k=2}^{4} \frac{(-1)^{k-1}}{k-1}$
43. Suppose that $\sum_{k=1}^{n} a_k = -5$ and $\sum_{k=1}^{n} b_k = 6$. Find the values of

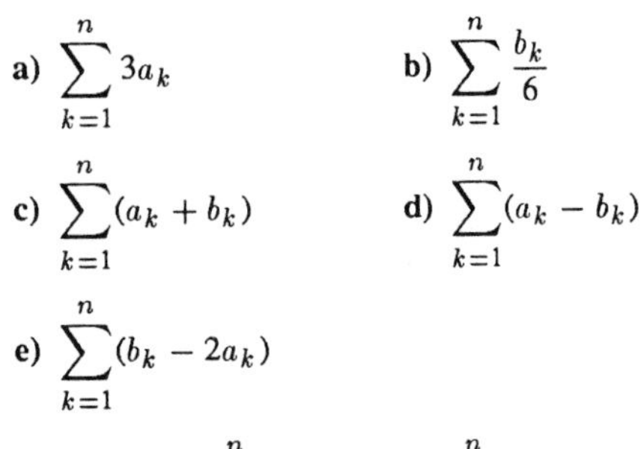

a) $\sum_{k=1}^{n} 3a_k$ b) $\sum_{k=1}^{n} \frac{b_k}{6}$

c) $\sum_{k=1}^{n} (a_k + b_k)$ d) $\sum_{k=1}^{n} (a_k - b_k)$

e) $\sum_{k=1}^{n} (b_k - 2a_k)$

44. Suppose that $\sum_{k=1}^{n} a_k = 0$ and $\sum_{k=1}^{n} b_k = 1$. Find the values of

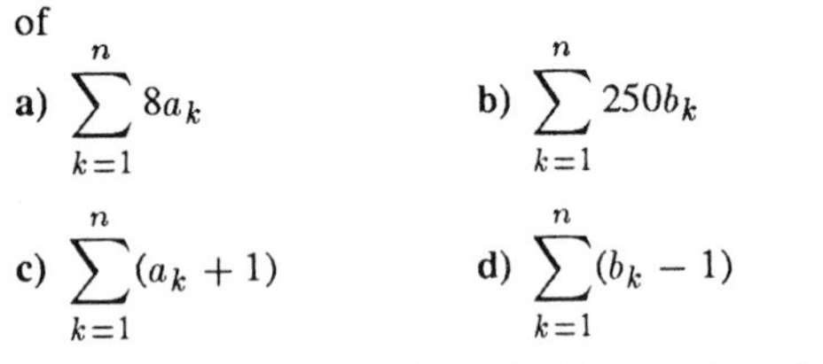

a) $\sum_{k=1}^{n} 8a_k$ b) $\sum_{k=1}^{n} 250b_k$

c) $\sum_{k=1}^{n} (a_k + 1)$ d) $\sum_{k=1}^{n} (b_k - 1)$

45. Use a summation formula to find the number of cans in this supermarket display.

Use arithmetic rules and the formula for the value of $\sum_{k=1}^{n} k$ to establish the formulas in Exercises 46 and 47.

46. $\sum_{k=1}^{n} (2k - 1) = n^2$ **47.** $\sum_{k=1}^{n} k + \sum_{k=1}^{n-1} k = n^2$

Proofs without Words

Here are three informal pictorial proofs of summation formulas. See if you can tell what is going on in each case.

48.

49.

50.

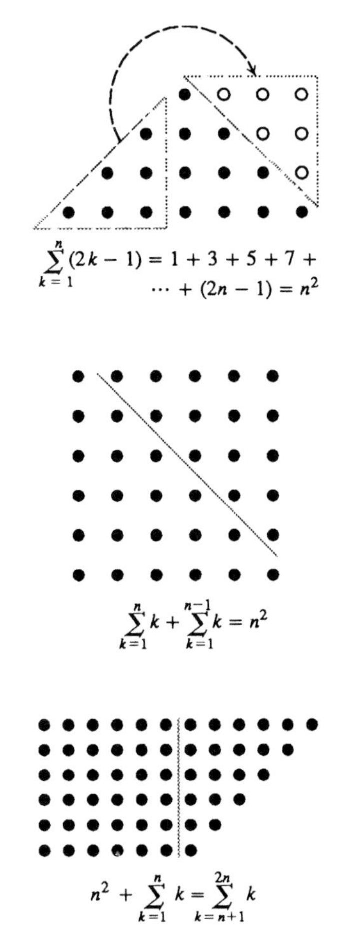

5.3 Definite Integrals

In this section we develop the mathematics that supports the area calculations in Sections 5.1 and 5.2. We do this by defining a limit of sums called the definite integral of a function $y = f(x)$ over an interval $[a, b]$, a limit that exists whenever f is continuous, regardless of the numerical signs of the values of f. In the special case where f is nonnegative, the definite integral of f from a to b is also the number we call the area under the curve $y = f(x)$ from a to b.

Definite integrals define and calculate many things besides areas. We use them to find the volumes of solids, the lengths of curves, the forces against dams, and the average values of periodic functions. The list is nearly endless, and it grows with every passing year.

The RAM for area determination of the previous two sections involved summing the areas of rectangles. These were special cases of a more general sum called a Riemann sum. The limit of a Riemann sum is central to the understanding of integral calculus.

Riemann Sums

We begin with an arbitrary continuous function $y = f(x)$ defined over a closed interval $a \leq x \leq b$. Like the function graphed in Fig. 5.11, it may have negative values as well as positive values.

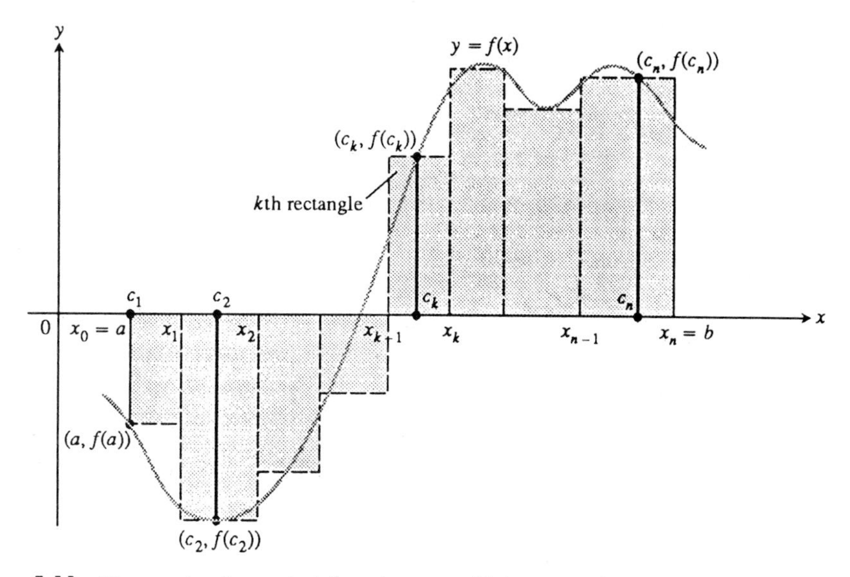

5.11 The graph of a typical function $y = f(x)$ over a closed interval $[a, b]$. The rectangles approximate the region between the graph of the function and the x-axis.

We partition the interval $[a, b]$ into n subintervals by choosing $n - 1$ points, say $x_1, x_2, \ldots, x_{n-1}$, between a and b subject only to the condition that

$$a < x_1 < x_2 < \cdots < x_{n-1} < b. \tag{1}$$

To make the notation consistent, we usually denote a by x_0 and b by x_n. The set

$$P = \{x_0, x_1, \ldots, x_n\} \tag{2}$$

is then called a **partition** of $[a, b]$.

The partition P defines n closed **subintervals**

$$[x_0, x_1], [x_1, x_2], \ldots, [x_{n-1}, x_n]. \tag{3}$$

The typical closed subinterval $[x_{k-1}, x_k]$ is called the **kth subinterval** of P.

kth subinterval

$x_0 = a$ x_1 x_2 ... x_{k-1} x_k ... x_{n-1} $x_n = b$ x

The length of the kth subinterval is $\Delta x_k = x_k - x_{k-1}$.

Δx_1 Δx_2 Δx_k Δx_n

$x_0 = a$ x_1 x_2 ... x_{k-1} x_k ... x_{n-1} $x_n = b$ x

On each subinterval we stand a vertical rectangle that reaches from the x-axis to the curve $y = f(x)$. The exact height of the rectangle does not matter as long as its top or base touches the curve at some point $(c_k, f(c_k))$. See Fig. 5.11 again.

If $f(c_k)$ is positive, the number $f(c_k)\Delta x_k$ = height × base is the area of the rectangle. If $f(c_k)$ is negative, then $f(c_k)\Delta x_k$ is the negative of the area. In any case, we add the n products $f(c_k)\Delta x_k$ to form the sum

$$S_P = \sum_{k=1}^{n} f(c_k)\Delta x_k. \tag{4}$$

This sum, which depends on P and the choice of the numbers c_k, is called a **Riemann sum for f on the interval $[a, b]$,** in honor of the German mathematician Georg Friedrich Bernhard Riemann (1826–1866), who studied the limits of such sums.

EXAMPLE 1 Find a Riemann sum for $f(x) = \sin \pi x$ on the interval $[0, 3/2]$.

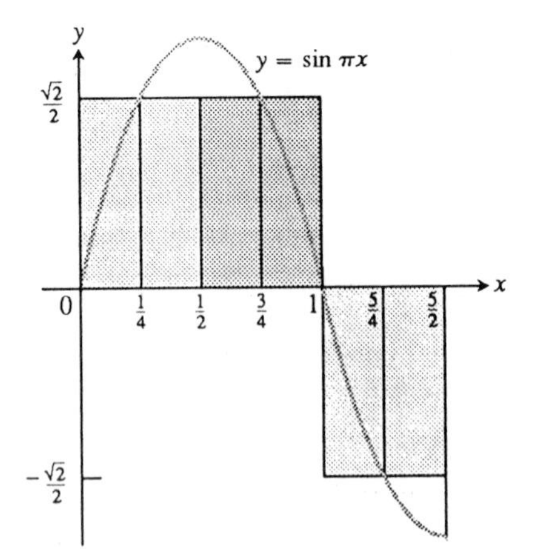

Solution We can choose the following partition and equal subinterval lengths.

Partition:

$$P = \left\{0, \frac{1}{2}, 1, \frac{3}{2}\right\}$$

Subinterval lengths:

$$\Delta x_k = \frac{1}{2}, \qquad k = 1,\ 2,\ 3$$

Choice of c_k: the interval midpoints

$$c_1 = \frac{1}{4}, \qquad c_2 = \frac{3}{4}, \qquad c_3 = \frac{5}{4}$$

The Riemann sum:

$$\sum_{k=1}^{3} f(c_k)\Delta x_k = \sum_{k=1}^{3} \sin(\pi c_k) \cdot \frac{1}{2}$$

$$= \frac{1}{2}\sum_{k=1}^{3} \sin(\pi c_k) = \frac{1}{2}\left(\sin\frac{\pi}{4} + \sin\frac{3\pi}{4} + \sin\frac{5\pi}{4}\right)$$

$$= \frac{1}{2}\left(\frac{\sqrt{2}}{2} + \frac{\sqrt{2}}{2} - \frac{\sqrt{2}}{2}\right) = \frac{\sqrt{2}}{4}$$

If instead of choosing the c_k's to be midpoints we choose each c_k to be the left-hand endpoint of its interval, then

$$c_1 = 0, c_2 = \frac{1}{2}, c_3 = 1,$$

and the Riemann sum is

$$\sum_{k=1}^{3} \sin(\pi c_k)\Delta x_k = \frac{1}{2}\sum_{k=1}^{3} \sin(\pi c_k)$$

$$= \frac{1}{2}\left(\sin 0 + \sin\frac{\pi}{2} + \sin\pi\right)$$

$$= \frac{1}{2}(0 + 1 + 0) = \frac{1}{2}.$$

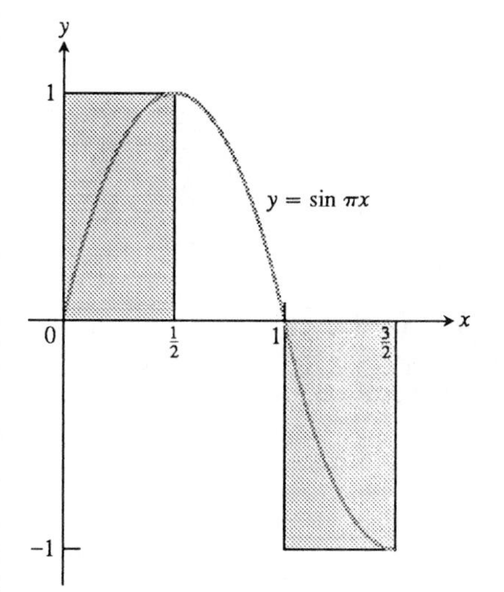

If we choose each c_k to be the right-hand endpoint of its subinterval, then

$$c_1 = \frac{1}{2}, c_2 = 1, c_3 = \frac{3}{2},$$

and the Riemann sum is

$$\sum_{k=1}^{3} \sin(\pi c_k)\Delta x_k = \frac{1}{2}\left(\sin\frac{\pi}{2} + \sin\pi + \sin\frac{3\pi}{2}\right) = \frac{1}{2}(1 + 0 - 1) = 0.$$

What happens to the Riemann sums as the number of points in the partition increases and the partition becomes finer? As Fig. 5.12 suggests, the rectangles involved overlap the region between the curve and the x-axis with increasing accuracy, and we should find the sums approaching a limiting value of some kind. To make this idea precise, we need to define what it means for partitions to become finer and for Riemann sums to have a limit. We accomplish this with the following definitions.

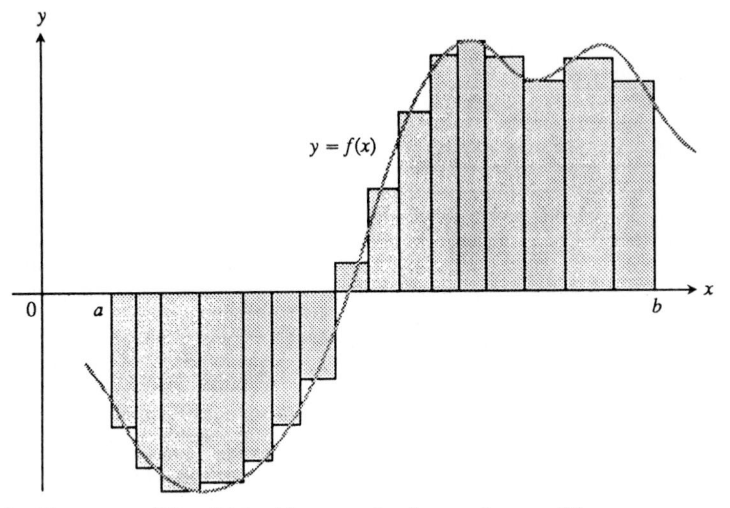

5.12 The curve of Fig. 5.11 with rectangles from a finer partition.

Definition

The **norm** of a partition is the length of the partition's longest subinterval. If the partition is denoted by P, its norm is denoted by putting double bars around the P:

$$||P||. \quad \text{(The norm of } P\text{)}$$

EXAMPLE 2 Find the norm of the partition $P = \left\{0, \frac{1}{4}, \frac{2}{3}, 1, \frac{3}{2}, 2\right\}$ of the interval $[0, 2]$.

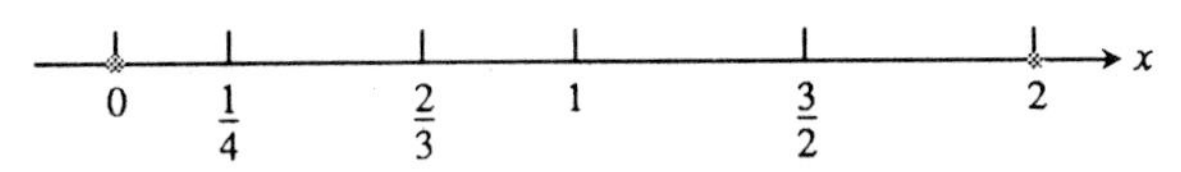

Solution The subintervals in the partition are

$$\left[0, \frac{1}{4}\right], \quad \left[\frac{1}{4}, \frac{2}{3}\right], \quad \left[\frac{2}{3}, 1\right], \quad \left[1, \frac{3}{2}\right], \quad \left[\frac{3}{2}, 2\right].$$

The lengths of the subintervals are

$$\Delta x_1 = \frac{1}{4}, \quad \Delta x_2 = \frac{2}{3} - \frac{1}{4} = \frac{5}{12}, \quad \Delta x_3 = \frac{1}{3}, \quad \Delta x_4 = \frac{1}{2}, \quad \Delta x_5 = \frac{1}{2}.$$

The norm of the partition is 1/2, the longest of these lengths. As you can see, there are two subintervals of length 1/2. There can be more than one longest subinterval. ≡

The way we say that successive partitions of an interval become finer is to say that their norms approach zero.

Definition

The Definite Integral as a Limit of Riemann Sums

Let $f(x)$ be a function defined on a closed interval $[a, b]$. We say that the **limit** of the Riemann sums $\sum_{k=1}^{n} f(c_k)\Delta x_k$ on $[a, b]$ as $||P|| \to 0$ is the number I if the following condition is satisfied:

Given any positive number ϵ, there exists a positive number δ such that for every partition P of $[a, b]$

$$||P|| < \delta \quad \text{implies} \quad \left|\sum_{k=1}^{n} f(c_k)\Delta x_k - I\right| < \epsilon \tag{5}$$

for any choice of the numbers c_k in the subintervals $[x_{k-1}, x_k]$.

If the limit exists, we write

$$\lim_{||P|| \to 0} \sum_{k=1}^{n} f(c_k)\Delta x_k = I. \tag{6}$$

We call I the **definite integral** of f over $[a, b]$, we say that f is **integrable** over $[a, b]$, and we say that Riemann sums of f on $[a, b]$ **converge** to I.

The amazing fact is that despite the potential for variety in the Riemann sums $\Sigma f(c_k)\Delta x_k$ as the partitions change and the c_k's are chosen at random in the intervals of each new partition, the sums always have a limit as $||P|| \to 0$ when f is continuous on $[a, b]$. The existence of this limit, of the definite integral of a continuous function on a closed interval, blithely assumed by the mathematicians of the seventeenth and eighteenth centuries, was finally established, once and for all, by Georg Riemann in 1854.

Explore Box

Use the formula for $\sum_{k=1}^{n} k^3$ given in Eq. (3) in Section 5.2 to compute a Riemann sum RR_n for $f(x) = x^3$ on (a) $[0, 1]$, (b) on $[0, 5]$, and (c) on $[0, a]$. What is $A_0^a(x^3)$ exactly?

Theorem 1

The Existence of Definite Integrals

All continuous functions are integrable. That is, if a function $y = f(x)$ is continuous on an interval $[a, b]$, then its definite integral over $[a, b]$ exists.

You can find a current version of Riemann's proof of this theorem in most advanced calculus books.

Theorem 1 says nothing about *how* to calculate definite integrals. Except for a few special cases, that takes another theorem, and we shall get to it in the next section. We do know how to find the limit of a particular Riemann sum and thus complete the definite integral for the special cases of $f(x) = ax$, $f(x) = ax^2$ and $f(x) = ax^3$. However, this process is tedious at best.

Finally, Theorem 1 speaks only about continuous functions. Many discontinuous functions are integrable as well, but we shall not deal with this here.

Terminology

There is a fair amount of terminology to learn in connection with definite integrals.

The definite integral of a function $f(x)$ over an interval $[a, b]$ is usually denoted by the symbol

$$\int_a^b f(x)\,dx, \tag{7}$$

which is read as "the integral of f of $x\,d\,x$ from a to b." The symbol $\int$ is an **integral sign**. Leibniz chose it because it resembled the S in the German word for *summation*.

When we find the value of $\int_a^b f(x)\,dx$, we say that we have **evaluated the integral** and that we have **integrated** f from a to b. We call $[a, b]$ the **interval of integration.** The numbers a and b are the **limits of integration,** a being the **lower limit of integration** and b the **upper limit of integration.** The function f is the **integrand** of the integral. The variable x is the **variable of integration.**

The value of the definite integral of a function over any particular interval depends on the function and not on the letter we choose to represent its independent variable. If we decide to use t or u instead of x, we simply write the integral as

$$\int_a^b f(t)\,dt \quad \text{or} \quad \int_a^b f(u)\,du \quad \text{instead of} \quad \int_a^b f(x)\,dx.$$

No matter how we write the integral, it is still the same number, defined as a limit of Riemann sums. Since it does not matter what letter we use, the variable of integration is called a **dummy variable.**

EXAMPLE 3 Express the limit of Riemann sums

$$\lim_{\|P\|\to 0} \sum_{k=1}^{n} (3c_k^2 - 2c_k + 5)\Delta x_k$$

as an integral if P denotes a partition of the interval $[-1, 3]$.

Solution The function being evaluated at c_k in each term of the sum is $f(x) = 3x^2 - 2x + 5$. The interval being partitioned is $[-1, 3]$. The limit is therefore the integral of f from -1 to 3:

$$\lim_{\|P\|\to 0} \sum_{k=1}^{n} (3c_k^2 - 2c_k + 5)\Delta x_k = \int_{-1}^{3} (3x^2 - 2x + 5)\, dx.$$

Constant Functions

Integrals of constant functions are always easy to evaluate.

Theorem 2

If $f(x) = c$ has the constant value c on the interval $[a, b]$, then

$$\int_a^b f(x)\, dx = \int_a^b c\, dx = c(b - a). \tag{8}$$

Theorem 2 says that the integral of a constant function over a closed interval is the constant times the length of the interval.

Proof of Theorem 2 The Riemann sums of f on $[a, b]$ all have the constant value

$$\begin{aligned} \sum f(c_k)\Delta x_k &= \sum c \cdot \Delta x_k \\ &= c \cdot \sum \Delta x_k && \text{(Constant multiple rule for sums)} \\ &= c(b - a). && \text{(The } \Delta x_k\text{'s add up to the length of } [a, b].) \end{aligned}$$

The limit of these sums, the integral they converge to, therefore has this value too.

EXAMPLE 4

$$\int_1^4 5\, dx = 5(4 - 1) = 5(3) = 15$$

Area Is Strictly a Special Case

If an integrable function $y = f(x)$

is nonnegative throughout an interval $[a, b]$, as in Fig. 5.13, each term $f(c_k)\Delta x_k$ is the area of a rectangle reaching from the x-axis up to the curve

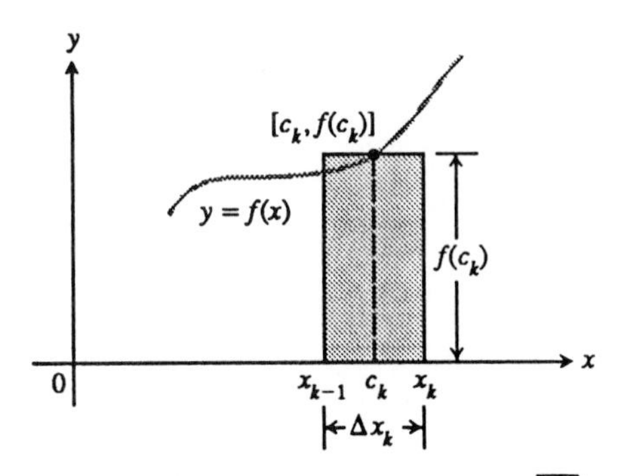

5.13 If $f(x) \geq 0$, then each term in the Riemann sum $\sum f(c_k)\Delta x_k$ is a rectangle area.

$y = f(x)$. The Riemann sum

$$\sum_{k=1}^{n} f(c_k)\Delta x_k,$$

which is the sum of the areas of these rectangles, gives an estimate of the area of the region between the curve and the x-axis from a to b. Since the rectangles give an increasingly good approximation of the region as we use subdivisions with smaller and smaller norms, we call the limiting value

$$\lim_{\|P\|\to 0} \sum f(c_k)\Delta x_k = \int_a^b f(x)\,dx$$

the area under the curve.

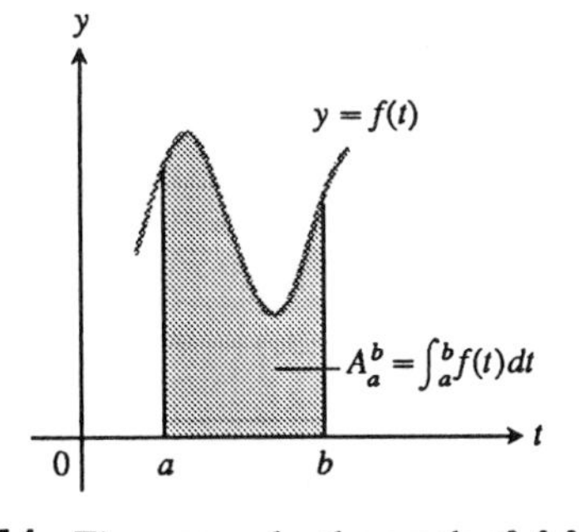

5.14 The area under the graph of f from a to x is defined and calculated as an integral.

Definition

> If $y = f(x)$ is nonnegative and integrable over a closed interval $[a, b]$, then the integral of f from a to b is the **area** of the region between the graph of f and the x-axis from a to b. We sometimes call this number the **area under the curve $y = f(x)$ from a to b.**

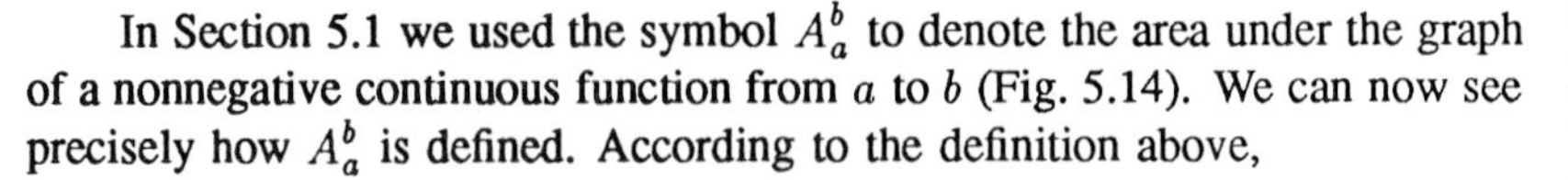

In Section 5.1 we used the symbol A_a^b to denote the area under the graph of a nonnegative continuous function from a to b (Fig. 5.14). We can now see precisely how A_a^b is defined. According to the definition above,

$$A_a^b f(x) = \int_a^b f(t)\,dt. \tag{9}$$

EXAMPLE 5 Find the value of the integral

$$\int_{-2}^{2} \sqrt{4 - x^2}\,dx$$

by regarding it as the area under the graph of an appropriately chosen function.

Solution We graph the integrand $f(x) = \sqrt{4 - x^2}$ over the interval of integration $[-2, 2]$ and see that the graph is a semicircle of radius 2. The area between the semicircle and the x-axis is

$$\text{Area} = \frac{1}{2} \cdot \pi r^2 = \frac{1}{2}\pi(2)^2 = 2\pi.$$

Because the area is also the value of the integral of f from -2 to 2,

$$\int_{-2}^{2} \sqrt{4 - x^2}\, dx = 2\pi.$$

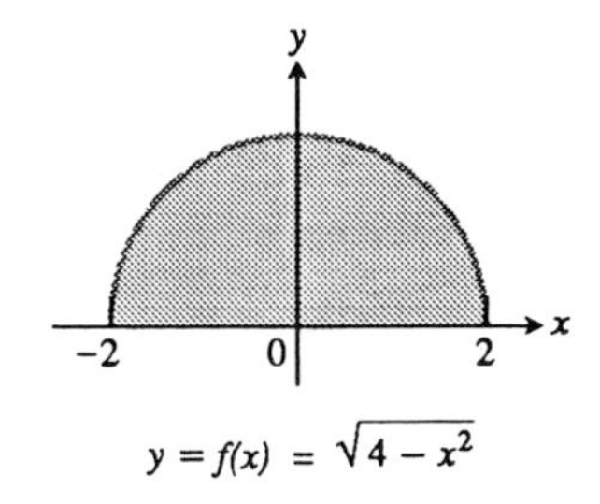

If an integrable function $y = f(x)$ is nonpositive, the terms $f(c_k)\Delta x_k$ in the Riemann sums for f over an interval $[a, b]$ are all negatives of rectangle areas. The limit of the Riemann sums, the integral of f from a to b, is therefore the negative of the area between the graph of f and the x-axis.

$$\int_a^b f(x)\, dx = -\text{ (the area) if } f(x) \leq 0. \qquad (10)$$

Or, turning this around,

$$\text{Area} = -\int_a^b f(x)\, dx \text{ when } f(x) \leq 0. \qquad (11)$$

EXAMPLE 6

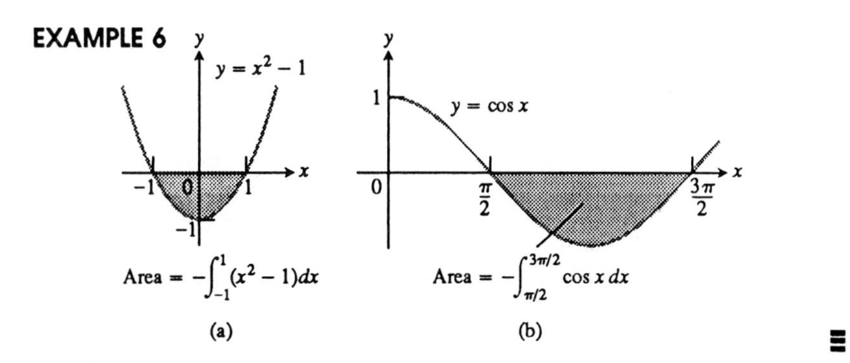

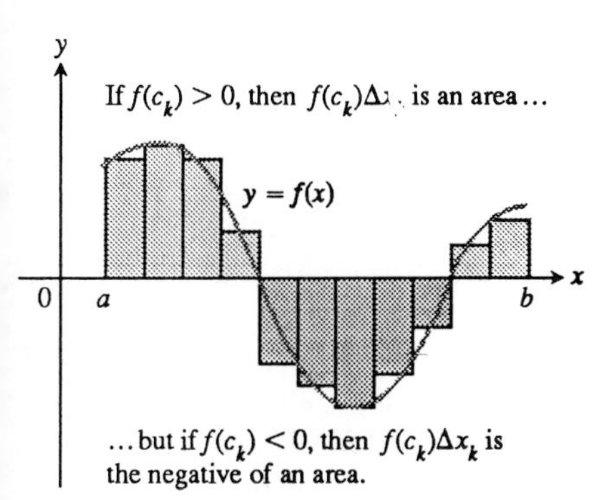

5.15 An integrable function f with negative as well as positive values.

If an integrable function $y = f(x)$ has both positive and negative values on an interval $[a, b]$, then the Riemann sums for f on $[a, b]$ add the areas of the rectangles that lie above the x-axis to the negatives of the areas of the rectangles that lie below the x-axis, as in Fig. 5.15. The resulting cancellation reduces the sums, so their limiting value is a number whose magnitude is less than the total area between the curve and the x-axis. The value of the integral is the area above the axis minus the area below the axis.

For any integrable function,

$$\int_a^b f(x)\,dx = (\text{area above } x\text{-axis}) - (\text{area below } x\text{-axis}). \quad (12)$$

NET AREA

Sometimes $\int_a^b f(x)$ is called the *net area* of the region determined by the curve $y = f(x)$ and the x-axis between $x = a$ and $x = b$.

EXAMPLE 7

$$\int_0^{2\pi} \sin x\,dx = 0$$

The integral is the area of region A above the x-axis minus the area of region B below the x-axis. The areas are the same, so the integral is zero.

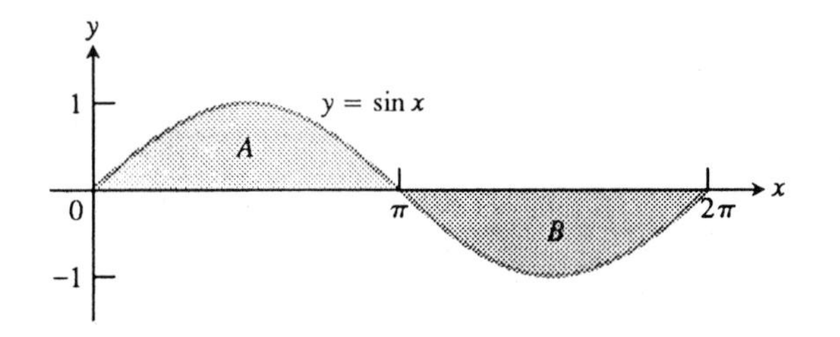

Evaluating Definite Integrals Numerically with Technology

In this textbook we assume you have a graphing calculator or computer software that has a simple method (usually a key) for approximating the value of any definite integral. We use the notation

$$\text{FnInt}(f(x), x, a, b)$$

to denote a calculator or computer approximation of $\int_a^b f(x)dx$. And we write

$$\int_a^b f(x)dx = \text{FnInt}(f(x), x, a, b)$$

with the understanding that the right side of the equation is an approximation to the left side.

There are many methods for numerically approximating $\int_a^b f(x)\,dx$. Thus $\text{FnInt}(f(x), x, a, b)$ values will vary depending on the machine you are using. However, most machines or software packages give results that are very accurate, usually 5 or 6 significant digits.

The accuracy issue is further complicated by the fact that some machines and software packages allow the user to specify an integration tolerance much like when you specify the value of δ in the "NDer" symmetric difference numerical derivative computation. You should become familiar with the calculator or computer software integration options that are available on your machine. In this textbook, unless stated otherwise, we will use TI-85 calculations with tol $= 10^{-5}$ for FnInt evaluation. We will usually report answers showing the first six significant digits or the first six decimal places.

EXAMPLE 8 Evaluate **a)** $\int_{-1}^{2} x \sin x \, dx$ and **b)** $\int_{0}^{5} e^{-x^2} dx$.

Solution $\text{FnInt}(x \sin x, x, -1, 2) = 2.042760$ and $\text{FnInt}(e^{-x^2}, x, 0, 5) = 0.8862269$. Remember your answers may vary slightly from these due to differences in machine computation. The use of "=" is understood to mean "approximately equal."

It is comforting to note that the *exact* solution to a) in Example 8 is $-2\cos 2 + \sin 2 - \cos 1 + \sin 1$ which is 2.04275977886 to 12 digits. You will learn how to determine this exact solution using analytic methods in Chapter 8. It is *not* comforting to note that there is no known *exact* solution to part (b) of Example 8. The best anyone can do is to numerically approximate this definite integral. Here technology is *essential*!

Exercises 5.3

Express the limits in Exercises 1–8 as definite integrals.

1. $\lim_{\|P\| \to 0} \sum_{k=1}^{n} c_k^2 \Delta x_k$, where P is a partition of $[0, 2]$
2. $\lim_{\|P\| \to 0} \sum_{k=1}^{n} 2c_k^3 \Delta x_k$, where P is a partition of $[-1, 0]$
3. $\lim_{\|P\| \to 0} \sum_{k=1}^{n} (c_k^2 - 3c_k) \Delta x_k$, where P is a partition of $[-7, 5]$
4. $\lim_{\|P\| \to 0} \sum_{k=1}^{n} \frac{1}{c_k} \Delta x_k$, where P is a partition of $[1, 4]$
5. $\lim_{\|P\| \to 0} \sum_{k=1}^{n} \frac{1}{1 - c_k} \Delta x_k$, where P is a partition of $[2, 3]$
6. $\lim_{\|P\| \to 0} \sum_{k=1}^{n} \sqrt{4 - c_k^2} \Delta x_k$, where P is a partition of $[0, 1]$
7. $\lim_{\|P\| \to 0} \sum_{k=1}^{n} \cos c_k \Delta x_k$, where P is a partition of $[0, 4]$
8. $\lim_{\|P\| \to 0} \sum_{k=1}^{n} \sin^3 c_k \Delta x_k$, where P is a partition of $[-\pi, \pi]$

Use $\text{FnInt}(f(x), x, a, b)$ in Exercises 9–12 to compute the definite integral $\int_a^b f(x) \, dx$. Graph each function $y = f(x)$ for $a \le x \le b$. Describe a region or regions for which the area of the region(s) is given by the definite integral.

9. $\int_{-1}^{3} (2 - x - 5x^2) \, dx$
10. $\int_{0}^{3} x^2 e^{-x^2} \, dx$
11. $\int_{0}^{2\pi} \sin(x^2) \, dx$
12. $\int_{0}^{10} \frac{\sin x}{x} \, dx$

In Exercises 13–18 express the area of the shaded region as an integral. Use $\text{FnInt}(f(x), x, a, b)$ to evaluate the integral.

13.

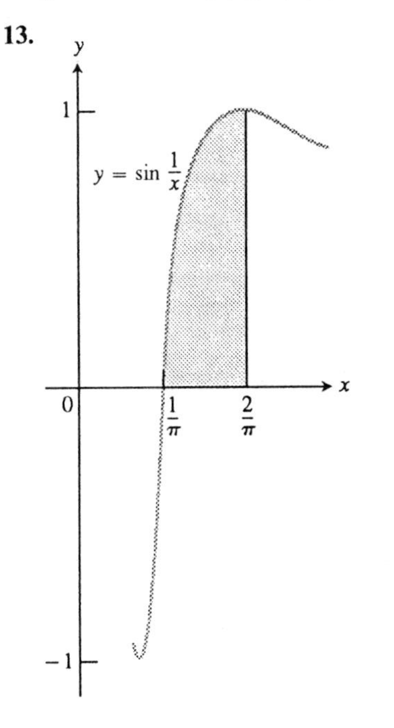

14.

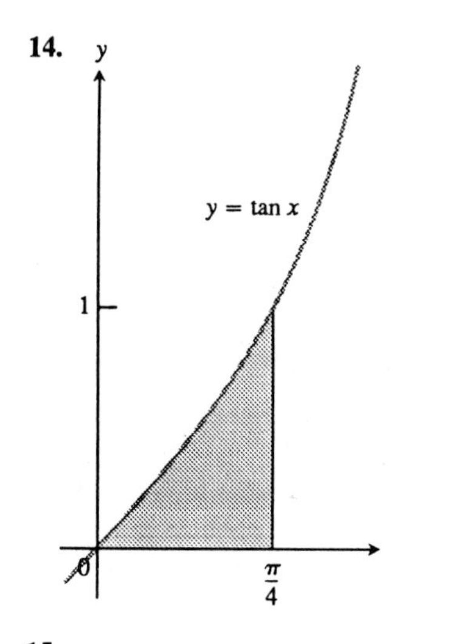

15.

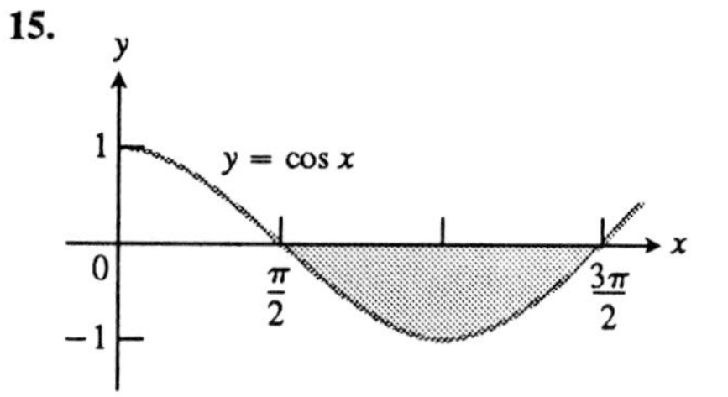

16.

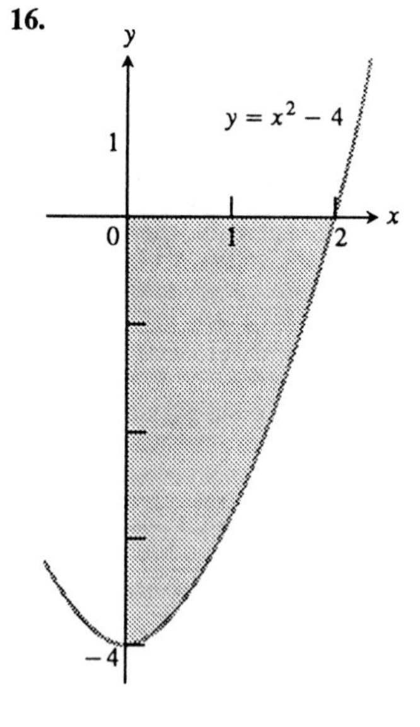

17.

18.

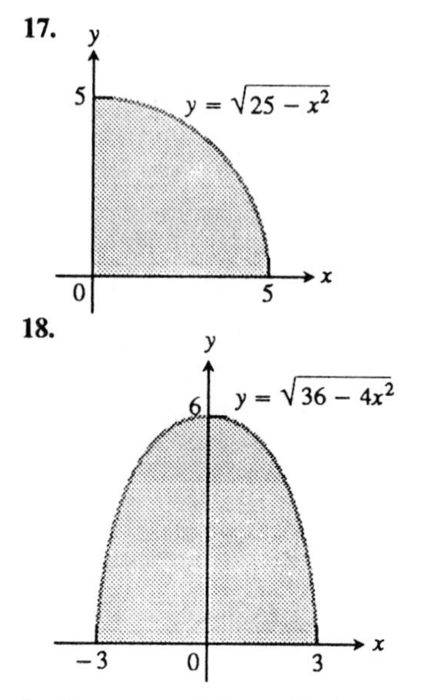

In Exercises 19–22, find the exact value of each integral by regarding it as the area under the graph of an appropriately chosen function and using an area formula from plane geometry. Compare the exact value with an FnInt computation.

19. $\int_{-1}^{1} \sqrt{1-x^2}\,dx$

20. $\int_{0}^{2} \sqrt{4-x^2}\,dx$

21. $\int_{-1}^{1} (1-|x|)\,dx$

22. $\int_{-1}^{1} (1+\sqrt{1-x^2})\,dx$

23. Explain why the sums $\text{LR}_n f$, $\text{RR}_n f$, and $\text{MR}_n f$ are Riemann sums.

24. Use the formula for $\sum_{k=1}^{n} (2k-1)^3$ given in Exercise 36(b) from Section 5.2 to compute a Riemann sum MR_n for $f(x) = x^3$ on $[0,5]$ and $[0,a]$. Find the limit of each Riemann sum.

25. Find the exact value of the area in Exercise 16 by finding the limit of a Riemann sum.

26. Use all three RAM's with $n = 100$ to support the FnInt computations in Example 8.

5.4 Antiderivatives and Definite Integrals

In this section we continue our discussion of definite integrals, introduce the concept of average value of a function, and conclude with a preview of the most remarkable result of calculus discovered independently by Newton and Liebniz.

Useful Rules for Working with Integrals

We often want to add and subtract definite integrals, multiply them by constants, and compare them with other definite integrals. We do this with the rules in Table 5.3 on the following page. All the rules except the first two follow from the way integrals are defined as limits of Riemann sums. The sums have these properties, so their limits do too. For example, Rule 3 says that the integral of k times a function is k times the integral of the function. This is true because

$$\begin{aligned}\int_a^b kf(x)dx &= \lim_{\|P\|\to 0} \sum_{i=1}^{n} kf(c_i)\Delta x_i \\ &= \lim_{\|P\|\to 0} k\sum_{i=1}^{n} f(c_i)\Delta x_i \qquad (1)\\ &= k \lim_{\|P\|\to 0} \sum_{i=1}^{n} f(c_i)\Delta x_i = k\int_a^b f(x)\,dx.\end{aligned}$$

Rule 1 is a definition. We want the integral of a function over an interval of zero length to be zero.

EXAMPLE 1 Suppose that f, g, and h are integrable, that

$$\int_{-1}^{1} f(x)\,dx = 5, \quad \int_{1}^{4} f(x)\,dx = -2, \quad \int_{-1}^{1} h(x)\,dx = 7,$$

and that $g(x) \geq f(x)$ on $[-1, 1]$. Then

1. $\displaystyle\int_4^1 f(x)\,dx = -\int_1^4 f(x)\,dx = -(-2) = 2,$ (Rule 2)

2. $\displaystyle\int_{-1}^{1} [2f(x) + 3h(x)]dx = 2\int_{-1}^{1} f(x)dx + 3\int_{-1}^{1} h(x)dx = 2(5) + 3(7) = 31,$ (Rules 3 and 5)

3. $\displaystyle\int_{-1}^{1} [f(x) - h(x)]dx = 5 - 7 = -2,$ (Rule 6)

4. $\displaystyle\int_{-1}^{1} g(x)dx \geq 5,$ (Because $g(x) \geq f(x)$ on $[-1, 1]$—Rule 8)

5. $\displaystyle\int_{-1}^{4} f(x)\,dx = \int_{-1}^{1} f(x)\,dx + \int_{1}^{4} f(x)\,dx = 5 + (-2) = 3.$ (Rule 10)

TABLE 5.3 Rules for Definite Integrals

1. The zero rule: $\int_a^a f(x)\,dx = 0.$ (A definition)
2. Reversing the order of integration changes an integral's sign:
$$\int_b^a f(x)\,dx = -\int_a^b f(x)\,dx. \quad \text{(Also a definition)}$$
3. The constant multiple rule: $\int_a^b kf(x)dx = k\int_a^b f(x)\,dx.$ (Any number k.)
4. Special case of Rule 3: $\int_a^b -f(x)\,dx = -\int_a^b f(x)\,dx.$ (Take $k = -1$ in Rule 3.)
5. The sum rule: If f and g are integrable on $[a, b]$, then $f + g$ is integrable on $[a, b]$ and
$$\int_a^b [f(x) + g(x)]dx = \int_a^b f(x)\,dx + \int_a^b g(x)dx.$$
6. The difference rule: If f and g are integrable on $[a, b]$, then $f - g$ is integrable on $[a, b]$ and
$$\int_a^b [f(x) - g(x)]dx = \int_a^b f(x)\,dx - \int_a^b g(x)dx.$$
7. Integrals of nonnegative functions are nonnegative:
$$f(x) \geq 0 \text{ on } [a, b] \quad \Rightarrow \quad \int_a^b f(x)dx \geq 0.$$
8. The domination rule:
$$g(x) \geq f(x) \text{ on } [a, b] \quad \Rightarrow \quad \int_a^b g(x)\,dx \geq \int_a^b f(x)\,dx.$$
9. The max-min rule: If $b > a$, then
$$\underbrace{\min f \cdot (b-a)}_{\text{Lower bound}} \leq \int_a^b f(x)\,dx \leq \underbrace{\max f \cdot (b-a)}_{\text{Upper bound}},$$
where max and min refer to the values of f on $[a, b]$.
10. The interval addition rule: $\int_a^b f(x)\,dx + \int_b^c f(x)\,dx = \int_a^c f(x)\,dx.$
The only requirement is that f be integrable on the intervals joining a, b, and c (Fig. 5.16).
11. The interval subtraction rule (Rule 10 in another form):
$$\int_b^c f(x)\,dx = \int_a^c f(x)\,dx - \int_a^b f(x)\,dx.$$

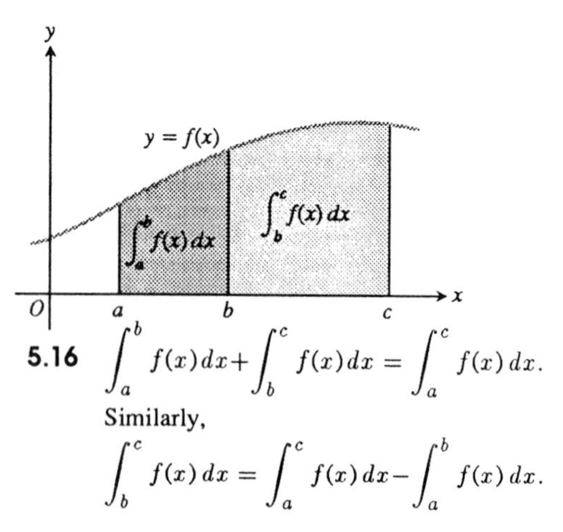

5.16 $\int_a^b f(x)\,dx + \int_b^c f(x)\,dx = \int_a^c f(x)\,dx.$ Similarly, $\int_b^c f(x)\,dx = \int_a^c f(x)\,dx - \int_a^b f(x)\,dx.$

The Average Value of a Function

If we divide the inequality in the max-min rule for definite integrals by $(b-a)$, we get the inequality

$$\min f \leq \frac{1}{b-a}\int_a^b f(x)\,dx \leq \max f. \tag{2}$$

If f is continuous, the Intermediate Value Theorem in Section 2.2 says that f must assume every value between $\min f$ and $\max f$. In particular, f must assume the value

$$\frac{1}{b-a}\int_a^b f(x)\,dx.$$

Theorem 2

The Mean Value Theorem for Definite Integrals

If f is continuous on the closed interval $[a, b]$, then, at some point c in the interval $[a, b]$,

$$f(c) = \frac{1}{b-a}\int_a^b f(x)\,dx. \tag{3}$$

The number on the right-hand side of Eq. (3) is called the mean value or average value of f on the interval $[a, b]$. We shall see some of the applications of average values in Section 5.5.

Definition

The **average value** of f on $[a, b]$ is $\dfrac{1}{b-a}\displaystyle\int_a^b f(x)\,dx.$ (4)

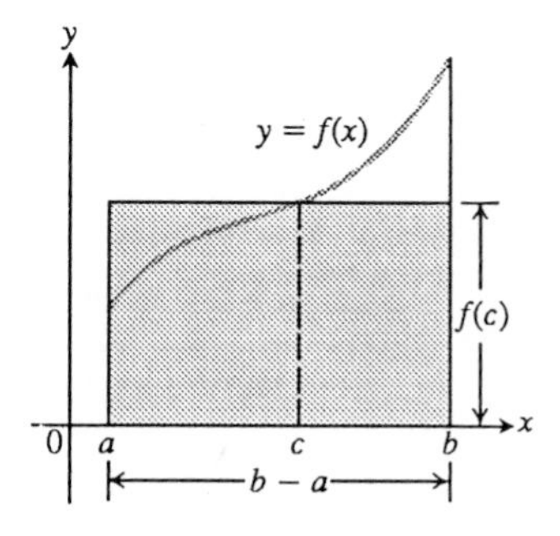

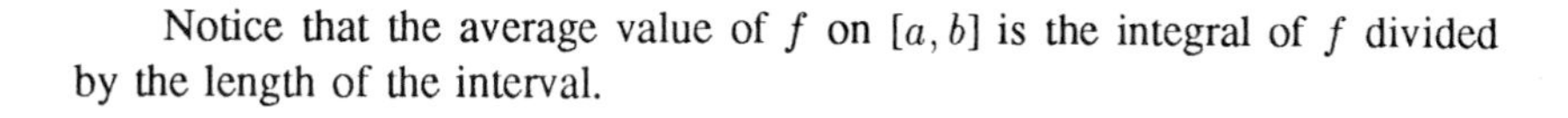

Notice that the average value of f on $[a, b]$ is the integral of f divided by the length of the interval.

EXAMPLE 2 If f is continuous and nonnegative on $[a, b]$, its average value is the height of a rectangle whose area,

$$f(c)(b-a) = \int_a^b f(x)\,dx,$$

is the area under the graph of f from a to b. ▬

EXAMPLE 3 Find the average value of $f(x) = \sqrt{4-x^2}\,dx$ on the interval $[-2, 2]$.

Solution

$$\text{Av. val. of } f \text{ on } [-2,2] = \frac{1}{2-(-2)}\int_{-2}^{2}\sqrt{4-x^2}\,dx \tag{4}$$

$$= \frac{1}{4}(2\pi) = \frac{\pi}{2} \qquad \left(\begin{array}{l}\text{Value from}\\ \text{Example 5, Section 5.3}\end{array}\right)$$

The average value of $f(x) = \sqrt{4-x^2}$ on $[-2,2]$ is $\pi/2$.

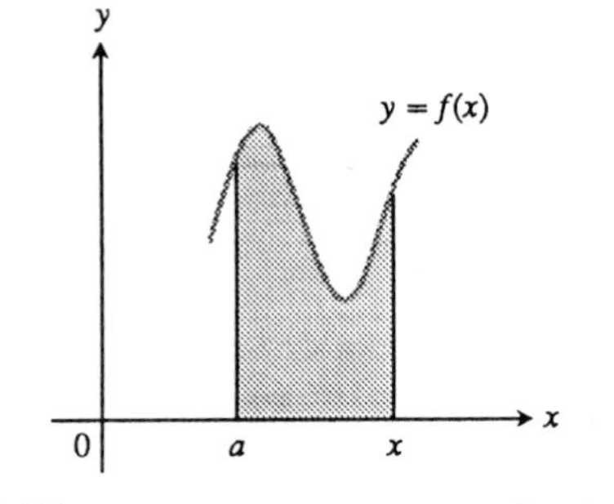

5.17 The area of the shaded region is $A_a^x f$.

Antiderivatives and their Connection to Integrals

We return to the discussion of area begun in Section 5.1. Let f be a nonnegative continuous function defined on an interval $[a,b]$. Recall $A_a^x f(x)$ denotes the area of the region that lies between the curve, the x-axis, and the vertical lines through the points a and x on the x-axis (Fig. 5.17). Here a is a constant and x a variable. The area A_a^x, whose value changes with each new value of x, is a function of x.

If we move from x to a nearby point $x+h$, the amount of area we add is A_x^{x+h}, the area under the curve from x to $x+h$. These two areas combine to give the area from a to $x+h$, so

$$A_a^x + A_x^{x+h} = A_a^{x+h} \tag{5}$$

and

$$A_x^{x+h} = A_a^{x+h} - A_a^x. \tag{6}$$

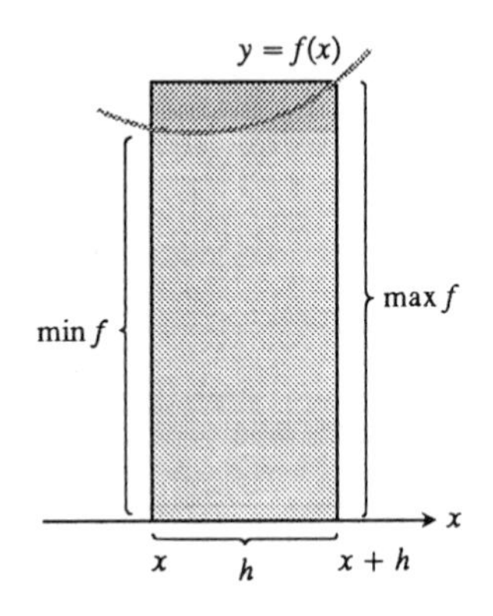

Now, the area under the curve from x to $x+h$ can be trapped between the areas of two rectangles of base length h, as shown here: The height of the shorter rectangle is min f, the minimum value of f on the interval from x to $x+h$. The area of this rectangle is therefore $h \cdot \min f$ (base times height). The height of the taller rectangle is max f, the maximum value of f on the interval from x to $x+h$. The area of this rectangle is therefore $h \cdot \max f$.

We record the observation that the area under the curve from x to $x+h$ lies between the areas of these rectangles by writing the following inequality:

$$h \cdot \min f \le A_a^{x+h} - A_a^x \le h \cdot \max f. \tag{7}$$

Dividing by h gives

$$\min f \le \frac{A_a^{x+h} - A_a^x}{h} \le \max f. \tag{8}$$

The fraction in the middle of the inequality in Eq. (8) is Fermat's difference quotient for the derivative of the area function A_a^x. It is the value of the function at $x+h$ minus the value of the function at x, all divided by h. We can therefore calculate the derivative of the area function at x by finding the limit of this quotient as h goes to zero.

As h goes to zero, the interval from x to $x+h$ gets shorter and shorter. As it does so, the values of max f and min f both approach the value of f at x (remember f is continuous). Hence, by the Sandwich Theorem of Section 2.3, the difference quotient approaches $f(x)$ as well. In symbols,

$$\lim_{h\to 0} \frac{A_a^{x+h} - A_a^x}{h} = f(x). \tag{9}$$

We are thus led to the astonishing conclusion that, when f is a nonnegative continuous function of x, the area under its graph from a to x is a differentiable function of x whose derivative at x is $f(x)$:

$$\frac{d}{dx} A_a^x = f(x). \tag{10}$$

Among other things, this means that we can find an explicit formula for A_a^x whenever we can solve the following initial value problem:

Differential equation: $\frac{d}{dx} A_a^x = f(x)$

Initial condition: $A_a^x = 0$ when $x = a$.

The initial condition comes from the observation that the area under the graph from a point a to the same point a is zero. There is no area under a graph of zero length.

We solve the initial value problem in the usual way. We find an antiderivative $F(x)$ of $f(x)$ to get

$$A_a^x = F(x) + C. \tag{11}$$

We then find the right value of C from the initial condition, by setting x equal to a:

$$\begin{aligned} A_a^a &= F(a) + C && (x = a \text{ in Eq. (11)}) \\ 0 &= F(a) + C && (A_a^x = 0 \text{ when } x = a) \\ C &= -F(a). && (\text{Solved for } C) \end{aligned}$$

The area under the curve $y = f(x)$ from a to x is therefore

$$A_a^x = F(x) - F(a). \tag{12}$$

Here, then, is the relation between area and calculus. We calculate area with antiderivatives.

How to Find the Area under the Graph of a Nonnegative Continuous Function $y = f(x)$ from $x = a$ to $x = b$

STEP 1: Find an antiderivative $F(x)$ of $f(x)$. (Any antiderivative will do.)
STEP 2: Calculate the number $F(b) - F(a)$. This number will be the area under the curve from a to b.

There are two practical questions to face here: How do we know f *has* an antiderivative, and how do we find one when f does? We shall take care of the existence when we get to Section 5.5. As for finding antiderivatives, there is no need to worry. We shall get better at that as we learn more calculus. Here, now, are some examples.

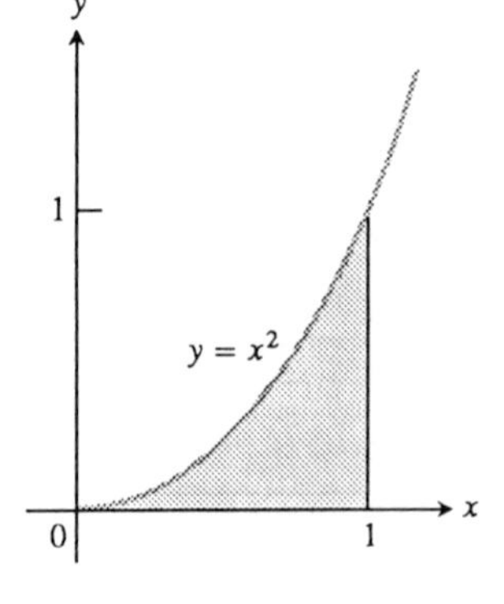

EXAMPLE 4 Find the area under the curve $y = x^2$ from $x = 0$ to $x = 1$.

Solution In this example $f(x) = x^2, a = 0$, and $b = 1$. We find the area in two steps.
STEP 1: Find an antiderivative $F(x)$ of $f(x) = x^2$. Any antiderivative will do, so we can choose the simplest one:

$$F(x) = \frac{x^3}{3}.$$

Does this look familiar?

STEP 2: Calculate $F(1) - F(0)$:

$$F(1) - F(0) = \frac{(1)^3}{3} - \frac{(0)^3}{3} = \frac{1}{3} - 0 = \frac{1}{3}.$$

The area is 1/3.

If we use a different antiderivative in Example 4, say

$$F(x) = \frac{x^3}{3} + 4,$$

we still get 1/3 for the answer:

$$F(1) - F(0) = \left(\frac{(1)^3}{3} + 4\right) - \left(\frac{(0)^3}{3} + 4\right) = \frac{1}{3} + 4 - 0 - 4 = \frac{1}{3}.$$

Notice how the 4 cancels out. Any constant we add to the $x^3/3$ will cancel out the same way when we subtract.

EXAMPLE 5 Find the area under the curve $y = x^3$ from $x = 0$ to $x = 5$.

Solution

STEP 1: Find an antiderivative $F(x)$ of $f(x) = x^3$: Since $D_x\left(x^4/4\right) = x^3$, any antiderivative is of the form $F[x] = x^4/4 + C$ for some real number C. Take $C = 0$ and $F[x] = x^4/4$.

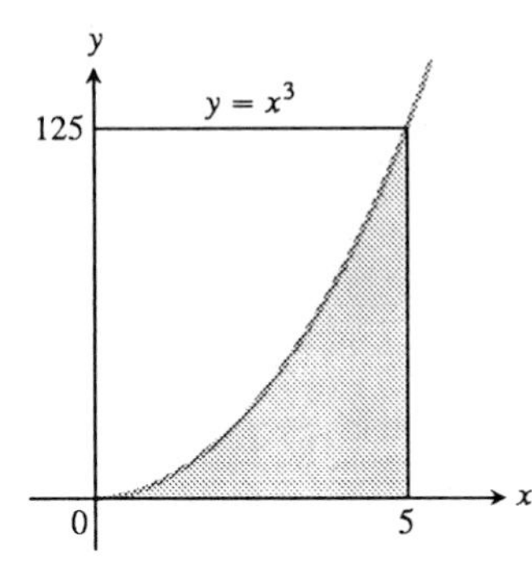

STEP 2: Compute $F(5) - F(0)$:

$$F(5) - F(0) = \frac{(5)^4}{4} - \frac{(0)^4}{4}$$
$$= 156.25.$$

Explore Box

Compute $\mathrm{MR}_{100}(x^3)$ for the interval from $x = 0$ to $x = 5$ and compare with the exact area computation in Example 5.

EXAMPLE 6 Find the area under one arch of the curve $y = \cos x$.

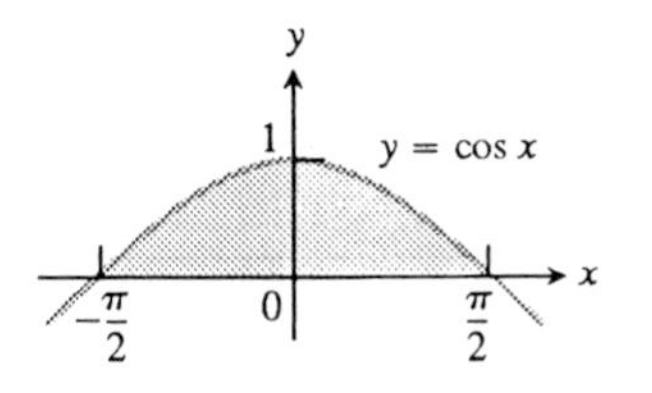

Solution

STEP 1: Find an antiderivative of $y = \cos x$:

$$F(x) = \sin x. \qquad \text{(Simplest one)}$$

STEP 2: Calculate $F\left(\frac{\pi}{2}\right) - F\left(-\frac{\pi}{2}\right)$:

$$F\left(\frac{\pi}{2}\right) - F\left(-\frac{\pi}{2}\right) = \sin\left(\frac{\pi}{2}\right) - \sin\left(-\frac{\pi}{2}\right) = 1 - (-1) = 2.$$

The area is 2 square units.

The reason we assumed f to be continuous in this development of antiderivatives, when integrability is all that is needed to define the integral $\int_a^x f(t)\,dt = A_a^x f(x)$, is that the continuity of f was used to show that $A_a^x f(x)$ was a differentiable function of x whose derivative at x was $f(x)$. The remarkable connection that $D_x(\int_a^x f(t)dt) = f(x)$ will be one of the highlights of the next section.

Exercises 5.4

Use antiderivatives to find the areas of the shaded regions in Exercises 1–16. Check using an FnInt computation.

1.

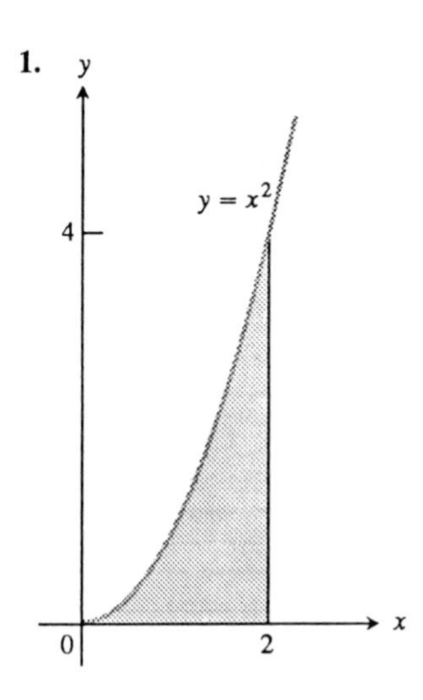

2.

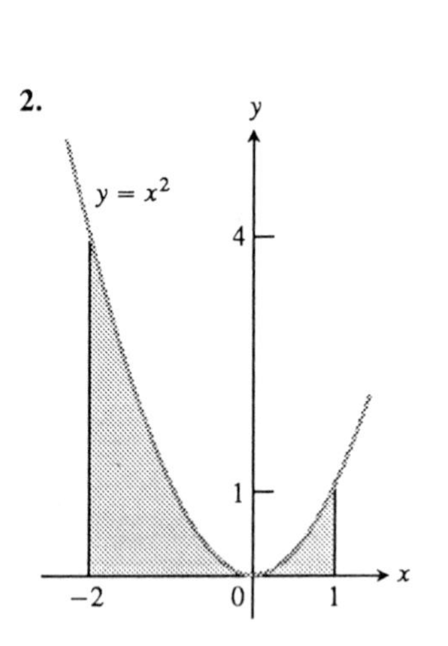

3.

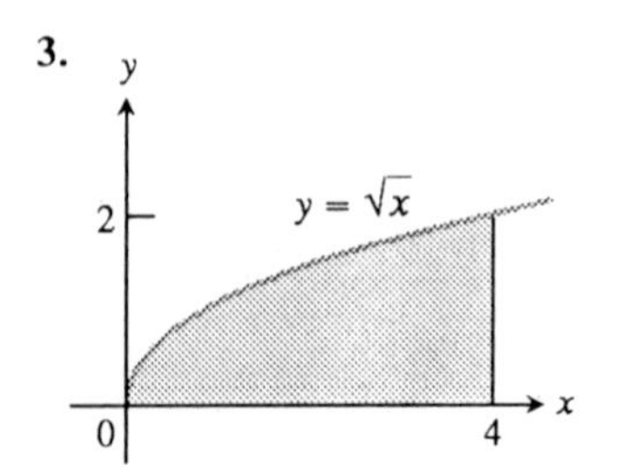

4.

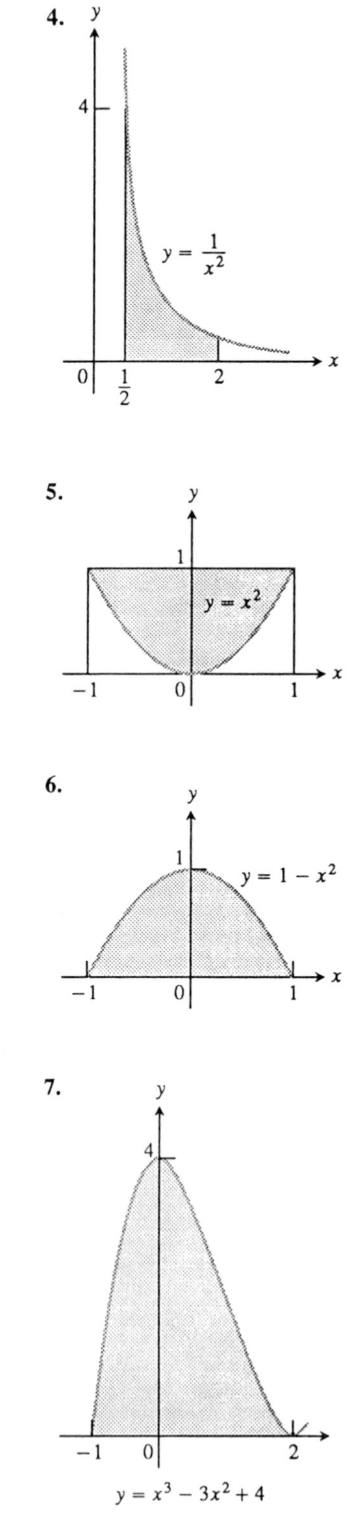

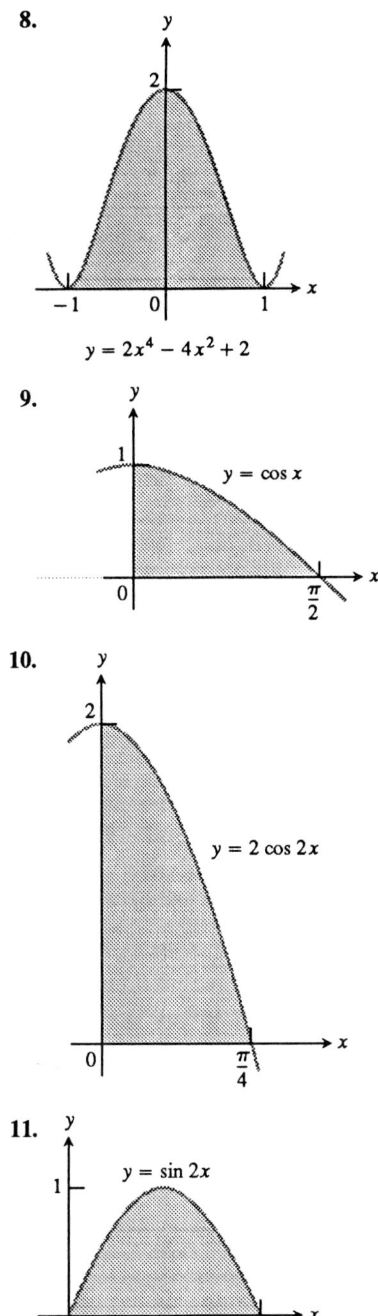
8.
$y = 2x^4 - 4x^2 + 2$
9.
$y = \cos x$
10.
$y = 2\cos 2x$
11.
$y = \sin 2x$

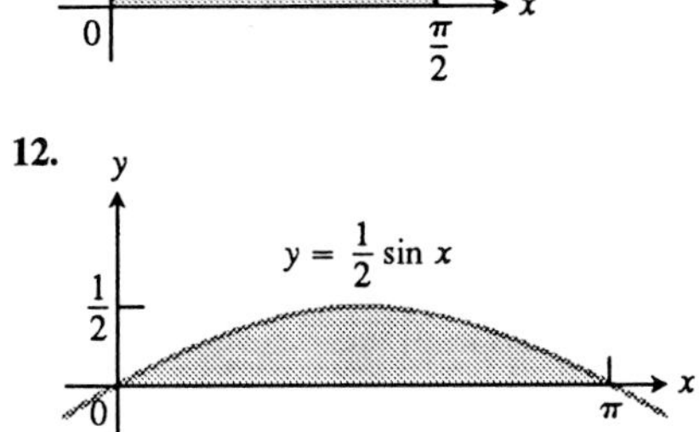
12.
$y = \frac{1}{2}\sin x$

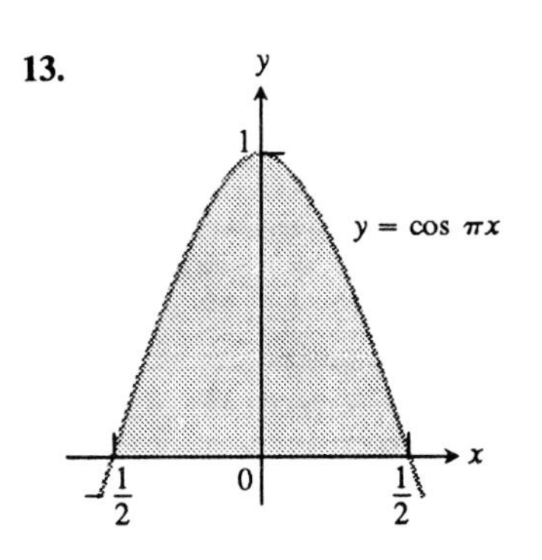
13.
$y = \cos \pi x$

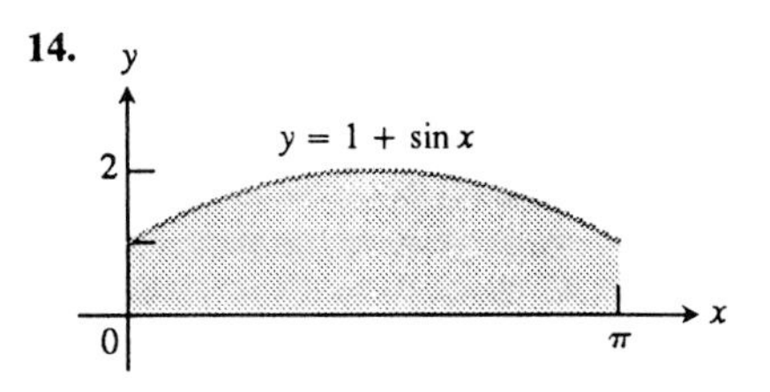
14.
$y = 1 + \sin x$

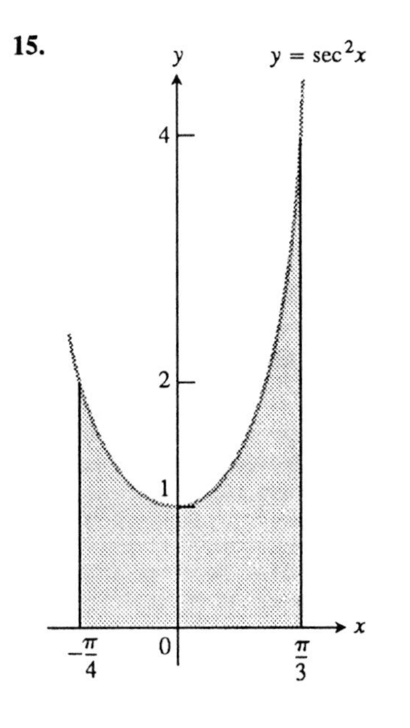
15.
$y = \sec^2 x$

16.

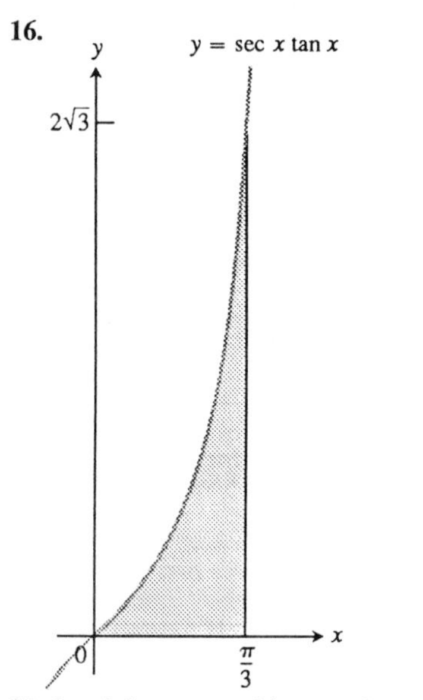

17. Let b be any positive number and n any rational number other than -1. Find a formula for the area under the curve $y = x^n$ from $x = 0$ to $x = b$.

18. Whenever we find a new way to calculate something, it is a good idea to be sure that the new and old ways agree on the objects to which they both apply. If you use an antiderivative to find the area of the triangle in Fig. 5.18, will you still get $A = (1/2)bh$? Try it and find out.

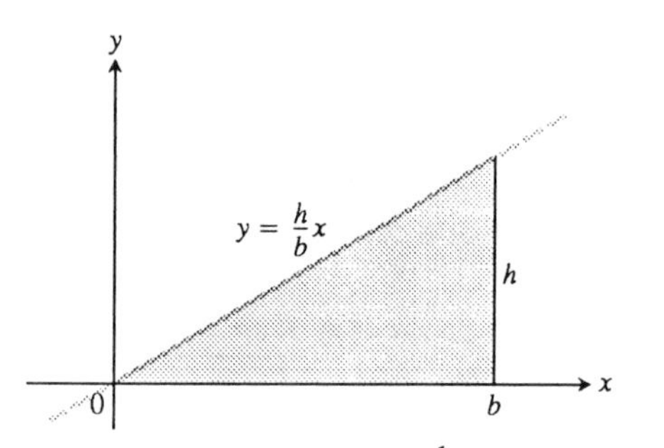

5.18 Is the area of this triangle still $\frac{1}{2}bh$? See Exercise 18.

19. Suppose f and g are continuous and that

$$\int_1^2 f(x)\,dx = -4, \quad \int_1^5 f(x)\,dx = 6, \quad \int_1^5 g(x)\,dx = 8.$$

Use the rules in Table 5.3 to find

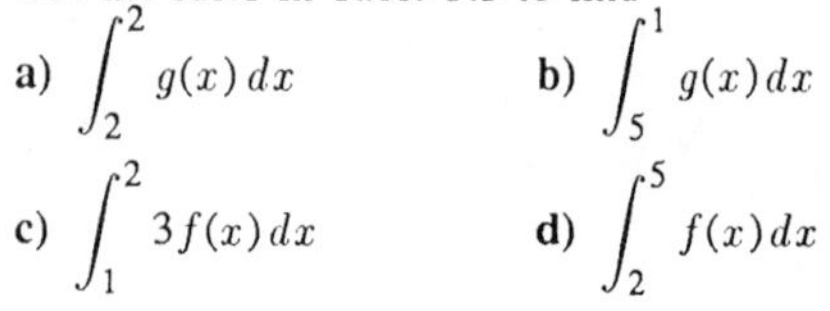

a) $\int_2^2 g(x)\,dx$ b) $\int_5^1 g(x)\,dx$

c) $\int_1^2 3f(x)\,dx$ d) $\int_2^5 f(x)\,dx$

e) $\int_1^5 [f(x) - g(x)]\,dx$ f) $\int_1^5 [4f(x) - g(x)]\,dx.$

20. Suppose f and h are continuous and that

$$\int_1^9 f(x)\,dx = -1, \quad \int_7^9 f(x)\,dx = 5, \quad \int_7^9 h(x)\,dx = 4.$$

Use the rules in Table 5.3 to find

a) $\int_1^9 -2f(x)\,dx$ b) $\int_7^9 [f(x) + h(x)]\,dx$

c) $\int_7^9 [2f(x) - 3h(x)]\,dx$ d) $\int_9^1 f(x)\,dx$

e) $\int_1^7 f(x)\,dx$ f) $\int_9^7 [h(x) - f(x)]\,dx.$

In Exercises 21–26, find the average value of f on $[a, b]$.

21. $f(x) = 2 - x^2$ on $[-3, 5]$ 22. $f(x) = x^3 - 1$ on $[-1, 3]$

23. $f(x) = \sqrt{x}$ on $[0, 3]$ 24. $f(x) = \cos^2 x$ on $[-\pi, \pi]$

25. $f(x) = x \sin x$ on $[0, 5]$ 26. $f(x) = 2e^{-x^2}$ on $[1, 4]$

27. Suppose $\int_1^2 f(x)\,dx = 5$. Find

a) $\int_1^2 f(u)\,du$, b) $\int_1^2 f(z)\,dz$, c) $\int_2^1 f(t)\,dt.$

28. Suppose f is continuous and that

$$\int_0^3 f(x)\,dx = 3, \quad \int_0^4 f(z)\,dz = 7.$$

Find

$$\int_3^4 f(y)\,dy.$$

29. Use Rule 9 in Table 5.3 to find upper and lower bounds for the value of

$$\int_0^1 \frac{1}{1+x^2}\,dx.$$

30. Continuation of Exercise 29. Use Rule 9 in Table 5.3 to find upper and lower bounds for the values of

$$\int_0^{1/2} \frac{1}{1+x^2}\,dx \quad \text{and} \quad \int_{1/2}^1 \frac{1}{1+x^2}\,dx.$$

Then add these to arrive at an improved estimate of

$$\int_0^1 \frac{1}{1+x^2}\,dx.$$

31. Suppose f is continuous and that

$$\int_1^2 f(x)\,dx = 4.$$

Show that $f(x) = 4$ at least once on the interval $[1, 2]$.

32. Show that the value of

$$\int_0^1 \sin^2 x\,dx$$

cannot possibly be 2.

5.5 The Fundamental Theorems of Integral Calculus

We know that definite integrals can be used to define area. We will see that integrals are also used to define arc length, volume, force and work. This list of applications of the definite integral is almost endless. This section presents the discovery by Newton and Leibniz of the astonishing connections between integration and differentiation which started the mathematical development that fueled the scientific revolution for the next 200 years and constitutes what is still regarded as the most important computational discovery in the history of the Western world. We introduce the two fundamental theorems of integral calculus. The first theorem says that the definite integral of a continuous function is a differentiable function of its upper limit of integration, and tells us what the value of that derivative is. The second theorem tells us that the definite integral of a continuous function from a to b can be found from any one of the function's antiderivatives F as the number $F(b) - F(a)$.

The First Fundamental Theorem

If $f(t)$ is an integrable function, its integral from any fixed number a to another number x defines a function F whose value at x is

$$F(x) = \int_a^x f(t)\,dt. \tag{1}$$

For example, if f is nonnegative and x lies to the right of a, the area under the graph of f from a to x is

$$A_a^x f = \int_a^x f(t)\,dt. \tag{2}$$

The variable x in the function A_a^x is the upper limit of integration of an integral, but the function is just like any other function. For each value of the input x there is an output A_a^x, in this case the value of the integral of f from a to x.

Notation and Agreements

In this textbook we use the notation

$$\text{FnInt }(f(t), t, a, x)$$

to denote a *computer* generated approximation to $F(x) = \int_a^x f(t)\,dt$ for each value of x. We use the *graph* of

$$y = \text{FnInt }(f(t), t, a, x)$$

to support the explicit analytic determination of $F(x) = \int_a^x f(t)\,dt$ much in the same way as we used $y = \text{NDer }(f(x), x)$ to support the explicit analytic determination of $D_x f(x)$. We will find visualizing $y = F(x) = \int_a^x f(t)\,dt$ by graphing $y = \text{FnInt }(f(t), t, a, x)$ a very powerful problem solving technique.

Explore Box

Graphing $y = \text{FnInt}(f(t), t, a, x)$ is sometimes affected by the tolerance setting. Try graphing $y = \text{FnInt}(\cos(t), t, 0, x)$ in $[-10, 10]$ by $[-2, 2]$ for tol $= 1, 0.001$, and 0.00001. Usually, tol $= 1$ is accurate enough for FnInt graphing.
Remark: Do you recognize the graph of $y = \text{FnInt }(\cos(t), t, 0, x)$? What function is it?

The formula

$$F(x) = \int_a^x f(t)\,dt$$

gives an important way to define new functions in science and provides an especially useful way to describe solutions of differential equations (more about this later). The reason for our mentioning the formula now is that this formula makes the connection between integrals and derivatives. For, if f is any continuous function whatever, F is a differentiable function of x and, even more important, its derivative, dF/dx, is f itself. At every value of x,

$$\frac{d}{dx}\int_a^x f(t)\,dt = f(x). \qquad (3)$$

If you were being sent to a desert island and could take only one equation with you, Eq. (3) might well be your choice. It says that the differential equation $dF/dx = f$ has a solution for any continuous function f. It says that every continuous function f is the derivative of some other function, namely, $\int_a^x f(t)\,dt$. It says that every continuous function has an antiderivative. Equation (3) is so important that we call it the First Fundamental Theorem of Calculus.

Theorem 3

The First Fundamental Theorem of Calculus

If f is continuous on $[a, b]$, then the function

$$F(x) = \int_a^x f(t)\, dt \tag{4}$$

has a derivative at every point in $[a, b]$ and

$$\frac{dF}{dx} = \frac{d}{dx}\int_a^x f(t)\, dt = f(x). \tag{5}$$

Corollary

The Existence of Antiderivatives of Continuous Functions

If f is continuous on $[a, b]$, then there exists a function F whose derivative on $[a, b]$ equals f.

Proof of the Corollary Take $F(x) = \int_a^x f(t)\, dt$. The integral exists by Theorem 1, Section 5.3, and $dF/dx = f$ by Theorem 3 above. ∎

Proof of the First Fundamental Theorem We prove Theorem 3 by applying the definition of derivative directly to the function $F(x)$. This means writing out Fermat's difference quotient,

$$\frac{F(x+h) - F(x)}{h}, \tag{6}$$

and showing that its limit as $h \to 0$ is the number $f(x)$.

When we replace $F(x + h)$ and $F(x)$ by their defining integrals, the numerator in Eq. (6) becomes

$$F(x+h) - F(x) = \int_a^{x+h} f(t)\, dt - \int_a^x f(t)\, dt. \tag{7}$$

The interval subtraction rule for integrals (Table 5.3, preceding section) simplifies this to

$$\int_x^{x+h} f(t)\, dt, \tag{8}$$

so that Eq. (6) becomes

$$\frac{F(x+h) - F(x)}{h} = \frac{1}{h}[F(x+h) - F(x)] = \frac{1}{h}\int_x^{x+h} f(t)\, dt. \tag{9}$$

According to the Mean Value Theorem for definite integrals, Theorem 2 in the preceding section, the value of the entire expression on the right-hand side of Eq. (9) is one of the values taken on by f in the interval joining x and $x+h$. That is, for some number c in this interval,

$$\frac{1}{h}\int_{x}^{x+h} f(t)\,dt = f(c). \tag{10}$$

We can therefore find out what happens to $(1/h)$ times the integral as $h \to 0$ by watching what happens to $f(c)$ as $h \to 0$.

What does happen to $f(c)$ as $h \to 0$? As $h \to 0$, the endpoint $x+h$ approaches x, taking c along with it like a bead on a wire:

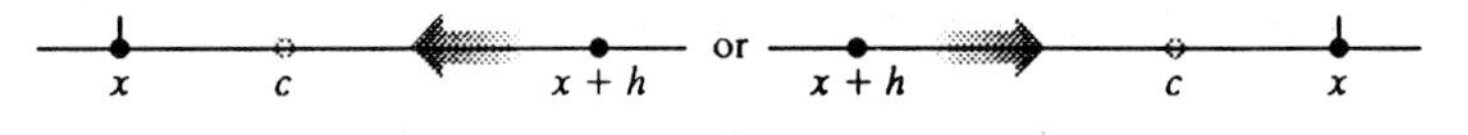

So c approaches x, and since f is continuous at x, $f(c)$ approaches $f(x)$:

$$\lim_{h\to 0} f(c) = f(x). \tag{11}$$

Going back to the beginning, then, we have

$$\begin{aligned}
\frac{dF}{dx} &= \lim_{h\to 0}\frac{F(x+h)-F(x)}{h} && \text{(The definition of derivative)}\\
&= \lim_{h\to 0}\frac{1}{h}\int_{x}^{x+h} f(t)\,dt && \text{(Eq. 9)}\\
&= \lim_{h\to 0} f(c) && \text{(Eq. 10)}\\
&= f(x). && \text{(Eq. 11)}
\end{aligned}$$

This concludes the proof. ≡

The above corollary to the First Fundamental Theorem is an *existence* statement. This means that we can be sure that *any* continuous function has an antiderivative, and we even have a definition for an antiderivative. However, there is a major difficulty with the corollary in that $F(x) = \int_a^x f(t)\,dt$ can be determined *explicitly as a function of x* only in relatively few cases. And these few special cases are what make up much of the traditional integral calculus you are studying. However, with computer technology assumed, we can graph the antiderivative of an integral and study its properties even though we can not find an explicit analytic formula for it. The graph is almost as good as the explicit formula! The next two examples demonstrate the power of this computer graphing approach.

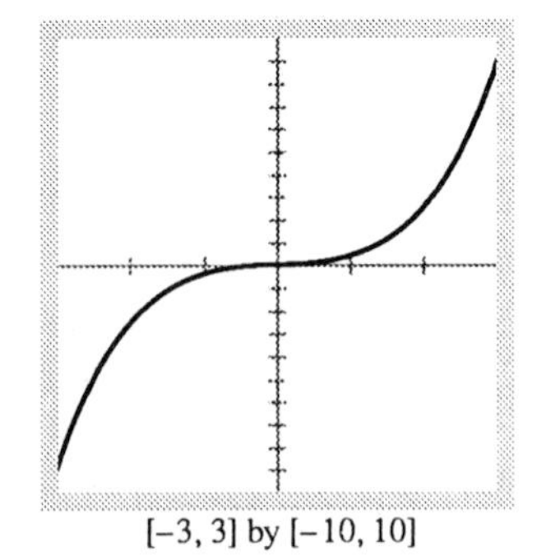

[−3, 3] by [−10, 10]

5.19 The graphs of $y = \text{FnInt}\,(t^2, t, 0, x)$ and $y = x^3/3$ are identical.

EXAMPLE 1 Compare the graph of $F(x) = \int_0^x t^2\,dt$ with the graph of $y = x^3/3$.

Solution Figure 5.19 suggests the graphs of $y = \text{FnInt}\,(t^2, t, 0, x)$ and $y = x^3/3$ are identical. This gives visual support for the corollary to Theorem 3. Because $D_x x^3/3 = x^2$, it follows that $y = x^3/3$ is one such function F whose existence is guaranteed by the corollary. Since the graphs appear identical, it is reasonable to conclude that $F(x) = \int_0^x t^2\,dt = x^3/3$. ≡

It would really be interesting to see what the graph of the *derivative* of $\int_0^x t^2 dt$ is.

EXAMPLE 2 Compare the graph of $y = D_x\left(\int_0^x t^2\,dt\right)$ with the graph of $y = x^2$.

Solution Figure 5.20 suggests the graphs of $y = \text{NDer}\,(\text{FnInt}\,(t^2, t, 0, x), x)$ and $y = x^2$ are identical. This is truly a remarkable result! ■

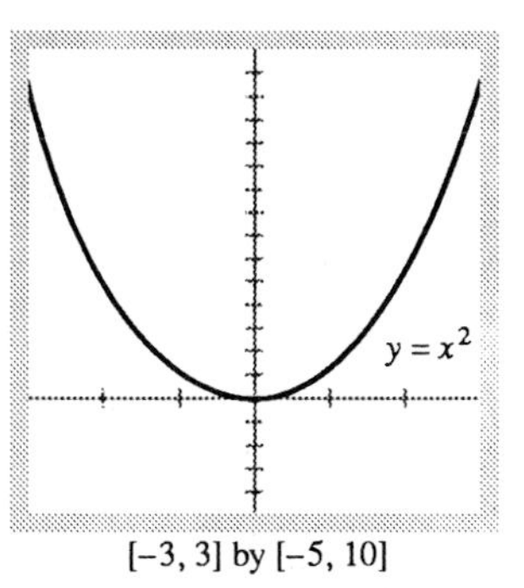

[−3, 3] by [−5, 10]

5.20 The graphs of $y = x^2$ and $y = \text{NDer}\,(\text{FnInt}\,(t^2, t, 0, x), x)$.

Think of what the computer software is doing to create the graph of $y = \text{NDer}(\text{FnInt}\,(t^2, t, 0, x), x)$. For each value of x (127 of them on the graphing calculator we use), *two* approximations of the definite integral are determined since, by definition,

$$\text{NDer}\,(\text{FnInt}\,(t^2, t, 0, x), x) = \frac{\text{FnInt}\,(t^2, t, 0, x, x+h) - \text{FnInt}\,(t^2, t, 0, x-h)}{2h}.$$

Example 2 visually supports the First Fundamental Theorem of Calculus. The next three examples are typical. Example 4 is most important.

EXAMPLE 3

$$\frac{d}{dx}\int_{-\pi}^{x} \cos t\,dt = \cos x \qquad \text{(Eq. (5) with } f(t) = \cos t)$$

$$\frac{d}{dx}\int_0^x \frac{1}{1+t^2}\,dt = \frac{1}{1+x^2} \qquad \left(\text{Eq. (5) with } f(t) = \frac{1}{1+t^2}\right)$$

■

EXAMPLE 4 Find dy/dx if

$$y = \int_1^{x^2} \cos t\,dt.$$

Solution Notice that the upper limit of integration is not x but x^2. To find dy/dx we must therefore treat y as the composite of

$$y = \int_1^u \cos t\,dt \quad \text{and} \quad u = x^2$$

and apply the Chain Rule:

$$\begin{aligned}
\frac{dy}{dx} &= \frac{dy}{du}\frac{du}{dx}. && \text{(Chain Rule)}\\
&= \frac{d}{du}\int_1^u \cos t\,dt \cdot \frac{du}{dx} && \text{(Substitute the formula for } y.)\\
&= \cos u \cdot \frac{du}{dx} && \text{(Eq. (5) with } f(t) = \cos t)\\
&= \cos x^2 \cdot 2x && (u = x^2)\\
&= 2x \cos x^2. && \text{(Usual form)}
\end{aligned}$$

■

EXAMPLE 5 Show that the function

$$y = \int_0^x \tan t \, dt + 5$$

solves the initial value problem

Differential equation: $\quad \dfrac{dy}{dx} = \tan x$

Initial condition: $\quad y = 5$ when $x = 0$.

Solution The function satisfies the differential equation because

$$\frac{d}{dx}\left(\int_0^x \tan t \, dt + 5\right) = \tan x + 0 = \tan x. \qquad \text{(Eq. (5) with } f(t) = \tan t\text{)}$$

It fulfills the initial condition because

$$y(0) = \int_0^0 \tan t \, dt + 5 = 0 + 5. \qquad \text{(Table 5.3, Rule 1)}$$

To evaluate $\int_a^b f(x)\,dx$ exactly:

1. Find an antiderivative F of f.
2. Calculate the number $F(b) - F(a)$.

This number will be the exact value of $\int_a^b f(x)\,dx$. Any antiderivative will do.

The Integral Evaluation Theorem

We now come to the theorem that tells how to find the exact value of definite integrals with antiderivatives, a theorem so useful it is called the Second Fundamental Theorem of Calculus.

Theorem 4

The Integral Evaluation Theorem (Second Fundamental Theorem of Calculus)

If f is continuous at every point of $[a, b]$ and F is any antiderivative of f on $[a, b]$, then

$$\int_a^b f(x)\,dx = F(b) - F(a). \qquad (12)$$

NOTATION

The usual notation for the number $F(b) - F(a)$ is $F(x)]_a^b$ or $[F(x)]_a^b$, depending on whether F has one or more terms. As you will see, this notation provides a compact "recipe" for the evaluation.

Write $F(x)]_a^b$ for $F(b) - F(a)$ when $F(x)$ has a single term.

Write $[F(x)]_a^b$ for $F(b) - F(a)$ when $F(x)$ has more than one term.

Theorem 4 says that to evaluate the definite integral of a continuous function f from a to b, all we need do is find an antiderivative F of f and calculate the number $F(b) - F(a)$. The existence of the antiderivative is assured by the First Fundamental Theorem.

Proof of the Integral Evaluation Theorem To prove the theorem, we use Corollary 3 of the Mean Value Theorem for derivatives (Section 4.7), which says that functions with identical derivatives can differ only by a constant. We already know one function whose derivative equals f, namely,

$$G(x) = \int_a^x f(t)\,dt.$$

Therefore, if F is any other such function, then

$$F(x) = G(x) + C \qquad (13)$$

throughout $[a, b]$ for some constant C. When we use Eq. (13) to calculate $F(b) - F(a)$, we find that

$$F(b) - F(a) = [G(b) + C] - [G(a) + C] = G(b) - G(a)$$
$$= \int_a^b f(t)\,dt - \int_a^a f(t)\,dt = \int_a^b f(t)\,dt - 0 = \int_a^b f(t)\,dt.$$

This establishes Eq. (12) and concludes the proof.

EXAMPLE 6

$$\int_{-1}^{3} (x^3 + 1)\,dx = \left[\frac{x^4}{4} + x\right]_{-1}^{3} \quad \left(D_x\left(\frac{x^4}{4} + x\right) = x^3 + 1\right)$$
$$= \left(\frac{81}{4} + 3\right) - \left(\frac{1}{4} - 1\right)$$
$$= \frac{80}{4} + 4 = 24$$

Approximating Definite Integrals

We assume, unless stated otherwise, that when evaluating definite integrals you will first try to find an explicit antiderivative and find the exact value, as in Example 6, and then support your exact solution with a calculator approximation using the definite integral approximation FnInt $(f(x), x, a, b)$. However, in many cases you will not be able to find an explicit antiderivative for reasons that will become clear later. In these cases we apply only the FnInt $(f(x), x, a, b)$ numerical approximation. For example, consider evaluating the definite integral:

$$\int_1^{10\pi} x \cos x\,dx.$$

Can you think of a function whose derivative is $x \cos x$? Probably not (eventually you will be able to find such a function). However, applying FnInt $(x \cos x, x, 1, 10\pi)$ yields -0.381773290682. The exact solution is

$$[x \sin x + \cos x]_1^{10\pi} = (10\pi \sin 10\pi + \cos 10\pi) - (\sin 1 + \cos 1)$$
$$= 1 - \sin 1 - \cos 1.$$

If you approximate this exact solution you obtain (to 12 decimal places) -0.381773290676, so our FnInt calculator approximation was *very* accurate indeed.

Explore Box

Verify that $x \sin x + \cos x$ is an antiderivative of $x \cos x$.

It is important to note that the *exact* solution is of no real value (just what do you do with $1 - \sin 1 - \cos 1$?). Approximations are very *useful.* Use of a calculator definite integral key is valuable *and* often appropriate.

> **Evaluating Integrals**
>
> To evaluate $\int_a^b f(x)\,dx$, first try to find an exact solution by determining an antiderivative $F(x)$ and computing $F(b) - F(a)$. If you can't find an antiderivative, then use a calculator or computer approximation FnInt $(f(x), x, a, b)$ for the value of $\int_a^b f(x)\,dx$.

Warning. Most calculator computations of FnInt $(f(x), x, a, b)$ are usually highly accurate approximations of $\int_a^b f(x)dx$ as the evaluation of $\int_1^{10\pi} x\cos x\,dx$ suggested. However, remember accuracy varies from machine to machine because different machines use different internal approximation algorithms. As mentioned in Section 5.3, you should become familiar with the method used and error produced by your machine. Furthermore, some machines will give a real number value to definite integrals which *do not exist.* For example, a graph of $y = 1/x$ for x in $[-1, 2]$ will convince you that $\int_{-1}^{2} 1/x\,dx$ does not exist. Can you explain why? However, some machines will, incorrectly, give a real number value to FnInt $(1/x, x, -1, 2)$. This is always a possibility when machines use discrete methods to analyze continuous functions.

A Big Advantage

For almost two hundred years we have focused our efforts to evaluate definite integrals in textbooks on applying Theorem 5 (the second Fundamental Theorem of Calculus) to a small set of *contrived* examples whose antiderivatives can be expressed as explicit functions. The problem is that most functions DO NOT HAVE antiderivatives that *anyone* can write down explicitly as functions. Here are two such important elementary functions,

$$f(x) = e^{-x^2} \text{ and } g(x) = \frac{\sin x}{x}.$$

There are many, many more. Today, happily, things are much different because of technology. Using computer graphing we can see (almost touch and feel) the antiderivative that we know exists for *any* continuous function. Let's look at an example.

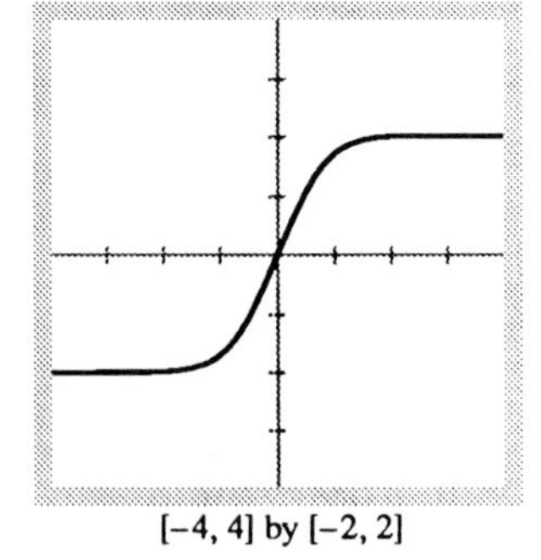

5.21 The graph of $y =$ FnInt $(e^{-t^2}, t, 0, x)$.

EXAMPLE 7 Find an explicit formula for an antiderivative of e^{-x^2}. This function is central to probability theory. Try as hard as you like, but you won't find one! None exists. However, we know $F(x) = \int_0^x e^{-x^2}$ is one antiderivative. And its graph is shown in Figure 5.21.

Now we really "know" it! That is, the graph is one of the antiderivatives of e^{-x^2}. We can examine it and understand its behavior. It is almost as good as having the explicit formula (which remember doesn't exist).

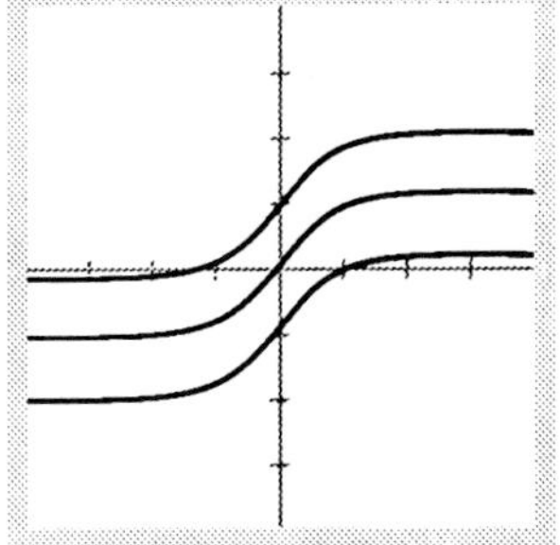

[−4, 4] by [−2, 2]

5.22 The graphs of $y = F_1(x) = $ FnInt $(e^{-t^2}, t, -1, x)$, $y = F_2(x) = $ FnInt $(e^{-t^2}, t, 0, x)$, and $y = F_3(x) = $ FnInt $(e^{-t^2}, t, 1, x)$.

> **Explore Box**
>
> Graph $f(x) = e^{-x^2}$ in $[-4, 4]$ by $[-2, 2]$. Explain the behavior of $F(x) = \int_0^x e^{-t^2} dt$ based solely on the graph of f. Why are there two horizontal asymptotes? Why is this graph increasing on any interval? Confirm your analysis by comparing with Figure 5.21. Why is the graph below the x-axis for $x < 0$?

There are many antiderivatives of $f(x) = e^{-x^2}$. Figure 5.22 displays the graphs of $F_1(x) = $ FnInt $(e^{-t^2}, t, -1, x)$, $F_2(x) = $ FnInt $(e^{-t^2}, t, 0, x)$ and $F_3(x) = $ FnInt $(e^{-t^2}, t, 1, x)$. Notice they are all related by a vertical shift. That is

$$F_1 = F_2 + C_1, F_1 = F_3 + C_2, F_2 = F_3 + C_3, \text{ etc.}$$

for constants $C_1, C_2, C_3, \ldots$.

We also know they *each* are antiderivatives of $f(x) = e^{-x^2}$ by the First Fundamental Theorem. Thus, $D_x F_1 = D_x F_2 = D_x F_3 = f(x) = e^{-x^2}$. We can provide dramatic visual support for the First Fundamental Theorem by graphing the numerical derivatives of F_1, F_2 and F_3. The result is shown in Figure 5.23, again providing a graphic demonstration of the First Fundamental Theorem of Calculus.

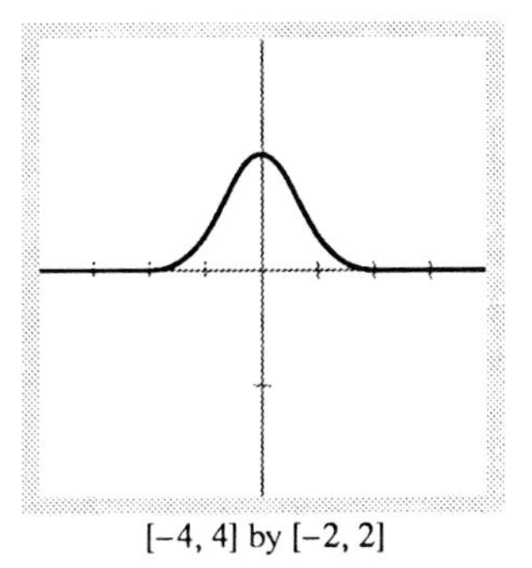

[−4, 4] by [−2, 2]

5.23 The graphs of $y = e^{-x^2}$, $y = $ NDer (F_1, x), $y = $ NDer (F_2, x), and $y = $ NDer (F_3, x), where F_1, F_2, and F_3 are shown in Fig. 5.22, are all the same!

Theorem 4 also justifies the way we calculated areas with antiderivatives in Section 5.4. The area under the graph of a nonnegative function $y = f(x)$ from $x = a$ to $x = b$ is $\int_a^b f(x)dx$ (from the definition of area in Section 5.3), and the value of this integral is $F(b) - F(a)$.

EXAMPLE 8 Find the area between the x-axis and the curves

a) $y = 4 - x^2$ **b)** $y = x^2 - 4$

for $-2 \le x \le 2$.

Solution We graph the curves over $[-2, 2]$ to see where the function values are positive and negative (Fig. 5.24).

a) Since $y = 4 - x^2 \ge 0$ on $[-2, 2]$, the area between the curve and the x-axis from -2 to 2 is

$$\text{Area} = \int_{-2}^{2} (4 - x^2)\, dx = \left[4x - \frac{x^3}{3}\right]_{-2}^{2} = \frac{32}{3}.$$

b) Since $y = x^2 - 4 \le 0$ on $[-2, 2]$, the area between the curve and the x-axis from -2 to 2 is the negative of the integral of $x^2 - 4$ from -2 to 2:

$$\text{Area} = -\int_{-2}^{2} (x^2 - 4)\, dx = \int_{-2}^{2} -(x^2 - 4)\, dx \qquad \text{(Rule 4, Table 5.3)}$$

$$= \int_{-2}^{2} (4 - x^2)\, dx = \frac{32}{3}. \qquad \text{(Value from above)}$$

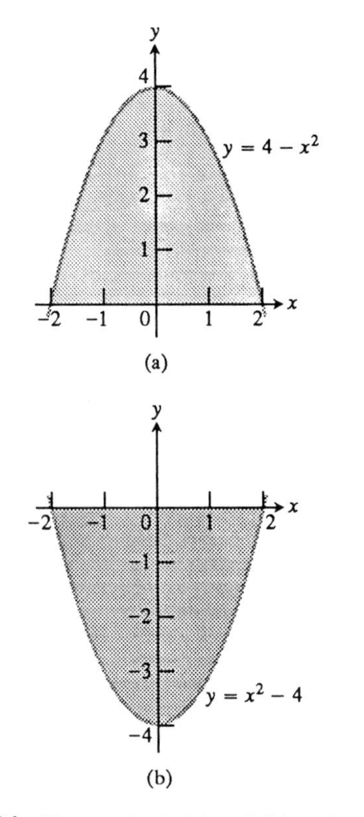

5.24 The graphs in (a) and (b) enclose the same amount of area with the x-axis, but the definite integrals of the functions from -2 to 2 differ in sign.

When the graph of $y = f(x)$ crosses the x-axis between $x = a$ and $x = b$, we find the area between the graph and the axis from a to b by taking the following steps.

Steps for Finding Area When f Has Both Positive and Negative Values on (a, b)

STEP 1: Find the points where $f = 0$.

STEP 2: Use the zeros of f to partition $[a, b]$ into subintervals.

STEP 3: Integrate f over each subinterval.

STEP 4: Add the absolute values of the results.

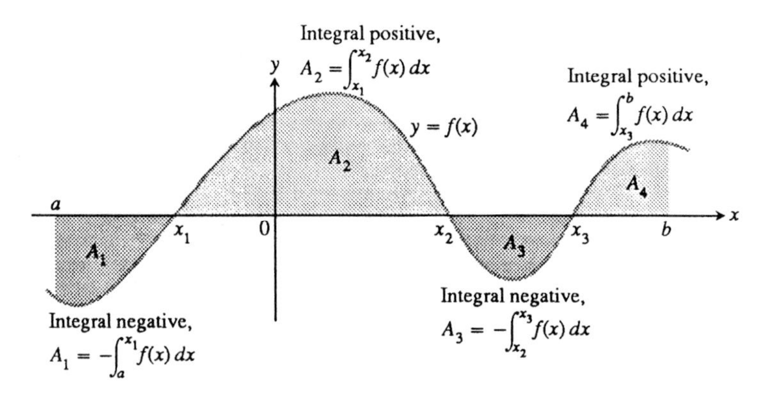

Here the area between the graph of f and the x-axis from a to b is found with four integrations. This is less work than it seems because we have to find the antiderivative of f only once.

We illustrate these steps in the next example.

EXAMPLE 9 Find the area of the region beween the x-axis and the curve

$$y = x^3 - 4x, \quad -2 \le x \le 2.$$

Explore Box

1. Compute FnInt $(x^3 - 4x, x, -2, 0)$ and FnInt $(x^3 - 4x, x, 0, 2)$. Do these results support the solutions in Step 3?
2. Explain why $\int_{-2}^{2} (x^3 - 4x)\,dx$ is not the desired area in Example 9.

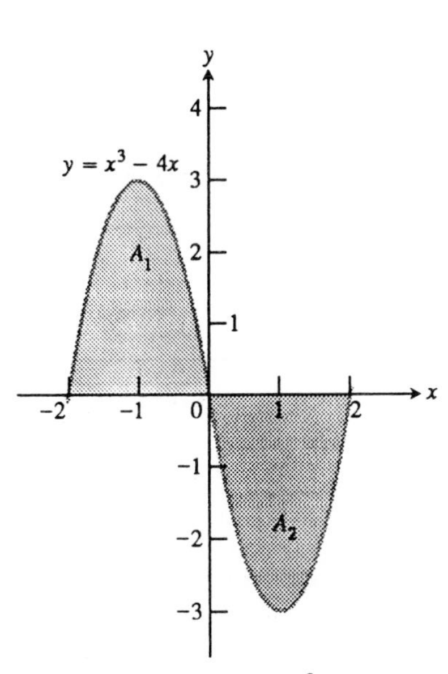

5.25 The graph of $y = x^3 - 4x$ from $x = -2$ to $x = 2$.

Solution

STEP 1: *The zeros of y.* We factor the formula for y to find where y is zero:

$$y = x^3 - 4x = x(x^2 - 4) = x(x-2)(x+2).$$

The zeros occur at $x = -2, 0$, and 2.

STEP 2: *The intervals of integration.* The points $x = -2, 0$, and 2 partition $[-2, 2]$ into two subintervals, $[-2, 0]$ and $[0, 2]$.

STEP 3: *The integrations.*

$$\int_{-2}^{0} (x^3 - 4x)\,dx = \left[\frac{x^4}{4} - 2x^2\right]_{-2}^{0} = [0] - [4 - 8] = 4$$

$$\int_{0}^{2} (x^3 - 4x)\,dx = \left[\frac{x^4}{4} - 2x^2\right]_{0}^{2} = [4 - 8] - [0] = -4$$

STEP 4: *The absolute values added.*

$$\text{Area of region } = |4| + |-4| = 4 + 4 = 8$$

Figure 5.25 shows the graph of $y = x^3 - 4x$ over $[-2, 2]$. The first integral in Step 3 gives the area A_1. The second integral gives the negative of the area A_2. The sum of the integrals' absolute values gives $A_1 + A_2$. ∎

EXAMPLE 10 Cost from marginal cost. The fixed cost of starting a manufacturing run and producing the first 10 units is \$200. After that the marginal cost at x units output is

$$\frac{dc}{dx} = \frac{1000}{x^2}.$$

Find the total cost of producing the first 100 units.

Solution If $c(x)$ is the cost of x units, then

$$\underbrace{c(100)}_{\substack{\text{cost of}\\ \text{100 units}}} = \underbrace{200}_{\substack{\text{startup}\\ \text{first 10}}} + \underbrace{c(100 - c(10)}_{\substack{\text{cost of units}\\ \text{11–100}}}$$

$$= 200 + \int_{10}^{100} \frac{dc}{dx}\,dx$$

$$= 200 + \int_{10}^{100} \frac{1000}{x^2}\,dx \qquad \left(\frac{dc}{dx} = \frac{1000}{x^2}\right)$$

$$= 200 + 1000\int_{10}^{100} \frac{1}{x^2}\,dx \qquad \text{(Rule 3, Table 5.3)}$$

$$= 200 + 1000\left[-\frac{1}{x}\right]_{10}^{100} = 200 + 1000\left[-\frac{1}{100} + \frac{1}{10}\right]$$

$$= 200 - 10 + 100 = 290.$$

The total cost of producing the first 100 units is \$290. ∎

Average Daily Inventory

The notion of a function's average value is used in economics to study things like average daily inventory. If $I(x)$ is the number of radios, tires, shoes, or whatever product a firm has on hand on day x (we call $I(x)$ an **inventory function**), the average value of I over a time period $a \leq x \leq b$ is the firm's average daily inventory for the period.

Definition

If $I(x)$ is the number of items on hand on day x, the **average daily inventory** of these items for the period $a \leq x \leq b$ is

$$I_{av} = \frac{1}{b-a}\int_a^b I(x)\,dx. \tag{14}$$

If h is the dollar cost of holding one item per day, the **average daily holding cost** for the period $a \leq x \leq b$ is $I_{av} \cdot h$.

EXAMPLE 11 Suppose a wholesaler receives a shipment of 1200 cases of chocolate bars every 30 days. The chocolate is sold to retailers at a steady rate, and x days after the shipment arrives, the inventory of cases still on hand is $I(x) = 1200 - 40x$. Find the average daily inventory. Also find the average daily holding cost for the chocolate if the cost of holding one case is 3¢ a day.

Solution The average daily inventory is

$$I_{av} = \frac{1}{30-0}\int_0^{30}(1200-40x)\,dx = \frac{1}{30}\left[1200x - 20x^2\right]_0^{30} = 600.$$

The average daily holding cost for the chocolate is the dollar cost of holding one case times the average daily inventory. This works out to \$18 a day:

$$\text{Average daily holding cost} = (0.03)(600) = 18.$$ ∎

Exercises 5.5

Use the Second Fundamental Theorem to evaluate the integrals in Exercises 1–14. Support with an FnInt computation.

1. $\int_0^3 (4 - x^2)\,dx$
2. $\int_0^1 (x^2 - 2x + 3)\,dx$
3. $\int_0^1 (x^2 + \sqrt{x})\,dx$
4. $\int_0^5 x^{3/2}\,dx$
5. $\int_1^{32} x^{-6/5}\,dx$
6. $\int_{-2}^{-1} \frac{2}{x^2}\,dx$
7. $\int_0^{\pi} \sin x\,dx$
8. $\int_0^{\pi} (1 + \cos x)\,dx$
9. $\int_0^{\pi/3} 2\sec^2 x\,dx$
10. $\int_{\pi/6}^{5\pi/6} \csc^2 x\,dx$

11. $\int_{\pi/4}^{3\pi/4} \csc x \cot x \, dx$

12. $\int_{0}^{\pi/3} 4 \sec x \tan x \, dx$

13. $\int_{-1}^{1} (r+1)^2 \, dr$

(*Hint:* Square first.)

14. $\int_{9}^{4} \frac{1-\sqrt{u}}{\sqrt{u}} \, du$

(*Hint:* Divide first.)

In Exercises 15–18, find the total area of the region between the curve and the x-axis.

15. $y = 2 - x, 0 \leq x \leq 3$

16. $y = 3x^2 - 3, -2 \leq x \leq 2$

17. $y = x^3 - 3x^2 + 2x, 0 \leq x \leq 2$

18. $y = x^3 - 4x, -2 \leq x \leq 2$

Find the areas of the shaded regions in Exercises 19–22.

19.

20.

21.

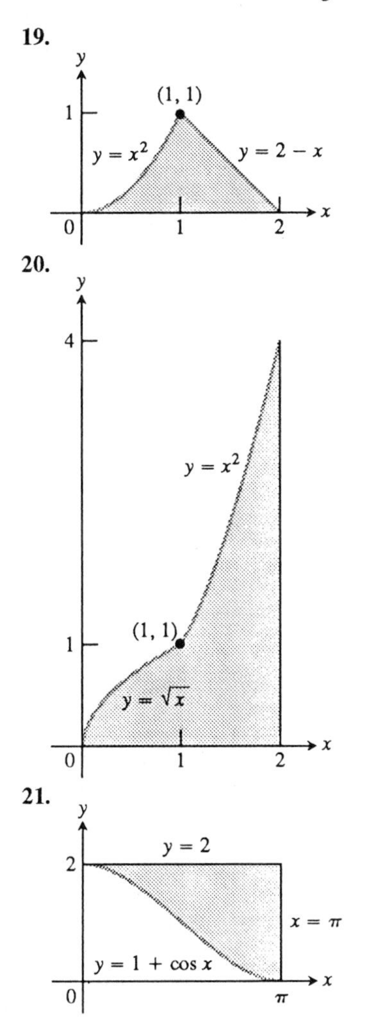

22.

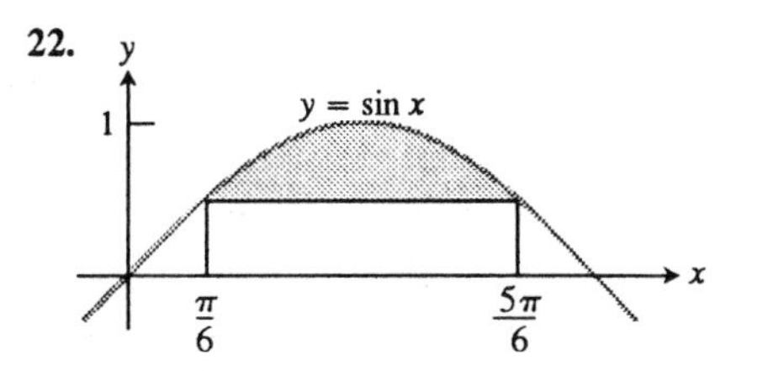

In Exercises 23–26, determine an explicit function for $F(x) = \int_a^k f(t)\,dt$ in terms of x. Graph $y = F(x)$ and $y =$ FnInt $(f(t), t, 0, x)$ in the same viewing window. Compare values at $x = 0.5, 1, 1.5, 2,$ and 5.

23. $\int_0^x (t-2)\,dt$

24. $\int_0^x (t^3+1)\,dt$

25. $\int_0^x (t^2 - 3t + 6)\,dt$

26. $\int_0^x 3 \sin t \, dt$

Graph in the interval specified in Exercises 27–30.

27. $\int_0^x t^2 \sin t \quad$ for $-3 \leq x \leq 3$

28. $\int_0^x \sqrt{1+t^2}\,dt \quad$ for $0 \leq x \leq 5$

29. $\int_0^x 5e^{-0.3t^2} \quad$ for $0 \leq x \leq 5$

30. $\int_0^x t \sin(t^3)\,dt \quad$ for $0 \leq x \leq \pi$

In Exercises 31 and 32 let $F(x) = \int_0^x f(t)\,dt$. Graph $y =$ NDer $(F(x), x)$ and compare with the graph of $y = f(x)$.

31. $f(x) = 4 - x^2 \quad$ for $-5 \leq x \leq 5$

32. $f(x) = x \sin x \quad$ for $0 \leq x \leq 2\pi$

In Exercises 33 and 34 find K so that $\int_a^x f(t)\,dt + K = \int_b^x f(t)\,dt$.

33. $f(x) = x^2 - 3x + 1; a = -1; b = 2$

34. $f(x) = \sin^2 x; a = 0; b = 2$

(The Power of Visualization) Solve for x in Exercises 35 and 36.

35. $\int_0^x e^{-t^2}\,dt = 0.9$

36. $\int_0^x \frac{\sin t}{t} = 1.8$

Find dy/dx in Exercises 37–40.

37. $y = \int_0^x \sqrt{1+t^2}\,dt$ 38. $y = \int_1^x \frac{1}{t}\,dt, x > 0$

39. $y = \int_0^{\sqrt{x}} \sin(t^2)\,dt$ 40. $y = \int_0^{2x} \cos t\,dt$

Each of the following functions solves one of the initial value problems in Exercises 41–44. Which function solves which problem?

a) $y = \int_1^x \frac{1}{t}\,dt - 3$ b) $y = \int_0^x \sec t\,dt + 4$

c) $y = \int_{-1}^x \sec t\,dt + 4$ d) $y = \int_\pi^x \frac{1}{t}\,dt - 3$

41. $\frac{dy}{dx} = \frac{1}{x}, y(\pi) = -3$ 42. $y' = \sec x, y(-1) = 4$

43. $y' = \sec x, y(0) = 4$ 44. $y' = \frac{1}{x}, y(1) = -3$

45. Show that if k is a positive constant, then the area between the x-axis and one arch of the curve $y = \sin kx$ is always $2/k$.

46. ARCHIMEDES' AREA FORMULA FOR PARABOLAS Archimedes (287–212 B.C.), inventor-military engineer, physicist, and the greatest mathematician of classical times, discovered that the area under a parabolic arch like the one shown here is always two-thirds the base times the height.

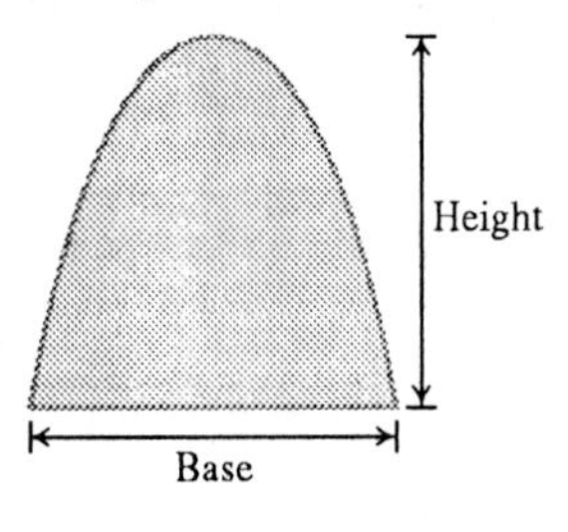

a) Find the area under the parabolic arch

$$y = 6 - x - x^2, -3 \le x \le 2.$$

b) Find the height of the arch. (Where does y have its maximum value?)

c) Show that the area is two-thirds the base times the height.

47. COST FROM MARGINAL COST The marginal cost of printing a poster when x posters have been printed is

$$\frac{dc}{dx} = \frac{1}{2\sqrt{x}}$$

dollars. Find (a) $c(100) - c(1)$, the cost of printing posters 2–100; (b) $c(400) - c(100)$, the cost of printing posters 101–400.

48. REVENUE FROM MARGINAL REVENUE Suppose that a company's marginal revenue from the manufacture and sale of egg beaters is

$$\frac{dr}{dx} = 2 - 2/(x+1)^2,$$

where r is measured in thousands of dollars and x in thousands of units. How much money should the company expect from a production run of $x = 3$ thousand egg beaters? To find out, integrate the marginal revenue from $x = 0$ to $x = 3$.

49. Solon Container receives 450 drums of plastic pellets every 30 days. The inventory function (drums on hand as a function of days) is $I(x) = 450 - x^2/2$. Find the average daily inventory. If the holding cost for one drum is 2¢ per day, find the average daily holding cost.

50. Mitchell Mailorder receives a shipment of 600 cases of athletic socks every 60 days. The number of cases on hand x days after the shipment arrives is $I(x) = 600 - 20\sqrt{15x}$. Find the average daily inventory. If the holding cost for one case is 1/2¢ per day, find the average daily holding cost.

51. For what value of x is

$$\int_a^x f(t)\,dt$$

sure to be zero?

52. Suppose $\int_1^x f(t)\,dt = x^2 - 2x + 1$. Find $f(x)$. (*Hint:* Differentiate both sides of the equation with respect to x.)

53. Find $f(4)$ if $\int_0^x f(t)\,dt = x \cos \pi x$.

54. Find the linearization of

$$f(x) = 2 + \int_0^x \frac{10}{1+t}\,dt$$

at $x = 0$.

55. Let $f(x) = \frac{\cos x}{x}$.

a) Draw a complete graph of f.

b) Draw a graph of $g(x) = \int_1^x f(t)\,dt, x > 0$.

c) Explain why $\int_0^x f(t)dt$ is undefined.

d) How are $h(x) = \int_{0.5}^x f(t)\,dt$ and $g(x)$ related? Graph both and support your answer.

56. Explain how you know $\int \frac{\sin 2x}{x}\,dx$ has no explicit analytic anti-derivative? Draw a graph of $f(x) = \int_0^x \frac{\sin 2t}{t}\,dt$. How is it related to the graph of $y = \int_0^x \frac{\sin t}{t}\,dt$?

5.6 Indefinite Integrals

Because antiderivatives make it possible to evaluate definite integrals with arithmetic, we shall be working with antiderivatives a lot. We therefore need a notation that will make antiderivatives easier to describe and work with. This section introduces the notation and shows how to use it.

The Indefinite Integral of a Function

We call the set of all antiderivatives of a function the indefinite integral of the function, according to the following definition.

Definition

If the function $f(x)$ is a derivative, then the set of all antiderivatives of f is called the **indefinite integral** of f, denoted by the symbols

$$\int f(x)\,dx.$$

As in definite integrals, the symbol $\int$ is called an **integral sign.** The function f is the **integrand** of the integral and x is the **variable of integration.**

Since every continuous function has an antiderivative (Corollary of the First Fundamental Theorem), every continuous function has an indefinite integral.

Once we have found an antiderivative $F(x)$ of a function $f(x)$, the other antiderivatives of f differ from F only by a constant (Corollary 3 of the Mean Value Theorem). We indicate this in the new notation by writing

$$\int f(x)\,dx = F(x) + C. \tag{1}$$

The constant C is called the **constant of integration** or the **arbitrary constant,** and Eq. (1) is read, "The indefinite integral of f with respect to x is $F(x)+C$." When we find $F(x)+C$, we say that we have **evaluated** the indefinite integral.

EXAMPLE 1 Evaluate $\int (x^2 - 2x + 5)\,dx$.

Solution

$$\int (x^2 - 2x + 5)\,dx = \underbrace{\frac{x^3}{3} - x^2 + 5x}_{\text{An antiderivative of } f(x) = x^2 - 2x + 5} \quad + \underbrace{C}_{\text{The constant of integration}}$$

■

TABLE 5.4 Integration Formulas

1. $\int x^n\,dx = \frac{x^{n+1}}{n+1} + C \quad (n \neq -1)$	
2. $\int \sin kx\,dx = -\frac{\cos kx}{k} + C$	3. $\int \cos kx\,dx = \frac{\sin kx}{k} + C$
4. $\int \sec^2 x\,dx = \tan x + C$	5. $\int \csc^2 x\,dx = -\cot x + C$
6. $\int \sec x \tan x\,dx = \sec x + C$	7. $\int \csc x \cot x\,dx = -\csc x + C$

To evaluate $\int f(x)\,dx$:

STEP 1: Find an antiderivative $F(x)$ of $f(x)$.

STEP 2: Add C (the constant of integration). Then

$$\int f(x)\,dx = f(x) + C.$$

The formulas for general antiderivatives in Section 4.7 translate into the formulas for evaluating indefinite integrals listed in Table 5.4.

In case you are wondering why the integrals of the tangent, cotangent, secant, and cosecant are not listed here, the answer is that the usual formulas for them require logarithms. We know that these functions do have indefinite integrals on intervals where they are continuous, but we shall have to wait until Chapters 7 and 8 to see what the integrals are.

EXAMPLE 2 Selected Integrals from Table 5.4

a) $$\int x^5\,dx = \frac{x^6}{6} + C \qquad \text{(Formula 1)}$$

b) $$\int \sin 2x\,dx = -\frac{\cos 2x}{2} + C \qquad \text{(Formula 2 with } k = 2\text{)}$$

c) $$\int \cos\frac{x}{2}\,dx = \int \cos\frac{1}{2}x\,dx = \frac{\sin(1/2)x}{1/2} + C \qquad \text{(Formula 3 with } k = 1/2\text{)}$$

$$= 2\sin\frac{x}{2} + C$$

The formulas in Table 5.4 hold because, in each case, the derivative of the function $F(x) + C$ on the right is the integrand $f(x)$ on the left. Finding an integral formula is often an impossible task, but *checking* an integral formula, once found, is relatively easy: Differentiate the right-hand side. If the derivative is the integrand, the formula is correct; otherwise it is wrong.

EXAMPLE 3

RIGHT: $$\int x\cos x\,dx = x\sin x + \cos x + C$$

Reason: The derivative of the right-hand side is the integrand:

$$\frac{d}{dx}(x\sin x + \cos x + C) = x\cos x + \sin x - \sin x + 0 = x\cos x.$$

WRONG: $$\int x \cos x \, dx = x \sin x + C$$

Reason: The derivative of the right-hand side is not the integrand:

$$\frac{d}{dx}(x \sin x + C) = x \cos x + \sin x + 0 \neq x \cos x.$$

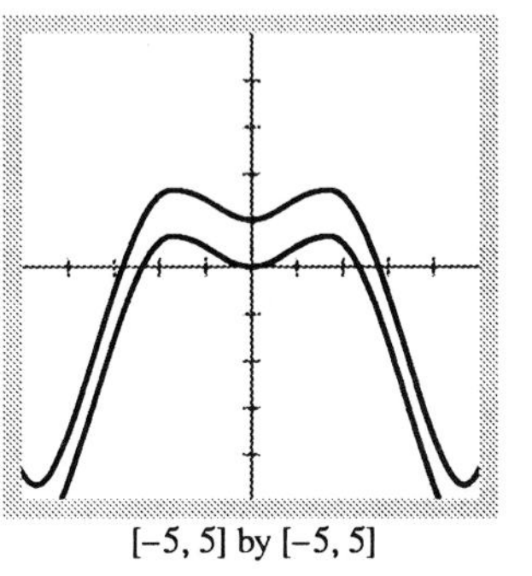

[-5, 5] by [-5, 5]

5.26 Graphs of $y = x \sin x + \cos x$ and y = FnInt($t \cos t, t, 0, x$) for $-5 \leq x \leq 5$.

Do not worry about where the (correct) integral formula in Example 3 comes from right now. There is a nice technique in Chapter 8 for evaluating integrals like this.

The Role of Graphing Integrals

Another effective way to check an integral formula is by graphing as illustrated in this example.

EXAMPLE 4 Support graphically: $\int x \cos x \, dx = x \sin x + \cos x + C$.

Solution Figure 5.26 displays a graph of $y = x \sin x + \cos x$ (*one* antiderivative of $x \cos x$) and the graph of $F(x) =$ FnInt($t \cos t, t, 0, x$) which is also an antiderivative $f(x) = x \cos x$.

Notice they appear to differ by a constant because the graphs should be related by a vertical shift. This is strong evidence that $\int x \cos x \, dx = x \sin x + \cos x + C$. Even stronger graphical evidence is provided by comparing the graphs of the numerical derivatives of both $y = x \sin x + \cos x$ and $F(x) =$ FnInt ($t \cos x, t, 0, x$) with the graph of $f(x) = x \cos x$. They all will be identical!

Explore Box

1. Figure 5.26 suggests that $x \sin x + \cos x + C =$ FnInt ($t \cos t, t, 0, x$) for some value of C. Determine the value of C in this particular case. Why are you sure?
2. Graph $y = x \cos x, y =$ NDer ($x \sin x + \cos x, x$) and $y =$ NDer (FnInt ($t \cos t, t, 0, x$), x) in the $[-5, 5]$ by $[-5, 5]$ window. What can you conclude? Why?

Rules of Algebra

Among the things we know about antiderivatives are these:

1. The general antiderivative of a derivative dF/dx is $F(x) + C$.
2. When we differentiate an antiderivative of a function $f(x)$ with respect to x, we get f back again.
3. A function is an antiderivative of a constant multiple kf of a function f if and only if it is k times an antiderivative of f.

4. In particular, a function is an antiderivative of $-f$ if and only if it is the negative of an antiderivative of f.
5. A function is an antiderivative of a sum $f(x)+g(x)$ if and only if it is the sum of an antiderivative of f and an antiderivative of g.
6. A function is an antiderivative of a difference $f(x)-g(x)$ if and only if it is an antiderivative of f minus an antiderivative of g.

These observations can be expressed very nicely in the notation of indefinite integrals (Table 5.5).

EXAMPLE 5

$$\int \cos x\,dx = \int \frac{d}{dx}(\sin x)\,dx = \sin x + C \qquad \text{(Rule 1 with } F(x)=\sin x)$$

EXAMPLE 6

$$\frac{d}{dx}\int \tan x\,dx = \tan x \qquad \text{(Rule 2 with } f(x)=\tan x)$$

We know this even without knowing how to evaluate the integral.

EXAMPLE 7 Rewriting the constant of integration.

$$\begin{aligned}
\int 5\sec x\tan x\,dx &= 5\int \sec x\tan x\,dx && \left(\begin{array}{l}\text{Table 5.5, Rule 3, with}\\ k=5,\ f(x)=\sec x\tan x\end{array}\right)\\
&= 5(\sec x + C) && \text{(Table 5.4, Rule 6)}\\
&= 5\sec x + 5C && \text{(First form)}\\
&= 5\sec x + C' && \text{(Shorter form)}\\
&= 5\sec x + C. && \text{(Usual form—no prime)}
\end{aligned}$$

TABLE 5.5 Rules for Indefinite Integrals

1. $\displaystyle\int \frac{dF}{dx}\,dx = F(x)+C$
2. $\displaystyle\frac{d}{dx}\int f(x)\,dx = f(x)$
3. $\displaystyle\int k\,f(x)\,dx = k\int f(x)\,dx$ (Provided k is a constant. Does not work if k varies with x.)
4. $\displaystyle\int -f(x)\,dx = -\int f(x)\,dx$ (Special case of Rule 3)
5. $\displaystyle\int [f(x)+g(x)]\,dx = \int f(x)\,dx + \int g(x)\,dx$
6. $\displaystyle\int [f(x)-g(x)]\,dx = \int f(x)\,dx - \int g(x)\,dx$

What about all the different forms in Example 7? Each one of them gives all the antiderivatives of $f(x) = 5 \sec x \tan x$, so each answer is correct. But the least complicated of the three, and the usual choice, is

$$\int 5 \sec x \tan x \, dx = 5 \sec x + C.$$

The general rule, in practice, is this:

> If $F'(x) = f(x)$ and k is a constant, then
> $$k \int f(x)\, dx = \int k\, f(x)\, dx = k\, F(x) + C. \qquad (2)$$

EXAMPLE 8

$$\int 8 \cos x \, dx = 8 \sin x + C \qquad \left(\begin{array}{l}\text{Eq. (2) with } k = 8,\ f(x) = \cos x, \\ F(x) = \sin x\end{array}\right)$$

EXAMPLE 9 When we integrate term by term, we combine the constants of integration into a single arbitrary constant. Evaluate $\int (x^2 - 2x + 5)\, dx$.

Solution This is the integral of Example 1, which we evaluated by finding an antiderivative of $(x^2 - 2x + 5)$ and adding an arbitrary constant:

$$\int (x^2 - 2x + 5)\, dx = \overbrace{\frac{x^3}{3} - x^2 + 5x}^{\text{Antiderivative}} + \underbrace{C}_{\text{Arbitrary constant}}.$$

Suppose, however, we evaluate the integral term by term by applying the sum and difference rules (Rules 5 and 6 in Table 5.5). Then, instead of getting one constant of integration, we get three:

$$\begin{aligned}\int (x^2 - 2x + 5)\, dx &= \int x^2\, dx - \int 2x\, dx + \int 5\, dx \\ &= \frac{x^3}{3} + C_1 - x^2 + C_2 + 5x + C_3.\end{aligned}$$

This formula certainly does give all the antiderivatives of $x^2 - 2x + 5$. But it is more complicated than it needs to be. If we were to combine C_1, C_2, and C_3 into a single constant $C = C_1 + C_2 + C_3$, the formula would simplify to

$$\frac{x^3}{3} - x^2 + 5x + C$$

and *still* give all the antiderivatives there are. For this reason we recommend that you go right to the final form even if you elect to integrate term by term. Write

$$\int (x^2 - 2x + 5)\, dx = \int x^2\, dx - \int 2x\, dx + \int 5\, dx = \frac{x^3}{3} - x^2 + 5x + C.$$

Find the simplest antiderivative you can for each part, then add the constant at the end. ∎

The Integrals of $\sin^2 x$ and $\cos^2 x$

We can sometimes use trigonometric identities to transform indefinite integrals we do not know how to evaluate into indefinite integrals we do know how to evaluate. Among the examples you should know about, important because of how frequently they arise in applications, are the integral formulas for $\sin^2 x$ and $\cos^2 x$. You need not remember the formulas themselves, but try to remember how they can be derived when you need them.

EXAMPLE 10

$$\int \sin^2 x \, dx = \int \frac{1 - \cos 2x}{2} dx \qquad \left(\text{Because } \sin^2 x = \frac{1 - \cos 2x}{2}\right)$$

$$= \frac{1}{2}\int (1 - \cos 2x)\, dx = \frac{1}{2}\int dx - \frac{1}{2}\int \cos 2x \, dx$$

$$= \frac{1}{2}x - \frac{1}{2}\frac{\sin 2x}{2} + C = \frac{x}{2} - \frac{\sin 2x}{4} + C$$

∎

We leave $\int \cos^2 x \, dx$ as an exercise. Another important trigonometric integral that occurs in applications is

$$\int \frac{\sin x}{x} dx.$$

We know $f(x) = (\sin x)/x$ has a removable discontinuity at 0 because $\lim_{x \to 0} (\sin x)/x = 1$. The function $\mathrm{Si}(x)$ defined by

$$\mathrm{Si}(x) = \begin{cases} 1 & \text{if } x = 0 \\ \dfrac{\sin x}{x} & \text{if } x \neq 0 \end{cases}$$

is, for all practical purposes, equal to $(\sin x)/x$. Henceforth in this chapter when we write $f(x) = (\sin x)/x$ we will mean the function $y = \mathrm{Si}(x)$. $y = \mathrm{Si}(x)$ is continuous so $F(x) = \int_a^x (\sin t)/t \, dt$ exists and is an antiderivative of $\mathrm{Si}(x)$. What is F? There is no explicit formula in terms of x for F!

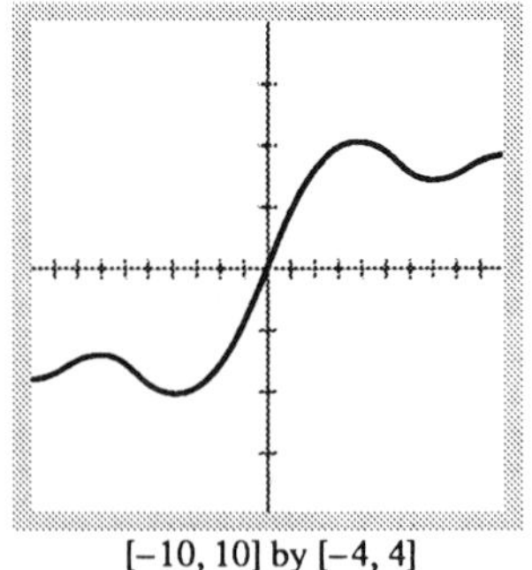

[−10, 10] by [−4, 4]

5.27 The graph of $y = \mathrm{FnInt}((\sin t)/t, t, 0, x)$.

EXAMPLE 11 Graphically determine $\displaystyle\int \frac{\sin x}{x} dx.$

Solution No formula exists in terms of x for this definite integral but we can graph one of the antiderivatives of $(\sin x)/x$ as shown in Figure 5.27.

Even though we don't have an explicit formula for the antiderivative, we can "see" its graph which is, in many cases, even better than having the exact analytic form. This is true because we can also apply a graphical method to solve equations involving integrals, as we will see in many examples later. ∎

Solving Initial Value Problems with Indefinite Integrals

As you know, we solve initial value problems with antiderivatives. We now look at the solutions in the language of indefinite integration. The first example is based on Example 7 in Section 4.7. You need not look back, however, unless you are interested in the modeling that gave rise to the problem. The solution we give here is mathematically self-contained.

EXAMPLE 12 Finding velocity from acceleration. As a function of elapsed time t, the velocity $v(t)$ of a body falling from rest in a vacuum near the surface of the earth satisfies

Differential equation: $\dfrac{dv}{dt} = 9.8$ (The acceleration is 9.8 m/sec^2.)

Initial condition: $v = 0$ when $t = 0$. (The velocity is 0 at the start.)

Find v as a function of t.

Solution We find the general solution of the differential equation by integrating both sides of it with respect to t:

$$\frac{dv}{dt} = 9.8 \qquad \text{(Differential equation)}$$

$$\int \frac{dv}{dt}\,dt = \int 9.8\,dt \qquad \text{(Integral equation)}$$

$$v + C_1 = 9.8t + C_2 \qquad \text{(Integrals evaluated)}$$

$$v = 9.8t + C_2 - C_1 \qquad \text{(Solved for } v\text{)}$$

$$v = 9.8t + C. \qquad \text{(Constants combined as one)}$$

This last equation tells us that the body's velocity t seconds into the fall is $9.8t + C$ m/sec for some value of C. What value? We find out from the initial condition:

$$v = 9.8t + C$$

$$0 = 9.8(0) + C \qquad (v = 0 \text{ when } t = 0)$$

$$C = 0.$$

Conclusion: The body's velocity t seconds into the fall is

$$v = 9.8t + 0 = 9.8t \text{ m/sec.}$$

In the next example we have to integrate a second derivative twice to find the function we are after. The first integration,

$$\int \frac{d^2y}{dx^2}\,dx = \frac{dy}{dx} + C, \tag{3}$$

gives the function's first derivative. The second integration gives the function.

EXAMPLE 13 Solve the following initial value problem for y as a function of x:

Differential equation: $\dfrac{d^2y}{dx^2} = 6x - 2$

Initial conditions: $\dfrac{dy}{dx} = 0$ and $y = 10$ when $x = 1$.

Solution We integrate the differential equation with respect to x to find dy/dx:

$$\int \frac{d^2y}{dx^2}\,dx = \int (6x - 2)\,dx$$

$$\frac{dy}{dx} = 3x^2 - 2x + C_1. \qquad \text{(Constants of integration combined as } C_1\text{)}$$

We apply the first initial condition to find C_1:

$$0 = 3(1)^2 - 2(1) + C_1 \qquad \left(\frac{dy}{dx} = 0 \text{ when } x = 1\right)$$

$$C_1 = -3 + 2 = -1.$$

This completes the formula for dy/dx :

$$\frac{dy}{dx} = 3x^2 - 2x - 1.$$

We integrate dy/dx with respect to x to find y:

$$\int \frac{dy}{dx}\,dx = \int (3x^2 - 2x - 1)\,dx$$

$$y = x^3 - x^2 - x + C_2. \qquad \text{(Constants of integration combined as } C_2\text{)}$$

We apply the second initial condition to find C_2:

$$10 = (1)^3 - (1)^2 - 1 + C_2 \qquad (y = 10 \text{ when } x = 1)$$

$$10 = -1 + C_2$$

$$C_2 = 11.$$

This completes the formula for y as a function of x:

$$y = x^3 - x^2 - x + 11.$$

When we find a function by integrating its first derivative, we have one constant of integration, as in Example 12. When we find a function from its second derivative, we have to deal with two constants of integration, one from each integration, as in Example 13. If we were to find a function from its third derivative, we would have to find the values of three constants of integration, and so on. In each case the values of the constants are determined by the problem's initial conditions. Each time we integrate, we need an initial condition to tell us the value of C.

EXAMPLE 14 Solve the following initial value problem.

Differential equation: $\dfrac{dy}{dx} = (0.6)2^x$

Initial condition: $y = -1$ when $x = 0$.

Solution We integrate the differential equation with respect to x to find y.

$$\int \frac{dy}{dx} = \int (0.6)2^x\,dx$$

$$y = \int (0.6)2^x\,dx$$

Now we don't yet have an explicit antiderivative of $(0.6)2^x$ in terms of x, but we do know that $F(x) = \int_a^x (0.6)2^t\,dt$ gives one antiderivative for each value of a. Figure 5.28 displays the graph of FnInt $((0.6)2^t, t, 1.1, x)$ in the window $[-3, 3]$ by $[-3, 3]$.

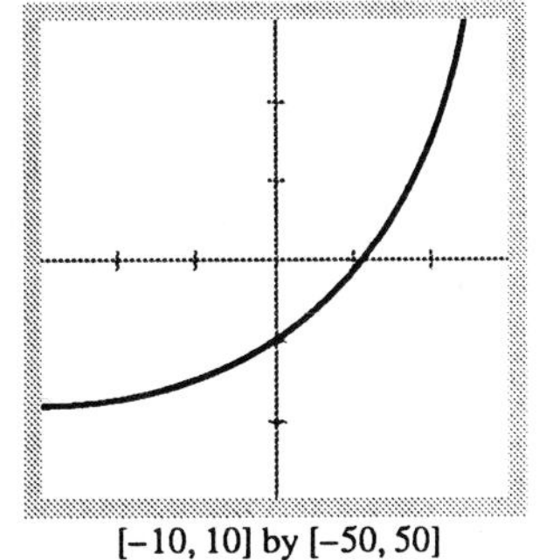

[−10, 10] by [−50, 50]

5.28 The graph of FnInt$((0.6)2^t, t, 1.1, x)$.

Notice the y-intercept appears to be -1. This means $F(x) = \int_{1.1}^x (0.6)2^t\,dt$ is the solution to the initial value problem. And we can "see" the solution. It should be noted that the value 1.1 is an approximation. The actual value can be found by solving the initial condition equation $F(0) = -1$, that is by solving the equation $F(0) = \int_a^0 (0.6)2^t dt = -1$ for a. It is amazing to discover that this can be done very accurately graphically as illustrated in the next Explore activity.

Explore Box

Graph $y =$ FnInt $((0.6)2^t, t, x, 0) + 1$ and zoom-in or use a ROOT finder to confirm that the solution to $\int_a^0 (0.6)2^t dt = -1$ is $a = 1.10785$ accurate to five decimal places. Explain why the graph gives the solution.

In a later section we will find that the explicit solution y as a function of x to this differential equation is based on the antiderivative $((0.6)2^x)/\ln 2$. Notice, $D_x\left[((0.6)2^x)/\ln 2\right] = (0.6)2^x$. So the initial value problem now boils down to determining C so that $((0.6)2^x)/\ln 2 + C = -1$ when $x = 0$. That is $0.6/\ln 2 + C = -1$ for C so that the initial condition will be satisfied. It is easy to compute that $C = -1.86562$.

In the next Explore we ask you to verify graphically that the two forms of the solution to the initial value problem of Example 14, namely,

1. $F(x) = \displaystyle\int_{1.10785}^{x} (0.6)2^t\,dt,$

and

2. $y = \dfrac{(0.6)2^x}{\ln 2} - 1.86561,$

are identical and thus really describe the same solution.

Explore Box

Graph $y = ((0.6)2^x)/\ln 2 - 1.86561$ and $y = \text{FnInt}\,((0.6)2^t, t, 1.10785, x)$ in the $[-4, 4]$ by $[-3, 3]$ window. Can you conclude they are identical? Use the TRACE feature of your grapher and compare values. How close are they?

As we have seen, the ideal way to evaluate a definite integral

$$\int_a^b f(x)\,dx$$

is to find a formula $F(x)$ for one of the antiderivatives of $f(x)$ and calculate the number $F(b) - F(a)$. But many antiderivatives are hard to find and still others, like the antiderivatives of $(\sin x)/x$ and $\sqrt{1+x^4}$, have no elementary formulas. We do not mean merely that no one has yet succeeded in finding simple expressions for evaluating the antiderivatives of $(\sin x)/x$ and $\sqrt{1+x^4}$; we mean it has been proved that no such expressions exist.

Whatever the reason, when we cannot evaluate a definite integral with an antiderivative, we turn to numerical and graphical methods like those illustrated in this section and in Sections 5.2 and 5.3. We will also illustrate numerical and graphical methods in Section 5.9.

Exercises 5.6

Evaluate the integrals in Exercises 1–30. Support visually by comparing the graphs of the analytic antiderivative and $y = \text{FnInt}\,(f(t), t, 0, x)$.

1. $\int x^3\,dx$
2. $\int 7\,dx$
3. $\int (x+1)\,dx$
4. $\int (6-6x)\,dx$
5. $\int 3\sqrt{x}\,dx$
6. $\int \frac{4}{x^2}\,dx$
7. $\int x^{-1/3}\,dx$
8. $\int (1-4x^{-3})\,dx$
9. $\int (5x^2+2x)\,dx$
10. $\int \left(\frac{x^2}{2}+\frac{x^3}{3}\right)\,dx$
11. $\int (2x^3-5x+7)\,dx$
12. $\int (1-x^2-3x^5)\,dx$
13. $\int 2\cos x\,dx$
14. $\int 5\sin\theta\,d\theta$
15. $\int \sin\frac{x}{3}\,dx$
16. $\int 3\cos 5x\,dx$
17. $\int 3\csc^2 x\,dx$
18. $\int \frac{\sec^2 x}{3}\,dx$
19. $\int \frac{\csc x\cot x}{2}\,dx$
20. $\int \frac{2}{5}\sec x\tan x\,dx$
21. $\int (4\sec x\tan x - 2\sec^2 x)\,dx$
22. $\int \frac{1}{2}(\csc^2 x - \csc x\cot x)\,dx$
23. $\int (\sin 2x - \csc^2 x)\,dx$
24. $\int (2\cos 2x - 3\sin 3x)\,dx$
25. $\int 4\sin^2 y\,dy$
26. $\int \frac{\cos^2 x}{7}\,dx$
27. $\int \sin x\cos x\,dx$
(Hint: $2\sin x\cos x = \sin 2x$)
28. $\int (1-\cos^2 t)\,dt$

29. $\int (1 + \tan^2 \theta)\, d\theta$

(Hint: $1 + \tan^2 \theta = \sec^2 \theta$)

30. $\int \frac{1 + \cot^2 x}{2}\, dx$

Show that the integral formulas in Exercises 31–34 are correct by showing that the derivatives of the right-hand sides are the integrands in the integrals on the left-hand sides. (In Section 5.7 we shall see where formulas like these come from.)

31. $\int (7x - 2)^3\, dx = \frac{(7x - 2)^4}{28} + C$

32. $\int \sec^2 5x\, dx = \frac{\tan 5x}{5} + C$

33. $\int \frac{1}{(x + 1)^2}\, dx = -\frac{1}{x + 1} + C$

34. $\int \frac{1}{(x + 1)^2}\, dx = \frac{x}{x + 1} + C$

35. Right, or wrong? Say which for each formula.

a) $\int x \sin x\, dx = \frac{x^2}{2} \sin x + C$

b) $\int x \sin x\, dx = -x \cos x + C$

c) $\int x \sin x\, dx = -x \cos x + \sin x + C$

36. Right, or wrong? Say which for each formula.

a) $\int (2x + 1)^2\, dx = \frac{(2x + 1)^3}{3} + C$

b) $\int 3(2x + 1)^2\, dx = (2x + 1)^3 + C$

c) $\int 6(2x + 1)^2\, dx = (2x + 1)^3 + C$

Solve the initial value problems in Exercises 37–44 for y as a function of x. Graph the solution $y = f(x)$ if analytic methods are unavailable.

37. Differential equation: $\frac{dy}{dx} = 3\sqrt{x}$

Initial condition: $y = 4$ when $x = 9$

38. Differential equation: $\frac{dy}{dx} = \frac{1}{2\sqrt{x}}$

Initial condition: $y = 0$ when $x = 4$

39. Differential equation: $\frac{dy}{dx} = 2^x$

Initial condition: $y = 2$ when $x = 0$

40. Differential equation: $\frac{dy}{dx} = \frac{\cos x}{x}$

Initial condition: $y = 0.5$ when $x = 0$

41. Differential equation: $\frac{d^2y}{dx^2} = 0$

Initial conditions: $\frac{dy}{dx} = 2$ and $y = 0$ when $x = 0$

42. Differential equation: $\frac{d^2y}{dx^2} = \frac{2}{x^3}$

Initial conditions: $\frac{dy}{dx} = 1$ and $y = 1$ when $x = 1$

43. Differential equation: $\frac{d^2y}{dx^2} = \frac{3x}{8}$

Initial conditions: $\frac{dy}{dx} = 3$ and $y = 4$ when $x = 4$

44. Differential equation: $\frac{d^3y}{dx^3} = 6$

Initial conditions: $\frac{d^2y}{dx^2} = -8$, $\frac{dy}{dx} = 0$, and $y = 5$ when $x = 0$

45. STOPPING A CAR IN TIME You are driving along a highway at a steady 60 mph (88 ft/sec) when you see an accident ahead and slam on the brakes. What constant deceleration is required to stop your car in 242 ft? To find out, carry out the following steps:

STEP 1: Solve the initial value problem

Differential equation: $\frac{d^2s}{dt^2} = -k$ (k constant)

Initial conditions: $\frac{ds}{dt} = 88$ and $s = 0$ when $t = 0$.

(Measuring time and distance from when the brakes are applied)

STEP 2: Find the value of t that makes $ds/dt = 0$. (The answer will involve k.)

STEP 3: Find the value of k that makes $s = 242$ for the value of t you found in Step 2.

46. MOTION ALONG A COORDINATE LINE A particle moves along a coordinate line with acceleration $a = d^2s/dt^2 = 15\sqrt{t} - (3/\sqrt{t})$, subject to the conditions that $ds/dt = 4$ and $s = 0$ when $t = 1$. Find

a) the velocity $v = ds/dt$ in terms of t,

b) the position s in terms of t.

47. THE HAMMER AND THE FEATHER When Apollo 15 astronaut David Scott dropped a hammer and a feather on the moon to demonstrate that in a vacuum all bodies fall with the same (constant) acceleration, he dropped them from about 4 ft above the ground. The television footage of the event shows the hammer and feather falling more slowly than on earth, where, in a vacuum, they would have taken only half a second to fall the four feet. How long did it take the hammer and feather to fall the four feet on the moon? To

find out, solve the following initial value problem for s as a function of t. Then find the value of t that makes s equal 4.

Differential equation: $\frac{d^2s}{dt^2} = 5.2 \text{ ft/sec}^2$

Initial conditions: $\frac{ds}{dt} = 0$ and $s = 0$ when $t = 0$

48. THE STANDARD EQUATION FOR FREE FALL The standard equation for free fall near the surface of every planet is

$$s(t) = \frac{1}{2}gt^2 + v_0t + s_0,$$

where $s(t)$ is the body's position on the line of fall, g is the planet's (constant) acceleration of gravity, v_0 is the body's initial velocity, and s_0 is the body's initial position. Derive this equation by solving the following initial value problem:

Differential equation: $\frac{d^2s}{dt^2} = g$

Initial conditions: $\frac{ds}{dt} = v_0$ and $s = s_0$ when $t = 0$.

49. Derive the integration formula

$$\int \cos^2 x \, dx = \frac{x}{2} + \frac{\sin 2x}{4} + C.$$

5.7 Integration by Substitution—Running the Chain Rule Backward

A change of variable can often turn an unfamiliar integral into one we can evaluate. The method for doing this is called the substitution method of integration. It is the principal method by which integrals are evaluated. This section shows how and why the method works.

The Generalized Power Rule in Integral Form

When u is a differentiable function of x and n is a rational number different from -1, the Chain Rule tells us that

$$\frac{d}{dx}\left(\frac{u^{n+1}}{n+1}\right) = u^n \frac{du}{dx}. \tag{1}$$

This same equation, from another point of view, says that $u^{n+1}/(n+1)$ is one of the antiderivatives of $u^n(du/dx)$. The set of all antiderivatives of $u^n(du/dx)$ is therefore

$$\int \left(u^n \frac{du}{dx}\right) dx = \frac{u^{n+1}}{n+1} + C. \tag{2}$$

The integral on the left-hand side of this equation is usually written in the simpler "differential" form,

$$\int u^n \, du, \tag{3}$$

obtained by treating the dx's as differentials that cancel. Combining Eqs. (2) and (3) then gives the following rule.

If u is any differentiable function of x,

$$\int u^n \, du = \frac{u^{n+1}}{n+1} + C. \tag{4}$$

Whenever we can cast an integral in the form

$$\int u^n \, du$$

with u a differentiable function of x and du the differential of u, we can integrate with respect to u in the usual way to evaluate the integral as $[u^{n+1}/(n+1)] + C$.

EXAMPLE 1 Evaluate $\int (x+2)^5 \, dx$.

Solution We can put the integral in the form

$$\int u^5 \, du$$

by substituting

$$u = x + 2, \quad du = d(x+2) = dx.$$

Then

$$\begin{aligned}
\int (x+2)^5 \, dx &= \int u^5 \, du && \text{(Substitute } u = x+2, du = dx.\text{)} \\
&= \frac{u^6}{6} + C && \text{(Integrate, using Eq. (4) with } n = 5.\text{)} \\
&= \frac{(x+2)^6}{6} + C. && \text{(Replace } u \text{ by } x+2.\text{)}
\end{aligned}$$

EXAMPLE 2

$$\begin{aligned}
\int \sqrt{1+x^2} \cdot 2x \, dx &= \int u^{1/2} du && \left(\begin{array}{l}\text{Substitute } u = 1+x^2, \\ du = 2x \, dx.\end{array}\right) \\
&= \frac{u^{(1/2)+1}}{(1/2)+1} + C && \left(\begin{array}{l}\text{Integrate using Eq. (4)} \\ \text{with } n = \frac{1}{2}.\end{array}\right) \\
&= \frac{2}{3} u^{3/2} + C && \text{(Simpler form)} \\
&= \frac{2}{3}(1+x^2)^{3/2} + C && \text{(Replace } u \text{ by } 1+x^2.\text{)}
\end{aligned}$$

EXAMPLE 3 Adjusting the integrand by a constant to put it in standard form.

$$\int \sqrt{4x-1}\,dx = \int u^{1/2}\cdot\frac{1}{4}du \qquad \begin{pmatrix}\text{Substitute } u = 4x-1,\\ du = 4\,dx,\ \frac{1}{4}du = dx.\end{pmatrix}$$

$$= \frac{1}{4}\int u^{1/2}du \qquad \begin{pmatrix}\text{With the } \frac{1}{4} \text{ out front, the integral}\\ \text{is now in standard form.}\end{pmatrix}$$

$$= \frac{1}{4}\frac{u^{3/2}}{3/2} + C \qquad \left(\text{Integrate, using Eq. (4) with } n = \frac{1}{2}.\right)$$

$$= \frac{1}{6}u^{3/2} + C \qquad (\text{Simpler form})$$

$$= \frac{1}{6}(4x-1)^{3/2} + C \qquad (\text{Replace } u \text{ by } 4x-1.)$$

If the "2x" had been left out of the integrand of Example 2, the corresponding indefinite integration problem would be nasty.

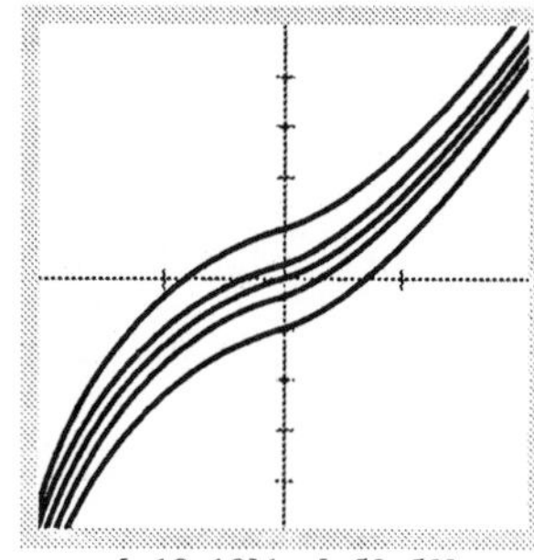

[−10, 10] by [−50, 50]

5.29 The graphs of $y = \text{FnInt}\,(\sqrt{1+t^2}, t, a, x)$ for $a = -4, -2, 0, 2,$ and 4. This is strong support for the fact that $\int \sqrt{1+x^2}\,dx = \int_0^x \sqrt{1+t^2}\,dt + C$ where C is any constant.

EXAMPLE 4 A picture is worth a thousand words.

We write

$$\int \sqrt{1+x^2}\,dx = \int_a^x \sqrt{1+t^2}\,dt \tag{5}$$

with the same understanding that we write $\int f(x)dx = F(x) + C$ where $D_x F(x) = f(x)$. Here the right side of Eq. (5) represents the set of *all* antiderivatives of $\sqrt{1+x^2}$. The graph in Figure 5.29 displays some members of this set. Again notice how each is a particular antiderivative, say $\int_0^x \sqrt{1+t^2}\,dt$, by a vertical translation.

Sines and Cosines

If u is a differentiable function of x, then $\sin u$ is a differentiable function of x. The Chain Rule gives the derivative of $\sin u$ as

$$\frac{d}{dx}\sin u = \cos u\frac{du}{dx}. \tag{6}$$

From another point of view, however, this same equation says that $\sin u$ is one of the antiderivatives of the product $\cos u \cdot (du/dx)$. The set of all antiderivatives of the product is therefore

$$\int \left(\cos u\frac{du}{dx}\right) dx = \sin u + C. \tag{7}$$

A formal cancellation of the dx's in the integral on the left leads to the following rule.

If u is a differentiable function of x, then

$$\int \cos u\, du = \sin u + C. \qquad (8)$$

Equation (8) says that whenever we can cast an integral in the form

$$\int \cos u\, du,$$

we can integrate with respect to u in the usual way to evaluate the integral as $\sin u + C$.

EXAMPLE 5

$$\int \cos(7x+5)\,dx = \int \cos u \cdot \frac{1}{7}du \qquad \left(\begin{array}{l}\text{Substitute } u = 7x+5,\\ du = 7\,dx,\ \frac{1}{7}\,du = dx.\end{array}\right)$$

$$= \frac{1}{7}\int \cos u\, du \qquad \left(\begin{array}{l}\text{With the } \frac{1}{7} \text{ out front, the inte-}\\ \text{gral is now in standard form.}\end{array}\right)$$

$$= \frac{1}{7}\sin u + C \qquad \text{(Integrate with respect to } u.)$$

$$= \frac{1}{7}\sin(7x+5) + C \qquad \text{(Replace } u \text{ by } 7x+5.)$$

∎

The companion formula for the integral of $\sin u$ when u is a differentiable function of x is

$$\int \sin u\, du = -\cos u + C. \qquad (9)$$

EXAMPLE 6

$$\int x^2 \sin(x^3)\,dx = \int \sin u \cdot \frac{1}{3}du \qquad \left(\begin{array}{l}\text{Substitute } u = x^3,\\ du = 3x^2\,dx,\ \frac{1}{3}\,du = x^2\,dx.\end{array}\right)$$

$$= \frac{1}{3}\int \sin u\, du$$

$$= \frac{1}{3}(-\cos u) + C \qquad \text{(Integrate with respect to } u.)$$

$$= -\frac{1}{3}\cos(x^3) + C \qquad \text{(Replace } u \text{ by } x^3.)$$

∎

The Chain-Rule formulas for the derivatives of the tangent, cotangent, secant, and cosecant of a differentiable function u of x lead to the following integrals.

Explore Box

Support the integration formula in Example 6

$$\int x^2 \sin(x^3)\,dx = -\frac{1}{3}\cos(x^3) + C$$

by graphing $y = -\frac{1}{3}\cos(x^3)$ and $y =$ FnInt $(t^2 \sin(t^3), t, 0, x)$ in the same viewing window. Use TRACE to compare values.

$$\int \sec^2 u\,du = \tan u + C \tag{10}$$

$$\int \csc^2 u\,du = -\cot u + C \tag{11}$$

$$\int \sec u \tan u\,du = \sec u + C \tag{12}$$

$$\int \csc u \cot u\,du = -\csc u + C \tag{13}$$

In each formula we assume u to be a differentiable function of x. Each formula can be checked by differentiating the right-hand side with respect to x. In each case the Chain Rule applies to produce the integrand on the left.

EXAMPLE 7

$$\int \frac{1}{\cos^2 2x}\,dx = \int \sec^2 2x\,dx \qquad \left(\sec 2x = \frac{1}{\cos 2x}\right)$$

$$= \int \sec^2 u \cdot \frac{1}{2}\,du \qquad \left(\text{Substitute } u = 2x,\ du = 2\,dx,\ dx = \frac{1}{2}\,du.\right)$$

$$= \frac{1}{2}\int \sec^2 u\,du$$

$$= \frac{1}{2}\tan u + C \qquad \text{(Integrate, using Eq. (10).)}$$

$$= \frac{1}{2}\tan 2x + C \qquad \text{(Replace } u \text{ by } 2x.\text{)}$$

The Substitution Method of Integration

The substitutions we have been using in the examples are all instances of a general rule:

$$\int f(g(x)) \cdot g'(x)\,dx = \int f(u)\,du \qquad \text{1. Substitute } u = g(x),\ du = g'(x)\,dx.$$

$$= F(u) + C \qquad \text{2. Evaluate by finding an antiderivative of } f(u). \text{ (Any one will do.)}$$

$$= F(g(x)) + C. \qquad \text{3. Substitute back.}$$

These three steps are the steps of the substitution method of integration.

The Substitution Method of Integration

To evaluate the integral

$$\int f(g(x))g'(x)\,dx$$

when f and g' are continuous functions, carry out the following steps:

STEP 1: Substitute $u = g(x)$ and $du = g'(x)\,dx$ to obtain the integral

$$\int f(u)\,du.$$

STEP 2: Integrate with respect to u.

STEP 3: Replace u by $g(x)$ in the result.

EXAMPLE 8

$$\int (x^2 + 2x - 3)^2(x+1)\,dx = \int u^2 \cdot \frac{1}{2}\,du \qquad \left(\begin{array}{l}\text{Substitute } u = x^2 + 2x - 3,\\ du = 2x\,dx + 2\,dx\\ \quad = 2(x+1)\,dx,\\ \frac{1}{2}du = (x+1)\,dx.\end{array}\right)$$

$$= \frac{1}{2}\int u^2\,du$$

$$= \frac{1}{2} \cdot \frac{u^3}{3} + C = \frac{1}{6}u^3 + C \qquad \left(\begin{array}{l}\text{Integrate with}\\ \text{respect to } u.\end{array}\right)$$

$$= \frac{1}{6}(x^2 + 2x - 3)^3 + C \qquad \left(\begin{array}{l}\text{Replace } u \text{ by}\\ x^2 + 2x - 3.\end{array}\right)$$

EXAMPLE 9

$$\begin{aligned}\int \sin^4 x \cos x\, dx &= \int u^4\, du && \text{(Substitute } u = \sin u\, du = \cos x\, dx.\text{)}\\ &= \frac{u^5}{5} + C && \text{(Integrate with respect to } u.\text{)}\\ &= \frac{\sin^5 x}{5} + C && \text{(Replace } u \text{ by } \sin x.\text{)}\end{aligned}$$

There is often more than one way to make a successful substitution, as the next example shows.

EXAMPLE 10 Evaluate $\int \frac{2z\, dz}{\sqrt[3]{z^2+1}}$.

Solution We can use the substitution method of integration as an exploratory tool: Substitute for the most troublesome part of the integrand and see how things work out. For the integral here, we might try $u = z^2 + 1$ or we might even press our luck and take u to be the entire cube root. Here is what happens in each case.

SOLUTION 1: Substitute $u = z^2 + 1$.

$$\begin{aligned}\int \frac{2z\, dz}{\sqrt[3]{z^2+1}} &= \int \frac{du}{u^{1/3}} && \text{(Substitute } u = z^2 + 1, du = 2z\, dz.\text{)}\\ &= \int u^{-1/3}\, du && \text{(In the form } \int u^n\, du.\text{)}\\ &= \frac{u^{2/3}}{2/3} + C && \text{(Integrate with respect to } u.\text{)}\\ &= \frac{3}{2}u^{2/3} + C\\ &= \frac{3}{2}(z^2+1)^{2/3} + C && \text{(Replace } u \text{ by } z^2 + 1.\text{)}\end{aligned}$$

SOLUTION 2: Substitute $u = \sqrt[3]{z^2+1}$ instead.

$$\begin{aligned}\int \frac{2z\, dz}{\sqrt[3]{z^2+1}} &= \int \frac{3u^2\, du}{u} && \left(\begin{array}{l}\text{Substitute } u = \sqrt[3]{z^2+1},\\ u^3 = z^2+1, 3u^2\, du = 2z\, dz.\end{array}\right)\\ &= 3\int u\, du\\ &= 3 \cdot \frac{u^2}{2} + C && \text{(Integrate with respect to } u.\text{)}\\ &= \frac{3}{2}(z^2+1)^{2/3} + C && \text{(Replace } u \text{ by } (z^2+1)^{1/3}.\text{)}\end{aligned}$$

Substitution in Definite Integrals

The formula for evaluating definite integrals by substitution first appeared in a book by Isaac Barrow (1630–1677), Newton's mathematics teacher at Cambridge University. It looks like this:

Substitution in Definite Integrals

The Formula	How to Use It	
$\int_a^b f(g(x)) \cdot g'(x)\,dx = \int_{g(a)}^{g(b)} f(u)\,du$	Substitute $u = g(x)$, $du = g'(x)\,dx$, and integrate from $g(a)$ to $g(b)$.	**(14)**

To use the formula, make the same u-substitution you would use to evaluate the corresponding indefinite integral. Then integrate with respect to u from the value u has at $x = a$ to the value u has at $x = b$.

EXAMPLE 11

$$\int_0^{\pi/4} \tan x \sec^2 x\,dx = \int_0^1 u\,du \qquad \left(\begin{array}{l}\text{Substitute } u = \tan x,\ du = \sec^2 x\,dx, \\ \text{and integrate from } \tan 0 = 0 \\ \text{to } \tan \frac{\pi}{4} = 1.\end{array}\right)$$

$$= \left.\frac{u^2}{2}\right]_0^1 \qquad \text{(Evaluate the definite integral.)}$$

$$= \frac{(1)^2}{2} - \frac{(0)^2}{2}$$

$$= \frac{1}{2}$$

We do not have to use Eq. (14) if we do not want to. We can always transform the integral as an indefinite integral, integrate, change back to x, and use the original x limits. In the next example we evaluate a definite integral both ways—with Eq. (14) and without.

EXAMPLE 12 Evaluate $\int_{-1}^{1} 3x^2\sqrt{x^3+1}\,dx$.

Solution We have two choices here:

METHOD 1: Transform the integral and evaluate it with transformed limits.

$$\int_{-1}^{1} 3x^2\sqrt{x^3+1}\,dx = \int_0^2 \sqrt{u}\,du \qquad \left(\begin{array}{l}\text{Substitute } u = x^3 + 1, \\ du = 3x^2\,dx \text{ and integrate} \\ \text{from } u(-1) = 0 \text{ to } u(1) = 2.\end{array}\right)$$

$$= \left.\frac{2}{3}u^{3/2}\right]_0^2 \qquad \text{(Integrate.)}$$

$$= \frac{2}{3}\left[2^{3/2} - 0^{3/2}\right] = \frac{2}{3}[2\sqrt{2}] = \frac{4\sqrt{2}}{3}$$

METHOD 2: Transform the integral as an indefinite integral, integrate, change back to x, and use the original x-limits.

$$\int 3x^2\sqrt{x^3+1}\,dx = \int \sqrt{u}\,du \qquad \begin{pmatrix}\text{Substitute } u = x^3+1,\\ du = 3x^2\,dx.\end{pmatrix}$$

$$= \frac{2}{3}u^{3/2} + C \qquad \text{(Integrate with respect to } u.)$$

$$= \frac{2}{3}(x^3+1)^{3/2} + C \qquad \text{(Replace } u \text{ by } x^3+1.)$$

$$\int_{-1}^{1} 3x^2\sqrt{x^3+1}\,dx = \frac{2}{3}(x^3+1)^{3/2}\Big]_{-1}^{1} \qquad \begin{pmatrix}\text{Using the indefinite}\\ \text{integral just found}\end{pmatrix}$$

$$= \frac{2}{3}\left[((1)^3+1)^{3/2} - ((-1)^3+1)^{3/2}\right]$$

$$= \frac{2}{3}\left[2^{3/2} - 0^{3/2}\right] = \frac{2}{3}[2\sqrt{2}] = \frac{4\sqrt{2}}{3}$$

Which method is better—evaluating the transformed integral with transformed limits or transforming back to use the original limits of integration? In Example 12 the first method seems easier, but that is not always the case. As a rule, it is best to know both methods and use whichever one seems better at the time.

In the last example it is particularly useful to confirm your solution using FnInt $(f(x), x, a, b)$. For example, FnInt $(3x^2\sqrt{x^3+1}, x, -1, 1) =$ 1.88561826351. Note that the exact answer $(4\sqrt{2})/3 = 1.88561808316$, is accurate to 12 digits.

It should be noted that definite integral problems in textbooks are usually *contrived* so that the resulting arithmetic is not too bad. How would you do this problem?

Evaluate $\int_{2.1}^{5.6} 3x^2\sqrt{x^3+1}$.

You could find an antiderivative analytically and then use a *calculator* with an antiderivative to compute $2/3(x^3+1)^{3/2}\big|_{2.1}^{5.6} = 1542.869$, or you could use a calculator directly and compute FnInt $(3x^2\sqrt{x^3+1}, x, 2.1, 5.6) = 1542.869$. Which method do you prefer? They *both* require calculators!

Exercises 5.7

Evaluate the indefinite integrals in Exercises 1–10 by using the given substitutions to reduce the integrals to standard form. Support by graphing the analytic antiderivative and $y =$ FnInt $(f(t), t, 0, x)$ in the same viewing window.

1. $\int \sin 3x\,dx,\ u = 3x$

2. $\int x\sin(2x^2)\,dx,\ u = 2x^2$

3. $\int \sec 2x \tan 2x\,dx,\ u = 2x$

4. $\int \left(1 - \cos\frac{t}{2}\right)^2 \sin\frac{t}{2}\,dt,\ u = 1 - \cos\frac{t}{2}$

5. $\int 28(7x-2)^3\,dx,\ u = 7x - 2$

6. $\int 4x^3(x^4-1)^2\,dx,\ u=x^4-1$

7. $\int \frac{9r^2\,dr}{\sqrt{1-r^3}},\ u=1-r^3$

8. $\int 12(y^4+4y^2+1)^2(y^3+2y)dy,\ u=y^4+4y^2+1$

9. $\int \csc^2 2\theta \cot 2\theta\, d\theta,$

a) Using $u=\cot 2\theta$ b) Using $u=\csc 2\theta$

10. $\int \frac{dx}{\sqrt{5x}},$

a) Using $u=5x$ b) Using $u=\sqrt{5x}$

Evaluate the definite integrals in Exercises 11–18 by using the given substitutions. Support with a numerical FnInt computation.

11. $\int_0^{1/2} \frac{dx}{(2x+1)^3},\ u=2x+1$

12. $\int_0^1 \sqrt{5x+4}\,dx,\ u=5x+4$

13. $\int_0^{\pi/6} \frac{\sin 2x}{\cos^2 2x}\,dx,\ u=\cos 2x$

14. $\int_{\pi/6}^{\pi/2} \sin^2\theta\cos\theta\,d\theta,\ u=\sin\theta$

15. $\int_{-1}^{1} x\sqrt{1-x^2}\,dx,\ u=1-x^2$

16. $\int_1^4 \frac{dy}{2\sqrt{y}(1+\sqrt{y})^2},\ u=1+\sqrt{y}$

17. $\int_{-\pi/2}^{\pi/2} \frac{\cos x}{(2+\sin x)^2}\,dx,\ u=2+\sin x$

18. $\int_{\pi^2/4}^{\pi^2} \frac{\sin\sqrt{x}}{\sqrt{x}}\,dx,\ u=\sqrt{x}$

Evaluate the indefinite integrals in Exercises 19–28. Support by graphing the analytic antiderivative and $y=\text{FnInt}\,(f(t),t,0,x)$ in the same viewing window.

19. $\int \frac{dx}{(1-x)^2}$

20. $\int \frac{4y}{\sqrt{2y^2+1}}\,dy$

21. $\int \sec^2(x+2)\,dx$

22. $\int \sec^2\left(\frac{x}{4}\right)\,dx$

23. $\int 8r(r^2-1)^{1/3}\,dr$

24. $\int x^4(7-x^5)^3\,dx$

25. $\int \sec\left(\theta+\frac{\pi}{2}\right)\tan\left(\theta+\frac{\pi}{2}\right)\,d\theta$

26. $\int \sqrt{\tan x}\sec^2 x\,dx$

27. $\int \frac{6x^3}{\sqrt[4]{1+x^4}}\,dx$

28. $\int (s^3+2s^2-5s+6)^2(3s^2+4s-5)\,ds$

Evaluate the definite integrals in Exercises 29–47 by analytic means.

29. a) $\int_0^3 \sqrt{y+1}\,dy$ b) $\int_{-1}^0 \sqrt{y+1}\,dy$

30. a) $\int_0^1 r\sqrt{1-r^2}\,dr$ b) $\int_{-1}^1 r\sqrt{1-r^2}\,dr$

31. a) $\int_0^{\pi/4} \tan x\sec^2 x\,dx$ b) $\int_{-\pi/4}^0 \tan x\sec^2 x\,dx$

32. a) $\int_0^1 x^3(1+x^4)^3\,dx$ b) $\int_{-1}^1 x^3(1+x^4)^3\,dx$

33. a) $\int_0^1 \frac{x^3}{\sqrt{x^4+9}}\,dx$ b) $\int_{-1}^0 \frac{x^3}{\sqrt{x^4+9}}\,dx$

34. a) $\int_{-1}^1 \frac{x}{(1+x^2)^2}\,dx$ b) $\int_0^1 \frac{x}{(1+x^2)^2}\,dx$

35. a) $\int_0^{\sqrt{7}} x(x^2+1)^{1/3}\,dx$ b) $\int_{-\sqrt{7}}^0 x(x^2+1)^{1/3}\,dx$

36. a) $\int_0^{\pi} 3\cos^2 x\sin x\,dx$ b) $\int_{2\pi}^{3\pi} 3\cos^2 x\sin x\,dx$

37. a) $\int_0^{\pi/6} (1-\cos 3x)\sin 3x\,dx$

b) $\int_{\pi/6}^{\pi/3} (1-\cos 3x)\sin 3x\,dx$

38. a) $\int_0^{\sqrt{3}} \frac{4x}{\sqrt{x^2+1}}\,dx$ b) $\int_{-\sqrt{3}}^{\sqrt{3}} \frac{4x}{\sqrt{x^2+1}}\,dx$

39. a) $\int_0^{2\pi} \frac{\cos x}{\sqrt{2+\sin x}}\,dx$ b) $\int_{-\pi}^{\pi} \frac{\cos x}{\sqrt{2+\sin x}}\,dx$

40. a) $\int_{-\pi/2}^0 \frac{\sin x}{(3+\cos x)^2}\,dx$ b) $\int_0^{\pi/2} \frac{\sin x}{(3+\cos x)^2}\,dx$

41. $\int_0^1 \sqrt{t^5+2t}(5t^4+2)\,dt$

42. $\int_1^4 \frac{dy}{2\sqrt{y}(1+\sqrt{y})^2}$

43. $\int_0^{\pi/2} \cos^3 2x\sin 2x\,dx$

44. $\int_{-\pi/4}^{\pi/4} \tan^2 x\sec^2 x\,dx$

45. $\int_0^{\pi} \frac{8\sin t}{\sqrt{5-4\cos t}}\,dt$

46. $\int_0^{\pi/4} (1-\sin 2t)^{3/2}\cos 2t\,dt$

47. $\displaystyle\int_0^1 15x^2\sqrt{5x^3+4}\,dx$

48. Do analytically and support numerically:

$$\int_0^1 (y^3+6y^2-12y+5)(y^2+4y-4)\,dy.$$

Find the total areas of the shaded regions in Exercises 49 and 50.

49.

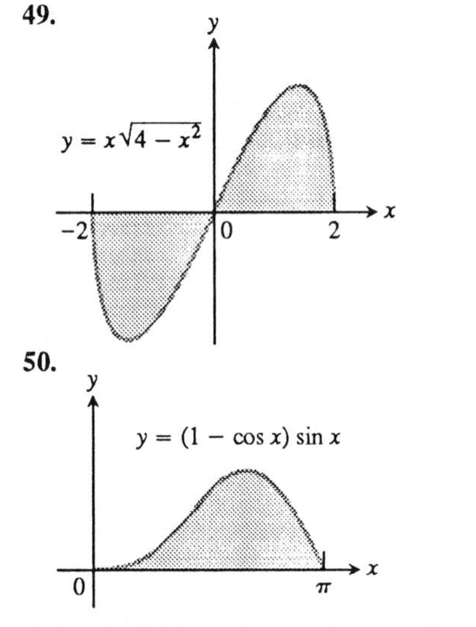

50.

Solve the initial value problems in Exercises 51–54.

51. $\dfrac{ds}{dt} = 24t(3t^2-1)^3$, $s = 0$ when $t = 0$

52. $\dfrac{dy}{dx} = 4x(x^2+8)^{-1/3}$, $y = 0$ when $x = 0$

53. $\dfrac{ds}{dt} = 6\sin(t+\pi)$, $s = 0$ when $t = 0$

54. $\dfrac{d^2s}{dt^2} = -4\sin\left(2t-\dfrac{\pi}{2}\right)$, $\dfrac{ds}{dt} = 100$ and $s = 0$ when $t = 0$

Sequences of Substitutions

If you do not know what substitution to make, try reducing the integral step by step, using a trial substitution to simplify the integral a bit, then another to simplify it some more. You will see what we mean if you try the sequences of substitutions in Exercises 55 and 56.

55. $\displaystyle\int_0^{\pi/4} \frac{18\tan^2 x\sec^2 x}{(2+\tan^3 x)^2}\,dx$

a) $u = \tan x$, followed by $v = u^3$, then by $w = 2+v$

b) $u = \tan^3 x$ followed by $v = 2+u$

c) $u = 2+\tan^3 x$

56. $\displaystyle\int \sqrt{1+\sin^2(x-1)}\,\sin(x-1)\cos(x-1)\,dx$

a) $u = x-1$, followed by $v = \sin u$, then by $w = 1+v^2$

b) $u = \sin(x-1)$ followed by $v = 1+u^2$

c) $u = 1+\sin^2(x-1)$

57. It looks as if we can integrate $2\sin x\cos x$ with respect to x in three different ways:

a) $\displaystyle\int 2\sin x\cos x\,dx = \int 2u\,du \qquad \begin{pmatrix} u = \sin x, \\ du = \cos x\,dx \end{pmatrix}$

$\qquad = u^2 + C_1 = \sin^2 x + C_1;$

b) $\displaystyle\int 2\sin x\cos x\,dx = \int -2u\,du \qquad \begin{pmatrix} u = \cos x, \\ du = -\sin x\,dx, \\ -du = \sin x\,dx \end{pmatrix}$

$\qquad = -u^2 + C_2 = -\cos^2 x + C_2;$

c) $\displaystyle\int 2\sin x\cos x\,dx = \int \sin 2x\,dx \qquad \begin{pmatrix} 2\sin x\cos x \\ = \sin 2x \end{pmatrix}$

$\qquad = -\dfrac{\cos 2x}{2} + C_3.$

Can all three integrations be correct? Graph each antiderivative with $C = 0$. Explain.

5.8 A Brief Introduction to Logarithms and Exponentials

You may recall our saying that the exponential model for bacterial growth, reproduced here from Section 4.7, is one of the really good models in science because it applies to so many phenomena people want to forecast and understand. Whenever you have a quantity y whose rate of change over time is proportional to the amount of y present, you have a function that satisfies the differential equation

$$\frac{dy}{dt} = ky.$$

If, in addition, $y = y_0$ when $t = 0$, the function is none other than the exponential function

$$y = y_0 e^{kt}.$$

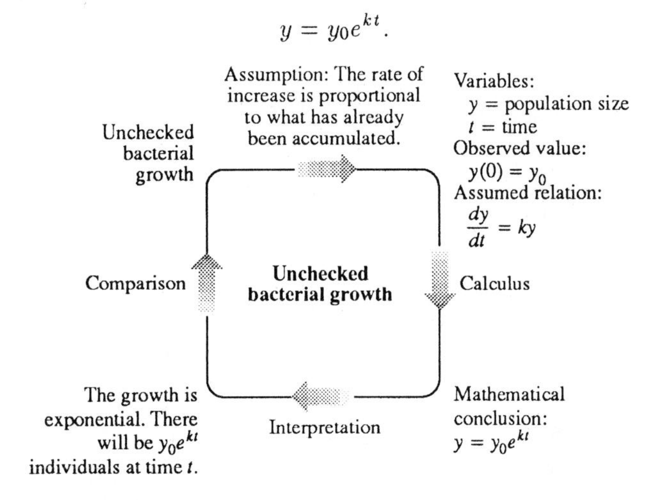

This section shows where the exponential function comes from and how it can be used to solve problems in the life sciences and in business and economics.

The key to understanding the exponential function is its inverse, the natural logarithm function, so we introduce that first. You will find detailed treatments of exponential and logarithmic functions, along with additional applications, in Chapter 7.

The Natural Logarithm Function

The function $f(x) = 1/x$ is continuous except at $x = 0$. It follows from the First Fundamental Theorem of Calculus that $f(x) = 1/x$ has antiderivatives of the form $F(x) = \int_a^x 1/t\,dt$ if both a and x are *positive* numbers. Figure 5.30 displays a family of antiderivatives of $1/x$.

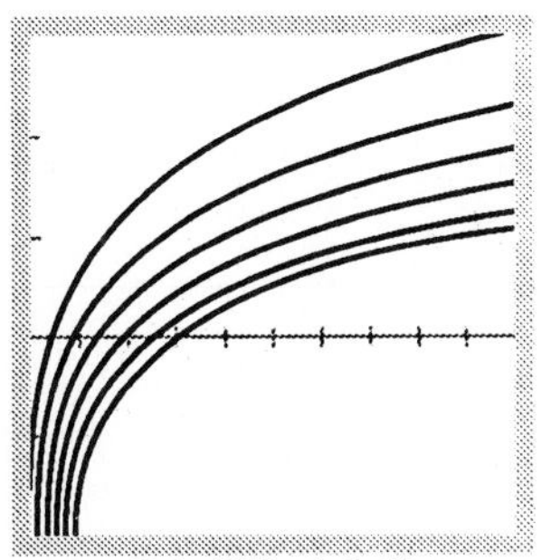

[0.1, 10] by [−2, 3]

5.30 The graphs of $y = \text{FnInt}\,(1/t, t, a, x)$ for $a = 0.5, 1, 1.5, 2, 2.5$, and 3.

> **Explore Box**
>
> Reproduce the six graphs in Fig. 5.30 by graphing $y = \text{FnInt}\,(1/t, t, a, x)$ for $a = 0.5, 1, 1.5, 2, 2.5$, and 3.
>
> **Note:** Because of the infinite discontinuity at $x = 0$, set $x\text{Min} = 0.1$ in RANGE before graphing.

For $x > 0$, the antiderivative $\int_1^x 1/t\,dt$ of $1/x$ shown in Fig. 5.30 has extremely special properties and is called the "natural logarithm" of the positive number x (written as $\ln x$).

Definition

The Natural Logarithm Function

$$\ln x = \int_1^x \frac{1}{t} dt, \qquad x > 0 \tag{1}$$

If $x > 1$, then $\ln x$ is the area under the curve $y = 1/t$ from $t = 1$ to $t = x$ (Fig. 5.31). If x is less than 1 (but still positive), $\ln x$ gives the negative of the area under the curve from $t = x$ to $t = 1$. The function is not defined for $x \leq 0$. The natural logarithm of 1 itself is zero because

$$\ln 1 = \int_1^1 \frac{1}{t}\, dt = 0. \qquad \text{(Upper and lower limits equal)} \tag{2}$$

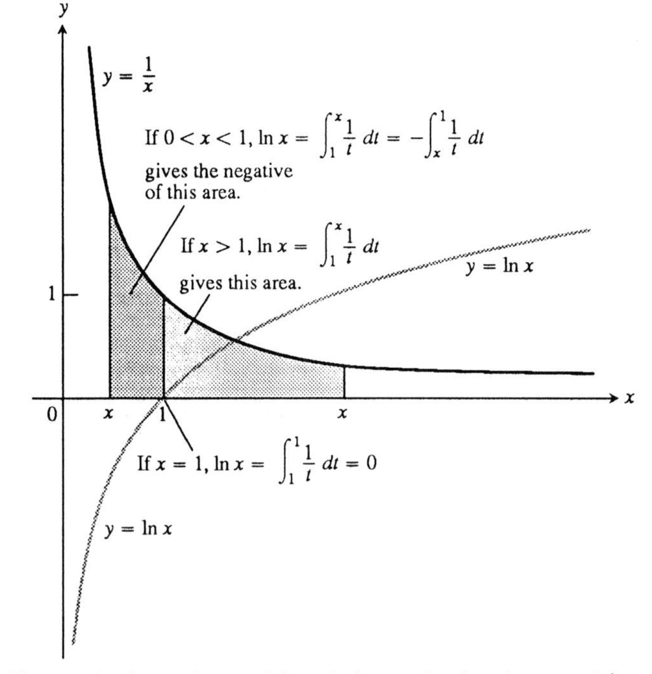

5.31 The graph of $y = \ln x$ and its relation to the function $y = 1/x$, $x > 0$. The graph rises above the x-axis as x moves from 1 to the right and falls below the axis as x moves from 1 to the left.

Notice that we show the graph of $y = 1/x$ in Fig. 5.31 but use the function $y = 1/t$ in the integral. Using x for everything would have us writing

$$\ln x = \int_1^x \frac{1}{x}\, dx. \tag{3}$$

This has too many x's to make sense, so we changed the dummy variable to t.

Historically the importance of logarithms came from the improvement they brought to arithmetic. The revolutionary properties of logarithms made possible the calculations behind the great seventeenth-century advances in off-shore navigation and celestial mechanics. Logarithms are still useful today, but mainly for reasons other than computations. Today calculators do the arithmetic of logarithms far better than we could in the past.

Rules for Logarithms

The following rules hold for any positive numbers a and x and for any exponent n.

1. $\ln ax = \ln a + \ln x$
2. $\ln \frac{a}{x} = \ln a - \ln x$
3. $\ln \frac{1}{x} = -\ln x$ (Rule 2 with $a = 1$)
4. $\ln x^n = n \ln x$

EXAMPLE 1

$$\ln 6 = \ln(2 \cdot 3) = \ln 2 + \ln 3 \qquad \text{(Rule 1)}$$

$$\ln 4 - \ln 5 = \ln\left(\frac{4}{5}\right) = \ln 0.8 \qquad \text{(Rule 2)}$$

$$\ln \frac{1}{8} = -\ln 8 \qquad \text{(Rule 3)}$$

$$= -\ln 2^3 = -3 \ln 2 \qquad \text{(Rule 4)}$$

The Derivative of $y = \ln x$

Explore Box

Let $F(x) = \text{FnInt}\ (1/t, t, 1, x)$. Graph $y = \text{NDer}\ (F(x), x)$ in the $[0.1, 5]$ by $[0, 5]$ window and compare it with the graph of $y = 1/x$. Use TRACE and compare function values. What can you conclude?

By the First Fundamental Theorem of Calculus, Section 5.5,

$$\frac{d}{dx} \ln x = \frac{d}{dx} \int_1^x \frac{1}{t}\, dt = \frac{1}{x}.$$

For every positive value of x, therefore,

$$\frac{d}{dx}\ln x = \frac{1}{x}. \tag{4}$$

If u is a differentiable function of x whose values are positive, so that $\ln u$ is defined, then applying the Chain Rule

$$\frac{dy}{dx} = \frac{dy}{du}\cdot\frac{du}{dx}$$

to the function $y = \ln u$ gives

$$\frac{d}{dx}\ln u = \frac{d}{du}\ln u \cdot \frac{du}{dx} = \frac{1}{u}\frac{du}{dx}. \tag{5}$$

$$\frac{d}{dx}\ln u = \frac{1}{u}\frac{du}{dx}, \qquad u > 0 \tag{6}$$

EXAMPLE 2

$$\frac{d}{dx}\ln 2x = \frac{1}{2x}\frac{d}{dx}(2x) = \frac{1}{2x}(2) = \frac{1}{x}$$

Notice the remarkable occurrence in Example 2. The function $y = \ln 2x$ has the same derivative as the function $y = \ln x$. This is true of $y = \ln kx$ for any constant k:

$$\frac{d}{dx}\ln kx = \frac{1}{kx}\cdot\frac{d}{dx}(kx) = \frac{1}{kx}(k) = \frac{1}{x}. \tag{7}$$

EXAMPLE 3 Equation (6) with $u = x^2 + 3$ gives

$$\frac{d}{dx}\ln(x^2+3) = \frac{1}{x^2+3}\cdot\frac{d}{dx}(x^2+3) = \frac{1}{x^2+3}\cdot 2x = \frac{2x}{x^2+3}.$$

The Integral $\int \frac{1}{u}\,du,\ u > 0$

Equation (6) leads to the integral formula

$$\int \frac{1}{u}\,du = \ln u + C, \qquad u > 0. \tag{8}$$

We shall see what to do when $u < 0$ when we get to Chapter 7. It is not really a problem—we just want to spend our time on something else right now.

EXAMPLE 4

$$\begin{aligned}
\int \frac{2x}{x^2+3}\,dx &= \int \frac{du}{u} && \text{(Substitute } u = x^2+3,\ du = 2x\,dx.\text{)} \\
&= \ln u + C && \text{(Eq. 8)} \\
&= \ln(x^2+3) + C && \text{(Replace } u \text{ by } x^2+3.\text{)}
\end{aligned}$$

EXAMPLE 5

$$\begin{aligned}
\int_0^3 \frac{2x}{x^2+1}\,dx &= \ln(x^2+3)\Big]_0^3 && \begin{pmatrix}\text{Antiderivative} \\ \text{from Example 4}\end{pmatrix} \\
&= \ln(3^2+3) - \ln(0+3) \\
&= \ln 12 - \ln 3 && \text{(OK answer)} \\
&= \ln \frac{12}{3} && \text{(Rule 2)} \\
&= \ln 4 && \text{(Equivalent to } \ln 12 - \ln 3\text{)}
\end{aligned}$$

The Exponential Function exp(x)

As you may recall, a function $y = f(x)$ is one-to-one if different input values for x always give different output values for y. The function $y = x$ is one-to-one because different x-values always give different y-values. The function $y = \sin x$ is not one-to-one because different x-values sometimes give the same y-value. The sines of 0 and π both have the value 0, even through 0 and and π are not the same number.

The function $y = \ln x$ is one-to-one because its derivative, $y' = 1/x$, is positive at every point in the function's domain (Corollary 1 of the Mean Value Theorem, Section 4.1). The natural logarithm function therefore has an inverse. We call the inverse the exponential function of x, abbreviated $\exp(x)$.

Definition

> The function $y = \exp(x)$, **the exponential function of** x, is the inverse of the natural logarithm function, $y = \ln x$.

When functions are inverses of one another, their composites in either order give the identity function. For the exponential and natural logarithm, this means that

$$\exp(\ln x) = x, \quad x > 0, \tag{9}$$

$$\ln(\exp(x)) = x, \quad \text{all } x. \tag{10}$$

We shall have more to say about these equations in a while.

The Graph of $y = \exp(x)$

In a sense, the graph of $y = \ln x$ is already a graph of $y = \exp(x)$, with the xy-pairs plotted in reverse. The usual graph of the exponential function, however, is obtained by reflecting the graph of $y = \ln x$ across the line $y = x$ to turn the pairs around the right way. The reflection interchanges the x and y to place the independent variable along the horizontal axis in the normal way (Fig. 5.32).

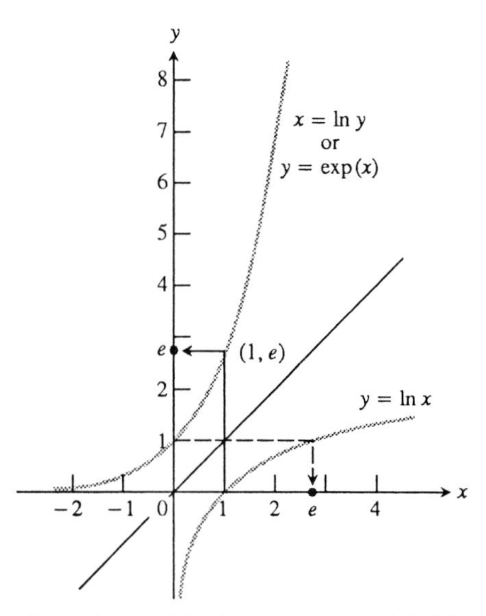

5.32 The graphs of $y = \ln x$ and its inverse $y = \exp(x)$. The number e is the one number whose natural logarithm is 1.

The Number *e*

The number e is the *one* number whose natural logarithm is 1. That is, e is the solution to the equation $\ln x = \int_1^x 1/t\, dt = 1$. Thus, $\ln e = 1$. This also means $\exp(1) = e$ by the inverse property (Figure 5.32). Although e is not a rational number, it is possible to find its value with a computer to as many decimal places as we please by using the formula

$$e = \lim_{n \to \infty} \left(2 + \frac{1}{2} + \frac{1}{6} + \cdots + \frac{1}{n!} \right). \tag{11}$$

To fifteen places,

$$e = 2.7\,1828\,1828\,45\,90\,45.$$

(This layout makes the number easier to remember.)

We will revisit this formula and its justification in a later chapter.

The number e is one of the most amazing numbers in all of mathematics. It occurs naturally in many different applications from business, science and pure mathematics. One of the most fundamental properties of the number e involves the derivative of the exponential function $y = a^x$. The exponential function ($a > 0$) is continuous, nonnegative, and has domain the set of real numbers. The domain issue is a bit tricky (e.g. just what is $3^{\sqrt{2}}$?).

We pose this basic question. Is there a solution to this differential equation

$$D_x(a^x) = a^x?$$

That is, is there a value of $a > 0$ so that $y' = y$ where $y = a^x$. We can approach this question graphically as suggested by the following Explore activity.

Explore Box

Let $y1 = a^x$ and $y2 =$ NDer $(y1, x)$. Graph both $y1$ and $y2$ for $a =$ 2.4, 2.6, 2.8, 3, 3.2 and 3.4 in the window $[-1, 3]$ by $[-1, 10]$. What value of a do you estimate to be the solution to $D_x(a^x) = a^x$?

By guess and graphical check you can verify that the graphs of $y = 2.72^x$ and $y =$ NDer $(2.27^x, x)$ are identical in all but very small windows. This strongly suggests that $e = 2.718281\ldots$ *is* the desired solution to $D_x(a^x) = a^x$. This means $y = e^x$ is a non-trivial function that is its own derivative. We will come back to this amazing property of $y = e^x$ soon in a more formal manner.

Equation (12) provides a way to complete the definition of e^x to include irrational values of x. The function $\exp(x)$ is defined for all x, so we can use it to give a value to e^x at points where e^x had no previous value.

Definition

For every real number x, $\quad e^x = \exp(x)$. **(12)**

EXAMPLE 6 You can find values of e^x on most scientific calculators to eight or ten digits by entering x and pressing [e^x] or [INV] [$\ln x$]. Typical values (rounded) are:

$$e^{-1} = 0.3679 \qquad e^2 = 7.3891$$

$$e^{10} = 22026 \qquad e^{100} = 2.6881 \times 10^{43}.$$

The values of e^x grow very rapidly as $x \to \infty$, another point we shall return to in Section 7.6.

Equations Involving $\ln x$ and e^x

When we replace $\exp(x)$ by e^x in Eqs. (9) and (10), we get the two most important rules for combining $\ln x$ and e^x.

For all positive x :	$e^{\ln x} = x$	**(13)**
For all x :	$\ln(e^x) = x$	**(14)**

The statements in Example 7 can be verified by applying the inverse properties (Eqs. (9) and (10)) of logarithmic and exponential functions.

EXAMPLE 7

a) $\ln e^2 = 2$ **b)** $\ln e^{-1} = -1$

c) $\ln \sqrt{e} = \frac{1}{2}$ **d)** $\ln e^{\sin x} = \sin x$

e) $e^{\ln 2} = 2$ **f)** $e^{\ln(x^2+1)} = x^2 + 1$

g) $e^{3 \ln 2} = e^{\ln 2^3} = e^{\ln 8} = 8$ (One way)

h) $e^{3 \ln 2} = (e^{\ln 2})^3 = 2^3 = 8$ (Another way)

EXAMPLE 8 Find y if $\ln y = 3t + 5$.

Solution Exponentiate:

$$e^{\ln y} = e^{3t+5} \qquad \text{(Eq. 13)}$$

$$y = e^{3t+5}$$

EXAMPLE 9 Find k if $e^{2k} = 10$.

Solution Take the logarithm of both sides:

$$e^{2k} = 10$$

$$\ln e^{2k} = \ln 10 \qquad \text{(Eq. 14)}$$

$$2k = \ln 10$$

$$k = \frac{1}{2} \ln 10$$

Two Useful Operating Rules

1. To remove logarithms from an equation, exponentiate both sides.
2. To remove exponentials, take the logarithm of both sides.

As we shall see in Chapter 7, the function e^x obeys the familiar rules of exponents from algebra.

Rules of Exponents

1. For any number x: $$e^{-x} = \frac{1}{e^x} \qquad (15)$$

2. For any x_1 and x_2: $$e^{x_1+x_2} = e^{x_1} \cdot e_2^x \qquad (16)$$

EXAMPLE 10

$$e^{-\ln 2} = \frac{1}{e^{\ln 2}} = \frac{1}{2} \qquad \text{(Eq. 15)}$$

$$e^{4t+\ln 5} = e^{4t} \cdot e^{\ln 5} \qquad \text{(Eqs. 16 and 13)}$$

$$= e^{4t} \cdot 5 = 5e^{4t}$$

The Derivative of $y = e^x$

As it turns out, the function $y = e^x$ is not only continuous at every value of x but also differentiable at every value of x. Although we shall have to wait until Chapter 7 to see exactly why this is so, we can calculate the derivative in a straightforward way right now. We assume it is differentiable.
Starting with $y = e^x$, we have, in order,

$$y = e^x$$

$$\ln y = \ln e^x \qquad \text{(Take logarithms of both sides)}$$

$$\ln y = x \qquad (\text{Because } \ln e^x = x)$$

$$\frac{d}{dx} \ln y = \frac{d}{dx} x \qquad \left(\begin{array}{l}\text{Implicit differentiation—assumes} \\ e^x \text{ to be differentiable}\end{array}\right)$$

$$\frac{1}{y}\frac{dy}{dx} = 1 \qquad (\text{Eq. (6) with } u = y)$$

$$\frac{dy}{dx} = y$$

$$\frac{dy}{dx} = e^x \qquad (\text{Replace } y \text{ by } e^x.)$$

The conclusion we draw from this sequence of equations is that the function $y = e^x$ is its own derivative. Never before have we encountered such a function. No matter how many times we differentiate it, we always get the function back. Constants times e^x are the only other functions to behave this way.

$$\frac{d}{dx} e^x = e^x \qquad (17)$$

EXAMPLE 11

$$\frac{d}{dx}(5e^x) = 5\frac{d}{dx}e^x = 5e^x$$

The Chain Rule extends Eq. (17) in the usual way to a more general form.

If u is any differentiable function of x, then

$$\frac{d}{dx}e^u = e^u\frac{du}{dx}. \qquad \textbf{(18)}$$

EXAMPLE 12 Equation (18) with $u = -x$:

$$\frac{d}{dx}e^{-x} = e^{-x}\frac{d}{dx}(-x) = e^{-x}(-1) = -e^{-x}.$$

EXAMPLE 13 Equation (18) with $u = \sin x$:

$$\frac{d}{dx}e^{\sin x} = e^{\sin x}\frac{d}{dx}(\sin x) = e^{\sin x} \cdot \cos x.$$

The integral-formula equivalent of Eq. (18) is

$$\int e^u\,du = e^u + C. \qquad \textbf{(19)}$$

EXAMPLE 14

$$\int_0^{\ln 2} e^{3x}\,dx = \int_0^{3\ln 2} e^u \cdot \frac{1}{3}du \qquad \left(\begin{array}{l} u = 3x,\ du = 3dx,\ \frac{1}{3}\,du = dx, \\ u(0) = 0,\ u(\ln 2) = 3\ln 2 \end{array}\right)$$

$$= \frac{1}{3}\int_0^{3\ln 2} e^u\,du = \frac{1}{3}e^u\Big]_0^{3\ln 2} = \frac{1}{3}[e^{3\ln 2} - e^0] = \frac{1}{3}[8-1] = \frac{7}{3}$$

EXAMPLE 15

$$\int_0^{\pi/2} e^{\sin x}\cos x\,dx = e^{\sin x}\Big]_0^{\pi/2} \qquad \text{(Antiderivative from Example 13)}$$

$$= e^{\sin(\pi/2)} - e^{\sin 0} = e^1 - e^0 = e - 1$$

The Law of Exponential Change

In many instances in biology and economics, some positive quantity y grows or decreases at a rate that at any given time t is proportional to the amount that is present. If we also know the initial amount y_0 at time $t = 0$, we can find y by solving the initial value problem

Differential equation: $\quad \dfrac{dy}{dt} = ky \qquad (20)$

Initial condition: $\quad y = y_0$ when $t = 0$.

The constant k is positive if y is increasing and negative if y is decreasing.

To solve Eq. (20), we divide through by y to get

$$\frac{1}{y}\frac{dy}{dt} = k$$

and integrate both sides with respect to t to get

$$\ln y = kt + C. \tag{21}$$

We then solve this for y by exponentiating:

$$e^{\ln y} = e^{kt+C}$$

$$y = e^{kt} \cdot e^{C}$$

$$y = Ae^{kt} \qquad \text{(Write } A \text{ for } e^{C}.)$$

We find the value of A from the initial condition:

$$y_0 = Ae^{k(0)} = A \cdot 1 = A.$$

The solution of the initial value problem is $y = y_0 e^{kt}$. We call this equation the Law of Exponential Change.

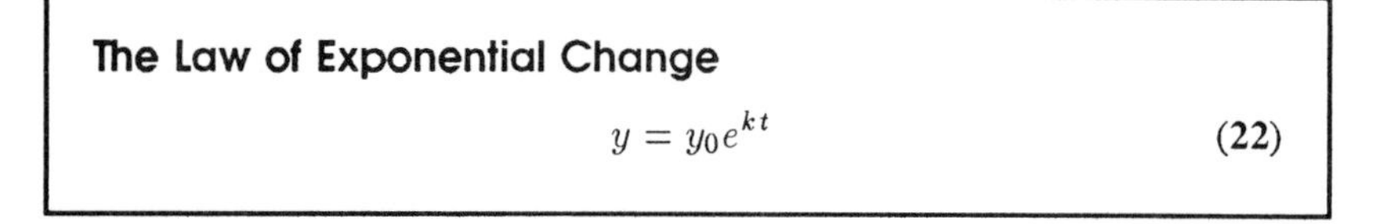

The Law of Exponential Change

$$y = y_0 e^{kt} \tag{22}$$

EXAMPLE 16 The growth of a cell. In an ideal environment, the mass m of a cell will grow exponentially, at least early on. Nutrients pass quickly through the cell wall, and growth is limited only by the metabolism within the cell, which in turn depends on the mass of participating molecules. If we make the reasonable assumption that, at each instant of time, the cell's growth rate dm/dt is proportional to the mass that has already been accumulated, then

$$\frac{dm}{dt} = km \qquad \text{and} \qquad m = m_0 e^{kt}.$$

There are limitations, of course, and in any particular case we would expect this equation to provide reliable information only for values of m below a certain size. ≡

EXAMPLE 17 Birth rates and population growth. Strictly speaking, the number of individuals in a population (of people, plants, foxes, or whatever) is a discontinuous function of time because it takes on discrete values. However, as soon as the number of individuals becomes large enough, it may safely be described with a continuous and even differentiable function. If we assume that the proportion of reproducing individuals remains constant and assume a constant fertility, then at any instant t the birth rate is proportional to the number $y(t)$ of individuals present. If, further, we neglect departures, arrivals, and deaths, the growth rate dy/dt will be the same as the birth rate ky. In other words,

$$\frac{dy}{dt} = ky.$$

Once again, we find that $y = y_0 e^{kt}$. ■

The Incidence of Disease

One model for the way diseases spread assumes that the rate dy/dt at which the number of infected people changes is proportional to the number y itself. The more infected people there are, the faster the disease will spread. The fewer there are, the slower it will spread. Once again,

$$y = y_0 e^{kt}.$$

EXAMPLE 18 In the course of any given year, the number y of cases of a disease is reduced by 20%. If there are 10,000 cases today, how many years will it take to reduce the number of cases to 1000?

Solution The equation we use is $y = y_0 e^{kt}$, and there are three things to find:

1. The value of y_0.
2. The value of k,
3. The value of t that makes $y = 1000$.

STEP 1: *The value of* y_0. We are free to count time from anywhere we want. If we start counting from today, then $y = 10,000$ when $t = 0$, so $y_0 = 10{,}000$. Our equation becomes

$$y = 10{,}000e^{kt}.$$

STEP 2: *The value of* k. When $t = 1$, the number of cases will be 80% of its present value, or 8000. Hence

$$\begin{aligned} 10{,}000e^{k(1)} &= 8000 \\ e^k &= 0.8 \\ \ln e^k &= \ln 0.8 \\ k &= \ln 0.8. \end{aligned}$$

At any given time t, therefore,

$$y = 10{,}000e^{(\ln 0.8)t}.$$

Two down and one to go.

STEP 3: *The value of t that makes* $y = 1000$. Set y equal to 1000 and solve for t:

$$10{,}000e^{(\ln 0.8)t} = 1000$$

$$e^{(\ln 0.8)t} = 0.1 \qquad \text{(Divide by 10,000.)}$$

$$(\ln 0.8)t = \ln 0.1 \qquad \text{(Take logs of both sides.)}$$

$$t = \frac{\ln 0.1}{\ln 0.8} = 10.32. \qquad \text{(With a calculator, rounded)}$$

It will take a little more than 10 years to reduce the number of cases to 1000. ≡

Continuously Compounded Interest

If you invest an amount A_0 of money at a fixed annual interest rate r and interest is added to your account k times a year, it turns out that the amount of money you will have at the end of t years is

$$A_1 = A_0\left(1 + \frac{r}{k}\right)^{kt}. \tag{23}$$

The money might be added ("compounded," bankers say) monthly ($k = 12$), weekly ($k = 52$), daily ($k = 365$), or even more frequently, say by the hour or by the minute. But there is still a limit to how much you will earn that way, and the limit is

$$\lim_{k\to\infty} A_t = \lim_{k\to\infty} A_0\left(1 + \frac{r}{k}\right)^{kt} = A_0e^{rt}. \tag{24}$$

(See Section 7.5.)

The resulting formula for the amount of money in your account after t years is $A(t) = A_0e^{rt}$.

> **The Continuous Compound Interest Formula**
>
> $$A(t) = A_0e^{rt} \tag{25}$$
>
> Interest paid according to this formula is said to be **compounded continuously.** The number r is called the **continuous interest rate.**

EXAMPLE 19 Suppose you deposit \$621 in a bank account that pays 6% compounded continuously. How much money will you have in the account 8 years later?

Solution We use Eq. (25) with $A_0 = 621, r = 0.06$, and $t = 8$:

$$A(8) = 621e^{(0.06)(8)} = 621e^{0.48} = 1003.58 \qquad \text{(Calculator, nearest cent)}$$

Had the bank paid interest quarterly ($k = 4$ in Eq. (23)), the amount in your account would have been \$1000.01. Thus the effect of continuous compounding, as compared with quarterly compounding, has been an addition of \$3.57. A bank might decide it would be worth this additional amount to be able to advertise, "We compound your money every second, night and day—better than that, we compound the interest continuously." ≡

The Rules of 70 and 72

How long does it take to double money at a continuous interest rate of 7%? To find out, we set $A_0e^{0.07t}$ equal to $2A_0$ and solve for t:

$$A_0e^{0.07t} = 2A_0$$

$$e^{0.07t} = 2 \qquad \text{(Divide out } A_0\text{.)}$$

$$\ln e^{0.07t} = \ln 2 \qquad \text{(Take logs of both sides.)}$$

$$0.07t = \ln 2$$

$$t = \frac{\ln 2}{0.07} \approx \frac{0.693}{0.07} = \frac{69.3}{7} \approx 10. \qquad \text{(Calculator)}$$

It takes about 10 years for money to double at 7%.

A similar calculation using i in place of 7 tells us that it takes about

$$t = \frac{69.3}{i}$$

years for money to double at i%.

The number 69.3 is not convenient for mental arithmetic, so most people use 70 instead. This gives rise to the so-called Rule of 70.

The Rule of 70

It takes about $y = 70/i$ years for money to double at i percent.

The Rule of 70 gives quick results that are usually accurate enough for the work at hand. Some people prefer to use a "Rule of 72" because 72 has more divisors than 70.

The Rule of 72

It takes about $y = 72/i$ years for money to double at i percent.

The Rules of 70 and 72 can be used "backward" to estimate the interest rate needed to double money in a given period of time.

EXAMPLE 20 What interest rate do you need to double your money in 6 years.

Solution

$$\text{Rule of 72}: \quad i = \frac{72}{y} = \frac{72}{6} = 12\%$$

$$\text{Rule of 70}: \quad i = \frac{70}{y} = \frac{70}{6} = 11\frac{2}{3}\%$$

Either way, it takes a rate of about 12% to double money in 6 years. ∎

Strictly speaking, the Rules of 70 and 72 refer to continuous compounding and not, say, to annual or quarterly compounding. As Example 19 shows, however, the results of the different frequencies of compounding are much the same, and the Rules of 70 and 72 have good predictive value for any of the frequencies currently in use.

The Consumer Price Index

As you know, prices often change (they usually go up) and it is important to know what things will cost in the years ahead. The economists in the U.S. Department of Labor measure the cost of living with a number called the **consumer price index (CPI).** This index is a weighted average of the costs of food, clothing, housing, transportation, medical care, personal care, and entertainment.

The current index, set at 100 (an arbitrary choice for convenient arithmetic) in 1984, was 112.1 in March of 1987 and 116.5 in March of 1988. What you could buy for \$1.00 in 1984 cost roughly \$1.12 in 1987 and \$1.17 in 1988.

One of the predictors of price change assumes that the rate dp/dt at which the CPI changes is proportional to p, so that

$$\frac{dp}{dt} = kp. \tag{26}$$

The constant k is called the **continuous rate of inflation.** In newspapers k is usually expressed as a percent. If $k = 0.04$, for example, the U.S. Bureau of Labor Statistics reports an inflation rate of 4%. The solution of Eq. (26) is

$$p(t) = p_0 e^{kt}. \tag{27}$$

If the CPI is p_0 at time $t = 0$, then t years later it will be $p(t) = p_0 e^{kt}$.

EXAMPLE 21 The CPI in March 1988 was 116.5. What will its value be 10 years later, in March 1998, if the inflation rate is a constant 4%?

Solution We take $p_0 = 116.5$ and $k = 0.04$ in Eq. (27) and find $p(10)$:

$$\begin{aligned} p(10) &= 116.5e^{(0.04)(10)} = 116.5e^{0.4} \\ &= (116.5)(1.49) = 173.6. \qquad \text{(Calculator, rounded)} \end{aligned}$$

What conclusions can we draw from this? One conclusion is that from 1988 to 1998 prices will rise

$$\frac{173.6 - 116.5}{116.5} \times 100 = 49\%.$$

Another is that by 1998, relative to March 1984, when the index was 100, prices will have risen

$$\frac{173.6 - 100}{100} \times 100 = 73.6\%.$$

(Now you see why the index is set at 100 in the base year. Subtracting 100 from any later value automatically gives the percentage change.)

Still another conclusion is that by 1998 it will cost nearly $1.74 to buy what $1.00 bought in 1984.

EXAMPLE 22 About how many years will it take the CPI to increase 50% if the inflation rate is a steady 4%?

Solution We want to know when $p_0e^{0.04t}$ will equal $1.5p_0$, so we set these equal and solve for t:

$$\begin{aligned} p_0e^{0.04t} &= 1.5p_0 \\ e^{0.04t} &= 1.5 \\ \ln e^{0.04t} &= \ln 1.5 \\ 0.04t &= \ln 1.5 \\ t &= \frac{\ln 1.5}{0.04} \approx \frac{0.405}{0.04} \approx 10.1. \end{aligned}$$

With 4% inflation, it will take a little more than 10 years for the cost of living to increase 50%.

The Purchasing Power of the Dollar

The higher the consumer price index, the less a dollar will buy, and economists call the number $100/p$ the **purchasing power of the dollar.**

In 1984, when the CPI was 100, the purchasing power of the dollar was $100/100 = 1.00$. Two years later, in 1986, it was $100/112.1 = 0.89$, and a year after that it was $100/116.5 = 0.86$.

EXAMPLE 23 Assuming the 4% inflation rate of Example 21, what will the purchasing power of the dollar be in 1998?

Solution We divide the CPI for 1998 into 100:

$$\frac{100}{1998 \text{ CPI}} = \frac{100}{173.6} = 0.58. \qquad \text{(CPI value from Example 21, rounded)}$$

This represents a change of -0.42, or a loss in purchasing power of 42% since 1984.

Exercises 5.8

The logarithms in Exercises 1–8 can all be expressed in terms of $\ln 2$ and $\ln 3$. For example, $\ln 1.5 = \ln(3/2) = \ln 3 - \ln 2$. See if you can do the others.

1. $\ln 4/9$ **2.** $\ln 12$

3. $\ln(1/2)$ **4.** $\ln(1/3)$

5. $\ln 4.5$ **6.** $\ln \sqrt[3]{9}$

7. $\ln 3\sqrt{2}$ **8.** $\ln \sqrt{13.5}$

Use the fact that the functions $y = \ln x$ and $y = e^x$ are inverses of each other to simplify the expressions in Exercises 9–14.

9. $e^{\ln 7}$ **10.** $e^{-\ln 7}$

11. $\ln e^2$ **12.** $e^{3\ln 2}$

13. $e^{2+\ln 3}$ **14.** $e^{-2\ln 3}$

Find dy/dx in Exercises 15–24. Support graphically.

15. $y = \ln(x^2)$ **16.** $y = (\ln x)^2$

17. $y = \ln(1/x)$ **18.** $y = \ln(10/x)$

19. $y = \ln(x+2)$ **20.** $y = \ln(2x+2)$

21. $y = \ln(2 - \cos x)$ **22.** $y = \ln(x^2+1)$

23. $y = \ln(\ln x)$ **24.** $y = x\ln x - x$

Find dy/dx in Exercises 25–34. Support graphically.

25. $y = 2e^x$ **26.** $y = e^{2x}$

27. $y = e^{-x}$ **28.** $y = e^{-5x}$

29. $y = e^{2x/3}$ **30.** $y = e^{-x/4}$

31. $y = xe^2 - e^x$ **32.** $y = x^2e^x - xe^x$

33. $y = e^{\sqrt{x}}$ **34.** $y = e^{(x^2)}$

Evaluate the integrals in Exercises 35–42 by using the given substitutions. Check using an FnInt computation.

35. $\int_0^3 \frac{1}{x+1}dx, u = x+1$

36. $\int_0^4 \frac{2x\,dx}{x^2+9}, u = x^2+9$

37. $\int_{\ln 3}^{\ln 5} e^{2x}\,dx, u = 2x$

38. $\int_0^{\ln 2} e^{-x}\,dx, u = -x$

39. $\int_0^1 (1+e^x)e^x\,dx, u = 1+e^x$

40. $\int_e^{e^2} \frac{dx}{x\ln x}, u = \ln x$

41. $\int_1^4 \frac{e^{\sqrt{x}}\,dx}{2\sqrt{x}}, u = \sqrt{x}$

42. $\int_0^\pi \frac{\sin x}{2-\cos x}dx, u = 2 - \cos x$

Evaluate the integrals in Exercises 43–50. Check using an FnInt computation.

43. $\int_1^{e^2} \frac{1}{x}\,dx$ **44.** $\int_1^e \frac{2}{x}\,dx$

45. $\int_{\ln 2}^{\ln 3} e^x\,dx$ **46.** $\int_{-1}^1 e^{(x+1)}\,dx$

47. $\int_2^4 \frac{dx}{x+2}$ **48.** $\int_{-1}^0 \frac{8\,dx}{2x+3}$

49. $\int_{-1}^1 2xe^{-x^2}\,dx$ **50.** $\int_0^1 \frac{x\,dx}{4x^2+1}$

Probability In Exercises 51–56 the function f is the Standard Normal Probability distribution and is given by the formula $f(x) = (1/\sqrt{2\pi})e^{(-x^2/2)}$ where x is measured in σ-units (standard deviation units) from the mean $\mu = 0$.

51. Draw a complete graph of f.

52. What is the domain of f?

53. Draw a complete graph of $F(x) = \int_{-5}^x f(t)\,dt$ for $-5 \le x \le 5$.

54. Draw a complete graph of $F(x) = \int_0^x f(t)\,dt$ for $x \ge 0$.

55. Draw a complete graph of $G(x) = \int_x^0 f(t)\,dt$ for $x \le 0$.

56. We will later formally define the *improper* integral

$$\int_{-\infty}^{\infty} f(x)\,dx = \lim_{x\to-\infty}\int_x^0 f(t)\,dt + \lim_{x\to\infty}\int_0^x f(t)\,dt.$$

Based on your graphs in Exercises 54 and 55, what is your conjecture for the value of $\int_{-\infty}^{\infty} f(x)\,dx$?

57. Assume the probabilities of a certain event are normally distributed. Then the probability of an observation being within one standard deviation from the mean is $\int_{-1}^1 f(t)\,dt$. Compute this probability.

58. Compute the probability of an observation being within two standard deviations from the mean.

59. Compute the probability of an observation being within three standard deviations from the mean.

In Exercises 60–65, solve for k.

60. $e^{2k} = 4$

61. $e^{5k} = \frac{1}{4}$

62. $100e^{10k} = 200$

63. $100e^{k} = 1$

64. $2^{k+1} = 3^k$

65. $e^{-k^2} = k^2$

In Exercises 66–71, solve for t.

66. $e^t = 1$

67. $e^{kt} = \frac{1}{2}$

68. $e^{-0.3t} = 27$

69. $e^{-0.01t} = 1000$

70. $2^{e^t} = 2 - t$

71. $e^{-2^t} = t + 2$

In Exercises 72–77, solve for y.

72. $\ln y = 2t + 4$

73. $\ln y = -t + 5$

74. $\ln(y - 40) = 5t$

75. $\ln(1 - 2y) = t$

76. $5 + \ln y = 2^{x^2+1}$

77. $\ln(2^y - 1) = x^2 - 3$

78. Solve for x: $\displaystyle\int_0^x \frac{1}{\sqrt{2\pi}} e^{(-t^2/2)}\, dt = 0.3$

79. Solve for x: $\displaystyle\int_1^x 2^t \ln t\, dt = 0.1$

80. CHOLERA BACTERIA Suppose that the bacteria in a colony can grow unchecked, by the law of exponential change. The colony starts with 1 bacterium and doubles every half hour. How many bacteria will the colony contain at the end of 24 hr? (Under favorable laboratory conditions, the number of cholera bacteria can double every 30 min. In an infected person, many bacteria are destroyed, but this example helps explain why a person who feels well in the morning may be dangerously ill by evening.)

81. GROWTH OF BACTERIA A colony of bacteria is grown under ideal conditions in a laboratory so that the population increases exponentially with time. At the end of 3 hr there are 10,000 bacteria. At the end of 5 hr there are 40,000. How many bacteria were present initially?

82. THE INCIDENCE OF A DISEASE (CONTINUATION OF EXAMPLE 18) Suppose that in any given year the number of cases can be reduced by 25% instead of 20%.

a) How long will it take to reduce the number of cases to 1000?

b) How long will it take to eradicate the disease, that is, reduce the number of cases to less than 1?

83. JOHN NAPIER'S QUESTION John Napier, who invented natural logarithms, was the first person to answer the question, what happens if you invest an amount of money at 100% interest, compounded continuously?

a) What does happen?

b) How long does it take to triple your money?

c) How much can you earn in a year?

84. BENJAMIN FRANKLIN'S WILL The Franklin Technical Institute of Boston owes its existence to a provision in a codicil to the will of Benjamin Franklin. In part it reads:

> I was born in Boston, New England and owe my first instruction in Literature to the free Grammar Schools established there: I have therefore already considered those schools in my Will. ... I have considered that among Artisans good Apprentices are most likely to make good citizens ... I wish to be useful even after my Death, if possible, in forming and advancing other young men that may be serviceable to their Country in both Boston and Philadelphia. To this end I devote Two thousand Pounds Sterling, which I give, one thousand thereof to the Inhabitants of the Town of Boston in Massachusetts, and the other thousand to the inhabitants of the City of Philadelphia, in Trust and for the Uses, Interests and Purposes hereinafter mentioned and declared.

Franklin's plan was to lend money to young apprentices at 5% interest with the provision that each borrower should pay each year

> ... with the yearly Interest, one tenth part of the Principal, which sums of Principal and Interest shall be again let to fresh Borrowers. ... If this plan is executed and succeeds as projected without interruption for one hundred Years, the Sum will then be one hundred and thirty-one thousand Pounds of which I would have the Managers of the Donation to the Inhabitants of the Town of Boston, then lay out at their discretion one hundred thousand Pounds in Public Works. ... The remaining thirty-one thousand Pounds, I would have continued to be let out on Interest in the manner above directed for another hundred Years. ... At the end of this second term if no unfortunate accident has prevented the operation the sum will be Four Millions and Sixty-one Thousand Pounds.

It was not always possible to find as many borrowers as Franklin had planned, but the managers of the trust did the best they could; they lent money to medical students as well as to others. At the end of 100 years from the reception of the Franklin gift, in January 1894, the fund had grown from 1000 pounds to almost exactly 90,000 pounds. In 100 years the original capital had multiplied about 90 times instead of the 131 times Franklin had imagined.

What rate of interest, compounded continuously for 100 years, would have multiplied Benjamin Franklin's original capital by 90?

85. In Benjamin Franklin's estimate that the original 1000 pounds would grow to 131,000 in 100 years, he was using an annual rate of 5% and compounding once each year. What rate of interest per year when compounded continuously for 100 years would multiply the original amount by 131?

86. THE U.S. POPULATION The Museum of Science in Boston, Massachusetts, displays the running total of the U.S. population. On December 27, 1988, it was increasing the total by 1 every 21 seconds, which works out to an instantaneous rate of 1,502,743 people a year. Working from 246,605,103, the display's population figure for 2:40 P.M. that day, this gives $k = 0.00609$, or 0.609%. Assuming this value of k for the next 10 years, what will the U.S. population be at 2:40 P.M. Boston time on December 27, 1998?

87. THE RULE OF 70 Use the Rule of 70 to answer these questions:

a) How long does it take to double money at 5% interest? At 7% interest?

b) What interest rate do you need to double money in 5 years? In 20 years?

88. The Consumer Price Index in March 1988 was 116.5. Assuming a constant inflation rate of 2.8% from then on, about when would you expect the index to double? (*Hint*: Use the Rule of 70.)

89. Repeat Exercise 88, but now use Eq. (26) directly instead of the Rule of 70.

90. THE PURCHASING POWER OF THE DOLLAR Assuming a constant inflation rate of 4.4% starting in March 1988, when the Consumer Price Index was 116.5, what will the purchasing power of the dollar be in March 1990? March 1992? March 1994?

91. INFLATION AND THE CPI You have just seen a newspaper headline saying that the consumer price index rose 4% last year. Assuming this was caused by a constant inflation rate, what was that rate?

92. RUNAWAY INFLATION At the end of 1988 the consumer price index in Brazil was increasing at the continuous annual rate of 800% (yes, it was). How many days does it take prices to double at this rate?
To find out, solve the equation

$$p_0 e^{8t} = 2p_0$$

for t and convert your answer to days, rounding your answer to the nearest day.

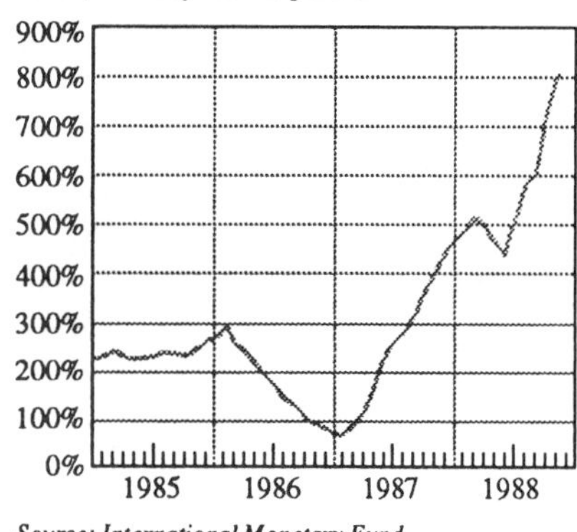

(Source: *Wall Street Journal*, Thursday, December 8, 1988, p. 1.)

93. THE PURCHASING POWER OF THE DOLLAR How long will it take the purchasing power of the dollar to fall to half its present value if the inflation rate holds at a steady 4%?

94. OIL DEPLETION Suppose the amount of oil pumped from one of the canyon wells in Whitter, California, decreases at the continuous rate of 10% a year. When will the well's output fall to a fifth of its present value?

95. CONTINUOUS PRICE DISCOUNTING To encourage buyers to place 100-unit orders, your firm's sales department applies a continuous discount that makes the unit price a function $p(x)$ of the number of units x ordered. The discount decreases the price at the rate of \$0.01 per unit ordered. The price per unit for a 100-unit order is $p(100) = \$20.09$.

a) Find $p(x)$ by solving the following initial value problem:

Differential equation: $\dfrac{dp}{dx} = -\dfrac{1}{100}p,$

Initial condition: $p(100) = 20.09.$

b) Find the unit price $p(10)$ for a 10-unit order and the unit price $p(90)$ for a 90-unit order.

c) The sales department has asked you to find out if it is discounting so much that the firm's revenue, $r(x) = x \cdot p(x)$, will actually be less for a 100-unit order than, say, for a 90-unit order. Reassure them by showing that r has its maximum value at $x = 100$.

d) Graph the revenue function $r(x) = xp(x)$ for $0 \leq x \leq 200$.

96. SUBINDICES OF THE CONSUMER PRICE INDEX There are many "subindices" of the Consumer Price Index—separate indices for food, rent, medical care, and so on. Each index has its own inflation rate. From March 1987 to March

1988 the costs of food, rent, and medical care rose at 3%, 5%, and 6.4%, respectively. At these rates

a) how long will it take costs to increase 50%?

b) how long will it take the costs to double? (Use the Rule of 70.)

Variable Inflation Rates

When the inflation rate k varies with time instead of being constant, the formula $p = p_0 e^{kt}$ in Eq. (27) no longer gives the solution of the equation $dp/dt = kp$. The corrected formula, as you will see in Chapter 16, is

$$p(t) = p_0 e^{\int_0^t k(t)dt} \tag{28}$$

Use this formula to answer the questions in Exercises 97 and 98.

97. Suppose that $p_0 = 100$ and that $k(t) = 0.04/(1+t)$, a rate that starts at 4% when $t = 0$ but decreases steadily as the years pass.

a) Find $\displaystyle\int_0^9 k(t)\,dt$.

b) Use Eq. (28) above to find the value of p when $t = 9$ yr.

98. Suppose that $p_0 = 100$ and that the inflation rate is $k(t) = 1 + 1.3t$ (as it was in Brazil during the first few months of 1987).

a) What does the formula in Eq. (28) look like in this case?

b) Find $p(1), p(2)$, and the 1-yr and 2-yr percentage increases in the associated consumer price index.

5.9 Numerical Integration: The Trapezoidal Rule and Simpson's Method

The Trapezoidal Rule and Simpson's Rule are described in this section. These rules enable us to estimate an integral's value to as many decimal places as we please whenever we want.

Today these numerical techniques are less valuable than in the past because most modern graphing calculators for calculus have a powerful and accurate numerical integration algorithm built-in. We will see the theory of calculus is central to effective implementation of these algorithms on computers. Furthermore the theory of calculus is central to understanding how and when it is appropriate to apply numerical integration techniques. This material is included mainly for its historical interest and as an introduction to *error* analysis. These rules also enable us to calculate integrals with reasonable accuracy from numerical tables of function values. This comes in handy when the only information we have about a function is a set of specific values measured in the laboratory or in the field as the last example of this section demonstrates.

The Trapezoidal Rule

The Trapezoidal Rule for estimating the value of

$$\int_a^b f(x)\,dx$$

is based on approximating the region between the graph of f and the x-axis with n trapezoids of equal width (Fig. 5.33). The trapezoids have the common base length $h = (b-a)/n$, and the side of each trapezoid runs from the x-axis up (or down) to the curve. The areas of the trapezoids add up to

$$\begin{aligned} T &= \frac{1}{2}(y_0 + y_1)h + \frac{1}{2}(y_1 + y_2)h + \cdots \\ &\quad + \frac{1}{2}(y_{n-2} + y_{n-1})h + \frac{1}{2}(y_{n-1} + y_n)h \\ &= h\left(\frac{1}{2}y_0 + y_1 + y_2 + \cdots + y_{n-1} + \frac{1}{2}y_n\right) \qquad (1) \\ &= \frac{h}{2}(y_0 + 2y_1 + 2y_2 + \cdots + 2y_{n-1} + y_n), \end{aligned}$$

where

$$y_0 = f(a), y_1 = f(x_1), \ \ldots, \ y_{n-1} = f(x_{n-1}), y_n = f(b).$$

The Trapezoidal Rule says: Use T to estimate the integral of f from a to b.

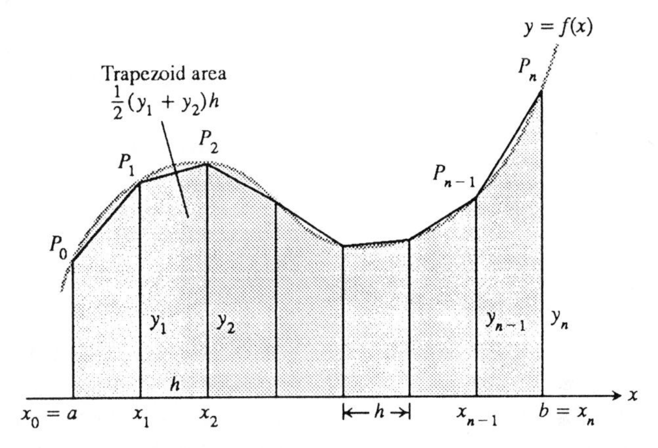

5.33 The Trapezoidal Rule approximates short stretches of curve with line segments. To estimate the shaded area, we add the areas of the trapezoids made by joining the ends of these segments to the x-axis.

> **The Trapezoidal Rule**
>
> To approximate
>
> $$\int_a^b f(x)\,dx,$$
>
> use
>
> $$T = \frac{h}{2}(y_0 + 2y_1 + 2y_2 + \cdots + 2y_{n-1} + y_n) \qquad (2)$$
>
> (n subintervals of length $h = (b - a)/n$).

EXAMPLE 1 Use the Trapezoidal Rule with $n = 4$ to estimate

$$\int_1^2 x^2\,dx.$$

Compare the estimate with the exact value of the integral.

Solution The exact value of the integral is

$$\int_1^2 x^2\,dx = \frac{x^3}{3}\bigg]_1^2 = \frac{8}{3} - \frac{1}{3} = \frac{7}{3}.$$

TABLE 5.6

x	$y = x^2$
1	1
$\frac{5}{4}$	$\frac{25}{16}$
$\frac{6}{4}$	$\frac{36}{16}$
$\frac{7}{4}$	$\frac{49}{16}$
2	4

To find the trapezoidal approximation, we divide the interval of integration into four subintervals of equal length and list the values of $y = x^2$ at the endpoints and subdivision points (see Table 5.6). We then evaluate Eq. (2) with $n = 4$ and $h = 1/4$:

$$T = \frac{h}{2}(y_0 + 2y_1 + 2y_2 + 2y_3 + y_4)$$

$$= \frac{1}{8}\left(1 + 2\left(\frac{25}{16}\right) + 2\left(\frac{36}{16}\right) + 2\left(\frac{49}{16}\right) + 4\right) = \frac{75}{32} = 2.34375.$$

The approximation overestimates the area by about half a percent of its true value. Each trapezoid contains slightly more than the corresponding strip under the curve (Fig. 5.34).

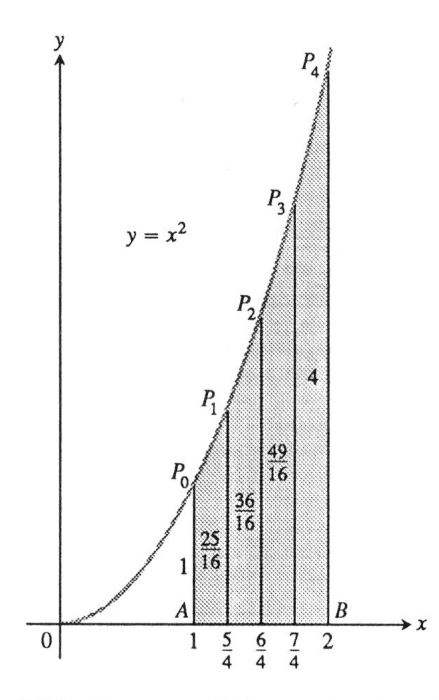

5.34 The trapezoidal approximation of the area under the graph of $y = x^2$ from $x = 1$ to $x = 2$ is a slight overestimate.

Controlling the Error in the Trapezoidal Approximation

Pictures suggest that the error

$$E_T = T - \int_a^b f(x)\,dx \tag{3}$$

in the trapezoidal approximation will go down as the **step size** h decreases, because the trapezoids fit the curve better as their number increases. A theorem from advanced calculus assures us that this will always be the case if f has a continuous second derivative.

The Error Estimate for the Trapezoidal Rule

If f'' is continuous and M is any upper bound for the values of $|f''|$ on $[a, b]$, then

$$|E_r| \le \frac{b-a}{12}h^2 M. \tag{4}$$

Although theory tells us there will always be a smallest safe value of M, in practice we can hardly ever find it. Instead, we find the best value we can and go on to estimate $|E_T|$ from there. This may seem sloppy, but it works. To make $|E_T|$ small for a given M, we just make h small.

EXAMPLE 2 Find an upper bound for the error in the approximation found in Example 1 for the value of

$$\int_1^2 x^2\,dx.$$

Solution We first find an upper bound M for the magnitude of the second derivative of $f(x) = x^2$ on the interval $1 \le x \le 2$. Since $f''(x) = 2$ for all x, we may safely take $M = 2$. With $b - a = 1$ and $h = 1/4$, Eq. (4) gives

$$|E_T| \le \frac{b-a}{12}h^2 M = \frac{1}{12}\left(\frac{1}{4}\right)^2 (2) = \frac{1}{96}.$$

This is precisely what we find when we subtract $T = 75/32$ from $\int_1^2 x^2\,dx = 7/3$, since $7/3 - 75/32 = -1/96$. Here we are able to give the error *exactly*, but this is exceptional.

EXAMPLE 3 The Trapezoidal Rule is used to estimate the value of

$$\int_0^1 x \sin x \, dx$$

when $n = 10$ steps. Find an upper bound for the error in the estimate.

Solution We use the formula

$$|E_T| \le \frac{b-a}{12}h^2 M$$

with $b = 1, a = 0$, and $h = 1/n = 1/10$. This gives

$$|E_T| \le \frac{1}{12}\left(\frac{1}{10}\right)^2 M = \frac{1}{1200}M.$$

The number M can be any upper bound for the values of $|f''|$ on $[0, 1]$. To choose a value for M, we calculate f'' to see how big it might be. A straightforward differentiation gives

$$f'' = 2\cos x - x\sin x.$$

Hence, by the triangle inequality,

$$|f''| \le 2|\cos x| + |x||\sin x| \le 2 + (1)(1) = 3$$

because $0 \le x \le 1$ and $|\cos x|$ and $|\sin x|$ never exceed 1. We can safely take $M = 3$. Therefore

$$|E_T| \le \frac{1}{1200}(3) = \frac{1}{400} = 0.0025.$$

The error is no greater than 2.5×10^{-3}.

For greater accuracy we would not try to improve M but would take more steps. With $n = 100$ steps, for example, $h = 1/100$ and

$$|E_T| \le \frac{1}{12}\left(\frac{1}{100}\right)^2 (3) < 2.5 \times 10^{-5}.$$

EXAMPLE 4 As we saw in Section 5.8, the value of $\ln 2$ ("log two"), the natural logarithm of 2, is given by the integral

$$\ln 2 = \int_1^2 \frac{1}{x}\,dx.$$

How many subintervals (steps) should be used in the Trapezoidal Rule to approximate the integral, and hence the value of ln 2, with an error of absolute value less than 10^{-4}?

Solution To determine n, the number of subintervals, we use Eq. (4) with

$$b-a=2-1=1, \quad h=\frac{b-a}{n}=\frac{1}{n},$$

$$f''(x)=\frac{d^2}{dx^2}(x^{-1})=2x^{-3}=\frac{2}{x^3}.$$

Then

$$|E_T|\le\frac{b-a}{12}h^2\max|f''(x)|=\frac{1}{12}\left(\frac{1}{n}\right)^2\max\left|\frac{2}{x^3}\right|,$$

where max refers to $[1,2]$.

This is one of the rare cases where we can find the exact value of max $|f''|$. On $[1,2]$, $y=2/x^3$ decreases steadily from a maximum of $y=2$ to a minimum of $y=1/4$ (Fig. 5.35). Therefore

$$|E_T|\le\frac{1}{12}\left(\frac{1}{n}\right)^2\cdot 2=\frac{1}{6n^2}.$$

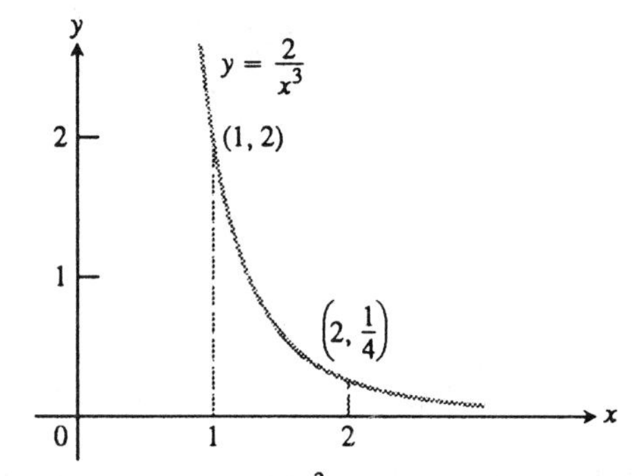

5.35 The continuous function $y=2/x^3$ has its maximum value on $[1,2]$ at $x=1$.

The error's absolute value will therefore be less than 10^{-4} if

$$\frac{1}{6n^2}<10^{-4},\quad \frac{10^4}{6}<n^2,\quad \frac{100}{\sqrt{6}}<n,\quad \text{or}\quad 40.83<n.$$

The first integer beyond 40.83 is $n=41$. With $n=41$ subdivisions we can guarantee calculating ln 2 with an error of magnitude less than 10^{-4}. Any larger n will work, too.

Simpson's Rule

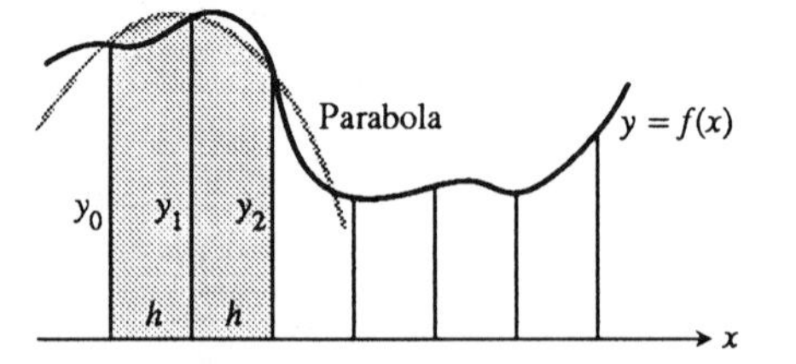

5.36 Simpson's Rule approximates short stretches of curve with parabolas.

Simpson's Rule is based on approximating curves with parabolas instead of line segments. The shaded area under the parabola in Fig. 5.36 is

$$A_p=\frac{h}{3}(y_0+4y_1+y_2). \tag{5}$$

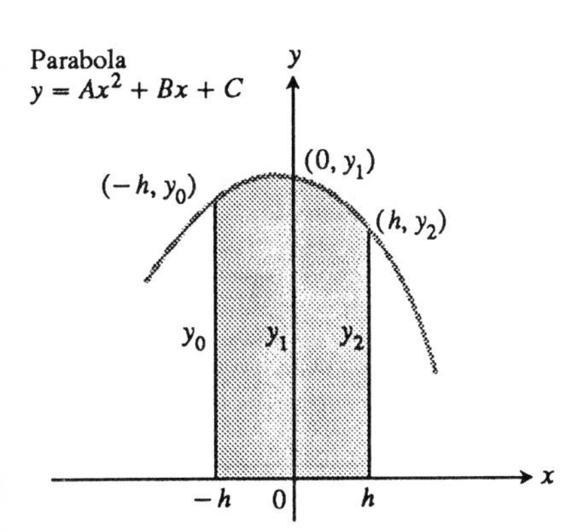

5.37 By integrating from $-h$ to h, the shaded area is found to be

$$A_p = \frac{h}{3}(y_0 + 4y_1 + y_2).$$

Applying this formula successively along a continuous curve $y = f(x)$ from $x = a$ to $x = b$ leads to an estimate of $\int_a^b f(x)\,dx$ that is generally more accurate than T for a given step size h.

We can derive the formula for A_p in the following way. To simplify the algebra, we use the coordinate system shown in Fig. 5.37. The area under the parabola is the same no matter where the y-axis is, as long as we preserve the vertical scale. The parabola has an equation of the form

$$y = Ax^2 + Bx + C,$$

so the area under it from $x = -h$ to $x = h$ is

$$\begin{aligned} A_p = \int_{-h}^{h} (Ax^2 + Bx + C)\,dx &= \frac{Ax^3}{3} + \frac{Bx^2}{2} + Cx\Big]_{-h}^{h} \\ &= \frac{2Ah^3}{3} + 2Ch = \frac{h}{3}(2Ah^2 + 6C). \end{aligned} \tag{6}$$

Since the curve passes through the three points $(-h, y_0), (0, y_1)$, and (h, y_2), we also have

$$y_0 = Ah^2 - Bh + C, \quad y_1 = C, \quad y_2 = Ah^2 + Bh + C,$$

from which we obtain

$$\begin{aligned} C &= y_1, \\ Ah^2 - Bh &= y_0 - y_1, \\ Ah^2 + Bh &= y_2 + y_1, \\ 2Ah^2 &= y_0 + y_2 - 2y_1. \end{aligned} \tag{7}$$

Hence expressing the area A_p in terms of y_0, y_1, and y_2, we have

$$A_p = \frac{h}{3}(2Ah^2 + 6C) = \frac{h}{3}((y_0 + y_2 - 2y_1) + 6y_1) = \frac{h}{3}(y_0 + 4y_1 + y_2). \tag{8}$$

Simpson's Rule follows from applying the formula for A_p to successive pieces of the curve $y = f(x)$ between $x = a$ and $x = b$. Each separate piece of the curve, covering an x-subinterval of width $2h$, is approximated by an arc of a parabola through its ends and midpoint. The areas under the parabolic arcs are then added to give Simpson's Rule.

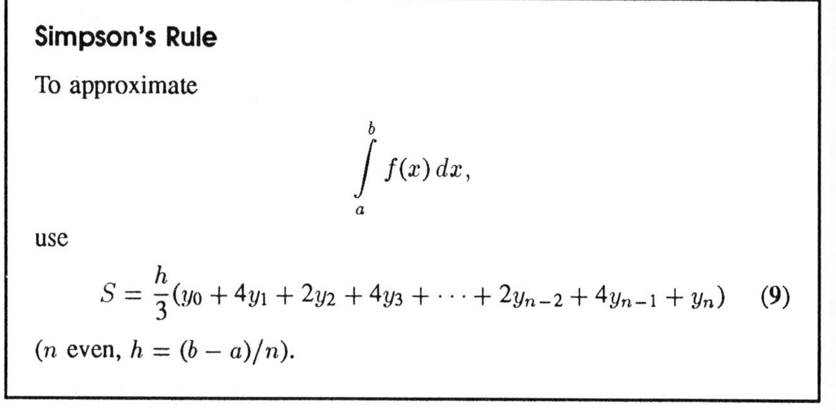

Simpson's Rule

To approximate

$$\int_a^b f(x)\,dx,$$

use

$$S = \frac{h}{3}(y_0 + 4y_1 + 2y_2 + 4y_3 + \cdots + 2y_{n-2} + 4y_{n-1} + y_n) \qquad (9)$$

(n even, $h = (b - a)/n$).

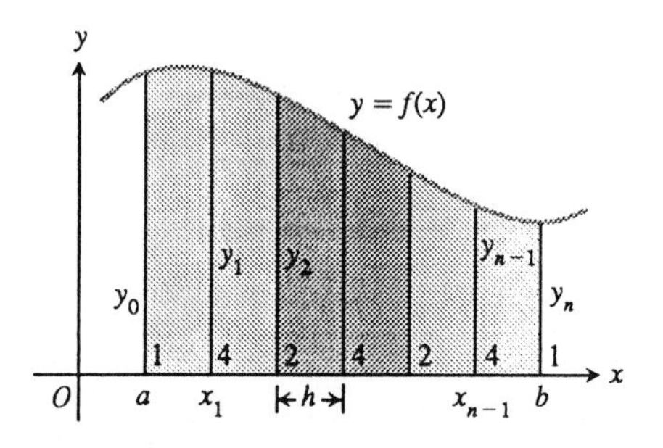

5.38 The y's in Eq. (9) are the values of f at the points of subdivision.

The y's in Eq. (9) are the values of $y = f(x)$ at the points

$$a, \quad x_1 = a + h, \quad x_2 = a + 2h, \quad \ldots, \quad x_{n-1} = a + (n-1)h, \quad b$$

that subdivide $[a, b]$ into n equal subintervals of length $h = (b-a)/n$ (Fig. 5.38). The number n must be even to apply the rule because each parabolic arc uses two subintervals.

Error Control for Simpson's Rule

The Simpson's Rule error,

$$E_S = S - \int_a^b f(x)\,dx, \qquad (10)$$

decreases with the step size too. The inequality for controlling it, however, assumes f to have a continuous fourth derivative instead of merely a continuous second derivative. The formula, once again from advanced calculus is this:

The Error Estimate for Simpson's Rule

If $f^{(4)}$ is continuous and M is any upper bound for the values of $|f^{(4)}|$ on $[a, b]$, then

$$|E_S| \le \frac{b-a}{180} h^4 M. \qquad (11)$$

As with the Trapezoidal Rule, we can almost never find the smallest possible value of M. We just find the best value we can and go on from there to estimate $|E_S|$.

EXAMPLE 5 Use Simpson's Rule with $n = 4$ to approximate

$$\int_0^1 5x^4\,dx.$$

SIMPSON'S ONE-THIRD RULE

The rule

$$\text{Area} = \frac{h}{3}(y_0 + 4y_1 + y_2)$$

for calculating area by replacing curves by parabolas, Eq. (5) in the text, was discovered long before Thomas Simpson (1720–1761) was born. It is another of history's beautiful quirks that one of the ablest mathematicians of eighteenth-century England is remembered not for his successful texts and his contributions to mathematical analysis but for a rule that was never his, that he never laid claim to, and that bears his name only because he happened to mention it in one of his books.

What estimate does Eq. (11) give for the error in the approximation?

Solution Again we have chosen an integral whose exact value we can calculate directly:

$$\int_0^1 5x^4\,dx = x^5\Big]_0^1 = 1.$$

TABLE 5.7

x	$y = 5x^4$
0	0
$\frac{1}{4}$	$\frac{5}{256}$
$\frac{2}{4}$	$\frac{80}{256}$
$\frac{3}{4}$	$\frac{405}{256}$
1	5

To find the Simpson approximation, we partition the interval of integration into four subintervals and evaluate $f(x) = 5x^4$ at the partition points (Table 5.7). We then evaluate Eq. (9) with $n = 4$ and $h = 1/4$:

$$S = \frac{h}{3}(y_0 + 4y_1 + 2y_2 + 4y_3 + y_4)$$

$$= \frac{1}{12}\left(0 + 4\left(\frac{5}{256}\right) + 2\left(\frac{80}{256}\right) + 4\left(\frac{405}{256}\right) + 5\right)$$

$$= 1.00260. \qquad \text{(Rounded)}$$

To estimate the error, we first find an upper bound M for the magnitude of the fourth derivative of $f(x) = 5x^4$ on the interval $0 \le x \le 1$. Since the fourth derivative has the constant value $f^{(4)}(x) = 120$, we may safely take $M = 120$. With $(b - a) = 1$ and $h = 1/4$, Eq. (11) then gives

$$|E_S| \le \frac{b-a}{180}h^4 M = \frac{1}{180}\left(\frac{1}{4}\right)^4(120) = \frac{1}{384} < 0.002604.$$

CALCULUS AND COMPUTERS

Here is another example of calculus having something important to say about computation. It is easy to implement the Trapezoidal Rule and Simpson's Rule on a computer. But that in itself is not enough. We need to know how many steps to take to achieve the accuracy we want, and the guidance for *that* comes from calculus.

A Graphical Analysis of Error

Let's evaluate $\int_{-1}^{1} e^{-x^2}dx$. First remember there is no explicit formula in terms of x for the antiderivative of e^{-x^2}—none exists! Take $n = 20$ and apply both the Trapezoidal Rule and Simpson's Method. Use a computer program (consult your *Graphing Calculator Laboratory Manual for Calculus*) to do the arithmetic. Applying each method yields

$$\int_{-1}^{1} e^{-x^2}dx = 1.4924216 \qquad \text{(Trapezoid, } n = 20\text{)}$$

and

$$\int_{-1}^{1} e^{-x^2}dx = 1.4936499 \qquad \text{(Simpson, } n = 20\text{)}$$

The question is how accurate are the results. Is it reasonable to conclude the answer is 1.49 with error of at most 0.01? We can answer the accuracy question by graphing each of the error *functions*,

$$ET(x) = \frac{b-a}{12}h^2\left|f''(x)\right| \qquad \text{(Trapezoid),}$$

and

$$ES(x) = \frac{b-a}{180}h^4\left|f^{(4)}(x)\right| \qquad \text{(Simpson),}$$

for $-1 \leq x \leq 1$ to determine the *maximum* value of each function. It is easy to compute the first two derivatives of f:

$$f'(x) = -2xe^{-x^2}$$

$$f''(x) = e^{-x^2}(4x^2 - 2).$$

We will use $\mathrm{N_2Der}\ (f''(x), x) = \mathrm{NDer}\ (\mathrm{NDer}\ (f''(x), x), x)$ to graph $y = ES(x)$ which involves $|f^{(4)}(x)|$ and not deal with the messy analytic form of $f^{(4)}$.

Since $b - a = 2$ and $h = 2/20 = 0.1$, the functions to be graphed are

$$y = ET(x) = \left(\frac{1}{6}\right)(0.1)^2 \left|(4x^2 - 2)e^{-x^2}\right| \qquad \text{and}$$

$$y = ES(x) = \left(\frac{1}{90}\right)(0.1)^4 \left|N_2 \text{ Der } ((4x^2 - 2)e^{-x^2}, x)\right|.$$

The maximum value of $y = ET(x)$ on $[-1, 1]$ is 3.333×10^{-3} and the maximum value of $y = ES(x)$ in $[-1, 1]$ is 1.333×10^{-5} (Fig. 5.39). Thus the error in using the trapezoid estimate ($n = 20$) is no greater than 3.333×10^{-3} and the error in using the Simpson's method estimate ($n = 20$) is no greater than 1.333×10^{-5}. So the $n = 20$ Simpson's result of 1.4936499 is very accurate indeed!

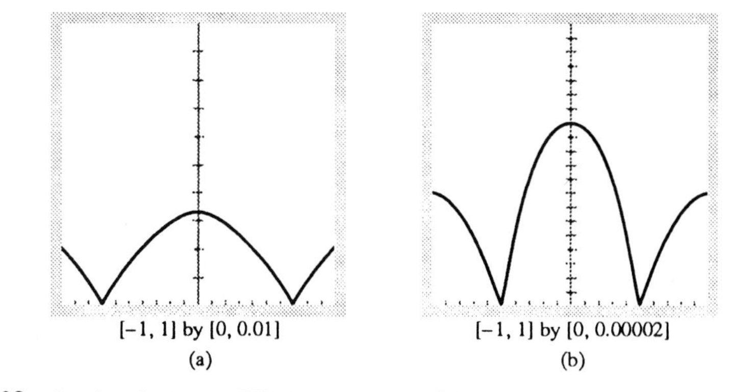

5.39 Graphs of (a) $y = ET(x)$ and (b) $y = ES(x)$.

In the exercises you will compute the analytic fourth derivative of $f(x) = e^{-x^2}$ and compare the graphs of $y = ES(x)$ using the exact form of $f^{(4)}$ versus $\mathrm{N_2Der}\ (f''(x))$ in the Simpson's Rule error analysis.

Which Rule Gives Better Results?

The answer lies in the error-control formulas for the two rules:

$$|E_T| \le \frac{b-a}{12}h^2M, \quad |E_S| \le \frac{b-a}{180}h^4M. \tag{12}$$

The M's of course mean different things, the first being an upper bound on $|f''|$ and the second an upper bound on $|f^{(4)}|$. But there is more than that going on here. The factor $(b-a)/180$ in the Simpson formula is one-fifteenth of the factor $(b-a)/12$ in the trapezoidal formula. More important still, the Simpson formula has an h^4 while the trapezoidal formula has only an h^2. If h is one-tenth, then h^2 is a hundredth but h^4 is only a ten-thousandth. If both M's are 1 for example, and $b-a=1$, then, with $h=1/10$,

$$|E_T| \le \frac{1}{12}\left(\frac{1}{10}\right)^2 \cdot 1 \le \frac{1}{1200}, \tag{13}$$

while

$$|E_S| \le \frac{1}{180}\left(\frac{1}{10}\right)^4 \cdot 1 \le \frac{1}{1{,}800{,}000} = \frac{1}{1500}\cdot\frac{1}{1200}. \tag{14}$$

For roughly the same amount of computational effort, we get better accuracy with Simpson's Rule—at least in this case.

The h^2 versus h^4 is the key. If h is less than 1, then h^4 can be significantly smaller than h^2. On the other hand, if h equals 1, there is no difference between h^2 and h^4. If h is greater than 1, the value of h^4 may be significantly larger than the value of h^2. In the latter two cases, the error-control formulas offer little help. We have to go back to the geometry of the curve $y=f(x)$ to see whether line segments or parabolas, if either, are going to give a better fit.

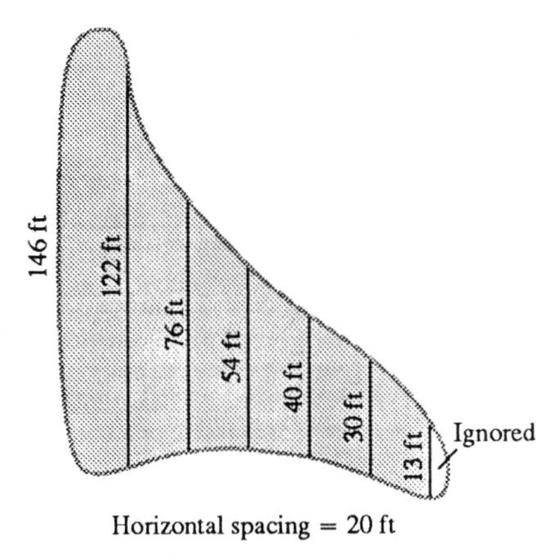

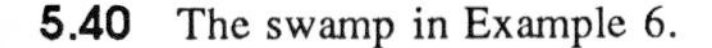

5.40 The swamp in Example 6.

Working with Numerical Data

The next example shows how to use Simpson's Rule to estimate the integral of a function from values measured in the laboratory or in the field even when we have no formula for the function. The Trapezoidal Rule can be used the same way.

EXAMPLE 6 A town wants to drain and fill the small swamp shown in Fig. 5.40. The swamp averages 5 ft deep. About how many cubic yards of dirt will it take to fill the area after the swamp is drained?

Solution To calculate the volume of the swamp, we estimate the surface area and multiply by 5. To estimate the area, we use Simpson's Rule with $h=20$ ft and the y's equal to the distances measured across the swamp, as shown in Fig. 5.40:

$$S = \frac{h}{3}(y_0 + 4y_1 + 2y_2 + 4y_3 + 2y_4 + 4y_5 + y_6)$$

$$= \frac{20}{3}(146 + 488 + 152 + 216 + 80 + 120 + 13) = 8100.$$

The volume is about (8100)(5) = 40,500 ft^3 or 1500 yd^3. ■

Round-off Errors

Although decreasing the step size h reduces the error in the Simpson and trapezoidal approximations in theory, it may fail to do so in practice. When h is very small, say $h = 10^{-5}$, the round-off errors in the arithmetic required to evalute S and T may accumulate to such an extent that the error formulas no longer describe what is going on. Shrinking h below a certain size can actually make things worse rather than better. While this will not be an issue in the present book, you should consult a text on numerical analysis for alternative methods if you are having problems with round-off.

Computer Programs

As we mentioned earlier, we assume you have a simple computer program available for both the Trapezoidal Rule and Simpson's Rule. Tables 5.8 and 5.9 are based on data using programs for a TI-81 graphing calculator. Consult the *Graphing Calculator Laboratory Manual for Calculus* that accompanies this textbook for program details for most brands of graphing calculators.

For comparison: ln 2 = 0.69314718056...

Table 5.8 shows the results for $f(x) = 1/x, a = 1$ and $b = 2$. The listed numbers are the trapezoidal approximations to $\ln 2 = \int_1^2 1/x\,dx$ for $n = 10$, 20, 30, 40, 50, and 100.

For comparison: ln 2 = 0.69314718056...

Table 5.9 shows the result of using the Simpson program with $f(x) = 1/x, a = 1$, and $b = 2$, to approximate $\ln 2 = \int_1^2 1/x\,dx$ with $n = 10$, 20, 30,

TABLE 5.8 Trapezoidal Approximations of $\ln 2 = \int_1^2 1/x\,dx$

n	$T(n)$	|Error| less than ...
10	.6937714	.0006242
20	.6933034	.0001562
30	.6932168	.0000694
40	.6931862	.0000391
50	.6931722	.0000250
100	.6931534	.0000062

TABLE 5.9 Simpson's Rule Approximations of $\ln 2 = \int_1^2 1/x\,dx$

n	$T(n)$	|Error| less than ...
10	.6931502	.0000031
20	.6931474	.0000003
30	.6931472	.0000001
40	.6931472	.0000001
50	.6931472	.0000001
100	.6931471809	3.4×10^{-10}

40, 50, and 100 subintervals. Notice the improvement over the Trapezoidal Rule: The approximation for $n = 50$ rounds accurately to seven places instead of three. The approximation for $n = 100$ is accurate almost to 10 decimal places!

We close with computing the value of $\int_1^5 (\sin x)/x\,dx$. There is no way to evaluate this definite integral other than to apply a numerical approximation. The results in Table 5.10 are based on programs for the TI-81.

It is known that $\int_1^5 (\sin x)/x = 0.603848$ is accurate to six decimal places. So both Simpson's method with 50 subintervals and a FnInt computation (tol $= 10^{-5}$) produce results accurate to 10^{-6}!

TABLE 5.10 Approximations of $\int_1^5 \sin x/x\,dx$.

Method	Subintervals	Value
(L)RAM	50	0.6453898
(R)RAM	50	0.5627293
(M)RAM	50	0.6037425
TRAP	50	0.6040595
SIMP	50	0.6038481
FnInt	tol = 0.00001	0.6038482

Explore Box

Some graphing calculators or computer software can generate sequences and sum them. Try this if your graphing calculator or software allows. Type the following on the Home screen:

$$10 \rightarrow N\text{: seq}(2/(1+K(1/10)), K, 0, N, 1)$$

Press ENTER and confirm the terms of the sequence are almost the terms in the $N = 10$ trapezoidal approximation of $\int_1^2 1/x\, dx$, namely

$$\frac{h}{2}(y_0 + 2y_1 + 2y_2 + \cdots + 2y_9 + y_{10}).$$

Now show that if the *sum* of

$$\text{seq }(2/(1+K*h), K, 0, N, 1)$$

is denoted by S, then,

$$\frac{h}{2}(y_0 + 2y_1 + 2y_2 + \cdots + 2y_{n-1} + y_n) = \frac{h}{2}(S - 1 - 0.5).$$

Next, enter this on the Home screen:

$10 \rightarrow N$: $1/N \rightarrow h$: $(h/2)(-1-0.5+\text{sum seq}(2/(1+K*h), K, 0, N, 1))$

Press ENTER and confirm you obtain the trapezoid estimate for $\int_1^2 1/x\, dx$ when $N = 10$. Repeat by editing and storing 20, 30, 40, and 50 to N, respectively. Compare with Table 5.8. Also compare with $\ln 2$. Explain.

Exercises 5.9

Use (a) the Trapezoidal Rule and (b) Simpson's Rule to approximate the integrals in Exercises 1–6 with $n = 4$. Then (c) find the integral's exact value for comparison.

1. $\int_0^2 x\, dx$ **2.** $\int_0^2 x^2\, dx$ **3.** $\int_0^2 x^3\, dx$

4. $\int_1^2 \frac{1}{x^2}\, dx$ **5.** $\int_0^4 \sqrt{x}\, dx$ **6.** $\int_0^{\pi} \sin x\, dx$

7. Use Eq. (4) to estimate the error in using the Trapezoidal Rule with $n = 10$ to estimate the value of

$$\ln 2 = \int_1^2 \frac{1}{x}\, dx.$$

Compare your answer with the result in Table 5.6.

8. Use Eq. (11) to estimate the error in using Simpson's Rule with $n = 10$ to estimate the value of

$$\ln 2 = \int_1^2 \frac{1}{x}\, dx.$$

Compare your answer with the result in Table 5.7.

In Exercises 9–14, estimate the minimum number of subdivisions needed to approximate the integrals with an error of absolute value less than 10^{-4} by (a) the Trapezoidal Rule and (b) Simpson's Rule.

9. $\int_0^2 x\, dx$ **10.** $\int_0^2 x^2\, dx$ **11.** $\int_0^2 x^3\, dx$

12. $\int_1^2 \frac{1}{x^2}\, dx$ **13.** $\int_1^4 \sqrt{x}\, dx$ **14.** $\int_0^{\pi} \sin x\, dx$

15. As the fish-and-game warden of your township, you are responsible for stocking the town pond with fish before fishing season. The average depth of the pond is 20 ft. You plan to start the season with one fish per 1000 ft^3. You intend to have at least 25% of the opening day's fish population left at the end of the season. What is the maximum number of licenses the town can sell if the average seasonal catch is 20 fish per license?

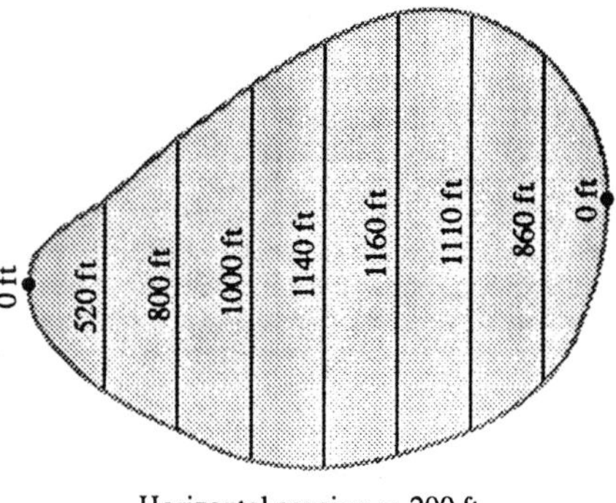

Horizontal spacing = 200 ft

16. The design of a new airplane requires a gasoline tank of constant cross-section area in each wing. A scale drawing of a cross section is shown here. The tank must hold 5000 lb of gasoline that has a density of 42 lb/ft^3. Estimate the length of the tank.

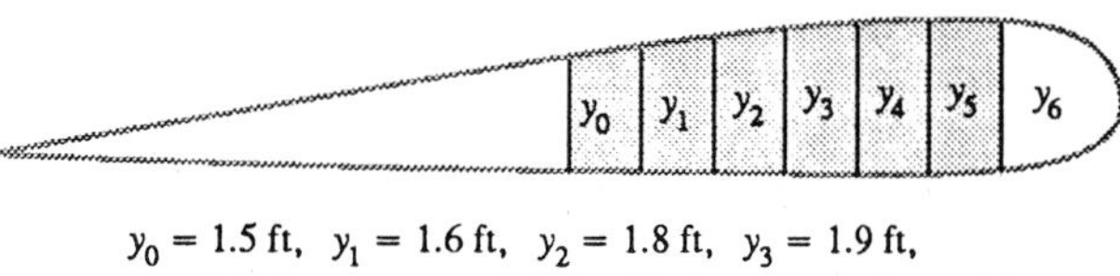

$y_0 = 1.5$ ft, $y_1 = 1.6$ ft, $y_2 = 1.8$ ft, $y_3 = 1.9$ ft,
$y_4 = 2.0$ ft, $y_5 = y_6 = 2.1$ ft, Horizontal spacing = 1 ft

17. A vehicle's aerodynamic drag is determined in part by its cross-section area and, all other things being equal, engineers try to make this area as small as possible. Use Simpson's Rule to estimate the cross-section area of James Worden's solar-powered Solectria car at M.I.T. from the diagram below.

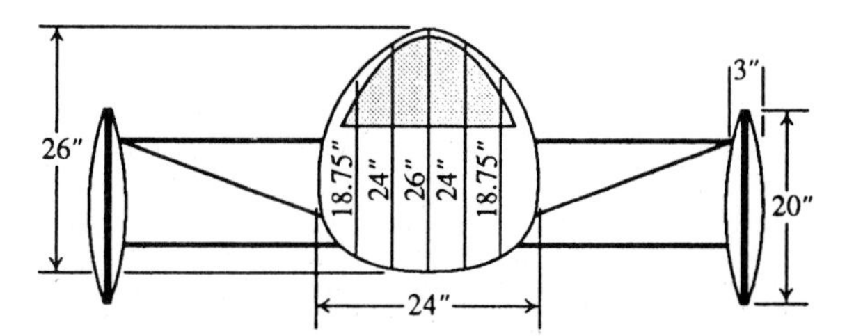

18. THE DYE-DILUTION TECHNIQUE FOR MEASURING CARDIAC OUTPUT Instead of measuring a patient's cardiac output with exhaled carbon dioxide, as in Exercise 21, Section 4.1, a doctor may prefer to use the dye-dilution technique described here. You start by injecting 5–10 mg of dye in a main vein near the heart. The dye is drawn into the right side of the heart and then pumped through the lungs and out the left side of the heart into the aorta, where its concentration is measured each second as the blood flows past. The data in Table 5.11 and the plot in Fig. 5.41 show the response of a healthy, resting patient to an injection of 5.6 mg of dye.

The patient's cardiac output is calculated by dividing the area under the concentration curve into the number of milligrams of dye and multiplying the result by 60:

$$\text{Cardiac output} = \frac{\text{milligrams of dye}}{\text{area under curve}} \times 60. \qquad (15)$$

You can see why if you check the units in which these quantities are measured. The dye is in milligrams, the area is in (milligrams/liter) × seconds, and

$$\frac{\text{mg}}{\frac{\text{mg}}{\text{L}} \cdot \text{sec}} \cdot 60 = \text{mg} \cdot \frac{\text{L}}{\text{mg} \cdot \text{sec}} \cdot 60 = \frac{\text{L}}{\text{sec}} \cdot 60 = \frac{\text{L}}{\text{min}}.$$

a) Use the Trapezoidal Rule and the data in Table 5.8 to calculate the area under the concentration curve in Fig. 5.41.

b) Then use Eq. (15) to calculate the patient's cardiac output.

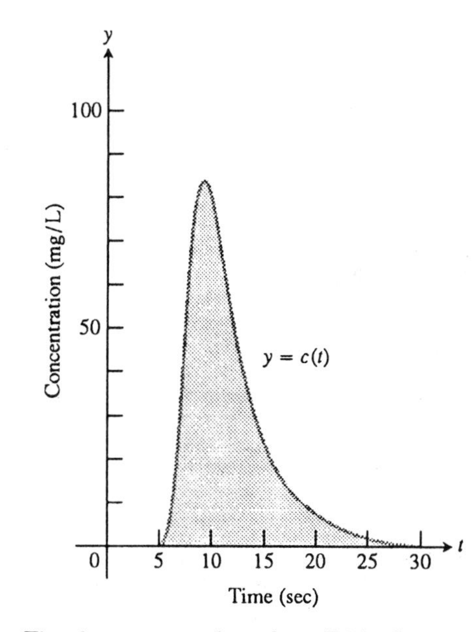

5.41 The dye concentrations from Table 5.11, plotted as a function of time t. The data points have been fitted with a smooth curve. Time is measured with $t = 0$ at the time of injection. The concentration readings are zero at the beginning, while the dye passes through the lungs. They then rise to a maximum at about $t = 9$ seconds and taper exponentially to zero at $t = 30$.

Electric Energy Consumption

The Louisiana Power and Light Company tries to forecast the demand for electricity throughout the day so that it can have enough generators on line at any given time to carry the load.

TABLE 5.11 Dye-Dilution Data for Exercise 18

Seconds after injection t	Dye concentration (adjusted for recirculation) $c(t)$	Seconds after injection t	Dye concentration (adjusted for recirculation) $c(t)$
1	0	16	18.5
2	0	17	14.5
3	0	18	11.5
4	0	19	9.1
5	0	20	7.3
6	1.5	21	5.7
7	38.0	22	4.5
8	67.0	23	3.6
9	80.0	24	2.8
10	73.0	25	2.3
11	61.0	26	1.8
12	48.0	27	1.4
13	36.0	28	1.1
14	29.0	29	0.9
15	23.0	30	0

Boilers take a while to fire up, so the company has to know in advance what the load is going to be if service is not to be interrupted. Like all power companies, Louisiana Power measures electrical demand in kilowatt-hours (KWH on your electric bill). A 1500-watt space heater, running for 10 hours, for instance, uses 15,000 watt-hours or 15 kilowatt-hours of electricity. A kilowatt, like a horsepower, is a unit of power. A kilowatt-hour is a unit of energy, and energy is what Louisiana Power, despite its name, sells.

Table 5.12 and the graph in Fig. 5.42 show the results of a 1984 residential-load study of Louisiana Power residential customers with electric heating, for typical weekdays and weekend days in January. A typical customer used 56.60 kwh on a weekday that January and 53.46 kwh on a weekend day. At the time of the study, 53.60 kwh cost a customer $3.40.

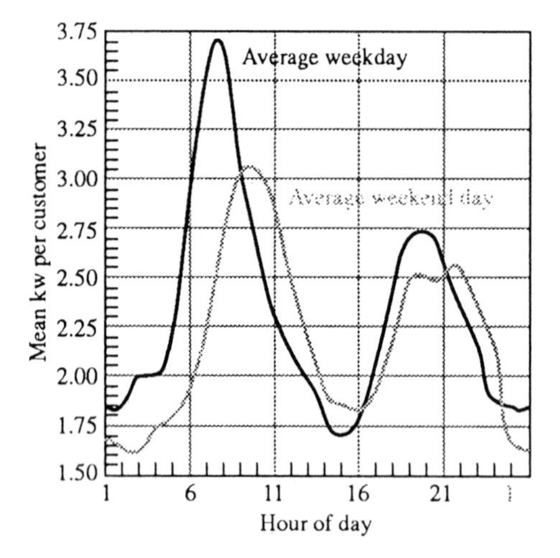

5.42 Louisiana Power and Light Company 1984 residential-load profile, residential customers with electric heating, averaged days for January. The curves plot kilowatts against hours. The areas under the curves give kilowatt-hours. See Table 5.12 for data.

TABLE 5.12 Residential Electric Loads (Exercises 19 and 20)

Hour of day	Weekday kw	Weekend kw
1	1.88	1.69
2	1.88	1.64
3	2.02	1.63
4	2.02	1.73
5	2.25	1.80
6	2.76	1.97
7	3.60	2.25
8	3.66	2.68
9	3.05	3.05
10	2.70	3.05
11	2.38	2.88
12	2.17	2.55
13	2.02	2.25
14	1.82	1.95
15	1.72	1.87
16	1.77	1.83
17	1.97	1.90
18	2.43	2.17
19	2.68	2.46
20	2.75	2.52
21	2.65	2.50
22	2.40	2.57
23	2.21	2.40
24	1.90	2.22

19. Find the weekday energy consumption of a typical residential customer by using the Trapezoidal Rule to estimate the area under the weekday power curve. To save time, you might try using only the data for the even-numbered hours or the data for the odd-numbered hours. How close do you come to Louisiana Power's own estimate if you do that? (If you use the even-numbered hours, be sure to count hour 2 twice so that you cover a complete 24-hr period. If you use odd-numbered hours, use hour 1 twice.)

20. CONTINUATION OF EXERCISE 19 Use the Trapezoidal Rule to find the daily weekend energy consumption.

Use Simpson's Rule with $N = 50$ and $N = 100$ to compute values of the integrals in Exercises 21–24.

21. $\int_{-1}^{1} 2\sqrt{1-x^2}\,dx$ (The exact value is π.)

22. $\int_{0}^{1} \sqrt{1+x^4}\,dx$ (A nonelementary integral that came up in Newton's research)

23. $\int_{0}^{\pi/2} \frac{\sin x}{x}\,dx$ (The integral from Exercise 23. To avoid division by zero, you may have to start the integration at a small positive number like 10^{-6} instead of 0.)

24. $\int_{0}^{\pi/2} \sin(x^2)\,dx$ (An integral associated with the diffraction of light)

25. a) Compute both the Trapezoid and Simpson's Rule estimate of $\int_{\frac{\pi}{2}}^{2\pi} (\sin x)/x\,dx$ for $n = 10$.

b) Graph the error functions $y = ET(x)$ and $y = ES(x)$ and establish the maximum error of each estimate.

c) Repeat part b) for $n = 20$.

d) Repeat part b) for $n = 50$.

e) What is the *exact* value of the definite integral?

26. Repeat Exercise 25 with $\int_{1}^{\pi} x \sin x^2\,dx$.

27. Repeat Exercise 25 with $\int_{0}^{2} \sqrt{1+x^4}\,dx$.

28. Let $f(x) = e^{-x^2}$. Graph $y_1 = f^{(4)}(x)$ where f'' is the *exact* analytic fourth derivative and $y_2 = N_2 \operatorname{Der}(f''(x), x)$ in the viewing window $[-1, 1]$ by $[0, 0.00002]$. Compare function values at $x = -0.2, 0.2, 0.4$, and 0.6. What is the maximum error in using y_2 to approximate y_1 in $[-1, 1]$? What does this imply about $ES(x)$ computations when y_2 is used rather than y_1?

Review Questions

1. How can you use antiderivatives to find areas? Give an example.
2. How are finite sums written in sigma notation? Give examples.
3. What is a partition of an interval? What is the norm of a partition? Give examples.
4. What is a Riemann sum? Give an example.
5. What does $\int_a^b f(x)\,dx$ mean? Is it a number? How is it defined? What is it called? When does it exist?
6. What is the relation between definite integrals and area? Does a definite integral have to represent an area?
7. State eleven rules for working with definite integrals (Table 5.1). Give a specific example of each rule.
8. What is the average value of a function $f(x)$ over an interval

$[a, b]$? Give an example.

9. State the Mean Value Theorem for Definite Integrals.

10. State the two fundamental theorems of integral calculus. What are they good for? Illustrate each theorem with an example.

11. What is an indefinite integral? How are indefinite integrals evaluated? What corollary of the Mean Value Theorem for derivatives makes the evaluation possible?

12. What specific formulas do you know for evaluating indefinite integrals (standard forms, so to speak)?

13. How does integration by substitution work? Does it apply to definite integrals as well as indefinite integrals? Give specific examples.

14. What numerical methods are available for estimating the values of definite integrals that cannot be evaluated directly with antiderivatives? What formulas sometimes help to determine the accuracy of these methods? How do you know if you are using a step size small enough to get the accuracy you want? Give an example.

15. How is the function $y = \ln x$ defined? What are the rules for doing arithmetic with natural logarithms?

16. What is the derivative of $y = \ln x$? What integrals involve logarithms? Give examples.

17. What is the function $y = \exp(x)$? What is number e? How is the function $y = e^x$ defined?

18. What derivatives and integrals are associated with the function $y = e^x$?

19. What initial value problem is solved by the function $y = y_0 e^{kt}$?

20. How does the exponential function arise in the life sciences and in business and economics? Give examples.

21. Explain how to visualize the First Fundamental Theorem of Calculus.

22. Explain how to use FnInt to support a definite integral computation. Why is FnInt often more useful than applying the Second Fundamental Theorem of Calculus to evaluate definite integrals?

23. a) Given that there is no explicit formula in terms of x for $\int (\sin 2x)/x\, dx$, draw a graph of one antiderivative of $(\sin 2x)/x$.

    b) How are all the antiderivatives of $(\sin 2x)/x$ related to the graph in part (a)?

Practice Exercises

For each of the following functions $y = f(x)$, consider the area above the x-axis and under the graph of $y = f(x)$ from $x = a$ to $x = b$.

a) Make a sketch illustrating the RAM for LR_5, RR_5, and MR_5 showing the five approximating rectangles.

b) Write out by hand LR_5, RR_5, and MR_5, and compute each sum.

1. $f(x) = 6 - x$ for $x = 0$ to $x = 5$
2. $f(x) = x^2 + 2x - 1$ for $x = 1$ to $x = 6$

For each of the following functions, estimate the area above the x-axis and under the graph of $y = f(x)$ from $x = a$ to $x = b$ using the RAM for $n = 10$, 100, and 1000. Do all three RAM's; $LR_n f$, $RR_n f$, and $MR_n f$. First verify that each function is non-negative on the specified interval $[a, b]$.

3. $f(x) = x^2 + 5$ for $x = 0$ to $x = 3$
4. $f(x) = x^3 - 2x + 1$ for $x = 0$ to $x = 2$
5. $f(x) = 2 \sin x$ from $x = 0$ to $x = \pi$
6. $f(x) = \frac{1}{2} e^{-x^2}$ for $x = -5$ to $x = 5$

Use standard formulas to evaluate the sums in Exercises 7–10.

7. a) $\sum_{k=1}^{10} (k + 2)$ b) $\sum_{k=1}^{10} (2k - 12)$

8. a) $\sum_{k=1}^{6} \left(k^2 - \frac{1}{6}\right)$ b) $\sum_{k=1}^{6} k(k + 1)$

9. a) $\sum_{k=1}^{5} (k^3 - 45)$ b) $\sum_{k=1}^{6} \left(\frac{k^3}{7} - \frac{k}{7}\right)$

10. Evaluate the following sums.

    a) $\sum_{k=1}^{3} 2^{k-1}$ b) $\sum_{k=0}^{4} (-1)^k \cos k\pi$

    c) $\sum_{k=-1}^{2} k(k + 1)$ d) $\sum_{k=1}^{4} \frac{(-1)^{k+1}}{k(k + 1)}$

11. Express the following sums in sigma notation.

    a) $1 + 2 + 4 + 8$ b) $1 + \frac{1}{3} + \frac{1}{9} + \frac{1}{27} + \frac{1}{81}$

    c) $1 - 2 + 3 - 4 + 5$ d) $\frac{5}{2} + \frac{5}{4} + \frac{5}{6}$

Express the limits in Exercises 12 and 13 as definite integrals.

12. $\lim_{\|P\|\to 0} \sum_{k=1}^{n} \frac{1}{c_k} \Delta x_k$, where P is a partition of $[1, 2]$.

13. $\lim_{\|P\|\to 0} \sum_{k=1}^{n} e^{c_k} \Delta x_k$, where P is a partition of $[0, 1]$.

Find a formula for $\text{RR}_n f$ in terms of n for each of the functions $y = f(x)$. Then determine $\lim_{x\to\infty} \text{RR}_n f$ to find the *exact* area under the curve $y = f(x)$ from $x = a$ to $x = b$.

14. $f(x) = x^2 + 2x + 3$ for $x = 0$ to $x = 4$

15. $f(x) = 2x^3 + 3x$ for $x = 0$ to $x = 5$

16. Suppose $\int_{-2}^{2} f(x)\,dx = 4$, $\int_{2}^{5} f(x)\,dx = 3$, $\int_{-2}^{5} g(x)\,dx = 2$. Which, if any, of the following statements are true, and which, if any, are false?

a) $\displaystyle\int_{5}^{2} f(x)\,dx = -3$

b) $\displaystyle\int_{-2}^{5} (f(x) + g(x)) = 9$

c) $f(x) \le g(x)$ on the interval $-2 \le x \le 5$

17. Suppose $\displaystyle\int_0^1 f(x)\,dx = \pi$. Find

a) $\displaystyle\int_0^1 f(t)\,dx$ b) $\displaystyle\int_1^0 f(y)\,dy$ c) $\displaystyle\int_0^1 -3f(z)\,dz$.

Find the areas of the shaded regions in Exercises 18–21. Support with an FnInt computation.

18.

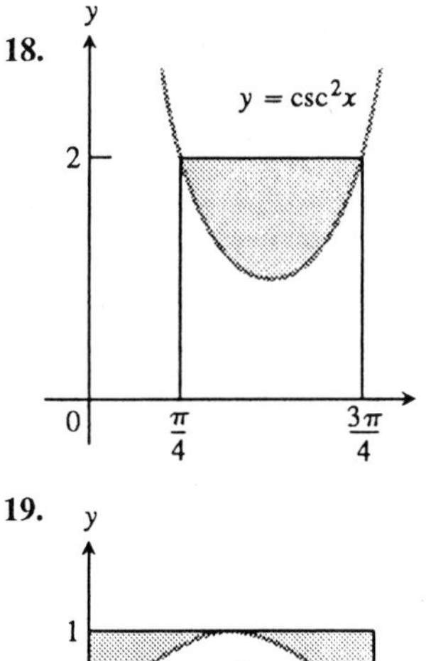

19.

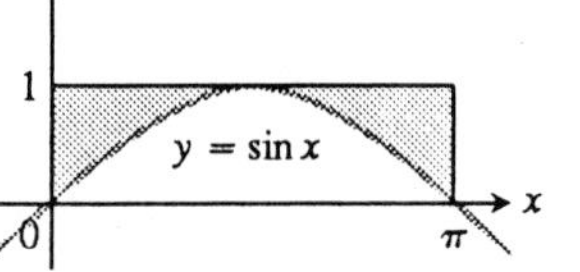

20.

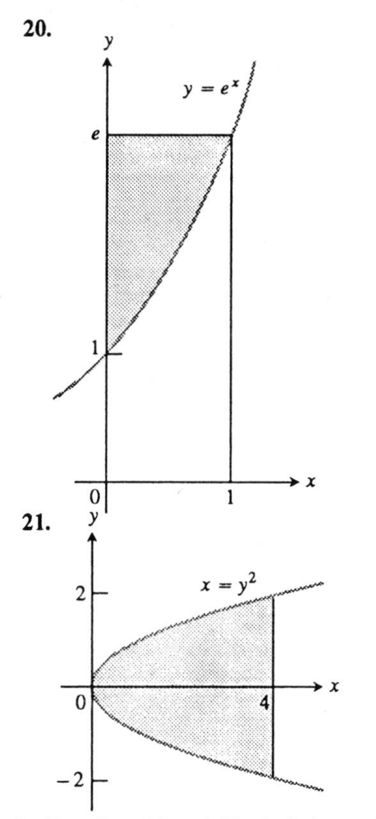

21.

In Exercises 22 and 23, find the total area between the curve and the x-axis.

22. $y = 4 - x, 0 \le x \le 6$

23. $y = \cos x, -\pi \le x \le \pi$

Evaluate the integrals in Exercises 24–49. Support with an FnInt computation.

24. $\displaystyle\int_{-1}^{1} (3x^2 - 4x + 7)\,dx$

25. $\displaystyle\int_0^1 (8s^3 - 12s^2 + 5)\,ds$

26. $\displaystyle\int_1^2 \frac{4}{x^2}\,dx$

27. $\displaystyle\int_1^{27} x^{-4/3}\,dx$

28. $\displaystyle\int_1^4 \frac{dt}{t\sqrt{t}}$

29. $\displaystyle\int_0^2 3\sqrt{4x+1}\,dx$

30. $\displaystyle\int_0^1 \frac{36\,dx}{(2x+1)^3}$

31. $\displaystyle\int_1^2 \left(x + \frac{1}{x^2}\right) dx$

32. $\displaystyle\int_0^{\pi} \sin 5\theta\,d\theta$

33. $\displaystyle\int_0^{\pi} \cos 5t\,dt$

34. $\displaystyle\int_0^{\pi/3} \sec^2\theta\,d\theta$

35. $\displaystyle\int_{\pi/4}^{3\pi/4} \csc^2 x\,dx$

36. $\displaystyle\int_{\pi}^{3\pi} \cot^2 \frac{x}{6}\,dx$

37. $\displaystyle\int_0^{\pi} \tan^2 \frac{\theta}{3}\,d\theta$

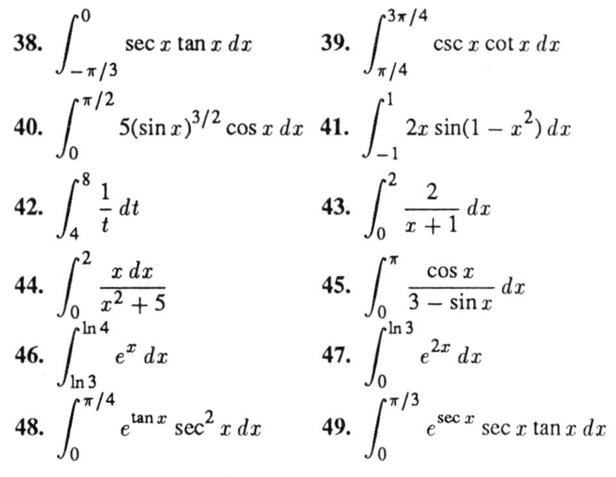

38. $\int_{-\pi/3}^{0} \sec x \tan x \, dx$

39. $\int_{\pi/4}^{3\pi/4} \csc x \cot x \, dx$

40. $\int_{0}^{\pi/2} 5(\sin x)^{3/2} \cos x \, dx$

41. $\int_{-1}^{1} 2x \sin(1 - x^2) \, dx$

42. $\int_{4}^{8} \frac{1}{t} \, dt$

43. $\int_{0}^{2} \frac{2}{x+1} \, dx$

44. $\int_{0}^{2} \frac{x \, dx}{x^2 + 5}$

45. $\int_{0}^{\pi} \frac{\cos x}{3 - \sin x} \, dx$

46. $\int_{\ln 3}^{\ln 4} e^x \, dx$

47. $\int_{0}^{\ln 3} e^{2x} \, dx$

48. $\int_{0}^{\pi/4} e^{\tan x} \sec^2 x \, dx$

49. $\int_{0}^{\pi/3} e^{\sec x} \sec x \tan x \, dx$

Evaluate the integrals in Exercises 50–53. Support with an FnInt computation.

50. $\int_{2}^{3} \left(t - \frac{2}{t}\right)\left(t + \frac{2}{t}\right) dt$ (*Hint:* Multiply first.)

51. $\int_{-1}^{0} (1 - 3w)^2 \, dw$ (*Hint:* Square first.)

52. $\int_{-4}^{0} |x| \, dx$ (*Hint:* Write it without absolute values first.)

53. $\int_{1}^{e} \frac{x+1}{x} \, dx$ (*Hint:* Divide first.)

Use the given substitutions to evaluate the integrals in Exercises 54–57. Support with an FnInt computation.

54. $\int_{-\pi/2}^{\pi/2} 15 \sin^4 3x \cos 3x \, dx, \quad u = \sin 3x$

55. $\int_{0}^{1} \frac{e^x}{1 + e^x} \, dx, \quad u = 1 + e^x$

56. $\int_{0}^{\pi/2} \frac{3 \sin x \cos x}{\sqrt{1 + 3\sin^2 x}} \, dx, \quad u = 1 + 3\sin^2 x$

57. $\int_{0}^{1} (2x - 1)e^{(x^2 - x)} \, dx, \quad u = x^2 - x$

In Exercises 58–65, find dy/dx. Support graphically.

58. $y = \ln \sqrt{x}$

59. $y = \ln\left(\frac{e^x}{2}\right)$

60. $y = \ln(3x^2 + 6)$

61. $y = \ln(1 + e^x)$

62. $y = \frac{1}{e^x}$

63. $y = xe^{-x}$

64. $y = e^{(1 + \ln x)}$

65. $y = \int_{1}^{x} \ln t \, dt$

Verify the integral formulas in Exercises 66–69 by differentiating their right-hand sides. Support with a graph involving FnInt.

66. $\int \frac{\ln 5x}{x} \, dx = \frac{1}{2}(\ln 5x)^2 + C$

67. $\int x^2 \ln x \, dx = \frac{x^3}{3} \ln x - \frac{x^3}{9} + C$

68. $\int e^x \sin x \, dx = \frac{e^x}{2}(\sin x - \cos x) + C$

69. $\int xe^x \, dx = xe^x - e^x + C$

Determine a graphical representation of the integrals in Exercises 70 and 71. Can you find an explicit antiderivative in terms of x?

70. $\int \frac{\sin 3x}{x} \, dx$

71. $\int \frac{1}{4} e^{-x^2/2} \, dx$

72. Solve for x: $\int_{0}^{x} (t^3 - 2t + 3) \, dt = 4$

73. Solve for x: $\int_{1}^{x} \frac{\cos t}{t} \, dt = 1$

74. What is the domain of $f(x) = \int_{1}^{x} \frac{\cos t}{t} \, dt$? Draw a graph of the function.

75. Simplify:

a) $\ln e^{2x}$ b) $\ln 2e$ c) $\ln \frac{1}{e}$

76. Simplify:

a) $e^{2 \ln 2}$ b) $e^{-\ln 4}$ c) $e^{\ln(\ln x)}$

Solve for y in Exercises 77–80.

77. $\ln(y^2 + y) - \ln y = x, \quad y > 0$

78. $\ln(y - 4) = -4t$

79. $e^{2y} = 4x^2, \quad x > 0$

80. $e^{-0.1y} = \frac{1}{2}$

81. Solve for t: $3^{t-1} = t^3 - 2t + 1$

82. Solve for x: $\int_{1}^{x} \frac{1}{2} e^{-t^2} \, dt = 0.25.$

83. Calculate the ratio $(\ln x)/(\log x)$ for various values of x, noting the results as you go along. What happens? When you think you know, try it for a few more values of x. Then calculate $\ln 10$. Section 7.3 will explain the connection.

84. Which of the following methods could be used successfully to prepare the integral

$$\int 3x^2(x^3 - 1)^5 \, dx$$

for evaluation?

a) Expand $(x^3 - 1)^5$ and multiply the result by $3x^2$ to get a polynomial to integrate term by term.

b) Factor $3x^2$ out front to get an integral of the form $3x^2 \int u^5\,du$.

c) Substitute $u = x^3 - 1$ to get an integral of the form $\int u^5 du$.

85. The substitution $u = \tan x$ gives

$$\int \sec^2 x \tan x\,dx = \int \tan x \cdot \sec^2 x\,dx$$

$$= \int u\,du = \frac{u^2}{2} + C = \frac{\tan^2 x}{2} + C.$$

The substitution $u = \sec x$ gives

$$\int \sec^2 x \tan x\,dx = \int \sec x \cdot \sec x \tan x\,dx$$

$$= \int u\,du = \frac{u^2}{2} + C = \frac{\sec^2 x}{2} + C.$$

Can both integrations be correct? Explain.

86. Suppose that $y = f(x)$ is continuous and positive throughout the interval $[0, 1]$. Suppose also that for every value of x in this interval the area between the graph of f and the subinterval $[0, x]$ is $\sin x$. Find $f(x)$.

87. Suppose that f has a positive derivative for all values of x and that $f(1) = 0$. Which of the following statements must be true of the function

$$g(x) = \int_0^x f(t)\,dt?$$

a) g is a differentiable function of x.

b) g is a continuous function of x.

c) The graph of g has a horizontal tangent at $x = 1$.

d) g has a local maximum at $x = 1$.

e) g has a local minimum at $x = 1$.

f) The graph of g has an inflection point at $x = 1$.

g) The graph of dg/dx crosses the x-axis at $x = 1$.

88. Suppose $F(x)$ is an antiderivative of $f(x) = \sqrt{1 + x^4}$. Express $\int_0^1 \sqrt{1 + x^4}\,dx$ in terms of F.

89. Show that $y = x^2 + \int_1^x 1/t\,dt + 1$ solves the initial value problem

Differential equations: $y'' = 2 - \dfrac{1}{x^2}$

Initial conditions: $y = 2$ and $y' = 3$ when $x = 1$.

90. The acceleration of a particle moving back and forth along a line is $d^2s/dt^2 = \pi^2 \cos \pi t$ m/sec^2. If $s = 0$ m and $v = 8$ m/sec when $t = 0$, find the value of s when $t = 1$.

91. Solve the following initial value problems.

a) Differential equation: $y^{(4)} = \cos x$
Initial conditions: $y = 3$, $y' = 2$, $y'' = 1$, and $y''' = 0$ when $x = 0$

b) Differential equation: $\dfrac{dy}{dx} = \dfrac{4x}{(1 + x^2)^2}$
Initial condition: $y = 0$ when $x = 0$

c) Differential equation: $\dfrac{dy}{dt} = -ky$
Initial condition: $y = y_0$ when $t = 0$

92. STOPPING A MOTORCYCLE The State of Illinois Cycle Rider Safety Program requires riders to be able to brake from 30 mph (44 ft/sec) to 0 in 45 ft. What constant deceleration does it take to do that? To find out, carry out these steps:

STEP 1: Solve the following initial value problem. The answer will involve k.

Differential equation: $\dfrac{d^2s}{dt^2} = -k$
Initial conditions: $ds/dt = 44$ and $s = 0$ when $t = 0$

STEP 2: Find the time t^* when $ds/dt = 0$. The answer will still involve k.

STEP 3: Solve the equation $s(t^*) = 45$ for k.

93. Which of the following graphs shows the solution of the initial value problem

$$\frac{dy}{dx} = 2x, \quad y = 4 \text{ when } x = 1?$$

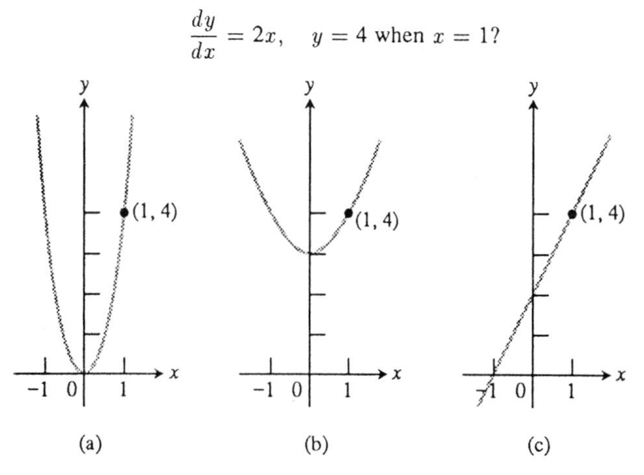

94. HOUSEHOLD ELECTRICITY We model the voltage V in our homes with the sine function

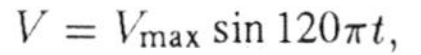

$$V = V_{\max} \sin 120\pi t,$$

which expresses V in volts as a function of time t in seconds. The function runs through 60 cycles each second. The number V_{max} is the **peak voltage**.

To measure the voltage effectively, we use an instrument that measures the square root of the average value of the square of the voltage over a 1-sec interval:

$$V_{rms} = \sqrt{(V^2)_{av}}.$$

The subscript "rms" (read the letters separately) stands for "root mean square." It turns out that

$$V_{rms} = V_{max}/\sqrt{2}. \quad (1)$$

The familiar phrase "115 volts ac" means that the rms voltage is 115. The peak value, obtained from Eq. (1) as $V_{max} = 115\sqrt{2}$, is about 163 volts.

a) Find the average value of V^2 over a 1-sec interval. Then find V_{rms} and verify Eq. (1).

b) The circuit that runs your electric stove is rated 240 volts rms. What is the peak value of the allowable voltage?

95. Compute the average value of the temperature function

$$f(x) = 37 \sin\left[\frac{2\pi}{365}(x - 101)\right] + 25$$

for a 365-day year. This is one way to estimate the annual mean air temperature in Fairbanks, Alaska. The National Weather Service's official figure, a numerical average of the daily normal mean air temperatures for the year, is 25.7°F, which is slightly higher than the average value of $f(x)$. Figure 1.148 shows why.

96. Let f be a function that is differentiable on $[a, b]$. In Chapter 1 we defined the average rate of change of f on $[a, b]$ to be

$$\frac{f(b) - f(a)}{b - a}$$

and the instantaneous rate of change of f at x to be $f'(x)$. In this chapter we defined the average value of a function. For the new definition of average to be consistent with the old one, we should have

$$\frac{f(b) - f(a)}{b - a} = \text{average value of } f' \text{ on } [a, b].$$

Show that this is the case.

97. Find the average value of

a) $y = \sqrt{3x}$ over the interval $0 \le x \le 3$,

b) $y = \sqrt{ax}$ over the interval $0 \le x \le a$.

98. What step size h would you use to be sure of estimating the value of

$$\ln 3 = \int_1^3 \frac{1}{x}\, dx$$

by Simpson's Rule with an error of no more than 10^{-4} in absolute value?

99. A brief calculation shows that if $0 \le x \le 1$, then the second derivative of $f(x) = \sqrt{1 + x^4}$ lies between 0 and 8. Based on this, about how many subdivisions would you need to estimate the integral of f from 0 to 1 with an error no greater than 10^{-3} in absolute value?

100. A direct calculation shows that

$$\int_0^{\pi} 2\sin^2 x\, dx = \pi.$$

How close do you come to this value by using the Trapezoidal Rule with $n = 6$? Simpson's Rule with $n = 6$? Try them and find out.

101. You are planning to use Simpson's Rule to estimate the value of the integral

$$\int_1^2 f(x)\, dx$$

with an error magnitude less than 10^{-5}. You have determined that $|f^{(4)}(x)| \le 3$ throughout the interval of integration. How many subdivisions should you use to assure the required accuracy? (Remember that for Simpson's Rule the number has to be even.)

102. A NEW PARKING LOT To meet the demand for parking, your town has allocated the area shown in Fig. 5.43. As the town engineer, you have been asked by the town council to find out if the lot can be built for \$11,000. The cost to clear the land will be \$0.10 a square foot, and the lot will cost \$2.00 a square foot to pave. Can the job be done for \$11,000?

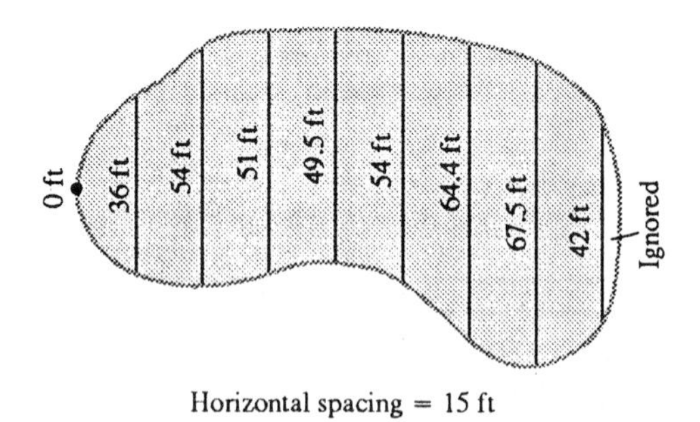

5.43

103. APPRECIATION A violin made in 1785 by John Betts, one of England's finest violin makers, cost \$250 in 1924 and sold for \$7500 in 1988. Assuming a constant rate of appreciation, what was that rate?

104. WORKING UNDER WATER The intensity $L(x)$ of light x feet beneath the surface of the ocean satisfies the differential equation

$$\frac{dL}{dx} = -kL.$$

As a diver, you know from experience that diving to 18 ft in the Caribbean Sea cuts the intensity in half. You cannot work without artificial light when the intensity falls below a tenth of the surface value. About how deep can you expect to work without artificial light?

105. THE PURCHASING POWER OF THE DOLLAR How long will it take the purchasing power of the dollar to decrease to three fourths of its present value if the annual inflation rate holds steady at 4%? (*Hint*: How long will it take the Consumer Price Index to increase to four thirds its present value at this inflation rate?)

106. INFLATION IN WEST GERMANY The consumer price index (1967 = 100) in West Germany was 175.8 in 1980 and 211.9 in 1986.

a) Assuming a constant rate of inflation, what was it?

b) What would you expect the index to be in 1992 if this rate continued?

107. INFLATION IN ITALY In contrast to West Germany (Exercise 95), the index in Italy was 295.5 in 1980 and 480.1 in 1986.

a) Assuming a constant rate of inflation, what was it?

b) What would you expect the index to be in 1992 if this rate were to continue?

108. THE RULE OF 70 FOR DEPLETION The Rule of 70 applies to deflation and depletion, too, to estimate how many years it will take something that decreases at a constant rate to reach one half its present value. To see why, solve the equation

$$A_0 e^{-kt} = \frac{1}{2} A_0$$

to show that $t = (\ln 2)/k$. Thus if $k = i/100$ is given as $i\%$,

$$t = \frac{\ln 2}{(i/100)} = \frac{100 \ln 2}{i} \approx \frac{69.3}{i} \approx \frac{70}{i}.$$

As a rule, it will take about $70/i$ years for the amount in question to decline from A_0 to $A_0/2$.

109. CONTINUATION OF EXERCISE 108 Use the Rule of 70 to tell about how long it will take

a) for prices to decline to half their present level at a constant deflation rate of 5%;

b) for oil reserves to deplete to half their present volume at a constant depletion rate of 7%.

Applications of Definite Integrals

OVERVIEW The importance of integral calculus stems from the fact that thousands of things we want to know can be calculated with integrals. This includes areas between curves, volumes and surface areas of solids, lengths of curves, the amount of work it takes to pump liquids up from below ground, the forces against flood gates, and the coordinates of the points where solid objects will balance. We can define all these things in natural ways as limits of Riemann sums of continuous functions on closed intervals. Then we can evaluate these limits by applying techniques of integral calculus including the use of technology.

There is a clear pattern to how we go about defining and calculating the integrals in our applications, a pattern that, once learned, enables us to define new integrals whenever we need them. In this chapter we look at specific applications first; then, in the concluding section, we examine the pattern and show how it leads to integrals in new situations.

6.1 Areas between Curves

This section shows how to find the area of a region in the coordinate plane by integrating the functions that define the region's boundaries.

The Basic Formula, Derived from Riemann Sums

Suppose that functions $f_1(x)$ and $f_2(x)$ are continuous and that $f_1(x) \geq f_2(x)$ throughout an interval $a \leq x \leq b$ (Fig. 6.1). The region between the curves $y = f_1(x)$ and $y = f_2(x)$ from a to b might accidentally have a shape that would let us find its area with a formula from geometry, but we usually have to find the area with an integral instead.

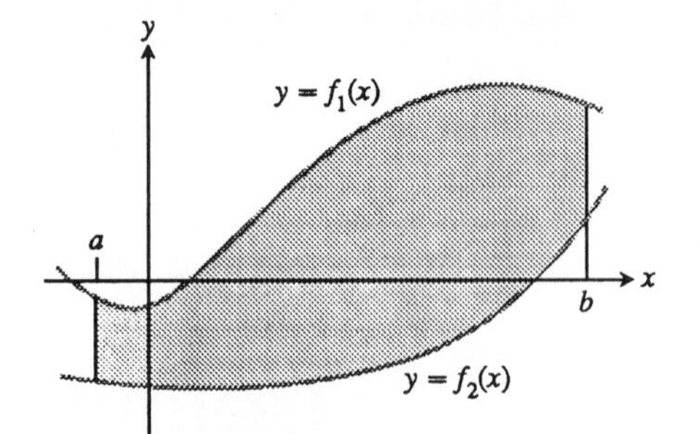

6.1 A typical region between two curves.

To see what that integral should be, we start by approximating the region with vertical rectangles. Figure 6.2 shows a typical approximation, based on a partition $P = \{x_0, x_1, \ldots, x_n\}$ of the interval $[a, b]$. The kth rectangle, shown in detail in Fig. 6.3, is Δx_k units wide and runs from the point $(c_k, f_2(c_k))$ on the lower curve to the point $(c_k, f_1(c_k))$ on the upper curve. Its area, height $\times$ width, is $(f_1(c_k) - f_2(c_k))\Delta x_k$.

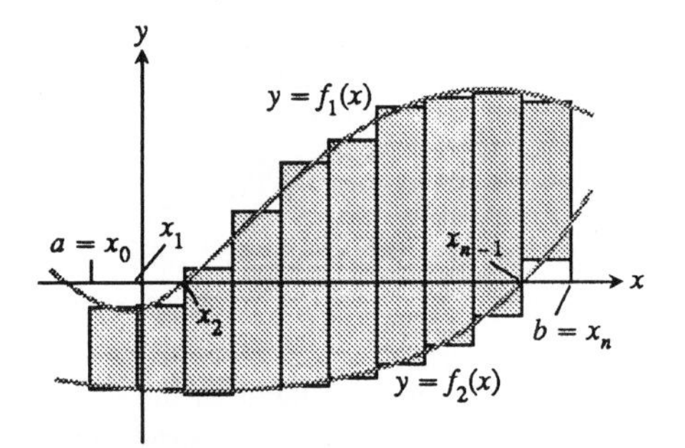

6.2 To derive a formula for the region's area, we think of approximating the region with rectangles perpendicular to the x-axis.

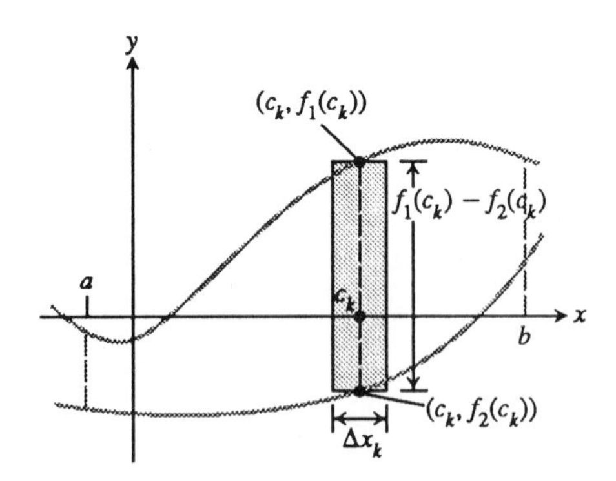

6.3 A typical rectangle is $f_1(c_k) - f_2(c_k)$ units high by Δx_k units wide.

We approximate the area of the region by adding the areas of the rectangles from a to b:

$$\text{Rectangle area sum} = \sum_{k=1}^{n} (f_1(c_k) - f_2(c_k))\Delta x_k. \tag{1}$$

This sum is a Riemann sum for the difference function $(f_1 - f_2)$ over the closed interval $[a, b]$.

Since f_1 and f_2 are continuous, two things will happen as we subdivide $[a, b]$ more finely and let the norm of the partition go to zero. The rectangles will approximate the region with increasing geometric accuracy and the Riemann sums in (1) will approach a limit. We therefore define this limit, the definite integral of $(f_1 - f_2)$ from a to b, to be the area of the region.

Definition

> If functions f_1 and f_2 are continuous and if $f_1(x) \geq f_2(x)$ throughout the interval $a \leq x \leq b$, then the **area of the region between the curves** $y = f_1(x)$ and $y = f_2(x)$ from a to b is the integral of $(f_1 - f_2)$ from a to b:
>
> $$\text{Area} = \int_a^b (f_1(x) - f_2(x))\, dx. \tag{2}$$

To apply Eq. (2), we take the following steps.

How to Find the Area between Two Curves

1. *Graph the curves together.* This tells you which is f_1 (upper curve) and which is f_2 (lower curve). It also helps find the limits of integration if you do not already know them.
2. *Find the limits of integration.*
3. *Write a formula for* $f_1(x) - f_2(x)$. Simplify it if you can.
4. *Integrate* $f_1(x) - f_2(x)$ *from* a *to* b. The number you get is the area.

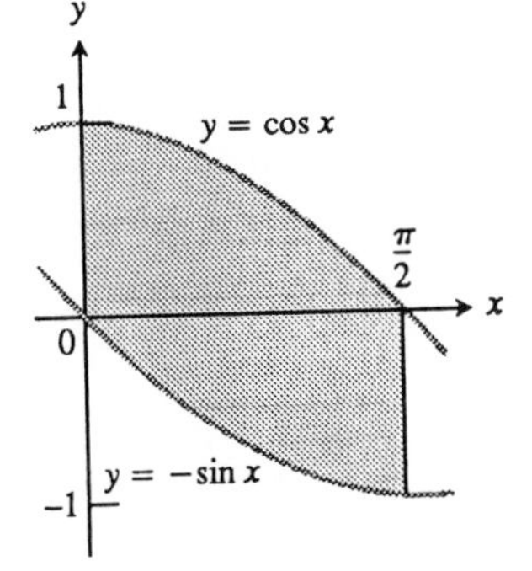

EXAMPLE 1 Find the area between the curves $y = \cos x$ and $y = -\sin x$ from 0 to $\pi/2$.

Solution

STEP 1: *The graphs.* We graph the curves together. The upper curve is $y = \cos x$, so we take $f_1(x) = \cos x$ in the area formula. The lower curve is $y = -\sin x$ so $f_2(x) = -\sin x$.

STEP 2: *The limits of integration.* They are already given: $a = 0$ and $b = \pi/2$.

STEP 3: *The formula for* $f_1(x) - f_2(x)$. From Step 1,

$$f_1(x) - f_2(x) = \cos x - (-\sin x) = \cos x + \sin x.$$

STEP 4: *Integrate* $f_1(x) - f_2(x)$ *from* $a = 0$ *to* $b = \pi/2$:

$$\int_0^{\pi/2} (\cos x + \sin x)\, dx = \Big[\sin x - \cos x \Big]_0^{\pi/2} = [1 - 0] - [0 - 1] = 2.$$

The area between the curves is 2.

Curves That Cross

When a region is determined by curves that cross, the crossing points give the limits of integration.

EXAMPLE 2 Find the area of the region enclosed by the parabola $y = 2 - x^2$ and the line $y = -x$.

Solution

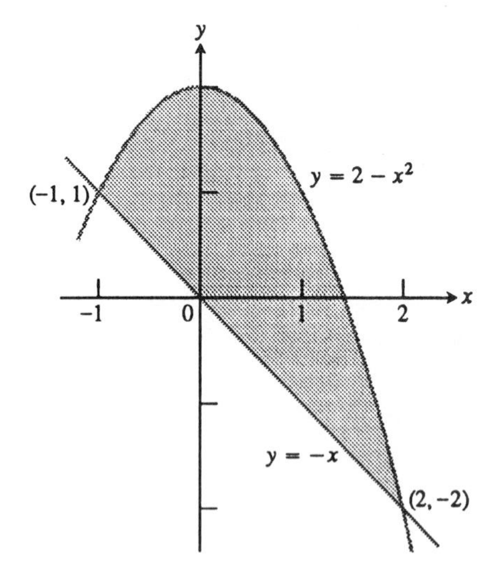

STEP 1: *The graphs.* We graph the curves together. The upper curve is $y = 2 - x^2$, so $f_1(x) = 2 - x^2$. The lower curve is $y = -x$, so $f_2(x) = -x$. The x-coordinates of the points where the parabola and line cross are the limits of integration. We find them in Step 2.

STEP 2: *The limits of integration.* We find the limits of integration by solving the equations $y = 2 - x^2$ and $y = -x$ simultaneously for x:

$$2 - x^2 = -x \qquad \text{(Equate } f_1(x) \text{ and } f_2(x))$$

$$x^2 - x - 2 = 0 \qquad \text{(Transpose)}$$

$$(x + 1)(x - 2) = 0 \qquad \text{(Factor)}$$

$$x = -1, x = 2. \qquad \text{(Solve)}$$

The region runs from $x = -1$ on the left to $x = 2$ on the right. The limits of integration are $a = -1$ and $b = 2$.

STEP 3: *The formula for* $f_1(x) - f_2(x)$.

$$f_1(x) - f_2(x) = (2 - x^2) - (-x) = 2 - x^2 + x$$

$$= 2 + x - x^2. \qquad \begin{pmatrix}\text{Rearranged—a} \\ \text{matter of taste}\end{pmatrix}$$

STEP 4: ***Integrate.***

$$\int_a^b [f_1(x) - f_2(x)]\,dx = \int_{-1}^{2} (2 + x - x^2)\,dx = \left[2x + \frac{x^2}{2} - \frac{x^3}{3}\right]_{-1}^{2}$$

$$= \left(4 + \frac{4}{2} - \frac{8}{3}\right) - \left(-2 + \frac{1}{2} + \frac{1}{3}\right)$$

$$= 6 + \frac{3}{2} - \frac{9}{3} = \frac{9}{2}.$$

The area of the region is $9/2$. ■

CHECKING ANSWERS

Use FnInt to support the computation in Example 2.

Example 2 is typical of contrived area problems. That is, problems where the limits of integration are easy to find (e.g., by factoring). The next example is not contrived and requires the use of technology.

EXAMPLE 3 Find the area of the region in the first quadrant above the curve $y = x$ and below the curve $y = 4\sin((1/2)x)$.

Solution The two graphs intersect at $x = 0$ and have only one point of intersection (near $x = 4$) in the first quadrant (Fig. 6.4).

If the x-coordinate of the point of intersection in the first quadrant of the two graphs is a then the area of the desired region is

$$\int_0^a \left(4\sin\left(\frac{1}{2}x\right) - x\right)\,dx.$$

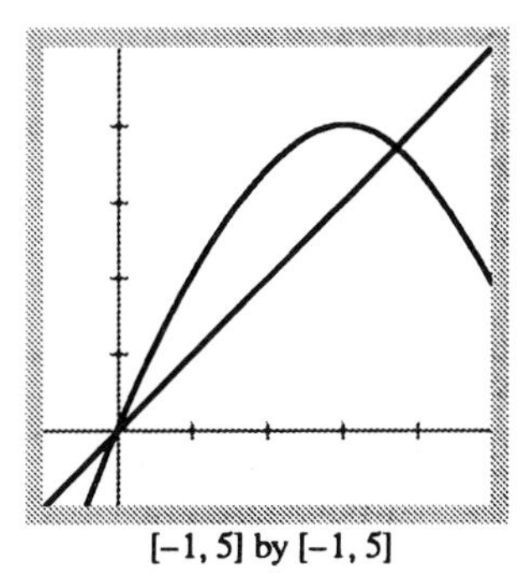

[−1, 5] by [−1, 5]

6.4 Graphs of $y = x$ and $y = 4\sin((1/2)x)$.

Next we need to find a by solving $x = 4\sin((1/2)x)$ for x. Algebra is no help! Using technology (graphing zoom-in or a root finder on $y = x - 4\sin((1/2)x)$), we find $a = 3.790988534$ (accurate to the digits shown). Finally we apply FnInt to compute

$$\int_0^a (4\sin(\frac{1}{2}x) - x)\,dx = 3.36638\ldots$$

You will confirm this in the exercises. Technology was essential in this example. ■

AGREEMENT

From now on, unless otherwise specified, we will report results of FnInt computations with 5 decimal place accuracy.

Boundaries with Changing Formulas

If the formula for one of the bounding curves changes at some point across the region, you may have to add two or more integrals to find the area.

EXAMPLE 4 Find the area of the region in the first quadrant bounded above by the curve $y = \sqrt{x}$ and below by the x-axis and the line $y = x - 2$.

Solution

STEP 1: *The graphs.* We graph the curves together. The entire upper boundary of the region consists of the curve $y = \sqrt{x}$, so $f_1(x) = \sqrt{x}$. The lower boundary consists of two curves, first $y = 0$ for $0 \le x \le 2$ and then $y = x - 2$ for $2 \le x \le 4$. Hence the formula for $f_2(x)$ changes from $f_2(x) = 0$ for $0 \le x \le 2$ to $f_2(x) = x - 2$ for $2 \le x \le 4$.

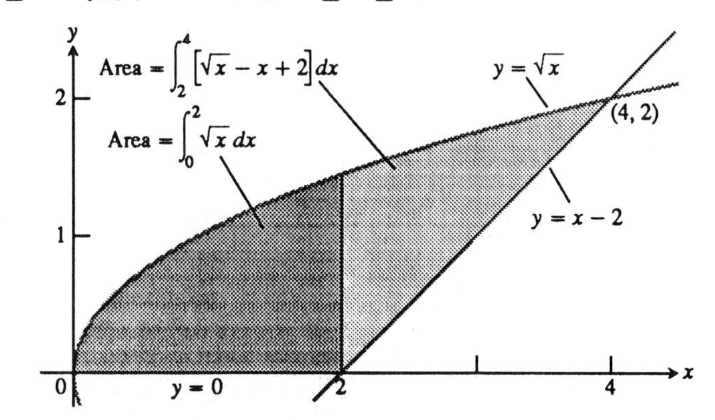

STEP 2: *The limits of integration.* The limits of integration for the pair $f_1(x) = \sqrt{x}$ and $f_2(x) = 0$ are $a = 0$ and $b = 2$.

For the pair $f_1(x) = \sqrt{x}$ and $f_2(x) = x - 2$, the left-hand limit is $a = 2$ and the right-hand limit is the x-coordinate of the upper point where the line crosses the parabola. To find it, we solve the equations $y = \sqrt{x}$ and $y = x - 2$ simultaneously for x:

$$\sqrt{x} = x - 2 \qquad \left(\text{Equate } f_1(x) = \sqrt{x} \text{ and } f_2(x) = x - 2\right)$$

$$x = (x-2)^2 = x^2 - 4x + 4 \qquad \text{(Square)}$$

$$x^2 - 5x + 4 = 0 \qquad \text{(Rearrange)}$$

$$(x-1)(x-4) = 0 \qquad \text{(Factor)}$$

$$x = 1, x = 4. \qquad \text{(Solve)}$$

The value $x = 1$ does not satisfy the equation $\sqrt{x} = x - 2$. It is an extraneous root introduced by squaring. The value $x = 4$ gives our upper limit of integration.

STEP 3: *The formulas for* $f_1(x) - f_2(x)$.

For $0 \le x \le 2$: $\quad f_1(x) - f_2(x) = \sqrt{x} - 0 = \sqrt{x}$,

For $2 \le x \le 4$: $\quad f_1(x) - f_2(x) = \sqrt{x} - (x - 2) = \sqrt{x} - x + 2.$

STEP 4: *Integrate.* We have two integrals to evaluate. Their sum is the area.

$$\text{Area} = \int_0^4 [f_1(x) - f_2(x)]\,dx = \int_0^2 \sqrt{x}\,dx + \int_2^4 (\sqrt{x} - x + 2)\,dx$$

$$= \left[\frac{2}{3}x^{3/2}\right]_0^2 + \left[\frac{2}{3}x^{3/2} - \frac{x^2}{2} + 2x\right]_2^4$$

$$= \frac{2}{3}(2)^{3/2} + \left(\frac{2}{3}(4)^{3/2} - \frac{16}{2} + 8\right) - \left(\frac{2}{3}(2)^{3/2} - \frac{4}{2} + 4\right)$$

$$= \frac{2}{3}(8) - 2 = \frac{10}{3}.$$

The area of the region is 10/3. ■

Integrating with Respect to *y*

When a region's bounding curves are described by giving x as a function of y, the basic formula changes.

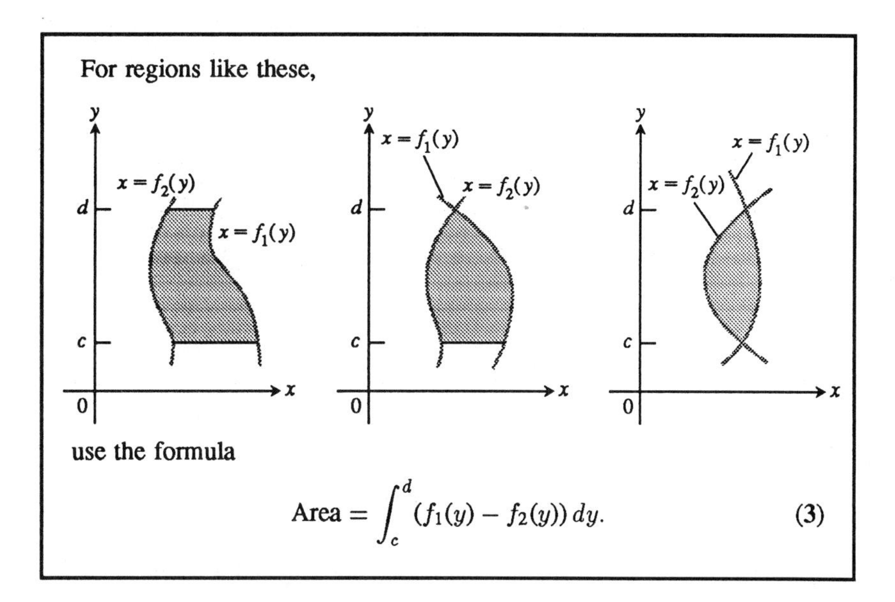

$$\text{Area} = \int_c^d (f_1(y) - f_2(y))\,dy. \qquad (3)$$

The only difference between this formula and the one in Eq. (2) is that we are now integrating with respect to y instead of x. We can sometimes save time by doing so. The basic steps are the same as before.

EXAMPLE 5 *The area in Example 4, found by a single integration with respect to y.* Find the area of the region between the curves $x = y^2$ and $x = y + 2$ in the first quadrant.

Solution

STEP 1: *The graphs.* We graph the curves together. The right-hand curve is $x = y + 2$, so $f_1(y) = y + 2$. The left-hand curve is $x = y^2$, so $f_2(y) = y^2$.

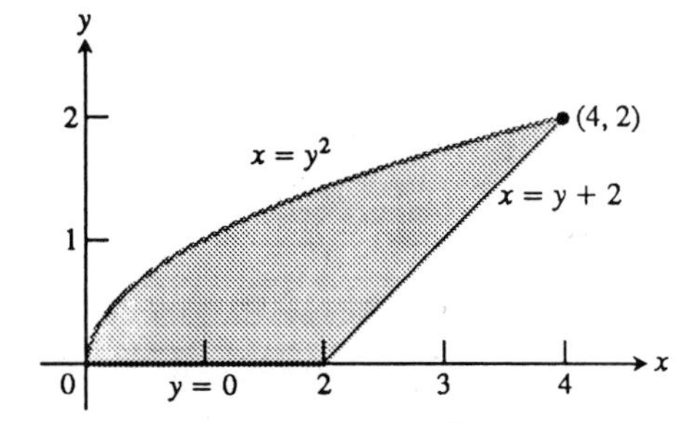

STEP 2: *The limits of integration.* The lower limit of integration is $y = 0$. The upper limit is the y-coordinate of the upper point where the line crosses the parabola. We find it by solving the equations $x = y + 2$ and $x = y^2$ simultaneously for y:

$$\begin{aligned} y + 2 &= y^2 && \text{(Equate } f_1(y) = y + 2 \text{ and } f_2(y) = y^2\text{)} \\ y^2 - y - 2 &= 0 && \text{(Rearrange)} \\ (y+1)(y-2) &= 0 && \text{(Factor)} \\ y = -1, \quad y &= 2. && \text{(Solve)} \end{aligned}$$

The upper limit of integration is 2. (The value $y = -1$ gives the point of intersection *below* the x-axis.)

STEP 3: *The formula for $f_1(y) - f_2(y)$.*

$$f_1(y) - f_2(y) = y + 2 - y^2 = 2 + y - y^2. \qquad \begin{pmatrix}\text{Rearrangement} \\ \text{a matter of taste}\end{pmatrix}$$

STEP 4: *Integrate.*

$$\text{Area} = \int_a^b (f_1(y) - f_2(y))\,dy = \int_0^2 (2 + y - y^2)\,dy$$

$$= \left[2y + \frac{y^2}{2} - \frac{y^3}{3}\right]_0^2 = 4 + \frac{4}{2} - \frac{8}{3} = \frac{10}{3}.$$

The area of the region is 10/3. This is the result of Example 4, found with less work.

Combining Integrals with Formulas from Geometry

Sometimes the fastest way to find the area of a region is to combine calculus and geometry.

EXAMPLE 6 The area of the region in Examples 4 and 5 found the fastest way. Find the area of the region in Examples 4 and 5.

Solution The area we want to find is the area between the curve $y = \sqrt{x}, 0 \leq x \leq 4$, and the x-axis, *minus* the area of a triangle with base 2 and height 2:

$$\text{Area} = \int_0^4 \sqrt{x}\,dx - \frac{1}{2}(2)(2) = \frac{2}{3}x^{3/2}\Big]_0^4 - 2$$

$$= \frac{2}{3}(8) - 0 - 2 = \frac{10}{3}.$$

Moral of Examples 4–6 It is sometimes easier to find the area between two curves by integrating with respect to y instead of x. Examine each region beforehand to determine which method, if either, is easier. Sketching the region may also reveal how to use geometry to simplify your work.

Exercises 6.1

Find the areas of the shaded regions in Exercises 1–6 analytically and support with technology.

1.

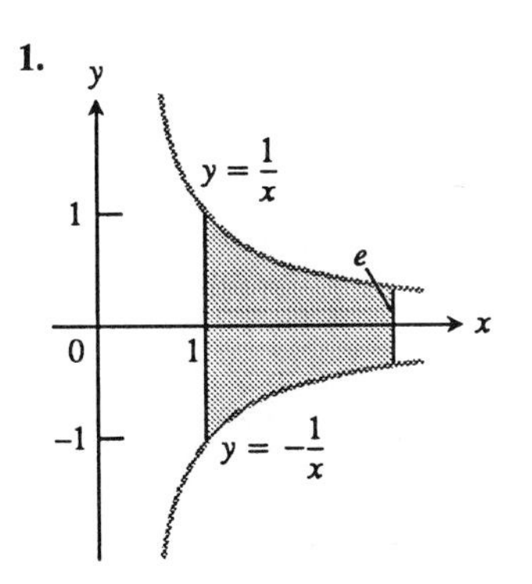

2.

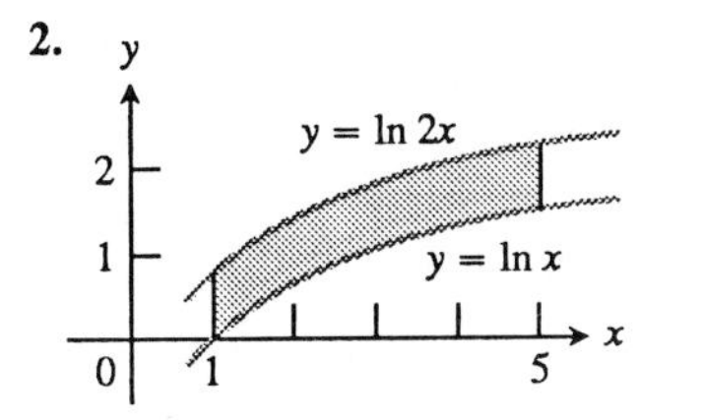

3. **4.**

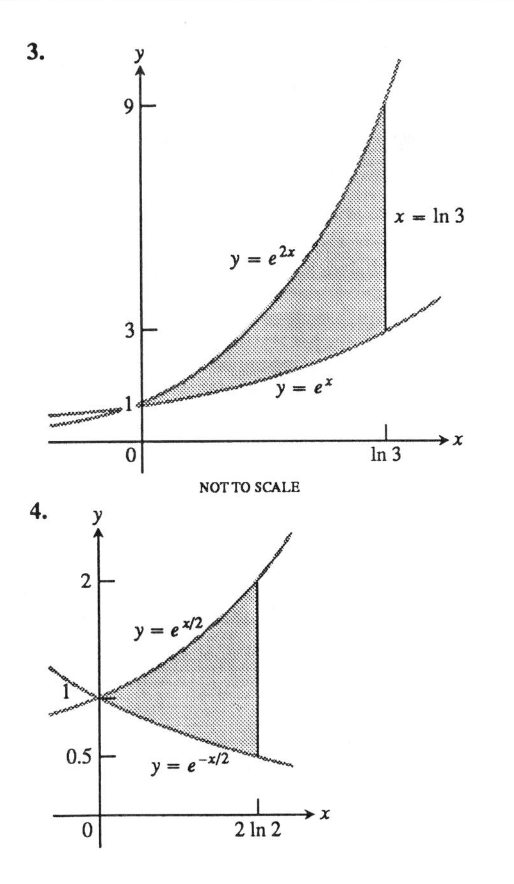

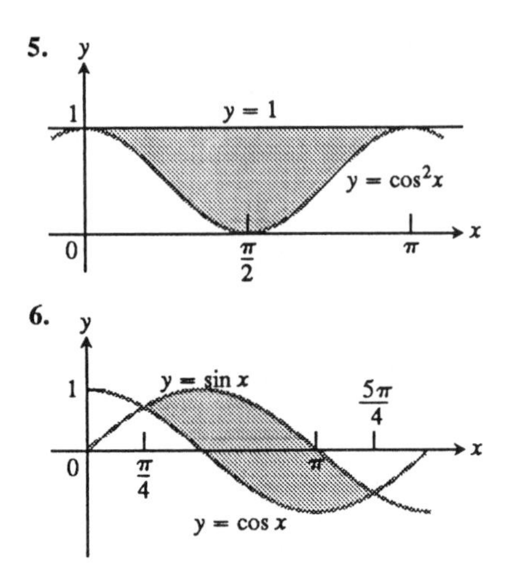

Whenever possible, find the area of the regions enclosed by the lines and curves in Exercises 7–22 analytically. Use technology where necessary.

7. The curve $y = x^2 - 2$ and the line $y = 2$
8. The x-axis and the curve $y = 2x - x^2$
9. The curve $y^2 = x$ and the line $x = 4$
10. The curve $y = 2x - x^2$ and the line $y = -3$
11. The curve $y = x^2$ and the line $y = x$
12. The curve $x = 3y - y^2$ and the line $x + y = 3$
13. The line $y = 2x$ and the curve $y = x^3 + 2x^2 - 3x + 1$
14. Below the line $y = 3-2x$ and above the curve $y = 2\cos 2x$ in the first quandrant
15. The curve $y = x^2 - 2x$ and the line $y = x$
16. The curve $x = 10 - y^2$ and the line $x = 1$
17. Above the line $y = 2x$ and below the curve $y = e^{-x^2}$ in the first quadrant
18. The curve $y = e^x$ and the lines $y = -x$ and $x = 2$
19. The line $y = x$ and the curve $y = 2 - (x - 2)^2$
20. The curves $y = 7 - 2x^2$ and $y = x^2 + 4$
21. The line $y = x$ and the curve $y = x^3 - 2x^2 - 3x + 1$ (*Hint:* There are two regions.)
22. The curve $y = 2 - x^2$ and $y = 2\cos 2x$ (*Hint:* There are two regions.)

Find the areas of the shaded regions in Exercises 23–26 analytically and support with technology.

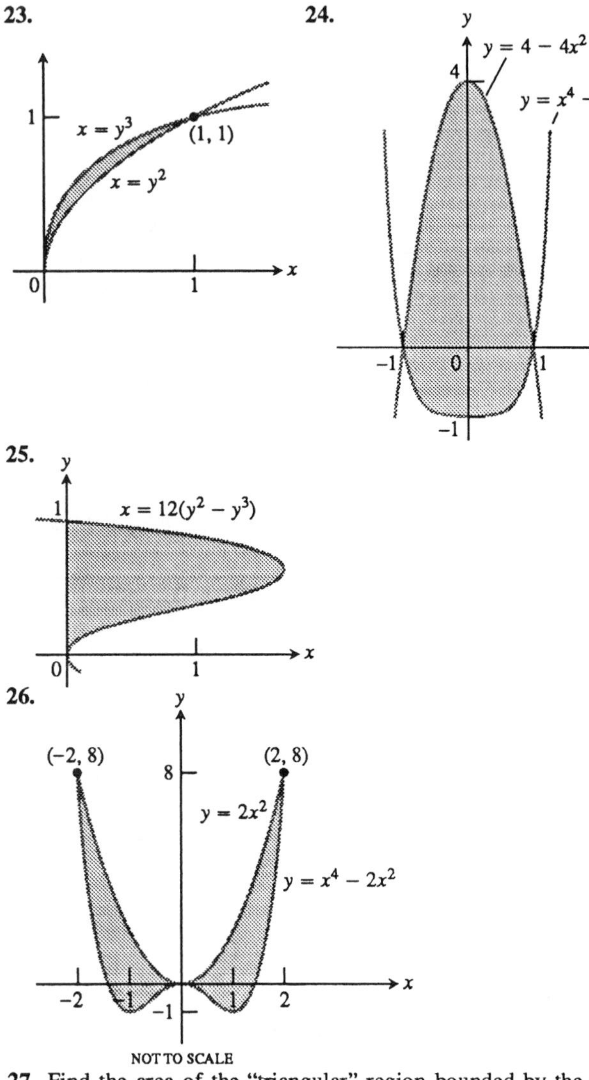

27. Find the area of the "triangular" region bounded by the y-axis and the curves $y = \sin x$ and $y = \cos x$ in the first quadrant.
28. Find the area of the region between the curve $y = 3 - x^2$ and the line $y = -1$ by integrating with respect to (a) x; (b) y.
29. The area of the region between the curve $y = x^2$ and the line $y = 4$ is divided into two equal portions by the line $y = c$.
    a) Find c by integrating with respect to y. (This puts c into the limits of integration.)
    b) Find c by integrating with respect to x. (This puts c into the integrand as well.)
30. Figure 6.5 shows triangle AOC inscribed in the region cut from the parabola $y = x^2$ by the line $y = a^2$. Find the

limit of the ratio of the area of the triangle to the area of the parabolic region as a approaches zero.

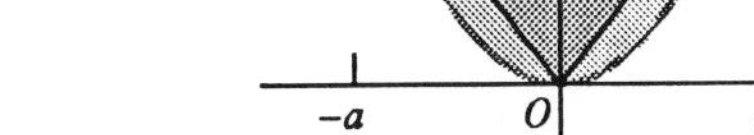

6.5 The Figure for Exercise 30.

31. Suppose that the area between the continuous curve $y = f(x)$ shown here and the x-axis from $x = a$ to $x = b$ is 4 square units. Find the area between the curves $y = f(x)$ and $y = 2f(x)$ from $x = a$ to $x = b$.

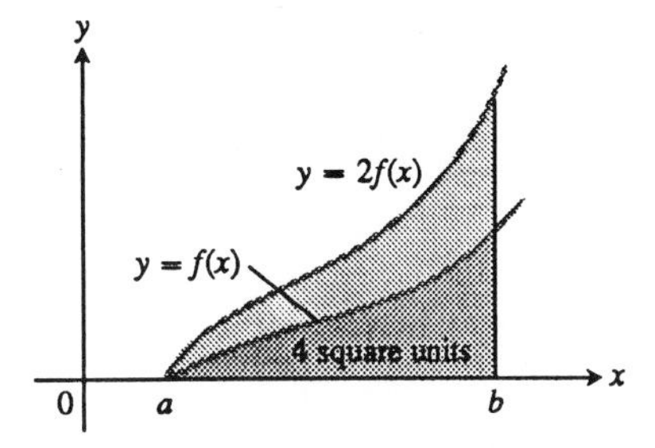

32. Determine the value of a and do the FnInt computation of Example 3.

33. Write a paragraph that explains why the first quadrant region determined by $y = 2x$, $y = e^{-x^2}$, and the x-axis is not bounded.

6.2 Volumes of Solids of Revolution—Disks and Washers

Solids of revolution are solids whose shapes can be generated by revolving plane regions about axes. Thread spools are solids of revolution; so are hand weights and billiard balls. Solids of revolution sometimes have volumes we can find with formulas from geometry, as we can the volume of the billiard ball. But, likely as not, we want to find the volume of a blimp instead, or to predict the weight of a part we are going to have turned on a lathe. In cases like these, formulas from geometry are of little help and we must turn to calculus for the answers. In this section and the next we show how to find these answers.

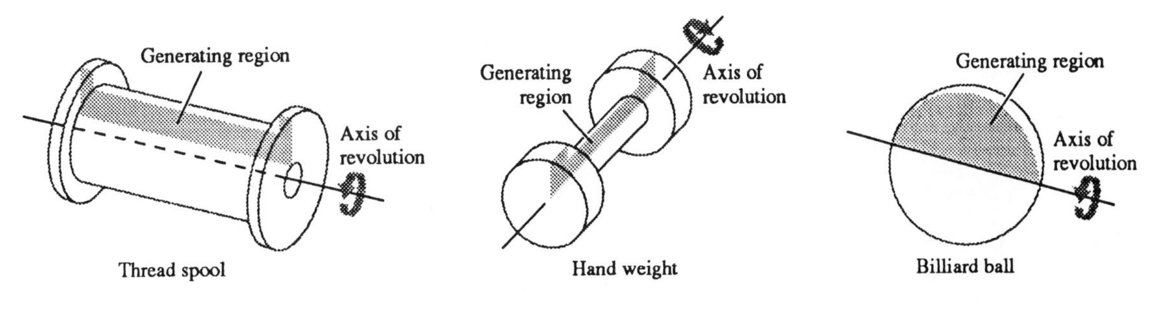

The Disk Method

As we said, a solid of revolution is generated by revolving a region about an axis. We usually assume the axis lies in the same plane as the region. If we can set things up so that the region is the region between the graph of a continuous function $y = f(x), a \leq x \leq b$, and the x-axis, and the axis of revolution is the x-axis, then we can define and calculate the resulting solid's volume in the following way.

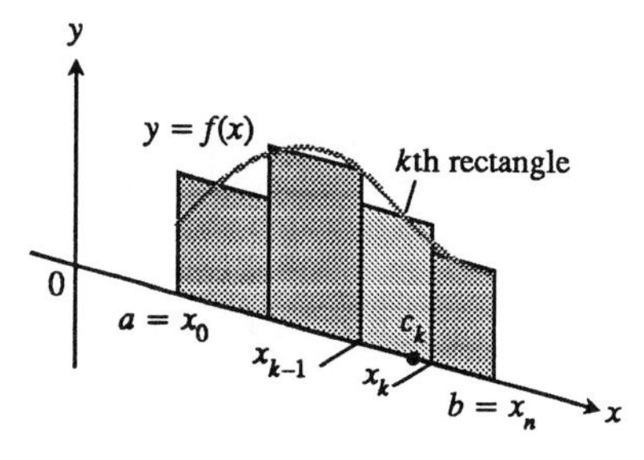

6.6 The first step in developing an integral to calculate the volume of a solid of revolution is to approximate the generating region with rectangles.

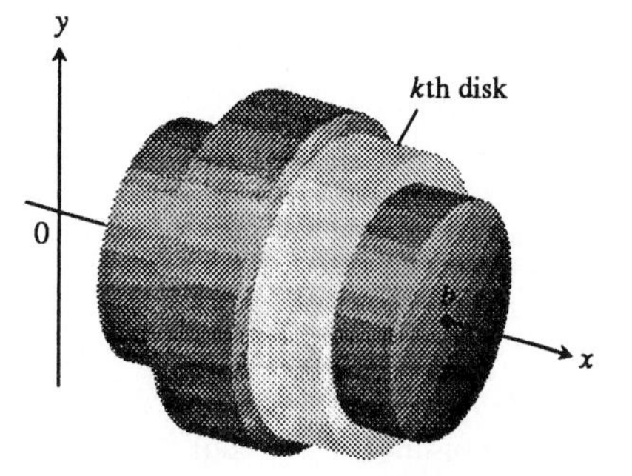

6.7 The rectangles revolve around the x-axis to generate solid disks whose volumes, when added, approximate the volume of the solid.

We begin by approximating the region itself with vertical rectangles based on a partition of the closed interval $[a, b]$ (Fig. 6.6). We then imagine the rectangles to be revolved about the x-axis along with the region. Each rectangle generates a solid disk (Fig. 6.7). The disks, taken together, approximate the solid of revolution. The approximation is similar to what we would get if we sliced up the original solid like a loaf of bread and reshaped each slice into a disk.

If we focus on a typical rectangle and the disk it generates (Fig. 6.8), we see that the disk is a right circular cylinder with height Δx_k and radius $f(c_k)$. We can therefore calculate the volume of the disk with the geometry formula $V = \pi r^2 h$:

$$\text{Disk volume} = \pi \times (\text{ radius})^2 \times \text{ height} = \pi(f(c_k))^2 \Delta x_k. \tag{1}$$

The volumes of all the disks together add up to

$$\text{Disk volume sum} = \sum_{k=1}^{n} \pi(f(c_k))^2 \Delta x_k. \tag{2}$$

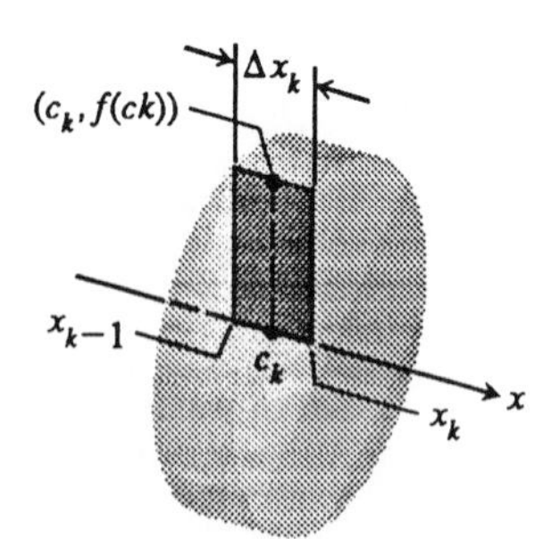

6.8 Enlarged view of the kth disk from Fig. 6.7. Its volume is base area × height $= \pi[f(c_k)]^2 \Delta x_k$.

The expression on the right-hand side of Eq. (2) is a Riemann sum for the function πf^2 on the closed interval $[a, b]$. As the norm of the partition of $[a, b]$ approaches zero, two things happen simultaneously. First, the rectangles approximating the revolved region fit the region with increasing accuracy and the disks they generate fit the solid of revolution with increasing accuracy. Second, the Riemann sums in Eq. (2) approach the integral of πf^2 from a to b. We therefore define the volume of the solid of revolution to be the value of this integral.

Definition

Volume of a Solid of Revolution (Rotation about the x-Axis) The volume of the solid generated by revolving the region between the graph of a continuous function $y = f(x)$ and the x-axis from $x = a$ to $x = b$ about the x-axis is

$$\text{Volume} = \int_a^b \pi(\text{radius})^2\,dx = \int_a^b \pi(f(x))^2\,dx. \tag{3}$$

TO APPLY EQ. (3)

1. Square the expression for the radius function $f(x)$.
2. Multiply by π.
3. Integrate from a to b.

EXAMPLE 1 The region between the curve $y = \sqrt{x}, 0 \le x \le 4$, and the x-axis is revolved about the x-axis to generate the shape in Fig. 6.9. Find its volume.

Solution

$$\begin{aligned}\text{Volume} &= \int_a^b \pi(\text{radius})^2\,dx && \text{(Eq. 3)}\\ &= \int_0^4 \pi(\sqrt{x})^2\,dx && \left(\begin{array}{l}\text{The radius function is}\\ f(x) = \sqrt{x}, 0 \le x \le 4.\end{array}\right)\\ &= \pi \int_0^4 x\,dx = \pi \frac{x^2}{2}\bigg]_0^4 = \pi \frac{(4)^2}{2} = 8\pi\end{aligned}$$

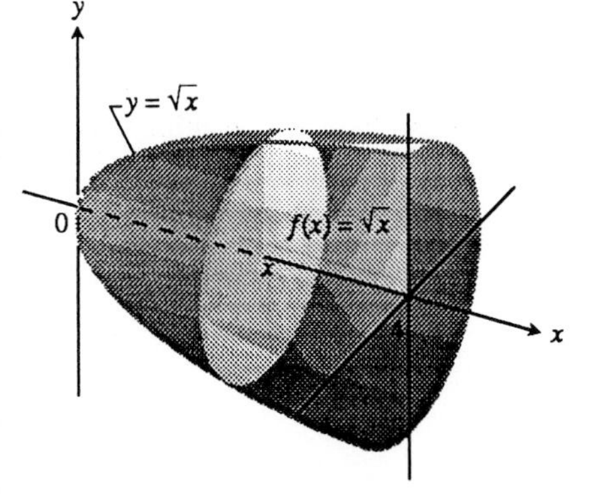

6.9 A slice perpendicular to the axis of the solid in Example 1. The radius at x is $f(x) = \sqrt{x}$.

The volume formula in Eq. (3) is consistent with all the standard formulas from geometry. If we use Eq. (3) to calculate the volume of a sphere of radius a, for instance, we get $(4/3)\pi a^3$, just as we should. The next example shows how the calculation goes.

EXAMPLE 2 The region enclosed by the semicircle $y = \sqrt{a^2 - x^2}$ and the x-axis is revolved about the x-axis to generate a sphere (Fig. 6.10). Find the volume of the sphere.

Solution

$$\begin{aligned}\text{Volume} &= \int_{-a}^a \pi(\text{radius})^2\,dx && \left(\begin{array}{l}\text{Eq. (3) with limits of}\\ \text{integration } -a \text{ and } a\end{array}\right)\\ &= \int_{-a}^a \pi(\sqrt{a^2 - x^2})^2\,dx && \left(\begin{array}{l}\text{The radius at } x \text{ is}\\ f(x) = \sqrt{a^2 - x^2}.\end{array}\right)\\ &= \pi \int_{-a}^a (a^2 - x^2)\,dx = \pi \left[a^2 x - \frac{x^3}{3}\right]_{-a}^a\\ &= \pi\left(a^3 - \frac{a^3}{3}\right) - \pi\left(-a^3 + \frac{a^3}{3}\right)\\ &= \pi\left(\frac{2a^3}{3}\right) - \pi\left(-\frac{2a^3}{3}\right) = \frac{4}{3}\pi a^3\end{aligned}$$

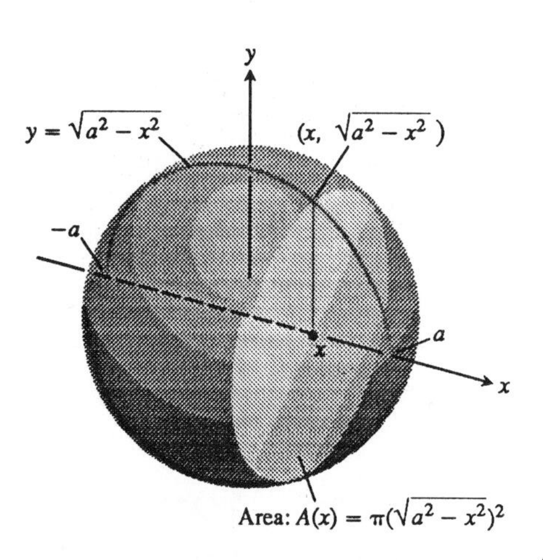

6.10 The sphere generated by revolving the semicircle $y = \sqrt{a^2 - x^2}$ about the x-axis. The radius at x is $f(x) = \sqrt{a^2 - x^2}$.

The axis of revolution in the next example is not the x-axis, but the rule for calculating the volume is the same: Integrate $\pi(\text{radius})^2$ between appropriate limits.

EXAMPLE 3 Find the volume generated by revolving the region bounded by $y = \sqrt{x}$ and the lines $y = 1$ and $x = 4$ about the line $y = 1$.

Solution We draw a figure that shows the region and the radius at a typical point on the axis of revolution (Fig. 6.11). The region runs from $x = 1$ to $x = 4$. At each x in the interval $1 \le x \le 4$, the cross-section radius is

$$\text{Radius at } x = \sqrt{x} - 1.$$

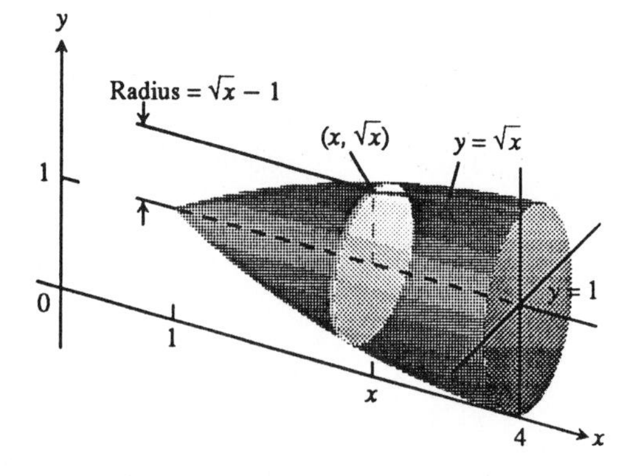

6.11 The solid swept out by revolving the region bounded by $y = \sqrt{x}$, $x = 4$, and $y = 1$ about the line $y = 1$.

Therefore

$$\begin{aligned}
\text{Volume} &= \int_1^4 \pi(\text{radius})^2\,dx = \int_1^4 \pi(\sqrt{x}-1)^2\,dx \\
&= \pi \int_1^4 (x - 2\sqrt{x} + 1)\,dx \\
&= \pi \left[\frac{x^2}{2} - 2\cdot\frac{2}{3}x^{3/2} + x\right]_1^4 \\
&= \frac{7\pi}{6}. \qquad \left(\begin{array}{l}\text{Arithmetic}\\ \text{omitted}\end{array}\right)
\end{aligned}$$

■

Explore with a Grapher

We assume you check your analytic definite integral evaluations with a calculator or computer. For instance, in Example 3,

$$\text{FnInt}\ (\pi(\sqrt{x}-1)^2, x, 1, 4) = 3.66519142919.$$

Compare this answer with the exact answer $7\pi/6$ on a calculator.

To find the volume of the solid generated by revolving the region between a curve $x = f(y)$ and the y-axis from $y = c$ to $y = d$ about the y-axis, use Eq. (3) with x replaced by y. (Fig. 6.12).

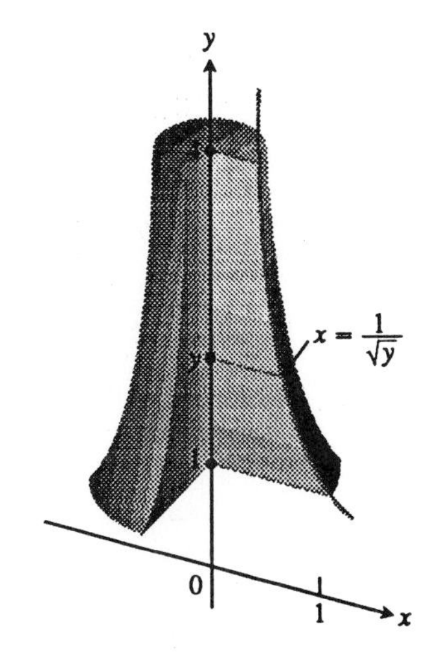

6.12 The solid in Example 4.

Volume of a Solid of Revolution (Rotation about the y-Axis)

$$\text{Volume} = \int_c^d \pi(\text{radius})^2\,dy = \int_c^d \pi(f(y))^2\,dy \qquad (4)$$

EXAMPLE 4 The region between the curve $x = 1/\sqrt{y}$, $1 \le y \le 4$, and the y-axis is revolved about the y-axis to generate a solid Find the volume of the solid.

Solution

$$\begin{aligned}
\text{Volume} &= \int_1^4 \pi(\text{radius})^2\,dy = \int_1^4 \pi\left(\frac{1}{\sqrt{y}}\right)^2 dy \qquad \left(\begin{array}{l}\text{The radius function}\\ \text{is } f(y) = 1/\sqrt{y}.\end{array}\right) \\
&= \pi \int_1^4 \frac{1}{y}\,dy = \pi \ln y\Big]_1^4 = \pi \ln 4 - 0 \qquad (\ln 1 = 0) \\
&= 2\pi \ln 2 \qquad (\ln 4 = \ln 2^2 = 2\ln 2)
\end{aligned}$$

■

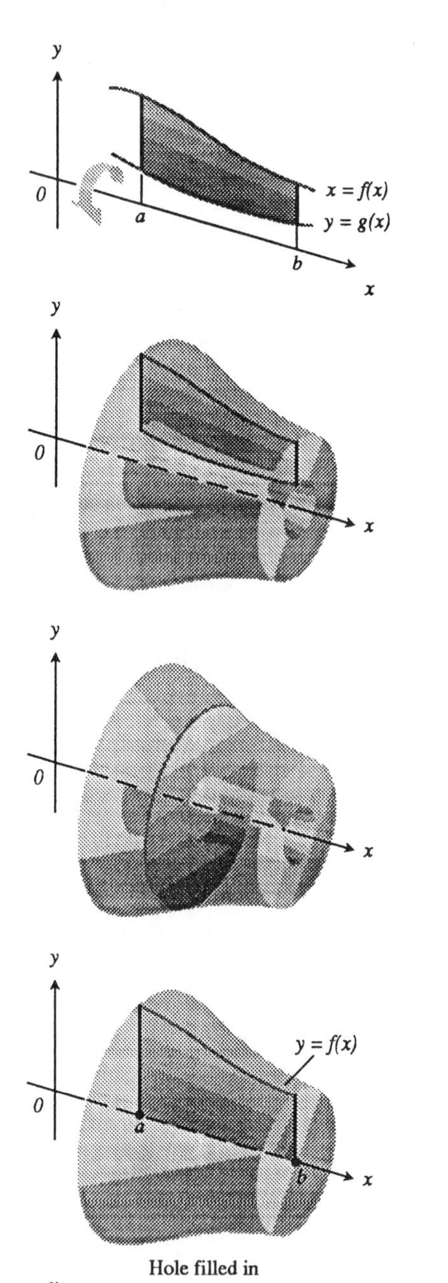

Hole filled in

Volume of the hole

The Washer Method

If the region we revolve to generate a solid does not border on the axis of revolution,

... the solid has a hole in it.

The cross sections perpendicular to the axis of revolution are washers instead of disks.

If we fill in the hole, the volume is

$$\int_a^b \pi(f(x))^2\,dx.$$

The volume of the hole itself is

$$\int_a^b \pi(g(x))^2\,dx.$$

The volume of the original solid is therefore

$$\begin{aligned}\text{Volume} &= (\text{volume with hole filled in}) - (\text{volume of hole})\\ &= \int_a^b \pi(f(x))^2\,dx - \int_a^b \pi(g(x))^2\,dx = \int_a^b \pi(f^2(x) - g^2(x))\,dx. \qquad (5)\end{aligned}$$

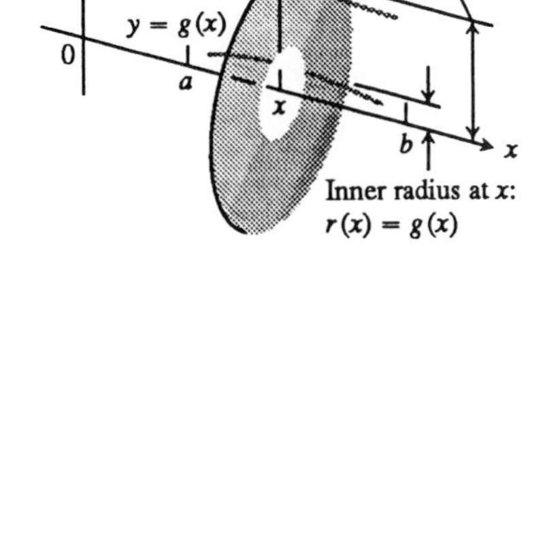

If we return to the cross sections to look at the typical washer at x, we see that it has two radii, an outer one and an inner one:

Outer washer radius: $\quad R(x) = f(x)$

Inner washer radius: $\quad r(x) = g(x)$.

Equation (5) can be described in terms of these radii as

$$\begin{aligned} \text{Volume} &= \int_a^b \pi((\text{outer radius})^2 - (\text{inner radius})^2)\,dx \\ &= \int_a^b \pi(R^2(x) - r^2(x))\,dx. \end{aligned} \tag{6}$$

Washer Method for Calculating Volumes

$$\text{Volume} = \int_a^b \pi(R^2(x) - r^2(x))\,dx \tag{7}$$

$R(x) =$ outer radius, $\qquad r(x) =$ inner radius

Notice that the function being integrated in Eq. (7) is $\pi(R^2 - r^2)$, not $\pi(R - r)^2$. Also notice that Eq. (7) turns into the disk-method formula if the inner radius $r(x)$ is zero throughout the interval $a \le x \le b$. The disk-method formula is a special case of what we have here.

We apply Eq. (7) with the steps shown in the following example. (We shall list the steps separately again after the example.)

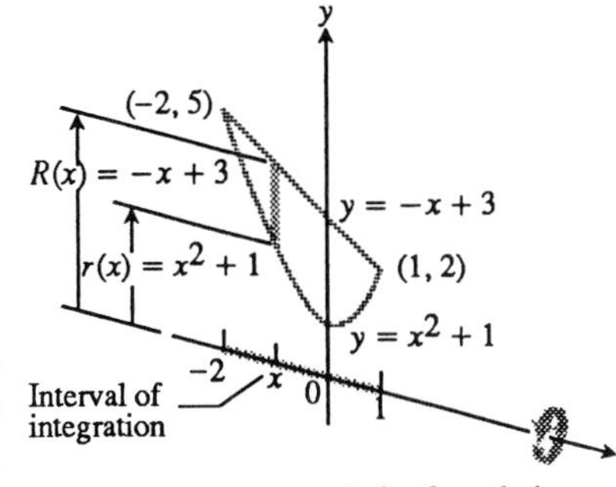

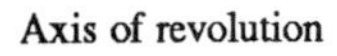

6.13 The region bounded by the curve $y = x^2 + 1$ and the line $y = -x + 3$ with a thin rectangle perpendicular to the axis of revolution.

EXAMPLE 5 The region bounded by the curve $y = x^2 + 1$ and the line $y = -x + 3$ is revolved about the x-axis to generate a solid. Find the volume of the solid.

Solution

STEP 1: Draw the region and find the limits of integration (Fig. 6.13). The limits of integration are the x-coordinates of the points where the parabola and line cross. We find them by solving the equations $y = x^2 + 1$ and $y = -x + 3$ simultaneously for x:

$$\begin{aligned} x^2 + 1 &= -x + 3 \\ x^2 + x - 2 &= 0 \\ (x + 2)(x - 1) &= 0 \\ x &= -2,\, x = 1. \end{aligned}$$

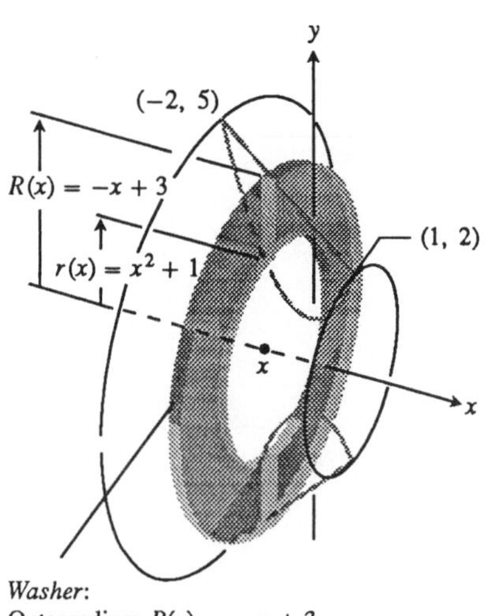

6.14 The inner and outer radii of the washer swept out by revolving the rectangle in Fig. 6.13 about the x-axis are the distances of the rectangle's two ends from the axis of revolution.

STEP 2: Draw a thin rectangle across the region perpendicular to the axis of revolution (the shaded rectangle in Fig. 6.13).

STEP 3: Find the outer and inner radii of the washer that would be swept out by the rectangle if it were revolved about the x-axis along with the region. (We drew the washer in Fig. 6.14, but in your own work you need not do that.) These radii are the distances of the two ends of the rectangle from the axis of revolution. You can read these distances from the formulas for the bounding curves:

Outer radius: $R(x) = -x + 3$

Inner radius: $r(x) = x^2 + 1.$

STEP 4: Evaluate the volume integral.

$$\text{Volume} = \int_a^b \pi(R^2(x) - r^2(x))\,dx \qquad \text{(Eq. 7)}$$

$$= \int_{-2}^{1} \pi((-x+3)^2 - (x^2+1)^2)\,dx \qquad \left(\begin{array}{l}\text{Values from}\\ \text{Steps 1 and 3}\end{array}\right)$$

$$= \int_{-2}^{1} \pi(8 - 6x - x^2 - x^4)\,dx \qquad \left(\begin{array}{l}\text{Expressions squared}\\ \text{and combined}\end{array}\right)$$

$$= \pi\left[8x - 3x^2 - \frac{x^3}{3} - \frac{x^5}{5}\right]_{-2}^{1} = \frac{117\pi}{5} \qquad \text{(Arithmetic omitted)}$$

The volume of the solid is $117\pi/5$.

As illustrated in Example 5, the steps we take to implement the washer method are these:

How to Find Volumes by the Washer Method

STEP 1: Draw the region and find the limits of integration.

STEP 2: Draw a thin rectangle across the region perpendicular to the axis of revolution.

STEP 3: Find the distances of the ends of the rectangle from the axis of revolution. These give the radius functions for the volume integral.

STEP 4: Integrate to find the volume.

To find the volume of a solid generated by revolving a region about the y-axis, we use the steps listed above and integrate with respect to y instead of x. The next example shows how to do this.

EXAMPLE 6 The region bounded by the parabola $y = x^2$ and the line $y = 2x$ in the first quadrant is revolved about the y-axis to generate a solid. Find the volume of the solid.

Solution

STEP 1: Draw the region (Fig. 6.15) and find the limits of integration. The curves $y = x^2$ and $y = 2x$ cross at $(0, 0)$ and $(2, 4)$, so we integrate from $y = 0$ to $y = 4$.

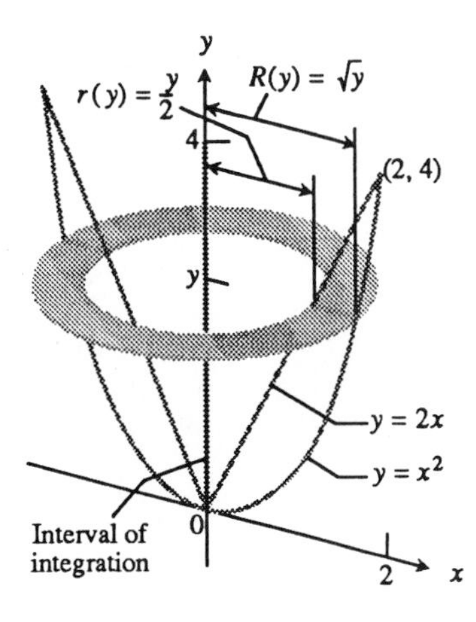

6.15 The region bounded by the parabola $y = x^2$ and the line $y = 2x$. The thin rectangle perpendicular to the axis of revolution sweeps out a washer whose radii are given by the formulas for the curves.

STEP 2: Draw a thin rectangle across the region perpendicular to the axis of revolution (the rectangle in Fig. 6.15).

STEP 3: Find the distances of the ends of the rectangle from the axis of revolution. From Fig. 6.15,

$$R(y) = \sqrt{y}, \qquad r(y) = \frac{y}{2}.$$

STEP 4: Integrate to find the volume.

$$\text{Volume} = \int_c^d \pi(R^2(y) - r^2(y))\,dy \qquad \text{(Eq. (7) with } y \text{ in place of } x)$$

$$= \int_0^4 \pi\left((\sqrt{y})^2 - \left(\frac{y}{2}\right)^2\right) dy \qquad \text{(Values from Steps 1 and 3)}$$

$$= \pi \int_0^4 \left(y - \frac{y^2}{4}\right) dy = \pi \left[\frac{y^2}{2} - \frac{y^3}{12}\right]_0^4 = \frac{8}{3}\pi$$

If the axis of revolution is not one of the coordinate axes, the rule for finding the volume is still the same: Find formulas for the solid's outer and innter radii and integrate $\pi(R^2 - r^2)$ between appropriate limits. In the next example we revolve a region about a line parallel to the y-axis. We would handle a region revolved about a line parallel to the x-axis in a similar way.

Solids of Revolution

We can generate a solid of revolution by revolving a plane region about an axis. To calculate the volume of the solid, we integrate the square of the radius of the solid along an interval on the axis and multiply by π.

If the axis of revolution is the x-axis, we integrate along an interval of the x-axis (Example 1).

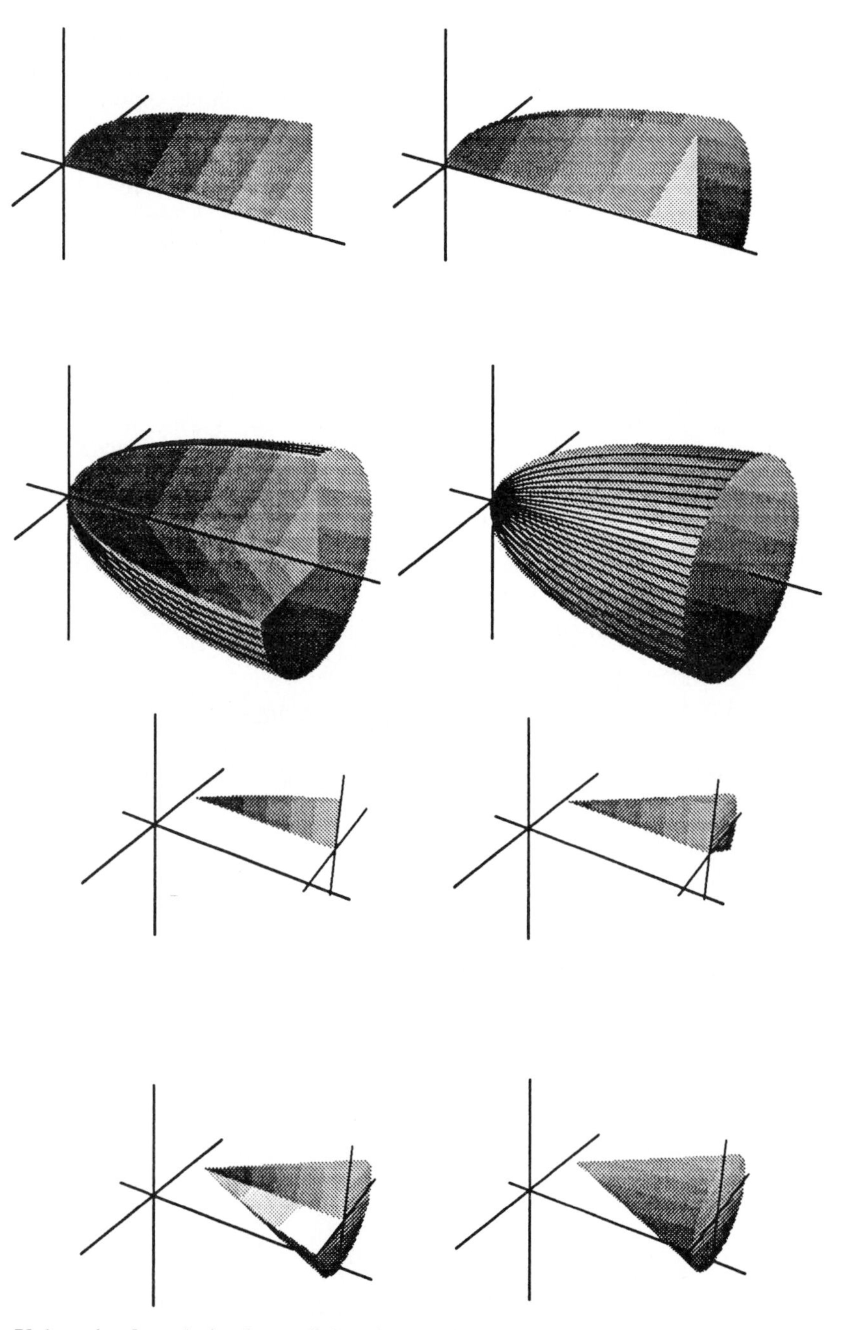

If the axis of revolution is parallel to the x-axis, we also integrate along an interval of the x-axis (Example 3). (Generated by Mathematica)

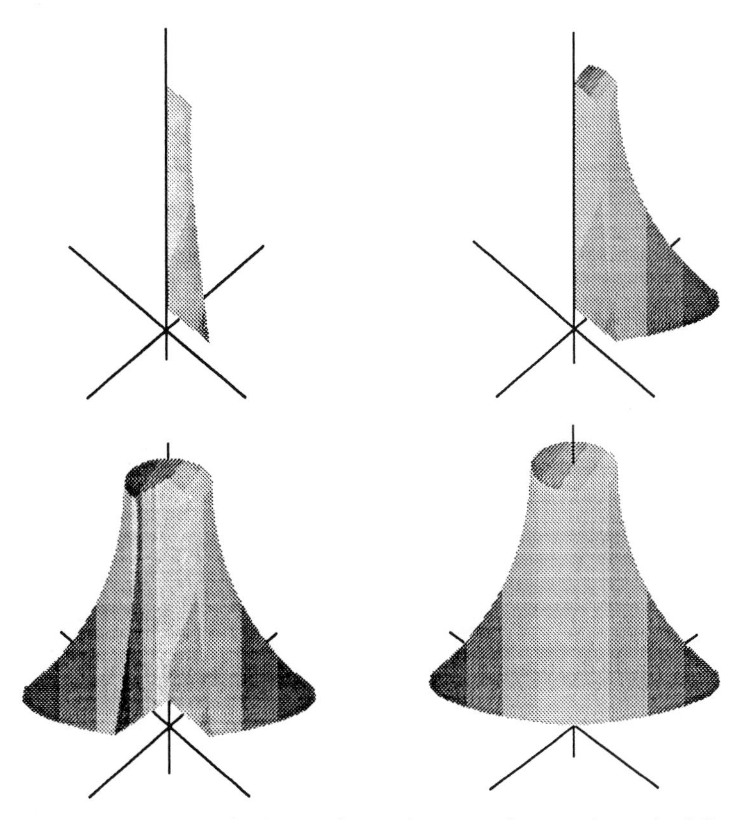

If the axis of revolution is the y-axis, we integrate along an interval of the y-axis (Example 4).

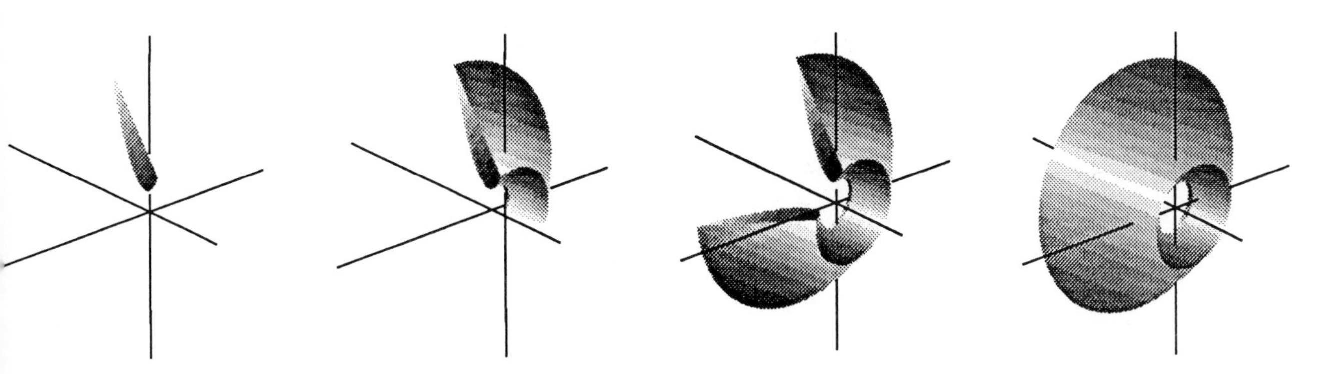

If the axis of revolution is not part of the region being revolved, the solid has a hole in it. We integrate along the coordinate axis parallel to the axis of revolution (Example 5).

EXAMPLE 7 The region between the parabola $y = x^2$ and the line $y = 2x$ is revolved about the line $x = 2$ parallel to the y-axis. Find the volume swept out.

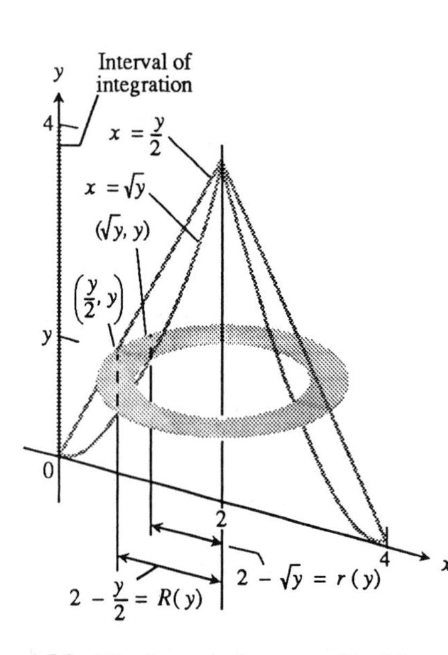

6.16 The inner and outer radii of the washer are measured, as always, as distances from the axis of revolution.

Solution

STEP 1: Draw the region (Fig. 6.16) and find the limits of integration. The line crosses the parabola at (0,0) and (2,4), so we integrate from $y = 0$ to $y = 4$.

STEP 2: Draw a thin rectangle across the region perpendicular to the axis of revolution (the rectangle in Fig. 6.16).

STEP 3: Find the distances of the ends of the rectangle from the axis of revolution. From Fig. 6.16,

$$R(y) = 2 - \frac{y}{2}, \quad r(y) = 2 - \sqrt{y}.$$

STEP 4: Integrate with respect to y to find the volume.

$$\text{Volume} = \int_c^d \pi(R^2(y) - r^2(y))\,dy \qquad \begin{pmatrix}\text{Eq. (7) with } y\\ \text{in place of } x\end{pmatrix}$$

$$= \int_0^4 \pi\left(\left(2 - \frac{y}{2}\right)^2 - (2 - \sqrt{y})^2\right) dy \qquad \begin{pmatrix}\text{Values from}\\ \text{Steps 1 and 3}\end{pmatrix}$$

$$= \pi \int_0^4 \left(\frac{y^2}{4} - 3y + 4\sqrt{y}\right) dy \qquad \begin{pmatrix}\text{Expressions squared}\\ \text{and combined}\end{pmatrix}$$

$$= \pi \left[\frac{y^3}{12} - \frac{3y^2}{2} + \frac{8}{3}y^{3/2}\right]_0^4 = \frac{8}{3}\pi \qquad \text{(Arithmetic omitted)}$$

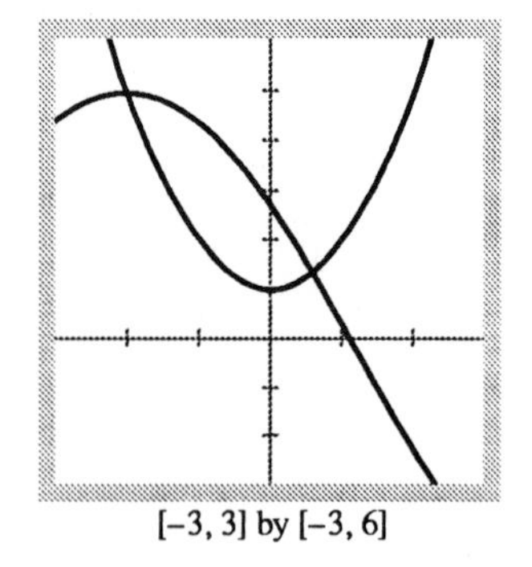

[−3, 3] by [−3, 6]

6.17 The graphs of $y = 5\cos(0.5x + 1)$ and $y = x^2 + 1$.

The examples in this section are contrived. They have been designed to keep the arithmetic simple. Consider this slight modification of Example 5. Replace the line $y = -x + 3$ by the sinusoid $y = 5\cos(0.5x + 1)$. In the exercises you will show that the volume of the solid determined by revolving the region bounded by the curves $y = 5\cos(0.5x + 1)$ and $y = x^2 + 1$ about the x-axis is given by

$$\int_a^b \pi\left[(5\cos(0.5x + 1))^2 - (x^2 + 1)^2\right] dx$$

where a and b are the x-coordinates of the points of intersection of the two curves. (See Figure 6.17.) In the exercises you will find a and b and determine the volume of the solid of revolution. You will need to use technology for this problem.

Exercises 6.2

Use the disk method to find the volumes of the solids generated by revolving the regions bounded by the lines and curves in Exercises 1–8 about the x-axis.

1. $x + y = 2, \; y = 0, \; x = 0$

2. $y = x^2, \; y = 0, \; x = 2$

3. $y = \sqrt{9 - x^2}, \; y = 0$

4. $y = x - x^2, \; y = 0$

5. $y = x^3, \; y = 0, \; x = 2$

6. $y = e^x, \; y = 0, \; x = 0, \; x = \ln 2$

7. $y = \sqrt{\cos x}, 0 \le x \le \pi/2, \; y = 0, \; x = 0$

8. $y = \sec x, \; y = 0, \; x = -\pi/4, \; x = \pi/4$

Use the disk method to find the volumes of the solids generated by revolving about the y-axis the regions bounded by the lines and curves in Exercises 9–16.

9. $y = x/2, \; y = 2, \; x = 0$

10. $x = \sqrt{4 - y}, \; x = 0, \; y = 0$

11. $x = \sqrt{5}y^2, \; x = 0, \; y = -1, \; y = 1$

12. $x = 1 - y^2, \; x = 0$

13. $x = y^{3/2}, \; x = 0, \; y = 2$

14. $x = \sqrt{2 \sin 2y}, \; 0 \le y \le \pi/2, \; x = 0$

15. $x = 2/\sqrt{y + 1}, \; x = 0, \; y = 0, \; y = 3$

16. $x = 2/(y + 1), \; x = 0, \; y = 0, \; y = 1$

Use the washer method to find the volumes of the solids generated by revolving about the x-axis the regions bounded by the lines and curves in Exercises 17–24.

17. $y = x, \; y = 1, \; x = 0$

18. $y = 2x, \; y = x, \; x = 1$

19. $y = x^2, \; y = 4, \; x = 0$

20. $y = x^2 + 3, \; y = 4$

21. $y = x^2 + 1, \; y = x + 3$

22. $y = 4 - x^2, \; y = 2 - x$

23. $y = \sec x, \; y = \sqrt{2}, \; -\pi/4 \le x \le \pi/4$

24. $y = 2/\sqrt{x}, \; y = 2, \; x = 4$

Use the washer method to find the volumes of the solids generated by revolving the regions bounded by the lines and curves in Exercises 25–30 about the y-axis.

25. $y = x - 1, \; y = 1, \; x = 1$

26. $y = x - 1, \; y = 0, \; x = 1$

27. $y = x^2, \; y = 0, \; x = 2$

28. $y = x, \; y = \sqrt{x}$

29. The semicircle $x = \sqrt{25 - y^2}$ and the y-axis

30. The semicircle $x = \sqrt{25 - y^2}$ and the line $x = 4$

Find the volume of the solid generated by revolving the shaded region about the indicated axis in Exercises 31–34.

31. The x-axis

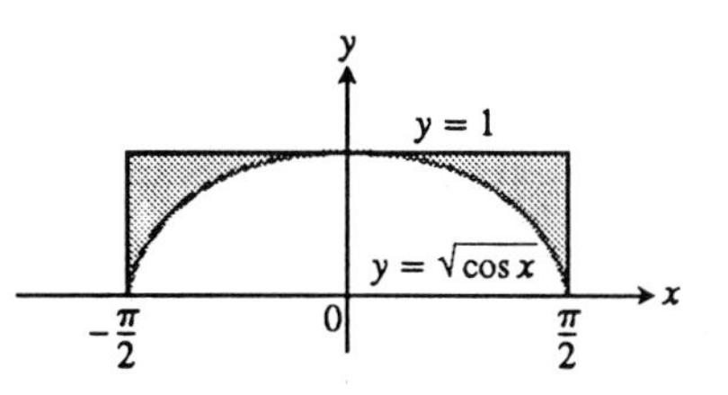

32. The x-axis

33. The y-axis

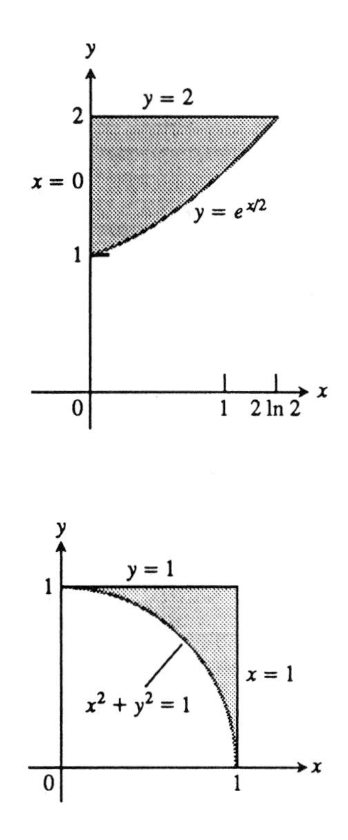

34. **a)** The x-axis

b) The y-axis

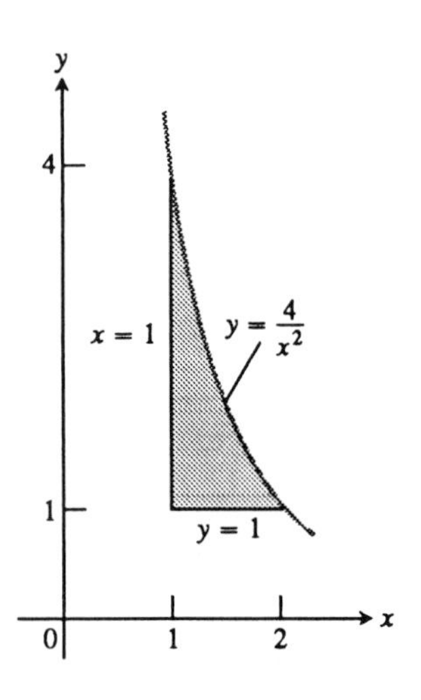

35. Find the volume of the solid generated by revolving the region bounded by $y = \sqrt{x}$ and the lines $y = 2$ and $x = 0$

a) about the x-axis

b) about the y-axis

c) about the line $y = 2$

d) about the line $x = 4$.

36. Find the volume of the solid generated by revolving the triangular region bounded by the lines $y = 2x$, $y = 0$, and $x = 1$

a) about the line $x = 1$

b) about the line $x = 2$.

37. Find the volume of the solid generated by revolving the region bounded by the parabola $y = x^2$ and the line $y = 1$

a) about the line $y = 1$

b) about the line $y = 2$

c) about the line $y = -1$.

38. By integration, find the volume of the solid generated by revolving the triangular region with vertices $(0, 0)$, $(b, 0)$, $(0, h)$

a) about the x-axis

b) about the y-axis.

39. **Minimizing a volume.** The arch $y = \sin x, 0 \leq x \leq \pi$, is revolved about the line $y = c$ to generate the solid shown in Fig. 6.18. Find the value of c that minimizes this volume. What is the minimum volume?

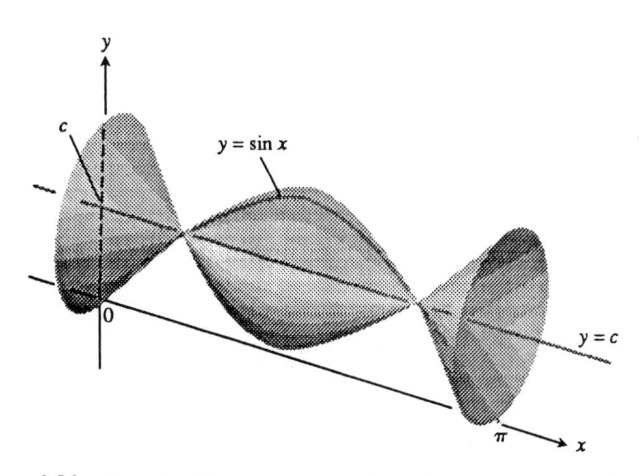

6.18 Exercise 39 asks for the value of c that minimizes this volume.

40. **Designing a plumb bob.** Having been asked to design a brass plumb bob that will weigh in the neighborhood of 190 gm, you decide to shape it like the solid of revolution shown here. What is the volume of the solid? If you specify a brass that weighs 8.5 gm/cm^3, about how much will the bob weigh?

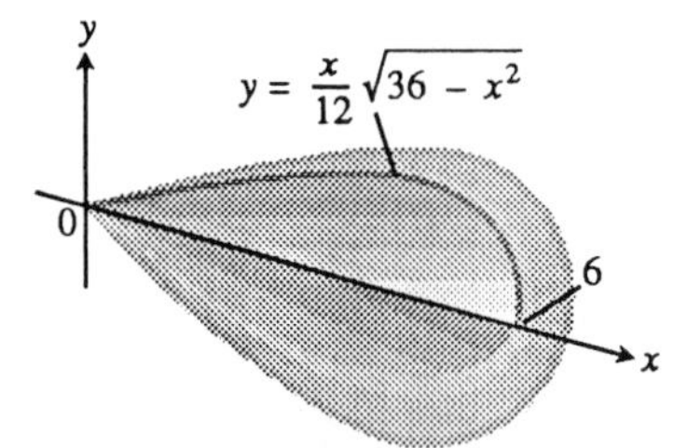

41. **Designing a wok.** You are designing a wok frying pan that will be shaped like a spherical bowl with handles.

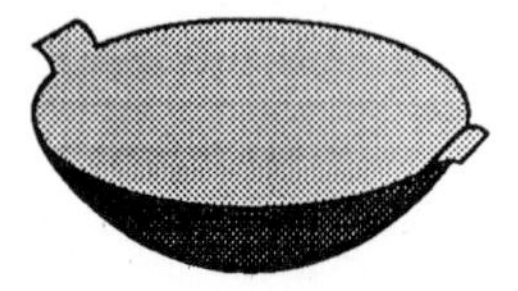

A bit of experimentation at home persuades you that you can get one that holds about 3 L if you make it 9 cm deep and give the sphere a radius of 16 cm. To be sure, you picture the wok as a solid of revolution and calculate the volume with an integral. Your picture looks like this:

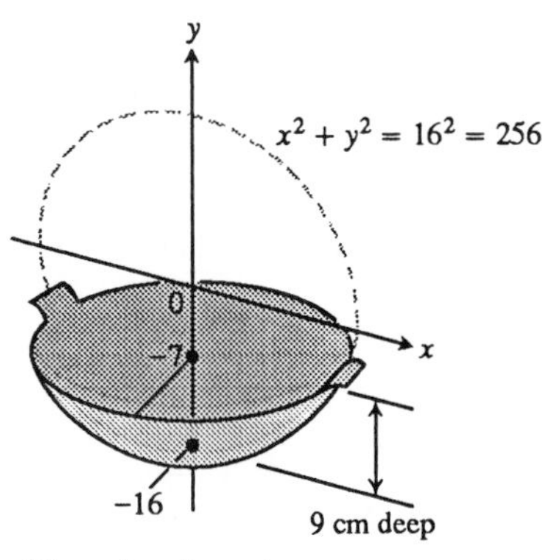

What volume do you really get?

42. **a)** Solve $5\cos(0.5x + 1) = x^2 + 1$ for x.

b) Find the volume of the solid of revolution determined by revolving the region bounded by the curves $y = 5\cos(0.5x + 1)$ and $y = x^2 + 1$ about the x-axis.

43. Solve Exercise 39 with the arch given by $y = \sin x\sqrt{x^2 + 3}$, $0 \le x \le \pi$.

6.3 Cylindrical Shells—An Alternative to Washers

If the rectangular strips that approximate a region being revolved about an axis lie parallel to the axis instead of perpendicular to it, they sweep out cylindrical shells. Cylindrical shells are sometimes easier to work with than washers because the formula they lead to does not require squaring.

The Basic Shell Formula

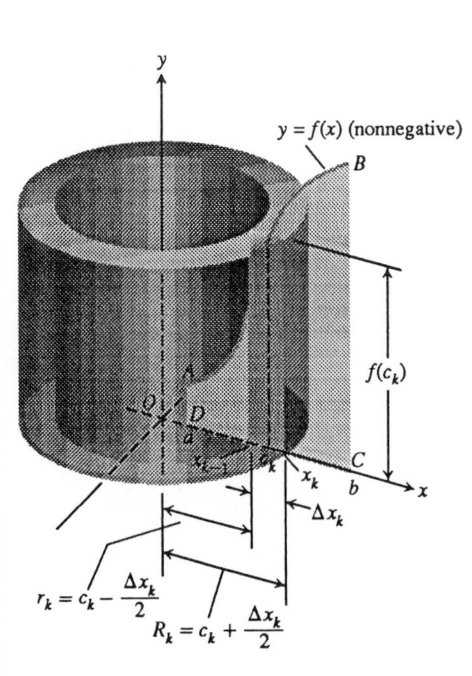

6.19 The solid swept out by revolving region $ABCD$ about the y-axis can be approximated with cylindrical shells like the one shown here.

We arrive at the cylindrical shell volume formula in the following way.

Suppose we revolve the tinted region $ABCD$ in Fig. 6.19 about the y-axis to generate a solid. To find the volume of the solid, we first approximate the region with vertical rectangles based on a partition of the closed interval $[a, b]$ on which the region stands. The rectangles run parallel to the y-axis, the axis of revolution.

Figure 6.19 also shows a typical approximating rectangle. Its dimensions are $f(c_k)$ by Δx_k. The point c_k is chosen to be the midpoint of the interval from x_{k-1} to x_k. Since it does not matter where the c_k's are chosen in their intervals when we find limits of Riemann sums, we are free to choose each c_k as we please. The resulting formula in this case will be less cumbersome if we stay with midpoints. You will see why in just a moment.

Again with reference to Fig. 6.19, the cylindrical shell swept out by revolving the rectangle about the y-axis is a solid cylinder with these dimensions:

Shell height: $f(c_k)$

Inner radius: $r_k = c_k - \dfrac{\Delta x_k}{2}$

Outer radius: $R_k = c_k + \dfrac{\Delta x_k}{2}$

Base ring area:
$$A_k = \pi R_k^2 - \pi r_k^2 = \pi(R_k + r_k)(R_k - r_k) \qquad (1)$$
$$= \pi(2c_k)\Delta x_k$$

$$\text{Shell volume:} \quad V_k = \text{base ring area} \times \text{shell height}$$
$$= 2\pi c_k f(c_k)\Delta x_k.$$

The advantage of choosing c_k to be the midpoint of its interval becomes clear in the formula for A_k. With this choice, $R_k + r_k$ equals $2c_k$; without this choice, it doesn't.

The volumes of all the shells generated by the partition of $[a, b]$ add up to

$$\text{Shell volume sum} = \sum_{k=1}^{n} 2\pi c_k f(c_k)\Delta x_k. \tag{2}$$

The volume of the solid swept out by revolving region $ABCD$ about the y-axis is taken to be the limit of the shell volume sums as the norm of the partition of $[a, b]$ goes to zero. If f is continuous, the limit exists and can be found by integrating the product $2\pi x f(x)$ from $x = a$ to $x = b$.

The Shell Method (Axis the y-Axis)

Suppose $y = f(x)$ is continuous throughout an interval $a \le x \le b$ that does not cross the y-axis. Then, the volume of the solid generated by revolving the region between the graph of f and the interval $a \le x \le b$ about the y-axis is found by integrating $2\pi x f(x)$ with respect to x from a to b.

$$\text{Volume} = \int_a^b 2\pi \begin{pmatrix}\text{shell}\\ \text{radius}\end{pmatrix} \begin{pmatrix}\text{shell}\\ \text{height}\end{pmatrix} dx = \int_a^b 2\pi x f(x)\, dx. \tag{3}$$

One way to remember Eq. (3) is to imagine that a cylindrical shell of average circumference $2\pi x$, height $f(x)$, and thickness dx has been cut along a generating rectangle and rolled flat like a sheet of tin (Fig. 6.20). The sheet is almost a rectangular solid of dimensions $2\pi x$ by $f(x)$ by dx. Hence the shell's volume is about $2\pi x f(x)\, dx$. Equation (3) says that the volume of the complete solid is the integral of $2\pi x f(x)\, dx$ from a to b.

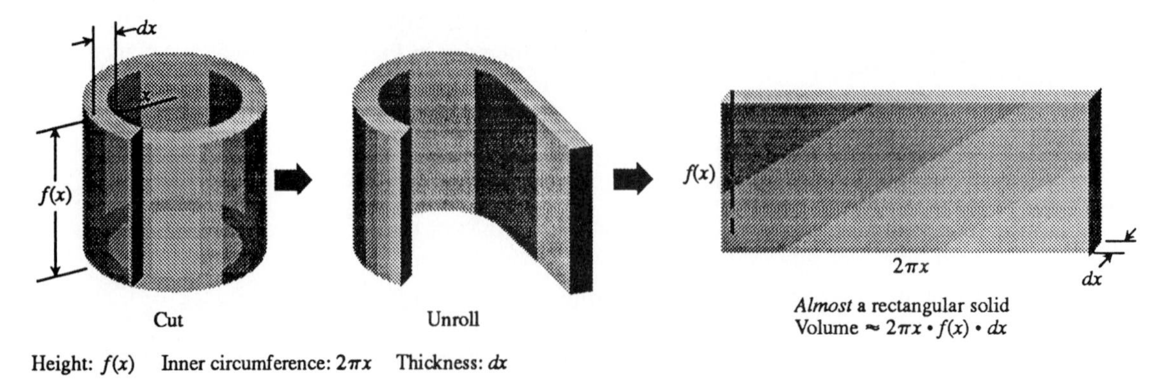

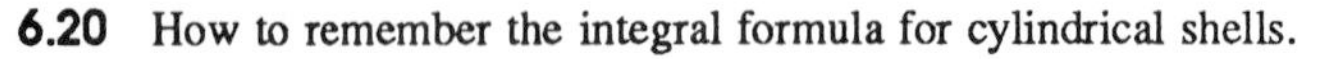

6.20 How to remember the integral formula for cylindrical shells.

How to Find a Volume by the Shell Method

To apply Eq. (3), take these steps:

STEP 1: Sketch the region.

STEP 2: Draw a thin rectangle across the region parallel to the y-axis (the axis of revolution). The radius of the cylindrical shell swept out by the rectangle is x. The height of the shell is $f(x)$, the height of the rectangle. The width of the rectangle is dx. Add this information to the picture. (We also sketched the shell, but you need not do that in your own work.)

STEP 3: Find limits of integration a and b that include all possible rectangles like this from one end of the region to the other.

STEP 4: Integrate the product $2\pi x f(x)$ with respect to x from a to b to find the volume.

EXAMPLE 1 The region bounded by the curve $y = \sqrt{x}$, the x-axis, and the line $x = 4$ is revolved about the y-axis to generate a solid. Find the volume of the solid.

Solution

STEP 1: Sketch the region (Fig. 6.21).

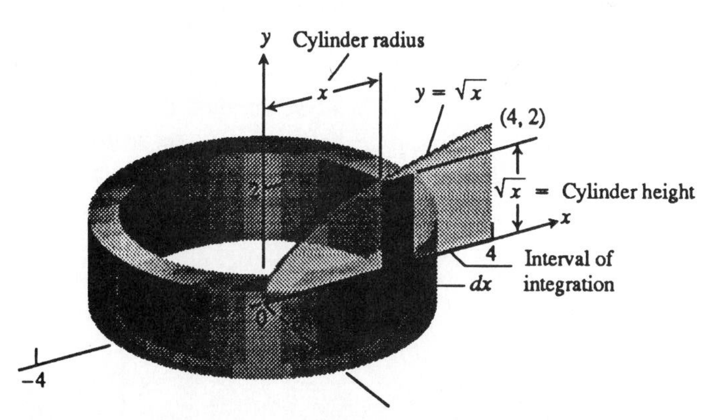

6.21 The region in Example 1, together with a thin rectangle parallel to the axis of revolution (the y-axis, in this case). The rectangle sweeps out a cylindrical shell.

STEP 2: Draw a thin rectangle across the region parallel to the y-axis. Label the rectangle's height $\sqrt{x}$, its width dx, and its distance from the y-axis x. (We added all this to Fig. 6.21.)

STEP 3: Find limits of integration that include all rectangles like the one in Step 2. The limits in this case are $x = 0$ and $x = 4$.

STEP 4: Integrate to find the volume.

$$\text{Volume} = \int_a^b 2\pi x f(x)\,dx = \int_0^4 2\pi x\sqrt{x}\,dx \qquad \begin{pmatrix}\text{Eq. (3) with values}\\ \text{from Steps 2 and 3}\end{pmatrix}$$

$$= 2\pi \int_0^4 x^{3/2}\,dx = 2\pi\left[\frac{2}{5}x^{5/2}\right]_0^4 = \frac{128\pi}{5} \qquad \begin{pmatrix}\text{Arithmetic}\\ \text{omitted}\end{pmatrix}$$

The volume is $128\pi/5$. ∎

To use shells to find the volume of a solid generated by revolving a region about the x-axis instead of the y-axis, use Eq. (3) with y in place of x. Except for changes in notation, the steps we take to implement the new formula are the same as before.

EXAMPLE 2 The region bounded by the curve $y = \sqrt{x}$, the x-axis, and the line $x = 4$ is revolved about the x-axis to generate a solid. Find the volume of the solid.

Solution

STEP 1: Sketch the region (Fig. 6.22).

STEP 2: Draw a thin rectangle across the region parallel to the x-axis (the axis of revolution). Describe the rectangle's height as a function of y. Label the rectangle's width dy and its distance from the x-axis y. (We added all this to Fig. 6.22.)

STEP 3: Find limits of integration that include all the rectangles like the one in Step 2. The rectangles run from $y = 0$, the smallest value of y in the region, to $y = 2$, the largest value of y in the region.

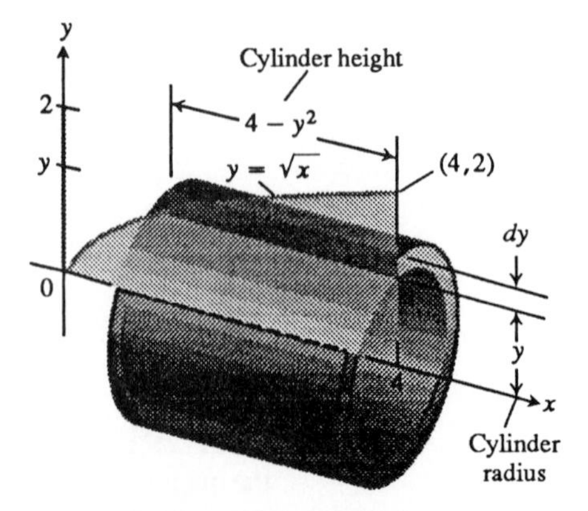

6.22 The region in Example 2, together with a thin rectangle parallel to the axis of revolution.

STEP 4: Integrate to find the volume.

$$\text{Volume} = \int_c^d 2\pi \begin{pmatrix}\text{shell}\\ \text{radius}\end{pmatrix} \begin{pmatrix}\text{shell}\\ \text{height}\end{pmatrix} dy = \int_0^2 2\pi y(4 - y^2)\, dy$$

$$= 2\pi \int_0^2 (4y - y^3)\, dy = 2\pi \left[2y^2 - \frac{y^4}{4}\right]_0^2 = 8\pi, \qquad \begin{pmatrix}\text{Arithmetic}\\ \text{omitted}\end{pmatrix}$$

in agreement with the disk-method calculation in Example 1, Section 6.2. ∎

If the axis of revolution is a line parallel to one of the coordinate axes, we use the same steps as before. The only added complication is that the expression for the radius of the typical cylinder is no longer simply x or y.

EXAMPLE 3 The region in the first quadrant bounded by the parabola $y = x^2$, the y-axis, and the line $y = 1$ is revolved about the line $x = 2$ to generate a solid. Find the volume of the solid.

Solution

STEP 1: Sketch the region (Fig. 6.23).

STEP 2: Draw a thin rectangle across the region parallel to the line $x = 2$ (the axis of revolution). Describe the rectangle's height as a function of x. Label the rectangle's width dx. Describe the rectangle's distance from the line $x = 2$ as a function of x. This is the radius of the typical cylindrical shell the rectangle sweeps out. (We added all this to Fig. 6.23.)

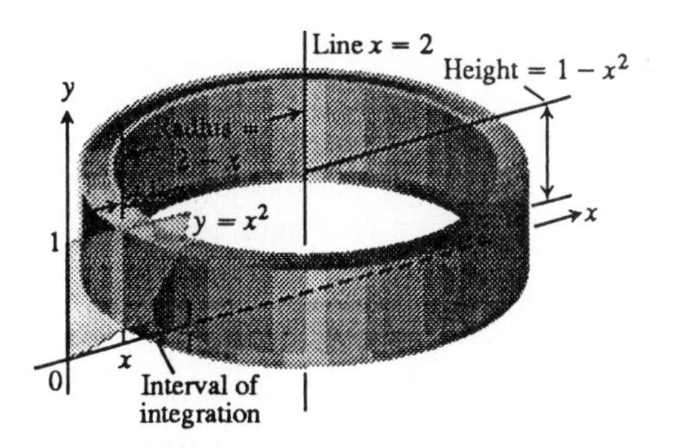

6.23 The region in Example 3. As always, the radius of the cylindrical shell is measured as a distance from the axis of revolution.

STEP 3: Find limits of integration that take in all rectangles like the one in Step 2. The limits are $x = 0$ and $x = 1$, the extreme values of x in the region.

STEP 4: Integrate to find the volume.

$$\text{Volume} = \int_a^b 2\pi \begin{pmatrix}\text{shell}\\ \text{radius}\end{pmatrix} \begin{pmatrix}\text{shell}\\ \text{height}\end{pmatrix} dx = \int_0^1 2\pi(2 - x)(1 - x^2)\, dx$$

$$= 2\pi \int_0^1 (2 - x - 2x^2 + x^3)\, dx = 2\pi \left[2x - \frac{x^2}{2} - \frac{2}{3}x^3 + \frac{x^4}{4}\right]_0^1 = \frac{13\pi}{6}$$

The disk, washer, and shell methods for calculating the volumes of solids of revolution always agree. We illustrate this agreement.

Explore with a Grapher

Consider this slight modification of Example 3. Replace the parabola $y = x^2$ by the curve $y = 1-(\sin 4x)/4x$. Show that the modified volume is given by $\int_0^b 2\pi(2 - x)\left((\sin 4x)/4x\right)\,dx$ where b is the x-coordinate of the first point of intersection of $y = 1$ and $y = 1 - (\sin 4x)/4x$ to the right of the y-axis. In the exercises you will find b and compute the volume. Will you need technology? Explain why.

EXAMPLE 4 The disk enclosed by the circle $x^2 + y^2 = 4$ is revolved about the y-axis to generate a solid sphere. A hole of diameter 2 is then bored through the sphere along the y-axis. Find the volume of the "cored" sphere.

Solution The cored sphere could have been generated by revolving the shaded region in Fig. 6.24(a) about the y-axis. Thus there are three methods we might use to find the volume: disks, washers, and shells.

METHOD 1: *Disks and subtraction.* Figure 6.24(b) shows the solid sphere with the core pulled out. The core is a circular cylinder with spherical end caps. Our plan is to subtract the volume of the core from the volume of the sphere.

We can simplify matters by imagining that the two caps have already been sliced off the sphere by planes perpendicular to the y-axis at $y = \sqrt{3}$ and $y = -\sqrt{3}$. With the caps removed, the truncated sphere has volume T, say. From this we subtract the volume of the hole, a right circular cylinder of radius 1 and height $2\sqrt{3}$.

The volume of the hole is

$$H = \pi(1)^2(2\sqrt{3}) = 2\pi\sqrt{3}.$$

The truncated sphere (before drilling) is a solid of revolution whose cross sections perpendicular to the y-axis are disks. The radius of a typical disk is $\sqrt{4-y^2}$. Therefore,

$$T = \int_{-\sqrt{3}}^{\sqrt{3}} \pi(\text{radius})^2\,dy = \int_{-\sqrt{3}}^{\sqrt{3}} \pi(4-y^2)\,dy = \pi\left[4y - \frac{y^3}{3}\right]_{-\sqrt{3}}^{\sqrt{3}} = 6\pi\sqrt{3}.$$

The volume of the cored sphere is

$$T - H = 6\pi\sqrt{3} - 2\pi\sqrt{3} = 4\pi\sqrt{3}.$$

METHOD 2: *Washers.* The cored sphere is a solid of revolution whose cross sections perpendicular to the y-axis are washers (Fig. 6.24c). The radii of a typical washer are

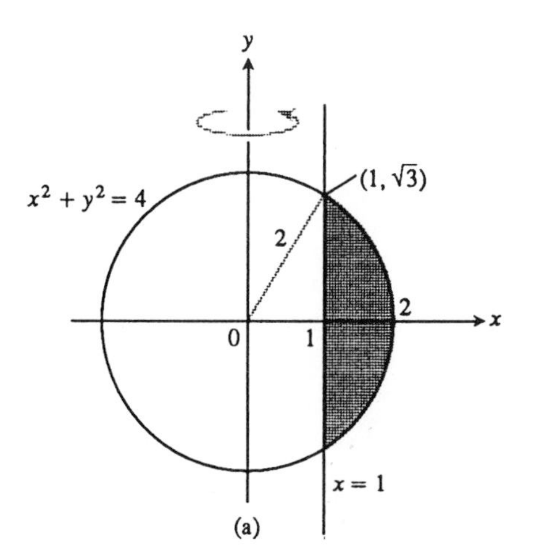

The volume of the cored sphere is the volume swept out by the shaded region as it revolves about the y-axis.

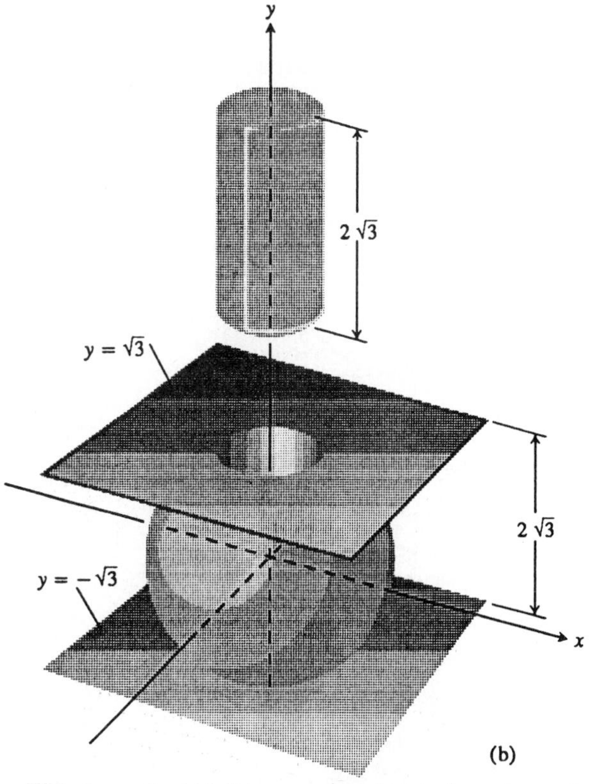

With the method of disks, we can calculate the volume of the hole left by the core and subtract it from the volume of the truncated sphere.

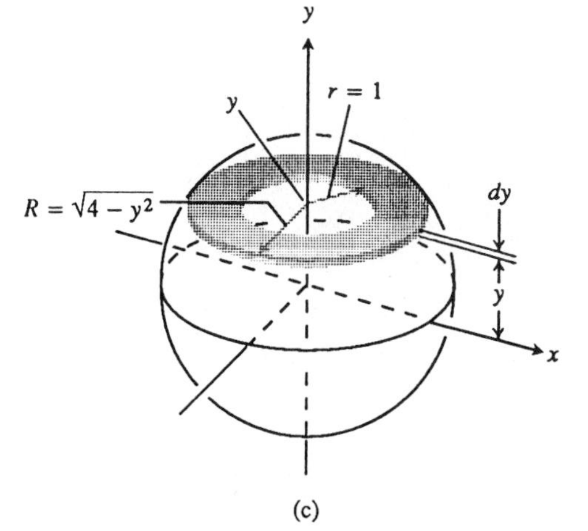

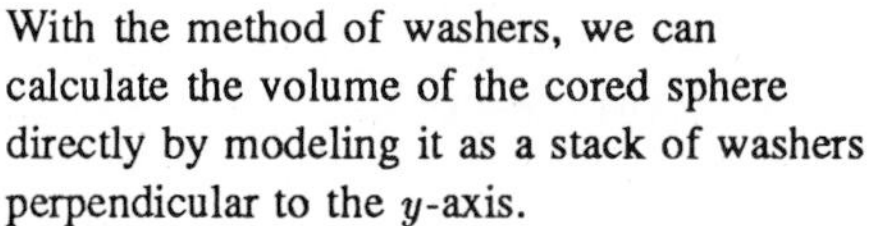

With the method of washers, we can calculate the volume of the cored sphere directly by modeling it as a stack of washers perpendicular to the y-axis.

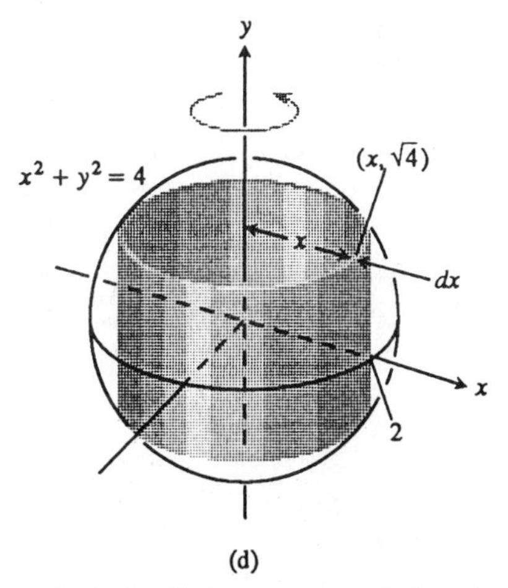

With the method of cylinders, we can calculate the volume of the cored sphere directly by modeling it as a union of cylindrical shells parallel to the y-axis.

6.24 (a) The tinted region generates the volume of revolution. (b) An exploded view showing the sphere with the core removed. (c) A phantom view showing a cross-section slice of the sphere with the core removed. (d) Filling the volume with cylindrical shells.

TABLE Washers versus Shells

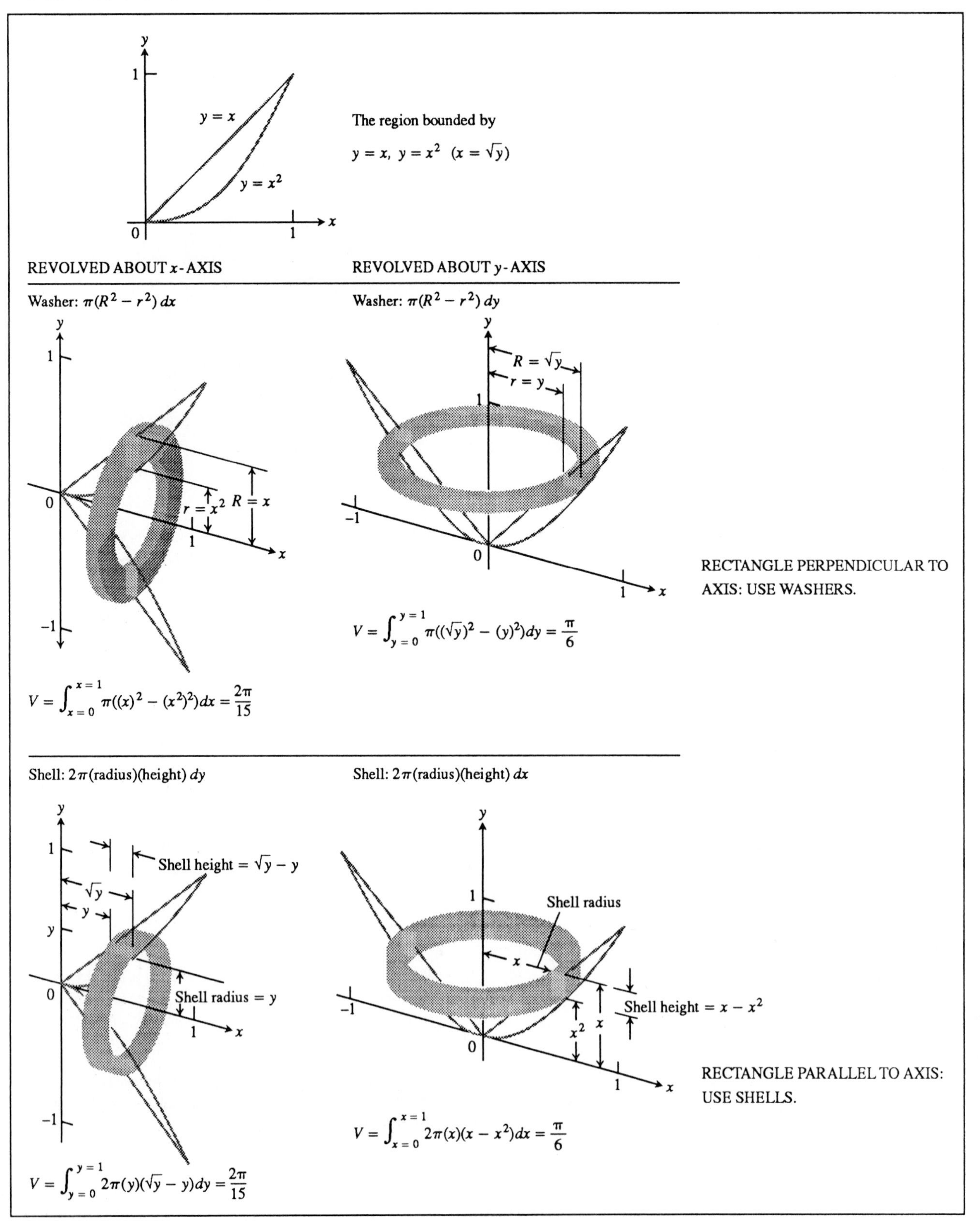

Outer radius: $R = \sqrt{4 - y^2}$

Inner radius: $r = 1.$

The volume of the cored sphere is therefore

$$V = \int_{-\sqrt{3}}^{\sqrt{3}} \pi(R^2 - r^2)\,dy = \int_{-\sqrt{3}}^{\sqrt{3}} \pi(4 - y^2 - 1)\,dy$$

$$= \pi\left[3y - \frac{y^3}{3}\right]_{-\sqrt{3}}^{\sqrt{3}} = 4\pi\sqrt{3}.$$

METHOD 3: *Cylindrical shells.* We model the volume of the cored sphere with cylindrical shells like the one in Fig. 6.24(d). The typical shell has radius x, height $2\sqrt{4 - x^2}$, and thickness dx. The volume of the cored sphere is

$$V = \int_1^2 2\pi \begin{pmatrix}\text{shell}\\ \text{radius}\end{pmatrix}\begin{pmatrix}\text{shell}\\ \text{height}\end{pmatrix} dx = \int_1^2 4\pi x\sqrt{4 - x^2}\,dx$$

$$= 4\pi\left[-\frac{1}{3}(4 - x^2)^{3/2}\right]_1^2 \qquad \begin{pmatrix}\text{After substituting } u = 4 - x^2,\\ \text{integrating, and substituting back}\end{pmatrix}$$

$$= 0 - 4\pi\left[-\frac{1}{3}(4 - 1)^{3/2}\right] = 4\pi\sqrt{3}.$$

Table 6.1 summarizes the methods of finding volumes with washers and shells.

Exercises 6.3

Use the shell method to find the volumes of the solids generated by revolving the regions bounded by the curves and lines in Exercises 1–6 about the y-axis.

1. $y = x$, $y = -x/2$, and $x = 2$

2. $y = \sqrt{x}$, $y = 0$, and $x = 4$

3. $y = x^2 + 1$, $y = 0$, $x = 0$, and $x = 1$

4. $y = 2x - 1$, $y = \sqrt{x}$, and $x = 0$

5. $y = 1/x$, $y = 0$, $x = 1/2$, $x = 2$

6. $y = 1/x^2$, $y = 0$, $x = 1/2$, $x = 2$

Use the shell method to find the volumes of the solids generated by revolving the regions bounded by the curves and lines in Exercises 7–12 about the x-axis.

7. $y = |x|$ and $y = 1$

8. $y = x$, $y = 1$, and $x = 2$

9. $y = \sqrt{x}$, $y = 0$, and $y = x - 2$

10. $y = 2$, $x = -y$, and $x = \sqrt{y}$

11. The parabola $x = 2y - y^2$ and the y-axis

12. The parabola $x = 2y - y^2$ and the line $y = x$

Find the volumes of the solids generated by revolving the shaded regions in Exercises 13–16 about the indicated axes.

13. The y-axis

14. The y-axis

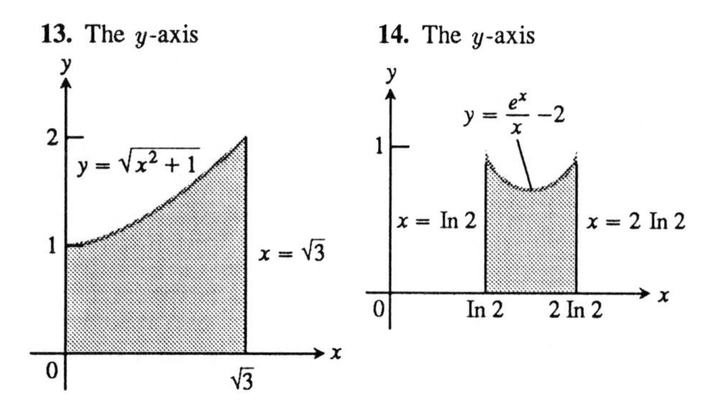

15. The x-axis

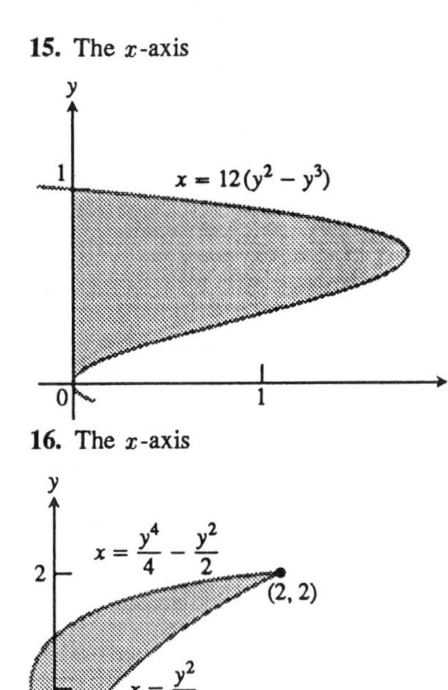

16. The x-axis

In Exercises 17–24, find the volumes of the solids generated by revolving the regions about the given axes.

17. The triangle with vertices $(1, 1)$, $(1, 2)$, and $(2, 2)$ about

a) the x-axis

b) the y-axis

18. The region in the first quadrant bounded by the curve $x = y - y^3$ and the y-axis about

a) the x-axis

b) the y-axis

19. The region in the first quadrant bounded by $x = y - y^3$, $x = 1$, and $y = 1$ about

a) the x-axis

b) the y-axis

c) the line $x = 1$

d) the line $y = 1$

20. The triangular region bounded by the lines $2y = x + 4$, $y = x$, and $x = 0$ about

a) the x-axis

b) the y-axis

c) the line $x = 4$

d) the line $y = 8$

21. The region in the first quadrant bounded by $y = x^3$ and $y = 4x$ about

a) the x-axis

b) the line $y = 8$

22. The region bounded by $y = \sqrt{x}$ and $y = x^2/8$ about

a) the x-axis

b) the y-axis

23. The region bounded by $y = 2x - x^2$ and $y = x$ about

a) the y-axis

b) the line $x = 1$

24. The region bounded by $y = \sqrt{x}$, $y = 2$, $x = 0$ about

a) the x-axis

b) the y-axis

c) the line $x = 4$

d) the line $y = 2$

25. a) Show that the volume of the solid determined by revolving the region in the first quadrant bounded by the curve $y = 1 - (\sin 4x)/4x$, the y-axis, and the line $y = 1$ about the line $x = 2$ is $\int_0^b 2\pi(2 - x)(\sin 4x)/4x\,dx$ where b is the x-coordinate of the first point of intersection of $y = 1$ and $y = 1 - (\sin 4x)/4x$ to the right of the y-axis.

b) Determine b given in part (a).

c) Find the volume of the solid determined in part (a).

6.4 Lengths of Curves in the Plane

We approximate the length of a curved path in the plane the way we use a ruler to estimate the length of a curved road on a map, by measuring from point to point with straight line segments and adding the results. There is a limit to the accuracy of such an estimate, however, imposed in part by how accurately we measure and in part by how many line segments we are willing to use.

With calculus we can usually do a better job because we can imagine using straight line segments as short as we please, each set of segments making a polygonal path that fits the curve more tightly than before. When we proceed this way, with a smooth enough curve, the lengths of the polygonal paths approach a limit we can calculate with an integral. In this section we shall see what that integral is. We shall also see what happens if, instead of being smooth, the curve is Helga von Koch's snowflake curve.

The Basic Formula

Suppose the curve whose length we want to find is the graph of the function $y = f(x)$ from $x = a$ to $x = b$. We partition the closed interval $[a, b]$ in the usual way and connect the corresponding points on the curve with line segments (Fig. 6.25). The line segments, taken together, form a polygonal path that approximates the curve.

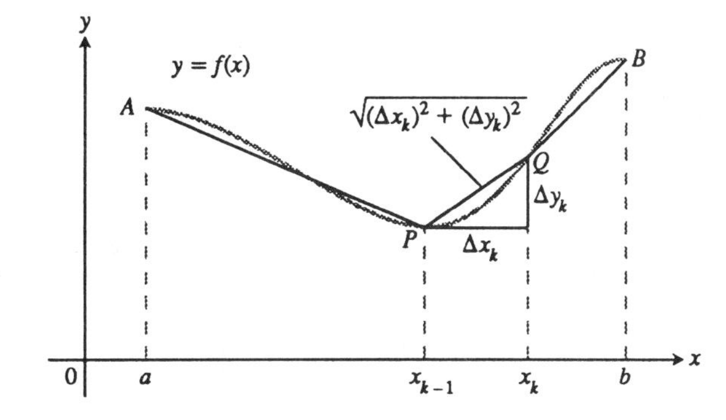

6.25 A typical segment PQ of a polygonal path approximating the curve AB.

The length of a typical line segment PQ (shown in the figure) is

$$\sqrt{(\Delta x_k)^2 + (\Delta y_k)^2}. \tag{1}$$

The length of the curve is therefore approximated by the sum

$$\sum_{k=1}^{n} \sqrt{(\Delta x_k)^2 + (\Delta y_k)^2}. \tag{2}$$

We expect the approximation to improve as the partition $[a, b]$ becomes finer, and we would like to show that the sums in (2) approach a calculable limit as the norm of the partition goes to zero. To show this, we rewrite the sum in (2) in a form to which we can apply the Integral Existence Theorem from Chapter 5. Our starting point, oddly enough, is the Mean Value Theorem for derivatives.

Suppose that f has a derivative that is continuous at every point of $[a, b]$. Then, by the Mean Value Theorem, there is a point $(c_k, f(c_k))$ on the curve between P and Q where the tangent is parallel to the segment PQ (Fig. 6.26). At this point

$$f'(c_k) = \frac{\Delta y_k}{\Delta x_k}, \tag{3}$$

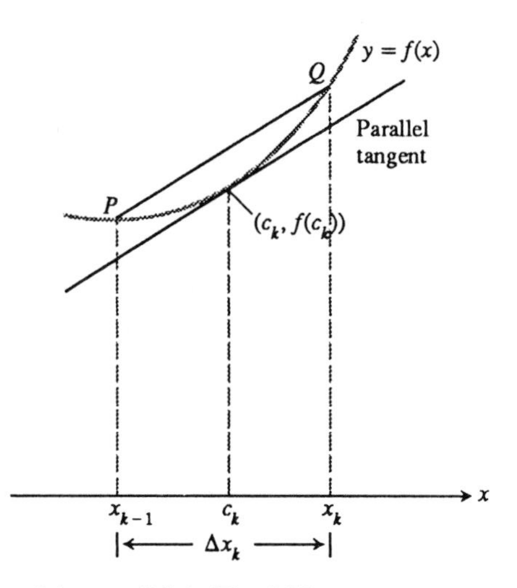

6.26 Enlargement of the arc PQ in Fig. 6.25.

so that

$$\Delta y_k = f'(c_k)\Delta x_k. \tag{4}$$

With this substitution for Δy_k, the sums in (2) take the form

$$\sum_{k=1}^{n} \sqrt{(\Delta x_k)^2 + (f'(c_k)\Delta x_k)^2} = \sum_{k=1}^{n} \sqrt{1 + (f'(c_k))^2}\,\Delta x_k. \tag{5}$$

The sums on the right are Riemann sums for the continuous function $\sqrt{1+(f'(x))^2}$ on the interval $[a, b]$. They therefore converge to the integral of this function as the norm of the partition of the interval goes to zero. We define this integral to be the length of the curve from a to b.

Definition

If the function f has a continuous first derivative throughout the interval $a \le x \le b$, the **length of the curve $y = f(x)$ from a to b** is the number

$$L = \int_a^b \sqrt{1 + \left(\frac{dy}{dx}\right)^2}\,dx. \tag{6}$$

EXAMPLE 1 Find the length of the curve

$$y = \frac{4\sqrt{2}}{3}x^{3/2} - 1, \quad 0 \le x \le 1.$$

Solution We use Eq. (6) with $a = 0$, $b = 1$, and

$$y = \frac{4\sqrt{2}}{3}x^{3/2} - 1$$

$$\frac{dy}{dx} = \frac{4\sqrt{2}}{3} \cdot \frac{3}{2}x^{1/2} = 2\sqrt{2}x^{1/2}$$

$$1 + \left(\frac{dy}{dx}\right)^2 = 1 + (2\sqrt{2}x^{1/2})^2 = 1 + 8x.$$

The length of the curve from $x = 0$ to $x = 1$ is

$$L = \int_0^1 \sqrt{1 + \left(\frac{dy}{dx}\right)^2}\, dx = \int_0^1 \sqrt{1 + 8x}\, dx \qquad \left(\begin{array}{l}\text{Eq. (6) with } a = 0,\\ b = 1\end{array}\right)$$

$$= \frac{2}{3} \cdot \frac{1}{8}(1 + 8x)^{3/2}\bigg]_0^1 = \frac{13}{6}. \qquad \left(\begin{array}{l}\text{Substitute } u = 1 + 8x\text{, integrate,}\\ \text{and replace } u \text{ by } 1 + 8x.\end{array}\right)$$ ■

This contrived example and other non-contrived curve length problems are obvious candidates for the use of technology. For example, FnInt$(\sqrt{1 + 8x}, x, 0, 1) = 2.16666666652$ provides an effective arithmetic check of Example 3. Compare this result with $13/6$. The fact is most curve-length problems produce a function whose antiderivative is impossible to find! Consider this non-contrived application.

EXAMPLE 2 A roller coaster track is modeled by the graph of $y = 100 + 30\sin(0.01x) + 40\cos(0.02x)$ from $x = 0$ to $x = 940$ feet. Think of a roller coaster track on the surface of a circular cylinder and then unwrap the cylinder. See Fig. 6.27. What is the length of the track?

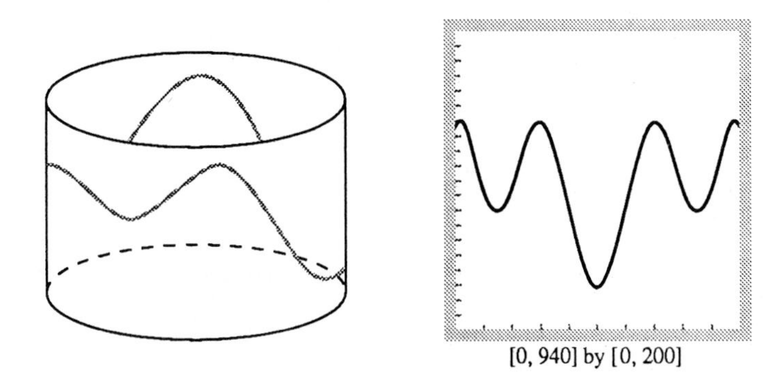

6.27 The roller coaster track of Example 2.

Solution The track length is modeled by the length of the curve $y = 100 + 30\sin(0.01x) + 40\cos(0.02x)$ from 0 to 940 and is given by

$$\int_0^{940} \sqrt{1 + (0.3\cos(0.01x) - 0.8\sin(0.02x))^2}\, dx.$$

This is an easy job for technology.

$$\text{FnInt}\ (\sqrt{1 + (0.3\cos 0.01x - 0.8\sin 0.02x)^2}, x, 0, 940) = 1064.47850.$$

So the length of the track is about 1064 feet. ∎

> **Explore with a Grapher**
>
> Can you explain why the FnInt computation takes a long time in Example 2?

Dealing with Discontinuities in dy/dx

At a point on a curve where dy/dx fails to exist, dx/dy may exist and we may be able to find the curve's length by interchanging x and y in Eq. (6). The revised formula looks like this:

$$L = \int_c^d \sqrt{1 + \left(\frac{dx}{dy}\right)^2}\, dy. \tag{7}$$

To use Eq. (7), we express x as a function of y, calculate dx/dy, and proceed as before to square, add 1, take the square root, and integrate.

EXAMPLE 3 Find the length of the curve $y = (x/2)^{2/3}$ from $x = 0$ to $x = 2$.

Solution The derivative

$$\frac{dy}{dx} = \frac{2}{3}\left(\frac{x}{2}\right)^{-1/3} = \frac{2}{3}\left(\frac{2}{x}\right)^{1/3}$$

is not defined at $x = 0$, so we cannot find the curve's length with Eq. (6).

We therefore rewrite the equation to express x in terms of y:

$$y = \left(\frac{x}{2}\right)^{2/3}$$

$$y^{3/2} = \frac{x}{2} \qquad \text{(Raise both sides to the power 3/2.)}$$

$$x = 2y^{3/2}. \qquad \text{(Solve for } x\text{.)}$$

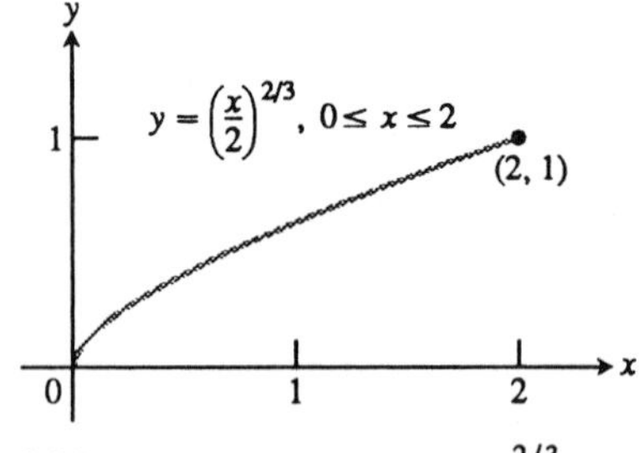

6.28 The graph of $y = (x/2)^{2/3}$ from $x = 0$ to $x = 2$ is also the graph of $x = 2y^{3/2}$ from $y = 0$ to $y = 1$, a function that has a continuous first derivative. We may therefore use Eq. (7) with $x = 2y^{3/2}$ to find the curve's length.

From this we see that the curve whose length we want is also the graph of $x = 2y^{3/2}$ from $y = 0$ to $y = 1$ (Fig. 6.28).

The derivative

$$\frac{dx}{dy} = 2 \cdot \frac{3}{2} y^{1/2} = 3y^{1/2}$$

is continuous throughout the interval $0 \le y \le 1$. We may therefore find the curve's length by setting

$$\left(\frac{dx}{dy}\right)^2 = (3y^{1/2})^2 = 9y$$

in Eq. (7) and integrating from $y = 0$ to $y = 1$:

$$\begin{aligned} L &= \int_c^d \sqrt{1 + \left(\frac{dx}{dy}\right)^2}\, dy = \int_0^1 \sqrt{1 + 9y}\, dy \qquad \text{(Eq. 7)} \\ &= \frac{1}{9} \cdot \frac{2}{3}(1 + 9y)^{3/2}\Big]_0^1 \qquad \left(\begin{array}{l}\text{Substitute } u = 1 + 9y,\ du/9 = dy, \\ \text{integrate, and substitute back.}\end{array}\right) \\ &= \frac{2}{27}(10\sqrt{10} - 1) = 2.27. \end{aligned}$$

The Short Differential Formula

The equations

$$L = \int_a^b \sqrt{1 + \left(\frac{dy}{dx}\right)^2}\, dx \quad \text{and} \quad L = \int_c^d \sqrt{1 + \left(\frac{dx}{dy}\right)^2}\, dy \tag{8}$$

are often written with differentials instead of derivatives. This is done formally by thinking of the derivatives as quotients of differentials and bringing the dx and dy inside the radicals to cancel the denominators. In the first integral we have

$$\sqrt{1 + \left(\frac{dy}{dx}\right)^2}\, dx = \sqrt{1 + \frac{dy^2}{dx^2}}\, dx = \sqrt{dx^2 + \frac{dy^2}{dx^2}\, dx^2} = \sqrt{dx^2 + dy^2}.$$

In the second integral we have

$$\sqrt{1 + \left(\frac{dx}{dy}\right)^2}\, dy = \sqrt{1 + \frac{dx^2}{dy^2}}\, dy = \sqrt{dy^2 + \frac{dx^2}{dy^2}\, dy^2} = \sqrt{dx^2 + dy^2}. \tag{9}$$

Thus the integrals in (8) reduce to a single differential formula:

$$L = \int_a^b \sqrt{dx^2 + dy^2}. \tag{10}$$

Of course, dx and dy must be expressed in terms of a common variable, and appropriate limits of integration must be found before the integration in Eq. (10) is performed.

We can shorten Eq. (10) still further. Think of dx and dy as two sides of a small triangle whose hypotenuse is

$$ds = \sqrt{dx^2 + dy^2} \tag{11}$$

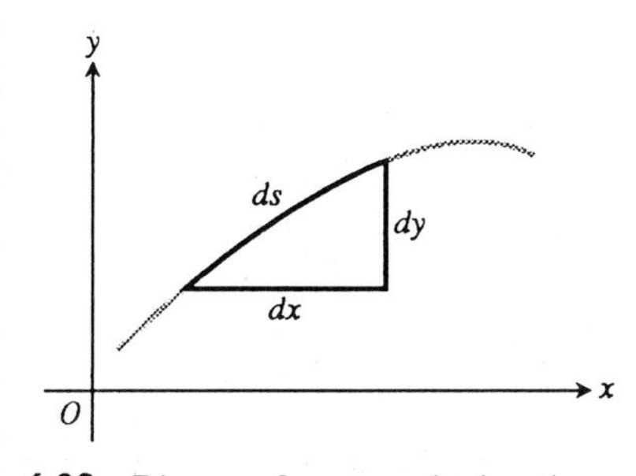

6.29 Diagram for remembering the equation $ds = \sqrt{dx^2 + dy^2}$.

(Fig. 6.29). The differential ds is then regarded as a differential of arc length that can be integrated between appropriate limits to give the length of the curve. With $\sqrt{dx^2 + dy^2}$ set equal to ds, the integral in Eq. (10) simply becomes the integral of ds.

Definition

The Arc Length Differential and the Differential Formula for Arc Length

$ds = \sqrt{dx^2 + dy^2}$	$L = \int ds$
Arc length differential	Differential formula for arc length

Curves with Infinite Length

As you may recall from Section 3.6, Helga von Koch's snowflake curve K is the limit curve of an infinite sequence $C_1, C_2, \ldots, C_n, \ldots$ of "triangular" polygonal curves. The first four curves in the sequence look like this:

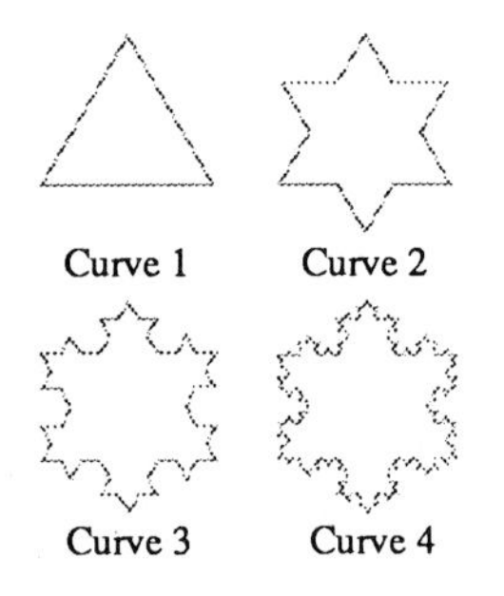

Each time we introduce a new vertex in the construction process, it remains as a vertex in all subsequent curves and becomes a point on the limit curve K. This means that each of the $C's$ is itself a polygonal approximation of K—the endpoints of its sides all belonging to K. The length of K should therefore be the limit of the lengths of the curves C_n. At least, that is what it should be if we stick to the definition of the length we developed for smooth curves.

What, then, is the limit of the lengths of the curves C_n? If the original equilateral triangle C_1 has sides of length 1, the total length of C_1 is 3. To make C_2 from C_1, we replace each side of C_1 by four segments, each of which is one-third as long as the original side.

To make C_2 from C_1, we do this to each side.

The total length of C_2 is therefore 3(4/3). To get the length of C_3, we multiply by 4/3 again. We do so again to get the length of C_4. By the time we get out to C_n, we have a curve of length $3(4/3)^{n-1}$.

Curve Number	1	2	3	...	n	...
Length	3	$3\left(\frac{4}{3}\right)$	$3\left(\frac{4}{3}\right)^2$	...	$3\left(\frac{4}{3}\right)^{n-1}$	...

The length of C_{10} is nearly 40 and the length of C_{100} is greater than 7,000,000,000,000. The lengths grow too rapidly to have any finite limit. Therefore the snowflake curve has no length, or, if you prefer, infinite length.

What went wrong? Nothing. The formulas we derived for length are for the graphs of functions with continuous first derivatives, curves that are smooth enough to have a continuously turning tangent at every point. Helga von Koch's snowflake curve is too rough for that, and our derivative-based formulas do not apply.

Exercises 6.4

Find the lengths of the curves in Exercises 1–10 analytically and support with technology.

1. $y = (1/3)(x^2 + 2)^{3/2}$ from $x = 0$ to $x = 3$

2. $y = x^{3/2}$ from $x = 0$ to $x = 4$

3. $9x^2 = 4y^3$ from $(0, 0)$ to $(2\sqrt{3}, 3)$

4. $y = x^{2/3}$ from $x = 0$ to $x = 4$

5. $y = (x^3/3) + 1/(4x)$ from $x = 1$ to $x = 3$
(Hint: $1 + (dy/dx)^2$ is a perfect square.)

6. $y = (x^{3/2}/3) - x^{1/2}$ from $x = 1$ to $x = 9$
(Hint: $1 + (dy/dx)^2$ is a perfect square.)

7. $x = (y^4/4) + 1/(8y^2)$ from $y = 1$ to $y = 2$
(Hint: $1 + (dx/dy)^2$ is a perfect square.)

8. $x = (y^3/6) + 1/(2y)$ from $y = 2$ to $y = 3$
(Hint: $1 + (dx/dy)^2$ is a perfect square.)

9. $y = (e^x + e^{-x})/2$ from $x = -\ln 2$ to $x = \ln 2$
(Hint: $1 + (dy/dx)^2$ is a perfect square.)

10. $x = y^2 - (1/8)\ln y$ from $y = 1$ to $y = 3$
(Hint: $1 + (dx/dy)^2$ is a perfect square.)

11. The length of an astroid. The graph of the equation $x^{2/3} + y^{2/3} = 1$ is one of a family of curves called *astroids* (not "asteroids") because of their starlike shapes. Find the length of this particular astroid by finding the length of half the first quadrant portion, $y = (1 - x^{2/3})^{3/2}$, $\sqrt{2}/4 \le x \le 1$, and multiplying by 8.

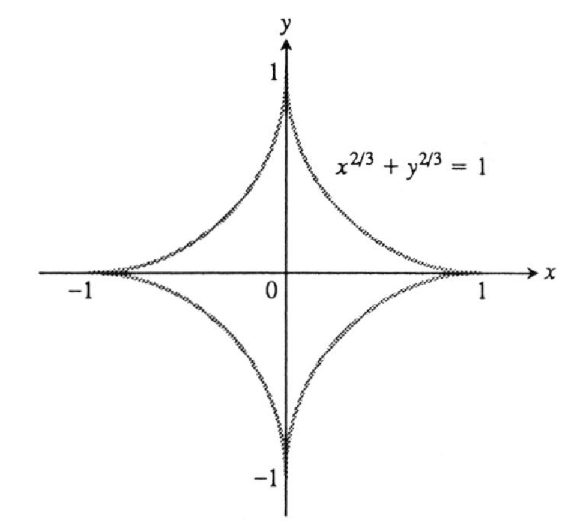

12. Find the length of the curve

$$y = \int_0^x \sqrt{\cos 2t}\, dt$$

from $x = 0$ to $x = \pi/4$ analytically and support with technology. *(Hint:* For $0 \le x \le \pi/4$, $1 + \cos 2x = 2\cos^2 x$.)

13. Find a curve through the origin whose length is

$$L = \int_0^4 \sqrt{1 + \frac{1}{4x}}\, dx.$$

(Hint: Solve the initial value problem $dy/dx = \sqrt{1/4x}$, $y(0) = 0$.)

14. Find a curve through the point $(0, 1)$ whose length from $x = 0$ to $x = 1$ is

$$L = \int_0^1 \sqrt{1 + e^{2x}}\, dx.$$

15. Find a curve through the point $(1, 0)$ whose length from $x = 1$ to $x = 2$ is

$$L = \int_1^2 \sqrt{1 + \frac{1}{x^2}}\, dx.$$

16. Without evaluating either integral, show why

$$2\int_{-1}^{1}\sqrt{1-x^2}\,dx = \int_{-1}^{1}\frac{1}{\sqrt{1-x^2}}\,dx,$$

and then support numerically. (*Hint:* Interpret one integral as an area and the other as a length.) (*Source:* Peter A. Lindstrom, *Mathematics Magazine*, Volume 45, Number 1, January 1972, page 47.)

You may have wondered why the curves we have been working with have such unusual formulas. They are contrived. The reason is that the square root $\sqrt{1+(dy/dx)^2}$ in the arc-length integral almost never produces a function whose anti-derivative we can find. In fact, this square root is a famous source of nonelementary integrals. Most arc-length integrals have to be evaluated numerically, as in the exercises that follow.

Determine the lengths of the curves in Exercises 17–18. Do not try to evaluate the integrals analytically right now . We shall see how to do that when we get to Chapter 8.

17. $y = \ln(1-x^2),\ \ 0 \le x \le 1/2$

18. $y = \ln(\cos x),\ \ 0 \le x \le \pi/3$

19. Your metal fabrication company is bidding for a contract to make sheets of corrugated iron roofing like the one shown in Fig. 6.30. The cross sections of the corrugated sheets are to conform to the curve

$$y = \sin\frac{3\pi}{20}x,\quad 0 \le x \le 20 \text{ in.}$$

If the roofing is to be stamped from flat sheets by a process that does not stretch the material, how wide should the original material be?

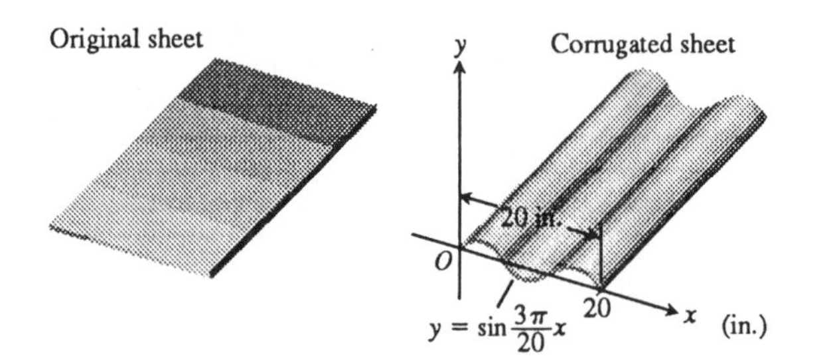

6.30 How wide does the original sheet have to be? See Exercise 19.

20. Your engineering firm is bidding for the contract to construct the tunnel shown in Fig. 6.31. The tunnel is 300 ft long and 50 ft wide at the base. The cross section is shaped like one arch of the curve $y = 25\cos(\pi x/50)$. Upon completion, the tunnel's inside surface (excluding the roadway) will be treated with a waterproof sealer that costs \$1.75 per ft^2 to apply. How much will it cost to apply the sealer?

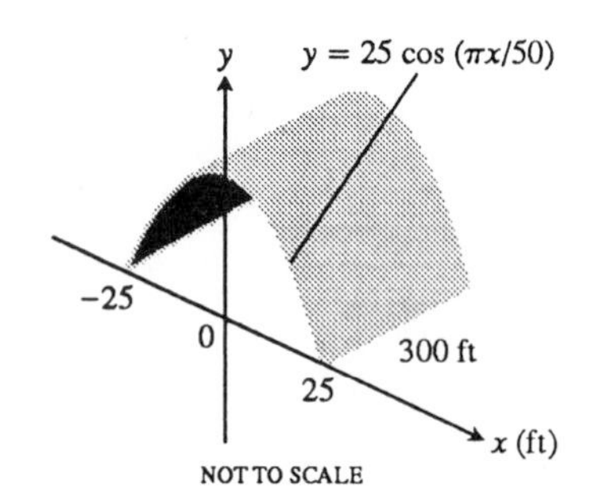

6.31 The tunnel in Exercise 20.

21. An asteroid follows the path modeled by $y^2 = 0.2x^2 - 1$ for $x > 0$ (x is in 10^6 miles units). Assume the earth is at the origin. How many miles will the asteroid travel starting from the point whose x-coordinate is $x = 10$ (million miles), until it reaches its closest point to earth?

22. Two lanes of a running track are modeled by semi-ellipses as shown in Fig. 6.32.

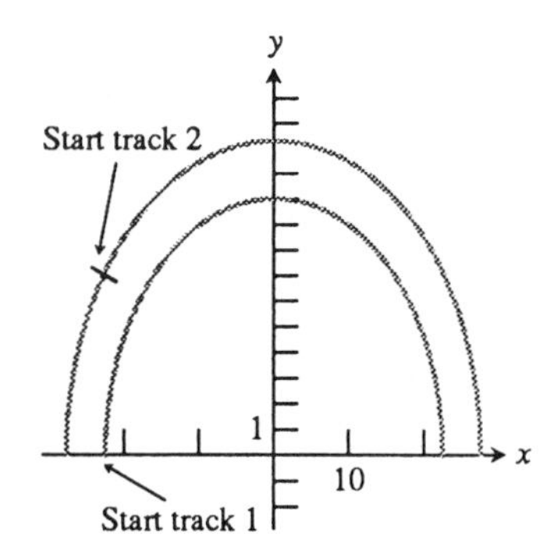

6.32 The two running tracks of Exercise 22.

The equation for track 1 is $y = \sqrt{100-0.2x^2}$ and the equation for track 2 is $y = \sqrt{150-0.2x^2}$. The starting point for track 1 is at the negative x-intercept $(-\sqrt{500}, 0)$. The finish points for both tracks are the positive x-intercepts. Where should the starting point be placed on track 2 so that the two track lengths will be equal?

23. The outside shape of a swimming pool is modeled by the equation $y^2 = 400(0.1x)^2(2-0.1x)$ for $x \ge 0$. Find the length of fence needed to completely enclose the pool.

24. Suppose the pool in Exercise 23 was modeled by the equation $y^2 = K(0.1x)^2(2-0.1x)x$ and only 80 feet of fencing was available to enclose the pool. Determine K.

6.5 Areas of Surfaces of Revolution

When you jump rope, the rope sweeps out a surface in the space around you, a surface called a surface of revolution. As you can imagine, the area of this surface depends on the rope's length and on how far away each segment of the rope swings. This section explores the relation between the area of a surface of revolution and the length and reach of the curve that generates it.

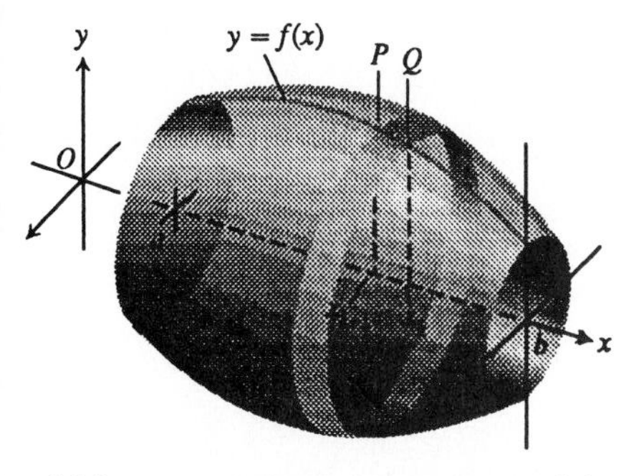

6.33 The surface swept out by revolving the curve $y = f(x), a \le x \le b$, about the x-axis is a union of bands like the one swept out by the arc PQ.

The Basic Formula

Suppose we want to find the area of the surface swept out by revolving the graph of the function $y = f(x), a \le x \le b$, about the x-axis. We partition the closed interval $[a, b]$ in the usual way and use the points in the partition to divide the graph into the short arcs. Figure 6.33 shows a typical arc PQ and the band it sweeps out as part of the graph of f.

As the arc PQ revolves about the x-axis, the line segment joining P and Q sweeps out part of a cone whose axis lies along the x-axis (magnified view in Fig. 6.34). A piece of a cone like this is called a frustum of the cone, *frustum* being Latin for "piece." The surface area of the frustum approximates the surface area of the band swept out by the arc PQ.

The surface area of the frustum of a cone (see Fig. 6.35) is 2π times the average of the base radii times the slant height:

$$\text{Frustum surface area } = 2\pi \cdot \frac{r_1 + r_2}{2} \cdot L = \pi(r_1 + r_2)L. \tag{1}$$

For the frustum swept out by the segment PQ (Fig. 6.36), this works out to be

$$\text{Frustum surface area } = \pi(f(x_{k-1}) + f(x_k))\sqrt{(\Delta x_k)^2 + (\Delta y_k)^2}. \tag{2}$$

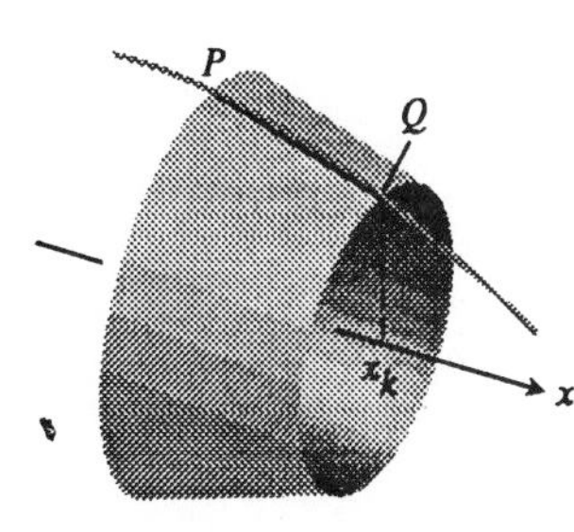

6.34 The straight line segment joining P and Q sweeps out a frustum of a cone.

The area of our original surface, being the sum of the areas of the bands swept out by arcs like arc PQ, is approximated by the frustum area sum

$$\sum_{k=1}^{n} \pi(f(x_{k-1}) + f(x_k))\sqrt{(\Delta x_k)^2 + (\Delta y_k)^2}. \tag{3}$$

We expect the approximation to improve as the partition of $[a, b]$ becomes finer, and we would like to show that the sums in (3) approach a calculable limit as the norm of the partition goes to zero.

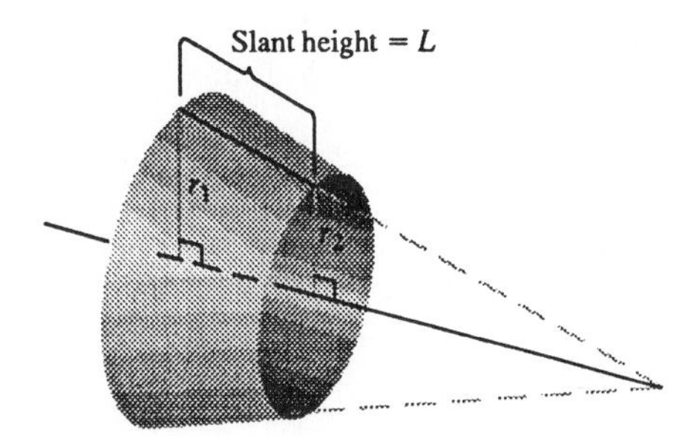

6.35 The important dimensions of the frustum in Fig. 6.34.

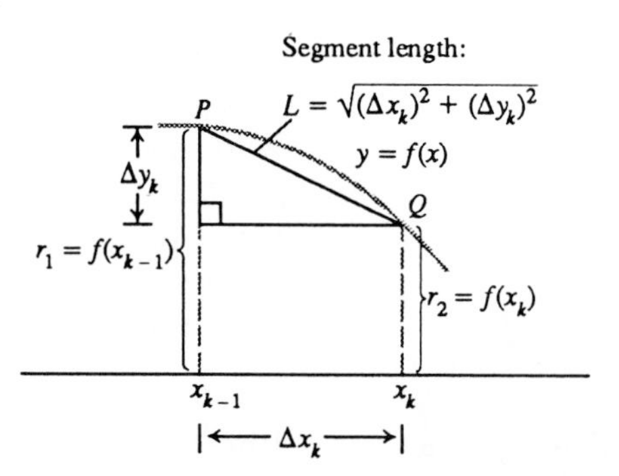

6.37 The important dimensions associated with the arc and segment PQ.

To show this, we try to rewrite the sum in (3) as the Riemann sum of some function over the interval from a to b. As in the calculation of arc length, we begin by appealing to the Mean Value Theorem for derivatives.

Suppose as before that f has a derivative that is continuous at every point of $[a, b]$. Then, by the Mean Value Theorem, there is a point $(c_k, f(c_k))$ on the curve between P and Q where the tangent is parallel to the segment PQ (Fig. 6.37). At this point,

$$f'(c_k) = \frac{\Delta y_k}{\Delta x_k}, \tag{4}$$

so

$$\Delta y_k = f'(c_k)\Delta x_k. \tag{5}$$

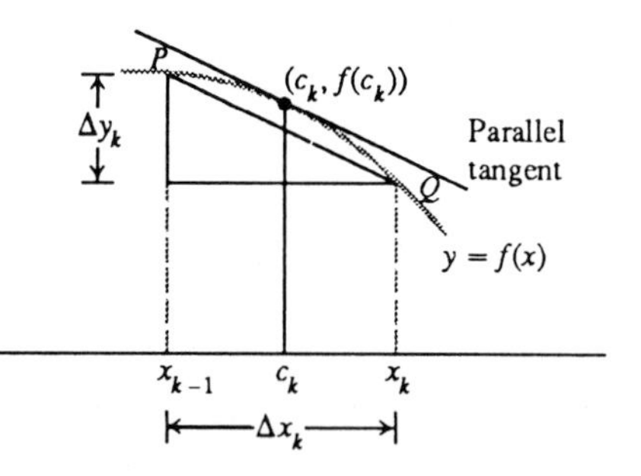

6.36 If f' is continuous, the Mean Value Theorem guarantees the existence of a point on arc PQ where the tangent is parallel to segment PQ.

With this substitution for Δy_k, the sums in (3) take the form

$$\sum_{k=1}^{n} \pi(f(x_{k-1}) + f(x_k))\sqrt{(\Delta x_k)^2 + (f'(c_k)\Delta x_k)^2}$$

$$= \sum_{k=1}^{n} \pi(f(x_{k-1}) + f(x_k))\sqrt{1 + (f'(c_k))^2}\Delta x_k. \tag{6}$$

At this point there is both good news and bad.

The bad news is that the sums in (6) are not the Riemann sums of any function because the points x_{k-1}, x_k, and c_k are not the same and there is no way to make them the same. The good news is that this does not matter. A theorem from advanced calculus assures us that as the norm of the subdivision of $[a, b]$ goes to zero the sums in (3) converge to

$$\int_a^b 2\pi f(x)\sqrt{1+(f'(x))^2}\,dx \tag{7}$$

just the way we want them to. We therefore define this integral to be the area of the surface swept out by the graph of f from a to b.

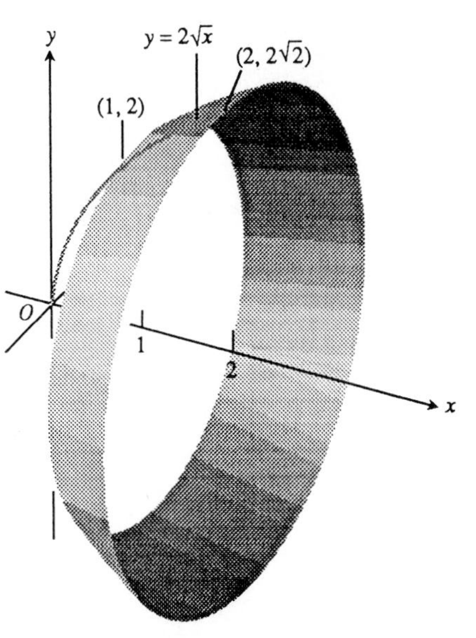

6.38 Example 1 calculates the area of this surface.

Definition

> If the function f has a continuous first derivative throughout the interval $a \le x \le b$, the **area of the surface** generated by revolving the curve $y = f(x)$ about the x-axis is the number
>
> $$S = \int_a^b 2\pi y\sqrt{1+\left(\frac{dy}{dx}\right)^2}\,dx. \tag{8}$$

Notice that the square root in Eq. (8) is the same one that appears in the formula for arc length. More about that later.

EXAMPLE 1 Find the area of the surface generated by revolving the curve $y = 2\sqrt{x}$, $1 \le x \le 2$, about the x-axis (Fig. 6.38).

Solution We evaluate the formula

$$\text{Surface area} = \int_a^b 2\pi y\sqrt{1+\left(\frac{dy}{dx}\right)^2}\,dx \qquad \text{(Eq. 8)}$$

with

$$a = 1, \quad b = 2, \quad y = 2\sqrt{x}, \quad \frac{dy}{dx} = \frac{1}{\sqrt{x}},$$

$$\sqrt{1+\left(\frac{dy}{dx}\right)^2} = \sqrt{1+\left(\frac{1}{\sqrt{x}}\right)^2} = \sqrt{1+\frac{1}{x}} = \sqrt{\frac{x+1}{x}} = \frac{\sqrt{x+1}}{\sqrt{x}}.$$

With these substitutions,

$$\text{Surface area} = \int_1^2 2\pi\cdot 2\sqrt{x}\frac{\sqrt{x+1}}{\sqrt{x}}\,dx = 4\pi\int_1^2 \sqrt{x+1}\,dx$$

$$= 4\pi\cdot\frac{2}{3}(x+1)^{3/2}\Big]_1^2 = \frac{8\pi}{3}(3\sqrt{3}-2\sqrt{2}).$$

Revolution about the y-Axis

If the axis of revolution is the y-axis, we use the formula we get from interchanging x and y in Eq. (8):

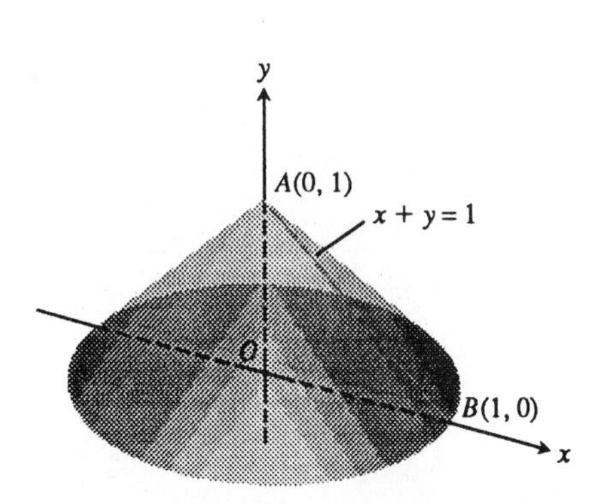

6.39 Revolving line segment AB about the y-axis generates a cone whose lateral surface area we can now calculate two different ways. See Example 2.

> **Revolution about the y-Axis**
>
> If $x = f(y)$ has a continuous first derivative throughout the interval $c \leq y \leq d$, the area of the surface S generated by revolving the curve $x = f(y), c \leq y \leq d$, about the y-axis is
>
> $$S = \int_c^d 2\pi x \sqrt{1 + \left(\frac{dx}{dy}\right)^2}\, dy. \tag{9}$$

EXAMPLE 2 The line segment $x = 1 - y, 0 \leq y \leq 1$, is revolved about the y-axis to generate the cone in Fig. 6.39. Find its lateral surface area.

Solution Here we have a calculation we can check with a formula from geometry:

$$\text{Lateral surface area} = \frac{\text{base circumference}}{2} \times \text{slant height} = \pi\sqrt{2}.$$

To see how Eq. (9) gives the same result, we take

$$c = 0, \quad d = 1, \quad x = 1 - y, \quad \frac{dx}{dy} = -1,$$

$$\sqrt{1 + \left(\frac{dx}{dy}\right)^2} = \sqrt{1 + (-1)^2} = \sqrt{2}$$

and calculate

$$\text{Surface area} = \int_c^d 2\pi x \sqrt{1 + \left(\frac{dx}{dy}\right)^2}\, dy = \int_0^1 2\pi(1 - y)\sqrt{2}\, dy \qquad \text{(Eq. 9)}$$

$$= 2\pi\sqrt{2}\left[y - \frac{y^2}{2}\right]_0^1 = 2\pi\sqrt{2}\left(1 - \frac{1}{2}\right) = \pi\sqrt{2}.$$

The results agree, as they should.

The Short Differential Form

The equations

$$S = \int_a^b 2\pi y \sqrt{1 + \left(\frac{dy}{dx}\right)^2}\, dx \quad \text{and} \quad S = \int_c^d 2\pi x \sqrt{1 + \left(\frac{dx}{dy}\right)^2}\, dy \tag{10}$$

are often written in terms of the arc length differential $ds = \sqrt{dx^2 + dy^2}$ as

$$S = \int_a^b 2\pi\, y\, ds \quad \text{and} \quad S = \int_c^d 2\pi\, x\, ds. \tag{11}$$

In the first of these, y is the distance from the x-axis to an element of arc length ds. In the second, x is the distance from the y-axis to an element of arc length ds. In both cases the integrals have the form

$$S = \int 2\pi(\text{radius})(\text{band width}) = \int 2\pi\, \rho\, ds, \tag{12}$$

where ρ is the radius from the axis of revolution to an element of arc length ds (Fig. 6.40).

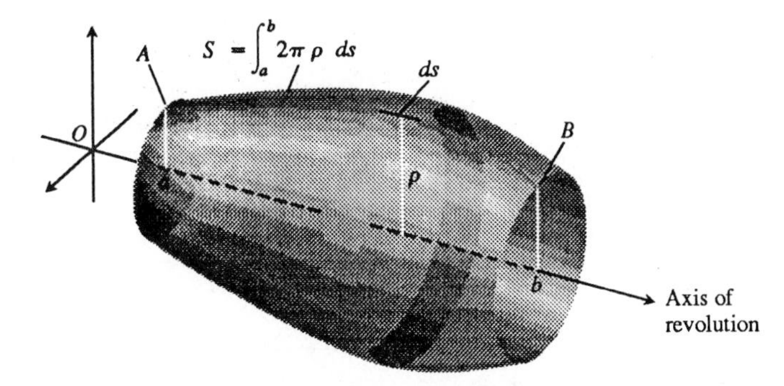

6.40 The area of the surface swept out by revolving arc AB about the axis shown here is $\int_a^b 2\pi\,\rho\,ds$. The exact expression depends on the formulas for ρ and ds.

If you wish to remember only one formula for surface area, you might make it the short differential form.

Short Differential Form

$$S = \int 2\pi\,\rho\,ds \tag{13}$$

In any particular problem you would then express the radius function ρ and the arc length differential ds in terms of a common variable and supply limits of integration for that variable.

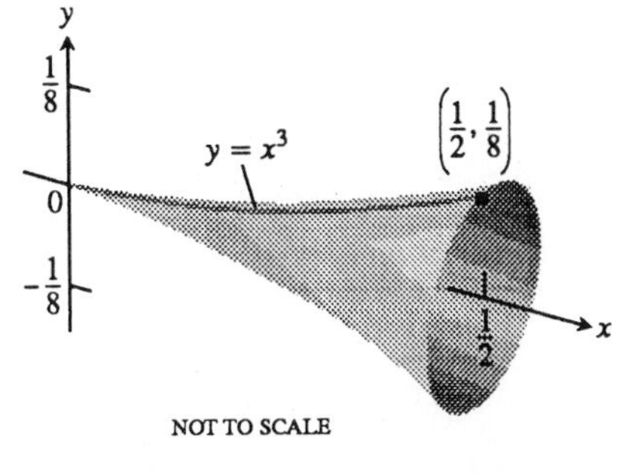

6.41 The surface generated by revolving the curve $y = x^3, 0 \le x \le 1/2$, about the x-axis could be the design for a champagne glass. The surface area is calculated in Example 3.

EXAMPLE 3 Find the area of the surface generated by revolving the curve $y = x^3, 0 \le x \le 1/2$, about the x-axis (Fig. 6.41).

Solution We start with the short differential form:

$$\begin{aligned} S &= \int 2\pi\,\rho\,ds \\ &= \int 2\pi\,y\,ds \qquad \left(\begin{array}{l}\text{For revolution about the } x\text{-axis,}\\ \text{the radius function is } \rho = y.\end{array}\right) \\ &= \int 2\pi\,y\sqrt{dx^2 + dy^2}. \qquad (ds = \sqrt{dx^2 + dy^2}) \end{aligned}$$

We then decide whether to express dy in terms of dx or dx in terms of dy. The original form of the equation, $y = x^3$, makes it easier to express dy in terms of dx, so we continue the calculation with

$$y = x^3, \qquad dy = 3x^2\,dx, \qquad \text{and} \qquad \sqrt{dx^2 + dy^2} = \sqrt{dx^2 + (3x^2\,dx)^2}$$
$$= \sqrt{1 + 9x^4}\,dx.$$

With these substitutions, x becomes the variable of integration, and we obtain

$$S = \int_{x=0}^{x=1/2} 2\pi\, y \sqrt{dx^2 + dy^2} = \int_0^{1/2} 2\pi\, x^3 \sqrt{1 + 9x^4}\, dx.$$

Using FnInt we obtain $S = 0.11090$.

In the exercises you will show analytically that the exact answer to Example 2 is $(61\pi)/1728$. Compare this answer on a calculator to the FnInt solution.

Exercises 6.5

Find the areas of the surfaces generated by revolving the curves in Exercises 1–14 about the axes indicated analytically and support with technology.

1. $y = x/2, 0 \le x \le 4$, about the x-axis. Check your result with a formula from geometry, as in Example 2.

2. $y = x/2, 0 \le x \le 4$, about the y-axis. Check your result with a formula from geometry, as in Example 2.

3. $y = (x/2) + (1/2), 1 \le x \le 3$, about the x-axis. Check your result with the geometry formula in Eq. (1).

4. $y = (x/2) + (1/2), 1 \le x \le 3$, about the y-axis. Check your result with the geometry formula in Eq. (1).

5. $y = x^3/9, 0 \le x \le 2$, about the x-axis

6. $y = \sqrt{x}, 3/4 \le x \le 15/4$, about the x-axis

7. $y = \sqrt{2x - x^2}, 0 \le x \le 2$, about the x-axis

8. $y = \sqrt{x + 1}, 1 \le x \le 5$, about the x-axis

9. $x = y^3/3, 0 \le y \le 1$, about the y-axis

10. $y = 2\sqrt{4 - x}, 1 \le x \le 4$, about the y-axis

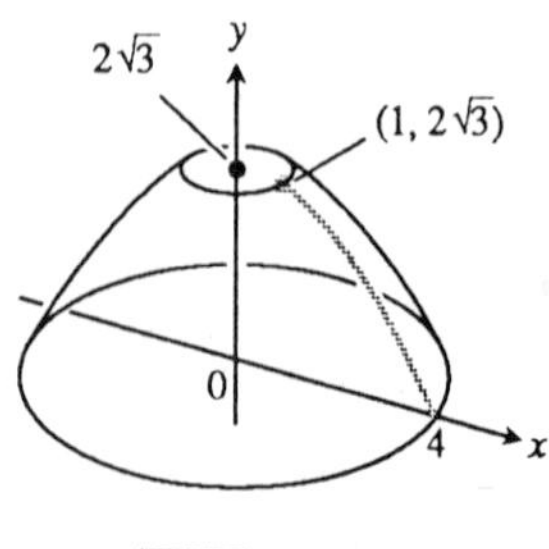

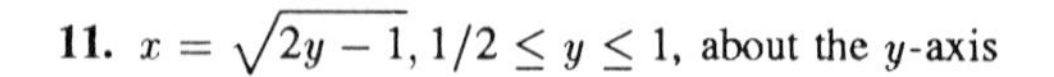

11. $x = \sqrt{2y - 1}, 1/2 \le y \le 1$, about the y-axis

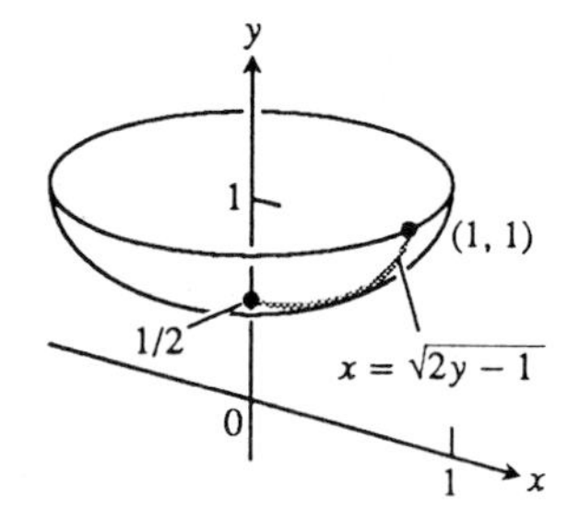

12. $x = (e^y + e^{-y})/2, 0 \le y \le \ln 2$, about the y-axis

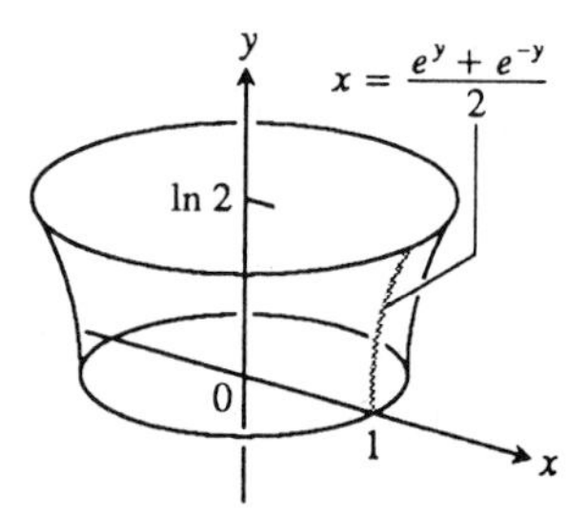

13. $x = (y^4/4) + 1/(8y^2), 1 \le y \le 2$, about the x-axis (*Hint:* Express $ds = \sqrt{dx^2 + dy^2}$ in terms of dy and evaluate the integral $S = \int 2\pi\, y\, ds$ with appropriate limits.)

14. $y = (1/3)(x^2 + 2)^{3/2}, 0 \le x \le 3$, about the y-axis (*Hint:* Express $ds = \sqrt{dx^2 + dy^2}$ in terms of dx and evaluate the integral $S = \int 2\pi\, x\, ds$ with appropriate limits.)

15. Use an integral to find the surface area of the sphere generated by revolving the semicircle $y = \sqrt{1 - x^2}, -1 \le x \le 1$, about the x-axis. Check your result with a formula from geometry.

16. Find the area of the surface generated by revolving the curve $y = \cos x, -\pi/2 \le x \le \pi/2$, about the x-axis.

17. **The surface of an astroid.** Find the area of the surface generated by revolving the portion of the astroid $x^{2/3} + y^{2/3} = 1$ shown below about the x-axis. (*Hint:* Revolve the first quadrant portion $y = (1 - x^{2/3})^{3/2}, 0 \le x \le 1$, about the x-axis and double your result.)

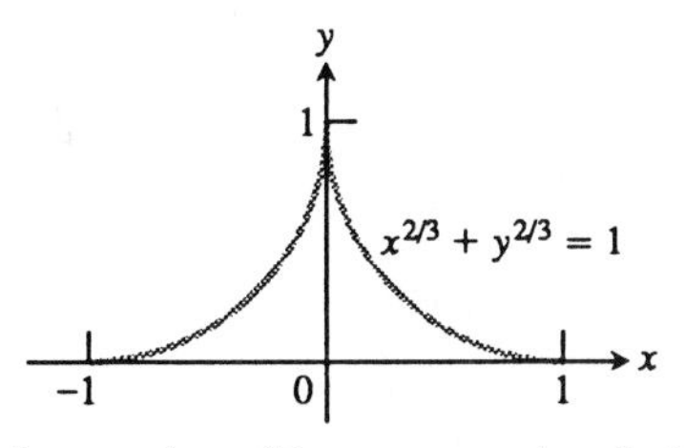

18. **Enamelling woks.** Your company has decided to put out a deluxe version of the successful wok you designed in Exercise 41 of Section 6.2. The plan is to coat it inside with white enamel and outside with blue enamel. Each enamel will be sprayed on a millimeter thick before baking. Manufacturing wants to know how much enamel it will take for a production run of 5,000 woks. What do you tell them? (Neglect waste and unused material. Answer in liters.)

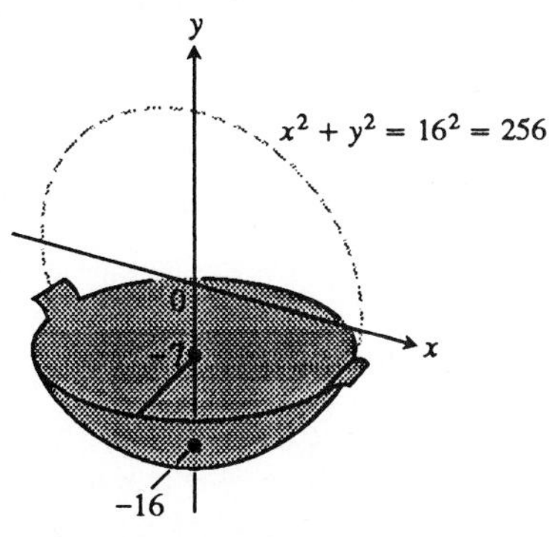

19. **Slicing bread.** Did you know that if you cut a spherical loaf of bread into slices of equal width, each slice will have the same amount of crust? To see why, suppose the semicircle $y = \sqrt{r^2 - x^2}$ in Fig. 6.42 is revolved about the x-axis to generate a sphere. Let AB be an arc of the semicircle that lies above an interval of length h on the x-axis. Show that the area swept out by AB does not depend on the location of the interval. (It does depend on the length of the interval.)

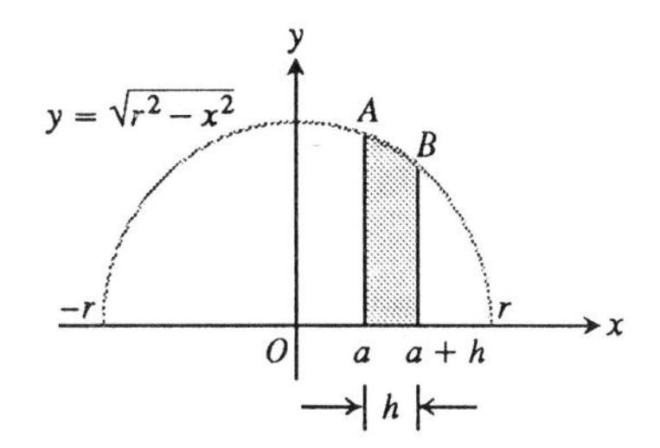

6.42 The semicircle in Exercise 19.

20. The nose cone of a space shuttle is modeled by the solid of revolution determined by revolving the region in the first quadrant given by the curve $y = 10x^{1.3/3}$ and the line $x = 40$ about the x-axis. Assume the special heat resistant tiles covering the nose cone are each 1 ft square. About how many tiles are necessary to cover the space shuttle's nose cone?

21. A parabolic mirror modeled by revolving the curve $y = 2\sqrt{x}$, $0 \le x \le c$ about the x-axis needs to have 50 square feet of surface area. Determine c.

22. Solve Example 3 using analytic methods of calculus.

6.6 Work

In everyday life, *work* describes any activity that takes muscular or mental effort. In science, however, the term is used in a narrower sense that involves the application of a force to a body and the body's subsequent displacement. This section shows how to calculate work. The applications run from stretching springs and pumping liquids from subterranean tanks to lifting satellites into orbit.

The Constant-Force Formula for Work

We begin with a definition.

Definition

When a body moves a distance d along a straight line as the result of being acted on by a force that has a constant magnitude F in the direction of the motion, the **work** W done by the force in moving the body is F times d.

$$W = Fd \tag{1}$$

We can see right away that there is a considerable difference between what we are used to calling work and what this formula says work is. If you push a car down a street, you are doing work, both by our own reckoning and according to Eq. (1). But if you push against the car and the car does not move, Eq. (1) says you are doing no work, no matter how hard or how long you push.

Work is measured in foot-pounds, newton-meters, or whatever force-distance unit is appropriate to the occasion.

EXAMPLE 1 It takes a force of about 1 newton (1 N) to lift an apple from a table. If you lift it 1 meter, you have done about 1 newton-meter (N·m) of work on the apple.

The Variable-Force Integral Formula for Work

If the force you apply varies along the way, as it will if you are lifting a leaking bucket or compressing a spring, the formula $W = Fd$ has to be replaced by an integral formula that takes the variation in F into account. It takes calculus to measure the work done by a variable force.

Suppose that the force performing the work varies continuously along a line that we can take to be the x-axis and that the force is represented by the function $F(x)$. We are interested in the work done along an interval from $x = a$ to $x = b$. We partition the closed interval $[a, b]$ in the usual way and choose an arbitrary point c_k in each subinterval $[x_{k-1}, x_k]$.

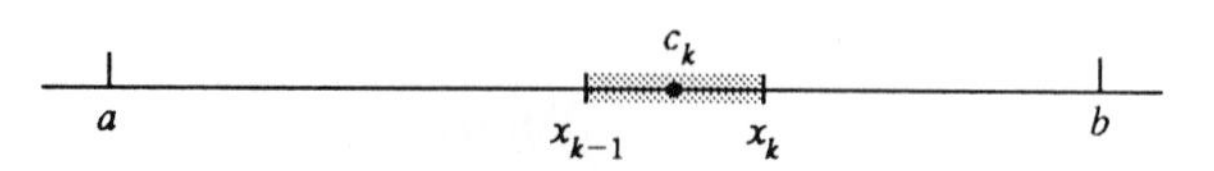

If the subinterval is short enough, F, being continuous, will not vary much from x_{k-1} to x_k. The amount of work done by the force from x_{k-1} to x_k will be nearly equal to $F(c_k)$ times distance Δx_k, as it would be if we could apply Eq. (1). The total work done from a to b is thus approximated by the Riemann sum

$$\sum_{k=1}^{n} F(c_k)\Delta x_k. \tag{2}$$

We expect the approximations to improve as the norm of the partition goes to zero, so we define the work done by the force from a to b to be the integral of F from a to b.

Definition

The **work** done by a continuous force $F(x)$ directed along the x-axis from $x = a$ to $x = b$ is

$$W = \int_a^b F(x)\,dx. \tag{3}$$

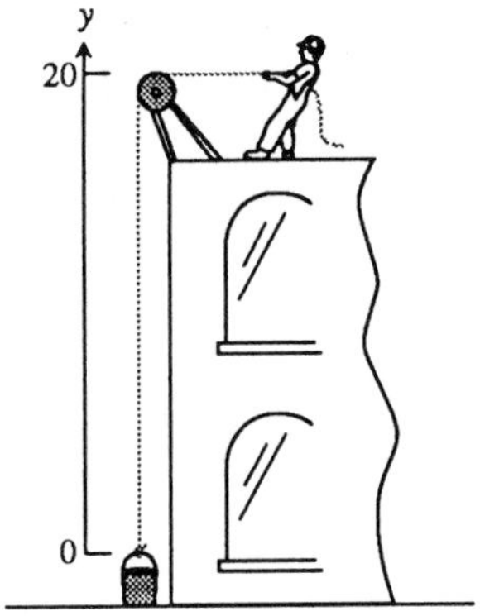

6.43 The leaky bucket in Example 2.

EXAMPLE 2 A leaky 5-lb bucket is lifted from the ground into the air by pulling in 20 ft of rope at a constant speed (Fig. 6.43). The rope weighs 0.08 lb/ft. The bucket starts with 2 gal of water (16 lb) and leaks at a constant rate. It finishes draining just as it reaches the top. How much work was spent

a) lifting the water alone?
b) lifting the water and bucket together?
c) lifting the water, bucket, and rope?

Solution

a) The water alone. The force required to lift the water is the water's weight, which varies steadily from 16 to 0 lb over the 20-ft lift. When the bucket is x ft off the ground, the water weighs

$$F(x) = \underbrace{16}_{\text{Original weight of water}} \underbrace{\left(\frac{20-x}{20}\right)}_{\text{Proportion left at elevation } x} = 16\left(1 - \frac{x}{20}\right) = 16 - \frac{4x}{5} \text{ lb.}$$

The work done is

$$W = \int_a^b F(x)\,dx \qquad \text{(Use Eq. (3) for variable forces.)}$$

$$= \int_0^{20}\left(16 - \frac{4x}{5}\right)dx = \left[16x - \frac{2x^2}{5}\right]_0^{20} = 320 - 160 = 160 \text{ ft} \cdot \text{lb.}$$

b) The water and bucket together. According to Eq. (1), it takes $5 \times 20 = 100$ ft · lb to lift a 5-lb weight 20 ft. Therefore

$$160 + 100 = 260 \text{ ft} \cdot \text{lb}$$

of work was spent lifting the water and bucket together.

c) The water, bucket, and rope. Now the total weight at level x is

$$F(x) = \underbrace{\left(16 - \frac{4x}{5}\right)}_{\text{Variable weight of water}} + \underbrace{5}_{\text{Constant weight of bucket}} + \underbrace{\overset{\text{lb/ft}}{(0.08)}\overset{\text{ft}}{(20 - x)}}_{\text{Weight of rope paid out at elevation } x}.$$

The work lifting the rope is

$$\text{Work on rope} = \int_0^{20} (0.08)(20 - x)\,dx = \int_0^{20} (1.6 - 0.08x)\,dx$$

$$= \left[1.6x - 0.04x^2\right]_0^{20} = 32 - 16 = 16 \text{ ft} \cdot \text{lb}.$$

The total work for the water, bucket, and rope combined is

$$160 + 100 + 16 = 276 \text{ ft} \cdot \text{lb}.$$

Hooke's Law for Springs: $F = kx$

Hooke's law says that the amount of force F it takes to stretch or compress a spring x length units from its natural length is proportional to x. In symbols,

$$F = kx. \tag{4}$$

The number k, measured in force units per unit length, is a constant characteristic of the spring, called the **spring constant.** Hooke's law (Eq. 4) holds as long as the force doesn't distort the metal in the spring. We shall assume that the forces in this section are too small to do that.

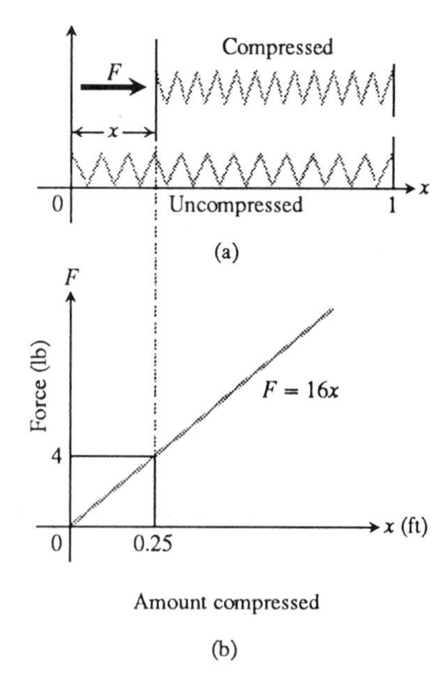

6.44 The force F required to hold a spring under compression increases linearly as the spring is compressed.

EXAMPLE 3 Find the work required to compress a spring from its natural length of 1 ft to a length of 0.75 ft if the spring constant is $k = 16$ lb/ft.

Solution We picture the uncompressed spring laid out along the x-axis with its movable end at the origin and its fixed end at $x = 1$ ft (Fig. 6.44). This enables us to describe the force required to compress the spring from 0 to x with the formula $F = 16x$. As the spring is compressed from 0 to 0.25 ft, the force varies from

$$F(0) = 16 \cdot 0 = 0 \text{ lb} \quad \text{to} \quad F(0.25) = 16 \cdot 0.25 = 4 \text{ lb}.$$

The work done by F over this interval is

$$W = \int_0^{0.25} 16x\,dx = 8x^2\Big]_0^{0.25} = 0.5 \text{ ft} \cdot \text{lb}. \qquad \begin{pmatrix}\text{Eq. (3) with } a = 0, \\ b = 0.25,\ F(x) = 16x\end{pmatrix}$$

EXAMPLE 4 A spring has a natural length of 1 m. A force of 24 N stretches the spring to a length of 1.8 m.

a) Find the spring constant k.

b) How much work will it take to stretch the spring 2 m beyond its natural length?

c) How far will a 45-N force stretch the spring?

Solution

a) The spring constant. We find it from Eq. (4). A force of 24 N stretches the spring 0.8 m, so

$$24 = k(0.8) \qquad \text{(Eq. (4) with } F = 24,\ x = 0.8)$$

$$k = 24/0.8 = 30 \text{ N/m}.$$

b) The work to stretch the spring 2 m. We imagine the unstressed spring hanging along the x-axis with its free end at $x = 0$ (Fig. 6.45). Then the force required to stretch the spring x m beyond its natural length is the force required to pull the free end of the spring x units from the origin. Hooke's law with $k = 30$ tells us this force is

$$F(x) = 30x.$$

The work required to apply this force from $x = 0$ m to $x = 2$ m is

$$W = \int_0^2 30x\,dx = 15x^2\Big]_0^2 = 60 \text{ N} \cdot \text{m}.$$

c) How far will a 45-N force stretch the spring? We substitute $F = 45$ in the equation $F = 30x$ to find

$$45 = 30x \quad \text{and} \quad x = 1.5 \text{ m}.$$

A 45-N force will stretch the spring 1.5 m. No calculus is required to find this.

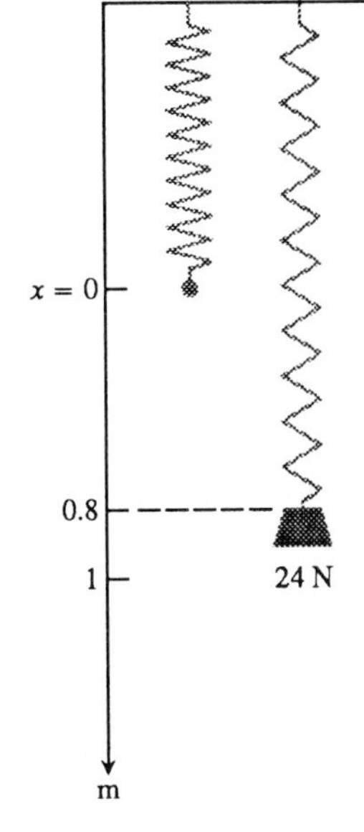

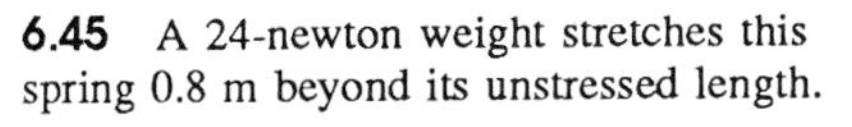

6.45 A 24-newton weight stretches this spring 0.8 m beyond its unstressed length.

Pumping Liquids from Containers—Do-it-yourself Integrals

To find how much work it takes to pump all or part of the liquid from a container, imagine lifting the liquid out one horizontal slab at a time and applying the equation $W = Fd$ to each slab. The integral we get each time depends on the weight of the liquid and the cross-section dimensions of the container, but the way we find the integral is the same for all containers. The next two examples show what to do.

EXAMPLE 5 How much work does it take to pump the water from a full upright right circular cylindrical tank of radius 5 ft and height 10 ft to a level 4 ft above the top of the tank?

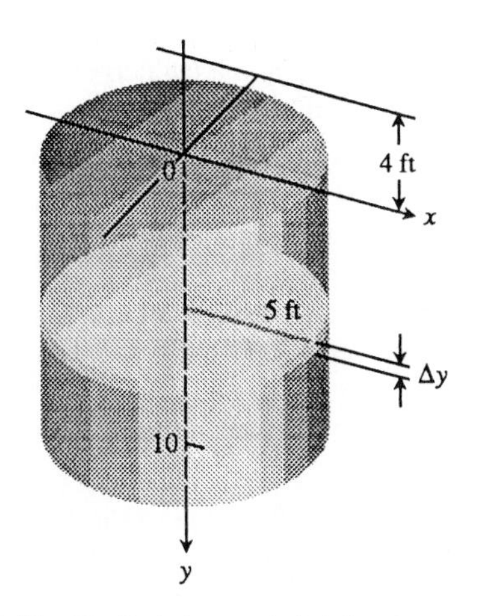

6.46 To find the work it takes to pump the water from a tank, think of lifting the water out one slab at a time. Pointing the y-axis down makes y measure the distance from the slab to the top of the tank.

Solution We draw coordinate axes (Fig. 6.46) and imagine the water divided into thin slabs by planes perpendicular to the y-axis at the points of a partition of the interval $[0, 10]$.

The typical slab between the planes at y and $y + \Delta y$ has a volume of approximately

$$\Delta V = \pi(\text{radius})^2(\text{thickness}) = \pi(5)^2 \Delta y = 25\pi \Delta y \text{ ft}^3.$$

The force $F(y)$ required to lift this slab is its weight,

$$F(y) = w\Delta V = 25\pi w \Delta y \text{ lb},$$

where w is the weight of a cubic foot of water (we can substitute for w later).

The distance through which $F(y)$ must act is about $(y+4)$ ft, so the work done in lifting this slab 4 ft above the top of the tank is about

$$\Delta W = 25\pi w(y+4)\Delta y \text{ ft} \cdot \text{lb}.$$

The work of lifting all the slabs of water is about

$$\sum_{0}^{10} \Delta W = \sum_{0}^{10} 25\pi w(y+4)\Delta y \text{ ft} \cdot \text{lb}.$$

This is a Riemann sum for the function $25\pi w(y+4)$ on the interval from $y=0$ to $y=10$. The work of pumping the tank dry is the limit of these sums as the norm of the partition goes to zero:

$$\begin{aligned}
\text{Work} &= \int_0^{10} 25\pi w(y+4)\,dy = 25\pi\, w \int_0^{10} (y+4)\,dy \\
&= 25\pi\, w \left[\frac{y^2}{2} + 4y\right]_0^{10} = 25\pi\, w(50+40) = 2250\pi\, w \\
&= 2250\pi(62.5) \qquad \text{(Water weighs 62.5 lb/ft}^3\text{.)} \\
&= 441{,}786 \text{ ft} \cdot \text{lb}. \qquad \text{(Nearest ft} \cdot \text{lb)}
\end{aligned}$$

A 1-horsepower pump, rated at 550 ft · lb per second, could empty the tank in a little less than 14 min.

WEIGHT-DENSITY

Weight per unit volume is called **weight-density**. Typical values (lb/ft^3) are

gasoline	42
mercury	849
milk	64.5
olive oil	57
seawater	64
water	62.5.

EXAMPLE 6 The inverted conical tank in Fig. 6.47 is filled to within 2 ft of the top with salad oil weighing 57 lb/ft^3. How much work does it take to pump the oil to the rim of the tank?

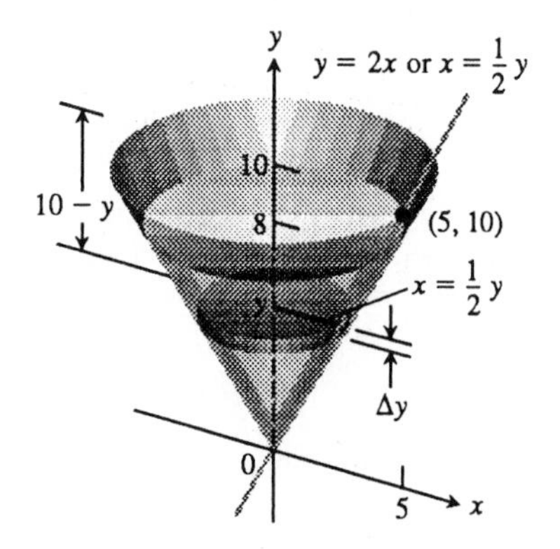

6.47 The salad oil in Example 6.

Solution We imagine the oil divided into thin slabs by planes perpendicular to the y-axis at the points of a partition of the interval $[0, 8]$.

The typical slab between the planes at y and $y + \Delta y$ has a volume of about

$$\Delta V = \pi(\text{radius})^2(\text{thickness}) = \pi\left(\frac{1}{2}y\right)^2 \Delta y = \frac{\pi}{4}y^2\Delta y \text{ ft}^3.$$

The force $F(y)$ required to lift this slab is its weight,

$$F(y) = 57\Delta V = \frac{57\pi}{4}y^2\Delta y \text{ lb}. \qquad \left(\begin{array}{l}\text{Weight} = \text{weight per} \\ \text{unit volume} \times \text{volume}\end{array}\right)$$

The distance through which $F(y)$ must act to lift this slab to the level of the rim of the cone is about $(10-y)$ ft, so the work done lifting the slab is about

$$\Delta W = \frac{57\pi}{4}(10-y)y^2\Delta y \text{ ft} \cdot \text{lb}.$$

The work done lifting all the slabs from $y = 0$ to $y = 8$ to the rim is about

$$\sum_{0}^{8} \frac{57\pi}{4}(10 - y)y^2 \Delta y \text{ ft} \cdot \text{lb}.$$

This is a Riemann sum for the function $(57\pi/4)(10-y)y^2$ on the interval from $y = 0$ to $y = 8$. The work of pumping the oil to the rim is the limit of these sums as the norm of the partition goes to zero.

$$\text{Work} = \int_0^8 \frac{57\pi}{4}(10 - y)y^2\, dy = \frac{57\pi}{4} \int_0^8 (10y^2 - y^3)\, dy$$

$$= \frac{57\pi}{4}\left[\frac{10y^3}{3} - \frac{y^4}{4}\right]_0^8 = 30{,}561 \text{ ft} \cdot \text{lb}. \qquad \text{(Nearest ft-lb)}$$

HOW TO FIND WORK DONE DURING PUMPING

1. Draw a figure with a coordinate system.
2. Find the weight of a thin horizontal slab of liquid.
3. Find the work to lift the slab to its destination.
4. Integrate the work expression from the top to the bottom of the liquid.

Exercises 6.6

1. The workers in Example 2 changed to a larger bucket that held 5 gal (40 lb) of water, but the new bucket had an even larger leak so that it, too, was empty by the time it reached the top. Assuming that the water leaked out at a steady rate, how much work was done lifting the water? (Do not include the rope and bucket.)

2. The bucket in Example 2 is hauled up twice as fast so that there is still 1 gal (8 lb) of water left when the bucket reaches the top. How much work is done lifting the water this time? (Do not include the rope and bucket.)

3. A mountain climber is about to haul up a 50-m length of hanging rope. How much work will it take if the rope weighs 0.74 newtons per meter?

4. A model rocket engine burned up its 2-oz fuel cartridge lifting the rocket to 170 ft. Assuming the fuel burned at a steady rate, how much work was spent just lifting fuel?

5. An electric elevator with a motor at the top has a multistrand cable weighing 4 lb/ft. One hundred eighty feet of cable are paid out when the car is at the first floor and effectively zero ft are out when the car is at the top floor. How much work does the motor do just lifting the cable when it takes the car from the first floor to the top?

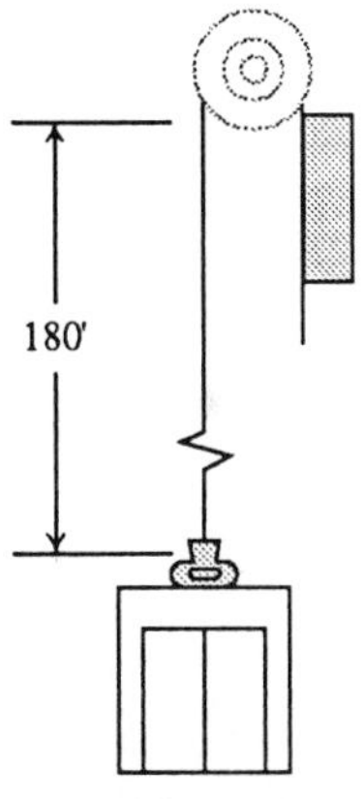

6. A bag of sand originally weighing 144 lb was lifted at a constant rate. The sand leaked out at a steady rate, and the sand was half gone by the time the bag had been lifted 18 ft. How much work was done lifting the sand this far? (Neglect the weight of the bag and lifting equipment.)

7. If a force of 6 N stretches a spring 0.4 m beyond its natural length, how much work is done?

8. If a force of 90 N stretches a spring 1 m beyond its natural length, how much work does it take to stretch the spring 5 m beyond its natural length?

9. A 10,000-lb force compressed a spring from its natural length of 12 in. to a length of 11 in. How much work did it do in compressing the spring
   a) the first half-inch?
   b) the second half-inch?

10. A bathroom scale is compressed 1/16 in. when a 150-lb

person stands on it. Assuming the scale behaves like a spring, how much does someone who compresses the scale 1/8 in. weigh? How much work does it take to compress the scale 1/8 in?

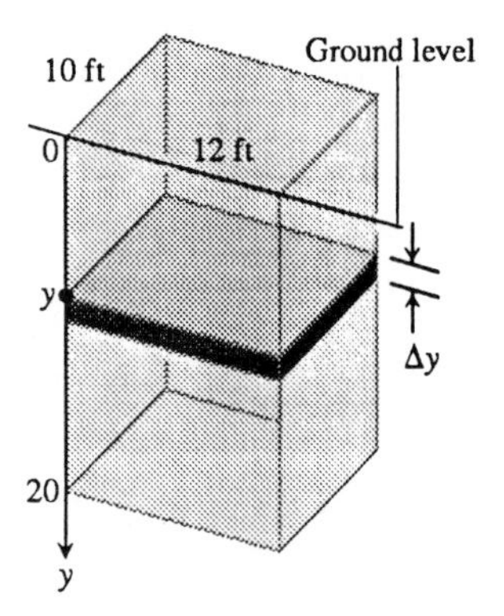

11. The rectangular tank shown here with its top at ground level is used to catch run-off water.

 a) How much work does it take to empty the tank by pumping the water back to ground level once the tank is full?

 b) If the water is pumped to ground level with a (5/11)-hp motor (work output 250 ft · lb/sec), how long will it take to empty the full tank?

 c) Show that the pump in part (b) will lower the water level 10 ft (half way) during the first 25 min of pumping.

12. The full rectangular cistern (rain water storage tank) shown here with its top 10 ft below ground level is to be emptied for inspection by pumping its contents to ground level.

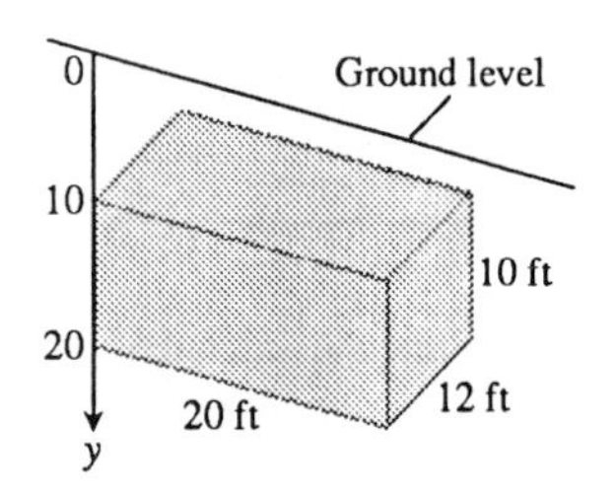

 a) How much work will it take to empty the cistern?

 b) How long will it take a (1/2)-hp pump, rated at 225 ft·lb/sec, to pump the tank dry?

 c) How long will it take the pump in part (b) to empty the tank half way? (It will be less than half the time required to empty the tank completely.)

13. How much work would it take to pump the water from the tank in Example 5 to the level of the top of the tank (instead of 4 ft higher)?

14. Suppose that instead of being completely full, the tank in Example 5 is only half full. How much work does it take to pump the water that's left to a level 4 ft above the top of the tank?

15. A vertical right circular cylindrical tank measures 30 ft high and 20 ft in diameter. It is full of kerosene weighing 51.2 lb/ft^3. How much work does it take to pump the kerosene to the level of the top of the tank?

16. The cylindrical tank shown here can be filled by pumping water from a lake 15 ft below the bottom of the tank. There are two ways to go about it. One is to pump the water through a hose to a valve in the bottom of the tank. The other is to attach the hose to the rim of the tank and let the water pour in. Which way will be faster?

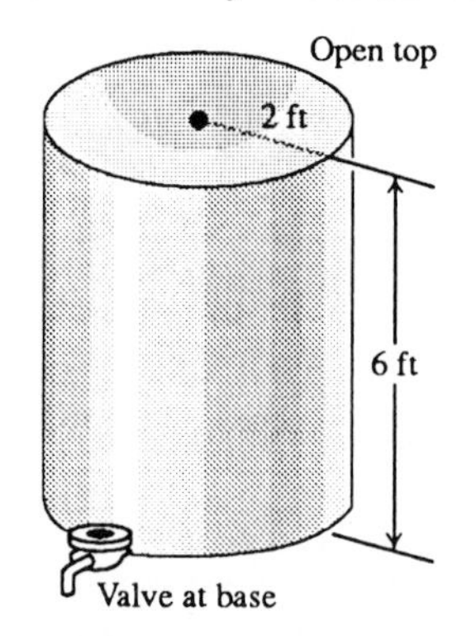

17. a) Suppose the cone in Example 6 contained milk (weight density 64.5 lb/ft^3) instead of salad oil. How much work would it have taken to pump the contents to the rim?

 b) How much work would it have taken to pump the oil in Example 6 to a level 3 ft above the cone's rim?

18. The truncated conical container shown here is full of strawberry milkshake that weighs (4/9) oz/in^3.

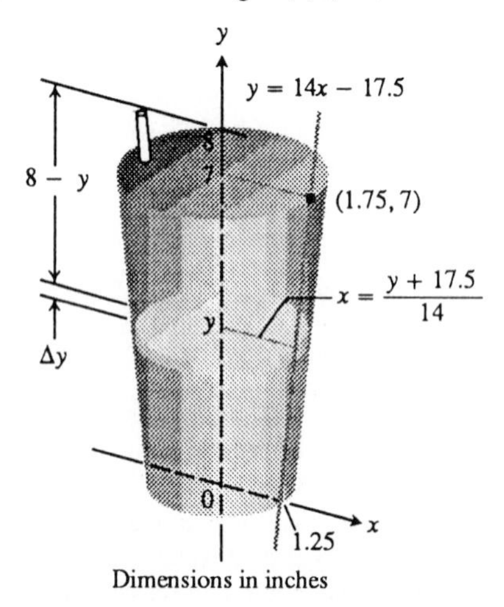

As you can see, the container is 7 in. deep, 2.5 in. across at the base, 3.5 in. across at the top (a standard size at

Brigham's in Boston). The straw sticks up an inch above the top. About how much work does it take to suck up the milkshake through the straw (neglecting friction)? Answer in inch-ounces.

19. To design the interior surface of a huge stainless steel tank, you revolve the curve $y = x^2, 0 \leq x \leq 4$, about the y-axis. The container, with dimensions in meters, is to be filled with seawater, which weighs 10,000 newtons per cubic meter. How much work will it take to empty the tank by pumping the water to the tank's top?

20. We model pumping from spherical containers the way we do from others, with the axis of integration along the vertical axis of the sphere. We drew Fig. 6.48 to help find out how much work it takes to pump the water from a full hemispherical bowl of radius 5 ft to a height 4 ft above the top of the bowl. How much work does it take?

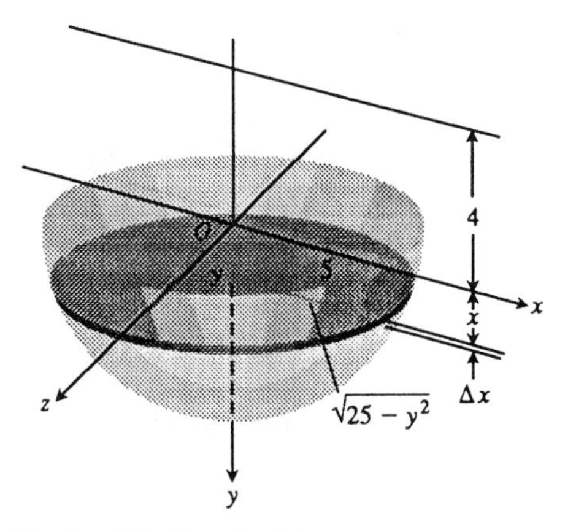

6.48 The bowl in Exercise 20.

21. You are in charge of the evacuation and repair of the storage tank shown in Fig. 6.49. The tank is a hemisphere of radius 10 ft and is full of benzene weighing 56 lb/ft^3. A firm you contacted says it can empty the tank for 1/2¢ per foot-pound of work. Find the work required to empty the tank by pumping the benzene to an outlet 2 ft above the top of the tank. If you have $5000 budgeted for the job, can you afford to hire the firm?

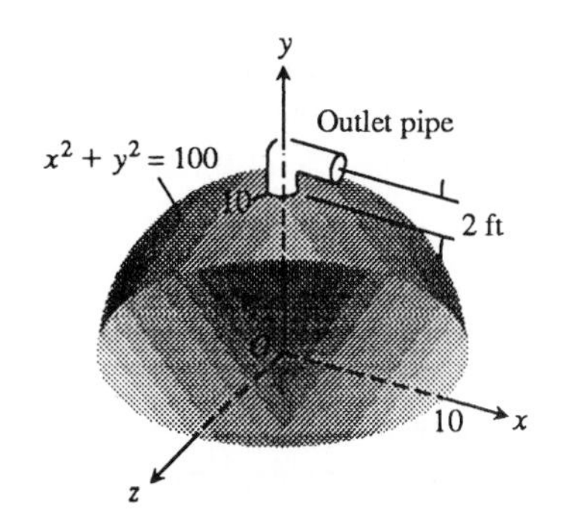

6.49 The tank in Exercise 21.

22. Your town has decided to drill a well to increase its water supply. As the town engineer, you have determined that a water tower will be necessary to provide the pressure needed for distribution, and you have designed the the system in Fig. 6.50. The water is to be pumped from a 300-ft well through a vertical 4-in. pipe into the base of a cylindrical tank 20 ft in diameter and 25 ft high. The base of the tank will be 60 ft above ground. The pump is a 3-hp pump, rated at 1650 ft · lb/sec. How long will it take to fill the tank the first time? (Include the time it takes to fill the pipe.)

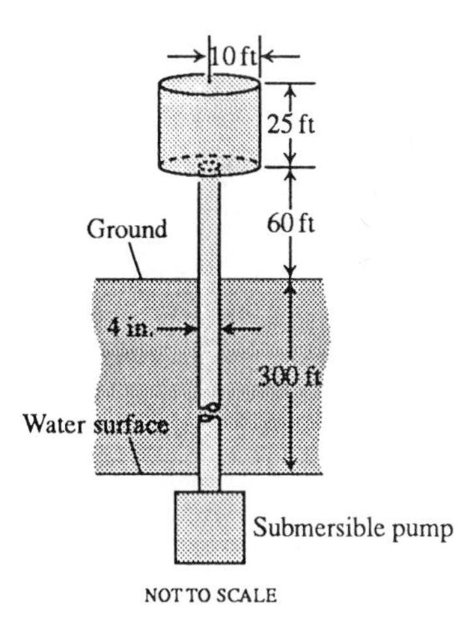

6.50 The water tower and well in Exercise 22.

23. Putting a satellite in orbit. The strength of the earth's gravitational field varies with the distance r from the earth's center, and the magnitude of the gravitational force experienced by a satellite of mass m during and after launch is

$$F(r) = \frac{m\,MG}{r^2}.$$

Here, $M = 5.975 \times 10^{24}$ kg is the earth's mass, $G = 6.6720 \times 10^{-11}$ Nm^{2}kg^{-2} is the universal gravitational constant, and r is measured in meters. The number of newton-meters of work it takes to lift a 1000-kg satellite from the earth's surface to a circular orbit 35,780 km above the earth's center is therefore given by the integral

$$\text{Work} = \int_{6,370,000}^{35,780,000} \frac{m\,MG}{r^2}\,dr.$$

Evaluate the integral. The lower limit of integration is the earth's radius in meters at the launch site. This calculation does not take into account energy spent by the launch vehicle or energy spent bringing the satellite to orbit velocity.

24. **Forcing electrons together.** Two electrons r meters apart repel each other with a force of

$$F = \frac{23 \times 10^{-29}}{r^2}$$

newtons.

a) Suppose one electron is held fixed at the point $(1, 0)$ on the x-axis (units in meters). How much work does it take to move a second electron along the x-axis from the point $(-1, 0)$ to the origin?

b) Suppose an electron is held fixed at each of the points $(-1, 0)$ and $(1, 0)$. How much work does it take to move a third electron along the x-axis from $(5, 0)$ to $(3, 0)$?

6.7 Fluid Pressures and Fluid Forces

We make dams thicker at the bottom than at the top (Fig. 6.51) because the pressure against them increases with depth. The deeper the water, the thicker the dam has to be.

6.51 To withstand the increasing pressure, dams are built thicker as they go down.

It is a remarkable fact that the pressure at any point on the dam depends only on how far below the surface the point is and not on how much the surface happens to be tilted at that point. The pressure, in pounds per square foot at a point h feet below the surface, is always $62.5\,h$. The number 62.5 is the weight-density of water in pounds per cubic foot.

The formula, pressure $= 62.5\,h$, makes sense when you think of the units involved: Pounds per square foot equals pounds per cubic foot times feet:

$$\frac{\text{lb}}{\text{ft}^2} = \frac{\text{lb}}{\text{ft}^3} \times \text{ft}. \tag{1}$$

As you can see, this equation depends only on units and not on what fluid is involved. The pressure h feet below the surface of any fluid is the fluid's weight-density times the depth.

WEIGHT-DENSITY

A fluid's weight-density is its weight per unit volume. Typical values (lb/ft^3) are

gasoline	42
mercury	849
milk	64.5
molasses	100
olive oil	57
seawater	64
water	62.5

The Pressure-Depth Equation

In a fluid that is standing still, the pressure at depth h is the fluid's weight-density times h:

$$p = wh. \tag{2}$$

In this section we use the equation $p = wh$ to derive a formula for the total force exerted by a fluid against all or part of a vertical or horizontal containing wall.

The Constant-Depth Formula for Force

In a container of fluid with a flat horizontal base, the total force exerted by the fluid against the base can be calculated by multiplying the area of the base by the pressure at that level. We can do this because total force equals force per unit area (pressure) times area. If F, p, and A are the total force, pressure, and area, then

$$\begin{aligned} F &= \text{total force} = \text{force per unit area} \times \text{area} \\ &= \text{pressure} \times \text{area} = p\,A \\ &= wh\,A. \qquad (p = wh \text{ from Eq. (2)}) \end{aligned}$$

Total Force on a Constant-Depth Surface

$$F = p\,A = wh\,A \tag{3}$$

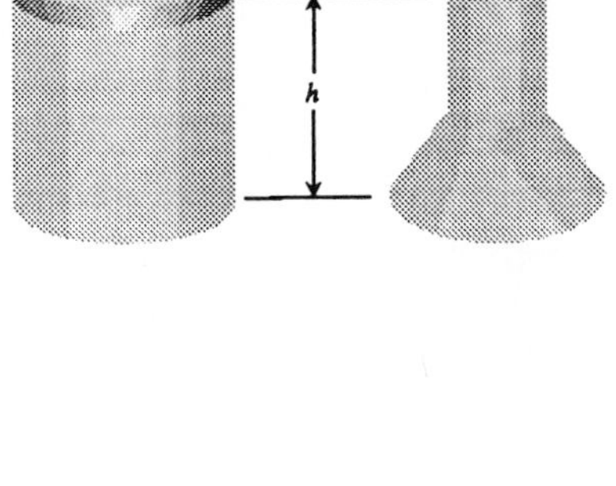

EXAMPLE 1 The two containers at the left are filled with water to the same depth and have the same base area. The total force is therefore the same on the bottom of each container. The containers' shapes do not matter here. The only things that count are depth and base area.

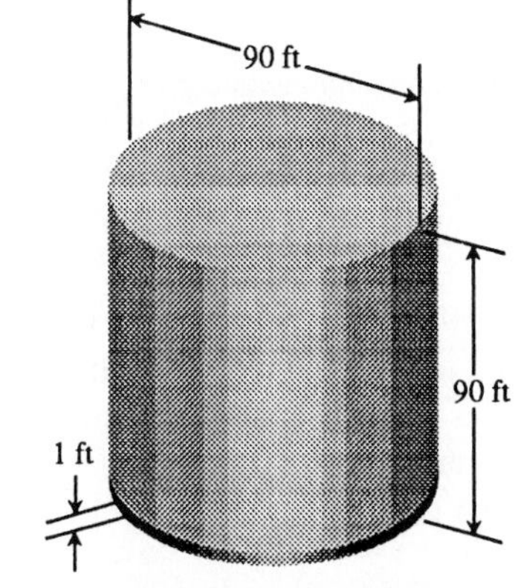

6.52 Schematic drawing of the molasses tank in Example 2. How much force did the bottom foot of wall have to withstand when the tank was full? It takes an integral to find out.

EXAMPLE 2 The Great Molasses Flood. At 1:00 P.M. on January 15, 1919, an unusually warm day, a 90-ft-high, 90-ft-diameter, cylindrical metal tank in which the Puritan Distilling Company was storing molasses at the corner of Foster and Commercial streets in Boston's North End exploded. The molasses flooded into the streets, 30 ft deep, trapping pedestrians and horses, knocking down buildings, and oozing into homes. It was eventually tracked all over town and even into the suburbs on people's shoes. The cleanup went on for weeks.

Given that the molasses weighed 100 lb/ft^3, what was the total force exerted by the molasses against the bottom of the tank at the time it blew? Assuming the tank was full, we can find out from Eq. (3):

$$\text{Total force} = wh\,A = (100)(90)(\pi(45)^2) = 57{,}255{,}526 \text{ lb.} \qquad \text{(Rounded)}$$

How about the force against the walls of the tank? For example, what was the total force against the bottom foot-wide band of tank wall (Fig. 6.52)?

The area of the band was

$$A = \pi\,dh = \pi(90)(1) = 90\pi \text{ ft}^2.$$

The tank was 90 ft deep, so the pressure near the bottom was about

$$p = wh = (100)(90) = 9000 \text{ lb/ft}^2.$$

Therefore the total force against the band was about

$$F = wh\ A = (9000)(90\pi) = 2{,}544{,}690 \text{ lb.} \qquad \text{(Rounded)}$$

But this is not exactly right. The top of the band was 89 ft below the surface, not 90, and the pressure there was less. To find out exactly what the force on the band was, we need to take the variation of the pressure across the band into account, and this means using calculus.

The Variable-Depth Integral for the Force against a Submerged Vertical Wall

Suppose we want to find the force against one side of a submerged vertical plate whose surface looks like the shaded region in Fig. 6.53. The region runs from a units below the surface to b units below the surface. We have chosen to measure depth with the y-axis, and the region's width at depth y is $L(y)$.

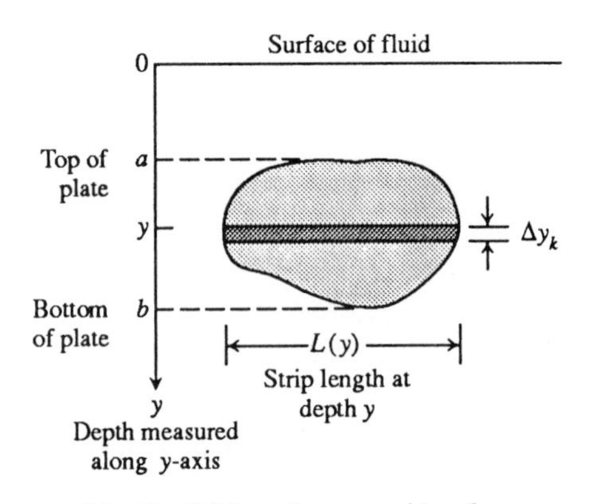

6.53 The force exerted by the fluid against one side of an approximating horizontal strip y units beneath the surface is about

$$\Delta F = (\text{pressure})(\text{area}) = (wy)(L(y)\Delta y).$$

We partition the closed interval $[a, b]$ in the usual way and imagine the region to be cut into thin horizontal strips by planes perpendicular to the y-axis at the points of the partition. The typical strip from y to $y + \Delta y$ is Δy units wide by $L(y)$ units long. We assume $L(y)$ to be continuous throughout the closed interval $[a, b]$.

The pressure varies across the strip from top to bottom, just as it did in the molasses tank. But, if the strip is narrow enough, the pressure will remain close to its top-edge value wy (w being the weight-density of the liquid we are working with, and y the depth of the top edge). The total force against one side of the strip will therefore be about

$$\begin{aligned}\Delta F &= (\text{pressure along top edge})(\text{area}) \\ &= (wy)(L(y)\Delta y) = wy\, L(y)\Delta y. \end{aligned} \qquad (4)$$

The force against the entire wall will be about

$$\sum_a^b \Delta F = \sum_a^b wy\, L(y)\Delta y. \tag{5}$$

The sum on the right-hand side of Eq. (5) is a Riemann sum for the continuous function $wy\, L(y)$ on the closed interval $[a, b]$. We expect the approximations to improve as the norm of the partition of $[a, b]$ goes to zero, so we define the total force against the wall to be the limit of these sums as the norm goes to zero.

Definition

The Integral for Fluid Force Suppose a submerged vertical plate running from depth $y = a$ to depth $y = b$ in a fluid of weight-density w is $L(y)$ units across at depth y, as measured along the plate. Then the total force of the fluid against one side of the plate is

$$F = \int_a^b wy\, L(y)\, dy. \tag{6}$$

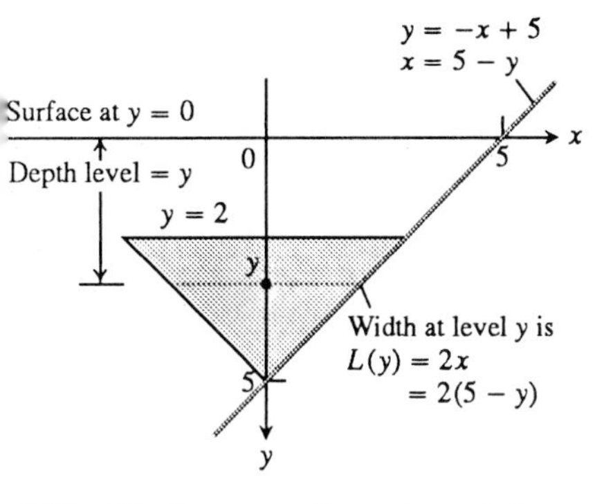

6.54 The important dimensions of the plate in Example 3.

EXAMPLE 3 A flat triangular plate is submerged vertically, base up, 2 ft below the surface of a swimming pool (Fig. 6.54). Find the fluid force against one side of the plate.

Solution We see from Fig. 6.54 that the plate runs from $y = 2$ to $y = 5$ and that its width at depth y is $L(y) = 2(5 - y)$. Therefore Eq. (6) gives

$$\text{Force} = \int_a^b wy\, L(y)\, dy = \int_2^5 (62.5)(y)(2)(5 - y)\, dy \qquad \text{(For water, } w = 62.5\text{)}$$

$$= 125 \int_2^5 (5y - y^2)\, dy = 125 \left[\frac{5}{2}y^2 - \frac{1}{3}y^3\right]_2^5 = 1687.5 \text{ lb.}$$

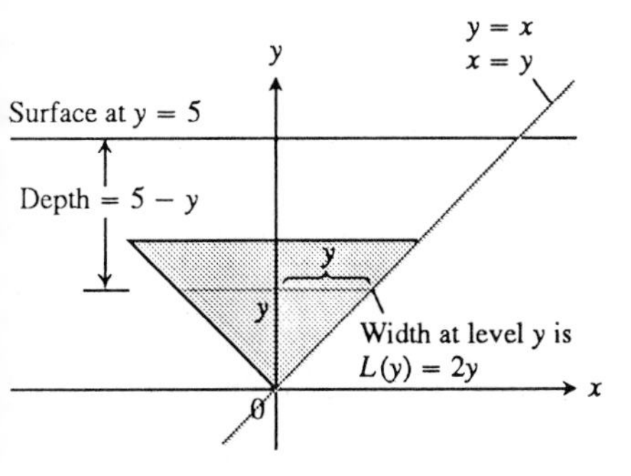

6.55 Example 4 calculates the force on the plate in Example 3 by using this more convenient coordinate system.

A coordinate system with the positive y-axis pointing down is not always the most convenient one to use. It is sometimes better to put the origin at the bottom of the plate instead of at the fluid's surface and have the y-axis point up. This changes the factor y in the integrand to some other expression for the depth, but the rest of the integral remains the same. Here is an example.

EXAMPLE 4 Find the force against one side of the plate in Example 3 by using the coordinate system in Fig. 6.55.

Solution The plate now runs from $y = 0$ to $y = 3$, and the length of a thin horizontal strip at level y is $L(y) = 2y$. The depth of this strip beneath the surface is $(5 - y)$. The force against the plate is therefore

$$\text{Force} = \int_a^b w \cdot \text{ depth } \cdot L(y)\,dy \qquad \text{(Modified Eq. 6)}$$

$$= \int_0^3 (62.5)(5 - y)(2y)\,dy = 125 \int_0^3 (5y - y^2)\,dy$$

$$= 125 \left[\frac{5}{2}y^2 - \frac{1}{3}y^3 \right]_0^3 = 1687.5 \text{ lb.} \qquad \text{(As before)}$$

Strategy for Finding Fluid Force

Whatever coordinate system you use, you can always find the fluid force against one side of a submerged vertical plate or wall by taking these steps:

1. Find expressions for the length and depth of a typical thin horizontal strip.
2. Multiply their product by the fluid's weight-density w and integrate over the interval of depths occupied by the plate or wall.

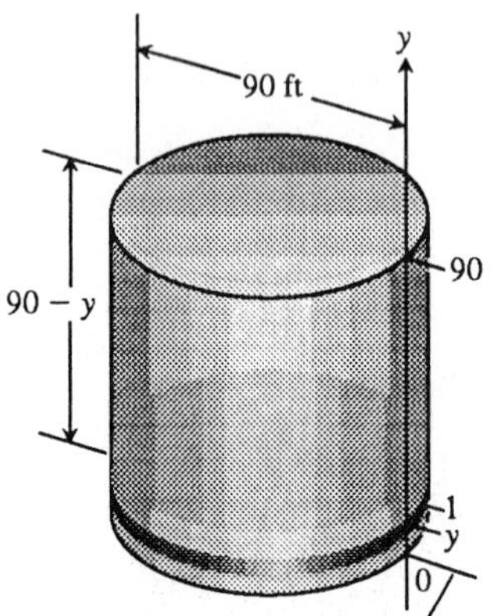

EXAMPLE 5 We can now calculate exactly the force exerted by the molasses against the bottom 1-ft band of the Puritan Distilling Company's storage tank when the tank was full.

The tank was a right circular cylindrical tank, 90 ft high and 90 ft in diameter. Using a coordinate system with the origin at the bottom of the tank and the y-axis pointing up, we find that the typical horizontal strip at level y has

Strip depth: $(90 - y)$

Strip length: $\pi \times \text{tank diameter} = 90\pi$.

The force against the band is therefore

$$\text{Force} = \int_0^1 w\,(\text{depth})(\text{length})\,dy$$

$$= \int_0^1 100(90 - y)(90\pi)\,dy \qquad \text{(For molasses, } w = 100\text{)}$$

$$= 9000\pi \int_0^1 (90 - y)\,dy = 9000\pi \left[90y - \frac{1}{2}y^2 \right]_0^1$$

$$= 9000\pi(89.5) = 2{,}530{,}553 \text{ lb.} \qquad \text{(Rounded)}$$

As expected, the force is slightly less than the constant-depth estimate following Example 2.

Exercises 6.7

1. Suppose the triangular plate in Fig. 6.55 is 4 ft beneath the surface instead of 2 ft. What is the force on one side of the plate now?

2. What was the total force against the side wall of the Puritan Distilling Company's molasses tank when the tank was full? Half full?

3. The vertical ends of a watering trough are inverted isosceles triangles like the one shown here. What is the force on each end of the trough when the trough is full? Does it matter how long the trough is?

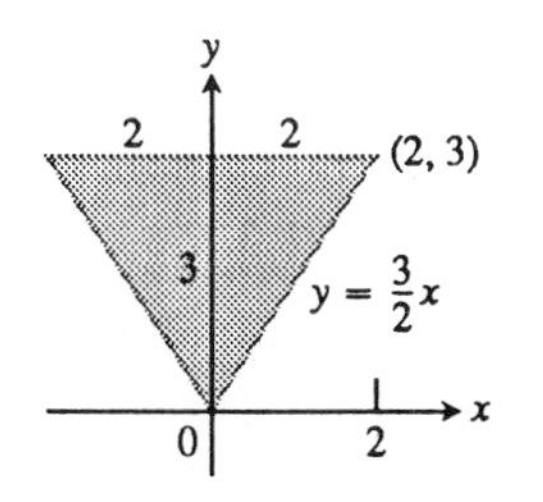

4. What is the force on each end of the trough in Exercise 3 if the water level is lowered 1 ft?

5. The triangular plate shown here is submerged vertically, 1 ft below the surface of the water.

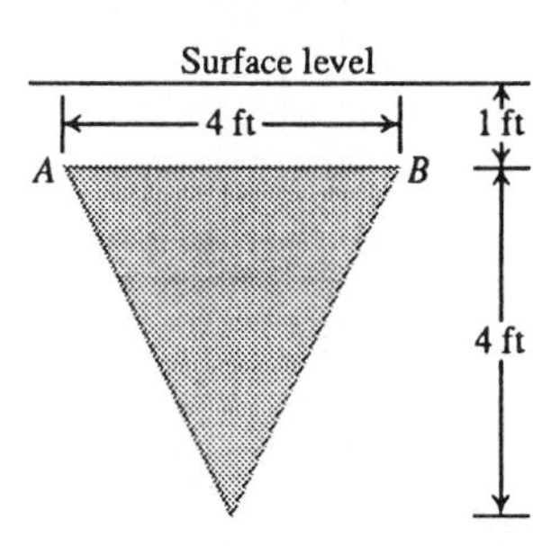

Find the force on one side of the plate.

6. The triangular plate in Exercise 5 is revolved 180° about the line AB so that part of it sticks up above the surface. What force does the water exert on one face of the plate now?

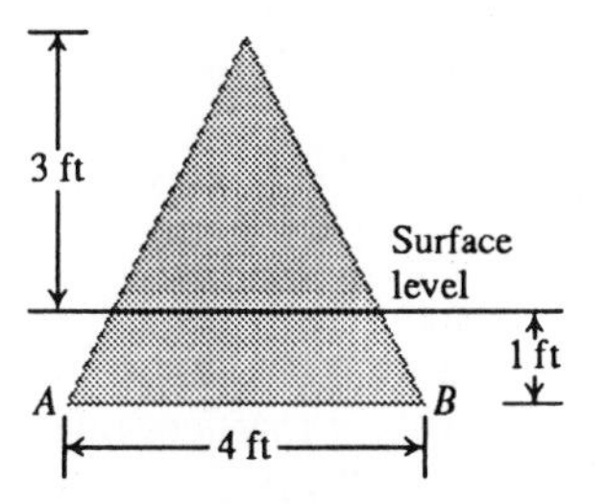

7. A semicircular plate is submerged straight down in the water with its diameter at the surface. Find the force exerted by the water on one side of the plate.

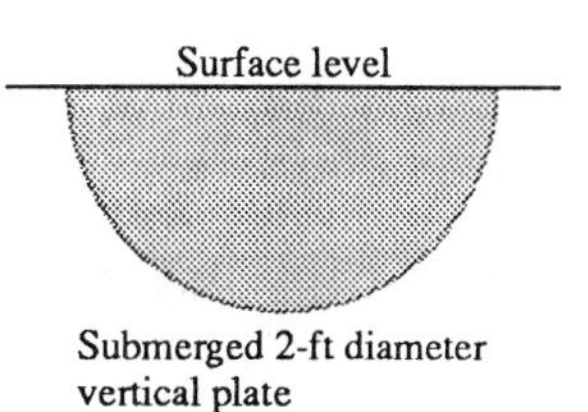

8. A rectangular fish tank of interior dimensions $2 \times 2 \times 4$ ft is filled to within 2 in. of the top with water. Find the force against the sides and ends of the tank.

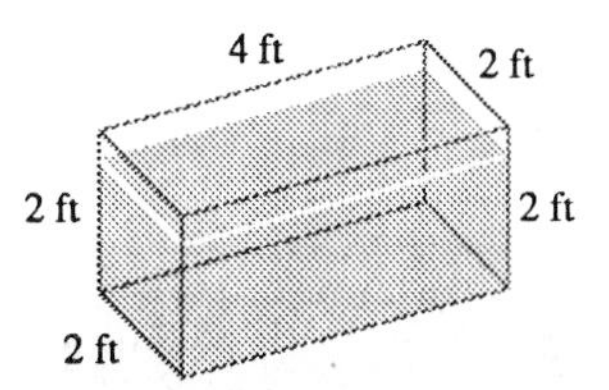

9. The viewing portion of the rectangular glass window in a typical fish tank at the New England Aquarium in Boston, Massachusetts, is 63 in. wide and runs from 0.5 in. below the water's surface to 33.5 in. below the surface. Find the force against this portion of the window. The weight-density of sea water is 64 lb/ft^3. (In case you were wondering, the glass is 3/4 in. thick and the tank walls extend 4 in. above the water to keep the fish from jumping out.)

10. A rectangular milk carton measures $3.75 \times 3.75 \times 7.75$ in. Find the force of the milk on one side when the carton is full.

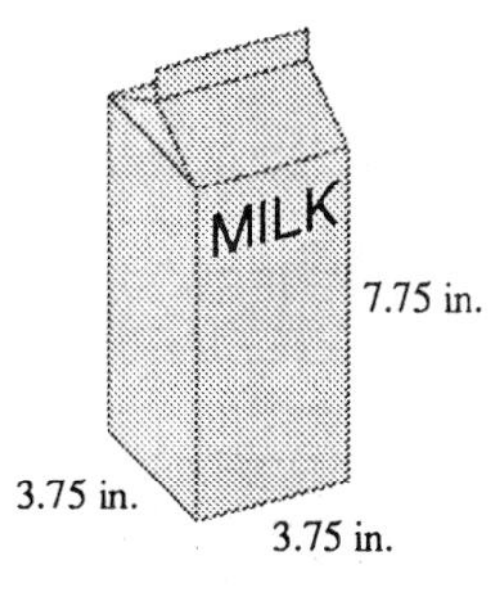

11. A tank truck hauls milk in a 6-ft diameter horizontal right circular cylindrical tank. What is the force on the end of the tank when the tank is half full?

12. The cubical metal tank shown here is used to store liquids. It has a parabolic gate, held in place by bolts and designed to withstand a force of 160 lb without rupturing. The liquid you plan to store has a weight-density of 50 lb/ft^3.

a) What will be the force on the gate when the liquid is 2 ft deep?

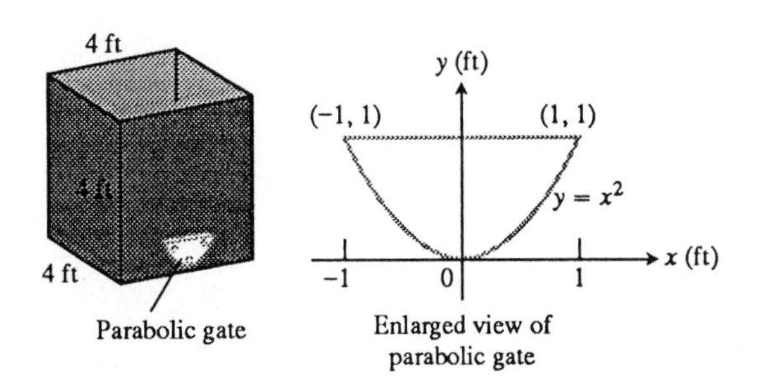

Parabolic gate

Enlarged view of parabolic gate

b) What is the maximum height to which the container can be filled without exceeding the design limitation on the gate?

13. The end plates in the trough shown here were designed to withstand a force of 6667 lb. How many cubic feet of water can the tank hold without exceeding design limitations?

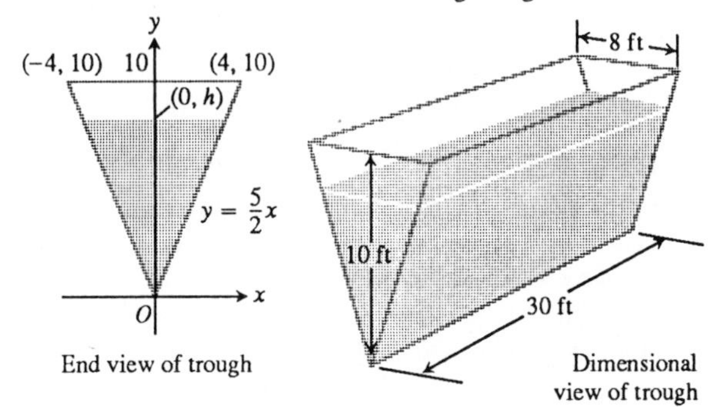

End view of trough

Dimensional view of trough

14. Water is running into the rectangular swimming pool shown here at the rate of 1000 ft^3/hr.

a) Find the force against the triangular drain plate after 9 hr of filling.

b) The plate is designed to withstand a force of 520 lb. How high can the pool be filled without exceeding this design limitation?

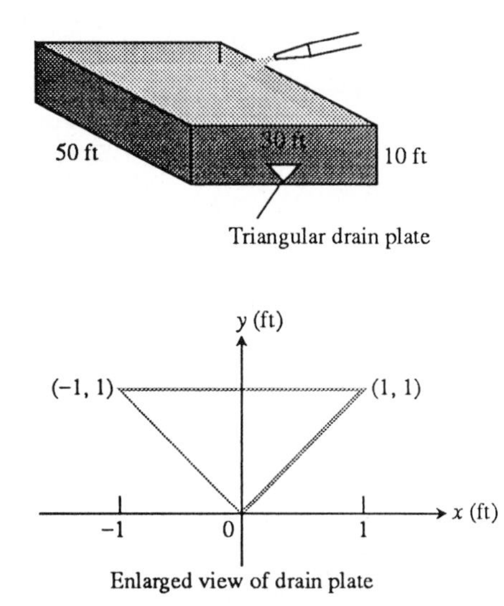

Triangular drain plate

Enlarged view of drain plate

15. A vertical rectangular plate is submerged in a fluid with its top edge parallel to the fluid's surface. Show that the force on one side of the plate is the average value of the pressure up and down the plate times the area of the plate.

6.8 Centers of Mass

Many structures and mechanical systems behave as if their masses were concentrated at a single point called the center of mass. It is important to know how to locate this point, and it turns out that doing so is basically a mathematical undertaking. We do it with calculus, and this section shows how. For the moment we shall deal only with one- and two-dimensional shapes. Three-dimensional shapes are best done with the so-called multiple integrals of Chapter 14.

Masses along a Line

If we imagine masses m_1, m_2, and m_3 placed on a rigid x-axis, and the axis supported by a fulcrum at the origin, the resulting system might or might not balance.

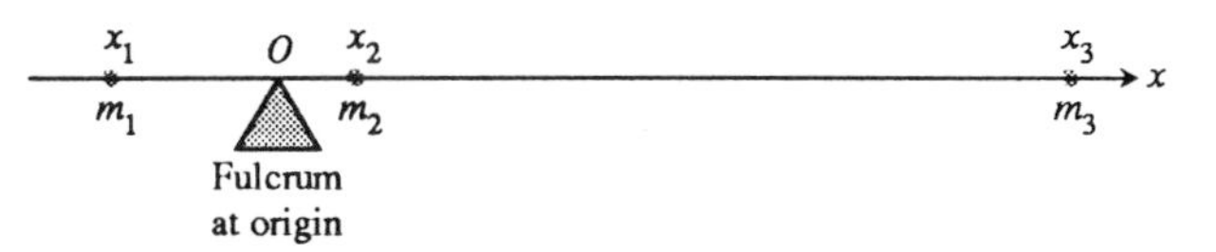

MASS VERSUS WEIGHT

Weight is the force that results from gravity pulling on a mass. If an object of mass m is placed in a location where the acceleration of gravity is g, the object's weight there is

$$F = mg$$

(a version of Newton's second law).

Each mass m_k exerts a downward force $m_k g$ equal to the magnitude of the mass times the acceleration of gravity. Each of these forces has a tendency to turn the axis about the origin, the way somebody's weight might turn a seesaw. This turning effect, called a **torque**, is measured by multiplying the force $m_k g$ by the signed distance x_k from the mass to the origin. Masses to the left of the origin exert a negative (counterclockwise) torque. Masses to the right of the origin exert a positive (clockwise) torque.

We use the sum of the torques to measure the tendency of a system to rotate about the origin. This sum is called the **system torque.**

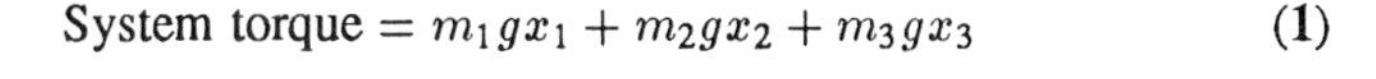

$$\text{System torque} = m_1 g x_1 + m_2 g x_2 + m_3 g x_3 \qquad (1)$$

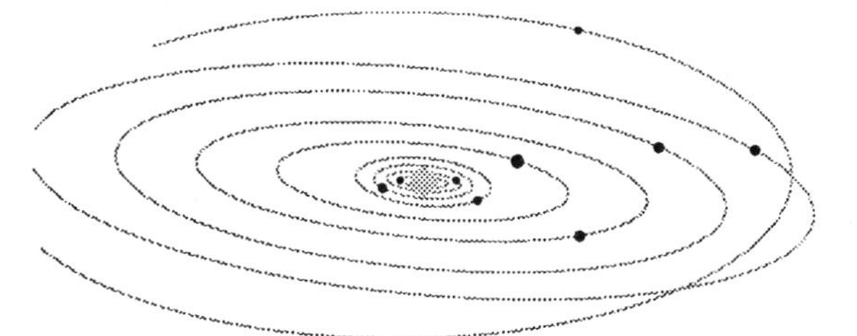

The planets, asteroids, and comets of our solar system revolve about their collective center of mass. (It lies inside the sun.)

The system will balance if and only if its net torque is zero.

If we factor out the g in Eq. (1), we see that the system torque is

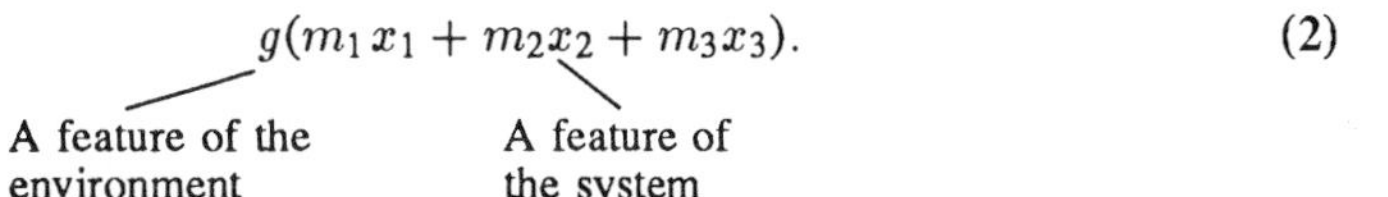

$$g(m_1 x_1 + m_2 x_2 + m_3 x_3). \qquad (2)$$

Thus the torque is the product of the gravitational acceleration g, which is a feature of the environment in which the system happens to reside, and the number $(m_1 x_1 + m_2 x_2 + m_3 x_3)$, which is a feature of the system itself, a constant that stays the same no matter where the system is placed.

The number $(m_1 x_1 + m_2 x_2 + m_3 x_3)$ is called the **moment of the system about the origin.**

$$M_O = \text{Moment of system about origin} = \sum m_k x_k \qquad (3)$$

(We shift to sigma notation here to allow for sums with more terms. If you want a quick way to read $\Sigma m_k x_k$, try "summation $m\ k\ x\ k$".)

We usually want to know where to place the fulcrum to make the system balance, that is, at what point $\bar{x}$ to place it to make the torque zero.

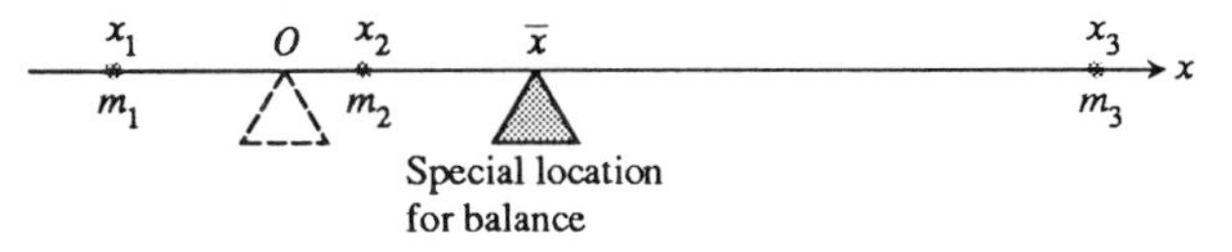

The torque of each mass about the fulcrum in this special location is

$$\text{Torque of } m_k \text{ about } \overline{x} = \begin{pmatrix}\text{signed distance}\\ \text{of } m_k \text{ from } \overline{x}\end{pmatrix} \cdot \begin{pmatrix}\text{downward}\\ \text{force}\end{pmatrix}$$

$$= (x_k - \overline{x}) \cdot m_k g. \tag{4}$$

When we write down the equation that says that the sum of these torques is zero, we get an equation we can solve for $\overline{x}$:

$$\sum (x_k - \overline{x}) m_k g = 0 \qquad \text{(Sum of the torques equals zero)}$$

$$g \sum (x_k - \overline{x}) m_k = 0 \qquad \text{(Constant multiple rule for sums)}$$

$$\sum (m_k x_k - \overline{x} m_k) = 0 \qquad (g \text{ divided out, } m_k \text{ distributed})$$

$$\sum m_k x_k - \sum \overline{x} m_k = 0 \qquad \text{(Difference rule for sums)}$$

$$\sum m_k x_k = \overline{x} \sum m_k \qquad \text{(Rearranged, constant multiple rule again)}$$

$$\overline{x} = \frac{\sum m_k x_k}{\sum m_k}. \qquad (\text{Solved for } \overline{x})$$

This last equation tells us to find $\overline{x}$ by dividing the system's moment about the origin by the system's total mass:

$$\overline{x} = \frac{\sum x_k m_k}{\sum m_k} = \frac{\text{System moment about origin}}{\text{System mass}}. \tag{5}$$

The point $\overline{x}$ is called the system's **center of mass.**

Wires and Thin Rods

In many applications, we want to know the center of mass of a rod or a thin strip of metal. In cases like these, where we can assume the distribution of mass is continuous, the summation signs in our formulas become integrals in a manner we shall now describe.

Imagine a long, thin strip lying along the x-axis from $x = a$ to $x = b$ and cut into small pieces of mass Δm_k by a partition of the interval $[a, b]$.

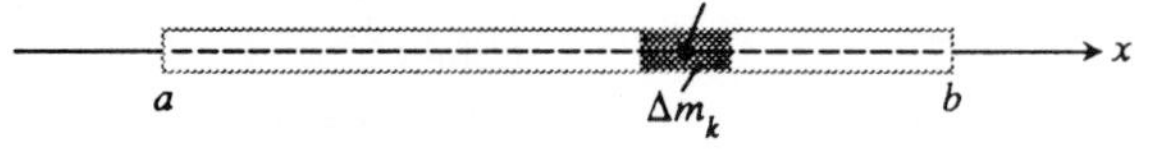

Each piece is Δx units long and lies approximately x_k units from the origin. Now observe three things.

First, the strip's center of mass $\overline{x}$ is nearly the same as the center of mass of the system of point masses we would get by attaching each mass Δm_k to the point x_k:

$$\overline{x} \approx \frac{\text{System moment}}{\text{System mass}}. \tag{6}$$

Second, the moment of each piece of the strip about the origin is approximately $x_k \Delta m_k$, so the system moment is approximately the sum of the $x_k \Delta m_k$:

$$\text{System moment} \approx \sum x_k \Delta m_k. \tag{7}$$

Third, if the density of the strip at x_k is $\delta(x_k)$, expressed in terms of mass per unit length, and δ is continuous, then Δm_k is approximately equal to $\delta(x_k)\Delta x$ (mass per unit length times length):

$$\Delta m_k \approx \delta(x_k)\Delta x. \tag{8}$$

DENSITY

A material's density is defined to be its mass per unit volume. In practice, however, we tend to use units we can conveniently measure. For wires, rods, and narrow strips we use mass per unit length. For flat sheets and plates we use mass per unit area.

Combining these three observations gives

$$\bar{x} \approx \frac{\text{System moment}}{\text{System mass}} \approx \frac{\sum x_k \Delta m_k}{\sum \Delta m_k} \approx \frac{\sum x_k \delta(x_k)\Delta x}{\sum \delta(x_k)\Delta x}. \tag{9}$$

The sum in the numerator of the last quotient in (9) is a Riemann sum for the continuous function $x\delta(x)$ over the closed interval $[a, b]$. The sum in the denominator is a Riemann sum for the function $\delta(x)$ over this interval. We expect the approximations in (9) to improve as the strip is partitioned ever more finely and are led to the equation

$$\bar{x} = \frac{\int_a^b x\delta(x)\,dx}{\int_a^b \delta(x)\,dx}. \tag{10}$$

This is the formula we use to calculate $\bar{x}$.

Moment, Mass, and Center of Mass of a Thin Rod or Strip along the *x*-Axis

Moment about the origin: $$M_O = \int_a^b x\delta(x)\,dx \tag{11a}$$

Mass: $$M = \int_a^b \delta(x)\,dx \tag{11b}$$

Center of mass: $$\bar{x} = \frac{M_O}{M} \tag{11c}$$

Equation (11c) says that to find the center of mass of a rod or a thin strip, we divide its moment about the origin by its mass.

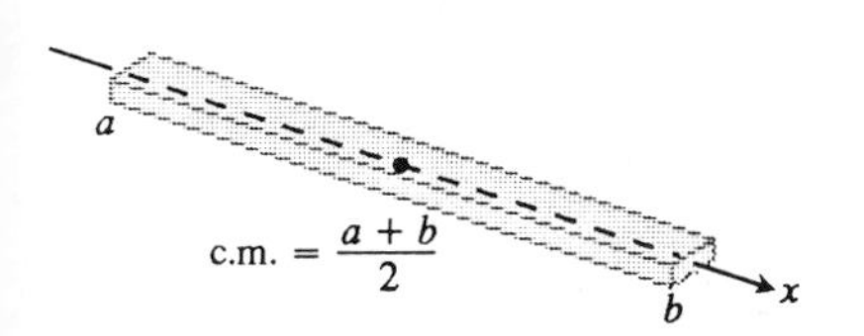

6.56 The center of mass of a straight, thin rod or strip of constant density lies half way between its ends.

EXAMPLE 1 A useful result. Show that the center of mass of a straight, thin strip or rod of constant density is always located half way between its two ends.

Solution We model the strip as a portion of the x-axis from $x = a$ to $x = b$ (Fig. 6.56). Our goal is to show that $\bar{x} = (a+b)/2$, the point half way between a and b.

The key is the density's having a constant value. This enables us to regard the function $\delta(x)$ in the integrals in Eqs. (11) as a constant (call it δ), with the

result that

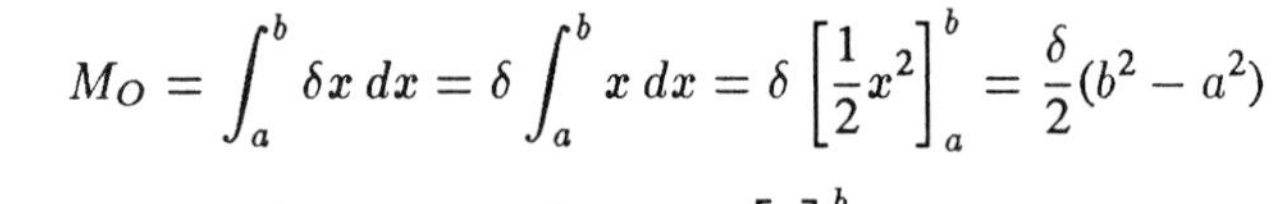

$$M_O = \int_a^b \delta x\,dx = \delta \int_a^b x\,dx = \delta \left[\frac{1}{2}x^2\right]_a^b = \frac{\delta}{2}(b^2 - a^2)$$

$$M = \int_a^b \delta\,dx = \delta \int_a^b dx = \delta \Big[x\Big]_a^b = \delta(b - a)$$

$$\bar{x} = \frac{M_o}{M} = \frac{\frac{\delta}{2}(b^2 - a^2)}{\delta(b - a)} = \frac{a + b}{2}. \qquad \text{(The } \delta\text{'s cancel.)}$$

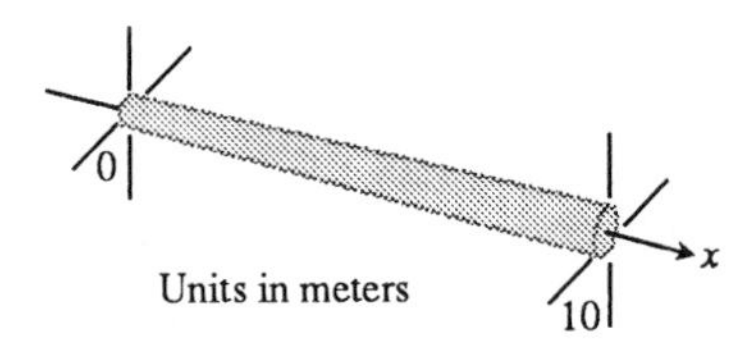

6.57 We can treat a rod of variable thickness as a rod of variable density. See Example 2.

EXAMPLE 2 A variable density. The 10-meter-long rod in Fig. 6.57 thickens from left to right so that its density, instead of being constant, is $\delta(x) = 1 + (x/10)$ kg/m. Find the rod's center of mass.

Solution The rod's moment about the origin (Eq. 11a) is

$$M_O = \int_0^{10} x\delta(x)\,dx = \int_0^{10} x\left(1 + \frac{x}{10}\right) dx = \int_0^{10} \left(x + \frac{x^2}{10}\right) dx$$

$$= \left[\frac{x^2}{2} + \frac{x^3}{30}\right]_0^{10} = 50 + \frac{100}{3} = \frac{250}{3}\text{kg} \cdot \text{m}. \qquad \left(\begin{matrix}\text{The units of a moment}\\ \text{are mass} \times \text{length.}\end{matrix}\right)$$

The rod's mass (Eq. 11b) is

$$M = \int_0^{10} \delta(x)\,dx = \int_0^{10} \left(1 + \frac{x}{10}\right) dx = \left[x + \frac{x^2}{20}\right]_0^{10} = 10 + 5 = 15 \text{ kg}.$$

The center of mass (Eq. 11c) is located at the point

$$\bar{x} = \frac{M_O}{M} = \frac{250}{3} \cdot \frac{1}{15} = \frac{50}{9} \approx 5.56 \text{ m}.$$

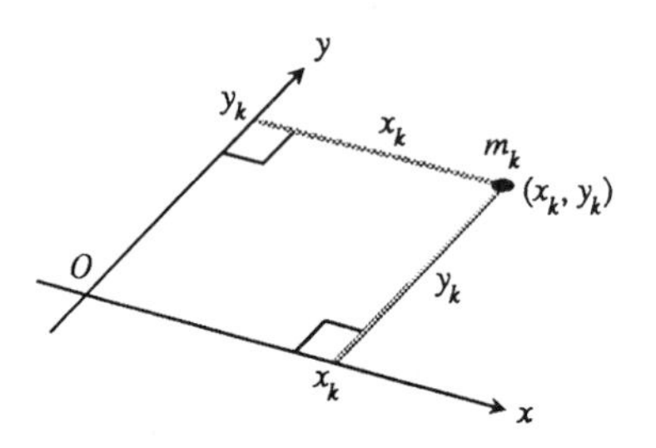

6.58 The mass m_k has a moment about each axis.

Masses Distributed over a Plane Region

Suppose we have a finite collection of masses located in the coordinate plane, the mass m_k being located at the point (x_k, y_k) (see Fig. 6.58). The total mass of the system is

$$\text{System mass:} \qquad M = \sum m_k.$$

Each mass m_k has a moment about each axis. Its moment about the x-axis is $m_k y_k$, and its moment about the y-axis is $m_k x_k$. The moments of the entire system about the two axes are

$$\text{Moment about } x\text{-axis:} \qquad M_x = \sum m_k y_k,$$

$$\text{Moment about } y\text{-axis:} \qquad M_y = \sum m_k x_k.$$

The x-coordinate of the system's center of mass is defined to be

$$\bar{x} = \frac{M_y}{M} = \sum m_k x_k \Big/ \sum m_k. \tag{12}$$

With this choice of $\overline{x}$, as in the one-dimensional case, the system balances about the line $x = \overline{x}$ (Fig. 6.59).

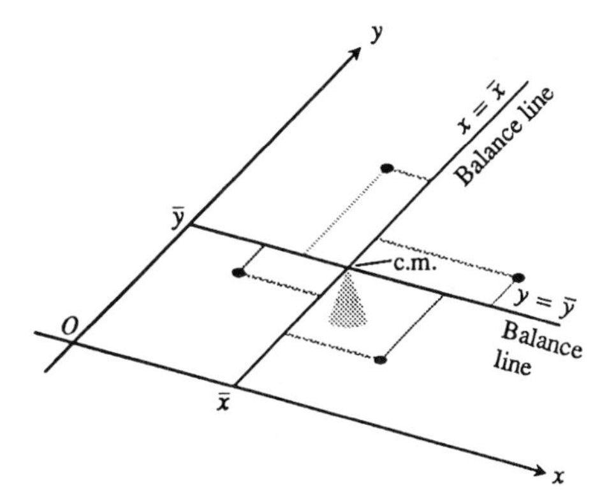

6.59 A two-dimensional array of masses balances on its center of mass.

The y-coordinate of the system's center of mass is defined to be

$$\overline{y} = \frac{M_x}{M} = \sum m_k y_k \Big/ \sum m_k. \tag{13}$$

With this choice of $\overline{y}$, the system balances about the line $y = \overline{y}$ as well. The torques exerted by the masses about the line $y = \overline{y}$ cancel out. Thus, as far as balance is concerned, the system behaves as if all its mass were at the single point $(\overline{x}, \overline{y})$. We call this point the system's center of mass.

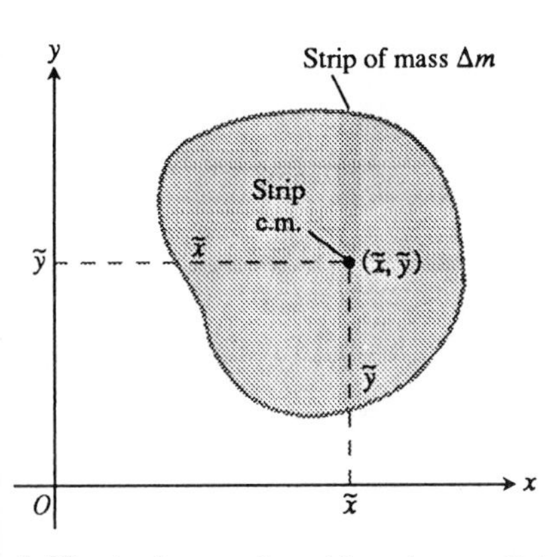

6.60 A plate cut into thin strips parallel to the y-axis. The moment exerted by a typical strip about each axis is the moment its mass Δm would exert if concentrated at the strip's center of mass $(\tilde{x}, \tilde{y})$.

Thin, Flat Plates

In many applications, we need to find the center of mass of a thin, flat plate: a disk of aluminum, say, or a triangular sheet of steel. In such cases we assume the distribution of mass to be continuous, and the formulas we use to calculate $\overline{x}$ and $\overline{y}$ contain integrals instead of finite sums. The integrals arise in the following way.

Imagine the plate occupying a region in the xy-plane, cut into thin strips parallel to one of the axes (in Fig. 6.60, the y-axis). The center of mass of a typical strip is $(\tilde{x}, \tilde{y})$. (The symbol ˜ over the x and y is a *tilde*, pronounced to rhyme with "Hilda." Thus $\tilde{x}$ is read "x tilde.") We treat the strip's mass Δm as if it were concentrated at $(\tilde{x}, \tilde{y})$. The moment of the strip about the y-axis is then $\tilde{x}\Delta m$, and the moment of the strip about the x-axis is $\tilde{y}\Delta m$. Equations (12) and (13) then become

$$\overline{x} = \frac{M_y}{M} = \frac{\sum \tilde{x}\Delta m}{\sum \Delta m}, \qquad \overline{y} = \frac{M_x}{M} = \frac{\sum \tilde{y}\Delta m}{\sum \Delta m}. \tag{14}$$

As in the one-dimensional case, the sums in the numerator and denominator are Riemann sums for integrals and approach these integrals as limiting values as the strips into which the plate is cut become narrower and narrower. We write these integrals symbolically as

$$\bar{x} = \frac{\int \tilde{x}\, dm}{\int dm} \quad \text{and} \quad \bar{y} = \frac{\int \tilde{y}\, dm}{\int dm}. \tag{15}$$

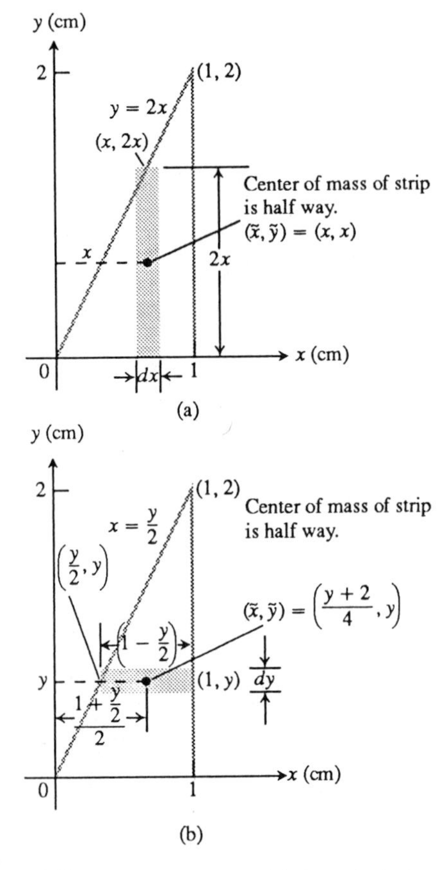

6.61 Two ways to model the calculation of the moment M_y of the triangular plate in Example 3.

Moments, Mass, and Center of Mass of a Thin Plate Covering a Region in the *xy*-Plane

Moment about the x-axis: $M_x = \int \tilde{y}\, dm$

Moment about the y-axis: $M_y = \int \tilde{x}\, dm$ (16)

Mass: $M = \int dm$

Center of mass: $\bar{x} = \dfrac{M_y}{M}, \quad \bar{y} = \dfrac{M_x}{M}$

To evaluate these integrals, we picture the plate in the coordinating plane and stretch a strip of mass parallel to one of the coordinate axes. We then express the strip's mass dm and the coordinates $(\tilde{x}, \tilde{y})$ of the strip's center of mass in terms of x or y. Finally, we integrate $\tilde{y}\, dm$, $\tilde{x}\, dm$, and dm between limits of integration determined by the plate's location in the plane.

EXAMPLE 3 The triangular plate shown in Fig. 6.61, bounded by the lines $y = 0, y = 2x$, and $x = 1$, has a constant density of $\delta = 3$ gm/cm^2. Find (a) the plate's moment M, about the y-axis, (b) the plate's mass M, and (c) the x-coordinate of the plate's center of mass.

Solution

METHOD 1: *Vertical strips* (Fig. 6.61a).

a) The moment M_y: The typical vertical strip has

center of mass (c.m.): $(\tilde{x}, \tilde{y}) = (x, x)$,

length: $2x$,

width: dx,

area: $dA = 2x\, dx$,

mass: $dm = \delta\, dA = 3 \cdot 2x\, dx = 6x\, dx$,

distance of c.m. from y-axis: $\tilde{x} = x$.

The moment of the strip about the y-axis is

$$\tilde{x}\, dm = x \cdot 6x\, dx = 6x^2\, dx.$$

The moment of the plate about the y-axis is therefore

$$M_y = \int \tilde{x}\, dm = \int_0^1 6x^2\, dx = 2x^3\Big]_0^1 = 2 \text{ gm} \cdot \text{cm}.$$

b) The plate's mass:

$$M = \int dm = \int_0^1 6x\, dx = 3x^2\Big]_0^1 = 3 \text{ gm}.$$

c) The x-coordinate of the plate's center of mass:

$$\bar{x} = \frac{M_y}{M} = \frac{2 \text{ gm} \cdot \text{cm}}{3 \text{ gm}} = \frac{2}{3} \text{ cm}.$$

By a similar computation we could find M_x and $\bar{y} = M_x/M$.

METHOD 2: *Horizontal strips* (Fig. 6.61b).

a) The moment M_y: The typical horizontal strip has

center of mass (c.m.): $(\tilde{x}, \tilde{y}) = \left(\frac{1}{2}\left(1 + \frac{y}{2}\right), y\right) = \left(\frac{y+2}{4}, y\right)$,

length: $1 - \frac{y}{2} = \frac{2-y}{2}$,

width: dy,

area: $dA = \frac{2-y}{2} dy$,

mass: $dm = \delta dA = 3 \cdot \frac{2-y}{2}\, dy$,

distance of c.m. to y-axis: $\tilde{x} = \frac{y+2}{4}$.

The moment of the strip about the y-axis is

$$\tilde{x}\, dm = \frac{y+2}{4} \cdot 3 \cdot \frac{2-y}{2}\, dy = \frac{3}{8}(4 - y^2)\, dy.$$

The moment of the plate about the y-axis is

$$M_y = \int \tilde{x}\, dm = \int_0^2 \frac{3}{8}(4 - y^2)\, dy = \frac{3}{8}\left[4y - \frac{y^3}{3}\right]_0^2 = \frac{3}{8}\left(\frac{16}{3}\right) = 2 \text{ gm} \cdot \text{cm}.$$

b) The plate's mass:

$$M = \int dm = \int_0^2 \frac{3}{2}(2 - y)\, dy = \frac{3}{2}\left[2y - \frac{y^2}{2}\right]_0^2 = \frac{3}{2}(4 - 2) = 3 \text{ gm}.$$

c) The x-coordinate of the plate's center of mass:

$$\bar{x} = \frac{M_y}{M} = \frac{2 \text{ gm} \cdot \text{cm}}{3 \text{ gm}} = \frac{2}{3} \text{ cm}.$$

By a similar computation, we could find M_x and $\bar{y}$.

How to Find a Plate's Center of Mass

1. Picture the plate in the xy-plane.
2. Sketch a strip of mass parallel to one of the coordinate axes and find its dimensions.
3. Find the strip's mass dm and center of mass $(\tilde{x}, \tilde{y})$.
4. Integrate $\tilde{y}\, dm$, $\tilde{x}\, dm$, and dm to find M_x, M_y, and M.
5. Divide the moments by the mass to calculate $\bar{x}$ and $\bar{y}$.

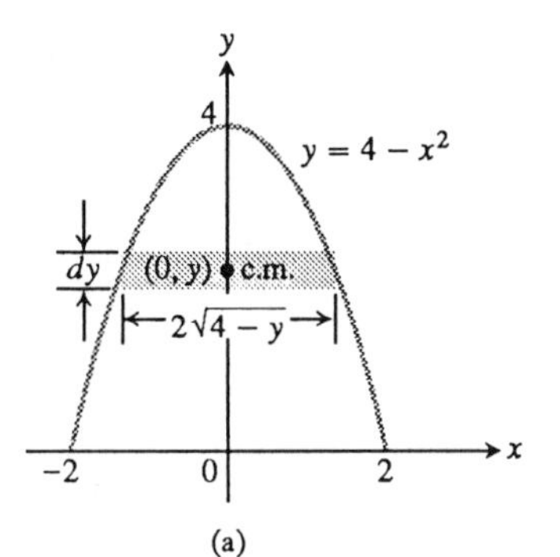

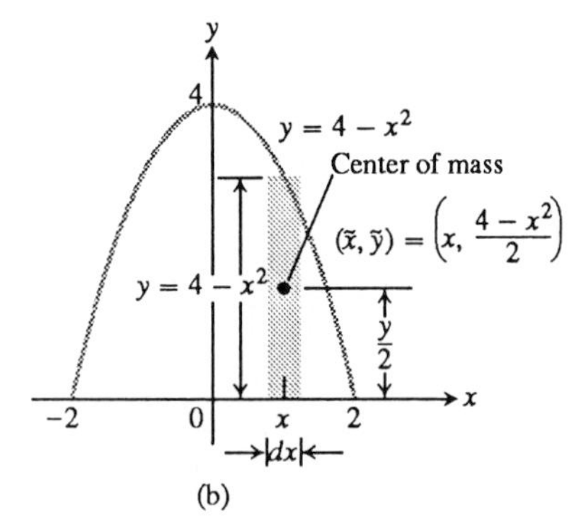

6.62 Modeling the problem in Example 4 with horizontal strips (a) leads to a difficult integration, so (b) we model with vertical strips instead.

EXAMPLE 4 Find the center of mass of a thin plate of constant density δ covering the region bounded above by the parabola $y = 4 - x^2$ and below by the x-axis (Fig. 6.62).

Solution Since the plate is symmetric about the y-axis and its density is constant, the center of mass lies on the y-axis. This means that $\bar{x} = 0$. It remains to find $\bar{y} = M_x/M$.

A trial calculation with horizontal strips (Fig. 6.62a) leads to a difficult integration using analytic methods:

$$M_x = \int_0^4 2\delta y\sqrt{4 - y}\, dy.$$

However, applying numerical methods, we easily obtain

$$\text{FnInt}\,(2y\sqrt{4 - y}, y, 0, 4) = 17.06666\ldots$$

Since $256/15 = 17.0666\ldots$ we suspect the exact value of M_x is $(256/15)\delta$. That is

$$M_x = \int_0^4 2\delta y\sqrt{4 - y}\, dy = \frac{256}{15}\delta.$$

We confirm this by modeling the distribution of mass with vertical strips (Fig. 6.62b). The typical vertical strip has

center of mass (c.m.): $(\tilde{x}, \tilde{y}) = \left(x, \dfrac{4 - x^2}{2}\right)$,

length: $4 - x^2$,

width: dx,

area: $dA = (4 - x^2)\, dx$,

mass: $dm = \delta dA = \delta(4 - x^2)\,dx$,

distance from c.m. to x-axis: $\tilde{y} = \dfrac{4 - x^2}{2}$.

The moment of the strip about the x-axis is

$$\tilde{y}\,dm = \frac{4 - x^2}{2} \cdot \delta(4 - x^2)\,dx = \frac{\delta}{2}(4 - x^2)^2\,dx.$$

The moment of the plate about the x-axis is

$$M_x = \int \tilde{y}\,dm = \int_{-2}^{2} \frac{\delta}{2}(4 - x^2)^2\,dx = \frac{256}{15}\delta. \tag{17}$$

The mass of the plate is

$$M = \int dm = \int_{-2}^{2} \delta(4 - x^2)\,dx = \frac{32}{3}\delta. \tag{18}$$

Therefore

$$\overline{y} = \frac{M_x}{M} = \frac{\frac{256}{15}\delta}{\frac{32}{3}\delta} = \frac{8}{5}.$$

The plate's center of mass is the point

$$(\overline{x}, \overline{y}) = \left(0, \frac{8}{5}\right).$$

EXAMPLE 5 Variable density. Find the center of mass of the plate in Example 4 if the density at any point (x, y) is $\delta = 2x^2$, twice the square of the distance from the point to the y-axis.

Solution The mass distribution is still symmetric about the y-axis, so

$$\overline{x} = 0.$$

With $\delta = 2x^2$, Eqs. (17) and (18) become

$$M_x = \int \tilde{y}\,dm = \int_{-2}^{2} \frac{\delta}{2}(4 - x^2)^2\,dx = \int_{-2}^{2} x^2(4 - x^2)^2\,dx = \frac{2048}{105} \tag{17'}$$

$$M = \int dm = \int_{-2}^{2} \delta(4 - x^2)\,dx = \int_{-2}^{2} 2x^2(4 - x^2)\,dx = \frac{256}{15}. \tag{18'}$$

Therefore

$$\overline{y} = \frac{M_x}{M} = \frac{2048}{105} \cdot \frac{15}{256} = \frac{8}{7}.$$

The plate's new center of mass is

$$(\overline{x}, \overline{y}) = \left(0, \frac{8}{7}\right).$$

Centers of Gravity, Homgeneity, Uniformity, and Centroids

As you read elsewhere, you will find some variety in the vocabulary used in connection with centers of mass.

When physicists discuss the effects of a constant gravitational force on a system of masses, they may call the center of mass the **center of gravity.**

Material that has a constant density δ is also said to be **homogeneous**, or to be **uniform**, or to have **uniform density.**

When the density function is constant, it cancels out of the numerator and denominator of the formulas for $\bar{x}$ and $\bar{y}$. This happened in nearly every example in this section. As far as $\bar{x}$ and $\bar{y}$ were concerned, δ might as well have been 1. Thus, when the density is constant, the location of the center of mass is a feature of the geometry of the object and not of the material from which it is made. In such cases engineers may call the center of mass the **centroid** of the shape, as in "Find the centroid of a triangle or a solid cone." To do so, we just set δ equal to 1 and proceed to find $\bar{x}$ and $\bar{y}$ as before, by dividing moments by masses.

Exercises 6.8

1. Two children are balancing on a seesaw. The 80-lb child is 5 ft from the fulcrum. How far from the fulcrum is the 100-lb child?

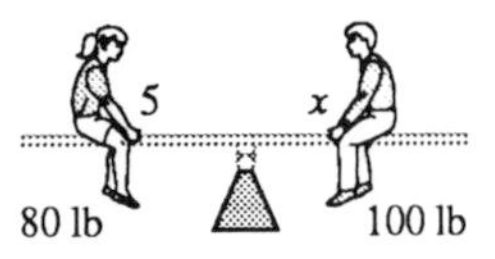

2. The ends of two thin steel rods of equal length are welded together to make a right-angled frame. Locate the frame's center of mass. (*Hint:* Where is the center of mass of each rod?)

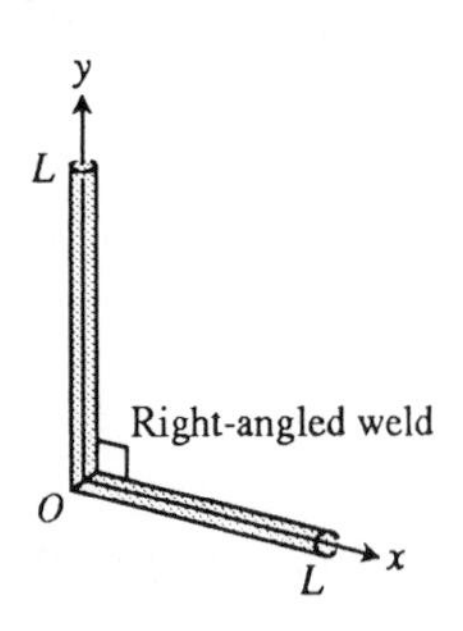

Exercises 3–6 give the density functions of thin rods lying along various intervals of the x-axis. Find each rod's moment about the origin and center of mass.

3. $\delta(x) = 4, 0 \le x \le 2$

4. $\delta(x) = 1 + (x/3), 0 \le x \le 3$

5. $\delta(x) = \left(1 + \frac{x}{4}\right)^2, 0 \le x \le 4$

6. $\delta(x) = \begin{cases} 2, & 0 \le x \le 3 \\ 1, & 3 \le x \le 6 \end{cases}$

In Exercises 7–18, find the center of mass of a thin plate of constant density δ covering the given region.

7. The triangular region bounded below by the x-axis and on the sides by the lines $y = 2x + 2$ and $y = -2x + 2$

8. The region bounded by the parabola $y = x^2$ and the line $y = 4$

9. The region bounded by the y-axis and the curve $x = y - y^3, 0 \le y \le 1$

10. The region bounded by the parabola $y = x - x^2$ and the line $y = -x$

11. The region bounded by the parabola $x = y^2 - y$ and the line $y = x$

12. The region bounded by the parabola $y = 25 - x^2$ and the x-axis

13. The region bounded by the x-axis and the curve $y = \cos x$, $-\pi/2 \le x \le \pi/2$

14. The region between the x-axis and the curve $y = \sec x$, $-\pi/4 \le x \le \pi/4$

15. The region bounded by the parabolas $y = 2x^2 - 4x$ and $y = 2x - x^2$

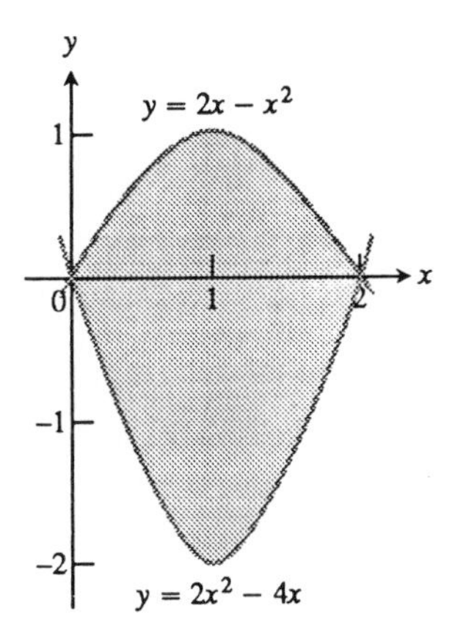

16. a) The region cut from the first quadrant by the circle $x^2 + y^2 = 9$

b) The region bounded by the x-axis and the semi-circle $y = \sqrt{9 - x^2}$

Compare your answer with the answer in (a).

17. The "triangular" region in the first quadrant between the circle $x^2 + y^2 = 9$ and the lines $x = 3$ and $y = 3$. (*Hint*: Use geometry to find the area.)

18. The region bounded above by the curve $y = 1/x^2$, below by the curve $y = -1/x^2$, and on the left and right by the lines $x = 1$ and $x = 2$.

It can be shown that the centroid of a triangle always lies at the intersection of the medians, one third of the way from the midpoint of each side toward the opposite vertex. Use this to find the centroids of the triangles whose vertices are given in Exercises 19–22. (*Hint*: Draw each triangle first.)

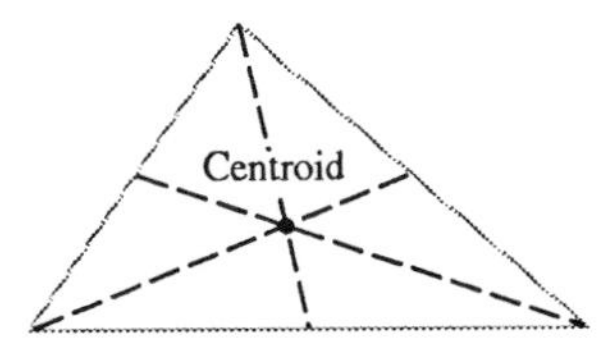

19. $(-1, 0), (1, 0), (0, 3)$

20. $(0, 0), (1, 0), (0, 1)$

21. $(0, 0), (a, 0), (0, a)$

22. $(0, 0), (a, 0), (0, b)$

23. Find the center of mass of a thin plate covering the region between the x-axis and the curve $y = 2/x^2, 1 \leq x \leq 2$, if the density is $\delta(x) = x^3$.

24. Find the center of mass of a thin plate covering the region bounded below by the parabola $y = x^2$ and above by the line $y = x$ if the density is $\delta(x) = 12x$.

The Theorems of Pappus

In the third century an Alexandrian Greek named Pappus discovered two formulas that relate centers of mass to surfaces and volumes of revolution. These formulas, easy to remember, provide useful shortcuts to a number of otherwise lengthy calculations.

Theorem 1

If a plane region is revolved once about an axis in the plane that does not pass through the region's interior, then the volume of the solid swept out by the region is equal to the region's area times the distance traveled by the region's center of mass. In symbols,

$$V = 2\pi\overline{y}A.$$

Theorem 2

If an arc of a plane curve is revolved once about a line in the plane that does not cut through the interior of the arc, then the area of the surface swept out by the arc is equal to the length of the arc times the distance traveled by the arc's center of mass. In symbols,

$$S = 2\pi\overline{y}L.$$

EXAMPLE 1 The volume of the torus (doughnut) generated by revolving a circle of radius a about an axis in its plane at a distance $b \geq a$ from its center (see Fig. 6.63) is

$$V = (2\pi b)(\pi a^2) = 2\pi^2 ba^2.$$

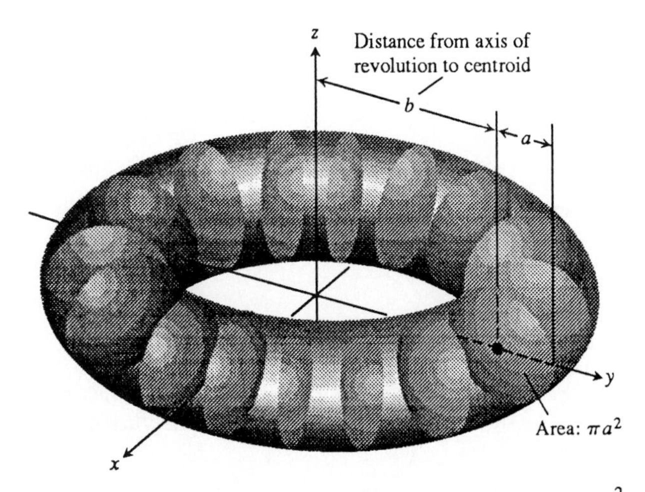

6.63 The volume swept out by the revolving disk is $(2\pi b)(\pi a^2)$.

EXAMPLE 2 The surface area of the torus in Example 1 is

$$S = (2\pi b)(2\pi a) = 4\pi^2 ba.$$

25. The square region with vertices $(0, 2)$, $(2, 0)$, $(4, 2)$ and $(2, 4)$ is revolved about the x-axis to generate a solid. Find the volume and surface area of the solid.

26. Use a theorem of Pappus to find the volume generated by revolving about the line $x = 5$ the triangular region bounded by the coordinate axes and the line $2x + y = 6$. (The centroid of a triangle lies at the intersection of the medians, one third of the way from the midpoint of each side toward the opposite vertex.)

27. Find the volume of the torus generated by revolving the circle $(x - 2)^2 + y^2 = 1$ about the y-axis.

28. Use the theorems of Pappus to find the lateral surface area and the volume of a right circular cone.

29. Use the second theorem of Pappus and the fact that the surface area of a sphere of radius a is $4\pi a^2$ to find the center of mass of the semicircle $y = \sqrt{a^2 - x^2}$.

30. As found in Exercise 29, the center of mass of the semicircle $y = \sqrt{a^2 - x^2}$ lies at the point $(0, 2a/\pi)$. Find the area of the surface swept out by revolving the semicircle about the line $y = a$.

31. Use the first theorem of Pappus and the fact that the volume of a sphere of radius a is $V = (4/3)\pi a^3$ to find the center of mass of the region enclosed by the x-axis and the semicircle $y = \sqrt{a^2 - x^2}$.

32. As found in Exercise 31, the center of mass of the region enclosed by the x-axis and the semicircle $y = \sqrt{a^2 - x^2}$ lies at the point $(0, 4a/3\pi)$. Find the volume of the solid generated by revolving this region about the line $y = -a$.

33. The region of Exercise 32 is revolved about the line $y = x - a$ to generate a solid. Find the volume of the solid.

34. As found in Exercise 29, the center of mass of the semicircle $y = \sqrt{a^2 - x^2}$ lies at the point $(0, 2a/\pi)$. Find the area of the surface generated by revolving the semicircle about the line $y = x - a$.

6.9 The Basic Idea. Other Modeling Applications

There is a pattern to what we have been doing in the preceding sections. In each section we wanted to measure something that was modeled or described by one or more continuous functions. In Section 6.1 it was the area between the graphs of two continuous functions. In Section 6.2 it was the volume of the solid defined by revolving the graph of a continuous function about an axis. In Section 6.6 it was the work done by a force directed along the x-axis whose magnitude was given by a continuous function, and so on. In each case we responded by partitioning the interval on which the function or functions were defined and approximating what we wanted to measure with Riemann sums over the interval. We then used the integral defined by the limit of the Riemann sums to define and calculate what we wanted to measure. You will see what we mean if you look at Table 6.2. Literally thousands of things in biology, chemistry, economics, engineering, finance, geology, medicine, and other fields (the list would fill pages) are modeled and calculated by exactly this process.

In this section we review the process and look at more of the important integrals it leads to.

Volumes of Arbitrary Solids—Slicing

Now that we can find the areas of regions bounded by smooth curves, we can find the volumes of a wide variety of cylinders. As we shall now see, this enables us to define the volumes of many new solids.

TABLE 6.2 The phases of developing an integral to calculate something

Phase 1	Phase 2	Phase 3
We describe or model something we want to measure in terms of one or more continuous functions defined on a closed interval $[a, b]$.	We partition $[a, b]$ into subintervals of length Δx_k and choose a point c_k in each subinterval. We approximate what we want to measure with a finite sum. We identify the sum as a Riemann sum of a continuous function over $[a, b]$.	The approximations improve as the norm of the partition goes to zero. The Riemann sums approach a limiting integral. We use the integral to define and calculate what we originally wanted to measure.
The area between the curves $y = f_1(x), y = f_2(x)$, on $[a, b]$ when $f_2(x) \leq f_1(x)$	$\sum (f_1(c_k) - f_2(c_k))\Delta x_k$	$\text{Area} = \int_a^b (f_1(x) - f_2(x))\,dx$
The volume of the solid defined by revolving the curve $y = f(x)$, $a \leq x \leq b$, about the x-axis	$\sum \pi f^2(c_k)\Delta x_k$	$\text{Volume} = \int_a^b \pi f^2(x)\,dx$
The length of a continuously differentiable curve $y = f(x)$, $a \leq x \leq b$	$\sum \sqrt{1 + (f'(c_k))^2}\Delta x_k$	$\text{Length} = \int_a^b \sqrt{1 + (f'(x))^2}\,dx$
The work done by a variable force $F(x)$ directed along the x-axis from a to b	$\sum F(c_k)\Delta x_k$	$\text{Work} = \int_a^b F(x)\,dx$

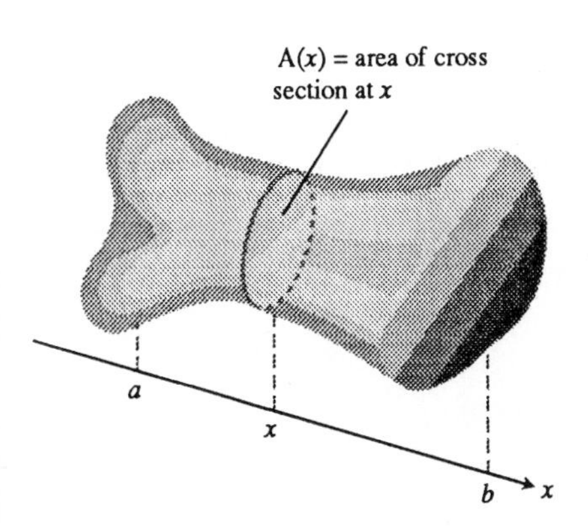

6.64 If the area of the cross section is a continuous function of x, we can find the volume of the solid in the way explained in the text.

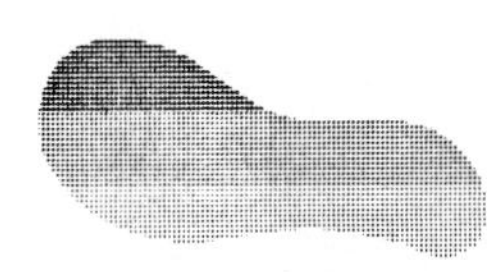
Plane region whose area we know

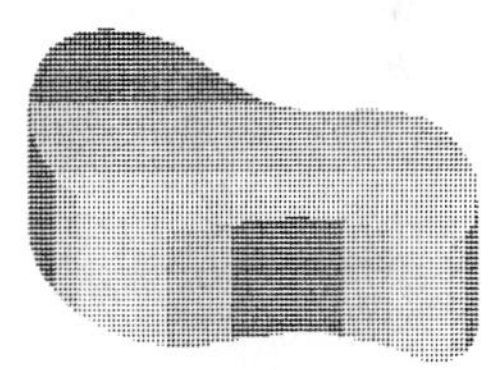
Cylinder based on region Volume = base area × height

Suppose we want to find the volume of a solid like the one shown in Fig. 6.64. The solid lies between planes perpendicular to the x-axis at $x = a$ and $x = b$. Each cross section of the solid by a plane perpendicular to the x-axis is a region whose area we know how to find. Specifically, at each point x in the closed interval $[a, b]$ the cross section of the solid is a region $R(x)$ whose area is $A(x)$. This makes A a real-valued function of x. If it is also a continuous function of x, we can use it to define and calculate the volume of the solid as an integral in the following way.

We partition the interval $[a, b]$ in the usual manner and slice the solid, as we would a loaf of bread, by planes perpendicular to the x-axis at the partition points. The kth slice, the one between the planes at x_{k-1} and x_k, has approximately the same volume as the cylinder between these two planes based on the region $R(x_k)$ (Fig. 6.65). The volume of this cylinder is

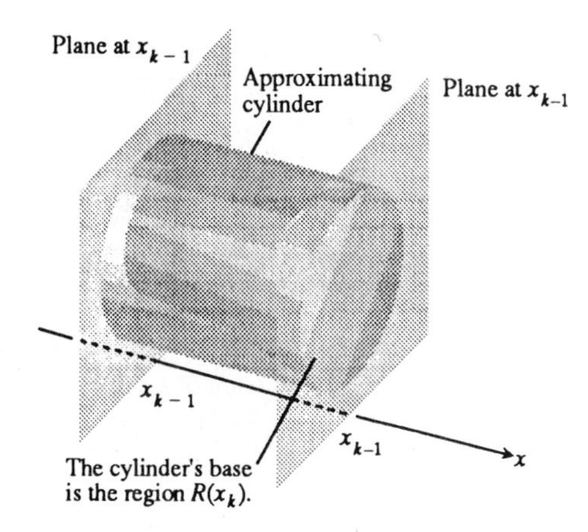

6.65 Enlarged view of the slice of the solid between the planes at x_{k-1} and x_k and its approximating cylinder.

$$\begin{aligned} V_k &= \text{base area} \times \text{height} \\ &= A(x_k) \times (\text{distance between the planes at } x_{k-1} \text{ and } x_k) \qquad (1) \\ &= A(x_k)\Delta x_k. \end{aligned}$$

The volume of the solid is therefore approximated by the cylinder volume sum

$$\sum_{k=1}^{n} A(x_k)\Delta x_k. \qquad (2)$$

This is a Riemann sum for the function $A(x)$ on the closed interval $[a, b]$. We expect the approximations we get from sums like these to improve as the norm of the partition of $[a, b]$ goes to zero, so we define their limiting integral to be the volume of the solid.

Definition

> The **volume** of a solid of known cross-section area $A(x)$ from $x = a$ to $x = b$ is the integral of A from a to b,
>
> $$\text{Volume} = \int_a^b A(x)\,dx. \qquad (3)$$

Notice that the new formula is consistent with the disk and washer formulas for solids of revolution (a good sign). In the disk formula,

$$V = \int_a^b \pi(f(x))^2\,dx,$$

the cross section at x is a disk of radius $f(x)$ whose area is

$$A(x) = \pi(f(x))^2.$$

In the washer formula,

$$V = \int_a^b \pi(R^2(x) - r^2(x))\,dx,$$

the cross section at x is a washer of inner radius $r(x)$ and outer radius $R(x)$ whose area is

$$A(x) = \pi R^2(x) - \pi r^2(x) = \pi(R^2(x) - r^2(x)).$$

To apply Eq. (3), we take the following steps.

The Method of Slicing

STEP 1: Sketch the solid and a typical cross section.

STEP 2: Find a formula for $A(x)$.

STEP 3: Find the limits of integration.

STEP 4: Integrate $A(x)$ to find the volume.

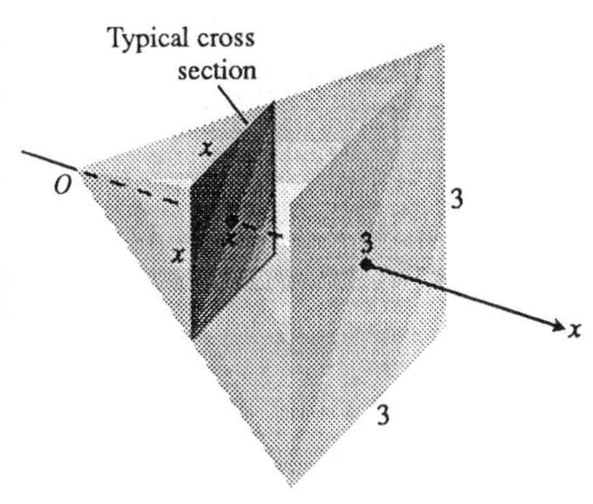

6.66 The cross sections of the pyramid in Example 1 are squares.

EXAMPLE 1 A pyramid 3 m high has a square base that is 3 m on a side. The cross section of the pyramid perpendicular to the altitude x units down from the vertex is a square x units on a side. Find the volume of the pyramid.

Solution

STEP 1: *A sketch.* We draw the pyramid with its altitude along the x-axis and its vertex at the origin and include a typical cross section (Fig. 6.66).

STEP 2: *A formula for A(x).* The cross section at x is a square x meters on a side, so its area is

$$A(x) = x^2.$$

STEP 3: *The limits of integration.* The squares go from $x = 0$ to $x = 3$.

STEP 4: *The volume:*

$$\begin{aligned}\text{Volume} &= \int_a^b A(x)\,dx \qquad \text{(Eq. 3)}\\ &= \int_0^3 x^2\,dx\\ &= \left.\frac{x^3}{3}\right]_0^3\\ &= 9.\end{aligned}$$

The volume is 9 m^3.

EXAMPLE 2 A curved wedge is cut from a cylinder of radius 3 by two planes. One plane is perpendicular to the axis of the cylinder. The second plane crosses the first plane at a 45° angle at the center of the cylinder. Find the volume of the wedge.

Solution

STEP 1: *A sketch.* We draw the wedge and sketch a typical cross section perpendicular to the x-axis (Fig. 6.67).

STEP 2: *The formula for $A(x)$.* The cross section at x is a rectangle of area

$$A(x) = (\text{height})(\text{width}) = (x)(2\sqrt{9-x^2}) = 2x\sqrt{9-x^2}.$$

STEP 3: *The limits of integration.* The rectangles run from $x = 0$ to $x = 3$.

STEP 4: *The volume:*

$$\text{Volume} = \int_a^b A(x)\,dx = \int_0^3 2x\sqrt{9-x^2}\,dx \qquad \text{(Eq. 3)}$$

$$= 18.0000001171. \qquad \text{(Using FnInt)}$$

In the Exercises, you will determine the exact solution is 18 using analytic methods of calculus.

EXAMPLE 3 Cavalieri's theorem. Cavalieri's theorem says that two solids with equal altitudes and identical parallel cross sections have the same volume (Fig. 6.68). We can see this immediately from Eq. (3) because the cross-section area function $A(x)$ is the same in each case.

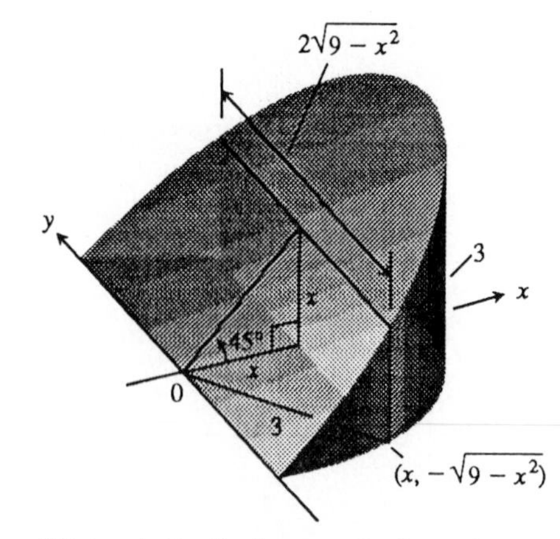

6.67 The wedge of Example 2, sliced perpendicular to the x-axis. The cross sections are rectangles.

BONAVENTURA CAVALIERI (1598–1647)

Cavalieri, a student of Galileo's whom Galileo told to study calculus, discovered that if two plane regions can be arranged to lie over the same interval of the x-axis in such a way that they have identical cross sections at every point, then the regions have the same area. The theorem was good enough to win Cavalieri a chair at the University of Bologna in 1629. The solid-geometry version in Example 3, which Cavalieri never proved, was given his name by later geometers.

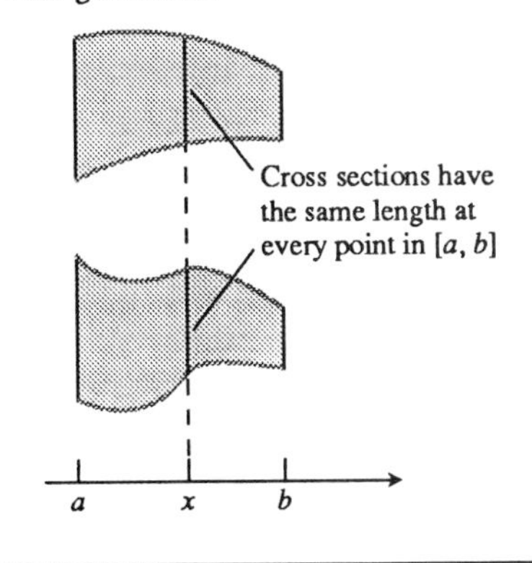

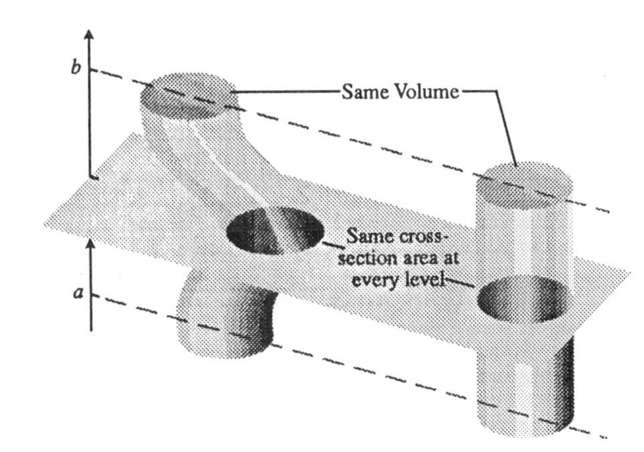

6.68 These two solids have the same volume even though the one on the left looks bigger. You can illustrate this yourself with two stacks of coins.

Position Shift and Distance Traveled

The total distance traveled by a body moving up and down a coordinate line is found by integrating the absolute value of the body's velocity (that is, the body's speed) over the time interval of the motion.

To see why, partition the time interval $a \leq t \leq b$ into subintervals in the usual way and let Δt_k denote the length of the kth interval. If Δt_k is small enough, the body's velocity $v(t)$ will not change much from t_{k-1} to t_k and the right-hand endpoint value $v(t_k)$ will give a good approximation of the velocity throughout the interval. Accordingly, the change in the body's position coordinate during the kth time interval will be about

$$v(t_k)\Delta t_k. \tag{4}$$

The change will be positive if $v(t_k)$ is positive and negative if $v(t_k)$ is negative.

If $s(t)$ is a body's position on a coordinate line at time t, then

$$\frac{ds}{dt} = v = \text{velocity}$$

and

$$\left|\frac{ds}{dt}\right| = |v| = \text{speed}.$$

In either case, the amount of distance traveled by the body during the kth interval will be about

$$|v(t_k)|\Delta t_k. \tag{5}$$

The total trip distance will be about

$$\sum_{k=1}^{n} |v(t_k)|\Delta t_k. \tag{6}$$

The sum in (6) is a Riemann sum for the speed $|v(t)|$ on the interval $[a, b]$. We expect the approximations we get from sums like these to improve as the norm of the partition of $[a, b]$ goes to zero. It therefore looks as if we should be able to calculate the total distance traveled by the body by integrating the body's speed from a to b. In practice, this turns out to be just the right thing to do. The mathematical model predicts the distance every time.

$$\text{Distance traveled} = \int_a^b |v(t)|\, dt \tag{7}$$

If we wish instead to predict how far up or down the line from its initial position a body will end up when a trip is over, we integrate v instead of its absolute value.

To understand why, let $s(t)$ be the body's position at time t and let F be an antiderivative of v. Then

$$s(t) = F(t) + C$$

for some constant C. The shift in the body's position caused by the trip from $t = a$ to $t = b$ is

$$s(b) - s(a) = (F(b) + C) - (F(a) + C) = F(b) - F(a) = \int_a^b v(t)\, dt. \tag{8}$$

CONCLUSIONS

1. To find distance traveled, integrate speed.
2. To find position shift, integrate velocity.

$$\text{Position shift} = \int_a^b v(t)\, dt \tag{9}$$

EXAMPLE 4 The velocity of a body moving along a line from $t = 0$ to $t = 3\pi/2$ seconds was

$$v(t) = 5\cos t \text{ m/sec.}$$

Find the total distance traveled and the shift in the body's position.

Solution

$$\text{Distance traveled} = \int_0^{3\pi/2} |5\cos t|\, dt \qquad \text{(Distance is the integral of speed.)}$$

$$= \int_0^{\pi/2} 5\cos t\, dt + \int_{\pi/2}^{3\pi/2} (-5\cos t)\, dt$$

$$= 5\sin t\Big]_0^{\pi/2} - 5\sin t\Big]_{\pi/2}^{3\pi/2} = 5(1-0) - 5(-1-1)$$

$$= 5 + 10 = 15 \text{ m}.$$

$$\text{Position shift} = \int_0^{3\pi/2} 5\cos t\, dt \qquad \text{(Shift is the integral of velocity.)}$$

$$= 5\sin t\Big]_0^{3\pi/2} = 5(-1) - 5(0) = -5 \text{ m}.$$

During the trip, the body traveled 5 m forward and 10 m backward for a total distance of 15 m. This shifted the body 5 m to the left (Fig. 6.69).

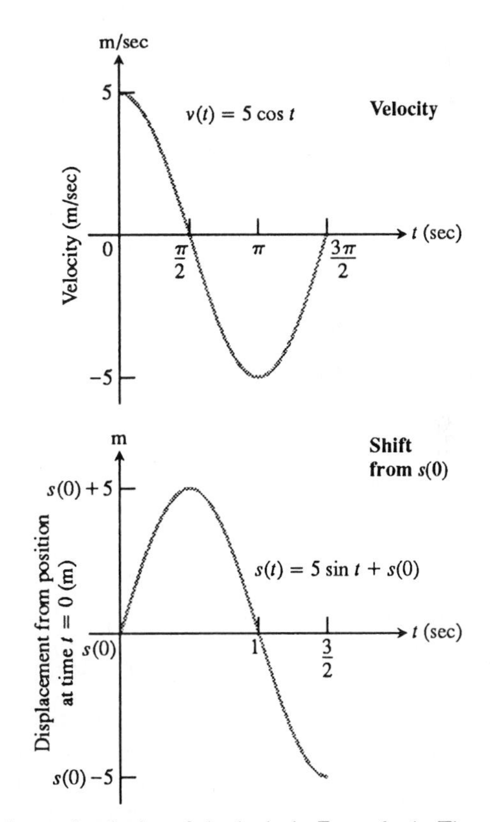

6.69 The position and velocity of the body in Example 4. The velocity is positive at first, and the corresponding displacement is positive. But the body stops at $t = \pi/2$ and reverses direction. By time $t = 3\pi/2$, the body lies 5 m to the left of where it started.

Explore with a Grapher

Example 4 can be nicely illustrated by graphing in parametric mode the following parametric equations

$$x(t) = s(t)$$

$$y(t) = 2.$$

Here $s(t)$ is the position function of the body moving along the line $y = 2$. The position function is the integral of the velocity function. That is, at a given time t, the position of the body is $(s(t), 2)$ or $(\int v(t)dt, 2)$. The catch is $\int v(t)dt = s(t) + C$ where $s(0) + C$ is the position at time $t = 0$ (the initial position). We assume $C = 0$, that is, the body starts out at the point $(s(0), 2)$.

Since $\int 5\cos t\, dt = 5\sin t + C$, graph $x(t) = 5\sin t$, and $y(t) = 2$ in the $[-10, 10]$ by $[0, 4]$ window. Set tMin $= 0$, tMax $= 3\pi/2$, and tStep $= \pi/24$. Use TRACE to simulate the path of the body (Fig. 6.70). Notice how easy it is to see that the total distance traveled is 15 m while the position shift is -5 m.

Achille Ernest Delesse was a mining engineer interested in determining the composition of rocks. To find out how much of a particular mineral a rock contained, he cut it through, polished an exposed face, and covered the face with transparent waxed paper, trimmed to size. He then traced on the paper the exposed portions of the mineral that interested him. After weighing the paper, he cut out the mineral traces and weighed them. The ratio of the weights gave not only the proportion of the surface occupied by the mineral but, more important, the proportion of the rock occupied by the mineral. Delesse described his method in an article entitled "A mechanical procedure for determining the composition of rocks," in the *Annales des Mines*, 13, 1848, pp. 379–388. His method is still used by petroleum geologists today. A two-dimensional analogue of it is used to determine the porosities of the ceramic filters that extract organic molecules in chemistry laboratories and screen out microbes in water purifiers.

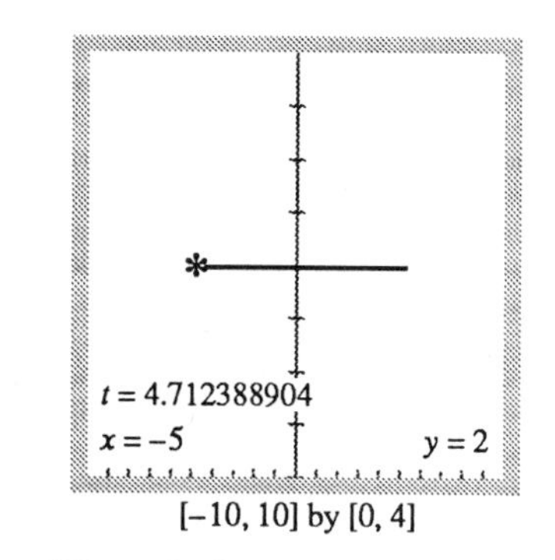

6.70 The moving body of Example 4.

Delesse's Rule

As you may know, the sugar in an apple starts turning into starch as soon as the apple is picked, and the longer the apple sits around, the starchier it becomes. You can tell fresh apples from stale by both flavor and consistency.

To find out how much starch is in a given apple, we can look at a thin slice under a microscope. The cross sections of the starch granules will show up clearly, and it is easy to estimate the proportion of the viewing area they occupy. This two-dimensional proportion will be the same as the three-dimensional proportion of uncut starch granules in the apple itself. The apparently magical quality of these proportions was first discovered by a French geologist, Achille Ernest Delesse, in the 1840s. Its explanation lies in the notion of average value.

Suppose we want to find the proportion of some granular material in a solid and that the sample we have chosen to analyze is a cube whose edges have length L. We picture the cube with an x-axis along one edge and imagine slicing the cube with planes perpendicular to points of the interval $[0, L]$. Call the proportion of the area of the slice at x occupied by the granular material of interest (starch, in our apple example) $r(x)$, and assume r is a continuous function of x.

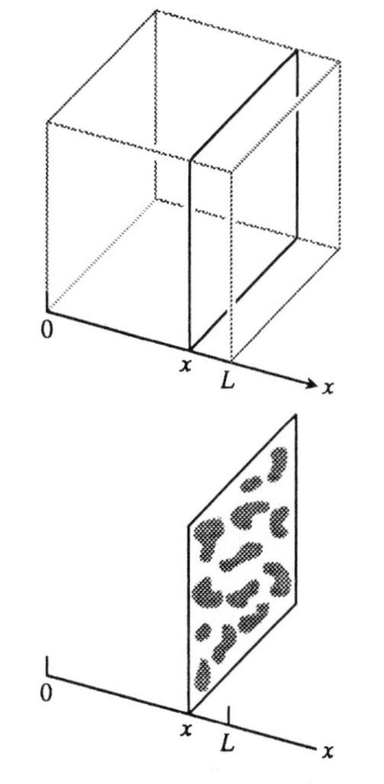

Now partition the interval $[0, L]$ into subintervals in the usual way. Imagine the cube sliced into thin slices by planes at the subdivision points. The length Δx_k of the kth subinterval is the distance between the planes at x_{k-1} and x_k. If the planes are close enough together, the sections cut from the grains by the planes will resemble cylinders with bases in the plane at x_k. The proportion of granular material between the planes will therefore be about the same as theproportion of cylinder base area in the plane at x_k, which in turn will be about $r(x_k)$. Thus the amount of granular material in the slab between the two planes will be about

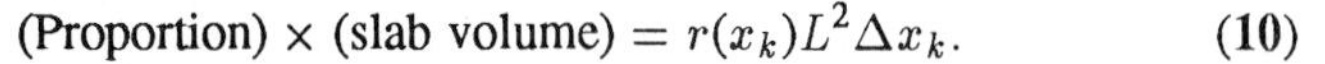

$$\text{(Proportion)} \times \text{(slab volume)} = r(x_k)L^2\Delta x_k. \tag{10}$$

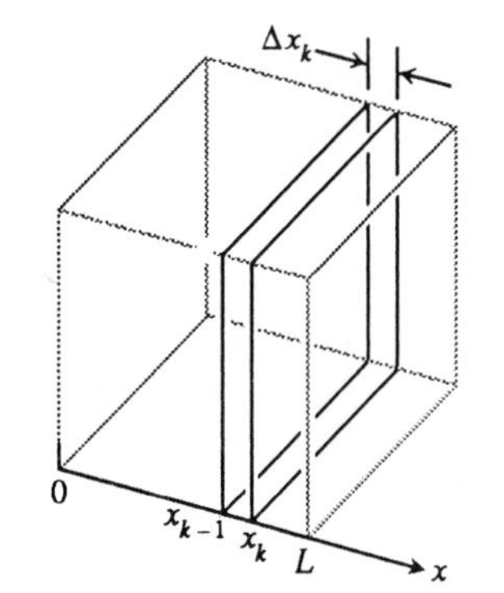

The amount of granular material in the entire sample cube will be about

$$\sum_{k=1}^{n} r(x_k)L^2\Delta x_k. \tag{11}$$

This sum is a Riemann sum for the function $r(x)L^2$ over the interval $[0, L]$. We expect the approximations by sums like these to improve as the norm of the subdivision of $[0, L]$ goes to zero and therefore expect the integral

$$\int_a^b r(x)L^2\,dx \tag{12}$$

to give the amount of granular material in the sample cube.

We can then obtain the proportion of granular material in the sample by dividing this amount by the cube's volume L^3. If we have chosen our sample well, this will also be the proportion of granular material in the solid from which the sample was taken. Putting it all together, we get

$$\begin{aligned}
\text{Proportion of granular material in solid} &= \text{Proportion of granular material in the sample cube} \\
&= \frac{\int_0^L r(x)L^2\,dx}{L^3} \\
&= \frac{L^2\int_0^L r(x)\,dx}{L^3} = \frac{1}{L}\int_0^L r(x)\,dx \qquad (13)\\
&= \text{average value of } r(x) \text{ over } [0, L] \\
&= \text{proportion of area occupied by granular material in a typical cross section.}
\end{aligned}$$

This is Delesse's rule. Once we have found $\bar{r}$, the average of $r(x)$ over $[0, L]$, we have found the proportion of granular material in the solid.

In practice, $\bar{r}$ is found by averaging over a number of cross sections. There are several things to watch out for in the process. In addition to the possibility that the granules cluster in ways that make representative samples difficult to find, there is the possibility we might not recognize a granule's trace for what it is. Some cross sections of normal red blood cells look like disks and ovals, but others look surprisingly like outlines of dumbbells. We do not want to dismiss the dumbbells as experimental error the way one research group we know of did a few years ago.

Useless Integrals

Some of the integrals we get from forming Riemann sums do what we want, but others do not. It all depends on how we choose to model the problems we want to solve. Some choices are good—others are not. Here is an example.

We use the surface area formula

$$\text{Surface area} = \int_a^b 2\pi f(x)\sqrt{1 + \left(\frac{df}{dx}\right)^2}\,dx \tag{14}$$

because it has predictive value and always gives results consistent with information from other sources. In other words, the model we used to derive the formula was a good one.

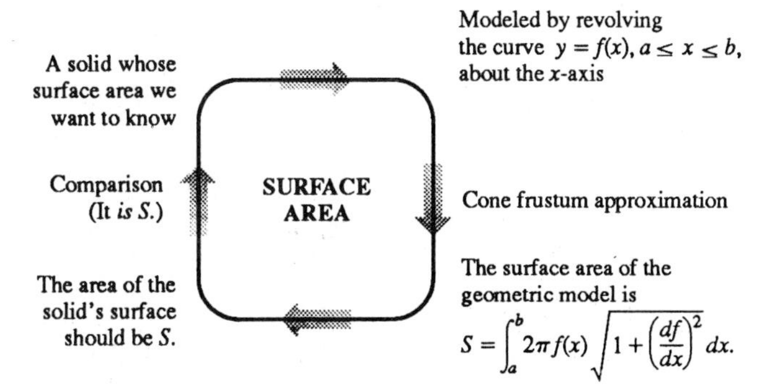

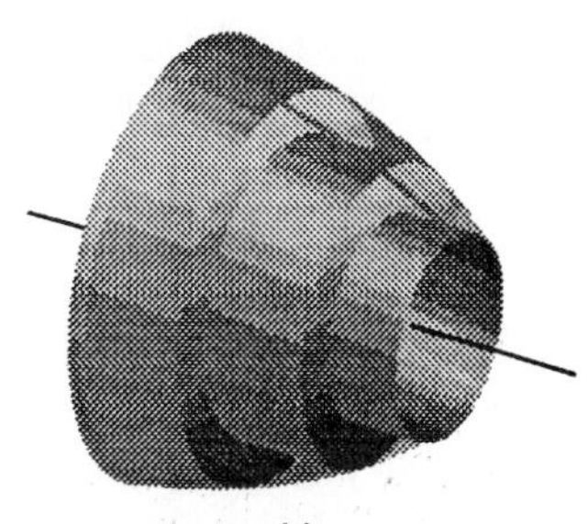

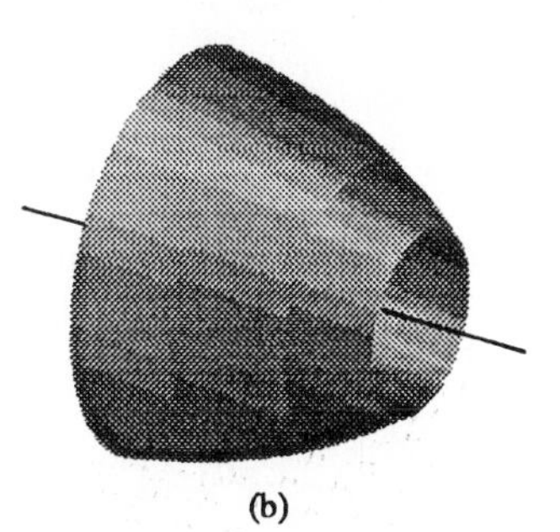

.71 Why not use cylindrical bands) instead of conical bands (b) to pproximate surface area?

Why not find the surface area by approximating with cylindrical bands instead of conical bands, as suggested in Fig. 6.71? The Riemann sums we get this way converge just as nicely as the ones based on conical bands, and the resulting integral is simpler. Instead of Eq. (14), we get

$$\text{Surface area candidate} = \int_a^b 2\pi f(x)\, dx. \tag{15}$$

After all, we might argue, we used cylinders to derive good volume formulas, so why not use them again to derive surface-area formulas?

The answer is that the formula in Eq. (15) has no predictive value and almost never gives results consistent with experience. The comparison step in the model fails for this formula.

There is a moral here: Just because we end up with a nice-looking integral does not mean it will do what we want. Constructing an integral is not enough—we have to test it too.

Exercises 6.9

Find the volumes of the solids in Exercises 1–8.

1. The solid lies between planes perpendicular to the x-axis at $x = 0$ and $x = 4$. The cross sections perpendicular to the axis on the interval $0 \le x \le 4$ are squares whose diagonals run from the parabola $y = -\sqrt{x}$ to the parabola $y = \sqrt{x}$.
2. The solid lies between planes perpendicular to the x-axis at $x = -1$ and $x = 1$. The cross sections perpendicular to the x-axis between these planes are squares whose diagonals run from the semicircle $y = -\sqrt{1 - x^2}$ to the semicircle $y = \sqrt{1 - x^2}$.
3. The solid lies between planes perpendicular to the x-axis at $x = -1$ and $x = 1$. The cross sections perpendicular to the axis between these planes are squares with edges running from the semicircle $y = -\sqrt{1 - x^2}$ to the semicircle $y = \sqrt{1 - x^2}$.

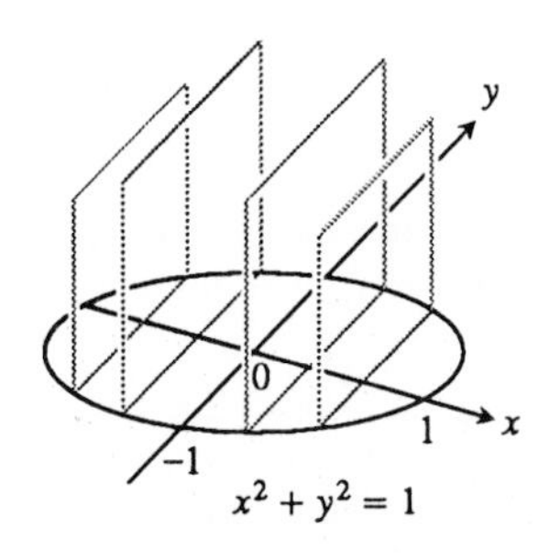

4. The solid lies between the planes perpendicular to the x-axis at $x = -1$ and $x = 1$. The cross sections perpendicular to the x-axis are circular disks whose diameters run from the parabola $y = x^2$ to the parabola $y = 2 - x^2$.

5. The solid lies between planes perpendicular to the x-axis at $x = 1$ and $x = 2$. The cross sections perpendicular to the x-axis are circular disks with diameters running from the x-axis up to the curve $y = 2/\sqrt{x}$.

6. The solid lies between planes perpendicular to the x-axis at $x = 0$ and $x = 2$. The cross sections perpendicular to the x-axis are circular disks with diameters running from the x-axis up to the parabola $y = \sqrt{5}x^2$.

7. The base of the solid is the disk $x^2 + y^2 \leq 1$. The cross sections by planes perpendicular to the y-axis between $y = -1$ and $y = 1$ are isosceles right triangles with one leg in the disk.

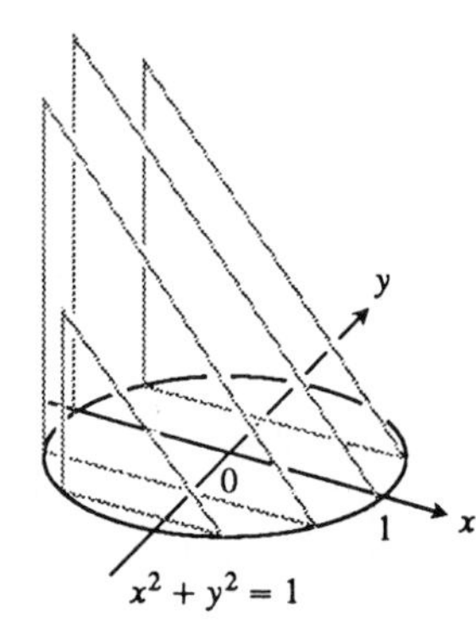

8. The base of the solid is the region between the curve $y = 2\sqrt{\sin x}$ and the interval $0 \leq x \leq \pi$ of the x-axis. The cross sections perpendicular to the x-axis are equilateral triangles.

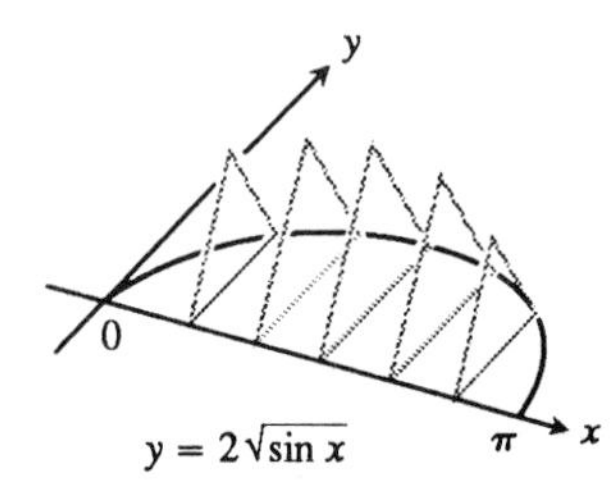

9. A twisted solid is generated as follows: We are given a line L in space and a square of side s in a plane perpendicular to L. One vertex of the square is on L. As this vertex moves a distance h along L, the square turns through a full revolution about L. Find the volume of the solid generated by this motion. What would the volume be if the square had turned two revolutions in moving the same distance along L?

10. Prove Cavalieri's original theorem (historical note, this section), assuming that each region is the region between the graphs of two continuous functions over the interval $a \leq x \leq b$.

In Exercises 11–18, the function $v(t)$ is the velocity in meters per second of a body moving along a coordinate line. (a) Graph v as a function of t to see where it is positive and negative. Then find (b) the total distance traveled by the body during the given time interval and (c) the shift in the body's position.

11. $v(t) = 5\cos t, 0 \leq t \leq 2\pi$

12. $v(t) = \sin \pi t, 0 \leq t \leq 2$

13. $v(t) = 6\sin 3t, 0 \leq t \leq \pi/2$

14. $v(t) = 4\cos 2t, 0 \leq t \leq \pi$

15. $v(t) = 49 - 9.8t, 0 \leq t \leq 10$

16. $v(t) = 8 - 1.6t, 0 \leq t \leq 10$

17. $v(t) = 6t^2 - 18t + 12 = 6(t-1)(t-2), 0 \leq t \leq 2$

18. $v(t) - 6t^2 - 18t + 12 = 6(t-1)(t-2), 0 \leq t \leq 3$

19. Figure 6.72 shows the velocity graphs of four bodies moving on coordinate lines. Find the distance traveled by each body and the position shift for the given time interval.

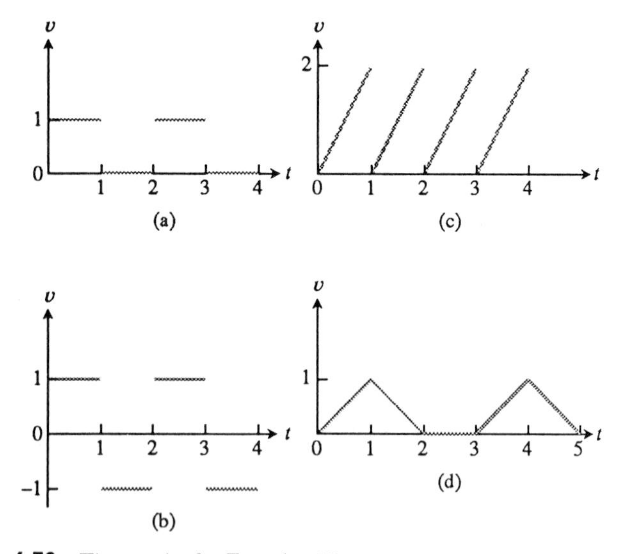

6.72 The graphs for Exercise 19.

20. Table 6.3 shows the velocity of a model train engine moving back and forth on a track for 10 sec. Use Simpson's rule to find the resulting position shift and total distance traveled.

TABLE 6.3 Selected velocities of the model train engine in Exercise 20

Time (sec)	Velocity (in. per sec)	Time (sec)	Velocity (in. per sec)
0	0	6	−11
1	12	7	−6
2	22	8	2
3	10	9	6
4	−5	10	0
5	−13		

21. **Modeling surface area.** The lateral surface area of the cone swept out by revolving the line segment $y = x/\sqrt{3}, 0 \leq x \leq \sqrt{3}$, about the x-axis should be (1/2) (base circumference)(slant height) $= (1/2)(2\pi)(2) = 2\pi$. What do you get if you use Eq. (15) with $f(x) = x/\sqrt{3}$?

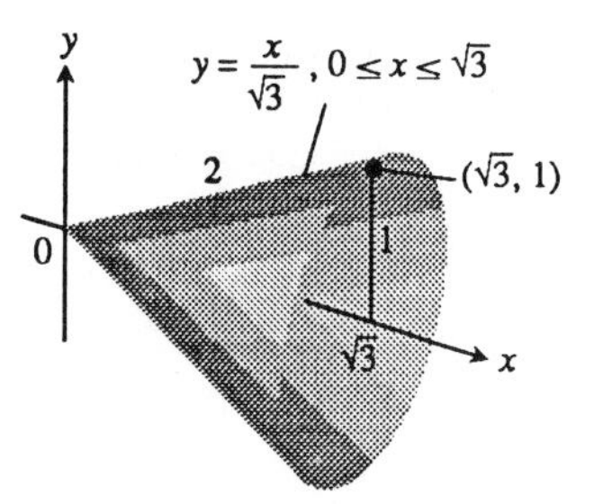

22. **Modeling surface area.** The only surface for which Eq. (15) gives the area we want is a cylinder. Show that Eq. (15) gives $S = 2\pi r h$ for the cylinder swept out by revolving the line segment $y = r, 0 \leq x \leq h$, about the x-axis.

23. Use analytic methods of calculus to determine that

$$\int_0^3 2x\sqrt{9 - x^2}\,dx = 18. \qquad \text{(See Example 2.)}$$

Review Questions

1. How do you define and calculate the area between the graphs of two continuous functions? Give an example.
2. How do you define and calculate the volume of a solid of revolution by
   a) the disk method?
   b) the washer method?
   c) the method of cylindrical shells? Give examples.
3. How do you define and calculate the length of the graph of a continuously differentiable function over a closed interval? Give an example. What about functions that aren't continuously differentiable?
4. How do you define and calculate the area of the surface swept out by revolving the graph of a continuously differentiable function $y = f(x), a \leq x \leq b$, about the x-axis? Give an example.
5. How do you define and calculate the work done by a force directed along a portion of the x-axis? How do you calculate the work it takes to pump liquid from a tank? Give examples.
6. How do you calculate the force exerted by a liquid against a portion of vertical wall? Give an example.
7. How do you locate the center of mass of a straight, narrow rod or strip of material? Given an example. If the density of the material is constant, you can tell right away where the center of mass is. Where is it?
8. How do you locate the center of mass of a thin, flat plate of material? Give an example.
9. Suppose that you know the velocity of a body that will move back and forth along a coordinate line tomorrow from time $t = a$ to time $t = b$. How can you calculate in advance how much the motion will shift the body's position? How do you predict the total distance the body will travel?
10. How do you define and calculate volumes of solids by the method of slicing? Give an example. How is the method of slicing related to the disk and washer methods?
11. What does Delesse's rule say? Give an example.
12. There is a basic pattern to the way we constructed integrals in this chapter. What is it?

Practice Exercises

Find the areas of the regions enclosed by the curves and lines in Exercises 1–14.

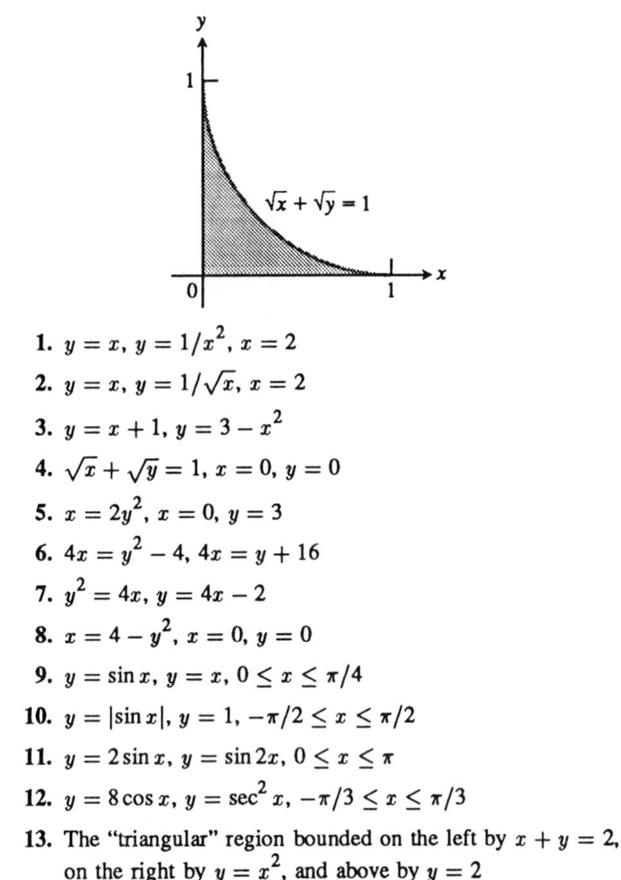

1. $y = x,\ y = 1/x^2,\ x = 2$
2. $y = x,\ y = 1/\sqrt{x},\ x = 2$
3. $y = x + 1,\ y = 3 - x^2$
4. $\sqrt{x} + \sqrt{y} = 1,\ x = 0,\ y = 0$
5. $x = 2y^2,\ x = 0,\ y = 3$
6. $4x = y^2 - 4,\ 4x = y + 16$
7. $y^2 = 4x,\ y = 4x - 2$
8. $x = 4 - y^2,\ x = 0,\ y = 0$
9. $y = \sin x,\ y = x,\ 0 \le x \le \pi/4$
10. $y = |\sin x|,\ y = 1,\ -\pi/2 \le x \le \pi/2$
11. $y = 2\sin x,\ y = \sin 2x,\ 0 \le x \le \pi$
12. $y = 8\cos x,\ y = \sec^2 x,\ -\pi/3 \le x \le \pi/3$
13. The "triangular" region bounded on the left by $x + y = 2$, on the right by $y = x^2$, and above by $y = 2$
14. The "triangular" region bounded on the left by $y = \sqrt{x}$, on the right by $y = 6 - x$, and below by $y = 1$

Find the areas of the shaded regions in Exercises 15 and 16.

15.

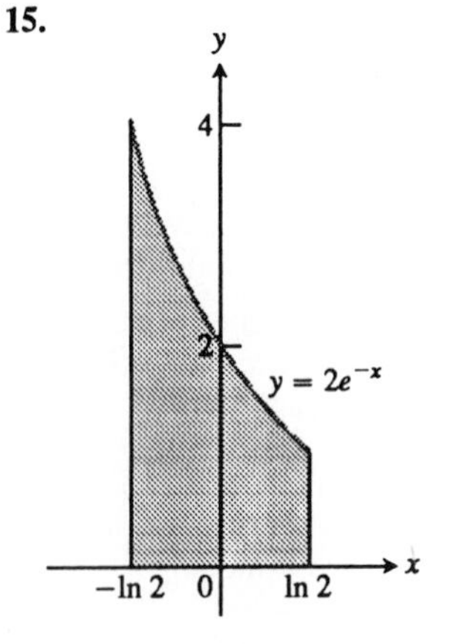

16.

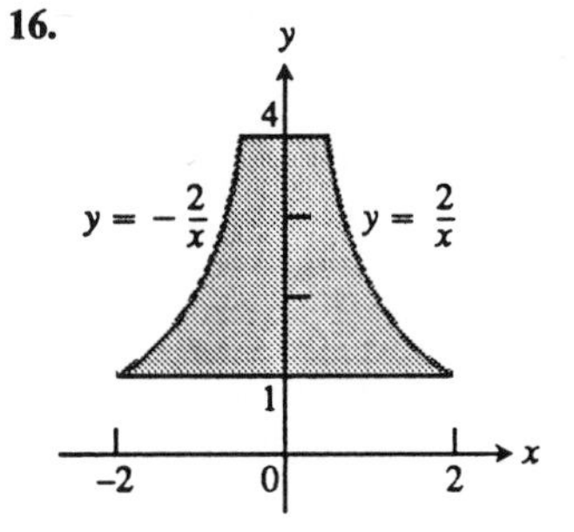

17. Find the volume of the solid generated by revolving the region bounded by the x-axis, the curve $y = 3x^4$, and the lines $x = 1$ and $x = -1$ about

    a) the x-axis

    b) the y-axis.

18. Find the volume of the solid generated by revolving the region in the first quadrant bounded by the x-axis, the parabola $y = x^2$, and the line $x = 3$ about (a) the x-axis; (b) the y-axis.

19. Find the volume of the solid generated by revolving the region bounded on the left by the parabola $x = y^2 + 1$ and on the right by the line $x = 5$ about (a) the x-axis; (b) the y-axis; (c) the line $x = 5$.

20. Find the volume of the solid generated by revolving the region bounded by the parabola $y^2 = 4x$ and the line $y = x$ about (a) the x-axis; (b) the y-axis; (c) the line $x = 4$; (d) the line $y = 4$.

21. Find the volume of the solid generated by revolving the "triangular" region bounded by the x-axis, the line $x = \pi/3$, and the curve $y = \tan x$ in the first quadrant about the x-axis.

22. Find the volume of the solid generated by revolving the region bounded by the curve $y = \sin x$ and the lines $x = 0$, $x = \pi$, and $y = 2$ about the line $y = 2$.

23. Find the volume of the solid spindle generated by revolving the region bounded by the x-axis, the curve $y = 1/\sqrt{x}$, and the lines $x = 1$ and $x = 16$ about the x-axis.

24. Find the volume of the solid generated by revolving the region bounded by the curve $y = e^{x/2}$ and the lines $x = \ln 3$ and $y = 1$ about the x-axis.

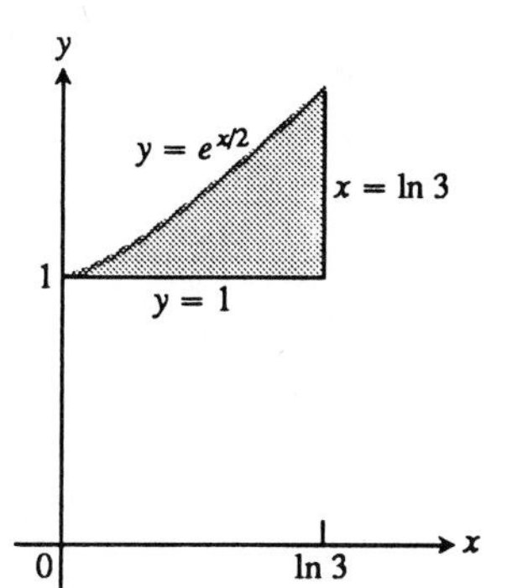

25. A round hole of radius $\sqrt{3}$ ft is bored through the center of a sphere of radius 2 ft. Find the volume cut out.

26. The profile of a football resembles the ellipse shown here. Find the volume of the football to the nearest cubic inch.

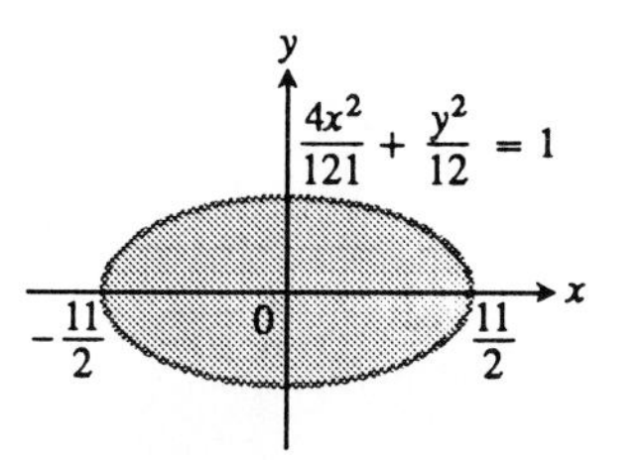

Find the lengths of the curves in Exercises 27–30.

27. $y = x^{1/2} - (1/3)x^{3/2}$, $0 \le x \le 3$
(*Hint*: $1 + (dy/dx)^2$ is a perfect square.)

28. $x = y^{2/3}$, $1 \le y \le 8$

29. $y = (x^2/8) - \ln x$, $4 \le x \le 8$

30. $y = e^x + (1/4)e^{-x}$, $0 \le x \le \ln 3$

In Exercises 31–36, find the areas of the surfaces generated by revolving the curves about the given axes.

31. $y = \sqrt{2x+1}$, $0 \le x \le 12$, x-axis

32. $y = x^3/9$, $-1 \le x \le 1$, x-axis

33. $y = (1/3)x^{3/2} - x^{1/2}$, $0 \le x \le 3$, x-axis

34. $y = e^x + (1/4)e^{-x}$, $-\ln 4 \le x \le \ln 2$, x-axis

35. $y = (1/3)(x^2 + 2)^{3/2}$, $0 \le x \le 1$, y-axis
(*Hint*: Express $ds = \sqrt{dx^2 + dy^2}$ in terms of dx and evaluate $\int 2\pi\, x\, ds$.)

36. $y = x^2$, $0 \le y \le 2$, y-axis

37. A rock climber is about to haul up 10 kg of equipment that has been hanging beneath her on 40 m of rope that weighs 0.8 newtons per meter. How much work will it take? (*Hint*: Solve for the rope and equipment separately; then add.)

38. You drove an 800-gal tank truck from the base to the summit of Mt. Washington and discovered on arrival that the tank was only half full. You started out with a full tank, climbed at a steady rate, and took 50 min to accomplish the 4750-ft elevation change. Assuming that the water leaked out at a steady rate, how much work was spent in carrying water to the top? Do not count the work done in getting yourself and the truck there. Water weighs 8 lb/U.S. gal.

39. If a force of 20 lb is required to hold a spring 1 ft beyond its unstressed length, how much work does it take to stretch the spring this far? How much work does it take to stretch the spring an additional foot?

40. A force of 2 N will stretch a rubber band 2 cm. Assuming Hooke's law applies, how far will a 4-N force stretch the rubber band? How much work does it take to stretch the rubber band this far?

41. A reservoir, shaped like an inverted right circular cone 20 ft across the top and 8 ft deep, is full of water. How much work does it take to pump the water to a level 6 ft above the top?

42. Continuation of Exercise 41. The reservoir is filled to a depth of 5 ft, and the water is to be pumped to the same level as the top. How much work does it take?

43. The vertical triangular plate shown here is the end plate of a triangular watering trough full of water. What is the force against the end of the plate?

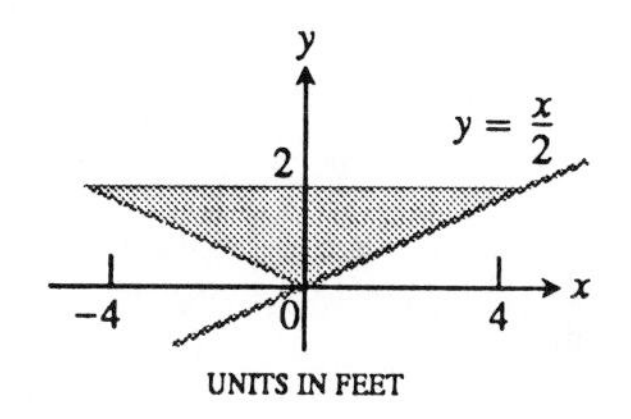

44. The vertical trapezoidal plate shown here is the end plate of a trough of maple syrup weighing 75 lb/ft^3. What is the force against the end of the trough when there are 10 in. of syrup in the trough?

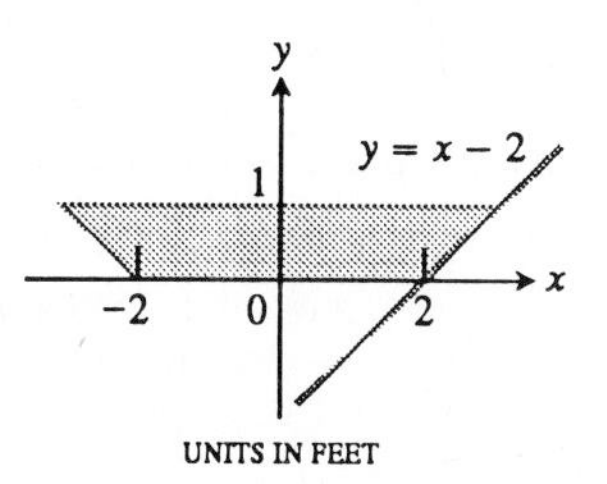

45. A flat vertical gate in the face of a dam is shaped like the parabolic region between the curve $y = 4x^2$ and the line $y = 4$, with measurements in feet. The top of the gate lies 5 ft below the surface of the water. Find the force against the gate.

46. A standard olive oil can measures 5.75 by 3.5 by 10 in. Find the fluid force against the base and each side of the can when the can is full.

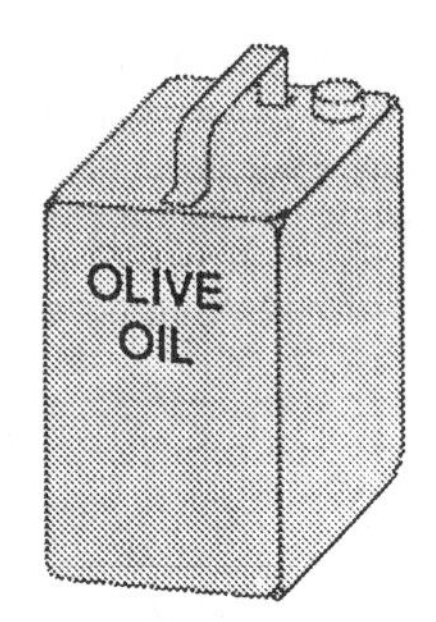

47. Find the center of mass of a thin, flat plate of constant density covering the region enclosed by the parabolas $y = 2x^2$ and $y = 3 - x^2$.

48. Find the center of mass of a thin, flat plate of constant density covering the region enclosed by the x-axis, the lines $x = 2$ and $x = -2$, and the parabola $y = x^2$.

49. Find the center of mass of a thin, flat plate of constant density covering the "triangular" region in the first quadrant bounded by the y-axis, the parabola $y = x^2/4$, and the line $y = 4$.

50. Find the center of mass of a thin, flat plate of density $\delta = 3$ covering the region enclosed by the parabola $y^2 = x$ and the line $x = 2y$.

51. Find the centroid of the trapezoid in Exercise 44.

52. Find the centroid of the triangle in Exercise 43.

53. Find the center of mass of a thin plate of constant density covering the region between the curve $y = 1/\sqrt{x}$ and the x-axis from $x = 1$ to $x = 16$.

54. Continuation of Exercise 53. Now find the center of mass, assuming that the density is $\delta(x) = 4/\sqrt{x}$.

Find the volumes of the solids in Exercises 55–58.

55. The solid lies between planes perpendicular to the x-axis at $x = 0$ and $x = 1$. The cross sections perpendicular to the x-axis between these planes are circular disks whose diameters run from the parabola $y = x^2$ to the parabola $y = \sqrt{x}$.

56. The base of the solid is the region in the first quadrant between the line $y = x$ and the parabola $y = 2\sqrt{x}$. The cross sections of the solid perpendicular to the x-axis are equilateral triangles whose bases stretch from the line to the curve.

57. The solid lies between planes perpendicular to the x-axis at $x = \pi/4$ and $x = 5\pi/4$. The cross sections between these planes are circular disks whose diameters run from the curve $y = 2\cos x$ to the curve $y = 2\sin x$.

58. The solid lies between planes perpendicular to the x-axis at $x = 0$ and $x = 6$. The cross sections between these planes are squares whose bases run from the x-axis up to the curve $x^{1/2} + y^{1/2} = \sqrt{6}$.

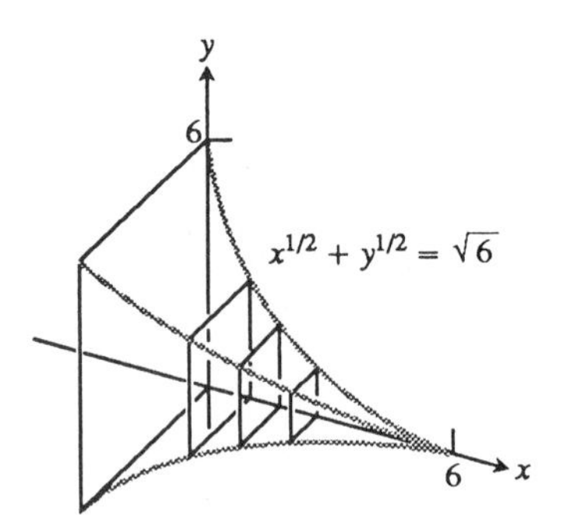

59. The volume of the wedge in Section 6.9, Example 2, could have been found by drawing a picture like the one in Fig. 6.73, taking cross sections perpendicular to the y-axis, and integrating with respect to y. Find the volume this way.

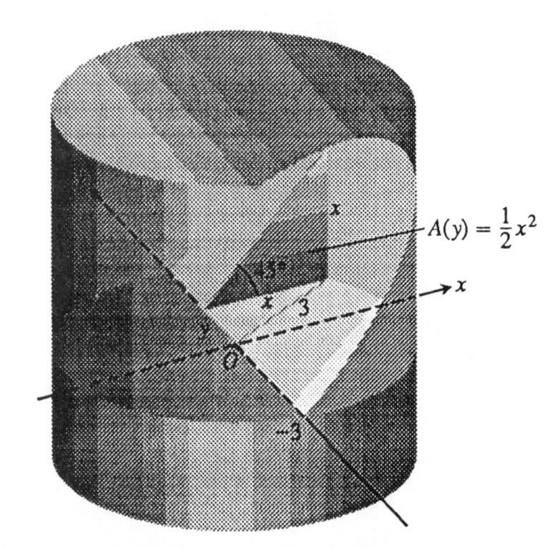

6.73 The cross sections of the wedge perpendicular to the y-axis are isosceles right triangles. Exercise 59 asks you to use their areas to find the volume of the wedge.

60. A solid lies between planes perpendicular to the x-axis at $x = 0$ and $x = 12$. The cross sections by planes perpendicular to the x-axis for $0 \le x \le 12$ are circular disks whose diameters run from the line $y = x/2$ to the line $y = x$. Use Cavalieri's theorem (Section 6.9, Example 3) to explain why the solid has the same volume as a right circular cone with base radius 3 and altitude 12.

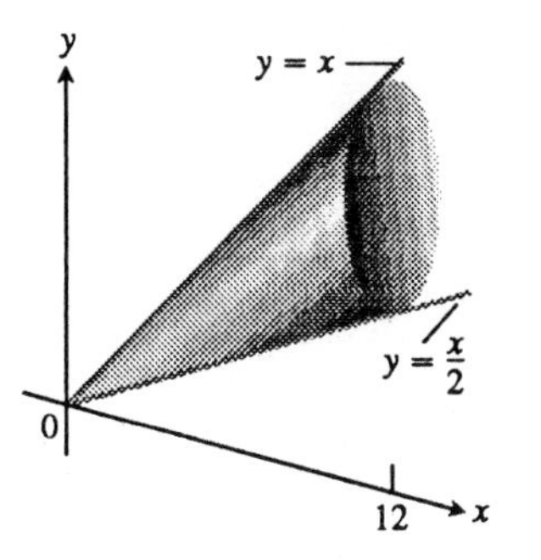

In Exercises 61–64, the function $v(t)$ is the velocity in feet per second of a body moving on a coordinate line. (a) Graph v as a function of t to see where it is positive and negative. Then find (b) the total distance traveled by the body during the given time interval and (c) how much the trip shifted the body's position.

61. $v(t) = (t/2) - 1,\ 0 \leq t \leq 6$

62. $v(t) = t^2 - 8t + 12 = (t - 2)(t - 6), 0 \leq t \leq 6$

63. $v(t) = 5\cos t,\ 0 \leq t \leq 3\pi/2$

64. $v(t) = \pi \sin \pi t,\ 0 \leq t \leq 3/2$

Appendices

A.1 Formulas from Precalculus Mathematics

Algebra

1. Laws of Exponents

$$a^m a^n = a^{m+n}, \quad (ab)^m = a^m b^m, \quad (a^m)^n = a^{mn}, \quad a^{m/n} = \sqrt[n]{a^m}$$

If $a \neq 0$,

$$\frac{a^m}{a^n} = a^{m-n}, \quad a^0 = 1, \quad a^{-m} = \frac{1}{a^m}.$$

2. Zero Division by zero is not defined.

If $a \neq 0$: $\quad \dfrac{0}{a} = 0, \quad a^0 = 1, \quad 0^a = 0$

For any number a: $\quad a \cdot 0 = 0 \cdot a = 0$

3. Fractions

$$\frac{a}{b} + \frac{c}{d} = \frac{ad + bc}{bd}, \qquad \frac{a}{b} \cdot \frac{c}{d} = \frac{ac}{bd}, \qquad \frac{a/b}{c/d} = \frac{a}{b} \cdot \frac{d}{c}, \qquad \frac{-a}{b} = -\frac{a}{b} = \frac{a}{-b},$$

$$\frac{(a/b) + (c/d)}{(e/f) + (g/h)} = \frac{(a/b) + (c/d)}{(e/f) + (g/h)} \cdot \frac{bdfh}{bdfh} = \frac{(ad + bc)fh}{(eh + fg)bd}$$

4. The Binomial Theorem

For any positive integer n,

$$(a + b)^n = a^n + na^{n-1}b + \frac{n(n-1)}{1 \cdot 2} a^{n-2}b^2 + \frac{n(n-1)(n-2)}{1 \cdot 2 \cdot 3} a^{n-3}b^3 + \cdots + nab^{n-1} + b^n.$$

For instance,

$$(a + b)^1 = a + b,$$

$$(a + b)^2 = a^2 + 2ab + b^2,$$

$$(a + b)^3 = a^3 + 3a^2b + 3ab^2 + b^3,$$

$$(a + b)^4 = a^4 + 4a^3b + 6a^2b^2 + 4ab^3 + b^4.$$

5. Difference of Like Integer Powers, $n > 1$

$$a^n - b^n = (a - b)(a^{n-1} + a^{n-2}b + a^{n-3}b^2 + \cdots + ab^{n-2} + b^{n-1})$$

For instance,

$$a^2 - b^2 = (a - b)(a + b),$$

$$a^3 - b^3 = (a - b)(a^2 + ab + b^2),$$

$$a^4 - b^4 = (a - b)(a^3 + a^2b + ab^2 + b^3).$$

6. Completing the Square

If $a \neq 0$, we can rewrite the quadratic $ax^2 + bx + c$ in the form $au^2 + C$ by a process called completing the square:

$$ax^2 + bx + c = a\left(x^2 + \frac{b}{a}x\right) + c \qquad \left(\text{Factor } a \text{ from the first two terms.}\right)$$

$$= a\left(x^2 + \frac{b}{a}x + \frac{b^2}{4a^2} - \frac{b^2}{4a^2}\right) + c \qquad \left(\text{Add and subtract the square of half the coefficient of } x.\right)$$

$$= a\left(x^2 + \frac{b}{a}x + \frac{b^2}{4a^2}\right) + a\left(-\frac{b^2}{4a^2}\right) + c \qquad \left(\text{Bring out the } -b^2/4a^2.\right)$$

$$= \underbrace{a\left(x^2 + \frac{b}{a}x + \frac{b^2}{4a^2}\right)}_{\text{This is } \left(x + \frac{b}{2a}\right)^2} + \underbrace{c - \frac{b^2}{4a}}_{\text{Call this part } C}$$

$$= au^2 + C \qquad (u = x + b/2a)$$

7. The Quadratic Formula

By completing the square on the first two terms of the equation

$$ax^2 + bx + c = 0$$

and solving the resulting equation for x (details omitted), we obtain the formula

$$x = \frac{-b \pm \sqrt{b^2 - 4ac}}{2a}.$$

This equation is called the **quadratic formula.**

The solutions of the equation $2x^2 + 3x - 1 = 0$ are

$$x = \frac{-3 \pm \sqrt{(3)^2 - 4(2)(-1)}}{2(2)} = \frac{-3 \pm \sqrt{9 + 8}}{4},$$

or

$$x = \frac{-3 + \sqrt{17}}{4} \quad \text{and} \quad x = \frac{-3 - \sqrt{17}}{4}.$$

The solutions of the equation $x^2 + 4x + 6 = 0$ are

$$x = \frac{-4 \pm \sqrt{(4)^2 - 4 \cdot 1 \cdot 6}}{2} = \frac{-4 \pm \sqrt{16 - 24}}{2}$$

$$= \frac{-4 \pm \sqrt{-8}}{2} = \frac{-4 \pm 2\sqrt{2}\sqrt{-1}}{2} = -2 \pm \sqrt{2}\,i.$$

The solutions are the complex numbers $-2 + \sqrt{2}i$ and $-2 - \sqrt{2}i$. Appendix A.6 has more on complex numbers.

Geometry

(A = area, B = area of base, C = circumference, S = lateral area or surface area, V = volume)

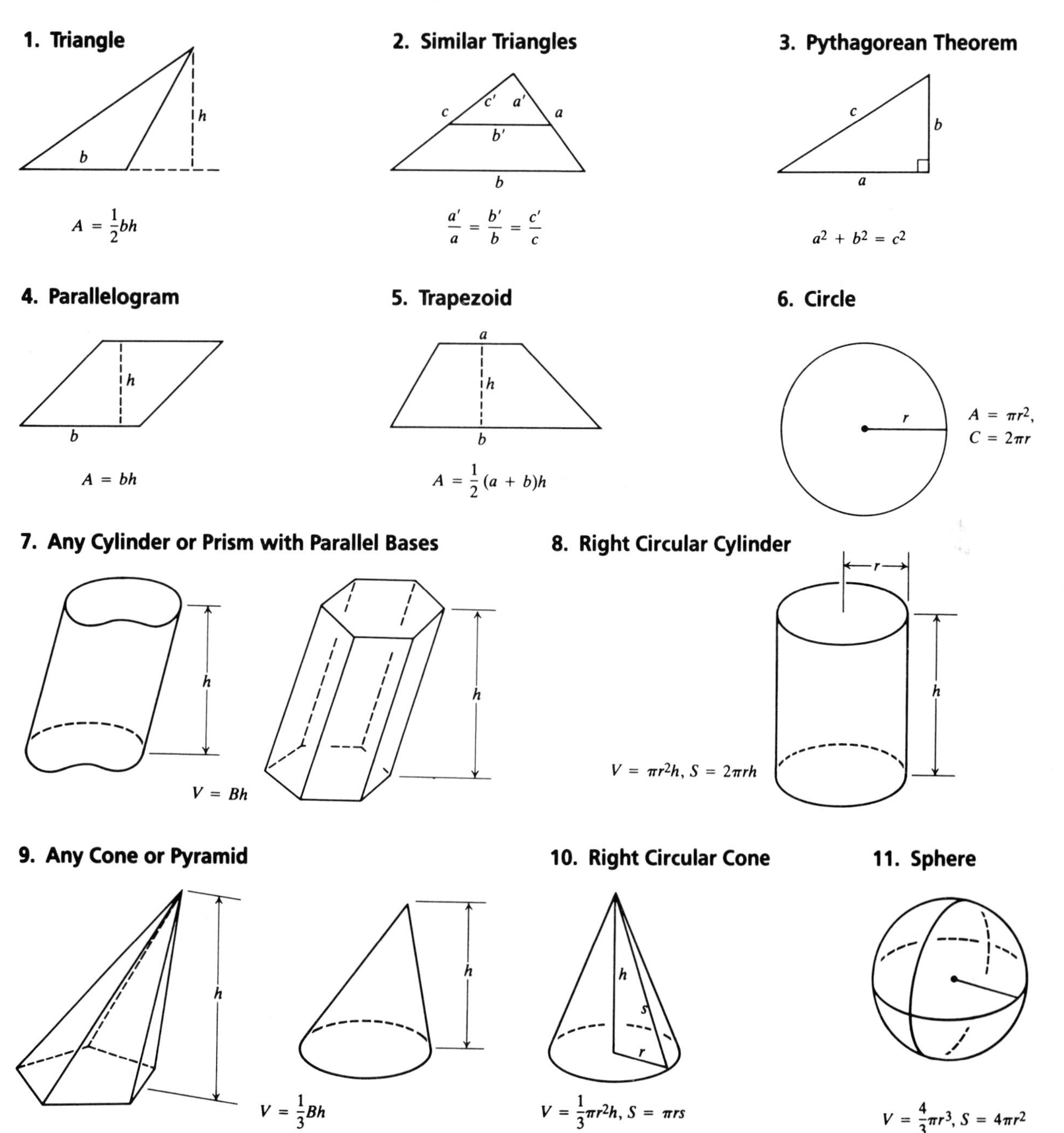

Trigonometry

1. Definitions and Fundamental Identities

Sine: $\sin\theta = \dfrac{y}{r} = \dfrac{1}{\csc\theta}$

Cosine: $\cos\theta = \dfrac{x}{r} = \dfrac{1}{\sec\theta}$

Tangent: $\tan\theta = \dfrac{y}{x} = \dfrac{1}{\cot\theta}$

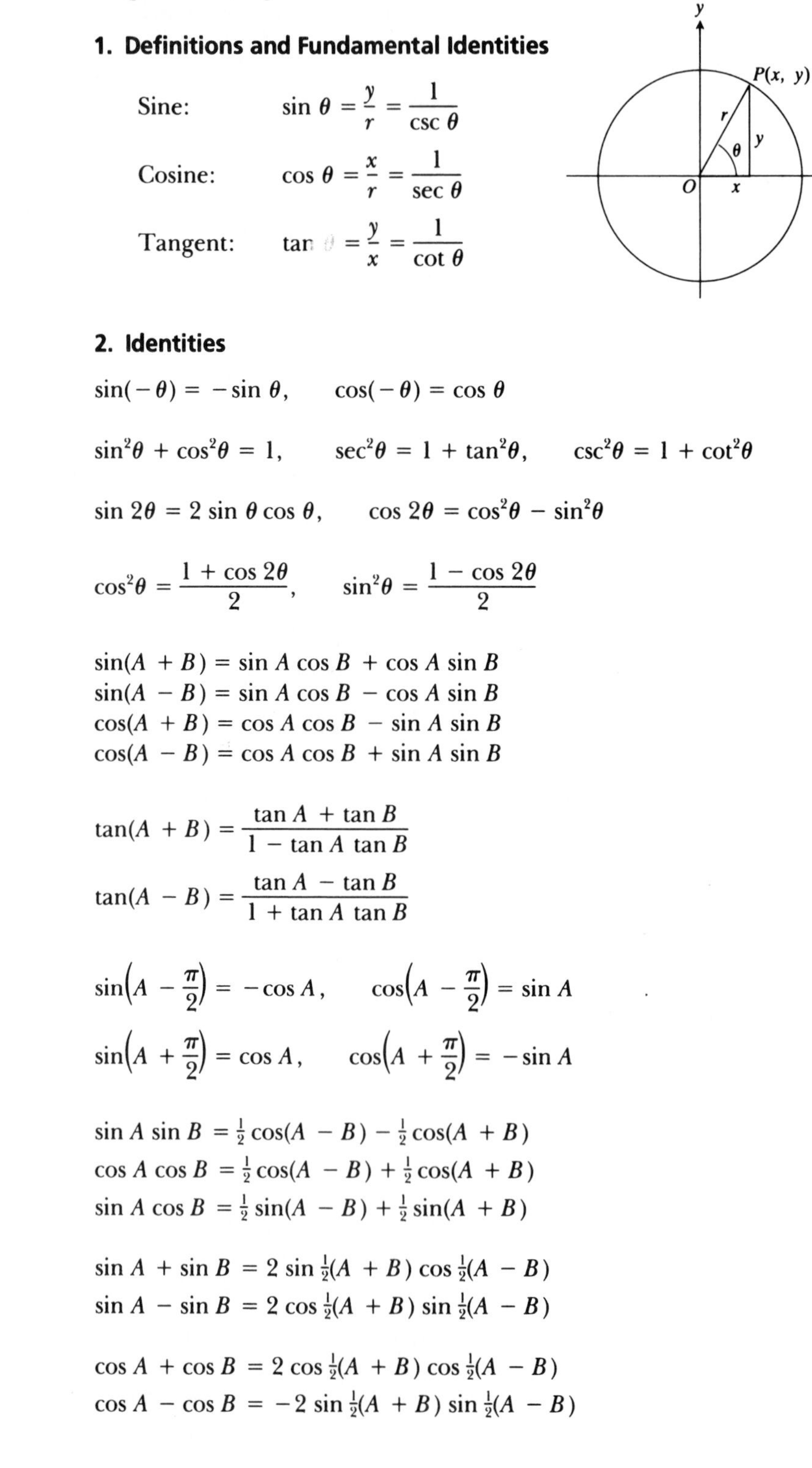

2. Identities

$\sin(-\theta) = -\sin\theta, \qquad \cos(-\theta) = \cos\theta$

$\sin^2\theta + \cos^2\theta = 1, \qquad \sec^2\theta = 1 + \tan^2\theta, \qquad \csc^2\theta = 1 + \cot^2\theta$

$\sin 2\theta = 2\sin\theta\cos\theta, \qquad \cos 2\theta = \cos^2\theta - \sin^2\theta$

$$\cos^2\theta = \frac{1+\cos 2\theta}{2}, \qquad \sin^2\theta = \frac{1-\cos 2\theta}{2}$$

$\sin(A+B) = \sin A\cos B + \cos A\sin B$
$\sin(A-B) = \sin A\cos B - \cos A\sin B$
$\cos(A+B) = \cos A\cos B - \sin A\sin B$
$\cos(A-B) = \cos A\cos B + \sin A\sin B$

$$\tan(A+B) = \frac{\tan A + \tan B}{1 - \tan A\tan B}$$

$$\tan(A-B) = \frac{\tan A - \tan B}{1 + \tan A\tan B}$$

$$\sin\left(A - \frac{\pi}{2}\right) = -\cos A, \qquad \cos\left(A - \frac{\pi}{2}\right) = \sin A$$

$$\sin\left(A + \frac{\pi}{2}\right) = \cos A, \qquad \cos\left(A + \frac{\pi}{2}\right) = -\sin A$$

$\sin A\sin B = \frac{1}{2}\cos(A-B) - \frac{1}{2}\cos(A+B)$
$\cos A\cos B = \frac{1}{2}\cos(A-B) + \frac{1}{2}\cos(A+B)$
$\sin A\cos B = \frac{1}{2}\sin(A-B) + \frac{1}{2}\sin(A+B)$

$\sin A + \sin B = 2\sin\frac{1}{2}(A+B)\cos\frac{1}{2}(A-B)$
$\sin A - \sin B = 2\cos\frac{1}{2}(A+B)\sin\frac{1}{2}(A-B)$

$\cos A + \cos B = 2\cos\frac{1}{2}(A+B)\cos\frac{1}{2}(A-B)$
$\cos A - \cos B = -2\sin\frac{1}{2}(A+B)\sin\frac{1}{2}(A-B)$

3. Common Reference Triangles

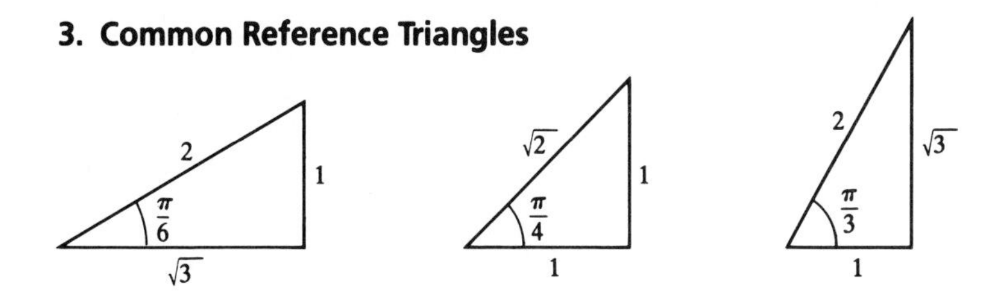

4. Angles and Sides of a Triangle

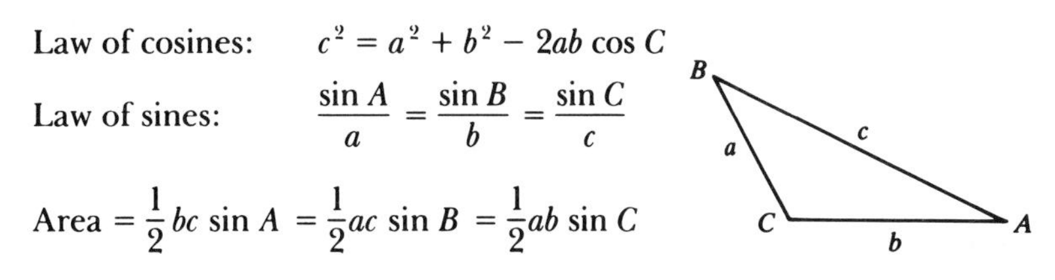

Law of cosines: $c^2 = a^2 + b^2 - 2ab\cos C$

Law of sines: $\dfrac{\sin A}{a} = \dfrac{\sin B}{b} = \dfrac{\sin C}{c}$

$$\text{Area} = \frac{1}{2}bc\sin A = \frac{1}{2}ac\sin B = \frac{1}{2}ab\sin C$$

A.2 Proofs of the Limit Theorems in Chapter 2

This appendix furnishes the ϵ–δ proofs of the limit theorems in Chapter 2.

THEOREM 1

> **If $\lim_{x\to x_0} f_1(x) = L_1$ and $\lim_{x\to x_0} f_2(x) = L_2$, then**
>
> 1. ***Sum Rule:*** $\lim\,(f_1(x) + f_2(x)) = L_1 + L_2$
> 2. ***Difference Rule:*** $\lim\,(f_1(x) - f_2(x)) = L_1 - L_2$
> 3. ***Product Rule:*** $\lim f_1(x)\cdot f_2(x) = L_1 \cdot L_2$
> 4. ***Constant Multiple Rule:*** $\lim k \cdot f_2(x) = k \cdot L_2$ **(any number k)**
> 5. ***Quotient Rule:*** $\lim \dfrac{f_1(x)}{f_2(x)} = \dfrac{L_1}{L_2}$ **if $L_2 \neq 0$.**
>
> **The limits are all to be taken as $x \to x_0$ in the appropriate domain (D), and L_1 and L_2 are to be real numbers.**

We proved the Sum Rule in Section 2.5, and we obtain the Difference Rule by replacing $f_2(x)$ by $(-f_2(x))$ and L_2 by $(-L_2)$ in the Sum Rule. The Constant Multiple Rule is the special case $f_1(x) = k$ of the Product Rule. This leaves only the Product and Quotient Rules to prove.

PROOF of the Limit Product Rule We need to show that for any $\epsilon > 0$ there exists a $\delta > 0$ such that for all x in D

$$0 < |x - x_0| < \delta \quad \Rightarrow \quad |f_1(x)f_2(x) - L_1L_2| < \epsilon. \tag{1}$$

Suppose then that ϵ is a positive number, and write $f_1(x)$ and $f_2(x)$ as

$$f_1(x) = L_1 + (f_1(x) - L_1) \qquad \text{and} \qquad f_2(x) = L_2 + (f_2(x) - L_2).$$

Multiply these expressions and subtract L_1L_2:

$$\begin{aligned} f_1(x) \cdot f_2(x) - L_1L_2 &= (L_1 + (f_1(x) - L_1))(L_2 + (f_2(x) - L_2)) - L_1L_2 \\ &= L_1L_2 + L_1(f_2(x) - L_2) + L_2(f_1(x) - L_1) \\ &\quad + (f_1(x) - L_1)(f_2(x) - L_2) - L_1L_2 \\ &= L_1(f_2(x) - L_2) + L_2(f_1(x) - L_1) + (f_1(x) - L_1)(f_2(x) - L_2). \end{aligned} \tag{2}$$

Since f_1 and f_2 have limits L_1 and L_2 as $x \to x_0$, there exist positive numbers δ_1, δ_2, δ_3, and δ_4 such that for all x

$$\begin{aligned} 0 < |x - x_0| < \delta_1 \quad &\Rightarrow \quad |f_1(x) - L_1| < \sqrt{\epsilon/3}, \\ 0 < |x - x_0| < \delta_2 \quad &\Rightarrow \quad |f_2(x) - L_2| < \sqrt{\epsilon/3}, \\ 0 < |x - x_0| < \delta_3 \quad &\Rightarrow \quad |f_1(x) - L_1| < \epsilon/(3(1 + |L_2|)), \\ 0 < |x - x_0| < \delta_4 \quad &\Rightarrow \quad |f_2(x) - L_2| < \epsilon/(3(1 + |L_1|)). \end{aligned} \tag{3}$$

All four of the inequalities on the right-hand side of (3) will hold for $0 < |x - x_0| < \delta$ if we take δ to be the smallest of the numbers δ_1 through δ_4. Therefore for all x, $0 < |x - x_0| < \delta$ implies

$$\begin{aligned} &|f_1(x) \cdot f_2(x) - L_1L_2| \\ &\qquad \le |L_1|\,|f_2(x) - L_2| + |L_2|\,|f_1(x) - L_1| + |f_1(x) - L_1|\,|f_2(x) - L_2| \\ &\qquad\qquad \text{(Triangle inequality applied to Eq. (2))} \\ &\le (1 + |L_1|)|f_2(x) - L_2| + (1 + |L_2|)|f_1(x) - L_1| + |f_1(x) - L_1|\,|f_2(x) - L_2| \\ &\le \frac{\epsilon}{3} + \frac{\epsilon}{3} + \sqrt{\frac{\epsilon}{3}}\sqrt{\frac{\epsilon}{3}} = \epsilon. \qquad \text{(Values from (3))} \end{aligned}$$

This completes the proof of the Limit Product Rule. ■

PROOF of the Limit Quotient Rule We show that

$$\lim_{x \to x_0} \frac{1}{f_2(x)} = \frac{1}{L_2}.$$

Then we can apply the Limit Product Rule to show that

$$\lim_{x \to x_0} \frac{f_1(x)}{f_2(x)} = \lim_{x \to x_0} f_1(x) \cdot \frac{1}{f_2(x)} = \lim_{x \to x_0} f_1(x) \cdot \lim_{x \to x_0} \frac{1}{f_2(x)} = L_1 \cdot \frac{1}{L_2} = \frac{L_1}{L_2}.$$

To show that $\lim_{x \to x_0}(1/f_2(x)) = 1/L_2$, we need to show that for any $\epsilon > 0$ there exists a $\delta > 0$ such that for all x

$$0 < |x - x_0| < \delta \quad \Rightarrow \quad \left|\frac{1}{f_2(x)} - \frac{1}{L_2}\right| < \epsilon.$$

Since $|L_2| > 0$, there exists a positive number δ_1 such that for all x

$$0 < |x - x_0| < \delta_1 \quad \Rightarrow \quad |f_2(x) - L_2| < \frac{|L_2|}{2}. \tag{4}$$

For any numbers A and B it can be shown that $|A| - |B| \leq |A - B|$ and $|B| - |A| \leq |A - B|$, from which it follows that

$$||A| - |B|| \leq |A - B|. \tag{5}$$

With $A = f_2(x)$ and $B = L_2$, this gives

$$||f_2(x)| - |L_2|| \leq |f_2(x) - L_2|,$$

which we can combine with the right-hand inequality in (4) to get, in turn,

$$||f_2(x)| - |L_2|| < \frac{|L_2|}{2},$$

$$-\frac{|L_2|}{2} < |f_2(x)| - |L_2| < \frac{|L_2|}{2},$$

$$\frac{|L_2|}{2} < |f_2(x)| < \frac{3|L_2|}{2},$$

$$|L_2| < 2|f_2(x)| < 3|L_2|,$$

$$\frac{1}{|f_2(x)|} < \frac{2}{|L_2|} < \frac{3}{|f_2(x)|}. \tag{6}$$

Therefore $0 < |x - x_0| < \delta_1$ implies that

$$\left|\frac{1}{f_2(x)} - \frac{1}{L_2}\right| = \left|\frac{L_2 - f_2(x)}{L_2 f_2(x)}\right| \leq \frac{1}{|L_2|} \cdot \frac{1}{|f_2(x)|} \cdot |L_2 - f_2(x)|$$

$$< \frac{1}{|L_2|} \cdot \frac{2}{|L_2|} \cdot |L_2 - f_2(x)|. \qquad \text{(Eq. 6)}$$

Suppose now that ϵ is an arbitrary positive number. Then $\frac{1}{2}|L_2|^2 \epsilon > 0$, so there exists a number $\delta_2 > 0$ such that for all x

$$0 < |x - x_0| < \delta_2 \quad \Rightarrow \quad |L_2 - f_2(x)| < \frac{\epsilon}{2}|L_2|^2. \tag{7}$$

The conclusions in (6) and (7) both hold for all x such that $0 < |x - x_0| < \delta$ if we take δ to be the smaller of the positive values δ_1 and δ_2. Combining (6) and (7) then gives

$$0 < |x - x_0| < \delta \quad \Rightarrow \quad \left|\frac{1}{f_2(x)} - \frac{1}{L_2}\right| < \epsilon. \tag{8}$$

This concludes the proof of the Limit Quotient Rule. ■

THEOREM 2

The Sandwich Theorem
Suppose that $g(x) \leq f(x) \leq h(x)$ for all $x \neq x_0$ in some interval about x_0 and that $\lim_{x\to x_0} g(x) = \lim_{x\to x_0} h(x) = L$. Then $\lim_{x\to x_0} f(x) = L$.

PROOF for Right-hand Limits Suppose $\lim_{x\to x_0^+} g(x) = \lim_{x\to x_0^+} h(x) = L$. Then for any $\epsilon > 0$ there exists a $\delta > 0$ such that for all x the inequality $x_0 < x < x_0 + \delta$ implies

$$L - \epsilon < g(x) < L + \epsilon \quad \text{and} \quad L - \epsilon < h(x) < L + \epsilon. \tag{9}$$

These inequalities combine with the inequality $g(x) \le f(x) \le h(x)$ to give

$$L - \epsilon < g(x) \le f(x) \le h(x) < L + \epsilon, \tag{10}$$

$$L - \epsilon < f(x) < L + \epsilon,$$

$$-\epsilon < f(x) - L < \epsilon.$$

Therefore, for all x the inequality $x_0 < x < x_0 + \delta$ implies $|f(x) - L| < \epsilon$.

PROOF for Left-hand Limits Suppose $\lim_{x\to x_0^-} g(x) = \lim_{x\to x_0^-} h(x) = L$. Then for any $\epsilon > 0$ there exists a $\delta > 0$ such that for all x the inequality $x_0 - \delta < x < x_0$ implies

$$L - \epsilon < g(x) < L + \epsilon \quad \text{and} \quad L - \epsilon < h(x) < L + \epsilon.$$

We conclude as before that for all x the inequality $x_0 - \delta < x < x_0$ implies $|f(x) - L| < \epsilon$.

PROOF for Two-sided Limits If $\lim_{x\to x_0} g(x) = \lim_{x\to x_0} h(x) = L$, then $g(x)$ and $h(x)$ both approach L as $x \to x_0^+$ and as $x \to x_0^-$; so $\lim_{x\to x_0^+} f(x) = L$ and $\lim_{x\to x_0^-} f(x) = L$. Hence $\lim_{x\to x_0} f(x)$ exists and equals L. ■

Exercises A.2

1. Suppose that functions $f_1(x)$, $f_2(x)$, and $f_3(x)$ have limits L_1, L_2, and L_3, respectively, as $x \to x_0$. Show that their sum has limit $L_1 + L_2 + L_3$. Use mathematical induction (Appendix A.4) to generalize this result to the sum of any finite number of functions.

2. Use mathematical induction and the Limit Product Rule in Theorem 1 to show that if functions $f_1(x)$, $f_2(x)$, $\ldots$, $f_n(x)$ have limits $L_1, L_2, \ldots, L_n$ as $x \to x_0$, then $\lim_{x\to x_0} f_1(x) f_2(x) \cdot \cdots \cdot f_n(x) = L_1 \cdot L_2 \cdot \cdots \cdot L_n$.

3. Use the fact that $\lim_{x\to x_0} x = x_0$ and the result of Exercise 2 to show that $\lim_{x\to x_0} x^n = x_0^n$ for any integer $n > 1$.

4. Use the fact that $\lim_{x\to x_0} (k) = k$ for any number k together with the results of Exercises 1 and 3 to show that $\lim_{x\to x_0} f(x) = f(x_0)$ for any polynomial function
$$f(x) = a_0 x^n + a_1 x^{n-1} + \cdots + a_{n-1} x + a_n.$$

5. Use Theorem 1 and the result of Exercise 4 to show that if $f(x)$ and $g(x)$ are polynomial functions and $g(x_0) \ne 0$ then
$$\lim_{x\to x_0} \frac{f(x)}{g(x)} = \frac{f(x_0)}{g(x_0)}.$$

6. Figure A.1 gives the diagram for a proof that the composite of two continuous functions is continuous. Reconstruct the proof from the diagram. The statement to be proved is this: If g is continuous at $x = x_0$ and f is continuous at $g(x_0)$ then $f \circ g$ is continuous at x_0.

   Assume that x_0 is an interior point of the domain of g and that $g(x_0)$ is an interior point of the domain of f. This will make the limits involved two-sided. (The arguments for the cases that involve one-sided limits are similar.)

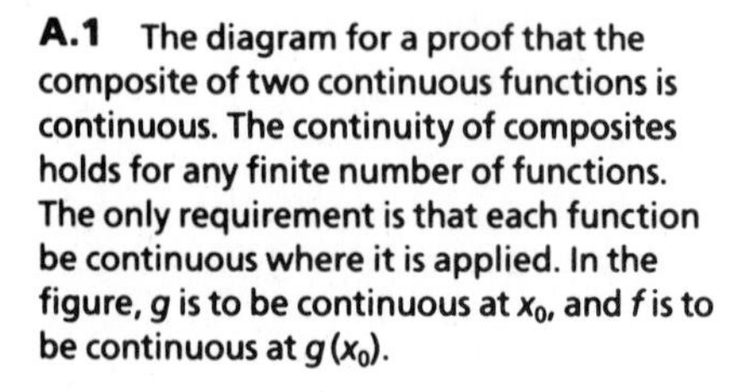

A.1 The diagram for a proof that the composite of two continuous functions is continuous. The continuity of composites holds for any finite number of functions. The only requirement is that each function be continuous where it is applied. In the figure, g is to be continuous at x_0, and f is to be continuous at $g(x_0)$.

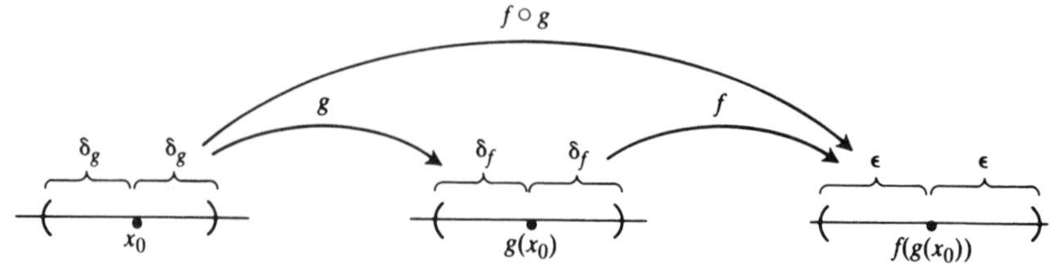

A.3 A Proof of the Chain Rule for Functions of a Single Variable

Our goal is to show that if $f(u)$ is a differentiable function of u and $u = g(x)$ is a differentiable function of x then the composite $y = f(g(x))$ is a differentiable function of x. More precisely, if g is differentiable at x_0 and f is differentiable at $g(x_0)$, then the composite is differentiable at x_0 and

$$\left.\frac{dy}{dx}\right|_{x=x_0} = f'(g(x_0)) \cdot g'(x_0). \tag{1}$$

Suppose that Δx is an increment in x and that Δu and Δy are the corresponding increments in u and y. The following figure is drawn with the increments all positive, but Δx could be negative and Δu and Δy could be negative or even zero.

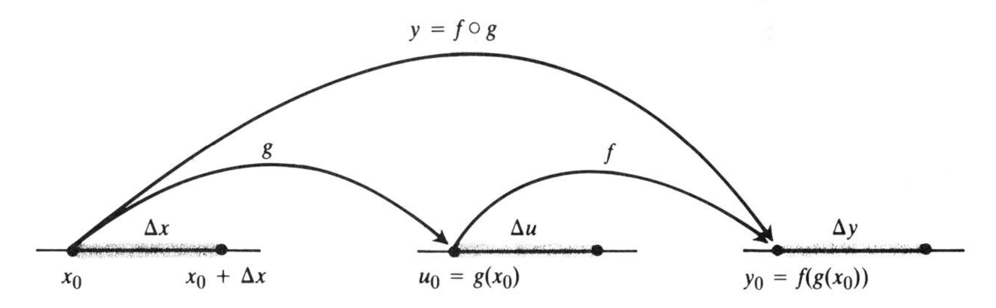

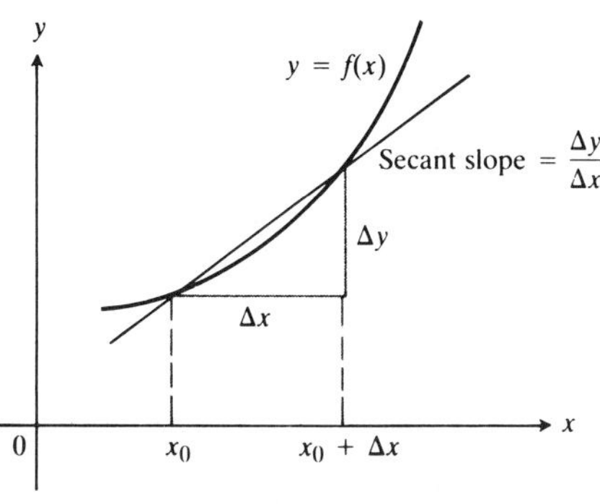

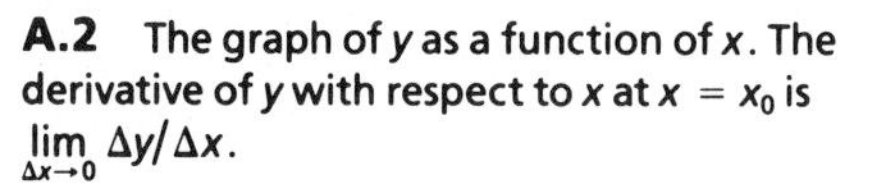
A.2 The graph of y as a function of x. The derivative of y with respect to x at $x = x_0$ is $\lim_{\Delta x \to 0} \Delta y/\Delta x$.

As you can see in Fig. A.2

$$\left.\frac{dy}{dx}\right|_{x=x_0} = \lim_{\Delta x \to 0} \frac{\Delta y}{\Delta x},$$

so our goal is to show that this limit is $f'(g(x_0)) \cdot g'(x_0)$.

By virtue of Eq. (12) in Section 3.7,

$$\Delta u = g'(x_0)\Delta x + \epsilon_1 \Delta x = [g'(x_0) + \epsilon_1]\,\Delta x, \tag{2}$$

where $\epsilon_1 \to 0$ as $\Delta x \to 0$. Similarly,

$$\Delta y = f'(u_0)\Delta u + \epsilon_2 \Delta u = [f'(u_0) + \epsilon_2]\,\Delta u, \tag{3}$$

where $\epsilon_2 \to 0$ as $\Delta u \to 0$. Combining the equation for Δu with the one for Δy gives

$$\Delta y = [f'(u_0) + \epsilon_2][g'(x_0) + \epsilon_1]\,\Delta x, \tag{4}$$

so

$$\frac{\Delta y}{\Delta x} = f'(u_0)g'(x_0) + \epsilon_2\, g'(x_0) + f'(u_0)\epsilon_1 + \epsilon_2\epsilon_1.$$

Since ϵ_1 and ϵ_2 go to zero as Δx goes to zero, the three terms on the right vanish in the limit and

$$\lim_{\Delta x \to 0} \frac{\Delta y}{\Delta x} = f'(u_0)g'(x_0) = f'(g(x_0)) \cdot g'(x_0).$$

This concludes the proof. ■

A.4 Mathematical Induction

Many formulas, like

$$1 + 2 + \cdots + n = \frac{n(n + 1)}{2},$$

can be shown to hold for every positive integer n by applying an axiom called the *mathematical induction principle.* A proof that uses this axiom is called a *proof by mathematical induction* or a *proof by induction.*

The steps in proving a formula by induction are

STEP 1: Check that it holds for $n = 1$.

STEP 2: Prove that if it holds for any positive integer $n = k$, then it also holds for $n = k + 1$.

Once these steps are completed (the axiom says), we know that the formula holds for all positive integers n. By Step 1 it holds for $n = 1$. By Step 2 it holds for $n = 2$, and therefore by Step 2 also for $n = 3$, and by Step 2 again for $n = 4$, and so on. If the first domino falls, and the kth domino always knocks over the $(k + 1)$st when it falls, all the dominoes fall.

From another point of view, suppose we have a sequence of statements

$$S_1, S_2, \ldots, S_n, \ldots,$$

one for each positive integer. Suppose we can show that assuming any one of the statements to be true implies that the next statement in line is true. Suppose that we can also show that S_1 is true. Then we may conclude that the statements are true from S_1 on.

EXAMPLE 1 Show that for every positive integer n

$$1 + 2 + \cdots + n = \frac{n(n + 1)}{2}.$$

Solution We accomplish the proof by carrying out the two steps of mathematical induction.

STEP 1: The formula holds for $n = 1$ because

$$1 = \frac{1(1 + 1)}{2}.$$

STEP 2: If the formula holds for $n = k$, does it also hold for $n = k + 1$? The answer is yes, and here's why: If

$$1 + 2 + \cdots + k = \frac{k(k + 1)}{2},$$

then

$$1 + 2 + \cdots + k + (k + 1) = \frac{k(k + 1)}{2} + (k + 1) = \frac{k^2 + k + 2k + 2}{2}$$

$$= \frac{(k + 1)(k + 2)}{2} = \frac{(k + 1)((k + 1) + 1)}{2}.$$

The last expression in this string of equalities is the expression $n(n + 1)/2$ for $n = (k + 1)$.

The mathematical induction principle now guarantees the original formula for all positive integers n.

Notice that all *we* have to do is carry out Steps 1 and 2. The mathematical induction principle does the rest. ■

EXAMPLE 2 Show that for all positive integers n

$$\frac{1}{2^1} + \frac{1}{2^2} + \cdots + \frac{1}{2^n} = 1 - \frac{1}{2^n}.$$

Solution We accomplish the proof by carrying out the two steps of mathematical induction.

STEP 1: The formula holds for $n = 1$ because

$$\frac{1}{2^1} = 1 - \frac{1}{2^1}.$$

STEP 2: If

$$\frac{1}{2^1} + \frac{1}{2^2} + \cdots + \frac{1}{2^k} = 1 - \frac{1}{2^k},$$

then

$$\frac{1}{2^1} + \frac{1}{2^2} + \cdots + \frac{1}{2^k} + \frac{1}{2^{k+1}} = 1 - \frac{1}{2^k} + \frac{1}{2^{k+1}} = 1 - \frac{1 \cdot 2}{2^k \cdot 2} + \frac{1}{2^{k+1}}$$

$$= 1 - \frac{2}{2^{k+1}} + \frac{1}{2^{k+1}} = 1 - \frac{1}{2^{k+1}}.$$

Thus, the original formula holds for $n = k + 1$ whenever it holds for $n = k$.

With these two steps verified, the mathematical induction principle now guarantees the formula for every positive integer n. ■

Other Starting Integers

Instead of starting at $n = 1$, some induction arguments start at another integer. The steps for such an argument are

STEP 1: Check that the formula holds for $n = n_1$ (whatever the appropriate first integer is).

STEP 2: Prove that if the formula holds for any integer $n = k \geq n_1$, then it also holds for $n = k + 1$.

Once these steps are completed, the mathematical induction principle will guarantee the formula for all $n \geq n_1$.

EXAMPLE 3 Show that $n! > 3^n$ if n is large enough.

Solution How large is large enough? We experiment:

n	1	2	3	4	5	6	7
$n!$	1	2	6	24	120	720	5040
3^n	3	9	27	81	243	729	2187

It looks as if $n! > 3^n$ for $n \geq 7$. To be sure, we apply mathematical induction. We take $n_1 = 7$ in Step 1 and try for Step 2.

Suppose $k! > 3^k$ for some $k \geq 7$. Then

$$(k+1)! = (k+1)(k!) > (k+1)3^k > 7 \cdot 3^k > 3^{k+1}.$$

Thus, for $k \geq 7$,

$$k! > 3^k \Rightarrow (k+1)! > 3^{k+1}.$$

The mathematical induction principle now guarantees $n! \geq 3^n$ for all $n \geq 7$. ■

Exercises A.4

1. Assuming that the triangle inequality $|a+b| \leq |a| + |b|$ holds for any two numbers a and b, show that
$$|x_1 + x_2 + \cdots + x_n| \leq |x_1| + |x_2| + \cdots + |x_n|$$
for any n numbers.

2. Show that if $r \neq 1$, then
$$1 + r + r^2 + \cdots + r^n = \frac{1 - r^{n+1}}{1 - r}$$
for all positive integers n.

3. Use the Product Rule
$$\frac{d}{dx}(uv) = u\frac{dv}{dx} + v\frac{du}{dx}$$
and the fact that
$$\frac{d}{dx}(x) = 1$$
to show that
$$\frac{d}{dx}(x^n) = nx^{n-1}$$
for all positive integers n.

4. Suppose that a function $f(x)$ has the property that $f(x_1x_2) = f(x_1) + f(x_2)$ for any two positive numbers x_1 and x_2. Show that
$$f(x_1x_2\cdots x_n) = f(x_1) + f(x_2) + \cdots + f(x_n)$$
for the product of any n positive numbers $x_1, x_2, \ldots, x_n$.

5. Show that
$$\frac{2}{3^1} + \frac{2}{3^2} + \cdots + \frac{2}{3^n} = 1 - \frac{1}{3^n}$$
for all positive integers n.

6. Show that $n! > n^3$ if n is large enough.

7. Show that $2^n > n^2$ if n is large enough.

8. Show that $2^n \geq 1/8$ for $n \geq -3$.

9. Show that the sum of the squares of the first n positive integers is $n(n+1)(2n+1)/6$.

10. Show that the sum of the cubes of the first n positive integers is $(n(n+1)/2)^2$.

11. Show that the following finite-sum rules hold for every positive integer n.

a) $\sum_{k=1}^{n} (a_k + b_k) = \sum_{k=1}^{n} a_k + \sum_{k=1}^{n} b_k$

b) $\sum_{k=1}^{n} (a_k - b_k) = \sum_{k=1}^{n} a_k - \sum_{k=1}^{n} b_k$

c) $\sum_{k=1}^{n} ca_k = c \cdot \sum_{k=1}^{n} a_k$ (Any number c)

d) $\sum_{k=1}^{n} a_k = n \cdot c$ if a_k has the constant value c.

A.5 Limits That Arise Frequently

This appendix verifies the limits in Table 9.1 of Section 9.1.

1. $\lim_{n\to\infty} \frac{\ln n}{n} = 0$	2. $\lim_{n\to\infty} \sqrt[n]{n} = 1$
3. $\lim_{n\to\infty} x^{1/n} = 1 \quad (x > 0)$	4. $\lim_{n\to\infty} x^n = 0 \quad (\lvert x\rvert < 1)$
5. $\lim_{n\to\infty} \left(1 + \frac{x}{n}\right)^n = e^x \quad (\text{Any } x)$	6. $\lim_{n\to\infty} \frac{x^n}{n!} = 0 \quad (\text{Any } x)$

In Eqs. (3)–(6), x remains fixed while n varies.

1. $\lim_{n\to\infty} \frac{\ln n}{n} = 0$ We proved this in Section 9.1, Example 9.

2. $\lim_{n\to\infty} \sqrt[n]{n} = 1$ Let $a_n = n^{1/n}$. Then

$$\ln a_n = \ln n^{1/n} = \frac{1}{n} \ln n \to 0. \tag{1}$$

Applying Theorem 3, Section 9.1, with $f(x) = e^x$ gives

$$n^{1/n} = a_n = e^{\ln a_n} = f(\ln a_n) \to f(0) = e^0 = 1. \tag{2}$$

3. If $x > 0$, $\lim_{n\to\infty} x^{1/n} = 1$ Let $a_n = x^{1/n}$. Then

$$\ln a_n = \ln x^{1/n} = \frac{1}{n} \ln x \to 0 \tag{3}$$

because x remains fixed as $n \to \infty$. Applying Theorem 3, Section 9.1, with $f(x) = e^x$ gives

$$x^{1/n} = a_n = e^{\ln a_n} \to e^0 = 1. \tag{4}$$

4. If $|x| < 1$, $\lim_{n\to\infty} x^n = 0$ We need to show that to each $\epsilon > 0$ there corresponds an integer N so large that $|x^n| < \epsilon$ for all n greater than N. Since $\epsilon^{1/n} \to 1$, while $|x| < 1$, there exists an integer N for which

$$\epsilon^{1/N} > |x|. \tag{5}$$

In other words,

$$|x^N| = |x|^N < \epsilon. \tag{6}$$

This is the integer we seek because, if $|x| < 1$, then

$$|x^n| < |x^N| \quad \text{for all } n > N. \tag{7}$$

Combining (6) and (7) produces

$$|x^n| < \epsilon \quad \text{for all } n > N, \tag{8}$$

and we're done.

5. For any number x, $\lim_{n\to\infty}\left(1+\frac{x}{n}\right)^n = e^x$ Let

$$a_n = \left(1+\frac{x}{n}\right)^n.$$

Then $$\ln a_n = \ln\left(1+\frac{x}{n}\right)^n = n\ln\left(1+\frac{x}{n}\right)\to x,$$

as we can see by the following application of l'Hôpital's rule, in which we differentiate with respect to n:

$$\lim_{n\to\infty} n\ln\left(1+\frac{x}{n}\right) = \lim_{n\to\infty}\frac{\ln(1+x/n)}{1/n}$$

$$= \lim_{n\to\infty}\frac{\left(\frac{1}{1+x/n}\right)\cdot\left(-\frac{x}{n^2}\right)}{-1/n^2} = \lim_{n\to\infty}\frac{x}{1+x/n} = x.$$

Apply Theorem 3, Section 9.1, with $f(x) = e^x$ to conclude that

$$\left(1+\frac{x}{n}\right)^n = a_n = e^{\ln a_n}\to e^x.$$

6. For any number x, $\lim_{n\to\infty}\frac{x^n}{n!} = 0$ Since

$$-\frac{|x|^n}{n!}\le\frac{x^n}{n!}\le\frac{|x|^n}{n!},$$

all we need to show is that $|x|^n/n!\to 0$. We can then apply the Sandwich Theorem for Sequences (Section 9.1, Theorem 2) to conclude that $x^n/n!\to 0$.

The first step in showing that $|x|^n/n!\to 0$ is to choose an integer $M > |x|$, so that

$$\frac{|x|}{M} < 1 \qquad\text{and}\qquad \left(\frac{|x|}{M}\right)^n\to 0.$$

We then restrict our attention to values of $n > M$. For these values of n, we can write

$$\frac{|x|^n}{n!} = \frac{|x|^n}{1\cdot 2\cdot\cdots\cdot M\cdot\underbrace{(M+1)(M+2)\cdot\cdots\cdot n}_{(n-M)\text{ factors}}}$$

$$\le\frac{|x|^n}{M!\,M^{n-M}} = \frac{|x|^n M^M}{M!\,M^n} = \frac{M^M}{M!}\left(\frac{|x|}{M}\right)^n.$$

Thus, $$0\le\frac{|x|^n}{n!}\le\frac{M^M}{M!}\left(\frac{|x|}{M}\right)^n.$$

Now, the constant $M^M/M!$ does not change as n increases. Thus the Sandwich Theorem tells us that

$$\frac{|x|^n}{n!}\to 0 \qquad\text{because}\ \left(\frac{|x|}{M}\right)^n\to 0.$$

Answers

Chapter 1

Section 1.1

1. (0, 6), (5, −5) **3.** [−17, 21] by [−12, 76] **5.** One choice: $x\,Scl = 5$, $y\,Scl = 5$ gives many scale marks at convenient points which are still distinguishable. **7.** $x\,Scl = 0.05$, $y\,Scl = 10$ **9.** (e) **11.** (e) **13.** (e)
15. Intercepts: $x = -1$, $y = 1$.

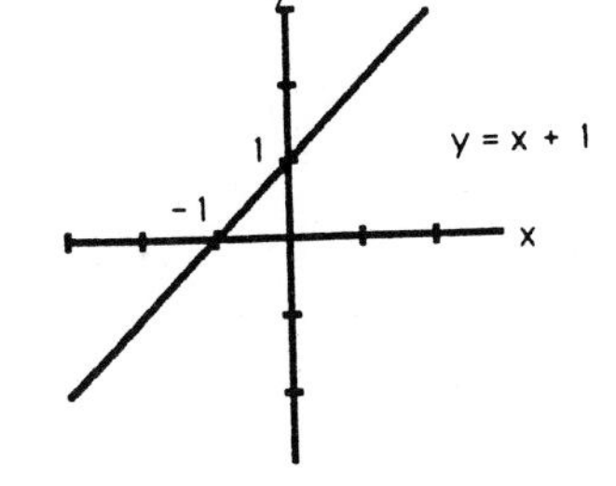

17. $y = -x^2$. (0, 0) gives the only intercepts.

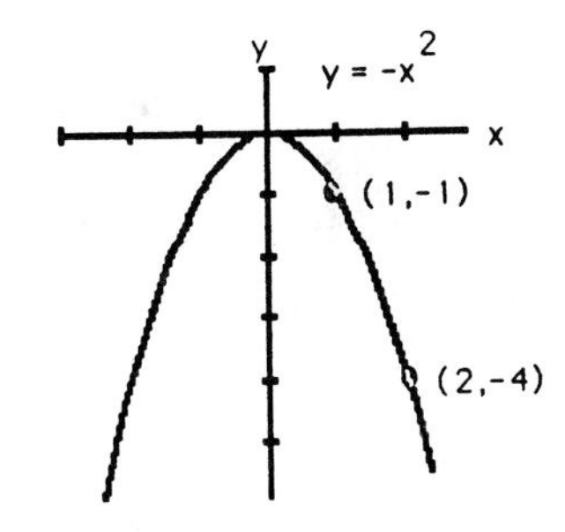

19. $x = -y^2$. (0, 0) gives the only intercepts.

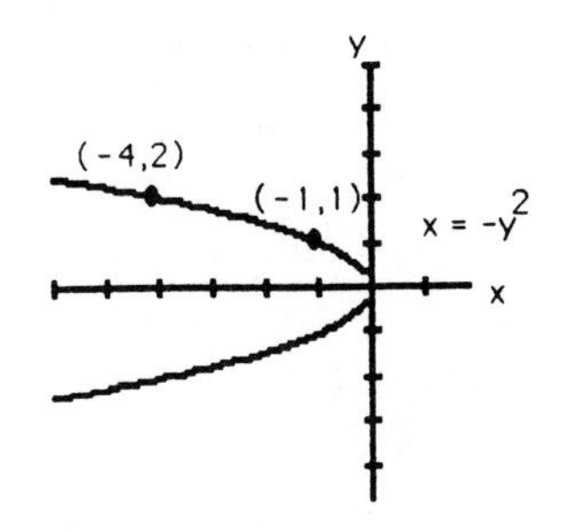

21. A complete graph of $y = 3x - 5$ may be viewed in the rectangle $[-2,4]$ by $[-10,5]$. x-intercept: $x = 5/3$. y-intercept: $y = -5$. **23.** A graph of $y = 10 + x - 2x^2$ may be viewed in the rectangle $[-6,6]$ by $[-30,15]$. Intercepts: $x = -2,\ 5/2$; $y = 10$. **25.** A graph of $y = 2x^2 - 8x + 3$ may be viewed in the rectangle $[-7,10]$ by $[-20,70]$. x-intercepts: $x \approx 0.42,\ 3.58$. y-intercept: $y = 3$. **27.** Using TRACE on the graph of $y = x^2 + 4x + 5 = (x+2)^2 + 1$ in the rectangle $[-8,4]$ by $[-2,20]$, we see there are no x-intercepts, $y = 5$ is the y-intercept and $(-2,1)$ is the low point. **29.** We graph $y = 12x - 3x^3$ in the rectangle $[-4,4]$ by $[-15,15]$. $x = \pm 2, 0$ are the x-intercepts, $y = 0$ is the y-intercept. Using TRACE, $(-1.14,-9.24)$ and $(1.14,9.24)$ are the approximate local low and high points, respectively. **31.** A complete graph of $y = -x^3 + 9x - 1$ can be obtained in the viewing rectangle $[-5,5]$ by $[-35,35]$. The y-intercept is -1. Use of TRACE yields the following approximations. x-intercepts: -3.02, 0.08, 2.93; low point: $(-1.75,-11.39)$; high point: $(1.75,9.39)$. **33.** An idea of a complete graph of $y = x^3 + 2x^2 + x + 5$ can be obtained by using the viewing rectangle $[-3,2]$ by $[-2,10]$ and $[-2,1]$ by $[4,6]$. The y-intercept is 5 and using TRACE in the first rectangle, we obtain -2.44 as the approximate x-intercept. In the second rectangle, we obtain $(-1,5)$ and $(-0.33,4.85)$ as the approximate local high and low points. **35.** With $(N,M) = (126,62)$, $a = 116$, $b = 52$ **37.** With $(N,M) = (126,62)$, $a = 53$, $b = 52$ **39.** a is closer to x_i than it is to $x_i \pm \Delta x$. Thus $x_i - \Delta x/2 \leqq a \leqq x_i + \Delta x/2$ so $-\Delta x/2 \leqq a - x_i \leqq \Delta x/2$. That is, $|a - x_i| \leqq \Delta x/2$. **41.** Yes

Chapter 1

Section 1.2

1. $\Delta x = -2,\ \Delta y = -3$ **3.** $\Delta x = -5,\ \Delta y = 0$ **5.** Slope is 3; slope of lines perpendicular is $-1/3$. **7.** $m = 0$. The perpendicular lines are vertical and have no slope. **9.** $\sqrt{2}$ **11.** 6 **13.** $\sqrt{a^2+b^2}$ **15.** 3 **17.** 5 **19.** 6 **21.** a) $x = 2$ b) $y = 3$ **23.** a) $x = 0$ b) $y = -\sqrt{2}$ **25.** $y = x$ **27.** $y = x+2$ **29.** $y = 2x+b$ **31.** $y = \frac{3}{2}x$ **33.** $x = 1$ **35.** $x = -2$ **37.** $y = 3x - 2$ **39.** $y = x + \sqrt{2}$ **41.** $y = -5x + 2.5$ **43.** x-intercept is 4, y-intercept is 3 **45.** x-intercept: $x = 3$, y-intercept: $y = -4$ **47.** x-intercept: $x = -2$, y-intercept: $y = 4$ **49.** Perpendicular: $y = x$. $y = -x + 2$ and $y = x$ may be viewed in the rectangle $[-6,6]$ by $[-4,4]$. Distance $= \sqrt{2}$ **51.** Perpendicular: $y = 2x$. $y = -\frac{1}{2}x + \frac{3}{2}$ and $y = 2x$ may be viewed in the rectangle $[-6,6]$ by $[-4,4]$. Distance $= 2\sqrt{5}/5$. **53.** Perpendicular: $y = x + 3$. $y = -x + 3$ and $y = x + 3$ may be viewed in the rectangle $[-7,8]$ by $[-3,7]$. Distance $= 3\sqrt{2}$ **55.** $y = x - 1$. $y = x + 2$ and $y = x - 1$ may graphed in $[-10,10]$ by $[-10,10]$. **57.** $y = -2x + 2$ **59.** $(3,-3)$

61. $x = -2$, $y = -9$ **63.** $\pi/4$ **65.** $A(-12, 2)$, $B(-12, -5)$, $C(9, -5)$ **67.** $b = 2$ **69.** a) $a = -2$ b) $a \geqq 0$ **71.** a) $|x - 3|$ b) $|x + 2|$ **73.** a) $-2.5°/\text{in}$ b) $-16.1°/\text{in}$ c) $-8\frac{1}{3}°/\text{in}$ **75.** 5.97 atmospheres **77.** $-40°$C is equivalent to $-40°$F

79. The possibilities for the third vertex are $(-1, 4)$, $(-1, -2)$, $(5, 2)$

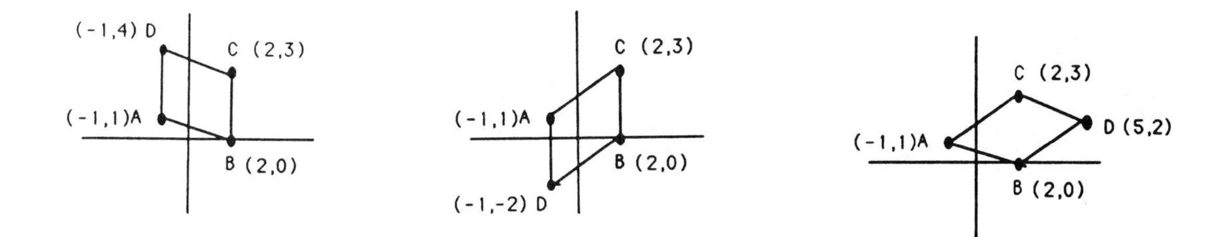

81. -2; 2

Chapter 1

Section 1.3

1. $(3, -1)$, $(-3, 1)$, $(-3, -1)$ **3.** $(-2, -1)$, $(2, 1)$, $(2, -1)$ **5.** $(1, \sqrt{2})$, $(-1, -\sqrt{2})$, $(-1, \sqrt{2})$ **7.** $(0, -\pi)$, $(0, \pi)$, $(0, -\pi)$ **9.** Domain: $[1, \infty)$, range: $[2, \infty)$ **11.** Domain: $(-\infty, 0]$, range: $(-\infty, 0]$ **13.** Domain: $(-\infty, 3]$, range: $[0, \infty)$ **15.** Domain: $(-\infty, 2) \cup (2, \infty)$, range: $(-\infty, 0) \cup (0, \infty)$ **17.** Odd **19.** Neither **21.** Even **23.** Even **25.** Odd **27.** Domain: $(-\infty, \infty)$, range: $[-9, \infty)$. Symmetric with respect to the y-axis. **29.** Domain = range = $(-\infty, \infty)$. No symmetry **31.** Domain = range = $(-\infty, \infty)$. No symmetry **33.** Domain = range = $(-\infty, 0) \cup (0, \infty)$. Symmetric with respect to the origin. **35.** Domain: $(-\infty, 0) \cup (0, \infty)$, range: $(-\infty, 1) \cup (1, \infty)$. No symmetry **37.** a) No b) No c) $(0, \infty)$ **39.** Symmetric with respect to the y-axis. Graph $y = -x^2$ in the viewing rectangle $[-10, 10]$ by $[-10, 0]$. **41.** Symmetric with respect to the y-axis. Graph $y = 1/x^2$ in the viewing rectangle $[-5, 5]$ by $[0, 3]$. **43.** Symmetric with respect to the origin. Graph $y = 1/x$ in the viewing rectangle $[-4, 4]$ by $[-4, 4]$. **45.** $x^2y^2 = 1$ has graph symmetric to both axes and the origin. Graph $y = 1/|x|$ and $y = -1/|x|$ in the viewing rectangle $[-4, 4]$ by $[-4, 4]$. **47.** Graph $y = abs(x + 3)$ in the viewing rectangle $[-7, 1]$ by $[0, 4]$. **49.** Graph $y = \frac{|x|}{x}$ in the viewing rectangle $[-2, 2]$ by $[-2, 2]$. There is no point on the graph when $x = 0$. **51.** $y = x$ when $x \leqq 0$ and $y = 0$ when $x \geqq 0$. **53.** Graph $y = 3 - x + 0\sqrt{1 - x}$ and $y = 2x + 0\sqrt{x - 1}$ in the viewing rectangle $[-10, 10]$ by $[0, 20]$. **55.** Graph $y = 1 + 0\sqrt{5 - x}$ in the viewing rectangle.

It is understood that the x-axis for $x \geqq 5$ is part of the graph and the point $(5, 1)$ is not. **57.** Graph $y = 4 - x^2 + 0\sqrt{1-x}$, $y = \frac{3}{2}x + \frac{3}{2} + 0\sqrt{x-1} + 0\sqrt{3-x}$ and $y = x + 3 + 0\sqrt{x-3}$ in the viewing rectangle $[-3, 7]$ by $[-5, 10]$. **59.** a) $0 \leqq x < 1$ b) $-1 < x \leqq 0$
61. a) Graph of $y = x - \lfloor x \rfloor$, $-3 \leqq x \leqq 3$ b) Graph of $y = \lfloor x \rfloor - \lceil x \rceil$, $-3 \leqq x \leqq 3$

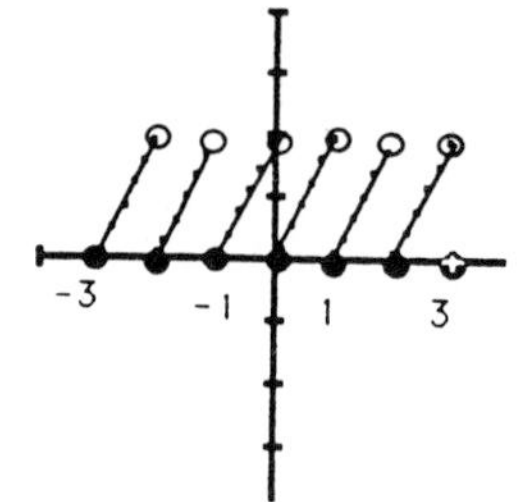

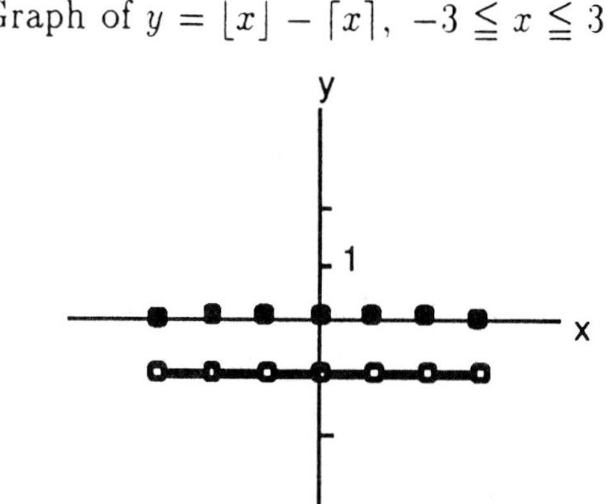

63. a) $1 - |x - 1|$, $0 \leqq x \leqq 2$ b) $y = 2$, $0 \leqq x < 1$, $2 \leqq x < 3$, $y = 0$, $1 \leqq x < 2$, $3 \leqq x \leqq 4$
65. Graph $y = abs(x+1) + 2abs(x-3)$ in the viewing rectangle $[-2, 5]$ by $[0, 12]$.

$$f(x) = \begin{cases} -3x + 5, & x \leqq -1 \\ -x + 7, & -1 < x \leqq 3 \\ 3x - 5, & x > 3 \end{cases}$$

67. Graph $y = |x| + |x - 1| + |x - 3|$ in the viewing rectangle $[-1, 4]$ by $[0, 7]$. $y = -3x + 4$, $x \leqq 0$, $y = -x + 4$, $0 < x \leqq 1$, $y = x + 2$, $1 < x \leqq 3$, $y = 3x - 4$, $x > 3$. **69.** $f(x) = x$, domain: $(-\infty, \infty)$. $g(x) = \sqrt{x-1}$, domain: $[1, \infty)$. $f(x) + g(x) = x + \sqrt{x-1}$, domain $(f+g) : [1, \infty)$, complete graph in $[0, 5]$ by $[0, 10]$. $f(x) - g(x) = x - \sqrt{x-1}$, domain $(f - g) : [1, \infty)$, complete graph in $[0, 5]$ by $[0, 4]$. $f \circ g(x) = f(g(x)) = f(\sqrt{x-1}) = \sqrt{x-1}$, domain $f \circ g : [1, \infty)$, complete graph in $[0, 5]$ by $[0, 2]$. $f(x)/g(x) = x/\sqrt{x-1}$, domain: $(1, \infty)$, complete graph in $[0, 10]$ by $[0, 5]$. $g(x)/f(x) = \sqrt{x-1}/x$, domain: $[1, \infty)$, complete graph in $[0, 10]$ by $[0, 0.5]$, graph starts at $(1, 0)$. **71.** a) 2 b) 22 c) $x^2 + 2$ d) $x^2 + 10x + 22$ e) 5 f) -2 g) $x + 10$ h) $x^4 - 6x^2 + 6$ **73.** a) $\sqrt{x-7}$ b) $3(x+2)$ c) x^2 d) x e) $\frac{1}{x-1}$ f) $\frac{1}{x}$ **75.** Domain $\sqrt{x^2}$ = domain $|x| = (-\infty, \infty)$, range $|x| = [0, \infty)$. Domain $(\sqrt{x})^2 = [0, \infty)$ = range $(\sqrt{x})^2$ **77.** $g(x) = \sqrt{x}$ is one possibility. **79.** Graph $y = abs(x+3) + abs(x-2) + abs(x-4)$ in the viewing rectangle $[-4, 5]$ by $[0, 15]$. We see that $d(x)$ is minimized when $x = 2$ so you would put the table next to Machine 2. **81.** $d(x)$ now has minimum value 17 when $x = 2$. The table should be placed next to Machine 3. **83.** Graph $y = (x^2)^{(1/3)} = (x^{(1/3)})^2$ in the viewing rectangle $[-5, 5]$ by $[-1, 5]$.

Our graphing utility does not give the complete graph with the instruction Graph $y = x^{(2/3)}$. (The graphing utility defines x^a as $e^{a \ln x}$ with domain $x > 0$ except when $a = \pm n^{\pm 1}$.)

Chapter 1

Section 1.4

1. a) $y = (x+4)^2$ b) $y = (x-7)^2$ **3.** a) Position 4 b) Position 1 c) Position 2 d) Position 3 **5.** Shift the graph of $|x|$ to the left 4 units. Then shift the resulting graph down 3 units. **7.** Reflect the graph of $y = \sqrt{x}$ over the y-axis and then stretch the resulting graph vertically by a factor of 3. **9.** Stretch the graph of $y = \frac{1}{x}$ vertically by a factor of 2 and shift the resulting graph down 3 units. **11.** Shift the graph of $y = x^2$ right 3 units. Shrink the resulting graph by a factor of 0.5. Reflect the last graph over the x-axis and shift the last graph up 1 unit. **13.** Shift the graph of $y = \frac{1}{x}$ to the left 2 units (obtaining the graph of $\frac{1}{x+2}$). Reflect the resulting graph over the y-axis (obtaining the graph of $y = \frac{1}{-x+2} = \frac{1}{2-x}$). Shift the resulting graph up 3 units. **15.** Graph the function in the viewing rectangle $[-8, 12]$ by $[-15, 5]$. Domain = range = $(-\infty, \infty)$ **17.** Sketch $y = \sqrt[3]{-x} - 1$ in the viewing rectangle $[-10, 10]$ by $[-4, 2]$. Domain = range = $(-\infty, \infty)$. **19.** Check your result by graphing $y = -(x+3)^{(-2)} + 2$ in the viewing rectangle $[-10, 5]$ by $[-3, 3]$. Domain = $(-\infty, -3) \cup (-3, \infty)$, range = $(-\infty, 2)$ **21.** Check your result by graphing $y = -2((x-1)^{(1/3)})^2 + 1$ in the viewing rectangle $[-1, 3]$ by $[-2, 1]$. Domain = $(-\infty, \infty)$, range = $(-\infty, 1]$. See comment in answer to 1.3 #83. **23.** Check your result by graphing $y = 2[1-x] = 2\,\text{Int}(1-x)$ in the viewing rectangle. (Graph this in Dot Mode if possible.) Domain = $(-\infty, \infty)$, range = $\{2n : n = 0, \pm 1, \pm 2, \ldots\}$. **25.** $y = -2(x-3)^2 + 7$, vertex: $(3, 7)$, axis of symmetry: $x = 3$. Check your result by graphing $y = -2x^2 + 12x - 11$ in the viewing rectangle $[-1, 7]$ by $[-16, 7]$. **27.** Vertex: $(-\frac{5}{2}, -6)$, axis of symmetry: $x = -\frac{5}{2}$. Check your graph by graphing the function in $[-5, 0]$ by $[-6, 10]$. **29.** $y = 3x^2 + 4$ **31.** $y = 0.2(\frac{1}{x} - 2)$ **33.** $y = 3|x+2| + 5$ **35.** $y = -0.8(x-1)^3 - 2$ **37.** $y = 5\sqrt{-(x+6)} + 5$ **39.** In #29 and #30 we obtain, respectively, $y = 3x^2 + 4$ and $y = 3(x^2 + 4) = 3x^2 + 12$. We obtain different geometric results by reversing the order of the transformations. The second graph turns out to be 8 units above the first graph. **41.** a) Reflect through the y-axis b) Reflect through the x-axis c) They are the same: $y = \sqrt[3]{-x} = -\sqrt[3]{x}$ **43.** $(2,3)$, $(3,2)$, $(4,3)$ **45.** $(-1,-2)$, $(0,0)$, $(1,-2)$

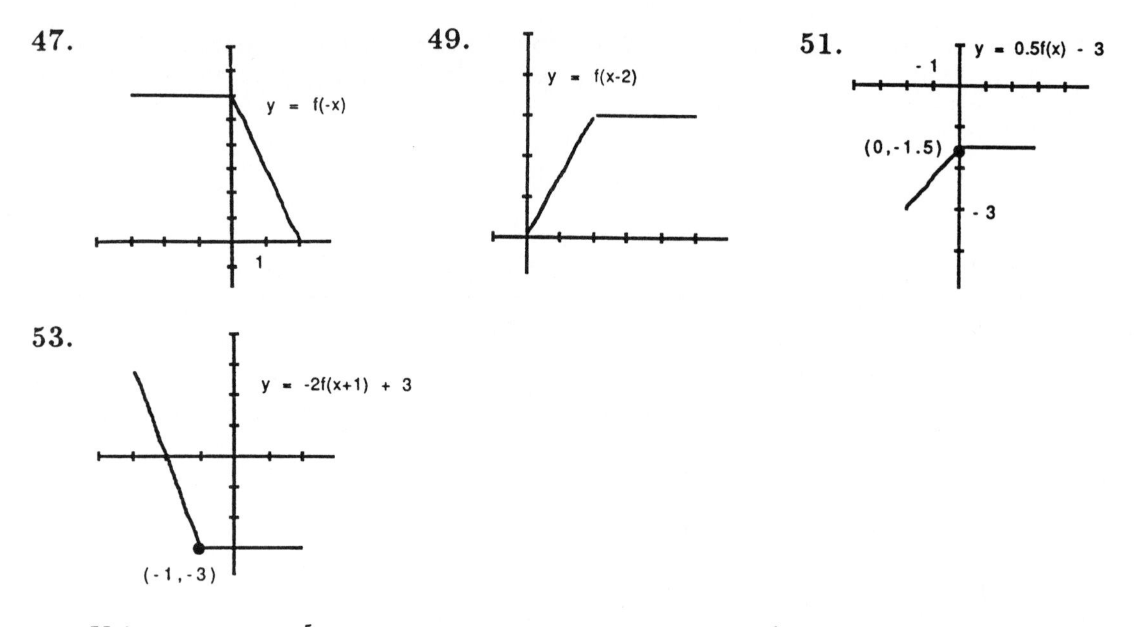

55. Using $y = 2 + \frac{5}{x-2}$, we start with the graph of $y = \frac{1}{x}$. Stretch vertically by a factor of 5, shift right 2 units, shift up 2.

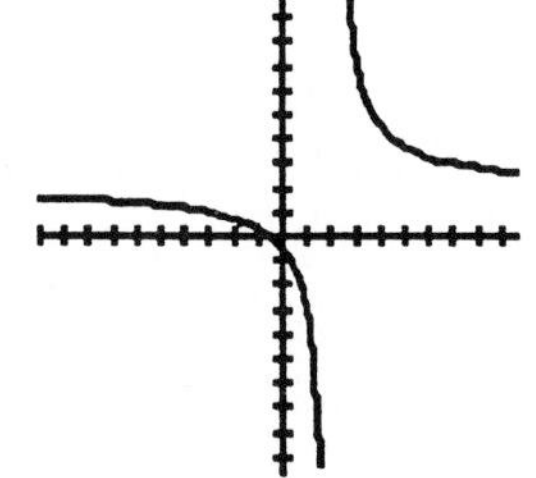

57. $x = (y-3)^2 + 2$. Shift up 3, shift right 2. Check your sketch by graphing $y = 3 + \sqrt{x-2}$ and $y = 3 - \sqrt{x-2}$ in $[2, 20]$ by $[-5, 11]$. **59.** $x = 2(y+1)^2 - 1$. Shift down 1, horizontal stretch by 2, shift left 1. Check your sketch by graphing $y = -1 + \sqrt{\frac{x+1}{2}}$ and $y = -1 - \sqrt{\frac{x+1}{2}}$ in $[-1, 10]$ by $[-4, 2]$. **61.** $x = -2(y-3)^2 + 5$. Shift up 3, horizontal stretch by 2, reflection through y-axis, shift right 5. Check your sketch by graphing $y = 3 + \sqrt{\frac{5-x}{2}}$ and $y = 3 - \sqrt{\frac{5-x}{2}}$ in $[-9, 5]$ by $[0, 7]$. **63.** Two **65.** $y = mx + b$

Section 1.5

1. a) and g) are false. The others are all equivalent to $2 < x < 6$. **3.** $\{-2, 2\}$ **5.** $\{-9/2, -1/2\}$ **7.** $\{-1/3, 17/3\}$ **9.** $\{-3/2, 2/3\}$ **11.** $\{1 - \sqrt{5}/2, 2, 1 + \sqrt{5}/2\}$ **13.** $\{-4, \frac{1}{2}\}$ **15.** $\{-0.5, -0.41, 1.08\}$ or $\{-\frac{1}{2}, \frac{1 \pm \sqrt{5}}{3}\}$ **17.** We give a sequence for the smallest solution only. $[-2, -1]$ by $[-1, 1]$, xscl $= 0.1$; $[-2, -1.9]$ by $[-0.1, 0.1]$, xscl $= 0.01$; $[-1.94, -1.93]$ by $[-0.01, 0.01]$, xscl $= 0.001$; $[-1.931, -1.930]$ by $[-0.001, 0.001]$, xscl $= 0.0001$ **19.** No real solution **21.** $\{3, \frac{1 \pm \sqrt{7}}{2}\}$ **23.** $\{0.74, 7.56, 12.70\}$ **25.** $-1 \leqq y \leqq 3$ **27.** $5/3 < y < 3$ **29.** $0.9 < y < 1.1$ **31.** $|x - 6| < 3$ **33.** $|x + 1| < 4$ **35.** $3 < x < 7$ **37.** $(-\infty, -1) \cup [3, \infty)$ **39.** $-8 \leqq x \leqq 2$ **41.** $(-\infty, -5) \cup (-1, \infty)$ **43.** $[-5, 2]$ **45.** $(-2/3, 2)$ **47.** $\{2, 2 \pm \sqrt{7}\}$

49. $(15, 225)$ **51.** b) If the y-range is too large, the graphing utility cannot distinguish between very close values of y. **53.** Non-real roots always occur in conjugate pairs and therefore the real cubic will have 0 or 2 non-real roots. Since there are 3 roots, there will be at least one real root. To find all real roots, first find a complete graph, determine the number of x-intercepts, and zoom in on each of them. The set of x-intercepts is the solution set of the cubic. **55.** a) $A(x) = x(50 - x)$ b) Check your sketch by graphing $y = x(50 - x)$ in $[-10, 60]$ by $[-100, 700]$ c) domain $= (-\infty, \infty)$, range $= (-\infty, 625)$ d) $0 < x < 50$ e) Using TRACE in b) leads to the approximation 13.6ft by 36.4ft. The exact dimensions are $25 - 5\sqrt{5}$ft by $25 + 5\sqrt{5}$ft f) $\{0 < x < 25 - 5\sqrt{5}\} \cup \{25 + 5\sqrt{5} < x < 50\}$ **57.** a) $A(x) = (8.5 - 2x)(11 - 2x)$ b) Graph $y = A(x)$ in $[0, 10]$ by $[-2, 50]$ c) Domain $= (-\infty, \infty)$, range $= [-1.5625, \infty)$ d) $0 < x < 4.25$ only makes sense. This part of the graph is the graph of the problem situation. e) 0.95in. **59.** a) $V = x(20-2x)(25-2x)$ b) Graph $y = V$ in $[-2, 17]$ by $[-100, 900]$ c) Domain = range $= (-\infty, \infty)$ d) $0 < x < 10$. Graphing $y = V$ in $[0, 10]$ by $[0, 900]$ gives a graph of the problem situation. e) $x = 3.68$in., maximum volume is 820.53in^3. **61.** a) $V(x) = .1x + .25(50 - x)$ b) Graph $y = V(x)$ in $[-20, 100]$ by $[-2, 20]$ c) Domain = range = $(-\infty, \infty)$ d) $0 \leqq x \leqq 50$, x and integer e) 22 dimes, 28 quarters f) There is no integral solution.

Section 1.6

1. Not one-to-one **3.** One-to-one **5.** Your graph should be the line through $(0, -3)$ and $(2, 0)$.

7.

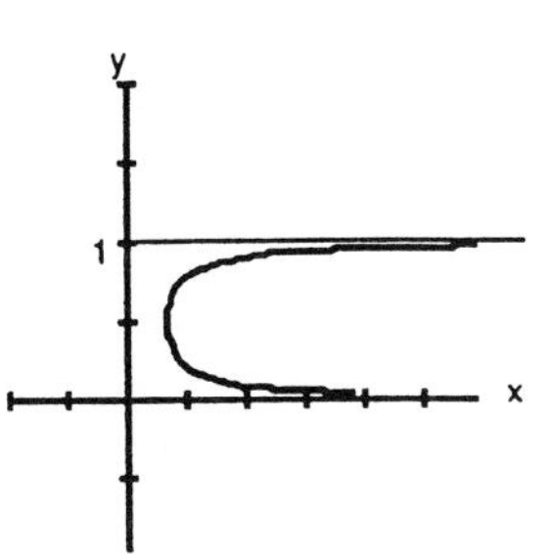

9. The graph of the inverse relation can be viewed by graphing $x_1 = 3/(t - 2) - 1$, $y_1 = t$, $-10 \leqq t \leqq 10$, Tstep $= .1$ in $[-10/10]$ by $[-10, 10]$. It is a function. **11.** Graph $x = t^3 - 4t + 6$, $y = t$, $-10 \leqq t \leqq 10$, Tstep $= .1$, in $[-1000, 1000]$ by $[-15, 15]$. This is a function. **13.** For the inverse graph $y = e^{x/2}$ and $y = -e^{x/2}$ in $[-2, 5]$ by $[-10, 10]$. This is not a function. **15.** Check your sketch by graphing $y = 2\,\ell n(x - 4)/\ell n\, 3 - 1$ in $[3, 14]$ by $[-10, 4]$ noting that $x = 4$ is a vertical asymptote. **17.** Check your sketch by graphing $y = -3\,\ell n(x + 2)/\ell n(.5) + 2$ in $[-2, 4]$ by $[-18, 10]$. **19.** The graph of $\log x = \log_{10} x$ has the same shape as that in Figure 1.122. We start with the graph of $y = \log x$, reflect it through the x-axis and stretch vertically by a factor of 3. (We now have the graph of $y = -3 \log x$.) We then shift left 2 and then up 1. The domain comes from $x + 2 > 0$ so is $(-2, \infty)$ and the range is $(-\infty, \infty)$. **21.** Start with the graph of $y = 3^x$, shift left 1 ($y = 3^{x+1}$), reflect through the y-axis ($y = 3^{-x+1}$), stretch vertically by a factor of 2 and shift up 1.5. Domain $= (-\infty, \infty)$, range $= (1.5, \infty)$ **23.** Shift the graph of $x^2 + y^2 = 9$ three units left and 5 units up. Domain $= [-6, 0]$, range $= [2, 8]$ **25.** $x^2 + (y - 2)^2 = 4$. Graph $y = 2 + \sqrt{4 - x^2}$ and $y = 2 - \sqrt{4 - x^2}$ in $[-3.4, 3.4]$ by $[0, 4]$. **27.** $(x - 3)^2 + (y + 4)^2 = 25$. Graph $y = -4 + \sqrt{25 - (x - 3)^2}$ and $y = -4 - \sqrt{25 - (x - 3)^2}$ in $[-5.5, 11.5]$ by $[-9, 1]$. **29.** $x^2 + y^2 = 4$ **31.** $(x-3)^2 + (y-3)^2 = 9$

33. a) All points outside the boundary of the unit circle b) All points in the interior of the circular disk with center $(0,0)$ and radius 2 c) The ring between the two circles **35.** $(x+2)^2+(y+1)^2 < 6$ **37.** $f^{-1}(x) = (x-3)/2$. **39.** $f^{-1}(x) = \sqrt[3]{x+1}$ **41.** $f^{-1}(x) = \sqrt{x-1}$ **43.** $f^{-1}(x) = 2 - \sqrt{-x},\ x \leqq 0$ **45.** $f^{-1}(x) = \frac{1}{\sqrt{x}}$ **47.** $f^{-1}(x) = \frac{1-3x}{x-2},\ x \neq 2$ **49.** $\{\ell n(\frac{3-\sqrt{5}}{2}), \ell n(\frac{3+\sqrt{5}}{2})\}$ **51.** $\{2-\sqrt{3}, 2+\sqrt{3}\}$ **53.** Graph $y = \frac{4}{3}\sqrt{x^2-9}$ and $y = -\frac{4}{3}\sqrt{x^2-9}$ in $[-10, 10]$ by $[-10, 10]$. **55.** Show that the line $y = x$ is the perpendicular bisector of the line segment with endpoints (a,b) and (b,a) **57.** In Theorem 6 again let $\log_a r = u$, $r = a^u$, $\log_a s = v$, $s = a^v$. (b) $\log_a \frac{r}{s} = \log_a \frac{a^u}{a^v} = \log_a a^{u-v} = u - v = \log_a r - \log_a s$. (c) $\log_a r^c = \log_a a^{uc} = cu = c \log_a r$ **59.** The graph of each relation does not pass the Vertical Line Test. For certain x's in the domain there correspond more than one y.

Section 1.7

1. 8.901rad **3.** $-\frac{5\pi}{6}$ rad **5.** 355.234° **7.** −114.592° **9.** a) $\sin\frac{\pi}{3} = \frac{\sqrt{3}}{2}$, $\cos\frac{\pi}{3} = \frac{1}{2}$, $\tan\frac{\pi}{3} = \sqrt{3}$, $\cot\frac{\pi}{3} = \frac{\sqrt{3}}{3}$, $\sec\frac{\pi}{3} = 2$, $\csc\frac{\pi}{3} = \frac{2\sqrt{3}}{3}$ b) $\sin(-\frac{\pi}{3}) = -\frac{\sqrt{3}}{2}$, $\cos(-\frac{\pi}{3}) = \frac{1}{2}$, $\tan(-\frac{\pi}{3}) = -\sqrt{3}$, $\cot(-\frac{\pi}{3}) = -\frac{\sqrt{3}}{3}$, $\sec(-\frac{\pi}{3}) = 2$, $\csc(-\frac{\pi}{3}) = -\frac{2\sqrt{3}}{3}$ **11.** a) $\sin(6.5) = 0.2151$, $\cos(6.5) = 0.9766$, $\tan(6.5) = 0.2203$, $\cot(6.5) = 4.5397$, $\sec(6.5) = 1.02450$, $\csc(6.5) = 4.6486$ b) $\sin(-6.5) = -0.2151$, $\cos(-6.5) = 0.9766$, $\tan(-6.5) = -0.2203$, $\cot(-6.5) = -4.5397$, $\sec(-6.5) = 1.0240$, $\csc(-6.5) = -4.6486$ **13.** a) $\sin\frac{\pi}{2} = 1$, $\cos\frac{\pi}{2} = 0$, $\tan\frac{\pi}{2}$ is undefined, $\cot(\frac{\pi}{2}) = 0$, $\sec\frac{\pi}{2}$ is undefined, $\csc\frac{\pi}{2} = 1$ b) $\sin\frac{3\pi}{2} = -1$, $\cos\frac{3\pi}{2} = 0$, $\tan\frac{3\pi}{2}$ is undefined, $\cot\frac{3\pi}{2} = 0$, $\sec\frac{3\pi}{2}$ is undefined, $\csc\frac{3\pi}{2} = -1$ **15.** $\frac{\pi}{6}$, 30° **17.** −1.3734, −78.6901° **19.** In parametric mode graph $x_1(t) = \cos t$, $y_1(t) = \sin t$, t Min $= 0$, t Max $= 2\pi$, t Step $= 0.1$. Then use TRACE. Then for a given t value, the displayed $x = \cos t$ and $y = \sin t$. **21.** $[-\pi, 2\pi]$ by $[-1,1]$, $[-\pi, 2\pi]$ by $[-1,1]$, $[-1.5\pi, 1.5\pi]$ by $[-2,2]$, respectively **23.** $[-270°, 450°]$ by $[-3,3]$, $[-360°, 360°]$ by $[-3,3]$, $[-180°, 180°]$ by $[-3,3]$, respectively **25.** Check by graphing the functions in $[-\pi, \pi]$ by $[-3,3]$. **27.** Graph the functions in $[0, 2\pi]$ by $[-1,1]$. **29.** Amplitude $= 2$, period $= 6\pi$, horizontal stretch by a factor of 3. To see one period of the function graph it in $[0, 6\pi]$ by $[-2,2]$. **31.** $y = \cot(2x + \frac{\pi}{2}) = \cot[2(x + \frac{\pi}{4})]$. A horizontal shrinking by a factor of $\frac{1}{2}$ is applied to the graph of $y = \cot x$ followed by a horizontal shift left $\frac{\pi}{4}$ units. The period is $\pi/2$. Graphing the function in $[-\frac{\pi}{4}, \frac{3\pi}{4}]$ by $[-2,2]$ shows two periods of the function. **33.** Period $= 2\pi/3$, domain is all real numbers except $n\pi/3$, n an integer, range $= (-\infty, -5) \cup (1, \infty)$. One period of the graph may be viewed in $[-\frac{\pi}{3}, \frac{\pi}{3}]$ by $[-11, 7]$. Start with the graph of $y = \csc x$, shrink horizontally by a factor of 1/3, shift horizontally left $\pi/3$ units, stretch vertically by a factor of 3, shift vertically downward 2 units. **35.** Period $= \pi/3$, domain is all real numbers except odd multiples of $\pi/6$, range is all real numbers. One period of the graph may be viewed in $[-\frac{\pi}{6}, \frac{\pi}{6}]$ by $[-11, 15]$. Start with the graph of $y = \tan x$, shrink horizontally by a factor of 1/3, shift horizontally left $\pi/3$ units, stretch vertically by a factor of 3, reflect through the x-axis, shift vertically upward 2 units. **37.** $\{\pm\cos^{-1}(-0.7) + 2n\pi\}$ **39.** $\{(\tan^{-1} 4) + n\pi\}$ **41.** Let $x_1 = \sqrt{5}\cos t$, $y_1 = \sqrt{5}\sin t$, $0 \leqq t \leqq 2\pi$ in $[-\sqrt{5}, \sqrt{5}]$ by $[-\sqrt{5}, \sqrt{5}]$. **43.** Graph $x_1 = 2 + 3\cos t$, $y_1 = -3 + 3\sin t$, $0 \leqq t \leqq 2\pi$, in $[-2, 6]$ by $[-7, 1]$. **45.** Let $A = \sqrt{a^2 + b^2}$ and let α be the unique solution in $[0, 2\pi)$ of $\cos\alpha = a/A$, $\sin\alpha = b/A$. $(a/A, b/A)$ is a point on the unit circle so there is such an angle α. Then $A\sin(x+\alpha) = A\sin x\cos\alpha + A\cos x\sin\alpha = a\sin x + b\cos x$. **47.** $y = \sqrt{13}\sin(x + \alpha)$ where $\alpha = \sin^{-1}(3/\sqrt{13}) = 0.9828$ **49.** $\sqrt{2}\sin(2x + \frac{\pi}{4})$ **51.** Four sets of equations with the same graphs: $\{a, j\}$, $\{b, d, g\}$, $\{c, e, i, m, n\}$, $\{f, h, k, \ell\}$ **53.** Graphs of cosine, sine and tangent are symmetric with respect to the y-axis, the origin and the origin, respectively. **55.** a) yes b) $-1 \leqq \cos 2x \leqq 1$ c) $0 \leqq \frac{1+\cos 2x}{2} \leqq 1$ d) The domain is the set of all reals; the range is the interval $[0, 1]$. **57.** a) 37 b) 365 c) 101 units to the right d) 25 units upward **59.** We obtain $\cos(A - A) = \cos A\cos A + \sin A\sin A$

or $1 = \cos^2 A + \sin^2 A$. **61.** $\cos(A + \frac{\pi}{2}) = \cos A \cos \frac{\pi}{2} - \sin A \sin \frac{\pi}{2} = -\sin A$. If we start with the cosine curve, reflect it through the x-axis and shift horizontally $\frac{\pi}{2}$ units to the left, we obtain the sine curve. **63.** $\frac{\sqrt{6}+\sqrt{2}}{4}$ **65.** $\frac{\sqrt{2}+\sqrt{6}}{4}$ **67.** $\frac{2+\sqrt{2}}{4}$ **69.** $\frac{2-\sqrt{3}}{4}$ **73.** a) Let $f(x) = \cot x$. $f(-x) = \frac{\cos(-x)}{\sin(-x)} = \frac{\cos x}{-\sin x} = -f(x)$, proving $f(x)$ is odd. b) Let $g(x) = \frac{h(x)}{k(x)}$ where $h(x)$ is even and $k(x)$ is odd. $g(-x) = \frac{h(-x)}{k(-x)} = \frac{h(x)}{-k(x)} = -g(x)$ proving $g(x)$ is odd where defined. c) The graph of $y = \cot(-x) = -\cot x$ can be obtained by reflecting the graph of $y = \cot x$ through the x-axis. **75.** Use the method indicated in the solution of Exercise 73.

Practice Exercises, Chapter 1

1. a) $(1,-4)$ b) $(-1,4)$ c) $(-1,-4)$ **3.** a) $(-4,-2)$ b) $(4,2)$ c) $(4,-2)$ **5.** a) origin b) y-axis **7.** a) both axes and the origin b) none of the mentioned symmetries **9.** $x = 1$, $y = 3$ **11.** $x = 0$, $y = -3$ **13.** $y = 2x - 1$. Intercepts: $x = \frac{1}{2}$, $y = -1$ **15.** $y = -x + 1$. Intercepts: $x = 1$, $y = 1$ **17.** $y + 6 = 3(x - 1)$ or $y = 3x - 9$. Intercepts: $x = 3$, $y = -9$ **19.** $y - 2 = -\frac{1}{2}(x + 1)$ or $y = -\frac{1}{2}x + \frac{3}{2}$. Intercepts: $x = -3$, $y = \frac{3}{2}$. **21.** $3y = 5x + 4$ **23.** $y = \frac{5}{2}x - 6$ **25.** $y = \frac{1}{2}x + 2$ **27.** $y = -2x - 1$ **29.** a) $2x - y = 12$ b) $x + 2y = 6$, distance $= 14\sqrt{5}/5$ **31.** a) $4x + 3y = -20$ b) $y + 12 = \frac{3}{4}(x - 4)$, distance $= \frac{32}{5}$ **33.** Domain = range = all real numbers **35.** Check your sketch by graphing $y = 2abs(x - 1) - 1$ in $[-2, 4]$ by $[-1, 5]$. Domain $= (-\infty, \infty)$, range $= [-1, \infty)$ **37.** Domain $= (-\infty, \infty)$, range $= [-1, 1]$ **39.** Domain $= (-\infty, 0]$, range $= (-\infty, \infty)$. Check your sketch by graphing $y = -\sqrt{-x}$ and $y = \sqrt{-x}$ in $[-9, 0]$ by $[-3, 3]$. **41.** Domain = range $= (-\infty, \infty)$. Graph $f(x)$ in $[-9, 4]$ by $[-100, 100]$. **43.** Domain $= (1, \infty)$, range $= (-\infty, \infty)$. Graph $y = \ell n(x - 1)/\ell n\, 7 + 1$ in $[1, 3]$ by $[-3, 3]$, recalling that $x = 1$ is a vertical asymptote. **45.** Domain $= (-\infty, \infty)$, range $= [5, \infty)$. Graph $y = abs(x-2) + abs(x+3)$ in $[-5, 4]$ by $[4, 9]$. **47.** Stretch vertically by a factor of 2, reflect through the x-axis, shift horizontally right one unit, shift vertically 5 units upward. **49.** Stretch vertically by a factor of 3, shrink horizontally by a factor of 1/3, shift horizontally $\pi/3$ units left. **51.** $y = -2(x - 2)^2 + 3$ **53.** $y = \frac{3}{x+2} + 5$
55. **57.**

59. Vertex is $(2, 3)$, $x = 2$ is the line of symmetry. Check your sketch by graphing y in $[-2, 5]$ by $[-6, 3]$. **61.** All are even **63.** a) even b) odd c) odd **65.** a) even b) odd c) odd **67.** Graph y in $[-2, 2]$ by $[0, 2]$. The function is periodic of period 1. **69.** Graph $y_1 = \sqrt{-x}$ and $y_2 = \sqrt{x}$ at the same time in $[-4, 4]$ by $[0, 2]$. **71.** The graph consists of one period of the sine function on $[0, 2\pi]$ together with all points on the x-axis larger than 2π. **73.** $y = \begin{cases} 1 - x, & 0 \leq x < 1 \\ 2 - x, & 1 \leq x \leq 2 \end{cases}$ **75.** For $f(x)$, domain = range = all real numbers except 0. For the remaining functions, domain = range = all positive real numbers. **77.** $(x - 1)^2 + (y - 1)^2 = 1$ **79.** $(x - 2)^2 + (y + 3)^2 = \frac{1}{4}$ **81.** $(3, 5)$, 4 **83.** $(-1, 7)$, 11 **85.** a) $x^2 + y^2 < 1$ b) $x^2 + y^2 \leq 1$ **87.** $\{\frac{1}{2}, \frac{3}{2}\}$ **89.** $\{-20, 15\}$ **91.** $-\frac{5}{2} \leq x \leq -\frac{3}{2}$

93. $-\frac{1}{5} < y < 1$ **95.** $\{0.95, 2.47, 4.34\}$ **97.** $\{\frac{15-\sqrt{5}}{6}, \frac{15+\sqrt{5}}{6}\}$ or $\{2.127, 2.873\}$ **99.** $-1 < x < 2$ **101.** $(-\infty, -1) \cup (5, \infty)$ **103.** $(-\infty, 0.95) \cup (2.47, 4.34)$ **105.** a) $\frac{\pi}{6}$ b) 0.122π c) -0.722π d) $-\frac{5\pi}{6}$ **107.** a) 0.891, 0.454, 1.965, 0.509, 2.205, 1.122 b) −0.891, 0.454, −1.965, −0.509, 2.205, −1.122 c) $\sqrt{3}/2$, $-1/2$, $-\sqrt{3}$, $-\sqrt{3}/3$, -2, $2\sqrt{3}/3$ d) $-\sqrt{3}/2$, $-1/2$, $\sqrt{3}$, $\sqrt{3}/3$, -2, $-2\sqrt{3}/3$ **109.** Graph the functions in $[0, 2\pi]$ by $[-1, 2]$. **111.** $\frac{3}{4}$ **113.** $f^{-1}(x) = \frac{2-x}{3}$ **115.** The inverse relation is not a function. Its graph may be obtained using $x_1 = t^3 - t$, $y_1 = t$, $-2 \leqq t \leqq 2$ in $[-6, 6]$ by $[-2, 2]$. **117.** 0.775, 44.427° **119.** Graph $y = |\cos x|$ in $[-\frac{\pi}{2}, \frac{\pi}{2}]$ by $[0, 1]$. The graph is complete because the function has period π. **121.** For x in the interval $[-\frac{\pi}{2}, \frac{3\pi}{2}]$, $y = \begin{cases} 0, & -\frac{\pi}{2} \leqq x < \frac{\pi}{2} \\ -\cos x, & \frac{\pi}{2} \leqq x \leqq \frac{3\pi}{2} \end{cases}$ Graphing this part gives a complete graph because the function has period 2π. **123.** a) $A(x) = (\frac{x}{4})^2 + (\frac{100-x}{4})^2$ b) Graph this function in $[-50, 150]$ by $[300, 1000]$ c) Domain $= (-\infty, \infty)$, range $= [312.5, \infty)$ d) $0 < x < 100$ e) $50 - 10\sqrt{7}$in. and $50 + 10\sqrt{7}$in. f) The maximum $(\frac{100}{4})^2 = 625\text{in}^2$ cannot be attained. The minimum of 312.5in^2 is attained if both pieces are 50in. **125.** The graph is the square with vertices $(1, 0)$, $(0, 1)$, $(-1, 0)$, $(0, -1)$.

Chapter 2

Section 2.1

1. 4 **3.** 2 **5.** 9 **7.** 1 **9.** -15 **11.** -2 **13.** 0 **15.** Limit does not exist. **17.** The limit does not exist. **19.** $\frac{1}{2}$ **21.** $-\frac{1}{2}$ **23.** $\frac{1}{2}$ **25.** $-\frac{1}{4}$ **27.** c) $f(990) - g(990) \approx -1.9301 \times 10^{-4}$; \$5181.03 **29.** There appear to be no points of the graph very near to $(0, 4)$. **31.** The graph of $f(x)$ in the very narrow, tall viewing rectangle $[1.99, 2.01]$ by $[-10,000, 10,000]$ strongly indicates $\lim_{x \to 2^+} f(x) = \infty$ and $\lim_{x \to 2^-} f(x) = -\infty$.

33. a)

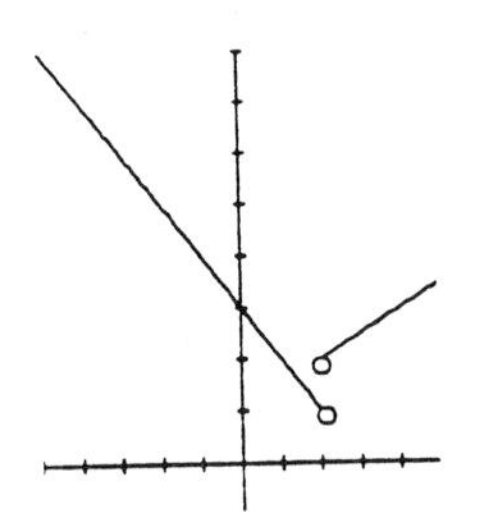

b) $\lim_{x \to 2^+} f(x) = 2, \lim_{x \to 2^-} f(x) = 1$

c) Does not exist because right-hand and left-hand limits are not equal.

35. a) A complete graph of $f(x)$ can be obtained on a graphing calculator by graphing both $y = (x-1)^{-1} + 0\sqrt{(1-x)}$ and $y = x^3 - 2x + 5 + 0\sqrt{(x-1)}$ in the viewing rectangle $[-3, 5]$ by $[-25, 25]$.

b) $\lim_{x \to 1^+} f(x) = 4$ and $\lim_{x \to 1^-} f(x) = \infty$

c) No. For this limit to exist, the two limits in b) must be equal to the same finite number.

37. a = 15 **39.** a), b), d), e), f)

41. a)

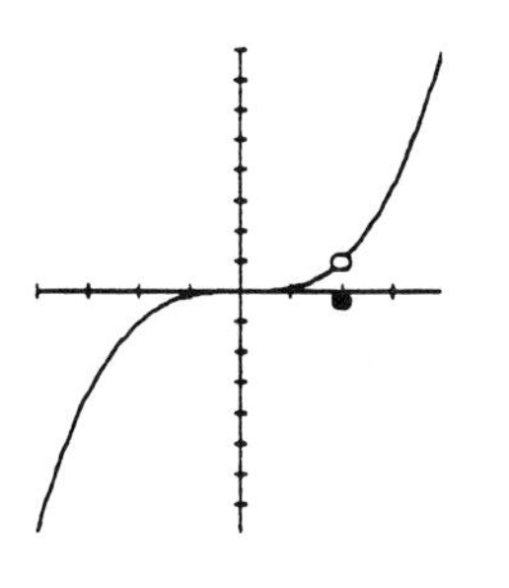

b) $\text{Lim}_{x \to 1^-} f(x) = 1 = \lim_{x \to 1^+} f(x)$
c) $\text{Lim}_{x \to 1} f(x) = 1$.

43.

a) All points c except $c = 0, 1, 2$ b) $x = 2$ c) $x = 0$

45. a) Viewing the graph of the function in the two viewing rectangles $[0, 5]$ by $[-2, 2]$ and $[0, 0.001]$ by $[-2, 2]$ will help give you a good idea of the complete graph of the function.

b) No c) $\lim_{x \to 0^-} f(x) = 0$ d) $\lim_{x \to 0} f(x)$ does not exist because the two one-sided limits are not equal to the same finite number.

47. 0 **49.** 0 **51.** 1 **53.** a) -10 b) -20 **55.** a) 4 b) -21 c) -12 d) -7/3 **57.** 0 **59.** 0 **61.** 0 **63.** $\text{Lim}_{x \to 0}(1 + x)^{3/x} = L$ where $L \approx 20.085$ **65.** 2

Section 2.2

1. a) Yes, $f(-1) = 0$ b) Yes. $\text{Lim}_{x \to -1^+} f(x) = 0$ c) Yes d) Yes **3.** a) No b) No **5.** a) 0 b) $f(2) = 0$ should be assigned. **7.** All points except $x = 2$ **9.** All points except $x = 1$ **11.** All points in $[-1, 2]$ except $x = 0$ and $x = 1$ **13.** All points except $x = 1$

15. a)

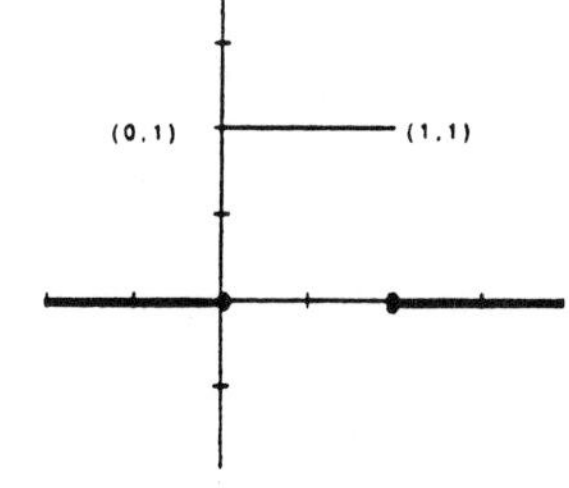

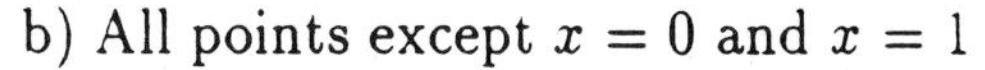
b) All points except $x = 0$ and $x = 1$

17. $x = 2$ **19.** 1 and 3 **21.** ± 1 **23.** There are no points of discontinuity **25.** $x = 0$ **27.** All x with $x < -3/2$. **29.** No points of discontinuity **31.** $\text{Lim}_{x \to 1} \frac{x^2-1}{x-1} = \lim_{x \to 1}(x + 1) = 2$. Hence $f(1) = \lim_{x \to 1} f(x)$ and f is continuous at $x = 1$. **33.** $h(2) = 7$ **35.** $g(4) = 8/5$

37. 4/3

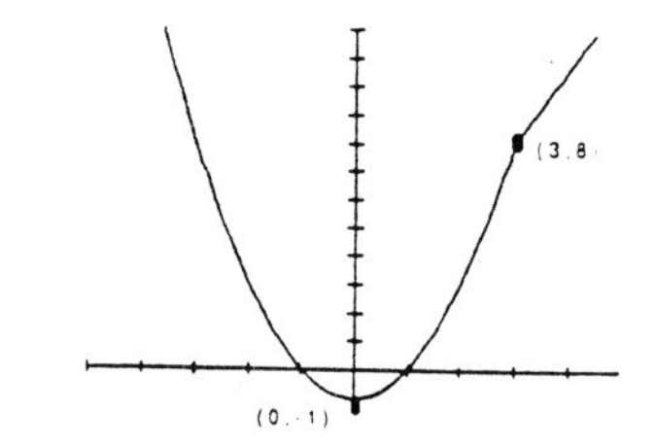

39. 1 **41.** 0 **43.** Both methods give 1.324717957
45. 2 and 3, f does not take on its minimum value 0 but only approaches it arbitrarily closely. Theorem 6 is not contradicted because of the discontinuities.
47. The maximum value 1 is not attained but is only approached as x approaches ± 1. The minimum value 0 is attained at $x = 0$. Theorem 6 is not contradicted because the interval $(-1, 1)$ is not closed.
49. We are given $f(0) < 0 < f(1)$. By Theorem 7 there exists some c in $[0, 1]$ such that $f(c) = 0$. A possible graph is

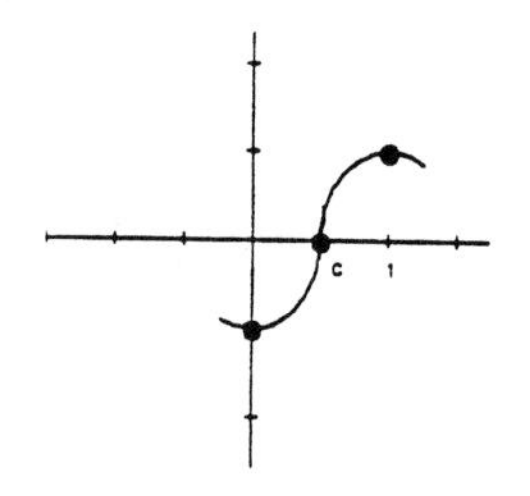

Section 2.3

1. 1 **3.** $\frac{1}{2}$ **5.** 1 **7.** 2 **9.** 1 **11.** -1 **13.** 0 **15.** 4 **17.** a) Approximately 0.6 b) very close c) 3/5 **19.** 1 **21.** a) 0.5 b) $\frac{1}{2}$ **23.** The numerator $\cos x$ approaches 1 as $x \to 0$ while the denominator approaches 0. Thus the fraction can be made arbitrarily large in absolute value if x is sufficiently close to 0. Therefore the fraction cannot approach any finite number and so the limit does not exist.

Section 2.4

1. a) $\frac{2}{5}$ b) $\frac{2}{5}$ **3.** a) 0 b) 0 **5.** a) ∞ b) $-\infty$ **7.** a) 0 b) 0
9. a) $-\frac{2}{3}$ b) $-\frac{2}{3}$ **11.** a) -1 b) -1 **13.** ∞ **15.** ∞ **17.** ∞ **19.** $-\infty$
21. $y = 0$ is the end behavior asymptote **23.** $y = 3$ **25.** $x - 4$ **27.** $x^2 + 2x + 2$
29. $y = 1$ **31.** $\infty, -\infty, -\infty, \infty$ **33.** $\infty, -\infty$ **35.** 0 **37.** 1 **39.** ∞
41. $-\infty$ **43.** $0, -\infty, -1, -1$ **45.** 2 **47.** 2 **49.** 0 **51.** Both limits are equal to 2. **53.** Each graph satisfies $y \to \infty$ as $x \to \infty$ and $y \to -\infty$

as $x \to -\infty$. As the power of x increases, the vertical steepness of the graph increases for $|x| > 1$. **55.** Carrying out the hint, proves that $y = -\frac{1}{7}$ is an end behavior model for $f(x)$ by definition of end behavior model. **57.** Let $f(x) = x^3, g(x) = \frac{1}{x^2}$. Then $\lim_{x\to 0}(fg) = \lim_{x\to 0} x = 0$. Let $f(x) = 5x^2, g(x) = \frac{1}{x^2}$. Then $\lim_{x\to 0}(fg) = \lim_{x\to 0} 5 = 5$. Let $f(x) = x^2, g(x) = \frac{1}{x^4}$. Then $\lim_{x\to 0}(fg) = \lim_{x\to 0} \frac{1}{x^2} = \infty$. In each case $\lim_{x\to 0} f(x) = 0$ and $\lim_{x\to 0} g(x) = \infty$. **59.** $\lim_{x\to\pm\infty} \frac{f(x)}{a_n x^n} = \lim_{x\to\pm\infty}(1 + \frac{a_{n-1}}{a_n}\frac{1}{x} + \cdots + \frac{a_1}{a_n}\frac{1}{x^{n-1}} + \frac{a_0}{a_n}\frac{1}{x^n}) = 1 + 0 + \cdots + 0 = 1$

Section 2.5

1. All are true except a) **3.** g) **5.** e) **7.** h) **9.** b) **11.** $-1 \leqq y \leqq 3$ **13.** $\frac{5}{3} < y < 3$ **15.** Answer: $0 \leqq y \leqq 4$ **17.** $\frac{9}{10} < y < \frac{11}{10}$ **19.** $|x - 6| < 3$ **21.** $|x + 1| < 4$ **23.** $-1.22 < x < -0.71$ **25.** $0.93 < x < 1.36$ **27.** $-1.36 < x < -0.93$ **29.** $9.995 < x < 10.004$ rounding to thousandths appropriately **31.** $22.21 < x < 23.81$ **33.** $20 < x < 30$ **35.** $-\frac{1}{9} < x < \frac{1}{11}$ **37.** $3.94 < x < 4.06$ **39.** $-2.68 < x < -2.66$ rounding to hundredths appropriately **41.** $\ell n\, 0.4 < x < \ell n\, 0.6$ or rounding to hundredths appropriately $-0.91 < x < -0.52$ **43.** $|x-3| < 0.5$ **45.** $|x-1| < 0.04$ **47.** $3.384 < x < 3.387$ or, in symmetric form, $|x - x_0| < 0.001$.

Section 2.6

1. $\delta = 2$ **3.** $\delta = \frac{1}{2}$ **5.** $\delta = 0.1$ **7.** $\delta = 0.23$ **9.** $\delta = \frac{7}{16}$ **11.** $L = 5$. $\delta = 0.005$ **13.** $L = 4$. $\delta = 0.05$ **15.** $L = 2$. $\delta = 0.0399$ or any other smaller positive number **17.** $L = 2$. $\delta = \frac{1}{3}$ **19.** $\delta = \varepsilon$ in each case **21.** $L = \sin 1 \approx 0.84$. $\delta = \varepsilon/0.54 = 1.85\varepsilon$ using the method of Example 6 and rounding down δ to hundredths to be safe. **23.** 1.17ε **25.** 0.30ε **27.** 1.78ε **29.** $I = (5, 5 + \varepsilon^2)$. $\lim_{x\to 5^+} \sqrt{x - 5} = 0$
31.

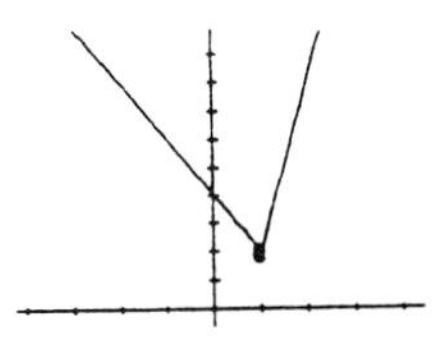

$\delta = \varepsilon/6$ and $I = (1 - \varepsilon/6, 1 + \varepsilon/6)$

33. $\lim_{x\to 2} f(x) = 5$ means corresponding to any radius $\varepsilon > 0$ about 5, there exists a radius $\delta > 0$ about 2 such that for all x

$$0 < |x - 2| < \delta \text{ implies } |f(x) - 5| < \varepsilon.$$

37. $\delta = \sqrt{4+\varepsilon} - 2$. $\lim_{x\to 2} x^2 = 4$ or $\lim_{x\to 2}(x^2 - 4) = 0$. $\delta \to 0$ as $\varepsilon \to 0$. The graph of δ as a function of ε can be viewed by graphing $y = \sqrt{4+x} - 2$ in the rectangle $[0,4]$ by $[0,1]$ and excluding the endpoints.

Practice Exercises, Chapter 2

1. -4 **3.** 0 **5.** -1 **7.** Does not exist **9.** 2 **11.** $\frac{1}{5}$ **13.** 3 **15.** $\frac{2}{5}$ **17.** 0 **19.** $-\infty$ **21.** ∞ **23.** ∞ **25.** $\frac{1}{2}$ **27.** 8 **29.** a) 0.78 b) all close to 0.78 c) f appears to have a minimal value at $x = 0$. d) 7/9 **31.** a) ∞ b) $-\infty$
33. a)

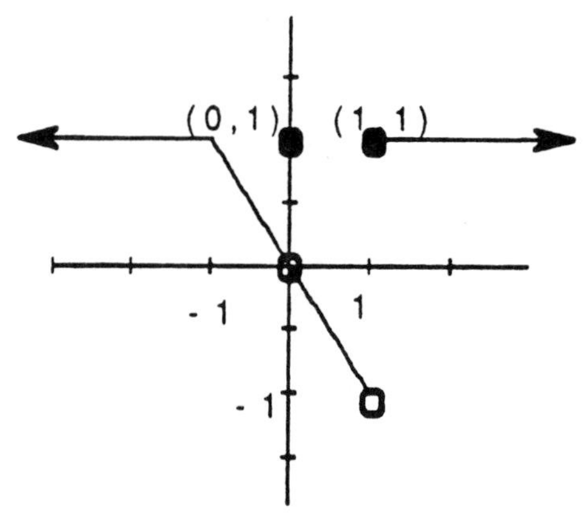

b) $\lim_{x\to -1^+} f(x) = 1, \lim_{x\to -1^-} f(x) = 1, \lim_{x\to 0^+} f(x) = 0, \lim_{x\to 0^-} f(x) = 0, \lim_{x\to 1^+} f(x) = 1, \lim_{x\to 1^-} f(x) = -1$

c) $\lim_{x\to -1} f(x) = 1, \lim_{x\to 0} f(x) = 0$ but $\lim_{x\to 1} f(x)$ does not exist because the right-hand and left-hand limits of f at 1 are not equal.

d) Only at $x = -1$

35. a) A graph of f may be obtained by graphing the functions $y = abs(x^3 - 4x) + 0\sqrt{1-x}$ and $y = x^2 - 2x - 2 + 0\sqrt{x-1}$ in the viewing rectangle $[-5, 7]$ by $[-4, 10]$.

b) $\lim_{x\to 1^+} f(x) = \lim_{x\to 1^+}(x^2 - 2x - 2) = -3$.
$\lim_{x\to 1^-} f(x) = \lim_{x\to 1^-} |x^3 - 4x| = 3$

c) f does not have a limit at $x = 1$ because the right-hand and left-hand limits at $x = 1$ are not equal.

d) $x^3 - 4x$ is continuous by 2.2 Example 5 and $|x|$ is continuous by 2.2 Example 8. Thus $|x^3 - 4x|$ is continuous by Theorem 5 and so f is continuous for $x < 1$. For $x > 1$, $f(x) = x^2 - 2x - 2$, a polynomial, is continuous. Thus $f(x)$ is continuous at all points except $x = 1$.

e) f is not continuous at $x = 1$ because the two limits in b) are not equal and so $\lim_{x\to 1} f(x)$ does not exist.

37. a) A graph of f is obtained by graphing $y = -x + 0\sqrt{1-x}$ and $y = x - 1 + 0\sqrt{x-1}$ in the rectangle $[-2,4]$ by $[-2,4]$.

b) $\lim_{x\to 1^+} f = \lim_{x\to 1^+} x - 1 = 0$. $\lim_{x\to 1^-} f = \lim_{x\to 1^-} -x = -1$.

c) No value assigned to $f(1)$ makes f continuous at $x = 1$.

39. $x = \pm 2$ **41.** $y = 0$ **43.** $x^2 - x$ **45.** a) -21 b) 49 c) 0 d) 1 e) 1 f) 7 **47.** 0 **49.** 0 **51.** 1 **53.** Set $k = 8$ **55.** This is not a contradiction because $0 < x < 1$ is not a *closed* interval. **57.** True because $0 = f(1) < 2.5 < f(2) = 3$ and so by Theorem 7, $2.5 = f(c)$ for some c in $[1,2]$. **59.** $\text{Lim}_{x\to 1} f(x) = 3$ means given any radius $\varepsilon > 0$ about 3 there exists a radius $\delta > 0$ about 1 such that for all x

$$0 < |x-1| < \delta \text{ implies } |f(x) - 3| < \varepsilon.$$

61. Let $f(x) = x^2$. $f(x)$ gets closer to -1 as x approaches 0 but -1 is not equal to $\lim_{x\to 0} f(x)$. **63.** $7 < x < 23$, $|x - 15| < 8$ **65.** $4.82 < x < 5.22$. Taking $x_0 = 5$, $|x-5| < 0.18$. **67.** $2.38 < x < 2.39$. If we take $x_0 = 2.383$ (near the root of $f(x) = 4$), we can say $|x - 2.383| < 0.003$ implies $|f(x) - 4| < 0.1$. **69.** $0 < \delta \le \varepsilon/2$ **71.** $L = 5$, $\delta = 0.01$ **73.** $L = 2$, $\delta = 3$ **75.** 1.13ε **77.** $3.8416 < x < 4.1616, 3.9204 < x < 4.0804$

Chapter 3

Section 3.1

1. $f'(x) = 4x$

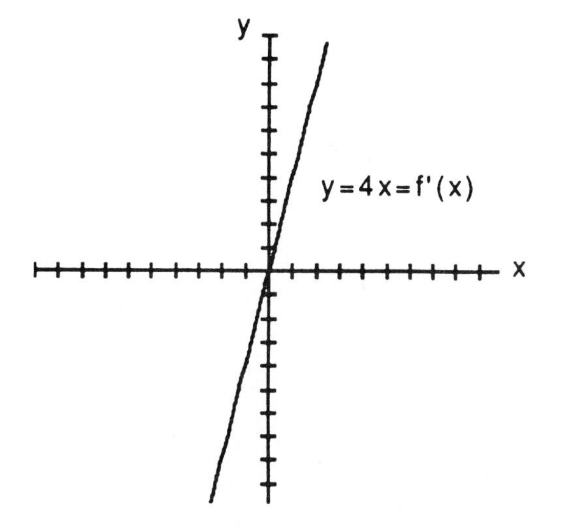

When $x = 3$, $m = 12$ and the tangent is $y = 12x - 23$.

3. $f'(x) = -2/x^2$. One can see a complete graph of this function in the rectangle $[-4,4]$ by $[-10,1]$. At $x = 3$ the slope is $-2/9$ and the tangent has equation $2x + 9y = 12$. **5.** $f'(x) = 1 - 9/x^2$. A complete graph of this function can be seen in the rectangle $[-10,10]$ by $[-10,1]$. At $x = 3$ the slope is 0 and the tangent has equation $y = 6$. **7.** $f'(x) = 1/\sqrt{2x}$. A complete graph of this function can be seen in the rectangle $[0,5]$ by $[0,4]$. At $x = 3$ the slope is $1/\sqrt{6}$ and the tangent has equation $\sqrt{6}\,y = x + 3$. **9.** $f'(x) = -1/2x^{3/2}$. A complete graph of this function can be seen in the rectangle $[0,3]$ by $[-10,1]$. At $x = 3$ the slope is $-1/2(3)^{3/2} \approx -0.096$ and the tangent has equation $y - 1/\sqrt{3} = -(1/6\sqrt{3})(x-3)$. **11.** $y = 4x - 3$. The graphs of $y = x^2 + 1$ and $y = 4x - 3$ can be viewed in the rectangle $[-10,10]$ by $[-10,20]$. **13.** $y = \sqrt{3} + 0.58(x+1)$. The graphs can be viewed in the rectangle $[-6,6]$ by $[-4,4]$. **15.** $y = 0.8(x-2)$. The graphs can be viewed in the rectangle $[-8,8]$ by $[-8,8]$. **17.** We can draw the graph without lifting our pencil so the function is continuous. But the function is not differentiable at each of the points which are peaks or low points. At these points the left-hand and right-hand derivatives are not equal (there is not a unique tangent line). **19.** $x = 0$ is not a point of the domain. At every other point there is a unique tangent line so the function both continuous and differentiable. **21.** The right-hand and left-hand derivatives at $x = 0$ are, respectively, 1 and 0. Since these are unequal, the function is not differentiable at $x = 0$.

23. $f(x) = \begin{cases} x^2, & x \leq 0 \\ x, & x > 0 \end{cases}$. $f(x)$ may be graphed by separately graphing $y = x^2 + 0\sqrt{-x}$ and $y = x + 0\sqrt{x}$. For the numerical derivative graph $\mathrm{NDer}(x^2, x) + 0\sqrt{-x}$ and $\mathrm{NDer}(x,x) + 0\sqrt{x}$. **25.** c) Conjectures: $f_1'(2) = 10$, $f_1'(0) = 2$. $S(h)$ is closer in both cases (in fact exact). $f_2'(2) = 28$, $S(h)$ is closer. $f_2'(0) = 4$, $D(h) = S(h)$. $f_3'(2) \approx -0.416147$ $S(h)$ is closer. $f_3'(0) = 1$, $D(h) = S(h)$. $f_4'(2) = 0.25$, $S(h)$ is closer. $f_4'(0)$ and $D(h)$ for $a = 0$ are not defined but $S(h) = 0$ for all h.

d) Even though the derivative may not exist, the values of $S(h)$ may be defined giving meaningless approximations of the derivative.

27. -0.12, -0.03, 0.03, -0.06 **29.** $[0.25895206, 0.25895208]$ by $[0.135, 0.145]$ is one possibility. **31.** b) The graph is a straight line according to what we see. It is the graph of $y = -2x$.

c) $\mathrm{NDer}(f(x), 0) = 200$. $f'(0)$ does not exist by Theorem 2.

33.

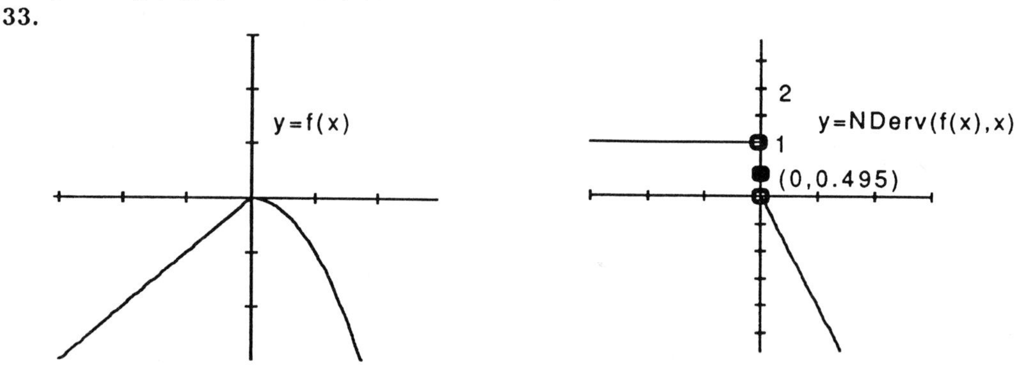

$f'(0)$ does not exist. However, $\mathrm{NDer}(f(x), 0) = 0.495$.

35. The graphs may be viewed by graphing the four functions $-\sqrt[3]{-x} + 0\sqrt{-x} = \sqrt[3]{x} + 0\sqrt{-x}$, $-\sqrt[3]{x} + 0\sqrt{x}$, $\mathrm{NDer}(\sqrt[3]{x}, x) + 0\sqrt{-x}$, $\mathrm{NDer}(-\sqrt[3]{x}, x) + 0\sqrt{x}$ in $[-3,3]$ by $[-5,3]$. $f'(0)$ does not exist. **37.** b) $D_x(\sin x) = \cos x$ **39.** By Example 11, $f'(0)$ does not exist. But $\mathrm{NDer}(|x|, 0) = 0$. $\mathrm{NDer}(f(x), a)$ may exist even if $f'(a)$ does not exist.

Section 3.2

1. $1, 0$ **3.** $-2x, -2$ **5.** $2, 0$ **7.** $x^2 + x,\ 2x + 1$ **9.** $4x^3 - 21x^2 + 4x,\ 12x^2 - 42x + 4$ **11.** $8x - 8,\ 8$ **13.** $y' = 2x - 1,\ y'' = 2$ **15.** $y' = 2x^3 - 3x - 1,\ y'' = 6x^2 - 3,\ y''' = 12x,\ y^{(4)} = 12$ **17.** $3x^2 + 2x + 1$ **19.** $3x^2$ **21.** $\frac{8}{(x+7)^2}$ **23.** $\frac{x^2-2x-1}{(1+x^2)^2}$ **25.** $\frac{x^4+2x}{(1-x^3)^2}$ **27.** $y' = -\frac{6}{x^3},\ y'' = \frac{18}{x^4}$

29. $y' = -\frac{20}{x^5}$, $y'' = \frac{100}{x^6}$ **31.** $y' = 1 - \frac{1}{x^2}$, $y'' = \frac{2}{x^3}$ **33.** $\frac{2x^3-7}{x^2}$ **35.** $3/x^4$ **37.** $y - 3^{0.2} = 0.63(x-1)$. The result is confirmed by viewing $y = x3^{-0.2x}$ and $y = 3^{-0.2} + 0.63(x-1)$ in the rectangle $[-10, 10]$ by $[-10, 10]$. **39.** $y = (3/5) + 0.44x$. The result is confirmed by graphing $y = f(x)$ and $y = (3/5) + 0.44x$ in the rectangle $[-1, 1]$ by $[0.3, 0.9]$. **41.** $y = f'(x)$ or $y = \text{NDer}(f(x))$ can be viewed in the rectangle $[-2, 8]$ by $[-4, 4]$. $y = f''(x)$ or $y = \text{N}_2\text{Der}(f(x))$ can be viewed in $[-2, 10]$ by $[-4, 10]$. **43.** $y = \text{NDer}(f(x))$ can be viewed in the rectangle $[-8, 8]$ by $[-2, 0]$. $y = \text{N}_2\text{Der}(f(x))$ can be viewed in $[-6, 8]$ by $[-1, 1]$. **45.** The graphs can be viewed in the two rectangles $[-10, 10]$ by $[-10, 10]$ and $[-50, 50]$ by $[-50, 50]$. **47.** We use zoom-in and the graphs of Exercise 41. $f'(x) = 0$ for $x = -0.31$ and $x = 3.20$. $f''(x) > 0$ for x in the set $(-\infty, -1) \cup (1, \infty)$. **49.** $f'(x) = 0$ has no solution. $f''(x) > 0$ has solution set $(-3, -0.33) \cup (5, \infty)$. **51.** The graph of $y = \text{NDer}(f(x))$ oscillates, appears to cross the x-axis infinitely often and to be symmetric with respect to the origin. The three solutions of $f'(x) = 0$ of smallest absolute value are $-2.03, 0, 2.03$. The graph of $y = \text{N}_2\text{Der}(f(x))$ appears to cross the x-axis infinitely often. The solution set of $f''(x) > 0$ consists of an infinite sequence of intervals. The two closest to the origin are $(-6.57, -3.65)$ and $(3.65, 6.57)$ rounding appropriately. **53.** $f''(x) = 6(6x^3 - 45x^2 - 24x + 20)/(3x^2 + 4)^3$. We cannot solve $f''(x) > 0$ exactly by a convenient method. The approximate solution set is $(-0.91, 0.46) \cup (7.95, \infty)$. **55.** a) 13 b) -7 c) $7/25$ d) 20 **57.** c) **59.** $y = 4x - 2$ and $y = 4x + 2$ are the tangents at $x = 1$ and $x = -1$, respectively. The smallest slope on the curve is 1 and it occurs at $x = 0$. **61.** $-4/3, 16$ **63.** $x + 2y = 4$ **65.** $\frac{ds}{dt} = 9.8t$, $\frac{d^2s}{dt^2} = 9.8$
67.

$$
\begin{aligned}
f'(-x) &= \lim_{h \to 0} \frac{f(-x+h) - f(-x)}{h} \\
&= \lim_{h \to 0} \frac{f[-(x-h)] - f(x)}{h} = \lim_{h \to 0} f(x-h) - f(x) \\
&= \lim_{h \to 0} -\frac{f(x-h) - f(x)}{-h} = -f'(x) \text{ so } f'(x) \text{ is odd.}
\end{aligned}
$$

Section 3.3

1. b) $(2, 3)$ at $t = 0$, $(0, 3)$ at $t = 1$, $(0, 3)$ at $t = 2$, $(2, 3)$ at $t = 3$ c) The particle changes direction at $(-0.25, 3)$ when $t = 1.5$. When $t = 1.5$, $v = 0$ and $a = 2$. d) 14.5 meters f) The particle is at rest when $t = 1.5$ sec. **3.** We use Tstep 0.05 in the viewing rectangle $[-10, 10]$ by $[-15, 25]$. a) The particle first moves to the right, then to the left and then to the right again. b) $(-3, 3)$, $(-1, 3)$, $(-5, 3)$, $(-9, 3)$ c) $(-0.70, 3)$ when $t = 0.7$, $v = 0.07$, $a = -7.8$; $(-9.30, 3)$ when $t = 3.3$, $v = 0.07$, $a = 7.8$ (TRACE approximations) d) 27.2 meters f) Approximately when $t = 0.7$ and $t = 3.3$ sec. **5.** In this problem we use $x_1(t) = t\sin t$, $y_1(t) = 3$, $0 \leq t \leq 15$ in the rectangle $[-6, 3]$ by $[-1, 5]$ with Tstep 0.05. Then in non-parametric mode we graphed $v = \text{NDer}(x \sin x)$ and $a = \text{N}_2\text{Der}(x \sin x) = \text{NDer}(v)$ in the rectangle $[0, 5]$ by $[-10, 10]$. a) The particle first moves right, then left and then right slightly. b) $(0, 3)$, $(0.84, 3)$, $(1.82, 3)$, $(0.42, 3)$ c) $(1.82, 3)$ when $t = 2.05$, $v = -0.07$, $a = -2.7$; $(-4.814, 3)$ when $t = 4.90$, $v = -0.10$, $a = 5.18$ d) The particle travels right from $(0, 3)$ to $(1.82, 3)$, from $(1.82, 3)$ to $(-4.81, 3)$ and from $(-4.81, 3)$ to $(-4.79, 3)$ when $t = 5$. Thus the total distance traveled is $1.82 + (1.82 + 4.81) + 0.02 = 8.48$ meters. f) $v = 0$ when $t = 0$, 2.05 and 4.90 sec. **7.** 4.46 sec on Mars, 0.73 sec on Jupiter **9.** One possibility: Graph $x_1(t) = t$, $y_1(t) = 24t - 0.8t^2$, $0 \leq t \leq 30$, in $[0, 30]$ by $[0, 180]$. Then use TRACE and zoom-in if more accuracy is desired. **11.** 320 sec, 52 sec **13.** a) 10^4 per hour b) 0 c) -10^4 per hour

15. a) The average cost of one washing machine when producing the first 100 washing machines is $c(100)/100 = \$110$. During production of the first 100 machines the average increase in producing one more machine is: average increase $= \frac{c(100)-c(0)}{100-0} = \$90$; the fixed cost $c(0) = \$2000$ is omitted with this method.

b) \$80 c) \$79.90

17. $a = -6m/\sec^2$ when $t = 1\sec$ and $a = 6m/\sec^2$ when $t = 3\sec$.

19. a) We use $-12 \leqq t \leqq 12$, Tstep 0.05 in the viewing rectangle $[-35, 35]$ by $[-3, 10]$.

b) For this graph we can use $0 \leqq t \leqq 6.29 \approx 2\pi$, Tstep 0.05 in $[-3, 8]$ by $[-3, 3]$.

c) This line segment can be viewed using $0 \leqq t \leqq 6.29$, Tstep 0.05 in $[-6, 10]$ by $[-6, 2]$

21. a) All have derivative $3x^2$. c) The result of a) suggests that the family consists of all functions of the form $x^3 + C$ where C can be any constant. d) Yes, $f(x) = x^3$. e) Yes, $g(x) = x^3 + 3$. **23.** a) 190ft/s b) 2 c) At 8s when $v = 0$ d) At 10.8s when it was falling at 90ft/s e) From t = 8s to t = 10.8s, i.e., 2.8s. f) Just before burnout, i.e., just before t = 2s. The acceleration was constant from t = 2s to t = 10.8s during free fall. **25.** a) 0, 0 b) 1700, 1400 **27.** (b) **29.** (d)

31. a)

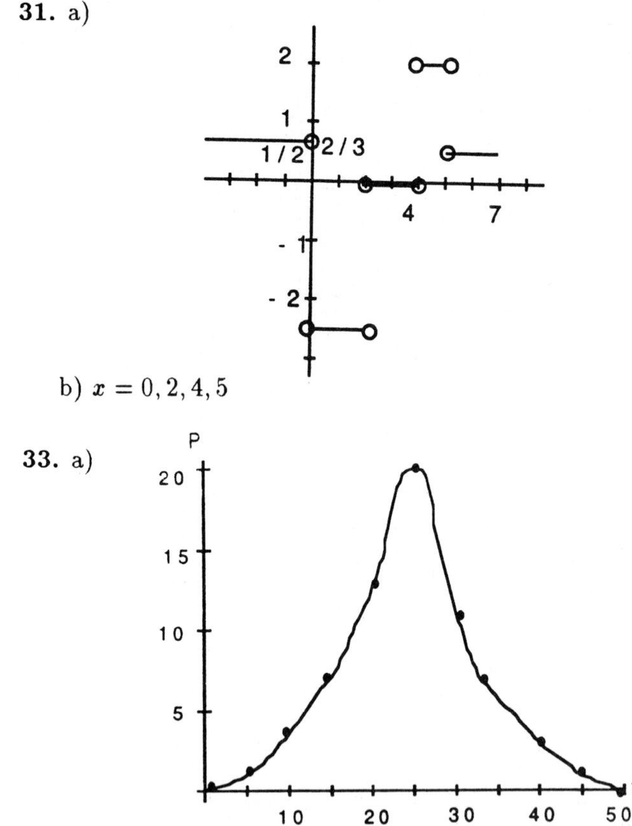

b) $x = 0, 2, 4, 5$

33. a)

The horizontal axis is measured in days, the vertical axis in fruit flies per day.

b) Growing fastest around day 25, slowest near the beginning day 0 and near the end day 50.

Section 3.4

1. $1+\sin x$ **3.** $-\frac{1}{x^2}+5\cos x$ **5.** $-\csc x\cot x-5$ **7.** $\sec x(x\tan x+1)$ **9.** $x(2\cot x - x\csc^2 x)$ **11.** $3 + x\sec^2 x + \tan x$ **13.** $\sec^2 x$ **15.** 0 **17.** $4\sec x\tan x$ **19.** $-\frac{x\sin x+\cos x}{x^2}$ **21.**

$\frac{1+\cos x+x\sin x}{(1+\cos x)^2}$ **23.** $-\frac{\csc^2 x}{(1+\cot x)^2}$ **25.** $\csc x(\csc^2 x+\cot^2 x)$ **27.** Tangent: $y=x$. Normal: $y=-x$. We may graph the three functions $y=\sin x$, $y=x$ and $y=-x$ is the viewing rectangle $[-3,3]$ by $[-2,2]$. **29.** Tangent: $y=2\sin^2 2+4\sin 2\cos 2(x-2)$. Normal: $y=2\sin^2 2-(4\sin 2\cos 2)^{-1}(x-2)$. We may view $y=2(\sin x)^2$, the tangent and the normal in $[0,4]$ by $[0,2.7]$. **33.** $(\tan x)'=\sec^2 x=\frac{1}{\cos^2 x}$ and $(\cot x)'=-\csc^2 x=-\frac{1}{\sin^2 x}$ cannot be 0 for any value of x. **35.** $y'=0$ is equivalent to $\cos x=-2$. Since the latter equation has no solution, the graph has no horizontal tangent. **37.** There are horizontal tangents at the points $(\pi/6,(\pi/6)+\sqrt{3})$, $(5\pi/6,(5\pi/6)-\sqrt{3})$. **39.** $s=A\cos(2\pi bt)$, $v=s'=-2\pi bA\sin(2\pi bt)$ and $a=v'=-4\pi^2b^2A\cos(2\pi bt)$. Now let $s_1=A\cos[2\pi(2b)t]=A\cos(4\pi bt)$. Then the new velocity and acceleration are given by $v_1=-4\pi bA\sin(4\pi bt)=2(-2\pi bA)\sin(4\pi bt)$ and $a_1=-16\pi^2b^2A\cos(4\pi bt)=4(-4\pi^2b^2A)\cos(4\pi bt)$. Thus the amplitude of v is doubled and the amplitude of a is quadrupled. **41.** $y=-x+1+\pi/4$, $y=x+1-\pi/4$ **43.** $y=-1$ is the tangent line at $(\pi/4,-1)$ the only point at which the tangent is horizontal. **45.** The graph of $\tan x$ and its derivative $\sec^2 x$, $-\pi/2<x<\pi/2$ may be viewed in the rectangle $[-1.57,1.57]$ by $[-5,5]$. **47.** 1/2

Section 3.5

1. $\cos(x+1)$ **3.** $-5\sin 5x$ **5.** $(2\pi/5)\cos(2\pi x/5)$ **7.** $-\sec^2(2-x)$ **9.** $2\sec(2x-1)\tan(2x-1)$ **11.** $-(2x+7)\csc(x^2+7x)\cot(x^2+7x)$ **13.** $-3\csc^2(3x+\pi)$ **15.** 0 **17.** $3\sin^2 x\cos x$ **19.** $10(2x+1)^4$ **21.** $-3(x+1)^{-4}$ **23.** $(1-\frac{x}{7})^{-8}$ **25.** $3(1+x-\frac{1}{x})^2(1+\frac{1}{x^2})$ **27.** $6(x^2+2x+3)^2(x+1)$ **29.** $-\sin(\sin x)(\cos x)$ **31.** $1+2\sec^2 x\tan x$ **33.** $\csc x(\csc x+\cot x)^{-1}$ or $\csc x(\csc x-\cot x)$ **35.** $\frac{5}{(x+3)^2}\cos\left(\frac{x-2}{x+3}\right)$ **37.** $6\sin(3x-2)\cos(3x-2)$ **39.** $-4(\sin 2x)(1+\cos 2x)$ **41.** $-2\sin(2x-5)\cos(\cos(2x-5))$ **43.** $2\sec^2 x\tan x$ **45.** $2\csc^2 x\cot x$ **47.** $\frac{5}{2}$ **49.** $-\frac{\pi}{4}$ **51.** 0 **53.** In both cases we get $-6\sin(6x+2)$ **55.** $\frac{dy}{dx}=1$ in both cases. **57.** 5 **59.** $\frac{1}{2}$ **61.** Tangent: $y-2=\pi(x-1)$. Normal: $y-2=-\frac{1}{\pi}(x-1)$ **63.** a) 2/3 b) $2\pi+5$ c) $15-8\pi$ d) 37/6 e) -1 f) $\sqrt{2}/24$ g) 5/32 h) $-5\sqrt{17}/51$

Section 3.6

1. $(9/4)x^{5/4}$ **3.** $(1/3)x^{-2/3}=\frac{1}{3\sqrt[3]{x^2}}$ **5.** $-(2x+5)^{-3/2}$ **7.** $\frac{2x^2+1}{\sqrt{x^2+1}}$ **9.** $-(2xy+y^2)/(x^2+2xy)$ **11.** $(1-2y)/(2x+2y-1)$ **13.** $x(1-y)/[y(x^2-1)]$ **15.** $1/[y(x+1)^2]$ **17.** $\frac{-1}{4\sqrt{x}\sqrt{1-\sqrt{x}}}$ **19.** $\frac{-\sin 2x}{\sqrt{1+\cos 2x}}$ **21.** $-(9/2)\csc^{3/2}x\cot x$ **23.** $\cos^2 y$ **25.** $-[y\sec^2(xy)+1]/x$ **27.** $-x/y$, $(y^2+x^2)/y^3$ **29.** $(x+1)/y$, $[y^2-(x+1)^2]/y^3$ **31.** $\frac{\sqrt{y}}{\sqrt{y}+1}$, $\frac{1}{2(\sqrt{y}+1)^3}$ **33.** (a) $7x-4y=2$ (b) $4x+7y=29$ **35.** (a) $y-3x=6$ (b) $x+3y=8$ **37.** x-intercepts are $\pm\sqrt{7}$. The tangents at $(\pm\sqrt{7},0)$ both have slope -2. **39.** $-\pi/2$ **41.** a) At $\left(\frac{\sqrt{3}}{4},\frac{\sqrt{3}}{2}\right)$ the slope is 2, at $\left(\frac{\sqrt{3}}{4},\frac{1}{2}\right)$ it is $\frac{2\sqrt{3}}{3}$. b) We graph $x=\pm\sqrt{t^2-t^4}$, $y=t$, $-1\leqq t\leqq 1$ in the viewing rectangle $[-0.5,0.5]$ by $[-1,1]$. **43.** b), c), d) **45.** (2/5)m/sec, $(-4/125)$m/sec^2 **47.** The graph of the relation $x=\pm\sqrt{2t-\sin t}$, $y=t$, $0\leqq t\leqq 10$, suggests that it passes the vertical line test and hence that the relation is a function. **49.** a) $x=-0.06251\ldots$ b) $\frac{dy}{dx}=\frac{-y\cos xy}{5y^4+x\cos xy}$; $1.88\ldots$ c) The tangent line has approximate equation $y=-0.5+1.88(x+0.06251)$

Section 3.7

1. $4x-3$ **3.** $2(x-1)$ **5.** $\frac{x}{4}+1$ **7.** $2x$ **9.** -5 **11.** $2+(1/12)(x-8)$ **13.** $L(x)=x$ **15.** $L(x)=\pi-x$ **17.** $L(x)=1+2(x-\pi/4)$ **19.** a) $1+2x$ b) $1-5x$ c) $2(1+x)$ d) $1-6x$ e) $3+x$ f) $1-(1/2)x$ **21.** $L(x)=1+(3/2)x$ is the sum of the linearizations of $\sqrt{x+1}$ and $\sin x$.

23. The limiting value of this sequence of numbers is 1 because $(\cdots(2^{\frac{1}{2}})^{\frac{1}{2}}\cdots)^{\frac{1}{2}} = 2^{\frac{1}{2^n}} \to 2^0 = 1$. **25.** a) 0.21 b) 0.2 c) 0.01 **27.** a) 0.231 b) 0.2 c) 0.031 **29.** a) $-1/3$ b) $-2/5$ c) $1/15$ **31.** $4\pi r_0^2 dr$ **33.** $3x_0^2 dx$ **35.** $2\pi r_0 h$ **37.** a) $0.08\pi \approx 0.2513m^2$ b) 2.000% **39.** 3% **41.** 3% **43.** 1/3 of 1% **45.** The variation of the radius should not exceed 1/2000 of its ideal value. **49.** b) $x = 28\sqrt{3} - 48$ c) $g(28\sqrt{3} - 48) \approx -0.009$ d) $x = 3.52$ with error at most 0.01 e) $\frac{225}{64}$ **51.** $3(x^2 - 1)dx$ **53.** $\frac{2(1-x^2)dx}{(1+x^2)^2}$ **55.** $y = \frac{x}{1-x}$, $dy = \frac{dx}{(1-x)^2}$ **57.** $5\cos(5x)dx$ **59.** $2\sec^2(x/2)dx$ **61.** $\csc(1-(x/3))\cot(1-(x/3))dx$

Practice Exercises, Chapter 3

1. $5x^4 - \frac{1}{4}x + \frac{1}{4}$ **3.** $2(x+1)(2x^2+4x+1)$ **5.** $2\cos 2x$ **7.** $\frac{1}{(x+1)^2}$ **9.** $-4x^2(x^3+1)^{-7/3}$ **11.** $2\sin(1-2x)$ **13.** $3(x^2+x+1)(2x+1)$ **15.** $\frac{2(1+u)}{\sqrt{2u+u^2}}$ **17.** $\frac{-y}{x+2y}$ **19.** $\frac{5-2x-y}{x+2y}$ **21.** $\frac{-1}{3(xy)^{1/5}}$ **23.** $\frac{1}{2y(x+1)^2}$ **25.** $(5x+1)(5x^2+2x)^{1/2}/(2y)$ **27.** $\frac{x-1}{2x\sqrt{x}}$ **29.** $3\sec(1+3x)\tan(1+3x)$ **31.** $-2x(\csc x^2)^2$ **33.** $\frac{1}{2}\sqrt{\frac{1+x^2}{1-x}}\,\frac{(x^2-1)}{(1+x^2)^2}$

35. a)

b) Yes

c) f is not differential at $x = 1$ because its left-hand derivative (1) is not equal to its right-hand derivative (-1) at $x = 1$.

37. $(-1, 27)$, $(2, 0)$ **39.** b) $5\sqrt{2}$ c) -10, 10 d) at -10, $v = 0$, $a = 10$; at 10, $v = 0$, $a = -10$. e) The particle first reaches the origin at $t = \pi/4$. At that time velocity $= -10$, speed $= 10$, acceleration $= 0$. **41.** a) (4/7) sec; 280 cm/ sec b) 560 cm/ sec; 980 cm/ sec^2 **43.** a) (iii) b) (i) c) (ii)

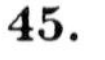

45.

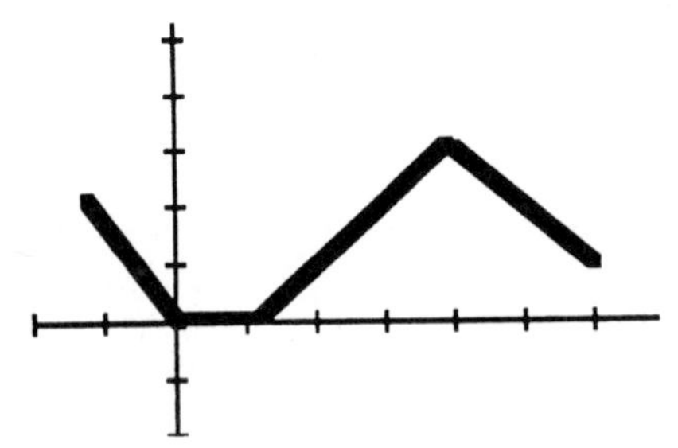

47. $\pi(20x - x^2)$ **49.** a) $y = 4 - \sqrt{3}$ b) $y - 2 = -(x - \pi/2)$ **51.** Yes, at the points $(2n\pi, 0)$ for all integers n **53.** a) 1 b) 6 c) 1 d) $-1/9$ e) $-40/3$ f) 2 g) $-4/9$ **55.** Differentiating both sides of the identity $\sin(x+a) = \sin x \cos a + \cos x \sin a$ with respect to x, we obtain the identity $\cos(x+a) = \cos x \cos a - \sin x \sin a$. We cannot do the same with $x^2 - 2x - 8 = 0$ because this is not an identity between two functions. **57.** 9/2 **59.** $(0, 1)$ and $(-4, 0)$ **61.** a) $x + 4y = 9$, $4x - y = 2$ b) $3x + 2y = 5$, $3y - 2x = 1$ c) $y = 2x - 4$, $x + 2y = 7$ **63.** 3ft

65. a) $y'' = \frac{-2x(x^3+y^3)}{y^5} = \frac{-2x}{y^5}$
b) $y'' = \frac{-(1+2xy^2)}{x^4y^3} = \frac{3-2x}{x^4y^3}$ replacing y^2 by its given value
67. a) $-\frac{1}{(2x+7)^{3/2}}$ b) $-\frac{1}{y^3}$ **69.** a) $L(x) = 2x + (\pi/2) - 1$. Graph $y = \tan x$ and $y = L(x)$ in the viewing rectangle $[-5,5]$ by $[-8,8]$. b) $L(x) = \sqrt{2} - \sqrt{2}(x + \pi/2)$. Graph $y = \sec x$ and $y = L(x)$ in the viewing rectangle $[-8,8]$ by $[-10,10]$. **71.** b) $L(x) = 0.5 + 1.5x = 0$ leads to $x = -1/3$ c) $f(-1/3) \approx -0.01$ d) By zoom-in, $x = -0.326\ldots$ **73.** $\frac{2}{3}\pi r_0 h dr$ **75.** a) With an error of no more than 1% b) With an estimated error of no more than 3% **77.** 2.33ft **79.** $E(x) = f(x) - g(x) = f(x) - m(x-a) - c$. $E(a) = 0$ implies $f(a) - 0 - c = 0$ and so $c = f(a)$. $E(x) = f(x) - f(a) - m(x-a)$. $0 = \lim_{x\to a} \frac{E(x)}{x-a} = \lim_{x\to a} \frac{f(x)-f(a)}{x-a} - m = f'(a) - m$. Thus $m = f'(a)$ and $g(x) = L(x)$.

Chapter 4

Section 4.1

1. $f(-2) = 11$, $f(-1) = 1$, $f'(x) \neq 0$ on $(-2,-1)$ **3.** $f(1) = -1$; $f(3) = 7/3$, $f'(x) = 1 + \frac{2}{x^2}$ which is never zero. On the interval $(1,3)$ the conditions of Rolle's Theorem are met. **5.** (b) Let $f(x) = x^n + a_{n-1}x^{n-1} + \cdots + a_1x + a_0$. Between every two zeros of $f(x)$ lies a zero of $f'(x) = nx^{n-1} + (n-1)a_{n-1}x^{n-2} + \cdots + a_1$. **7.** Between zeros of $f(x) = \sin x$ there is a zero of $f'(x) = \cos x$. **9.** a) Use the window $[-5,5]$ by $[-1,15]$. b) no, c) no, d) local maxima at $(\pm1.73, 10.39)$; local minima at $(\pm3, 0)$, $(0,0)$. f) increasing on $(-3,-1.73) \cup (0,1.73) \cup (3,\infty)$ decreasing on $(-\infty,-3) \cup (-1.73,0) \cup (1.73,3)$ **11.** Local maximum at $(5/2, 6.25)$; increasing on $(0, 5/2)$, decreasing on $(5/2, 6)$; absolute maximum at $(5/2, 6.25)$, absolute minimum at $(6,-6)$. **13.** No local extrema; absolute maximum at $(7,\sqrt{3})$, absolute minimum at $(4,0)$, increasing on $(4,7)$. **15.** Local and absolute minima at $(\pm\sqrt{5}, -16)$; local and absolute maximum at $(0,9)$ increasing on $(-\sqrt{5},0) \cup (\sqrt{5},3)$ decreasing on $(-3,-\sqrt{5}) \cup (0,\sqrt{5})$. **17.** Local minimum at $(3/4, -27/256)$. Absolute maximum : $(-1,2)$, absolute minimum $(3/4,-27/256)$. Decreasing on $(-1, 3/4)$, increasing on $(3/4, 1)$. **19.** 1/2 **21.** 1 **23.** $s'(c) = v(c) = 79.5$ **25.** $v(c) = 7.66$ knot **27.** $|\sin b - \sin a| = |\cos c||b-a| \leq |b-a|$ **29.** $y' = -1/x^2 < 0$. By Cor 1, y is decreasing. **31.** $1.099990 \leq f(0.1) \leq 1.10000$ **33.** Select $b \neq 3$. By the MVT $f(b) - 3 = 0 \cdot (b-0) = 0$.

Section 4.2

1. $f' > 0$ on $(-\infty,-1) \cup (1,\infty)$; $f' < 0$ on $(-1,1)$ **3.** Rising on $(-1,1)$; falling on $(-\infty,-1)$, $(1,\infty)$; local minimum $(-1,-3)$; local maximum $(1,3)$. **5.** Falling $(-\infty,1/2)$, rising $(1/2,\infty)$; concave up everywhere, local minimum at $(1/2,-5/4)$. **7.** Local maximum at $(1,5)$, local minimum at $(3,1)$, inflection point at $(2,3)$. **9.** Local minima $(\pm1,-1)$, local maximum $(0,1)$, inflection points $(\pm1/\sqrt{3}, -1/9)$. **11.** Decreasing between $(1,11)$ and $(1.5,11.5)$. Inflection point at $(5/6, 11.02)$. **13.** Local minima at $(\pm1/\sqrt{6}, -121/12)$, local maximum at $(0,-10)$. Inflection points occur when $x = \pm1/\sqrt{18}$. **15.** Inlfection point (π,π) **17.** Local minima at $(-2.13,-22.23)$ and $(1.88,-6.26)$; local maximum at $(0.24, 2.50)$; inflection points at $(-2/\sqrt{3}, -11.50)$ and $(2/\sqrt{3}, -2.27)$. **19.** Local maximum at $(-1.263, 2.435)$; inflection points at $(-\sqrt{1/3}, 0.2876)$ and $(\sqrt{1/3}, -3.176)$. **21.** The graph is always rising, always concave up. **23.** The particle starts at 3 and moves to left until $t = 2$. Thereafter it moves to the right. **25.** Moves left from $t = 0$ until $t = 1$; for $t > 1$ it moves right. **27.** a) $t = 2, 6, 9.5$ b) $t = 4, 8, 11.5$

29.

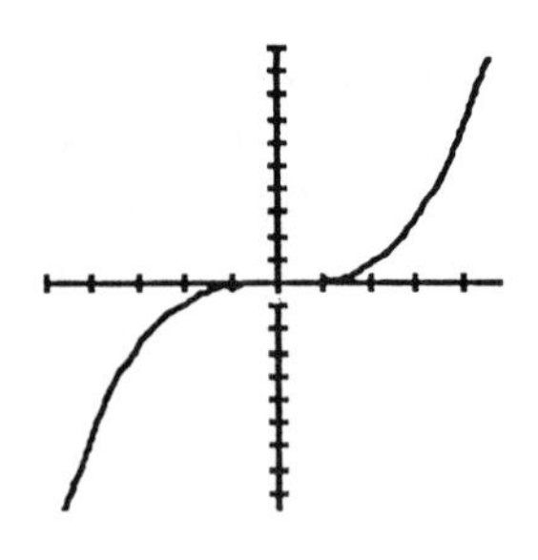

31.

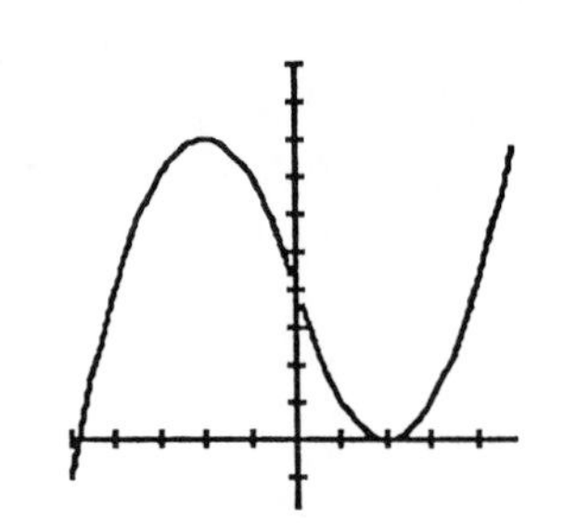

33. Local minimum at $x = 2$; inflection points at $x = 1$, $x = 5/3$. **35.** Local maximum at $(-1, -2)$; local minimum at $(1, 2)$. **37.** No; f might have an inflection point. **39.** True **41.** Viewing rectangle: $[-1, 1]$ by $[-1, 1]$ **43.** Viewing rectangle: $[-0.5, 0.5]$ by $[-5, 5]$

Section 4.3

1. 0.618033988, −1.618033989 **3.** 1, 16403514, −1.452626879 **5.** 0.6301153962, 2.5732719864 **7.** 3.216451347, −1.564587289 **9.** 1.1935 **11.** If $f'(x_0) \neq 0$, all $x_n = x_0$. **13.** $x_0 = h > 0 \Rightarrow x_1 = x_0 - \sqrt{x_0}/(\frac{1}{2\sqrt{x_0}}) = -h$.

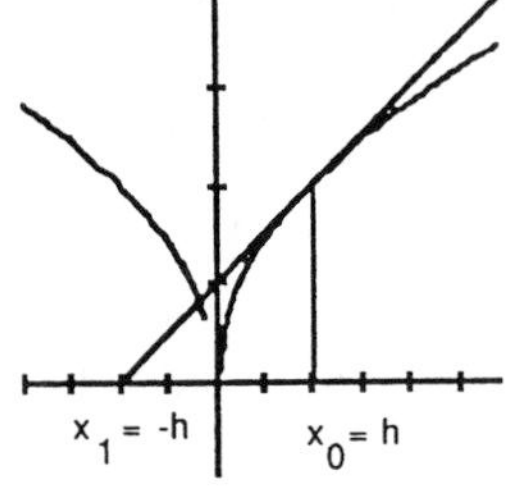

15. Always rising; concave down $x < 1$, concave up $x > 1$. Inflection point $(1, -1)$. One real root. **17.** Local maximum $(\frac{2-\sqrt{3}}{3}, 8.879)$; local minimum $(\frac{2+\sqrt{3}}{3}, 1.9354)$ inflection point $(\frac{2}{3}, 5.407)$, one real root. **19.** Local maximum $(-1/\sqrt{12}, 11.808)$; local minimum $(1/\sqrt{12}, 12.192)$, inflection point $(0, 12)$; one real root. **21.** Use $[-10, 10]$ by $[-60, 60]$. Graph is always falling; inflection point $(0, 20)$. **23.** Local minimum at $(-0.38, 7.8)$; concave up everywhere; no zeros. **25.** Use $[-1, 6]$ by $[-30, 10]$. Local maximum $(0.258, -0.001)$; inflection point $(1.417, -12.417)$; local minimum $(2.574, -24.825)$; one zero. **27.** Use $[-5, 5]$ by $[-17, 17]$. Graph rises to a local maximum at $(2, 15.33)$, then falls. Concave down for $x < 1.33$, then concave up. Inflection point $(1.33, 13.14)$; two zeros. **29.** Use $[-3, 3]$ by $[0, 34]$. Falling to a local minimum at $(-1.106, 17.944)$; rising to a local maximum at $(0.270, 20.130)$; falling to a local minimum at $(0.835, 19.927)$; rising thereafter. Concave down between inflection points $(-0.578, 18.865)$ and $(0.578, 20.021)$; concave up elsewhere. **31.** Use $[-8, 4]$ by $[-20, 35]$. Rising to a local maximum at $(-4, 32)$; falling to a local minimum at $(-1.52, -0.18)$; rising to a local maximum at $(0.29, 13.10)$; falling thereafter. Concave up between inflection points $(-2.86, 15.16)$ and $(-0.50, 7.03)$; concave down elsewhere. **33.** Falling to local minimum at $(-2.60, -7.58)$; rising to local maximum at $(-1.08, 21.19)$; falling to local minimum at $(0.51, -0.48)$; rising to $(2.35, 52.99)$; falling thereafter. Concavity: up until $(-2.03, 4.05)$, down until $(-0.25, 9.97)$, up until $(1.65, 29.61)$, down thereafter. **35.** Use $[-2, 2]$ by $[-50, 50]$. Always rising. Inflection point at $(-0.286, -2.887)$ where concavity changes from concave down to concave up. **37.** 0 and 20 **39.** 1/2 **41.** 5/3 × 14/3 × 35/3 **43.** a) $v = -32t + 100$; b) 170.122 feet; c) −150.00 feet/sec **45.** 18 × 18 × 36 inches **47.** a) $x = 12$, $y = 6$; b) $x = 12$, $y = 6$

49. 0.653 **51.** 0.774 **51.** a) $a = 3$; $b = -9$; b) $a = 3$, $b = 12$ **53.** $4\sqrt{3}$ wide $\times 4\sqrt{6}$ inches deep. **55.** $v' = 0$ when $x(t) = a/2$ **57.** Maximum v at $r = r_0/2$ **59.** Degree 7; local maximum at $x = -1$, local minimum at $x = 5$; concave up on $(-0.48, 1)$ and $(2.81, \infty)$. **61.** Answer varies, 1.003 is common. **63.** c) 1.13716, −0.39493, −0.74223 **65.** Use $x - (\tan x) \times (\cos x)^2 \rightarrow x$.

Section 4.4

1. No extrema or inflection points; concave up if $x < 0$, concave down if $x > 0$. $[-5, 5]$ by $[-5, 5]$ shows a complete graph. **3.** Local minima at $(\pm 1, 2)$, $[-5, 5]$ by $[0, 5]$ shows a complete graph. **5.** Always falling. Concavity: down$(-\infty, -2)$, up$(-2, 0)$, down$(0, 2)$, up$(2, \infty)$. $(0, 0)$ is inflection point. **7.** $[-5, 5]$ by $[-5, 5]$ shows a complete graph. $(0, -1)$ is a local maximum. No inflection points. **9.** Local minimum at $(0, 2)$, $[-5, 5]$ by $[-5, 5]$ shows complete graph. **11.** $[-10, 10] \times [10, 10]$ shows a complete graph. **13.** $[-5, 4]$ by $[-10, 10]$ shows a complete graph. Local minimum at $(0.575, 0.144)$ **15.** Use $[-4, 4]$ by $[-0.5, 1.5]$. $(-2.41, -0, 21)$ local minimum, $(-0.27, 0.69)$ inflection point, $(0.41, 1.21)$ local maximum, $(1.33, 0.84)$ inflection point. **17.** $[-4, 4]$ by $[-10, 10]$ shows a complete graph. $(-0.47, -3.33)$ local maximum, $(0.49, 0.80)$ local minimum. **19.** Graph needs both $[-2, 6]$ by $[-10, 30]$ and $[-2, 2]$ by $[-2, 2]$ to be seen completely. Local minima at $(0.22, -1.59)$ and $(2.54, 18.46)$; local maximum at $(1.20, -0.89)$; inflection point at $(0.86, -1.16)$. **21.** $[-5, 5]$ by $[-15, 15]$ shows a complete graph. Inflection points at $(0, -0.5)$ and $(-3.02, 10.88)$; local minimum at $(1.67, 2.88)$. **23.** Three views are necessary: $[-5, 5]$ by $[-100, 100]$, $[-3.2, -2.8]$ by $[-100, 100]$ and $[2.8, 3.2]$ by $[-100, 100]$. Local maximum at $(2.91, 45.57)$, local minumum at $(3.08, 62.54)$. **25.** $[-4, 4]$ by $[-1, 2]$ shows a complete graph. Inflection points at $(\pm 1.17, 1.49)$. **27.** 16 **29.** 12×18m., 72m **31.** $x = 15$, $y = 5$ **33.** $h = r = 10/\sqrt[3]{\pi}$ **35.** a) 16 b) −1 **37.** $y \geq 147/9 > 0$ **39.** Minimum value occurs when $x = 51/8$; minimum $L = 11.04$ inches. **41.** $50 + c/2$ **43.** $(h/2km)^{1/2}$ **45.** maximum of p is $p(3) = 0$ **47.** The denominators of f, f', f'' have the same zeros.

Section 4.5

1. Always rising; concave up for $x < 0$, concave down for $x > 0$. **3.** For $x > -3/2$, y is rising and concave down. **5.** The interval $(-0.144, 1.999)$ between $3x + 5 = 3\pi/2$ and $3x + 5 = 7\pi/2$ is one period; $y = 1$, a minimum at the endpoints. $y = 5$ a maximum at $x = 0.951$. Inflection points at $(.432, 3.01)$, $(1.147, 3.01)$. **7.** Local maximum at $(\pi/12, \sqrt{2})$, inflection point $(3\pi/12, 0)$, local minimum $(5\pi/12, -\sqrt{2})$, inflection point $(7\pi/12, 0)$, local maximum $(9\pi/12)$. **9.** Always decreasing; concave down $x < 3$, concave up $x > 3$. **11.** Defined for $x \geq 0$; always increasing, always concave down. **13.** Minimum at $(3\pi/4, -2)$; always concave up. **15.** Inflection points at $(0, 0)$ and $(0.27, 1.34)$. Local minimum at $(-0.35, -0.58)$ **17.** $x > 2 \Rightarrow y$ increasing and concave down. **19.** Graph on $[-\pi, \pi]$ by $[-2, 2]$. Minimum at $(0, 1)$, maximum at $(\pi/2, -1)$. **21.** Use $[-5, 5]$ by $[0, 100]$; always rising and concave up. **23.** Graph $y = 3(\ell n(x + 1))/\ell n\, 2$ in $[-5, 5]$ by $[-10, 10]$. **25.** Use $[-5, 5]$ by $[0, 4]$; concave down where defined. **27.** Use $[-5, 5]$ by $[-4, 4]$; always rising; inflection point $(0, 0)$. **29.** Use $[-3, 3]$ by $[-1, 1]$; inflection points at $(\mp 0.90, \pm 0.70)$. **31.** Use $[-3, 3]$ by $[-1, 10]$; minimum at $(0, 1)$. **33.** Inflection point at $(1, 0)$. **35.** Use $[-\pi/2, 7\pi/2]$ by $[-8, 8]$; $(0, 0)$ and $(\pi, 0)$ are inflection points. **37.** Use $[-\pi/2, \pi/2]$ by $[-3, 3]$; minimum at $(-3\pi/8, -1.41)$; maximum at $(\pi/8, 1.41)$; inflection points at $(-\pi/8, 0)$ and $(3\pi/8, 0)$. **39.** Use $[-\pi, \pi]$ by $[-6, 6]$; $(-1.30, -4.13)$ minimum, $(-0.85, -3.88)$ inflection point, $(-0.57, -2.73)$ maximum, $(0.58, 3.72)$ minimum, $(0.86, 3.90)$ inflection point, $(1.30, -8.67)$ maximum. **41.** Minimum at $(0.36, -0.37)$, analytically at $(1/e, -1/e)$. **43.** Local minima at $(\pm 0.60, -0.18) =$

$(\pm\sqrt{(1/e)}, -0.18)$. **45.** Minimum at $(0, 1/2)$ **47.** Minimum/maximum at $(\mp 0.86, \mp 0.52)$; inflection points at $(\mp 1.48, \mp 0.33)$ and $(0, 0)$. **49.** Maxima at $(\pm 4.49, -4.60)$, $(\pm 10.92, -10.96)$; minima at $(\pm 7.73, 7.80)$. **51.** Inflection point at $(1.088, 0.009)$. **53.** $\min y = (-5)^{3/5} = -2.63$; $\max y = 2.63$ **55.** maximum $= 7.42$, minimum $= -172.64$ **57.** $79/4$ and $1/4$ **59.** $\pi/2$ **61.** Maximum area occurs when $a^2 = 10$. **63.** $\sqrt{1.25}$ **65.** Minimum value is $f(\pi) = 0$; $f \geq 0$ is never negative. **67.** Reflect B in the mirror, obtaining B', AOB' must be a line. **69.** $r = 10\sqrt{2}/\sqrt{3}$, $h = 20/\sqrt{3}$

Section 4.6

1. $dA/dt = 2\pi r\, dr/dt$ **3.** $dV/dt = 3x^2 dx/dt$ **5.** $dV/dt = (2/3)\pi rh\, dr/dt$ **7.** $\pi\, cm^2/\min$ **9.** a) $14\, cm^2/\sec$, increasing, b) $0\, cm/\sec$, constant, c) $-14/13\, cm/\sec$ decreasing. **11.** -12 ft/sec, $-119/2$ ft^2/sec **13.** 0.0239 in^3/min **15.** $dV/dt = K(4\pi r^2) = 4\pi r^2\, dV/dt \Rightarrow dr/dt = K$ **17.** 5/2 ft/sec **19.** $10/9\pi$ in/min, $10/9\pi$ in/min **21.** 0.2772 units/min **23.** 1 rad/sec, 0 rad/sec **25.** 8 ft/sec, decreasing 3 ft/sec **27.** 20 ft/sec **29.** 80 mph **31.** 29.5 knots

Section 4.7

1. a) x^2+C b) $3x+C$, c) x^2+3x+C **3.** a) x^3+C b) $x^3/3+C$ c) $x^3/3+x^2+x+C$ **5.** a) $x^{-3}+C$ b) $-x^{-3}/3+C$ c) $-x^{-3}/3+x^2+3x+C$ **7.** a) $x^{3/2}+C$ b) $(8/3)x^{3/2}+C$ c) $(1/3)x^3-(8/3)x^{3/2}+C$ **9.** a) $x^{2/3}+C$ b) $x^{1/3}+C$ c) $x^{-1/3}+C$ **11.** a) $(1/3)\cos 3x+C$ b) $-3\cos x+C$ c) $-3\cos x+(1/3)\cos 3x+C$ **13.** a) $\tan x+C$ b) $\tan 5x+C$ c) $(1/5)\tan 5x+C$ **15.** a) $\sec x+C$ b) $\sec 2x+C$ c) $2\sec 2x+C$ **17.** $x+(1/2)\cos 2x+C$ **19.** a) $-\sqrt{x}+C$ b) $x+C$ c) $\sqrt{x}+C$ d) $-x+C$ e) $x-\sqrt{x}+C$ f) $-3\sqrt{x}-2x+C$ g) $x^2/2-\sqrt{x}+C$ h) $-3x+C$ **21.** $y=x^2-7x+10$ **23.** $y=x^3/3+x+1$ **25.** $y=5/x+2$ **27.** $y=x^3+x^2+x-3$ **29.** $y=x+\sin x+4$ **31.** $y=-x^3+x^2+4x+1$ **33.** $s=4.9t^2+10$ **35.** $s=16t^2+20t$ **37.** $y=2x^{3/2}-50$ **39.** $x=x^3-3x^2+12x$ **41.** 48 m/sec downwards **43.** 14 m/sec **45.** $y=(c-kt)^2/4$ **47.** d

Practice Exercises, Chapter 4

1. $-40\ m^2$/sec **3.** 1 cm/min **5.** b) $-125/144\pi$ ft/min **7.** -2 rad/sec **9.** Let $y = x^4+2x^2-2$, $y(0)y(1) < 0$; $y' > 0$ on $(0,1)$. **11.** $y' = 1/(x+1)^2 > 0$ **13.** Use $[-6,6]$ by $[-4,4]$; inflection point at $(-1,0)$. **15.** Use $[-6.6]$ by $[-4,4]$; inflection point at $(2,0)$. **17.** Use $[-6,6]$ by $[-4,4]$ **19.** Use $[-6.6]$ by $[-50,50]$. Since $y' \leq 0$ everywhere, there are no hidden extrema. **21.** Use $[-6,6]$ by $[-8,3]$ and $[-0.5,0.5]$ by $[0.99,1.01]$ to show the behavior of the graph. **23.** Use $[-5,5]$ by $[-15,25]$ **25.** Use $[-5,5]$ by $[-15,20]$; inflection point at $(3.71,-3.41)$. **27.** Graph $y = (\ell n\,|x|)/\ell n\, 3$ in $[-1,4]$ by $[-5,5]$. **29.** Use $[1.9,4]$ by $[-5,5]$ **31.** Use $[-1,2]$ by $[-1,2]$. $(0.50, 0.707)$ is a local maximum. **33.** Local minimum at $x=-1$, inflection points at $x=0$ and $x=2$. **35.** -3 **37.** a) T b) P **39.** $(-6,-74)$, $\left(\frac{-8\pm\sqrt{104}}{4}, 16.25\right)$ are the minimum and maximum points. **41.** Particle starts at $(5,0)$, moves left until it reaches $(0.94,0)$ then moves right. **43.** $r=25$, $s=50$ **45.** Height $=3$ft, side of base $=6$ft. **47.** $h=2$, $r=\sqrt{2}$ **49.** $x=48/\sqrt{7}$ miles, $y=36/\sqrt{7}$ miles. **51.** 276 Grade A tires, 553 Grade B tires **53.** All such functions are represented by the formula. **55.** a) C b) $x+C$ c) $x^2/2+C$ d) $x^3/3+C$ e) $x^{11}/11+C$ f) $-x^{-1}+C$ g) $-x^{-4}/4+C$ h) $(2/7)x^{7/2}+C$ i) $(3/7)x^{7/3}+C$ j) $(4/7)x^{7/4}+C$ k) $(2/3)x^{3/2}+C$ l) $2x^{1/2}+C$ m) $(7/4)x^{4/7}+C$ n) $(-3/4)x^{-4/3}+C$ **57.** $x^3+(5/2)x^2-7x+C$ **59.** $(2/3)x^{3/2}+2x^{1/2}+C$ **61.** $(3/5)\sin 5x+C$ **63.** $\tan 3x+C$ **65.** $(1/2)x-\sin x+C$ **67.** $3\sec(x/3)+5x+C$ **69.** $\tan x-x+C$ **71.** $x-(1/2)\sin 2x+C$ **73.** $y=1+x+x^2/2+x^3/6$ **75.** $y=x-1/x-1$

77. $y = \sin x$ **79.** $y = x$ **81.** Duck! **83.** a) $f(0)f(1) < 0,\ f'(x) > 0 \Rightarrow f = 0$ has exactly one solution; b) 0.682 **85.** 0.5604 **87.** 2.989

Chapter 5

Section 5.1

1. a)

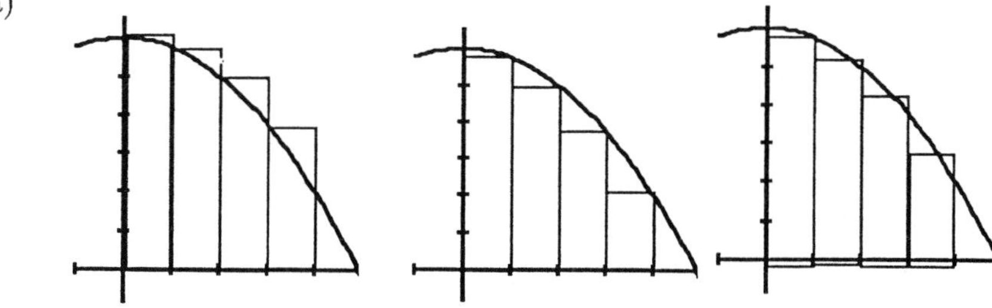

b) $LR_5(6-x^2) = (6-0^2)\Delta x+(6-(.4)^2)\Delta x+(6-(.8)^2)\Delta x+(6-(1.2)^2)\Delta x+(6-(1.6)^2)\Delta x = 10.08(\Delta x = 0.4)$. $RR_5(6-x^2) = (6-.4^2)\Delta x+(6-.8^2)\Delta x+(6-1.2^2)\Delta x+(6-1.6^2)\Delta x+(6-2^2)\Delta x = 8.48$. $MR_5(6-x^2) = (6-.2^2)\Delta x+(6-.6^2)\Delta x+(6-1^2)\Delta x+(6-1.4^2)\Delta x+(6-1.8^2)\Delta x = 9.36$.

3. a)

b) $LR_5(x+1) = [(0+1)+(1+1)+(2+1)+(3+1)+(4+1)]\Delta x = 15(\Delta x = 1)$. $RR_5(x+1) = [(1+1)+(2+1)+(3+1)+(4+1)+(5+1)]\Delta x = 20$. $MR_5(x+1) = [(.5+1)+(1.5+1)+(2.5+1)+(3.5+1)+(4.5+1)]\Delta x = 17.5$.

5. a)

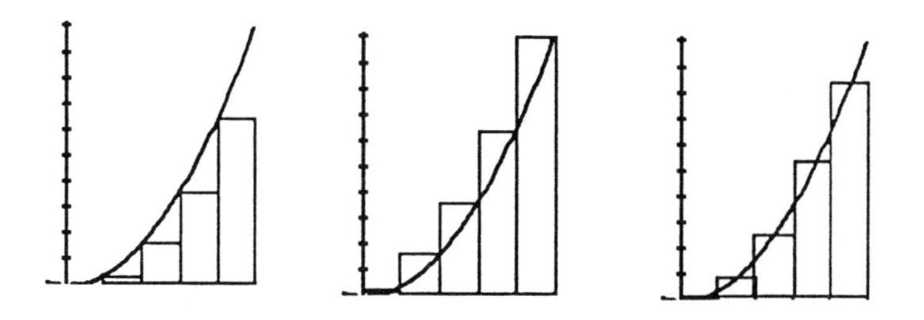

b) $LR_5(2x^2) = [2(0)^2+2(1)^2+2(2^2)+2(3^2)+2(4^2)]\Delta x = 60(\Delta x = 1)$. $RR_5(2x^2) = [2(1^2)+2(2^2)+2(3^2)+2(4^2)+2(5^2)]\Delta x = 110$. $MR_5(2x^2) = [2(.5^2)+2(1.5^2)+2(2.5^2)+2(3.5^2)+2(4.5^2)]\Delta x = 82.5$.

7.

n	$LR_n f$	$RR_n f$	$MR_n f$
10	12.645	14.445	13.4775
100	13.41045	13.59045	13.499775
1000	13.4910045	13.5090045	13.49999775

11.

n	$LR_n f$	$RR_n f$	$MR_n f$
10	1.98352..	1.98352..	2.00825..
100	1.99984..	1.99984	2.00008..
1000	1.99999..	1.99999..	2.00000..

9.

n	$LR_n f$	$RR_n f$	$MR_n f$
10	268.125	393.125	325.9375
100	333.78125	333.78125	327.48438..
1000	326.87531..	328.12531..	327.49984..

13.

n	$LR_n f$	$RR_n f$	$MR_n f$
10	1.77264..	1.77264..	1.77227..
100	1.77245..	1.77245..	1.77245..
1000	1.77245..	1.77245..	1.77245

15. 17.5, 83, 13.5, 327.5, 2, respectively. **19.** In $\text{LR}_n x^2$ each of the rectangles used is an inscribed rectangle and underestimates the area in its interval under the curve $y = x^2$. In $\text{RR}_n x^2$ each of the rectangles used is a superscribed rectangle and overestimates the area in its interval under curve $y = x^2$. **21.** Let the division points of the interval be $-2 = x_0, x_1, x_2, \ldots, x_n = 2$. Let $S = \sum_{k=1}^{n-1} f(x_k)\Delta x$. Then $\text{LR}_n f = f(x_0)\Delta x + S$ and $\text{RR}_n f = S + f(x_n)\Delta x$. Since $f(-2) = 0 = f(2)$, $\text{LR}_n f = \text{RR}_n f$. This can easily be generalized to prove the statement in Exercise 24 because there $f(a) = f(b)$ by symmetry. **23.** The graph of $y = \sin x$, $0 \leqq x \leqq \pi$, has $x = \pi/2$ as a line of symmetry. Now apply the method of Exercise 21.

Section 5.2

1. $\sum_{k=1}^4 \frac{1}{k} = \frac{1}{1} + \frac{1}{2} + \frac{1}{3} + \frac{1}{4} = \frac{25}{12}$ **3.** $\sum_{k=1}^3 (k+2) = (1+2) + (2+2) + (3+2) = 12$ **5.** $\sum_{k=0}^4 \frac{k}{4} = \frac{0}{4} + \frac{1}{4} + \frac{2}{4} + \frac{3}{4} + \frac{4}{4} = \frac{5}{2}$ **7.** $\sum_{k=1}^2 \frac{6k}{k+1} = \frac{6\cdot 1}{1+1} + \frac{6\cdot 2}{2+1} = 7$ **9.** $\sum_{k=1}^5 k(k-1)(k-2) = 1(1-1)(1-2) + 2(2-1)(2-2) + 3(3-1)(3-2) + 4(4-1)(4-2) + 5(5-1)(5-2) = 90$ **11.** $\sum_{k=1}^4 \cos k\pi = \cos(1\cdot\pi) + \cos 2\pi + \cos 3\pi + \cos 4\pi = 0$ **13.** $\sum_{k=1}^4 (-1)^k = (-1)^1 + (-1)^2 + (-1)^3 + (-1)^4 = 0$ **15.** $\sum_{k=1}^6 k$ **17.** $\sum_{k=1}^4 \frac{1}{2^k}$ **19.** $\sum_{k=1}^5 (-1)^{k+1}\frac{k}{5}$ **21.** 55 **23.** −91 **25.** −20 **27.** 1,000,000 **29.** $RR_n(x+1) = 5\left[\frac{5}{2}\frac{n+1}{n} + 1\right]$. $\text{Lim}_{n\to\infty} RR_n(x+1) = \frac{35}{2} = A_0^5(x+1)$ **31.** $RR_n(x^2+2) = 9\left[1 + \frac{n+1}{n} + \frac{n+1}{n}\frac{(2n+1)}{2n}\right]$. $A_1^4(x^2+2) = \lim_{n\to\infty} RR_n(x^2+2) = 27$ **33.** $RR_n(2x^3+3) = 5\left[\frac{125}{2}\left(\frac{n+1}{n}\right)^2 + 3\right]$. $\lim_{n\to\infty} RR_n(2x^3+3) = 327.5$ **35.** $LR_n x^2 = \frac{125}{n^3}\frac{(n-1)n[2(n-1)+1]}{6}$ **37.** $MR_n x^2 = \frac{125}{4n^3}\frac{(2n-1)n(2n+1)}{3}$ **39.** $MR_n x^3 = \frac{625}{8}(2 - \frac{1}{n^2})$ **41.** All **43.** a) −15 b) 1 c) 1 d) −11 e) 16 **45.** 78

Section 5.3

1. $\int_0^2 x^2 dx$ **3.** $\int_{-7}^5 (x^2 - 3x)dx$ **5.** $\int_2^3 \frac{1}{1-x}dx$ **7.** $\int_0^4 \cos x\, dx$
9. $\text{FnInt}(2 - x - 5x^2, x, -1, 3) = -42.666...$

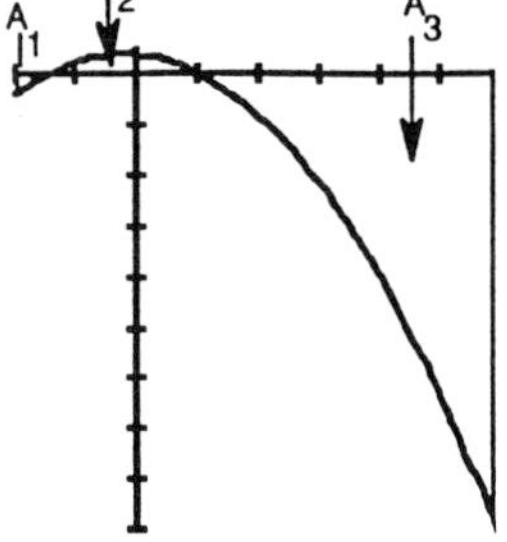

[-1,3] by [-46,2.5]

The definite integral has the value $A_2 - A_1 - A_3$. Since A_1 and A_3 are below the x-axis, they each contribute a negative value to the integral.
11. $\text{FnInt}(\sin(x^2), x, 0, 2\pi) = .642...$. Graph $y = \sin(x^2)$ in $[0, 2\pi]$ by $[-1, 1]$. The integral is

the sum of the signed areas (positive if above the x-axis, negative if below the x-axis) between the x-axis and the curve. **13.** $\int_{1/\pi}^{2/\pi} \sin(\frac{1}{x})dx = 0.238287$ **15.** $-\int_{\pi/2}^{3\pi/2}(\cos x)dx = 2$ **17.** $\int_0^5 \sqrt{25-x^2}dx = 19.63496$ **19.** $\int_{-1}^{1}\sqrt{1-x^2}dx = \frac{\pi}{2}$. FnInt$(\sqrt{1-x^2}, x, -1, 1) = 1.57080$ **21.** $\int_{-1}^{1}(1-|x|)dx = 1 =$ FnInt$(1-|x|, x, -1, 1)$ **23.** Each has the form $\sum_{k=1}^{n} f(c_k)\Delta x_k$ with $\Delta x_k = (b-a)/n$ for all n. In $\text{LR}_n f, \text{RR}_n f$, and $\text{MR}_n f, c_k$ is, respectively, the left-hand endpoint, the right-hand endpoint and the midpoint of the interval $[x_{k-1}, x_k]$. **25.** $\frac{16}{3}$

Section 5.4

1. $\frac{8}{3}$ **3.** $\frac{16}{3}$ **5.** $\frac{4}{3}$ **7.** $\frac{27}{4}$ **9.** 1 **11.** 1 **13.** $\frac{2}{\pi}$ **15.** $1+\sqrt{3}$ **17.** $\frac{b^{n+1}}{n+1}$ **19.** a) 0 b) -8 c) -12 d) 10 e) -2 f) 16 **21.** $\frac{-13}{3}$ **23.** $\frac{2}{\sqrt{3}}$ **25.** $-.475...$ **27.** a) 5 b) 5 c) -5 **29.** $\frac{1}{2} \leq \int_0^1 \frac{1}{1+x^2}dx \leq 1$ **31.** This is an immediate consequence of the Mean Value Theorem for Definite Integrals.

Section 5.5

1. 3 **3.** 1 **5.** $\frac{5}{2}$ **7.** 2 **9.** $2\sqrt{3}$ **11.** 0 **13.** $\frac{8}{3}$ **15.** $\frac{5}{2}$ **17.** $\frac{1}{2}$ **19.** $\frac{5}{6}$ **21.** π **23.** $F(x) = \frac{x^2}{2} - 2x$. The two graphs appear to be the same in $[-10, 10]$ by $[-10, 10]$. $F(0.5) = -0.875$ and we get the same value after zooming in. $F(1) = -1.5$ compared to -1.516 as one approximation. $F(1.5) = -1.875$ compared to -1.879 as one approximation. $F(2) = -2$ compared to -2. $F(5) = 2.5$ compared to 2.53 as one approximation. **25.** $F(x) = \frac{x^3}{3} - \frac{3}{2}x^2 + 6x$. The two graphs are indistinguishable in the viewing rectangle $[-15, 15]$ by $[-1000, 1000]$. The values of $F(x)$ and FnInt$(f(t), t, 0, x)$ agree when accurately calculated. **27.** Graph $y =$ FnInt$(t^2 \sin t, t, 0, x)$ in the viewing rectangle $[-3, 3]$ by $[0, 9]$. **29.** Graph $y =$ FnInt$(5e^{-0.3t^2}, t, 0, x)$ in the viewing rectangle $[0, 5]$ by $[0, 10]$. **31.** The two graphs are identical. **33.** $K = -\frac{3}{2}$ **35.** $x = 0.70$ **37.** $\sqrt{1+x^2}$ **39.** $\frac{\sin x}{2\sqrt{x}}$ **41.** d) **43.** b) **47.** a) 9 dollars b) 10 dollars **49.** $I_{av} = 300$. Average daily holding cost $= 6$ dollars per day. **51.** $x = a$ **53.** $f(4) = 1$ **55.** a) Compare your drawing with the result of graphing $y = (\cos x)/x$ in $[-15, 15]$ by $[-1, 1]$. One $x-$ and $y-$ axes are asymptotes. b) Graph $y =$ fnInt$((\cos t)/t, t, 1, x)$ in $[0, 15]$ by $[-1, 1]$. c) Because $f(0)$ is undefined d) For $x > 0$, $g(x)$ and $h(x)$ have the same derivative $f(x)$ and so they differ by an additive constant. This is confirmed if one graph can be obtained from the other by a vertical shift. Along with the function in b), graph $y =$ fnInt$((\cos t)/t, t, 0.5, x)$ in $[0.01, 3]$ by $[-3, 3]$ to see that this is the case. Alternatively, $\int_{0.5}^{x} f(t)dt = \int_{0.5}^{1} f(t)dt + \int_1^x f(t)dt \approx 0.5 + \int_1^x f(t)dt$.

Section 5.6

1. $\int x^3 dx = \frac{x^4}{4} + C$. The graphs of $\frac{x^4}{4}$ and FnInt$(t^3, t, 0, x)$ in $[-3, 3]$ by $[0, 9]$ see to be identical. **3.** $\frac{x^2}{2} + x + C$ **5.** $2x^{3/2} + C$ **7.** $\frac{3}{2}x^{2/3} + C$ **9.** $\frac{5}{3}x^3 + x^2 + C$ **11.** $\frac{1}{2}x^4 - \frac{5}{2}x^2 + 7x + C$ **13.** $2\sin x + C$ **15.** $-3\cos\frac{x}{3} + C$ **17.** $-3\cot x + C$ **19.** $-\frac{1}{2}\csc x + C$ **21.** $4\sec x + -2\tan x + C$ **23.** $-\frac{1}{2}\cos 2x + \cot x + C$ **25.** $2y - \sin 2y + C$ **27.** $-\frac{1}{4}\cos 2x + C$ **29.** $\tan\theta + C$ **37.** $y = 2x^{3/2} - 50$ **39.** $y = \int_0^x 2^x dx + 2$ **41.** $y = 2x$ **43.** $y = \frac{x^3}{16}$ **45.** 16 ft/sec^2 **47.** 1.24sec

Section 5.7

1. $-\frac{1}{3}\cos 3x + C$ **3.** $\frac{1}{2}\sec 2x + C$ **5.** $(7x-2)^4 + C$ **7.** $-6\sqrt{1-r^3} + C$ **9.** a) $-\frac{\cot^2 2\theta}{4} + C$ b) $-\frac{\csc^2 2\theta}{4} + C$ **11.** $\frac{3}{16}$ **13.** $\frac{1}{2}$ **15.** 0 **17.** $\frac{2}{3}$ **19.** $\frac{1}{1-x} + C$ **21.**

$\tan(x+2)+C$ **23.** $3(r^2-1)^{4/3}+C$ **25.** $\sec(\theta+\frac{\pi}{2})+C$ **27.** $2(1+x^4)^{3/4}+C$ **29.** a) $\frac{14}{3}$ b) $\frac{2}{3}$ **31.** a) $\frac{1}{2}$ b) $-\frac{1}{2}$ **33.** a) $\frac{1}{2}(\sqrt{10}-3)$ b) $\frac{1}{2}(3-\sqrt{10})$ **35.** a) $\frac{45}{8}$ b) $-\frac{45}{8}$ **37.** a) $\frac{1}{6}$ b) $\frac{1}{2}$ **39.** a) 0 b) 0 **41.** $2\sqrt{3}$ **43.** 0 **45.** 8 **47.** $\frac{38}{3}$ **49.** $\frac{16}{3}$ **51.** $s=(3t^2-1)^4-1$ **53.** $s=-6\cos(t+\pi)-6$ **55.** 1 **57.** The answers can be seen to be equivalent: $\sin^2 x+C_1=(1-\cos^2 x)+C_1=-\cos^2 x+(1+C_1)=-\cos^2 x+C_2=-\left(\frac{1+\cos 2x}{2}\right)+C_2=-\frac{\cos 2x}{2}+(-\frac{1}{2}+C_2)=-\frac{\cos 2x}{2}+C_3$. The graph of any one of the antiderivatives can be obtained from the graph of any other antiderivative by a vertical shift verifying that they differ by an additive constant.

Section 5.8

1. $2(\ell n\,2-\ell n\,3)$ **3.** $-\ell n\,2$ **5.** $2\,\ell n\,3-\ell n\,2$ **7.** $\ell n\,3+\frac{1}{2}\,\ell n\,2$ **9.** 7 **11.** 2 **13.** $3e^2$ **15.** $\frac{2}{x}$. The result is supported by graphing $2/x$ and $\text{der1}(\ell n(x^2),x,x)$ in $[-5,5]$ by $[-5,5]$. **17.** $-\frac{1}{x}$ **19.** $\frac{1}{x+2}$ **21.** $\frac{\sin x}{2-\cos x}$ **23.** $\frac{1}{x\,\ell n\,n}$ **25.** $2e^x$ **27.** $-e^{-x}$ **29.** $\frac{2}{3}e^{2x/3}$ **31.** xe^x **33.** $\frac{e^{\sqrt{x}}}{2\sqrt{x}}$ **35.** $\ell n\,4$ **37.** 8 **39.** $\frac{1}{2}(e^2+2e-3)$ **41.** e^2-e **43.** 2 **45.** 1 **47.** $\ell n(3/2)$ **49.** 0 **51.** Graph $y=f(x)$ in the viewing rectangle $[-5,5]$ by $[0,0.5]$. **53.** Graph $y=\text{fnInt}(f(t),t,-t,x)$ in $[-5,5]$ by $[0,1.5]$. **55.** Graph $y=\text{fnInt}(f(t),t,x,0)$ in $[-5,0]$ by $[0,1]$ **57.** 0.682689 **59.** 0.99730 **61.** $-\frac{\ell n\,4}{5}$ **63.** $\ell n\,0.01$ **65.** $\pm0.75\ldots$ **67.** $-\frac{\ell n\,2}{k}$ **69.** $-300\,\ell n\,10$ **71.** $-0.27\ldots$ **73.** e^{-t+5} **75.** $\frac{1-e^5}{2}$ **77.** $\frac{\ell n(e^{x^2-3}+1)}{\ell n\,2}$ **79.** $\{0.679\ldots,1.3086\}$ **81.** 1250 **83.** a) The amount in the amount after t years is $A(t)=A_0e^t$. b) 1.0986 years (rounded) c) e times the original amount **85.** $4.875\ldots\%$ **87.** a) 14 years, 10 years b) 14%, 3.5% **89.** About three quarters of the way through 2012 **91.** About 3.9% **93.** About 17.3 years **95.** a) $p(x)=20.09e^{1-0.01x}$ b) (rounded) \$49.41, \$22.20 d) Graph $y=20.09xe^{1-0.01x}$ in $[0,200]$ by $[0,2100]$ **97.** a) $0.04\,\ell n\,10$ b) 109.65 (rounded)

Section 5.9

1. a) 2 b) 2 c) 2 **3.** a) 4.25 b) 4 c) 4 **5.** a) $5.146\ldots$ b) $5.252\ldots$ c) $16/3=5.333\ldots$ **7.** $|E_T|\leq\frac{1}{600}=0.0016666\ldots$ **9.** a) $n=1$ b) $n=2$ **11.** a) $n=283$ b) $n=2$ **13.** a) $n=76$ b) $n=12$ **15.** 1013 **17.** $466.66\ldots\text{in}^2$ **19.** Using the odd-numbered hours, we get 56.86 kwh per customer. **21.** $3.1379\ldots$, 3.14029 **23.** $1.3669\ldots$, $1.3688\ldots$ **25.** a) $0.057\ldots$ and $0.0472\ldots$ b) Let $y_1=\sin x/x$, $y_2=\text{der}\,2(y_1,x,x)$, $y_3=\left(\frac{1.5\pi}{12}\right)\left(\frac{1.5\pi}{10}\right)^2 abs\,y_2$, $y_4=\text{nDer}(\text{nDer}(y_2,x,x),x,x)$, $y_5=\frac{1.5\pi}{180}\left(\frac{1.5\pi}{10}\right)^4 abs\,y_4$. We graph y_3 in $[\frac{\pi}{2},\frac{3\pi}{2}]$ by $[0,0.03]$ and y_5 in $[\frac{\pi}{2},\frac{3\pi}{2}]$ by $[0,2\times10^{-4}]$. $\max y_3=2.166\ldots\times10^{-2}$, $\max y_5=1.71\ldots\times10^{-4}$ c) $\max ET(x)=5.41\ldots\times10^{-3}$, $\max ES(x)=1.07\ldots\times10^{-5}$ d) $\max ET(x)=8.66\times10^{-4}$, $\max ES(x)=2.74\ldots\times10^{-7}$ e) We cannot find the exact value, but we can approximate the integral as closely as we like by increasing n. With $n=50$, Simpson's Rule gives the value $0.0473894\ldots$. By d) the error is at most 2.744×10^{-7}. **27.** Refer to the method of Exercise 25. a) $3.6664\ldots$ and $3.65348218\ldots$ b) $\max ET=0.2466\ldots$, $\max ES=2.55\ldots\times10^{-4}$ c) $\max ET=6.16\ldots\times10^{-2}$, $\max ES=1.59\ldots\times10^{-5}$ d) $\max ET=9.86\ldots\times10^{-3}$, $\max ES=4.08\ldots\times10^{-7}$ e) Simpson's Rule with $n=50$ yields $3.6534844\ldots$ with error at most $4.08\ldots\times10^{-7}$.

Practice Exercises, Chapter 5

1. a)

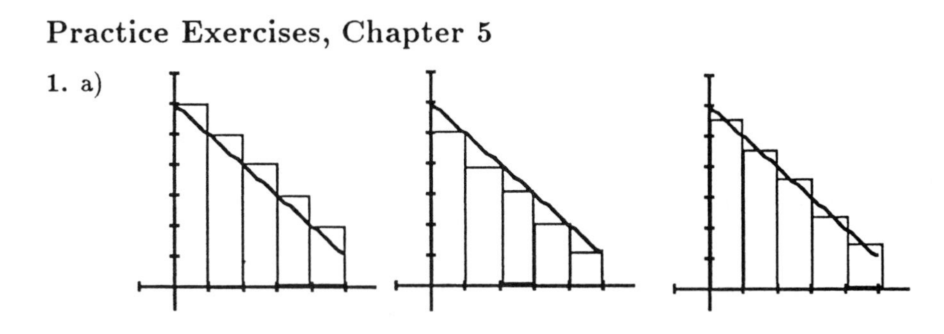

b) $20, 15, 17.5$

3.

	n = 10	n = 100	n = 1000
LR_n	22.695	23.86545	23.9865045
RR_n	25.395	24.13545	24.0135045
MR_n	23.9775	23.999775	23.99999775

5.

	n = 10	n = 100	n = 1000
LR_n	3.9670	3.99967	3.9999967
RR_n	3.9670	3.99967	3.9999967
MR_n	4.0165	4.00016	4.0000016

7. a) 75 b) -10 **9.** a) 0 b) 60 **11.** a) $\sum_{k=0}^{3} 2^k$ b) $\sum_{k=0}^{4} \frac{1}{3^k}$ c) $\sum_{k=1}^{5}(-1)^{k+1}k$ d) $\sum_{k=1}^{3} \frac{5}{2k}$ **13.** $\int_0^1 e^x\,dx$ **15.** $RR_n f = \frac{625}{2}(1+\frac{1}{n}) + \frac{75}{2}(1+\frac{1}{n})$. $\lim_{n\to\infty} RR_n f = 350$ **17.** a) π b) $-\pi$ c) -3π **19.** Approximately 1.14159 **21.** Area $= \frac{32}{3} = 10.666\ldots$. $2\,\text{fnInt}(\sqrt{x}, x, 0, 4) = 10.6667\ldots$ with tol $= 0.001$ **23.** 4 **25.** 3 **27.** 2 **29.** 13, $\text{FnInt}(3\sqrt{(4x+1)}, x, 0, 2) = 12.999\ldots$ **31.** 2 **33.** 0 **35.** 2 **37.** $3\sqrt{3} - \pi = 2.054559\ldots$ **39.** 0; $\text{fnInt}((\sin x \tan x)^{-1}, x, \frac{\pi}{4}, \frac{3\pi}{4}) = 1.9 \times 10^{-13}$ **41.** 0 **43.** $2\,\ell n\, 3$ **45.** 0 **47.** 4 **49.** $e^2 - e$ **51.** 7 **53.** e **55.** $\ell n(\frac{1+e}{2})$ **57.** 0 **59.** $\frac{dy}{dx} = 1$. Confirm by graphing $\text{nDer}(\ell n(e^x/2), x, x)$. **61.** $\frac{dy}{dx} = \frac{e^x}{1+e^x}$. We may confirm this graphically by graphing $\frac{e^x}{1+e^x} + 2$ and $\text{nDer}(\ell n(1+e^x), x, x)$ and seeing that the first curve may be obtained from the second by vertically shifting the latter curve 2 units upward. **63.** $\frac{dy}{dx} = e^{-x}(1-x)$ **65.** $\frac{dy}{dx} = \ell n\, x$. This can be confirmed by graphing $\ell n\, x + 2$ and $\text{nDer}(\text{fnInt}(\ell n\, t, t, 1, x), x)$ in $[1, 10]$ by $[-5, 5]$ and observing that one graph can be obtained from the other by a vertical shift. **71.** Graph $\text{fnInt}(0.25e^{-(0.5t^2)}, t, 0, x)$ in $[-10, 10]$ by $[-1, 1]$. There is no explicit elementary formula for this integral. **73.** No solution **75.** a) $2x$ b) $1 + \ell n\, 2$ c) -1 **77.** $y = e^x - 1$ **79.** $y = \ell n\, 2x$ **81.** $\{-1.61, 0.28, 1.61\}$ **85.** Both integrations are correct. $\frac{\sec^2 x}{2} = \frac{1+\tan^2 x}{2} = \frac{\tan^2 x}{2} + \frac{1}{2}$ shows that the two results differ by a constant. **87.** All are true except d) and f) **91.** a) $y = \cos x + x^2 + 2x + 2$ b) $y = \frac{2x^2}{1+x^2}$ c) $y = y_0 e^{-kt}$ **93.** b) **95.** 25° **97.** a) 2 b) $\frac{2a}{3}$ **99.** 26 subdivisions using the Trapezoidal Rule **101.** 8 **103.** About 5.3% **105.** About 7.19 years **107.** a) About 8% b) About 771.8 using the 8% of a) **109.** a) 14 years b) 10 years

Chapter 6

Section 6.1

1. 2 **3.** 2 **5.** $\pi/2$ **7.** 32/3 **9.** 32/3 **11.** 1/6 **13.** 26.15341 (use technology) **15.** 9/2 **17.** 0.22016 **19.** 1/6 **21.** 15.68376 **23.** 1/12 **25.** 1 **27.** $\sqrt{2}-1$ **29.** $C = 2^{4/3}$ **31.** 4

Section 6.2

1. $8\pi/3$ **3.** 36π **5.** $128\pi/7$ **7.** π **9.** $32\pi/3$ **11.** 2π **13.** 4π **15.** $4\pi \ln 4$ **17.** $2\pi/3$ **19.** $128\pi/5$ **21.** $117\pi/5$ **23.** $\pi^2 - 2\pi$ **25.** $4\pi/3$ **27.** 8π **29.** $500\pi/3$ **31.** $\pi^2 - 2\pi$ **33.** $\pi/3$ **35.** a. 8π, b. $32\pi/5$, c. $8\pi/3$, d. $224\pi/15$ **37.** a. $16\pi/15$, b. $56\pi/15$, c. $64\pi/15$ **39.** $c = 2\pi$, minimum value is 15.7392 **41.** 1053π **43.** $c = 0.1761$, $V = 4.656$

Section 6.3

1. 8π **3.** $3\pi/2$ **5.** 3π **7.** $4\pi/3$ **9.** $16\pi/3$ **11.** $8\pi/3$ **13.** $14\pi/3$ **15.** $6\pi/5$ **17.** a. $5\pi/3$, b. $4\pi/3$ **19.** a. $11\pi/15$, b. $97\pi/105$, c. $121\pi/210$, d. $23\pi/30$ **21.** a. $512\pi/21$ b. $832\pi/21$ **23.** a. $\pi/6$, b. $\pi/6$ **25.** $\pi/4$, 1.601937π

Section 6.4

1. 12 **3.** 14/3 **5.** 53/6 **7.** 123/32 **9.** 3/2 **11.** 6 **13.** $y = x^{1/2}$ **15.** $y = \ln x$ **17.** $L = \int_0^{1/2} \frac{1+x^2}{1-x^2}dx$ **19.** 21.0684 inches **21.** 204.3561 million miles **23.** 100.9080

Section 6.5

1. $4\pi\sqrt{5}$ **3.** $3\pi\sqrt{5}$ **5.** $98\pi/81$ **7.** 4π **9.** $(\sqrt{8}-1)\pi/9$ **11.** $2\pi(2\sqrt{2}-1)/3$ **13.** $253\pi/20$ **15.** 4π **17.** $12\pi/5$ **19.** $S_{AB} = 2\pi rh$ which is independent of a. **21.** 3.178

Section 6.6

1. 400 ft·lb **3.** 925 N·m **5.** 64800 ft·lb **7.** 1.2 N·m **9.** a. $104\frac{1}{6}$ ft·lb, b. 312.5 ft·lb **11.** a. 1,500,000 ft·lb, b. 100 minutes **13.** 245436.926 ft·lb **15.** 7238229.473 ft·lb **17.** a. 34582.652 ft·lb, b. 53482.473 ft·lb **19.** 2144605.85 N·m **21.** 967610.537 ft·lb **23.** 5.1398×10^{10} N·m

Section 6.7

1. 2812.5 lb **3.** 375 lb., No. **5.** 1166.67 lb **7.** 41.67 lb **9.** $F = 2618$ lb/in^3 **11.** 1161 lb **13.** 1034.16 ft^3 **15.** A plate, $h \times W$ with top d below the surface experiences a force of $\omega W\left[\frac{h^2+2dh}{2}\right] = \frac{\omega d+\omega(d+h)}{2}hW$

Section 6.8

1. 4 **3.** $M_0 = 8$, $\bar{x} = 1$ **5.** $M_0 = 68/3$, $\bar{x} = 17/7$ **7.** $(0, 2/3)$ **9.** $(16/105, 8/15)$ **11.** $(3/5, 1)$ **13.** $(0, \pi/8)$ **15.** $(1, -2/5)$ **17.** $(\frac{2}{4-\pi}, \frac{2}{4-\pi})$ **19.** $(0, 1)$ **21.** $(\frac{a}{3}, \frac{a}{3})$ **23.** $(14/9, (\ln 4)/3)$ **25.** $V = 32\pi$; $S = 32\sqrt{2}\pi$ **27.** $4\pi^2$ **29.** $(0, 2a/\pi)$ **31.** $(0, 4a/3\pi)$ **33.** $\frac{\pi a^3}{6}(4+3\pi)\sqrt{2}$

Section 6.9

1. 16 **3.** 16/3 **5.** $\pi \ln 2$ **7.** 8/3 **9.** s^2h, s^2h **11.** (b) 20 meters (c) 0 meters **13.** (b) 6 meters, (c) 2 meters **15.** (a) 245m, (b) 0 meters **17.** (b) 6 meters, (c) 4 meters **19.** (a) 2, 2 (b) 4, 4 (c) 4, 0 (d) 2, 2 **21.** about 68% **23.** $\sqrt{3}\pi \approx 1.7\pi$ **25.** 18

Practice Exercises, Chapter 6

1. 1 **3.** 9/2 **5.** 18 **7.** 9/8 **9.** $\frac{\pi^2}{32} + \frac{\sqrt{2}}{2} - 1$ **11.** 4 **13.** $(8\sqrt{2}-7)/6$ **15.** 3 **17.** (a) 2π, (b) π **19.** (a) 8π, (b) $544\pi/15$, (c) $256\pi/15$ **21.** $9\pi/2 - 8$ **23.** $\pi \ln 16$ **25.** $\frac{32\pi}{3} - \frac{13\sqrt{13}}{12}$ **27.** $2\sqrt{3}$ **29.** $6 + \ln 2$ **31.** $\frac{2\pi}{3}[26^{3/2} - 2^{3/2}] \approx 86.5\pi$ **33.** 3π **35.** $3\pi/2$ **37.** 4560 N·m **39.** 10 ft·lb, 30 ft·lb **41.** $6400\pi\omega/3$ ft·lb where $\omega = 62.5$ **43.** 333.3 lb **45.** 2200 lb **47.** $(0, 8/5)$ **49.** $(3/2, 12/5)$ **51.** $(0, 8/15)$ **53.** $(7, \frac{1}{3}\ln 2)$ **55.** $9\pi/280$ **57.** π^2 **59.** 18 **61.** (a) 5 ft, (b) 3 ft **63.**(b) 15 ft, (c) −5 ft

Index

A
absolute change, 3-85
absolute maximum, 4-2
absolute minimum, 4-2
absolute value, 1-23
  arithmetic with, 1-24
  equations and inequalities, 1-68
  inequalities, 1-25, 1-72–75
  with distance, 1-25
  with intervals, 1-69
absolute value finctions, 1-43–44
acceleration, 3-40
  finding velocity from, 4-80
accuracy agreement, 1-65, 3-12, 6-5
algebra,
  rules for antiderivatives, 5-55
  rules for finite sums, 5-12
algebraic properties, 2-17
  polynomial, 1-65
algebraic representation, 1-72
  complete graph, 1-10, 1-12, 4-21
algorithm,
  definition, 4-26
analysis of error,
  graphical solution of equations, 1-64
angiography, 3-87
angioplasty, 3-87
angle,
  double-angle formulas, 1-113
  half-angle formulas, 1-113
  sum and difference, 1-111
angles,
  of incidence and reflection, 4-65
  of inclination, 1-17
antiderivatives, 4-74
  definite integrals, 5-29–35
  finding, 4-77–79
applications,
  definite integrals, 6-1
approximations,
  analysis of error, 1-64
  definite integrals, 5-45
  error, 3-87
  Simpson's Rule, 5-105
  standard linear, 3-80
  Trapezoidal Rule, 5-104
arc length,
  of a curve, 6-36
  differential formula, 6-40
Archimedes Area Formula, 5-52
area, 5-1
  between curves, 6-1–9
  evaluation using subintervals, 5-48
  computing, 5-13–14
  of a surface of revolution, 6-45
  under a curve, 5-23–24
  under a graph,
    definition, 5-24
    of a nonnegative continuous
      function, 5-34
astroids, 6-41
asymptotes,
  end behavior, 2-43
  end behavior, finding, 2-35
  horizontal, 2-35
  vertical, 2-39, 2-41
average daily inventory, 5-50
average rate of change, 3-2, 3-41
average value of a function, 5-31
average velocity, 3-37

B
bacteria,
  cholera bacteria growth, 5-92
Barrow, Isaac, 5-70
behavior,
  end, 2-35, 2-39, 2-42
best location problems, 1-49

biology,
mathematical models in, 3-34
relief from a heart attack, 4-70
body's reaction to medicine, 3-34
boundaries,
with changing formulas, 6-5

C
calculators,
computing trig values, 1-101
graphing, 3-11, 4-28
Calculus and Computers, 5-101
cardiac output, 5-107–108
Cartesian coordinate system, 1-1
origin, 1-1
x-axis, 1-1
y-axis, 1-1
x-coordinate, 1-2
y-coordinate, 1-2
Cavalieri, Bonaventura, 6-81
Cavalieri's Theorem, 6-80
cell growth problem, 5-85
Celsius, 1-22, 1-30
center of gravity, 6-74
center of homogeneity, 6-74
center of mass, 6-64–74
along a line, 6-64
over a plane region, 6-68
thin, flat plates, 6-69
center of mass, 6-70
mass, 6-71
moments, 6-70
wires and thin rods, 6-66
center of mass, 6-67
mass, 6-67
moments, 6-67
center of uniformity, 6-74
centroids, 6-74
Chain Rule, 3-58–67
first form, 3-60
integration by substitution, 5-64–72
definition, 5-69
sines and cosines, 5-66
shorter form, 3-60
shortest form, 3-61
change,
absolute, relative, and percentage, 3-85
circles, 1-78
area of, 3-84
centered at the origin, 1-90
in a plane, 1-89
standard equation for, 1-89
unit, 1-91
using graphers to draw, 1-92
Cissoid of Diocles, 3-78
close,
sufficiently close to, 2-5
arbitrarily close to, 2-5
closed interval, 3-15
continuous functions on, 4-36
complete graph, 1-10
algebraic representation, 1-72
complex numbers, 1-34, 4-30
composites,
of continuous functions, 2-19
continuous compound interest,
formula, 5-87
computer generated graphs, 4-18
computer programs, 5-104–106
computer simulation, 4-84
computer visualization, 1-6
dot mode, 1-6
connected mode, 1-6
concavity, 4-15
second derivative test for, 4-16
confirming computer-generated graphs, 4-18–21
conic sections,
graphing, 1-93
connectivity, 2-23
Constant Multiple Rule, 2-34
differentiation, 3-22
constant of integration, 5-53
constant functions,
integrals of, 5-23
Consumer Price Index, 5-89
Purchasing Power of the Dollar, 5-90
continuity, 2-1,
at an interior point, 2-15
definition of, 2-13–15
test, 2-15
continuous extension, 2-18
continuous functions, 2-13
algebraic properties of, 2-17
composites of, 2-19
Intermediate Value Theorem, 2-23
Max-Min Theorem, 2-21
nonnegative,
area under the graph of, 5-4–7, 5-34
on closed intervals, 4-36
properties of, 2-20
zeros of, 5-48
contrapositive, 3-18

controlling function outputs,
a linear function example, 2-46
a quadratic function example, 2-47
area of a circle example, 2-51
rational function, 2-49
square root function example, 2-48
target values, 2-46
trigonometric function example, 2-49
coordinates,
Cartesian coordinate system, 1-1
pair, 1-2
coordinate plane,
distance between points, 1-23
cosecant (*see also* secant), 1-98
derivative of, 3-55
cosine, 1-98
calculating, 1-99
defining the inverse, 1-105
derivative of, 3-53
even function, 1-104
cotangent (*see also* tangent), 1-98
derivative of, 3-55
critical points, 4-36
curves
area under, 5-23–24
with infinite length, 6-40
planar,
length of, 6-34–37
cylindrical shells, 6-25–33

D

decimal numbers,
expression of, 1-67
definite integrals,
antiderivatives, 5-29
applications of, 6-1
approximation of, 5-45
as a limit of Riemann Sums, 5-21
existance of, 5-22
Mean Value Theorem for, 5-31, 6-44
numerical evaluation, 5-26–27
rules for, 5-29–30
substitution in, 5-70–72
Delesse, Achille Ernest, 6-84
Delesse's Rule, 6-84–86
density, 6-67
dependent variables, 1-34
derivatives, 3-1
Chain Rule, 3-58–67
differential notation, 3-88
exact, 3-12
First Derivative Theorem, 4-3–5
first derivative, 4-14
of a constant, 3-20
of e^x, 5-83
of functions, 3-8
graphs of, 3-11
higher order, 3-25
implicit differentiation, 3-74
implicit, 3-69–77
in economics, 3-42
Inside-Outside Rule, 3-63
Mean Value Theorem, 4-7
First Corollary, 4-9
not exists, 3-17
numerical, 3-9, 3-31
of products, 3-26–27
one-sided, 3-15
Product Rule, 3-27, 3-30
Quotient Rule, 3-28
Rolle's Theorem, 4-5
second order, 3-25
slopes, 3-1
trigonometric functions, 3-51
Descartes, Rene, 1-2
Difference Rule, 2-34
differentiable functions, 3-18
at a point, 3-8
closed intervals, 3-15
continuous, 3-14
fractional powers of, 3-74–77
integer powers of 3-62
modeling discrete phenomena, 4-51
differential equations, 5-33
initial value problems, 4-79
differential notation, 3-88
differentials,
arc length formula, 6-40
linear approximations, 3-79
short formula, 6-39
differentiation,
Power Rule, 3-21
rules, 3-20
Sum and Difference Rule, 3-23
directions, 1-2
discontinuous function, 3-8
in dy/dx, 6-38
points of discontinuity, 2-14, 2-15
disks,
solid of revolution, 6-11–13
volume calculation, 6-12
distance,
between points, 1-23
and a coordinate plane, 1-23
from a point to a line, 1-26

traveled, 6-81
division,
by zero, 1-34
domain,
endpoints, 4-36
of function, 1-34–36, 1-78
double-angle formulas, 1-113

E
economics,
cost and revenue, 4-48
derivatives in, 3-42
Eight Curve, 3-78
ellipse, 1-93, 3-73
end behavior, 2-35
asymptote, 2-43
model, 2-35
polynomial, 2-38
rational function, 2-42
endpoints,
continuity at, 2-15
maxima and minima at, 4-36
equations,
of circles in the plane, 1-89–93
of lines, 1-14, 1-18, 1-28
linear, 1-27, 1-28
point-slope, 1-18
polynomial, 1-66
solutions of, 4-6
error,
approximation, 3-87
graphical analysis of, 5-101–102
of a solution, 1-65
round-off, 5-104
Simpson's Rule, 5-100
Trapezoidal Rule, 5-6
estimating change,
with differentials, 3-83
estimating derivatives,
from functions, 3-45
estimating functions,
from derivatives, 3-44
Euler, Leonhard, 1-31
even functions, 1-38
exact derivative calculations, 3-12
existance,
of definite integrals, 5-22
of a function, 2-24
explore activity,
lines, 1-21
explore box,
area under a graph, 5-6, 5-35, 5-40
conic sections, 1-93
exponential functions, 1-85, 5-47, 5-81
geometric transformations, 1-55
graphing utility, 4-3
integrals, 5-40, 5-55, 5-61, 5-62, 5-68, 5-77
natural logarithms, 5-75
parabolas, 1-47
parametric equations, 1-103, 3-55, 3-70
Riemann Sums, 5-21
sequences and sums, 5-106
sines and cosines, 1-100, 1-112, 3-29, 5-45
standard viewing rectangle, 1-5
tangent lines, 3-4, 4-16
explore with a grapher,
analytic definite integral, 6-15
chain rule, 3-61
continuous extension, 2-19
derivatives, 3-12, 3-14 3-21, 3-76, 3-79, 4-3, 4-10
existance of derivatives, 3-76
Mean Value Theorem, 4-76
Newton's Method, 4-28
parabolas, 3-4
parametric equations, 3-55, 3-70, 6-84
sines and cosines, 3-29
tangent lines, 4-16
volume of rotation, 6-30, 6-38
volumes, 3-67
Exponential Change,
Law of, 5-85
exponential functions, 1-84–88, 4-59–64, 5-74–75
base, *e*, 5-80
defined, 1-85, 5-79
equations involving, 5-81
graph of, 4-59–62, 5-80
exponents,
rules of, 5-83
extrema, 4-31–35

F
Fahrenheit, 1-22, 1-30
Fermat, Pierre de, 4-55
finite limits,
calculation rules, 2-33–34
finite sums, 5-9–14
algebra rules for, 5-12
sigma notation, 5-9
first derivative, 3-25, 4-14

concavity, 4-15
First Derivative Theorem, 4-3
Fixx, Frank, 1-3
fluid,
fluid force integral, 6-61
pressures and forces, 6-58
pressure-depth equation, 6-58
force,
constant-depth formula, 6-59
fluid force integral, 6-61
variable-depth formula, 6-60
fractional powers,
of differentiable functions, 3-74, 4-78
Power Rule for, 3-75
free fall, 3-34, 3-35
standard equation for, 5-64
fruit fly problem, 3-1, 3-12
frustrum of a cone, 6-43
functions, 1-30–34
absolute value function, 1-43
average value of, 5-31
complete graph, 1-10
composition of, 1-46–47
continuous, 2-13
defined in pieces, 1-41
derivative of, 3-7–8
domain of, 1-32
even, 1-38, 1-104
graphs of, 1-32–34
integer-valued, 1-40
odd, 1-38, 1-104
periodic, 1-101
piecewise defined, 1-41
range of, 1-32
real-valued, 1-31
step, 1-42
sums, differences, quotients, 1-44–45
vertical line test, 1-79
Fundamental Theorem of Algebra, 4-30
Fundamental Theorems of Integral Calculus, 5-39–50
first fundamental theorem, 5-39–41
notation, 5-40

G
Galileo, 4-55
Generalized Power Rule,
in integral form, 5-64
geometry,
area of a rectangle, 4-56
combining integrals with formulas, 6-9
grade of a roadbed, 1-16
graphical method,
for solving equations, 1-63–75
for solving inequalities, 1-67–75
graphical representation, 1-72
graphing, 1-35
completeness, 1-9
integrals, 5-55
intercepts, 1-8
parametric, 1-83, 3-36, 3-69
with zoom-in, 1-63
graphing utility, 1-3, 1-35
explore box, 1-112, 4-3
screen coordinates, 1-4
viewing window, 1-3
graphs,
geometric transformations, 1–54
of derivatives, 3-11
of equations, 1-5
shifts, reflections, stretches, and shrinks, 1-50–54
Great Molasses Flood, 6-59

H
Hamilton, Sir William, 4-55
hammer and feather problem, 5-63
harmonic motion, 3-54
height function, 3-39
Helga von Koch's Snowflake Curve, 3-75
higher order derivatives,
implicit differentiation, 3-74
Hooke's Law,
for springs, 6-52
horizontal asymptotes, 2-39
horizontal line test, 1-80
horizontal tangents, 3-24
horizontal stretch and shrink, 1-108
hyperbola, 1-93
hyperbolic functions, 4-63–64

I
identity function, 1-83
implicit differentiation, 3-69–77
implicit functions,
differentiation, 3-70
graphing, 3-69
incidence of disease, 5-92
model, 5-86
increments, 1-14
indefinite integrals, 5-53–62

rules for, 5-56
independent variables, 1-34
inequality, 3, 69–73
absolute value, 1-68, 1-72
graphical solution, 1-67
representations,
algebraic, 1-72–74
graphical, 1-72–74
triangle, 1-25
infinite limits, 2-42
infinity, 2-31–32
limits involving, 2-37–38
inflection,
linearization at, 4-26
points of, 4-16, 4-18
initial value problems, 4-74, 5-60
differential equations, 4-79
solving with indefinite integrals, 5-59
Inside-Outside Rule, 3-63
instantaneous velocity, 3-37
Integral Evaluation Theorem, 5-44
integrals,
graphing of, 5-55
indefinite, 5-53–55
of sines and cosines, 5-58
useless, 6-86
variable of, 5-53
with respect to y, 6-7–9
integral sign 5-53
integrand, 5-22, 5-53
integration, 5-1
by substitution, 5-64
limits of, 6-4
intercepts, 1-7
integer powers,
differentiable functions, 3-62
differentiation, 3-20–22
integer-valued functions, 1-40, 2-16
intercept, 1-7
interest,
continuous compounded, 5-87
rules of 70 and 72, 5-88–89
interior point, 4-36
continuity at, 2-15
Intermediate Value Theorem, 2-23
for continuous functions, 2-23
intervals,
absolute values, 1-69–71
of integration, 5-49
inverse functions,
function text, 1-81
one-to-one, 1-81
trigonometric functions, 1-104
inverse relations, 1-78
definition, 1-80
functions, 1-80
inverse trigonometric functions, 1-105
irrational numbers, 3-25
iteration, 4-26

L

leading coefficient, 2-37
left-hand limits, 2-8, 2-62
limits, 2-1
definition, 2-3
examples, 2-2
formal definition, 2-53–56
informal definition, 2-3
of integration, 5-22, 6-4, 6-6
involving infinity, 2-31, 2-37
infinite, 2-42
left-hand, 2-8, 2-28
one-sided, 2-9, 2-61
properties of, 2-5–10
proving theorems, 2-60–62
of rational functions, 2-36
right-hand limits, 2-8, 2-28
two-sided limits, 2-9, 2-61
limit theorems,
proving, 2-60
linear approximations,
differentials, 3-79
linearization, 3-97
linear equations,
forms of, 1-28
general, 1-27
solving (*see also* equation), 44–47
lines,
equations of, 1-18–21
point-slope equations, 1-18–19
slope-intercept equations, 1-20
slopes, 3-9
local extrema, 4-19, 4-33
local maximum and minimum, 4-2, 4-14, 4-23
defined, 4-2
second derivative test, 4-23
locally straight functions, 2-59
locating a planet, 4-42
logarithmic functions, 1-84–88, 4-59–63
common, 291
defined, 1-86
derivative of, 5-77
equations involving, 5-81

natural, 5-75
rules for, 5-77
logarithms,
properties of, 1-87–88, 5-74
special properties, 1-88

M
marginal revenue, 3-44
mass,
distributed over a plane, 6-68
thin plates, 6-70
versus weight, 6-65
wires or thin rods, 6-67
mathematical modeling, 4-74
in biology, 3-34, 4-85
surface area example, 6-89
Max-Min problems,
strategy for solving, 4-38
Max-Min Theorem,
for Continuous Functions, 2-21
maxima, 4-2, 4-55
relative vs. absolute, 4-2
maxima and minima, 4-55
Mean Value Theorem, 4-7, 6-44
first corollary, 4-9
for definite integrals, 5-31
second and third corollary, 4-75–77
medicine,
body's reaction to, 3-34
how we cough problem, 4-41
melting ice cubes problem, 3-65
Mendel, Gregor Johann, 3-41
metal fabrication example, 4-37
minima, 4-2, 4-55
relative vs. absolute, 4-2
modeling applications, 6-86
position shift and distance traveled, 6-81
volumes of arbitrary solids, 6-76
moments about the origin,
thin plates, 6-70–72
wires or thin rods, 6-67–69

N
Napier, John, 5-92
natural logarithm function, 1-88, 5-75,
NDer function, 3-10–12, 3-21, 3-25, 3-51
net area, 5-26
Newton's Method, 4-26–38
Newton's Second Law, 6-65
nonnegative continuous function,
area under graph of, 5-4–5, 5-34
norm,
of a partition, 5-20
normal,
definition, 3-56
lines, 3-72
to a surface, 3-72
numerical derivatives, 3-9, 3-30
numerical evaluation,
definite integrals, 5-26
number line,
distance between points, 1-23

O
odd functions, 1-39
one-to-oneness,
graphical test, 1-81
one-sided derivatives, 3-15, 3-17
one-sided limits, 2-9
optimization,
of mathematical models, 4-35
ordered pair, 1-2
orthogonal curves, 3-78

P
Pappus, Theorems of, 6-75
parabola, 1-8, 1-56
graphing, 1-58
open to the left, 1-58–60
open to the right, 1-58–60
parametric equations, 6-84
conic sections, 1-93
graphing of, 1-103
parametric graphing, 1-82, 3-36, 3-69
parametric range setting, 1-83
particle motion problem, 4-21
partition,
norm of, 5-20
Riemann Sums, 5-17
percentage change, 3-85
periodic functions, 1-101
periodicity, 1-101
plane,
coordinates and graphs in a, 1-1
point-slope form, 1-18
points of inflection, 4-16
polynomial functions,
degree of, 1-65, 2-37
end behavior of, 2-38
limits, 2-6
zeros of, 4-30
polynomials,
continuity, 2-16
functions, 4-30

end behavior models, 2-38–39
equations, 1-65
limits of, 2-6
number of extrema, 4-31
roots of, 4-30
zeros of, 5-48
position shift, 6-81–83
Power Rule,
fractional exponents, 3-75–77
negative integers, 3-30
positive integers, 3-21
pressure under water problem, 1-29
Product Rule, 2-34, 3-37
products,
of numbers, 4-36
purchasing power of the dollar, 5-90
Pythagorean Theorem, 1-23

Q
quadrants, 1-2
Quotient Rule, 2-34, 3-28

R
radian measure, 1-95, 1-101
conversion, 1-96
radical functions, 4-54–63
RAM Computations, 5-5–8
rates of change, 3-34, 4-65–71
rational functions, 2-42–44
and asymptotes, 2-39
controlling a, 2-49
economic applications, 4-42–45
end behavior for, 2-42
limits of, 2-7, 2-36, 2-44
sketching the graph of, 4-46
Rational Zeros Theorem, 1-66
real variable, 1-31
rectangle approximation method, 5-5–8
recursion, 4-26
reflection,
angle of, 4-65
of a graph, 1-50–54
refraction,
of a light ray, 3-72
related rate problems, 4-69
relations,
definition, 1-78
graphs of, 91–99
inverse, 1-78
relative change, 3-85
removable discontinuity, 2-26
Riemann Sums, 5-17–20, 6-44, 6-54, 6-69, 6-77
right-hand limits, 2-8
rise, 1-15
rising balloon problem, 4-68
Rolle, Michal, 4-5
Rolle's Theorem, 4-5
root,
finding, 2-23
Fundamental Theorem of Algebra, 4-30
Newton's Method, 4-27
rounding, 1-97
Rules of 70 and 72, 5-88
run, 1-15

S
Sandwich Theorem, 2-26, 2-34, 5-33
screen coordinates, 1-4
secant, 1-98, 3-2
derivative of, 3-55
secant slope, 3-7, 3-38
second derivative,
test for local maxima and minima, 4-23
Second Fundamental Theorem of Calculus, 5-44
second order derivatives, 3-25
sensitivity, 3-87
shells, 6-25–33
cylindrical, 6-25, 6-33
method, 6-26
versus washers, 6-32
volume computation, 6-26–29
shift formulas, 1-109
shift of a graph, 1-50–54
shrink of a graph, 1-50–54
sigma notation,
finite sums, 5-9–11
simple harmonic motion, 3-54
Simpson's Method, 5-94–105
controlling error, 5-100
one-third rule, 5-100
Simpson's Rule, 5-98–100
table of approximations, 5-105
with numerical data, 5-103
Simpson, Thomas, 5-100
sine, 1-98
calculating, 1-98
defining the inverse, 1-105
derivative of, 3-52
general, 1-109
odd function, 1-104
sinusoid, 1-111
definition, 1-111

graphing, 1-112
sketching, 1-7, 1-44
slicing,
volumes of arbitrary solids, 6-76–80
sliding ladder problem, 4-72
slope, 1-14, 1-27, 3-8
calculating, 1-15
definition, 1-15, 1-27, 3-4
lines, 3-9
of nonvertical lines, 1-15
undefined, 3-16
slope-intercept equations, 1-20–21
solids of revolution, 6-11–22
definition, 6-13
generation of, 6-19–22
volume of, 6-15
solution methods, 1-75
soap bubble problem, 4-65
sonobouy problem, 4-40
speed, 3-34, 3-39
sphere, 3-86
standard formula for sums, 5-12
integers, 5-13
squares, 5-13
cubes, 5-13
standard linear approximation, 3-80
stiffness of a beam, 4-41
strength of a beam, 4-41
stretch of a graph. 1-50–54
horizontal, 1-107
subintervals,
Riemann Sums, 5-17
substitution,
definite integrals, 5-70
Substitution Method of Integration, 5-69
summation,
limits of, 5-10
Sum and Difference Rule, 3-23
Sum Rule, 2-34
sums,
Riemann, 5-17
standard formula for, 5-12
surfaces, of revolution,
areas of, 6-43–48
about the y-axis, 6-45–46
symmetric difference, 3-10
symmetry, 1-37–38
about the origin, 1-37

T

tangent, 1-98, 3-8
defining the inverse, 1-105
derivative of, 3-55
to a curve, 3-3
vertical, 3-17
tangent lines, 3-1, 3-80, 4-29
definition, 3-4
horizontal, 3-24
implicit differentiation, 3-72
trancendental functions, 4-54–63
Trapezoidal Rule, 5-94–105
table of approximations, 5-104
controlling error, 5-96
triangle,
area of, 5-2
triangle inequality, 1-25, 2-61
trigonometric functions, 1-95–112, 4-56–59
controlling a, 2-49
derivatives, 3-51–57
evaluating inverse, 1-105
graphs of, 1-102–104
inverse, 1-104
review of, 1-95–112
truck convoy problem, 4-69
two-sided limits, 2-9

U

under the curve,
area, 5-24
unit circle, 1-91
useless integrals, 6-86

V

value of integration, 5-53
variables, 1-34
velocity, 3-34, 3-36
finding from acceleration, 4-80–83
vertical asymptotes,
definition, 2-39, 2-41, 2-43
vertical line test,
for a function, 1-79
viewing rectangle, 1-4
square, 1-19
visualization, 1-9–12
volume,
of arbitrary solids, 6-76–79
of a disk, 6-12
solids of revolution, 6-11–22

W

water, bucket, and rope example, 6-51
washers,
solid of revolution, 6-11
versus shells, 6-32

volume calculation, 6-16–19
Weierstrass functions, 3-17
Weierstrass, Karl, 3-17
weight-density, 6-54, 6-58
work, 6-49–54
constant-force formula, 6-49
pumping liquids, 6-53
variable-force integral, 6-50–52

X

x-axis, 1-1
x-coordinate, 1-2
x-intercepts, 1-7

Y

y-axis, 1-1
y-coordinate, 1-2
y-intercepts, 1-7

Z

zero of a function,
finding rational, 195–196
zoom-in, 1-67, 4-20

1. $\int u\,dv = uv - \int v\,du$

2. $\int a^u\,du = \dfrac{a^u}{\ln a} + C, \quad a \neq 1, \quad a > 0$

3. $\int \cos u\,du = \sin u + C$

4. $\int \sin u\,du = -\cos u + C$

5. $\int (ax+b)^n\,dx = \dfrac{(ax+b)^{n+1}}{a(n+1)} + C, \quad n \neq -1$

6. $\int (ax+b)^{-1}\,dx = \dfrac{1}{a}\ln|ax+b| + C$

7. $\int x(ax+b)^n\,dx = \dfrac{(ax+b)^{n+1}}{a^2}\left[\dfrac{ax+b}{n+2} - \dfrac{b}{n+1}\right] + C, \quad n \neq -1, -2$

8. $\int x(ax+b)^{-1}\,dx = \dfrac{x}{a} - \dfrac{b}{a^2}\ln|ax+b| + C$

9. $\int x(ax+b)^{-2}\,dx = \dfrac{1}{a^2}\left[\ln|ax+b| + \dfrac{b}{ax+b}\right] + C$

10. $\int \dfrac{dx}{x(ax+b)} = \dfrac{1}{b}\ln\left|\dfrac{x}{ax+b}\right| + C$

11. $\int (\sqrt{ax+b})^n\,dx = \dfrac{2}{a}\dfrac{(\sqrt{ax+b})^{n+2}}{n+2} + C, \quad n \neq -2$

12. $\int \dfrac{\sqrt{ax+b}}{x}\,dx = 2\sqrt{ax+b} + b\int \dfrac{dx}{x\sqrt{ax+b}}$

13. (a) $\int \dfrac{dx}{x\sqrt{ax+b}} = \dfrac{2}{\sqrt{-b}}\tan^{-1}\sqrt{\dfrac{ax+b}{-b}} + C, \quad \text{if } b < 0$

(b) $\int \dfrac{dx}{x\sqrt{ax+b}} = \dfrac{1}{\sqrt{b}}\ln\left|\dfrac{\sqrt{ax+b}-\sqrt{b}}{\sqrt{ax+b}+\sqrt{b}}\right| + C, \quad \text{if } b > 0$

14. $\int \dfrac{\sqrt{ax+b}}{x^2}\,dx = -\dfrac{\sqrt{ax+b}}{x} + \dfrac{a}{2}\int \dfrac{dx}{x\sqrt{ax+b}} + C$

15. $\int \dfrac{dx}{x^2\sqrt{ax+b}} = -\dfrac{\sqrt{ax+b}}{bx} - \dfrac{a}{2b}\int \dfrac{dx}{x\sqrt{ax+b}} + C$

16. $\int \dfrac{dx}{a^2+x^2} = \dfrac{1}{a}\tan^{-1}\dfrac{x}{a} + C$

17. $\int \dfrac{dx}{(a^2+x^2)^2} = \dfrac{x}{2a^2(a^2+x^2)} + \dfrac{1}{2a^3}\tan^{-1}\dfrac{x}{a} + C$

18. $\int \dfrac{dx}{a^2-x^2} = \dfrac{1}{2a}\ln\left|\dfrac{x+a}{x-a}\right| + C$

19. $\int \dfrac{dx}{(a^2-x^2)^2} = \dfrac{x}{2a^2(a^2-x^2)} + \dfrac{1}{2a^2}\int \dfrac{dx}{a^2-x^2}$

20. $\int \dfrac{dx}{\sqrt{a^2+x^2}} = \sinh^{-1}\dfrac{x}{a} + C = \ln|x + \sqrt{a^2+x^2}| + C$

Continued

21. $\int \sqrt{a^2+x^2}\,dx = \frac{x}{2}\sqrt{a^2+x^2} + \frac{a^2}{2}\sinh^{-1}\frac{x}{a} + C$

22. $\int x^2\sqrt{a^2+x^2}\,dx = \frac{x(a^2+2x^2)\sqrt{a^2+x^2}}{8} - \frac{a^4}{8}\sinh^{-1}\frac{x}{a} + C$

23. $\int \frac{\sqrt{a^2+x^2}}{x}\,dx = \sqrt{a^2+x^2} - a\sinh^{-1}\left|\frac{a}{x}\right| + C$

24. $\int \frac{\sqrt{a^2+x^2}}{x^2}\,dx = \sinh^{-1}\frac{x}{a} - \frac{\sqrt{a^2+x^2}}{x} + C$

25. $\int \frac{x^2}{\sqrt{a^2+x^2}}\,dx = -\frac{a^2}{2}\sinh^{-1}\frac{x}{a} + \frac{x\sqrt{a^2+x^2}}{2} + C$

26. $\int \frac{dx}{x\sqrt{a^2+x^2}} = -\frac{1}{a}\ln\left|\frac{a+\sqrt{a^2+x^2}}{x}\right| + C$

27. $\int \frac{dx}{x^2\sqrt{a^2+x^2}} = -\frac{\sqrt{a^2+x^2}}{a^2x} + C$

28. $\int \frac{dx}{\sqrt{a^2-x^2}} = \sin^{-1}\frac{x}{a} + C$

29. $\int \sqrt{a^2-x^2}\,dx = \frac{x}{2}\sqrt{a^2-x^2} + \frac{a^2}{2}\sin^{-1}\frac{x}{a} + C$

30. $\int x^2\sqrt{a^2-x^2}\,dx = \frac{a^4}{8}\sin^{-1}\frac{x}{a} - \frac{1}{8}x\sqrt{a^2-x^2}\,(a^2-2x^2) + C$

31. $\int \frac{\sqrt{a^2-x^2}}{x}\,dx = \sqrt{a^2-x^2} - a\ln\left|\frac{a+\sqrt{a^2-x^2}}{x}\right| + C$

32. $\int \frac{\sqrt{a^2-x^2}}{x^2}\,dx = -\sin^{-1}\frac{x}{a} - \frac{\sqrt{a^2-x^2}}{x} + C$

33. $\int \frac{x^2}{\sqrt{a^2-x^2}}\,dx = \frac{a^2}{2}\sin^{-1}\frac{x}{a} - \frac{1}{2}x\sqrt{a^2-x^2} + C$

34. $\int \frac{dx}{x\sqrt{a^2-x^2}} = -\frac{1}{a}\ln\left|\frac{a+\sqrt{a^2-x^2}}{x}\right| + C$

35. $\int \frac{dx}{x^2\sqrt{a^2-x^2}} = -\frac{\sqrt{a^2-x^2}}{a^2x} + C$

36. $\int \frac{dx}{\sqrt{x^2-a^2}} = \cosh^{-1}\frac{x}{a} + C = \ln\left|x+\sqrt{x^2-a^2}\right| + C$

37. $\int \sqrt{x^2-a^2}\,dx = \frac{x}{2}\sqrt{x^2-a^2} - \frac{a^2}{2}\cosh^{-1}\frac{x}{a} + C$

38. $\int (\sqrt{x^2-a^2})^n\,dx = \frac{x(\sqrt{x^2-a^2})^n}{n+1} - \frac{na^2}{n+1}\int (\sqrt{x^2-a^2})^{n-2}\,dx, \quad n \neq -1$

39. $\int \frac{dx}{(\sqrt{x^2-a^2})^n} = \frac{x(\sqrt{x^2-a^2})^{2-n}}{(2-n)a^2} - \frac{n-3}{(n-2)a^2}\int \frac{dx}{(\sqrt{x^2-a^2})^{n-2}}, \quad n \neq 2$

40. $\int x(\sqrt{x^2-a^2})^n\,dx = \frac{(\sqrt{x^2-a^2})^{n+2}}{n+2} + C, \quad n \neq -2$

41. $\int x^2\sqrt{x^2-a^2}\,dx = \frac{x}{8}(2x^2-a^2)\sqrt{x^2-a^2} - \frac{a^4}{8}\cosh^{-1}\frac{x}{a} + C$

42. $\int \frac{\sqrt{x^2-a^2}}{x}\,dx = \sqrt{x^2-a^2} - a\sec^{-1}\left|\frac{x}{a}\right| + C$

43. $\int \frac{\sqrt{x^2-a^2}}{x^2}\,dx = \cosh^{-1}\frac{x}{a} - \frac{\sqrt{x^2-a^2}}{x} + C$

44. $\int \frac{x^2}{\sqrt{x^2-a^2}}\,dx = \frac{a^2}{2}\cosh^{-1}\frac{x}{a} + \frac{x}{2}\sqrt{x^2-a^2} + C$

45. $\int \frac{dx}{x\sqrt{x^2-a^2}} = \frac{1}{a}\sec^{-1}\left|\frac{x}{a}\right| + C = \frac{1}{a}\cos^{-1}\left|\frac{a}{x}\right| + C$

46. $\int \frac{dx}{x^2\sqrt{x^2-a^2}} = \frac{\sqrt{x^2-a^2}}{a^2x} + C$

47. $\int \frac{dx}{\sqrt{2ax-x^2}} = \sin^{-1}\left(\frac{x-a}{a}\right) + C$

48. $\int \sqrt{2ax-x^2}\,dx = \frac{x-a}{2}\sqrt{2ax-x^2} + \frac{a^2}{2}\sin^{-1}\left(\frac{x-a}{a}\right) + C$

49. $\int (\sqrt{2ax-x^2})^n\,dx = \frac{(x-a)(\sqrt{2ax-x^2})^n}{n+1} + \frac{na^2}{n+1}\int (\sqrt{2ax-x^2})^{n-2}\,dx,$

50. $\int \frac{dx}{(\sqrt{2ax-x^2})^n} = \frac{(x-a)(\sqrt{2ax-x^2})^{2-n}}{(n-2)a^2} + \frac{(n-3)}{(n-2)a^2}\int \frac{dx}{(\sqrt{2ax-x^2})^{n-2}}$

51. $\int x\sqrt{2ax-x^2}\,dx = \frac{(x+a)(2x-3a)\sqrt{2ax-x^2}}{6} + \frac{a^3}{2}\sin^{-1}\frac{x-a}{a} + C$

52. $\int \frac{\sqrt{2ax-x^2}}{x}\,dx = \sqrt{2ax-x^2} + a\sin^{-1}\frac{x-a}{a} + C$

53. $\int \frac{\sqrt{2ax-x^2}}{x^2}\,dx = -2\sqrt{\frac{2a-x}{x}} - \sin^{-1}\left(\frac{x-a}{a}\right) + C$

54. $\int \frac{x\,dx}{\sqrt{2ax-x^2}} = a\sin^{-1}\frac{x-a}{a} - \sqrt{2ax-x^2} + C$

55. $\int \frac{dx}{x\sqrt{2ax-x^2}} = -\frac{1}{a}\sqrt{\frac{2a-x}{x}} + C$

56. $\int \sin ax\,dx = -\frac{1}{a}\cos ax + C$

57. $\int \cos ax\,dx = \frac{1}{a}\sin ax + C$

58. $\int \sin^2 ax\,dx = \frac{x}{2} - \frac{\sin 2ax}{4a} + C$

59. $\int \cos^2 ax\,dx = \frac{x}{2} + \frac{\sin 2ax}{4a} + C$

60. $\int \sin^n ax\,dx = \frac{-\sin^{n-1} ax\cos ax}{na} + \frac{n-1}{n}\int \sin^{n-2} ax\,dx$

61. $\int \cos^n ax\,dx = \frac{\cos^{n-1} ax\sin ax}{na} + \frac{n-1}{n}\int \cos^{n-2} ax\,dx$

62. (a) $\int \sin ax\cos bx\,dx = -\frac{\cos(a+b)x}{2(a+b)} - \frac{\cos(a-b)x}{2(a-b)} + C, \quad a^2 \neq b^2$

(b) $\int \sin ax\sin bx\,dx = \frac{\sin(a-b)x}{2(a-b)} - \frac{\sin(a+b)x}{2(a+b)}, \quad a^2 \neq b^2$

(c) $\int \cos ax\cos bx\,dx = \frac{\sin(a-b)x}{2(a-b)} + \frac{\sin(a+b)x}{2(a+b)}, \quad a^2 \neq b^2$

Continued

63. $\int \sin ax \cos ax\, dx = -\frac{\cos 2ax}{4a} + C$

64. $\int \sin^n ax \cos ax\, dx = \frac{\sin^{n+1} ax}{(n+1)a} + C, \quad n \neq -1$

65. $\int \frac{\cos ax}{\sin ax}\, dx = \frac{1}{a} \ln |\sin ax| + C$

66. $\int \cos^n ax \sin ax\, dx = -\frac{\cos^{n+1} ax}{(n+1)a} + C, \quad n \neq -1$

67. $\int \frac{\sin ax}{\cos ax}\, dx = -\frac{1}{a} \ln |\cos ax| + C$

68. $\int \sin^n ax \cos^m ax\, dx = -\frac{\sin^{n-1} ax \cos^{m+1} ax}{a(m+n)} + \frac{n-1}{m+n} \int \sin^{n-2} ax \cos^m ax\, dx,$

$n \neq -m \quad$ (If $n = -m$, use No. 86.)

69. $\int \sin^n ax \cos^m ax\, dx = \frac{\sin^{n+1} ax \cos^{m-1} ax}{a(m+n)} + \frac{m-1}{m+n} \int \sin^n ax \cos^{m-2} ax\, dx,$

$m \neq -n \quad$ (If $m = -n$, use No. 87.)

70. $\int \frac{dx}{b + c \sin ax} = \frac{-2}{a\sqrt{b^2 - c^2}} \tan^{-1}\left[\sqrt{\frac{b-c}{b+c}} \tan\left(\frac{\pi}{4} - \frac{ax}{2}\right)\right] + C, \quad b^2 > c^2$

71. $\int \frac{dx}{b + c \sin ax} = \frac{-1}{a\sqrt{c^2 - b^2}} \ln \left|\frac{c + b \sin ax + \sqrt{c^2 - b^2} \cos ax}{b + c \sin ax}\right| + C, \quad b^2 < c^2$

72. $\int \frac{dx}{1 + \sin ax} = -\frac{1}{a} \tan\left(\frac{\pi}{4} - \frac{ax}{2}\right) + C$

73. $\int \frac{dx}{1 - \sin ax} = \frac{1}{a} \tan\left(\frac{\pi}{4} + \frac{ax}{2}\right) + C$

74. $\int \frac{dx}{b + c \cos ax} = \frac{2}{a\sqrt{b^2 - c^2}} \tan^{-1}\left[\sqrt{\frac{b-c}{b+c}} \tan \frac{ax}{2}\right] + C, \quad b^2 > c^2$

75. $\int \frac{dx}{b + c \cos ax} = \frac{1}{a\sqrt{c^2 - b^2}} \ln \left|\frac{c + b \cos ax + \sqrt{c^2 - b^2} \sin ax}{b + c \cos ax}\right| + C, \quad b^2 < c^2$

76. $\int \frac{dx}{1 + \cos ax} = \frac{1}{a} \tan \frac{ax}{2} + C$

77. $\int \frac{dx}{1 - \cos ax} = -\frac{1}{a} \cot \frac{ax}{2} + C$

78. $\int x \sin ax\, dx = \frac{1}{a^2} \sin ax - \frac{x}{a} \cos ax + C$

79. $\int x \cos ax\, dx = \frac{1}{a^2} \cos ax + \frac{x}{a} \sin ax + C$

80. $\int x^n \sin ax\, dx = -\frac{x^n}{a} \cos ax + \frac{n}{a} \int x^{n-1} \cos ax\, dx$

81. $\int x^n \cos ax\, dx = \frac{x^n}{a} \sin ax - \frac{n}{a} \int x^{n-1} \sin ax\, dx$

82. $\int \tan ax\, dx = \frac{1}{a} \ln |\sec ax| + C$

83. $\int \cot ax\, dx = \frac{1}{a} \ln |\sin ax| + C$

84. $\int \tan^2 ax\, dx = \frac{1}{a} \tan ax - x + C$

85. $\int \cot^2 ax\, dx = -\frac{1}{a} \cot ax - x + C$

86. $\int \tan^n ax\, dx = \frac{\tan^{n-1} ax}{a(n-1)} - \int \tan^{n-2} ax\, dx, \quad n \neq 1$

87. $\int \cot^n ax\, dx = -\frac{\cot^{n-1} ax}{a(n-1)} - \int \cot^{n-2} ax\, dx, \quad n \neq 1$

88. $\int \sec ax\, dx = \frac{1}{a} \ln |\sec ax + \tan ax| + C$

89. $\int \csc ax\, dx = -\frac{1}{a} \ln |\csc ax + \cot ax| + C$

90. $\int \sec^2 ax\,dx = \frac{1}{a}\tan ax + C$

91. $\int \csc^2 ax\,dx = -\frac{1}{a}\cot ax + C$

92. $\int \sec^n ax\,dx = \frac{\sec^{n-2} ax \tan ax}{a(n-1)} + \frac{n-2}{n-1}\int \sec^{n-2} ax\,dx, \quad n \neq 1$

93. $\int \csc^n ax\,dx = -\frac{\csc^{n-2} ax \cot ax}{a(n-1)} + \frac{n-2}{n-1}\int \csc^{n-2} ax\,dx, \quad n \neq 1$

94. $\int \sec^n ax \tan ax\,dx = \frac{\sec^n ax}{na} + C, \quad n \neq 0$

95. $\int \csc^n ax \cot ax\,dx = -\frac{\csc^n ax}{na} + C, \quad n \neq 0$

96. $\int \sin^{-1} ax\,dx = x\sin^{-1} ax + \frac{1}{a}\sqrt{1-a^2x^2} + C$

97. $\int \cos^{-1} ax\,dx = x\cos^{-1} ax - \frac{1}{a}\sqrt{1-a^2x^2} + C$

98. $\int \tan^{-1} ax\,dx = x\tan^{-1} ax - \frac{1}{2a}\ln(1+a^2x^2) + C$

99. $\int x^n \sin^{-1} ax\,dx = \frac{x^{n+1}}{n+1}\sin^{-1} ax - \frac{a}{n+1}\int \frac{x^{n+1}\,dx}{\sqrt{1-a^2x^2}}, \quad n \neq -1$

100. $\int x^n \cos^{-1} ax\,dx = \frac{x^{n+1}}{n+1}\cos^{-1} ax + \frac{a}{n+1}\int \frac{x^{n+1}\,dx}{\sqrt{1-a^2x^2}}, \quad n \neq -1$

101. $\int x^n \tan^{-1} ax\,dx = \frac{x^{n+1}}{n+1}\tan^{-1} ax - \frac{a}{n+1}\int \frac{x^{n+1}\,dx}{1+a^2x^2}, \quad n \neq -1$

102. $\int e^{ax}\,dx = \frac{1}{a}e^{ax} + C$

103. $\int b^{ax}\,dx = \frac{1}{a}\frac{b^{ax}}{\ln b} + C, \quad b > 0,\ b \neq 1$

104. $\int xe^{ax}\,dx = \frac{e^{ax}}{a^2}(ax-1) + C$

105. $\int x^n e^{ax}\,dx = \frac{1}{a}x^n e^{ax} - \frac{n}{a}\int x^{n-1}e^{ax}\,dx$

106. $\int x^n b^{ax}\,dx = \frac{x^n b^{ax}}{a\ln b} - \frac{n}{a\ln b}\int x^{n-1}b^{ax}\,dx, \quad b > 0,\ b \neq 1$

107. $\int e^{ax}\sin bx\,dx = \frac{e^{ax}}{a^2+b^2}(a\sin bx - b\cos bx) + C$

108. $\int e^{ax}\cos bx\,dx = \frac{e^{ax}}{a^2+b^2}(a\cos bx + b\sin bx) + C$

109. $\int \ln ax\,dx = x\ln ax - x + C$

110. $\int x^n(\ln ax)^m\,dx = \frac{x^{n+1}(\ln ax)^m}{n+1} - \frac{m}{n+1}\int x^n(\ln ax)^{m-1}\,dx, \quad n \neq -1$

111. $\int x^{-1}(\ln ax)^m = \frac{(\ln ax)^{m+1}}{m+1} + C, \quad m \neq -1$

112. $\int \frac{dx}{x\ln ax} = \ln|\ln ax| + C$

113. $\int \sinh ax\,dx = \frac{1}{a}\cosh ax + C$

114. $\int \cosh ax\,dx = \frac{1}{a}\sinh ax + C$

115. $\int \sinh^2 ax\,dx = \frac{\sinh 2ax}{4a} - \frac{x}{2} + C$

116. $\int \cosh^2 ax\,dx = \frac{\sinh 2ax}{4a} + \frac{x}{2} + C$

117. $\int \sinh^n ax\,dx = \frac{\sinh^{n-1} ax \cosh ax}{na} - \frac{n-1}{n}\int \sinh^{n-2} ax\,dx, \quad n \neq 0$

Continued

118. $\displaystyle\int \cosh^n ax\,dx = \frac{\cosh^{n-1} ax \sinh ax}{na} + \frac{n-1}{n}\int \cosh^{n-2} ax\,dx, \qquad n \neq 0$

119. $\displaystyle\int x \sinh ax\,dx = \frac{x}{a}\cosh ax - \frac{1}{a^2}\sinh ax + C$

120. $\displaystyle\int x \cosh ax\,dx = \frac{x}{a}\sinh ax - \frac{1}{a^2}\cosh ax + C$

121. $\displaystyle\int x^n \sinh ax\,dx = \frac{x^n}{a}\cosh ax - \frac{n}{a}\int x^{n-1}\cosh ax\,dx$

122. $\displaystyle\int x^n \cosh ax\,dx = \frac{x^n}{a}\sinh ax - \frac{n}{a}\int x^{n-1}\sinh ax\,dx$

123. $\displaystyle\int \tanh ax\,dx = \frac{1}{a}\ln(\cosh ax) + C$

124. $\displaystyle\int \coth ax\,dx = \frac{1}{a}\ln|\sinh ax| + C$

125. $\displaystyle\int \tanh^2 ax\,dx = x - \frac{1}{a}\tanh ax + C$

126. $\displaystyle\int \coth^2 ax\,dx = x - \frac{1}{a}\coth ax + C$

127. $\displaystyle\int \tanh^n ax\,dx = -\frac{\tanh^{n-1} ax}{(n-1)a} + \int \tanh^{n-2} ax\,dx, \qquad n \neq 1$

128. $\displaystyle\int \coth^n ax\,dx = -\frac{\coth^{n-1} ax}{(n-1)a} + \int \coth^{n-2} ax\,dx, \qquad n \neq 1$

129. $\displaystyle\int \operatorname{sech} ax\,dx = \frac{1}{a}\sin^{-1}(\tanh ax) + C$

130. $\displaystyle\int \operatorname{csch} ax\,dx = \frac{1}{a}\ln\left|\tanh\frac{ax}{2}\right| + C$

131. $\displaystyle\int \operatorname{sech}^2 ax\,dx = \frac{1}{a}\tanh ax + C$

132. $\displaystyle\int \operatorname{csch}^2 ax\,dx = -\frac{1}{a}\coth ax + C$

133. $\displaystyle\int \operatorname{sech}^n ax\,dx = \frac{\operatorname{sech}^{n-2} ax \tanh ax}{(n-1)a} + \frac{n-2}{n-1}\int \operatorname{sech}^{n-2} ax\,dx, \qquad n \neq 1$

134. $\displaystyle\int \operatorname{csch}^n ax\,dx = -\frac{\operatorname{csch}^{n-2} ax \coth ax}{(n-1)a} - \frac{n-2}{n-1}\int \operatorname{csch}^{n-2} ax\,dx, \qquad n \neq 1$

135. $\displaystyle\int \operatorname{sech}^n ax \tanh ax\,dx = -\frac{\operatorname{sech}^n ax}{na} + C, \qquad n \neq 0$

136. $\displaystyle\int \operatorname{csch}^n ax \coth ax\,dx = -\frac{\operatorname{csch}^n ax}{na} + C, \qquad n \neq 0$

137. $\displaystyle\int e^{ax}\sinh bx\,dx = \frac{e^{ax}}{2}\left[\frac{e^{bx}}{a+b} - \frac{e^{-bx}}{a-b}\right] + C, \qquad a^2 \neq b^2$

138. $\displaystyle\int e^{ax}\cosh bx\,dx = \frac{e^{ax}}{2}\left[\frac{e^{bx}}{a+b} + \frac{e^{-bx}}{a-b}\right] + C, \qquad a^2 \neq b^2$

139. $\displaystyle\int_0^\infty x^{n-1}e^{-x}\,dx = \Gamma(n) = (n-1)!, \qquad n > 0.$

140. $\displaystyle\int_0^\infty e^{-ax^2}\,dx = \frac{1}{2}\sqrt{\frac{\pi}{a}}, \qquad a > 0$

141. $$\int_0^{\pi/2} \sin^n x\,dx = \int_0^{\pi/2} \cos^n x\,dx = \begin{cases} \dfrac{1\cdot 3\cdot 5\cdots(n-1)}{2\cdot 4\cdot 6\cdots n}\cdot\dfrac{\pi}{2}, & \text{if } n \text{ is an even integer} \geq 2, \\ \dfrac{2\cdot 4\cdot 6\cdots(n-1)}{3\cdot 5\cdot 7\cdots n}, & \text{if } n \text{ is an odd integer} \geq 3 \end{cases}$$